Handbook of Laser Technology and Applications

Handbook of Laser Technology and Applications

Lasers: Principles and Operations (Volume One)
Second Edition

Lasers Design and Laser Systems (Volume Two)
Second Edition

Lasers Application: Material Processing and Spectroscopy (Volume Three)
Second Edition

Laser Applications: Medical, Metrology and Communication (Volume Four)
Second Edition

Handbook of Laser Technology and Applications

Lasers: Principles and Operations (Volume One)

Second Edition

Edited by

Chunlei Guo

Subhash Chandra Singh

CRC Press is an imprint of the
Taylor & Francis Group, an **informa** business

Second edition published 2021
by CRC Press
6000 Broken Sound Parkway NW, Suite 300, Boca Raton, FL 33487-2742

and by CRC Press
2 Park Square, Milton Park, Abingdon, Oxon, OX14 4RN

© 2021 Taylor & Francis Group, LLC

First edition published by IOP Publishing 2004

CRC Press is an imprint of Taylor & Francis Group, LLC

The right of Chunlei Guo and Subhash Chandra Singh to be identified as the authors of the editorial material, and of the authors for their individual chapters, has been asserted in accordance with sections 77 and 78 of the Copyright, Designs and Patents Act 1988.

Reasonable efforts have been made to publish reliable data and information, but the author and publisher cannot assume responsibility for the validity of all materials or the consequences of their use. The authors and publishers have attempted to trace the copyright holders of all material reproduced in this publication and apologize to copyright holders if permission to publish in this form has not been obtained. If any copyright material has not been acknowledged please write and let us know so we may rectify in any future reprint.

Except as permitted under U.S. Copyright Law, no part of this book may be reprinted, reproduced, transmitted, or utilized in any form by any electronic, mechanical, or other means, now known or hereafter invented, including photocopying, microfilming, and recording, or in any information storage or retrieval system, without written permission from the publishers.

For permission to photocopy or use material electronically from this work, access www.copyright.com or contact the Copyright Clearance Center, Inc. (CCC), 222 Rosewood Drive, Danvers, MA 01923, 978-750-8400. For works that are not available on CCC please contact mpkbookspermissions@tandf.co.uk

Trademark notice: Product or corporate names may be trademarks or registered trademarks and are used only for identification and explanation without intent to infringe.

Library of Congress Cataloging-in-Publication Data
Names: Guo, Chunlei, editor. | Singh, Subhash Chandra, editor.
Title: Handbook of laser technology and applications : four volume set /
[edited by] Chunlei Guo and Subhash Chandra Singh.
Description: 2nd edition. | Boca Raton : CRC Press, 2021- |
Series: Handbook of laser technology and applications | Includes bibliographical references and index. | Contents: v. 1. Lasers: principles and operations —
v. 2. Laser design and laser systems — v. 3. Lasers applications: materials processing —
v. 4. Laser applications: medical, metrology a [?].
Identifiers: LCCN 2020037189 (print) | LCCN 2020037190 (ebook) |
ISBN 9781138032613 (v. 1 ; hardback) | ISBN 9781138032620 (v. 2 ; hardback) |
ISBN 9781138033320 (v. 3 ; hardback) | ISBN 9780367649173 (v. 4 ; hardback) |
ISBN 9781138196575 (hardback) | ISBN 9781315389561 (v. 1 ; ebook) |
ISBN 9781003127130 (v. 2 ; ebook) | ISBN 9781315310855 (v. 3 ; ebook) |
ISBN 9781003130123 (v. 4 ; ebook)
Subjects: LCSH: Lasers.
Classification: LCC TK7871.3 .H25 2021 (print) | LCC TK7871.3 (ebook) |
DDC 621.36/6—dc23
LC record available at https://lccn.loc.gov/2020037189
LC ebook record available at https://lccn.loc.gov/2020037190

ISBN: 9781138032613 (hbk)
ISBN: 9780367649692 (pbk)
ISBN: 9781315389561 (ebk)

Typeset in Times
by codeMantra

Contents

Preface ix
Editors xi
Contributors xiii

1. **Laser Principle: Section Introduction** 1
   Richard Shoemaker

2. **Basic Laser Principles** 3
   Christopher C. Davis

3. **Interference and Polarization** 51
   Alan Rogers

4. **Introduction to Numerical Analysis for Laser Systems** 77
   George Lawrence

5. **Optical Cavities: Free-Space Laser Resonators** 97
   Robert C. Eckardt

6. **Optical Cavities: Waveguide Laser Resonators** 121
   Chris Hill

7. **Nonlinear Optics** 135
   Orad Reshef and Robert W. Boyd

8. **Laser Beam Control** 153
   Jacky Byatt

9. **Optical Detection and Noise** 171
   Gerald Buller and Jason Smith

10. **Laser Safety** 189
    J. Michael Green and Karl Schulmeister

11. **Optical Components: Section Introduction** 205
    Julian Jones

12. **Optical Components** 207
    Leo H. J. F. Beckmann

13. **Optical Control Elements** 219
    Alan Greenaway

14. **Adaptive Optics and Phase Conjugate Reflectors** 227
    Michael J. Damzen and Carl Paterson

15. **Opto-mechanical Parts** 233
    Frank Luecke

16. **Optical Pulse Generation: Section Introduction** 239
    Clive Ireland

17. **Quasi-cw and Modulated Beams** 241
    K. Washio

18. **Short Pulses** 247
    Andreas Ostendorf

19. **Ultrashort Pulses** 259
    Derryck T. Reid

20. **Mode-locking Techniques and Principles** 289
    Rüdiger Paschotta

21. **Attosecond Metrology** 307
    Pierre Agostini, Andrew J. Piper, and Louis F. DiMauro

22. **Chirped Pulse Amplification** 321
    Donna Strickland

23. **Optical Parametric Devices** 331
    M. Ebrahimzadeh

24. **Optical Parametric Chirped-Pulse Amplification (OPCPA)** 363
    László Veisz

25. **Laser Beam Delivery: Section Introduction** 383
    Julian Jones

26. **Basic Principles** 385
    D. P. Hand

27. **Free-space Optics** 391
    Leo H. J. F. Beckmann

28. **Optical Waveguide Theory** 417
    George Stewart

29. **Fibre Optic Beam Delivery** 435
    D. P. Hand

30. **Positioning and Scanning Systems** 445
    Jürgen Koch

31. **Laser Beam Measurement: Section Introduction** 459
    Julian Jones

32. **Beam Propagation** 461
    B. A. Ward

33. **Laser Beam Management Detectors** 467
    Alexander O. Goushcha and Bernd Tabbert

34. **Laser Energy and Power Measurement** 479
    Robert K. Tyson

35. Irradiance and Phase Distribution Measurement 483
  B. Schäfer

36. The Measurement of Ultrashort Laser Pulses 487
  *Rick Trebino, Rana Jafari, Peeter Piksarv, Pamela Bowlan, Heli Valtna-Lukner, Peeter Saari,
  Zhe Guang, and Günter Steinmeyer*

Index 537

Preface

This updated Handbook comes at the time when the world just celebrated the 60th anniversary of the laser. Compared to most fields in science and technology, the laser is still a relatively young one, but its developments have been astonishing. Today, hardly any area of modern life is left untouched by lasers, so it is almost impossible to provide a complete account of this subject.

As challenging as it is, this updated Handbook attempts to provide a comprehensive coverage on modern laser technology and applications, including recent advancements and state-of-the-art research and developments. The main goal of developing this Handbook is to provide both an overview and details of ever-expanding technologies and applications in lasers.

We want this Handbook to be useful for both newcomers and experts in lasers. To meet these goals, the chapters in this Handbook are typically developed in a style that does not require advanced mathematical tools. On the other hand, they are written by the experts in each area so that the most important concepts and developments are covered.

The first edition of the Handbook was released in 2003. It has been hugely popular and ranked as one of the top ten most referenced materials by the publisher. Eighteen years later, although a relatively short period for many more established scientific fields, the Handbook has become outdated, and an update is overdue. The rapid changes in lasers are certainly reinforced by my own experience of teaching and researching the subject in the Institute of Optics at University of Rochester. Flipping through my old lecture notes on lasers, I am often amazed at how much progress we have witnessed in this field over the years.

I am indebted to the editors of the first edition, Colin Webb and Julian Jones, who brought this original Handbook into existence. When I was asked to take over this second edition, it laid before me a daunting task of how to rejuvenate the Handbook while keeping its original flavour. Since many of the fundamental principles of the laser are well established, we tried to honour the original authors by keeping the chapters on fundamental concepts where possible. If a revision is needed, we usually started by asking the original authors for the revision but if impossible, we brought in new authors to revise these chapters.

As the laser shines in modern applications, we added a large number of new chapters reflecting the most recent advancements in laser technologies. Throughout the Handbook, entirely new sections were added, including sections on materials processing, laser spectroscopy and lasers in imaging and communications. Nearly all chapters in these sections are either entirely new or substantially revised. On the other hand, some of the topics previously included have seen dwindling relevance today. We had to make the hard decision to let go of some of these outdated chapters from the first edition. Despite these deletions, this new Handbook still grows significantly from the original three volumes to the current four volumes.

Bringing this large project to its conclusion is the collective efforts of many individuals. It began with the encouragement and guidance of Lu Han, the then managing editor of this Handbook. I know how much Lu cared about this project. I still remember an initial phone call with Lu, we finished it at a late afternoon past 5 p.m. Over the phone, I was told that I would receive the first edition of this Handbook. To my surprise, I had the handbooks in my hand the next morning. At CRC press, this project was later passed onto Carolina Antunes and finally to Lara Spieker, who has been essential in bringing this project to its conclusion.

Many people have provided me with indispensable help. My co-editor, Subhash Chandra Singh, at the University of Rochester, helped chart the layout of this new edition and worked along with me throughout this project. Ying Zhang, who was a senior editor at Changchun Institute of Optics, Fine Mechanics, and Physics (CIOMP) in China, spent a half year with us in Rochester, where his years of professional editorial experience helped move this project forward significantly. Lastly, my thanks go to Pavel Redkin of CIOMP, who made significant contributions in communicating with the chapter authors and guiding them throughout the project. Additionally, my appreciation goes to Kai Davies, Sandeep K. Maurya, Xin Wei, and Wenting Sun for their help in this Handbook project.

Chunlei Guo
Editor-in-Chief
University of Rochester

Editors

Chunlei Guo is a Professor in The Institute of Optics and Physics at the University of Rochester. Before joining the Rochester faculty in 2001, he earned a Ph.D. in Physics from the University of Connecticut and did his postdoctoral training at Los Alamos National Laboratory. His research is in studying femtosecond laser interactions with matter, spanning from atoms and molecules to solid materials. His research at University of Rochester has led to the discoveries of a range of highly functionalized materials through femtosecond laser processing, including the so-called black and colored metals and superhydrophillic and superhydrophobic surfaces. These innovations may find a broad range of applications, and have also been extensively featured by the media, including multiple New York Times articles. Lately, he devoted a significant amount of efforts to developing technologies for global sanitation by working with the Bill & Melinda Gates Foundation. Through this mission, he visited Africa multiple times to understand humanitarian issues. To further expand global collaboration under the Gates project, he helped establish an international laboratory at Changchun Institute of Optics, Fine Mechanics, and Physics in China. He is a Fellow of the American Physical Society, Optical Society of America, and International Academy of Photonics & Laser Engineering. He has authored about 300 referred journal articles.

Subhash Chandra Singh is a scientist at the Institute of Optics, University of Rochester and an Associate Professor at Changchun Institute of Optics, Fine Mechanics, and Physics. Dr. Singh earned a Ph.D. in Physics from University of Allahabad, India in 2009. Prior to working with the Guo Lab, he was IRCSET-EMPOWER Postdoctoral Research Fellow at Dublin City University, Ireland for 2 years and a DST-SERB Young Scientist at University of Allahabad for 3 years. He has more than 10 years of research experience in the fields of laser-matter interaction, plasma, nanomaterial processing, spectroscopy, energy applications, plasmonics, and photonics. He has published more than 100 research articles in reputable refereed journals and conference proceedings. His past editor experience includes serving as the main editor for Wiley-VCH book *Nanomaterials: Processing and Characterization with Lasers* and guest editor for special issues of a number of journals.

Contributors

Charles Adams
Pierre Agostini
Markus-Christian Amann
S. Anders
A.J. Annala
Graham Appleby
Sergey Babin
B.D. Barmashenko
Norman P. Barnes
G.P. Barwood
T.T. Basiev
Gary Beane
Leo H.J.F. Beckmann
David Binks
Peter Blood
Brendan S. Brown
Stephen Brown
Georges Boulon
Pamela Bowlan
Stephen Bown
Ian W. Boyd
Robert W. Boyd
Igor Bufetov
Gerald Buller
Jacky Byatt
Cheng Chen
Douglas B. Chrisey
Christophe Codemard
Michael Copeland
Kevin Cordina
Geoffrey Cranch
Daoxin Dai
Michael J. Damzen
Bill Davies
Christopher Davis
Tuphan Devkota
Evgeny Dianov
Louis F. DiMauro
Malcolm Dunn
Frank Duschek
Niloy K. Dutta
Robert C. Eckardt
Qiang Fang
Daniel Farkas
Maria Farsari
Lea Fellner
Peter Fendel
Sergei Firstov
Jens Flugge
Alan Fry
Shijie Fu
K Gao

Xinya Gao
John O. Gerguis
P Gill
T.C. Sabari Girisun
E. Gornik
Alexander O. Goushcha
J. Michael Green
Alan Greenaway
Philippe Grelu
Karin Grünewald
Zhe Guang
Tony Gutierrez
Harald Haas
Denis Hall
Neil Halliwell
Byoung S. Ham
Yu Han
D. P. Hand
Gregory V. Hartland
Hamid Hemmati
Chris Hill
Jinxin Huang
Y. Huang
Ifan Hughes
C. Indumathi
Clive Ireland
Mohamed Islim
Steven Jacques
Rana Jafari
E. Duco Jansen
Y. Jiang
Julian Jones
Haruhisa Kato
Hitoshi Kawaguchi
Diaa Khalil
Terence A. King
Randall Knize
Jürgen Koch
Shimon Kolkowitz
K. Naga Krishnakanth
Ashok Krishnamoorthy
Stefanie Kroker
Sean Lanigan
George Lawrence
Gary Lewis
Qiang Li
Jinyang Liang
Wang Lihong
Thomas Lippert
Frank Luecke
J.I. Mackenzie
Mohammad Malekzadeh

John Marsh
Robert Martin
Leonardo Mastropasqua
Terry McKee
Vasileia Melissinaki
Mikhail Melkumov
V.A. Mikhailov
S.B. Mirov
Seyedeh Zahra Mortazavi
Harry Moseley
Peter Moulton
J.B. Murphy
Philip Nash
Deepak Ranjan Nayak
Beat Neuenschwander
P.G. O'Shea
Shinji Okazaki
Haitham Omran
Andreas Ostendorf
Taiichi Otsuji
Mahesh Pai
Daniel Palanker
Alexandra Palla Papavlu
Harold V. Parks
Parviz Parvin
Rüdiger Paschotta
Carl Paterson
Michael Patterson
Stephen A. Payne
Fabienne Pellé
Dirk Petring
Peeter Piksarv
Andrew J. Piper
John Powell
R.C. Powell
Wilhelm Prettl
Venkata S. Puli
Gregory J. Quarles
Soma Venugopal Rao
Pavel Redkin
Derryck T. Reid
Orad Reshef
Jorge J. Rocca
Jose Rodriguez
Alan Rogers
Parham Rohani
Jannick Rolland
S. Rosenwaks
Khokan Roy
Peeter Saari
Yasser M. Sabry
B. Schäfer
Harald Schnatz
Karl Schulmeister

Lee Sentman
I.A. Shcherbakov
D.P. Shepherd
Bei Shi
Wei Shi
Yaocheng Shi
Richard Shoemaker
Sanchita Sil
William T. Silfvast
Jason Smith
Martin Sparkes
W.M. Steen
Günter Steinmeyer
George Stewart
G. Strasser
Donna Strickland
Binod Subedi
Mark T. Swihart
Sándor Szatmári
Bernd Tabbert
Lance Thomas
David Titterton
Mary Tobin
Rick Trebino
Cameron Tropea
Lisa Tsufura
Robert K. Tyson
Siva Umapathi
Peter Unger
Heli Valtna-Lukner
Z. Vangelatos
Peter P. Vasil'ev
László Veisz
Wachsmann-Hogiu
B.A. Ward
K. Washio
Colin Webb
Alan D. White
Ian White
Adam Whybrew
Garth Williams
Brian C. Wilson
Peter J. Winzer
W.J. Wittman
Peng Xi
Yiwei Xie
Jianjun Yang
Lianxiang Yang
Jun Ye
M. Ebrahimzadeh
A.I. Zagumennyi
Michalis N. Zervas
Boris Zhdanov
P.G. Zverev

1

Laser Principle: Section Introduction

Richard Shoemaker

Since the operation of the first laser in 1960, literally hundreds of different laser varieties have been developed and the light that they produce is being used in thousands of applications ranging from precision measurement to materials processing to medicine. Underlying all these varieties, however, is a small set of basic physical principles upon which laser operation, laser beam propagation and the interaction of laser beams with matter depend. The explanation of these principles is the subject of this section. Chapter 2 begins by explaining the basic physics that allows one to construct optical amplifiers, including discussions of energy levels and level populations, stimulated and spontaneous emission, optical lineshapes and gain saturation. It then discusses the principles that allow an optical amplifier to be turned into a laser (i.e. an optical oscillator) by the addition of feedback in the form of an optical resonator. The article closes with a discussion of the physics that determines the linewidth, coherence properties and power of the laser output. The frequency and spatial distribution of a beam produced by a laser are largely determined by the laser resonator, and as a result, an understanding of optical resonators and their modes is key to understanding the properties of laser beams.

Many applications of lasers rely upon the fact that the light produced by most lasers is coherent and thus can exhibit strong interference effects. Usually, the light is also highly polarized, and this polarization can be utilized to good effect in many other applications. Chapter 3 covers the basic principles of coherent wave interference, Mach–Zehnder interferometers, Michelson interferometers, Fabry–Pérot interferometers and partial coherence. The discussion then moves to polarization concepts including the polarization ellipse, crystal optics, retarding wave-plates, polarizing prisms, circular birefringence, polarization analysis, and applications of polarization optics, including electro-optic and magneto-optic effects.

Chapter 4 discusses the numerical modelling of laser beam propagation within and outside the laser resonator. These models are important tools used by optical engineers in designing laser systems and laser applications. The article begins by discussing the representation of optical beams for numerical work, followed by descriptions of specific methods for handling beam propagation: the split step method, finite difference propagation and angular spectrum propagation. Numerical calculations of propagation in homogeneous media including issues of sampling and propagation control are then presented, followed by an elementary discussion of propagation through gain and non-linear media including the use of Beer's law, rate equations, the Franz–Nodvik solution, refractive index effects and the inclusion of spontaneous emission. The last section of the article discusses the selection and validation of laser-modelling software packages. Excluded from the discussion of numerical modelling in Chapter 4 is a treatment of numerical modelling for semiconductor lasers. Although obviously important, these lasers are by far the most difficult laser systems to model, and the development of software that can do such modelling is currently an active research topic at a number of universities and companies. The essential physics needed to model these lasers properly includes the complex non-linear interactions between the multi-component electron-hole plasma that produces the laser radiation, the intense laser radiation within the waveguide resonator, and the several layers of semiconductor materials that form the laser. As a consequence, the gain and refractive index cannot be represented in a parametric form using the laser rate equations discussed in Chapter 4. The gain peak and the gain lineshape both change on the fly with changes in internal carrier density and temperature, and electrical and heat transport from the external contacts into the active region of the p-i-n structure also critically influence the optical properties by modifying the optical gain and refractive index.

Chapter 5 presents the principles of Gaussian beams, stable resonators, stable resonator axial and transverse modes, beam quality, mode matching, plane parallel resonators, unstable resonators and frequency selection. Chapter 6 supplements this material by discussing the principles governing hollow waveguide optical resonators, widely used for carbon dioxide lasers. The purpose for which most lasers are purchased or built is to make use of the laser beam that it produces. In many applications, making effective use of this beam requires that it be properly controlled in time (e.g. pulsed lasers), space (e.g. beam focusing), frequency or amplitude.

Chapter 7 forms the theoretical basis for the non-linear optical phenomena.

Chapter 8 covers the principles used in laser beam control, including beam focusing with lenses, beam transmission through apertures, the M value, transverse and axial mode control, frequency stabilization, frequency selection, astigmatic beam shaping, Q-switching, mode locking, cavity dumping and spatial filtering. One of the key features that make lasers so useful is their ability to produce optical fields having very high intensity. When these fields interact with matter, a great variety of non-linear optical effects can occur. Some of these, such as optical frequency doubling, can be very useful, while others, such as optical damage, cause problems that need to be controlled. Some of the most economically important applications of lasers rely upon our ability to confine laser beams within optical waveguides where they can be modulated, amplified, split, switched and recombined in ways similar to those used to manipulate currents in electronic circuits. These capabilities together with the ability to transmit the light over long distances through optical fibres with very low loss make optical communication systems possible. Chapter 7 covers the theory of optical waveguides and fibres. The chapter first introduces the primary types of waveguides and their fabrication, and then presents the basic theory of planar and 2D waveguides. The second part of the article turns to optical fibres, beginning with basic fibre propagation theory and then turning to a variety of important propagation effects including attenuation, dispersion, birefringence and polarization, non-linear effects and mode coupling. Many applications of lasers would be severely limited or impossible if we were unable to accurately and sensitively detect the energy or intensity of the beam with some type of optical detector.

Chapter 9 presents basic descriptions and operating principles of photomultipliers, p-n photodiodes, Schottky and avalanche diode detectors, photoconductive detectors and thermal detectors, including bolometers and pyroelectric detectors. The final sections of the article discuss noise in photodetection, including detector figures of merit, noise sources, and methods of minimizing detector noise.

Chapter 10 describes the principles of laser.

2

Basic Laser Principles[1]

Christopher C. Davis

CONTENTS

2.1 Introduction 4
2.2 The Amplifier–Oscillator Connection 4
2.3 The Energy Levels of Atoms, Molecules and Condensed Matter 5
2.4 Spontaneous and Stimulated Transitions 6
    2.4.1 Spontaneous Emission 6
    2.4.2 The Lineshape Function 7
    2.4.3 Stimulated Emission 8
    2.4.4 The Relation between Energy Density and Intensity 8
    2.4.5 Stimulated Absorption 10
2.5 Transitions between Energy Levels for a Collection of Particles in Thermal Equilibrium 11
2.6 The Relationship between the Einstein A and B Coefficients 11
    2.6.1 The Effect of Level Degeneracy 12
    2.6.2 Ratio of Spontaneous and Stimulated Transitions 13
2.7 Optical Frequency Amplifiers and Line Broadening 14
    2.7.1 Homogeneous Line Broadening 14
    2.7.2 Natural Broadening 14
    2.7.3 Other Homogeneous Broadening Mechanisms 15
2.8 Inhomogeneous Broadening 16
    2.8.1 Doppler Broadening 16
    2.8.2 Energy Bands in Condensed Matter 18
2.9 Optical Frequency Amplification with a Homogeneously Broadened Transition 18
    2.9.1 The Stimulated Emission Rate in a Homogeneously Broadened System 21
    2.9.2 Optical Frequency Amplification with Inhomogeneous Broadening Included 21
2.10 Optical Frequency Oscillation—Saturation 22
    2.10.1 Homogeneous Systems 22
    2.10.2 Inhomogeneous Systems 24
2.11 Power Output from a Laser Amplifier 28
2.12 The Electron Oscillator Model of a Radiative Transition 28
    2.12.1 The Connection between the Complex Susceptibility, Gain and Absorption 30
    2.12.2 The Classical Oscillator Explanation for Stimulated Emission 31
2.13 From Amplifier to Oscillator—the Feedback Structure 33
2.14 Optical Resonators Containing an Amplifying Media 34
2.15 The Oscillation Frequency 37
    2.15.1 Multi-mode Laser Oscillation 37
    2.15.2 Mode Beating 42
2.16 The Characteristics of Laser Radiation 43
    2.16.1 Laser Modes 44
    2.16.2 Beam Divergence 46
    2.16.3 Linewidth of Laser Radiation 46
2.17 Coherence Properties 47
    2.17.1 Temporal Coherence 47

---

[1] This chapter is based on a longer and more detailed exposition of these principles in Ref. [1].

2.17.2 Laser Speckle 47
2.17.3 Spatial Coherence..48
2.18 The Power Output of a Laser 48
2.18.1 Optimum Coupling..49
Acknowledgement 50
References..50

2.1 Introduction

A laser is an oscillator that operates at *optical* frequencies. These frequencies of operation lie within a spectral region that extends from the very far infrared to the *vacuum ultraviolet* (VUV) or soft X-ray region. At the lowest frequencies at which they operate, lasers overlap with the frequency coverage of masers, to which they are closely related, and millimetre wave sources using solid-state or vacuum tube electronics, such as TRAPATT, IMPATT and Gunn diodes, klystrons, gyroklystrons and travelling wave tube oscillators, whose principles of operation are quite different [1]. In common with electronic circuit oscillators, a laser is constructed using an amplifier with an appropriate amount of positive feedback. The acronym LASER, which stands for *light amplification by stimulated emission of radiation*, is in reality, therefore, a slight misnomer.

Of central importance are the fundamental processes that allow amplification at optical frequencies to be obtained. These processes use the energy that is involved when the discrete particles making up matter, specifically atoms, ions and molecules, move from one energy level to another. These particles have energies that can have only certain discrete values. This discreteness, or *quantization*, of energy is intimately connected with the duality that exists in nature. Light sometimes behaves as if it were a wave and in other circumstances it behaves as if it were composed of particles. These particles, called *photons*, carry the discrete packets of energy associated with the wave. For light of frequency ν, the energy of each photon is $h\nu$, where h is Planck's constant—6.6×10^{-34} J s.

The energy $h\nu$ is the *quantum* of energy associated with the frequency ν. On an atomic scale, the amplification of light within a laser involves the emission of such quanta. Thus, the term *quantum electronics* is often used to describe the branch of science that has grown from the development of the maser in 1954 and the laser in 1960. The widespread use of lasers and other optical devices in practical applications such as communications, signal processing, imaging and data storage has also led to the use of the term *photonics*. Whereas electronics uses electrons in various devices to perform analogue and digital functions, photonics aims to replace the electrons with photons. Because photons have zero mass, do not interact with each other to any significant extent and travel at the speed of light, photonic devices promise small size and high speed.

2.2 The Amplifier–Oscillator Connection

In 'conventional' electronics, whereby the word 'conventional' for the present purposes, we mean frequencies where solid-state devices such as transistors or diodes will operate, say below 10^{11} Hz, an oscillator is conveniently constructed by applying an appropriate amount of positive feedback to an amplifier. Such an arrangement is shown schematically in Figure 2.1. The input and output voltages of the amplifier are V_i and V_o, respectively. The voltage gain of the amplifier is A_0 where, in the absence of feedback, $A_0 = V_o/V_i$. The feedback circuit returns part of the amplifier output to the input. The feedback factor $\beta = |\beta|e^{j\phi}$ is, in general, a complex number with amplitude $|\beta| \leq 1$ and phase ϕ.

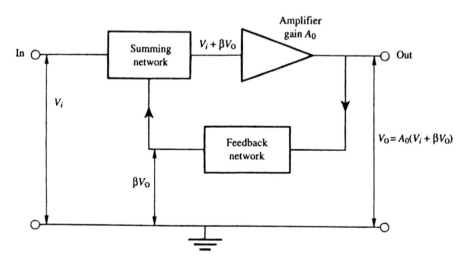

FIGURE 2.1 Circuit diagram of a simple amplifier with feedback.

Basic Laser Principles

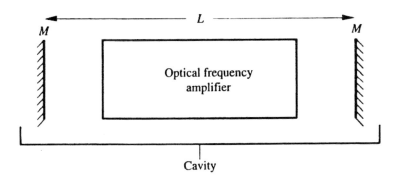

FIGURE 2.2 Schematic diagram of a basic laser structure with an amplifying medium in a resonant cavity formed by two feedback mirrors, M.

$$V_o = A_0(V_i + \beta V_o) \quad (2.1)$$

so

$$V_o = \frac{A_0 V_i}{1 - \beta A_0} \quad (2.2)$$

and the overall voltage gain is

$$A = \frac{A_0}{1 - \beta A_0}. \quad (2.3)$$

As βA_0 approaches +1, the overall gain of the circuit goes to infinity, and the circuit would generate a finite output without any input. In practice, electrical 'noise', which is a random oscillatory voltage present to a greater or lesser extent in all electrical components in any amplifier system, provides a finite input. Because βA_0 is generally a function of frequency, the condition $\beta A_0 = +1$ is usually satisfied only at one frequency. The circuit oscillates at this frequency by amplifying the noise at this frequency present at its input. The output does not grow infinitely large, because as the signal grows, A_0 falls—this process is called saturation. This phenomenon is fundamental to all oscillator systems. A laser (or maser) is an optical (microwave) frequency oscillator constructed from an optical (microwave) frequency amplifier with positive feedback, shown schematically in Figure 2.2. Light waves, which are amplified in passing through the amplifier, are returned through the amplifier by the reflectors and grow in intensity, but this intensity growth does not continue indefinitely because the amplifier saturates. The arrangement of mirrors (and sometimes other components) that provides the feedback is generally referred to as the laser cavity or resonator.

The characteristics of the device consisting of amplifying medium and resonator will be covered later; for the moment we must concern ourselves with the problem of how to construct an amplifier at optical frequencies, which range from 10^{11} Hz to beyond 10^{16} Hz. The operating frequencies of masers overlap this frequency range at the low-frequency end; the fundamental difference between the two devices is primarily one of scale. If the length of the resonant cavity which provides feedback is L, then for $L \gg \lambda$, where λ is the wavelength at which oscillation occurs, we generally have a laser: for $L \simeq \lambda$, we usually have a maser, although the development of *microlasers*, which have small cavity lengths, has removed this easy way of distinguishing lasers from masers.

2.3 The Energy Levels of Atoms, Molecules and Condensed Matter

All particles in nature have distinct states[2] that they can occupy. These states in general have different energies, although it is possible for particles in different states to have the same energy. The term 'energy level' is used to describe a particle with a specific, distinct energy, without implying any particular information about its (quantum) state. The lowest energy state in which a particle is stable is called the *ground state*. All higher energy states are called *excited* states. Excited states are intrinsically unstable, and a particle occupying one will eventually lose energy and fall to lower energy states. When a particle falls from a higher energy state to a lower, energy is conserved. The energy ΔE lost by the particle can be emitted as a photon with energy $h\nu = \Delta E$: this is radiative energy loss. The particle can also lose energy *non-radiatively*, in which case the energy is dissipated into heating. Atomic systems have only electronic states, which in the simple Bohr model of the atom correspond to different configurations of electron orbits. The types of energy state that exist in a molecular system are more varied and include electronic, vibrational and rotational states.

In a molecule, changes in the inter-nuclear separation of the constituent atoms give rise to *vibrational* energy states, which have quantized energies. The various characteristic vibrational motions of a molecule are called its normal modes, which for a molecule with N atoms number $3N-6$, unless the molecule is linear, in which case they number $3N-5$. The quantized energies of a normal mode can be written as [2]

$$E_{vib} \simeq \left(n + \frac{1}{2}\right)h\nu_{vib} \quad (2.4)$$

and form a ladder of (almost) equally spaced energy levels.

Molecules also have quantized rotational energy levels, whose energies can be written as

$$E_{rot} \simeq BJ(J+1), \quad (2.5)$$

where B is a rotational energy constant, and J is called the rotational quantum number. The overall energy state of a

---

[2] The term 'state' in quantum mechanics corresponds to a configuration with a particular 'state function', which often corresponds to a specific set of quantum numbers that identify the state.

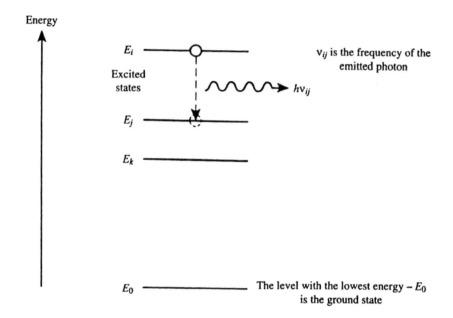

FIGURE 2.3 Simple energy-level diagram for a particle.

molecule thus has electronic, vibrational and rotational components. A molecule in a particular combination of electronic and vibrational states is described as being in a *vibronic* state. A state with a specific combination of vibrational and rotational energies would be described as being in a *vibrot* state.

As a rough rule of thumb, transitions between different vibronic states where the electronic state changes lie in the visible spectrum with energy spacings[3] ~20 000 cm^{-1} and correspond to an energy spacing of $3 \times 10^{10} h/\lambda$ J. Transitions between vibrot states where the electronic energy does not change but the vibrational state does are typically ~1000 cm^{-1}. Transitions between different rotational states where the electronic and vibrational states do not change are typically ~100 cm^{-1}. In practical terms, vibrational transitions are typically in the 3–20 μm range and rotational transitions are typically in the 50–1000 μm range.

In the gas phase, the energy levels of atoms or molecules are quite sharp and distinct, as shown schematically in Figure 2.3, although we shall see later that even these precise energies are 'broadened'. This broadening occurs for several reasons but perhaps most importantly because of the interactions between neighbouring particles. In condensed matter, whether this be in the solid or liquid state, there are so many particles close to any individual particle of interest, and the inter-particle interactions are strong. Consequently, the allowed energies of particles in the medium occupy broad, continuous ranges of energy called energy 'bands'. The lowest-lying energy band, which is analogous to the ground state of an isolated particle, is called the *valence* band. The next highest band of allowed energies is called the *conduction* band. An energy band can be thought of as the result of very many sharp isolated energy states having their energies 'smeared' out so that they overlap. We will reserve further discussion of the energy bands in condensed matter until a little later and, for the moment, will consider the energy levels of particles as relatively sharp and not strongly influenced by inter-particle interactions.

2.4 Spontaneous and Stimulated Transitions

To build an amplifier that operates at optical frequencies, we use the energy delivered as the particles that constitute the amplifying medium make jumps between their different energy levels. The medium may be gaseous, liquid, a crystalline or glassy solid, an insulating material or a semi-conductor. The particles of the amplifying medium, whether these are atoms, molecules or ions, can occupy only certain discrete energy levels. Consider such a system of energy levels, shown schematically in Figure 2.3, particles can make jumps between these levels in three ways. In the case of an atomic amplifier, these energy jumps involve electrons moving from one energy level to another.

2.4.1 Spontaneous Emission

When a particle spontaneously falls from a higher energy level to a lower one, as shown in Figure 2.4, the emitted photon has frequency

$$v_{ij} = \frac{E_i - E_j}{h}. \qquad (2.6)$$

This photon is emitted in a random direction with arbitrary polarization (except in the presence of magnetic fields but this need not concern us here). The photon carries away momentum $h/\lambda = hv/c$, and the emitting particle (atom, molecule or ion) recoils in the opposite direction. The probability of a spontaneous jump within a small time interval is given quantitatively by the Einstein A coefficient defined by $A_{ij}\Delta t$ = 'probability' of

---

[3] The cm^{-1} unit is often used to describe energy spacings. A transition at wavelength λ (cm) between two levels has an energy spacing characterized by $1/\lambda$ cm^{-1}.

Basic Laser Principles

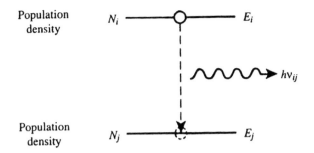

FIGURE 2.4 Representation of the spontaneous emission process for two levels of energy E_i and E_j.

a spontaneous jump from level i to level j during a short time interval $\Delta t \cdot A_{ij}$ has units of s^{-1}. To preserve the concept of $A_{ij}\Delta t$ as a true measure of the probability of a spontaneous emission, the time interval must be chosen so that $A_{ij}\Delta t \ll 1$.

For example, if there are N_i particles per unit volume in level i, then $N_i A_{ij} \Delta t$ make jumps to level j in a short time interval. The total rate at which jumps are made between the two levels is

$$\frac{dN_{ij}}{dt} = -N_i A_{ij}. \tag{2.7}$$

There is a negative sign because the population of level i is decreasing.

Generally a particle can make jumps to more than one lower level, unless it is already in the first (lowest) excited level. The total probability that the particle will make a spontaneous jump to any lower level in a small time interval is $A_i \Delta t$, where

$$A_i = \sum_j A_{ij}. \tag{2.8}$$

The summation runs over all levels j lower in energy than level i. The total rate at which the population of level i changes by spontaneous emission is

$$\frac{dN_i}{dt} = -N_i A_i \tag{2.9}$$

which has the solution

$$N_i = N_i^0 e^{-A_i t} \tag{2.10}$$

where N_i^0 is the population density of level i at time $t = 0$.

The population of level i falls exponentially with time as particles leave that level by spontaneous emission. The time in which the population falls to $1/e$ of its initial value is called the natural lifetime of level i, τ_i, where $\tau_i = 1/A_i$. The magnitude of this lifetime is determined by the actual probabilities of jumps from level i by spontaneous emission. Jumps which are likely to occur are called *allowed* transitions, those which are unlikely are said to be *forbidden*. Allowed transitions in the visible region typically have A_{ij} coefficients in the range 10^6–$10^8 s^{-1}$. Forbidden transitions in this region have A_{ij} coefficients below $10^4 s^{-1}$. These probabilities decrease as the wavelength of the transition increases. Consequently, levels that can decay by allowed transitions in the visible have lifetimes generally shorter than 1 μs; similar forbidden transitions have lifetimes in excess of 10–100 μs. Although no jump turns out to be absolutely forbidden, some jumps are so unlikely that levels whose electrons can only fall to lower levels by such jumps are very long lived. Levels with lifetimes in excess of 1 h have been observed under laboratory conditions. Levels which can only decay slowly, and usually only by forbidden transitions, are said to be *metastable*.

2.4.2 The Lineshape Function

When a particle loses energy spontaneously the emitted radiation is not, as might perhaps be expected, all at the same frequency. Real energy levels are not infinitely sharp; they are smeared out or *broadened*. A particle in a given energy level can actually have any energy within a finite range. The frequency spectrum of the spontaneously emitted radiation is described by the *lineshape function*, $g(\nu_0, \nu)$, where ν_0 is a reference frequency, usually the frequency where $g(\nu_0, \nu)$ has a maximum. The lineshape function is normalized so that

$$\int_{-\infty}^{\infty} g(\nu_0, \nu) d\nu = 1. \tag{2.11}$$

$g(\nu_0, \nu) d\nu$ represents the probability that a photon will be emitted spontaneously in the frequency range $\nu + d\nu$. The lineshape function $g(\nu_0, \nu)$ is a true probability function for the spectrum of emitted radiation and is usually sharply peaked near the frequency ν_0, as shown in Figure 2.5. Since negative frequencies do not exist in reality, the question might properly be asked: 'Why does the integral have a lower limit of minus infinity?' This is done because $g(\nu_0, \nu)$ can be viewed as the Fourier transform of a real function of time, so negative frequencies have to be permitted mathematically. In practice, $g(\nu_0, \nu_0)$ is only of significant value around a large value of ν_0 so

$$\int_0^{\infty} g(\nu_0, \nu) d\nu \simeq 1. \tag{2.12}$$

The amount of radiation emitted spontaneously by a collection of particles can be described quantitatively by their *spectral radiant intensity* $I_e(\nu)$. The units of spectral radiant intensity are watts per hertz per steradian.[4] The total power (watts) emitted in a given frequency interval $d\nu$ is

$$W(\nu) = \int_s I_e(\nu) d\nu \, d\Omega, \tag{2.13}$$

---

[4] The steradian is the unit of solid angle, Ω. The surface of a sphere encompasses a solid angle of 4π steradians.

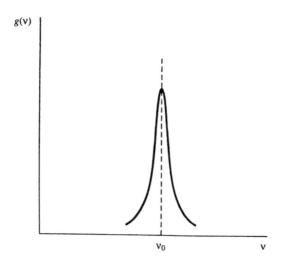

FIGURE 2.5 A lineshape function $g(v_0, v)$.

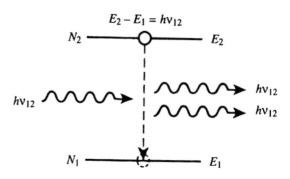

FIGURE 2.6 Representation of the stimulated emission process for two levels of energy E_2 and E_1.

where the integral is taken over a closed surface S surrounding the emitting particles. The total power emitted is

$$W_0 = \int_{-\infty}^{\infty} W(v)\,dv. \quad (2.14)$$

$W(v)$ is closely related to the lineshape function

$$W(v) = W_0 g(v_0, v). \quad (2.15)$$

For a collection of N_i identical particles, the total spontaneously emitted power per frequency interval (Hz) is

$$W(v) = N_i A_i h v g(v). \quad (2.16)$$

Clearly, this power decreases with time if the number of excited particles decreases.

For a plane electromagnetic wave, we can introduce the concept of *intensity*, which has units of W m^{-2}. The intensity is the average amount of energy per second transported across a unit area in the direction of travel of the wave. The spectral distribution of intensity, $I(v)$, is related to the total intensity, I_0, by

$$I(v) = I_0 g(v_0, v). \quad (2.17)$$

Although perfect plane waves do not exist, because such waves would have a unique propagation direction and infinite radiant intensity, they represent a useful, simple idealization. To a very good degree of approximation, we can treat the light from a small source as a plane wave if we are far enough away from the source. The light coming from a star viewed outside the Earth's atmosphere is a good example.[5]

2.4.3 Stimulated Emission

As well as being able to make transitions from a higher level to a lower one by spontaneous emission, particles can also be stimulated to make these jumps by the action of an externally applied radiation field, as shown in Figure 2.6.

The probability of the external radiation field causing stimulated emission depends on its energy density, which is written as $\rho(v)$ and is measured in J m^{-3}Hz^{-1}. The rate for stimulated emissions to occur within a small band of frequencies dv is

$$\frac{dN_2}{dt}(v)\,dv = N_2 B'_{21}(v)\rho(v)\,dv\ s^{-1}m^{-3}, \quad (2.18)$$

where $B'_{21}(v)$ is a function specific to the transition between levels 2 and 1, and N_2 is the number of particles per unit volume in the upper level of the transition. Stimulated emission will occur if the external radiation field contains energy in a frequency range that overlaps the lineshape function. The frequency dependence of $B'_{21}(v)$ is the same as the lineshape function:

$$B'_{21}(v) = B_{21} g(v_0, v). \quad (2.19)$$

B_{21} is called the Einstein B coefficient for stimulated emission. The total rate of change of particle concentration in level 2 by stimulated emission is

$$\begin{aligned}\frac{dN_2}{dt} &= -N_2 \int_{-\infty}^{\infty} B'_{21}(v)\rho(v)\,dv \\ &= -N_2 B_{21} \int_{-\infty}^{\infty} g(v)\rho(v)\,dv.\end{aligned} \quad (2.20)$$

Note that, for the dimensions of both sides of equation (2.20) to balance, B_{21} must have units m^{3}J^{-1}s^{-2}. To evaluate the integral in equation (2.20) we must consider how energy density is related to intensity and varies with frequency.

2.4.4 The Relation between Energy Density and Intensity

The energy density of a radiation field $\rho(v)$ can be most easily related to its spectral intensity by examining the case of a plane electromagnetic wave. In Figure 2.7 a plane wave propagating along carries energy across an area A oriented perpendicular to the direction of propagation. If the intensity of the wave is $I(v)$ W m^{-2}Hz^{-1}, then in one second the energy crossing A occupies a volume cA, where c is the velocity of light in the medium.[6] Clearly,

---

[5] The plane waves coming from a star are distorted by the atmosphere because of density and refractive index fluctuations referred to as *atmospheric turbulence*.

[6] $c = c_0/n$, where c_0 is the velocity of light in a vacuum and n is the *refractive index*.

Basic Laser Principles

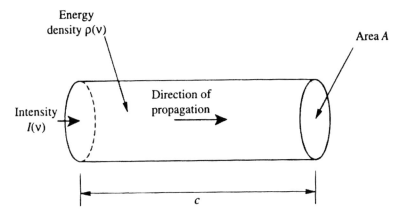

FIGURE 2.7 A volume of space swept through per second by part of a plane wave.

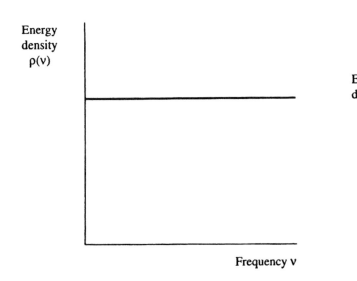

FIGURE 2.8 A 'white' energy density spectrum.

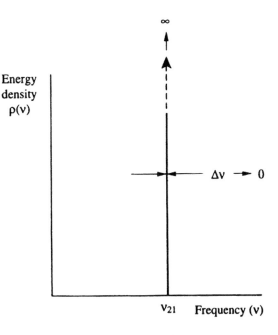

FIGURE 2.9 A monochromatic energy density spectrum.

$$\rho(v) = \frac{I(v)}{c}. \quad (2.21)$$

The energy density of a general radiation field $\rho(v)$ is a function of frequency v. If $\rho(v)$ is independent of frequency, the radiation field is said to be *white*, as shown in Figure 2.8. If the radiation field is *monochromatic* at frequency v_{21}, its spectrum is as shown in Figure 2.9. The ideal *monochromatic* radiation field has a δ-function energy density profile at frequency v_{21}.

$$\rho(v) = \rho_{21}\delta(v - v_0). \quad (2.22)$$

The δ-function has the property

$$\delta(v - v_{21}) = 0 \quad \text{for } v \neq v_{21} \quad (2.23)$$

and

$$\int_{-\infty}^{\infty} \delta(v - v_{21}) \, dv = 1. \quad (2.24)$$

For a general radiation field, the total energy stored per unit volume between frequencies v_1 and v_2 is $\int_{v_1}^{v_2} \rho(v_2) \, dv$.

For a monochromatic radiation field, the total stored energy per unit volume is

$$\int_{-\infty}^{\infty} \rho(v) \, dv = \int_{-\infty}^{\infty} \rho_{21}\delta(v - v_{21}) \, dv = \rho_{21}. \quad (2.25)$$

The rate of stimulated emissions caused by a monochromatic radiation field can be calculated by using equation (2.20) and is given by

$$\frac{dN_2}{dt} = -N_2 B_{21} \int_{-\infty}^{\infty} g(v_0, v)\rho_{21}\delta(v - v_{21}) \, dv$$

$$= -N_2 B_{21} \int_{-\infty}^{\infty} g(v_0, v)\rho_{21}\delta(v - v_{21}) \, dv \quad (2.26)$$

It is very important to note that the rate of stimulated emissions produced by this input monochromatic radiation is directly proportional to the value of the lineshape function at the input frequency. The maximum rate of stimulated emission is produced, all else being equal, if the input radiation is at the line-centre frequency v_0.

If the stimulating radiation field has a spectrum that is broad, we can assume that the energy density $\rho(v)$ is constant over the narrow range of frequencies where $g(v_0, v)$ is significant. In this case, equation (2.20) gives

$$\frac{dN_2}{dt} = -N_2 B_{21} \rho(v) \quad (2.27)$$

where $\rho(v) \simeq \rho(v_0)$ is the energy density in the frequency range where transitions take place.

2.4.5 Stimulated Absorption

As well as making stimulated transitions in a downward direction, particles may make transitions in an upward direction between their energy levels by absorbing energy from an electromagnetic field, as shown in Figure 2.10. The rate of such absorptions and the rate at which particles leave the lower level are $N_1 \rho(v) B_{12} g(v_0, v)$ s^{-1}Hz^{-1}m^{-3}, which yields a result similar to equation (2.20)

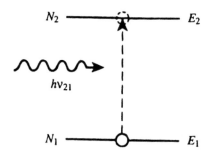

FIGURE 2.10 Representation of the stimulated absorption process for two levels of energy E_1 and E_2.

$$\frac{dN_1}{dt} = -N_1 B_{12} \int_{-\infty}^{\infty} g(v_0, v) \rho(v) \, dv. \quad (2.28)$$

Once again B_{12} is a constant specific to the transition between levels 1 and 2 and is called the Einstein coefficient for stimulated absorption. Here, again, $\rho(v)$ is the energy density of the stimulating field. There is no analogue in the absorption process to spontaneous emission. A particle cannot spontaneously *gain* energy without an external energy supply. Thus, it is unnecessary for us to continue to describe the absorption process as stimulated absorption.

It is interesting to view both stimulated emission and absorption as photon–particle collision processes. In *stimulated* emission, the incident photon produces an identical photon by 'colliding' with the particle in an excited level, as shown in Figure 2.11a. After the stimulated emission process, both photons are travelling in the same direction and with the same polarization as the incident photon originally had. When light is described in particle terms, the polarization state describes the angular motion or spin of individual photons. Left- and right-hand circularly polarized light corresponds in this particle picture to beams of photons that spin clockwise and counterclockwise, respectively, about their direction of propagation. Linearly polarized light corresponds to a beam of photons that has no net angular momentum about an axis parallel to their direction of propagation. In stimulated emission, the stimulated photon has exactly the same frequency as the stimulating photon. In absorption, the incident photon disappears, as shown in Figure 2.11b. In both stimulated emission and absorption, the particle recoils to conserve linear momentum.

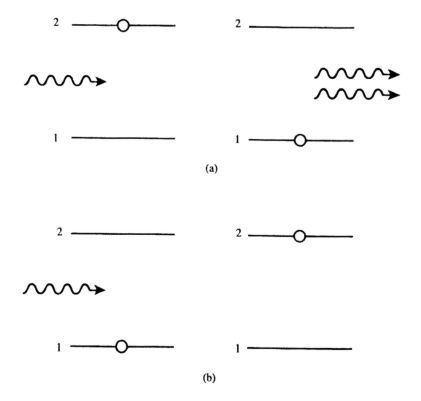

FIGURE 2.11 Photon–particle 'collision' pictures of the stimulated emission and absorption processes: (a) stimulated emission and (b) absorption.

Basic Laser Principles

2.5 Transitions between Energy Levels for a Collection of Particles in Thermal Equilibrium

A collection of particles in thermal equilibrium is described by a common temperature T K. Although the collection of particles is described as being in 'equilibrium' this is a dynamic equilibrium. The processes of spontaneous emission, stimulated emission and absorption continuously occur because there is always a radiation field present. Even though no external radiation field is supplied, the thermal background radiation is always present. This radiation is called *black body* radiation and constantly interacts with each particle. Black-body radiation is so called because of the special characteristics of the radiation emitted and absorbed by a *black body*. Such a body absorbs with 100% efficiency all the radiation falling on it, irrespective of the radiation frequency. A close approximation to a black body (absorber and emitter) is an enclosed cavity containing a small hole. Radiation that enters the hole has very little chance of escaping. If the inside of this cavity is in thermal equilibrium, it must lose as much energy as it absorbs, and the emission from the hole is therefore characteristic of the equilibrium temperature T inside the cavity. Thus, this type of radiation is often called 'thermal' or 'cavity' radiation. Black-body radiation has a spectral distribution as shown in Figure 2.12.

Thermodynamically, the shape of the cavity should not influence the characteristics of the radiation; otherwise, we could make a heat engine by connecting together cavities of different shapes. If, for example, two cavities of different shapes, but at the same temperature, were connected together with a reflective hollow pipe, we could imagine placing filters having different narrow frequency bandpass characteristics in the pipe. Unless the radiation emitted in each elemental frequency band from both cavities was identical, one cavity could be made to heat up and the other cool down, thereby violating the second law of thermodynamics.

In the latter part of the 19th century, experimental measurements of the spectral profile of black-body radiation had already been obtained and the data fitted to an empirical formula. Attempts had been made to explain the form of the data by treating the electromagnetic radiation as a collection of oscillators, each oscillator with its own characteristic frequency; however, these efforts had failed. It was a striking success of the new quantum theory and Planck's hypothesis is that the radiation field was quantized that led to a theoretical description of the energy density of black-body radiation. Planck's hypothesis was that an oscillator at frequency v could only have energies

$$E_{n_v} = \left(n + \frac{1}{2}\right)hv \qquad n = 0, 1, 2, 3 \ldots \qquad (2.29)$$

where the term $\frac{1}{2}hv$ is called the *zero-point energy*. Planck's hypothesis led to a theoretical prediction of the energy density of black-body radiation, which was

$$\rho(v) = \frac{8\pi h v^3}{c^3}\left(\frac{1}{2} + \frac{1}{e^{hv/kT}-1}\right). \qquad (2.30)$$

The term arising from zero-point energy corresponds to energy that cannot be released, so the available stored energy in the field is

$$\rho(v) = \frac{8\pi h v^3}{c^3}\left(\frac{1}{e^{hv/kT}-1}\right). \qquad (2.31)$$

This formula predicts exactly the observed spectral character of black-body radiation. The term $8\pi h v^3/c^3$ is called the *density of states* or the number of modes of the radiation field per frequency interval. The term $\left(\frac{1}{e^{hv/kT}-1}\right)$ is called the *occupation number*. It represents the average number of photons occupying a 'mode' of the radiation field at frequency v.

2.6 The Relationship between the Einstein A and B Coefficients

We can derive a useful relationship between Einstein's A and B coefficients by considering a collection of particles in thermal equilibrium inside a cavity at temperature T. The energy density of the radiation within the cavity is given by

$$\rho(v) = \frac{8\pi h v^3}{c^3}\left(\frac{1}{e^{hv/kT}-1}\right) \qquad (2.32)$$

since in thermal equilibrium, the radiation in the cavity will be black-body radiation. Although real particles in such a cavity possess many energy levels, we can restrict ourselves to considering the dynamic equilibrium between any two of them, as shown in Figure 2.13. The transitions that occur between two such levels as a result of interaction with radiation essentially occur independently of the energy levels of the system, which are not themselves involved in the transition.

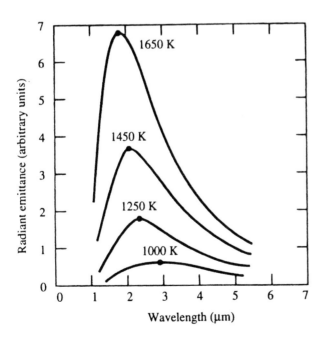

FIGURE 2.12 Spectral distribution of black-body radiation at different temperatures.

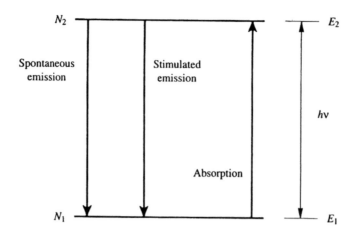

FIGURE 2.13 Radiative processes connecting two energy levels in thermal equilibrium at temperature T.

In thermal equilibrium, the populations N_2 and N_1 of these two levels are constant, so

$$\frac{dN_2}{dt} = \frac{dN_1}{dt} = 0 \quad (2.33)$$

and the rates of transfer between the levels are equal. Since the energy density of a black-body radiation field varies very little over the range of frequencies where transitions between levels 2 and 1 take place, we can use equations (2.7) and (2.27) and write

$$\frac{dN_2}{dt} = -N_2 B_{21}\rho(\nu) - A_{21}N_2 + N_1 B_{12}\rho(\nu). \quad (2.34)$$

Therefore, substituting from equation (2.32)

$$N_2\left[B_{21}\frac{8\pi h\nu^3}{c^3(e^{h\nu/kT}-1)} + A_{21}\right] = N_1\left[B_{12}\frac{8\pi h\nu^3}{c^3\left(e^{h\nu/kT}-1\right)}\right]. \quad (2.35)$$

For a collection of particles that obeys Maxwell–Boltzmann statistics, in thermal equilibrium, energy levels of high energy are less likely to be occupied than levels of low energy. In exact terms, the ratio of the population densities of two levels whose energy difference is $h\nu$ is

$$\frac{N_2}{N_1} = e^{-h\nu/kT}. \quad (2.36)$$

So,

$$\frac{8\pi h\nu^3}{c^3\left(e^{h\nu/kT}-1\right)} = \frac{A_{21}}{B_{12}e^{h\nu/kT}-B_{21}}. \quad (2.37)$$

This equality can only be satisfied if

$$B_{12} = B_{21} \quad (2.38)$$

and

$$\frac{A_{21}}{B_{21}} = \frac{8\pi h\nu^3}{c^3} \quad (2.39)$$

so a single coefficient A_{21} (say) will describe both stimulated emission and absorption. Equations (2.38) and (2.39) are called the *Einstein relations*. The stimulated emission rate is W_{21}, where

$$W_{21} = B_{21}\rho(\nu) = \frac{c^3 A_{21}}{8\pi h\nu^3}\rho(\nu) \quad (2.40)$$

which is proportional to energy density. The spontaneous emission rate is A_{21}, which is independent of external radiation.

Although spontaneous emission would appear to be a different kind of radiative process from stimulated emission, in fact, that is not really the case. Spontaneous emission can be shown to result from the zero-point energy of the radiation field, which was described in equation (2.30).

2.6.1 The Effect of Level Degeneracy

In real systems containing atoms, molecules or ions, it frequently happens that different configurations of the system can have exactly the same energy. If a given energy level corresponds to a number of different arrangements specified by an integer g, we call g the *degeneracy* of the level. We call the separate states of the system with the same energy *sub-levels*. The levels 2 and 1 that we have been considering may consist of a number of degenerate sub-levels, where each sub-level has the same energy, as shown in Figure 2.14, with g_2 sub-levels making up level 2 and g_1 sub-levels making up level 1. For each of the sub-levels of levels 1 and 2 with population n_1, n_2, respectively, the ratio of populations is

$$\frac{n_2}{n_1} = e^{-h\nu/kT} \quad (2.41)$$

and

$$N_1 = g_1 n_1 \quad N_2 = g_2 n_2. \quad (2.42)$$

Therefore,

$$\frac{n_2}{n_1} = \frac{g_1 N_2}{g_1 N_2} \quad (2.43)$$

Basic Laser Principles

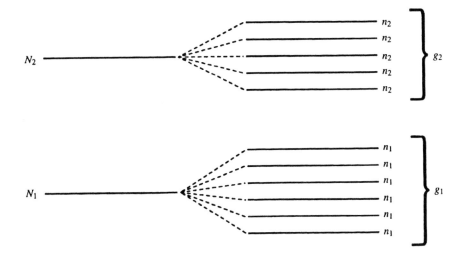

FIGURE 2.14 Two energy levels, each of which has a number of sub-levels of the same energy.

and

$$\frac{N_2}{N_1} = \frac{g_2}{g_1} e^{-h\nu/kt}. \qquad (2.44)$$

From equations (2.35) and (2.44), it follows in this case, where degenerate levels are involved, that the Einstein relations become

$$g_1 B_{12} = g_2 B_{21} \qquad (2.45)$$

and, as before,

$$\frac{A_{21}}{B_{21}} = \frac{8\pi h\nu^3}{c^3}. \qquad (2.46)$$

Note that

$$A_{21} = B_{21} \frac{8\pi h\nu^3}{c^3} = B_{21} \frac{8\pi \nu^2}{c^3} h\nu \qquad (2.47)$$

which can be described as $B_{21} \times$ no. of modes per unit volume per frequency interval $\times$ photon energy.

If there were only one photon in each mode of the radiation field, then the resulting energy density would be

$$\rho(\nu) = \frac{8\pi \nu^2}{c^3} h\nu. \qquad (2.48)$$

The resulting number of stimulated transitions would be

$$W_{21} = B_{21} \frac{8\pi \nu^2}{c^3} h\nu = A_{21} \qquad (2.49)$$

Thus, the number of spontaneous transitions per second is equal to the number of stimulated transitions per second that would take place if there was just one photon excited in each mode.

2.6.2 Ratio of Spontaneous and Stimulated Transitions

It is instructive to examine the relative rates at which spontaneous and stimulated processes occur in a system in equilibrium at temperature T. This ratio is

$$R = \frac{A_{21}}{B_{21} \rho(\nu)}. \qquad (2.50)$$

We choose the $\rho(\nu)$ appropriate to a black-body radiation field, since such radiation is always present, to interact with an excited particle that is contained within an enclosure at temperature T.

$$R = \frac{A}{B\rho(\nu)} = (e^{h\nu/kT} - 1). \qquad (2.51)$$

If we use $T = 300$ K and examine the *microwave region*, $\nu = 10^{10}$ (say), then

$$\frac{h\nu}{kT} = \frac{6.626 \times 10^{-34} \times 10^{10}}{1.38 \times 10^{-23} \times 300} = 1.6 \times 10^{-3}$$

so

$$R = e^{0.0016} - 1 \approx 0.0016$$

and stimulated emission dominates over spontaneous. Particularly, in any microwave laboratory experiment

$$\rho(\nu)_{\text{laboratory created}} > \rho(\nu)_{\text{black-body}}$$

and spontaneous emission is negligible. However, spontaneous emission is still observable as a source of noise—the randomly varying component of the optical signal.

In the *visible region*,

$$\nu \approx 10^{15} \quad \frac{h\nu}{kT} \approx 160 \quad \text{and} \quad A \gg B\rho(\nu)$$

So, in the visible and near-infrared region, spontaneous emission generally dominates unless we can arrange for there to be several photons in a mode. The average number of photons in a mode for black-body radiation is very small in the visible and infrared.

2.7 Optical Frequency Amplifiers and Line Broadening

When an electromagnetic wave propagates through a medium, stimulated emissions increase the intensity of the wave, while absorptions diminish it. The overall intensity will increase if the number of stimulated emissions can be made larger than the number of absorptions. If we can create such a situation, then we have built an amplifier that operates through the mechanism of stimulated emission. This *laser amplifier*, in common with electronic amplifiers, only has useful gain over a particular frequency bandwidth. Its operating frequency range will be determined by the lineshape of the transition and we expect the frequency width of its useful operating range to be of the same order as the width of the lineshape. It is very important to consider how this frequency width is related to the various mechanisms by which transitions between different energy states of a particle are smeared out over a range of frequencies. This *line broadening* affects in a fundamental way not only the frequency bandwidth of the amplifier but also its gain.

A laser amplifier can be turned into an oscillator by supplying an appropriate amount of positive feedback. The level of oscillation will stabilize because the amplifier saturates. Laser amplifiers fall into two categories, which saturate in different ways. The *homogeneously* broadened amplifier consists of a number of amplifying particles that are essentially equivalent while the *inhomogeneously* broadened amplifier contains amplifying particles with a distribution of amplification characteristics.

2.7.1 Homogeneous Line Broadening

All energy states, except the lowest energy state of a particle (the ground state), cover a range of possible energies. This is reflected in the lineshape function, which shows the range of energies for a transition between one or two broadened energy states. At the fundamental level, this smearing out of the energy is caused by the uncertainty involved in the energy measurement process. This gives rise to an intrinsic and unavoidable amount of line broadening called *natural broadening*.

2.7.2 Natural Broadening

This most fundamental source of line broadening arises, as just mentioned, because of uncertainty in the exact energy of the states involved. This uncertainty in measured energy, ΔE, arises from the time uncertainty, Δt, involved in making such a measurement. The product of these uncertainties is [3–5] $\Delta E \Delta t \sim \hbar$.[7] Because an excited particle can only be observed for a time that is of the order of its lifetime, the measurement time uncertainty, Δt, is roughly the same as the lifetime, so

$$\Delta E \sim \hbar/\tau = A\hbar. \quad (2.52)$$

The uncertainty in emitted frequency Δv is $\Delta E/h$, so

$$\Delta v \sim A/2\pi. \quad (2.53)$$

---

[7] $\hbar = h/2\pi$.

When the decay of an excited particle is viewed as a photon emission process, we can think of the particle, initially placed in the excited state at time $t' = 0$, emitting a photon at time t. The distribution of these times t among many such particles varies as $e^{-t/\tau}$. Our knowledge of when the photon is likely to be emitted with respect to the time origin restricts our ability to be sure of its frequency. For example, if a photon is observed at time t and is known to have come from a state with lifetime τ, we know that the *probable* time t' at which the atom became excited was $t - \tau < t' \leq t$. The longer the lifetime of the state is, the greater the uncertainty about when the particle acquired its original excitation becomes. In the limit as $\tau \to \infty$, our knowledge of the time of excitation becomes infinitely uncertain and we can ascribe a very well-defined frequency to the emitted photon; in this limit the electromagnetic waveform emitted by the atom approaches infinite length and is undamped.

We can put this approximate determination of Δv on a more exact basis by considering the exponential intensity decay of a group of excited particles: The decay of each individual excited particle is modelled as an exponentially decaying (damped) co-sinusoidal oscillation, as shown schematically in Figure 2.15. It must be stressed that this is only a convenient way of *picturing* how an excited particle decays and emits electromagnetic radiation. It would not be possible in practice to observe such an electromagnetic field by watching single excited particle decay. We can only observe a classical field by watching many excited particles simultaneously. Within the framework of our model, we can represent the electric field of a decaying excited particle as

$$e(t) = E_0 e^{-t/\tau_c} \cos \omega_0 t. \quad (2.54)$$

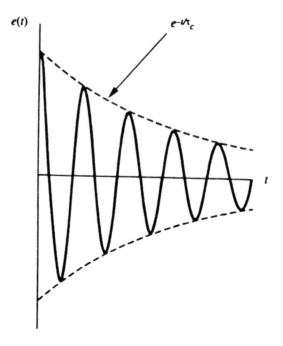

FIGURE 2.15 A damped oscillation used to represent the electric field produced by an excited particle as it decays.

Basic Laser Principles

We need to determine the time constant τ_c that applies to this damped oscillation. The instantaneous intensity $i(t)$ emitted by an individual excited atom is

$$i(t) \propto |e(t)|^2 = E_0^2 e^{-2t/\tau_c} \cos^2 \omega_0 t. \quad (2.55)$$

If we observe many such atoms, the total observed intensity is

$$I(t) = \sum_{\text{particles}} i(t) = \sum_i E_0^2 e^{-2t/\tau_c} \cos^2(\omega_0 t + \varepsilon_i)$$
$$= \sum_i \frac{E_0^2}{2} e^{-2t/\tau_c} [1 + \cos 2(\omega_0 t + \varepsilon_i)] \quad (2.56)$$

where ε_i is the phase of the wave emitted by atom i. In the summation the cosine term gets smeared out because individual atoms are emitting with random phases. So, $I(t) \propto e^{-2t/\tau_c}$. However, we know that $I(t) \propto e^{-2t/\tau}$, where τ is the lifetime of the emitting state, so the time constant τ_c is in fact $= 2\tau$. Thus,

$$e(t) = E_0 e^{-t/2\tau} \cos \omega_0 t. \quad (2.57)$$

To find the frequency distribution of this signal, we take its Fourier transform

$$E(\omega) = \frac{1}{2\pi} \int_{-\infty}^{\infty} e(t) e^{-i\omega t} \, dt \quad (2.58)$$

where

$$e(t) = \tfrac{1}{2} E_0 \left(e^{i(\omega_0 + i/2\tau)t} + e^{-i(\omega_0 - i/2\tau)t} \right) \quad \text{for } t > 0$$
$$= 0 \quad \text{for } t < 0. \quad (2.59)$$

The start of the period of observation at $t = 0$, taken for example at an instant when all the particles are pushed into the excited state, allows the lower limit of integration to be changed to 0, so

$$E(\omega) = \frac{1}{2\pi} \int_0^{\infty} e(t) e^{-i\omega t} \, dt$$
$$= \frac{E_0}{4\pi} \left[\frac{i}{(\omega_0 - \omega + i/2\tau)} - \frac{i}{(\omega_0 - \omega + i/2\tau)} \right] \quad (2.60)$$

The intensity of emitted radiation is

$$I(\omega) \propto |E(\omega)|^2 = E(\omega) E^*(\omega) \propto \frac{1}{(\omega - \omega_0)^2 + (1/2\tau)^2}. \quad (2.61)$$

Or, in terms of ordinary frequency

$$I(\nu) \propto \frac{1}{(\nu - \nu_0)^2 + (1/4\pi\tau)^2}. \quad (2.62)$$

The full width at half maximum height (FWHM) of this function is found from the half-intensity points of $I(\nu)$, which occur at frequencies $\nu_{\pm\frac{1}{2}}$ as shown in Figure 2.16. This occurs where

$$\left(\frac{1}{4\pi\tau}\right)^2 = (\nu_{\pm\frac{1}{2}} - \nu_0)^2. \quad (2.63)$$

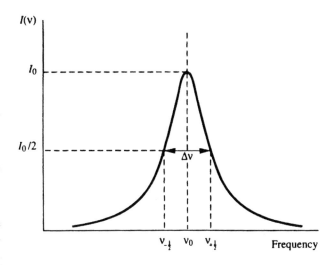

FIGURE 2.16 Lorentzian lineshape function for natural broadening.

The FWHM is $\Delta\nu = \nu_{+\frac{1}{2}} - \nu_{-\frac{1}{2}}$, which gives[8]

$$\Delta\nu = \frac{1}{2\pi\tau} = \frac{A}{2\pi} \quad (2.64)$$

So, from equation (2.62),

$$I(\nu) \propto \frac{1}{(\nu - \nu_0)^2 + (\Delta\nu/2)^2}. \quad (2.65)$$

The normalized form of this function is the lineshape function for natural broadening:

$$g(\nu_0, \nu)_N = \frac{(2/\pi\Delta\nu)}{1 + [2(\nu - \nu_0)/\Delta\nu]^2}. \quad (2.66)$$

This type of function is called a Lorentzian. Since natural broadening is the same for each particle, it is said to be a *homogeneous* broadening mechanism.

2.7.3 Other Homogeneous Broadening Mechanisms

Besides natural broadening other mechanisms of homogeneous broadening exist, for example:

i. In a crystal, the constituent particles of the lattice are in constant vibrational motion. This collective vibration can be treated as being equivalent to sound waves bouncing around inside the crystal. These sound waves, just like electromagnetic waves, can only carry energy in quantized amounts. These packets of acoustic energy are called *phonons* and are analogous in many ways to photons. The principal differences between them are that phonons travel at the speed of sound and can only exist in a material medium. Collisions of phonons with the particles of

---
[8] A more exact treatment gives $\Delta\nu = (A_1 + A_2)/2\pi$ where A_2 and A_1 are the Einstein coefficients of the upper and lower levels of the transition.

the lattice perturb the phase of any excited, emitting particles present. This type of collision, which does not abruptly terminate the lifetime of the particle in its emitting state, is called a *soft* collision.

ii. By *pressure* broadening, particularly in the gaseous and liquid phases, interaction of an emitting particle with its neighbours causes perturbation of its emitting frequency and subsequent broadening of the transition. This interaction may arise in a number of ways:

a. Collisions with neutral particles, which may be *soft* or *hard*. A hard collision causes abrupt decay of the emitting species.

b. Collisions with charged particles. These collisions need not be very direct but may involve a very small interaction that occurs when the charged particle passes relatively near, but perhaps as far as several tens of atomic diameters away from, the excited particle. In any case, the relative motion of the charged and excited particles leads to a time-varying electric field at the excited particle that perturbs its energy states. This general effect in which an external electric field perturbs the energy levels of an atom (molecule or ion) is called the *Stark* effect; hence, line broadening caused by charged particles (ions or electrons) is called Stark broadening.

c. By van der Waals and resonance interactions (usually small effects). Resonance interactions occur when an excited particle can easily exchange energy with like neighbours; the effect is most important for transitions involving the ground state since, in this case, there are generally many particles near an excited particle for which the possibility of energy exchange exists. Broadening occurs because the *possibility* of energy exchange exists, not because an actual emission/reabsorption process occurs.

2.8 Inhomogeneous Broadening

When the environment or properties of particles in an emitting sample are non-identical, *inhomogeneous* broadening can occur. In this type of broadening, the shifts and perturbations of emission frequencies differ from particle to particle.

For example, in a real crystal, the presence of imperfections and impurities in the crystal structure alters the physical environment of atoms from one lattice site to another. The random distribution of lattice point environments leads to a distribution of particles whose centre frequencies are shifted in a random way throughout the crystal.

2.8.1 Doppler Broadening

In a gas, the random distribution of particle velocities leads to a distribution in the emission centre frequencies of different emitting particles seen by a stationary observer. For an atom whose component of velocity towards the observer is v_x, the observed frequency of the transition, whose stationary centre frequency is v_0, is

$$v = v_0 + \frac{v_x}{c} v_0 \qquad (2.67)$$

where c is the velocity of light in the gas.

The Maxwell–Boltzmann distribution of atomic velocities for particles of mass M at absolute temperature T is [6,7]

$$f(v_x, v_y, v_z) = \left(\frac{M}{2\pi kT}\right)^{3/2} \exp\left[-\frac{M}{2kT}(v_x^2, v_y^2, v_z^2)\right]. \qquad (2.68)$$

The number of particles per unit volume that have velocities simultaneously in the range $v_x \to v_x + dv_x$, $v_y \to v_y + dv_y$, $v_z \to v_z + dv_z$ is $Nf(v_x, v_y, v_z)dv_x dv_y dv_z$, where N is the total number of particles per unit volume. The $(M/2\pi kT)^{3/2}$ factor is a normalization constant that ensures that the integral of $f(v_x, v_y, v_z)$ over all velocities is equal to unity, i.e.

$$\iiint_{-\infty}^{\infty} f(v_x, v_y, v_z)\, dv_x\, dv_d\, dv_z = 1. \qquad (2.69)$$

The normalized one-dimensional velocity distribution is

$$f(v_x) = \sqrt{\frac{M}{2\pi kT}}\, e^{-Mv_x^2/2kT}. \qquad (2.70)$$

This Gaussian-shaped function is shown in Figure 2.17. It represents the probability that the velocity of a particle towards an observer is in the range $v_x \to v_x + dv_x$. This is the same as the probability that the frequency be in the range

$$v_0 + \frac{v_x}{c}v_0 \to v_0 + \left(\frac{v_x + dv_x}{c}\right) = v_0 + \frac{v_x}{c}v_0 + \frac{dv_x}{c}v_0. \qquad (2.71)$$

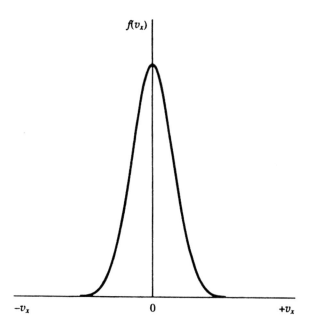

FIGURE 2.17 The Gaussian distribution of velocities in the x-direction for particles in a gas.

Basic Laser Principles

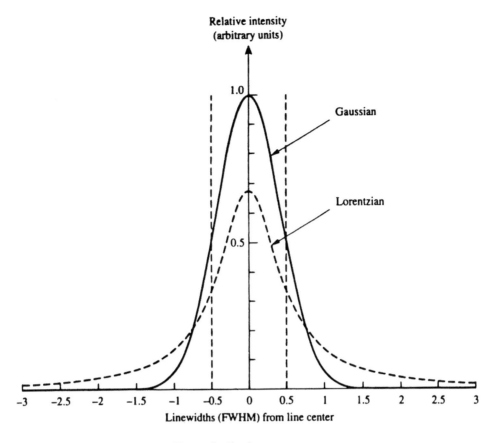

FIGURE 2.18 Comparison of normalized Gaussian and Lorentzian lineshapes.

The probability that the frequency lies in the range $v \to v + dv$ is the same as the probability of finding the velocity in the range $(v-v_0)c/v_0 \to (v-v_0)c/v_0 + c\ dv/v_0$, so the distribution of the emitted frequencies is

$$g(v) = \frac{c}{v_0}\sqrt{\frac{M}{2\pi kT}} \exp\left[\left(-\frac{M}{2kT}\right)\left(\frac{c^2}{v_0^2}\right)(v-v_0)^2\right]. \quad (2.72)$$

This is already normalized (since $f(v_x)$ was normalized). It is called the normalized Doppler-broadened lineshape function. Its FWHM is

$$\Delta v_D = 2v_0\sqrt{\frac{2kT\ln 2}{Mc^2}}. \quad (2.73)$$

It can be seen that this increases with $\sqrt{T}$ and falls with atomic mass as $1/\sqrt{M}$. In terms of the Doppler-broadened linewidth Δv_D, the normalized Doppler lineshape function is

$$g(v_0, v) = \frac{2}{\Delta v_D}\sqrt{\frac{\ln 2}{\pi}} e^{-[2(v-v_0)/\Delta v_D]^2 \ln 2}. \quad (2.74)$$

This is a Gaussian function. It is shown in Figure 2.18 compared with a Lorentzian function of the same FWHM. The Gaussian function is much more sharply peaked while the Lorentzian has considerable intensity far away from its centre frequency, in its *wings*.

Example. The 632.8 nm transition of neon is the most important transition to show laser oscillation in the HeNe laser (see Chapter 27). The atomic mass of neon is 20. Therefore, using

$$M = 20 \times 1.67 \times 10^{-27}\,\text{kg}$$
$$v_0 = 3 \times 10^8 / 632.8 \times 10^{-9}\,\text{Hz}$$
$$T = 400\,\text{K}$$

which is an appropriate temperature to use for the gas in a HeNe laser, the Doppler width is $\Delta v_D \sim 1.5\,\text{GHz}$.[9] Doppler broadening usually dominates over all other sources of broadening in gaseous systems, except occasionally in very heavy gases at high pressures and in highly ionized plasmas of light gases; in the latter case, Stark broadening frequently dominates.

Doppler broadening gives rise to a Gaussian lineshape and is a common source of inhomogeneous broadening. The inhomogeneous lineshape covers a range of frequencies because many *different* particles are being observed. The particles are *different* in the sense that they have different velocities and, consequently, different centre frequencies. Homogeneous broadening also always occurs at the same time as inhomogeneous broadening, to a greater or lesser degree. To illustrate this, imagine a hypothetical experiment in which only those particles in a gas within a certain narrow velocity range are observed. The centre frequencies of these particles are confined to a narrow frequency band and, in this sense, there is

---

[9] The gigahertz (GHz) is a unit of frequency $\equiv 10^9$ Hz, another designation of high frequency is the terahertz (THz) $\equiv 10^{12}$ Hz.

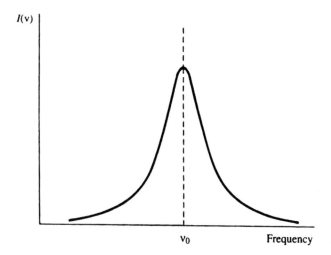

FIGURE 2.19 Homogeneous broadening of a group of particles in a gas that share the same velocity.

no inhomogeneous broadening—all the *observed* particles are the same. However, a broadened lineshape would still be observed—the homogeneous lineshape resulting from natural and pressure broadening. This is illustrated in Figure 2.19. When all particles are observed, irrespective of their velocity, an *overall* lineshape called a *Voigt* profile is observed. This overall lineshape results from the superposition of Lorentzian lineshapes spread across the Gaussian distribution of Doppler-shifted centre frequencies, as shown in Figure 2.20. If the constituent Lorentzians have FWHM $\Delta v_L \ll \Delta v_D$, then the overall lineshape remains Gaussian and the system is properly said to be *inhomogeneously* broadened. However if all observed particles are identical, or almost identical, so that $\Delta v_D \ll \Delta v_L$, then the system as a whole is *homogeneously* broadened.

In solid materials, inhomogeneous broadening, when it is important, results from lattice imperfections and impurities that cause the local environment of individual excited particles to differ in a random way. We shall assume that the broadening that thereby results also gives rise to a Gaussian lineshape of appropriate FWHM, which we shall also designate as Δv_D.[10]

2.8.2 Energy Bands in Condensed Matter

Energy bands are the result of extensive line broadening in condensed matter and also occur because the energy states of electrons in such material reflect the de-localization of electrons throughout the material. The outermost electrons of atoms in condensed matter, the *valence* electrons, can no longer be thought of as being bound to a particular atom, rather they 'wander' throughout the material. Because of the Pauli's exclusion principle, which states that no two electrons can occupy the exact same energy state; even in the ground state these valence electrons separate themselves into very many, closely spaced energy states. These closely spaced states overlap and form the valence band. At absolute zero, all energy states in this band are filled with electrons.

---

[10] The designation Δv_d originates from Doppler broadening but is used generally to designate inhomogeneous linewidth.

The next highest band of energies results from the electrons in the lowest excited states of the individual atoms becoming delocalized, interacting strongly and forming a band, called the conduction band. At absolute zero, all the energy states in the conduction band are empty. At any temperature above absolute zero, some electrons are thermally excited into the conduction band. It is the accessibility of the energy states in the conduction band that characterizes the difference between *insulators*, *metals* and *semiconductors*. This is illustrated in Figure 2.21. The gap in energy between the top of the valence band and the bottom of the conduction band is called the *energy gap*, E_g. If $E_g \gg kT$, where kT is the characteristic thermal energy at temperature T, then electrons do not readily jump across the bandgap into the conduction band. The probability of such a jump depends on the Boltzmann factor $e^{-E_g/kT}$. In a metal, however, E_g is either a lot less than kT or zero. In a semiconductor, $E_g \simeq kT$. It is the presence of electrons in the conduction band, and also the absence of some electrons from the valence band (the 'holes'), which gives rise to electrical conductivity. In a simple sense, a material conducts electricity if there are available energy states for electrons to move into, since conduction corresponds to directed electron motion, which implies that an electron has moved into a different energy state. There are higher-lying energy bands above the conduction band, which are analogous to the higher-lying energy states of the isolated particles. They result from these states interacting, spreading over a range of energies and forming a band.

Some energy levels in condensed matter can remain relatively sharp if inter-particle interactions do not lead to extensive broadening. For example, for some levels of an impurity in a solid material, the spacing between adjacent impurity atoms can, on average, be as large as it might be in the gas phase, and broadening of the energy levels of these impurities may not lead to broad energy bands. In a laser using condensed matter, the laser transition involves a jump between energy bands or sometimes between relatively sharper levels that have not been extensively broadened. In a semiconductor laser, the laser transition generally involves electrons jumping across the energy gap.

2.9 Optical Frequency Amplification with a Homogeneously Broadened Transition

In an optical frequency amplifier, we are generally concerned with the interaction of a monochromatic radiation field with a transition between two energy states whose centre frequency is at, or near, the frequency of the monochromatic field. The magnitude of this interaction with each particle is controlled by the homogeneous lineshape function of the transition.

In the general case, the monochromatic radiation field and the centre frequency of the transition are not the same. This situation is shown schematically in Figure 2.22. The stimulating radiation field is taken to be at frequency v whilst the centre frequency of the transition is at v'. The closer v is to v', the greater the number of transitions that can be stimulated. The stimulated transitions occur at frequency v, since this is the frequency of the stimulating radiation. The number

Basic Laser Principles

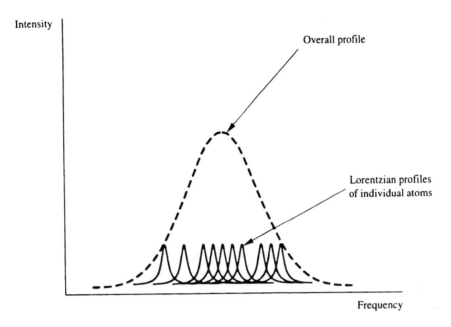

FIGURE 2.20 A Doppler-broadened distribution of Lorentzian lineshapes.

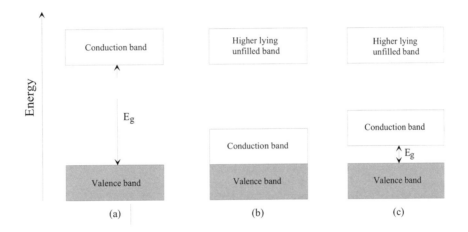

FIGURE 2.21 Schematic energy band diagram for (a) an insulator, (b) a metal and (c) a semiconductor. E_g is the energy gap.

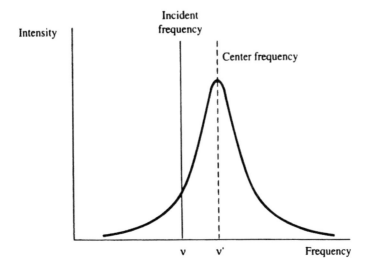

FIGURE 2.22 A monochromatic field interacting with a homogeneously broadened lineshape.

of stimulated transitions is proportional to the homogeneous lineshape function $g(v', v)$ and can be written as

$$N_S = N_2 B_{21} \rho(v) g(v', v). \quad (2.75)$$

We have written $g(v', v)$ to indicate that this lineshape function has its centre frequency at v' but is being evaluated at frequency v. It is important to stress that the lineshape function $g(v', v)$ that is used here is the *homogeneous* lineshape function of the individual particles in the system, even though the contribution of homogeneous broadening to the overall broadening in the system may be small, for example, when the overall broadening is predominantly inhomogeneous. The important point about the interaction of a particle with radiation is that an excited atom, molecule or ion can only interact with a monochromatic radiation field that overlaps its homogeneous (usually Lorentzian-shaped) lineshape profile. For example, consider the case of two excited atoms with different centre frequencies, these may be the different centre frequencies of atoms with different velocities relative to a fixed observer. The homogeneous lineshapes of these two atoms are shown in Figure 2.23 together with a monochromatic radiation field at frequency v. Particle A with centre frequency v_A and *homogeneous width* Δv can interact strongly with the field while the interaction of particle B is negligible.

We can analyse the interaction between a plane monochromatic wave and a collection of homogeneously broadened particles with reference to Figure 2.24. As the wave passes through the medium it grows in intensity if the number of stimulated emissions exceeds the number of absorptions. The change in intensity of the wave in travelling a small distance dz through the medium is

dI_v = (number of stimulated emissions
 − number of absorptions)/vol $\times$ hv $\times$ dz (2.76)

$$= \left(N_2 B_{21} g(v', v) \frac{I_v}{c} - N_2 B_{12} g(v', v) \frac{I_v}{c} \right) hv \, dz.$$

Use of the Einstein relations, equations (2.38) and (2.39), gives

$$dI_v = \frac{I_v}{c} \left(N_2 - \frac{g_2}{g_1} N_1 \right) \frac{c^3 A_{21}}{8\pi hv^3} hv g(v', v) \, dz. \quad (2.77)$$

Therefore,

$$\frac{dI_v}{dz} = \left(N_2 - \frac{g_2}{g_1} N_1 \right) \frac{c^2 A_{21}}{8\pi v^2} g(v', v) I_v \quad (2.78)$$

which has the solution

$$I_v = I_v(0) e^{\gamma(v) z} \quad (2.79)$$

where $I_v(0)$ is the initial intensity at $z = 0$ and

$$\gamma(v) = \left(N_2 - \frac{g_2}{g_1} N_1 \right) \frac{c^2 A_{21}}{8\pi v^2} g(v', v). \quad (2.80)$$

$\gamma(v)$ is called the gain coefficient of the medium and has the same frequency dependence as $g(v', v)$. If $N_2 > (g_2/g_1) N_1$, then $\gamma(v) > 0$ and we have an optical frequency amplifier. If $N_2 < (g_2/g_1) N_1$, then $\gamma(v) < 0$ and net absorption of the incident radiation occurs.

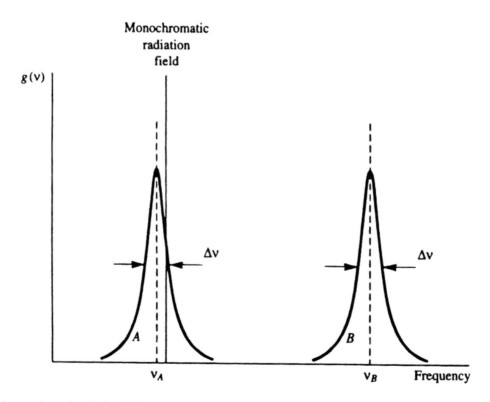

FIGURE 2.23 A monochromatic radiation field interacting with two homogeneously broadened lineshapes whose centre frequencies are different.

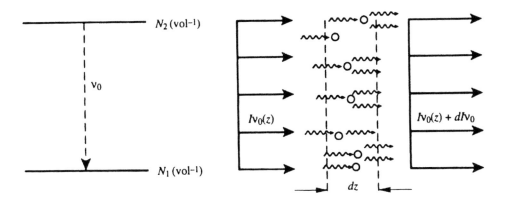

FIGURE 2.24 A plane wave travelling through and interacting with a collection of homogeneously broadened particles.

For a system in thermal equilibrium,

$$\frac{N_2}{N_1} = \frac{g_2}{g_1} e^{-h\nu/kT} \quad (2.81)$$

and for $T > 0$, $e^{-h\nu/kT} < 1$, which implies that, in thermal equilibrium at positive temperatures, we cannot have positive gain. If we allow the formal existence of a negative temperature, at least for our system of two levels, then we can have $N_2 > (g_2/g_1)N_1$. Such a situation, which is essential for the construction of an optical frequency amplifier, is called a state of *population inversion* or *negative temperature*. This is not a true state of thermal equilibrium and can only be maintained by feeding energy into the system.

In the previous discussion, we have neglected the occurrence of spontaneous emission: this is reasonable for a truly plane wave, as the total number of spontaneous emissions into the zero solid angle subtended by the wave is zero. If the waves being amplified were diverging into a small solid angle $\delta\omega$, then $N_2 A_{21} \delta\omega/4\pi$ spontaneous emissions per second per unit volume would contribute to the increase in intensity of the wave. However, such emissions, being independent of the incident wave, are not in a constant phase relationship with this wave as are the stimulated emissions. These spontaneous emissions constitute a kind of 'noise' superimposed on the beam of identical photons created by stimulated emission.

2.9.1 The Stimulated Emission Rate in a Homogeneously Broadened System

The stimulated emission rate $W_{21}(\nu)$ is the number of stimulated emissions per particle per second caused by a monochromatic input wave at frequency ν:

$$W_{21}(\nu) = B_{21} g(\nu', \nu) \rho_\nu. \quad (2.82)$$

ρ_ν (J m^{-3}) is the energy density of the stimulating radiation. Equation (2.82) can be rewritten in terms of more practical parameters as

$$W_{21}(\nu) = \frac{A_{21} c^2 I_\nu}{8\pi h \nu^3} g(\nu', \nu). \quad (2.83)$$

$W_{21}(\nu)$ has units s^{-1} per particle. Note that the frequency variation of $W_{21}(\nu)$ follows the lineshape function. The total number of stimulated emissions is

$$N_s = N_2 W_{21}(\nu). \quad (2.84)$$

2.9.2 Optical Frequency Amplification with Inhomogeneous Broadening Included

Although in our discussion so far we have been restricting our attention to homogeneous systems, we can show that equations (2.76)–(2.80) hold generally, even in a system with inhomogeneous broadening, if we take $g(\nu', \nu)$ as the *total* lineshape function.

In an inhomogeneously broadened system, we can divide the atoms up into classes, each class consisting of particles with a certain range of centre emission frequencies and the same homogeneous lineshape. For example, the class with centre frequency ν'' in the frequency range $d\nu''$ has $N g_D(\nu', \nu'') d\nu''$ particles in it, where $g_D(\nu', \nu'')$ is the normalized inhomogeneous distribution of centre frequencies—the inhomogeneous lineshape function centred at ν'. This class of particles contributes to the change in intensity of a monochromatic wave at frequency ν as

$$\Delta(dI_\nu)(\text{from the group of particles in the band } d\nu'')$$

$$= \left(N_2 B_{21} g_D(\nu', \nu'') \, d\nu'' g_L(\nu'', \nu) \frac{I_\nu}{c} \right. \quad (2.85)$$
$$\left. - N_1 B_{12} g_D(\nu', \nu'') \, d\nu'' g_L(\nu'', \nu) \frac{I_\nu}{c} \right) h\nu \, dz$$

where $g_L(\nu'', \nu)$ is the homogeneous lineshape function of a particle at centre frequency ν''. Equation (2.85) is equivalent to equation (2.76). The increase in intensity from all the classes of particles is found by integrating over these classes, that is, over the range of centre frequencies ν'', so equation (2.85) becomes

$$dI_\nu = \frac{I_\nu}{c}(N_2 B_{21} - N_1 B_{12})\left[\int_{-\infty}^{\infty} g_D(\nu', \nu'') g_L(\nu'', \nu) d\nu''\right] h\nu \, dz$$

(2.86)

which, in a similar fashion to equations (2.76)–(2.80), gives

$$\gamma(v) = \left(N_2 - \frac{g_2}{g_1}N_1\right)\frac{c^2 A_{21}}{8\pi v^2} g(v', v) \quad (2.87)$$

where $g(v', v)$ is now the overall lineshape function defined by the equation

$$g(v', v) = \int_{-\infty}^{\infty} g_D(v', v'') g_L(v'', v) \, dv''. \quad (2.88)$$

In other words, the overall lineshape function is the convolution [8] of the homogeneous and inhomogeneous lineshape functions. The convolution integral in equation (2.88) can be put in more familiar form if we measure frequency relative to the centre frequency of the overall lineshape, that is, we put $v' = 0$, and equation (2.88) becomes

$$g(0, v) = \int_{-\infty}^{\infty} g_D(0, v'') g_L(v'', v) \, dv''$$

$$= \int_{-\infty}^{\infty} g_D(0, v'') g_L(0, v - v'') \, dv'' \quad (2.89)$$

which can be written in simple form as

$$g(v) = \int_{-\infty}^{\infty} g_D(v'') g_L(v - v'') \, dv''. \quad (2.90)$$

This is recognizable as the standard convolution integral of two functions $g_D(v)$ and $g_L(v)$.

If $g_D(v', v'')$ is indeed a normalized Gaussian lineshape as in equation (2.74) and $g_L(v'', v)$ is a Lorentzian, then equation (2.88) can be written in the form

$$g(v', v) = \frac{2}{\Delta v_D}\sqrt{\frac{\ln 2}{\pi}} \frac{y}{\pi} \int_{-\infty}^{\infty} \frac{e^{-t^2}}{y^2 + (x-t)^2} \, dt, \quad (2.91)$$

where $y = \Delta v_L \sqrt{\ln 2}/\Delta v_D$ and $x = 2(v - v')\sqrt{\ln 2}/\Delta v_D$.

This is one way of writing a normalized Voigt profile [9,10]. The integral in equation (2.91) cannot be evaluated analytically but must be evaluated numerically. For this purpose the Voigt profile is often written in terms of the error function for complex argument $W(z)$, which is available in tabulated form [11]

$$g(v', v) = \frac{2}{\Delta v_D}\sqrt{\frac{\ln 2}{\pi}} R[W(z)] \quad (2.92)$$

where $z = x + iy$ and $R[W(z)]$ denotes the real part of the function.

2.10 Optical Frequency Oscillation—Saturation

If we can prepare a medium in a state of population inversion for a pair of its energy levels, then the transition between these levels can be used to make an optical frequency amplifier. To produce an oscillator, we need to apply appropriate feedback by inserting the amplifying medium between a pair of suitable mirrors. If the overall gain of the medium exceeds the losses of the mirror cavity and ancillary optics then oscillation will result. The level at which this oscillation stabilizes is set by the way in which the amplifier saturates.

2.10.1 Homogeneous Systems

Consider an amplifying transition at centre frequency v_0 between two energy levels of a particle. We maintain this pair of levels in population inversion by feeding in energy. In equilibrium in the absence of an external radiation field, the rates R_2 and R_1 at which particles are fed into these levels must be balanced by spontaneous emission and non-radiative loss processes (such as collisions). The population densities and effective lifetimes of the two levels are N_2, τ_2 and N_1, τ_1, respectively, as shown in Figure 2.25. These effective lifetimes include the effect of non-radiative deactivation. If X_{2j} is the rate per particle per unit volume by which collisions depopulate level 2 and cause the particle to end up in a lower state j, we can write

$$\frac{1}{\tau_2} = \sum_j (A_{2j} + X_{2j}). \quad (2.93)$$

In equilibrium,

$$\frac{dN_2^o}{dt} = R_2 - \frac{N_2^o}{\tau_2} = 0 \quad (2.94)$$

where the term N_2^o/τ_2 is the total loss rate per unit volume from spontaneous emission and other deactivation processes, so

$$N_2^o = R_2 \tau_2 \quad (2.95)$$

where the superscript o indicates that the population is being calculated in the absence of a radiation field. Similarly, for the lower level of the transition, in equilibrium

$$\frac{dN_1^o}{dt} = R_1 + N_2^o A_{21} - \frac{N_1^o}{\tau_1} = 0 \quad (2.96)$$

The term N_1^o/τ_1 is the total loss rate per unit volume from the level by spontaneous emission and other deactivation processes, while the term $N_2^o A_{21}$ is the rate at which atoms are feeding into level 1 by spontaneous emission from level 2. So from equations (2.95) and (2.96)

$$N_1^o = (R_1 + N_2^o A_{21})\tau_1 = (R_1 + R_2 \tau_2 A_{21})\tau_1. \quad (2.97)$$

The population inversion is

$$\left(N_2^o - \frac{g_2}{g_1}N_1^o\right) = \Delta N^o = R_2\tau_2 - \frac{g_2}{g_1}\tau_1(R_1 + R_2\tau_2 A_{21}) \quad (2.98)$$

or, when the level degeneracies are equal,

$$\Delta N^o = R_2\tau_2 - \tau_1(R_1 + R_2\tau_2 A_{21}). \quad (2.99)$$

If we now feed in a monochromatic (or other) signal, then stimulated emission and absorption processes will occur. We

Basic Laser Principles

take the energy density of this signal at some point within the medium to be $\rho(v) = I(v)/c$. The rate at which this signal causes stimulated emissions is

$$W_{21}(v) = \int_{-\infty}^{\infty} B_{21} g(v_0, v) \rho(v)\, dv \quad \text{per particle per second,} \tag{2.100}$$

where $g(v_0, v)$ is the *homogeneous* lineshape function. If the input radiation was *white*, which in this context means that $\rho(v)$ is a constant over the range of frequencies spanned by the lineshape function, then

$$W_{21}(v) = B_{21} \rho(v) \int_{-\infty}^{\infty} g(v_0, v)\, dv = B_{21} \rho(v). \tag{2.101}$$

The total rate at which a monochromatic plane wave causes stimulated transitions is

$$\begin{aligned} W_{21}(v) &= \int_{-\infty}^{\infty} B_{21} g(v_0, v) \rho(v)\, dv \\ &= \frac{I_v}{c} B_{21} g(v_0, v) \int_{-\infty}^{\infty} \delta(v - v'')\, dv'' \\ &= B_{21} g(v_0, v) \frac{I_v}{c}. \end{aligned} \tag{2.102}$$

In the presence of a radiation field, in equilibrium, the population densities of the pair of energy levels shown in Figure 2.25 are

$$\frac{dN_2}{dt} = R_2 - \frac{N_2}{\tau_2} - N_2 B_{21} g(v_0, v) \rho(v) + N_1 B_{12} g(v_0, v) \rho(v) = 0 \tag{2.103}$$

and

$$\frac{dN_1}{dt} = R_1 + N_2 A_{21} - \frac{N_1}{\tau_1} + N_2 B_{21} g(v_0, v) \rho(v) \\ - N_1 B_{12} g(v_0, v) \rho(v) = 0. \tag{2.104}$$

If we write $B_{21} g(v_0, v) \rho(v) = W_{12}(v)$, the stimulated emission rate at frequency v per particle, and neglect degeneracy factors so that we can assume that $W_{12}(v) = W_{21}(v) = W$ (equivalent to $B_{12} = B_{21}$), then we have

$$R_2 - \frac{N_2}{\tau_2} - N_2 W + N_1 W = 0 \tag{2.105}$$

and

$$R_1 + N_2 A_{21} - \frac{N_1}{\tau_1} + N_2 W - N_1 W = 0. \tag{2.106}$$

From equation (2.109), this gives

$$N_2 = \frac{N_1 W + R_2}{1/\tau_2 + W} \tag{2.107}$$

and from equation (2.106),

$$N_2 = \frac{N_1/\tau_1 + N_1 W - R_1}{A_{21} + W} \tag{2.108}$$

so

$$N_1 = \frac{R_1/\tau_2 + R_2 A_{21} + W(R_1 + R_2)}{1/\tau_1 \tau_2 + W(1/\tau_1 + 1/\tau_2 - A_{21})}. \tag{2.109}$$

The population inversion in the system is now

$$N_2 - N_1 = \frac{N_1 W + R_2}{1/\tau_2 + W} - N_1 = \frac{R_2 - N_1/\tau_2}{1/\tau_2 + W}. \tag{2.110}$$

From equation (2.109), this gives

$$\begin{aligned} N_2 &- N_1 \\ &= \frac{1/\tau_2 (R_2/\tau_1 - R_2 A_{21} - R_1/\tau_2) + W(R_2/\tau_1 - R_2 A_{21} - R_1/\tau_2)}{(1/\tau_2 + W)(1/\tau_1 \tau_2 + W/\tau_2 + W/\tau_1 - W A_{21})} \\ &= \frac{R_2/\tau_1 - R_1/\tau_2 - R_2 A_{21}}{1/\tau_1 \tau_2 + W(1/\tau_2 + 1/\tau_1 - A_{21})}. \end{aligned} \tag{2.111}$$

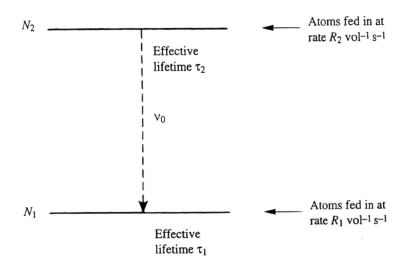

FIGURE 2.25 Pair of energy levels used in discussion of amplifier saturation.

Multiplying the numerator and denominator of equation (2.111) by $\tau_1 \tau_2$, we get

$$N_2 - N_1 = \frac{R_2\tau_2 - R_1\tau_1 - R_2\tau_1\tau_2 A_{21}}{1 + W\tau_2(1 + \tau_1/\tau_2 - A_{21}\tau_1)}. \quad (2.112)$$

The numerator of this expression is just the population inversion in the absence of any light signal, ΔN^o. Thus,

$$N_2 - N_1 = \frac{\Delta N^o}{1 + W\tau_2(1 + \tau_1/\tau_2 - A_{21}\tau_1)} \quad (2.113)$$

or with the substitution

$$\phi = A_{21}\tau_2\left[1 + (1 - A_{21}\tau_2)\frac{\tau_1}{\tau_2}\right] \quad (2.114)$$

$$N_2 - N_1 = \frac{\Delta N^o}{1 + \phi W/A_{21}}. \quad (2.115)$$

Now from equation (2.83),

$$W = \frac{c^2 A_{21}}{8\pi h v^3} I_v g(v_0, v). \quad (2.116)$$

If we define

$$I_s(v) = \frac{8\pi h v^3}{c^2 \phi g(v_0, v)} \quad (2.117)$$

then

$$N_2 - N_1 = \frac{\Delta N^o}{1 + I_v/I_s(v)} \quad (2.118)$$

and $I_s(v)$, called the saturation intensity, is the intensity of an incident light signal (power area^{-1}) that reduces the population inversion to half its value when no signal is present. Note that the value of the saturation intensity depends on the frequency of the input signal relative to the line centre.

Returning to our expression for the gain constant of a laser amplifier,

$$\gamma(v) = (N_2 - N_1)\frac{c^2 A_{21}}{8\pi v^2} g(v_0, v); \quad (2.119)$$

the gain as a function of intensity is, in a *homogeneously* broadened system,

$$\gamma(v) = \frac{\Delta N^o}{[1 + I_v/I_s(v)]} \frac{c^2 A_{21}}{8\pi v^2} g(v_0, v) \quad (2.120)$$

which is reduced, that is, *saturates* as the strength of the amplified signal increases. A good optical amplifier should have a large value of saturation intensity; from equation (2.117), this implies that ϕ should be a minimum. In such systems often $A_{21} \simeq 1/\tau_2$ so $\phi \simeq 1$.

2.10.2 Inhomogeneous Systems

The problem of gain saturation in inhomogeneous media is more complex. For example, in a gas, a plane monochromatic wave at frequency v interacts with a medium whose individual particles have Lorentzian homogeneous lineshapes with FWHM Δv_N, but whose centre frequencies are distributed over an inhomogeneous (Doppler) broadened profile of width (FWHM) Δv_D. The Lorentzian contribution to the overall lineshape is

$$g_L(v', v) = \frac{(2/\pi \Delta v_N)}{1 + [2(v - v'/\Delta v_N]^2} \quad (2.121)$$

where v' is the centre frequency of a particle set by its velocity relative to the observer. The Doppler-broadened profile of all the particles is

$$g_D(v_0, v') = \frac{2}{\Delta v_D}\sqrt{\frac{\ln 2}{\pi}} e^{-\left[(2(v'-v_0/\Delta v_D)^2 \ln 2\right]} \quad (2.122)$$

where v_0 is the centre frequency of a particle at rest. The overall lineshape (from all the particles) is a sum of Lorentzian profiles spread across the particle velocity distribution, as shown in Figure 2.20.

If $\Delta v_N \gg \Delta v_D$, the overall profile remains approximately Lorentzian and the observed behaviour of the system will correspond to *homogeneous* broadening. Such a situation is likely to arise for long-wavelength transitions in a gas, particularly if this has a high atomic or molecular weight, at pressures where pressure broadening (which is a homogeneous process) is important, and frequently in solid materials. If $\Delta v_D \gg \Delta v_N$ (as is often the case in gases), the overall lineshape remains Gaussian and the system is *inhomogeneously* broadened.

Once again, we reduce the problem to a consideration of the interaction of a plane electromagnetic wave with a two-level system as shown in Figure 2.25. As the wave passes through the system, its intensity changes according to whether the medium is amplifying or absorbing. We take the intensity of the monochromatic wave to be $I(v, z)$ at plane z within the medium. The individual particles of the medium have a distribution of emission centre frequencies (or absorption centre frequencies) because of their random velocities (or, for example, their different crystal environments). We take the population density functions (particles vol^{-1}Hz^{-1}) in the upper and lower levels whose centre frequency is at v' to be $N_2(v', z)$ and $N_1(v', z)$, respectively, at plane z. Particles are fed into levels 2 and 1 at rates $R_2(v')$ and $R_1(v')$. These rates are assumed to be uniform throughout the medium. $N_2(v', z)$, $N_1(v', z)$, $R_2(v')$ and $R_1(v')$ are assumed to follow the Gaussian frequency dependence set by the particle velocity distribution and are normalized so that, for example, the total pumping rate of level 2 is

$$R_2 = \int_{-\infty}^{\infty} R_2(v')\,dv' = R_{20}\int_{-\infty}^{\infty} e^{-\left[2(v'-v_0/\Delta v_D)^2 \ln 2\right]}\,dv'. \quad (2.123)$$

In practice, the primary pumping process may not have this Gaussian dependence, but, even when this is the case, the effect of collisions among particles that have been excited will be to *smear out* any non-Gaussian pumping process into a near-Gaussian form. This conclusion is justified by observations of

Basic Laser Principles

Doppler-broadened lines under various excitation conditions where deviations from a true Gaussian lineshape are found to be minimal. From equation (2.123)

$$R_2 = \sqrt{\frac{\pi}{\ln 2}} \frac{\Delta \nu_D}{2} R_{20} \qquad (2.124)$$

where R_{20} is a pumping rate constant and the total population density of level at plane z is

$$N_2 = \int_{-\infty}^{\infty} N_2(\nu', z) \, d\nu'. \qquad (2.125)$$

The rate equations for the atoms whose centre frequencies are at ν' are

$$\frac{dN_2}{dt}(\nu', z) = R_2(\nu') + N_2(\nu', z)\left[\frac{1}{\tau_2} + B'_{21}(\nu', \nu)\frac{I(\nu, z)}{c}\right]$$
$$+ N_1(\nu', z) B'_{12}(\nu', \nu)\frac{I(\nu, z)}{c} \qquad (2.126)$$

and

$$\frac{dN_1}{dt}(\nu', z) = R_1(\nu') + N_2(\nu', z)\left[A_{21} + B'_{21}(\nu', \nu)\frac{I(\nu, z)}{c}\right]$$
$$- N_1(\nu', z)\left[\frac{1}{\tau_1} + B'_{12}(\nu', \nu)\frac{I(\nu, z)}{c}\right]. \qquad (2.127)$$

Here we have used the modified Einstein coefficients $B'_{21}(\nu', \nu)$, $B'_{12}(\nu', \nu)$ that describe stimulated emission processes when the stimulating radiation is at frequency ν and the particle's centre emission frequency is at ν'. Written out in full,

$$B'_{21}(\nu', \nu) = B_{21} g(\nu', \nu) \qquad (2.128)$$

where $g(\nu', \nu)$ is the *homogeneous* lineshape function. The rate of change of intensity of the incident wave due to atoms with centre frequencies in a small range $d\nu'$ at ν' is

$$\left[\frac{dI}{dz}(\nu, z)\right]_{d\nu'} = h\nu \frac{I(\nu, z)}{c}$$
$$\times \left[B'_{21}(\nu', \nu) N_2(\nu', z) - B'_{12}(\nu', \nu) N_1(\nu', z)\right] d\nu'. \qquad (2.129)$$

The total rate of change of intensity due to all the particles, that is from all possible centre frequencies ν', is

$$\frac{dI(\nu, z)}{dz} = \frac{h\nu I(\nu, z)}{c}$$
$$\times \int_{-\infty}^{\infty} \left[B'_{21}(\nu', \nu) N_2(\nu', z) - B'_{12}(\nu', \nu) N_1(\nu', z)\right] d\nu'. \qquad (2.130)$$

In the steady state,

$$\frac{dN_2}{dt}(\nu', z) = \frac{dN_1}{dt}(\nu', z) = 0 \qquad (2.131)$$

so from equations (2.126) and (2.127)

$$B'_{21}(\nu', \nu) N_2(\nu') - B'_{12}(\nu', \nu) N_1(\nu')$$
$$= \frac{B'_{21}(\nu', \nu)\left[\frac{R_2(\nu')}{1/\tau_2} - \left(\frac{g_2}{g_1}\right)\frac{(R_2(\nu')A_{21}+R_1(\nu')/\tau_2)}{1/\tau_1\tau_2}\right]}{1 + \left[\frac{(g_1/g_2)(1/\tau_2 - A_{21})}{1/\tau_1\tau_2} + \tau_2\right] B'_{21}(\nu', \nu) \frac{I(\nu, z)}{c}}. \qquad (2.132)$$

We note that

$$R_2(\nu') = R_{20} e^{-[2(\nu'-\nu_0)/\Delta\nu_D]^2 \ln 2}. \qquad (2.133)$$

Substituting in equation (2.130) from (2.132) and (2.133) and bearing in mind that has a Lorentzian form,

$$B'_{21}(\nu', \nu) = \frac{B_{21} \frac{2}{\pi \Delta \nu_N}}{1 + \left[\frac{2(\nu - \nu')}{\Delta \nu_N}\right]^2} \qquad (2.134)$$

where $\Delta \nu_N$ is the *homogeneous* FWHM of the transition, gives

$$\frac{1}{I(\nu, z)} \frac{dI(\nu, z)}{dz} = \gamma(\nu)$$
$$= \frac{\gamma_0 \int_{-\infty}^{\infty} d\nu' \left(\frac{2}{\pi \Delta \nu_N}\right)\{1 + [2(\nu - \nu'/\Delta\nu_N)]^2\}^{-1} \exp\left(-\left[\frac{2(\nu'-\nu)}{\Delta\nu_D}\right]^2 \ln 2\right)}{1 + \eta I(\nu, z)\left(\frac{2}{\pi \Delta \nu_N}\right)[1 + [2(\nu - \nu')/\Delta\nu_N]^2]^{-1}}. \qquad (2.135)$$

We have made the substitutions

$$\gamma_0 = \frac{h\nu}{c} B_{21} \left[R_{20} \tau_2 - \frac{g_2}{g_1}\left(\frac{R_{20} A_{21} + R_{10} A_2}{1/\tau_1 \tau_2}\right)\right] \qquad (2.136)$$

$$\eta = \left[\frac{g_2}{g_1} \frac{1/(\tau_2 - A_{21})}{1/\tau_1 \tau_2} + \tau_2\right] \frac{B_{21}}{c}. \qquad (2.137)$$

Equation (2.135) can be written as

$$\gamma(\nu) = \frac{2\gamma_0 \Delta \nu_N}{\pi} \int_{-\infty}^{\infty} \frac{\exp(-[2(\nu' - \nu_0/\Delta\nu_D]^2 \ln 2) \, d\nu'}{4(\nu - \nu')^2 + \Delta\nu_N^2 [1 + 2\eta I(\nu, z)/\pi \Delta \nu_N]}. \qquad (2.138)$$

Although it is fairly clear from equation (2.138) that the gain of the amplifier falls as $I(\nu, z)$ increases, it is not easy to see from the integral exactly how this occurs. If the intensity is small, the gain approaches its *unsaturated* value

$$\gamma_0(\nu) = \frac{2\gamma_0 \Delta\nu_N}{\pi} \int_{-\infty}^{\infty} \frac{e^{-[2(\nu'-\nu_0/\Delta\nu_D]^2 \ln 2} \, d\nu'}{4(\nu - \nu')^2 + \Delta\nu_N^2}. \qquad (2.139)$$

If equation (2.138) is examined closely, for frequencies ν' close to the input frequency ν, the integrand can be written approximately as

$$\frac{e^{-[2(\nu'-\nu_0)/\Delta\nu_D]^2 \ln 2}}{\Delta\nu_N^2[1 + 2\eta I(\nu, z)/\pi \Delta \nu_N]}.$$

However, for frequencies ν' far from the input frequency, the integrand can be written approximately as

$$\frac{e^{-[2(v'-v_0)/\Delta v_D]^2 \ln 2}}{4(v-v')^2 + \Delta v_N^2}$$

which can be seen to be identical to the integrand in the unsaturated gain expression, equation (2.139). Thus, we conclude that, in making their contribution to the overall gain, particles whose frequencies are far from the input frequency are relatively unaffected by the input radiation, whereas particles whose frequencies are close to that of the input show strong saturation effects. The gain in the system comes largely from those particles whose frequencies are within (roughly) a homogeneous linewidth of the input radiation frequency. The consequences of this are best illustrated by considering a hypothetical experiment, shown schematically in Figure 2.26, in which the small-signal gain of a predominantly inhomogeneously broadened amplifier is measured with and without a strong saturating signal simultaneously present. Without the presence of a strong signal at a fixed frequency v_S, the observed (small-signal) gain follows the Gaussian curve of the overall line profile of the amplifier, as shown in Figure 2.27a. However, if we perform this experiment again when a strong fixed frequency field is also present, which causes saturation of the gain, we find the gain is reduced locally by the saturating effect of the strong field as shown in Figure 2.27b. This phenomenon is called *hole burning* [12]. The width of the *hole* that is thus produced is determined by the quantity

$$\Delta v_N^2 \left[1 + \frac{2\eta I(v,z)}{\pi \Delta v_N}\right].$$

If

$$\Delta v_N \sqrt{1 + \frac{2\eta I(v,z)}{\pi \Delta v_N}} \ll \Delta v_D$$

for example, in a gaseous system where Doppler broadening is the largest contribution to the total observed line broadening, equation (2.138) can be integrated by bringing the much less sharply peaked exponential factor outside the integral. In this case, the sharply peaked Lorentzian lineshape makes the integrand largest for frequencies v' near to v; over a small range of frequencies v' near v, the exponential factor remains approximately constant, so equation (2.138) can be written

$$\gamma(v) = \frac{2\gamma_0 \Delta v_N}{\pi} e^{-[2(v-v_0)/\Delta v_D]^2 \ln 2}$$
$$\times \int_{-\infty}^{\infty} \frac{dv'}{4(v-v')^2 + \Delta v_N^2 [1 + 2\eta I(v,z)/\pi \Delta v_N]}. \quad (2.140)$$

Now the integral can be evaluated to give

$$\gamma(v) = \gamma_0 \left[1 + \frac{2\eta I(v,z)}{\pi \Delta v_N}\right]^{-\frac{1}{2}} e^{-[2(v-v_0)/\Delta v_D]^2 \ln 2} \quad (2.141)$$

which gives

$$\gamma(v) = \gamma_0 \left[1 + \frac{I(v,z)}{I_s'(v)}\right]^{-\frac{1}{2}} e^{-[2(v-v_0)/\Delta v_D]^2 \ln 2} \quad (2.142)$$

where $I_s' = \pi \Delta v_N / 2\eta$ is called the saturation intensity for inhomogeneous broadening. Note that γ_0 is the small-signal gain at line centre of the inhomogeneously broadened line, which can be written in the form

$$\gamma_0 = \frac{1}{4\pi} \sqrt{\frac{\ln 2}{\pi}} \frac{\lambda^2 A_{21}}{\Delta v_D} \left(N_2 - \frac{g_2}{g_1} N_1\right). \quad (2.143)$$

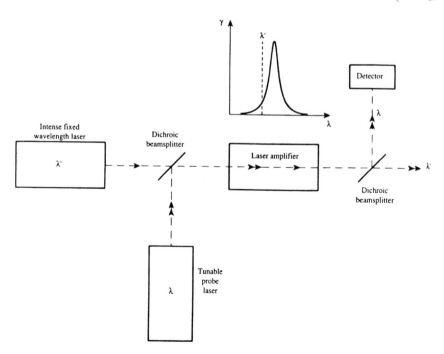

FIGURE 2.26 Schematic experimental arrangement for measuring the frequency dependence of the gain of an amplifier that is experiencing saturation from a strong fixed monochromatic signal.

Basic Laser Principles

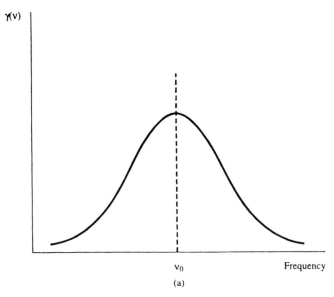

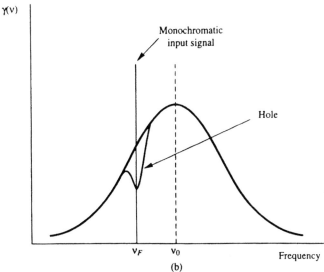

FIGURE 2.27 Gain as a function of frequency in an inhomogeneously broadened amplifier. (a) Small-signal situation when no saturation has occurred. (b) Showing the production of a 'hole' in the gain curve by a strong monochromatic input at frequency v_F.

When Doppler broadening dominates in a system, incident radiation at frequency v cannot interact with those particles whose Doppler-shifted frequency is different from v by much more than Δv_N.

If the amplifier is homogeneously broadened, that is, if

$$\Delta v_N \sqrt{1 + \frac{\eta I(v,z)}{\pi \Delta v_N}} \gg \Delta v_D.$$

equation (2.138) can be integrated by bringing the less-sharply peaked Lorentzian factor outside the integral to give

$$\gamma(v) = \frac{\gamma_0 \Delta v_D}{\Delta v_N \sqrt{\pi \ln 2}} \left\{ \left[\frac{2(v-v_0)}{\Delta v_N}\right]^2 + 1 + \frac{I(v,z)}{I'_s(v)} \right\}^{-1}$$

$$= \frac{\Delta v_D}{2} \sqrt{\pi/\ln 2} \frac{\gamma_0 g(v_0, v)}{[1 + I(v,z)/I_s(v)]}$$
(2.144)

where $g(v_0, v)$ is the homogeneous lineshape function

$$g(v_0, v) = \frac{(2/\pi \Delta v_N)}{1 + [2(v-v_0)\Delta v_N]^2}$$
(2.145)

and $I_s(v)$ is the saturation intensity for homogeneous broadening given by

$$I_s(v) = \frac{2I'_s(v)}{\pi \Delta v_N g(v_0, v)} = \frac{1}{\eta g(v_0, v)}.$$
(2.146)

It can be seen from equation (2.137) that for $g_2 = g_1$, η reduces to the expression

$$\eta = \left[\left(\frac{(1/\tau_2 - A_{21})}{1/\tau_1 \tau_2} \right) + \tau_2 \right] \frac{B_{21}}{c}$$
(2.147)

and $I_S(v)$ reduces to the expression obtained previously as the saturation intensity for homogeneous broadening. Namely,

$$I_S(v) = \frac{8\pi h v^3}{c^2 \phi g(v_0, v)} \quad (2.148)$$

where

$$\phi = A_{21}\tau_2 \left[1 + (1 - A_{21}\tau_2) \frac{\tau_1}{\tau_2} \right]. \quad (2.149)$$

2.11 Power Output from a Laser Amplifier

For a laser amplifier of length ℓ and gain coefficient $\gamma(v)$, the output intensity for a monochromatic input intensity of I_0(W m^{-2}) at frequency v is

$$I = I_0 e^{\gamma(v)\ell} \quad (2.150)$$

if saturation effects are neglected. If saturation effects cannot be neglected, then the differential equation that describes how intensity increases must be re-examined. This is

$$\gamma(v) = \frac{1}{I}\frac{dI}{dz}. \quad (2.151)$$

For a homogeneously broadened amplifier with saturation, an explicit solution to this equation can be found. In this case, if $\gamma_0(v)$ is the small-signal gain,

$$\gamma(v) = \frac{\gamma_0(v)}{1 + I/I_s(v)} = \frac{1}{I}\frac{dI}{dz} \quad (2.152)$$

which can be rewritten in the form

$$\frac{dI}{I} + \frac{dI}{I_s(v)} = \gamma_0(v)dz. \quad (2.153)$$

The solution to this equation is

$$I = I_0 e^{\gamma_0(v)\ell - (I - I_0)/I_s(v)}. \quad (2.154)$$

equation (2.154) must be solved iteratively. The solution will be somewhere between the input intensity and the output intensity when saturation is neglected.

2.12 The Electron Oscillator Model of a Radiative Transition

When a particle decays from a higher energy level to a lower, we can model the resultant electric field as a damped oscillation. There is an analogy between this decay of an excited particle and the damped oscillation of an electric circuit. For example, for the RLC circuit shown in Figure 2.28, the resonant frequency is $v_0 = 1/2\pi\sqrt{LC}$. If the sinusoidal driving voltage is disconnected from the circuit, then the oscillation of the circuit decays exponentially—provided the circuit is underdamped. The power spectrum of the decaying electric current is Lorentzian, just as it is for a spontaneous transition.

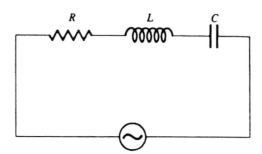

FIGURE 2.28 RLC circuit.

The FWHM of the circuit resonance is v_0/Q, where Q is called the *quality factor* of the circuit, analogous to the homogeneously broadened linewidth Δv_N. A transition between levels almost always has $\Delta v_N \ll v_0$ so clearly has a very high Q.

In the classical theory of how a particle responds to electromagnetic radiation, each of the n electrons attached to the particle is treated as a damped harmonic oscillator. For example, when an electric field acts on an atom, the nucleus, which is positively charged, moves in the direction of the field while the electron cloud, which is negatively charged, moves in the opposite direction to the field. The resultant separation of the centres of positive and negative charge causes the atom to become an elemental dipole. If the separation of the nucleus and electron cloud is d, then the resultant dipole has magnitude ed and points from the negative towards the positive charge.[11] As the frequency of the electric field that acts on the atom increases, the amount of nuclear motion decreases much more rapidly than that of the electrons. At optical frequencies, we generally neglect the motion of the nucleus; its great inertia compared to the electron cloud prevents it following the rapidly oscillating applied electric field. If the vector displacement of the ith electron on the atom from its equilibrium position is x_i then at any instant, the atom has acquired a dipole moment

$$\mu = -\sum_{i=1}^{n} e x_i \quad (2.155)$$

where the summation runs over all the n electrons on the atom. The magnitude of the displacement of each electron depends on the value of the electric field $\mathbf{E}_i$ at the electron

$$k_i x_i = -e\mathbf{E}_i \quad (2.156)$$

where k_i is a force constant. A time-varying field E leads to a time-varying dipole moment. This dipole moment can become large if there is a resonance between the applied field and a particular electron on the atom. This happens if the frequency of the field is near the natural oscillation frequency of a particular electron. Classically, the resonance frequency of electron i is, by analogy with a mass attached to a spring, $\omega_i = \sqrt{k_i/m}$.

If the applied electric field is near this frequency, one electron in the summation in equation (2.155) makes a dominant contribution to the dipole moment, and we can treat the atom

---
[11] The magnitude of the electronic charge is $e \simeq 1.6 \times 10^{-19}$ C, the charge on an electron is $-e$.

Basic Laser Principles

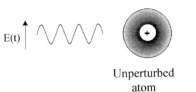

FIGURE 2.29 An atom exposed to an external electric field. The electron cloud and nucleus are displaced in opposite directions.

as a single electron oscillator. The physical significance of the resonant frequencies of the electrons is that they correspond to the frequencies of transitions that the electrons of the atom can make from one energy level to another. If we confine our attention to one of these resonances, then we can treat an *n*-electron atom as a one-electron classical oscillator. A time-varying electric field $E(t)$, which we assume *a priori* to be at a frequency near to an atomic resonance, perturbs only a single electron to a significant degree, thereby inducing a dipole moment which varies at the same frequency as the applied field. The electron and nucleus are perturbed in opposite directions by the field as shown in Figure 2.29.

The motion of the electrons on each of the particles in the medium, which for simplicity can be assumed to be identical, obeys the differential equation

$$\frac{d^2x}{dt^2} + 2\Gamma\frac{dx}{dt} + \frac{k}{m}x = -\frac{e}{m}E(t). \quad (2.157)$$

The terms on the left-hand side of the equation represent, reading from left to right: the acceleration of the electron, a damping term proportional to the electron velocity and a restoring force. These terms are balanced by the effect of the applied electric field $E(t)$. The restoring force is analogous to the restoring force acting on a mass suspended from a spring and given a small displacement from its equilibrium position. The damping can be regarded as a viscous drag that the moving electron experiences because of its interaction with the other electrons on the particle.

We take $E(t) = R(Ee^{i\omega t})$ and $x(t) = R[X(\omega)e^{i\omega t}]$. The possibility that $E(t)$ and $x(t)$ are not in phase is taken into account by allowing the function $X(\omega)$ to include a phase factor. If we define a resonant frequency by $\omega_0 = \sqrt{k/m}$, then the differential equation (2.157) becomes

$$(\omega_0^2 - \omega^2)X + 2i\omega\Gamma X = -\frac{e}{m}E \quad (2.158)$$

giving

$$X(\omega) = \frac{-(e/m)E}{\omega_0^2 - \omega^2 + 2i\omega\Gamma}. \quad (2.159)$$

This is the amplitude of the displacement of the electron from its equilibrium position as a function of the frequency of the applied field.

Near resonance $\omega \simeq \omega_0$, so

$$X(\omega \simeq \omega_0) = \frac{-(e/m)E}{2\omega_0(\omega_0 - \omega) + 2i\omega_0\Gamma}. \quad (2.160)$$

The dipole moment of a single particle is $\mu(t) = -e[x(t)]$, which arises from the separation of the electron charge cloud and the nucleus.

If there are N electron oscillators per unit volume, there results a net polarization (dipole moment per unit volume) of

$$P(t) = -Nex(t) = R[P(\omega)e^{i\omega t}] \quad (2.161)$$

where $P(\omega)$ is the complex amplitude of the polarization

$$P(\omega) = -NeX(\omega) = \frac{(Ne^2/m)E_0}{2\omega_0(\omega_0 - \omega) + 2i\omega_0\Gamma}$$

$$= \frac{-i(Ne^2/(2m\omega_0\Gamma))}{1 + i(\omega - \omega_0)/\Gamma}E_0 \quad (2.162)$$

The electronic susceptibility $\chi(\omega)$ is defined by the equation

$$P(\omega) = \varepsilon_0\chi(\omega)E_0 \quad (2.163)$$

where ε_0 is the permittivity of free space.[12] The susceptibility $\chi(\omega)$ is complex and can be written in terms of its real and imaginary parts as

$$\chi(\omega) = \chi'(\omega) - i\chi''(\omega) \quad (2.164)$$

where the use of the negative sign is a common convention.

The polarization is

$$P(t) = R[\varepsilon_0\chi(\omega)E_0e^{i\omega t}] = \varepsilon_0E_0\chi'(\omega)\cos\omega t$$
$$+ \varepsilon_0E_0\chi''\sin\omega t. \quad (2.165)$$

Therefore, $\chi'(\omega)$, the real part of the susceptibility, is related to the in-phase polarization, while $\chi''(\omega)$, the complex part, is related to the out-of-phase (quadrature) component. From equations (2.162) and (2.163),

$$\chi(\omega) = -i\left(\frac{Ne^2}{2m\omega_0\Gamma\varepsilon_0}\right)\frac{1}{1 + i(\omega - \omega_0)/\Gamma} \quad (2.166)$$

so

$$\chi'(\omega) = \left(\frac{Ne^2}{2m\omega_0\Gamma\varepsilon_0}\right)\frac{(\omega_0 - \omega)/\Gamma}{1 + (\omega - \omega_0)^2/\Gamma^2} \quad (2.167)$$

$$\chi''(\omega) = \left(\frac{Ne^2}{2m\omega_0\Gamma\varepsilon_0}\right)\frac{1}{1 + (\omega - \omega_0)^2/\Gamma^2}. \quad (2.168)$$

Changing to conventional frequency, $v = \omega/2\pi$, and putting $\Delta v = \Gamma/\pi$, which is the FWHM of the Lorentzian shape that describes $\chi''(\omega)$, we obtain

$$\chi''(v) = \left(\frac{Ne^2}{16\pi^2mv_0\varepsilon_0}\right)\frac{\Delta v}{(\Delta v/2)^2 + (v - v_0)^2} \quad (2.169)$$

---

[12] $\varepsilon_0 = 8.854 \times 10^{-12}$ Fm^{-1}

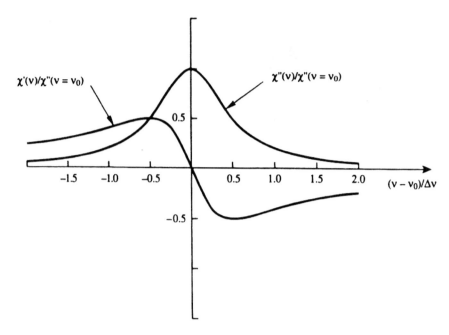

FIGURE 2.30 Frequency variation of the real, $\chi'(v)$, and imaginary, $\chi''(v)$, parts of the susceptibility calculated using the electron oscillator model.

$$\chi'(v) = \frac{2(v_0 - v)}{\Delta v}\chi''(v) = \left(\frac{Ne^2}{8\pi^2 m v_0 \varepsilon_0}\right)\frac{v_0 - v}{(\Delta v/2)^2 + (v - v_0)^2}$$

(2.170)

Figure 2.30 is a plot of χ'' and χ' normalized to the peak value χ_0'' of χ''. Note that χ'' has the characteristic Lorentzian shape common to the frequency response of RLC circuits and homogeneously broadened spectral lines.[13]

2.12.1 The Connection between the Complex Susceptibility, Gain and Absorption

The complex susceptibility of a particle, or a medium containing a collection of particles, is closely related to the change in amplitude of an electromagnetic wave passing through such a medium. We can make this connection by considering some fundamental concepts in classical electromagnetic theory.

The relationship between the applied electric field $\boldsymbol{E}$ and the electron displacement vector $\boldsymbol{D}$ is

$$\boldsymbol{D} = \varepsilon_0 \boldsymbol{E} + \boldsymbol{P}$$
$$= \varepsilon_0 (1 + \chi) \boldsymbol{E}$$

(2.171)

which, by introducing the dielectric constant $\varepsilon_r = 1 + \chi$, can be written as

$$\boldsymbol{D} = \varepsilon_0 \varepsilon_r \boldsymbol{E}.$$

(2.172)

The refractive index n of the medium is related to ε_r by $n = \sqrt{\varepsilon_r}$.[14]

When an external electric field interacts with a group of particles, there are two contributions to the induced polarization, a macroscopic contribution $\boldsymbol{P}_m$ from the collective properties of the particles, for example, their arrangement in a crystal lattice, and a contribution from the polarization $\boldsymbol{P}_t$ associated with transitions in the medium, so, in general

$$\boldsymbol{P} = \boldsymbol{P}_m + \boldsymbol{P}_t.$$

(2.173)

Usually, there are many transitions possible for the particles of the medium but only one will be in near resonance with the frequency of an applied field. $\boldsymbol{P}_t$ is dominated by the contribution of this single transition near resonance. Far from any such resonance $\boldsymbol{P}_t$ is negligible and $\boldsymbol{P} = \boldsymbol{P}_m$, which allows us to define the macroscopic dielectric constant ε_r (far from resonance) from

$$\boldsymbol{P} = \varepsilon_0 \boldsymbol{E} + \boldsymbol{P}_m = \varepsilon_0 \boldsymbol{E} + \chi_m \varepsilon_0 \boldsymbol{E} = \varepsilon_r \varepsilon_0 \boldsymbol{E}.$$

(2.174)

where χ_m is the macroscopic, non-resonant susceptibility. However, if the frequency of the electric field is near the frequency of a possible transition within the medium, then there is a significant contribution to $\boldsymbol{P}$ from this transition. Other possible transitions far from resonance do not contribute, and we can write

$$\boldsymbol{D} = \varepsilon_0 \boldsymbol{E} + \boldsymbol{P}_m + \boldsymbol{P}_t = \varepsilon_r \varepsilon_0 \boldsymbol{E} + \boldsymbol{P}_t.$$

(2.175)

$\boldsymbol{P}_t$ is related to the complex susceptibility that results from the transition according to $\boldsymbol{P}_t = \varepsilon_0 \chi(\omega)\boldsymbol{E}$. Therefore, we can rewrite equation (2.171) as

---

[13] It can be shown that although equations (A1.1.169) and (A1.1.170) have the correct frequency dependence for the susceptibility of a real particle, they have incorrect magnitudes. This arises because of inadequacies in the classical electron oscillator model and these inadequacies should be borne in mind before the classical model is used to make detailed predictions about the behaviour of a particle.

[14] In a magnetic material with relative magnetic permeability $\mu_r \neq 1$, $n = \sqrt{\mu_r \varepsilon_r}$.

Basic Laser Principles

$$\boldsymbol{D} = \varepsilon_0[\varepsilon_r + \chi(\omega)]\,\boldsymbol{E} = \varepsilon_0 \varepsilon_r^* \boldsymbol{E} \quad (2.176)$$

so the complex susceptibility modifies the effective dielectric constant from ε_r to ε_r^*.

When an electromagnetic wave propagates through a medium with a complex susceptibility, both the amplitude and phase velocity of the wave are affected. This can be illustrated easily for a plane wave propagating in the z-direction with a field variation $\sim e^{i(\omega t - kz)}$, where the propagation constant, k, is given by the expression

$$k = \omega\sqrt{\mu\varepsilon} = \omega\sqrt{\mu_r\mu_0\varepsilon_r\varepsilon_0}. \quad (2.177)$$

The relative permeability μ_r is generally unity for optical materials. For a complex dielectric constant, equation (2.177) can be rewritten as

$$k' = \omega\sqrt{\mu_0\varepsilon_r\varepsilon_0}\sqrt{1 + \frac{\chi(\omega)}{\varepsilon_r}} = k\sqrt{1 + \frac{\chi(\omega)}{\varepsilon_r}} \quad (2.178)$$

where k' is now the new propagation constant, which differs from the non-resonant propagation constant k because of the complex susceptibility resulting from a transition. If $|\chi(\omega)| \ll \varepsilon_r$, equation (2.178) can be simplified by the use of the binomial theorem to give

$$k' = k\left[1 + \frac{\chi(\omega)}{2\varepsilon_r}\right] = k\left[1 + \frac{\chi'(\omega)}{2\varepsilon_r} - \frac{i\chi''(\omega)}{2\varepsilon_r}\right] \quad (2.179)$$

The wave now propagates through the medium as $e^{-ik'z}$. Written out in full, the electric field varies as

$$E = E_0 \exp\left(i\left\{\omega t - k\left[1 + \frac{\chi'(\omega)}{2\varepsilon_r} - \frac{i\chi''(\omega)}{2\varepsilon_r}\right]z\right\}\right)$$

$$= E_0 \exp\left(i\left\{\omega t - k\left[1 + \frac{\chi'(\omega)}{2\varepsilon_r}\right]\right\}\right)\exp\left[-\frac{k\chi''(\omega)}{2\varepsilon_r}z\right]. \quad (2.180)$$

Clearly, this is a wave whose phase velocity is

$$c' = \frac{\omega}{k[1 + \chi'(\omega)/2\varepsilon_r]} = \frac{\omega}{k + \Delta k} \quad (2.181)$$

and whose field amplitude changes exponentially with distance. If we write

$$\gamma = -\frac{k\chi''(\omega)}{\varepsilon_r} \quad (2.182)$$

then the wave changes its electric field amplitude with distance according to $e^{(\gamma/2)z}$.

The intensity of the wave is $I \propto E(z,t)E^*(z,t)$, which changes as the wave passes through the medium as $I \propto e^{\gamma z}$. We can identify γ as the familiar gain coefficient of the medium, which was calculated previously by considering the spontaneous and stimulated radiative jumps between two energy levels.

Now from equation (2.182)

$$\gamma(\nu) = -\frac{k\chi''(\nu)}{n^2} = -\left(\frac{2\pi\nu_0 n}{c_0}\right)\frac{\chi''(\nu)}{n^2} \quad (2.183)$$

which, from equation (2.184), obtained by consideration of the system as a collection of classical oscillators, is

$$\gamma(\nu) = -\left(\frac{2\pi\nu}{nc_0}\right)\left(\frac{Ne^2}{16\pi^2 m\nu_0\varepsilon_0}\right)\frac{\Delta\nu}{(\Delta\nu/2)^2 + (\nu - \nu_0)^2} \quad (2.184)$$

which is *always negative*. Clearly, this is an incorrect result, since we have previously shown that

$$\gamma(\nu) = \left(N_2 - \frac{g_2}{g_1}N_1\right)\frac{c^2 A_{21}}{8\pi\nu^2}g(\nu_0, \nu) \quad (2.185)$$

which can be positive or negative depending on the sign of $N_2 - (g_2/g_1)N_1$. Thus, the classical electron oscillator model appears to predict only absorption of incident radiation. It is possible, however, within the framework of the classical electron oscillator model to show that, in certain conditions, stimulated emission can occur. Although the classical electron oscillator model is instructive, it is not entirely adequate in describing the interaction between particles and radiation. It is better to accept that $\gamma(\nu) = -k\chi''(\nu)/n^2$ and use equation (2.185) as the expression for $\gamma(\nu)$. In this case, we find that the imaginary part of the complex susceptibility of the medium is

$$\chi''(\nu) = -\frac{n^2\gamma(\nu)}{k} = -\left(\frac{n^2 c}{2\pi\nu}\right)\gamma(\nu) \quad (2.186)$$

which, from equation (2.183), gives

$$\chi''(\nu) = -\frac{[N_2 - (g_2/g_1)N_1]\,n^2 c^3 A_{21}}{8\pi^3\nu^3\Delta\nu}\cdot\frac{1}{1 + [2(\nu - \nu_0)/\Delta\nu]^2}. \quad (2.187)$$

This quantum mechanical susceptibility is negative or positive depending on whether $N_2 - (g_2/g_1)N_1$ is positive or not. A *negative* value of $\chi''(\nu)$ corresponds to a system in population inversion.

2.12.2 The Classical Oscillator Explanation for Stimulated Emission

If there were no applied electric field acting on the electron, then, from equation (2.157), the position of the electron would satisfy the equation

$$x(t) = x_0 e^{-\Gamma t}\cos(\omega_0 t + \phi_0) \quad (2.188)$$

where ω_0 is the resonant frequency and ϕ_0 is a phase factor set by the initial conditions. If at time $t = 0$ the position and velocity of the electron are a_0, v_0, respectively, then

$$a_0 = x_0 \cos\phi_0$$
$$v_0 = -\Gamma x_0\cos\phi_0 - \omega_0 x_0 \sin\phi_0. \quad (2.189)$$

When the electric field is applied, we have already seen that, in the steady state, energy is apparently only absorbed. However, this impression is erroneous. It neglects that, in reality, no electromagnetic field interacts indefinitely with an electron. Therefore, we must consider what happens when an electron, which can already be regarded as oscillating if it is in an excited state, is suddenly subjected to the additional perturbation of an applied field. We are going to be interested in the behaviour of the electron over the first few cycles of the applied field, so we neglect damping and write

$$\frac{d^2x}{dt^2} + \omega_0^2 = -\frac{eE_0}{2m}\left(e^{i\omega t} + e^{-i\omega t}\right) \quad (2.190)$$

where the applied field is $E_0 \cos \omega t$ and has been written in its complex exponential form. By introducing a new variable $z = \dot{X} + i\omega_0 x$, where the dot indicates differentiation with respect to time, equation (2.190) can be rewritten in the form

$$\frac{dz}{dt} - i\omega_0 z = -\frac{eE_0}{2m}\left(e^{i\omega t} + e^{-i\omega t}\right). \quad (2.191)$$

This equation can be solved by multiplying each term by $e^{-i\omega_0 t}$ and then integrating to give

$$ze^{-i\omega_0 t} = -\frac{eE_0}{2m}\int \left(e^{i(\omega-\omega_0)t} + e^{-i(\omega+\omega_0)t}\right)dt. \quad (2.192)$$

This gives

$$\frac{dx}{dt} + i\omega_0 x = -\frac{eE_0}{2m}\left[-\frac{ie^{i\omega t}}{(\omega-\omega_0)} + \frac{ie^{-i\omega t}}{(\omega+\omega_0)} + Ae_0^{i\omega t}\right] \quad (2.193)$$

where A is a constant of integration. By integrating a second time in a similar manner and introducing the initial values of position and velocity, it is straightforward to show that the final solution is

$$x(t) = -\frac{eE_0}{m}\left[\cos\omega_0 t - \frac{\cos\omega t}{(\omega^2 - \omega_0^2)}\right] + \sqrt{\left(\frac{v_0}{\omega_0}\right)^2 + x_0^2}\cos(\omega_0 t + \phi)$$

(2.194)

where $\tan\phi = -v_0/\omega_0 x_0$.

By the use of the trigonometrical identity,

$$\cos X - \cos Y = -2\sin\left(\frac{X+Y}{2}\right)\sin\left(\frac{X-Y}{2}\right) \quad (2.195)$$

and assuming that the applied frequency is close to resonance, equation (2.194) can be written

$$x(t) = -\frac{eE_0}{2m\omega_0}t\sin\omega_0 t + \sqrt{\left(\frac{v_0}{\omega_0}\right)^2 + x_0^2}\cos(\omega_0 t + \phi). \quad (2.196)$$

Thus, near resonance, the amplitude of oscillation will increase linearly with time, which is, of course, a consequence of our neglect of damping. It is more interesting, however, to use equation (2.196) to calculate the work done during the first n cycles of the applied field. This work is calculated as the work done by the electric field in polarizing the medium: the polarization P is proportional to electron displacement. The work done is $E \cdot \partial P/\partial t$ [2]. During the first n cycles, the total work done by the field is

$$W = \int_0^{2n\pi/\omega_0} E \cdot \frac{\partial P}{\partial t}dt = -NeE_0\int_0^{2n\pi/\omega_0}(\cos\omega_0 t)\dot{x}(t)dt \quad (2.197)$$

where N is the total number of particles per unit volume. Writing $(v_0/\omega_0)^2 + x_0^2 = a^2$ and substituting from equation (2.196) gives

$$W = -NeE_0\int_0^{2n\pi/\omega_0}\left[-\frac{eE_0}{m\omega_0}\sin 2\omega_0 t - \frac{eE_0 t}{4}(1+\cos 2\omega_0 t) - \frac{a\omega_0}{2}\sin\phi + \frac{a\omega_0}{2}\sin(2\omega_0 t + \phi)\right]dt$$

(2.198)

Clearly, the first and last terms of the integrand average to zero over a whole number of cycles. The remaining terms can be integrated to give

$$W = \frac{Ne^2 E_0^2}{m}\left(\frac{n^2\pi^2}{2\omega_0^2} + \frac{n\pi ma}{eE_0}\sin\phi\right). \quad (2.199)$$

This work done by the applied field is negative, implying that the oscillating electrons supply energy to the field if $\sin\phi < 0$ and $|\sin\phi| > eE_0 n\pi/2ma\omega_0^2$. This is the condition set by classical theory for stimulated emission to occur. Because the charge on the electron is negative, stimulated emission can only result if the electron velocity when the applied field is turned on is in the direction of the field. If the electron velocity is in the same direction as the field, the electron is decelerated by the field and, consequently, radiates energy. If the electrons were accelerated by the field then absorption of energy from the field would occur.

There are a maximum number of cycles of the applied field after which the oscillating electrons start, and continue indefinitely, to absorb energy. This is set by the condition

$$n < 2ma\omega_0^2/eE_0\pi. \quad (2.200)$$

After a long enough time, the motion of the electron is dominated by the first term in equation (2.196) and can be written

$$x(t) = -\frac{eE_0}{2m\omega_0}t\sin\omega_0 t \quad (2.201)$$

and the electron velocity is, for large enough t,

$$\dot{x}(t) \simeq -\frac{eE_0}{2m}t\cos\omega_0 t. \quad (2.202)$$

The electron now has a velocity that is oppositely directed from the applied field, and the electron is being accelerated and absorbs energy from the field.

We can conclude by saying that when an electric field near resonance is applied to an already oscillating electron, stimulated emission can occur at early times provided the initial velocity of the electron is in the same direction as the field.

Basic Laser Principles

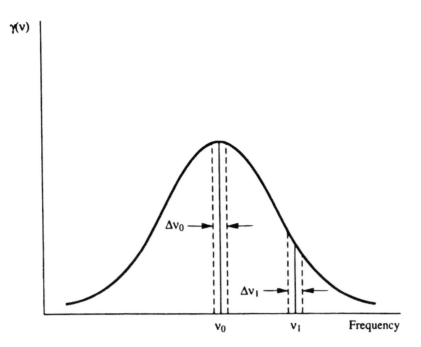

FIGURE 2.31 To illustrate how laser oscillation might build up at two different frequencies in a non-interactive laser cavity.

2.13 From Amplifier to Oscillator— the Feedback Structure

A feedback structure is used to channel output emissions from a laser amplifier back through the amplifier. The simplest example consists of a pair of plane mirrors placed parallel to each other but at opposite ends of the amplifying medium. The structure is called the laser resonator or optical cavity (see Section 2.2). It is often referred to as a Fabry–Pérot resonator.[15] The start of laser oscillation begins with spontaneous emission that occurs close to the axis of the resonator system: the direction perpendicular to both mirrors. This spontaneous emission is redirected by the mirrors back through the amplifying medium and its intensity grows. The fact that only photons that are travelling close to the axis direction can make many passes through the amplifying medium, especially if this is of small width, explains the directional character of laser beams.

The frequency spectrum of laser radiation is controlled by the interaction between the lineshape function of the amplifying transition and the properties of the optical resonator. A plane wave bouncing back and forth between the two flat mirrors of an empty cavity will 'resonate' if the wavelength of the radiation satisfies the simple condition, with m integer,

$$\frac{m\lambda}{2} = \ell \quad (2.203)$$

where ℓ is the axial spacing between the two parallel mirrors. This is the simple condition for the bouncing waves to be in phase and develop a maximum amplitude standing wave between the two mirrors, much like the acoustic resonances of an open or closed pipe. When an amplifying medium is present, this resonance condition must still be satisfied and determines the precise frequency (or frequencies) at which laser oscillation can occur. In this case, the effective 'length' ℓ of the cavity is modified by the presence of the amplifier.

Before exploring this further, let us consider at what frequency a laser would oscillate if the resonator did not interact with the gain profile of the amplifying medium in any ways. Suppose the amplifying medium has a gain profile (gain/frequency response) of a Gaussian form, as shown in Figure 2.31. Such a gain profile occurs in a gaseous amplifying medium where the individual homogeneous lineshapes of the particles are significantly narrower than the overall Doppler width of the spontaneous transition.

The maximum gain of the medium is at frequency ν_0, the line centre, so it is perhaps logical to expect that oscillation will build up at this frequency rather than at any other. If we view the build-up of oscillation as a process triggered by spontaneous emission, we can see why this is so. A photon travelling in a direction that keeps it bouncing back and forth within the resonator is more likely to be emitted in a narrow band of frequencies $\Delta\nu_0$ near ν_0 than in some other band $\Delta\nu_1$ at frequency ν_1. As oscillation builds up, one can imagine photons spontaneously emitted at all points of the lineshape being amplified to some extent but oscillation at ν_0 builds up fastest. As its intensity grows, it depletes the atomic population by causing sufficient stimulated emission that the medium ceases to be amplifying at frequencies near ν_0 (within a few homogeneous widths, say). If the medium has a homogeneous (Lorentzian) gain profile, then since photons oscillating at ν_0 can stimulate emission from all the atoms in the medium, it is easy to see that oscillation at frequency ν_0 can suppress oscillation at any other frequencies under the gain profile. The possibility of additional oscillation at frequencies far away from

---

[15] Named for Marie P A C Fabry and Jean B G G A Pérot, the inventors of the plane-parallel mirror structure developed originally as an optical instrument for interferometry and first described in 1899.

ν_0 in an inhomogeneously broadened gaseous amplifier is not precluded by this discussion: this, in fact, often happens, as we shall see later.

The monochromatic character of the oscillation can be predicted by a simple consideration of the shape of the gain profile of the amplifier. In the early stages of oscillation, photons with a frequency distribution $g(\nu_0, \nu)$ (the total lineshape) are being amplified in a material whose gain/frequency response is $\gamma(\nu)$ (proportional to $g(\nu_0, \nu)$). The amplification process changes the lineshape of the emitted photons circulating in the cavity by a process that is dependent on the product of $g(\nu_0, \nu)$ and $\gamma(\nu)$, that is, on $[g(\nu_0, \nu)]^2$. The resulting profile of the laser radiation is dependent on higher powers of $[g(\nu_0, \nu)]^2$ as the oscillation is dependent on many passes of photons back and forth through the amplifying medium. For Gaussian lineshapes like $e^{-[2(\nu-\nu_0)/\Delta\nu_D]^2 \ln 2}$, which is like e^{-x^2/σ^2}, the product of two lineshapes produces a narrower profile, for example

$$\left[e^{-x^2/\sigma^2} \right]^2 = e^{-2x^2/\sigma^2} = e^{-x^2/(\sigma/\sqrt{2})^2}, \quad (2.204)$$

a function that has a width $1/\sqrt{2}$ of than the original. The same can also be shown to be true for Lorentzian profiles. In both cases, the gain of the medium causes a narrowing of the original, spontaneously emitted lineshape. Thus, we can see that in a non-interactive laser cavity, the laser oscillation will be highly monochromatic and at the line centre.

2.14 Optical Resonators Containing an Amplifying Media

When an optical resonator is filled with an amplifying medium, laser oscillation will occur at specific frequencies if the gain of the medium is large enough to overcome the loss of energy through the mirrors and by other loss mechanisms within the laser medium. The onset of laser oscillation and the frequency, or frequencies, at which it occurs is governed by threshold amplitude and phase conditions. Once laser oscillation is established, it stabilizes at a level that depends on the saturation intensity of the amplifying medium and the reflectance of the laser mirrors.

Figure 2.32 represents an optical (Fabry–Pérot) resonator, whose interior is filled with an amplifying medium and which has plane mirrors. We consider the complex amplitudes of the waves bouncing backwards and forwards normally between the resonator mirrors. These waves result from an incident beam with electric vector E_0 at the first mirror, as shown in Figure 2.32, where E_0 is the complex amplitude at some reference point. The reflection and transmission coefficients in the various directions at the mirrors are as shown in the figure. For example, t is the transmission coefficient for electromagnetic fields passing through the left-hand reflector and r_1 is the reflection coefficient for fields striking the left-hand reflector from inside the resonator. A, A_1 and A_2 are absorption coefficients, which lead to energy dissipation in the resonator reflectors. These coefficients are not used explicitly in the analysis that follows, but they modify the values of the reflection and transmission coefficients. If the mirrors are lossless, then their reflectances R_1, R_2 and transmittances T_1, T_2 obey $R_1 + T = 1$, etc., and $|r_1|^2 = R_1$, $|r_2|^2 = R_2$, $|t|^2 = |t_1|^2 = T_1$, $|t_2|^2 = T_2$. The absorption coefficients, A, A_1, A_2, at the two mirrors include any reflection losses or scattering that send energy out of the resonator: we do not at this stage include diffraction losses, which result from the finite lateral dimensions of the mirrors or medium.

If there were nothing inside the resonator, then a wave propagating between the mirrors would propagate as $E_0 e^{i(\omega t - kz)}$ to the right and $E_0 e^{i(\omega t - kz)}$ to the left. The presence of a gain medium changes the otherwise passive propagation factor k to

$$k'(\omega) = k + \Delta k = k + \frac{k\chi'(\omega)}{2n^2} \quad (2.205)$$

and the gain coefficient $\gamma(\omega) = -k\chi''(\omega)/n^2$ causes the amplitude of the wave to change with distance as $e^{(\gamma/2)z}$. We allow for the possible existence of a distributed loss per pass given by an absorption coefficient α. Such absorption causes a fractional change in intensity for a single pass through the medium of $e^{-\alpha \ell}$. Such a distributed loss could, for example, arise from scattering by imperfections in a solid laser medium. This distributed loss modifies the complex amplitude by a factor $e^{-i\alpha\ell/2}$

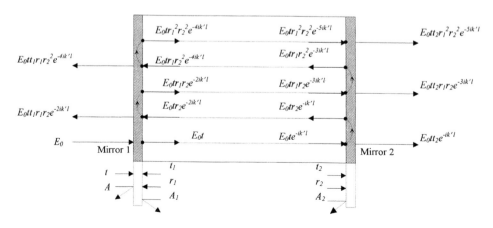

FIGURE 2.32 The amplitude of the electric field vectors of the successively transmitted, amplified and reflected waves in a Fabry–Pérot resonator system containing an amplifying (or absorbing) medium. The absorption factors A, A_1 and A_2 are not used explicitly in the analysis given in the text, but they modify the values of t, t_1 and, t_2.

Basic Laser Principles

per pass. Therefore, the full propagation constant of the wave in the presence of both gain and loss is

$$k'(\omega) = k + k\frac{\chi'(\omega)}{2n^2} - \frac{ik\chi''(\omega)}{2n^2} - \frac{i\alpha}{2} \quad (2.206)$$

and the wave propagates as $e^{i(\omega t \pm k'z)}$.

A wave travelling to the right with complex amplitude E_0 at plane $z = 0$ in the resonator, the left-hand mirror, has at plane ℓ, the right-hand mirror, become

$$E = E_0 e^{i(\omega t - k\ell)} = E_0 e^{-ik\ell} e^{i\omega t} = E_0' e^{i\omega t}.$$

This wave then begins to propagate to the left as

$$E = E_0' e^{i(\omega t + kz)}.$$

At plane $-\ell$, the left-hand mirror, with the right-hand mirror now taken as the origin, has become once more a wave travelling to the right

$$E = E_0 e^{ik\ell} e^{i(\omega t - k\ell)} = E_0 e^{-2ik\ell} e^{i\omega t}.$$

In this way, we can write down the complex amplitudes of successive rays travelling at normal incidence between the two reflectors, as shown in Figure 2.32.

The output beam through the right-hand mirror arises from the transmission of waves travelling to the right: its total electric field amplitude is

$$E_t = E_0 tt_2 e^{-ik'\ell} + E_0 tt_2 r_1 r_2 e^{-3ik'\ell} + \cdots$$

$$= E_0 tt_2 e^{-ik'\ell} + (1 + r_1 r_2 e^{-2ik'\ell} + r_1^2 r_2^2 e^{-4ik'\ell} + \cdots)$$

$$= \frac{E_0 tt_2 e^{-ik'\ell}}{1 - r_1 r_2 e^{-2ik'\ell}} \quad (2.207)$$

$$= \frac{E_0 tt_2 e^{-1(k+\Delta k)\ell} e^{(\gamma-\alpha)\ell/2}}{1 - r_1 r_2 e^{-2i(k+\Delta k)\ell} e^{(\gamma-\alpha)\ell}}$$

where

$$\gamma(\nu) = \left[N_2 - \left(\frac{g_2}{g_1}\right) N_1 \right] \left(\frac{c^2 A_{21}}{8\pi\nu^2}\right) g(\nu_0, \nu). \quad (2.208)$$

The ratio of input-to-output intensities is

$$\left(\frac{E_t}{E_0}\right)^2 = \frac{I_t}{I_0} = \frac{t^2 t_2^2 e^{(\gamma-\alpha)\ell}}{\left(1 - r_1 r_2 e^{-2i(k+\Delta k)\ell} e^{(\gamma-\alpha)\ell}\right)\left(1 - r_1 r_2 e^{2i(k+\Delta k)\ell} e^{(\gamma-\alpha)\ell}\right)} \quad (2.209)$$

which becomes

$$\frac{I_t}{I_0} = \frac{t^2 t_2^2 e^{(\gamma-\alpha)\ell}}{(1 - r_1^2 r_2^2 e^{2(\gamma-\alpha)\ell} - 2r_1 r_2 e^{(\gamma-\alpha)\ell}[\cos 2(k+\Delta k)\ell]}. \quad (2.210)$$

In a passive resonator, which has no gain γ or loss α, $\Delta k = 0$, and if $|r_1|^2 = |r_2|^2 = R$

$$\frac{I_t}{I_0} = \frac{T^2}{1 + R^2 - 2R\cos 2k\ell}. \quad (2.211)$$

This function is shown in Figure 2.33 for a few values of the variable $2k\ell = \delta$ and for three different values of the mirror reflectance R. Resonances in the transmittance (I_t/I_0) of this structure satisfy $2k\ell = m\pi$ and correspond to wavelengths that satisfy the condition $m\lambda/2 = \ell$. These wavelengths also correspond to the largest standing wave fields between the two mirrors. If R is close to 1, then the resonances in Figure 2.33 are very sharp. It is easy to show [2] that they approximate a Lorentzian shape with an FWHM defined as

$$\Delta\nu_{\frac{1}{2}} = \Delta\nu/F \quad (2.212)$$

where $\Delta\nu = c/2\ell$ and $F = \pi\sqrt{R}/(1-R)$. $\Delta\nu$ is called *the free spectral range* of the resonator, and F is called its *finesse*.

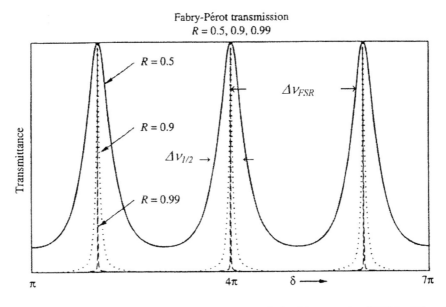

FIGURE 2.33 Theoretical transmittance (I_t/I_0) characteristic of a laser resonator calculated from equation (2.211) for different values of (equal) mirror reflectance R. The theoretical reflection characteristic that corresponds to these curves can be viewed by turning the picture upside down.

In a resonator containing an active medium, as $\gamma - \alpha$ increases from zero, the denominator of equation (2.207) approaches zero and the whole expression blows up when

$$r_1 r_2 e^{-2i(k+\Delta k)\ell} e^{(\gamma-\alpha)\ell} = 1. \quad (2.213)$$

When this happens, we have an infinite amplitude transmitted wave for a finite amplitude incident wave. In other words, a finite amplitude transmitted wave for zero incident wave—*oscillation*. Physically, equation (2.213) is the condition that must be satisfied for a wave to make a complete round trip inside the resonator and return to its starting point with the same amplitude and, apart from a multiple of 2π, with the same phase.

Equation (2.213) provides an amplitude condition for oscillation that gives an expression for the threshold gain constant, $\gamma_t(\nu)$,

$$r_1 r_2 e^{[\gamma_t(\omega)-\alpha]\ell} = 1. \quad (2.214)$$

To satisfy equation (2.213), $e^{-2i(k+\Delta k)\ell}$ must be real, which provides us with the phase condition

$$2[k + \Delta k(\nu)]\ell = 2\pi m \qquad m = 1, 2, 3 \ldots \quad (2.215)$$

The threshold gain coefficient can be written

$$\gamma_t(\nu) = \alpha - \frac{1}{\ell} \ln r_1 r_2 \quad (2.216)$$

which from the gain equation (2.185) gives the population inversion needed for oscillation

$$\left(N_2 - \frac{g_2}{g_1} N_1\right)_t = \frac{8\pi}{g(\nu_0, \nu) A_{21} \lambda^2} \left(\alpha - \frac{1}{\ell} \ln r_1 r_2\right). \quad (2.217)$$

For a homogeneously broadened transition, the parametric variation of equation (2.217) that depends on the gain medium can be written as

$$\left(N_2 - \frac{g_2}{g_1} N_1\right)_t \propto \frac{\Delta \nu}{A_{21} \lambda^2}. \quad (2.218)$$

Whereas, for an inhomogeneously broadened transition, since $\Delta \nu_D \propto 1/\lambda$,

$$\left(N_2 - \frac{g_2}{g_1} N_1\right)_t \propto \frac{1}{A_{21} \lambda^3}. \quad (2.219)$$

Clearly, lower inversions are needed to achieve laser oscillation at longer wavelengths. It is much easier to build lasers that oscillate in the infrared than at visible, ultraviolet or X-ray wavelengths. For example, in an inhomogeneously broadened laser, a population inversion 10^6 times greater would be required for oscillation at 200 nm than at 20 µm (all other factors such as A_{21} being equal). In practice, since A_{21} factors generally increase at shorter wavelengths, the difference in population inversion may not need to be as great as this.

In a resonator as shown in Figure 2.32, if $R_1 = r_1^2 \simeq 1$, $R_2 = r_2^2 \simeq 1$ and distributed losses are small, a wave starting with intensity I inside the resonator will, after one complete round trip, have intensity $I R_1 R_2 e^{-2\alpha \ell}$, the change in intra-cavity intensity after one round trip is

$$dI = (R_1 R_2 e^{-2\alpha \ell} - 1) I. \quad (2.220)$$

This loss occurs in a time $dt = 2\ell/c$. So,

$$\frac{dI}{dt} = cI[R_1 R_2 e^{-2\alpha \ell} - 1]/2\ell. \quad (2.221)$$

This equation has the solution

$$I = I_0 \exp\{-[1 - R_1 R_2 e^{-\alpha \ell}] ct/2\ell\} \quad (2.222)$$

where I_0 is the intensity at time $t = 0$. The time constant for intensity (energy) loss is

$$\tau_0 = \frac{2\ell}{c\left(1 - R_1 R_2 e^{-2\alpha \ell}\right)}. \quad (2.223)$$

Now if $R_1 R_2 e^{-2\alpha \ell} \simeq 1$, with α small as we have assumed here, then

$$\left(1 - R_1 R_2 e^{-2\alpha \ell}\right) \simeq -\ln\left(R_1 R_2 e^{-2\alpha \ell}\right) = -\ln(R_1 R_2) + 2\alpha \ell \quad (2.224)$$

and we get

$$\tau_0 = \frac{2\ell}{c(2\alpha \ell - \ln R_1 R_2)} = \frac{1}{c[\alpha - (1/\ell) \ln r_1 r_2]}. \quad (2.225)$$

Thus, the threshold population inversion can be written

$$N_t = \frac{8\pi}{A_{21} \lambda^2 g(\nu) c \tau_0}. \quad (2.226)$$

Threshold population inversion—numerical example. For the 488 nm transition in the argon ion laser (see Chapter 30)

$$\lambda = 488 \text{ nm} \quad c = 3 \times 10^8 \text{ ms}^{-1} \quad A_{21} \simeq 10^9 \text{s}^{-1} \quad \Delta \nu_D \sim 3 \text{ GHz}.$$

Take $\ell = 1$ m, $R_1 = 100\%$, $R_2 = 90\%$ (typical values for a practical device). Since this is a gas laser, internal losses are easily kept small so $\alpha \simeq 0$. In this case

$$\tau_0 = 2\ell/c(1 - R_1 R_2)$$
$$= 66.67 \text{ ns}.$$

For oscillation at or near line centre,

$$g(\nu_0, \nu_0) = \frac{2}{\Delta \nu_D} \sqrt{\frac{\ln 2}{\pi}} = \frac{0.94}{\Delta \nu_D} \sim \frac{1}{\Delta \nu_D}.$$

The threshold inversion is, from equation (2.226),

$$N_t = \frac{8\pi \times 3 \times 10^9}{10^9 \times (488 \times 10^{-9})^2 \times 3 \times 10^8 \times 66.67 \times 10^{-9}}$$
$$= 1.58 \times 10^{13} \text{ m}^{-3}.$$

Basic Laser Principles

2.15 The Oscillation Frequency

To determine the frequency at which laser oscillation can occur, we return to the phase condition, equation (2.215). This phase condition was

$$(k + \Delta k)\ell = m\pi \quad (2.227)$$

which, from equation (2.205), gives

$$k\ell\left[1 + \frac{\chi'(v)}{2n^2}\right] = m\pi. \quad (2.228)$$

Now, from equation (2.170),

$$\chi'(v) = \frac{2(v_0 - v)}{\Delta v}\chi''(v) \quad (2.229)$$

where v_0 is the line-centre frequency and Δv is its *homogeneous* FWHM and

$$\gamma(v) = -\frac{k\chi''(v)}{n^2}. \quad (2.230)$$

So we must have

$$\frac{2\pi v\ell}{c}\left[1 - \frac{(v_0 - v)}{\Delta v}\frac{\gamma(v)}{k}\right] = m\pi \quad (2.231)$$

and on rearranging,

$$v\left[1 - \frac{(v_0 - v)}{\Delta v}\frac{\gamma(v)}{k}\right] = \frac{mc}{2\ell} = v_m \quad (2.232)$$

where v_m is the mth resonance of the passive laser resonator in normal incidence as calculated previously. Equation (2.232) can be rewritten as

$$v = v_m - (v - v_0)\frac{\gamma(v)c}{2\pi\Delta v}. \quad (2.233)$$

We expect the actual oscillation frequency v to be close to v_m, so we can write $(v - v_0) \simeq (v_m - v_0)$ and $\gamma(v) \simeq \gamma(v_m)$ to give

$$v = v_m - (v_m - v_0)\frac{\gamma(v_m)c}{2\pi\Delta v}. \quad (2.234)$$

At threshold

$$\gamma_t(v_m) = \alpha - \frac{1}{\ell}\ln r_1 r_2 \quad (2.235)$$

and if $\alpha \simeq 0, r_1 = r_2 = \sqrt{R}$

$$\gamma_t(v_m) = \frac{(1 - R)}{\ell}. \quad (2.236)$$

Now, the FWHM of the passive resonances (the transmission intensity maxima of the Fabry–Pérot resonator) is

$$\Delta v_{1/2} = \frac{\Delta v_{FSR}}{F} = \frac{c(1 - R)}{2\pi\ell\sqrt{R}} \quad (2.237)$$

which, with $R \simeq 1$, gives

$$\Delta v_{1/2} = \frac{c(1 - R)}{2\pi\ell} \quad (2.238)$$

and, finally,

$$v = v_m - (v_m - v_0)\frac{\Delta v_{1/2}}{\Delta v}. \quad (2.239)$$

Thus, if v_m coincides with the line centre, oscillation occurs at the line centre. If $v_m \neq v_0$, oscillation takes place near v_m but is shifted slightly towards v_0. This phenomenon is called 'mode-pulling' and is illustrated in Figure 2.34.

2.15.1 Multi-mode Laser Oscillation

We have seen that for oscillation to occur in a laser system, the gain must reach a threshold value $\gamma_t(v) = \alpha - (1/\ell)\ln r_1 r_2$. For gain coefficients greater than this, oscillation can occur at,

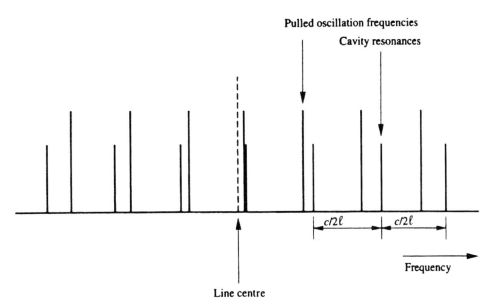

FIGURE 2.34 Relative position of line centre, cavity resonances and pulled oscillation frequencies that satisfy the phase condition (2.227).

or near (because of mode-pulling effects), one or more of the passive resonance frequencies of the Fabry–Pérot laser cavity. The resulting oscillations of the system are called longitudinal modes. As oscillation at a particular one of these mode frequencies builds up, the growing intra-cavity energy density depletes the inverted population and gain saturation sets in. The reduction in gain continues until

$$\gamma(v) = \gamma_t(v) = \alpha - \frac{1}{\ell}\ln r_1 r_2. \qquad (2.240)$$

Further reduction of $\gamma(v)$ below $\gamma_t(v)$ does not occur; otherwise, the oscillation would cease. Therefore, the gain is stabilized at the loss

$$\alpha - \frac{1}{\ell}\ln r_1 r_2.$$

Usually, α, r_1 and r_2 are nearly constant over the frequency range covered by typical amplifying transitions, so, over such moderate frequency ranges, 10^{11} Hz say, $\alpha - (1/\ell)\ln r_1 r_2$ as a function of frequency is a straight line parallel to the frequency axis. This line is called the *loss line*. At markedly different frequencies α, r_1 and r_2 can be expected to change: for example, a laser mirror with high reflectivity in the red region of the spectrum could have quite low reflectivity in the blue.

In a homogeneously broadened laser, because the reduction in gain caused by a monochromatic field is uniform across the whole gain profile, the clamping of the gain at $\gamma_t(v)$ leads to final oscillation at only one of the cavity resonance frequencies, the one where the original unsaturated gain was highest. We can show this schematically by plotting $\gamma(v)$ at various stages as oscillation builds up. Remember first the effect on $\gamma(v)$ produced by a monochromatic light signal of increasing intensity as shown in Figure 2.35. Note that the gain profile is depressed uniformly even though the saturating signal is not at the line centre, as predicted by equation (2.120).

In a laser, as oscillation begins, several such monochromatic fields start to build up at those cavity resonances where gain exceeds loss, as shown in Figure 2.36. The oscillation stabilizes when the highest (small-signal) gain has been reduced to the loss line by saturation as shown in Figures 2.37 and 2.38. Thus, in a *homogeneously* broadened laser, oscillation only occurs at *one* longitudinal mode frequency.

In an inhomogeneously broadened laser, the onset of gain saturation due to a monochromatic signal only reduces the gain locally over a region which is of the order of a homogeneous width. Only particles whose velocities (or environments in a crystal) make their centre emission frequencies lie within a homogeneous width of the monochromatic field can interact strongly with it. Schematically, the effect of an increasing intensity monochromatic field on the gain profile is as shown in Figure 2.39. A localized dip, or *hole*, in the gain profile occurs. If only one cavity resonance has a small-signal gain above the loss line then only this longitudinal mode oscillates. The stabilization of the oscillation might be expected to occur schematically as shown in Figure 2.40. However, the situation is not quite as simple as this. Oscillation at this single longitudinal mode frequency implies waves travelling in both directions inside the laser cavity. These waves can be represented by

a. the wave travelling to the right $\sim E_0\,\mathrm{e}^{i(\omega t - kz)}$ and
b. the wave travelling to the left $\sim E_0\,\mathrm{e}^{i(\omega t + kz)}$

where we choose for convenience that $\omega = 2\pi v < 2\pi v_0$. Wave (a) can interact with particles whose centre frequency is near v. These particles are, as far as their Doppler shifts are concerned, moving away from an observer looking into the laser from right to left. Their centre frequencies satisfy $v = v_0 - |v|v_0/c$,

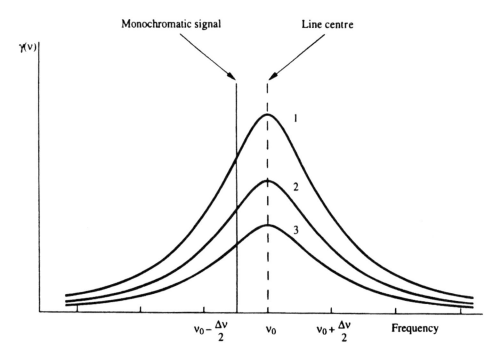

FIGURE 2.35 Saturation of gain of a homogeneously broadened transition produced by a monochromatic signal whose intensity increases from 1→2→3.

Basic Laser Principles

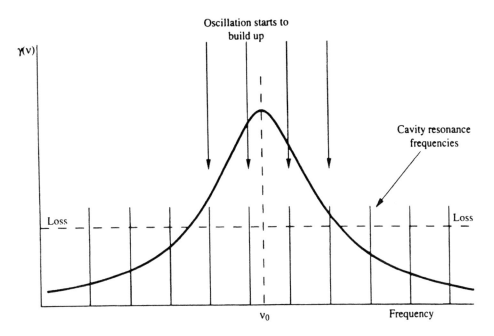

FIGURE 2.36 Schematic illustration of the onset of oscillation at cavity resonances that lie above the loss line in a homogeneously broadened laser.

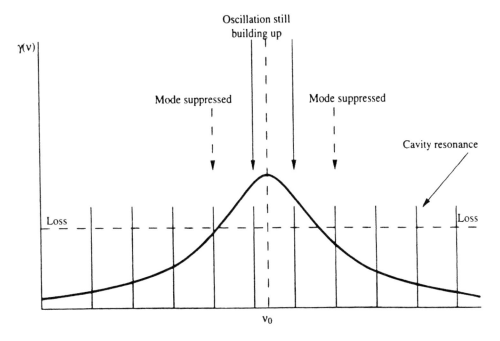

FIGURE 2.37 Oscillation building up in a homogeneously broadened laser. Gain saturation has already suppressed oscillation at two of the cavity modes that were above the loss line in figure 2.36.

where positive atom velocities correspond to particles moving from left to right. Wave (b) which is travelling in the opposite direction (to the left) and is monitored, still at frequency $v(< v_0)$, by a second observer looking into the laser from left to right cannot interact with the same velocity group of particles as wave (a). The particles which interacted with wave (a) were moving away from the first observer and were Doppler shifted to lower frequencies so as to satisfy

$$v = v_0 - \frac{|v|}{c} v_0. \quad (2.241)$$

The second observer sees these particles approaching and their centre frequency as

$$v = v_0 + \frac{|v|}{c} v_0 \quad (2.242)$$

so they cannot interact with wave (b). Wave (b) interacts with particles moving away from the second observer so that their velocity would be the solution of

$$v = v_0 - \frac{|v|}{c} v_0. \quad (2.243)$$

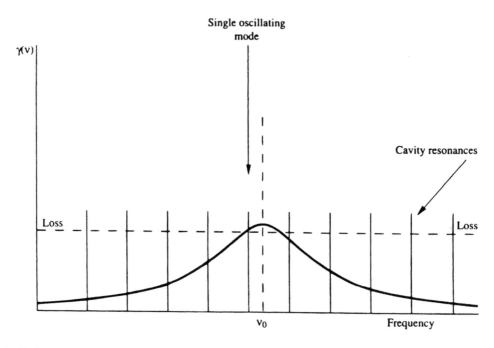

FIGURE 2.38 Oscillation stabilized in a homogeneously broadened laser. The gain has been uniformly reduced by saturation until only one mode remains at the loss line.

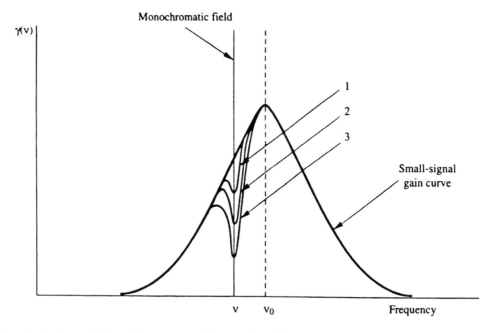

FIGURE 2.39 Localized gain saturation in an inhomogeneously broadened amplifier produced by a monochromatic signal whose intensity increases from 1→2→3.

These particles would be monitored by the first observer at centre frequency

$$v = v_0 + \frac{|v|}{c} v_0. \qquad (2.244)$$

So the oscillating waves interact with two velocity groups of particles as shown in Figure 2.41. This leads to saturation of the gain by a single laser mode in an inhomogeneously broadened laser both at the frequency of the mode v and at a frequency $v_0 + (v_0 - v)$, which is equally spaced on the opposite side of the line centre, as shown in Figure 2.42). The power output of the laser (strictly the intra-cavity power) comes from those groups of particles that have gone into stimulated emission and left the two holes. The combined area of these two holes gives a measure of the laser power.

If the frequency of the oscillating mode is moved in towards the line centre, the main hole and image hole begin to overlap. This corresponds physically to the left and right travelling waves within the laser cavity beginning to interact with the

Basic Laser Principles

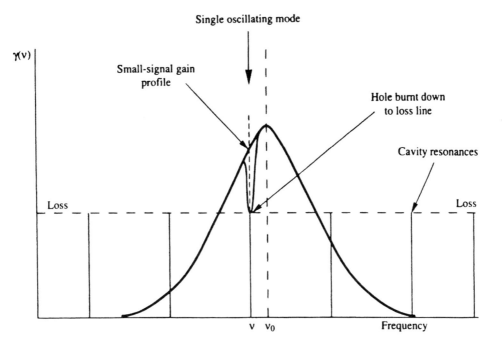

FIGURE 2.40 Simplified illustration of how saturation stabilizes oscillation at a single longitudinal mode in an inhomogeneously broadened laser.

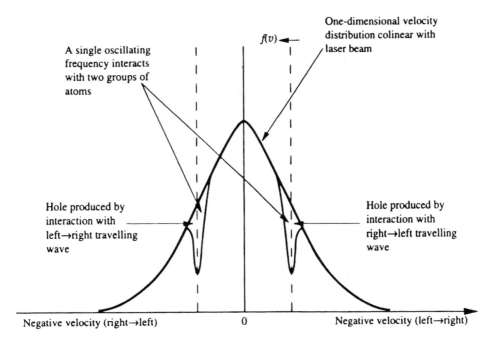

FIGURE 2.41 Production of two holes in the velocity distribution of a collection of amplifying particles by single-cavity mode.

same velocity group of particles. As the oscillating mode moves in towards the line centre, the holes overlap further, the combined area decreases and the laser output power falls, reaching a minimum at the line centre. This phenomenon is called the Lamb dip, named after Willis E Lamb, Jr, who first predicted the effect [13], and is illustrated in Figure 2.43. When the cavity resonance is at the line centre frequency v_0, both travelling waves are interacting with the same group of atoms—those with near-zero directed velocity along the laser resonator axis.

Because hole burning in gain saturation in inhomogeneously broadened lasers is localized near the frequency of a cavity mode, one oscillating mode does not reduce the gain at other cavity modes, so simultaneous oscillation at several longitudinal modes is possible. If several such modes have small-signal gains above the loss line, the oscillation stabilizes in the manner shown in Figure 2.44a. The output frequency spectrum from the laser would appear as is shown in Figure 2.44b. This simultaneous oscillation at several closely spaced frequencies ($c/2\ell$ apart) can be observed with a high-resolution spectrometer—for example, a scanning Fabry–Pérot interferometer. The multiple modes are almost exactly $c/2\ell$ in frequency apart but are not exactly equally spaced because of

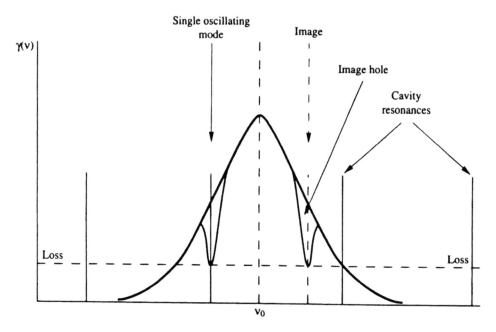

FIGURE 2.42 Stable saturation of a single longitudinal mode in an inhomogeneously broadened laser.

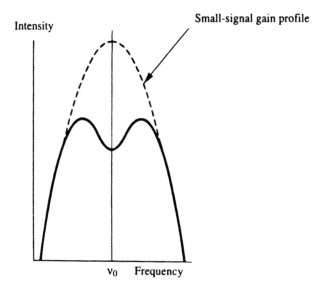

FIGURE 2.43 The Lamb dip—a reduction in the intensity of a single oscillating longitudinal mode in an inhomogeneously broadened laser as its frequency is scanned through line centre.

mode pulling. This effect can be observed in the beat spectrum observed with a square-law optical detector (which means most optical detectors). Such a detector responds to the intensity, not the electric field of an incident light signal.

2.15.2 Mode Beating

Suppose we shine the light from a two-mode laser on a square-law detector. The incident electric field is

$$E_i = R(E_1 e^{i\omega t} + E_2 e^{i(\omega + \Delta\omega)t}) \quad (2.245)$$

where E_1 and E_2 are the complex amplitudes of the two modes and $\Delta\omega$ is the frequency spacing between them.

Using real notation for these fields, the output current i from the detector is

$$\begin{aligned}
i &\propto \left\{ |E_1|\cos(\omega t + \phi_1) + |E_2|\cos[(\omega + \Delta\omega)t + \phi_2] \right\}^2 \\
&\propto |E_1|^2 \cos^2(\omega t + \phi_1) + |E_2|^2 \cos^2[(\omega + \Delta\omega)t + \phi_2] \\
&\quad + 2|E_1||E_2|\cos(\omega t + \phi_1)\cos[\omega + \Delta\omega t + \phi_2] \\
&\propto |E_1|^2 \cos^2(\omega t + \phi_1) + |E_2|^2 \cos^2[(\omega + \Delta\omega)t + \phi_2] \\
&\quad + |E_1||E_2|\left[\cos(2\omega + \Delta\omega)t + \phi_1 + \phi_2\right] \\
&\quad + |E_1||E_2|\cos(\Delta\omega t + \phi_2 - \phi_1).
\end{aligned} \quad (2.246)$$

Basic Laser Principles

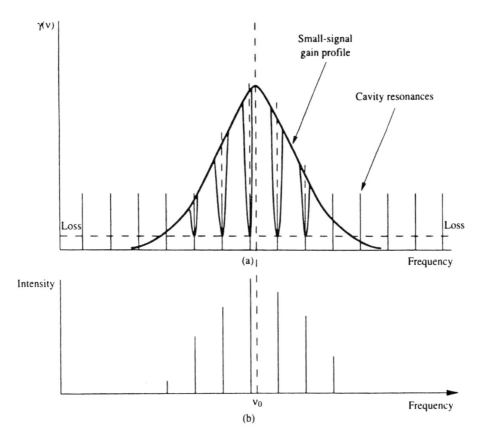

FIGURE 2.44 Multi-longitudinal-mode oscillation in an inhomogeneously broadened laser. (a) Only the primary holes are shown burnt down to the loss line. The image holes are not shown. (b) Schematic laser output spectrum.

And since, for example,

$$|E_1|^2 \cos^2(\omega t + \phi_1) = \tfrac{1}{2}|E_1|^2 [1 + \cos 2(\omega t + \phi_1)] \quad (2.247)$$

the output frequency spectrum of the detector appears to contain the frequencies 2ω, $2(\omega + \Delta\omega)$, $2\omega + \Delta\omega$ and $\Delta\omega$. However, the first three of these frequencies are very high, particularly for light in the visible and infrared regions of the spectrum, and do not appear in the output of the detector. It is as if the high-frequency terms are averaged to zero by the detector time response to give

$$i \propto \frac{|E_1|^2}{2} + \frac{|E_2|^2}{2} + |E_1||E_2|\cos(\Delta\omega t + \phi_2 - \phi_1) \quad (2.248)$$

so only the difference frequency beat $\Delta\omega$ is observed.

If the output from the square-law detector is analysed with a radio frequency spectrum analyser (because it is in this frequency range where the difference frequencies between longitudinal laser modes are usually observed), different displays are obtained according to how many longitudinal modes of a multi-mode laser are simultaneously oscillating. Figure 2.45 gives some examples. Because equation (2.239) is not quite exact, the beat frequencies can split as shown because of non-linear mode-pulling. This splitting will only be observed if non-linear mode-pulling is large and the spectrum analyser that analyses the output of the photo-detector has high resolution.

If a predominantly inhomogeneously broadened laser also has a significant amount of homogeneous broadening, the holes burnt in the gain curve can start to overlap, for example, when $\Delta v \gtrsim c/2\ell$. If Δv is large enough, this causes neighbouring oscillating modes to compete and may lead to oscillation on a strong mode suppressing its weaker neighbours, as shown in Figure 2.46. This effect has been observed in several laser systems, for example, in the argon ion laser, where an increase in the strength of the oscillation can lead to the successive disappearance, first of every other mode, then two modes out of every three, and so on.

2.16 The Characteristics of Laser Radiation

Laser radiation has special properties that distinguish it from ordinary light. We have already seen that laser radiation should be very directional.[16] Laser radiation is, in general, very monochromatic: the spectrum of a laser longitudinal mode is confined to a narrow spectral range. Laser radiation is generally very *coherent* compared to the light from traditional light sources. This coherence is a measure of the temporal and spatial phase relationships that exist for the fields associated with laser radiation (see Chapter 3, Section 3.2.2).

---

[16] Diffraction effects cause the laser beams from lasers with small lateral dimensions, such as semiconductor lasers, to diverge substantially after they leave the laser resonator.

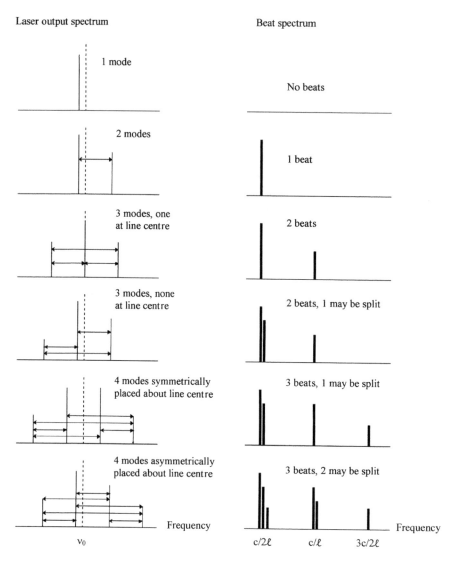

FIGURE 2.45 Schematic mode-beating spectra observed with a square-law detector and a multi-mode laser.

The special nature of laser radiation is graphically illustrated by the ease with which the important optical phenomena of interference and diffraction are demonstrated using it. Interference effects demonstrate the coherence properties of laser radiation, while diffraction effects are intimately connected with the beam-like properties that make this radiation special.

2.16.1 Laser Modes

When a laser oscillates, it emits radiation at one or more frequencies that lie close to passive resonant frequencies of the cavity. These frequencies are called *longitudinal* modes (see Chapter 5, Section 5.9). In our initial discussion of these modes we treated them as plane waves reflecting back and forth between two plane laser mirrors. In practice, laser mirrors are not always plane. Usually, at least one of the laser mirrors will have concave spherical curvature. The use of spherical mirrors relaxes the alignment tolerance that must be maintained for adequate feedback to be achieved. Even if the laser mirrors are plane, the waves reflecting between them cannot be plane, as true plane waves can only exist if there is no lateral restriction of the wave fronts. Practical laser mirrors are of finite size so any waves reflecting from them will spread out because of diffraction [14]. Diffraction results whenever a wave is restricted laterally, for example by passing it through an aperture. Reflection from a finite-size mirror produces equivalent effects. We can explain this phenomenon qualitatively by introducing the concept of Huygens secondary wavelets.[17] If a plane mirror is illuminated by a plane wave then each point on the mirror can be treated as a source of a spherical wave called a secondary wavelet. The overall reflected wave is the envelope of the sum total of secondary wavelets originating from every point on the mirror surface, as shown in Figure 2.47. This construction shows that the reflected wave from a finite-size mirror is not a plane wave.

The existence of diffraction in a laser resonator places restrictions on the minimum size of mirrors that can be used at a given wavelength λ and spacing ℓ. If the diameters of the

---

[17] Christian Huygens (1629–1695) was a Dutch astronomer who first suggested the concept of secondary wavelets.

Basic Laser Principles

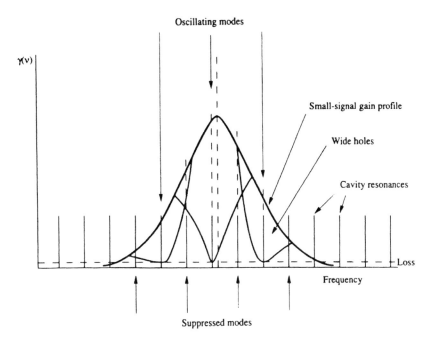

FIGURE 2.46 Schematic illustration of mode competition in an inhomogeneously broadened laser in which there is significant homogeneous broadening.

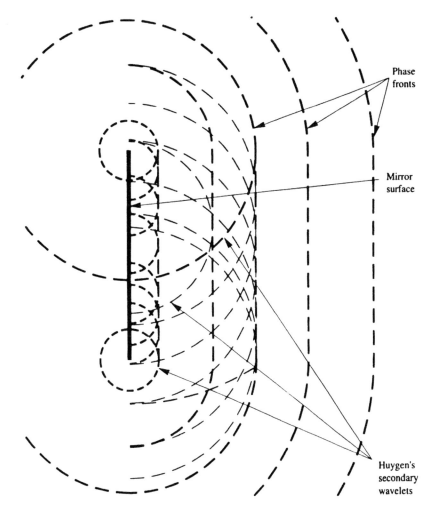

FIGURE 2.47 Secondary wavelets originating from a finite plane mirror.

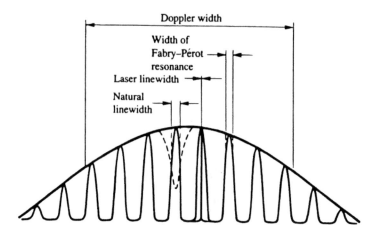

FIGURE 2.48 Linewidth factors in a laser.

resonator mirrors are d_1, d_2, respectively, and the resonator has length ℓ, then the resonator will have high loss unless the *Fresnel condition* is satisfied [2], namely,

$$\frac{d_2 d_2}{\lambda \ell} \geq 1. \quad (2.249)$$

The actual waves that reflect back and forth between the mirrors of a laser resonator are not plane waves. They have characteristic spatial patterns of electric (and magnetic) field amplitude and are called *transverse modes* (see Chapter 5, Section 5.4). To be amplified effectively, such modes must correspond to rays which make substantial numbers of specular reflections before being lost from the cavity. A transverse mode is a field configuration on the surface of one reflector that propagates to the other reflector and back, returning in the same pattern, apart from a complex amplitude factor that gives the total phase shift and loss of the round trip. To each of these transverse modes, there corresponds a set of longitudinal modes spaced by approximately $c/2\ell$. A more detailed treatment of these transverse modes is given in Chapter 5.

2.16.2 Beam Divergence

Since the oscillating field distributions inside a laser are not plane waves, when they propagate through the mirrors as output beams, they spread by diffraction. The semi-vertical angle of the cone into which the output beam diverges is[18]

$$\theta_{\text{beam}} = \tan^{-1}\left(\frac{\lambda}{\pi \omega_0}\right) \approx \frac{\lambda}{\pi \omega_0} \quad (2.250)$$

where λ is the wavelength of the output beam and ω_0 is a parameter called the 'minimum spot size' that characterizes the transverse mode.

Let us take the specific example of a 530-nm laser beam with $\omega_0 = 1$ mm for which $\theta_{\text{beam}} = 169$ μrad $\simeq 1$ millidegree. This is a highly directional beam but the beam does become wider the further it goes away from the laser. Such a beam is, however, highly useful in providing the perfect straight line reference. Over a 100 m distance, the laser beam just described

---

[18] For a derivation of this result see Chapter 5, Section 5.2.

would have expanded to a diameter of 34 mm. After travelling the distance to the moon (~ 390,000 km), the beam would be $\simeq$ 132 km in diameter.

2.16.3 Linewidth of Laser Radiation

A single longitudinal mode of a laser is an oscillation resulting from the interaction of a broadened gain curve with a passive resonance of the Fabry–Pérot laser cavity. The frequency width of the gain curve is Δv, and the frequency width of the passive cavity resonance is $\Delta v_{1/2} = \Delta v_{\text{FSR}}/F$. We expect the linewidth of the resulting oscillation to be narrower than either of these widths, as shown schematically in Figure 2.48. It can be shown that the frequency width of the laser oscillation itself is [10,15–17]

$$\Delta v_{\text{laser}} = \frac{\pi h v_0 (\Delta v_1/2)^2}{P} = \frac{N_2}{[N_2 - (g_2/g_1)N_1]_{\text{threshold}}} \quad (2.251)$$

here P is the output power.

Equation (2.252) predicts very low linewidths for many lasers. For a typical He–Ne laser with 99% reflectance mirrors and a cavity 30 cm long,

$$\Delta v_{1/2} = \frac{c(1-R)}{2\pi\ell} = 1.59 \text{ MHz}.$$

The factor $N_2/[N_2 - (g_2/g_1) N_1]_{\text{threshold}}$ is close to unity for a typical low power, say 1 mW, laser. Consequently,

$$\Delta v_{\text{laser}} = \frac{\pi \times 6.626 \times 10^{-34} \times 3 \times 10^8 \times 1.59^2 \times 10^{12}}{10^{-3} \times 632.8 \times 10^{-9}}$$

$$= 2.5 \times 10^{-3} \text{ Hz}.$$

Such a small linewidth is never observed in practice because thermal instabilities and acoustic vibrations lead to variations in resonator length that further broaden the output radiation lineshape. The best observed minimum linewidths for highly stabilized gas lasers operating in the visible region of the spectrum are around 10^3 Hz. Even if macroscopic thermal and acoustic vibrations could be eliminated from the system, a fundamental limit to the resonator length stability would be

Basic Laser Principles

set by Brownian motion of the mirror assemblies. For example, consider two laser mirrors mounted on a rigid bar. The mean stored energy in the Brownian motion of the whole bar is $\overline{E} = kT$.

The frequency spread of the laser output that thereby results is

$$\Delta v_{\text{Brownian}} = v\sqrt{\frac{2kT}{YV}} \quad (2.252)$$

where Y is Young's modulus of the bar material and V is the volume of the mounting bar. Typical values of $\Delta v_{\text{Brownian}}$ are $\sim 2\,\text{Hz}$.

2.17 Coherence Properties

2.17.1 Temporal Coherence

Because of its extremely narrow output linewidth, the output beam from a laser exhibits considerable *temporal coherence* (longitudinal coherence). To illustrate this concept, consider two points A and B a distance L apart in the direction of propagation of a laser beam, as shown in Figure 2.49. If a definite and fixed phase relationship exists between the wave amplitudes at A and B, then the wave shows temporal coherence for a time c/L. The further apart A and B can be, while still maintaining a fixed phase relation with each other, the greater is the temporal coherence of the output beam. The maximum separation at which the fixed phase relationship is retained is called the *coherence length*, L_c, which is a measure of the length of the continuous uninterrupted wave trains emitted by the laser. The coherence length is related to the *coherence time* τ_c by $L_c = c\tau_c$. The coherence time itself is a direct measure of the monochromaticity of the laser, since by Fourier transformation, done in an analogous manner to the treatment of natural broadening,

$$\tau_c \simeq \frac{1}{\Delta v_L} \quad \text{and} \quad L_c \simeq \frac{c}{\Delta v_L}. \quad (2.253)$$

The coherence length and time of a laser source are considerably better than a conventional monochromatic source (a spontaneously emitted line source). The greatly increased coherence can be demonstrated in a Michelson interferometer experiment, which allows interference between waves at longitudinally different positions in a wavefront to be studied [14,18,19].

2.17.2 Laser Speckle

Perhaps, the most easily observable consequence of the high temporal coherence of most lasers is the occurrence of *speckle* [20–22]. When a laser beam strikes an object, unless the object is *very* flat (to better than a faction of a wavelength over the illuminated area) light that is scattered from the object will be observed to have a 'grainy' texture. If the scattered light falls directly onto a screen, then an *objective* speckle pattern will be observed. This manifests itself as a pattern of bright and dark patches with a 'salt and pepper' appearance. If the speckle pattern is collected by a lens, or enters the eye, then a *subjective* speckle pattern is observed. Speckle results from the roughness of the illuminated object, and on a scale of visible wavelengths, most objects are intrinsically very rough. A piece of paper provides a good object to observe the production of a speckle pattern. When a rough surface is illuminated with temporally coherent light, each point on the surface serves as a scattering centre for the production of an outgoing spherical wave. At a point in space, these many scattered waves arrive with different phases, because they have come from the 'hills and valleys' of the surface roughness. If these waves add substantially in phase, then bright illumination results at that point. However, if destructive interference occurs, a dark region will result.

The spatial structure of the speckle pattern results largely from the size of the illuminated region on the object. If this region is large, an objective speckle pattern with fine spatial scale results. If the region illuminated is small, then a speckle pattern with a coarser spatial scale is observed. This is related to the diffraction angle associated with the diameter D of the illuminated region. At a dark spot in the speckle pattern, the waves that have come from the two halves of the illuminated

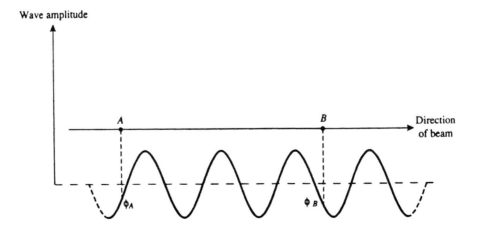

FIGURE 2.49 Illustration of the concept of temporal coherence. If an unbroken wave train connects the points A and B, then the phase difference $(\phi_B - \phi_A)$ will have a constant value.

region can be thought of as being π radians out of phase. At an adjacent bright spot, the waves can be thought of as being in phase. It is easy to see that the angular separation of adjacent bright and dark spot is $\sim \lambda/D$.

For a fixed object, the objective speckle pattern is stationary in space. If this pattern of light and dark enters the eye, then because of the natural and continuous small motions of the eye, an apparent twinkling effect will be perceived as the eye moves through the regions of bright and dark illumination of the speckle pattern.

2.17.3 Spatial Coherence

A laser also possesses *spatial* (lateral) coherence, which implies a definite fixed-phase relationship between points separated by a distance L transverse to the direction of beam propagation. The transverse coherence length, which has similar physical meaning to the longitudinal coherence length, is

$$L_{tc} \sim \frac{\Delta}{\theta_{beam}} \sim \pi \omega_0 \qquad (2.254)$$

for a laser source.

The existence of spatial coherence in a wavefront and the limit of its extent can be demonstrated in a classic Young's slits interference experiment [14,18,19]. A pair of thin, parallel slits or a pair of pinholes spaced a distance d apart is illuminated normally with a spatially coherent monochromatic plane wave. If an interference pattern of bright and dark bands is observed on the other side of the slits, then the illumination laser beam is spatially coherent over at least the distance d; if the slit variation separations were increased to beyond the lateral coherence length, so that $d > L_{tc}$, then the fringe pattern would disappear.

2.18 The Power Output of a Laser

When a laser oscillates, the intra-cavity field grows in amplitude until saturation reduces the gain to the loss line for each oscillating mode. What this means in practice can be best illustrated with reference to Figure 2.50.

For an asymmetrical resonator, whose mirror reflectances are not equal, the distribution of standing-wave energy within the resonator is not symmetrical. For example, in Figure 2.50, if $R_2 > R_1$, the distribution of intra-cavity travelling wave intensity will be schematically as shown, and

$$\frac{I_3}{I_2} = R_2 \qquad \frac{I_1}{I_4} = R_1. \qquad (2.255)$$

The left travelling wave, of intensity I_-, grows in intensity from I_3 to I_4 on a single pass. The right travelling wave, of intensity I_+, grows in intensity from I_1 to I_2 on a single pass. The total output intensity is

$$I_{out} = T_2 I_2 + T_1 I_4. \qquad (2.256)$$

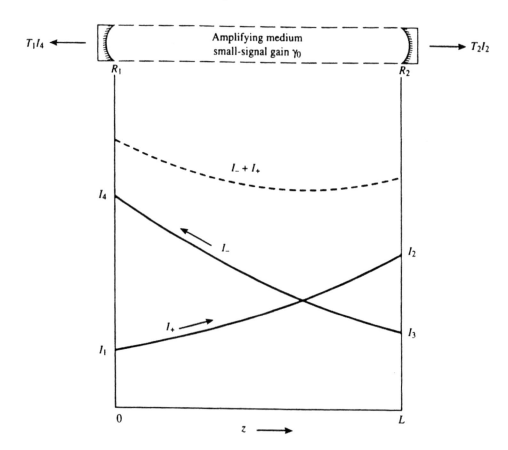

FIGURE 2.50 Distribution of bi-directional propagating wave intensities in an asymmetric laser cavity.

Basic Laser Principles

However, calculation of I_2 and I_4 is not straightforward in general case. I_2 grows from I_1 through a gain process that depends in a complex way on $I_+ + I_-$ as does the growth of I_3–I_4. We can identify at least three scenarios in which the calculation proceeds differently:

a. homogeneously broadened amplifier and single-mode operation,
b. an inhomogeneously broadened amplifier and single-mode operation and
c. an inhomogeneously broadened amplifier and multi-mode operation.

In both cases (b) and (c), the calculation of the output power becomes more complicated as the homogeneous contribution to the broadening grows more significant compared to $\Delta\nu_D$. This additional complexity arises because each oscillating mode burns both a primary and an image hole in the gain curve. The resultant distribution of overlapping holes makes the gain for each mode dependent not only on its own intensity but also on the intensity of the other simultaneously oscillating modes. The presence of distributed intra-cavity loss presents additional complications. We shall not attempt to deal with these complex situations here but will follow Rigrod [23] in dealing with a homogeneously broadened amplifier in which the primary intensity loss occurs at the mirrors. Inhomogeneously broadened systems and multi-mode operation have been discussed elsewhere by Smith [24].

In a purely homogeneously broadened system, the saturated gain in Figure 2.50 is

$$\gamma(z) = \frac{\gamma_0}{1 + (I_+ + I_-)/I_s}. \tag{2.257}$$

Both I_- and I_+ grow according to $\gamma(z)$

$$\frac{1}{I_+}\frac{dI_+}{dz} = \frac{1}{I_-}\frac{dI_-}{dz} = \gamma(z). \tag{2.258}$$

Consequently,

$$I_+ I_- = \text{constant} = C. \tag{2.259}$$

From equation (2.260)

$$I_4 I_1 = I_2 I_3 = C \tag{2.260}$$

and, therefore, from equation (2.255)

$$I_2/I_4 = \sqrt{R_1/R_2}. \tag{2.261}$$

For the right travelling wave, using equations (2.258) and (2.259) gives

$$\frac{1}{I_+}\frac{dI_+}{dz} = \frac{\gamma_0}{1 + (I_+ + C/I_+)/I_s} \tag{2.262}$$

which can be integrated to give

$$\gamma_0 L = \ln\left(\frac{I_2}{I_1}\right) + \frac{(I_2 - I_1)}{I_s} - \frac{C}{I_s}\left(\frac{1}{I_2} - \frac{1}{I_1}\right). \tag{2.263}$$

In a similar way, for the left travelling wave

$$\gamma_0 L = \ln\left(\frac{I_4}{I_3}\right) + \frac{(I_4 - I_3)}{I_s} - \frac{C}{I_s}\left(\frac{1}{I_4} - \frac{1}{I_3}\right). \tag{2.264}$$

Adding equations (2.263) and (2.264) and using equations (2.255), (2.260) and (2.261) give

$$I_2 = \frac{I_s\sqrt{R_1}\,(\gamma_0 L + \ln\sqrt{R_2 R_1})}{(\sqrt{R_1} + \sqrt{R_1})(1 - \sqrt{R_2 R_1})}. \tag{2.265}$$

From equation (2.261)

$$I_4 = I_2 \sqrt{\frac{R_2}{R_1}}. \tag{2.266}$$

Now

$$T_1 = 1 - R_1 - A_1 \tag{2.267}$$

$$T_2 = 1 - R_2 - A_2 \tag{2.268}$$

so, from equations (2.256) and (2.265), if $A_1 = A_2 = A$

$$I_{\text{out}} = I_s = \frac{\left(1 - A - \sqrt{R_1 R_2}\right)}{1 - \sqrt{R_1 R_2}}\left(\gamma_0 L + \ln\sqrt{R_1 R_2}\right). \tag{2.269}$$

If one mirror is made perfectly reflecting, say $T_1 = 0$, $R_1 = 1$, then

$$I_{\text{out}} = T_2 I_2 = \frac{T_2 I_s[\gamma_0 L + \frac{1}{2}\ln(1 - A_2 - T_2)]}{(A_2 + T_2)}. \tag{2.270}$$

For a symmetrical resonator, defined by $R_1 R_2 = R^2$,

$$R = 1 - A - T \tag{2.271}$$

the output intensity at each mirror is

$$\frac{I_{\text{out}}}{2} = \frac{I_s}{2}\frac{(1 - A - R)}{1 - R}(\gamma_0 L + \ln R). \tag{2.272}$$

2.18.1 Optimum Coupling

To maximize the output intensity from the symmetrical resonator, we must find the value of R such that $\partial I_{\text{out}}/\partial R = 0$ which gives

$$\frac{T_{\text{opt}}}{A} = \left(\frac{1 - A - T_{\text{opt}}}{A + T_{\text{opt}}}\right)[\gamma_0 L + \ln(1 - A - T_{\text{opt}})]. \tag{2.273}$$

For small losses, such that $A + T_{\text{opt}} \ll 1$, equation (2.273) gives

$$\frac{T_{\text{opt}}}{A} = \sqrt{\frac{\gamma_0 L}{A}} - 1. \tag{2.274}$$

Figure 2.51 shows the calculated optimum coupling for various values of the loss parameter A and the unsaturated gain in dB (4.343 $\gamma_0 L$).

In practice, it should be pointed out that the optimum mirror transmittance in a laser system is generally determined empirically. For example, for the cw CO_2 laser, whose unsaturated gain varies roughly inversely with the tube diameter d, the optimum mirror transmittance has been determined to be $T \simeq L/500d$.

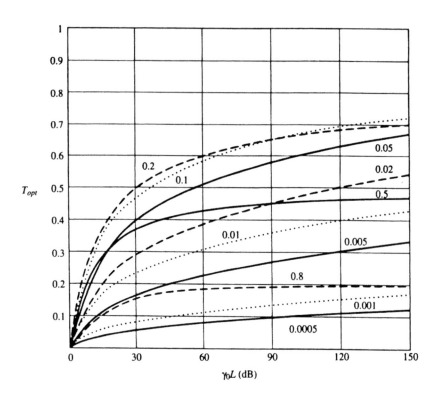

FIGURE 2.51 Calculated optimum coupling for a symmetrical resonator for various values of the loss parameter A and the unsaturated gain.

Acknowledgement

All figures except 2.21, 2.29, 2.32, 2.33 and 2.45 are reprinted by permission of Cambridge University Press.

REFERENCES

1. Liao S Y 1985 *Microwave Solid-State Devices* (Englewood Cliffs, NJ: Prentice-Hall).
2. Davis C C 1996 *Lasers and Electro-Optics* (Cambridge: Cambridge University Press).
3. Dicke R H and Wittke J P 1960 *Introduction to Quantum Mechanics* (Reading, MA: Addison-Wesley).
4. Liboff L 2003 *Introductory Quantum Mechanics* 4th edn (San Francisco, CA: Benjamin-Cumings).
5. White R L 1966 *Basic Quantum Mechanics* (New York: McGraw-Hill).
6. Jeans Sir J H 1940 *An Introduction to the Kinetic Theory of Gases* (Cambridge: Cambridge University Press).
7. Present R D 1958 *Kinetic Theory of Gases* (New York: McGraw-Hill).
8. Champeney D C 1973 *Fourier Transforms and Their Physical Applications* (London: Academic).
9. Mitchell A G C and Zemansky M W 1971 *Resonance Radiation and Excited Atoms* (Cambridge: Cambridge University Press)
10. Milonni P W and Eberly J H 1988 *Lasers* (New York: Wiley).
11. Abramowitz M and Stegun I A 1968 *Handbook of Mathematical Functions* (New York: Dover).
12. Bennett W R Jr 1962 Hole-burning effects in a He–Ne optical maser *Phys. Rev.* **126** 580–93.
13. Lamb W E 1964 Theory of an optical maser *Phys. Rev. A* **134** 1429–50.
14. Born M and Wolf E 1999 *Principles of Optics Electromagnetic Theory of Propagation, Interference and Diffraction of Light* 7th edn (Cambridge: Cambridge University Press).
15. Siegman A E 1986 *Lasers* (Mill Valley, CA: University Science Books).
16. Yariv A 1991 *Introduction to Optical Electronics* 4th edn (New York: Holt, Rinehart and Winston).
17. Yariv A 1989 *Quantum Electronics* 3rd edn (New York: Wiley).
18. Ditchburn R W 1976 *Light* 3rd edn (New York: Academic).
19. Hecht E 1998 *Optics* 3rd edn (Reading, MA: Addison-Wesley).
20. Dainty J C (ed) 1975 *Laser Speckle and Related Phenomena* (*Topics in Applied Physics* 9) (Berlin: Springer).
21. Francon M 1979 *Laser Speckle and Applications in Optics* (New York: Academic).
22. Jones R and Wykes C 1989 *Holographic and Speckle Interferometry* 2nd edn (Cambridge: Cambridge University Press).
23. Rigrod W W 1965 Saturation effects in high-gain lasers *J. Appl. Phys.* 36 2487–90 See also Rigrod W W 1963 Gain saturation and output power of optical masers *J. Appl. Phys.* **34** 2602–9. Rigrod W W 1978 Homogeneously broadened CW lasers with uniform distributed loss *IEEE J. Quantum Electron.* QE-**14** 377–81.
24. Smith P W 1966 The output power of a 6328 Å He–Ne gas laser *IEEE J. Quantum Electron.* QE-**2** 62–8.

3

Interference and Polarization

Alan Rogers

CONTENTS

3.1 Introduction 51
3.2 Interference 51
    3.2.1 Wave Coherence 51
    3.2.2 Coherent-wave Interference 52
    3.2.3 Interferometers 53
    3.2.4 Interference between Partially Coherent Waves 57
    3.2.5 Practical Examples 59
3.3 Polarization 62
    3.3.1 Introduction 62
    3.3.2 The Polarization Ellipse 62
    3.3.3 Material Interactions 63
    3.3.4 Crystal Optics 63
    3.3.5 Retarding Waveplates 65
    3.3.6 Polarizing Prisms 66
    3.3.7 Circular Birefringence 68
    3.3.8 Polarization Analysis 68
    3.3.9 Applications of Polarization Optics 71
3.4 Conclusions 74
Acknowledgements 75
References 75

3.1 Introduction

Light is a form of electromagnetic radiation. It is distinguished from other forms only by its particular wavelength range: 0.4–0.7 μm for the visible range, with regions below 0.4 μm (ultraviolet) and above 0.7 μm (infrared) also conventionally classified as lying within the 'optical' range.

Electromagnetic radiation exhibits both wave and particle (i.e. photon) properties. Wave properties are usually more appropriate for description of behaviour at the longer wavelengths (radio and microwave wavelengths, for example) where photon energies are small, and therefore, the number of photons per energy range is large; particle properties are more appropriate at the shorter wavelengths (X-rays, γ-rays, for example), where there are very few photons per energy range and the discrete, particulate nature of the radiation is more evident in the, now, relatively rare occurrence of photon arrival at a detector. The optical range is intermediate between these two regimes, so that it is sometimes more useful to work with the wave description and sometimes with the photon description. Examples of the former are the subject of the present chapter, interference and polarization; examples of the latter are photodetection and photoemission processes.

Interference and polarization are both concerned with wave interaction. Interference is concerned with the interaction between waves, whilst polarization is concerned with the interaction of a wave within itself. They are not independent: the interference between waves depends upon their relative polarization states, and both interference and polarization phenomena depend upon the 'purity' of the wave—on its 'coherence'. We shall, however, deal with each topic separately and, in turn, cross-linking the two where appropriate.

3.2 Interference

3.2.1 Wave Coherence

We shall begin by considering the interference between 'pure' waves. 'Pure' in this sense means that the wave is idealized as a sinusoid in all time and space, so that the way in which such waves interact is constant for all time and space. This is a useful idealization in that many of the light sources with which we deal in practice (including most lasers) are 'pure' for the observation times and experimental volumes which are used. A stricter definition of this 'purity' is that the wave within itself, or two (or more) interfering waves, maintains a constant phase relationship over these times and spaces—they are then said to be self- or mutually-'coherent', respectively. We shall deal more fully with the concept of coherence and its quantitative effect on interference phenomena in Section 3.2.4.

3.2.2 Coherent-wave Interference

Light, in the wave description, comprises electric and magnetic fields oscillating in phase and mutually at right angles in space (Figure 3.1). We know that these fields are vector fields since they represent forces (on unit charge and unit magnetic pole, respectively). The fields will thus add vectorially. Consequently, when two light waves are superimposed on each other we obtain the resultant by constructing their vector sum at each point in time and space.

It is clear that the amplitude of such a wave in a given direction can only be affected by another if they both have components in that direction. Two waves oscillating mutually at right angles cannot, therefore, influence each other, and to find the effect at other angles, one wave must be resolved in the direction of the other. The simplest case to deal with is that where the two waves of the same frequency are both oscillating in the same direction and propagating in the same (orthogonal) direction, so that their mutual interaction is maximized.

If two such sinusoids are added, the result is another sinusoid. Suppose that two such light waves given, via their electric fields, as:

$$e_1 = E_1 \cos(\omega t + \phi_1)$$
$$e_2 = E_2 \cos(\omega t + \phi_2)$$

have the same oscillation direction and are superimposed at a point in space (Figure 3.2). We know that the resultant field at the point will be given, using elementary trigonometry, by

$$e_t = E_T \cos(\omega t + \phi_T)$$

where

$$E_T^2 = E_1^2 + E_2^2 + 2E_1 E_2 \cos(\phi_2 - \phi_1)$$

and

$$\tan \phi_T = \frac{(E_1 \sin \phi_1 + E_2 \sin \phi_2)}{(E_1 \cos \phi_1 + E_2 \cos \phi_2)}.$$

For the important case where $E_1 = E_2 = E$, say, we have:

$$E_T^2 = 4E^2 \cos^2 \frac{(\phi_2 - \phi_1)}{2} \qquad (3.1)$$

and

$$\tan \phi_T = \tan \frac{(\phi_2 + \phi_1)}{2}.$$

The intensity of the wave will be proportional to E_T^2, so that, from (3.1), it can be seen to vary from $4E^2$ to 0, as $(\phi_2 + \phi_1)/2$ varies from 0 to $\pi/2$.

Consider now the arrangement shown in Figure 3.3. Here two slits, separated by a distance p, are illuminated by a plane wave. The portions of the wave which pass through the slits will interfere on the screen S, a distance d away. Now each of the slits will act as a source of cylindrical waves, from Huygens' principle. Moreover, since they originate from the same plane wave, they will start in phase. On a line displaced a distance s from the line of symmetry on the screen, the waves from the two slits will differ in phase by:

$$\delta = \frac{2\pi}{\lambda} \cdot \frac{sp}{d} = k \cdot \frac{sp}{d} (d \gg s, p) \text{ with } k = \frac{2\pi}{\lambda}.$$

Thus, as s increases, the intensity will vary between a maximum and zero, in accordance with equation (3.1). These variations will be viewed as fringes, i.e. lines of constant intensity parallel with slits. They are known as Young's fringes, after

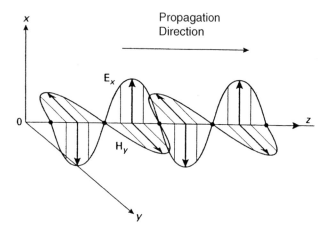

FIGURE 3.1 Electromagnetic wave. (From Rogers A 1997 essentials of optoelectronics (Cheltenham: Nelson Thornes) with permission.)

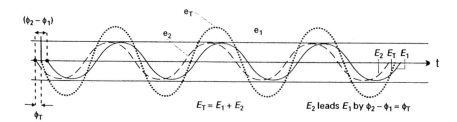

FIGURE 3.2 Addition of two waves of the same frequency. (From Rogers A 1997 essentials of optoelectronics (Cheltenham: Nelson Thornes) with permission.)

Interference and Polarization

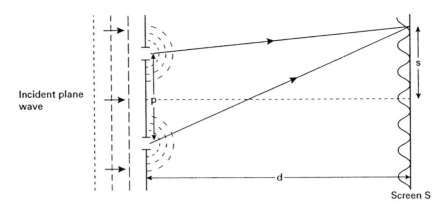

FIGURE 3.3 Two-slit interference. (From Rogers A 1997 essentials of optoelectronics (Cheltenham: Nelson Thornes) with permission.)

their discoverer, and are the simplest example of light interference. We shall now consider some important measuring instruments which depend upon these interference principles.

3.2.3 Interferometers

In Section 3.2.2, the essentials of dual-beam interference were discussed. Although very simple in concept, the phenomenon is extremely useful in practice. The reason for this is that the maxima of the resulting fringe pattern appear where the phase difference between the interfering light beams is a multiple of 2π. Any quite small perturbation in the phase of one of the beams will thus cause a transverse shift in the position of the fringe pattern which, using opto-electronic techniques, is readily observed to about 10^{-4} of the fringe spacing. Such a shift is caused by, for example, an increase in path length of one of the beams by one-hundredth of a wavelength, or about 5×10^{-9} m for visible light. This means that differential distances of this order can be measured, leading to obvious applications in, for example, sensitive strain monitoring on mechanical structures.

Another example of a dual-beam interferometer is shown in Figure 3.4. Here, the two beams are produced from the partial reflection and transmission at a dielectric, or partially silvered, mirror M_1. Another such mirror, M_4, recombines the two beams after their separate passages. Such an arrangement is known as a Mach–Zehnder interferometer and is used extensively to monitor changes in the phase differences between two optical paths. An optical-fibre version of a Mach–Zehnder interferometer is shown in Figure 3.5.

In this case, the 'mirrors' are optical couplings between the cores of the two fibres. The 'fringe pattern' consists effectively of just one fringe, since the fibre core acts as an efficient spatial filter. However, the light which emerges from the fibre end (E) clearly will depend on the phase relationship between the two optical paths when the light beams recombine at R, and thus, it will depend critically on propagation conditions within the two arms. If one of the arms varies in temperature, strain, density, etc., compared with the other, then the light output will also vary. Hence, the latter can be used as a sensitive measure of any physical parameters which are capable of modifying the phase propagation properties of the fibre.

Finally, Figure 3.6a shows another, rather more sophisticated variation of the Mach–Zehnder idea. In this case, the beams are again separated by means of a beam-splitting mirror but are returned to the same point by fully silvered mirrors placed at the ends of the two respective optical paths. (The plate P is necessary to provide equal optical paths for the two beams in the absence of any perturbation.) This arrangement is called the Michelson interferometer after the experimenter who, in the late 19th century, used optical interferometry with great skill to make many physical advances [1]. His interferometer (not to be confused with his 'stellar' interferometer, of which more later) allows for a greater accuracy of fine adjustment by control of the reflecting mirrors but uses, of course, just the same basic interferometric principles as before. The optical-fibre version of this device is shown in Figure 3.6b.

For completeness and because of its historical importance, mention must be made of the use of Michelson's interferometer

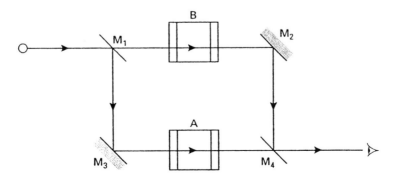

FIGURE 3.4 Basic Mach–Zehnder interferometer. (From Rogers A 1997 essentials of optoelectronics (Cheltenham: Nelson Thornes) with permission.)

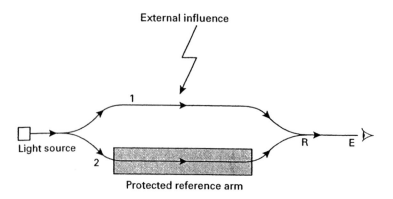

FIGURE 3.5 An optical-fibre Mach–Zehnder interferometer. (From Rogers A 1997 essentials of optoelectronics (Cheltenham: Nelson Thornes) with permission.)

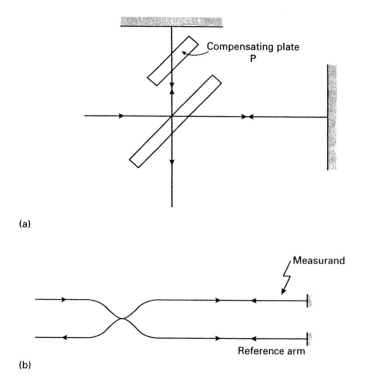

FIGURE 3.6 Michelson interferometers. (a) Bulk version; (b) optical-fibre version. (From Rogers A 1997 essentials of optoelectronics (Cheltenham: Nelson Thornes) with permission.)

in the famous Michelson–Morley experiment of 1887 [2]. This demonstrated that light travelled with the same velocity in each of two orthogonal paths, no matter what was the orientation of the interferometer with respect to the earth's 'proper' motion through space. This result was crucial to Einstein's formulation of special relativity in 1905 and is thus certainly one of the most important results in the history of experimental physics.

Valuable as dual-beam interferometry is, it suffers from the limitation that its accuracy depends upon the location of the maxima (or minima) of a sinusoidal variation. For very accurate work, such as precision spectroscopy, this limitation is severe. By using the interference amongst many beams, rather than just two, we find that we can improve the accuracy very considerably. We can see this by considering the arrangement of Figure 3.7. Light from a single source gives a large number of phase-related, separate beams by means of multiple reflections and transmissions within a dielectric (e.g. glass) plate. For a given angle of incidence (θ), there will be fixed values for the amplitude transmission (T, T') and amplitude reflection (R) coefficients, as shown. If we start with a wave of amplitude 'a', the waves on successive reflections will suffer attenuation by a constant factor and will increase in phase by a constant amount. If we consider the transmitted light only, then the total amplitude which arrives at the focus of the lens L is given by the sum

$$A_T = aTT' \exp(i\omega t) + aTT'R^2 \exp i(\omega t - ks)$$
$$+ aTT'R^4 \exp i(\omega t - 2ks) + \cdots$$

Interference and Polarization

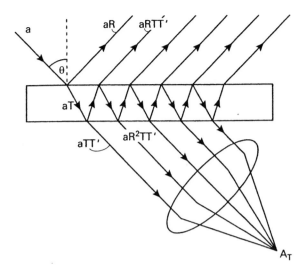

FIGURE 3.7 Multiple-wave interference. (From Rogers A 1997 Essentials of Optoelectronics (Cheltenham: Nelson Thornes) with permission.)

where, again,

$$k = \frac{2\pi}{\lambda}$$

and where s is the optical path difference between successive reflections at the lower surface (including the phase changes on reflection and transmission). The sum can be expressed as

$$A_T = aTT' \sum_{p=0}^{\infty} R^{2p} \exp i(\omega t - pks)$$

which is a geometric series whose sum value is:

$$A_T = \frac{aTT' \exp(i\omega t)}{(1 - R^2 \exp(-iks))}.$$

Hence, the intensity (I) of the light is given by:

$$I \propto |A_T|^2 = \frac{(aTT')^2}{(1 + R^4 - 2R^2 \cos ks)}. \quad (3.2)$$

We note from this equation the ratio of maximum and minimum intensities

$$\frac{I_{max}}{I_{min}} = \frac{(1+R^2)^2}{(1-R^2)^2}$$

so that the fringe contrast increases with R. However, as **R** increases so does the attenuation between the successive reflections. Hence, the total transmitted light power will fall. Figure 3.8 shows how I varies with ks for different values of R. We note that the fringes become very sharp for large values of R. Hence, the position of the maxima may now be accurately determined. Further, since the spacing of the maxima specifies ks, this information can be used to determine either k or s, if the other is known. Consequently, multiple interferences may be used either to select (or measure) a very specific wavelength or to measure very small changes in optical path length.

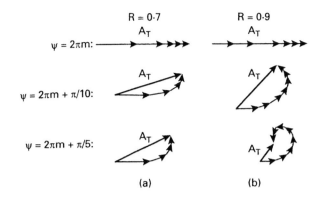

FIGURE 3.8 Variation of intensity with optical path, for various reflectivities, in a multiple-interference plate. (From Rogers A 1997 essentials of optoelectronics (Cheltenham: Nelson Thornes) with permission.)

The physical reason for the sharpening of the fringes as the reflectivity increases is indicated in Figure 3.9. The addition of the multiplicity of waves is equivalent to the addition of vectors with progressively decreasing amplitude and increasing relative phase. For small reflectivity (Figure 3.9a), the wave amplitudes decrease rapidly, so that the phase increase has a relatively small effect on the resultant wave amplitude. In the case of high reflectivity (Figure 3.9b), the reverse is the case and a small successive phase change rapidly reduces the resultant.

Two important devices based on these ideas of multiple reflections are the Fabry–Pérot interferometer and the Fabry–Pérot etalon [3]. In the former case the distance between the two surfaces is finely variable for fringe control; in the case of the etalon the surfaces are fixed. In both cases, the flatness and parallelism of the surfaces must be accurate to $\sim \lambda/100$ for good-quality fringes. This is difficult to achieve in a variable device, and the etalon is preferred for most practical purposes.

The Fabry–Pérot interferometer is extremely important in opto-electronics. We have already noted its wavelength selectivity, but we should also note its ability to store optical energy by continually bouncing light between two parallel mirrors. For this reason, it is often called a Fabry–Pérot 'cavity' and is, roughly speaking, the optical equivalent of an electronic oscillator. The optical term is 'resonator', and it is this property which makes it an integral feature in all lasers. Because of its importance, it is useful to be aware of the parameters which characterize the performance of the Fabry–Pérot resonator. There are three main parameters: (i) finesse, (ii) resolving power, (iii) free spectral range.

These parameters relate, as is to be expected, to the instrument's ability to separate closely spaced optical wavelengths. The first is a measure of the sharpness of the fringes. This measure is normalized to the separation of the fringes for a single wavelength, since, clearly, there is no advantage in having narrow fringes if they are all crowded together, so that the orders of different wavelengths overlap. We hence define a quantity:

$$\Phi = \frac{\text{separation of successive fringes}}{\text{width at half maximum of a single fringe}}$$

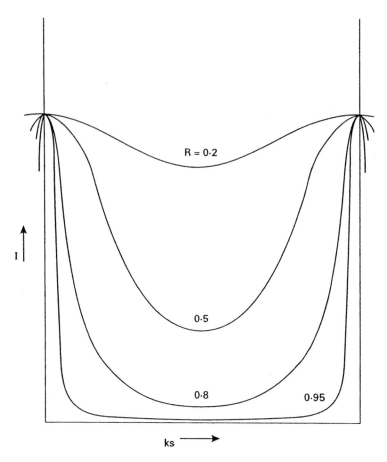

FIGURE 3.9 Graphical illustration of the dependence of fringe sharpness on reflectivity (R). (From Rogers A 1997 essentials of optoelectronics (Cheltenham: Nelson Thornes) with permission.)

Φ is called the 'finesse' and is roughly equivalent to the 'Q' ('quality' factor measuring the sharpness of the resonance) of an electronic oscillator.

It is easy to derive an expression for Φ from equation (3.2) as follows. Equation (3.2) may be written in the form:

$$I = \frac{I_{max}}{1 + F \sin^2\left(\frac{1}{2}\psi\right)}$$

where

$$F = \frac{4R^2}{(1-R^2)^2}$$

and

$$\Psi = ks.$$

F is sometimes known as the 'coefficient of finesse'. From this, it is clear that $I = I_{max}/2$ when:

$$\psi_h = \frac{2}{\sqrt{F}}.$$

Hence the width at half maximum $= 2\psi_h = 4/\sqrt{F}$. The 'ψ distance' between successive maxima is just 2π, and thus, the finesse is given by:

$$\Phi = \frac{2\pi}{2\psi_h} = \frac{\pi\sqrt{F}}{2} = \frac{\pi R}{(1-R^2)}.$$

This quantity has a value of two for a dual beam interferometer. For a Fabry–Pérot etalon with $R = 0.9$ its value is 15. Clearly, the higher the value of R the sharper are the fringes for a given fringe separation and the more wavelength-selective is the device.

The next quantity we need to look at is the resolving power. This is a measure of the smallest detectable wavelength separation ($\delta\lambda$) at a given wavelength (λ) and is defined as:

$$\rho = \frac{\lambda}{\delta\lambda}.$$

If we take λ to be that which corresponds to a ψ difference equal to the width of the half maximum, we find that:

$$\rho = \frac{\lambda}{\delta\lambda} = p \times \text{finesse}$$

i.e. $\rho = p\Phi$, where p is the 'order' of the maximum. If the etalon is being viewed close to normal incidence, then p will be effectively just the number of wavelengths in a double passage across the etalon. If the etalon has optical thickness t we have $p = 2t/\lambda$ and with:

we have:
$$F = \frac{4R^2}{(1-R^2)^2}$$

$$\rho = \pi t \frac{\sqrt{F}}{\lambda}.$$

Resolving power, ρ, is typically of the order of 10^6, compared with a figure $\sim 10^4$ for a dual-beam interferometer such as the Michelson (see Section 3.2.5(i)). The ratio of these figures thus represents the improvement in accuracy afforded by multiple-beam interferometry over dual-beam techniques.

Finally, we define a quantity concerned with the overlapping of orders. If the range of wavelengths ($\Delta \lambda$) under investigation is such that the $(p+1)$th maximum of λ is to coincide with the pth maximum of $(\lambda + \Delta \lambda)$, then, clearly, there is an unresolvable confusion. For this to be so:

$$(p+1)k = p(k + \Delta k)$$

so that

$$\frac{\Delta k}{k} = \frac{\Delta \lambda}{\lambda} = \frac{1}{p}.$$

Again, close to normal incidence, we may write, with $p = 2t/\lambda$:

$$\Delta \lambda = \frac{\lambda}{p} = \frac{\lambda^2}{2t}.$$

$\Delta \lambda$ is called the 'free spectral range' of the etalon and represents the maximum usable wavelength range without recourse to prior separation of the confusable wavelengths.

For a more detailed discussion of the Fabry–Pérot interferometer, see [3].

3.2.4 Interference between Partially Coherent Waves

In dealing with the subjects of interference it has been assumed that each of the interfering waves bears a constant phase relationship to the others in both time and space. Such an assumption cannot be valid for all time and space intervals since the atomic emission processes which give rise to light are largely uncorrelated, except for the special case of laser emission.

In this section, we shall look at interference between waves which are not fully coherent but only 'partially coherent'. These are waves which are mutually related in phase to only a limited extent, which must be quantified. This can conveniently be done by considering again the problem of dual-beam interference, but this time with partially coherent beams.

Consider the two-beam interference diagram of Figure 3.10. It is clear, from our previous look at this topic, that interference fringes will be formed if the two waves bear a constant phase relationship to each other, but we must now consider the form of the interference pattern for varying degrees of mutual coherence. In particular, we must consider the 'visibility' of the pattern, in other words, the extent to which it contains measurable structure and contrast.

At the point 0 (in Figure 3.10) the (complex) amplitude resulting from the two sources P_1 and P_2 is given by:

$$A = f_1(t'') + f_2(t'')$$

where $t'' = t' + \tau_0$; τ_0 is the time taken for light to travel from P_1 or P_2 to 0. If f_1, f_2 represent the electric field amplitudes of the waves, the observed intensity at 0 will be given by the square of the modulus of this complex number. Hence, in this case, the optical intensity is given by:

$$I_0 = \langle AA^* \rangle = \langle (f_1(t'') + f_2(t''))(f_1^*(t'') + f_2^*(t'')) \rangle$$

where the triangular brackets indicate an average taken over the response time of the detector (e.g. the human eye) and we assume that f_1 and f_2 contain the required constant of proportionality ($K^{1/2}$) to relate optical intensity with electric field strength, i.e. $I = KE^2$.

At point Q the amplitudes will be:

$$f_1(t'' - \tau/2), f_2(t'' + \tau/2)$$

τ being the time difference between paths P_2Q and P_1Q. Writing $t = t'' - \tau/2$, we have the intensity at Q:

$$I_Q = \langle (f_1(t) + f_2(t + \tau))(f_1^*(t) + f_2^*(t + \tau)) \rangle$$

i.e.

$$I_Q = \langle f_1(t)f_1^*(t) \rangle + \langle f_2(t)f_2^*(t) \rangle$$
$$+ \langle f_2(t + \tau)f_1^*(t) \rangle + \langle f_1(t)f_2^*(t + \tau) \rangle. \quad (3.3)$$

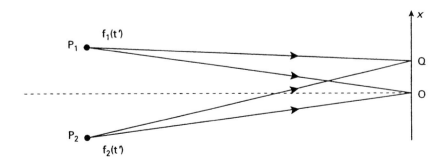

FIGURE 3.10 Interference between partially coherent sources. (From Rogers A 1997 essentials of optoelectronics (Cheltenham: Nelson Thornes) with permission.)

The first two terms are clearly the independent intensities, I_1, I_2, of the two sources at Q. The second two terms will have values that depend upon the extent to which f_1 and f_2 correlate in phase and amplitude when displaced in time by τ. We may, in fact, define a 'mutual correlation function', $c_{12}(t)$:

$$c_{12}(t) = \langle f_1(t) f_2^*(t+\tau) \rangle$$
$$c_{12}^*(t) = \langle f_1^*(t) f_2(t+\tau) \rangle.$$

We may note, in passing, that each of these terms will be zero if f_1 and f_2 have orthogonal polarizations, since, in that case, neither field amplitude has a component in the direction of the other, there can be no superposition, and the two cannot interfere. Hence, the average value of their product is again just the product of their averages, each of which is zero, being a sinusoid.

If $c_{12}(\tau)$ is now written in the form:

$$c_{12}(\tau) = |c_{12}(\tau)| \exp(i\omega\tau)$$

(which is valid provided that f_1 and f_2 are sinusoids in ωt), we have:

$$c_{12}(\tau) + c_{12}^*(\tau) = 2|c_{12}(\tau)|\cos\omega\tau.$$

Hence, provided that we observe the light intensity at Q with a detector which has a response time very much greater than the coherence times (self and mutual) of the sources (so that the time averages are valid), then we may write the intensity at Q as (equation (3.3)):

$$I_Q = I_1 + I_2 + 2|c_{12}(\tau)|\cos\omega\tau. \quad (3.4)$$

As we move along x, we shall effectively increase τ, so we shall see a variation in intensity whose amplitude will be $2|c_{12}(\tau)|$ (i.e. twice the modulus of the mutual correlation function) and which varies about a mean value equal to the sum of the two intensities (Figure 3.11). Thus, we have an experimental method by which the mutual correlation of the sources, $c_{12}(\tau)$, can be measured.

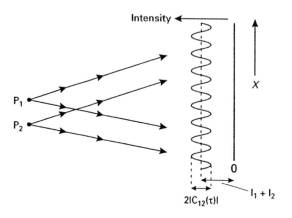

FIGURE 3.11 Mutual coherence function ($|C12(\tau)|$) from the two-source interference pattern. (From Rogers A 1997 essentials of optoelectronics (Cheltenham: Nelson Thornes) with permission.)

If we now define a fringe visibility for this interference pattern by:

$$V = \frac{(I_{\max} - I_{\min})}{(I_{\max} + I_{\min})}$$

which quantifies the contrast in the pattern, i.e. the difference between maxima and minima as a fraction of the mean level, then from equation (3.4):

$$V(\tau) = \frac{2|c_{12}(\tau)|}{(I_1 + I_2)}$$

so that the visibility of the fringes is seen to be directly related to the mutual correlation of the sources. We further define a 'coherence' function, $\gamma(t)$, which is just the mutual correlation function normalized to its value when $\tau = 0$, so that it now only depends upon differences between the phases at τ. The value of the mutual correlation function at $\tau = 0$ is given by:

$$c_{12}(0) = \langle f_1(t) f_2^*(t) \rangle = K \langle E_1 E_2 \rangle = (I_1 I_2)^{1/2}$$

so with:

$$\gamma(\tau) = \frac{|c_{12}(\tau)|}{|c_{12}(0)|}$$

we have:

$$\gamma(\tau) = \frac{|c_{12}(\tau)|}{(I_1 I_2)^{1/2}}.$$

Hence the visibility function $V(\tau)$ is related to the coherence function $\gamma(\tau)$ by:

$$V(\tau) = \frac{2(I_1 I_2)^{1/2}}{(I_1 + I_2)} \cdot \gamma(\tau)$$

and, if the two intensities are equal, we have:

$$V(\tau) = \gamma(\tau)$$

i.e. the visibility and coherence functions are identical.

From this, we may conclude that, for equal intensity coherent sources, the visibility is 100% ($\gamma = 1$); for incoherent sources it is zero, and for partially coherent sources, the visibility gives a direct measure of the actual coherence.

If we arrange that the points P_1 and P_2 are pinholes equidistant from and illuminated by a single source S, then the visibility function clearly measures the self-coherence of S. Suppose now that the two holes are placed in front of an extended source, S, as shown in Figure 3.12, and that their separation is variable.

The interference pattern produced by these sources of light now measures the correlation between the two corresponding points on the extended source. If the separation is initially zero and is increased until the visibility first falls to zero, the value of the separation at which this occurs defines a spatial coherence dimension for the extended source. Also, if the source is isotropic, a coherence area is correspondingly defined. In other words, in this case, any given source point has no phase

Interference and Polarization

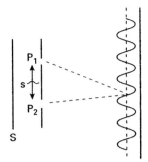

FIGURE 3.12 Extended-source interference. (From Rogers A 1997 essentials of optoelectronics (Cheltenham: Nelson Thornes) with permission.)

correlation with any point that lies outside the circular area of which it is the centre point.

3.2.5 Practical Examples

In order to appreciate fully the practical importance of the concept of optical coherence, we shall conclude with four examples of the concept in action.

   i. *The Michelson and Twyman–green interferometers.* The Michelson [1] interferometer is a very sensitive device for measuring optical path length differences. It is also important for its crucial role in the original formulation of special relativity. The basic arrangement is shown in Figure 3.13. Light from a collimated, extended source is split into two beams by a partial mirror S. The two light signals are then returned to S by the two plane mirrors M_1 and M_2. The partial mirror then returns parts of each light signal in the direction E, to a screen where they are allowed to interfere. The plate P is included to compensate that path (i.e. via mirror 2) for the extra distance travelled along the other path (via mirror 1) as a result of the triple passage of the partial mirror S. Hence, in this pristine state, the two signals in direction E are precisely matched in the path of travel. (This is important to ensure that high-visibility interference is obtained over a broad range of wavelengths, in the face of dispersion in the mirror material). With an extended source the field, looking into the direction E, will appear as a set of interference rings corresponding to the rays from different positions on the source aperture. The interference pattern can be made to consist of straight lines by setting M_1 and M_2 at a small angle to each other, thus creating a linear phase difference (which then dominates) across the field. If an optical component is now introduced into one of the arms, its added optical path length will displace the fringes. By measuring this displacement, accurate measurements can be made of, for example, length and refractive index. This interferometer was used to measure the number of wavelengths of cadmium light in 1 m, thus leading to a new definition of that length unit.

The Michelson interferometer's contribution to the formulation of special relativity derives from the fact that, if the interferometer is moving physically in the direction of one of the light paths, the double passage of the light in that direction should be shorter, according to classical electromagnetic theory, than in the orthogonal direction, thus resulting in a fringe displacement. It thus should have been possible to detect the motion of the earth through the aether in its passage around the sun. The fact that Michelson and Morley, who performed this experiment in 1887, could detect no such motion, led Einstein (in 1905) to abandon altogether the notion of the aether and to formulate entirely new ideas about the nature of space and time, in his Special Theory of Relativity.

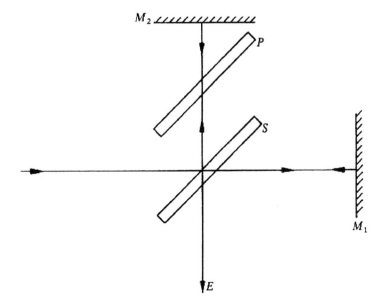

FIGURE 3.13 Arrangement for the Michelson interferometer.

An important variation of the Michelson interferometer idea is shown in Figure 3.14. It is known as the Twyman–Green interferometer [4]. In this case the illuminating light originates from a point source and is rendered plane by a lens, so that a uniform phase front results. If an optical component is now inserted into one of the arms and care is taken to ensure that the light returning to P is again plane, after having twice traversed the inserted component, the interference pattern at S will be distorted by any imperfections in the component, such as variations in refractive index. This allows the imperfections to be accurately quantified. This interferometer is thus a powerful tool in the preparation of high-quality optical elements, even to the extent of arranging for its output to control polishing functions, in order to equalize path lengths to less than a fraction of a wavelength.

ii. *Michelson's stellar interferometer.* The concept of the spatial coherence of a light source is used in an instrument known as Michelson's stellar interferometer to measure the angular diameter of (nearer) stars (Figure 3.15). If the star subtends an angle α at the two mirrors, spaced at distance d, then the two monochromatic (with the aid of an optical filter) rays A, A' (essentially parallel due to the very large distance of the star), from one point at the edge of the star, will be coherent and will produce an interference pattern with visibility = 1. Similarly, so also will the two rays B, B' from a diametrically opposite point at the other edge of the star. If the distance d between the mirrors is such that the ray B' is just one wavelength closer to M_2 than B is to M_1 then the second interference pattern, from BB', will coincide with the first from AA'. All the intermediate points across the star produce interference patterns between these two to give a total resultant visibility of zero. Hence, the value of d for which the fringe visibility first disappears provides the angular diameter of the star as λ/d (in fact, due to the circular rather than rectangular area, it is $1.22\,\lambda/d$). This method was first used by Michelson in 1920 [5] to determine the angular diameter of the star Betelgeuse as 0.047 seconds of arc. Distances between mirrors (d) of up to 10 m have been used. (Betelgeuse is a large star in the constellation of Orion and is quite close to Earth (~4 light years).) The vast majority of stars are too distant even for this very sensitive method to be of any use.

iii. *The Mach–Zehnder interferometer.* Consider now the two-arm, optical-fibre, Mach–Zehnder arrangement of Figure 3.16. A measurand (quantity to be measured) M in a Mach–Zehnder interferometer causes a phase change in arm 1 which is detected by means of a change in the position of the interference pattern resulting from the recombination of the light at point R. Interference can only occur if the recombining beams have components of the same polarization and if the difference in path length between the two arms is less than the source coherence length. This is not practicable with an LED, which has a coherence length ~0.02 mm, but even a modest semiconductor laser has a coherence length ~1 m (coherence time ~5 ns) and can easily be used in this application. A single mode He–Ne laser has a coherence length of several kilometres. It is clear that, in order to make an accurate measurement of M, it is necessary to choose a source with a fairly large coherence length. However, if the coherence length is too large, every reflection in the system interferes with every other, and an unwanted interference 'noise' results. This is an important problem for the opto-electronic designer: the coherence of the source must be optimized for the system in question.

This Mach–Zehnder arrangement is widely used in optical measurement technology. A special case of the Mach–Zehnder interferometer is described in the next section.

iv. *The optical-fibre gyroscope.* A rather more sophisticated example of the effect of coherence occurs in the optical-fibre gyroscope (Figure 3.17) [6]. The principle of the gyroscope is essentially the same as for the Mach–Zehnder interferometer we have just considered, the only important difference being

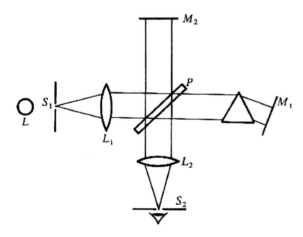

FIGURE 3.14 The Twyman–Green interferometer.

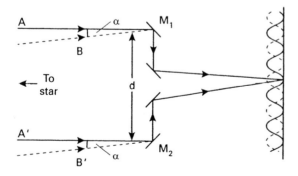

FIGURE 3.15 Michelson's stellar interferometer. (From Rogers A 1997 essentials of optoelectronics (Cheltenham: Nelson Thornes) with permission.)

Interference and Polarization

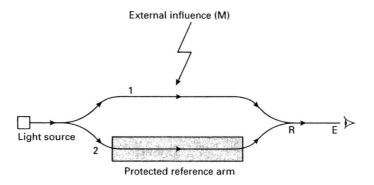

FIGURE 3.16 An optical-fibre Mach–Zehnder interferometer. (From Rogers A 1997 essentials of optoelectronics (Cheltenham: Nelson Thornes) with permission.)

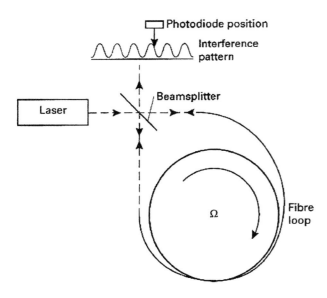

FIGURE 3.17 A Sagnac interferometer: the optical-fibre gyroscope.

that the two interfering arms now exist in the same fibre; the distinction between them lies in the fact that the light propagates in opposite directions in the two arms. The single source launches light in each of the two directions around a fibre loop, and the light which emerges from the two ends of the loop interferes to produce a pattern on the screen. If the loop now rotates about an axis normal to its plane in, say, a clockwise direction, the light travelling in that direction will see its exit end receding from it, while the counter-propagating light will see its end approaching. Consequently, the two components will traverse different optical paths and will emerge with a different relative phase relationship compared with that when the loop is stationary. The effect of the rotation is thus to cause a shift in the interference pattern, the magnitude of which is a measure of the rotation. Clearly, the greater the rotational velocity the greater will be the path length shift, and the coherence length of the source must be large enough to embrace the range of rotation which is to be measured. However, if the coherence length is too great, another problem arises. Some of the light will be backscattered, by the fibre material, as it propagates, and the light which is backscattered from a region around the half-way point will itself interfere for the two directions (Figure 3.17) and construct its own interference pattern. This will generate a noise level which will degrade the device performance. Clearly, the greater the coherence length of the source the greater will be the region around the midpoint from which this can occur and thus the greater will be the noise level. Hence a compromise or 'trade-off' has to be struck, as it always does in device and system design. Gyroscopes are very important devices for navigation and automatic flight control. The conventional gyroscope based on the conservation of angular momentum in a spinning metal disc is highly developed but contains parts which take time to be set in motion ('spin-up' time) and which wear. The device is also relatively expensive both to install and to maintain. The optical-fibre gyroscope overcomes all these problems (but, inevitably, has some of its own).

The phase difference between the counter-propagating signals is given by [7]:

$$\Phi = \frac{8\pi A \Omega}{c_0 \lambda_0}$$

or

$$\Phi = \frac{2\pi L D \Omega}{c_0 \lambda_0}$$

where A is the total effective area of the coil (i.e. the total area enclosed by N turns), L is the total length of the fibre, D is the diameter of the coil, c_0 is the velocity of light in free space, λ_0 is the wavelength of the light in free space and Ω is the rotation rate. Let us now insert some numbers into these expressions. Suppose that we use a wavelength of 1 µm with a coil of length 1 km and a diameter of 0.1 m. This gives:

$$\Phi = 2.1 \Omega$$

For the earth's rotation of $15°$ h^{-1} (7.3×10^{-5} radians s^{-1}), we must therefore be able to measure $\sim 1.5 \times 10^{-4}$ radian of phase shift. This can quite readily be done. In fact it is possible, using this device, to measure ~10^{-6} radian of phase shift, corresponding to ~5×10^{-7} radians s^{-1} of rotation rate.

3.3 Polarization

3.3.1 Introduction

We know that the electric and magnetic fields, for a freely propagating light wave, lie transversely to the propagation direction and orthogonally to each other. Normally, when discussing polarization phenomena, we fix our attention on the electric field, since it is this that has the most direct effect when the wave interacts with matter. In saying that an optical wave is 'polarized', we are implying that the direction of the electric field is either constant or is changing in an ordered, prescribable manner. In general, the tip of the electric vector circumscribes an ellipse, performing a complete circuit in a time equal to the period of the wave, or in a distance of one wavelength. Clearly, the two parameters are equivalent in this respect.

As is well known, linearly polarized light can conveniently be produced by passing any light beam through a sheet of 'polaroid'. This is a material which absorbs light of one linear polarization (the 'acceptance' direction) to a much smaller extent (~1000 times) than the orthogonal polarization, thus effectively allowing just one linear polarization state to pass. The material's properties result from the fact that it consists of long-chain polymeric molecules aligned in one direction (the acceptance direction) by stretching a plastic and then stabilizing it. Electrons can move more easily along the chains than transversely to them, and thus, the optical wave transmits easily only when its electric field lies along this acceptance direction. The material is cheap and allows the use of large optical apertures. It thus provides a convenient means whereby, for example, a specific linear polarization state can be defined; this state then provides a ready polarization reference which can be used as a starting point for other manipulations.

In order to study these manipulations and other aspects of polarization optics, we shall begin by looking more closely at the polarization ellipse.

3.3.2 The Polarization Ellipse

The most general form of polarized light wave propagating in the O_z direction is derived from the two linearly polarized components in the O_x and O_y directions (Figure 3.18):

$$E_x = e_x \cos(\omega t - kz + \delta_x)$$
$$E_y = e_y \cos(\omega t - kz + \delta_y). \quad (3.5a)$$

If we eliminate $(\omega t - kz)$ from these equations, we obtain the expression:

$$\frac{E_x^2}{e_x^2} + \frac{E_y^2}{e_y^2} + \frac{2 E_x E_y}{e_x e_y} \cos(\delta_y - \delta_x) = \sin^2(\delta_y - \delta_x) \quad (3.5b)$$

which is the ellipse (in the variables E_x, E_y) circumscribed by the tip of the resultant electric vector at any one point in space over one period of the combined wave. This can only be true; however, if the phase difference $(\delta_y - \delta_x)$ is constant in time or, at least, changes only slowly when compared with the speed of response of detector. In other words, we say that the two waves must have a large mutual 'coherence'. If this were not so then relative phases and, hence, resultant field vectors would vary randomly within the detector response time, giving no ordered pattern to the behaviour of the resultant field and thus presenting to the detector what would be, essentially, unpolarized light. Assuming that the mutual coherence is good, we may investigate further the properties of the polarization ellipse.

Note, first, that the ellipse always lies in the rectangle shown in Figure 3.19, but that the axes of the ellipse are not parallel with the original x, y directions. The ellipse is specified as follows: with e_x, e_y, $\delta(= \delta_y - \delta_x)$ known, then we define $\tan \beta = e_y/e_x$. The orientation of the ellipse, α, is given by:

$$\tan 2\alpha = \tan 2\beta \cos \delta.$$

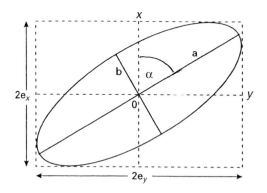

FIGURE 3.19 The polarization ellipse. (From Rogers A 1997 essentials of optoelectronics (Cheltenham: Nelson Thornes) with permission.)

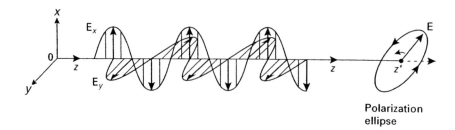

FIGURE 3.18 Electric field components for an elliptically polarized wave. (From Rogers A 1997 Essentials of Optoelectronics (Cheltenham: Nelson Thornes) with permission.)

Semi-major and semi-minor axes a, b are given by:

$$e_x^2 + e_y^2 = a^2 + b^2 \sim I \text{ (the wave intensity)}.$$

Also, the ellipticity, e, is given by:

$$e = \tan \chi = \pm \frac{b}{a} \text{ (the sign determines the sense of the rotation)}$$

where

$$\sin 2\chi = \sin 2\beta \sin \delta.$$

We should note also that the electric field components along the major and minor axes are always in quadrature (i.e. $\pi/2$ phase difference, the sign of the difference depending on the sense of the rotation).

Linear and circular states of polarization may be regarded as special cases where the polarization ellipse degenerates into a straight line or a circle, respectively. A linear state is obtained with the components in equations (3.5a) when either:

$$\left.\begin{array}{l} e_x = 0 \\ e_y \neq 0 \end{array}\right\} \text{linearly polarized in } O_y \text{ direction}$$

$$\left.\begin{array}{l} e_x \neq 0 \\ e_y = 0 \end{array}\right\} \text{linearly polarized in } O_x \text{ direction}$$

or,

$$\delta_y - \delta_x = m\pi$$

where m is an integer. In this latter case, the direction of polarization will be at an angle:

$$+\tan^{-1}(e_y/e_x) \quad m \text{ even}$$
$$-\tan^{-1}(e_y/e_x) \quad m \text{ odd}$$

with respect to the O_x axis. A circular state is obtained when

$$e_x = e_y$$

and

$$(\delta_y - \delta_x) = (2m+1)\pi/2$$

i.e. in this case the two waves have equal amplitudes and are in phase quadrature. The waves will be right-hand circularly polarized when m is even and left-hand circularly polarized when m is odd.

3.3.3 Material Interactions

Light can become polarized as a result of the intrinsic directional properties of matter: either the matter which is the original source of the light or the matter through which the light passes. These intrinsic material directional properties are the result of directionality in the bonding which holds together the atoms of which the material is made. This directionality leads to variations in the response of the material according to the direction of an imposed force, be it electric, magnetic or mechanical. The best-known manifestation of directionality in solid materials is the crystal, with the large variety of crystallographic forms, some symmetrical and some asymmetrical. The characteristic shapes which we associate with certain crystals result from the fact that they tend to break preferentially along certain planes known as cleavage planes, which are those planes between which atomic forces are weakest.

It is not surprising, then, to find that directionality in a crystalline material is also evident in the light which it produces or is impressed upon the light which passes through it.

In order to understand the ways in which we may produce polarized light, control it and use it, we must make a gentle incursion into the subject of crystal optics.

3.3.4 Crystal Optics

Light propagates through a material by stimulating the elementary atomic dipoles to oscillate and thus to radiate. In our previous discussions, the forced oscillation was assumed to take place in the direction of the driving electric field, but in the case of a medium whose physical properties vary with direction, an anisotropic medium, this is not necessarily the case. If an electron in an atom or molecule can move more easily in one direction than another, then an electric field at some arbitrary angle to the preferred direction will move the electron in a direction which is not parallel with the field direction (Figure 3.20). As a result, the direction in which the oscillating dipole's radiation is maximized (i.e. normal to its oscillation direction) is not the same as that of the driving wave.

The consequences of this simple piece of physics for the optics of anisotropic media are complex. One consequence is that the refractive index varies with the direction of the electric field of the wave, E. If we have a wave travelling in direction O_z, its velocity now will depend upon its polarization state: if the wave is linearly polarized in the O_x direction it will travel

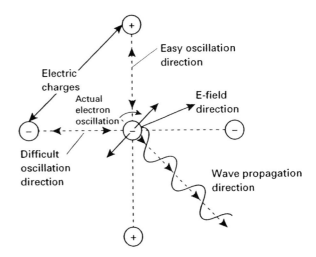

FIGURE 3.20 Electron response to electric field in an anisotropic medium. (From Rogers A 1997 essentials of optoelectronics (Cheltenham: Nelson Thornes) with permission.)

with velocity c_0/n_x, while if it is linearly polarized in the O_y direction its velocity will be c_0/n_y. Hence, the medium is offering two refractive indices to the wave travelling in this direction: we have the phenomenon known as double refraction or 'birefringence'. A wave which is linearly polarized in a direction at 45° to O_x will split into two equal-amplitude components, linearly polarized in directions O_x and O_y, respectively, the two components travelling at different velocities. Hence, the phase difference between the two components will steadily increase, and the composite polarization state of the wave will vary progressively from linear to circular back to linear again. The two special directions, O_x and O_y, for which there is no resolution of amplitude, are referred to as the birefringence axes.

This behaviour is, of course, a direct consequence of the basic physics which was discussed earlier: it is easier, in the anisotropic crystal, for the electric field to move the atomic electrons in one direction than in another. Hence, for the direction of easy movement, the light polarized in this direction can travel faster than when it is polarized in the direction for which the movement is more sluggish.

It follows from these discussions that an anisotropic medium may be characterized by means of three refractive indices, corresponding to polarization directions along O_x, O_y, O_z, and that these will have values n_x, n_y, n_z, respectively. We can use this information to determine the refractive index (and thus the velocity) for a wave in any direction with any given linear polarization state. To do this, we construct an 'index ellipsoid' or 'indicatrix', as it is sometimes called. This ellipsoid has the following important properties [8].

Suppose that we wish to investigate the propagation of light at an arbitrary angle to the crystal axes (polarization as yet unspecified). We draw a line, OP, corresponding to this direction within the index ellipsoid, passing through its centre O (Figure 3.21). Now we construct the plane, also passing through O, for which OP is its normal. This plane will cut the ellipsoid in an ellipse. This ellipse has the property that the directions of its major and minor axes define the directions of the birefringence axes for this propagation direction, and the lengths of these axes OA and OB are equal to the refractive indices for these polarizations. Since these two linear polarization states are the only ones which propagate without the change in polarization form for this crystal direction, they are sometimes referred to as the 'eigenstates' or 'polarization eigenmodes' for this direction, conforming to the matrix terminology of eigenvectors and eigenvalues.

The propagation direction we first considered, along O_z, corresponds, of course, to one of the axes of the ellipsoid, and the two refractive indices n_x, n_y are the lengths of the other two axes in the central plane normal to O_z. The refractive indices n_x, n_y, n_z are referred to as the principal refractive indices. Several other points are very well worth noting. Suppose, first, that

$$n_x > n_y > n_z$$

It follows that there will be a plane which contains O_z for which the two axes of interception with the ellipsoid are equal (Figure 3.22). This plane will be at some angle to the O_yO_z plane and will thus intersect the ellipsoid in a circle. This means that, for the light propagation direction corresponding to the normal to this plane, all polarization directions have the same velocity; there is no double refraction for this direction. This direction is an optic axis of the crystal, and there will, in general, be two such axes, since there must also be such a plane at an equal angle to the O_yO_z plane on the other side (see Figure 3.22). Such a crystal, with two optic axes, is said to be biaxial. Suppose now that:

$$n_x = n_y = n_o \quad \text{(say), the 'ordinary' index}$$

and

$$n_z = n_e \quad \text{(say), the 'extraordinary' index}$$

In this case, one of the principal planes is a circle and it is the only circular section (containing the origin) which exists. Hence, in this case, there is only one optic axis, along the O_z direction. Such crystals are said to be uniaxial. The crystal is said to be positive when $n_e > n_o$ and negative when $n_e < n_o$. For example, quartz is a positive uniaxial crystal and calcite a

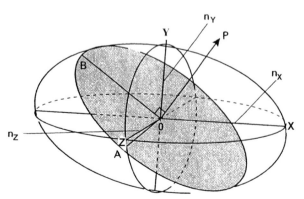

OA and OB represent the linearly polarized eigenstates for the direction OP

FIGURE 3.21 The index ellipsoid. (From Rogers A 1997 essentials of optoelectronics (Cheltenham: Nelson Thornes) with permission.)

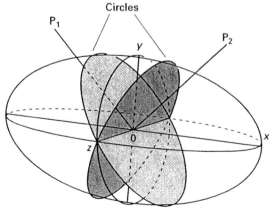

$n_x > n_y > n_z$
P_1 and P_2 are the optic axes of the crystal

FIGURE 3.22 Ellipsoid for a biaxial crystal. (From Rogers A 1997 essentials of optoelectronics (Cheltenham: Thornes) with permission.)

Interference and Polarization

negative uniaxial crystal. These features are, of course, determined by the crystal class to which these materials belong. It is clear that the index ellipsoid is a very useful device for determining the polarization behaviour of anisotropic media. Let us now consider some practical consequences of all of this.

3.3.5 Retarding Waveplates

Consider a positive uniaxial crystal plate (e.g. quartz) cut in such a way (Figure 3.23) as to set the optic axis parallel with one of the faces. Suppose a wave is incident normally on to this face. If the wave is linearly polarized with its electric field parallel with the optic axis, it will travel with refractive index n_e as we have described; if it has the orthogonal polarization, normal to the optic axis, it will travel with refractive index n_o.

The two waves travel in the same direction through the crystal but with different velocities. For a positive uniaxial crystal $n_e > n_o$ and, thus, the light linearly polarized parallel with the optic axis will be a 'slow' wave, whilst the one at right angles to the axis will be 'fast'. For this reason, the two crystal directions are often referred to as the 'slow' and 'fast' axes.

Suppose that the wave is linearly polarized at 45° to the optic axis. The phase difference between the components parallel with and orthogonal to the optic axis will now increase with distance, l, into the crystal according to:

$$\phi = \frac{2\pi}{\lambda}(n_e - n_o)l.$$

Hence, if, for a given wavelength λ,

$$l = \frac{\lambda}{4(n_e - n_o)}$$

then

$$\phi = \frac{\pi}{2}$$

and the light emerges from the plate circularly polarized. We have inserted a phase difference of π/2 between the components, equivalent to a distance shift of λ/4, and the crystal plate,

when of this thickness, is called a 'quarter-wave' plate. It will (for an input polarization direction at 45° to the axes) convert linearly polarized light into circularly polarized light or *vice versa*. If the input linear polarization direction lies at some arbitrary angle α to the optic axis then the two components:

$$E \cos\alpha$$
$$E \sin\alpha$$

will emerge with a phase difference of π/2. We noted in Section 3.3.3 that the electric field components along the two axes of a polarization ellipse were always in phase quadrature. It follows that these two components are now the major and minor axes of the elliptical polarization state which emerges from the plate. Thus, the ellipticity of the ellipse (i.e. the ratio of the major and minor axes) is just tan α, and by varying the input polarization direction α, we have a means by which we can generate an ellipse of any ellipticity. The orientation of the ellipse will be defined relative to the direction of the optic axis of the waveplate (Figure 3.23a).

Suppose now that the crystal plate has twice the previous thickness and is used at the same wavelength, it becomes a 'half-wave' plate. A phase difference of π is inserted between the components (linear eigenstates). The result of this is that an input wave which is linearly polarized at angle α to the optic axis will emerge still linearly polarized but with its direction now at −α to the axis. The plate has rotated the polarization direction through an angle −2α. Indeed, any input polarization ellipse will emerge with the same ellipticity but with its orientation rotated through −2α (Figure 3.23b).

It follows that, with the aid of these two simple plates, we can generate elliptical polarization of any prescribed ellipticity and orientation from linearly polarized light, which can itself be generated from any light source plus a simple polaroid sheet. Equally valuable is the reverse process: that of the analysis of an arbitrary elliptical polarization state or its conversion to a linear state. Suppose we have light of unknown elliptical polarization. By inserting an analysing polarizer and rotating it around the axis parallel to the propagation direction, we shall find a position of maximum transmission and an

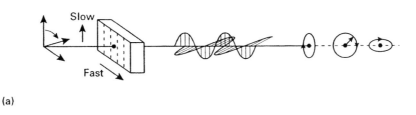

(a)

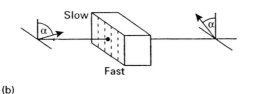

(b)

FIGURE 3.23 Polarization control with waveplates. (a) Quarter-wave plate; (b) half-wave plate. (From Rogers A 1997 essentials of optoelectronics (Cheltenham: Nelson Thornes) with permission.)

orthogonal position of minimum transmission. These are the major and minor axes of the ellipse (respectively), and the ratio of the two intensities at these positions will give the square of the ellipticity of the ellipse, i.e.

$$e = \frac{b}{a} = \frac{E_b}{E_a} = \left(\frac{I_b}{I_a}\right)^{1/2}.$$

Clearly, the orientation of the ellipse is also known since this is, by definition, just the direction of the major axis and is given by the position at which the maximum occurs. In order to convert the elliptical state into a linear one, all we need is a quarter-wave plate (appropriate to the wavelength of the light used, of course). Since the components of electric field along the major and minor axis of the ellipse are always in phase quadrature (see Section 3.3.2), the insertion of a quarter-wave plate with its axes aligned with the axes of the polarization ellipse brings the components into phase or into antiphase, and the light thus becomes linearly polarized. The quarter-wave plate is used in conjunction with a following polaroid sheet (or prism polarizer), and the two are rotated (independently) about the propagation axis until the light is extinguished. The quarter-wave plate must then have the required orientation in line with the ellipse axes, since only when the light has become linearly polarized can the polarizer extinguish it completely. (If there are no positions for which the light is extinguished, then it is not fully polarized.)

Such are the quite powerful manipulations and analyses which can be performed with very simple devices. However, manual human intervention for the rotation of plates is not always convenient or even possible. In many cases, polarization analysis and control must be done very quickly (perhaps in nanoseconds) and automatically, using electronic processing. For these cases, more advanced polarization devices must be used, and in order to understand and use these, a more advanced theoretical framework is necessary. We shall introduce this in Section 3.3.8. Before doing so, however, we shall first look at another very important component for polarization control and then at another crucial polarization parameter.

3.3.6 Polarizing Prisms

The same ideas as those just described are also useful in devices that produce linearly polarized light with a higher degree of polarization than a polaroid sheet is capable of and without its intrinsic loss (even for the 'acceptance' direction there is a significant loss). We shall look at just two of these devices, in order to illustrate the application of the ideas, but there are several others (these are described in most standard optics texts).

The first device is the Nicol prism, illustrated in Figure 3.24. Two wedges of calcite crystal are cut as shown, with their optic axes in the same direction (in the plane of the page) and cemented together with 'Canada balsam', a material whose refractive index at visible wavelengths lies midway between n_e and n_o.

When unpolarized light enters parallel to the axis of the prism (as shown) and at an angle to the front face, it splits, as always, into the e and o components, each with its own refractive index, and thus each with its own refractive angle according to Snell's law. (Calcite is a negative uniaxial crystal so $n_o > n_e$.) When the light reaches the Canada balsam interface between the two wedges, it finds that the geometry and refractive indices have been arranged such that the ordinary (o) ray, with the larger deflection angle, strikes this interface at an angle greater than the total internal reflection (TIR) angle and is thus not passed into the second wedge, whereas the extraordinary (e) ray is so passed. Hence, only the e ray emerges from the prism and this is linearly polarized. Thus, we have an effective prism polarizer, albeit one of limited angular acceptance (~14°) since the TIR condition is quite critical in respect of angle of incidence.

The second prism we shall discuss is widely used in practical polarization optics: it is called the Wollaston prism and is shown in Figure 3.25. Again we have two wedges of positive (say) uniaxial crystal. They are equal in size, placed together to form a rectangular block (sometimes a cube) and have their optic axes orthogonal, as shown. Consider a wave entering

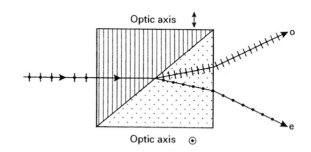

FIGURE 3.25 Action of the Wollaston prism. (From Rogers A 1997 essentials of optoelectronics (Cheltenham: Nelson Thornes) with permission.)

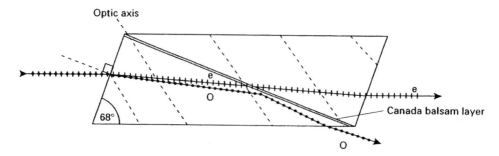

FIGURE 3.24 Action of the Nicol prism. (From Rogers A 1997 essentials of optoelectronics (Cheltenham: Nelson Thornes) with permission.)

normally from the left. The e and o waves travel with differing velocities and strike the boundary between the wedges at the same angle. On striking the boundary, one of the waves sees a positive change in refractive index $(n_e - n_o)$, the other a negative change $(n_e - n_o)$, so that they are deflected, respectively, up and down (Figure 3.25) through equal angles. The e and o rays thus diverge as they emerge from the prism allowing either to be isolated, or the two to be observed (or detected) simultaneously but separately. Also it is clear that, by rotating this prism around the propagation axis, we may reverse the positions of the two components.

It is extremely useful to be able to separate the two orthogonally polarized components in this controllable way. For example, consider the problem of the measurement of the rotation of the direction of a linearly polarized optical wave. Such a problem arises, for example, when measuring an electric current using the magneto-optic effect in the single-mode optical fibre (see Section 3.3.9(b)).

How can we actually measure such a rotation, ρ? Suppose that the emerging linearly polarized light falls on to a linear polarizer which is set with its polarization direction parallel with that of the light's *input* polarization direction (Figure 3.26). In the absence of a magnetic field ($\rho = 0$), all the light will be passed by the polarizer (ignoring its intrinsic attenuation).

Let us assume that the electric field amplitude of the propagating light is E_0, so that an intensity proportional to E_0^2 is passed, in the absence of current. When current flows, the polarization is rotated through an angle ρ and only a field component $E_0 \cos \rho$ will now be passed by the polarizer, giving a measurable intensity proportional to $E_0^2 \cos^2 \rho$. This intensity, in principle, allows ρ to be deduced.

However, there is a more convenient way to measure ρ. Suppose that, instead of a simple polarizer, we use a Wollaston prism, with its polarization axes set at $\pm 45°$ to the input polarization direction. We now have two intensity outputs from the Wollaston prism (see Figure 3.26):

$$I_1 = K E_0^2 \cos^2\left(\frac{1}{4}\pi - \rho\right)$$

$$I_2 = K E_0^2 \cos^2\left(\frac{1}{4}\pi + \rho\right)$$

where K is the usual universal constant.

If we detect these two intensities separately (by measuring the optical powers falling on two separate photodiodes: remember that power = intensity × area), then we can readily arrange for the electronics to construct the function

$$S = \frac{I_1 - I_2}{I_1 + I_2} = \frac{\cos^2\left(\frac{1}{4}\pi - \rho\right) - \cos^2\left(\frac{1}{4}\pi + \rho\right)}{\cos^2\left(\frac{1}{4}\pi - \rho\right) + \cos^2\left(\frac{1}{4}\pi + \rho\right)}$$

which gives, on manipulation of the functions,

$$S = \sin 2\rho$$

and, if 2ρ is small ($\ll /2$):

$$S \approx 2\rho$$

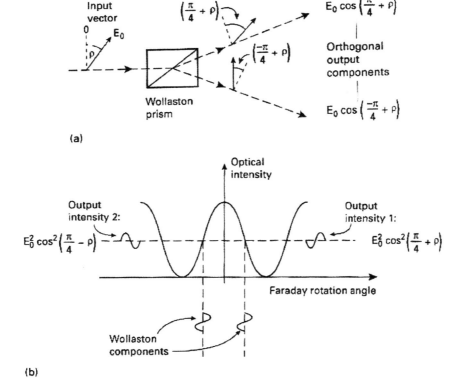

FIGURE 3.26 Measurement of polarization rotation. (a) Wollaston E-field components; (b) output intensities for Wollaston components.

Hence, with the aid of the polarizing beam-splitter, we have succeeded in measuring the polarization rotation independently of the light intensity and, thus, free from any variations in the source output, or variations in the attenuation along the optical path.

3.3.7 Circular Birefringence

So far we have considered only linear birefringence, where two orthogonal linear polarization eigenstates propagate, each remaining linear, but with different velocities. Some crystals also exhibit circular birefringence. Quartz (again) is one such crystal and its circular birefringence derives from the fact that the crystal structure spirals around the optic axis in a right-handed (dextro-rotatory) or left-handed (laevo-rotatory) sense depending on the crystal specimen: both forms exist in nature.

It is not surprising to find, in view of this knowledge and our understanding of easy motions of electrons, that light which is right-hand circularly polarized (clockwise rotation of the tip of the electric vector as viewed by a receiver of the light) will travel faster down the axis of a matching right-hand spiralled crystal structure then left-hand circularly polarized light. We now have circular birefringence: the two circular polarization components propagate without change of form (i.e. they remain circularly polarized) but at different velocities. They are the circular polarization eigenstates for this case.

The term 'optical activity' has been traditionally applied to this phenomenon, and it is usually described in terms of the rotation of the polarization direction of a linearly polarized wave as it passes down the optic axis of an 'optically active' crystal. This fact is exactly equivalent to the interpretation in terms of circular birefringence, since a linear polarization state can be resolved into two oppositely rotating circular components (Figure 3.27). If these travel at different velocities, a phase difference is inserted between them. As a result of this, when recombined, they again form a resultant which is linearly polarized but rotated with respect to the original direction (Figure 3.27). Hence 'optical activity' is equivalent to circular birefringence.

In general, both linear and circular birefringence might be present simultaneously in a material (such as quartz). In this case, the polarization eigenstates which propagate without change of form (and at different velocities) will be elliptical states, the ellipticity and orientation depending upon the ratio of the magnitudes of the linear and circular birefringences, and on the direction of the linear birefringence eigen-axes within the crystal.

It should, again, be emphasized that only the polarization eigenstates propagate without change of form. All other polarization states will be changed into different polarization states by the action of the polarization element (e.g. a crystal component). These changes of polarization state are very useful in opto-electronics. They allow us to control, analyse, modulate and demodulate polarization information impressed upon a light beam and to measure important directional properties relating to the medium through which the light has passed. We must now explore a more rigorous formalism to handle these more general polarization processes.

3.3.8 Polarization Analysis

As has been stated, with both linear and circular birefringence present, the polarization eigenstates (i.e. the states which propagate without change of form) for a given optical element are elliptical states, and the element is said to exhibit elliptical birefringence, since these eigenstates propagate with different velocities.

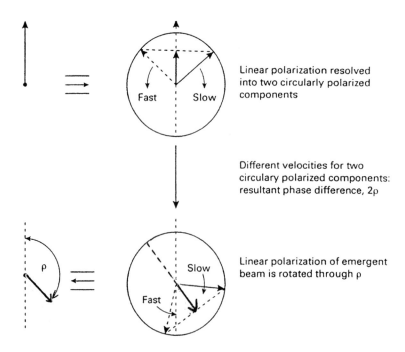

FIGURE 3.27 Resolution of linear polarization into circularly polarized components in circular birefringence (2ρ). (From Rogers A 1997 essentials of optoelectronics (Cheltenham: Nelson Thornes) with permission.)

In general, if we have, as an input to a polarization-optical element, light of one elliptical polarization state, it will be converted, on emergence, into a different elliptical polarization state (the only exceptions being, of course, when the input state is itself an eigenstate). We know that any elliptical polarization state can always be expressed in terms of two orthogonal electric field components defined with respect to chosen axes O_x, O_y, i.e.

$$E_x = e_x \cos(\omega t - kz + \delta_z)$$
$$E_y = e_y \cos(\omega t - kz + \delta_y)$$

or, in complex exponential notation:

$$E_x = |E_x|\exp(i\phi_x) \quad (\phi_x = \omega t - kz + \delta_x)$$
$$E_y = |E_y|\exp(i\phi_y) \quad (\phi_y = \omega t - kz + \delta_y).$$

When this ellipse is converted into another by the action of a lossless polarization element, the new ellipse will be formed from components which are linear combinations of the old, since it results from directional resolutions and rotations of the original fields. Thus, these new components can be written:

$$E'_x = m_1 E_x + m_4 E_y$$
$$E'_y = m_3 E_x + m_2 E_y$$

or, in matrix notation:

$$\mathbf{E'} = \mathbf{M} \cdot \mathbf{E}$$

where

$$\mathbf{M} = \begin{pmatrix} m_1 & m_4 \\ m_3 & m_2 \end{pmatrix} \quad (3.6)$$

and the m_n are, in general, complex numbers. **M** is known as a 'Jones' matrix after the mathematician who developed an extremely useful 'Jones calculus' for manipulations in polarization optics [9]. Now, in order to make measurements of the input and output states in practice, we need a quick and convenient experimental method. In Section 3.3.6 there was described a method for doing this which involved the manual rotation of a quarter wave plate and/or a polarizer, but the method we seek now must lend itself to automatic operation.

A convenient method for this practical determination is again to use the linear polarizer and the quarter-wave plate but to measure the light intensities for a series of fixed orientations of these elements.

Suppose that $I(\theta, \varepsilon)$ denotes the intensity of the incident light passed by the linear polarizer set at angle θ to O_x, after the O_y component has been retarded by angle ε as a result of the insertion of the quarter-wave plate with its axes parallel with O_x, O_y. We measure what are called the four Stokes parameters, as follows:

$$S_0 = I(0°, 0) + I(90°, 0) = e_x^2 + e_y^2$$
$$S_1 = I(0°, 0) - I(90°, 0) = e_x^2 - e_y^2$$
$$S_2 = I(45°, 0) - I(135°, 0) = 2e_x e_y \cos\delta$$
$$S_3 = I\left(45°, \frac{\pi}{2}\right) - I\left(135°, \frac{\pi}{2}\right) = 2e_x e_y \sin\delta$$
$$(\delta = \delta_y - \delta_x).$$

These parameters can be measured directly, with the aid of a photodetector. If the light is 100% polarized, only three of these parameters are independent, since:

$$S_0^2 = S_1^2 + S_2^2 + S_3^2$$

S_0 being the total light intensity.

If the light is only partially polarized, the fraction:

$$\eta = \frac{(S_1^2 + S_2^2 + S_3^2)}{S_0^2}$$

defines the degree of polarization. In what follows, we shall assume that the light is fully polarized ($\eta = 1$). It is easy to show that measurement of the S_n provides the ellipticity, e, and the orientation, α, of the polarization ellipse according to the relations:

$$e = \tan\chi$$
$$\sin 2\chi = \frac{S_3}{S_0}$$
$$\tan 2\alpha = \frac{S_2}{S_1}.$$

Now, these relations suggest a geometrical construction which provides a powerful and elegant means for description and analysis of polarization-optical phenomena. The Stokes parameters S_1, S_2, S_3 may be regarded as the Cartesian coordinates of a point referred to axes Ox_1, Ox_2, Ox_3. Thus, every elliptical polarization state corresponds to a unique point in three-dimensional space. For a constant S_0 (lossless medium), it follows that all such points lie on a sphere of radius S_0—the Poincaré sphere (Figure 3.28). The properties of the sphere are quite well known (see, for example [10]). We can see that the equator comprises the continuum of linearly polarized states, whilst the two poles correspond to the two oppositely handed states of circular polarization.

It is clear that any change resulting from the passage of light through a lossless element, from one polarization state to another, corresponds to a rotation of the sphere about a diameter. Now, any such rotation of the sphere may be expressed as a unitary 2×2 matrix, **M**. Thus, the conversion from one polarization state, **E**, to another **E**' may also be expressed in the form:

$$\mathbf{E'} = \mathbf{M} \cdot \mathbf{E}$$

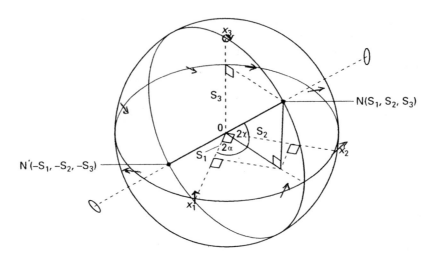

FIGURE 3.28 The Poincaré sphere: the eigenmode diameter (NN'). (From Rogers A 1997 essentials of optoelectronics (Cheltenham: Nelson Thornes) with permission.)

or

$$\begin{pmatrix} E'_x \\ E'_y \end{pmatrix} = \begin{pmatrix} m_1 & m_4 \\ m_3 & m_2 \end{pmatrix} \begin{pmatrix} E_x \\ E_y \end{pmatrix}$$

$$E'_x = m_1 E_x + m_4 E_y$$

$$E'_y = m_3 E_x + m_2 E_y$$

where

$$\mathbf{M} = \begin{pmatrix} m_1 & m_4 \\ m_3 & m_2 \end{pmatrix}$$

and this **M** may be immediately identified with our previous **M**, in equation (3.6). **M** is a Jones matrix [9] which completely characterizes the polarization action of the element and is also equivalent to a rotation of the Poincaré sphere. The two eigenvectors of the matrix correspond to the eigenmodes (or eigenstates) of the element (i.e. those polarization states which can propagate through the element without change of form). These two polarization eigenstates lie at opposite ends of a diameter of the Poincaré sphere, and the polarization effect of the element is to rotate the sphere about this diameter (Figure 3.29) through an angle which is equal to the phase which the polarization element inserts between its eigenstates.

The polarization action of the element may thus be regarded as that of resolving the input polarization state into the two eigenstates with appropriate amplitudes, and then inserting a phase difference between them before recombining to obtain the emergent state. Thus, a pure rotator (e.g. optically active crystal) is equivalent to a rotation about the polar axis, with the two oppositely handed circular polarizations as eigenstates. The phase velocity difference between these two eigenstates is a measure of the circular birefringence. Analogously, a pure linear retarder (such as a wave plate) inserts a phase difference between orthogonal linear polarizations which measures the linear birefringence. The linear retarder's eigenstates lie at opposite ends of an equatorial diameter.

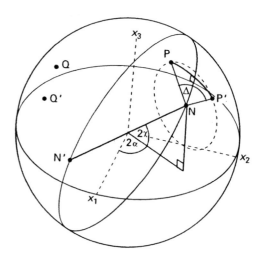

FIGURE 3.29 Rotation of the Poincaré sphere about the eigenmode diameter (NN'). (From Rogers A 1997 essentials of optoelectronics (Cheltenham: Nelson Thornes) with permission.)

It is useful for many purposes to resolve the polarization action of any given element into its linear and circular birefringence components. The Poincaré sphere makes it clear that this may always be done, since any rotation of the sphere can always be resolved into two sub-rotations: one about the polar diameter and the other about an equatorial diameter.

From this brief discussion we can begin to understand the importance of the Poincaré sphere. It is a construction which converts all polarization actions into visualisable relationships in three-dimensional space.

To illustrate this point graphically, let us consider a particular problem. Suppose that we ask what is the smallest number of measurements necessary to define completely the polarization properties of a given lossless polarization element, about which we know nothing in advance. Clearly, we must provide known polarization input states and measure their corresponding output states, but how many input/output pairs are necessary: one, two or more?

Interference and Polarization

The Poincaré sphere answers this question easily. The element in question will possess two polarization eigenmodes, and these will be at opposite ends of a diameter. We need to identify this diameter. We know that the action of the element is equivalent to a rotation of the sphere about this diameter and through an angle equal to the phase difference which the element inserts between its eigenmodes. Hence, if we know one input/output pair of polarization states, we know that the rotation from the input to the output state must have taken place about a diameter lying in the plane which perpendicularly bisects the line joining the two states (see Figure 3.29). Two other input/output states will similarly define another such plane and, thus, the required diameter is clearly seen as the common line of intersection of these planes.

Further, the phase difference inserted between the eigenstates (i.e. the sphere's rotation angle) is easily calculated from either pair of states, once the diameter is known.

Hence, the answer is that two pairs of input/output states will define completely the polarization properties of the element. Simple geometry has provided the answer. A good general approach is to use the Poincaré sphere to determine or visualize the nature of the solution to a problem and then to revert to the Jones matrices to perform the precise calculations. Alternatively, some simple results in spherical trigonometry will usually suffice.

Having dealt with the theoretical tools by which polarization characteristics are manipulated and analysed, we shall now turn to some practical applications of the ideas. We shall look at ways in which directional properties within materials can impose polarizations on light waves passing through them.

3.3.9 Applications of Polarization Optics

A polarized optical wave is essentially one which is asymmetrical with regard to its transverse vibrations. In other words, there is a preference for oscillation in some directions when compared with others.

When an optical wave passes through a material medium, it does so by stimulating the elementary atomic dipoles to radiate. These secondary radiations combine vectorially with the primary wave to give rise to the resultant propagation through the medium, thus defining the latter's (complex) refractive index.

In such circumstances, any directionality inherent in the medium itself, resulting either from its crystal structure or from externally applied asymmetrical forces, will be impressed also on the propagating wave. Consequently, carefully chosen materials can be used to control polarization state, and the polarization analysis of the resultant wave can be used sensitively to probe material structures.

Clearly then, polarization effects may arise naturally or may be induced deliberately. Of those which occur naturally, the most common are the ones which are a consequence of an anisotropic material, an asymmetrical material strain or asymmetrical waveguide geometries.

If an optical medium is compressed in a particular direction, there results the same kind of directional restriction on the atomic or molecular electrons as in the case of crystals, and hence, the optical polarization directions parallel and orthogonal to these imposed forces (for isotropic materials) will encounter different refractive indices.

Somewhat similarly, if an optical wave is being guided in a channel, or other type of guide, with a refractive index greater than its surroundings, we have to be aware of the effect of any asymmetry in the geometry of the guide's cross section. Clearly, if the cross section is a perfect circle, as in the case of an ideal optical fibre, all linear polarization directions must propagate with the same velocity. If, however, the cross section is elliptical, then it is not difficult to appreciate that a linear polarization direction parallel with the minor axis propagates at a different velocity from that parallel with the major axis.

The optical fibre is, in fact, a good medium for illustrating these passive polarization effects, since all real fibres possess the same directional asymmetry due to one or more of the following: non-circularity of core cross section, linear strain in the core and twist strain in the core. Bending will introduce linear strain and twisting will introduce circular strain (Figure 3.30). Linear strain leads to linear birefringence and circular (twist) strain to circular birefringence.

The linear birefringence in 'standard' telecommunications optical fibre can be quite troublesome for high-performance links since it introduces velocity differences between the two orthogonal linear polarization states, which lead to relative time lags of the order of 1–10 ps km^{-1}. Clearly, this distorts the modulating signal: a pulse in a digital system, for example, will be broadened, and thus degraded, by this amount. This so-called 'polarization-mode dispersion' can be reduced by spinning the preform from which the fibre is being drawn, whilst it is being drawn, so as to average out the cross-sectional anisotropies. This 'spun preform' technique [11] reduces this form of dispersion to ~0.01 ps km^{-1}, i.e. by two orders of magnitude.

It is sometimes valuable deliberately to introduce linear or circular birefringence into a fibre. In order to introduce linear

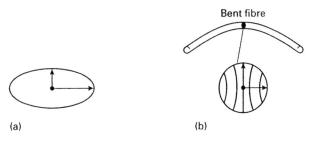

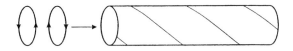

(c) Circularly birefringent fibres

FIGURE 3.30 Birefringence in optical fibres. (a) Geometrical 'form'; (b) bending 'strain'; and (c) twist-strain circularly birefringent fibre. (From Rogers A 1997 essentials of optoelectronics (Cheltenham: Nelson Thornes) with permission.)

birefringence the fibre core may be made elliptical (with consequences previously discussed) or stress may be introduced by asymmetric doping of the cladding material which surrounds the core (Figure 3.31) [12]. The stress results from asymmetric contraction as the fibre cools from the melt.

Circular birefringence may be introduced by twisting and then clamping the fibre or by spinning an asymmetric preform (from which the fibre is being pulled). One important application of fibre with a high value of linear birefringence ('hi-bi' fibre) is that linearly polarized light launched into one of the two linear eigenmodes will tend to remain in that state, thus providing a convenient means for conveying linearly polarized light between two points. The reason for this 'polarization holding' property is that light, when coupled (i.e. transferred) to the other eigenmode, will be coupled to a mode with a different velocity and will not, in general, be in phase with other previous light couplings into the mode; thus the various couplings will interfere destructively overall and only a small amplitude will result. There is said to be a 'phase mismatch'. (This is yet another example of wave interference!) Clearly, however, if a deliberate attempt is made to couple light only at those points where the two modes are in phase, then constructive interference can occur and the coupling will be strong. This is known as 'resonant' coupling and has a number of important applications.

An extremely convenient way of inducing polarization anisotropies into materials is by subjecting them to electric and/or magnetic fields. As we know very well, these fields can exert forces on electrons, so it is not surprising to learn that, through their effects on atomic electrons, the fields can influence the polarization properties of media, just as the chemical bond restrictions on these electrons in crystals were able to do. The use of electric and magnetic fields thus allows us to build convenient polarization controllers and modulators. Some examples of the effects which can be used will help to establish these ideas.

FIGURE 3.31 Asymmetrically doped, linearly birefringent optical fibre ('bow-tie'). (From Rogers A 1997 essentials of optoelectronics (Cheltenham: Nelson Thornes) with permission.)

a. *The electro-optic effect.* When an electric field is applied to an optical medium, the electrons suffer restricted motion in the direction of the field, when compared with that orthogonal to it. Thus, the material becomes linearly birefringent in response to the field. This is known as the electro-optic effect.

Consider the arrangement of Figure 3.32. Here, we have incident light which is linearly polarized at 45° to an electric field and the field acts on a medium transversely to the propagation direction of the light. The field-induced linear birefringence will cause a phase displacement between components of the incident light which lie, respectively, parallel and orthogonal to the field; hence, the light will emerge elliptically polarized.

A (perfect) polarizer placed with its acceptance direction parallel with the input polarization direction will, of course, pass all the light in the absence of a field. When the field is applied, the fraction of light power which is passed will depend upon the form of the ellipse, which in turn depends upon the phase delay introduced by the field. Consequently, the field can be used to modulate the intensity of the light, and the electro-optic effect is, indeed, very useful for the modulation of light.

The phase delay introduced may be proportional either to the field (Pockels effect) or to the square of the field (Kerr effect). All materials manifest a transverse Kerr effect. Only crystalline materials can manifest any kind of Pockels effect or longitudinal (E field parallel with propagation direction) Kerr effect. The reason for this is physically quite clear. If a material is to respond linearly to an electric field, the effect of the field must change sign when the field changes sign. This means that the medium must be able to distinguish (for example) between 'up' (positive field) and 'down' (negative field). But it can only do this if it possesses some kind of directionality in itself; otherwise all field directions must be equivalent in their physical effects. Hence, in order to make the necessary distinction between up and down, the material must possess an intrinsic asymmetry and, hence, must be crystalline. By a similar argument, a longitudinal E field can only produce a directional effect orthogonally to itself (i.e. in the direction of the optical electric field) if the medium is anisotropic (i.e. crystalline), for otherwise, all transverse directions will be equivalent.

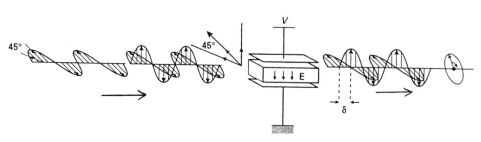

FIGURE 3.32 The electro-optic effect. (From Rogers A 1997 essentials of optoelectronics (Cheltenham: Nelson Thornes) with permission.)

In addition to the modulation of light (phase or intensity/power), it is clear that the electro-optic effect could be used to measure an electric field and/or the voltage which gives rise to it. Several modulation and sensors are based on this idea.

b. *The magneto-optic effect.* If a magnetic field is applied to a medium in a direction parallel to the direction in which light is passing through the medium, the result is a rotation of the polarization direction of whatever is the light's polarization state: in general, the polarization ellipse is rotated. The phenomenon, known as the Faraday (after its discoverer, in 1845) magneto-optic effect, is normally used with a linearly polarized input, so that there is a straightforward rotation of a single polarization direction (Figure 3.33). The magnitude of the rotation due to a field, H, over a path length, L, is given by:

$$\rho = V \int_0^L H \cdot \mathrm{d}l$$

where V is a constant known as the Verdet constant: V is a constant for any given material but is wavelength-dependent. Clearly, if H is constant over the optical path, we have:

$$\rho = VHL$$

From the discussion in Section 3.3.7, we see that this is a magnetic-field-induced circular birefringence.

The physical reason for the effect is easy to understand in qualitative terms. When a magnetic field is applied to a medium, the atomic electrons find it easier to rotate in one direction around the field than in the other: the Lorentz force acts on a moving charge in a magnetic field and this will act radially on the electron as it circles the field. The force will be outwards for one direction of rotation and inwards for the other. The consequent electron displacement will lead to two different radii of rotation and thus two different rotational frequencies and electric permittivities. Hence, the field will result in two different refractive indices and thus to circular birefringence. Light which is circularly polarized in the 'easy' (say clockwise) direction will travel faster than that polarized in the 'hard' direction (anti-clockwise), leading to the observed effect (Figure 3.33b). Another important aspect of the Faraday magneto-optic effect is that it is 'non-reciprocal'. This means that linearly polarized light (for example) is always rotated in the same absolute direction in space, independently of the direction of propagation of the light (Figure 3.34a). For an optically active crystal, this is not the case: if the polarization direction is rotated from right to left (say) on forward passage (as viewed by a fixed observer), it will be rotated from left to right on backward passage (as viewed by the same observer), so that back-reflection of light through an optically active crystal will result in light with zero final rotation, the two rotations having cancelled out (Figure 3.34c). This is called as reciprocal rotation because the rotation looks the same for an observer who always looks in the direction of propagation of the light (Figure 3.34c).

For the Faraday magneto-optic case, however, the rotation always takes place in the same direction with respect to the magnetic field (not the propagation direction) since it is this which determines 'easy' and 'hard' directions. Hence, an observer always looking in the direction of light propagation will see different directions of rotation since he/she is, in one case, looking along the field and, in the other, against it. It is a non-reciprocal effect. The Faraday effect has a number of practical applications. It can be used to modulate light, although it is less convenient for this than the electro-optic effect. This is a result of the greater difficulty of producing and manipulating large and rapidly varying (for high-modulation bandwidth) magnetic fields when compared with electric fields (large solenoids have large inductance!).

The Faraday magneto-optic effect can valuably be used in optical isolators, however. In these devices, light from a source passes through a linear polarizer and then through a magneto-optic element which rotates the polarization direction through 45°.

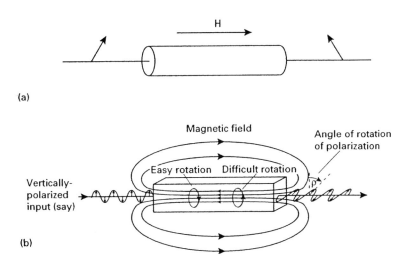

FIGURE 3.33 The Faraday magneto-optic effect. (From Rogers A 1997 essentials of optoelectronics (Cheltenham: Nelson Thornes) with permission.)

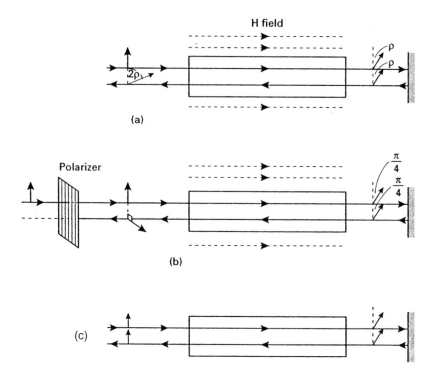

FIGURE 3.34 Reciprocal and non-reciprocal polarization rotation. (a) Non-identical rotation (Faraday effect). Rotation in same direction in relation to the magnetic field. (b) Optical isolator action. Total rotation of π/2 for polarization blocking. (c) Reciprocal rotation (optical activity). Rotation in same direction in relation to propagation direction. (From Rogers A 1997 Essentials of Optoelectronics (Cheltenham: Nelson Thornes) with permission.)

Any light which is back-reflected by the ensuing optical system suffers a further 45° rotation during the backward passage, in the same rotational direction, thus arriving back at the polarizer having been rotated through 90°; it is therefore blocked by the polarizer (Figure 3.34b). Hence the source is isolated from back-reflections by the magneto-optic element/polarizer combination which is thus known as a Faraday magneto-optic isolator. This is very valuable for use with devices whose stability is sensitive to back-reflection, such as lasers and optical amplifiers, and it effectively protects them from feedback effects. The Faraday magneto-optic effect can also be used to measure magnetic fields and the electric currents which give rise to them [13]. There are other magneto-optic effects (e.g. Kerr, Cotton–Mouton, Voigt) but the Faraday effect is by far the most important for opto-electronics.

3.4 Conclusions

In the first part of this chapter we began by noting that the light wave comprises field vibrations which take place transversely to the propagation direction. This makes it possible to explain satisfactorily the phenomenon of optical interference. With this understanding, we saw also how to design useful interferometers for the analysis and control of light.

We have also looked at the conditions necessary for optical waves to interfere in a consistent and measurable way, with themselves and with other waves. We have seen that the conditions relate to the extent to which the properties such as amplitude, phase, frequency and polarization remain constant in time and space, i.e. the extent to which knowledge of the properties at one point in time or space tells us about these properties at other points.

Any interference pattern will remain detectable only as long as coherence persists, and by studying the rise and fall of interference patterns, much can be learned about the sources themselves and about the processes which act upon the light from them.

Coherence also relates critically to the information-carrying capacity of light and to our ability to control and manipulate it sensibly. The design and performance of any device or system that relies on interference or diffraction phenomena must take into account the coherence properties of the sources to be used; some of these works to the designer's disadvantage, but others do not.

In the second part of this chapter we have looked closely at the directionality possessed by the optical transverse electric field, i.e. we have looked at optical polarization. We have seen how to describe it, to characterize it, to control it, to analyse it and how, in some ways, to use it.

We have also looked at the ways in which the transverse electric and magnetic fields interact with directionalities (anisotropies) in material media through which the light propagates; we looked briefly at the ways in which these material interactions allow us to control light: to modulate it and, perhaps, to analyse it.

All of these ideas bear upon more advanced phenomena such as those which allow light to switch light and to process light, opening up a new range of possibilities in the world of very fast (femtosecond, ~ 10^{-15} s) phenomena.

Acknowledgements

Much of the material in this chapter was first presented in 'Essentials of Optoelectronics' (Rogers), published by Chapman and Hall, 1997, and is included here with permission.

REFERENCES

1. Michelson A A 1882 Interference phenomena in a new form of refractometer *Am. J. Sci.* 23 395–400.
2. Michelson A A and Morley E W 1887 On the relative motion of the earth and the luminiferous aether *Phil. Mag.* 24 449–63.
3. Born M and Wolf E 1975 The Fabry–Pérot Interferometer *Principles of Optics* 5th edn (Oxford: Pergamon) section 7.6.2, pp 329–33.
4. Twyman F and Green A 1916 British Patent No 103832.
5. Michelson A A 1920 An interferometer for measurement of stellar diameters *Astrophys. J.* 51 263.
6. Vali V and Shorthill R W 1976 Fiber ring interferometer *Appl. Opt.* 15 1099–100.
7. Lefevre H 1993 *The Fiber-Optic Gyroscope* (Boston: Artech House).
8. Nye J F 1976 *Physical Properties of Crystals* (Oxford: Clarendon).
9. Clark Jones R 1941 A new calculus for the treatment of optical systems *J. Opt. Soc. Am.* 38 671–85.
10. Jerrard H G 1954 Transmission of light through birefringent and optically-active media *J. Opt. Soc. Am.* 44 634–40.
11. Barlow A J, Ramskov-Hansen J J and Payne D N 1982 Anisotropy in spun single-mode fibres *Electron Lett.* 18 200–2.
12. Varnham P et al 1983 Single polarization operation of highly-Birefringent 'bow-tie' optical fibres *Electron. Lett.* 19 246–7.
13. Rogers A J 1988 Optical-fibre current measurement *Int. J. Optoelectron.* 3 391–407.

4

Introduction to Numerical Analysis for Laser Systems

George Lawrence

CONTENTS

4.1 Introduction 77
    4.1.1 Representation of the Optical Beams 78
    4.1.2 Split-step Method 79
    4.1.3 Solving the Diffraction Part of the Split-step Method 79
    4.1.4 Finite-difference Propagation 79
    4.1.5 Angular Spectrum Propagation 79
4.2 Propagation in Homogeneous Media 81
    4.2.1 Sampling 83
    4.2.2 Propagation Control 83
4.3 Gain and Non-linear Media 84
    4.3.1 Saturated Beer's Law Gain 84
    4.3.2 Rate Equation Model 84
        4.3.2.1 Frantz–Nodvik Solution 85
        4.3.2.2 Offline Effects 86
        4.3.2.3 Spontaneous Emission 86
4.4 Integration of Geometrical and Physical Optics 86
4.5 Dielectric Waveguides 87
4.6 Reflecting Wall Waveguides 90
4.7 Laser Modelling Software 91
    4.7.1 Traditional Methods of Modelling 91
    4.7.2 Selecting Commercial Numerical Modelling Software 95
    4.7.3 Validation of Software 95
References 95

4.1 Introduction

The detailed design of laser systems often requires numerical modelling to include all aspects of the optical system: diffraction propagation, non-linear gain, lenses and mirrors, apertures and other optical elements. As the complexity and variety of laser systems have expanded over the years, the need for powerful analytical methods has become increasingly important. The optical engineer or scientist can determine the end-to-end performance of a complex system based on the characteristics of the lenses and mirrors, propagation distances, apertures, aberrations, laser gain and other effects.

Optical rays were the first type of optical model developed and continue to be of great use in optical design where an optical system is sufficiently short that, from front surface to rear surface, there is little diffraction spreading. In such optics, rays may be used to calculate the optical aberrations with good precision. Optical rays serve well when used in their original role as predictors of aberration error in 'well-behaved' systems, providing many digits of precision in optical path difference calculations. However, rays neglect both near-field diffraction and the interaction between intensity and wavefront gradients. It may be said that geometric ray models are precise but not accurate and that physical optics theory is accurate but not precise (due to sampling limitations).

Among the effects that are difficult to treat with rays are diffraction ripples near edge boundaries, spontaneous emission and speckle, laser modes with discontinuous phase, strong non-linear gain, non-linear optics, waveguides and optical fibres and wavefront discontinuities. In particular, speckles cannot be represented well by rays. In many lasers, light originates from spontaneous emission and evolves through various stages of speckle size: initially fine-structured speckle, and after passing through a resonator multiple times, the high spatial frequencies are scraped out of the beam and the speckles become larger.

By far the most difficult types of laser systems to model are semiconductor diode lasers. Although the resonators for these lasers can be quite simple, consisting of a rectangular waveguide structure with plane mirrors formed by the surfaces at the ends of the waveguide, the interactions between the charge carriers that produce the laser radiation, the intense laser radiation within the waveguide and the semiconductor materials that form the waveguide are extremely complex. As a result, even fairly sophisticated software is not capable of accurately modelling many aspects of semiconductor diode laser behaviour. The development of software that can do such modelling is currently an active research topic at a number of universities and companies.

The earliest work in resonator analysis codes was done for optical communications in the 1960s by Fox and Li [1]. The military interest in high-energy lasers and laser fusion stimulated intense development of physical optics modelling codes in the mid-1970s. The work by Siegman in 1973 and Sziklas and Siegman in 1974 studied gas dynamic lasers including diffraction, the active gain medium, apertures and aberration. The first paper by Siegman and Sziklas used a Hermite Gaussian expansion for propagation [2]. The second paper by Sziklas and Siegman used a fast Fourier transform (FFT) method for propagation [3]. A third method based on finite difference propagation—a direct solution to the differential equation of diffraction—was used by Rench and Chester in 1974 [4]. Over time, the FFT method has become the mainstay of optical propagation codes for system analysis, as much for its modest and well-understood sensitivity to error as for its computational efficiency for many types of problems.

4.1.1 Representation of the Optical Beams

For numerical solution of diffraction calculations, it is convenient to use the complex amplitude, designated a, which is related to irradiance and to the electric field by $I = |a|^2 = (nc\varepsilon_0/2)|E|^2$, where c is the speed of light in vacuum and $n = \sqrt{\varepsilon/\varepsilon_0}$. In this form, the complex amplitude may take the form of square root watts per unit length.

For a full vector treatment, it is necessary to have the complex amplitude fields in all three directions: the two transverse directions as well as the propagation direction. The most common cases that require a full vector treatment are strongly converging or diverging beams, scattering from features comparable to or smaller than the wavelength or dielectric waveguides with high core-to-cladding differences—that is, situations in which the diffraction angles are greater than about 10 or 15° for which the cosine of the diffraction angle is no longer close to unit.

Smaller diffraction angles allow a scalar treatment. This may be described as Fresnel diffraction. A small-angle scalar treatment may be generalized to consider the two transverse fields which, for propagation in the -direction, would be a_x and a_y. Use of the two transverse fields allows representation of any polarization state. This could be considered a small-angle vector treatment but may be better described as Fresnel diffraction with polarization effects considered. By defining the relative amplitudes and phase differences between a_x and a_y, various states of polarization can be defined: linear, circular and general elliptical polarizations. Where different polarization states are not required, calculations may be performed using only one array. Most physical optics code defines two-dimensional computer arrays that represent the transverse distribution of the optical beam at a specific axial point.

The dependence of the optical beam on time may be neglected in many cases: either because the optical beam is so slowly varying that only the steady-state solution is needed or because the pulse is so short that all physical processes in the system see only the integrated effects of the optical pulse. In either case, the time dependence may be dropped for the purpose of diffraction calculations. Time dependence may be added, if necessary, by breaking a temporal waveform or pulse into discrete time samples and propagating each time sample through the system using diffraction propagation. For example, a Q-switched laser system might be sampled over nanosecond intervals. By including time-varying phase between the temporal samples, finite temporal coherence can be modelled. For broad-band signals, it may be necessary to sample the wavelength spectrum, propagate each wavelength sample through the system independently and add them incoherently. Both temporal and wavelength sampling depend upon proper numerical calculation of strictly coherent propagation.

For small-angle scalar propagation, a coherent field may be represented as

$$a(x,y;z) = a_x(x,y;z)\hat{i} + a_y(x,y;z)\hat{j}. \qquad (4.1)$$

The evolution of the optical fields is a function of diffraction and the gain and loss mechanisms in the beam train. For a detailed derivation from Maxwell's equations, the reader is referred to one of the many excellent texts on laser physics such as Sargent et al. [5]. For many lasers, the differential equation for the optical field may be written as follows:

$$\frac{\partial a}{\partial z} = -j\frac{1}{2k}\nabla_\perp^2 a - j\frac{\mu\omega^2}{2k}p \qquad (4.2)$$

where p represents the effect of the medium and ignoring the time variation of a and dropping $e^{-j\omega t}$. For non-linear optical effects, the medium polarization may take a more complex form, according to Bloembergen [6]:

$$p \propto \varepsilon_0 \chi a + \chi^{(2)} aa + \chi^{(3)} aaa + \cdots \qquad (4.3)$$

where the superscripts indicate the linear and various higher-order non-linear susceptibilities. For example, the polarization term for four-wave mixing takes the form

$$p \propto \chi_{ijk} a_i a_j a_k^* \exp[j(k_i + k_j - k_k \cdot z)] \qquad (4.4)$$

For linear media, equation (4.2) takes the form

$$\frac{\partial a}{\partial z} = -j\frac{1}{2k}\nabla_\perp^2 a - j\frac{k\chi}{2n^2}a \qquad (4.5)$$

where n is the index of refraction in the medium. Equation (4.5) describes the propagation of a laser beam in gain media and may also be applied to waveguide and many other forms of material. The first term on the right is the diffraction term; the second is the effect of the medium. In general, this equation cannot be solved in closed form. Numerical methods are well understood for solving each of the terms on the right-hand side of equation (4.5) if taken separately [7].

4.1.2 Split-step Method

For equation (4.5), the field after a small propagation z is given by

$$a(z + \Delta z) = a(z) + \Delta a. \quad (4.6)$$

For small steps, the term may be separated into a diffraction and medium term:

$$\Delta a = \Delta a_{\text{diff}} + \Delta a_{\text{medium}} \quad (4.7)$$

$$\Delta a_{\text{diff}} = -j\frac{1}{2k}\nabla_\perp^2 a \Delta z \quad (4.8)$$

and

$$\Delta a_{\text{medium}} = -j\frac{k\chi}{2n^2} a \Delta z. \quad (4.9)$$

The solution of equation (4.8) is performed in separate steps

Even when the medium has non-linear gain or absorption, the effect of the non-linearities on diffraction effects is often relatively modest. The errors may be reduced to an acceptable level by taking short steps through the medium. This method is often referred to as the split-step method.

4.1.3 Solving the Diffraction Part of the Split-step Method

The two primary means of solving the diffraction part of the split-step method are the finite difference method and angular spectrum decomposition.

4.1.4 Finite-difference Propagation

The finite difference propagator was developed by Rench [4]. The parabolic wave equation may be solved directly:

$$\frac{\partial a}{\partial z} = \frac{1}{2jk}\left(\frac{\partial^2}{\partial^2 x} + \frac{\partial^2}{\partial^2 y}\right)a. \quad (4.10)$$

The second derivative is taken by considering only immediately neighbouring points, by

$$\frac{\partial^2 a}{\partial x^2} \approx \frac{a(I+1,J) + a(I-1,J) - 2a(I,J)}{\Delta x^2} \quad (4.11)$$

$$\frac{\partial^2 a}{\partial y^2} \approx \frac{a(I,J+1) + a(I,J-1) - 2a(I,J)}{\Delta y^2} \quad (4.12)$$

where $a(I, J)$ is an element of the complex amplitude array, and I and J are the indices. To calculate a single point in two dimensions, four sums are needed: two subtractions and two divisions. The method is extremely fast—its principal virtue.

The longest single allowable propagation step was calculated by Rench to be

$$\Delta z < \frac{k}{2}\Delta x^2. \quad (4.13)$$

To propagate any significant distance, the algorithm must be repeated many times. For a strong non-linear gain, it may be necessary to calculate diffraction and gain intermittently at many points along the axis, taking steps no longer than the characteristic length. In that case, the requirement for short steps with the finite difference propagator is not a problem.

The finite-difference method, as it is based on the calculation of second derivatives, may prove unstable at apertures and other sources of discontinuities in the optical field. In the case of waveguides, where the optical fields may have no true discontinuities, the method proves to be very valuable.

High Fresnel diffraction and high spatial frequencies require high sampling rates and, necessarily, large arrays. FFT computations are time-consuming and the question naturally arises as to whether alternative methods have advantages. The method most frequently considered is the finite-difference propagator (FDP). The advantages of the FDP are that it is very fast for short steps [4]. The disadvantages are as follows:

1. one must take short steps (repeated application of the algorithm is required for long propagation),
2. the algorithm is numerically unstable at discontinuities and
3. certain diffraction effects are washed out.

This length is called the characteristic diffraction length. Since the iterative solution of the propagation problem in a non-linear active medium requires re-calculation of the diffraction effects for every characteristic length, it may be advantageous to use the FDP.

The amplitude–wavefront representation provides an alternative method of representing the optical beam: complex amplitude $a \exp(jw)$, where a is the amplitude and w is the wavefront error. In the amplitude–wavefront representations, equation (4.11) takes the form

$$\frac{\partial a}{\partial z} = \frac{a}{2}\left(\frac{\partial^2 w}{\partial x^2} + \frac{\partial^2 w}{\partial y^2}\right) + \frac{\partial a}{\partial x}\frac{\partial w}{\partial y} + \frac{\partial a}{\partial y}\frac{\partial w}{\partial x} \quad (4.14)$$

$$\frac{\partial w}{\partial z} = \frac{1}{2}\left[\left(\frac{\partial w}{\partial x}\right)^2 + \left(\frac{\partial w}{\partial y}\right)^2\right] - \frac{1}{2ak^2}\left(\frac{\partial^2 a}{\partial x^2} + \frac{\partial^2 a}{\partial y^2}\right). \quad (4.15)$$

Equations (4.14) and (4.15) allow direct propagation of the amplitude–wavefront representation of the beam. It is capable of representing large wavefront errors without the problem of higher order intrinsic to the complex amplitude form. It is, however, subject to the same difficulties of finite-difference propagation as equations (4.11) and (4.12).

4.1.5 Angular Spectrum Propagation

One very effective method of calculating diffraction propagation of an arbitrary complex amplitude distribution is to decompose the distribution into a summation of plane waves, propagate the plane waves individually using the eigenvalues and re-sum the plane waves. This procedure is called the angular spectrum decomposition method [8].

The geometrical representation of the wavefront and propagation may be compared with the complex amplitude and angular spectrum propagation. Geometrical rays are normals to the wavefront. Enough rays are needed to sample the wavefront thoroughly. For example, an optical system may be traced using hundreds of rays. The ray direction is defined by wavenumber unit vector, $\hat{k}$, with a direction perpendicular to the wavefront. For free-space propagation along the ray a distance q, the ray position vector is transformed:

$$r_2 = r_1 + q\hat{k}. \quad (4.16)$$

Propagation of a plane wave is very similar to geometric propagation. A plane wave of amplitude $A(k)$ is propagated by the equation,

$$a(k;z) = a(k;0)e^{jk \cdot z}. \quad (4.17)$$

The propagation distance depends on the direction of the plane wave. Evaluating the phase along the z-axis,

$$e^{jk \cdot z} = e^{jk_z z} \quad (4.18)$$

where k_x, k_y and k_z are the components of the wavenumber vector:

$$k_x^2 + k_y^2 + k_z^2 = |k|^2 = k^2. \quad (4.19)$$

We now make the approximation [8]

$$\exp(jk_z z) = \exp\left[jz\sqrt{k^2(1-\alpha^2-\beta^2)}\right]$$
$$\approx \exp(jkz)\exp\left[\frac{1}{2}jkz(\alpha^2+\beta^2)\right]. \quad (4.20)$$

where k_x/k and k_y/k are the direction cosines in the transverse direction. Equation (4.20) is the transfer function for a plane wave in homogeneous, isotropic media. The term $\exp(jkz)$ is generally dropped although it may be important in phased array and coupled resonator studies.

The direction cosines may be associated with spatial frequency variables ξ and η,

$$\xi\lambda = \alpha \text{ and } \eta\lambda = \beta \quad (4.21)$$

and the transfer function for a plane wave described in terms of spatial frequency variables is

$$e^{jk_z z} \approx e^{jkz}e^{-j\pi\lambda z\rho^2} \quad (4.22)$$

where $\rho^2 = \xi^2 + \eta^2$.

Any well-behaved function may be written as a summation of spatial frequency components:

$$a(x,y;0) = \int_{-\infty}^{\infty}\int_{-\infty}^{\infty} A(\xi,\eta;0)e^{j2\pi(x\xi+y\eta)}\,d\xi\,d\eta \quad (4.23)$$

where $A(\xi, \eta;0)$ is the spatial frequency component at the zeroth axial position and spatial frequency coordinates (ξ, η). From equation (4.22),

$$A(\xi,\eta;z) = A(\xi,\eta;0)e^{-j\pi\lambda z\rho^2}. \quad (4.24)$$

Using equation (4.24), propagation in homogeneous media can be written in the operator notation:

$$a(x,y;z) = \mathrm{FF}^{-1}[T(z)\mathrm{FF}[a(x,y;0)]] \quad (4.25)$$

where

$$T(z) = e^{-j\pi\lambda z\rho^2}. \quad (4.26)$$

is the transfer function of diffraction propagation. The forward and inverse Fourier transforms, FF and FF^{-1}, are defined by [9]

$$\mathrm{FF}[\ ] = \int_{-\infty}^{\infty}\int_{-\infty}^{\infty} [\ ]\exp[-j2\pi(x\xi+y\eta)]\,dx\,dy \quad (4.27)$$

$$\mathrm{FF}[\ ]^{-1} = \int_{-\infty}^{\infty}\int_{-\infty}^{\infty} [\ ]\exp[j2\pi(x\xi+y\eta)]\,d\xi\,d\eta \quad (4.28)$$

Propagation may be written as a convolution by taking the Fourier transform of equation (4.24) [10].

$$a(x_2,y_2;z_2)$$
$$= \int_{-\infty}^{\infty}\int_{-\infty}^{\infty} a(x_1,y_1;z_1)t(x_1-x_2,y_1-y_2,z_2-z_1)\,dx_1\,dy_1 \quad (4.29)$$

$$t(x,y;\Delta z) = \frac{1}{j\lambda z}\exp[j(kr^2/2\Delta z)]. \quad (4.30)$$

The quantity $t(x, y; \Delta z)$ is the point spread function or impulse response function at position Δz. Phase factors which are constant over the field have been dropped. The quadratic phase factor of equation (4.19) can be factored to give equation (4.31).

$$a(x_2,y_2;z_2) = \frac{1}{j\lambda z}q(r_2,\Delta z)$$
$$\times \int_{-\infty}^{\infty}\int_{-\infty}^{\infty} a(x_1,y_1;z_1)q(r_1;\Delta z)\exp\left[-j\frac{2\pi}{\lambda z}(x_1 x_2+y_1 y_2)\right]dx_1\,dy_1$$
$$(4.31)$$

where $q(r;z) = \exp[jk(r^2/2z)]$ is a quadratic phase factor and simplifies many of the diffraction equations. In operator notation,

$$a(x_2,y_2;z_2) = \frac{1}{j\lambda\Delta z}q(r_2,\Delta z)FF^s[a(x_1,y_1;z_1)q(r_1;\Delta z)] \quad (4.32)$$

where $s = \Delta z/|\Delta z|$.

Equations (4.25) and (4.32) are the near- and far-field propagation expressions. In the continuous mathematical formulation, there is no difference between the two expressions. In discrete formulation for numerical calculations, errors are reduced if the correct selection of a near- or far-field propagator is made [11]. This arises from the quadratic phase factors that must be evaluated. In the near field, the phase factor is found from equation (4.26) and in the far field, the phase factor is found from equation (4.30):

$$T(\Delta z) = e^{-j\pi\lambda\Delta z p^2} \text{ and } t(x, y; \Delta z) = \frac{1}{j\pi\lambda\Delta z} e^{j(kr^2/2\Delta z)}. \quad (4.33)$$

The phase factor, $T(\Delta z)$, for the near field varies rapidly as $\Delta z \to \infty$ to but slowly as $\Delta z \to 0$. However, $t(\Delta z)$ varies slowly as $\Delta z \to \infty$ to and rapidly as $\Delta z \to 0$. Rapidly varying phase factors create numerical errors called aliasing, which are described later. The near-field propagator aliases at large propagation distances but is well behaved at short propagation distances. These relationships are summarized in Table 4.1.

By using the far-field expression at long propagation distances and the near-field expression at short distances, aliasing can be reduced to a tolerable level in most cases.

In the case of strictly rotationally symmetric functions, Hankel transforms may, in principle, be used to solve the diffraction integrals. The Hankel transform pair may be written as follows:

$$A(\rho) = 2\pi \int_0^\infty a(r')J_0(2\pi\rho r')r'\,dr' \quad (4.34)$$

$$a(\rho) = 2\pi \int_0^\infty A(r')J_0(2\pi r\rho')\rho'\,d\rho'. \quad (4.35)$$

Direct solution in terms of the Bessel function representation is relatively slow. The fast-Hankel transform was devised by Siegman to provide a faster method [15–18]. This method has proved successful in many cases but suffers in numerical implementation from a singularity at zero radius in both spatial and frequency domains and non-linear sample spacing. The zero radius singularity may be minimized by good programming but may still have a tendency to exhibit 'edge droop' or 'sloping shoulders'. The non-linear sampling results in higher sampling densities at the edge of the array which is an advantage in some cases. Given the high speed and inexpensive memory of modern computers, one may find that the higher accuracy of the methods based on square arrays makes these methods more attractive than rotational propagation methods in spite of slower speed.

An efficient circular propagator with uniform sample spacing and no zero radius singularity results in excellent energy conservation. The method is based on a degenerate form of two-dimensional Fourier transform. In the general case, a two-dimensional Fourier transform is used to calculate a two-dimensional frequency spectrum of the form $A(\xi, \eta)$. If, however, only the frequency spectrum is needed along a single row where $\eta=0$, the two-dimensional Fourier transform may be simplified into a sum along the y-direction and a one-dimensional transform

TABLE 4.1

A Time Slice of the Pulse Length Interacts with a Gain Region of Length L

	Far Field	Near Field
$\Delta z \to 0$	Rapid	Slow
$\Delta z \to \infty$	Slow	Rapid

$$A(\xi, 0) = \iint a(x, y)e^{-j2\pi(x\xi+y\eta)}\,dx\,dy\Big|_{\eta=0} = \int a(x)e^{-j2\pi x\xi}\,dx. \quad (4.36)$$

where $a(x) = \int a(x)\,dy \cdot a(x)$ is the sum along the y-direction. The y-sum $a(x)$ can be quickly computed from the centre row of the square array $a(x, 0)$ by interpolation to find the values at the various (x, y) points. The projection method based on y-sums may be used to calculate a Hankel transform pair between $a(x, 0)$ and $A(\xi, 0)$. For an array size of 1024×1024, an improvement in speed of between 20 and 40 times may be realized doing the one-dimensional transformation of $a(x)$ rather than the two-dimensional transforms of $a(x, y)$ but with diminished accuracy. Modern computers are so fast that for most problems, the higher accuracy of the two-dimensional method makes it the most convenient choice.

4.2 Propagation in Homogeneous Media

The propagation through any well-behaved system can be separated into geometrical aberration calculations and propagation in homogeneous media. The mathematically equivalent expressions of equations (4.25) and (4.32) provide a complete description of diffraction propagation in homogeneous media in the Fresnel approximation [11].

In numerical calculations, only discrete points may be represented. Also, only a limited region of space may be considered because of computer memory limitations. Consider a two-dimensional function represented in a rectangular computer array of $M \times N$ points. The sampling intervals for the x- and y-directions are Δx and Δy. In the general case, $M \neq N$ and $x \neq y$. The width of the computer array representation is $M\Delta x$ by $N\Delta y$. Information exists in the computer only at the discrete points defined by the rectangular grid. Any functions to be represented must be truncated by the finite width of the computer array.

The computer points in the spatial domain may be counted with the indices k and l. The indices have the ranges

$$\frac{-M}{2} \leq k \leq \frac{M}{2} - 1 \quad \frac{-N}{2} \leq l \leq \frac{N}{2} - 1. \quad (4.37)$$

Note that the centre of the distribution has been chosen to be at $(M/2+1, N/2+1)$. Many FFT routines based on arrays with dimensions which are powers of two are implemented with natural centres either at $(1, 1)$ or $(M/2+1, N/2+1)$ by shifting the array one-half cycle in each direction. The natural centre of the array is defined to be the point at which a delta function will give a perfectly constant real Fourier transform [12]. The physical limits are obtained by multiplying equation (4.37) by Δx and Δy:

$$-\frac{M\Delta x}{2} \leq k\Delta x \leq \frac{M\Delta x}{2} - \Delta x - \frac{N\Delta y}{2} \leq l\Delta y \leq \frac{N\Delta y}{2} - \Delta y. \quad (4.38)$$

Sampling can be represented as multiplication by a special function called the comb function. The comb function is an infinite array of delta functions spaced apart by Δx and Δy [9]:

$$\text{comb}\left(\frac{x}{\Delta x},\frac{y}{\Delta y}\right)=|\Delta x||\Delta y|\sum_k\sum_l\delta(x-k\Delta x,y-l\Delta y). \quad (4.39)$$

The comb function is useful in transforming a continuous function into a discrete representation:

$$a(x,y)\to a(x,y)\text{comb}\left(\frac{x}{\Delta x},\frac{y}{\Delta y}\right). \quad (4.40)$$

where $a(x, y)$ is the continuous function to be sampled. The arrow indicates transformation from continuous to discrete form.

The discrete nature of the spatial domain causes the frequency domain to be periodic (and necessarily of infinite extent). The continuous function $A(\xi, \eta)$, the Fourier transform of $a(x, y)$, is replicated with a period of $(1/\Delta x, 1/\Delta y)$.

The Fourier transform domain functions must also be discrete. The most common (and most efficient) form of the FFT has the same dimensions for the spatial and frequency domains. The frequency domain indices m and n have the ranges

$$\frac{-M}{2}\leq m\leq\frac{M}{2}-1\quad\frac{-N}{2}\leq n\leq\frac{N}{2}-1. \quad (4.41)$$

Multiplication of equation (4.41) by $\Delta\xi = 1/(M\Delta x)$ and $\Delta\eta = 1/(N\Delta y)$ gives the frequency range

$$-\frac{1}{2\Delta x}\leq\xi\frac{1}{2\Delta x}\leq\left(1-\frac{2}{M}\right)-\frac{1}{2\Delta y}\leq\eta\leq\frac{1}{2\Delta y}\left(1-\frac{2}{N}\right). \quad (4.42)$$

The frequency domain bounds are the Nyquist sampling frequencies. The FFT algorithm is occasionally blamed for this restriction, but it is more accurately attributed to the discrete sampling process and will exist for any form of propagation of sampled data.

The continuous frequency domain function is also transformed to discrete representation by means of the comb function,

$$A(\xi,\eta)\to A(\xi,\eta)\text{comb}\left(\frac{\xi}{\Delta\xi},\frac{\eta}{\Delta\eta}\right)=a(m\Delta\xi,n\Delta\eta). \quad (4.43)$$

The discrete nature of the frequency domain forces the spatial domain also to be periodic with period $(1/(\Delta\xi), 1/(\Delta\eta))$.

A Fourier transform pair can be defined for modified spatial and frequency functions $a(k\Delta x, l\Delta y)$ and $A(m\Delta\xi, n\Delta\eta)$ such that an exact Fourier relationship is obtained:

$$a(k\Delta x,l\Delta y)\Leftrightarrow A(m\Delta\xi,n\Delta\eta) \quad (4.44\text{a})$$

$$a(k\Delta x,l\Delta y)\Leftrightarrow a(x,y)\text{comb}\left(\frac{x}{\Delta x},\frac{y}{\Delta y}\right)$$
$$**|\Delta\xi\Delta\eta|\text{comb}(x\Delta\xi,y\Delta\eta) \quad (4.44\text{b})$$

$$A(m\Delta\xi,n\Delta\eta)\Leftrightarrow A(\xi,\eta)\text{comb}\left(\frac{\xi}{\Delta\xi},\frac{\eta}{\Delta\eta}\right)$$
$$**|\Delta x\Delta y|\text{comb}(\xi\Delta x,\eta\Delta y) \quad (4.44\text{c})$$

where ** indicates two-dimensional convolution and $\Leftrightarrow$ indicates two-dimensional Fourier transformation pairs.

The function, $\text{comb}(x\Delta\xi, y\Delta\eta)$, causes the spatial domain to be periodic with minimum periods of $M\Delta x$ and $N\Delta y$ in the x- and y-directions. There is, in effect, an infinite rectangular array of functions separated by $M\Delta x$ and $N\Delta y$. Therefore, the frequency domain sampling periods are

$$\Delta\xi=\frac{1}{M\Delta x}\Delta\eta=\frac{1}{N\Delta y}. \quad (4.45)$$

The Fourier transform operator can be written in discrete form:

$$\text{FF}[\ ]=\sum_k\sum_l[\ ]\exp\left(-sj2\pi\left(\frac{km}{M}+\frac{ln}{N}\right)\right) \quad (4.46)$$

where s is $+1$ for forward transformation and -1 for inverse. Various forms of algorithms are used in FFTs and some have a normalization step for the forward or inverse transformation.

Evaluation of the far-field expression, equation (4.32), in discrete terms causes a redefinition of the sampling period,

$$A(m\Delta\xi,n\Delta\eta)=\text{FF}[a(k\Delta x,l\Delta y)] \quad (4.47\text{a})$$

$$\Delta\xi=\frac{1}{M\Delta x_1}\Delta n=\frac{1}{N\Delta y_1} \quad (4.47\text{b})$$

The coordinates x_2, y_2 are related to $\Delta\xi, \Delta\eta$ by,

$$\Delta\xi=\frac{\Delta x_2}{\lambda|\Delta z|}\Delta\eta=\frac{\Delta y_2}{\lambda|\Delta z|} \quad (4.48)$$

based on equation (4.45). The discrete far-field calculation is, therefore,

$$a(k\Delta z_2,l\Delta z_2)=\frac{1}{j\lambda\Delta z}q(r_2,\Delta z)\text{FF}^s[a(k\Delta x_1,l\Delta y_1)q(r_1,\Delta z)] \quad (4.49\text{a})$$

$$\Delta x_2=\frac{\lambda|\Delta z|}{M\Delta x_1}\quad\Delta y_2=\frac{\lambda|\Delta z|}{N\Delta y_1} \quad (4.49\text{b})$$

$$r^2=(k\Delta x)^2+(l\Delta)^2 \quad (4.49\text{c})$$

$$s=\frac{z}{|\Delta z|}. \quad (4.49\text{d})$$

Note the scale change of the new sampling periods, Δx_2 and Δy_2. The discrete near-field propagation equation is

$$a(k\Delta x, l\Delta y; z_2) = \text{FF}^{-1}[T(\Delta z)\text{FF}[a(k\Delta x, l\Delta y; z_1)]]. \quad (4.50)$$

4.2.1 Sampling

There are two important and related issues in determining the numerical sampling. The highest spatial frequency which can be represented in the computer is determined by the sample spacing Δx and Δy. The region of space which can be represented is determined by the width of the computer array $M\Delta x$ and $N\Delta y$. First, the diffraction phenomenology, which can be observed with a given sample spacing, will be considered.

The Nyquist sampling frequency—the highest frequency which can be represented—is

$$f_{\text{Nyquist}} = \frac{1}{2\Delta x}. \quad (4.51)$$

Failure to resolve the highest spatial frequencies may not result in an unacceptable representation of the function. In particular, using the near-field propagator for very short distances will show the distribution to be largely unchanged (the correct answer) even though the high spatial frequencies may not be correctly sampled.

Aliasing has a much more serious effect on the accuracy of the information. If the distribution grows outside the bounds of the array, severe aliasing will result which may render the calculation unusable. These errors arise from the finite size of the computer array. With propagation, a collimated beam expands and the complex amplitude grows beyond the bounds of the array and is folded back on itself. This folded amplitude is the source of aliasing errors. The folded amplitude causes high spatial frequency errors in the intensity pattern.

Often the most severe errors tend to be where the distribution has the highest amplitude, not near the edge of the distribution. This is because the amplitude of the signal and error add rather than the intensities. Consider a nominally top-hat function of unit amplitude and an aliasing contribution of ε. Assume that ε is slowly varying across the array. Near the edge of the array, the intensity is of the order of magnitude of ε^2. In the centre, the intensity is roughly $1 + 2\varepsilon + \varepsilon^2$. The error in the centre will be of order 2ε—much larger than the error at the edge of order ε^2. It is important not to be deceived into believing that the aliasing errors are negligible by seeing an intensity distribution roll-off at the edge of the array. Approximate guidelines for the magnitude of aliasing errors may be determined for top-hat functions, i.e. uniformly filled circular apertures. The results will be generally characteristic of distributions with strong discontinuities. For top-hat functions, there is an exact solution based on Lommel functions [13]. The Lommel functions may be approximated by an asymptotic solution [14]. This approximation enables the calculation of aliasing errors to be made for the bright region inside the geometric aperture area and the dark region in the shadow. Because the amplitude values add—not the irradiance values—that the aliasing errors are affected by the signal level. Let ε_b and ε_d be the errors in the bright and dark regions, then approximate expressions for the errors are

$$\varepsilon_b = 8\sqrt{\frac{1}{\pi^2 F_n} \frac{a^3/r^3}{(1-a^2/r^2)}} \quad \varepsilon_d = \frac{3}{\pi^2 F_n} \frac{a^3/r^3}{(1-a^2/r^2)} \quad (4.52)$$

These expressions give the order of magnitude of the effects. Aliasing errors are not always immediately distinguishable from diffraction ripples. Generally, high spatial frequency ripples will be manifested in the immediate vicinity of an aperture, but high spatial frequency aliasing errors will be present all over the distribution with the largest errors where the distribution has high intensity.

Some experimentation with different size arrays and sampling may be required to gain an understanding of the appearance of the two phenomena. Consider a case of a beam of 5 mm diameter, wavelength 1.6 μm and propagation of 100 cm. An array of 128 × 128 points is selected, and in the first case, the array is almost completely filled by the aperture (78%). The aperture is less than half the size of the array (39%). Both beams are propagated a distance of 100 cm, which is a Fresnel number of 3.9.

Close examination of the degree of aliasing either by performing numerical experiments or using equation (4.52) may, at first, be discouraging since most near-field diffraction calculations have significant amounts of aliasing. In practice, many calculations are not adversely affected by significant levels of aliasing. For specific problems, one can try various guard-band values to determine whether results are affected. Ideally, one increases the array size and the guard-band width—keeping the same number of sample points across the distribution—until no appreciable change in the results is observed. In practice, one may choose the array size based on the computational time that is consistent with one's own level of patience.

4.2.2 Propagation Control

The beam spreads due to diffraction and may, therefore, overfill the computer array. Fortunately, the near-field and far-field propagators may be used to control the size of the array so that the beam aliasing does not change much from the initial state. The sampling period of the near field is constant. The sampling period of the far field is

$$\Delta x_2 = \frac{\lambda |\Delta z|}{M\Delta x}. \quad (4.53)$$

By use of a combination of near-field and far-field propagators, the sampling period may be set to any required value. The far-field propagation has an expanding coordinate system of the form of equation (4.53) and the near-field propagation has a constant coordinates system of the form, $\Delta x_2 = \Delta x_1$.

A function $f(x, y)$ may be defined as the complex amplitude with respect to the curved reference surface of radius, z, such that

$$f(x, y) = a(x, y)e^{(jkr^2/2z)} \qquad (4.54)$$

where z is still the distance from the waist. Either $a(x, y)$, referenced to a plane surface, or $f(x, y)$, referenced to the curved surface, may be propagated using the equations to be presented. Either $a(x, y)$ or $f(x, y)$ may be selected depending on which has the smaller residual phase. The surrogate Gaussian beam again proves to be useful. At any point in space, the Gaussian beam is

$$a(r) = e^{(-r^2/\omega^2)} e^{(-jkr^2/2R)}. \qquad (4.55)$$

The function $f(r)$, using the curved reference, is

$$f(r) = e^{(-r^2/\omega^2)} e^{(-jkr^2/2R)} e^{(jkr^2/2z)}. \qquad (4.56)$$

The critical question is whether the residual phase of equations (4.55) or (4.56) is less. Consider the phase error of the actual wavefront, with respect to either a planar or spherical reference surface, evaluated at the 1/e point of the Gaussian amplitude:

$$\Delta W_{\text{plane}} = -\frac{h^2}{2}\frac{1}{R}\bigg|_{h=\omega(z)} \quad \Delta W_{\text{sphere}} = -\frac{h^2}{2}\left(\frac{1}{R} - \frac{1}{z}\right)\bigg|_{h=\omega(z)}.$$

$$(4.57)$$

Consider the phase error for a representative Gaussian beam for the two choices of reference surface:

$$\Delta W_{\text{plane}} = -\frac{\omega_0^2}{2}\frac{z}{z_R^2} \quad \Delta W_{\text{sphere}} = -\frac{\omega_0^2}{2}\frac{1}{z} \qquad (4.58)$$

where the Gaussian beam propagation equations:

$$\omega(z) = \omega_0\sqrt{1 + \frac{z^2}{z_R^2}} \text{ and } R = z + \frac{z^2}{zR} \qquad (4.59)$$

have been used.

The phase error ΔW is minimized by choosing a plane reference inside the Rayleigh distance and spherical reference outside the Rayleigh distance. A system for propagation between any combinations, near- or far-field positions to any other near- or far-field positions, was developed by Lawrence [11].

4.3 Gain and Non-linear Media

Propagation through active media involves both diffraction and gain or absorption. The numerical approach to solution is the split-step method, described in previous sections. In this section, the gain part of the inhomogeneous wave equation is described.

In general, gain is described as a function of the density of the active medium and the intensity of the optical field. Medium density influences the small-signal gain and, in general, has some spatial variation. Because of saturation of the medium, gain is a non-linear function of the intensity of the optical field.

4.3.1 Saturated Beer's Law Gain

A simple model of gain using Beer's law (with a saturation intensity) may be used. The saturated form of Beer's law may be represented by

$$I(z + \Delta z) = I(z) \exp\left(\frac{g_0 \Delta z}{1 + I(z)/I_{\text{sat}}}\right)^q \qquad (4.60)$$

where g_0 is the small signal gain, I_{sat} is the saturation intensity and $q = 1/2$ for inhomogeneously and $q = 1$ for homogeneously broadened gain.

The gain grows exponentially at low values,

$$\frac{dI}{dz} \approx g_0 I. \qquad (4.61)$$

The characteristic gain length is $1/g_0$. When I is comparable to I_{sat}, the homogeneously broadened gain takes the form

$$\frac{dI}{dz} \approx g_{\text{sat}} I_{\text{sat}} \qquad (4.62)$$

which is a linear increase in intensity.

4.3.2 Rate Equation Model

A rate equation model for the two-level atom provides a more detailed model capable of treating laser start-up and transient effects such as Q-switching. The state of an active medium may be characterized by the density of the medium and the population inversion. Computer arrays may be used to store the population density of the upper and lower level for a collection of x-, y- and z-points representing samples of the gain region. A series of transverse arrays at different axial positions may represent the gain volume. The constituent transverse arrays may be referred to as gain sheets.

A four-level treatment of gain is one of the most commonly used models. The rate equations are [20,21]

$$\Delta N_2 = \left[R_2 - \frac{N_2}{t_2} - (N_2 - N_1)W_i(\nu)\right]\Delta t \qquad (4.63)$$

$$\Delta N_1 = \left[R_1 - \frac{N_2}{t_{10}} + \frac{N_2}{t_{\text{spont}}} - (N_2 - N_1)W_i(\nu)\right]\Delta t \qquad (4.64)$$

where ΔN_1 is the change in population of lower level (atoms cm^{-3}), ΔN_2 is the change in population of upper level (atoms cm^{-3}), R_2 is the pump rate for upper level (excitations s^{-1}cm^{-3}), R_1 is the pump rate for lower level (excitations s^{-1}cm^{-3}), t_{spont} is the spontaneous decay lifetime (s), t_{20} is the decay time from upper level to ground (s), t_2 is the total decay time from upper level to ground (s)($1/t_2 = 1/t_{20} + 1/t_{\text{spont}}$), t_{10} is the decay time from lower level to ground (s), $W_i(\nu)$ is the transition probability density (probability s^{-1}cm^{-3}) and Δt is the elapsed time.

The transition probability density is

$$W_i(\nu) = \frac{\lambda^2 f(\nu)}{8\pi n^2 h \nu t_{\text{spont}}} I \qquad (4.65)$$

where λ is the wavelength, $f(v)$ is the normalized lineshape, n is the index of refraction, h is Planck's constant, v is the frequency of the radiation and I is the irradiance of the radiation.

The transition probability of equation (4.65) may be written in terms of the Einstein B-coefficient: $W_i(v) = B(v)\dfrac{I}{hv}$, where

$$B(v) = \frac{\lambda^2 f(v)}{8\pi n^2 h v t_{spont}} I \tag{4.66}$$

The small-signal amplification takes the form

$$I(z) = I(0)e^{B\Delta Nz}. \tag{4.67}$$

Under steady-state conditions, the irradiance of the optical field is constant, and equations (4.63) and (4.64) lead to the steady-state solution for the population inversion:

$$\Delta N^0 R_2 t_2 - \left(R_1 + \frac{t}{t_{spont}}\right) R_2 t_{10}. \tag{4.68}$$

The small-signal gain coefficient is

$$g_0(v) = B(v)\Delta N^0. \tag{4.69}$$

The gain coefficient for homogeneous broadening and for arbitrary irradiance magnitude is

$$g(v) = \frac{g_0(v)}{1 + \dfrac{I}{I_s}} \tag{4.70}$$

where,

$$I_s = \frac{8\pi n^2 hv}{\left(\dfrac{t_2}{t_{spont}}\right)\lambda^2 g(v)} = \frac{hv}{B(v)t_2}. \tag{4.71}$$

In the case of strong saturation, equation (4.70) is well approximated by

$$\frac{dI}{dz} \approx g_0(v)I_s = (B\Delta N^0)\left(\frac{hv}{B(v)t_2}\right) = \frac{\Delta N^0 hv}{t_2}. \tag{4.72}$$

where the pumping rate into the upper level R_2 dominates the process; equations (4.68) and (4.72) give the saturated gain coefficient as

$$\frac{dI}{dz} = R_2 hv, g(v) = R_2 hv \tag{4.73}$$

showing, in the case of saturated steady-state gain, a linear growth of irradiance with distance based on the pumping flux density.

4.3.2.1 Frantz–Nodvik Solution

The laser may be treated as consisting of a discrete amplifier and efficiency loss due to out-coupling and other factors. The equation of optical amplification in a laser rod may be represented as [19]

$$\frac{\partial I(z)}{\partial z} = B\Delta N(z)I(z). \tag{4.74}$$

The population inversion at each point is driven by the transition probability:

$$\frac{\partial \Delta N}{\partial z} = -2\Delta N W_i(v) = -2\Delta N \frac{BI}{hv} \tag{4.75}$$

In a coordinate system moving with the optical field, the change of variables $t = zn/c$ may be simplified to

$$\frac{\partial \Delta N(z)}{\partial z} = \frac{-2n}{hvc} B\Delta N(z)I(z) \tag{4.76}$$

where $I(z)$ is the irradiance, $N(z)$ is the population inversion and n is the index of the medium. The constant B is the cross section and has the value

$$B = \frac{\lambda^2}{8\pi t_{spont}} f(v) \tag{4.77}$$

where λ is the wavelength in the medium (not the vacuum wavelength) and $f(v)$ is area-normalized spectral lineshape function,

$$f(v) = \frac{\Delta v}{2\pi\left[(V - V_0)^2 + \left(\dfrac{\Delta v}{2}\right)^2\right]}. \tag{4.78}$$

A method that is both fast and robust is possible by re-evaluating the problem. Equations (4.74) and (4.76) are appropriate for a small temporal sample of a beam travelling through an optical amplifier. In a resonator, the gain medium interacts with the entire optical field in the device. A single computer array may be used (or at most two, for the two polarizations) to represent the entire optical field in the resonator. One cannot, therefore, distinguish temporal events which occur on a scale less than the round-trip time of the resonator. One could, in principle, use multiple temporal samples to resolve time events shorter than the round-trip time, but this is not necessary for the Q-switch study.

If one considers the optical field to be of intensity I and of duration, Δt, the round-trip time, then the optical field contains a well-defined photon flux. The potential photon flux increase due to the population inversion is

$$\frac{1}{2}\Delta N(0)L \tag{4.79}$$

where L is the length of the gain region. The net energy of an incident square pulse of irradiance $I(z)$ and temporal length Δt, giving the energy density as $I(0)\Delta t$. The energy density in a gain sheet representing a length of L is $\Delta N(0)hvL/2$. The sum of these two energy densities is a constant by conservation of energy:

$$\text{total energy density} = I(0)\Delta t + \frac{1}{2}\Delta N(0)h\nu L. \quad (4.80)$$

By dividing by Δt, one can calculate the maximum possible irradiance if all the population inversions were transformed into light:

$$I_{\max} = I(0) + \frac{\Delta N(0)h\nu L}{2\Delta t} \quad (4.81)$$

and by dividing by $h\nu L/2$, one has the maximum population inversion if all the lights were subsumed by stimulated absorption,

$$\Delta N_{\text{total}} = \Delta N(0) + \frac{2I(0)\Delta t}{h\nu L}. \quad (4.82)$$

One can use ΔN_{total} to calculate $N(z)$:

$$\Delta N(z) = \Delta N_{\text{total}} - \frac{2I(z)\Delta t}{h\nu L}. \quad (4.83)$$

Equation (4.74) now takes the form:

$$\frac{\partial I(z)}{\partial z} = B\left[\Delta N_{\text{total}} - \frac{2I(z)\Delta t}{h\nu L}\right]I(z). \quad (4.84)$$

Equation (4.84) has the exact solution:

$$I(L) = \frac{I_{\max}I(0)}{I(0) + (I_{\max} - I(0))e^{-B\Delta N_{\text{total}}L}}. \quad (4.85)$$

At low saturation, equation (4.85) approaches the expected simple exponential gain. At high saturation, equation (4.85) approaches $I_{\max}$. Equation (4.85) works well for both high- and low-energy amplifiers, and the rate equation algorithms have been modified to use this expression.

4.3.2.2 Offline Effects

The gain and offline index of refraction effects may be represented by a complex index of refraction using χ'_m and χ''_m, such that

$$n \to n\left(1 + \frac{\chi'_m}{2n^2} + \frac{\chi''_m}{2n^2}\right) \quad (4.86)$$

$$\chi''_m = (N_1 - N_2)\frac{\lambda^3}{16\pi^3 t_{\text{spont}} n}f(\nu), \chi'_m = \frac{2(\nu_{\text{off}} + m\Delta\nu_c)}{\Delta\nu}\chi''_m \quad (4.87)$$

$$f(\nu_m) = \frac{\Delta\nu}{2\pi\left[\nu_{\text{off}} + m\Delta\nu_c)^2 + \left(\frac{\Delta\nu}{2}\right)\right]} \quad (4.88)$$

where $\nu_m - \nu_{\text{cen}} = \nu_{\text{off}} + m\Delta\nu_c$, and m is the mode number. The optical field, under steady-state conditions, varies as

$$a_m(x,y;\Delta t) = a_m(x,y,0)\exp[(jk\chi'_m + k\chi''_m L/2n^2)] \quad (4.89)$$

where $\chi_m = \chi'_m - j\chi''_m$ is the electric susceptibility and n is the index of refraction.

4.3.2.3 Spontaneous Emission

Spontaneous emission arises from the decay of the population inversion. This spontaneous emission is a noise source for many laser processes. The noise power injected into each mode at a distance Δz is

$$\Delta I_{\text{noise}} = \frac{(N_2 - N_1)h\Delta z}{2t_{\text{spont}}}\frac{\lambda^2}{4\pi\Delta x\Delta y} \quad (4.90)$$

where the solid angle subtended by the computer array is $\Delta\Omega = \lambda^2/4\pi\Delta x\Delta y$ when using sampling intervals of Δx and Δy. This noise is introduced as a delta-correlated, normally distributed random phasor.

4.4 Integration of Geometrical and Physical Optics

This section outlines a method of combining geometrical and physical optics. We shall be primarily concerned with reasonably well-behaved, i.e. imaging optical systems. Such systems image a region in object space into image space. Generally, the best imaging is found for a limited range of object to lens distances and a limited range of object angles. The imaging is never perfect, suffering from truncation of the beam by the apertures and aberrations of the optical system.

The approach for modelling well-behaved optical systems derives from their geometrical optics behaviour. If the imaging was perfect, then the input beam would be geometrically modified into an output beam in image space. Considered between the conjugate planes, the ideal system simply applies magnification, change of direction and change of phase radius of curvature—no diffraction effects, no aberration and no aperture clipping. The real optical system—necessarily imperfect—is a perturbation from the ideal system.

In many lens designs, there is a well-defined aperture stop (at least for the on-axis optical beam), unlike general physical optics systems where clipping apertures may be more arbitrarily located. In addition to the aperture stop, an off-axis beam may be clipped by apertures ahead of or following the aperture stop. Figure 4.1 illustrates a representative system. Generally, there is a well-defined region where the beam is expanded and the elements are relatively close, such that the elements are effectively in the near field with respect to each other. This is an important condition for geometrical design to be valid. Apertures at different axial points in the expanded region are effectively co-located in terms of diffraction calculations. The various apertures in the expanded region may be collapsed into the entrance pupil. The aperture stop may be accurately included by using diffraction propagation from the object to the entrance pupil, applying the entrance pupil aperture as the aperture stop and including phase aberrations, and

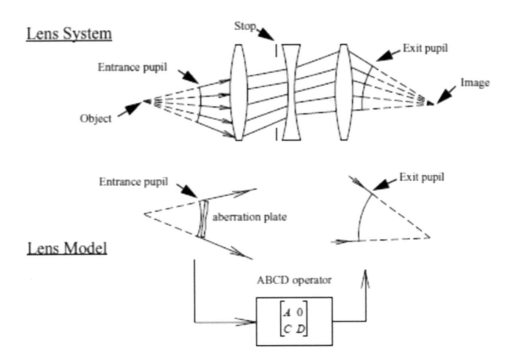

FIGURE 4.1 The object point may be transformed to its conjugate in image space by a simple ABCD matrix. Aberrations determined by geometric ray tracing may be used to add an aberration plate at the entrance pupil which is relayed to its image in the exit pupil.

propagating back to the object point. The object point may be transformed to its conjugate in image space by a simple ABCD matrix of the form:

$$ABCD = \begin{bmatrix} A & 0 \\ C & D \end{bmatrix} \quad (4.91)$$

where $A = M$ (M is the magnification), $D = 1/M$, and C represents the optical power. $B = 0$ indicates there is no diffraction propagation in going between conjugate points.

The aberrations may be determined by probing the system with rays. The rays are constructed to be normal to the reference surface. Ideally, the rays are constructed to be normal to the wavefront, but in order to probe the system with rays, it is generally not necessary to make a distinction between the wavefront normals and the reference surface normals.

4.5 Dielectric Waveguides

Optical fibres, dielectric waveguides and general gradient index materials may be modelled in diffraction codes by using the split-step method where the effect of the material is to alter the index of refraction. The result is that with propagation through the material, a phase change results from propagation through the material. Let us consider a three-dimensional case where propagation is proceeding in z, and x and y are the transverse dimensions. For an incremental step in z, the phase changes due to Δn, which is the difference between the local index and that of the bulk material; the complex amplitude changes according to

$$a_1(x, y, z + \Delta z) = a_0(x, y, z + \Delta z) e^{i 2\pi \Delta n(x,y) \Delta z} \quad (4.92)$$

Combining the effect of variable index of refraction with diffraction may be solved with the split-step method illustrated in Figure 4.2, which is adapted from the general split-step method to specifically address the index of refraction combined with diffraction.

In our three-dimensional case, we consider a circular core of higher index embedded in a cladding of lower index. Ultimately, the cladding is surrounded by air, but to simplify the discussion, we will neglect the cladding-to-air interface. However, in a numerical analysis, it is generally necessary to use an absorbing boundary (sometimes called a transparent boundary) that is implemented with an absorbing region near the edge of the array describes the (x, y) region of the calculation.

To make the case interesting, we will make the core initially straight, then go through an S-curve and finally exit with a second straight section. We will inject a Gaussian beam into the upper-left side of the waveguide and follow the evolution of the beam as it proceeds towards the lower left.

If the Gaussian beam has a narrow waist then, initially, the beam will diverge. If the starting Gaussian is larger, then the beam converges initially. For any inserted Gaussian beam, the beam will not exactly match the waveguide mode. As this waveguide was set to support only the lowest order waveguide mode and in due course higher modes due to the Gaussian being mismatched and also due to the bends in the waveguide, we see transient structure as the higher-order modes beat with each other. In the initial straight section, the beam exhibits cycles of blooming and pinching. These cycles damp out in as the higher order modes are radiated out of the waveguide.

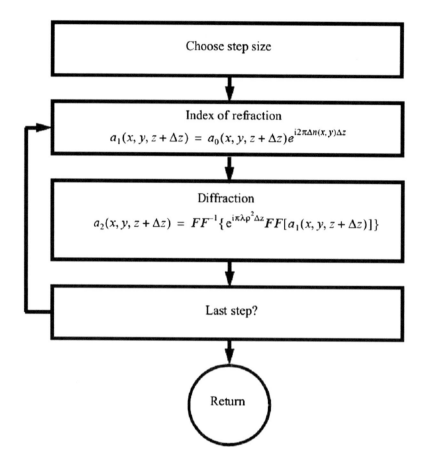

FIGURE 4.2 Flow chart for split-step method of treating diffraction and the refractive index function $\Delta n(x, y)$. For a small step Δz, the effect of the refractive index is implemented as a phase screen of the form $e^{i2\pi \Delta n(x,y)\Delta z}$ to change the initial complex amplitude $a_0(x, y, z + \Delta z)$ into the intermediate result $a_1(x, y, z + \Delta z)$. A diffraction step is applied to the intermediate result, implemented by FFT methods, to create the full split-step procedure.

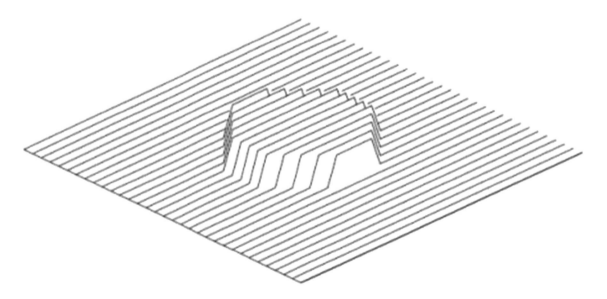

FIGURE 4.3 Circular shape of higher index core which is extended along the optical path.

We will use a circular core as shown in Figure 4.3. The S-curve of the fibre is implemented simply by moving the centre position of the core transversely as shown schematically in Figure 4.4. The irradiance profile is shown

The position of the core versus propagation length through the waveguide is shown in Figure 4.5. The length of the propagation is greatly compressed to better display the effects (Figure 4.6).

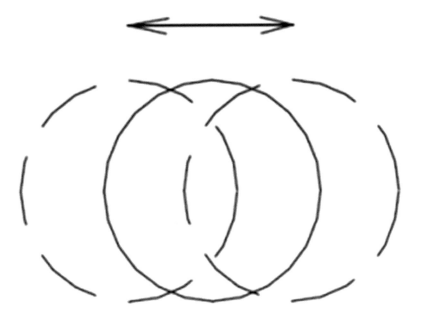

FIGURE 4.4 Transverse motion of core versus length of the core to generate the S-curve.

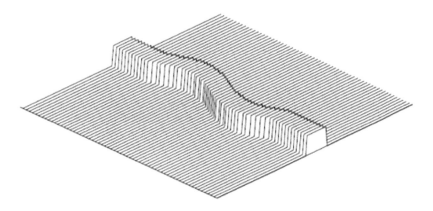

FIGURE 4.5 Plot of index difference versus distance from upper left to lower right showing initial straight section, S-curve and a second straight section. The propagation length is shown greatly compressed for better illustration.

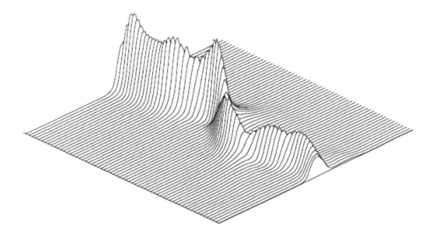

FIGURE 4.6 History of irradiance profiles after the injection of Gaussian mode at upper left. Mode beating is shown in straight section, then more mode beating in the S-curve, followed by another straight section exiting at the lower left. Note that output is considerably reduced from input because of radiative losses.

4.6 Reflecting Wall Waveguides

Reflecting wall waveguides are frequently used as optical integrators. Diffraction propagation in these devices can be done using a simple modification of free-space methods previously discussed.

Aliasing is usually considered an unwanted effect of numerical propagation, but in the special case of reflecting wall waveguides, aliasing allows us to calculate reflecting wall waveguides very accurately.

When doing numerical calculations with a fixed size array for representation of the optical distribution, we must take care to handle errors of aliasing due to light tending to scatter outside the boundary of the array. For the reflecting wall waveguides, aliasing may be used to model reflections using a simple method.

Aliasing and reflecting from a wall are different. A reflecting wall reflects the light back into the beam path, but aliasing "wraps" the distribution back into the array on the opposite side. Figure 4.7 illustrates a simple triangle as an object inside a reflecting wall configuration.

We represent the pattern of reflecting images produced by the reflecting walls in Figure 4.8. If we set the size of the array to exactly match the reflection walls, aliasing will occur but it

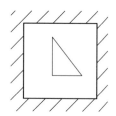

FIGURE 4.7 Consider a reflecting wall waveguide with four sides. We wish to calculate the diffraction propagation of an object along the axis with effects from the reflecting side walls. This figure shows a simple triangle as object.

will appear as though the ghost images are all identical so that we would have images of the object that are all exactly identical to the object for all image positions. In the general case, reflection and aliasing are similar but not identical.

However, for bilaterally symmetric objects, the images due to reflection and those due to aliasing are identical. We can force arbitrary distribution to be bilaterally symmetric by putting the array of interest into the upper-left quadrant of an array of twice the width. We then fill the other three quadrants with mirror images of the upper-left quadrant about their respective common boundaries with an appropriate phase change due to reflection. We now have a super cell of the object and three images as shown in Figure 4.9. As the super cell, which is now acting as the object, is bilaterally symmetric, all images of the super cell are identical.

To perform the diffraction calculation for any arbitrary distribution, we copy the object distribution into the upper-left quadrant of a super cell of twice the size and populate the other three quadrants with the images properly oriented and adjusted for phase associated with reflection. We now propagate the super cell arrangement the desired distance and extract the upper-left quadrant of the super cell which is our desired diffraction pattern for the reflecting wall configuration. The calculation is essentially perfect to the level of Fresnel diffraction. The only disadvantages are that the diffraction calculation takes longer because the array for the super cell is twice as large and there is a small calculation time required for copying the complex amplitude distributions.

To illustrate this effect, a circular beam was injected into the waveguide at an offset position and with a tilt, so that the beam will reflect off all four walls in sequence with the diffraction effects growing continually for the length of the reflecting wall device. When the beam is near a reflecting wall, strong self-interference is observed as expected. Here, we will show only reflection from the left and then the top walls as indicated in Figure 4.10. We choose to start with a circular distribution near the left wall and headed into the left wall. Figure 4.11 shows

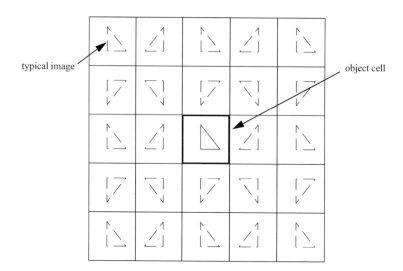

FIGURE 4.8 The object cell contains the distribution to be propagated in a reflecting wall waveguide having the same size as the object cell. Due to the reflecting walls, many image cells are formed around the object cell with their respective orientation. For aliasing the images would always have the same orientation as the object. For reflection, light that goes out the left side is reflected back from that side. For aliasing, light leaving from the left reappears on the right. Reflection and aliasing are not, in the general case, the same.

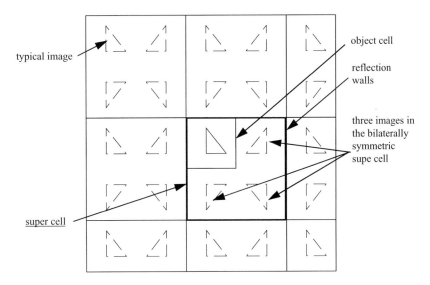

FIGURE 4.9 This figure shows the object cell embedded in the upper-left quadrant of an array of twice the size of object cell. The other three quadrants are populated with images of the object cell that exist if there were reflecting walls between the object and image cells. The object cell and the three image cells form a super cell which is bilaterally symmetric. For the bilaterally symmetric super cell, aliasing and reflection are identical. We make the array exactly fit the super cell and rely on diffraction propagation with the inevitable aliasing associated with a finite size array to create spillover between the super cell and its images to correctly represent a reflecting wall waveguide on the boundary of the object cell.

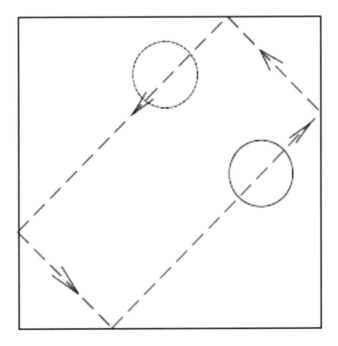

FIGURE 4.10 Consider a beam in a hollow waveguide with reflecting walls. The beam is given a tilt which sends it towards the upper right. The beam will reflect around the walls while expanding because of diffraction.

the appearance of the super cell at the start of the calculation. Figures 4.12–4.18 show the evolution of diffraction in a quarter of the first full round trip.

4.7 Laser Modelling Software

The complexity of diffraction, laser gain and non-linear optical calculations are often best handled by computer programs. The consumer will find a great variety of programs available with the choice expanding year-to-year.

4.7.1 Traditional Methods of Modelling

By tradition, optical system analysis falls into the general categories of ray-based, hybrid ray-based method using Gaussian beamlets that evolve along optical rays and complex amplitude representations. Ray-based methods have proved to be the method of choice for conventional optical design where the optical designer is altering radius, thickness, glass type, aspheric coefficients, etc., to minimize and balance optical aberrations. Hybrid ray-tracing methods are similar to traditional ray tracing in being capable of calculating optical

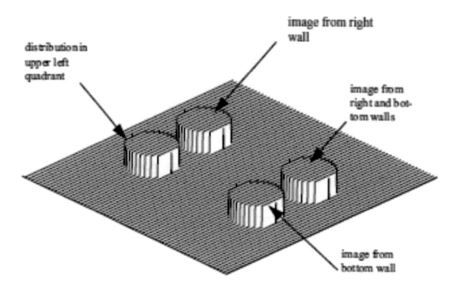

FIGURE 4.11 A hollow waveguide may be represented by placing the distribution in one quadrant (here the upper-left quadrant) and placing images as formed by the walls in the other three quadrants. The starting distribution is offset and has a tilt which directs the beam initially toward the upper right.

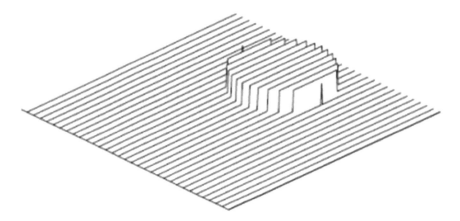

FIGURE 4.12 Start with tilt aberration, showing only upper-left quadrant, as with the rest of figures.

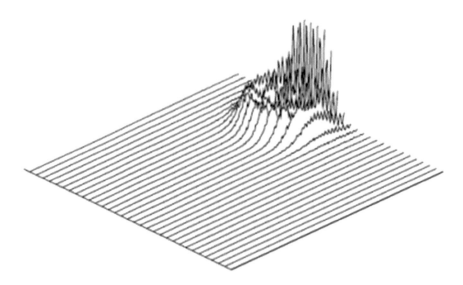

FIGURE 4.13 Beam is tilted to upper right and hits the right wall. Fine structure is due to self-interference at reflection from the wall.

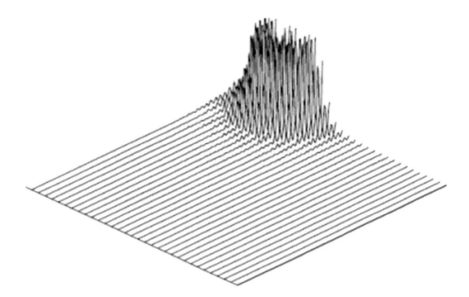

FIGURE 4.14 Beam collides with right wall and is deflected toward top wall.

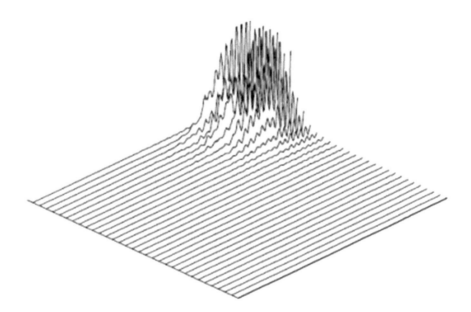

FIGURE 4.15 Beam is passing from right to top wall.

path differences but have the additional capability of carrying intensity along the optical rays so that surface scattering, multi-faceted optical integrators and other non-imaging applications may be analysed. Both the traditional and hybrid ray methods have the virtue of being able to treat large aberration values without the aliasing difficulties of sampled, complex amplitude descriptions. Rays could be used to represent a flashlight beam. Rays have difficulty in representing the coupling of intensity and phase such as occurs most prominently near a focus region, near-field diffraction, laser gain and non-linear optics.

Beam propagation methods (BPM) represent the complex amplitude on a point-by-point basis. The research literature shows an overwhelming preference for BPM for analysis of lasers and laser beam trains. BPM may be implemented with propagation by plane-wave decomposition or by finite-difference methods. The point-by-point representation of the optical fields facilitates the calculation of laser gain and non-linear effects. More difficult photonic applications, such as fibres and waveguides with high core–cladding index differences, sharp bends, holey fibres and photonic crystals, may require finite difference or finite-element methods. The primary limitation of BPM is that the spread of angles associated with strong aberration requires fine sample spacing and angles above about 10 or 15 degrees invalidate the simpler scalar theory.

Many commercial programs are not limited to a single method of analysis, although most programs still have their principal strength in either ray tracing or BPM. The complexity of analysis required for lasers and laser beam trains is vastly greater than for traditional optical design. Most ray-tracing

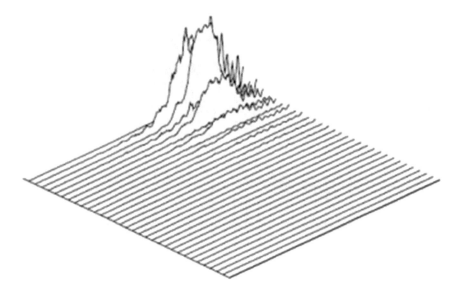

FIGURE 4.16 Beam is now reflecting off top wall.

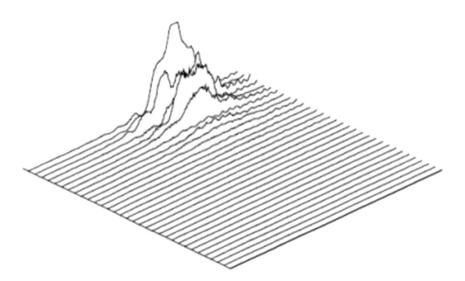

FIGURE 4.17 Beam is now headed for left wall after reflection from top wall.

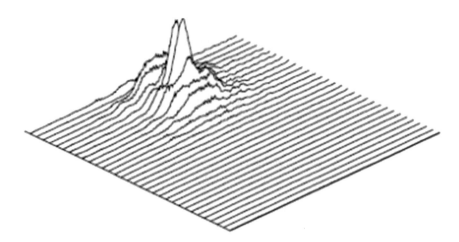

FIGURE 4.18 Beam is coming off top wall.

analysis is a repeated application of Snell's law or the law of reflection. One needs little theoretical explanation and just a few examples may suffice.

4.7.2 Selecting Commercial Numerical Modelling Software

Most software vendors will provide a detailed list of features and capabilities. One should consider both the overall emphasis of the software, e.g. ray tracing, BPM, etc., and the detailed list of features. The software vendor should back up claimed capabilities with numerous demonstration examples and associated explanations. A broad range of examples is also a great aid to learning how to use the software. Examples should demonstrate critical phenomena as part of the validation of the program as well as illustrating the use of program features. The thoroughness and clarity of the documentation seems to be a good predictor of overall program quality. Before buying, it is a good idea to evaluate the technical support by posing a question by email or phone.

4.7.3 Validation of Software

It is appropriate for the user to expect the software vendor to provide proof of accuracy, and the user should make an effort to be familiar with such validation material. Virtually, all models rely on various assumptions to simplify the calculations and such approximate models will have a finite range of parameters for which accuracy is satisfactory. While one will hear it said that certain numerical models may be trusted because they have been 'anchored to experiment', it is unwise to rely on such testimonials (with the exception of measurement of material properties). It is quite possible for unsound computer models to agree with certain data points of certain experiments but to fail for other conditions. Sound computer models will be based on theory and validation exercises, which should be designed to illustrate agreement with critical points of the theory. For example, a gain model may be validated by showing both correct small-signal gain and correct operation under strong saturation. The software builder may rely on the availability of sound and complete theory in virtually all areas of laser technology and diffraction propagation theory. Such a sound theoretical basis is the best way of ensuring accuracy for the broadest range of conditions.

The user is well advised to perform his or her own validation studies to gain familiarity and confidence with the software. For example, one might check near-field diffraction of a circular aperture at even Fresnel number to observe the zero at the centre and the centre structure. It is valuable to perform numerical experiments to observe the detail in the calculated pattern and degree of aliasing for different choices of sampling density and guard band. Laser gain may be checked for a short section of gain media (to minimize effects of diffraction) for weak input intensity to check small-signal gain and high intensity to check saturated gain. The results can be checked against hand calculations from the basic equations. If necessary, any functional feature in physical optics modelling may be checked in isolation against hand calculations.

REFERENCES

1. Fox A G and Li T 1961 Resonant modes in a maser interferometer *Bell Syst. Tech. J.* 46 453.
2. Siegman A E 1973 Hermite Gaussian functions of complex argument as optical beam eigenfunctions *J. Opt. Soc. Am.* 63 1093.
3. Sziklas E A and Siegman A E 1974 Diffraction calculations using fast Fourier transform methods *Proc. IEEE* 62 410–12.
4. Rench D B and Chester 1974 Three dimensional unstable resonators with laser medium *Appl. Opt.* 13 2546–61.
5. Sargent M, Scully M and Lamb W 1974 *Laser Physics* (Reading, MA: Addison-Wesley).
6. Bloembergen N 1965 *Nonlinear Optics* (Reading, MA: Benjamin).
7. Hardin R H and Tappert F D 1973 Applications of the split step Fourier method to the numerical solution of nonlinear and variable coefficient wave equations *SIAM Rev.* 15 423.
8. Goodman J W 1968 *Introduction to Fourier Optics* (New York: McGraw-Hill).
9. Gaskill J 1976 *Linear Systems, Transforms, and Optics* (New York: Academic) p. 139.
10. Kraus H 1989 Huygens Fresnel Kirchoff wave front diffraction formulation: spherical waves *J. Opt. Soc. Am.* A 6 1196.
11. Lawrence George N 1992 *Optical Modelling (Applied Optics and Optical Engineering XI)* eds R Shannon and J Wyant (New York: Academic) pp. 125–200.
12. Hayes J 1992 *Fast Fourier Transforms and their Applications (Applied Optics and Optical Engineering 11)* eds R Shannon and J Wyant (New York: Academic).
13. Born M and Wolf E 1965 *Principles of Optics* (New York: Pergammon).
14. Lawrence G 1980 *Optical Performance Analysis of CO_2 Laser Fusion Systems* Doctoral Dissertation University of Arizona.
15. Siegman A E 1977 Quasi fast Hankel transform *Opt. Lett.* 1 13–15.
16. Sheng S-C 1980 *Studies of Laser Resonators and Beam Propagation Using Fast Transfrom Methods* PhD Dissertation ch 3, Department of Applied Physics, Stanford University.
17. Sheng S-C and Siegman A E 1980 Nonlinear optical calculations using fast transform methods: Second harmonic generation with depletion and diffraction *Phys. Rev.* A 21 599–606.
18. Oppenheim A V, Frisk G V and Martinez G R 1980 Computation of the Hankel transform using projections *J. Acoust. Soc. Am.* 68 523–9.
19. Frantz L M and Nodvik J S 1963 Theory of pulse propagation in a laser amplifier *J. Appl. Phys.* 34 2346–9.
20. Yariv A 1976 *Introduction to Optical Electronics* (New York: Holt, Rinehart, and Winston).
21. Hader J et al 2002 Semiconductor quantum-well designer active materials *Opt. Photon. News* special issue 'Photonics in 2002'

5
Optical Cavities: Free-Space Laser Resonators

Robert C. Eckardt

CONTENTS

5.1 Introduction 97
5.2 Gaussian Beams 99
    5.2.1 Conventions and Notation 99
    5.2.2 Description of Gaussian Beams 99
    5.2.3 Ray Transfer Matrices 101
    5.2.4 Gaussian Resonant Modes 103
5.3 Stable Resonators 104
    5.3.1 Two Mirror Resonators 104
5.4 Higher-order Modes of Stable Resonators 105
    5.4.1 Cartesian Coordinates 106
    5.4.2 Cylindrical Coordinates 106
    5.4.3 Beam Quality 107
5.5 Mode-Matching 109
    5.5.1 One-lens Approach 109
    5.5.2 Two-lens Mode-matching 109
5.6 Plane Parallel Resonators 110
5.7 Unstable Resonators 111
    5.7.1 Hard-Edged Apertures 112
    5.7.2 Soft-edged Apertures 115
5.8 Distortion Effects 115
5.9 Axial Modes 116
    5.9.1 Stable-resonator Axial-mode Spectral Separation 116
5.10 Frequency Selection and Frequency Stability 117
5.11 Temporal Resonator Characteristics 118
5.12 Fibre Laser Resonators 118
5.13 Conclusion 119
References 119

5.1 Introduction

Resonators provide the optical structure in which laser oscillations are established. Passive optical resonators can also be used to increase locally the power of coherent optical radiation or to filter optical radiation. An understanding of laser resonators is necessary in analysing the spatial beam characteristics and temporal coherence properties of the light output of laser systems. Optimizing the design of a laser system requires resonator analysis. In addition to the operation and design of lasers, an understanding of optical resonators is important in coupling the laser output to the application in which it is to be used. This chapter reviews the fundamental aspects of laser resonators and discusses some of the techniques of laser operation.

Optical resonators have resonant modes and these modes are important to the analysis of laser operation. An optical resonator mode is a field distribution that is resonant with the structure and that is reproduced in phase and in relative intensity after a round-trip transit of the resonator. The intensity of a mode may decrease due to resonator losses, or it may be increased by amplification in an active laser material or by light introduced from outside the resonator. Mode competition and selection usually occur when an oscillation builds up in a resonator. Typically, laser oscillations start from random quantum fluctuations and build to high intensity in a single mode or many simultaneous modes. After many resonator transits the mode or modes with minimum loss tend to dominate and to be reproduced after each cavity transit. An iterative numerical analysis that simulated this process was used in the seminal paper by Fox and Li [1] to analyse the field distributions of laser resonators. Numerical methods of resonator analysis augmented by fast Fourier transform techniques and computers of increasing capability are currently in wide use. The analysis of resonators that have large dynamic changes, non-linearities or significant diffraction typically requires numerical methods rather than analytical techniques. However, certain analytical methods for resonator analysis also have wide application and are extensively used.

Many of the analytical techniques for describing stable open resonators originated in the work of Boyd and Gordon [2]. The Gaussian beams of stable open resonators can be characterized with analytical expressions and provide a widely used approach to the discussion of laser resonators. A Gaussian beam is the fundamental mode of a set of Hermite–Gaussian modes in rectangular coordinates, and the same Gaussian mode is also the fundamental mode of a set of Laguerre–Gaussian modes in cylindrical coordinates. Either of these sets of modes forms complete sets of orthogonal functions, which can be used in a series expansion to describe a transverse field distribution to an arbitrary degree of accuracy. These modes will usually be a very good approximation, although still an approximation, to the modes of any real laser resonator. Real resonators will always have some limiting aperture, which introduces a small loss, and possibly other perturbations that will slightly distort the modes and make them non-orthogonal.

A resonator is called 'stable' if a ray of a geometrical optical analysis always remains within the resonator rather than wandering off an increasing distance from the resonator centre with an increasing number of cavity transits. Unstable resonators, which do not satisfy this condition, nonetheless have modes that are reproduced through diffraction from one cavity transit to the next. The term 'unstable resonator' is somewhat misleading because these resonators can be quite stable in operation, and they are useful for efficient energy extraction in high-gain systems that can support the higher diffraction losses typical of unstable resonators. Siegman [3] prefers the descriptive name 'geometrically unstable' as being more accurate for these resonators. Stable resonators typically have small diffraction losses and support Gaussian transverse fundamental modes. They may also support many higher-order transverse modes giving a great deal of structure to the beam. Unstable resonators usually offer large loss discrimination between the lowest-loss transverse mode and other transverse modes. For this reason, they often provide good spatial beam quality in large-diameter laser output beams. Stable and unstable resonators are illustrated schematically in Figure 5.1.

Resonators simultaneously have axial modes in addition to their transverse modes. The condition that the phase distribution be reproduced after a round-trip cavity transit allows axial modes that cycle through a phase change of integer multiples of 2π on a round-trip transit. Higher-order transverse modes have more complex spatial amplitude distributions, which are also reproduced from one cavity transit to the next. The higher-order transverse modes have additional phase shifts that place them, in frequency, between the fundamental transverse modes. The axial or longitudinal resonator modes with a fundamental transverse distribution will be nearly equally spaced in frequency or wavenumber. The modes will only be precisely spaced if care is taken to compensate for dispersion and frequency-dependent phase shifts of the optical elements of the resonator. Such compensation becomes important for

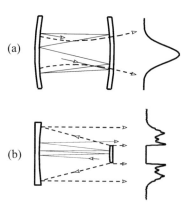

FIGURE 5.1 Schematic representation of a stable resonator (a) and an unstable resonator (b). A ray path (full line) is confined to remain on the mirrors in the stable resonator, but the geometrical ray walks off the mirrors in the unstable resonator. The curves on the right-hand side and the broken lines represent the intensity distributions of the fundamental resonator modes.

mode locking, in which case many axial modes are locked in phase to synthesize a single short pulse that propagates back and forth in the resonator. Other cavity control techniques include frequency selection or narrowing to restrict axial modes and spatial filtering to restrict transverse modes. Laser resonator losses can be controlled to hold off oscillation and then suddenly reduced in a technique called Q-switching, to produce an energetic pulse of only a few resonator transits duration.

The objective of this chapter is to present a basic description of laser resonators at a level that will allow the reader to apply the material to practical systems. Derivations are not given in this review but are covered in detail in the references. There have been many excellent papers and reviews on this subject. The books by Siegman [3], by Hall and Jackson [4] and by Hodgson and Weber [5] cover the topic thoroughly. Other more general books dealing with lasers also have good discussions of the topic [6–8]. Two review papers by Siegman [9,10] detail the development of the analysis of laser resonators and laser beams and include extensive bibliographies. Many of the results, terminology and notation developed in the earlier papers have become standard and are followed in this chapter. The basic types of resonators that will be discussed include geometrically stable two-mirror cavities, simple unstable resonators and resonators with plane-parallel mirrors.

The next section of this chapter reviews the optics of Gaussian light beams. These beams are solutions of the wave equation for propagation in a homogeneous, isotropic medium and are the fundamental modes of geometrically stable open resonators. The diffraction of Gaussian beams can be treated directly with analytical methods, simplifying calculations of laser-beam propagation and mode-matching between resonators. The minimum diffraction of a Gaussian beam is used to define the diffraction-limited beam quality factor M^2 of one. The fundamental mode of a stable two-mirror cavity can be determined by matching the Gaussian-beam wavefront radii to the mirror radii. More complicated cavities are typically analysed with ABCD or ray-transfer-matrix techniques.

Optical Cavities

Such analysis and application of the ray transfer matrices to the propagation of Gaussian beams are discussed here. A description of higher-order transverse modes of stable resonators follows and is used for a further description of beam quality.

Unstable resonators are discussed here both from the geometrical optics and diffraction points of view. An illustrative example of numerical resonator analysis is used to discuss transverse-mode selection in several types of unstable resonators. The calculation is seeded with a transverse distribution obtained by randomizing the phases of the initial Fourier-spatial-frequency components to simulate development from quantum fluctuations. The comparison includes plane-parallel-mirror resonators and collimated-output unstable resonators with hard-edged and soft mirrors. The use of variable-reflectivity laser mirrors is important to mitigate diffraction effects in unstable resonators. Gain-guided laser resonators, common in laser-pumped lasers, offer another technique for controlling the diffraction effects in lasers with plane-parallel resonators and in unstable resonators.

The discussion of axial modes starts with the frequency spacing of fundamental and higher-order transverse modes. The topic of cavity finesse deals with the frequency width of individual resonator modes. The bandwidth of the active gain medium of the resonator limits the number of axial modes or the frequency bandwidth of the oscillation. Components such as prisms, gratings and etalons are added to resonators to further narrow the bandwidth. Frequency stability is achieved first through mechanical and thermal stabilization of resonators. Higher levels of frequency stabilization require active stabilization to a frequency standard and careful control of the pumping stability. Modulation within the resonator is used to achieve mode locking and Q-switching. Techniques for injection seeding and injection locking can be used to transfer the properties of highly coherent low-power resonators to high-power systems. It is necessary to match the properties of a beam to the cavity into which it is injected. Examples of mode-matching are presented with one lens and two lenses used to relay the laser beam. The optics of Gaussian beams is fundamental to all of these topics and that is where we begin.

5.2 Gaussian Beams

5.2.1 Conventions and Notation

The discussion here will be restricted to paraxial optics. In this limit, the sine of an angle is approximated by that angle expressed in radians. Spherical surfaces can be represented by a parabolic approximation: $z = (x^2+y^2)/2R$ is used to describe a spherical surface of radius of curvature R. A positive wavefront radius of curvature indicates a beam that is convex in the direction of propagation or diverging. A negative wavefront radius of curvature indicates a beam that is concave in the direction of propagation or converging (Figure 5.2). Similarly, the radius of curvature of a concave spherical mirror surface is positive and that of a convex surface is negative. Coordinate rotations or reflections are assumed for refraction or reflection at arbitrary angles such that the beam remains centred on the z-axis and propagating in the +z-direction. There is no coordinate translation or scaling in the direction of propagation, and z remains the cumulative physical distance of propagation. A plane wave travelling in the +z-direction is expressed as $\exp\{-i(kz-\omega t)\}$. If the complex conjugate of this expression is used to describe this plane wave, appropriate changes of sign may be required in some expressions.

5.2.2 Description of Gaussian Beams

The derivation of the properties of a Gaussian beam is discussed in detail in texts such as those by Siegman [3] and Hodgson and Weber [5]. The early review article by Kogelnik and Li [11] standardized much of the notation for the treatment of Gaussian beams, and the article remains relevant. The results presented in these sources are reproduced here. Gaussian beams are solutions of the paraxial wave equation

$$\frac{\partial^2 E}{\partial x^2} + \frac{\partial^2 E}{\partial y^2} - 2ik\frac{\partial E}{\partial z} = 0 \tag{5.1}$$

and, equivalently, Fresnel's approximation to Huygen's integral

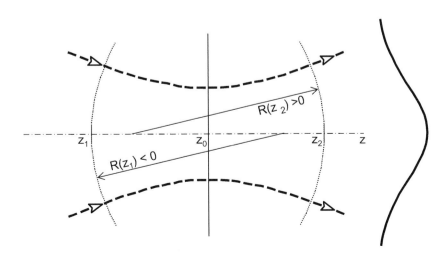

FIGURE 5.2 The sign convention for wavefront radii of curvature is positive when the beam is divergent and negative when the beam is convergent.

$$E(x, y, z) = \frac{i e^{-ik(z-z_0)}}{(z-z_0)} \iint E(x_0, y_0, z_0)$$
$$\times \exp\left[-ik\frac{(x-x_0)^2 + (y-y_0)^2}{2(z-z_0)}\right] dx_0 dy_0. \quad (5.2)$$

Here E is the electric field, which is specified at coordinates (x_0, y_0, z_0) and calculated at (x, y, z), $i = \sqrt{-1}$ and $k = 2\pi/\lambda$, where λ is the wavelength. For these equations to be valid, it is necessary that there is no sharp discontinuity, such as an aperture edge or a phase step, near the plane at which the field is evaluated.

A Gaussian beam has transverse amplitude distributions proportional to $\exp\{-(r/\omega)^2\}$, and the transverse intensity distribution is proportional to $\exp\{-2(r/\omega)^2\}$. The radial coordinate is the distance from the centre of the beam: $r = \sqrt{x^2 + y^2}$. The parameter ω is called the spot size and changes with distance as the beam propagates. Circularly symmetric beams are initially discussed here, but later, the discussion is expanded to include astigmatic beams with amplitude distributions expressed as $\exp\{-(x^2/\omega_x^2) - (y^2/\omega_y^2)\}$. Wavefronts of constant phase of circularly symmetric Gaussian beams form spherical surfaces, and the wavefronts of the astigmatic beams are circular in cross section along the principal axes of the astigmatic components, which are assumed here to be aligned for all components in the beam.

The mathematical description of the electric field of a Gaussian light beam or TEM_{00} mode is a real function expressed as the sum of a complex expression and its complex conjugate:

$$E(x,y,z,t) = \frac{1}{2} E_0 \frac{\omega_0}{\omega(z)} \exp\left\{-\frac{r^2}{\omega^2(z)} - i\left(\frac{r^2 k}{2R(z)} - \Phi(z)\right)\right\}$$
$$\times \exp\{-i(kz - \omega t)\} + c.c. \quad (5.3)$$

This is a solution to the wave equation representing a nearly collimated beam propagating in the $+z$-direction. The complex conjugate is not carried in the discussion, and it is sufficient to treat the field as a complex parameter. Only in a few special cases, not encountered here, is it necessary to retain the real number representation of the electric field. It is assumed that the transverse dimension of the beam is much larger than the wavelength, that is, $\omega_0 \gg \lambda$, a condition which is typical of most laser beams and laser resonators. The factor $\exp\{-i(kz - \omega t)\}$, with $k = 2\pi n/\lambda_0$, n is the index of refraction of the medium in which the beam is propagating, λ_0 is the free-space wavelength, ω is the angular frequency and t is the time, represents the plane wave component of the distribution. The remaining portion of the expression describes differences in field distribution and phase from that of the plane wave. The parameter $\omega(z)$ is the spot size and $R(z)$ is the wavefront radius of curvature. The change due to propagation with diffraction for these two parameters is

$$\omega^2(z) = \omega_0^2\left[1 + \left\{\frac{\lambda(z-z_0)}{\pi\omega_0^2}\right\}^2\right] = \omega_0^2[1 + \{(z-z_0)/z_R\}^2] \quad (5.4)$$

and

$$R(z) = (z-z_0)\left[1 + \left\{\frac{\pi\omega_0^2}{\lambda(z-z_0)}\right\}^2\right] = (z-z_0)[1 + \{z_R/(z-z_0)\}^2]. \quad (5.5)$$

The beam waist is located at z_0 where $\omega(z_0) = \omega_0$ has a minimum value. The parameter z_R is the Rayleigh range or Rayleigh length:

$$z_R = \pi\omega_0^2/\lambda. \quad (5.6)$$

The Rayleigh length is the distance necessary to travel from the beam waist for the spot size to increase by a factor of $\sqrt{2}$. The confocal parameter b of a Gaussian beam, a commonly used parameter, is twice the Rayleigh range:

$$b = k\omega_0^2 = 2\pi n \omega_0^2/\lambda_0. \quad (5.7)$$

The parameter $\Phi(z)$ is the Gouy phase shift and gives the departure of the on-axis phase from that of a plane wave:

$$\Phi(z) = \arctan\{\lambda(z-z_0)/(\pi\omega_0^2)\}. \quad (5.8)$$

When propagated into the far field, a Gaussian beam will expand with a divergence half-angle of $\theta = \lambda/(\pi\omega_0)$ (Figure 5.3). The constant E_0 in equation (5.3) is the peak electric field at the beam waist and is expressed in SI units in terms of the total power in the beam P or peak intensity I_0 as

$$E_0 = \sqrt{2I_0/(n\varepsilon_0 c)} = \sqrt{4P/(\pi\omega_0^2 n\varepsilon_0 c)} \quad (5.9)$$

where E_0 is in units of V m^{-1}, P is in W, I_0 is in W m^{-2}, $\varepsilon_0 \simeq 8.854 \times 10^{-12}$ CN m^{-2} is the permittivity of free space and $c \simeq 2.997 \times 10^8$ m s^{-1} is the speed of light.

Expressing a Gaussian beam in terms of ω_0, the beam waist spot size, and z_0, the position of the beam waist, offers the advantage of explicitly stating the spot size and wavefront radius of curvature as a function of position. An alternate description is given in terms of $q(z)$, the 'complex beam parameter' or 'complex radius of curvature',

$$1/q(z) = 1/R(z) - i\lambda/\{\pi\omega^2(z)\}. \quad (5.10)$$

Using $q(z)$, the expression (5.3) for the Gaussian beam, becomes

$$E(x,y,z,t) = \frac{1}{2} E_0 \frac{q_0}{q(z)} \exp\left\{-i\frac{kr^2}{2q(z)}\right\} \exp\{-i(kz-\omega t)\} + c.c. \quad (5.11)$$

where $q_0 = -i\lambda/(\pi\omega_0^2)$. It follows that $q_0/q(z) = \exp(i\Phi(z)) \times \omega_0/\omega(z)$, and the two forms equations (5.3) and (5.11) are equivalent. Using the complex beam parameter offers advantages in terms of mathematical simplicity. For example, the change of q with propagation is simply

Optical Cavities

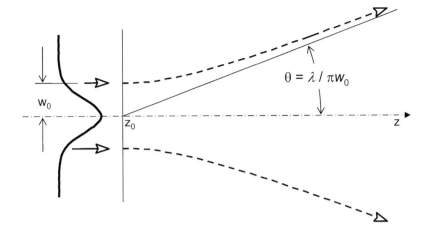

FIGURE 5.3 Gaussian beam diffracts into a half-angle of $\theta = \lambda/\pi\omega_0$ as it propagates into the far field.

$$q(z) = q_0 + z - z_0. \tag{5.12}$$

The use of the complex beam parameter and the 2×2 ABCD or ray transfer matrices greatly simplifies the analysis of complex resonators. Before discussing the ray transfer matrices, some additional relationships involving spot sizes and wavefront radii are given.

The parameters ω_0 and z_0, along with the direction of propagation, wavelength, index of refraction and beam power, specify a Gaussian beam. Usually, the characterization of a beam depends on the determination of ω_0 and z_0 with the other parameters known. Table 5.1 lists relationships that can be used to characterize a beam. Equations (5.14) can be used to determine the fundamental mode of a two-mirror laser cavity by matching the wavefront radius of curvature to the radii of curvature of the cavity mirrors (Figure 5.4). The stability condition for a two-mirror oscillator is implicit in equation (5.14) but difficult to extract in a simple form. Use of the complex beam parameter equation (5.10) and the ray transfer matrices provides this simplicity.

5.2.3 Ray Transfer Matrices

Ray transfer matrices or ABCD matrices provide concise and useful representations of both the geometrical propagation of paraxial rays and the propagation with diffraction of Gaussian beams. The propagation of rays through simple optical components is described by

$$\begin{bmatrix} x_2 \\ x_2' \end{bmatrix} = \begin{bmatrix} A & B \\ C & D \end{bmatrix} \begin{bmatrix} x_1 \\ x_1' \end{bmatrix} \text{ and }$$

$$\begin{bmatrix} y_2 \\ y_2' \end{bmatrix} = \begin{bmatrix} A & B \\ C & D \end{bmatrix} \begin{bmatrix} y_1 \\ y_1' \end{bmatrix}. \tag{5.17}$$

TABLE 5.1

Equations for Gaussian beam characterization

if $\omega(z)$ and $R(z)$ are known at some position z: (5.13)
$$\omega_0^2 = \omega^2(z)\{1 + \pi\omega^2(z)/(\lambda R(z))\}, \quad z_0 = z - R(z)\{1 + \lambda R(z)/(\pi\omega^2(z))\}$$

if $R_1 = R(z_1)$ and $R_2 = R(z_2)$ are known, $d = z_2 - z_1$: (5.14)
$$\omega_0^4 = \frac{\lambda^2 d(R_2 - d)(R_1 + d)(R_1 - R_2 + d)}{\pi^2(R_1 - R_2 + 2d)^2}, \quad z_0 = \frac{z_2 - d(R_1 + d)}{R_1 - R_2 + 2d}$$

if $R_2 = R(z_2)$ and $\omega_1 = \omega(z_1)$ are known, $d = z_2 - z_1$: (5.15)
$$z_0 = z_1 + \left(b \pm \sqrt{b^2 - 4ac}\right)/(2a), \quad \omega_0^2 = \lambda\sqrt{(z_2 - z_0)(R_2 - z_2 + z_0)}/\pi$$
where $a = (\lambda(R_2 - 2d)/\pi)^2 + \omega_1^2$, $b = \omega_1^4(2d - R_2) + 2\lambda^2 d(R_2 - 2d)(R_2 - d)/\pi^2$, and
$$c = \omega_1^4 d(d - R_2) + (\lambda d(R_2 - d)/\pi)^2$$

if $\omega_1 = \omega(z_1)$ and $\omega_2 = \omega(z_2)$ are known at positions z_1 and z_2, $d = z_2 - z_1$: (5.16)
$$\omega_0^2 = \left(-b \pm \sqrt{b^2 - 4ac}\right)/(2a), \quad z_0 = z_1 - \{(\omega_1^2 - \omega_1^2)(\pi\omega_0/\lambda)^2 - d^2\}/(2d)$$
where $a = 1 + \{(\omega_2^2 - \omega_1^2)\pi/(2d\lambda)\}^2$, $b = -(\omega_2^2 + \omega_1^2)/2$, and $c = (\lambda d/2\pi)^2$.

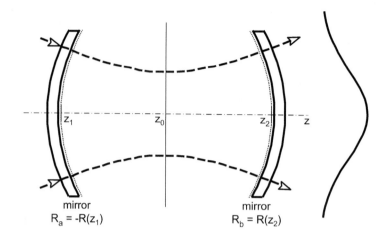

FIGURE 5.4 The modes of a two-mirror, stable resonator can be determined by matching wavefront radii of curvature to the mirror surfaces.

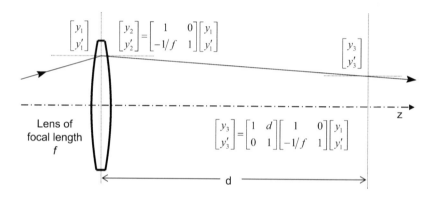

FIGURE 5.5 Application of ray transfer matrices. Subscript 1 refers to height and slope immediately before a thin lens, 2 immediately after and 3 after propagating a distance d beyond the lens.

The ray is assumed to be propagating nearly parallel to the z-axis. The distance of the ray from the z-axis is given by (x, y), and the projections of the slope of the ray are $x' = dx/dz$ and $y' = dy/dz$ (Figure 5.5).

The ray position and slope before the optical element are identified by the subscript 1 and after the element by the subscript 2. Ray transfer matrices for six simple elements are given in table 5.2: propagation over a distance d; a thin lens of focal length f; propagation through a spherical interface from a medium of index n_1 to one of n_2; an element with flat parallel surfaces and an index distribution of $n(z) = n_0 - n_2(x^2 + y^2)/2$ inside and $n = 1$ outside; a spherical mirror and a flat interface from index n_1 to index n_2. Ray transfer matrices can be combined by matrix multiplication to describe multiple-component systems:

$$\begin{bmatrix} A & B \\ C & D \end{bmatrix} = \begin{bmatrix} A_n & B_n \\ C_n & D_n \end{bmatrix} \begin{bmatrix} A_{n-1} & B_{n-1} \\ C_{n-1} & D_{n-1} \end{bmatrix}$$
$$\cdots \begin{bmatrix} A_1 & B_1 \\ C_1 & D_1 \end{bmatrix}$$

(5.18)

The multiplication must be performed as shown in equation (5.18) with subscript 1, representing the first element encountered and n the last. When the indices of refraction before the first element and after the last element of a system of components are equal, the determinant of the resultant matrix is unity:

$$AB - CD = 1.$$ (5.19)

There is a generalization of the ray matrix formulation, namely, multiplying the slope by the local index of refraction, that results in the unity determinant even if the index changes [3]. For optical resonators and propagation of one round-trip of the resonator, the index of refraction at the start and finish must be the same. For an arbitrary number of cavity transits of the resonator represented by an ABCD matrix, the distance of the ray from the z-axis is

$$r_s = r_{\max} \sin(s\theta + \delta)$$ (5.20)

where $\theta = \cos^{-1}\{(A+D)/2\}$, δ and $r_{\max}$ are determined by initial conditions and s is the number of cavity transits. If the quantity $(A+D)/2$ is in the range

TABLE 5.2

Ray transfer matrices for circularly symmetric components

Propagate distance d	Thin lens of focal length f
$\begin{bmatrix} 1 & d \\ 0 & 1 \end{bmatrix}$	$\begin{bmatrix} 1 & 0 \\ -1/f & 1 \end{bmatrix}$
Spherical interface of radius of curvature R going from index n_1 to n_2	Thick slab of graded index $n(z) = n_A - n_B(x^2 + y^2)/2$ with $n = 1$ outside
$\begin{bmatrix} 1 & 0 \\ \dfrac{n_2 - n_1}{n_2 R} & \dfrac{n_1}{n_2} \end{bmatrix}$	$\begin{bmatrix} \cos(d\sqrt{n_B/n_A}) & \dfrac{\sin(d\sqrt{n_B/n_A})}{\sqrt{n_A n_B}} \\ -\sqrt{n_A n_B}\sin(d\sqrt{n_B/n_A}) & \cos(d\sqrt{n_B/n_A}) \end{bmatrix}$
Spherical mirror with radius of curvature R_A	Flat interface going from index n_1 to n_2
$\begin{bmatrix} 1 & 0 \\ \dfrac{-2}{R_A} & 1 \end{bmatrix}$	$\begin{bmatrix} 1 & 0 \\ 0 & \dfrac{n_1}{n_2} \end{bmatrix}$

$$-1 < (A+D)/2 < 1 \quad (5.21)$$

the ray will be bound within the resonator never exceeding some maximum distance from the z-axis. If this condition is not met, the relationship describing the distance from the axis becomes a hyperbolic trigonometric function and the distance eventually expands without limit. Resonators that satisfy condition (5.21) are geometrically stable.

The ray transfer matrix also describes the propagation of Gaussian beams by

$$q_2 = (Aq_1 + B)/(Cq_1 + D). \quad (5.22)$$

Here, A, B, C and D are the components of a ray transfer matrix for a single component or a combination of components. The complex beam parameter (5.10) before the element or combination of elements is q_1, and after, it is q_2. Equation 5.22 is called the ABCD law. It is easily confirmed using matrices from Table 5.1 for propagation over a distance d or for a thin lens of focal length f; the respective results are $q_2 = q_1 + d$ and $1/R_2 = 1/R_1 - 1/f$.

5.2.4 Gaussian Resonant Modes

Modes of stable two-mirror resonators can be found by fitting a Gaussian beam to the curvature and spacing of the resonator mirrors. More complex multiple-element resonators require the use of ABCD matrix techniques to determine the modes of a stable cavity. The ray transfer matrix for a resonator satisfies the condition that the initial index is the same as the final index, and the determinant is unity, that is, equation (5.19) applies. If a resonator is to have a Gaussian beam as the fundamental mode, the complex beam parameter must be reproduced after each cavity round-trip transit. Starting at z_1, where $q_1 = q(z_1)$, and propagating through all n resonator components and returning to the same position requires

$$q_n = (Aq_1 + B)/(Cq_1 + D) = q_1 \quad (5.23)$$

The ABCD matrix elements are obtained by ordered matrix multiplication as performed in equation (5.18).

The complex beam parameter at the position z_1 is obtained from equation (5.22) using equation (5.19):

$$\frac{1}{q_1} = \frac{D-A}{2B} - \frac{i}{2|B|}\sqrt{4-(A+D)^2} = \frac{1}{R(z_1)} - \frac{i\lambda}{\pi\omega^2(z_1)}. \quad (5.24)$$

The sign of the square root is chosen to obtain a positive value of ω^2. The condition for ω to be real is the same as equation (5.21), i.e. that required for a geometrically stable resonator. It is also possible to write

$$q_1 = \frac{A-D}{2C} + \frac{i\sqrt{(A+D)^2}}{2|C|} \quad (5.25)$$

and obtain

$$z_0 = z_1 + (D-A)/(2C) \quad \text{and} \quad \omega_0^2 = \lambda\sqrt{4-(A+D)^2}/(2\pi|C|). \quad (5.26)$$

There are other cavity configurations that can support Gaussian modes, such as gain-guided resonators and those with apodized apertures but these are not treated with the simple ray transfer matrices used here. Siegman [3] describes higher-order matrices for generalized astigmatism and complex matrices to treat the apodized apertures. One generalization that is described later is an orthogonal astigmatic system in which the axes of the astigmatic components remain parallel and perpendicular. In this case, the sagittal and tangential characteristics of the resonator modes are treated separately with appropriate ray transfer matrices. This is illustrated schematically for a mirror in Figure 5.6. A tabulation of ray transfer matrices for astigmatic elements is given in Table 5.3.

5.3 Stable Resonators

5.3.1 Two Mirror Resonators

The previous section contains the analytic techniques necessary to characterize stable resonators. The simplest stable resonators have two mirrors. In a first approximation, the active laser material and other intra-cavity components can be accommodated by an effective cavity length change resulting from flat parallel plates of uniform refractive index. A graded index guide that approximates the effects of thermal loading in a laser rod could be used for a more accurate approximation. The resonator considered here simply has two spherical mirrors of radius of curvature R_A and R_B separated by distance d. The mirror radii are positive for concave mirrors and negative for convex mirrors. The first technique is to match the mirror curvatures to the wavefront curvatures by considering a Gaussian beam that passes through the two mirrors as shown

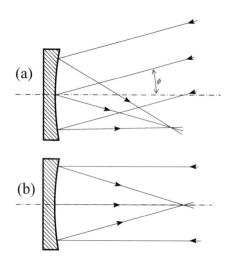

FIGURE 5.6 Top view showing the tangential focus (a) and side view showing the sagittal focus (b) of a spherical mirror. The axis of the mirror and incident beam are in a horizontal plane.

TABLE 5.3

Ray transfer matrices for astigmatic components at an angle of incidence φ

Sagittal	Tangential

Thin lens of focal length f with index n_2 surrounded by medium index n_1:

$$\begin{bmatrix} 1 & 0 \\ -\dfrac{\sqrt{n_2^2 - n_1^2 \sin^2 \phi} - n_1 \cos\phi}{(n_2 - n_1)f} & 1 \end{bmatrix} \qquad \begin{bmatrix} 1 & 0 \\ -\dfrac{\sqrt{n_2^2 - n_1^2 \sin^2 \phi} - n_1 \cos\phi}{(n_2 - n_1)f \cos^2 \phi} & 1 \end{bmatrix}$$

Thin cylindrical lens of tangential focal length f and index n_2 in a medium of index n_1:

$$\begin{bmatrix} 1 & 0 \\ 0 & 1 \end{bmatrix} \qquad \begin{bmatrix} 1 & 0 \\ -\dfrac{\sqrt{n_2^2 - n_1^2 \sin^2 \phi} - n_1 \cos\phi}{(n_2 - n_1)f \cos^2 \phi} & 1 \end{bmatrix}$$

Flat interface going from index n_1 to index n_2:

$$\begin{bmatrix} 1 & 0 \\ 0 & \dfrac{n_1}{n_2} \end{bmatrix} \qquad \begin{bmatrix} \dfrac{\sqrt{n_2^2 - n_1^2 \cos\phi}}{n_2 \cos\phi} & 0 \\ 0 & \dfrac{n_1 \cos\phi}{\sqrt{n_2^2 - n_1^2 \cos\phi}} \end{bmatrix}$$

Concave spherical mirror of radius of curvature R:

$$\begin{bmatrix} 1 & 0 \\ -\dfrac{2\cos\phi}{R} & 1 \end{bmatrix} \qquad \begin{bmatrix} 1 & 0 \\ -\dfrac{2}{R\cos\phi} & 1 \end{bmatrix}$$

Spherical interface of radius of curvature R going from index n_1 to index n_2:

$$\begin{bmatrix} 1 & 0 \\ \dfrac{\sqrt{n_2^2 - n_1^2 \sin^2 \phi} - n_1 \cos\phi}{n_2 R} & \dfrac{n_1}{n_2} \end{bmatrix} \qquad \begin{bmatrix} \dfrac{\sqrt{n_2^2 - n_1^2 \cos\phi}}{n_2 \cos\phi} & 0 \\ \dfrac{1}{R\cos\phi} - \dfrac{n_1/(n_2 R)}{\sqrt{n_2^2 - n_1^2 \cos^2 \phi}} & \dfrac{n_1 \cos\phi}{\sqrt{n_2^2 - n_1^2 \cos^2 \phi}} \end{bmatrix}$$

Thick plate of thickness d and index n_2 with parallel surfaces and index n_1 outside:

$$\begin{bmatrix} 1 & \dfrac{n_1 d}{\sqrt{n_2^2 - n_1^2 \cos^2 \phi}} \\ 0 & 1 \end{bmatrix} \qquad \begin{bmatrix} 1 & \dfrac{d n_1 n_2 \cos\phi}{n_2^2 - n_1^2 \sin^2 \phi} \\ 0 & 1 \end{bmatrix}$$

Optical Cavities

in Figure 5.4. Use equation (5.14) with $R_1 = -R_A$ and $R_2 = R_B$ to obtain

$$\omega_0^4 = \frac{\lambda^2 d(R_B - d)(R_A - d)(d - R_A - R_B)}{\pi^2 (2d - R_A - R_B)^2}$$

$$\text{and} \quad z_0 = \frac{z_2 - d(d - R_A)}{2d - R_A - R_B}. \tag{5.27}$$

Here, z_1 is the position of the first mirror, z_2 is the position of the second mirror, $d = z_2 - z_1$ and z_0 is the location of the beam waist, which has spot size ω_0. There will be a real value for the beam waist only when the right-hand side of the first equation of equation (5.27) is positive. This is the condition for a stable cavity.

To perform a ray-transfer-matrix analysis of the same resonator, pick a starting position, for example, just before mirror R_B. Then, the appropriate matrices from Table 5.2 are used for reflection from the mirror R_B, propagation over the resonator length d, reflection from mirror R_A and propagation back to just before the right-hand mirror.

$$\begin{bmatrix} A & B \\ C & D \end{bmatrix}$$

$$= \begin{bmatrix} 1 & d \\ 0 & 1 \end{bmatrix} \begin{bmatrix} 1 & 0 \\ -2/R_A & 1 \end{bmatrix} \begin{bmatrix} 1 & d \\ 0 & 1 \end{bmatrix} \begin{bmatrix} 1 & 0 \\ -2/R_B & 1 \end{bmatrix}$$

$$= \begin{bmatrix} \left(1 - \frac{2d}{R_A} - \frac{4d}{R_B} + \frac{4d^2}{R_A R_B}\right) & 2d\left(1 - \frac{d}{R_A}\right) \\ \left\{\frac{4d}{R_A R_B} - 2\left(\frac{1}{R_A} + \frac{1}{R_B}\right)\right\} & \left(1 - \frac{2d}{R_A}\right) \end{bmatrix}. \tag{5.28}$$

The manipulation of the completed matrix multiplications in the second line of equation (5.28) could be continued to obtain equation (5.27). When the condition for a stable resonator (5.21) is applied to equation (5.28),

$$0 < (1 - d/R_A)(1 - d/R_B) < 1 \tag{5.29}$$

is obtained. The quantities $(1 - d/R_A)$ and $(1 - d/R_B)$ are commonly referred to as the resonator g parameters g_1 and g_2, and the stability condition (5.29) in terms of these quantities becomes

$$0 < g_1 g_2 < 1. \tag{5.30}$$

The familiar stability diagram (shown in Figure 5.7) first used by Boyd and Gardner [2] is based on equation (5.29).

It is possible to include a large amount of information in the stability diagram. Resonators for which the $g_1 g_2$ product is located in the unshaded area are stable, and those in the shaded area are unstable. The straight line from point a ($g_1 = 1, g_2 = 1$) to point b ($g_1 = -1, g_2 = -1$) represents symmetric, stable cavities. The end points of this line, the plane-parallel

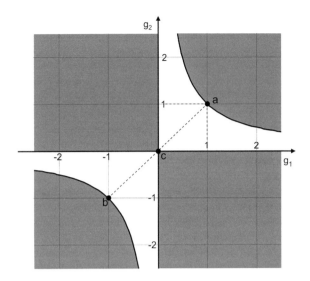

FIGURE 5.7 Resonator stability diagram. The two-mirror resonators for which $g_1 g_2$ falls in the unshaded area are stable. Point a represents the plane-parallel resonator, c a confocal resonator and b a concentric one.

cavity at point a and the concentric cavity at point b, are on the border of the stability region. It is not possible to implement these geometries as stable cavities because an infinite spot size is required at the cavity mirrors. The centre point c ($g_1 = 0, g_2 = 0$) representing the confocal cavity is also on the border of instability. A change from a symmetric geometry could make the confocal resonator unstable. The horizontal and vertical broken lines starting at point a ($g_1 = 1, g_2 = 1$) represent stable resonators with one flat mirror. If a resonator is chosen for stability, it is useful to have $g_1 g_2 \approx 1/2$. Other resonator choices could be based on other considerations such as the desired spot size at specific locations in the resonator. If a resonator is to accommodate a lensing that develops due to thermal loading of the active laser material, a good choice of resonator geometry might be to have the cold-cavity $g_1 g_2$ product near the appropriate end of the stable-symmetric-cavity line of Figure 5.7 and have the thermal lensing bring the cavity towards the other end of the line. It is a geometrical property of stable resonators that the on-axis line from the surface of one mirror to its centre of curvature will partially overlap the corresponding line for the other mirror. If this condition is not met and there is no overlap or one line is completely contained within the other, the cavity is unstable.

5.4 Higher-order Modes of Stable Resonators

Stable lasers often run in multiple stable modes leading to mode beats. For laser operation on multiple longitudinal modes but restricted to a single transverse mode, the mode beating will result in temporal intensity and phase modulation that repeats with the cavity round-trip-transit frequency. If there are multiple transverse modes as well as multiple longitudinal modes, the mode beating becomes more complex. The transverse intensity and phase distribution will fluctuate as well. The average or time-integrated intensity distribution, however, may appear uniform, obscuring the instantaneous spatial structure.

Typically, special apertures or obstructions such as wires are required to force a laser resonator to operate in a single higher-order mode. Modelling of higher-order modes is useful because actual laser output is often closely represented by a superposition of several resonator transverse modes, and therefore, it offers a technique for dealing with actual laser oscillations and beams. Higher-order modes for open stable resonators or free-space propagation usually are derived mathematically by substitution of a trial solution in the paraxial wave (equation 5.1) or Fresnel's paraxial approximation to Huygen's integral (equation 5.2). Depending on the form of trial solution, the result can be equations that describe the Hermite–Gaussian modes for rectangular symmetry or the Laguerre–Gaussian modes for cylindrical symmetry. This procedure determines the scaling factors and relationship between modes of different order. A fundamental-mode spot size $\omega(z)$ with propagation dependence given by equation (5.4) and a wavefront radius of curvature $R(z)$ described by equation (5.5) will carry over unchanged as parameters of both sets of modes. The higher-order modes retain their shape but expand in proportion to $\omega(z)$ with propagation. The Gouy phase shift changes with the order of the modes. The different phase shifts will cause the superposition of a series of modes to synthesize amplitude and phase distributions that change with propagation. Only individual transverse modes are guaranteed to retain their relative distribution with propagation. The more commonly used Hermite–Gaussian modes are described next, followed by the Laguerre–Gaussian modes.

5.4.1 Cartesian Coordinates

The Hermite–Gaussian modes, often called the transverse electromagnetic modes of order m and n or TEM$_{mn}$, have the form

$$E_{mn}(x,y,z,t) = \frac{1}{2} E_{mn0} \sqrt{\frac{\omega_{x0}\omega_{y0}}{\omega_x(z)\omega_y(z)}}$$

$$\times \exp\{-i(kz-\omega t)\} H_m\left(\frac{\sqrt{2}x}{\omega_x(z)}\right) H_n\left(\frac{\sqrt{2}y}{\omega_y(z)}\right)$$

$$\exp\left\{-x^2\left(\frac{1}{\omega_x^2(z)} + \frac{ik}{2R_x(z)}\right) - y^2\left(\frac{1}{\omega_y^2(z)} + \frac{ik}{2R_y(z)}\right) + i\Phi_{mn}(z)\right\}$$

$$+ \text{c.c.}$$

(5.31)

The constant E_{mn0} in the above equation is given by

$$\sqrt{\frac{2P_{mn}}{n\varepsilon_0 c}} \sqrt{\frac{1}{\omega_{x0}\omega_{y0}\pi 2^{m+n-1} m! n!}}$$

(5.32)

where P_{mn} is the power of the TEM$_{mn}$ mode in watts. The Gouy phase shifts of the Hermite–Gaussian modes are given by

$$\Phi_{mn}(z) = (m+1/2)\arctan\{\lambda(z-z_{x0})/(\pi\omega_{x0}^2)\}$$

$$+ (n+1/2)\arctan\{\lambda(z-z_{y0})/(\pi\omega_{y0}^2)\}.$$

(5.33)

Ellipticity and astigmatism are included in equations (5.31)–(5.33) by using separate spot sizes $\omega_x(z)$ and $\omega_y(z)$ and wavefront radii of curvature $R_x(z)$ and $R_y(z)$ for the orthogonal transverse directions. Implicit in the separate parameters for the two transverse directions is the possibility of different beam-waist positions z_{x0} and z_{y0}. The functions H_m and H_n are Hermite polynomials [12] and, for order up to four, are

$$H_0 = 1$$
$$H_1(s) = 2s$$
$$H_3(s) = 8s^3 - 12s$$
$$H_4(s) = 16s^4 - 48s^2 + 12.$$

(5.34)

Higher-order Hermite polynomials can be obtained with the recurrence relation

$$H_{n+1}(s) = 2sH_n(s) - 2nH_{n-1}(s)$$

(5.35)

The Hermite–Gaussian-polynomial orthogonality integrals are

$$\int_{-\infty}^{+\infty} \exp(-s^2) H_n(s) H_m(s) \, ds = 0 \quad (n \neq m)$$

(5.36)

and

$$\int_{-\infty}^{+\infty} \exp(-s^2) H_n^2(s) \, ds = \sqrt{\pi} 2^n n!.$$

(5.37)

The orthogonality integrals permit the evaluation of the constant E_{mn0} in equation (5.31), describing the TEM$_{mn}$ mode amplitude. Another integral that is required shortly is the second moment.

$$\langle s^2 \rangle = \int_{-\infty}^{+\infty} s^2 \exp(-s^2) H_n^2(s) \, ds \Big/ \int_{-\infty}^{+\infty} \exp(-s^2) H_n^2(s) \, ds$$

$$= (2n+1)/2.$$

(5.38)

This is obtained using the recurrence relation and orthogonality integrals.

The Hermite–Gaussian modes form the familiar rectangular array of transverse distribution intensity peaks that are demonstrated in laser outputs with cavity perturbations that favour a single mode. Normalized mode amplitudes and relative intensities are plotted for one dimension in Figure 5.8. The plots are for a beam waist or point where the beam is collimated to simplify the figure.

5.4.2 Cylindrical Coordinates

The modes of a resonator with strict circular symmetry can be described in cylindrical coordinates using generalized Laguerre polynomials. The fundamental mode in cylindrical coordinates is identical to the TEM$_{00}$ Hermite–Gaussian mode with no astigmatism. Higher-order modes are given by

Optical Cavities

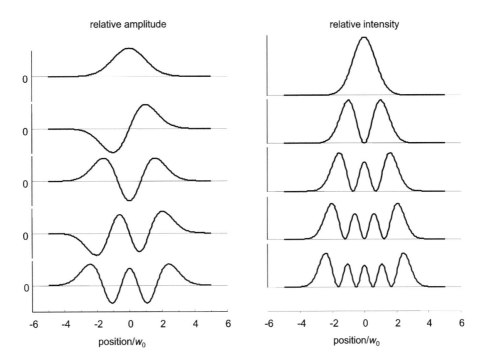

FIGURE 5.8 Relative amplitudes and intensities in one dimension of the five lowest-order Hermite–Gaussian modes.

$$E_{pl}(r,\phi,z,t) = \frac{1}{2} E_{pl0} \frac{\omega_0}{\omega(z)} \exp\{-\mathrm{i}(kz-\omega t)\}\left(\sqrt{2}\,\frac{r}{\omega(z)}\right)^{|l|}$$

$$\times L_p^{|l|}(2r^2/\omega^2(z))\,\mathrm{e}^{-\mathrm{i}l\phi}$$

$$\times \exp\left\{-r^2\left(\frac{1}{\omega^2(z)} + \frac{\mathrm{i}k}{2R(z)}\right) + \Phi_{pl}(z)\right\} + \text{c.c.}$$

(5.39)

where (r, φ) are cylindrical coordinates and $L_p^{|l|}(\rho)$ is a generalized Laguerre polynomial with l an integer or zero and p zero or a positive integer. The parameters ω_0, $\omega(z)$, k, ω and $R(z)$ are the same as those used with the Hermite–Gaussian modes. The first two generalized Laguerre polynomials are

$$L_0^{|l|}(\rho) = 1 \qquad L_0^{|l|}(\rho) = |l|+1-\rho. \tag{5.40}$$

Higher-order generalized Laguerre polynomials can be obtained from the recurrence relation

$$(p+1)L_{p+1}^{|l|}(\rho) = (2p+|l|+1-\rho)L_p^{|l|}(\rho) - (p+|l|)L_{p-1}^{|l|}(\rho). \tag{5.41}$$

The Gouy phase shifts for the cylindrical modes are given by

$$\Phi_{pl} = (2p+|l|+1)\tan^{-1}(\lambda z/\pi\omega_0^2). \tag{5.42}$$

The orthogonality integral for the generalized Laguerre polynomials is

$$\int_0^\infty \exp(-\rho)\rho^{\{(|l_1|+|l_2|)/2\}} L_{p_1}^{|l_1|}(\rho) L_{p_2}^{|l_2|}(\rho)\,\mathrm{d}\rho = \delta_{|l_1||l_2|}\delta_{p_1 p_2}(|l|+p)!/p! \tag{5.43}$$

where the Kronecker delta $\delta_{\alpha\beta}$ is 1 for $\alpha=\beta$ and 0 if $\alpha\neq\beta$. The coefficient E_{pl0} in equation (5.39) is given in terms of P_{pl}, the power of the p, l mode in watts, by

$$E_{pl0} = \sqrt{\frac{2P_{pl}}{n\varepsilon_0 c}\frac{2}{\pi\omega_0}\frac{p!}{(|l|+p)!}}. \tag{5.44}$$

In practice, it is difficult to eliminate astigmatism completely and attain the degree of circular symmetry necessary to produce cylindrical modes.

5.4.3 Beam Quality

Actual resonators will have imperfections that distort a laser oscillation from the ideal Hermite–Gaussian modes, even though the distortion may be small. Laser oscillations may also consist of multiple transverse modes. Siegman's 'M squared' or M^2 parameter has been established as a measure of beam quality [13,14]. M^2 is the ratio of the product of the square root of the second moment of the time-averaged transverse spatial distribution and the square root of the second moment of the angular distribution of a beam and the corresponding value for an ideal Gaussian beam. The transverse spatial distribution second moment or variance changes with propagation, and the minimum values are used in the measurement of M^2. The M^2 value can be specified separately for the x and y coordinates as M_x^2 and M_y^2, or a combined value for the total beam can be specified. The variance for the x coordinate of the fundamental or TEM$_{00}$ Gaussian mode is $\sigma_{x,m=0}^2(z)=\omega_x^2(z)/4$, which has a minimum value at the beam waist of $\sigma_{x,m=0}^2(z_0)=\omega_{x,0}^2/4$. The angular distribution can be obtained by Fourier transform or, equivalently, by propagating to the far field, where the angular half width in the x-direction at the $1/e$ maximum amplitude is $\theta_{x,m=0} = \lambda/(\pi w x, 0)$. It follows that the angular variance of the fundamental Gaussian mode is $\sigma_{\theta x,m=0}^2 = \lambda^2/(4\pi^2\omega_{x,0}^2)$ and the product that defines an M_x^2 of 1 is $\sigma_{x,m=0}(z_0)\sigma_{\theta x,m=0} = \lambda/(4\pi)$. For an arbitrary Hermite–Gaussian mode of order m, n,

$$\sigma_{x,m}(z_0)\sigma_{\theta x,m} = (2m+1)\lambda/(4\pi) \quad \text{and} \quad M_{x,m}^2 = 2m+1. \tag{5.45}$$

Similar expressions apply for the y coordinate.

In numerical calculations, it is common to describe an arbitrary transverse distribution by the field amplitudes at an array of uniformly spaced sampling points. It is also possible to describe a transverse amplitude distribution as the sum of the orthogonal modes. The summation over a set of normalized Hermite–Gaussian modes can be expressed as

$$E(x,y,z,t) = \sum_m \sum_n c_{mn} E_{mn}(x,y,z,t) \tag{5.46}$$

where the c_{mn} are the complex amplitudes of the normalized modes $E_{mn}(x, y, z, t)$. If the phases of the modes are random, a statistical average will yield

$$M_x^2 = \sum_m \sum_n (2m+1)|c_{mn}|^2 \Big/ \sum_m \sum_n |c_{mn}|^2 \tag{5.47}$$

and

$$M_y^2 = \sum_m \sum_n (2n+1)|c_{mn}|^2 \Big/ \sum_m \sum_n |c_{mn}|^2. \tag{5.48}$$

When Fourier transform techniques are used to describe beam propagation, the transverse amplitude distribution is typically described by its value at a rectangular array of $N_1 \times N_2$ discrete sampling points:

$$E_{n_1,n_2} = E(x_{n_1}, y_{n_2}, z, t) = E(n_1 \Delta x/N_1, n_2 \Delta y/N_2, z, t). \tag{5.49}$$

The indices n_1 and n_2 are integers and range in value: $0 \leq n_1 < N_1$ and $0 \leq n_2 < N_2$. The array size is Δx in the x-direction and Δy in the y-direction. A discrete Fourier transform pair such as [15,16]

$$\tilde{E}_{q_1,q_2} = \sum_{n_1}^{N_1-1} \sum_{n_2}^{N_2-1} \exp\{-i2\pi(q_1 n_1/N_1 + q_2 n_2/N_2)\} E_{n_1,n_2} \tag{5.50}$$

$$\tilde{E}_{n_1,n_2} = (1/N_1 N_2) \sum_{q_1}^{N_1-1} \sum_{q_2}^{N_2-1} \exp\{i2\pi(q_1 n_1/N_1 + q_2 n_2/N_2)\} \tilde{E}_{q_1,q_2} \tag{5.51}$$

is used to change from the spatial representation to the angular representation and back again. In the angular representation, the individual components propagate at angles to the yz plane given by

$$\theta_{x,q_1} = \lambda q_1/\Delta x \quad \text{when } 0 \leq q_1 < N_1/2,$$

and $\quad \theta_{x,q_1} = \lambda(q_1 - N_1)/\Delta x \quad \text{when } N_1/2 \leq q_1 < N. \tag{5.52}$

A similar expression describes the angle with respect to the xz plane. The angular variance in the x-direction is

$$\sigma_{\theta x}^2 = \sum_{q_1}^{N_1-1} \sum_{q_2}^{N_2-1} (\theta_{x,q_l} - \bar{\theta}_x)^2 \tilde{E}_{q_1,q_2} \tilde{E}_{q_1,q_2}^* \Big/ \sum_{q_1}^{N_1-1} \sum_{q_2}^{N_2-1} \tilde{E}_{q_1,q_2} \tilde{E}_{q_1,q_2}^*. \tag{5.53}$$

The x spatial variance is

$$\sigma_x^2 = \sum_{n_1}^{N_1-1} \sum_{n_2}^{N_2-1} \left(\frac{n_1 \Delta x}{N_1} - \bar{x}\right)^2 E_{n_1,n_2} E_{n_1,n_2}^* \Big/ \sum_{n_1}^{N_1-1} \sum_{n_2}^{N_2-1} E_{n_1,n_2} E_{n_1,n_2}^*. \tag{5.54}$$

It is necessary to propagate to a z position where the spatial variance is minimum to obtain M^2 or, equivalently, to remove the spherical curvature [17] from the wave front before calculation of the spatial and angular variances. When this is done, the M^2 values are given by

$$M_x^2 = 4\pi \sigma_{\theta x} \sigma_{x,\min}/\lambda \quad \text{and} \quad M_y^2 = 4\pi \sigma_{\theta y} \sigma_{y,\min}/\lambda. \tag{5.55}$$

There are as many ways to obtain the M^2 of an actual beam experimentally as there are for numerically modelled beams. One experimental technique involves estimating a Gaussian spot size W_n of the beam at many positions z_n along the beam and fitting these to functions of the form:

$$W_x^2(z) = M_x^2 \{\omega_{x,0}^2 + (z - z_{x,0})^2 (\lambda/\pi \omega_{x,0})^2\}$$
$$= W_{x,0}^2 \{1 + M_x^4 (z - z_{x,0})^2 (\lambda/\pi W_{x,0}^2 s)^2\} \tag{5.56}$$

to obtain the M^2 value, the embedded Gaussian beam-waist spot size w_0 and the beam-waist position z_0. The capital letter W is used to indicate the spot of an actual beam, which is larger than the spot size of the embedded Gaussian beam. In performing the least-squares fit to a set of experimental data, it is helpful to weight the individual measurements of W^2 by $1/W^2$ to place more significance on the measurements near the beam waist and reduce the possibility of a fit that predicts meaningless negative values of W^2. The spot sizes W_n may be estimated by taking the difference of knife-edge positions that transmit 16% and 84% of the total beam. A refinement could involve obtaining best-fit Gaussian distributions from several knife-edge positions at each of many propagation distances, z_n. Scanning pinholes, scanning slits and array detectors also can be used to measure beam distributions and second moments. It is necessary to consider the properties of the measurement and application. For example, a small amount of energy or error in measuring at a large distance from the central lobe of a beam will increase M^2. However, the small amount of energy at a large distance from the central lobe may or may not be significant in the application. The M^2 parameter provides both a measure of beam quality and a mechanism to use Gaussian beam propagation methods to deal with the propagation of actual laser beams.

5.5 Mode-Matching

In many applications it is necessary to be precise in the spatial distribution of a laser beam delivered on a target. Knowledge of the beam's intensity distribution at the target is obtained from the beam power and the beam propagation parameters. The size and beam quality parameters are critically important when coupling laser beams into optical fibres. It is desirable to match the transverse distribution of a pump beam to the mode size of a resonator for laser-pumped lasers. In some applications, mode-matching is extended to matching the confocal parameters of a pump beam and an external resonant cavity. With a single lens or spherical mirror, it is possible to control either the size of the beam waist or its location. Two lenses adjustable in position are required to control both the beam-waist position and size simultaneously.

Configurations for single-lens mode-matching with an ideal Gaussian beam are described next. The application of the beam quality parameter and the concept of an embedded Gaussian beam allow these results to be extended to actual beams that are far from perfect. The results for single-lens mode-matching can be cascaded to make predictions for two-lens mode-matching.

5.5.1 One-lens Approach

The standard approach is to consider the ABCD matrix that results when starting with a Gaussian beam waist of spot size $w_{0,1}$, propagating over a distance d_1 to a lens of focal length f and finally propagating an additional distance d_2 to the new beam waist of spot size $w_{0,2}$ formed by the lens (Figure 5.9). The ABCD law (5.22) is applied to the complex beam parameter (5.10). The real and imaginary parts of the resulting equation are separated to yield [11]

$$(d_1 - f) b_2 = (d_2 - f) b_1 \tag{5.57}$$

and

$$(d_1 - f)(d_2 - f) = f^2 - b_1 b_2 / 4. \tag{5.58}$$

Here, $b_1 = 2\pi\omega_{0,1}^2/\lambda$ and $b_2 = 2\pi\omega_{0,2}^2/\lambda$ are the confocal parameters of the beam of wavelength λ before and after the lens. The quantity $b_1 b_2 / 4$ is sometimes labelled f_0^2. It is necessary that $f^2 \geq b_1 b_2 / 4 = f_0^2$ for there to exist distances d_1 and d_2 that will yield a confocal parameter b_2 from an initial beam with confocal parameter b_1; that is, the absolute value of the lens' focal length must be longer than a minimum value. In this case, when $\omega_{0,1}$, $\omega_{0,2}$ and f are specified, the distance from the first waist to the lens and the distance from the lens to the second waist are given by

$$d_1 = f \pm (\omega_{0,1}/\omega_{0,2})\sqrt{f^2 - b_1 b_2 / 4} \tag{5.59}$$

and

$$d_2 = f \pm (\omega_{0,2}/\omega_{0,2})\sqrt{f^2 - b_1 b_2 / 4}. \tag{5.60}$$

Here, either + or − signs should be used in both equations (5.59) and (5.60).

Another set of useful equations comes from combining the last two equations. What is the focal length of the required lens when the beam waists $\omega_{0,1}$ and $\omega_{0,2}$ and their separation $d = d_1 + d_2$ are specified? This will aid in the choice of a lens from available focal lengths to provide approximately the desired waist separation. The resulting equation is quadratic in f:

$$\{4 - (\omega_{0,1}^2 + \omega_{0,2}^2)^2 / \omega_{0,1}^2 \omega_{0,2}^2\}f^2 - 4df + (\omega_{0,1}^2 + \omega_{0,2}^2)^2/4 + d^2 = 0 \tag{5.61}$$

which has either two or no real solutions. Even with real solutions, it is necessary to check that the lens positions are physically realizable, e.g. not beyond the target. The position of a lens of focal length f determined in equation (5.61) is

$$d_1 = (\omega_{0,1}^2(d - f) + \omega_{0,2}^2 f)/(\omega_{0,1}^2 + \omega_{0,2}^2). \tag{5.62}$$

5.5.2 Two-lens Mode-matching

One way to approach calculations of two-lens mode-matching is to step through a series of lens positions until the desired mode-matching is found. Each iteration involves a different placement of the first lens, for which the position of the waist and the confocal parameter of the beam formed by the first lens of focal length f_1 are given by

$$d_b = f_1\{b_1^2 + 4(d_a - f_1)d_a\}/\{b_1^2 + 4(d_a - f_1)^2\} \tag{5.63}$$

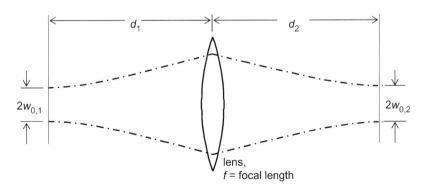

FIGURE 5.9 Parameters for mode-matching with one lens.

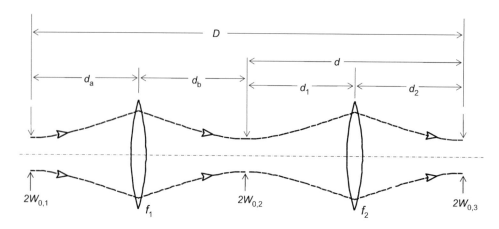

FIGURE 5.10 Mode-matching with two lenses may be necessary when there is a fixed distance D between the initial and final beam waists.

and

$$b_2 = b_1(d_b - f_1)/(d_a - f_1). \quad (5.64)$$

This leaves a known distance d from the position of the waist formed by the first lens to the desired position of the final waist. The second lens is then numerically placed at a position that produces a third beam waist at the desired position. The numerical analysis is stepped through a range of positions for the first lens. The parameters are illustrated in Figure 5.10. A cubic equation is obtained for the position of the second lens when the distance between beam waists $d(=d_1+d_2)$, $w_{0,2}$, the size of the intermediate beam waist and f_2, the focal length of the second lens, are specified:

$$d_2^3 - (f_2 + d)d_2^2 + (2f_2 d + b_2^2/4)d_2 - (d - f_2)b_2^2/4 - f_2^2 d = 0 \quad (5.65)$$

There are either one or three real roots to equation (5.65) and the confocal parameter of each of the resulting beams is given by

$$b_3 = b_2(f_2 - d_2)/(f_2 - d_1) \quad (5.66)$$

Single-lens mode-matching is simpler than two-lens mode-matching. In cases where the distance between the initial and final beam waists can be adjusted, a single lens will work well. However, if that distance is fixed, two-lens mode-matching may be required. It is possible to add two more adjustable parameters by tilting the lenses, but any astigmatism or ellipticity in the beam might best be removed before mode-matching.

5.6 Plane Parallel Resonators

The analysis of ideal stable resonators provides a useful background for understanding practical laser resonators including plane-parallel-mirror resonators and unstable resonators. We begin with plane-parallel resonators in this section and continue with unstable resonators in the next section. Plane-parallel resonators are characterized by their Fresnel number:

$$N_F = a^2/(L\lambda) \quad (5.67)$$

where a is the radius of the limiting aperture of the resonator, L is the propagation distance from one encounter of the limiting aperture to the next and λ is the wavelength of the resonated light. The limiting aperture could be a laser rod, a cavity mirror or an actual aperture placed in the resonator. Resonators with the same Fresnel number will have equivalent diffraction properties. A circular aperture is used here, and the resonator has circular symmetry. The circular symmetry is lost, however, when the Fourier components of the initial amplitude distribution are given a random phase.

The diffraction for various apertures and beams provides some insight into the significance of the Fresnel number. In many cases, there is a diffraction spread of a collimated beam that is approximately λ/a rad. For example, the diffraction half-angle of a Gaussian beam is $\lambda/(\pi w_0)$. The centre to first minimum angle of the Airy diffraction pattern of a uniformly illuminated circular aperture is $1.22\lambda/(2a)$, and the full width at half maximum of the diffraction pattern from a slit of width $2a$ illuminated by a plane wave is $1.39\lambda/(\pi a)$. With a Fresnel number of $N_F = 1$, a nearly collimated beam will spread by diffraction in a resonator length L to slightly overfill the aperture for significant loss on the next encounter with the aperture. Numerical calculations show that a resonator with a Fresnel number of 1 will have about 18% loss for each cavity transit from aperture to aperture, whereas the loss will be approximately 0.88% for a Fresnel number of 10. The Fresnel number is a useful and general concept. For example, the N_F of a stable resonator is approximately the number of Hermite–Gaussian modes that the resonator will support.

Iterative computer techniques are commonly used to determine the cavity modes of plane-parallel resonators. In the original paper by Fox and Li [1], iterative solutions to Huygen's integral in cylindrical symmetry starting with a uniform plane wave were used. It is more common now to use Hankel transform techniques for cylindrical symmetry and Fourier transform techniques for rectangular symmetry [15,16]. A discrete Fourier transform pair such as equations (5.50) and (5.51) is used. The spatial amplitude and phase distribution transmitted through the aperture are transformed into a sum of plane waves propagating at regularly spaced directions with respect

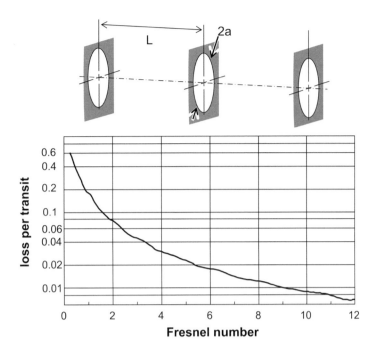

FIGURE 5.11 Calculated loss per transit in a plane-parallel resonator with circular mirrors as a function of Fresnel number $N_F = a^2/(L\lambda)$.

to the central direction of propagation. The propagation of the plane waves to the next encounter with the aperture is straightforward, each having a relative phase shift dependent on the direction of propagation. The plane waves are then summed to synthesize the spatial distribution, which is modified by transmission through the aperture. Routines for calculating the Fourier transformations are available [18,19].

Calculated diffraction losses as a function of Fresnel number for plane-parallel resonators of circular symmetry are shown in Figure 5.11. Starting with a uniform distribution and propagating through 300 cavity transits yielded the plotted values. The power in the beam was re-normalized after each cavity transit. The plotted line shows transmission on the 300th transit. The process is more slowly converging for Fresnel numbers greater than ten, typically requiring more than 300 transits to converge. Virtual source techniques [20] are useful for the analysis of resonators with large Fresnel numbers.

A modification of the Fox and Li iterative cavity transit technique was used for the calculations illustrated in Figures 5.12–5.14. A two-dimensional FFT technique was used to calculate beam propagation. In this case, the initial spatial frequency or angular components all had equal power but the phases were random. A Fresnel number of 10 was used for the calculation. An initial intensity distribution transmitted through an aperture is shown in Figure 5.12a. This is intended to simulate a laser oscillation that is growing out of zero-point quantum fluctuations. The initial distribution is limited by the number of spatial sampling points used in the calculation, and the distribution changes for a new calculation with different random phases. After each transit, the intensity was again re-normalized. After 10 cavity transits (Figure 5.12b), the high spatial frequencies are greatly attenuated but the beam is still strongly structured with an M^2 of 9.5. Spatial filtering continues with fewer high-frequency components and an M^2 of 5.3 after 20 transits (Figure 5.12c). Many high-gain Q-switched lasers reach their peak output power in 10 or 20 cavity transits, and this type of distribution could be present in an instantaneous sampling of the output. Averaging over the total Q-switched pulse could give the appearance of a uniform intensity beam with an M^2 between 5 and 10. The loss per transit behaviour and the evolution of M^2 during the iterative calculation are illustrated in Figure 5.13.

After several hundred cavity transits, the intensity and phase distribution become constant, as shown in Figure 5.14. At this point, the beam has an M^2 of 1.5 and the intensity loss per transit is 0.88%. In practice, such a distribution would be difficult to obtain, even with hundreds of cavity transits, due to the sensitivity of the plane-parallel resonator to misalignment. Injection seeding of a field distribution close to that of the fundamental mode of the resonator would quickly produce a dominant oscillation of that mode in the resonator. The use of injection seeding, however, is more commonly used to achieve single-frequency oscillation in resonators with large gain. Selection of the fundamental transverse mode is easily achieved in unstable resonators.

5.7 Unstable Resonators

Unstable resonators offer advantages of good energy extraction efficiency from a large-volume active laser material and reasonably good spatial beam quality in the laser output. Typically, unstable resonators have large loss or output coupling that must be offset by high laser gain. The high-gain requirement usually restricts operation to the pulsed mode because of the difficulty in maintaining high can gain. Frequency control and narrow spectral bandwidth operation are more difficult with high laser gain. Unstable resonators have transverse modes that reproduce from one cavity transit to the next. These modes, however, are not orthogonal or even nearly orthogonal

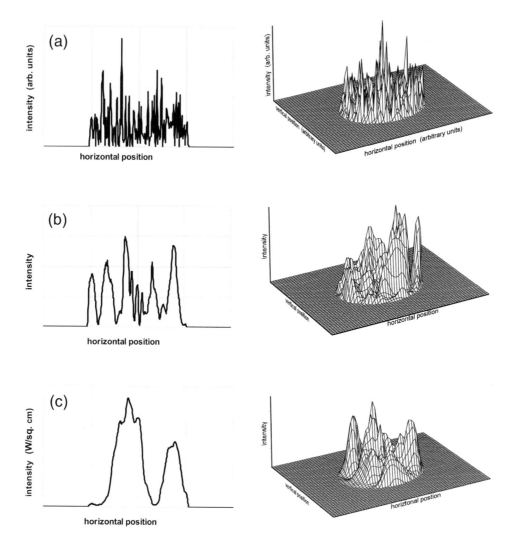

FIGURE 5.12 Intensity on a trace through the centre of the transverse distribution of a beam propagating in a plane-parallel circular-mirror resonator of Fresnel number 10. The calculation starts with randomly phased Fourier components (a) and is shown after 10 transits (b) and 20 transits (c).

as in the case of stable resonators. The non-orthogonality of the modes leads to some perhaps surprising effects, such as a mode is most efficiently seeded with a conjugate beam; that is, the backward propagating beam, converging where the output beam was diverging, will most effectively seed the oscillation. Transverse mode discrimination is usually large, and oscillations usually resolve to a single transverse mode in a small number of cavity transits.

5.7.1 Hard-Edged Apertures

Unstable resonators can be classified as either hard- or soft-edged. In a hard-edged resonator, output coupling is typically by expansion of the resonated beam beyond the edge of some limiting aperture. This could be a hole cut in the centre of a mirror used for output coupling, a high-reflectivity mirror smaller than the expanded beam or a high-reflectivity spot on a substrate used to reflect and output couple the resonated beam. Soft-edged resonators can be created with apodized apertures, variable reflectivity mirrors (VRMs) and gain guiding such as is common in laser-pumped lasers. Unstable resonators are

also classified as either negative or positive branch according to the sign of the product $g_1 g_2 = (1-d/R_A)(1-d/R_B)$. Two-mirror negative-branch unstable resonators have a beam focus inside the resonator, whereas a two-mirror positive-branch resonator can have either none or two. The absence of an intra-cavity beam focus is an advantage for high-power laser oscillations. Confocal unstable resonators are useful because the resonated beam is collimated in the output part of the resonator round trip [21]. This can be useful in producing a collimated output or region of collimated propagation inside the resonator.

Unstable resonators are usually analysed on two levels. Geometrical analysis of unstable resonators provides a fair first approximation. Some soft-edge unstable resonators can be analysed using Gaussian optics and an extension of the ABCD matrix techniques to include complex matrix elements. Numerical beam propagation calculations, however, are usually required for a detailed understanding of the diffraction effects. Fourier transform techniques are commonly used to perform these numerical calculations.

When equation (5.24), $q = (Aq+B)/(Cq+D)$, is solved for $1/q$ in the case of an unstable resonator, the result is real:

Optical Cavities

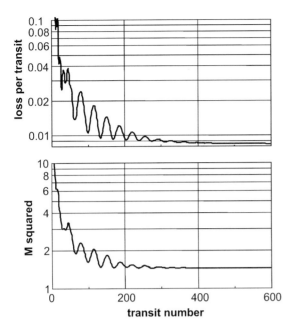

FIGURE 5.13 Both loss per transit and M^2 become smaller on approaching limiting values in the calculation of beam development. The results for a calculation starting with randomly phased Fourier components in a plane-parallel resonator of $N_F = 10$ are shown.

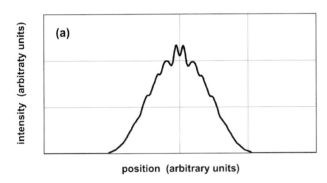

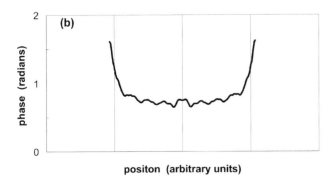

FIGURE 5.14 Intensity and phase on a trace through the centre of the transverse distribution. This is the calculated distribution of the same Fresnel number 10 resonator used in Figures 5.12 and 5.13 after several hundred transits.

$$\frac{1}{q} = \frac{D-A}{2B} \pm \frac{1}{B}\sqrt{\left(\frac{A+D}{2}\right)^2 - 1} = \frac{1}{R}. \qquad (5.68)$$

This result gives a non-physical interpretation of spherical waves since $1/\omega^2 = 0$. When the accurate solutions with limiting apertures and diffraction are considered, it is found that the two solutions for R represent a stable solution of a divergent beam and an unstable solution for a convergent beam. The convergent beam becomes smaller only for a few cavity transits until diffraction begins to dominate and it quickly changes to become a divergent beam.

Geometrical magnification of the unstable resonator is also obtained from the ABCD matrices. For positive-branch unstable resonators, $m = (A+B)/2 > 1$; the values for magnification are

$$M = m + \sqrt{m^2 - 1} \qquad (5.69a)$$

for the expanding beam and

$$1/M = m - \sqrt{m^2 - 1} \qquad (5.69b)$$

for the convergent beam. For negative-branch unstable resonators, $m = (A+B)/2 < -1$; the values for magnification are

$$M = m - \sqrt{m^2 - 1} < -1 \qquad (5.70a)$$

and

$$-1 < 1/M = m + \sqrt{m^2 - 1} < 0. \qquad (5.70b)$$

A parameter important for characterization of hard-edged unstable resonators is the equivalent Fresnel number:

$$N_{eq} = \left(\frac{M^2 - 1}{MB}\right)\frac{a^2}{2\lambda}. \qquad (5.71)$$

Here, M is the magnification of the resonator as given earlier; B is the component of the ABCD matrix of the resonator; a is the radius of the limiting circular aperture and λ is the wavelength of the light circulating in the resonator.

We now restrict the discussion to an illustrative example of a positive-branch, confocal, unstable resonator as shown in Figure 5.1b. This resonator is formed by a concave mirror of radius of curvature $R_A > 0$ and a convex mirror of radius of curvature $R_B < 0$ separated by distance d. The output mirror has a central highly reflecting spot of radius a and is transmitting for a radius greater than a. Such an output mirror could be a small suspended mirror, as shown, or a reflecting spot deposited on a meniscus substrate to preserve the collimation of the output beam. The confocal property of the resonator specifies that $R_A = 2d - R_B$. In the special case of a confocal unstable resonator, the magnification is given by $M = -R_A/R_B$ and the equivalent Fresnel number is $N_{eq} = a^2/(\lambda|R_B|)$. The magnification is a positive value because $R_B < 0$.

A calculation of loss per round-trip transit as a function of N_{eq} for a positive-branch confocal resonator with magnification $M = 3$ is shown in Figure 5.15. At larger values of N_{eq}, the loss converges on the geometrical value of $(1-1/M^2) = 0.89$. The broken line in Figure 5.15 represents this value. At smaller values, loss is minimum at half-integer values of N_{eq}, a

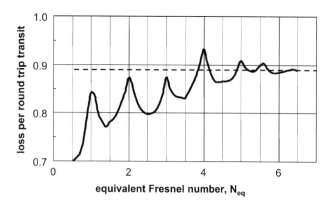

FIGURE 5.15 Loss as a function of equivalent Fresnel number for a positive-branch, confocal, unstable resonator with magnification $M = 3$. The broken line is the geometrical value, and the full curve is obtained with a diffraction calculation.

feature common to hard-edged unstable resonators. This feature is due to the transverse-mode properties of the resonator. At half-integer values of N_{eq}, the loss difference between modes is large, and the oscillation of a single transverse mode dominates after a few cavity transits. At integer values of N_{eq}, there are two transverse modes of nearly equal loss and many cavity transits are required to obtain a reproducible loss and transverse field distribution.

The initial distribution and distribution after eight transits for a resonator with magnification $M = 3$ and equivalent Fresnel number $N_{eq} = 5.5$ (Figure 5.16) illustrate that the lowest loss mode can be resolved quickly both in calculations and in the actual build-up of oscillation in an unstable resonator laser. The calculation used to generate Figure 5.16 used random phasing of the initial Fourier components all of equal power. The total power was normalized on each cavity transit, preserving the phase and relative intensity of the individual components in the method of the Fox and Li calculation. The lowest loss distribution of this resonator is numerically propagated outside the cavity in Figure 5.17. In a distance equal to four times the cavity length, a strong spot of Arago develops in the centre of the beam. The development of the spot of Arago, with a peak intensity many times that of the other portions of the beam, is a disadvantage of hard-edged unstable resonators.

A short description of the numerical techniques used in the calculations is appropriate. A direct application of Fourier beam propagation methods to the unstable resonator would require prohibitively large arrays of sampling points to handle the diverging portions of beam propagation in the cavity. This problem is avoided with a simple transformation that reduces the problem of calculating the propagation of the collimated beam. The technique is illustrated with propagation in a Galilean telescope equivalent to reflections from the convex mirror followed by reflection from the concave mirror in our resonator. The beam, just before the convex mirror, is first magnified by setting

$$E_{mag}(x, y) = (1/M)E_{in}(x/M, y/M) \quad (5.72)$$

where E_{in} is the incident electric field, E_{mag} is the magnified field, $M = -R_A/R_B$ is the magnification and the factor $1/M$ is needed for conservation of energy. Next, the field is propagated over an effective distance of the magnification times the mirror separation:

$$d_{eff} = M \times d. \quad (5.73)$$

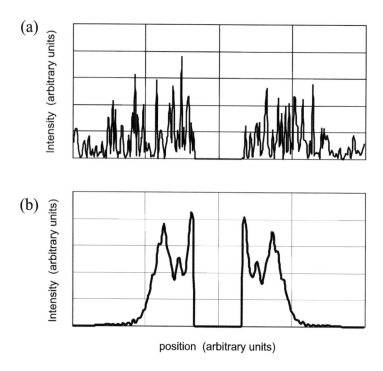

FIGURE 5.16 Calculated development of transverse intensity distribution in a positive-branch, confocal, unstable resonator with equivalent Fresnel number, $N_{eq} = 5.5$, and magnification $M = 3$: (a) initial distribution; (b) After eight cavity transits the distribution is close to the final form.

Optical Cavities

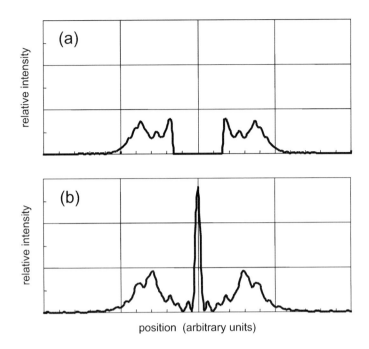

FIGURE 5.17 Development of the spot of Arago for the distribution described in Figure 5.16: (a) at resonator output; (b) four cavity lengths from the output mirror.

Finally, it is necessary to retrieve the spacing of the original array sampling points by some method such as interpolation using first, second and cross-derivatives. Siegman [3] provides a justification of this transformation based on Fermat's principle. The technique is more general than that described here and can be applied to an open optical system described by an ABCD matrix.

The value of the calculated beam quality parameter M^2 is dependent on the sharpness of the edge of the aperture and on the number of sampling points used in the calculation. For example, the distributions shown in Figures 5.16 and 5.17 were obtained using a rather sharp edge of intensity reflectivity on the output mirror given by $R_M = \exp\{-(r^2/a^2)^{64}\}$, and the value $M^2 = 7.2$ was obtained for the $N_{eq} = 5.5$ resonator. When the edge is softened to $R = \exp\{-(r^2/a^2)^{32}\}$, $M^2 = 5.6$ results. In the limit of an infinitely sharp edge, the paraxial approximation will fail. The number of sampling points also limits the resolution of the sharpness of the edge. The improvement of beam quality with softening of the reflector or aperture edge leads to VRMs.

5.7.2 Soft-edged Apertures

An example of a VRM with an intensity reflectivity of $R = 0.34 \exp(-r^2/a^2)$ is described here. The magnification of the resonator is reduced to 1.75 to yield a loss per transit of 0.89, the same as the resonator with the sharp-edged reflector discussed earlier. The reflectivity profiles of the hard-edged reflectors and the variable reflector used in these examples are shown in Figure 5.18. A positive-branch confocal resonator with an equivalent Fresnel number of $N_{eq} = 2.3$ is used with the VRM and a magnification of $M = 1.75$. There is nothing unique about these values. They were chosen for comparison because the cavity transit losses were the same as the hard-edged aperture

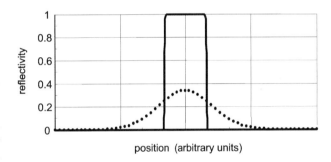

FIGURE 5.18 Reflectivity distributions used for hard-aperture (full line) and soft-aperture (dotted curve) calculations.

example and the output distribution was reasonable. Again, the phases of the Fourier components are initially random and the initial amplitudes are equal; the total power is renormalized after each cavity transit. After only eight resonator transits, the calculated intensity distribution is near the final form with a beam quality of $M^2 = 1.4$ (Figure 5.19). Good beam quality can develop in relatively few cavity transits in such a resonator.

These examples are for ideal conditions. An actual laser resonator would have a gain saturation that would make the distribution more 'top-hat' like and less like a Gaussian distribution. Laser-pumped laser resonators can also produce good beam quality. Some laser-pumped lasers have a transverse gain distribution that is nearly Gaussian.

5.8 Distortion Effects

There are a number of practical problems in actual lasers that lead to beam distortion. Heat deposition in the laser gain medium resulting from pumping can cause thermal lensing

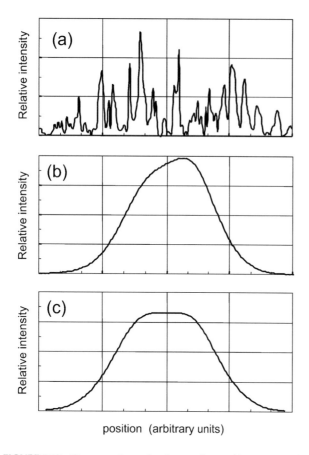

FIGURE 5.19 Transverse beam development in unstable resonator with VRM shown in Figure 5.18 (a) after first resonator transit, (b) after eight transits and (c) after 20 transits.

and thermally induced stress birefringence. Master oscillator power amplifier techniques may be useful. A high-quality beam is generated in a low-power oscillator where beam quality is more easily controlled. Power amplification is then obtained with a single or double pass through a high-power laser amplifier. This avoids multiple passes in a high-power resonator where greater beam distortion could accumulate. A technique that is finding wide use in these systems is phase-conjugate reflection. After the first pass through a laser amplifier, the beam acquires sufficient intensity that relatively efficient phase-conjugate reflection in a Brillouin cell is possible. The reflection of the conjugate mirror traverses the path through the laser amplifier in reverse, cancelling the distortion acquired on the first transit. These topics are beyond the scope of this discussion. Next, we turn to an overview of axial modes and temporal properties of laser resonators.

5.9 Axial Modes

5.9.1 Stable-resonator Axial-mode Spectral Separation

Up to this point, we have only discussed the phase difference between transverse resonator modes and plane waves propagating in the same direction. For a discussion of axial modes, it is necessary to add the additional constraint that the resonator mode must reproduce itself both in amplitude and phase except for uniform amplification or attenuation. This means the wavelength of a resonator mode must satisfy the condition that the optical length of a round-trip cavity transit is an integer multiple of that wavelength with adjustment for Gouy and other possible phase shifts.

Often it is the case that the optical length of a resonator is not known precisely on the scale of a fraction of a wavelength of light. This is commonly the case for resonators used as interferometers to determine the relative spectral distribution of an optical beam. It may be sufficient to consider only the spectral spacing of modes. For example, the free spectral range of fundamental modes of an open, stable resonator is given to a high degree of accuracy by

$$\Delta\tilde{v}_{FSR} = \Delta\lambda_{FSR}/\lambda^2 = 1/\lambda_{n+1} - 1/\lambda_n \approx 1/(2d). \quad (5.74)$$

Here, λ_{n+1} and λ_n are the wavelengths of adjacent modes. The free spectral range is given as a wavenumber separation $\Delta\tilde{v}_{FSR}$ and as a wavelength separation $\Delta\lambda_{FSR}$ in equation (5.74). The mirror separation of the resonator is d, and λ is the central or average wavelength. A symmetric confocal interferometer is commonly used in the spectral analysis of laser beams. In typical use, there are four reflections before the beam path inside the resonator is closed (Figure 5.20), and the free spectral range is $\Delta\tilde{v}_{FSR} = 1/(4d)$. Usually, the accuracy of these expressions is limited by the precision to which the mirror separation is known.

The optical length is the integral of the refractive index over the round-trip transit path followed by the centre of the transverse field distribution:

$$\text{optical length} = \oint_{\text{round trip}} n(\lambda, z) \, dz. \quad (5.75)$$

The condition that the phase of a resonator mode change by an integer multiple on a cavity round trip is

$$\frac{\text{optical length}}{\lambda} - \frac{\Phi}{2\pi} = \text{integer} \quad (5.76)$$

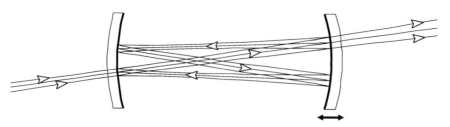

FIGURE 5.20 Schematic representation of the light path in a confocal interferometer.

where Φ represents the sum of the Gouy phase shifts on the complete cavity transit. The axial or longitudinal cavity modes associated with a transverse mode will be nearly equally spaced in wavenumber, but the spacing will not be exactly equal because of the dispersion of the refractive index. Transverse modes of different order will also usually be spectrally positioned between the fundamental transverse modes. Secondary changes such as additional wavelength-dependent phase shifts in optical components and wavelength-dependent path differences must be considered in critical applications such as the propagation of femtosecond-duration pulses.

The spectral width of a resonator mode depends on several factors. Losses typically determine the spectral width in a resonator used passively as a multi-pass interferometer. It is useful to consider an ideal case of a two-mirror stable resonator with identical mirror reflectivity R_m for lossless mirrors that have transmissions and reflections that sum to one: $T_m + R_m = 1$. The transmission of a monochromatic Gaussian beam mode matched to the resonator is

$$\frac{P_t}{P_i} = \frac{1}{1 + 4R_m \sin^2(\phi/2)/(1-R_m)^2}. \quad (5.77)$$

Here, P_i is the incident power in the beam, P_t the transmitted power and ϕ is the phase shift encountered on one round-trip transit of the resonator. When the reflectivity is high, the shape of the interferometer transmission peaks can be obtained by using the approximation $\sin(\Delta\phi/2) \approx \Delta\phi/2$, where $\Delta\phi$ is the small difference between ϕ and a multiple of 2π, and ϕ is nearly equal to a multiple of 2π. In this approximation, the full width at half maximum of the resonance is given by

$$\phi_{FWHM} = 2(1-R_m)/\sqrt{R_m}. \quad (5.78)$$

The finesse of the resonator is the ratio of the free spectral range divided by the resonance width:

$$\text{finesse} = 2\pi/\phi_{FWHM} = \pi\sqrt{R_m}/(1-R_m). \quad (5.79)$$

For less than ideal cavity mirrors, the finesse can be stated as a function of total resonator loss

$$\text{finesse} \approx \pi/(1-\text{loss}/2). \quad (5.80)$$

Such resonators, when illuminated by a single-frequency laser, can provide accurate measurements of low levels of loss in components inserted in the resonator.

5.10 Frequency Selection and Frequency Stability

The spectral properties of laser resonators are determined by the laser gain medium and the temporal character of the laser as well as by the resonator. The gain bandwidth of the active laser medium can be many times wider than the free spectral range of the resonator and support many longitudinal modes. As the laser oscillations build up, there is a frequency narrowing or frequency selection similar to transverse mode selection. High-gain pulsed lasers, however, typically have insufficient time to resolve single-mode operation. Effects such as spectral and spatial 'hole burning' can also limit frequency selection.

In mode-locked lasers, many modes are locked in phase to synthesize a single short pulse. The time–bandwidth limit specifies the minimum pulse width attainable for a given bandwidth or the minimum bandwidth required to support a given pulse duration. The minimum time–bandwidth product is obtained for pulses that have a Gaussian shape in time with no further amplitude or phase modulation. Frequently, the time–bandwidth product is given in terms of pulse width in seconds, Δt_{FWHM}, and spectral width, $\Delta \upsilon_{FWHM}$, in hertz:

$$\left(\Delta t_{FWHM} \Delta \upsilon_{FWHM}\right)_{min} = 0.44. \quad (5.81)$$

In terms of wavenumber, it is $\left(\Delta t_{FWHM} \Delta \tilde{\upsilon}_{FWHM}\right)_{min} = 14.72 \text{ ps cm}^{-1}$, and in terms of wavelength,

$$\left(\Delta t_{FWHM} \Delta \lambda_{FWHM}/\lambda^2\right)_{min} = 14.72 \times 10^{-6} \text{ ps nm}^{-1}$$

where λ is the central wavelength. The time–bandwidth product can also be stated as an uncertainty principle $(\Delta t_\sigma \Delta E_\sigma)_{min} = \hbar/2$ given in terms of the variance.

The properties of the laser gain have an effect on frequency selection. When the gain is inhomogeneously broadened, gain in a narrow spectral region can be depleted without depleting the gain in neighbouring regions. An example is a gas discharge where gain broadening is due to Doppler shifting in a distribution of velocities. Multiple modes can oscillate, each drawing gain from a different velocity population. The familiar helium–neon laser exhibits this type of behaviour with typically two or three modes oscillating simultaneously. The depletion of gain in a narrow spectral region of an inhomogeneously broadened laser is called spectral hole burning. It is also possible to have spatial hole burning in a standing-wave laser resonator. The gain is not depleted at the nodes of the standing-wave laser oscillation and remains to provide gain for modes of a different frequency with different node locations. Spatial hole burning can occur with either inhomogeneous or homogeneous gain broadening. When the laser gain is homogeneously broadened, energy extraction in a narrow spectral region will uniformly decrease gain over the entire gain bandwidth. Gain broadening by lifetime-limiting collisions or thermal vibrations in solids are examples of homogeneous broadening.

Techniques for frequency selection include the addition of dispersive elements such as prisms and gratings in the resonator. Etalons within the resonator are frequently used for gain narrowing as are multiple-element birefringent filters. Individual elements of birefringent filters are wavelength-dependent high-order waveplates providing a retardation between orthogonal polarizations. These elements are placed between polarizers or Brewster-angle surfaces to provide favoured transmission for wavelengths that have integer orders of retardation. The order or retardation is adjusted by rotating the plate: using plates of different thickness can provide a single region of high transmission in a wider spectral range. Etalons can be as simple as an uncoated plane-parallel plate of a transmitting material or a precise assembly of two closely spaced surfaces with multi-layer-dielectric reflective coatings to provide higher finesse. It is more difficult to control multiple etalons, and it is common to use only a single etalon. It is also common to use combinations of these techniques.

To obtain frequency-stable output from a laser, it is first necessary to make the laser as stable as possible before proceeding to active control techniques. Mechanical stability and the reduction of mechanical vibrations are a first essential step. Temperature stability must be considered next. At this point, it is necessary to consider the stability of the pump source. This is true even for a cw laser pumped by another cw laser. Active control of laser frequency requires an external reference and feedback control of the resonator. This is usually done with piezoelectric control of the cavity length. Frequency reference standards can be a stable external etalon for relative frequency stabilization or an atomic transition reference for absolute stabilization. Sub-Doppler spectroscopy techniques may be appropriate for the absolute frequency standard. Even greater absolute frequency accuracy is obtained by reference to cryogenically cooled atoms held in a trap or transiting through an atomic fountain.

Stable single-frequency operation is significantly easier to attain in low-power lasers. Injection locking is a technique for transferring the frequency characteristics of the low-power laser to a higher-power laser. To accomplish this, the resonances of the locking laser and locked laser must be actively controlled and locked together. One technique for accomplishing this is called Pound–Driever locking. The seeding laser beam is modulated to produce fm sidebands along with the central frequency. The sidebands are outside the resonance of the higher-power laser and are reflected. The central-frequency portion of the beam is partially transmitted into the second laser and drives the oscillation of that laser. The phase of the combined reflected locking beam and transmitted oscillation of the second laser change slightly if the two resonances are not exactly matched in frequency. This produces an amplitude modulation in the combined beam with the fm sidebands. The phase and amplitude of this amplitude modulation provide the error signal used to control the cavity length of the second laser. Ring resonators are suited to this type of locking. A unidirectional oscillation is established in the second laser. The combined transmitted and reflected beams from the second laser are directed away at an angle from the incident locking beam. This greatly helps in the isolation of the laser generating the locking radiation.

5.11 Temporal Resonator Characteristics

Mode-locking many cavity modes requires some modulation technique that couples the modes in phase. Acousto-optic modulators are used for active mode-locking of cw laser systems. An acoustic modulation driven at a frequency that matches the resonator mode spacing is established in a transparent material. This modulation produces sidebands on the modes, which couple to the adjacent modes, locking many modes in phase. Saturable absorbers are normally used with higher-peak-power pulsed mode-locked lasers. Saturable absorbers appropriate for mode locking must have short relaxation times much less than the round-trip cavity transit times. As the laser oscillation builds up with random fluctuations, the strongest fluctuation will begin to saturate the absorption. This fluctuation will build up more rapidly and come to dominate the laser oscillation. Gain will be depleted by the strong fluctuation reducing the amplitude of secondary fluctuations. The rapid saturation of the absorption on each cavity transit is a modulation that couples additional modes. The depth of modulation, gain, gain bandwidth and dispersion in the resonator determine the width on the mode-locked pulses. Kerr-lens-mode locking is an additional technique for producing very short mode-locked pulses. This technique uses the optical Kerr effect or change in refraction index that is produced by very high-intensity pulses.

In a Q-switched laser, oscillation is held off by introducing high losses in the resonator, while an inverted population is built up by some means of laser pumping. When the gain has reached a high level, the loss is abruptly removed. Laser oscillation develops in an intense short pulse that may be as short as a few cavity transits of the resonator. Electro-optic and acousto-optic Q-switches are frequently used to control cavity losses for the purpose of Q-switched operation. An acoustic wave deflects a beam by Bragg diffraction in the acousto-optic modulator. When the rf signal to a piezo-electric transducer is turned off, the resonator returns to the low-loss condition. Pockels cells, and waveplates and polarizers are used for electro-optic modulation. A double-pass through a quarter-wave-retardation waveplate rotates the polarization of the resonator beam, and the beam is deflected out of the resonator at the polarizer. When voltage is applied to the Pockels cell, a polarization retardation is produced that cancels the retardation of the waveplate, loss is minimized and a Q-switched laser oscillation rapidly develops.

5.12 Fibre Laser Resonators

Brief mention is made here of fibre laser resonators. In fibre lasers, the resonator beam is primarily a guided wave inside a fibre. Pump radiation, often generated by semiconductor diode lasers, is coupled into the cladding of the optical fibre. The pump radiation is absorbed over a relatively great length in a fibre core that is doped with the active laser ion. The fibre core is manufactured with a higher refractive index than that of the cladding, and the core is usually chosen to be of a size that will support only a single-fibre mode. Bragg reflectors can be established in the fibre by techniques such as processing with ultraviolet radiation. It is necessary to engineer the coupling of the pump radiation into the fibre carefully, predict the free-space propagation of the fibre laser radiation outside the fibre and to optimize the placement of discrete elements of the fibre laser resonator not incorporated in the fibre itself. Gaussian optics is useful in each of these areas. Fibre lasers have remarkable properties such as high efficiency and simplicity of operation. Optical properties are also remarkable, with high powers exceeding 100W having been demonstrated. Broad wavelength availability and high coherence are also being achieved.

5.13 Conclusion

The goal of this chapter has been to present an overview of laser resonators and the techniques of resonator analysis and design. Most of the detail of this discussion has focused on the fundamental aspects of Gaussian optics and stable resonators. Unstable resonators were discussed with illustrative examples. Other topics were briefly mentioned. It is hoped that the information presented is sufficient to address basic issues in resonator design and the management of laser beams. There has been a substantial amount of work performed on the topics of optical resonators since the first demonstrations of lasers in the early 1960s, and the investigation on optical resonators and beam propagation still continues. Several of the references have more extensive presentations and detailed bibliographies.

REFERENCES

1. Fox A G and Li T 1960 Resonant modes in an optical maser *Proc. IRE* **48** 18 904–5; Fox A G and Li T 1961 Resonant modes in maser interferometers *Bell Syst. Tech. J.* **40** 453–8.
2. Boyd G D and Gordon J P 1961 Confocal multi-mode resonator for millimeter through optical-wavelength masers *Bell Syst. Tech. J.* **40** 489–508.
3. Siegman A E 1986 *Lasers* (Mill Valley, CA: University Science Books).
4. Hall D R and Jackson P E (ed) 1990 *The Physics and Technology of Laser Resonators* (Philadelphia, PA: Institute of Physics Publishing).
5. Hodgson and Weber 1997 *Optical Resonators, Fundamentals, Advanced Concepts and Applications* (Berlin: Springer).
6. Selveto O 1998 *Principles of Lasers* 4th edn (New York: Plenum).
7. Kochner W 1976 *Solid-State Laser Engineering* (New York: Springer).
8. Yariv A 1989 *Quantum Electronics* 3rd edn (New York: Wiley).
9. Siegman A E 2000 Laser beams and resonators: the 1960s *IEEE J. Selected Topics Quantum Electron.* **6** 1380–8.
10. Siegman A E 2000 Laser beams and resonators: beyond the 1960s *IEEE J. Selected Topics Quantum Electron.* **6** 1389–99.
11. Kogelnik H and Li T 1966 Laser beams and resonators *Appl. Opt.* **5** 1550–67.
12. Abramowitz M and Stegun I A 1972 *Handbook of Mathematical Functions with Formulas, Graphs, and Mathematical Tables* (Washington, DC: National Bureau of Standards).
13. Siegman A E 1990 New developments in laser resonators *Proc. SPIE* **1224** 2–14.
14. Siegman A E and Townsend S W 1993 Output beam propagation and beam quality from a multimode stable-cavity laser *IEEE J. Quantum Electron.* **29** 1212–17.
15. Goodman J W 1968 *Introduction to Fourier Optics* (San Francisco, CA: McGraw-Hill).
16. Oughstun E E 1987 *Unstable Resonator Modes (Progress in Optics 24)* ed E Wolf (Amsterdam: North-Holland) pp 165–387.
17. Siegman A E 1991 Defining the effective radius of curvature for a non-ideal optical beam *IEEE J. Quantum Electron.* **27** 1146–8.
18. Press W H, Teukolsky S A, Vetterling W T and Flannery B P 1992 *Numerical Recipes in C: The Art of Scientific Computing* 2nd edn (New York: Cambridge University Press).
19. Frigo M and Johnson S G *FFTW, A C Subroutine Library for Computing the Discrete Fourier Transform* (Cambridge, MA: MIT) http://www.fftw.org.
20. Horwitz 1974 Asymptotic analysis of unstable resonator modes *J. Opt. Soc. Am.* **63** 1528–43.
21. Krupke W F and Sooy W R 1969 Properties of an unstable confocal resonator CO_2 laser system *IEEE J. Quantum Electron.* **QE-5** 575–86.

6

Optical Cavities: Waveguide Laser Resonators

Chris Hill

CONTENTS

6.1 Introduction 121
6.2 Propagation in Hollow Dielectric Waveguides 122
    6.2.1 Waveguide Mode Expressions 122
6.3 Waveguide Resonator Analysis 124
    6.3.1 The Concept of Resonator Modes 124
    6.3.2 Waveguide Modes 125
    6.3.3 Mode Coupling, Coupling Losses and Mode Losses 126
    6.3.4 Single-mode, Few-mode and Multi-mode Theory 127
6.4 First-Order Theory and Its Limits 127
    6.4.1 Coupling Loss Theory of Single-Mode Waveguide Resonators 127
    6.4.2 Dual Case I Waveguide Lasers 129
    6.4.3 Rigrod Analysis for Waveguide Lasers 129
6.5 Real Waveguide Resonators: Experiment and Theory 130
    6.5.1 Distant Mirrors 130
    6.5.2 Tilted Mirrors and Folded Lasers 131
    6.5.3 Tunability and Line Selection 132
    6.5.4 Resonator Mode Degeneracies: Hopping and Hooting 132
6.6 Summary 133
References 133
Reviews 133
Other Reading 133

6.1 Introduction

This chapter links the two topics of optical waveguide theory and free-space laser resonators. An optical resonator can contain any sort of aperture, lens, mirror, prism and obstruction. The total resonator thus formed will have its own self-repeating field patterns. The principle that a 'resonator mode' is self-repeating in phase and amplitude (so that it is an *eigenvector* for the resonator round trip) is rather general and powerful, extending even to apparently complicated structures. The presence of a *waveguide*, as either a minor or a major 'obstruction', does not affect this principle and need not cause alarm. Waveguides are, from one popular and useful viewpoint, only 'apertures'—albeit extended 3D ones. This chapter tries to make readers familiar with their presence and effects.

The optical resonator properties deliberately ignore any interaction of light with the side walls of the laser cavities. Apertures and mirror edges may be used for transverse-mode control but otherwise the light is assumed to propagate freely between the resonator mirrors. We now consider a class of resonator with intentional wall effects: the resonator path includes a *waveguide*, for example, a hollow dielectric tube in which light is guided by a series of Fresnel reflections from the tube walls. We will see how some resonator properties of waveguide lasers may be modelled theoretically and how such models compare with actual laser devices.

A general waveguide resonator is shown in Figure 6.1. The total influence of all the optical elements, including the guide, will determine what self-consistent field patterns may exist. Our modelling problem is to find these *resonator modes*, with their round-trip losses and resonant frequencies. The important point here is that, inside a sufficiently narrow guide, the Gaussian-beam modes and the corresponding propagation equations are inappropriate. Instead, we must recognize guide wall effects by using another set of functions (modes) which obey boundary conditions at those walls. We must also treat the coupling of optical radiation at the guide apertures and the free-space propagation to and from the mirrors.

Now, if we properly account for the coupling in, the propagation through and the coupling out again from the guide, so that the waveguide output for *any* Gaussian-beam mode (at the input plane) is again expressed in terms of Gaussian-beam modes (at the output plane), we have a description as full as can be desired. The mathematical quantities involved may be called waveguide coupling matrices, sets of overlap integrals or 'transfer functions' (in terms of transverse modes of

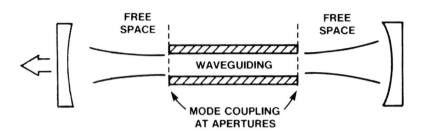

FIGURE 6.1 Sketch of a general linear waveguide resonator.

different spatial frequencies, rather than the more usual electrical waves of different radian frequencies).

Early workers on waveguide lasers sought definite *gain/bandwidth* advantages. In particular, because of favourable wall interactions, narrow-bore waveguide lasers with CO_2 as the active medium could run at much higher pressures than conventional Gaussian-resonator lasers, and thus offered increased frequency tunability. Here, we concentrate on the optical properties of a 'waveguide resonator', where *the presence of the waveguide significantly affects the mode structure of the optical resonator*. This seems the only reasonable answer to the question, when does a resonator become a waveguide resonator?

We restrict our discussions (although this is not essential) to infrared hollow dielectric guides. These are hollow guides as opposed to solid or clad optical fibre guides. They are assumed to be uniform, straight and of rectangular or circular cross-section, with half-widths or radii much greater than the radiation wavelength ($a \gg \lambda$). Such structures possess sets of *waveguide modes*. These are field patterns which can propagate along the guides, with tolerably low loss, maintaining their transverse shape. Formally, they are solutions of Maxwell's equations obeying the appropriate boundary conditions at the guide walls. With our assumptions, and freely using $a \gg \lambda$ to discard terms, we can derive fairly simple mode fields and propagation constants (and in any case, we omit most of the mathematical details here). The lowest-order linearly polarized mode of a square-bore guide, for instance, is called EH_{11} and looks rather like the lowest-order linearly polarized free-space mode TEM_{00}. As we shall see later, the fact that guide attenuation losses are strongly mode-dependent may be used to our advantage in controlling the transverse mode pattern actually emitted by the laser. When talking of mode fields, coupling losses and resonators, we will not use specific guide dimensions or wavelengths, since the theory is fairly general. The most common real devices, and most of our examples in this chapter, involve square guides and the familiar CO_2 laser medium ($\lambda \simeq 10\,\mu m$). There is no direct treatment here of laser media, the effects of guide walls on discharges or their consequences for laser design and engineering—no discussion, for example, of *non*-uniform structures which are tapered or otherwise shaped so that they encourage the desired spatial mode(s).

This chapter has four further sections. In Section 6.2, we briefly discuss light propagation and modes in hollow dielectric waveguide structures. Section 6.3 is a rather general account of *waveguide resonator analysis*. Section 6.4 presents some detail of first-order (single-mode) waveguide resonator theory. Finally, Section 6.5 is concerned with the resonator properties of some real waveguide lasers and how we interpret them in terms of available resonator theory.

6.2 Propagation in Hollow Dielectric Waveguides

In 1964, Marcatili and Schmeltzer proposed the use of hollow dielectric waveguides as low-loss components for laser amplifiers [1]. They calculated the allowed *modes of propagation* of hollow dielectric waveguides by solving the standard wave equation in free space but with dielectric boundary conditions; field components were matched at the guide walls instead of falling away to zero at infinity. These modes are field patterns which keep their shape, but experience a mode-dependent phase shift, as they propagate along the guide. Hollow dielectric waveguides used in typical gas waveguide lasers have half-widths or radii $a \simeq 1\,mm$, so the assumption $a \gg \lambda$ is reasonable. They cannot be made very much smaller without severe attenuation ($\propto a^{-3}$) and are inherently *multi-mode* or *over-moded*. When terms of first or higher order in λ/a are removed, the rather complex expressions for the field amplitudes of the waveguide modes reduce to simple 'first-order' ones. But the corresponding simplified propagation constants include essential terms of order λ/a and $(\lambda/a)^2$. For each waveguide mode the propagation constant can be written in the form $k = \beta + i\alpha$, where β is the phase constant and α is the amplitude attenuation constant. Propagation along the long axis of the waveguide is described by a term $\exp(ikz)$.

6.2.1 Waveguide Mode Expressions

In this section, we describe the simplified fields and propagation constants for the square and circular waveguide geometries. Inconsistent notation for waveguide modes is usually the first thing that irritates workers in this subject. A brief study of Degnan [2] or other reviews may help.

For circular cross-section dielectric guides (radius a, polar coordinates r and φ as in Figure 6.2a), we usually have a single dielectric constant $\varepsilon = (n_a - ik_a)^2$ for the wall material. Circular guides may support transverse circular electric modes (TE_{0m}), transverse circular magnetic modes (TM_{0m}) and sets of higher-order hybrid modes (EH_{nm}) which have both radial and tangential electric and magnetic fields. Figure 6.2b shows the first-order field amplitudes and transverse patterns for the lowest-order mode of each type. In practice, linear polarization with a fixed axis is usually desired and often enforced

Optical Cavities

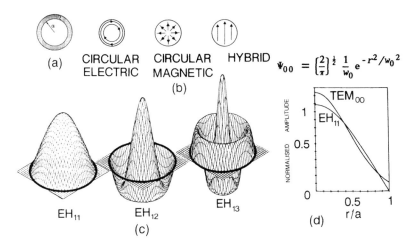

FIGURE 6.2 Characteristics of circular waveguide: (a) circular waveguide geometry, (b) electric field patterns of circular waveguide modes, (c) amplitude profiles of low-order EH_{1m} modes and (d) amplitude profile of EH_{11} and its best-fit Gaussian Ψ_{00} ($\omega_0 = 0.64a$). EH_{11} is normalized according to equations (6.2) and (6.3) with $a = 1$.

with a Brewster window or diffraction grating. We can express an arbitrary linearly polarized transverse field pattern within a circular guide as a combination of LP_{nm} modes, where n is the azimuthal or angular mode number and m is the radial mode number. The LP_{0m} modes are identical with the EH_{1m} modes of Marcatili and Schmeltzer [1] or the HE_{1m} modes of Snitzer [3]. This important set of zero-angular-order modes is sufficient to describe any circularly symmetric field pattern and to model many experiments. Figure 6.2c shows the amplitude distribution of the fundamental hybrid mode EH_{11} (or HE_{11} or LP_{01}). The higher-order modes must be included for a full description of even well-aligned lasers and are essential for any account of misalignment effects.

The first-order fields are given by

$$E_{nm} = f_{nm}\, J_n(\rho_{nm}r/a)\cos(n\varphi) \qquad 0 \le r \le a \qquad (6.1)$$

where ρ_{nm} is the mth root of the nth-order Bessel function, i.e. $J_n(\rho_{nm}) = 0$, and

$$f_{mn} = \begin{cases} \sqrt{2}[\sqrt{\pi}a J_{n+1}(\rho_{nm})]^{-1} & \text{if } n=2, 3, 4,\ldots \\ [\sqrt{\pi}a J_1(\rho_{0m})]^{-1} & \text{if } n=0 \end{cases} \qquad (6.2)$$

for the even LP_{nm} ($n \ne 1$) modes. The axis of polarization is arbitrary: neither the field strengths nor the propagation constants depend on it. The field strengths for the even linearly polarized ($n = 1$) modes obey the same form:

$$E_{1m} = \sqrt{2}[\sqrt{\pi}a J_2(\rho_{1m})]^{-1} J_1(\rho_{1m}r/a)\cos\varphi \qquad (6.3)$$

but the axis of polarization is not arbitrary. This choice of even (cosine) modes forces the choice of circular magnetic TM_{0m} modes (plus HE_{2m}) for x-polarized light, and circular electric TE_{0m} modes (plus HE_{2m}) for y-polarized light. The normalization and the propagation constants are given by

$$\int_0^{2\pi}\int_0^a E_{nm}(r,\varphi)E_{n'm'}(r,\varphi)\, r\, dr\, d\varphi = \delta_{nn'}\,\delta_{mm'} \qquad (6.4)$$

$$\beta_{nm} \simeq \frac{2\pi}{\lambda}\left[1 - \frac{1}{2}\left[\frac{\lambda\rho_{nm}}{2\pi a}\right]^2\right] \qquad (6.5)$$

$$\alpha_{nm} \simeq \frac{1}{a}\left[\frac{\lambda\rho_{nm}}{2\pi a}\right]^2 \text{Re}\left[\frac{1}{2}(\varepsilon+1)(\varepsilon-1)^{-\frac{1}{2}}\right] \qquad (6.6)$$

Figure 6.3 shows a general rectangular cross-section dielectric guide (width $2a$ and height $2b$) with two different dielectric materials. In practice, many guides use a single material but many others are 'hybrid' with one or two metal walls (which may be electrodes for transverse rf excitation). Common guide materials are aluminium, alumina (Al_2O_3) and beryllia (BeO). Figure 6.3 shows two complex relative dielectric constants:

$$\varepsilon_a = (n_a - i\, k_a)^2 \qquad \varepsilon_b = (n_b - i\, k_b)^2$$

where n_a and n_b are refractive indices and k_a and k_b are extinction coefficients; all four are positive real numbers. These rectangular guides support two sets of linearly polarized EH_{mn} hybrid modes but do not support circularly polarized modes. In a ceramic–metal hybrid guide with one or two horizontal metal walls, the horizontally polarized $E^x H_{mn}$ modes are usually favoured over the vertically polarized $E^y H_{mn}$ modes, chiefly because of the loss factors associated with the metal.

The first-order fields for the linearly polarized EH_{mn} modes are

$$E_{m,n}(x, y) = ab^{-\frac{1}{2}}\left[\begin{array}{c}\cos\\ \sin\end{array}\left(\frac{m\pi x}{2a}\right)\right]\left[\begin{array}{c}\cos\\ \sin\end{array}\left(\frac{n\pi y}{2b}\right)\right];$$

$$m, n = \begin{array}{c}\text{odd}\\ \text{even}\end{array} \qquad (6.7)$$

with normalization

$$\int_{-a}^{a}\int_{-b}^{b} E_{mn}(x,y)E_{m'n'}(x,y)\, dx\, dy = \delta_{mm'}\,\delta_{nn'} \qquad (6.8)$$

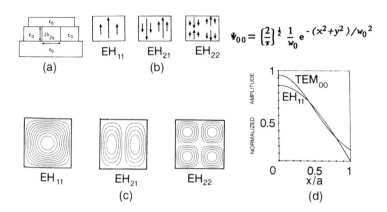

FIGURE 6.3 (a) Rectangular waveguide geometry, (b) electric field lines of low-order EH_{mn} modes, (c) low-order amplitude contours for a square-bore guide and (d) amplitude profile of EH_{11} and its best-fit Gaussian Ψ_{00} ($\omega_0 = 0.70\,a$). EH_{11} is normalized according to equation (6.7) with $a = b = 1$.

The propagation constants $k_{mn} = \beta_{mn} + i\alpha_{mn}$ are given by

$$\beta_{mn} \simeq \frac{2\pi}{\lambda}\left[1 - \frac{1}{2}\left[\frac{m\lambda}{4a}\right]^2 - \frac{1}{2}\left[\frac{n\lambda}{4b}\right]^2\right] \quad (6.9)$$

and (for x-polarized modes)

$$\alpha_{mn} \simeq \frac{m^2}{a}\left[\frac{\lambda}{4a}\right]^2 \mathrm{Re}\!\left(\varepsilon_a(\varepsilon_a - 1)^{-\frac{1}{2}}\right) + \frac{n^2}{b}\left[\frac{\lambda}{4b}\right]^2 \mathrm{Re}\!\left((\varepsilon_b - 1)^{-\frac{1}{2}}\right) \quad (6.10)$$

Figure 6.3 also has some contour maps for the lowest-order modes in a square waveguide, and a comparison of the EH_{11} field amplitude with that of the fundamental Gaussian beam which best approximates it (see Section 6.4.1).

Remember the benefits of this modal approach to propagating light. The appropriate (Maxwell) equations are solved once for all, and thereafter, little effort is needed to find the field (at any point in the guide) due to a specified initial field (at any previous point). Instead of solving a fresh set of wave equations, we need only to perform a set of multiplications by the complex numbers $\exp(ik_{mn}z)$. This is a great benefit and immediately invites a *matrix* treatment. Any possible field (consistent with our simplified model) will be represented by a linear combination of EH_{mn} modes; any possible propagation, reflection, scattering, etc., will produce another linear combination of the same modes. We can calculate once for all the self-coupling and cross-coupling coefficients to describe this, arrange them in suitably ordered propagation and coupling matrices and manipulate them easily in one of several popular matrix-based software packages.

6.3 Waveguide Resonator Analysis

This section gives a rather general account of resonator analysis as applied to waveguide lasers. A waveguide resonator (Figure 6.1) may be viewed as a free-space Gaussian resonator perturbed by an aperturing waveguide tube, or as a tube resonator perturbed by the addition of free-space sections. We summarize the usual methods of calculating the properties of waveguide resonators and later examine some of the geometric constraints and principles in the design of a resonator. Readers should be familiar with the main features of *optical resonators* and free-space Gaussian beams, so that concepts such as round-trip eigenvalues and their associated *resonator modes* may be introduced and examined without difficulty. Even so, large amounts of time and effort can be saved if such terms as 'mode', 'resonator mode', 'multi-mode' and 'mode loss' are clearly understood. At the risk of some duplication, we discuss these key concepts as they arise.

6.3.1 The Concept of Resonator Modes

The real waveguide lasers used in the factory or on the bench are more or less complicated assemblies with electrical, mechanical and optical elements including active media. We wish to model important laser properties such as output power, beam shape, frequency tunability and frequency purity. We discard all specific real-life elements and consider the very simple, and fairly abstract, idea of an optical resonator which possesses some *resonator modes*. The link with real-life devices will reappear shortly. A resonator mode is a field distribution which repeats itself in shape and in phase after one round trip of the resonator.

Let us start with a well-aligned Gaussian-beam resonator with two large-aperture, spherically curved mirrors. We know that a standard free-space scalar wave equation, when we require the light to stay near the z-axis with the dominant variation being simply the axial propagation term $\exp(ikz)$, yields Gaussian beam functions. These are the TEM_{pq} free-space *modes* of propagation: their properties change very slowly with wavelength and their transverse shape is given by a simple Gaussian $\exp(-r^2/w^2)$, multiplied by Laguerre polynomials (in cylindrical coordinates) or Hermite polynomials (in Cartesian coordinates). We can define a common beam-waist position z_0 and a common beam-waist radius w_0 for the whole orthonormal set of beams. Then, formally, we can express the free-space field E_{fs} in any plane along the resonator axis as a linear combination of Gaussian beams:

$$E_{fs} = \sum b_{pq} \Psi_{pq}(z - z_0, w_0) \quad (6.11)$$

$$b_{pq} = \int E_{\text{fs}} \Psi_{pq}^* \, dA \qquad (6.12)$$

where Ψ_{pq} are the TEM$_{pq}$ Hermite–Gaussian or Laguerre–Gaussian functions *including* the z-dependent amplitude and phase factors. The '*' indicates complex conjugation and the integral is performed over the infinite cross-section (though only the area near the z-axis contributes significantly). The b_{pq} are complex coefficients which, during lossless free-space propagation, do not depend on axial position z. Again we see great benefits from the modal approach, as apparently complicated wave equations or diffraction integrals are replaced by the simple Gaussian beam transformations.

Thus, equations (6.11) and (6.12) show a 'decomposition' of some arbitrary function E_{fs} into an orthonormal set of functions Ψ_{pq} representing various spatial frequencies, just as in Fourier analysis, we split some function $f(x)$ into sines and cosines. If the original field E_{fs} is associated with a precise temporal frequency, then so is each term of the sum: the precise frequency information is extra and not contained in the spatial form of the modes of propagation.

By contrast, resonator modes are *modes of oscillation*. This concept has two essentials: the mode shape must be one of the self-repeating *transverse modes* of the cavity, and the mode frequency must yield a precise *axial* resonance. Conventionally, we assign two transverse-mode integers r and s, and one longitudinal-mode integer j, where j is the number of full 2π propagation phase shifts per round trip. Thus, a general resonator mode frequency is $\nu_{j,rs}$. For most real waveguide lasers of relevance here, $j > 10\,000$ and $r, s < 10$, with r and s referring to the transverse-mode order in the x and y (or radial and azimuthal) directions. Usually, a change of 1 part in 10^4 in λ will make no important difference to the transverse part (the shapes and amplitudes of the Gaussian beams), whereas the change $j \to j+1$ (or a single-pass phase change $\simeq \pi$) is crucial in laser frequency studies. This means that laser axial modes and transverse modes are physically decoupled and can be treated separately. They are both present in the idea of an oscillating or resonating mode. It often helps to imagine a laser emitting some beam of arbitrary transverse shape with a single well-defined frequency; this shape, and this frequency $\nu_{j,rs}$, constitutes a resonator mode by definition.

Formally, the self-consistency condition translates into the complex eigenvalue equation:

$$M E_{\text{fs}} = \gamma E_{\text{fs}} \qquad (6.13)$$

where each self-consistent field E has its complex eigenvalue γ. We will define a *round-trip loss* $1-|\gamma|^2$ and *relative phase shift* $\arg(\gamma)$. Since the *round-trip matrix* M is complex and may be large, we may be no further forward. However, we see that if enough of M can be specified with reasonable accuracy, equation (6.13) can be solved in principle. For stable open resonators, the resonator transverse modes are accurately given by pure Gaussian beams with appropriate phase shifts: that is, by choosing w_0 and z_0 correctly, we can make M diagonal. Each eigenvector can then be written as a single term:

$$E_{\text{fs}} = \Psi_{pq}(z-z_0, w_0). \qquad (6.14)$$

The Ψ_{pq} relative phases are given by $2(p+q+1)\cos^{-1}[\pm(g_1 g_2)^{1/2}]$, where the 2 refers to a *round trip* or double pass between the mirrors. But, in general, with misalignments, obstructions or perturbations, M is not diagonal and each eigenvector is a mixture or linear combination as in our equation (6.11); the b_{pq} may now depend on z. We are still free to choose any w_0 and z_0 but no choice will yield pure Ψ_{pq} as the self-consistent fields, and equation (6.13) will generally be tedious to solve without a computer. Applying this empty resonator theory to active lasers, we *assume* that the various γ and E are unaffected by the laser medium—except that the overall cavity loss $1-|\gamma|^2$ is exactly balanced by the round-trip gain. This neglect of the active medium is convenient but (not surprisingly) may lead to inaccurate results.

If we understand all this, then the 'free-space' part of our problem is solved. To analyse a given resonator, we break one round trip into a sequence of well-defined segments, such as free-space propagation between the mirrors or reflection from a mirror. We choose a base set of free-space Gaussian beams Ψ_{pq}. In practice, with limited computing resources, p and q cannot take infinitely many values, so our base set is judiciously truncated. By ABCD matrices or by explicit Huygens–Fresnel diffraction calculations, we mathematically describe the effect of each segment on our arbitrary field (= linear combination of Ψ_{pq}) by a matrix. The ordered product $M = M_n \ldots M_3 M_2 M_1$ of all these matrices is the grand round-trip matrix, whose eigenvectors are self-consistent linear combinations and whose eigenvalues define the round-trip losses and phase shifts.

We often hope that the largest eigenvalue (lowest loss) refers to some field:

$$E = b_{00}\Psi_{00} + b_{01}\Psi_{01} + b_{10}\Psi_{10} + \ldots \qquad (6.15)$$

(truncated at some $p_{\max}, q_{\max}$) which is nearly pure TEM$_{00}$; that is, if we normalize to $\Sigma|b_{pq}|^2 = 1$, then we want $1-|b_{00}|^2 \ll 1$. This seems less likely as larger perturbations are introduced and is usually untrue of unstable resonators. *The perturbation that most concerns us here is a hollow dielectric waveguide, lying along the z-axis and acting as a 'three-dimensional aperture'*. From our present point of view, some obvious questions are: How do we form a matrix to describe the effect of this guide on our Gaussian beams? That is, how does radiation couple into the guide, propagate along it and couple out again? Is it easy to decide whether a given waveguide will significantly perturb the first few resonator modes? (If it will not, we may ignore it for most purposes.) And, importantly, does the eventual waveguide resonator matrix theory show any agreement with experiment?

6.3.2 Waveguide Modes

Suppose, by analogy with equation (6.11), we have an orthonormal set of guide modes and can express the field E_{wg} anywhere inside the guide as

$$E_{\text{wg}} = \Sigma a_{mn} E_{mn} \qquad (6.16)$$

$$a_{mn} = \int E_{\text{wg}} E_{mn}^* \, dA \qquad (6.17)$$

where we integrate over the guide cross-section. These functions E_{mn} will usually stand for the linearly polarized EH_{mn} modes, just as the Ψ_{pq} functions stand for the TEM_{pq} modes. Because waveguides are lossy, the complex coefficients a_{mn} depend on z. There is no analogous mode waist radius w_0 or waist plane at some z_0; the guide mode has the same spatial shape at every z within the guide.

For Gaussian beams, an 'ideal' reflector is a large-aperture mirror whose curvature exactly matches that of the incident beams. Each beam is reflected back on itself, with no coupling to any other beam: that is, the Gaussian beam coupling matrix of the mirror is diagonal. For waveguide modes, the corresponding reflector is a plane mirror aligned perpendicular to the guide axis and placed immediately against the guide end. Here, again, the modes are reflected back along the z-axis without cross-coupling, so we have amplitude self-consistency. It is then fairly clear that, with two such mirrors, we have a simple waveguide resonator whose modes of oscillation, or resonator modes, are pure waveguide modes—with, again, the important extra condition of phase self-consistency. The resonator matrix M can be written diagonally; its non-zero elements are the propagation constants $\exp(i2k_{mn}L)$. This diagonal feature is unique to this (in principle) simple tube resonator, called dual case I (see Section 6.4.1).

Later, we will meet other designs which offer nearly pure modes. We will also see that, in practice, even this simplest resonator does not always follow 'first-order' theory. In general, with perturbations present, M must be non-diagonal and the eigenvectors will be linear combinations of the E_{mn} with several significant non-zero terms a_{mn}. Note that we can order the values of m and n, so that $\Sigma a_{mn} E_{mn}$ is a column vector and M is a square matrix. If we understand this, the waveguide part of our problem will be solved also.

6.3.3 Mode Coupling, Coupling Losses and Mode Losses

The missing step so far is the coupling between free space and the waveguide. In mathematical terms, this is a change of basis; given the two orthonormal sets of field patterns, we express the coupling coefficients as overlap integrals across the guide aperture and form these coefficients into a coupling matrix. When describing mode-coupling loss, we must specify the modes from which and into which radiation is coupled. For example, let us consider the fundamental waveguide mode EH_{11}, together with a general curved-mirror reflector (Figure 6.4).

The *amplitude coupling coefficient* between EH_{11} and TEM_{pq} is given by $\int E_{11} \Psi_{pq}^* \, dA$, that is, the integral of the product of the waveguide mode field and the complex conjugate of the free-space field. The modulus squared of this is the *coupling efficiency*, which differs from unity by the *coupling loss*. Similar coefficients, efficiencies and losses could be defined for any EH_{mn} mode or for any sum of modes. However, of immediate interest is the amplitude *self-*coupling coefficient of the fundamental mode, that is, the amount of EH_{11} which reappears as EH_{11} after propagation to and from the reflector. This is defined as the overlap integral of EH_{11} with the

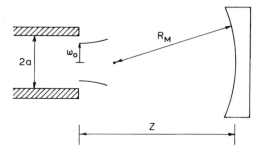

FIGURE 6.4 Sketch of general curved mirror–waveguide combination.

returned (conjugate) field due to a 'launched' EH_{11}. The modulus squared will be the EH_{11} self-coupling efficiency $|c_{11,11}|^2$, and the self-coupling *loss* will be

$$\Gamma_{11} = 1 - |c_{11,11}|^2 \qquad (6.18)$$

Γ_{11} is commonly called the 'EH_{11} coupling loss'. We can find these fields and integrals by diffraction theory or by assembling and tracking our linear combination of Ψ_{pq}, but the task may not be easy. There is much more on this subject in the literature, and Section 6.4 summarizes some results.

If in doubt about the conjugate (*) operations in these coupling-coefficient definitions, we may note that the 'change of basis' is possible at any plane: we can express a mode E as a sum of modes Ψ by defining appropriate E-to-Ψ coefficients $\int E \Psi^* \, dA$, and then at once (without any intervening propagation of the Ψ) change back to E with appropriate Ψ-to-E coefficients. This (correctly) achieves nothing, so long as the result is the original E, and not E* or anything else. Such a check of the mathematics, including the software conventions for complex arithmetic and arctangents, is recommended: errors of say $\pi/2$ or π in phase, while unimportant for single-mode propagation losses, may seriously affect the resonator mode content and hence the resonator mode losses and frequencies.

A particular combination of waveguide modes $\Sigma a_{mn} E_{mn}$ at the guide exit, associated with a given resonator mode, will experience its own self-coupling efficiency and loss. But this is a coherent assembly of guide modes, and the resonator mode-coupling efficiency is not generally the weighted sum of the individual EH_{mn} coupling efficiencies. Instead, it depends crucially on the relative phases of the waveguide modes. This is the simple but vital reason why a multi-mode approach is needed for all but the simplest resonators or all but the crudest estimates of laser loss and transverse-mode quality. Unless great pains are taken to force single-mode operation, significant higher-order mode amplitudes (a_{mn}) may creep in and cause unpredictable spatial interference, and small changes of only 1%–2% in round-trip loss can significantly alter the available laser power. The relative higher–order mode powers $|a_{mn}|^2$ may seem very small, but interference involves complex amplitudes.

Similarly, the power lost per unit length through guide attenuation of waveguide mode combinations is not generally $\Sigma 2\alpha_{mn}|a_{mn}|^2$. In first-order theory (Section 6.4), where terms of order λ/a are neglected, the EH_{mn} mode fields form a complete orthonormal set across the guide aperture and all the fields vanish at the walls. But to describe attenuation, we must treat the small non-zero fields at the walls, and when two or more modes are present, these fields include interference terms. Many resonator models tend to neglect this fact, either ignoring it totally or assuming that, with several modes travelling up and down the guide, the interference terms will rapidly 'wash out' to leave a smooth average loss. We should be aware of such neat but inaccurate assumptions and realize that the distribution of loss over the round-trip path in real lasers is seldom easy to calculate.

6.3.4 Single-mode, Few-mode and Multi-mode Theory

Once our method of resonator analysis is settled, and we can describe each resonator segment, the question usually arises: How many modes should, or must, we include? Broadly speaking, there have been three answers: one, a few and many; that is, single-mode theory, few-mode theory and multi-mode theory.

In *single-mode* theory, we assume that the lasing resonator mode is pure EH_{11} (or, rarely, another pure EH_{mn} mode); therefore, the resonator loss is formed by pure EH_{11} attenuation and coupling losses, and no interference effects are considered. This may be a good approximation for some lasers with well-aligned near-field plane mirrors or strong in-built mode selection; it is usually hopeless for lasers with gratings or misaligned mirrors or almost any perturbation. It requires all higher-order modes to be either very faintly excited or very strongly damped. But higher-order modes tend not to be suppressed very effectively by attenuation in typical guides: if we begin with equal amounts of EH_{11} and EH_{12}, it may take many metres of guide propagation before the ratio is even 2:1 in favour of EH_{11}. (To check this, we evaluate $L \simeq 0.5(\alpha_{12}-\alpha_{11})^{-1}$ ln 2.) Also, by making guides with low EH_{11} losses, we inevitably reduce any mode discrimination due to guide loss: the inference is that an 'ideal' laser of minimum EH_{11} loss, with an excellent straight smooth guide and plane mirrors near the guide ends, has negligible in-built transverse-mode discrimination and may emit beams whose transverse shapes are quite unlike E_{11} or Ψ_{00}.

To model real lasers in any detail, we must consider at least the *first few* modes. It is generally considered that about five or ten will do: the lowest-loss resonator mode and its loss will be modelled with fair accuracy, so that adding extra modes does not greatly change the results; beyond five or ten modes, the first-order theory must be suspect, and since with N modes, we generally have an $N \times N$ complex matrix, the computing load for $N > 10$ may be too heavy.

Multi-mode resonator models, when compared with real lasers (see Section 6.5), tend to give encouraging but not *very* accurate results. They are far better than single-mode models, and considerably better than three- or five-mode models, but yield diminishing returns as our initial approximations become strained and then broken.

In fact, these approximations are likely to collapse before we achieve numerical convergence to 'final' resonator predictions. This is, again, simply because typical waveguides are short and fat. Although the capillary waveguides used in lasers appear relatively long and slender, in terms of transverse-mode discrimination, they are not. The influence of interfering high-order modes can be complicated, and resonator losses can seem to depend (in theory) very sharply on precise guide geometry (see Section 6.5). Some caution is needed when we interpret rapidly varying loss curves, derived from idealized matrix equations, in terms of real laser behaviour. Real lasers have imperfections—roughnesses, random mirror defects and scattering centres—which may be very difficult to model accurately but which, on the whole, tend to smooth out such 'ideal' sharp variations or spikes.

In practice, we must draw a line at some reasonable level of theoretical complexity; it becomes unproductive to consider hundreds of modes in search of a small increase in numerical accuracy. Extra caution is perhaps needed at this point, because the number-crunching power of cheap computers continues to advance. It becomes ever easier to evaluate hundreds or thousands of matrix coefficients by numerical integration or FFT-based diffraction integrals, pass them through matrix inversion routines and plot the 'results' cheerfully. Section 6.4 offers some rough limits to common waveguide laser assumptions. The only general lesson is that we should keep clearly in mind the type of results and the degree of accuracy that will satisfy us, use the necessary number of modes to that end and lower our expectations if we find that number too high.

6.4 First-Order Theory and Its Limits

This section discusses the *single-mode* theory of waveguide resonators. This theory, built on the waveguide mode expressions in Section 6.2 and the coupling coefficients in Section 6.3, gives our first view of the impact of a wave-guiding region within the resonator. We quote some important results from the literature on coupling theory, the dual case I resonator and Rigrod modelling, but we stress that the resonator properties of most waveguide lasers are poorly described by single-mode theory.

6.4.1 Coupling Loss Theory of Single-Mode Waveguide Resonators

This topic introduces the essential concepts in resonator analysis and remains interesting for three main reasons. First, if we trust that our resonator transverse mode is almost a pure waveguide mode, we want to estimate coupling losses and judge their effect on laser power (see Section 6.5 where the Rigrod equation is discussed). Second, in resonator design, we want to exploit mode-dependent coupling losses and force near-single-mode operation; here, the relative losses of several modes are important. The third reason is that coupling losses are important in experiments and applications involving waveguide transmission, even if there is no resonator.

We consider the general reflector (waveguide plus spherical mirror) of Figure 6.4. A guide mode is launched from the aperture and, after reflection, couples to itself and to other modes. Researchers have found the returned field distribution in two main ways. One is to express the initial field as a linear combination of Gaussian modes and propagate it to and from the mirror according to the Gaussian beam equations of Kogelnik and Li [4], assuming that the mirror aperture is effectively infinite. The other is to use scalar diffraction integrals. The two approaches give identical results [5]. Early work [6–8] predicted the existence of three reflector configurations which result in low EH_{11} coupling loss. For a circular guide of radius a, with $E(r)$ the launched field and $E'(r)$ the returned field, equation (6.18) may be written:

$$\Gamma_{11} = 1 - \left| \int_0^a E(r)E'(r) 2\pi r \, dr \right|^2. \qquad (6.19)$$

Evaluating this for a range of mirror-guide distances and mirror curvatures yields the set of curves in Figure 6.5. The three low-loss reflector geometries (Figure 6.6) were named by Degnan and Hall [7]:

- Case I: large R mirrors very near the guide ($z \simeq 0$, $z/R \simeq 0$),
- Case II: large R mirrors centred near the guide entrance ($z \simeq R$),
- Case III: mirrors with $R \simeq 2b$ and $z \simeq b$, where $b = \pi \omega_0^2 / \lambda$.

Here, w_0 and b are the beam-waist radius and Rayleigh range of the EH_{11} mode's *approximating Gaussian*, i.e. the TEM_{00} beam having maximum overlap with EH_{11} across the guide aperture ($z = 0$). This occurs for $\omega_0/a \simeq 0.64$ (circular) or $\omega_0/a \simeq 0.70$ (square), and the power overlap is $\simeq 98\%$ in both geometries. We expect that any reflector which efficiently recouples this TEM_{00} to itself will do the same for EH_{11}. Thus, the simplest possible model of a general waveguide resonator, with only EH_{11} inside the guide and only TEM_{00} outside, suggests the choice of *phase-matched mirrors*, whose curved surfaces coincide with the phase fronts of this special TEM_{00} at z_1 and z_2, according to $R = z + b^2/z$. For such mirrors, Abrams [9] found that Γ_{11} is about 1.48% for case III and always less than 7% (circular bore). For case III, Degnan and Hall [7] revised this estimate to $\simeq 1.38\%$ and found that $\Gamma_{12} \simeq 78\%$. A reasonably well-built device with one or two case III mirrors usually offers the best chance of guaranteed TEM_{00}-like operation. On the whole, the approximating Gaussian concept is neat and intuitive and predicts Γ_{11} minima with fair accuracy, but with short fat guides, its practical value has not always been realized. Note: this definition $b = \pi\omega_0^2 / \lambda = z_R$ is common in the waveguide laser literature. It is half the 'b' of Kogelnik and Li [4].

In the far field, any launched field pattern will revert to a roughly spherical wave centred near the guide aperture. Case II is a far-field phase-matched reflector, for which any guide mode has low coupling loss. The poor mode discrimination and large z make it unpopular, unless the extra space is needed for intra-cavity elements.

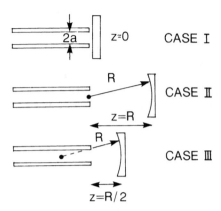

FIGURE 6.6 Low-loss coupling configurations for a fundamental waveguide mode.

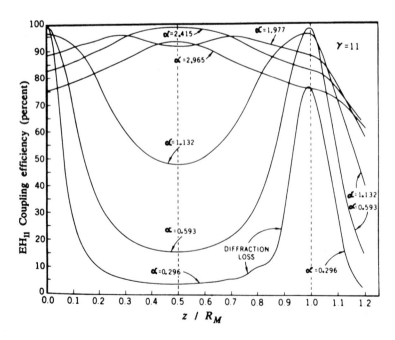

FIGURE 6.5 EH_{11} coupling losses for circular guides. Curves are shown for various values of $\alpha = (ka^2/R_M)$ and a constant mirror half width of $11a$. (From Degnan and Hall [7].)

Case I is used in most of today's commercial and scientific gas waveguide lasers: just a plane mirror at each end of the guide, placed against, or within a couple of millimetres of, the exit. There has been much study of the EH_{11} coupling loss for small (realistic) departures from perfect case I ($z = 0$, $R = \infty$, zero tilt), with surprising disagreements. Some simple and useful results are as follows:

EH_{11} coupling losses Γ_{11} ($z \ll b$). Degnan and Hall [7] give $38.4 N^{-1.5}$% for circular guides, where $N = a^2/\lambda z$ is the reflector Fresnel number (numerical approximation to diffraction integral results). Boulnois and Agrawal [10] give $= 33 N^{-1.5}$% for square guides (asymptotic approximation); the same answer is obtained from a Hermite-Gaussian expansion, and an earlier mistake [5] is corrected in Ref. [11].

EH_{mn} self-coupling efficiencies for square guides ($N > 1$). Calculations by Boulnois and Agrawal [10] yield the 1-D result:

$$|c_{mm}|^2 \simeq \left(1 - \frac{m^2}{6N^{1.5}} - \frac{\pi m^4}{240 N^{2.5}} + \frac{m^4}{72 N^3}\right)^2 \quad (6.20)$$

from which (keeping two terms) $\Gamma_{mn} = (1/6)(m^2 + n^2) N^{-1.5}$. Their analytical expressions, not given here, greatly reduce the computer time needed for square-bore coupling-loss calculations, if misalignment is not involved, and coupling amplitudes and cross-coupling coefficients are not required.

6.4.2 Dual Case I Waveguide Lasers

We recall that in a dual case I resonator, a single round trip is very nearly equivalent to propagation along an undistorted guide of length $2L$, and the resonator transverse modes are essentially the transverse modes of the waveguide itself. The mode frequencies are found from a standard phase equation:

$$2\beta_{mn} n(\nu_{j,mn}) L = 2j\pi \quad (6.21)$$

where the laser medium refractive index $n(\nu)$ includes a small anomalous dispersion term. By taking the difference between two mode frequencies, we can remove the axial mode integer $j \simeq 10^5$. In the absence of mode pulling ($n(\nu) = 1$), we have:

$$\nu_{j,mn} \simeq \frac{jc}{2L} + \frac{c\lambda}{32}\left(\frac{m^2}{a^2} + \frac{n^2}{b^2}\right) \quad \text{(rectangular)} \quad (6.22)$$

$$\nu_{j,mn} \simeq \frac{jc}{2L} + \frac{c\lambda \rho_{nm}^2}{8\pi^2 a^2} \quad \text{(circular)} \quad (6.23)$$

Equations (6.22) and (6.23) show roughly how the mode frequencies depend on waveguide geometry, although, in practice, the exact frequencies will depend on many other small corrections. In particular, the difference in frequency between two different transverse modes depends strongly on waveguide bore but weakly on waveguide length, in terms of typical manufacturing tolerances; the derivatives $\partial \nu / \partial a$ vary as a^{-3}.

By contrast with stable open resonators, these waveguide laser transverse-mode spacings vary as m^2 and n^2 or ρ_{nm}^2 and can exceed the axial mode spacing $c/2L$. Homogeneously broadened lasers tend, with some mostly unwelcome exceptions, to operate on the one mode (of all the available modes) with the highest ratio of small-signal gain to threshold loss. Now, if we fix $|\nu_{j,11} - \nu_{j,12}|$ at an integer multiple of $c/2L$, then the EH_{12} mode always coincides with a lower-loss EH_{11} mode and, *other things being equal*, EH_{11} must be preferred. This 'coincidence' of transverse-mode frequencies, if it can be achieved, will hold very well for different axial modes under the same line and fairly well for different CO_2 lines. Formally, this implies:

$$L = 16 s a^2 / \lambda \Delta \quad s = 1, 2, 3, 4, \ldots \quad \text{(rectangular)} \quad (6.24)$$

where $\Delta = |m^2 - m'^2 + n^2 - n'^2|$. This is for a square-bore dual case I laser and for any two modes EH_{mn} and $EH_{m'n'}$. For the circular-bore dual case I device, we obtain

$$L = 4\pi^2 s a^2 / \lambda \left(|\rho_{nm}^2 - \rho_{n'm'}^2|\right) \quad \text{(circular)} \quad (6.25)$$

for any two modes LP_{nm} and $LP_{n'm'}$. There is some evidence that such choices of guide geometry can improve the average laser mode quality by increasing the proportion of the active signature over which the fundamental mode is preferred.

It may be asked 'When is a case I?' We can show from equations (6.10) and (6.20) that at $\lambda = 10 \mu m$, the plane-mirror coupling loss Γ_{mn} ($N \gg 1$) becomes comparable with the first-order single-pass alumina guide loss $2\alpha_{mn} L$ when $z \simeq 0.18 L^{-2/3}$ (z and L in mm), independent of guide width and nearly independent of mode number [12]. It seems reasonable to propose, as a working definition of a case I mirror, that z should be considerably less than this value, so that the coupling loss can be neglected. However, this typically requires $z < 1$ mm and rules out many existing case I mirrors. Another reasonable definition is that a dual case I laser, where the mirror-guide distances z are gradually increased, ceases to be dual case I when the fundamental resonator mode has appreciable non-EH_{11} content.

There are no universally accepted definitions of mode quality. Formally, we would wish to know all the significant terms in the decompositions $\Sigma a_{mn} E_{mn}$ or $\Sigma b_{pq} \Psi_{pq}$, over the full laser signature and for a range of cavity perturbations. This is a most unrealistic goal. Many users are satisfied if a laser, after a brief warm-up, yields a 'reasonable' single-dot transverse intensity pattern with long-term power variations of a few per cent. In this case, extensive but simple beam measurements are adequate. Other users may set fierce limits on short-term frequency fluctuations, unwanted mode beating and departure from pure EH_{11} or pure TEM_{00}. They may also demand wide tunability and no line-hopping or multi-lining.

6.4.3 Rigrod Analysis for Waveguide Lasers

A useful and deliberately simple model for the output power of a homogeneously broadened gas laser has been widely adopted in the guise of one or other form of the Rigrod equation. It allows comparisons of waveguide resonator theory with experiment by expressing a measurable quantity, the laser output power, in terms of resonator losses. The gain is assumed constant along the length L of the laser and does not depend on direction; when homogeneous broadening is dominant, it is fully described by

$$g(z) = \frac{g_0(v_0)}{1+I_++I_-} - 2\alpha = \frac{1}{I_+}\frac{dI_+}{dz} = -\frac{1}{I_-}\frac{dI_-}{dz}. \quad (6.26)$$

This $g(z)$ is the saturated gain in intensity per unit length experienced by both the *forward* and *backward* travelling plane waves in the amplifying section of the resonator. These plane-wave intensities are normalized to the line-centre saturation intensity $I_s(v_0)$, the intensity in whose presence the available gain drops to $g_0(v_0)/2$ (half the small-signal value). Thus,

$$I_+ = I(\text{forward})/I_s(v_0)\, I_-\equiv I(\text{backward})/I_s(v_0).$$

There is a uniform loss constant 2α per unit length. By considering the boundary conditions defined by the mirror reflectances $R_1 = 1 - A_1 - T_1$ and $R_2 = 1 - A_2 - T_2$ (where A is the mirror loss and T is the transmittance) and manipulating this gain equation, Rigrod obtained essentially this expression for line-centre output power when $\alpha = 0$:

$$P = \frac{I_s(v_0)A_b T_2 \sqrt{R_1}}{(\sqrt{R_1}+\sqrt{R_2})(1-\sqrt{R_1 R_2})}[g_0(v_0)L + \ln(\sqrt{R_1 R_2})]. \quad (6.27)$$

This represents the normalized intensity incident on the outcoupler, multiplied by (i) the normalization factor $I_s(v_0)$, (ii) the area of the incident beam A_b and (iii) the outcoupler transmittance T_2. Often, there is only one outcoupler ($T_1 = 0$) and all the dissipative loss is lumped into one term A; then $R_1 = 1 - A$ and $R_2 = 1 - T$. To account for a small uniform guide loss, we can put $R_1 = 1-A-4\alpha L$. The full equation (6.26) cannot be integrated explicitly for a resonator but Rigrod [13] derived a useful approximate form. For a homogeneously broadened laser with a Lorentzian linewidth Δv and $f \equiv (v-v_0)/\Delta v$,

$$P_v = \frac{I_s(v_0)A_b T_2 \sqrt{R_1}[(g_0(v_0)-2\alpha(1+4f^2))L+(1+4f^2)\ln(\sqrt{R_1 R_2})]}{(\sqrt{R_1}+\sqrt{R_2})(1-\sqrt{R_1 R_2})(1-2\alpha L/\ln(\sqrt{R_1 R_2}))}$$
(6.28)

The waveguide beam area is usually defined as $A_b = \pi\omega_e^2$, where $\omega_e \equiv \omega_0/\sqrt{2}$ is the $1/e$-power radius of the EH_{11} approximating Gaussian (see Section 6.4.1). For a square guide, $A_b \approx \pi(0.70/\sqrt{2})^2 \approx 0.78a^2$. If the resonator mode *is* pure EH_{11}, the assumption of uniform A_b (and, hence, uniform active medium 'filling factor') will hold well—better than for open resonators with Gaussian beams of varying $\omega(z)$.

Note that we use a distributed loss term 2α, referring to "intensity" or magnitude-squared, not to magnitude: this point should be carefully checked in any literature. Unluckily, if the guide is narrow enough to force EH_{11}-like operation, then $2\alpha_{11}$ is often not greatly less than the *saturated* gain; but if it is wide enough to make $2\alpha_{11}$ very small, perturbations and lack of discrimination may introduce higher-order modes whose losses are not very small. Common dual case I CO_2 waveguide lasers tend to have lumped losses A of 1%–3%; higher values naturally occur in more complex resonators. The best obtainable alumina waveguides seem to have losses of $2\alpha_{11} \geq 1\%$ m^{-1} for $2a = 1.5$ mm, and the actual laser mode is never pure EH_{11}. Thus, it is usually unwise to ignore guide losses. Equation (6.28) is simple and useful; it pays at least lip service to distributed losses, and in many cases, it can be fitted fairly well to measured power.

In a single-mode model, the coupling loss is considered as part of the lumped loss A. In a multi-mode model, or in non-trivial resonators with several sources of loss, the easiest solution is often to insert an average distributed loss $-(1/L)\ln|\gamma|$ in equation (6.28). As mentioned in Section 6.3.3, we hope that loss variations will 'wash out' along the resonator path, but few or no real lasers have had their dissipative loss distributions examined in detail.

Rigrod analysis is often used to extract best-fit values of laser parameters, such as the g_0, I_s and A. One parameter, such as T_2 or active gain length, is varied while the others are kept constant, and output power is recorded for various pressures (or gas mixes, input powers and so on). Even after much curve-fitting, the error bars are almost always quite large ($\pm 10\%$ or worse). The equations are so simple that this is not surprising. What use is the idea of constant A_b, when we know that changing the pressure or over-pumping the discharge will seriously alter the output mode (and hence A_b, and hence 2α, and probably the effective g_0 and I_s too)? Why suppose that the loss is uniform in a waveguide laser, when we know that injecting an EH_{11} matched beam into a typical guide gives clear periodic variations in loss? The answer is that Rigrod analysis is *simple* and, though obviously deficient, *accurate enough to be useful* in understanding and designing waveguide laser resonators.

6.5 Real Waveguide Resonators: Experiment and Theory

6.5.1 Distant Mirrors

From time to time, the versatile toolbox of waveguide resonator theory has been refined on real-life lasers. Some problems, such as a spatially resolved treatment of gain and saturation, or accurate determination of waveguide E_{mn} and k_{mn}, are very hard. A few are both interesting and not too difficult, at least initially, and are summarized here: plane mirrors, then tilted plane mirrors, intra-line tunability and line selection; finally, unpleasant behaviour such as line hopping and resonator mode 'hooting'. Multi-mode predictions and detailed power/mode/frequency studies have rarely been made for real devices.

If we begin with a square-bore dual case I resonator, in theory, the fundamental transverse mode is pure EH_{11} and the others are pure higher-order EH_{mn} (which can be ranked in order of increasing loss). As one mirror is moved away, the round-trip efficiency for a pure EH_{mn} mode is given by the product of guide transmission and mirror-coupling efficiencies; if these are both near unity, we can write the round-trip loss as the sum of guide loss and coupling loss, or roughly $4\alpha_{mn}L + (1/6)(m^2+n^2)N^{-1.5}$ (see the similar small-term approximations, such as $R_1 = 1-A-4\alpha L$, introduced for the Rigrod treatment). But, however slightly at first, modes of the same parity now begin to couple among themselves. The fundamental resonator mode at a fixed wavelength is now whatever field minimizes the sum of guide loss and coupling loss. Thus, the resonator may do better than $4\alpha_{11}L + \Gamma_{11}$ by mixing in (say) some EH_{13}, EH_{15} and so on. But it may also do worse, and the fundamental mode loss may be higher or lower than the pure EH_{11} loss.

This is an important illustration of the central issue. For a given wavelength and geometry, the resonator can 'choose' only from the available set of self-consistent fields. The resonator cannot in this case choose pure EH_{11}, because pure EH_{11} is no longer a resonator mode—it ceased to be one when the mirror was moved.

If we fix L and vary z, the fundamental loss curve often shows non-monotonic behaviour (wiggles). This is confirmed by experiments where laser power fades when z is increased initially but revives when the plane mirror is pushed yet further away. If we fix z and vary L (easier to model than do), the fundamental loss curve often shows rather sharp periodic 'spiky' behaviour. The periods correspond with the 'beat wavelengths' over which pairs of important modes (notably EH_{11} and EH_{13}) change their relative guide propagation phase shifts by 2π. The published predictions of loss spikes (for example, [11] for rectangular geometry or [14] for circular) are in no doubt unrealistically sharp (see Section 6.3.4), but experiments confirm that phase-period effects are important in real lasers—often more important than attenuation effects. This is a further proof of shortness and fatness; if guides were very long and thin, high-order modes would be damped out efficiently.

There is another large literature based on beat wavelengths or phase periods in multi-mode waveguide structures. Simple relationships between lengths and mode-dependent phase shifts as in equation (6.24), if they can be assumed to hold for multiple modes in long, straight and low-loss rectangular guides, lead to 'degenerate' total phase shifts (modulo 2π). For propagation, the spatial pattern formed from a coherent sum of modes repeats itself at regular intervals; for resonators, the different transverse-mode frequencies coincide. There are also many sub-interval effects that scale with the guide Fresnel number $a^2/\lambda L$. These 'Talbot' or 'kaleidoscope' relations were (re)invented in the 1980s and applied to infrared hollow dielectric waveguides and slabs. They are probably more important nowadays in lasers, amplifiers, compact splitters and recombiners at shorter wavelengths around 1–2 μm, where the investments in technology are greater and the length scales $L \simeq a^2/\lambda$ are more convenient. The repetitions and coincidences, which would be exact in ideal guides that were lossless and otherwise obeyed first-order theory, are still impressive in imperfect real guides.

6.5.2 Tilted Mirrors and Folded Lasers

Some people, presented with a 'well-aligned' waveguide laser and an EH_{11}-like output, cannot resist 'tweaking' a mirror to be sure that it *is* well-aligned. The immediate result is familiar: the power drops sharply, the mode becomes asymmetric or multi-dot and things are never quite as good again. Tilt experiments are interesting for several reasons. First, of course, they are relatively easy, often easier than varying z or L. Figure 6.7 shows a typical arrangement [12]. A tilted plane mirror is a fair first approximation to a laser diffraction grating (see Section 6.5.3). Interesting things tend to happen with tilts of order λ/a, which is typically several milliradians for CO_2 guides: microradian precision is not needed, and the important dispersive tilts in CO_2 lasers happen to be 2–8 mrad. A curved or tapered guide can readily be modelled as a sequence of short segments with small tilts between them. Moreover, tilting is a simple and controllable way to introduce and study non-EH_{11} guide modes.

Suppose we tilt an ideal case I plane mirror ($z = 0$) around its vertical diameter: call this a pure x-tilt φ_x. To first order, we get a linear phase shift $\exp(i2\pi\varphi_x/\lambda)$ across the aperture width. The EH_{mn} fields are sine and cosine functions and their tilt-coupling coefficients emerge as sums of sinc functions. There is no 'y-coupling' between modes of different n. With circular guides (Bessel functions and polar coordinates), any tilt couples together modes with different n and different m.

For non-zero z, the tilt causes both a linear phase shift and a displacement Δx of the recoupling field, and a numerical solution is generally required. As φ_x and Δx grow to (typically) several mrad and a significant fraction of the guide width, the number of modes needed for a given accuracy tends to increase sharply. The theoretical dependence of resonator loss on z or L, for modest and relevant tilt values such as 6–8 mrad, can be surprisingly acute, and 'case I' assumptions may be improper even for $z = 1$–2 mm [12]. Results for a small square-bore rf-excited laser are also shown in Figure 6.7. The measured and predicted powers $P(\varphi_x)$ agree very well, and the Rigrod values for g_0 and I_s fall within the usual (rather wide) limits.

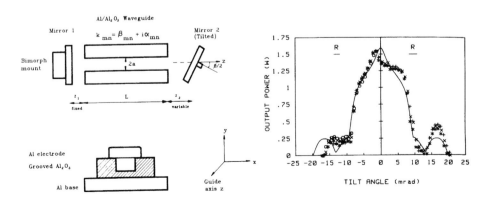

FIGURE 6.7 Square-bore rf-excited CO_2 waveguide laser with adjustable mirror mounts. (From Hill and Colley [12].)

Folded-waveguide CO_2 designs are popular because they squeeze long active lengths into compact devices and are well suited to slab-like rf excitation structures [15]. A V-fold or U-fold with plane mirrors is no harder to model than a distant plane mirror in a linear resonator. Curved folding mirrors can improve output mode quality, but they cause astigmatism and the mode-coupling coefficients need heavy computation [16]. Folds with 'part waveguiding' or Brewster-cut prisms reduce the losses and increase the theoretical difficulty. In one example [17], the CO_2 resonator had two waveguides, a fold prism, a plane mirror and a curved mirror, a Brewster window *and* an electro-optic modulator. Measured and predicted powers are in fair agreement but this is not even a true multi-transverse-mode model, simply an iterative lumped-element approach to a real device where no closed-form solution exists.

6.5.3 Tunability and Line Selection

For each available laser transition, a resonator has a set (perhaps small, perhaps very large) of potentially oscillating resonator modes. Many CO_2 devices can choose from several tens of distinct lines, each with 1–10 potential resonator modes. Well-behaved lasers ought to select the one mode/line combination offering the highest ratio of small-signal gain to threshold loss; the competitive processes characteristic of homogeneous broadening should suppress all other laser frequencies. This is often a reasonable description of real lasers. The main problem of CO_2 tunability theory is then to calculate the relevant gains and losses and find how far away from line centre we may move our chosen mode (usually the fundamental mode) before it slips below threshold or another mode takes over (There is a separate literature asking whether one can *formally* show that self-consistent fields exist and that if they exist they are orthogonal, but we need not explore this here).

Setting $P(\nu) = 0$ in equation (6.28) gives the maximum possible continuous tuning range:

$$2|\nu - \nu_0|_{\text{zero power}} = \Delta\nu \left(\frac{g_0(\nu_0)L}{2\alpha L - \ln(\sqrt{R_1 R_2})} - 1 \right)^{\frac{1}{2}}. \quad (6.29)$$

In these simple terms, the modelling problem is just to find the average loss $2\alpha = -(1/L)\ln|\gamma|$. However, no systematic check of laser tuning ranges against multi-mode theory seems to have been performed. To model 'hopping' to other resonator modes under the same CO_2 line, we need to know their losses and relative frequencies. These depend on guide material constants (poorly known) and precise guide geometry (hard to measure accurately and, if distorted, hard to model). To model hopping to another line, we need to know the relative gains of the two lines and the precise positions of the resonator modes within the line shapes.

In principle, we can probe the laser cavity for modes which, though still below threshold, are 'bubbling under' and threatening to lase, and this can be done neatly by reinjecting some of the laser's own output ([18,19]; note that UK authors, but not usually US authors, will call this an 'autodyne' configuration). The total problem of mode selection and tunability still looks very difficult. Nevertheless it is important to see, at least qualitatively, why waveguide CO_2 lasers are especially prone to line hopping. Their ability to run cw at high pressures (200–400 torr) offers a linewidth $\Delta\nu$ of 1–2 GHz. It is possible, but rather difficult, to obtain reliable dual case I single-mode CO_2 tuning ranges of $\simeq 1$ GHz [20,21]. Complex cavities can raise this to $\simeq 1.5$ GHz. Usually, unwanted lines or modes break in after a few hundred MHz. With narrow guides, only 100–300 grating lines are typically illuminated, but 'loss of resolution' is an incomplete explanation.

Lasing on a desired line, on the fundamental mode in a dual case I grating-tuned laser, will cease for one of three main reasons. We cannot exceed $|\nu - \nu_0| = c/4L$ because an adjacent axial mode will take over. We may fall below threshold according to equation (6.29). Or, another mode may achieve a higher gain/loss ratio. If this is a mode under the same line, we can infer the mode frequency separation from the laser signature or from heterodyne measurements, estimate the mode losses and hope that the experimental mode-hopping point coincides with the theoretical point where the gain/loss ratios are equal. This seems a straightforward test of theory. It is not and is seldom if ever tried. Real lasers do not obey ideal theory, and unwanted modes tend to be seen as nuisances to be eliminated, not as resonator features to be studied.

6.5.4 Resonator Mode Degeneracies: Hopping and Hooting

So far, we have accepted that a cw CO_2 waveguide laser chooses, from instant to instant, the one resonator mode with the highest ratio of small-signal gain to threshold loss. All other potential modes are frozen out by rapid collisional-broadening effects in the high-pressure active medium. At many times and for many devices, this is not true. Real gas discharge waveguide lasers are imperfect and cannot suppress other modes unless the gain/loss ratios are *clearly* larger than that of the fundamental mode, but what *clearly* means for a specific laser depends on molecular excitation rates and transverse mode spatial overlaps, and is hard to quantify. The rate equations can be readily solved only for cases well removed from real life. In very simple terms, the laser cannot be expected to run reliably single-frequency when two equal-loss modes lie within a few MHz or when mode hopping is imminent and the (gain/loss) ratios are very nearly equal; or, similarly, when line hopping is imminent. Thus, we may see more than one frequency emitted on one line, or more than one line emitted, or both; this is highly irritating to many users.

If, with a multi-mode resonator model, we plot the losses for the first few resonator modes as functions of z or φ_x, comparing curves for several known CO_2 wavelengths, it is tempting to identify regions of curve-crossing or near-coincidence with observations of multi-frequency output. This cannot be pushed too far, because even the multi-mode approaches outlined here are highly idealized and restricted to empty resonators, whereas multi-frequency waveguide CO_2 lasers are (by our definitions) non-ideal and badly behaved. For instance, we have not tried to model any effect that the two or more resonating modes have on each other; that is, a non-linear problem needing much more work. Despite these reservations, studies of cw mode beating ('hooting') and line-hopping provide encouraging support for multi-mode methods.

6.6 Summary

The general aim of this chapter has been to offer readers a brief and not mathematically detailed review of waveguide laser resonators, so that the literature can be approached with a knowledge of the theory's present limitations, and without a few common misconceptions. We began by introducing *modes of propagation* (both free-space and waveguide). Resonator modes are self-consistent field patterns, with certain definite round-trip phase shifts, which are potentially available to the laser cavity as *modes of oscillation*, and each is associated with a *single* frequency. The presence in the laser field of several guide modes does not imply several different oscillation frequencies.

From our point of view, there is nothing special about *waveguide* resonators as such. It is possible, and may sometimes help, to see them as conventional open resonators with more or less strongly perturbing 3D apertures. The transition from free-space propagation to wave-guiding, as the perturbation strengthens, is not necessarily sharp. And, we stress that many existing commercial and scientific 'waveguide' devices are not safely clear of this transition region; still less are they safely within the domain of single-mode theory. They are *short and fat*. This forces us towards a multi-mode treatment if we want reasonably accurate performance comparisons and predictions.

Current theory cannot explain all our experimental observations. Nevertheless, the steady expansion of waveguide laser technology in industrial, medical and remote-sensing applications should prompt continuing improvements in the depth and accuracy of waveguide resonator theory.

REFERENCES

1. Marcatili E A J and Schmeltzer R A 1964 Hollow metallic and dielectric waveguides for long distance optical transmission and lasers *Bell Syst. Tech. J.* **43** 1783–809.
2. Degnan J J 1976 The waveguide laser: a review *Appl. Phys.* **11** 1–33.
3. Snitzer E 1961 Cylindrical dielectric waveguide modes *J. Opt. Soc. Am.* **51** 491–8.
4. Kogelnik H and Li T 1966 Laser beams and resonators *Appl. Opt.* **15** 1550–67.
5. Hill C A and Hall D R 1985 Coupling loss theory of single-mode waveguide resonators *Appl. Opt.* **24** 1283–90.
6. Abrams R L and Chester A N 1974 Resonator theory for hollow waveguide lasers *Appl. Opt.* **13** 2117–25.
7. Degnan J J and Hall D R 1973 Finite-aperture waveguide-laser resonators *IEEE J. Quantum Electron.* **QE-9** 901–10.
8. Abrams R L and Bridges W B 1973 Characteristics of sealed-off waveguide CO_2 lasers *IEEE J. Quantum Electron.* **QE-9** 940–6.
9. Abrams R L 1972 Coupling losses in a hollow waveguide laser resonator *IEEE J. Quantum Electron.* **QE-8** 838–43.
10. Boulnois J-L and Agrawal G P 1982 Mode discrimination and coupling losses in rectangular waveguide resonators with conventional and phase-conjugate mirrors *J. Opt. Soc. Am.* **72** 853–60.
11. Hill C A 1988 Transverse modes of plane-mirror waveguide lasers *IEEE J. Quantum Electron.* **QE-24** 1936–46.
12. Hill C A and Colley A D 1990 Misalignment effects in a CO_2 waveguide laser *IEEE J. Quantum Electron.* **QE-26** 323–8.
13. Rigrod W W 1978 Homogeneously broadened cw lasers with uniform distributed loss *IEEE J. Quantum Electron.* **QE-14** 377–81.
14. Gerlach R, Wei D and Amer N M 1984 Coupling efficiency of waveguide laser resonators formed by flat mirrors: analysis and experiment *IEEE J. Quantum Electron.* **QE-20** 948–63.
15. Newman L A and Hart R A 1987 Recent R&D advances in sealed-off CO_2 lasers *Laser Focus/Electro-opt.* 80–96.
16. Banerji J, Davies A R, Hill C A, Jenkins R M and Redding J R 1995 Effects of curved mirrors in waveguide resonators *Appl. Opt.* **34** 3000–8.
17. Hill C A, Pearson G N, Tapster P, Vaughan J M and Miller G M 1996 Polarization states and output powers of a CO_2 laser with an electro-optic phase retarder *Appl. Opt.* **35** 5381–5.
18. Pearson G N, Harris M, Hill C A, Vaughan J M and Homby A M 1995 Inter-transverse-mode injection locking and subthreshold gain measurements in a CO_2 waveguide laser *IEEE J. Quantum Electron.* **QE-31** 1064–8.
19. Shackleton C J, Loudon R, Hill C A, Shepherd T J, Harris M and Vaughan J M 1995 Transverse modes above and below threshold in a single-frequency laser *Phys. Rev. A* **52** 4908–20.
20. Abrams R L 1974 Gigahertz tunable waveguide CO_2 laser *Appl. Phys. Lett.* **25** 304–6.
21. Gonchukov S A, Kornilov S T and Protsenko E D 1978 Tunable waveguide laser *Sov. Phys.—Tech. Phys.* **23** 1084–6.

REVIEWS

Abrams R L 1979 *Laser Handbook* vol 3, ed M L Stitch (Amsterdam: North-Holland) pp 41–88.

Hall D R and Hill C A 1987 *Handbook of Molecular Lasers* ed P K Cheo (New York and London: Marcel Dekker) pp 165–258.

Hill C A 1989 Theory of waveguide laser resonators *The Physics and Technology of Laser Resonators* ed D R Hall and P E Jackson (Bristol: Adam Hilger) (This forms the basis of the present chapter.)

Smith P W, Wood O R II, Maloney P J and Adams C R 1981 Transversely excited waveguide gas lasers *IEEE J. Quantum Electron.* **17** 1166–81.

OTHER READING

Bel'tyugov V N, Gracheva E V, Kuznetsov A A, Ochkin V N, Sobolev N N, Troitskii Yu V and Udalov Yu B 1988 Frequency selectivity of a multimode waveguide gas laser with a diffraction grating *Sov. J. Quantum Electron.* **18** 599–604.

Chester A N and Abrams R L 1972 Mode losses in hollow waveguide lasers *Appl. Phys. Lett.* **21** 576–8.

Degnan J J 1973 Waveguide laser mode patterns in the near and far field *Appl. Opt.* **12** 1026–30.

Henderson D M 1976 Waveguide lasers with intracavity electro-optic modulators: misalignment loss *Appl. Opt.* **15** 1066–70.

Hill C A and Hall D R 1986 Waveguide resonators with a tilted mirror *IEEE J. Quantum Electron.* **QE-22** 1078–87.

Hill C A, Redding J R and Colley A D 1990 Multimode treatment of misaligned CO_2 waveguide lasers *J. Mod. Opt.* **37** 473–81.

Laakmann K D and Steier W H 1976 Waveguides: characteristic modes of hollow rectangular dielectric waveguides *Appl. Opt.* **15** 1334–40.

Merkle G and Heppner J 1983 CO_2 waveguide laser with Fox-Smith mode selector *IEEE J. Quantum Electron.* **QE-19** 1663–7.

Roullard F P III and Bass M 1977 Transverse mode control in high gain, millimeter bore, waveguide lasers *IEEE J. Quantum Electron.* **QE-13** 813–19.

Smith P W 1971 A waveguide gas laser *Appl. Phys. Lett.* **19** 132–4.

Tang F and Henningsen J O 1987 Conditions for single-line and single-mode tuning of a CO_2 waveguide laser *Appl. Phys. B* **44** 93–8.

7
Nonlinear Optics

Orad Reshef and Robert W. Boyd

CONTENTS

- 7.1 Basic Concepts 135
- 7.2 Mechanisms of Optical Nonlinearity 136
  - 7.2.1 Influence of Inversion Symmetry on Second-order Nonlinear Optical Processes 137
  - 7.2.2 Influence of Time Response on Nonlinear Optical Processes 137
  - 7.2.3 Non-resonant Electronic Response 137
  - 7.2.4 Molecular Orientation 137
  - 7.2.5 Electrostriction 137
  - 7.2.6 Photorefractive Effect 137
- 7.3 Nonlinear Optical Materials 138
- 7.4 Optics in Plasmonic Materials 140
  - 7.4.1 Linear Optical Properties 140
  - 7.4.2 Plasmonic Mechanisms of Optical Nonlinearity 140
  - 7.4.3 Epsilon-Near-Zero Nonlinearities 142
- 7.5 Second- and Third-harmonic Generation 142
- 7.6 Optical Parametric Oscillation 143
- 7.7 Optical Phase Conjugation 144
- 7.8 Self-focusing of Light 145
- 7.9 Optical Solitons 145
- 7.10 Optical Bistability 146
- 7.11 Optical Switching 147
- 7.12 Stimulated Light Scattering 147
  - 7.12.1 Stimulated Raman Scattering 147
  - 7.12.2 Stimulated Brillouin Scattering 148
- 7.13 Multi-photon Absorption 148
- 7.14 Optically Induced Damage 149
- 7.15 Strong-field Effects and High-order Harmonic Generation 149
- References 150

7.1 Basic Concepts

Nonlinear optics is the study of the interaction of light with matter under conditions such that the linear superposition principle is not valid. The origin of this breakdown of the linear superposition principle can usually be traced to a modification of the optical properties of the material medium induced by the presence of an intense optical field. With a few exceptions [1], only laser light is sufficiently strong to lead to a significant modification of the optical properties of a material system and, for this reason, the field of nonlinear optics is basically the study of the interaction of laser light with matter. In this context, it is important to distinguish two different sorts of nonlinear optical effects: (i) effects associated with the nonlinear optical response of the material contained within the laser cavity itself; and (ii) effects induced by a prescribed laser beam outside of the laser cavity. In this chapter, we are concerned primarily with the second possibility, which constitutes the traditional field of nonlinear optics. Nonlinear optical processes occurring within the laser cavity itself constitute a central aspect of laser physics, as described in Chapter 1, and lead to important effects such as laser instabilities and chaos [2] and self-mode-locking of lasers [3]. The treatment of nonlinear optics presented in this chapter is necessarily limited in scope. More detailed treatments can be found in various monographs on the subject [4–10] as well as in the research literature. The present approach follows most closely the notational conventions of Ref. [5].

Nonlinear optical effects can often be described by assuming that the response of the material system can be expressed as a power series expansion in the strength $\tilde{E}(t)$ of the applied laser field:

$$\tilde{P}(t) = \chi^{(1)}\tilde{E}(t) + \chi^{(2)}\tilde{E}^2(t) + \chi^{(3)}\tilde{E}^3(t) + \ldots$$
$$\equiv \tilde{P}^{(1)}(t) + \tilde{P}^{(2)}(t) + \tilde{P}^{(3)}(t) + \ldots, \quad (7.1)$$

where $\tilde{P}(t)$ is the induced dipole moment per unit volume, *i.e.* the dielectric polarization. Here, the first term describes ordinary linear optics and includes the linear susceptibility $\chi^{(1)}$, the second term describes the second-order nonlinear optical effects and includes the second-order susceptibility $\chi^{(2)}$, *etc.* We shall see later that there is a significant qualitative difference between even- and odd-order nonlinear optical effects. To summarize these differences briefly, the crystal symmetry determines whether even orders are present within a material, and odd-order nonlinearities allow for processes where the output frequency is identical to an input frequency. We note that second-order nonlinear optical effects involve processes involving the simultaneous interaction of three photons, whereas third-order processes involve the interaction of four photons. Thus, second-order nonlinear optics includes processes such as second-harmonic generation (*i.e.* where two waves at a frequency ω combine to form a wave at a frequency of 2ω), sum- and difference-frequency generation and optical rectification (*i.e.* where a static field is generated under intense illumination); in contrast, third-order nonlinear optical effects include processes such as third-harmonic generation, the intensity dependence of the refractive index and four-wave mixing processes.

Typically, a nonlinear polarization of the type described in equation (7.1) is used as a source term in the driven wave equation:

$$\nabla^2 \tilde{E} - \frac{\varepsilon}{c^2}\frac{\partial^2 \tilde{E}}{\partial t^2} = \frac{1}{\varepsilon_0 c^2}\frac{\partial^2 \tilde{P}^{NL}}{\partial t^2}. \quad (7.2)$$

Here, we are assuming that the material is lossless and dispersionless and that the slowly varying approximation holds. For different nonlinear processes, $\tilde{P}^{NL}$ is replaced with the appropriate field terms and nonlinear susceptibility tensor elements, which generates the corresponding nonlinear signal.

Equation (7.1) has been written in a highly simplified form. In general, the relation between the polarization and the applied laser field must treat the tensor nature of the nonlinear coupling and any possible frequency dependence of the nonlinear susceptibility elements. One particularly useful way of generalizing equation (7.1) to deal with such issues is to express $\tilde{P}(t)$ and $\tilde{E}(t)$ in terms of their frequency components as

$$\tilde{P}(r,t) = \sum_n P(\omega_n)e^{-i\omega_n t} \quad \tilde{E}(r,t) = \sum_n E(\omega_n)e^{-i\omega_n t}, \quad (7.3)$$

where the summation extends over all positive and negative frequency components of the field. We then define the second-order susceptibility to be the coefficient relating the amplitude of the nonlinear polarization to the product of two field amplitudes according to

$$P_i(\omega_n + \omega_m) = \sum_{jk}\sum_{(mn)} \chi_{ijk}^{(2)}(\omega_n + \omega_m, \omega_n, \omega_m) E_j(\omega_n) E_k(\omega_m). \quad (7.4)$$

Here i, j and k refer to the Cartesian components of the fields, and the notation (nm) indicates that we are to sum over n and m while holding the sum $\omega_n + \omega_m$ fixed. By way of illustration, second-harmonic generation is described using these conventions by the susceptibility $\chi_{ijk}^{(2)}(2\omega, \omega, \omega)$, sum-frequency generation by the susceptibility $\chi_{ijk}^{(2)}(\omega_1 + \omega_2, \omega_1, \omega_2)$ and difference-frequency generation by the susceptibility $\chi_{ijk}^{(2)}(\omega_1 - \omega_2, \omega_1, -\omega_2)$. We similarly define the third-order susceptibility through the relation

$$P_i(\omega_0 + \omega_n + \omega_m) = \sum_{jkl}\sum_{mno} \chi_{ijkl}^{(3)}(\omega_0 + \omega_n + \omega_m, \omega_0, \omega_n, \omega_m)$$
$$\times E_j(\omega_0) E_k(\omega_n) E_l(\omega_m). \quad (7.5)$$

Third-harmonic generation is then described by the susceptibility $\chi_{ijkl}^{(3)}(3\omega, \omega, \omega, \omega)$, and the intensity-dependent refractive index is described by $\chi_{ijkl}^{(3)}(\omega, \omega, \omega, -\omega)$. The intensity dependence of the refractive index is alternatively described in terms of the nonlinear refractive index coefficient n_2, defined by the relation

$$n = n_0 + n_2 I, \quad (7.6)$$

where I is the laser intensity, which is related to the nonlinear susceptibility through

$$n_2 = \frac{3}{4n_0 \varepsilon_0 c \,\mathrm{Re}(n_0)} \chi^{(3)}. \quad (7.7)$$

It is often convenient to measure I in units of W m^{-2}, in which case n_2 is measured in units of m^2 W^{-1}. We then find that numerically

$$n_2\left(\frac{m^2}{W}\right) = \frac{283}{n_0 \,\mathrm{Re}(n_0)} \chi^{(3)}\left(\frac{m^2}{v^2}\right). \quad (7.8)$$

In certain cases, n_2 is ill-defined, such as in low-index media where $n_2 I > n_0$. In these rare instances, $\chi^{(3)}$ becomes the preferred quantity with which nonlinear responses should be characterized [85].

7.2 Mechanisms of Optical Nonlinearity

In this section, we present a brief summary of the various physical mechanisms that can lead to a nonlinear optical response of a material system. We first make some general comments regarding the conditions under which various types of optical nonlinearities can occur.

7.2.1 Influence of Inversion Symmetry on Second-order Nonlinear Optical Processes

A well-known result states that the second-order susceptibility $\chi^{(2)}$ necessarily vanishes for a material possessing inversion symmetry. Thus, the second-order effects neither can occur in liquids, gases or glasses nor can they occur in any of the crystal classes that possess inversion symmetry.

7.2.2 Influence of Time Response on Nonlinear Optical Processes

It should be noted that only very fast physical mechanisms can lead to an appreciable response for processes in which the output frequency is different from the input frequencies, because, in order for such processes to occur, the material has to be able to respond at the difference frequencies of the various interacting fields. In contrast, processes such as the intensity-dependent refractive index can occur even as the consequence of sluggish mechanisms, because in this case, the average intensity of the incident light field can lead to a change of the refractive index. We thus conclude that only very fast processes can lead to processes such as harmonic generation.

7.2.3 Non-resonant Electronic Response

Perhaps, the most important source of optical nonlinearity is the response of bound electrons to an applied laser field. The electronic response can lead to both second- and third-order nonlinear optical processes. For the important case of non-resonant excitation, this mechanism has a very short response time. This response time can be estimated as the time required for the electron cloud surrounding the atomic nucleus to move in response to an applied laser field; this time is of the order of magnitude of the period associated with the motion of an electron in a Bohr orbit about the nucleus, which is of the order of magnitude of 10^{-16} s.

Non-resonant electronic response can be described theoretically in one of several ways. One is to solve Schrödinger's equation for an atom in the presence of an intense laser field and extract that part of the induced response which is second or third order in the amplitude of the applied field. Another is to develop a totally classical model of the optical response based, for instance, on adding nonlinear contributions to the restoring force introduced into the equation of motion used in the Lorentz model of the atom. These approaches lead to consistent predictions in relevant limits. At an even more elementary level, one can make order-of-magnitude estimates [4,11,12] of the size of the nonlinear optical response by arguing that the ratio of linear to nonlinear optical response will be of the order of $(E/E_{at})^n$, where E_{at} is the characteristic atomic electric field strength and n is the order of the nonlinearity. Since $E_{at} = m^2 e^5/\hbar^4 = 5.14 \times 10^{11}$ V m^{-1}, this argument leads to the prediction that

$$\chi^{(2)} \sim \hbar^4/m^2 e^5 = 2 \times 10^{-12} \, \text{m/V} \quad (7.9)$$

$$\chi^{(3)} \sim \hbar^8/m^4 e^{10} = 4 \times 10^{-24} \, \text{m}^2/\text{V}^2 \quad (7.10)$$

These values are in good order-of-magnitude agreement with the measured values for typical nonlinear optical materials.

7.2.4 Molecular Orientation

The molecular orientation effect occurs in anisotropic molecules and leads to a nonlinear optical response as a consequence of the tendency of molecules to become aligned along the electric field vector of the incident laser field. This process is illustrated in Figure 7.1. This alignment tends to increase the refractive index of the material, that is, it leads to a positive value of $\chi^{(3)}$. This process typically has a response time of the order of 1 ps and produces a nonlinear optical response of the order of 10^{-20} m^2 V^{-2}. The contribution to the third-order susceptibility resulting from molecular orientation can be expressed as

$$\chi^{(3)} = \frac{4N}{135} \frac{(\alpha_3 - \alpha_1)^2}{\kappa T}, \quad (7.11)$$

where N is the number density of molecules and $(\alpha_3 - \alpha_1)$ is the difference in polarizabilities along the principal dielectric axes of the molecule.

7.2.5 Electrostriction

Electrostriction is the tendency of materials to become compressed in the presence of a static or oscillating electric field. Since for most materials the refractive index increases with material density, this process leads to a positive value of $\chi^{(3)}$, typically of the order of 10^{-20} m^2 V^{-2}. The response time of electrostriction is typically of the order of 1 ns. The contribution to the third-order susceptibility resulting from electrostriction can be expressed as

$$\chi^{(3)} = \frac{1}{3} \varepsilon_0 C_T \gamma_e^2 \quad (7.12)$$

where C_T is the isothermal compressibility and where $\gamma_e \equiv \rho \partial \varepsilon / \partial \rho$ is the electrostrictive constant.

7.2.6 Photorefractive Effect

The photorefractive effect [13,14] leads to a large nonlinear optical response but one that cannot usually be described in terms of a third-order (or any order) nonlinear susceptibility. The photorefractive effect occurs as a consequence of the

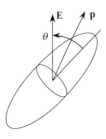

FIGURE 7.1 Origin of the molecular orientation effect, illustrating the tendency of an anisotropic molecule to become oriented in an electric field.

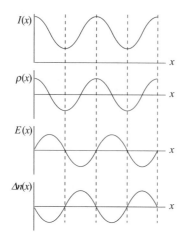

FIGURE 7.2 Origin of the photorefractive effect. $I(x)$ represents the spatially modulated laser intensity, $\rho(x)$ represents the free-charge distribution, $E(x)$ the static electric field created by this charge distribution and $\Delta n(x)$ is the resulting change in refractive index.

tendency of weakly bound electric charges within an optical material to migrate from regions of high intensity to regions of low intensity. This charge imbalance leads to the establishment of an electric field within the material, which modifies the refractive index of the material by means of the linear electro-optic (Pockels) effect. This basic process is illustrated in Figure 7.2. The photorefractive effect cannot be described in terms of a nonlinear susceptibility because the resulting change in refractive index tends to be independent of the strength of the incident laser field. Stronger laser fields tend to speed up the process of charge redistribution but do not change the final charge distribution. Typically, a laser beam of intensity $10\,\mathrm{kW\,m^{-2}}$ will produce a photorefractive response with a response time of the order of 1 s.

7.3 Nonlinear Optical Materials

The development of applications of nonlinear optics has historically been limited by the availability of materials with the required optical and environmental properties, and much effort has gone into the development of superior materials for use in nonlinear optics [15–17]. A brief representative sample of some materials of interest in second- and third-order nonlinear optics are given in Tables 7.1 and 7.2. More complete listings of material properties are to be found in various references [9,18–20]. A particularly useful approach towards the development of superior materials for nonlinear optics has been the development of nanocomposite materials or metamaterials [21–23,89–91].

TABLE 7.1

Properties of Several Second-Order Nonlinear Optical Materials

Crystal (class)	Transmission range (μm)	Refractive index (at 1.06 μm)	Nonlinear coefficient (pm V^{-1})
Silver gallium selenide	0.78–18	$n_o = 2.7010$	$d_{36} = 33$
		$n_e = 2.6792$	(at 10.6 μm)
AgGaSe$_2$ ($\bar{4}2m$)			
β-barium borate	0.21–2.1	$n_o = 1.6551$	$d_{22} = 2.3$
BBO (3m)		$n_e = 1.5425$	$d_{24} = d_{15} \leq 0.1$
Lithium iodate	0.31–5	$n_o = 1.8517$	$d_{31} = -7.11$
LiIO$_3$ (6)		$n_e = 1.7168$	$d_{33} = -7.02$
			$d_{14} = 0.31$
Lithium niobate		$n_o = 2.234$	$d_{31} = -5.95$
LiNbO$_3$ (3m)		$n_e = 2.155$	$d_{33} = -34.4$
Potassium dihydrogen phosphate	0.18–1.55	$n_o = 1.4944$	$d_{36} = 0.63$
		$n_e = 1.4604$	
KH$_2$PO$_4$ (KDP)			
KTiOPO$_4$	0.35–4.5	$n_x = 1.7367$	$d_{31} = 6.5$
KTP		$n_y = 1.7395$	$d_{32} = 5.0$
(mm^2)		$n_z = 1.8305$	$d_{33} = 13.7$
			$d_{24} = 6.6$
			$d_{15} = 6.1$

From a variety of sources including [19]. By convention, $d = \dfrac{1}{2}\chi^{(2)}$. The tensor nature of the nonlinear coefficients is expressed in contracted notation, in which the first index of d_{il} represents any of the three Cartesian indices and the second index l represents the product of two Cartesian indices according to the rule $l = 1$ implies xx, 2 implies yy, 3 implies zz, 4 implies yz or zy, 5 implies zx or xz and 6 implies xy or yx.

TABLE 7.2

Third-Order Nonlinear Optical Coefficients of Various Materials

Material	n_0	$\chi^{(3)}$ (m² V⁻²)	n_2 (m² W⁻¹)	Comments
Crystals				
Al_2O_3	1.8	3.1×10^{-22}	2.9×10^{-20}	
CdS	2.34	9.8×10^{-20}	5.1×10^{-18}	1.06 μm
Diamond	2.42	2.5×10^{-21}	1.3×10^{-19}	
GaAs	3.47	1.4×10^{-18}	3.3×10^{-17}	1, 1.06 μm
Ge	4.0	5.6×10^{-19}	9.9×10^{-18}	THG $|\chi^{(3)}|$
LiF	1.4	6.1×10^{-23}	9.0×10^{-21}	
Si	3.4	2.8×10^{-18}	2.7×10^{-18}	THG $|\chi^{(3)}|$
TiO_2	2.48	2.1×10^{-20}	9.4×10^{-19}	
ZnSe	2.7	6.1×10^{-20}	3.0×10^{-18}	1.06 μm
Glasses				
Fused silica	1.47	2.5×10^{-22}	3.2×10^{-20}	
As_2S_3 glass	2.4	4.0×10^{-19}	2.0×10^{-17}	
BK-7	1.52	2.8×10^{-22}	3.4×10^{-20}	
BSC	1.51	5.0×10^{-22}	6.4×10^{-20}	
Pb Bi gallate	2.3	2.2×10^{-20}	1.3×10^{-18}	
SF-55	1.73	2.1×10^{-21}	2.0×10^{-19}	
SF-59	1.953	4.3×10^{-21}	3.3×10^{-19}	
Nanoparticles				
CdSSe in glass	1.5	1.4×10^{-20}	1.8×10^{-18}	non-res.
CS 3-68 glass	1.5	1.8×10^{-16}	2.3×10^{-14}	res.
gold in glass	1.5	2.1×10^{-16}	2.6×10^{-14}	res.
Polymers				
Polydiacetylenes				
PTS		8×10^{-18}	3×10^{-16}	non-res.
PTS		-6×10^{-16}	-2×10^{-14}	res.
9BCMU			1.9×10^{-14}	$|n_2|$, res.
4BCMU	1.56	-1.3×10^{-19}	-1.5×10^{-17}	non-res, $\beta = 1 \times 10^{-12}$ m W⁻¹
Liquids				
Acetone	1.36	1.5×10^{-21}	2.4×10^{-19}	
Benzene	1.5	9.5×10^{-22}	1.2×10^{-19}	
Carbon disulphide	1.63	3.1×10^{-20}	3.2×10^{-18}	$\tau = 2$ ps
CCl_4	1.45	1.1×10^{-21}	1.5×10^{-19}	
Diiodomethane	1.69	1.5×10^{-20}	1.5×10^{-18}	
Ethanol	1.36	5.0×10^{-22}	7.7×10^{-20}	
Methanol	1.33	4.3×10^{-22}	6.9×10^{-20}	
Nitrobenzene	1.56	5.7×10^{-20}	6.7×10^{-18}	
Water	1.33	2.5×10^{-22}	4.1×10^{-20}	
Other materials				
Air	1.0003	1.7×10^{-25}	5.0×10^{-23}	
Vacuum	1	3.4×10^{-41}	1.0×10^{-38}	
Cold atoms	1.0	7.1×10^{-8}	2×10^{-3}	(EIT BEC)
Fluorescein dye in glass	1.5	$2.8(1+i) \times 10^{-8}$	$3.5(1+i) \times 10^{-6}$	$\tau = 0.1$ s

Here n_0 is the linear refractive index. The third-order susceptibility $\chi^{(3)}$ is defined by equation (7.1). This definition is consistent with that introduced by Bloembergen [4]. In compiling this table, we have converted the literature values when necessary to the definition of equation (7.1). The quantity β is the coefficient describing two-photon absorption. Reference [24] provides an extensive tabulation of third-order nonlinear optical susceptibilities. Other references used are [25–36].

7.4 Optics in Plasmonic Materials

Plasmonic materials (e.g. metals) are materials where unbound electrons in the conduction band make a significant contribution to the optical properties of the material [86–88]. Though the motion of electrons is usually linked to dissipative losses, metals have been empirically shown to exhibit stronger nonlinearities than insulating or dielectric materials. Plasmonic materials also possess many other favourable properties for nonlinear effects, such as the possibility to confine light to sub-wavelength scales. This effect is associated with a large local field enhancement, known as the 'lightning rod effect', and under the correct circumstances, it can be quite dramatic. Consider, for example, a sphere with a dielectric constant ε embedded in a background of permittivity ε_{BG}. If the sphere is placed in a uniform electric field of field strength E_0, the field within the sphere takes the value

$$E = \frac{3\varepsilon_{BG}}{\varepsilon + 2\varepsilon_{BG}} E_0. \quad (7.13)$$

This relation is valid in the quasi-static regime where the dimensions of the particle are smaller than the wavelength of an incoming light wave. The sphere enhances the electric field significantly if the real part of its permittivity is given by $\varepsilon = -2\varepsilon_{BG}$, known as the Fröhlich criterion [92,93]. Since the real part of the permittivity of plasmonic materials is strongly frequency-dependent and can take negative values, this material can show large field-enhancement effects at the frequency where the real part of the denominator vanishes. The enhancement is ultimately limited by the imaginary component of the permittivity of the plasmonic medium; however, it remains substantial, and a field enhancement $|E/E_0|$ of one to two orders of magnitude is routinely achieved for gold nanoparticles of various shapes.

7.4.1 Linear Optical Properties

For photon energies below the threshold for inter-band transitions, the plasmonic materials may be accurately modelled using the Drude model, which yields a dielectric function of the form

$$\varepsilon(\omega) = \varepsilon_\infty - \frac{\omega_D^2}{\omega^2 + i\gamma_D \omega}$$

$$= \left(\varepsilon_\infty - \frac{\omega_D^2}{\omega^2 + \gamma_D^2}\right) + i\left(\frac{\omega_D^2 \gamma_D}{\omega(\omega^2 + \gamma_D^2)}\right) \quad (7.14)$$

Here, ε_∞ is known as the high-frequency permittivity, γ_D is the electron damping term and ω_D is the plasma frequency. ε_∞ includes the residual polarization due to the positive background of the ion cores ($\varepsilon_\infty = 1$ for an ideal, undamped, free-electron gas, and usually $\varepsilon_\infty \lesssim 10$). The plasma frequency is the characteristic frequency at which a free-electron gas oscillates. It is given by

$$\omega_D \equiv \sqrt{\frac{Ne^2}{m^* \varepsilon_0}}, \quad (7.15)$$

where N is the free-electron volume density and m^* is the effective mass of the electron.

For a free-electron metal, this frequency also marks the metal/dielectric transition: as shown in equation (7.14), for frequencies significantly smaller than ω_D, the real part of ε becomes negative. Also, as the imaginary part scales with $1/\omega^3$, in this regime, it becomes large. Combined, these properties give the Drude material its metallic character. Near the plasma frequency, the real part of the permittivity takes small values, even crossing zero. Thus, the wavelength at which ω is equal to the plasma frequency is known as the epsilon-near-zero wavelength λ_{ENZ}. For solid conductors, the frequency of this zero-crossing is shifted to $\omega_D/\sqrt{\varepsilon_\infty}$, neglecting terms of order $(\gamma_D)^2$. This new frequency is known as the *shielded plasma frequency*.

At optical frequencies, the optical response of plasmonic materials exhibits significant deviations from the Drude model due to the onset of band-to-band transitions, even for what are traditionally considered 'good' metals (*e.g.* gold). These deviations can be accounted for by adding to the permittivity a series of Lorentz-oscillator terms, which are typically used to model bound electron effects [94]:

$$\varepsilon_L = -\sum_j f_j \frac{\omega_{L,j}^2}{\omega^2 + i\gamma_j \omega - \omega_{L,j}^2}. \quad (7.16)$$

Here, as for the Drude model in equation (7.14), each oscillator of index j represents the response of electrons harmonically oscillating at a resonance frequency of $\omega_{L,j}$ and a damping coefficient $\gamma_{L,j}$. The oscillator strength f_j is a unitless positive quantity bounded by unity. Though each of these coefficients holds a physical meaning, in the literature, they are typically treated as fitting parameters. Of note to our discussion earlier is that this contribution to the permittivity effectively shifts the epsilon-near-zero region for a given material so that it no longer depends solely on Drude coefficients.

7.4.2 Plasmonic Mechanisms of Optical Nonlinearity

Plasmonic materials exhibit some of the strongest observed ultrafast optical nonlinearities. The third-order nonlinear coefficients for a representative sample of these materials are given in Table 7.3. In addition to the aforementioned nonlinear mechanisms (*e.g.* molecular orientation, electrostriction, *etc.*), the plasmonic materials feature a few other important mechanisms of optical nonlinearity [95–98]:

- *Hot-electron nonlinearities.* Electrons can absorb heat from intense laser excitation. This absorption raises the free-electron temperature, changing the distribution of electrons in the band structure (also known as "Fermi smearing"), consequently modifying the effective mass of the electrons. Because these nonlinearities are thermal in origin, they are not instantaneous; however, they are typically the strongest and may still be ultrafast (*i.e.* possessing sub-picosecond rise and relaxation timescales).

TABLE 7.3

Third-Order Nonlinear Optical Coefficients of Selected Plasmonic Materials

Material	$\chi^{(3)}$ (m² V⁻²)	n_2 (m² W⁻¹)	λ (nm)	Pulse width	Comments
Metals					
Ag	3.4×10^{-17}		396	28 ps	DFWM
Ag	2.8×10^{-19}		1060	ps	THG
Au	2.1×10^{-16}	2.6×10^{-14}	532	28 ps	DFWM
Au	$(-1.4 + 5i) \times 10^{-16}$		532	30 ps	
Au	$(4.67 + 3.03i) \times 10^{-19}$		796	100 fs	Kretschmann–Raether
Au	7.6×10^{-19}		1060	ps	THG
Nanoparticles					
Ag	7×10^{-15}		532	4.5 ps	
Au	1.7×10^{-15}		532	7 ns	DFWM
Au	9.1×10^{-15}		532	4.5 ps	
Cu		2×10^{-11}	532	100 ps	
Cu	$(1.4 - 3.5) \times 10^{-16}$	$(2.0 - 4.2) \times 10^{-14}$	570 – 600	6 ps	
Cu	3.8×10^{-14}		532	4.5 ps	
Cu	$(1.9 - 6.0) \times 10^{-19}$		532	7 ns	
Cu		1.7×10^{-14}	770	130 fs	
Ni		5×10^{-15}	770	130 fs	
Pb		3×10^{-10}	532	100 ps	
Sn	2.1×10^{-14}		532	4.5 ps	
Transparent conducting oxides					
AZO	$(4 + 1i) \times 10^{-20}$	3.5×10^{-17}	1310	100 fs	FWM ENZ region
ITO		6×10^{-18}	970	150 fs	
ITO		2.6×10^{-16}	1240	150 fs	AOI = 0° ENZ region
ITO	$(1.60 + 0.50i) \times 10^{-18}$	1.1×10^{-14}	1240	150 fs	AOI = 60° ENZ region

Unless otherwise stated, the reported values were measured using a z-scan measurement [116,117]. Effective values for nanoparticle composites are distinguished from bulk material values, as various enhancement phenomena in small nanoparticles could account for many orders of magnitude difference in the reported values. References used are [85,101–115].

- *Conduction band filling.* Photons with energies larger than the inter-band transition are absorbed, promoting electrons from the valence band to the conduction band. For intense excitation, the conduction band gets filled, contributing to the nonlinear susceptibility largely in the form of saturable absorption.
- *Quantum-size effects.* Typically, intra-band transitions (*i.e.* due to electrons already in the conduction band) do not contribute to the nonlinear susceptibility; as free electrons experience no restoring forces, these bands contribute to the purely linear Drude response expressed in the equation (7.14) above. However, in small particles (*e.g.* <50 nm in diameter), nano-scale confinement leads to quantum-size nonlinear effects attributable to unbound electrons [99]. This nonlinearity is strongly size-dependent, scaling with the inverse of the particle volume.
- *Ponderomotive nonlinearities.* Certain metals (such as silver) also possess a ponderomotive nonlinearity. Here, charge carriers are repelled from high-intensity regions in the metal, depleting the electron density [100]. This effect manifests as a contribution to $\chi^{(3)}$ which looks like a highly dispersive (~ $1/\omega_4$) Kerr nonlinearity.

As a whole, this combination of nonlinear mechanisms exhibits a large dependence on laser parameters, such as the pulse duration and operating wavelength. This dependence is demonstrated explicitly in the large range of values reported for any single material listed in Table 7.3 and is discussed in greater detail for the case of gold in Ref. [97].

7.4.3 Epsilon-Near-Zero Nonlinearities

At the appropriate wavelength regime, hot-electron nonlinearities in particular can become quite important. Near their epsilon-near-zero wavelength, plasmonic materials possess intrinsic resonant properties that give rise to significant nonlinear optical effects. The origin of these large nonlinearities can be understood heuristically through the definition of the nonlinear refractive index coefficient n_2 defined in equation (7.7), where $n_2 \propto 1/(n \, \text{Re}(n))$. As the real part of the refractive index is typically smallest when the permittivity vanishes, n_2 is expected to diverge in the epsilon-near-zero regime.

Upon intense laser excitation near the epsilon-near-zero wavelength, the unbound electrons in a plasmonic material are excited resonantly; the change in the effective electron mass is so significant that the plasma frequency is dramatically red-shifted, modifying the permittivity throughout the nearby spectrum, as described by equation 7.14. This frequency shift yields a broadband refractive index change Δ_n, which is largest where ε is the smallest once losses are taken into account (Figure 7.3).

The small magnitude of the permittivity in the ENZ region gives rise to another unique field enhancement mechanism. In the absence of a surface charge, the interface conditions ensure the continuity of the perpendicular component of the electric displacement field. Thus, the magnitude of the electric field $\vec{E}$ within a medium is proportional to the external field $\vec{E}_0$ and to the inverse of its permittivity:

$$|\vec{E}_\perp| \propto \frac{1}{\varepsilon} |\vec{E}_{0,\perp}| \tag{7.17}$$

Equation (7.17) leads to the following expression for the total field within a medium of permittivity ε for a given angle of incidence (AOI) θ:

$$|\vec{E}| = |\vec{E}_0| \sqrt{\cos^2 \theta + \frac{\sin^2 \theta}{\varepsilon}} \tag{7.18}$$

Therefore, for a small permittivity and at an oblique angle, the electric field within the medium can be much larger than the incident field. This additional enhancement mechanism helps to explain the angle-dependence of the nonlinear refractive index n_2 reported in Table 7.3.

Combining the effects of resonantly exciting unbound electrons, and the multiple-field enhancement mechanisms, plasmonic materials have made accessible an impressive new regime of ultrafast third-order nonlinear effects. The total change in refractive index Δn has even been shown to exceed the linear refractive index n [101]. Because of their large nonlinearities, these materials have once again become an active area of research and hold considerable promise towards making major technological advances in nonlinear photonic devices.

7.5 Second- and Third-harmonic Generation

Second-harmonic generation is the process in which an incident field at frequency ω_1 is converted to an output field at frequency $\omega_2 = 2\omega_1$ by means of the second-order response of the material system. This was, in fact, one of the first nonlinear optical processes to be studied in detail [37] and was discovered shortly after the invention of the laser. Second-harmonic generation can be a very efficient process, leading to conversion efficiencies approaching 100%. This process is described pictorially in Figure 7.4.

Second-harmonic generation can be described mathematically by introducing coupled-amplitude equations that describe the propagation of the fundamental and second-harmonic waves. We take the fundamental wave to have amplitude $A_1(z) \exp(ik_1 z)$, where $k_1 = n_1 \omega_1 / c$ is its wavevector magnitude, and take the second-harmonic wave to have amplitude $A_2(z) \exp(ik_2 z)$, where $k_2 = n_2 \omega_2 / c$ is its wavevector magnitude. The coupled amplitude equations are derived by introducing the nonlinear polarizations $P(2\omega) = \chi^{(2)} A_1^2 \exp(2ik_1)$ and $P(\omega) = 2\chi^{(2)} A_2 A_1^* \exp[i(k_2 - k_1)z]$ of equation (7.4) into the driven wave equation (equation 7.2) and then making the slowly varying amplitude approximation. The resulting equations have the form

$$\frac{dA_1}{dz} = \frac{i\omega_1^2 \chi^{(2)}}{k_1 c^2} A_2 A_1^* e^{-i\Delta kz} \tag{7.19}$$

and

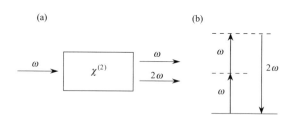

FIGURE 7.4 (a) The geometry of second-harmonic generation and (b) its description in terms of an energy-level diagram.

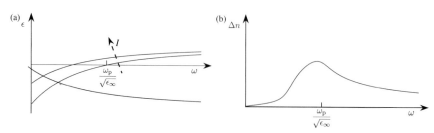

FIGURE 7.3 Hot-electron nonlinearities can lead to significant changes in optical properties. (a) Under intense laser excitation at the epsilon-near-zero wavelength, the plasma frequency is red-shifted as a function of pump intensity. (b) This red-shift may lead to a dramatic change in refractive index Δn.

$$\frac{dA_2}{dz} = \frac{i\omega_2^2 \chi^{(2)}}{k_2 c^2} A_1^2 e^{i\Delta kz}, \quad (7.20)$$

where $\Delta k = 2k_1 - k_2$. These equations express the fact that the amplitude of the second-harmonic wave is driven by the A_1^2 and that the generated second-harmonic wave acts back on the fundamental wave through the factor $A_2 A_1^*$. Coupled amplitudes for other nonlinear optical processes are derived using analogous procedures.

Second-harmonic generation (and in fact all nonlinear processes) can only occur with good efficiency only if the wave-vector mismatch factor Δk that appears in equations (7.12) and (7.13) is much smaller than the inverse of the length L of the interaction region. When this condition is met, the interaction is said to be phase-matched. Phase matching is typically achieved by making use of the natural birefringence of standard second-order nonlinear optical crystals and propagating the fundamental and second-order fields with orthogonal polarizations [38]. For $\Delta k = 0$, and assuming that only a fundamental field is present at the input to the medium, these equations can be solved exactly to find that

$$A_2(z) = \sqrt{n_1/n_2} A_1(0) \tanh(z/l) \quad (7.21)$$

where

$$l = \frac{\sqrt{n_1 n_2} c}{\omega_1 \chi^{(2)} |A_1(0)|}. \quad (7.22)$$

gives the characteristic distance over which the interaction occurs. Note that this model predicts that asymptotically the conversion efficiency can approach 100%. Second-harmonic generation in the plane-wave limit has been described more completely by Armstrong *et al.* [39], and the effects of laser-beam focusing on this process have been described by Boyd and Kleinman [40].

Radiation at the third-harmonic frequency can be created in one of two ways. One procedure is to create the third harmonic directly by means of a third-order interaction in which the amplitude of the nonlinear polarization is given by

$$P(3\omega) = \chi^{(3)} E(\omega) E(\omega) E(\omega). \quad (7.23)$$

Third-order interactions of this sort (and higher-order interactions, which give rise, for instance, to fifth- and seventh-harmonic generation) tend to be less efficient than second-order interactions but have the advantage that they can be used even at short wavelengths where standard nonlinear optical crystals are not transmitting. This approach to the generation of third-harmonic radiation has been described in detail by Ward and New [41] and by Miles and Harris [42].

The other approach to the generation of radiation at the third-harmonic frequency is to first generate a field at frequency 2ω through the process of second-harmonic generation followed by sum-frequency generation of the fields at frequencies ω and 2ω to produce an output at frequency 3ω. This approach can often be considerably more efficient than direct third-harmonic generation because lower-order processes tend to be stronger than third-order processes. In fact, through a judicious choice of experimental conditions, it is possible to produce radiation at the third-harmonic frequencies with efficiency exceeding 80% [43].

7.6 Optical Parametric Oscillation

An important technological application of nonlinear optics is the construction of parametric oscillators, which can produce tunable radiation over broad spectral regions spanning the infrared, visible and ultraviolet.

To understand the operation of an optical parametric oscillator (OPO), let us first examine the nature of the amplification that accompanies the process of difference frequency generation, which is illustrated in Figure 7.4. The left-hand side of this figure shows input waves at frequencies ω_3 and ω_2 with $\omega_3 > \omega_2$ incident on a second-order nonlinear optical material, within which the difference frequency wave at frequency $\omega_3 - \omega_2$ is generated. The energy-level diagram on the right-hand side of this figure reveals that one photon must be added to the field at frequency ω_2 for every photon that is created at frequency ω_1. The process of difference-frequency generation thus automatically leads to amplification of the lower-frequency input field (Figure 7.5).

This conclusion can be reached more rigorously by considering the coupled-waves equations describing the interaction of the two low-frequency waves in the presence of an undepleted pump wave at a frequency of ω_3,

$$\frac{dA_1}{dz} = \frac{i\omega_1^2 \chi^{(2)}}{k_1 c^2} A_3 A_2^* e^{-i\Delta kz} \quad (7.24)$$

$$\frac{dA_2}{dz} = \frac{i\omega_2^2 \chi^{(2)}}{k_2 c^2} A_3 A_1^* e^{i\Delta kz}, \quad (7.25)$$

where $\Delta k = k_3 - k_2 - k_1$. These equations can readily be solved for arbitrary boundary conditions. The solution for the special case of perfect phase matching ($\Delta k = 0$) and for no input at one of the lower frequencies (*i.e.* $A_2(0) = 0$) is given by

$$A_1(z) = A_1(0) \cosh \kappa z \Rightarrow \frac{1}{2} A_1(0) \exp(gz) \quad (7.26)$$

$$A_2(z) = i \left(\frac{n_1 \omega_2}{n_2 \omega_1} \right)^{1/2} \frac{A_3}{|A_3|} A_1^*(0) \sinh \kappa z \Rightarrow O(1) A_1^*(0) \exp(gz)$$

$$(7.27)$$

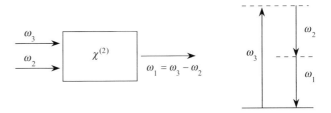

FIGURE 7.5 Illustration of the relation between difference frequency generation and optical parametric amplification. Note that amplification of the lower frequency input field ω_2 accompanies the creation of the difference frequency field ω_1.

where

$$g = \sqrt{k_1 k_2 k_j} = \frac{i\omega_j^2 \chi^{(2)} A_3}{k_j c^2}. \quad (7.28)$$

In these equations, the symbol $\Rightarrow$ denotes the asymptotic behaviour at large z and the symbol $O(1)$ denotes a number of the order of unity. Clearly, both waves asymptotically experience exponential growth.

The optical layout of an OPO is shown in Figure 7.6. Here, a pump wave of frequency ω_3 is incident on a second-order nonlinear optical crystal located inside an optical resonator. The end mirrors are assumed to be identical and to have reflectivities R_1 and R_2 at frequencies ω_1 and ω_2, respectively. The oscillator is said to be singly resonant if the end mirror reflectivity is large at either ω_1 or ω_2 and is said to be doubly resonant if the end mirror reflectivity is large at both ω_1 and ω_2. Generally speaking, doubly resonant oscillators have lower threshold pump intensities, but singly resonant oscillators are more readily operated in a stable manner because they do not require the independent establishment of a cavity resonance condition for the two separate frequencies ω_1 and ω_2.

Let us next consider the threshold condition for the establishment of oscillation in an OPO. For simplicity, we consider a simple model that applies to the doubly resonant oscillator. We assume that $R_1 = R_2 \approx 1$, that $\Delta k = 0$ and that the frequencies exactly meet the cavity resonance condition. The threshold condition can then be expressed as

$$\left(e^{2gL} - 1\right) = 2(1 - R). \quad (7.29)$$

Here, the left-hand side of the equation can be interpreted as the fractional energy gain per pass, and the right-hand side of the equation can be interpreted as the fractional energy loss per pass. The factor of two appears in the exponential because g is defined to be the amplitude gain coefficient. By expanding the exponential on the left-hand side to first order in gL, we find that the threshold condition can be expressed [44] as

$$gL = (1 - R). \quad (7.30)$$

Through the use of equation (7.28), we can use this result to determine the laser intensity required to reach the threshold for parametric oscillation.

The output frequencies of an OPO are usually controlled by adjusting the orientation of the nonlinear mixing crystal to determine which set of frequencies ω_1 and ω_2 (with $\omega_1 + \omega_2 = \omega_3$) satisfy the phase-matching condition ($\Delta k = 0$). OPOs tend to be broadly tunable because the tuning range is limited only by the limits of transparency of the crystal and by the limits over which the phase-matching relation can be established. Optical parametric oscillation was first observed experimentally by Giordmaine and Miller [44]. Continuous-wave OPO operation was first achieved by Smith *et al.* [45]. Early work on OPOs has been reviewed by Byer and Herbst [47]. An important material for the construction of OPOs is beta-barium borate [46].

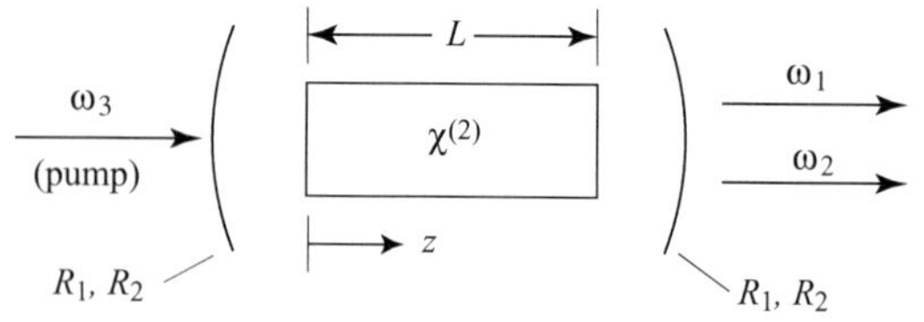

7.7 Optical Phase Conjugation

Optical phase conjugation [48–51] is a nonlinear optical process that has applications such as aberration correction, image processing and novel forms of interferometry [52]. The name phase conjugation derives from the fact that certain nonlinear optical processes have the ability to transform a field of the form

$$\tilde{E}(r,t) = A(r)e^{ikz - i\omega t} + c.c. \quad (7.31)$$

into the form

$$\tilde{E}_{pc}(r,t) = A^*(r)e^{-ikz - i\omega t} + c.c. \quad (7.32)$$

In addition to propagating in a direction opposite to that of the incident field, the wavefront of the phase-conjugate wave is changed from A to A^*. The nature of the phase-conjugation process is illustrated in Figure 7.7a, which shows an optical field falling onto a phase conjugating device which is often referred to as a phase conjugate mirror. The 'phase-conjugate' nature of this reversed wavefront allows it to remove, in double pass, the influence of aberrations in optical systems. The quantum statistical properties of the phase conjugation process have been described by Gaeta and Boyd [53].

The two primary means for forming a phase conjugate wavefront are degenerate four-wave mixing, which is illustrated in Figure 7.7b, and stimulated Brillouin scattering (SBS), which is illustrated in Figure 7.7c and will be discussed in further

FIGURE 7.6 Layout of the OPO.

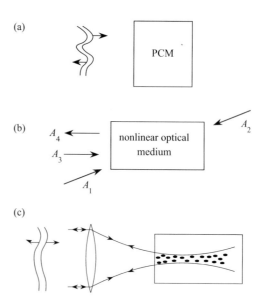

FIGURE 7.7 (a) Illustration of the nature of the phase-conjugation process, (b) phase conjugation by degenerate four-wave mixing and (c) phase conjugation by SBS.

detail later in the chapter. In the four-wave mixing interaction, a signal beam of amplitude A_3 interacts with two counter-propagating plane-wave pump beams of amplitudes A_1 and A_2 in a third-order nonlinear optical medium. Under these conditions, the dominant phase-matched contribution to the nonlinear optical susceptibility is of the form

$$P_{NL}(\omega_1) = 3\varepsilon_0 \chi^{(3)}(\omega_1, \omega_1, \omega_2, -\omega_2) A_1(\omega_1) A_2(\omega_2) A_3^*(\omega_2). \quad (7.33)$$

Here, ω_1 may be at the same frequency as ω_2. Since the nonlinear polarization is proportional to A_3^*, it will generate an output field that is the phase conjugate of the input field, that is a field proportional to A_3^*. The mutual interaction of the signal and conjugate beams can be described by the coupled amplitude equations

$$\frac{dA_3}{dz} = i\kappa A_4^* \quad \frac{dA_4}{dz} = i\kappa A_3^* \quad (7.34)$$

where the solution to these equations for the boundary conditions appropriate to the situation illustrated shows that the amplitude of the generated conjugate field is given by

$$A_4(0) = A_3^*(0) \frac{i\kappa}{|\kappa|} \tan|\kappa| L. \quad (7.35)$$

We see that the generated field is indeed proportional to the complex conjugate of the input field. We also see that a phase-conjugate mirror can have a reflectivity greater than 100%, because the pump waves provide energy to the phase-conjugate wave.

The other standard configuration for forming a phase-conjugate wavefront is through SBS, as illustrated in Figure 7.6c. SBS is described in more detail in Section 7.12 of this chapter. This process leads to phase conjugation because an aberrated input wave will produce a highly non-uniform volume intensity distribution in the focal region. The gain coefficient of the SBS process is proportional to the laser intensity, and the resulting non-uniform gain distribution will tend to generate an output wave whose wavefronts match those of the input wave.

7.8 Self-focusing of Light

Self-focusing is an example of a self-action effect of light. Other examples of self-action effects include self-trapping of light and the break-up of a beam of light into multiple filaments. These effects are illustrated schematically in Figure 7.8. These particular self-action effects can occur only if the nonlinear refractive index coefficient n_2 is positive. Self-focusing (Figure 7.8a) occurs because the refractive index at the centre of the laser beam is larger than in the wings of the laser beam. This effect causes the material medium to act like a positive lens, bringing the light to a focus within the material medium. Self-trapping (Figure 7.8b) occurs when the tendency of a beam to converge because of self-focusing precisely compensates for the tendency of the beam to diverge due to

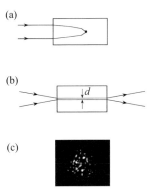

FIGURE 7.8 Several self-action effects of light are illustrated: (a) self-focusing, (b) self-trapping of light and (c) the break-up of a beam of light into multiple filaments.

diffraction effects. Simple arguments [5,54] show that this balance can occur only if the laser beam carries the critical power

$$P_{cr} = \lambda^2 / 8 n_0 n_2. \quad (7.36)$$

As a point of reference, P_{cr} has the value 30 kW for carbon disulphide at a wavelength of 700 nm. Self-trapped filaments are also known as spatial solitons. The use of spatial solitons has been proposed for optical switching applications [55]. According to the simple model leading to equation (7.36), self-trapped filaments can have any diameter d, as long as the total power contained in the beam has the value P_{cr}. Only if the power of the laser beam exceeds P_{cr}, self-focusing can occur. It is readily shown [5], on the basis of Fermat's principle, that the distance from the entrance face of the nonlinear material to the self-focus is given by

$$z_f = \frac{2n_0}{0.61} \frac{\omega_0^2}{\lambda} \frac{1}{(P/P_{cr} - 1)^{1/2}} \quad (7.37)$$

where ω_0 is the beam diameter. If the laser power is much greater than P_{cr}, another process known as filamentation (Figure 7.7c) can occur. In this process, the beam breaks up into multiple small filaments, each of which carries power P_{cr}. The origin of this process is that small perturbations on the incident laser wavefront experience exponential spatial growth as the consequence of near-forward four-wave mixing processes [56]. The maximum value of this growth rate is given by $g = (\omega/c)n_2 I$ and it occurs at the characteristic filamentation angle $\theta_{max} = \sqrt{2g/k}$. Filamentation is an undesirable process, and methods to suppress filamentation include the use of spatial filtering to remove aberrations from the laser wavefront, the use of specially structured beams [57] and the use of quantum interference effects [58] to eliminate the nonlinear response leading to filamentation.

7.9 Optical Solitons

Optical solitons are beams of light that propagate without changing their form, that is, they propagate as a self-similar solution to the wave equation [59]. By a *temporal soliton*, one means a pulse of light that propagates through a dispersive medium with no change in shape because of a balancing of dispersive and nonlinear effects. By a *spatial soliton*, one means a beam of light that propagates with a constant transverse profile because of a balance between diffraction and self-focusing effects. Similarly, a *spatio-temporal soliton* is a pulse that propagates without spreading in time or in the transverse directions.

The basic equation describing the propagation of optical solitons is a generalization of the so-called nonlinear Schrödinger equation and has the form

$$\frac{\partial A}{\partial z} + \frac{1}{v_g}\frac{\partial A}{\partial t} + \frac{i\beta}{2}\frac{\partial^2 A}{\partial t^2} + \frac{1}{2ik}\nabla_T^2 A = i2n_0\varepsilon_0 n_2 \omega_0 |A|^2 A \tag{7.38}$$

where $\vartheta_g \equiv \beta_1 = (\partial k/\partial \omega)^{-1}$ is the group velocity, $\beta_2 = \partial^2 k/\partial \omega^2$ is a measure of the dispersion in the group velocity, ∇_T^2 is the transverse Laplacian and ω_0 is the central frequency of the pulse. The term involving β_2 describes the tendency of the pulse to spread in time due to dispersive effects, the term involving ∇_T^2 describes diffraction and the term involving n_2 describes self-phase modulation and self-focusing effects. The propagation of temporal solitons can be described by this equation by discarding the transverse Laplacian, and the propagation of spatial solitons can be described by this equation by discarding the time derivatives. This equation possesses solutions relevant to many different physical situations. For instance, it possesses solutions in the form of bright solitons (*e.g.* a bright pulse on a zero background) or dark solitons (a decrease in intensity in an otherwise uniform non-zero background). Optical solitons can occur only for certain values of the material parameters. For instance, bright temporal solitons can occur only if n_2 and β_2 have opposite signs, and dark temporal solitons can occur only if these quantities have the same sign. Similarly, bright spatial solitons can occur only if n_2 is positive, whereas dark spatial solitons can occur only if n_2 is negative. Only certain solutions to equation (7.38) are stable to small perturbations. For instance, bright spatial solitons are stable in one transverse dimension but are unstable in two transverse dimensions.

A particularly important example of optical solitons is bright temporal solitons propagating through an optical fibre. The solution to equation (7.38), with the transverse terms discarded, that describes this occurrence is given by

$$A_s(z,\tau) = A_s^0 \mathrm{sech}(\tau/\tau_0)e^{i\kappa z}, \tag{7.39}$$

where $\tau = t - z/v_g$ and where the pulse amplitude A_s^0 and pulse width τ_0 must be related according to

$$|A_s^0|^2 = \frac{-\beta_2}{2n_0\varepsilon_0 n_2 \omega_0 \tau_0^2} \tag{7.40}$$

and where $k = -\beta_2/2\tau_0^2$ represents the phase shift experienced by the pulse upon propagation. Note that the condition (7.40) shows that β_2 and n_2 must have opposite signs in order for equation (7.39) to represent a physical pulse in which the intensity $|A_s^0|^2$ and the square of the pulse width τ_0^2 are both positive. We can see from equation (7.38) that, in fact, β_2 and γ must have opposite signs in order for group velocity dispersion to compensate for self-phase modulation.

For the case of an optical fiber, n_2 is positive with a value of approximately $3.2 \times 10^{-20} \mathrm{m}^2 \mathrm{W}^{-1}$. Bright optical solitons can then occur only if β_2 is negative, and consideration of the dispersion of the refractive index of silica glass shows that β_2 is negative only at wavelengths longer than 1.3 μm. Optical solitons of this sort have been observed experimentally [60].

7.10 Optical Bistability

Optical bistability refers to the possibility that a given optical system may possess two (or more) outputs for a given input. This possibility was first described theoretically by Szöke et al. [64] and first observed experimentally by Gibbs et al. [62]. Extensive treatments of bistability can be found in Ref. [61,63]. The realization that optical bistability can occur is important because it suggests that nonlinear optical techniques can be used to perform logical operations similar to those of electronic digital computers.

A standard design for a bistable optical device and its typical operating characteristics are shown in Figure 7.9. Here, a wave of amplitude A_1 is shown falling onto a device in the form of a Fabry–Pérot interferometer filled with a third-order nonlinear optical material. Such a device can be bistable in the sense that, under certain situations, there can be more than one output intensity for a given input intensity. The theoretical analysis of such a device proceeds by deriving a relation between the input amplitude A_1 and the output amplitude A_3. The result is the well-known Airy equation

$$A_3 = \frac{T^2 A_1}{1 - R^2 e^{2ikL-\alpha L}}, \tag{7.41}$$

where T is the amplitude transmittance of either end mirror and R is the amplitude reflectivity. Here k is the total propagation constant and α is the total intensity absorption coefficient of the material within the resonator, including both their linear and nonlinear contributions. A nonlinear contribution to k or α or both can lead to bistable behaviour. As an illustration, if

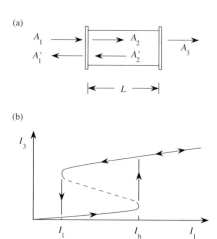

FIGURE 7.9 (a) Standard design for a bistable optical device and (b) typical operating characteristics.

we assume that the material within the resonator is a saturable absorber, for which the absorption coefficient α changes with intensity according to

$$\alpha = \frac{\alpha_0}{1 + I/I_s} \tag{7.42}$$

where I_s is the saturation intensity, we find that the output intensity $I_3 = |A_3|^2$ is related to the input intensity $I_1 = |A_1|^2$ according to

$$I_1 = I_3 \left(1 + \frac{C_0}{1 + 2I_3/TI_s}\right)^2 \tag{7.43}$$

where $C_0 = R\alpha_0 L/(1-R)$. Under certain conditions (in particular, for $C_0 > 8$), this equation predicts the occurrence of optical bistability. The input–output characteristics of the device under these conditions are illustrated schematically in Figure 7.8b. This curve has the form of a standard hysteresis loop and shows that, over a considerable range of input intensities, more than one output intensities can occur.

7.11 Optical Switching

Optical switching refers to the use of nonlinear optical methods to control the amplitude or propagation direction of one beam of light using a second beam of light. Early reviews that tended to define this field are given in Refs. [65] and [66].

A prototypical design for an all-optical switch is shown in Figure 7.10. In this design, a third-order nonlinear optical material is placed in one arm of a Mach–Zehnder interferometer. Let us first analyse the operation of such a device in the

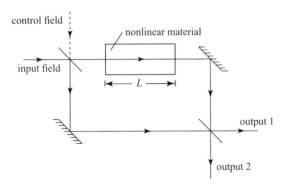

FIGURE 7.10 Typical design of an all-optical switching device, in the form of a nonlinear Mach–Zehnder interferometer.

absence of an applied control field. The relative phase of the two pathways through the interferometer changes with signal wave input intensity according to $\phi_{NL} = n_2(\omega/c)IL$, and thus, the input can be directed either towards output port 1 or port 2 depending upon the input intensity. The threshold intensity for switching from one output port to the other is given by the condition that the nonlinear phase shift ϕ_{NL} is equal to π radians. In this configuration, the nonlinear interferometer could be used to separate low-intensity pulses from high-intensity pulses in a pulse train of variable intensity. More sophisticated types of switching behaviour can be obtained by applying an additional input field to the control port of the interferometer. For instance, the presence or absence of a control field can be used to direct the signal field either to output port 1 or port 2, so that the device would operate as a router.

7.12 Stimulated Light Scattering

The scattering of light can occur either as a consequence of spontaneous or stimulated processes. The distinction between these two types of light-scattering processes can be understood by noting [67] that all light scattering occurs as a consequence of localized fluctuations in the optical properties of the material medium. From this perspective, spontaneous light scattering occurs as a consequence of fluctuations induced by thermal or by quantum mechanical zero-point fluctuations, and stimulated light scattering occurs as a consequence of fluctuations that are induced by the presence of the laser field.

It is believed that all spontaneous light-scattering processes possess a stimulated analogue, and quantitative models have been presented that relate the optical coefficients of these two types of processes [68]. Some of the important light-scattering processes are summarized in Table 7.4.

7.12.1 Stimulated Raman Scattering

Stimulated Raman scattering (SRS) [69] is characterized by exponential growth of a light wave at the Stokes sideband of the laser field. In particular, in this process, the intensity $I_S(t)$ of a beam of light at the Stokes frequency $\omega_S = \omega_L - \omega_v$, where ω_L is the laser frequency and ω_v is the vibrational frequency of the molecule, increases with propagation distance z according to

$$I_S(z) = I_S(0)e^{gI_L z} \tag{7.44}$$

TABLE 7.4

Summary of Light-Scattering Processes

Process	Spectral Shift (m^{-1})	Line Width (m^{-1})	Gain (m W^{-1})	Scattering Mechanism
Rayleigh scattering	None	5×10^{-2}	10^{-12}	Non-propagating density fluctuations
Rayleigh-wing scattering	None	10^3	10^{-11}	Fluctuations in the orientation of anisotropic molecules
Brillouin scattering	30	0.5	10^{-10}	Propagating sound waves
Raman scattering	10^5	500	5×10^{-11}	Vibrational modes of the molecules that constitute the scattering medium

where g is the Raman gain and I_L is the laser intensity. The Raman gain coefficient for various materials is given in Table 7.5. The Stokes shift for SRS is sufficiently large that SRS is an important technique for laser frequency shifting [70]. This nonlinear process was the one employed to create the first integrated continuous wave silicon lasers, since silicon cannot show optical gain based on normal stimulated emission in that it does not have a direct bandgap [118].

SRS can be understood by assuming that the optical polarizability $\alpha(t)$ of a molecule depends on the inter-atomic separation $q(t)$ according to

$$\alpha(t) = \alpha_0 + (\partial\alpha/\partial q)_0 \left[q(t) - q_0\right] \quad (7.45)$$

where q_0 is the equilibrium value of the inter-atomic separation [71]. Note that a periodic oscillation of $q(t)$ will produce a periodic oscillation of $\alpha(t)$ and consequently of the refractive index $n(t)$. Such a periodic modulation of $n(t)$ will tend to amplify light at a frequency detuned from the laser frequency by the vibrational frequency. Moreover, the simultaneous presence of the laser and Stokes beams will tend to reinforce the molecular vibration at the beat frequency of the waves. This process leads to exponential growth of the Stokes sideband, with a gain factor given by

$$g = \frac{i \epsilon_0 N_{\omega S}}{4 m \omega_\vartheta n S^c} \frac{(\partial a/\partial q)_0^2}{[\omega_s - (\omega_L - \omega_\vartheta)] + i\gamma} \quad (7.46)$$

where m is the reduced nuclear mass, γ is the vibrational damping rate and n_S is the refractive index at the Stokes frequency.

7.12.2 Stimulated Brillouin Scattering

The analysis of SBS shares many features in common with that of SRS but differs in the significant manner that SBS involves a collective excitation of the material medium. Consequently, the properties of SBS can be quite different in different directions. Our analysis here is restricted to geometries in which the laser and Stokes fields propagate in opposite directions.

The nature of the gain of the SBS process can be understood from the following perspective. The laser and counter-propagating Stokes waves beat together and, by means of the

TABLE 7.5

Properties of SRS for Several Materials

Substance	Frequency Shift (m^{-1})	Gain Factor g (m GW^{-1})
Liquids		
Benzene	99 200	0.03
Water	329 000	0.0014
Nitrogen	232 600	0.17
Oxygen	155 500	0.16
Gases		
Methane	291 600	0.0066 at 10 atm
Hydrogen	415 500 (vibrational)	0.015 (10 atm and above)
	45 000 (rotational)	0.005 (0.5 atm and above)
Deuterium	299 100 (vibrational)	0.011 (10 atm and above)
Nitrogen	232 600	7.1×10^{-4} (at 10 atm)
Oxygen	155 500	1.6×10^{-4} (at 10 atm)

FIGURE 7.11 Illustration of the nature of SBS.

TABLE 7.6

Properties of SBS for a Variety of Materials[a]

Substance	$\Omega_B/2\pi$ (MHz)	$\Gamma_B/2\pi$ (MHz)	g_0 (m GW^{-1})
CS$_2$	5850	52.3	1.5
Acetone	4600	224	0.2
Toluene	5910	579	0.13
CCl$_4$	4390	520	0.06
Methanol	4250	250	0.13
Ethanol	4550	353	0.12
Benzene	6470	289	0.18
H$_2$O	5690	317	0.048
Cyclohexane	5550	774	0.068
CH$_4$ (1400 atm)	150	10	1
Optical glasses	15 000–26 000	10–106	0.04–0.25
SiO$_2$	25 800	78	0.045

[a] Values are quoted for a wavelength of 0.694 μm. To convert to other laser frequencies ω, recall that Ω_B is proportional to ω, Γ_B is proportional to ω^2 and g_0 is independent of ω.

electrostrictive response of the material, produce a sound wave at the beat frequency which travels in the direction of the laser field. Some of the laser light then scatters from this sound wave and, in doing so, becomes Stokes-shifted and consequently reinforces the Stokes wave. But since the Stokes wave is now stronger, it tends to produce a stronger sound wave, and in this manner, the growth of the sound and Stokes wave mutually reinforce each other. These phenomena are illustrated in Figure 7.11. A consistent analysis of this situation shows that the Stokes wave experiences exponential amplification according to

$$I_S(z) = I_S(0) e^{g_0 I_L} \quad (7.47)$$

where the Brillouin gain factor is given by

$$g_0 = \frac{\gamma_e^2 \omega^2}{n v c^3 \rho_0 \Gamma_B} \quad (7.48)$$

and where γ_e is the electrostrictive constant introduced earlier in equation (7.12), ω is the laser frequency, n is the refractive index, v is the velocity of sound, ρ_0 is the mean material density and Γ_B is the phonon-damping rate. The properties of SBS for several materials are summarized in Table 7.6.

7.13 Multi-photon Absorption

Multi-photon absorption refers to optical processes in which more than one photon is removed from the optical field in a single optical transition. Some typical multi-photon absorption processes as well as normal one-photon absorption are shown

Nonlinear Optics

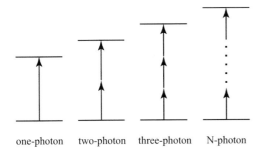

FIGURE 7.12 Illustration of one-photon and multi-photon absorption processes.

in Figure 7.12. Multiphoton absorption processes are important for a number of reasons, including the fact that multiphoton absorption constitutes a nonlinear loss mechanism that can limit the efficiency of certain optical interactions and also because of applications of multi-photon absorption such as the use of two-photon microscopy for biological applications [72] and 3D direct-laser writing [119–121].

Multiphoton absorption can be described theoretically by calculating the transition rate from the ground state to the final state through the use of time-dependent quantum-mechanical perturbation theory. The resulting expression is often referred to as Fermi's golden rule. This method applied to the case of two-photon absorption leads to a prediction for the value of the two-photon absorption cross section $\sigma^{(2)}_{ng}(\omega)$, which is defined such that the transition rate for transitions from level g to level n is given by

$$R^{(2)}_{ng} = \sigma^{(2)}_{ng}(\omega) I^2. \tag{7.49}$$

One finds that the two-photon cross section has the form

$$\sigma^{(2)}_{ng}(\omega) = \frac{1}{4n^2 \varepsilon_0^2 c^2} \left| \sum_m \frac{\mu_{nm} \mu_{mg}}{\hbar^2 (\omega_{mg} - \omega)} \right|^2 2\pi \rho_f (\omega_{ng} = 2\omega) \tag{7.50}$$

In this expression, μ_{nm} is the electric-dipole matrix element connecting levels n and m and $\rho_f(\omega_{ng} = 2\omega)$ is the density of states for the g to n transition evaluated at the laser frequency ω. Experimentally, two-photon cross sections are often quoted with intensities measured in units of photons m^{-2}s^{-1}. One finds, either from laboratory measurement [73] or from evaluation of equation (7.50), that a typical value of the two-photon cross section is

$$\bar{\sigma}^{(2)}_{ng} \approx 2.5 \times 10^{-58} \frac{m^4 s}{photon^2} \tag{7.51}$$

These predictions can readily be extended to higher-order multi-photon transition rates.

7.14 Optically Induced Damage

A topic of great practical importance is laser-induced damage of optical materials. Laser-induced damage is important because this process limits the maximum amount of optical power that can be transmitted through a given material and, consequently, limits the efficiency of many nonlinear optical interactions. The field of optical damage has been described in several review articles [74–77]; several investigations of optical damage include the following references [78,79]. There are several different mechanisms that lead to optical damage. In brief summary, these mechanisms are as follows:

- Linear absorption, leading to localized heating and cracking of the optical material. This is the dominant damage mechanism for continuous-wave and long-pulse ($\gtrsim 1$ μs) laser beams.
- Avalanche breakdown, which is the dominant mechanism for pulsed lasers (shorter then $\lesssim 1$ μs) for intensities in the range 10^{13}–10^{16}/W m^{-2}.
- Multiphoton ionization or dissociation of the optical material, which is the dominant mechanism for intensities in the range 10^{16}–10^{20} W m^{-2}.
- Direct (single cycle) field ionization, which is the dominant mechanism for intensities $> 10^{20}$ W m^{-2}.

This summary suggests that the avalanche breakdown mechanism is the dominant optical damage mechanism for laser pulses most often encountered in the laboratory and in applications. The nature of this mechanism is that a small number of free electrons initially present within the optical material are accelerated to high energies through their interaction with the laser field. These electrons can then impact-ionize other atoms within the material, thereby producing additional electrons which are subsequently accelerated by the laser field and eventually producing still more electrons. Some fraction of the energy imparted to each electron will lead to a localized heating of the material, which can eventually lead to damage of the material due to cracking or melting. A small number of electrons initially present within the material are created by one of several processes, including thermal excitation multi-photon excitation, or free electrons resulting from crystal defects.

Empirical evidence shows that, for many materials and for laser pulse lengths τ in the range of 10 ps to 100 ns, the threshold fluence for laser damage increases with pulse duration as $\tau^{1/2}$, and consequently, the threshold intensity decreases with pulse duration as $\tau^{-1/2}$. Results for the case of fused silica are shown in Table 7.7.

7.15 Strong-field Effects and High-order Harmonic Generation

Recent advances have led to the development of lasers that can produce pulses of a fraction of a femtosecond in duration. New nonlinear optical phenomena become accessible with these short laser pulses for two different reasons: (i) ultrashort laser pulses of only modest energy can produce super-intense fields; and (ii) nonlinear optical self-action effects are qualitatively different when excited by such short pulses, because of the dominance of dispersive and space-time coupling effects. Some of these new features are reviewed in the present section.

TABLE 7.7
Optical Damage Threshold of Fused Silica[a]

Pulse Duration	Threshold Fluence (kJ m^{-2})	Threshold Intensity (GW m^{-2})	Comments
1 ps	13	1.3×10^7	Deviation from $\tau^{1/2}$ scaling
10 ps	41	4.1×10^6	
100 ps	130	1.3×10^6	
1 ns	410	4.1×10^5	
10 ns	1300	1.3×10^5	
100 ns	4100	4.1×10^4	

[a] From Stuart et al. [78] and other sources.

Let us first consider how nonlinear optical effects are modified when excited by a super-intense pulse. nonlinear optical effects have historically been modelled using the power-series expansion of equation (7.1), but this series is not expected to converge if the laser field strength E exceeds the atomic unit of field strength $E_{at} = m^2 e^5/\hbar^4 = 5.14 \times 10^{11}$ V m^{-1}. This field strength corresponds to a laser intensity of $I_{at} = 4 \times 10^{20}$ W m^{-2}, which constitutes the threshold field-strength for exciting nonperturbative nonlinear optical response.

One of the dramatic consequences of excitation with intensities comparable to the atomic unit of intensity I_{at} is the occurrence of high-harmonic generation [80,81]. In brief, if an atomic gas jet is irradiated by high-intensity laser radiation, it is observed that all odd harmonics of the laser frequency, up to some maximum value N_{max}, are emitted. The various harmonics below N_{max} are typically emitted with approximately equal intensity; such an observation is incompatible with a perturbative explanation of this phenomenon. Harmonic generation with N_{max} as large as 341 has been demonstrated.

This phenomenon can be understood in terms of a simple physical model [82]. One imagines an atomic electron that has received kinetic energy from the laser field and is excited to a highly elliptical orbit. The positively charged atomic nucleus is at one focus of this ellipse, and each time the electron passes near the nucleus, it undergoes violent acceleration and emits a short pulse of radiation. This radiation is in the form of a train of short pulses; the spectrum of the radiation is the square of the Fourier transform of this pulse train, which contains the harmonics of the oscillation period up to some maximum frequency that is approximately the inverse of the time the electron spends near the atomic core. This argument can be made qualitative to show that the maximum harmonic number is given by

$$N_{max}\hbar\omega = 3.17K + U_p \quad (7.52)$$

where $K = e^2 E^2/m\omega^2$ is the 'ponderomotive energy' (the kinetic energy of an electron in a laser field) and U_p is the ionization energy of the atom.

Nonlinear optical self-action effects are also profoundly modified through excitation with ultrashort laser pulses. New phenomena come into play, including self-steepening of the laser pulse and space time focusing effects, in which different spectral components of the laser pulse undergo differing amounts of self-focusing. These effects have been described by a nonlinear envelope equation [83] which is a generalization of equation (7.38) and have been studied extensively by Gaeta [84] in the context of super-continuum generation.

REFERENCES

1. Lewis G N, Lipkin D and Magel T T 1941 *J. Am. Chem. Soc.* **63** 3005.
2. Boyd R W, Raymer M G and Narducci L M (ed) 1986 *Optical Instabilities* (Cambridge: Cambridge University Press).
3. Spence D E, Kean P N and Sibbett W 1991 *Opt. Lett.* **16** 42. Sibbett W Gran R S and Spence D E 1994 *Appl. Phys.— Lasers Opt. B* **58** 171.
4. Bloembergen N 1964 *Nonlinear Optics* (New York: Benjamin).
5. Boyd R W 2020 *Nonlinear Optics*, 4th ed. (San Diego, CA: Academic Press).
6. Butcher P N and Cotter D 1990 *The Elements of Nonlinear Optics* (Cambridge: Cambridge University Press).
7. Hannah D C, Yuratich M A and Cotter D 1979 *Nonlinear Optics of Free Atoms and Molecules* (Berlin: Springer).
8. Shen Y R 1984 *The Principles of Nonlinear Optics* (New York: Wiley).
9. Sutherland R L 1996 *Handbook of Nonlinear Optics* (New York: Dekker).
10. Agrawal G P and Boyd R W (ed) 1992 *Contemporary Nonlinear Optics* (Boston, MA: Academic).
11. Boyd R W 1996 *Laser Sources and Applications* ed A Miller and D M Finlayson (Scottish Universities Summer School in Physics) (Bristol: IOPP).
12. Boyd R W 1999 *J. Mod. Opt.* **46** 367.
13. Günter P and Huignard J-P (ed) 1988 *Photorefractive Materials and Their Applications Vols 1 and 2* (Topics in Applied Physics 61 and 62) (New York: Springer).
14. D D Nolte (ed) 1995 *Photorefractive Effects and Materials* (Dordrecht: Kluwer).
15. Prasad P N and Williams D J 1991 *Introduction of Nonlinear Optical Effects in Molecules and Polymers* (New York: Wiley).
16. Chemla D S and Zyss J 1987 *Nonlinear Optical Properties of Organic Molecules and Crystals Vols 1 and 2* (Orlando, FL: Academic).
17. Boyd R W and Fischer G L 2001 Nonlinear optical materials *Encyclopedia of Materials, Science and Technology* (Amsterdam: Pergamon) 6237–44.

18. Cleveland Crystals, Inc, 19306 Redwood Road, Cleveland, Ohio 44110 USA provides a large number of useful data sheets which may also be obtained at http://www.cleveland-crystals.com.
19. Smith A V maintains a public domain nonlinear optics data base SNLO, which can be obtained at http://www.sandia.gov/imrl/XWEB1128/xxtal.htm
20. Chase L L and van Stryland E W 1995 *CRC Handbook of Laser Science and Technology* (Boca Raton, FL: Chemical Rubber Company) section 8.1 (This reference provides an extensive tabulation of third-order nonlinear optical susceptibilities. The values of $\chi^{(3)}$ given in this reference need to be multiplied by a factor of four to conform with the standard convention of Bloembergen, which is the convention used in the present article.)
21. Hache F, Ricard D, Flytzanis C and Kreibig U 1988 *Appl. Phys. A* **47** 347.
22. Sipe J E and Boyd R W 1992 *Phys. Rev. A* **46** 1614; Boyd R W and Sipe J E 1994 *J. Opt. Soc. Am. B* **11** 297; Fischer G L, Boyd R W, Gehr R J, Jenekhe S A, J E Osaheni J A, Sipe and Weller-Brophy L A 1995 *Phys. Rev. Lett.* **74** 1871.
23. Nelson R L and Boyd R W 1999 *Appl. Phys. Lett.* **74** 2417.
24. Chase L L and van Stryland E W 1995 *CRC Handbook of Laser Science and Technology* (Boca Raton, FL: Chemical Rubber Company) section 8.1
25. Bloembergen N et al. 1969 *Opt. Commun.* **1** 195.
26. Vogel EM et al. 1991 *Phys. Chem. Glasses* **32** 231.
27. Hall D W et al. 1989 *Appl. Phys. Lett.* **54** 1293.
28. Lawrence B L et al. 1994 *Electron. Lett.* **30** 447.
29. Carter G M et al. 1985 *Appl. Phys. Lett.* **47** 457.
30. Molyneux et al. S 1993 *Opt. Lett.* **18** 2093.
31. Erlich J E et al. 1993 *J. Mod. Opt.* **40** 2151.
32. Sutherland R L 1996 *Handbook of Nonlinear Optics* (New York: Dekker) ch 8.
33. Pennington D M et al. 1989 *Phys. Rev. A* **39** 3003.
34. Euler H and Kockel B 1935 *Naturwiss.* **23** 246.
35. Hau LV et al. 1999 *Nature* **397** 594.
36. Kramer M A, Tompkin W R and Boyd R W 1986 *Phys. Rev. A* **34** 2026.
37. Franken P A, Hill A E, Peters C W and Weinreich G 1961 *Phys. Rev. Lett.* **7** 118.
38. Midwinter J E and Warner J 1965 *Brit. J. Appl. Phys.* **16** 1135
39. Armstrong J A, Bloembergen N, Ducuing J and Pershan P S *Phys. Rev.* **127** 1918.
40. Boyd G D and Kleinman D A 1968 *J. Appl. Phys.* **39** 3597.
41. Ward J F and New G H C 1969 *Phys. Rev.* **185** 57.
42. Miles R B and Harris S E 1973 *IEEE J. Quantum Electron.* **QE-9** 470.
43. Craxton R S 1980 *Opt. Commun.* **34** 474; Seka W, Jacobs S D, Rizzo J E, Boni R and Craxton R S 1980 *Opt. Commun.* **34** 469.
44. Giordmaine J A and Miller R C 1965 *Phys. Rev. Lett.* **14** 973; Giordmaine J A and Miller R C 1966 *Appl. Phys. Lett.* **9** 298.
45. Smith R G et al. 1968 *Appl. Phys. Lett.* **12** 308.
46. Bosenberg W R, Pelouch W S and Tang C L 1989 *Appl. Phys. Lett.* **55** 1952.
47. Byer R L and Herbst R L 1977 *Tunable Infrared Generation* ed Y R Shen (Berlin: Springer).
48. Zel'dovich B Ya, Popovichev V I, Ragulsky V V and Faizullov F S 1972 *JETP Lett.* **15** 109.
49. Zel'dovich B Ya, Pilipetsky N F and Shkunov V V 1985 *Principles of Phase Conjugation* (Berlin: Springer).
50. Fisher R A 1983 (ed) *Optical Phase Conjugation* (Orlando, FL: Academic).
51. Boyd R W and Grynberg G 1992 *Contemporary Nonlinear Optics* ed G P Agrawal and R W Boyd (Boston, MA: Academic).
52. Gauthier D J, Boyd R W, Jungquist R K, Lisson J B and Voci L L 1989 *Opt. Lett.* **14** 325.
53. Gaeta A L and Boyd R W 1988 *Phys. Rev. Lett.* **60** 2618.
54. Svelto O 1974 *Progress in Optics* vol XII, ed E Wolf (Amsterdam: North-Holland).
55. Blair S, Wagner K and McLeod R 1994 *Opt. Lett.* **19** 1943.
56. Bespalov V I and Talanov V I 1966 *JETP Lett.* **3** 307.
57. Maillotte H, Monneret J and Froehly C 1990 *Opt. Commun.* **77** 241.
58. Jain M, Xia H, Yin G Y, Merriam A J and Harris S E 1996 *Phys. Rev. Lett.* **77** 4326.
59. Zakharov V E and Shabat A B 1972 *Sov. Phys.–JETP* **34** 62.
60. Mollenauer L F, Stolen R H and Gordon J P 1980 *Phys. Rev. Lett.* **45** 1095.
61. Gibbs H M 1985 *Optical Bistability* (Orlando, FL: Academic).
62. Gibbs H M, McCall S L and Venkatesan T N 1976 *Phys. Rev. Lett.* **36** 113.
63. Lugiato L A 1984 'Theory of optical bistability' *Progress in Optics* vol XXI, ed E Wolf (Amsterdam: North-Holland).
64. Szöke A, Daneu V, Goldhar J and Kurnit N A 1969 *Appl. Phys. Lett.* **15** 376.
65. Stegeman G I and Miller A 1993 *Photonics in Switching* (San Diego, CA: Academic).
66. Gibbs H M, Khitrova G and Peyghambarian N (ed) 1990 *Nonlinear Photonics* (Berlin: Springer).
67. Fabelinskii I L 1986 *Molecular Scattering of Light* (New York: Plenum).
68. Hellwarth R W 1963 *Phys. Rev.* **130** 1850.
69. Kaiser W and Maier M 1972 *Laser Handbook* ed F T Arechi and E O Schulz-DuBois (Amsterdam: North-Holland).
70. Simon U and Tittel F K 1994 *Methods of Experimental Physics* vol III, ed R G Hulet and F B Dunning (Orlando, FL: Academic).
71. Garmire E, Pandarese F and Townes C H 1963 *Phys. Rev. Lett.* **11** 160.
72. Denk W, Strickler J H and Webb W W 1990 *Science* **248** 73; Xu C and Webb W W 1997 *Topics in Fluorescence Spectroscopy, Volume 5: Nonlinear and Two-Photon-Induced Fluorescence* ed J Lakowicz (New York: Plenum) ch 11.
73. Xu C and Webb W W 1996 *J. Opt. Soc. Am. B* **13** 481.
74. Bloembergen N 1974 *IEEE J. Quantum Electron.* **10** 375.
75. Lowdermilk W H and Milam D 1981 *IEEE J. Quantum Electron.* **17** 1888.
76. Raizer Y P 1965 *Sov. Phys.–JETP* **21** 1009.
77. Manenkov A A and Prokhorov A M 1986 *Sov. Phys.–Usp.* **29** 104.
78. Stuart B C et al. 1995 *Phys. Rev. Lett.* **74** 2248; Stuart B C et al. 1996 *Phys. Rev. B* **53** 1749.
79. Du D et al. 1994 *Appl. Phys. Lett.* **64** 3071.

80. Ferray M et al. *1988 J. Phys. B* **21** L31.
81. Chang Z et al. 1997 *Phys. Rev. Lett.* **79** 2967.
82. Corkum P B 1993 *Phys. Rev. Lett.* **71** 1994.
83. Brabec T and Krausz F 1997 *Phys. Rev. Lett.* **78** 3282.
84. Gaeta A L 2000 *Phys. Rev. Lett.* **84** 3583.
85. Reshef O 2017 *Opt. Lett.* **42** 3225.
86. Maier S A 2007 *Plasmonics: Fundamentals and Applications* (New York, NY: Springer).
87. Ozbay E 2006 *Science* **311** 189.
88. Berini P 2009 *Adv. Opt. Photonics* **1** 484.
89. Lee J et al. 2014 *Nature* **511** 65.
90. Li G et al. 2017 *Nat. Rev. Mater.* **2** 17010.
91. Alam M Z et al. 2018 *Nat. Photonics* **12** 79.
92. Fröhlich H 1968 *Theory of Dielectrics: Dielectric Constant and Dielectric Loss* (Oxford: Oxford University Press).
93. Bohren C E and Huffman D R 1998 *Absorption and Scattering of Light by Small Particles* (Weinheim, Germany: Wiley-VCH Verlag GmbH).
94. Vial A et al. 2005 *Phys. Rev. B* **71** 085416.
95. Hache F et al. 1986 *J. Opt. Soc. Am. B* **3** 1647.
96. Hache F et al. 1988 *App. Phys. A* **47** 347.
97. Boyd R W, Shi Z and De Leon I 2014 *Optics Comm.* **326** 74.
98. Clerici M et al. 2017 *Nat. Commun.* **8** 15829.
99. Qian H, Xiao Y and Liu Z 2016 *Nat. Commun.* **7** 13153.
100. Ginzburg P et al. 2010 *Opt. Lett.* **35** 1551.
101. Alam M Z, De Leon I and Boyd R W 2016 *Nat. Photonics* **352** 795.
102. Ricard D, Roussignol P and Flytzanis C 1985 *Opt. Lett.* **10** 511.
103. Vogel E M, Weber M J and Krol D 1991 *Phys. Chem. Glasses* **32** 231.
104. Bloembergen N, Burns W K and Matsuoka M 1969 *Optics Comm.* **1** 195.
105. Smith D D et al. 1999 *J. Appl. Phys.* **86** 6200.
106. De Leon I et al. 2014 *Opt. Lett.* **39** 2274.
107. Ila D et al. 1998 *Nucl. Instr. Meth. Phys. Res. B* **141** 289.
108. Fukumi K et al. 1991 *Jpn J. Appl. Phys.* **30** L742.
109. Haglund R F et al. 1992 *Nucl. Instr. Meth. Phys. Res. B* **65** 405.
110. Haglund R F et al. 1993 *Opt. Lett.* **18** 373.
111. Mazzoldi P et al. 1996 *J. Nonlinear Opt. Phys. Mater.* **5** 285.
112. Huang H H et al. 1997 *Langmuir* **13** 172.
113. Falconieri M et al. 1998 *Appl. Phys. Lett.* **73** 288
114. Caspani L et al. 2016 *Phys. Rev. Lett.* **116** 233901.
115. Elim H I, Ji W and Zhu F 2006 *Appl. Phys. B* **82** 439.
116. Sheik-Bahae M, Said A A and Van Stryland E W 1989 *Optics Lett.* **14** 955.
117. Sheik-Bahae M et al. 1990 *IEEE J. Quantum Electron* **26** 760.
118. Rong H et al. 2005 *Nature* **433** 725.
119. Maruo S, Nakamura O and Kawata S 1997 *Opt. Lett.* **22** 132.
120. Kawata S et al. 2001 *Nature* **412** 697.
121. Vora K et al. 2012 *Appl. Phys. Lett.* **100** 063120.

8

Laser Beam Control

Jacky Byatt

CONTENTS

8.1 Transforming a Gaussian Beam with Simple Lenses 154
    8.1.1 Beam Concentration 155
        8.1.1.1 Calculating a Correcting Surface 155
        8.1.1.2 Depth of Focus 155
    8.1.2 Truncation 155
    8.1.3 Non-Gaussian Laser Beams 156
8.2 Transverse Modes and Mode Control 157
    8.2.1 Mode Control 158
    8.2.2 Injection Locking 158
    8.2.3 Mode Control with Phase-conjugate Mirrors 158
8.3 Single Axial Mode Operation 159
    8.3.1 Theory of Longitudinal Modes 159
    8.3.2 Selecting a Single Longitudinal Mode 160
        8.3.2.1 The Ring Laser 160
    8.3.3 Frequency Stabilization 160
8.4 Tunable Operation 162
8.5 Beam Shape and Astigmatism in Diode Lasers 163
    8.5.1 Correcting Astigmatism in Collimators 163
    8.5.2 Circularizing a Diode Laser 163
8.6 Q-switching, Mode-locking and Cavity Dumping 164
    8.6.1 Q-switching 164
        8.6.1.1 Rotating Mirrors 165
        8.6.1.2 Electro-optic and Acousto-optic Q-switching 165
        8.6.1.3 Passive Q-switching 165
    8.6.2 Cavity Dumping 165
    8.6.3 Mode-locking 166
        8.6.3.1 Active Mode-locking 167
        8.6.3.2 Passive Mode-locking 167
        8.6.3.3 Synchronous Pumping 167
8.7 Beam Quality—Limits and Measurement 167
    8.7.1 Frequency and Amplitude Stabilization 167
    8.7.2 Methods for Suppressing Amplitude Noise and Drift 168
8.8 Spatial Filtering 169
References 169
Further Reading 169

In most laser applications, it is necessary to focus, modify or shape the laser beam by using lenses and other optical elements. In general, laser-beam propagation can be approximated by assuming that the laser beam has an ideal Gaussian intensity profile, corresponding to the theoretical TEM_{00} mode. Coherent Gaussian beams have transformation properties that require special consideration. The output from single-mode (TEM_{00}) lasers is highly Gaussian (helium–neon lasers and argon–ion lasers are very good examples). In contrast, high-power lasers often operate with many modes and are highly non-Gaussian. The propagation of non-Gaussian lasers is discussed later in this chapter.

In the fundamental TEM_{00} mode, the beam emitted from a laser forms as a Gaussian transverse irradiance profile as shown in Figure 8.1. The Gaussian shape is truncated at some diameter either by the internal dimensions of the laser or by some limiting aperture in the optical train.

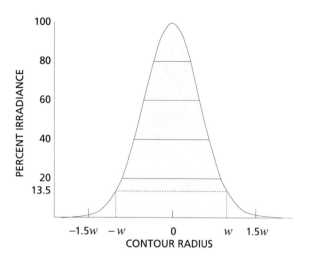

FIGURE 8.1 Irradiance profile of the fundamental TEM$_{00}$ mode.

8.1 Transforming a Gaussian Beam with Simple Lenses

Gaussian beams transform differently to non-Gaussian beams. Siegman [1] uses matrix transformations to treat the general problem of Gaussian beam propagation with lenses and mirrors. A less rigorous, but in many ways more insightful, approach to this problem has been developed by Self [2]. Self shows a method to model transformations of a laser beam through simple optics, under paraxial conditions, by calculating the Rayleigh range and beam waist location following each individual optical element. These parameters are calculated using a formula analogous to the well-known standard lens formula.

The standard lens equation is written as

$$\frac{1}{s} + \frac{1}{s''} = \frac{1}{f} \quad (8.1)$$

where s is the object distance, s'' is the image distance and f is the focal length of the lens. For Gaussian beams, Self has derived an analogous formula by assuming that the waist of the input beam represents the object, and the waist of the output beam represents the image. The formula is expressed in terms of the Rayleigh range of the input beam.

In the regular form

$$\frac{1}{s + z_R^2/(s-f)} + \frac{1}{s''} = \frac{1}{f} \quad (8.2)$$

or, in dimensionless form

$$\frac{1}{(s/f) + (z_R/f)^2/(s/f - 1)} + \frac{1}{s''/f} = 1. \quad (8.3)$$

In the far-field limit as $z_R \to 0$, this reduces to the geometric optics equation. A plot of s/f versus s''/f for various values of z_R/f is shown in Figure 8.2. For a positive thin lens, the three distinct regions of interest correspond to real object and real image, real object and virtual image, and virtual object and real image.

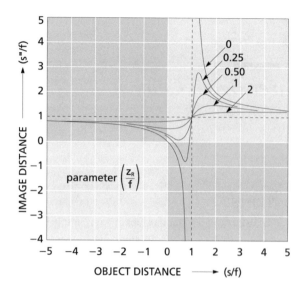

FIGURE 8.2 Plot of lens formula for Gaussian beams with normalized Rayleigh range of the input beam as the parameter.

The main differences between Gaussian beam optics and geometric optics, highlighted in such a plot, can be summarized as follows:

- There are both a maximum and a minimum image distance for Gaussian beams.
- The maximum image distance occurs at $s = f + z_R$, rather than at $s = f$.
- There is a common point in the Gaussian beam expression at $s/f = s''/f = 1$. For a simple positive lens, this is the point at which the incident beam has a waist at the front focus and the emerging beam has a waist at the rear focus.
- A lens appears to have a shorter focal length as z_R/f increases from zero (i.e. there is a Gaussian focal shift).

Self recommends calculating z_R, w_0 and the position of w_0 for each optical element in the system, in turn, so that the overall transformation of the beam can be calculated. To carry this out, it is also necessary to consider magnification: ω_0''/ω_0. The magnification is given by

$$m = \frac{\omega_0''}{\omega_0} = \frac{1}{\sqrt{\{[1-(s/f)]^2 + (z_R/f)^2\}}}. \quad (8.4)$$

The Rayleigh range of the output beam is then given by

$$z_R'' = m^2 z_R. \quad (8.5)$$

All the formulae are written in terms of the Rayleigh range of the input beam. Unlike the geometric case, the formulae are not symmetric with respect to input and output beam parameters. For back-tracing beams, it is useful to know the Gaussian beam formula in terms of the Rayleigh range of the output beam:

Laser Beam Control

$$\frac{1}{s} + \frac{1}{s'' + z_R{''}^2 / (s'' - f)} = \frac{1}{f}. \quad (8.6)$$

8.1.1 Beam Concentration

The spot size and focal position of a Gaussian beam can be determined from the previous equations. Two cases of particular interest occur when $s = 0$ (the input waist is at the first principal surface of the lens system) and $s = f$ (the input waist is at the front focal point of the optical system). For $s = 0$, we get

$$s'' = \frac{f}{1 + (\lambda f / \pi \omega_0^2)^2} \quad (8.7)$$

and

$$\omega = \frac{\lambda f / \omega_0}{[1 + (\lambda f / \pi \omega_0^2)]^{1/2}}. \quad (8.8)$$

For the case of $s = f$, the equations for image distance and waist size reduce to the following:

$$s'' = f$$

and

$$\omega = \lambda f / \pi \omega_0.$$

Substituting typical values into these equations yields nearly identical results, and for most applications, the simpler, second set of equations can be used.

8.1.1.1 Calculating a Correcting Surface

In cases where the laser uses a partially transmitting output mirror, the first lens seen by the laser beam is the output mirror itself. The beam is refracted as it passes through the second surface of the output mirror. If the mirror has a flat second surface, the apparent beam waist moves closer to the mirror and the divergence is increased. To counteract this, laser manufacturers often put a radius on the second surface to collimate the beam by making a waist at the output mirror, as shown in the case of a typical helium–neon laser cavity consisting of a flat high reflector and an output mirror with a radius of curvature of 20 cm separated by 15 cm. If the laser is operating at 633 nm, the intrinsic beam waist radius (ω_0), beam output radius (ω_{200}) and beam half-angle divergence θ are

$$\omega_0 = 0.13 \text{ mm} \quad \omega_{200} = 0.26 \text{ mm and } \theta = 1.5 \text{ mrad}$$

however, with a flat second surface, the divergence nearly doubles to 2.8 mrad. By solving equation (8.7) for f, with $s'' = 15$ cm, we see that the focal length of the correcting output coupler should be 15.1 cm. Using the lens-makers formula

$$\frac{1}{f} = (n-1)\left(\frac{1}{R_1} - \frac{1}{R_2}\right) \quad (8.9)$$

with the appropriate sign convention and assuming that $n = 1.5$, we get a convex correcting curvature of approximately 5.5 cm. At this point, the beam waist has been transferred to the output coupler, with a radius of 0.26 mm, and the far-field half-angle divergence is reduced to 0.76 mrad, a factor of nearly four.

Correcting surfaces are used primarily on output couplers whose radius of curvature is a metre or less. For longer radius output couplers, the effects are much less dramatic.

8.1.1.2 Depth of Focus

Depth of focus ($\pm \Delta z$), that is, the range in image space over which the focused spot diameter remains below an arbitrary limit, can be derived using the following equation:

$$\omega(z) = \omega_0 \left[1 + \left(\frac{\lambda z}{\pi \omega_0^2}\right)^2\right]^{1/2}. \quad (8.10)$$

The first step in performing a depth-of-focus calculation is to set the allowable degree of spot size variation. If we choose a typical value of 5%, or $\omega(z) = 1.05\, w_0$, and solve for $z = \Delta z$, the result is

$$\Delta z = \pm \frac{0.32 \pi \omega_0^2}{\lambda}.$$

Since the depth of focus is proportional to the square of focal spot size, and the focal spot size is directly related to *f*-number, the depth of focus is proportional to the square of the *f*-number of the focusing system.

8.1.2 Truncation

In a diffraction-limited lens, the diameter of the image spot is

$$d = K \times \lambda \times f/\# \quad (8.11)$$

where K is a constant dependent on truncation ratio (the ratio of the Gaussian beam diameter to the limiting aperture diameter) and pupil illumination, λ is the wavelength of light and *f*/# is the speed of the lens at truncation. The intensity profile of the spot is strongly dependent on the intensity profile of the radiation filling the entrance pupil of the lens. For uniform pupil illumination, the image spot takes on an Airy disc intensity profile as shown in Figure 8.3. If the pupil illumination is Gaussian in profile, an image spot of Gaussian profile results as shown in Figure 8.4. When the pupil illumination is between these two extremes, a hybrid intensity profile results.

In the case of the Airy disc, the intensity falls to zero at the point $d_{zero} = 2.44 \times \lambda \times f/\#$, defining the diameter of the spot (see Figure 8.3). When the pupil illumination is not uniform, the image spot intensity never falls to zero, making it necessary to define the diameter at some other point. This is commonly done for two points:

$$d_{FWHM} = 50\% \text{ intensity point}$$

and

$$d_{1/e^2} = 13.5\% \text{ intensity point}.$$

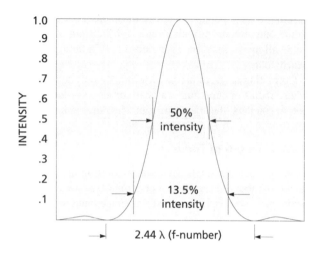

FIGURE 8.3 Airy disc intensity distribution at the image plane.

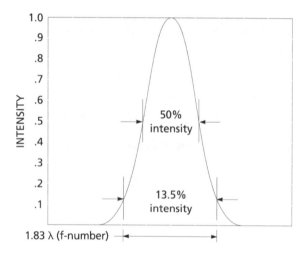

FIGURE 8.4 Gaussian intensity distribution at the image plane.

It is helpful to introduce the truncation ratio

$$T = \frac{D_b}{D_t} \quad (8.12)$$

where D_b is the Gaussian beam diameter measured at the $1/e^2$ intensity point and D_t is the limiting aperture diameter of the lens. If $T = 2$, which approximates uniform illumination, the image spot intensity profile approaches that of the classic Airy disc. When $T = 1$, the Gaussian profile is truncated at the $1/e^2$ diameter, and the spot profile is clearly a hybrid between an Airy pattern and a Gaussian distribution. When $T = 0.5$, which approximates the case for an untruncated Gaussian input beam, the spot intensity profile approaches a Gaussian distribution.

Calculation of spot diameter for these or other truncation ratios requires that K is evaluated. This is accomplished with the formulae

$$K_{FWHM} = 1.029 + \frac{0.7125}{(T-0.2161)^{2.179}} - \frac{0.6455}{(T-0.2161)^{2.221}} \quad (8.13)$$

and

$$K_{1/e^2} = 1.6449 + \frac{0.6460}{(T-0.2816)^{1.821}} - \frac{0.6455}{(T-0.2816)^{1.891}}. \quad (8.14)$$

The K function, plotted in Figure 8.5, permits calculation of on-axis spot diameter for any beam truncation ratio.

The optimal choice for truncation ratio depends on the relative importance of spot size, peak spot intensity and total power in the spot as demonstrated in Table 8.1. The total power loss, P_L, from the spot after propagating through an aperture can be calculated by using

$$P_L = e^{-2(D_t/D_b)^2} \quad (8.15)$$

for a truncated Gaussian beam. A good compromise between power loss and spot size is often a truncation ratio of one. When $T = 2$ (approximately uniform illumination), fractional power loss is 60%. When $T = 1$, d_{1/e^2} is just 8.0% larger than when $T = 2$, while fractional power loss is down to 13.5%. Because of this large saving in power with relatively little growth in the spot diameter, truncation ratios of 0.7–1.0 are typically used. Ratios as low as 0.5 might be employed when laser power must be conserved. However, this low value often wastes too much of the available clear aperture of the lens.

The mathematics of the effects of truncation on a real-world laser beam is beyond the scope of this chapter. For an in-depth treatment of this problem, please refer to Ref. [3].

8.1.3 Non-Gaussian Laser Beams

In the real world, perfectly Gaussian laser beams are very hard to find. Low-power beams from helium–neon lasers can be a close approximation, but the higher the power of the laser and the more complex the excitation mechanism (e.g. transverse discharges, flash-lamp pumping), or the higher the order of the mode, the more the beam deviates from the ideal.

The M^2 factor has come into general use to address the issue of non-Gaussian beams. For a theoretical Gaussian beam, the value of the radius-divergence product is

$$\omega_0 \theta = \lambda/\pi. \quad (8.16)$$

For a real laser beam, we have

$$\omega_{0M} \theta_M = M^2 \lambda/\pi > \lambda/\pi. \quad (8.17)$$

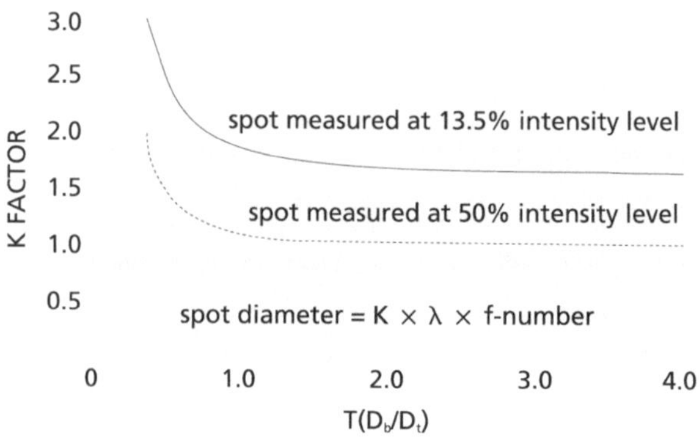

FIGURE 8.5 K factors as a function of truncation ratio.

Laser Beam Control

TABLE 8.1
Spot Diameters and Fractional Power Loss for Three Values of Truncation

Truncation Ratio	d_{FWHM}	d_{1/e^2}	d_0	$P_L(\%)$
∞	1.03	1.64	2.44	100
2.0	1.05	1.69	-	60
1.0	1.13	1.83	-	13.5
0.5	1.54	2.51	-	0.03

where ω_{0M} and θ_M are the $1/e^2$ intensity waist radius and the far-field half-divergence angle of the real laser beam, respectively. For a typical helium–neon laser operating in TEM_{00} mode, $M^2 < 1.1$. Ion lasers typically have an M^2 factor ranging from 1.1 to 1.7. For high-energy multi-mode lasers, the M^2 factor can be as high as 20 or 30. In all cases, the M^2 factor affects the characteristics of a laser beam and cannot be neglected in optical designs, and truncation, in general, increases the M^2 factor of the beam. M^2 factors into the equations for beam diameter and wavefront radius as follows:

$$\omega_M(z) = \omega_{0M}[1 + (z\lambda M^2/\pi\omega_{0M}^2)^2]^{1/2} \quad (8.18)$$

and

$$R_M(z) = z[1 + (\pi\omega_{0M}^2/z\lambda M^2)^2]. \quad (8.19)$$

The definition for the Rayleigh range remains the same for a real laser beam and becomes

$$z_R = \pi\omega_{0M}^2/\lambda \quad (8.20)$$

and the lens equation (8.6) becomes

$$\frac{1}{s + (z_R/M^2)/(s-f)} + \frac{1}{s''} = \frac{1}{f} \quad (8.21)$$

or, in normalized fashion [4]

$$\frac{1}{(s/f) + (z_R/M^2 f)^2/(s/f-1)} + \frac{1}{s''/f} = 1. \quad (8.22)$$

8.2 Transverse Modes and Mode Control

The fundamental TEM_{00} mode is only one of many transverse modes that satisfy the round-trip propagation criteria shown in equation (8.2). Figure 8.6 shows examples of the primary lower-order Hermite–Gaussian (rectangular) modes. Note that the subscripts m and n in the eigenmode TEM_{mn} are correlated to the number of nodes in the x- and y-directions. In each case, adjacent lobes of the mode are 180° out-of-phase.

The propagation equation can also be written in cylindrical form in terms of radius (ρ) and angle (ϕ). The eigenmodes ($E_{\rho\phi}$) for this equation are a series of axially symmetric modes, which, for stable resonators, are closely approximated by Laguerre–Gaussian functions, denoted by $TEM_{\rho\phi}$. For the lowest order mode, TEM_{00}, the Hermite–Gaussian and Laguerre–Gaussian functions are identical, but for higher-order modes, they differ significantly, as shown in Figure 8.7.

The mode, TEM_{01}^*, also known as the 'bagel' or 'doughnut' mode, is considered to be a superposition of the Hermite–Gaussian TEM_{10} and TEM_{01} modes, locked in phase quadrature [5].

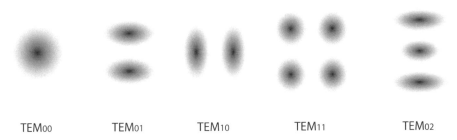

FIGURE 8.6 Low-order Hermite–Gaussian resonator modes.

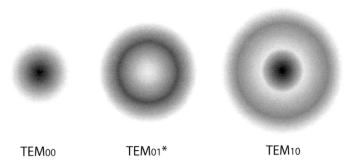

FIGURE 8.7 Low-order axisymmetric Laguerre–Gaussian resonator modes.

In real-world lasers, the Hermite–Gaussian modes predominate since strain, slight misalignment or contamination on the optics tend to drive the system towards rectangular coordinates. Nonetheless, the Laguerre–Gaussian TEM_{10} 'target' or 'bulls-eye' mode is clearly observed in well-aligned gas-ion and helium–neon lasers with the appropriate limiting apertures.

8.2.1 Mode Control

The transverse modes for a given stable resonator each have different beam diameters and divergences. The lower order the mode, the smaller the beam diameter, the lower the M^2 value and, although it is not intuitively obvious, the narrower the far-field divergence angle (the divergence is determined by the size of the uniphase lobes making up the higher-order mode, not the size of the mode itself). For example, the TEM_{01}^* doughnut mode is approximately 1.5 times greater than the diameter of the fundamental TEM_{00} mode, and the Laguerre TEM_{10} target mode is twice the diameter of the TEM_{00} mode. The theoretical M^2 values for the TEM_{00}, TEM_{01}^* and TEM_{10} modes are 1.0, 2.3 and 3.6, respectively [6]. Because of its smooth intensity profile, low divergence and ability to be focused to a diffraction-limited spot, it is usually desirable to operate in the lowest order mode possible, TEM_{00}. Lasers, however, tend to operate at the highest-order mode possible, either in addition to or instead of TEM_{00}, because the larger beam diameter may allow them to extract more energy from the lasing medium.

The primary method for reducing the order of the lasing mode is to add sufficient loss to the higher-order modes so that they cannot oscillate, without significantly increasing the losses at the desired, lower-order mode. In most lasers, this is accomplished by placing a fixed or variable aperture inside the laser cavity. Because of the significant differences in beam diameter, the aperture can cause significant diffraction losses for the higher-order modes without impacting the lower-order modes. As an example, consider the case of a typical argon-ion laser with a long-radius cavity and a variable mode-selecting aperture.

When the aperture is fully open, the laser oscillates in the TEM_{10} target mode. As the aperture is slowly reduced, the output changes smoothly to the TEM_{01}^* doughnut mode and, finally, to the TEM_{00} fundamental mode. (Many gas lasers will support only one mode at a time due to the nature of the discharge. Other lasers can have several modes operating simultaneously.)

In many lasers, the limiting aperture is provided by the geometry of the laser itself. For example, by designing the cavity of a helium–neon laser so that the diameter of the fundamental mode at the end of the laser bore is approximately 60% of the bore diameter, the laser will naturally operate in the TEM_{00} mode.

8.2.2 Injection Locking

In high-power high-gain lasers, suppressing higher-order modes with an aperture can be very difficult, if not impossible. One technique developed to address this problem is 'injection locking'.

An injection-locking scheme is shown in Figure 8.8. The relatively weak output of a well-controlled 'master' laser, operating at frequency ω_1 and intensity I_1, is fed into the cavity of a freely oscillating, higher-power, uncontrolled 'slave' laser, operating at frequency ω_0 and intensity I_0. If the frequency ω_1 is tuned very close to ω_0, the coherent photons from the master laser will compete with and overwhelm the ω_0 'noise build up' in the slave for the oscillator gain, 'locking' the output of the slave to that of the master, with output at frequency ω_1 and intensity I_0.

According to Siegman, the full locking range for the oscillator is given by

$$\Delta \omega_{\text{lock}} = 2(\omega_1 - \omega_0) \approx \frac{2\omega_0}{Q_e}\sqrt{\frac{I_1}{I_0}} \qquad (8.23)$$

where Q_e is a quality factor describing the losses from the resonator mirror and cavity configuration (for more information see Q-switching, later). As can be seen from the equation, the higher the intensity of the master laser beam, the wider the locking range.

A similar technique, called injection seeding, is used with high-energy pulsed lasers. In this case, the beam from a well-controlled cw laser replaces the 'noise' around which the pulsed laser builds up its output.

8.2.3 Mode Control with Phase-conjugate Mirrors

A major problem encountered in high-power and high-energy lasers and laser amplifiers is the wavefront distortion caused by inhomogeneities in the laser medium and optical cavity. These inhomogeneities can result from high-speed turbulence, naturally occurring impurities in the lasing medium, or thermal effects caused by the excitation method or the laser beam itself. Wavefront distortion and mode purity can be dramatically reduced by using phase-conjugate mirrors.

Optical phase conjugation is the generic name for a variety of non-linear optical processes that are capable of 'reflecting' waves in such a manner that both the direction of propagation

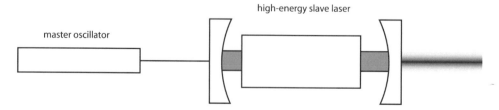

FIGURE 8.8 Configuration for laser injection locking.

Laser Beam Control

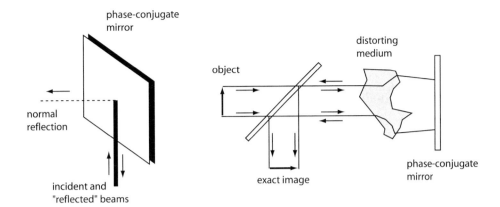

FIGURE 8.9 'Reflection' properties of a phase-conjugate mirror.

and the phase for each component of the wave are exactly reversed. This is illustrated in Figure 8.9. Whereas, with a regular mirror, only a ray normal to the mirror is reflected back upon itself, with a phase-conjugate mirror, *all* rays are reflected back upon themselves. Furthermore, if the wave passes through a distorting element on its way to the phase-conjugate mirror, the 'reflected' wave, after passing back through the distorting element, has exactly the same phase characteristics as the incident wave.

This has profound implications for high-energy laser applications, because, by using a phase-conjugate mirror as part of the laser cavity, or as a reflector in a laser amplifier, the distorting effects of the laser media can be effectively eliminated.

Two common techniques for generating a phase-conjugate mirror are stimulated Brillouin scattering (SBS), which uses acoustic waves to generate the reflection, and degenerate four-wave mixing, wherein two counter-propagating probe beams interact with the incident and reflected main beams to generate the reflection [7].

8.3 Single Axial Mode Operation

8.3.1 Theory of Longitudinal Modes

In a laser cavity, the requirement that the field exactly reproduce itself in relative amplitude and phase means that the only allowable laser wavelengths or frequencies are given by the formulae

$$\lambda = \frac{P}{N} \text{ or } v = \frac{Nc}{P} \quad (8.24)$$

where λ is the laser wavelength, v is the laser frequency, c is the speed of light in a vacuum, N is an integer whose value is determined by the lasing wavelength and P is the effective perimeter of the beam as it makes one round trip, taking into account the effects of the index of refraction, *etc*. For a conventional two-mirror cavity where the mirrors are separated by optical length L, these formulae revert to the familiar

$$\lambda = \frac{2L}{N} \text{ or } v = \frac{Nc}{2L}.$$

These allowable frequencies are referred to as longitudinal modes. The spacing between the modes, in terms of frequency, is given by

$$\Delta v = \frac{c}{P}. \quad (8.25)$$

As can be seen from equation (8.25), the shorter the laser cavity, the greater the cavity mode spacing. By differentiating the expression for v with respect to P we arrive at

$$\delta v = -\frac{Nc}{P^2}\delta P \text{ or } \delta v = -\frac{Nc}{2L^2}\delta L. \quad (8.26)$$

Consequently, for a helium–neon laser operating at 632.8 nm, with an effective cavity length of 25 cm, the mode spacing is approximately 600 MHz, and a 100 nm change in cavity length will cause a given longitudinal mode to shift by approximately 190 MHz.

The number of longitudinal laser modes that are present in a laser depends primarily on two factors: the length of the laser cavity and the width of the gain envelope of the lasing medium. For example, the gain of the red helium–neon laser is centred at 632.8 nm and has a width (FWHM) of approximately 1.4 GHz, meaning that, with a 25 cm laser cavity length, only two or three longitudinal modes can be present simultaneously, and a change in cavity length of less than 1 μm will cause a given mode to 'sweep' completely through the gain. By doubling the cavity length, the number of oscillating longitudinal modes that can fit under the gain curve doubles.

The gain of a gas-ion laser (e.g. argon or krypton) is approximately five times broader than that of a helium–neon laser, and the cavity spacing is typically much greater, allowing many more modes to oscillate simultaneously.

In most cases involving conventional lasers, only the portion of energy in the laser gain that is very close in frequency to that of a given longitudinal mode can contribute to the output energy in that mode; consequently, the greater the number of longitudinal modes present, the higher the total output energy. Likewise, a mode oscillating at a frequency near the peak of the gain will have higher energy than one oscillating at the fringes. This has a significant impact on the performance of a laser system because, as vibration and temperature changes cause small changes in the cavity length, modes sweep back and forth through the gain. A laser operating with only two or three longitudinal modes can experience power fluctuations of 10% or more, whereas a laser with ten or more longitudinal modes will see mode-sweeping fluctuations of 2% or less.

8.3.2 Selecting a Single Longitudinal Mode

A laser that operates with a single longitudinal mode is called a single-frequency laser. There are two ways to force a conventional two-mirror laser to operate with a single longitudinal mode. The first is to design the laser with a short enough cavity that only a single mode can be sustained. For example, in the helium–neon laser described before, a 10 cm cavity would allow only one mode to oscillate. This is not a practical approach for most gas lasers because, by the time the cavity is short enough to suppress additional modes, there is insufficient energy in the lasing medium to sustain any lasing action at all.

The second method is to introduce an element, typically a low-finesse Fabry–Pérot etalon, into the laser cavity. The free spectral range of the etalon should be several times the width of the gain curve, and the reflectivity of the surfaces should be sufficient to provide 10% or greater loss at frequencies half a laser mode spacing away from the etalon peak. The etalon is mounted at a slight angle to the optical axis of the laser to prevent parasitic oscillations between the etalon surfaces and the laser cavity.

Once the mode is selected, the challenge is to optimize and maintain its output power. Since the laser mode moves if the cavity length changes slightly, and the etalon pass band shifts if the etalon spacing varies slightly, it is important that both should be stabilized. Various mechanisms are used. Etalons can be passively stabilized by using zero-expansion spacers and athermalized designs, or they can be thermally stabilized by placing the etalon in a precisely controlled oven. Likewise, the overall laser cavity can be passively stabilized, or alternatively, the laser cavity can be actively stabilized by providing a mechanism to control cavity length.

8.3.2.1 The Ring Laser

Lasers with more than two resonator mirrors, as shown in Figure 8.10, are commonly used to produce a single longitudinal mode. These lasers have a mode spacing of c/P where P is the length of the perimeter of the laser cavity. Normally, these lasers produce counter-propagating standing waves, but by introducing a device such as a Faraday rotator into the laser cavity, or injecting leakage from a partially reflecting mirror back into the laser cavity, the laser beam can be constrained to propagate in a single direction as a travelling wave. The travelling wave sweeps through the laser gain, eliminating the spatial hole burning found with standing wave lasers, effectively increasing the homogeneity of the laser gain. When an etalon (or other frequency-selecting element) is introduced into the laser beam, more of the available energy is coupled into the single longitudinal mode than with a standing-wave laser. In addition, since adjacent modes are suppressed, the laser can be pumped with higher energy, further increasing single-mode output. For example, a ring dye laser can produce ten times the single frequency output of an equivalent standing-wave dye laser.

8.3.3 Frequency Stabilization

The frequency output of a single-longitudinal-mode laser can be stabilized by precisely controlling the laser cavity length. A moderate level of stability can be accomplished passively by building an athermalized resonator structure and carefully controlling the laser environment to eliminate expansion, contraction and vibration; or actively, by using a mechanism to determine the frequency (either relatively or absolutely) and quickly adjusting the laser cavity length to maintain the frequency within the desired parameters.

A typical stabilization scheme is shown in Figure 8.11. A portion of the laser output beam is directed into a low-finesse Fabry–Pérot etalon and tuned to the side of the transmission band. The throughput is compared with an unattenuated reference beam, as shown in the figure. If the laser frequency increases, the ratio of attenuated power to unattenuated power increases. If the laser frequency decreases, the ratio decreases. In other words, the etalon is used to create a frequency discriminant that converts changes in frequency to changes in power. By 'locking' the discriminant ratio at a specific value (e.g. 50%) and providing negative feedback to the device used to control cavity length, output frequency can be controlled. If the frequency increases from the pre-set value, the length of the laser cavity is increased to drive the frequency back to the set point. If the frequency decreases, the cavity length is decreased. The response time of the control electronics is

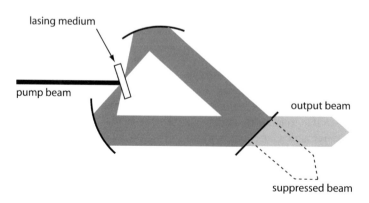

FIGURE 8.10 Ring laser cavity.

Laser Beam Control

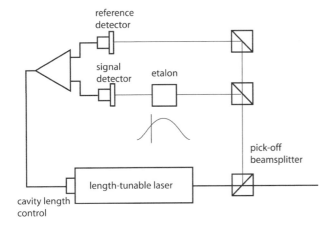

FIGURE 8.11 Laser frequency stabilization scheme.

determined by the characteristics of the laser system being stabilized. This system can, in general, be made as stable as the reference etalon. Mechanical, acoustic and thermal stabilization of the etalon can improve the absolute stability of this set-up.

Other techniques can be used to provide a discriminant. One common method used to provide an ultrastable, long-term reference is to replace the etalon or add to the etalon set-up, an absorption cell or atomic beam with a spectral absorption line within the laser's tuning band. The stabilization system can then be used to maintain the laser frequency at the centre of the appropriate transition. This method is the basis for modern atomic clocks.

A major problem encountered with tightly controlled stabilized lasers is unwanted frequency shifts caused by cavity length perturbations that are beyond the handling capability of the control electronics. This is a particular problem with dye lasers, where bubbles or thickness changes in the dye jet can change the effective cavity length sufficiently to move the frequency completely off of the discriminant. This is handled in commercial systems by providing two correction loops: a narrow-range high-frequency correction loop (typically with a piezoelectric actuator) that controls the cavity length when the system is in lock and a lower-frequency, longer-range correction loop (e.g. using a pair of galvanometer-driven tilting plates) to control the overall cavity length to within the range of the piezoelectric actuator. The discriminant must also be capable of providing the appropriate error signal over the maximum possible frequency jump. One method is to use two control etalons: one to provide a narrow perturbations and the second to provide a broader discriminant for long-range perturbations (see Figure 8.12).

Another stabilization method, shown in Figure 8.13, is used with commercial helium–neon lasers. It takes advantage of the fact that, for an internal mirror tube, the adjacent modes are orthogonally polarized. The cavity length is designed so that two modes can oscillate under the gain curve, as shown in Figure 8.14. The two modes are separated outside the laser by a polarization-sensitive beam splitter. Stabilizing the relative amplitude of the two beams stabilizes the frequency of both beams.

The cavity length changes needed to stabilize the laser cavity are very small. In principle, the maximum adjustment

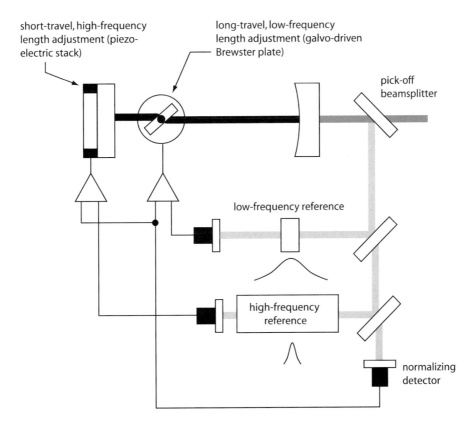

FIGURE 8.12 Correcting for long-range perturbations.

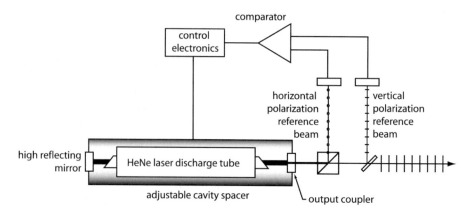

FIGURE 8.13 Frequency stabilization for a helium–neon laser.

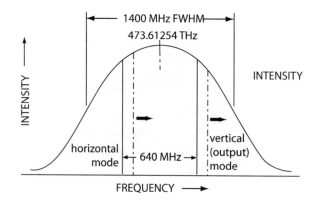

FIGURE 8.14 Two orthogonal modes oscillating in a helium–neon gain envelope.

needed is that required to sweep the frequency through one free spectral range of the laser cavity (the cavity-mode spacing). For the helium–neon laser cavity described earlier, the required change is only 320 nm, well within the capability of piezoelectric actuators.

Commercially available systems can stabilize frequency output to 1 MHz or less. Laboratory systems that stabilize the frequency to within tens of kilohertz have been developed.

8.4 Tunable Operation

Many lasers can operate at more than one wavelength. Argon and krypton lasers can operate at discrete wavelengths ranging from the ultraviolet to the near infrared. Dye lasers can be continuously tuned over a spectrum of wavelengths determined by the fluorescence bandwidths of the specific dyes (typically about 150 nm). Alexandrite and titanium sapphire lasers can be tuned continuously over specific spectral regions.

To create a tunable laser, the coatings on the cavity optics must be sufficiently broadband to accommodate the entire tuning range, and a variable-wavelength tuning element must be introduced into the cavity, either between the cavity optics or replacing the high-reflecting optic, to introduce loss at undesired wavelengths.

Three tuning mechanisms are in general use: Littrow prisms, diffraction gratings and birefringent filters. Littrow prisms (see Figure 8.15) and their close relative, the full-dispersing prism, are used extensively with gas lasers that operate at discrete wavelengths. In its simplest form, the Littrow prism is a 30°–60°–90° prism with the surface opposite the 60° coated with a broadband high-reflecting coating. The prism, which replaces the end mirror in the laser, is oriented so that the desired wavelength is reflected back along the optical axis and the other wavelengths are dispersed off axis. By rotating the prism, individual lines can be chosen. To improve performance, the prism's angles can be modified so that the beam enters the prism exactly at Brewster's angle, thereby reducing

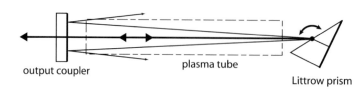

FIGURE 8.15 Littrow prism used to select a single wavelength.

intra-cavity losses. For higher power lasers that require greater dispersion to separate closely spaced lines, the Littrow prism can be replaced by a full-dispersing prism coupled with a high reflecting mirror.

Gratings are used for laser systems that require a higher degree of dispersion than that of a full-dispersing prism. They are usually placed inside the laser cavity.

Birefringent filters have come into general use for continuously tunable dye and Ti:sapphire lasers, since they introduce significantly lower loss than do gratings. The filter is made from a thin, crystalline-quartz plate with its fast axis oriented in the plane of the plate. The filter, placed at Brewster's angle in the laser beam, acts like a weak etalon with a free spectral range wider than the gain curve of the lasing medium. Rotating the filter around the normal to its face shifts the transmission bands, tuning the laser. Since there are no coatings and the filter is at Brewster's angle (thereby polarizing the laser), there are no inherent cavity losses at the peak of the transmission band. A single filter does not have as significant a line-narrowing effect as does a grating, but this can be overcome by stacking multiple filter plates together, with each successive plate having a smaller free spectral range.

8.5 Beam Shape and Astigmatism in Diode Lasers

Two characteristics of the output beams of semiconductor (diode) lasers are astigmatism (caused when the vertical and horizontal beam waists are not coincident) and ellipticity (caused when the vertical and horizontal beam waists are different sizes). In applications requiring collimation or transformation of the diode laser's beam, these characteristics must be considered, and often corrected, if the wavefront is to approach the diffraction limit.

8.5.1 Correcting Astigmatism in Collimators

Index-guided diode lasers typically exhibit a small amount of astigmatism (between 2 and 8 μm). Gain-guided diode lasers usually have between 30 and 60 μm of astigmatism.

In many applications, it is important to have as little wavefront distortion as possible in the final collimated or focused beam. The amount of astigmatic wavefront distortion, W, caused by the axial astigmatism of the diode laser can be given by the following expression:

$$W \approx (NA^2 \times z)/2\lambda \quad (8.27)$$

where NA is the numerical aperture of the diode, λ is the laser wavelength and z is the axial astigmatism. The divergence can be expressed as a function of the wavefront or the axial astigmatism:

$$\text{Divergence} = 8W/\varphi \quad (8.28)$$

where φ is the diameter of the clear aperture and W is expressed in the same units as φ.

The simplest way to correct astigmatism in a diode laser collimator is to place a plano-concave cylinder lens in front of the collimator. If the collimating lens is focused on the front facet of the diode, the cylinder should have negative refracting power with the power oriented parallel to the junction. If the collimating lens is focused on the waist behind the exit facet of the diode laser, then the cylinder lens should have positive power and be oriented perpendicular to the plane of the junction. If the cylinder lens is placed before the collimating lens, the cylinder radius R is given (approximately) by

$$R = \frac{\varphi^2}{8nD(1-\cos u)} \quad (8.29)$$

where φ is the clear aperture of the lens, n is the refractive index of the lens and u is the smaller of the half acceptance angle of the collimator or the divergence of the laser [8].

If the cylinder is placed after the collimating lens and we assume that the two lenses are thin, then we can treat the astigmatism as defocus and calculate the necessary focal lengths using

$$\frac{1}{f_{cyl}} = \frac{1}{f_{coll}} - \frac{1}{f_{ast}} \quad (8.30)$$

where f_{ast} is the focal length of the collimator and the amount of astigmatism present.

A tilted plate, placed between the negative and positive elements of a beam expander, will add longitudinal astigmatism. The amount of astigmatism added depends on the angle of tilt and the thickness of the plate. The added astigmatism, D, can be calculated using the following expression:

$$D = \frac{1}{\sqrt{n^2 - \sin^2(U_p)}} \left[\frac{n^2 \cos^2(U_p)}{n^2 - \sin^2(U_p)} - 1 \right] \quad (8.31)$$

where t is the thickness of the plate, U_p is the tilt angle and n is the index of refraction of the tilt plate.

8.5.2 Circularizing a Diode Laser

The elliptical output of a diode laser can be circularized in a variety of ways. The most common technique is to use an anamorphic beam expander. There are two general classes of anamorphic beam expanders currently used with diode lasers: beam expanders that use cylinder lenses and beam expanders that use prisms.

Beam expanders that use cylinder lenses to circularize the beam usually do so in the form of a Galilean beam-expanding telescope. This has two advantages over a prism telescope: the beam is not displaced from the original centre line and the cylindrical elements that make up the beam expander can be adjusted to correct for any natural astigmatism in the diode output. Unfortunately, for large magnifications, the length of the telescope becomes excessively long. Furthermore, the telescope is not easily adjustable—an important factor since the ellipticity ratio can vary dramatically from diode to diode.

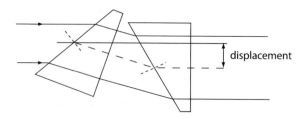

FIGURE 8.16 The Brewster telescope. Polarization must be in the *p*-plane for low reflective loss if the Brewster surface is not AR coated.

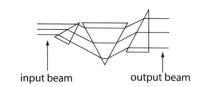

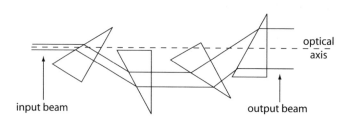

FIGURE 8.17 Anamorphic beam expanders that eliminate beam offset.

Because of the inherent disadvantages of the cylindrical beam expander, most designers opt to go with prism magnification for beam circularization. The prisms are relatively easy to manufacture with good transmitted wavefront, and they are easy to align. The most common configuration is the Brewster telescope shown in Figure 8.16. It is compact and it produces an exit beam parallel to the incoming beam. Differences in elliptical ratios can be accommodated by rotating the prism pair to different angles. The chief disadvantage is that the exit beam is displaced. This can be corrected by adding additional prisms, as shown in Figure 8.17.

8.6 Q-switching, Mode-locking and Cavity Dumping

Various techniques are used to generate laser pulses with much higher peak powers than for the same laser operating continuously. The most common are Q-switching, mode-locking and cavity dumping.

8.6.1 Q-switching

In a lasing system, the output power is proportional to the population inversion between the upper and lower energy levels of the lasing transition. The greater the inversion, the higher the output power. In an oscillating laser, the upper laser level is constantly being depopulated by emission stimulated by the photons circulating in the laser cavity, as well as by spontaneous emission and by non-radiating relaxation mechanisms. At equilibrium (i.e. cw operation), the excitation mechanism is repopulating the upper laser level at the same rate as the other mechanisms are depopulating it.

The Q (or quality factor) of a laser cavity is defined by the equation

$$Q = \frac{2\pi P}{\lambda \delta} \quad (8.32)$$

where P is the perimeter of the laser cavity ($2L$ for a linear laser) and δ is the round-trip cavity loss. Increasing the round-trip cavity loss (i.e. 'spoiling' the Q of the cavity) reduces, or effectively eliminates, the main source of stimulated emission depopulation—photons directed back into the lasing medium by the cavity mirrors. Assuming that the rate of population of the upper lasing level remains constant, the population inversion will then increase dramatically until a new equilibrium is reached, essentially storing up energy in the lasing medium. If the cavity losses are suddenly reduced (Q is 'switched' to its normal value), the round-trip gain will be much larger than the cavity loss, and the energy within the laser cavity will build up at an unusually rapid rate, resulting in a 'giant pulse'. The peak power in the Q-switched pulse can be three-to-four orders of magnitude higher than the power obtained from a non-switched laser using the same excitation mechanism.

For Q-switching to be most effective, the time needed to depopulate the upper laser level *via* spontaneous emission and non-radiating mechanisms must be much greater than the time needed to populate the level, as is the case for many solid-state materials (e.g. ruby, Nd:YAG). It is not true, however, for many gas-discharge lasers (e.g. helium–neon, argon–ion),

Laser Beam Control

Q-switching these lasers typically increases peak power by only 50%. Consequently, Q-switching is rarely used with gas lasers based on atomic transitions. Q-switching is, however, used on some molecular gas lasers (e.g. CO_2 lasers), because the vibrational and rotational transitions typically have a longer relaxation time than atomic transitions.

Four techniques routinely used to Q-switch a laser are shown in Figure 8.18.

8.6.1.1 Rotating Mirrors

One of the earliest methods was to use a rotating high-reflecting mirror or multifaceted prism as the high reflector. As the mirror spins, it alternately goes in and out of alignment changing the cavity Q. This technique has several weaknesses: slow switching speed, mechanical complexity, vibration and alignment problems, high motor wear and the inability to precisely time or trigger the pulses.

8.6.1.2 Electro-optic and Acousto-optic Q-switching

A second method is to insert an electro-optic (EO) or acousto-optic (AO) shutter inside the laser cavity. The EO shutter operates by rotating the polarization of the recirculating light in the cavity by 90° and directing it out of the cavity with a polarization-selecting optical element. The AO shutter creates an rf-induced Bragg grating which deflects light out of the cavity. The EO shutter is faster, with high hold-off (insertion loss in the low-Q state), but it is more expensive, requires a fast-switching power supply and has, in general, a higher insertion loss. The AO shutter is simpler and has very low insertion loss, but it is switching slower and should not be used with high-gain lasers. Both techniques can provide precise triggering and synchronization of the laser pulses.

8.6.1.3 Passive Q-switching

Lasers can also be Q-switched by placing an easily saturable absorbing medium inside the laser cavity. The absorber lowers the cavity gain but not sufficiently to eliminate totally the ability of the laser to oscillate. When the population inversion builds up sufficiently to overcome the additional cavity loss, the laser begins to oscillate weakly. The absorber quickly saturates and becomes transparent, restoring the cavity Q. Saturable absorbers are widely used in commercial lasers because they are simple and require no electronics. The main drawback of passive Q-switching is the inability to trigger pulses externally, and the systems exhibit much greater pulse-to-pulse timing jitter than is typically observed with other techniques.

8.6.2 Cavity Dumping

In a cw laser, the power circulating inside the laser cavity is much greater than the power escaping through the output coupler. For example, the circulating power in a 10-mW helium–neon laser cavity is approximately 2 W; for a 20 W ion laser cavity, approximately 400 W. By replacing the output couplers on these lasers with non-transmitting, high-reflecting mirrors,

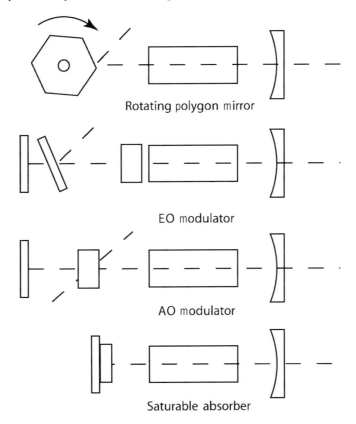

FIGURE 8.18 Techniques used to Q-switch a laser.

this circulating power can be increased by nearly an order of magnitude. Cavity dumping is used to access (dump) the circulating energy inside the laser cavity. The effect is the same as if the output coupler was suddenly pulled away, letting all of the circulating power escape in a pulse whose width, essentially, is the round-trip time of the laser cavity.

Cavity-dumping schemes employing AO and EO modulators are shown in Figure 8.19. As was the case with AO Q-switching, an rf signal creates a Bragg diffraction grating in the modulator, which deflects the beam out of the cavity, except that in this case, the laser is oscillating fully when the deflection occurs (Q-switching in reverse). The switching speed is determined by the time it takes the acoustic signal to travel across the beam; consequently, the beam is often focused at the modulator, as shown in the figure, to speed up the process. EO cavity dumping occurs when the Pockels cell rotates the polarization.

8.6.3 Mode-locking

Mode-locking is a method of producing a train of very narrow, extremely high-peak-power pulses from a cw or long-pulse laser. A complete understanding of mode-locking is beyond the scope of this article, and for detailed information, the author recommends the text by Anthony Siegman [1].

As with most other processes involving laser manipulation, mode-locking begins with the introduction of a selective loss into the laser cavity, forcing the laser to react to the effects of that loss. In the case of mode-locking, a periodic loss is introduced, which is timed to coincide with the round-trip time of the laser cavity. In one approach, the aperture at the laser mirror is widened and narrowed periodically at a frequency equal to the cavity round-trip time ($c/2L$). While the aperture is narrowed, fewer photons are transmitted to the gain medium. After the first round trip, there will be a non-uniform spatial distribution of photons throughout the cavity, with a higher density of photons in the group that passed through the open aperture (the favoured group). Consequently, as the favoured group passes through the gain medium, it adds more stimulated emission photons to its group than the other photons in the cavity, accentuating the non-uniformity. The process is repeated, with the round-trip time and the opening and closing of the aperture perfectly synchronized. The favoured group always sees the open aperture, while the rest of the circulating photons experience losses. As the energy in the favoured group increases, the width of the group narrows because there is somewhat more loss in the wings of the group than in the centre. Ultimately, the energy in the favoured group is so high that it absorbs all the remaining gain in the lasing medium, and the output becomes a train of very narrow pulses exactly spaced by the round-trip time of the laser cavity.

Three main techniques are used to mode-lock a laser: active mode-locking, passive mode-locking and synchronous pumping.

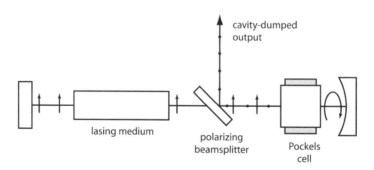

ELECTRO-OPTIC CAVITY DUMPING

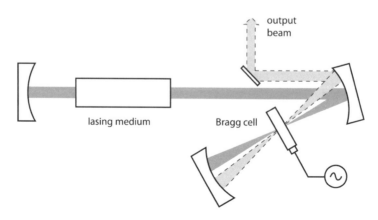

ACOUSTO-OPTIC CAVITY DUMPING

FIGURE 8.19 EO and AO cavity dumping configurations.

8.6.3.1 Active Mode-locking

Active mode-locking is achieved by placing a modulation element inside the laser cavity and operating it at a frequency very close to the round-trip time of the laser cavity. Because lasers can be effectively mode-locked with only 20% to 30% amplitude modulation, hold-off is not an issue, and AO modulators are typically used because of their low insertion loss. This technique is used to mode-lock a wide variety of gas and solid-state lasers.

8.6.3.2 Passive Mode-locking

It is possible to mode-lock a cw laser by placing a saturable absorber in the laser cavity. On the surface, this set-up is very similar to the passive Q-switching scheme discussed before, but in practice, the parameters of the absorber, laser cavity, gain medium and excitation mechanism are substantially different.

As a population inversion builds up in the lasing medium, the precursor of laser oscillation is noise in the form of stimulated and spontaneous emission. As the noise builds up, there will be one noise spike with sufficient energy to saturate the absorber and make its way around the resonator and back through the gain medium where it can be amplified. Although eventually other noise spikes will have sufficient energy to saturate the absorber, this initial spike will be preferentially amplified until it absorbs all the available gain. This technique has been used to create the shortest optical pulses with durations of less than 30 fs.

8.6.3.3 Synchronous Pumping

It is possible to mode-lock a laser by modulating the gain medium instead of modulating the cavity losses. This is done extensively with laser-pumped laser systems, such as dye lasers, by mode-locking the pump source (e.g. an argon laser) with an AO modulator and then pumping the dye laser with the mode-locked train of pulses. The key is to have the round-trip time of the dye laser cavity closely match that of the pump laser. This can lead to some very large systems. For optimal mode-locking, it is critical that the cavity lengths of both lasers, as well as the modulation frequency of the pump laser, be maintained precisely.

8.7 Beam Quality—Limits and Measurement

Defining the quality of a laser beam is often a difficult proposition because most lasers do not operate in a pure Gaussian mode. Indeed, in many cases, a Gaussian mode is undesirable (e.g. for photolithography applications, a 'top-hat' beam with uniform intensity over the transverse area is most desirable). The main parameters that go into determining beam quality are beam diameter and divergence, beam intensity profile, beam wavefront, mode uniformity, noise and frequency stability. No one instrument can measure all of these parameters, but there are instruments available that can measure each of the parameters.

Power and energy meters can measure peak power, average power, pulse energy and, in many cases, beam noise (rapid intensity fluctuations). Slit, knife-edge and pinhole scanning profilometers can provide accurate two-dimensional intensity profiles for cw laser beams as small as a few microns in diameter; CCD beam profilometers can give accurate two- and three-dimensional intensity profiles of both pulsed and cw beams, but with reduced resolution. Scanning interferometers are used to detect longitudinal modes and the presence of multiple transverses modes. They can also be used to determine the characteristics of the wavefront at a given point. Wavemetres can compare the laser output to a stabilized source and measure both relative and absolute frequency drift. Finally, the M^2 meter automatically measures beam diameter at multiple points on a focused beam and compares the observed beam waist and far-field divergence with that of a theoretical Gaussian, providing important information about how the beam propagates through an optical system.

8.7.1 Frequency and Amplitude Stabilization

The output of a freely oscillating laser will fluctuate in both amplitude and frequency. Fluctuations of less than 0.1 Hz are commonly referred to as 'drift'; faster fluctuations are termed 'noise' or, when talking about sudden frequency shifts, 'jitter'.

The major sources of noise in a laser are fluctuations in the pumping source and changes in length or alignment caused by vibration, stress and changes in temperature. For example, unfiltered line ripple can cause output fluctuations of 5%–10% or more.

Likewise, a 10-μrad change in alignment can cause a 10% variation in output power, and depending upon the laser, a 1-μm change in length can cause amplitude fluctuations of up to 50% (or more) and frequency fluctuations of several gigahertz.

High-frequency noise (>1 MHz) is caused primarily by 'mode beating'. For example, in a helium–neon laser, transverse Laguerre–Gaussian modes of adjacent order are separated by a few tens of MHz. If multiple transverse modes oscillate simultaneously, heterodyne interference effects, or 'beats', will be observed at the difference frequencies. Likewise, mode beating can occur between longitudinal modes at frequencies of

$$\Delta v_{\text{longitudinal}} = \frac{c}{2L} \text{ or } \frac{c}{P} \tag{8.33}$$

where L is the mirror separation of a linear laser cavity and P is the perimeter of a ring laser cavity. Mode beating can cause peak-to-peak power fluctuations of several percent. The only way to eliminate this noise component is to limit the laser output to a single transverse and single longitudinal mode.

Finally, when all other sources of noise have been eliminated, we are left with quantum noise, the noise generated by the spontaneous emission of photons from the upper laser level in the lasing medium. The randomness of this emission is proportional to the square root of the number of photons being detected. Thus, the signal-to-noise ratio is proportional to the square root of the number of photons being detected. The laser beam signal quality increases as the laser intensity increases. It is impossible to suppress spontaneous noise, but in most applications, it is inconsequential.

8.7.2 Methods for Suppressing Amplitude Noise and Drift

Two primary methods are used to stabilize amplitude fluctuations in commercial lasers: automatic current control (ACC), also known as current regulation, and automatic power control (APC), also known as light regulation. In ACC, the current driving the pumping process passes through a stable sensing resistor, as shown in Figure 8.20, and the voltage across this resistor is monitored. If the current through the resistor increases, the voltage drop across the resistor increases proportionately. Sensing circuitry compares this voltage to a reference and generates an error signal that causes the power supply to reduce the output current appropriately. If the current decreases, the inverse process occurs. ACC is an effective way to reduce noise generated by the power supply, including line ripple and fluctuations.

With APC, instead of monitoring the voltage across a sensing resistor, a small portion of the output power in the beam is diverted to a photodetector, as shown in Figure 8.21, and the voltage generated by the detector circuitry is compared to a reference. As output power fluctuates, the sensing circuitry generates an error signal that is used to make the appropriate corrections to maintain constant output.

ACC effectively reduces amplitude fluctuations caused by the driving electronics, but it has no effect on amplitude fluctuations caused by vibration or misalignment. APC can effectively reduce power fluctuations from all sources. Neither of these control mechanisms has any impact on frequency stability.

Not all cw lasers are amenable to APC as described. For the technique to be effective, there must be a monotonic relationship between output power and a controllable parameter (typically current or voltage). For example, throughout the typical operating range of a gas-ion laser, an increase in current will increase the output power and *vice versa*. This is not the case for some lasers. The output of a helium–neon laser is very insensitive to discharge current, and an increase in current may increase or decrease laser output. In a helium cadmium laser, where electrophoresis determines the density and uniformity of cadmium ions throughout the discharge, a slight change in discharge current in either direction can effectively kill lasing action.

If traditional means of APC are not suitable, the same result can be obtained by placing an AO modulator inside the laser cavity and using the error signal to control the amount of circulating power ejected from the cavity.

One consideration that is often overlooked in an APC system is the geometry of the light pickoff mechanism itself. One's first instinct is to insert the pick-off optic into the main beam at a 45° angle, so that the reference beam exits at a 90° angle. However, as shown in Figure 8.22, for uncoated glass, there is almost a 10% difference in reflectivity for s and p polarization. In a randomly polarized laser, the ratio of the s and p components is not necessarily stable, and using a 90° reference beam can actually increase amplitude fluctuations. This is of much less concern in a laser that has a high degree of linear polarization (e.g. 500:1 or better), but even then there is a slight presence of the orthogonal polarization. Good practice dictates that the pick-off element is inserted at an angle of 25° or less.

Never use the reflection off a Brewster window, used to polarize a laser, as the reference beam, because this is, in general, not representative of the actual laser output.

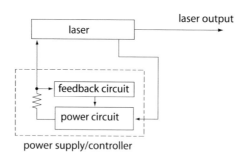

FIGURE 8.20 ACC schematic diagram.

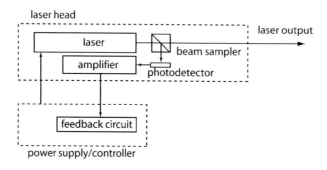

FIGURE 8.21 APC schematic diagram.

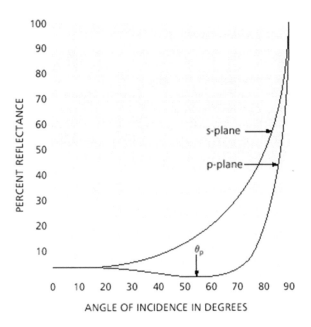

FIGURE 8.22 Reflectivity of the glass surface as a function of incidence angle.

8.8 Spatial Filtering

Laser light scattered from dust particles residing on optical surfaces may produce interference patterns resembling holographic zone planes. Such patterns can cause difficulties in interferometric and holographic applications where they form a highly detailed, contrasting and confusing background which interferes with desired information. In other cases, stray coherent light from satellite beams (beams generated by reflections from the front and back surfaces of the output coupler) or incoherent light from a laser discharge may interfere with a measurement. Spatial filtering is a simple way of suppressing this interference and maintaining a very smooth beam irradiance distribution. The scattered light propagates in different directions from the laser light and hence is spatially separated at a lens focal plane. By centring a small aperture around the focal spot of the direct beam, as shown in Figure 8.23, it is possible to block scattered light while allowing the direct beam to pass unscathed. The result is a cone of light with a smooth irradiance distribution which can be refocused to form a collimated beam that is almost equally smooth.

As a compromise between ease of alignment and complete spatial filtering, it is best that the aperture diameter is about two times the $1/e^2$ beam contour at the focus, or about 1.33 times the 99% throughput contour diameter.

REFERENCES

1. Siegman A 1986 *Lasers* (Sausalito, CA: University Science Books).
2. Self S A 1983 Focusing of spherical Gaussian beams *Appl. Opt.* **22** 658.
3. Belland P and Crenn J 1982 Changes in characteristics of a Gaussian beam weakly diffracted by a circular aperture *Appl. Opto.* **21** 522–7.
4. Sun H 1998 Thin lens equation for a real laser beam with weak lens aperture truncation *Opt. Eng.* **37** 2906–13.
5. Freiberg R J and Halsted A S 1969 Properties of low order transverse modes in argon ion lasers *Appl. Opt.* **8** 355–62.
6. Rigrod W W 1963 Isolation of axi-symmetric optical-resonator modes *Appl. Phys. Lett.* **2** 51–3.
7. Kuttner 1985 Laser beam scanning *Optical Engineering* vol 8 (New York: Dekker) p. 352.
8. Unsbo P 1995 Phase conjugation and four-wave mixing Doctoral Thesis Royal Institute of Technology, Stockholm.

FURTHER READING

Born M and Wolf E 1999 *Principles of Optics* 7th edn (Cambridge: Cambridge University Press); Koechner W 1999 *Solid-State Laser Engineering* 5th edn (Berlin: Springer).

Q-Switching and Cavity Dumping:

Chesler R B, Karr M A and Geusic J E 1970 An experimental and theoretical study of high repetition rate Q-switched Nd:YAG lasers *Proc. IEEE* **58** 1899–914.

Truntna R and Siegman A E 1977 Laser cavity dumping using an antiresonant ring *IEEE J. Quantum Electron.* **13** 955–62.

Chesler R B and Maydan D 1971 Q–Switching and cavity dumping of Nd:YAG lasers *J. Appl. Phys.* **42** 1028–34.

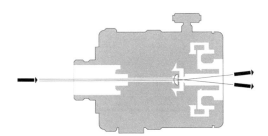

FIGURE 8.23 Spatial filtering.

Phase conjugation:

Bloom D M and Bjorklund G C 1977 Conjugate wave-front generation and image reconstruction by four-wave mixing *Appl. Phys. Lett.* **31** 592–4.

Hellwarth R W 1977 Generation of time-reversed wave fronts by nonlinear refraction *J. Opt. Soc. Am.* **67** 1–3.

Lera G and Nieto-Vesperinas M 1990 Phase conjugation by four-wavemixing of statistical beams *Phys. Rev. A* **41** 6400–5.

Rockwell D A 1988 A review of phase-conjugate solid-state lasers *IEEE J. Quantum Electron.* **24** 1124–40.

Frequency stabilization:

Hall J L 1986 Stabilizing lasers for applications in quantum optics *Quantum Optics IV, Proc. 4th Int. Symp.* (February 10–15, Hamilton, New Zealand) ed J D Harvey and D F Walls (Berlin: Springer) pp 273–84.

Hall J L, Hils D, Salomon C and Chartier J-M 1987 Towards the ultimate laser resolution *Laser Spectroscopy* vol VIII, ed W Persson and S Svanberg (Heidelberg: Springer) pp 376–80.

Hils D and Hall J L 1989 Ultra-stable cavity-stabilized lasers with subhertz linewidth *Proc. 4th Int. Symp. on Frequency Standards and Metrology* ed A De Marchi (Heidelberg: Springer) pp 162–73.

9

Optical Detection and Noise

Gerald Buller and Jason Smith

CONTENTS

- 9.1 Introduction 172
  - 9.1.1 Nomenclature and Figures of Merit 172
    - 9.1.1.1 Signal-to-noise Ratio 172
    - 9.1.1.2 Noise-equivalent Power 172
    - 9.1.1.3 Detectivity (D) and Specific Detectivity (D^*) 173
    - 9.1.1.4 Responsivity 173
- 9.2 Photoemissive Detectors 173
  - 9.2.1 The Photoemissive Effect 173
  - 9.2.2 Photomultipliers 174
- 9.3 Semiconductor Detectors 176
  - 9.3.1 Photoelectric Absorption 176
  - 9.3.2 pn and Pin Photodiodes 176
    - 9.3.2.1 Photovoltaic Mode 178
  - 9.3.3 Schottky Diode Detectors 178
  - 9.3.4 Avalanche Photodiodes 179
  - 9.3.5 Photoconductive Detectors 180
  - 9.3.6 Intra-band Detectors or QWIPs 181
- 9.4 Thermal Detectors 181
  - 9.4.1 Thermocouples and Thermopiles 182
  - 9.4.2 Bolometers and Thermistors 182
  - 9.4.3 Pyroelectric Detectors 182
- 9.5 Noise in Photodetection 183
  - 9.5.1 Noise in the Optical Signal 183
    - 9.5.1.1 Background Noise (Blackbody Radiation) 183
    - 9.5.1.2 Photon Noise 184
  - 9.5.2 Noise in the Photodetector 184
    - 9.5.2.1 Photoelectron Noise 184
    - 9.5.2.2 Shot Noise 184
    - 9.5.2.3 Generation–Recombination (G–R) Noise 185
    - 9.5.2.4 Gain Noise 185
    - 9.5.2.5 1/f or Flicker 186
    - 9.5.2.6 Temperature Noise 186
  - 9.5.3 Noise in the Measurement Circuit 186
    - 9.5.3.1 Thermal Noise (Aka Johnson or Nyquist Noise) 186
    - 9.5.3.2 Amplifier Noise and Impedance Matching 186
  - 9.5.4 Combining Noise Sources 187
  - 9.5.5 Bandwidth-related Noise Reduction Methods 187
- References 188
- Further Reading 188

9.1 Introduction

The reliable detection of optical radiation is an essential element of laser technology. Indeed, almost everywhere a laser is used, a photodetector will also be employed either to sample the laser output itself or to detect radiation produced by some other system as a result of laser excitation.

So far, the most common method by which the intensity of laser light is measured is by conversion to a real-time electrical signal. There are three processes commonly used in photodetection, by which this conversion takes place:

i. photoelectric emission,
ii. internal photoelectric (photoconductive, photocurrent) and
iii. photothermal/thermoelectric conversion.

Each of these processes will be discussed in detail in the following sections. For ease of recognition, the section on internal photoelectric effect detectors is entitled 'semiconductor detectors'; the devices described therein rely explicitly on excitation of electrons either across the semiconductor bandgap or between levels in a quantum well heterostructure. It should be noted that semiconductors are also used both in photoemission and in photothermal detection.

Each process has a range of optical wavelengths in which it is suitable for photodetection, as depicted in Figure 9.1. Broadly speaking, photoemissive detectors are best suited to wavelengths from the UV to the near-infrared. In the UV and visible, photoemissive and internal photoelectric detectors are comparable in performance and selection will depend on the particular application. Beyond about 0.9 μm, internal photoelectric detectors gain the upper hand. Photothermal detectors, whilst sensitive throughout the visible, currently only offer superior detectivity at wavelengths longer than about 20 μm. With semiconductor quantum well infrared photodetector (QWIP) technology advancing at a rapid pace, it is likely that this cross-over wavelength will grow ever longer in the coming years.

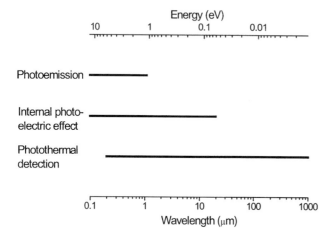

FIGURE 9.1 Sensitive spectral ranges of the three principal photodetection processes.

Photodetection in all its manifestations and detail is a vast subject, and here, we intend only to provide a brief introduction to the science, an overview of the existing technology and some basic theory aimed at the user (rather than the designer) of photodetection systems. For this reason, quite a lot of space is dedicated to the subject of noise—an area in which a clear understanding by the user can lead to real improvements in performance.

Most standard optoelectronics textbooks cover the subject of photodetection to some degree. The further reading list at the end of this chapter provides a selection of such references, along with some more specialist texts.

9.1.1 Nomenclature and Figures of Merit

In order to discuss the relative performance of different types of photodetector in Sections 9.2 through 9.4, it will be helpful first to provide definitions of some widely used figures of merit. A full discussion of the sources of noise and their relevance to the different detector types is left until Section 9.5.

9.1.1.1 Signal-to-noise Ratio

The output from a real-time photodetector subject to an optical signal of constant mean intensity can be represented as a time-dependent current i with a mean value $\langle i \rangle$ and a variance $\langle \delta i^2 \rangle = \langle i^2 \rangle - \langle i \rangle^2$. It is the variance that is most commonly used as the quantitative assessment of the absolute noise level. The often-quoted root-mean-square (rms) noise current, i_{rms}, is equal to the square root of the current variance as defined above.

The total output current, i_{tot}, is the sum of the current generated by the optical signal i_{sig}, and any background current, i_{bkgnd}, caused by non-signal light or leakage currents in the detector. These currents both contribute to the noise in the total current:

$$\langle \delta i_{tot}^2 \rangle = \langle \delta(i_{sig} + i_{bkgnd})^2 \rangle. \quad (9.1)$$

The simplest figure of merit for the quality of the photodetection signal is the signal-to-noise ratio (SNR) defined as the ratio of the square of the mean signal current to the variance in the total current,

$$\text{SNR} = \frac{\langle i_{sig} \rangle^2}{\langle \delta i_{tot}^2 \rangle}. \quad (9.2)$$

9.1.1.2 Noise-equivalent Power

The sensitivity of a photodetector can be defined in terms of the incident optical power required to achieve a given SNR. A figure of merit commonly used is the *noise-equivalent power* (NEP), defined as the mean optical power required to be incident on the detector to generate an SNR of unity. NEP in its most simple form, therefore, has units of watts and is specific to a particular detector, taking into account such parameters as the sensitive detection area A and the detection bandwidth Δf.

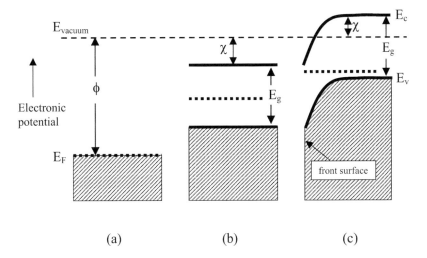

FIGURE 9.2 Work functions, ϕ, and electron affinities, χ, for (a) a metal and for a semiconductor with (b) positive and (c) NEA. The semiconductor bandgap is labelled E_g.

9.1.1.3 Detectivity (D) and Specific Detectivity (D*)

The detectivity D is the reciprocal of the NEP as previously defined. It is, therefore, specific to a particular detector operating with a particular bandwidth and has units of W^{-1}.

In order to generate a figure of merit for detector sensitivity that is independent of detection bandwidth, it is often assumed that the detector noise is independent of frequency or is white noise (cf white light). To this end, the NEP is often quoted as having units of $W\,Hz^{-1/2}$.

The specific or normalized detectivity (D^*) is defined by assuming also that the background signal is proportional to the device area, so that $D = D^*/\sqrt{A\Delta f}$. Specific detectivity is, therefore, usually quoted in units of $cm\,Hz^{1/2}\,W^{-1}$.

We shall see in the following sections that the assumption that detector noise is white is valid in many cases but that there are also common situations in which non-white noise sources must be taken into account. In all situations, however, the bandwidth is an important consideration when determining noise characteristics.

9.1.1.4 Responsivity

The responsivity of a detector indicates the change in the output of the detector per unit change in the incident optical power and is, therefore, unrelated to noise considerations. In a photodiode, for instance, the output is a photocurrent and so the responsivity will be quoted in $A\,W^{-1}$. In a pyroelectric detector, the output is a voltage and units of $V\,W^{-1}$ are appropriate.

9.2 Photoemissive Detectors

9.2.1 The Photoemissive Effect

When photons of sufficient energy are incident on a solid, electrons are emitted from the surface. This phenomenon is known as the photoemissive or photoelectric effect. Figure 9.2 represents this effect schematically in terms of energy levels for (a) a metal, (b) a semiconductor with positive electron affinity and (c) a semiconductor with negative electron affinity (NEA). We shall discuss each of these briefly in turn.

In metals, the Fermi energy E_F is situated within an energy band of allowed electron states, and so electrons occupy a quasi-continuum of states up to that energy.[1] The minimum energy that a photon must provide in order to eject an electron into the vacuum is

$$h\nu_{min} = e\phi \quad (9.3)$$

where $e\phi = E_{vacuum} - E_F$ is known as the work function of the material.

Since the majority of electrons exist well below the Fermi energy, and since energy can be lost by inelastic scattering of the excited electron prior to emission, a photon energy higher than $h\nu_{min}$ will usually be required.

The probability that an incident photon will cause an electron to be emitted (known as the quantum efficiency, or quantum yield, and represented by η) is, therefore, a strong function of the photon energy, falling off rapidly for $h\nu < e\phi$. The function $\eta(h\nu, T)$ is the most important consideration when selecting a photoemissive material for a detector. In an intrinsic semiconductor, depicted in Figure 9.2b, the highest energy electrons exist not at the Fermi energy, which lies within the bandgap, but at the valence band edge. Instead of being given by the work function, the minimum energy that a photon must provide to emit an electron from a semiconductor is given by

$$h\nu_{min} = E_g + \chi \quad (9.4)$$

where E_g is the bandgap energy and $\chi = E_{vacuum} - E_c$ is the electron affinity. χ is positive for intrinsic semiconductors. Direct-gap semiconductors such as GaAs and $In_xGa_{1-x}As$ are particularly attractive as photoemitters since their higher

---

[1] This is only strictly true at low temperatures. The energy distribution of electron state occupancy at finite temperatures is given by the Fermi–Dirac function $p(E, T) = 1/[\exp[(E - E_f)/kT] + 1]$.

optical absorption coefficients allow greater absorption near to the surface and so higher photoemission efficiencies.

NEA photocathodes are created by applying a layer of n-type impurities (commonly Cs or Cs_2O) to the surface of a heavily doped p-type semiconductor. Charge redistribution leads to bending of the electronic bands near to the surface, so that the conduction band minimum in the bulk material lies above the vacuum energy, as shown in Figure 9.2c. If the region of band bending is thinner than the electron mean-free path through the crystal, efficient photoemission can, therefore, occur for any photon energy above $hv = Eg$. Cs_2O provides the largest shift in bulk band energies but also forms a potential barrier at the surface through which the electrons must tunnel in order to be emitted. This barrier limits the quantum efficiency in narrow-gap NEA semiconductors. Note that NEA semiconductors work only as reflective and not as transmissive photocathodes.

Some common photocathode materials are listed in Table 9.1 and their responsivity spectra are shown in Figure 9.3. The internationally agreed S code identifying a particular photocathode design is also included in the table.

TABLE 9.1

Common Photodiode Materials with International S Codes and Sensitive Spectral Ranges (See Also Figure 9.3)

Photocathode Material	Code	Wavelength Range (μm)
Cs_3Sb	S-11	0.3–0.6
$[Cs]Na_2K\,Sb$	S-20	0.3–0.8
Ag–O–Cs	S-1	0.25–1.2
$GaAs(Cs_2O)$		0.2–0.9
$In_{0.18}Ga_{0.82}As(Cs_2O)$		0.2–1.1
$In_{0.52}Ga_{0.48}As/InP$		0.3–1.7

9.2.2 Photomultipliers

To create a simple photodetection circuit, the photoemissive material is negatively biased (and called a photocathode) and emitted electrons are accelerated towards and collected by a positively biased electrode (anode) whereupon a measurable current is generated. In order to improve detectivity from this basic principle, a multiplication process is used, the result being known as a photomultiplier. A generic diagram of a photomultiplier tube (PMT) is shown in Figure 9.4.

In a PMT, a series of electrodes, known as dynodes, are held at progressively higher potentials under high-vacuum conditions. Electrons emitted by the photocathode are accelerated towards the first dynode by a large potential difference (typically > 100 V). Each 'primary' electron gains sufficient kinetic energy to eject a number of 'secondary' electrons from the first dynode. These secondary electrons are then accelerated towards the second dynode where the gain process is repeated. A typical PMT contains eight to twelve dynodes and offers current gains of 50–70 dB.

Dynode materials are chosen for their low work functions to enable efficient secondary emission. To this end, many of the materials that are used for photocathodes are also suitable for dynodes, and in particular, CsSb is commonly used.

The design of the PMT and, in particular, the dynode configuration are crucial in ensuring the efficient transfer of electrons between dynodes and in optimizing the time response of the device by minimizing the range of electron path lengths—the transit time spread—during the multiplication process. Figure 9.5 shows three popular designs. In each case, a semi-transparent photocathode is employed and radiation is incident from the left of the diagram. Design (a) has dynodes in a 'venetian blind' configuration, with each dynode consisting of a number of small plates facing at 45° to the tube axis. Alternate dynodes face in opposite directions to maximize

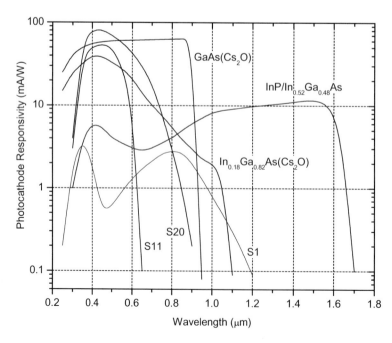

FIGURE 9.3 Responsivity spectra for some common photocathodes (see also Table 9.1).

Optical Detection and Noise 175

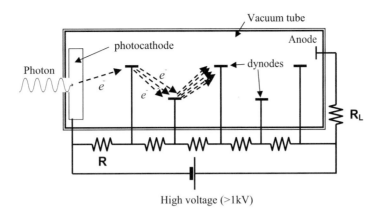

FIGURE 9.4 Schematic diagram of a photomultipier. Electrons emitted from the photocathode are accelerated in turn to a series of dynodes at progressively higher potential. At each dynode, a single incident electron causes emission of several secondary electrons and so the electron current increases exponentially through the device. The signal voltage is measured across load resistor R_L.

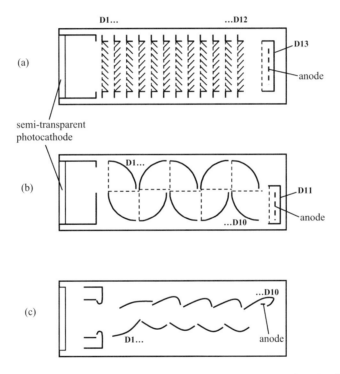

FIGURE 9.5 Three common photomultiplier dynode configurations: (a) venetian blind, (b) box and grid and (c) focused dynode chain.

collection efficiency. This design tends to have high stability in gain but poor time response. Design (b), the 'box and grid' design, has very high efficiency of transfer between dynodes, whilst in design (c), each dynode is curved so as to focus the secondary electrons onto the centre of the subsequent dynode, resulting in lessened transit time spreading and, therefore, achieving a fast time response. 'Focused dynode chain' PMTs with rise times of a few nanoseconds are readily available.

The high gains achieved using photomultipliers make them suitable for single-photon detection. In this configuration, the anode current is monitored and a 'count' is recorded whenever it exceeds a specified threshold value.

An adaptation of the standard photomultiplier design that achieves faster response times and allows imaging is the multichannel plate (MCP). A MCP is a thin disc consisting of many thin glass channels (~10 μm in diameter) fused in parallel in a 2D array. The interior of each capillary is coated with a photoemissive material and biased at each end, thus acting as a continuous dynode. Each channel operates as an independent electron multiplier, hence reducing the effects of transit time spreading and improving the rise time of the detector to <200 ps. A typical MCP contains 10^6–10^8 channels. Imaging is facilitated in some designs by the inclusion of independent anodes for the array of capillaries.

With no light incident, thermionic emission at the photocathode will give rise to a 'dark current'. For a photocathode of area A at temperature T in a photomultiplier of gain G, the dark current is given by the expression

$$i_T = G\alpha A T^2 \exp\left(-\frac{h\nu_{min}}{kT}\right) \qquad (9.5)$$

where α is a constant (for pure metals, $\alpha = 1.2 \times 10^{-6}$ A m^{-2} K^{-1}) and k is Boltzmann's constant. It is clear that, for minimized dark current, low-temperature operation is necessary—especially for low work function materials employed for detection at longer optical wavelengths. For example, the PMT that incorporates the InGaAs/InP photocathode in Figure 9.2 is operated at −80°C. 'Shot' and 'gain' noise in the dark current (see Section 9.5.2) and Johnson noise in the anode (Section 9.5.3) are the key factors which limit the detectivity of photomultipliers.

9.3 Semiconductor Detectors

9.3.1 Photoelectric Absorption

Absorption of a photon by an electron in a semiconductor can radically alter its conductive properties and enable photodetection. Assessing the relative suitability of different semiconductors for photodetection (or any other optoelectronic application) relies on a thorough understanding of the electronic band structure of the materials. The reader is referred to the bibliography for introductory texts.

The most commonly exploited process is that in which an electron is excited from the valence band, where it is bound to an atom, to the conduction band, where it is free to move around the crystal and act as a conducting 'carrier'. The 'hole' left behind in the valence band also acts as a carrier, and we speak of an 'electron–hole pair' being created upon absorption of a photon. This process is depicted in Figure 9.6a. In a photoconductor, the resulting change in conductivity is measured, whilst in a photodiode, the excited carriers are accelerated by an electric field to generate a photocurrent. These methods are effective for optical wavelengths from the UV to the mid-infrared, the long wavelength limit of sensitivity being determined by the band gap E_g of the absorptive material. Several semiconductors commonly used for inter-band absorption in photodetectors, and their electronic band gaps, are listed in Table 9.2. Optimum detectivity will often be achieved by selecting a material with a bandgap only slightly smaller than the photon energy to be detected, since the narrower-gap materials suffer much higher background signals due to thermal carrier generation.

Intra-band absorption—in which a carrier electron (hole) is excited to a higher energy state within the conduction (valence)

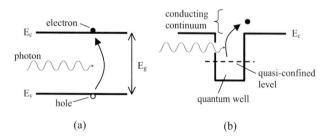

FIGURE 9.6 (a) Inter-band absorption of a photon to create an electron–hole pair and (b) intra-band absorption from a quasi-confined state into a conducting continuum. The conduction and valence band edges are labelled E_c and E_v, respectively.

TABLE 9.2

Some Semiconductor Photodiode Materials and Their Electronic Bandgaps at 300 K. Both the Bandgap Energy and the Corresponding Optical Wavelength are Shown

	Bandgap at 300 K	
	(eV)	(μm)
Si	1.14[a]	1.09
Ge	0.67[a]	1.86
GaAs	1.43	0.87
InP	1.35	0.92
In$_{0.53}$Ga$_{0.47}$As[b]	0.75	1.66
InAs	0.35	3.56
InSb	0.18	6.93
Hg$_{1-x}$Cd$_x$Te	$0 < E_g < 1.44$	$0.86 < \lambda < \infty$

[a] Indirect bandgap.
[b] Lattice matched to InP.

band—can also be exploited for photodetection. In QWIPs, carriers confined in quantum wells must be excited either into a higher quantum well sub-band or into a continuum state above the potential barrier, in order to contribute to the electrical current (Figure 9.6b). They can be designed to detect light at any wavelength longer than about 1 μm, whilst avoiding the need to use narrow-gap materials for many of which the processing technology is relatively immature. QWIPs will be discussed further in Section 9.3.6.

9.3.2 pn and Pin Photodiodes

The simplest design for a photodiode is a pn junction, an example of which is shown in Figure 9.7a. Here the n-type surface has been coated with an opaque metal contact, while the p-type surface has an annular ohmic contact with a transparent dielectric window—usually coated to reduce surface reflection—through which the light may enter. The device is shown under a reverse bias VA.

The potential profile of the conduction and valence band edges and of the electric field ξ through the device are shown in Figures 9.7b and c, respectively. The region of the device that is under electric field (i.e. $dV/dz \neq 0$) is known as the depletion region and within it, electrons (holes) in the conduction (valence) band drift towards the n-type (p-type) contact. The total electrical current flow is then the sum of these electron and hole currents. In the device shown, the p-side of the junction is much more heavily doped than the n-side, so the majority of the depletion region is in the n-type layer.

A typical current–voltage characteristic for a pn photodiode under dark and illuminated conditions is shown in Figure 9.8. Four regions are identified on the 'bias' axis, corresponding to different modes of photodetection using pn junction-based devices. Bias region I provides the most straightforward photodetection, offering low dark currents and linear photocurrent response. We shall discuss operation in this bias region first.

The photocurrent—the difference in current between the 'dark' and 'illuminated' curves in Figure 9.8—is approximately proportional to the rate of inter-band absorption in the depletion region. Carriers generated in the regions of the

Optical Detection and Noise

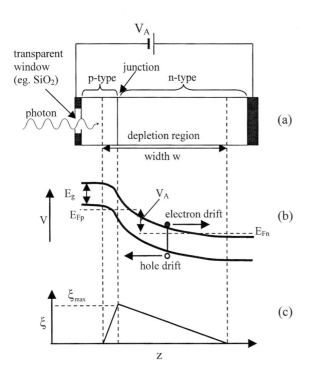

FIGURE 9.7 The pn junction semiconductor photodiode. (a) Schematic diagram of a typical device structure. The n-type material is less heavily doped and so contains most of the depletion region. (b) Profile of the electronic potential through the device. E_g is the bandgap energy, while the applied reverse bias V_A equals the difference between Fermi levels in the p and n-type contact regions, E_{Fp} and E_{Fn}. Photogenerated electrons (holes) drift towards the cathode (anode). (c) Electric field profile through the device. At each position on the z-axis, the doping type and density determine the electric field gradient.

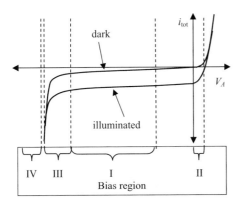

FIGURE 9.8 Typical current–voltage characteristic for a pn junction photodiode, both dark and illuminated. Four bias regions are identified for different modes of photodetection (see text for details).

device not under field can contribute to the photocurrent by diffusing into the depletion region, but this process is by far both less efficient and slower and should be kept to a minimum. Since the optical absorption length of any given material reduces with increased photon energy, the detectivity falls away at short wavelengths as more of the signal is absorbed in the top contact region.

Thermal carrier generation in the depletion region is a primary source of dark current in these devices. It depends upon the material bandgap and the operating temperature according to the expression

$$i_{therm} \propto e^{-E_g/2kT}. \qquad (9.6)$$

Noise in the dark current limits the detectivity and so cooling of photodiodes—especially those utilizing narrow-gap materials—is commonplace. For example, in a germanium pn photodiode near to ambient temperature, a reduction in dark current by a factor of two (and a corresponding increase in detectivity by a factor of $\sqrt{2}$) is obtained for each 7°C reduction in detector temperature.

The responsivity of the photodiode can be adjusted somewhat by changing the bias VA across the device, which changes the depletion width W. For an abrupt pn junction, the relationship is given by

$$W = \sqrt{\frac{2\varepsilon(V_0 + V_A)}{e}\left(\frac{N_A - N_D}{N_A N_D}\right)} \qquad (9.7)$$

where ε is the permittivity of the detector material, V_0 is the built-in potential difference that remains within each band when $V_A = 0$ (substituting E_g in place of V_0 is often a reasonable approximation) and N_A and N_D are the acceptor and donor doping concentration in the p- and n-type regions, respectively. Note that the detectivity is—to a first approximation—less affected by small changes in bias, since the dark current due to thermal carrier generation scales with W as does the photocurrent.

Slight increases in the detectivity of photodiodes can often be realized by tilting the detector by a small angle relative to the incident signal, whereupon the optical path length through the depletion layer becomes somewhat greater than W. An increase in photocurrent of several percent can often be obtained by this method, with no associated noise penalty.

The response time of a pn photodiode is determined by two important factors: (i) the transit time of carriers across the junction; and (ii) the RC time constant of the complete detection circuit.

A rough estimate of the transit time can be made using the expression

$$\tau_{drift} = \frac{W^2}{\mu(V_0 + V_A)} \qquad (9.8)$$

where μ is the carrier mobility. The quadratic dependence of equation (9.8) on the depletion thickness shows that a compromise has to be reached between high responsivity and high speed, and it is common for pn photodiodes to be designed either for one or the other. Comparing equations (9.7) and (9.8), we can see that, to a first approximation, the transit time is independent of the bias voltage, so the user should consider the detection speed required when selecting a device.

The appropriate RC time constant is the product of the device capacitance C_{pn} and the load resistance R_L:

$$\tau_{RC} = R_L \cdot C_{pn} = R_L \frac{\varepsilon A}{W} \qquad (9.9)$$

where A is the device area. Fast devices, therefore, tend to have narrow depletion widths, or 'shallow junctions', to minimize τ_{drift} but consequently rather small active areas to keep τ_{RC} from limiting the bandwidth.

To achieve greater control over the depletion depth and to maximize the mobility, an undoped layer is often grown between the p-type and n-type layers (which are then usually both highly doped). This is known as a pin photodiode and is shown in Figure 9.9. It operates in exactly the same way as a pn photodiode but the region of the device under field is determined by the thickness of the intrinsic layer rather than the depletion region. The electric field is uniform in the intrinsic region, and so equation (9.8) is a better approximation than for a pn structure. Variation of the reverse bias, to a first approximation, modifies only the electric field in these devices. The capacitance of a pin structure is given approximately by $C_{pin} \approx \varepsilon A/\omega_i$, where ω_i is the width of the intrinsic layer. The lack of dopant centres in the intrinsic region results in significantly increased carrier mobilities and, therefore, smaller transit times. Pin photodiodes operating at tens of GHz are now widely available, their development having been driven in large part by the telecommunications industry.

9.3.2.1 Photovoltaic Mode

A special case occurs in a circuit containing just a pn (or pin) photodiode and a load resistor. The built-in electric field in the diode means that a photocurrent is still generated in the reverse direction, causing a potential drop across the load. This manifests itself as a forward bias across the photodiode (bias region II in Figure 9.8), creating negative feedback in the photocurrent, with equilibrium occurring when $V_{forward} = i_{tot}R_L$. Electrical power is generated in the load resistor equal to $P_{PV} = V_{forward} i_{tot}$. This is the principle of operation of a solar, or photovoltaic, cell.

9.3.3 Schottky Diode Detectors

One of the problems in pn photodiode operation is that, with high absorption coefficients (often $\alpha > 10^4 \text{cm}^{-1}$), photons are absorbed close to the semiconductor surface, potentially lowering the photodiode quantum efficiency due to high surface recombination rates. One solution to this problem is the Schottky photodiode, shown in Figure 9.10, which consists of a thin metal layer deposited on to a lightly doped semiconductor. The materials are chosen such that the work function of the metal is greater than the electron affinity in the semiconductor,

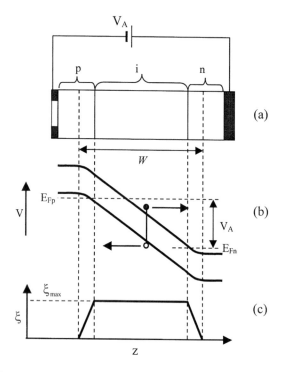

FIGURE 9.9 The pin photodiode, shown for comparison with figure A1.7.7: (a) schematic diagram of a typical device structure; (b) profile of the electronic potential and (c) electric field profile through the device. Since most of the bias is across the intrinsic region, modification of V_A primarily changes ξ_{max} and has little effect on W.

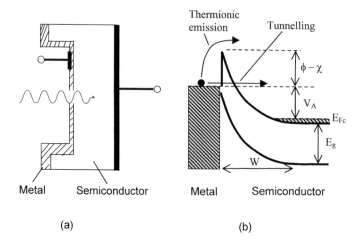

FIGURE 9.10 The Schottky photodiode: (a) schematic diagram of a typical device structure and (b) the profile of the electronic potential through the device. Photocurrent is generated by absorption in the semiconductor depletion region W. Electron transport is inhibited by a triangular potential barrier of height $\phi - \chi$. The two principal mechanisms leading to the dark current—thermionic emission over and tunnelling through the potential barrier—are also indicated.

Optical Detection and Noise

with the result that band bending occurs in the semiconductor similar to that on the lightly doped side of a pn junction. At the semiconductor–metal interface, a triangular potential barrier—the Schottky barrier—inhibits the conduction of majority carriers through the device and gives the device its diode-like electrical characteristics. The height and width of this barrier are fundamental to the device performance, since thermionic emission over and tunnelling through the barrier are the principal origins of dark current in the device. Optical access is through the deposited metal layer, which is typically only a few tens of nanometres thick. For modelling purposes, a Schottky photodiode can be treated like an asymmetric pn junction (equations (9.6)–(9.9)), the semiconductor representing the 'lesser doped' side of the junction. Like pin photodiodes, Schottky barrier detectors can be made to operate at very high speeds.

9.3.4 Avalanche Photodiodes

Avalanche gain is a convenient means by which a photocurrent can be amplified within the detector itself and can offer better noise performance than amplification by external electronics (see Section 9.5).

The avalanche process in semiconductors relies on the multiplication of carriers by impact ionization. If a carrier can obtain sufficient kinetic energy when drifting in an electric field, it can undergo intra-band relaxation whilst using the released kinetic energy to excite a further electron–hole pair. Figure 9.11 shows this process schematically as it occurs when instigated by an electron. The three resulting carriers are then accelerated in the field and have the opportunity to instigate similar processes, and so on. In this way, an 'avalanche' of carriers may be generated.

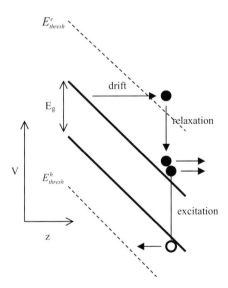

FIGURE 9.11 The (electron-initiated) impact ionization process. An electron obtains sufficient kinetic energy from the electric field to exceed the threshold energy, E^e_{thresh}. It can then relax to the conduction band edge, whilst exciting a further electron across the bandgap. The resulting two electrons and one hole then repeat the process, the hole needing to exceed kinetic energy E^h_{thresh} to instigate ionization.

Electric fields in excess of 10^5 V cm^{-1} are generally required to instigate impact ionization in semiconductors, since the saturation velocity $\mu\xi$ must be high enough that carriers reach the threshold kinetic energy required to satisfy the energy and momentum conservation requirements of the microscopic process. The threshold kinetic energy depends on the effective masses of the participant electrons and holes, and other bandstructure details, and is generally in the range $E_g < E_{thresh} < 3E_g/2$. The key parameters commonly used in describing the macroscopic multiplication are the empirically determined electron and hole impact ionization coefficients, α_e and α_h, which specify the ionization probability per cm of carrier drift, and are expressed as functions of the local electric field.

The net effect of the avalanche process in a device is the multiplication of the photocurrent by the factor

$$M = \frac{1}{1 - f_M} \quad (9.10)$$

where f_M is the McIntyre function which is determined by an integral of the electric-field-dependent ionization coefficients throughout the device [1–3]. Most usage of avalanche photodiodes (APDs) is in 'analogue' mode whereby $0 < f_M < 1$.

The singularity in equation (9.10) at $f_M = 1$ corresponds to avalanche breakdown, where the avalanche process becomes self-sustaining. In analogue-mode operation, APDs are usually biased at around a volt below this breakdown condition (bias region III in Figure 9.8), where M in excess of 100 can be achieved in high-quality devices.

Avalanche gain adversely affects both the bandwidth and the noise of a photodiode. The former is a result of the time taken for the avalanche current to build and to decay, approximately equal to the gain multiplied by the single carrier transit time (as such, the gain–bandwidth product is an important figure of merit for APDs). The latter is due to statistical fluctuations in the avalanche gain and will be discussed further in Section 9.5.3.

The desire for high-absorption efficiency combined with wide bandwidths has led to APDs of the 'reach through' design, in which gain is limited to a narrow region of the device by careful control of the doping profile (Figure 9.12).

A common modification of this design is employed for construction of APDs with sensitivity at optical wavelengths well into the near-infrared. In semiconductors with bandgaps narrower than about 1 eV, the electric fields needed for avalanche multiplication also result in significant increases in dark current due to inter-band tunnelling. For this reason, devices have been developed with separate absorption and multiplication regions (known as SAM devices). The structure is the same as a reach-through device but with a narrow-gap semiconductor occupying the intrinsic region. Absorption at photon energies between those of the two bandgaps occurs only in the narrow-band material, which is kept under low-enough electric field to prevent inter-band tunnelling of carriers. The principal exponent of the SAM design is the InGaAs/InP APD, commonly used for optical fibre-based telecommunications applications. Here, photons of wavelengths between 1.3 and 1.6 μm are absorbed in a low-field InGaAs layer, and the photogenerated

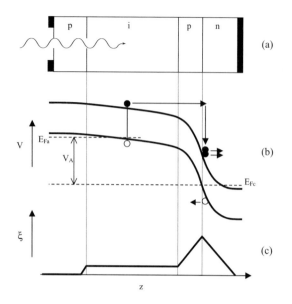

FIGURE 9.12 The reach-through APD (with electron injection into the gain region): (a) typical device structure and (b) electrical potential profile through the device. Optical absorption occurs throughout the device but avalanche multiplication only occurs in the high-field region near the pn junction. (c) Electric field profile.

holes are swept into a high-field InP layer where they then multiply by impact ionization. Note the role of holes as the primary ionization species (holes cause ionization more readily than electrons in InP), so that the high-field region of the device is near to the p-type contact, opposite to that shown in Figure 9.12.

Avalanche breakdown offers a particularly effective means of single-photon detection. If the high-electric-field region of a device is free of conducting carriers when biased above the reverse bias breakdown voltage, no avalanche can occur and no current flows. By injecting a single carrier optically, however, an exponentially growing avalanche current is generated, which can easily be recorded. This current must then be switched off, or 'quenched', before the device overheats and the carriers allowed to drain from the active region before the bias is restored to the above-breakdown value ready for the next photon to arrive. Operated in this mode, APDs are commonly referred to as single-photon avalanche diodes or SPADs.

For photon-counting applications in which the signal can be focused on to a small detector area (spot diameter $\simeq$ 100 µm or less) and the photon flux is less than about $10_6 s^{-1}$, commercially available SPADs offer detectivity about a 100 times higher than the best photon-counting photomultipliers. A thermoelectrically cooled silicon SPAD can offer a dark count rate as low as a few counts per second coupled with a quantum efficiency of 70%, equating to a NEP of the order of 10^{-18} W Hz$^{-1/2}$. Photomultipliers make better large-area devices, however, and are still standard equipment for sensitive fluorescence measurements in which the signal cannot be focused. Photomultipliers' lack of need for quenching also means that the maximum measurable photon flux is limited only by the discrimination of distinct current pulses and 10^7 photons s^{-1} or more is readily achievable.

9.3.5 Photoconductive Detectors

Photoconductive detectors rely on the fact that the conductivity σ of a semiconductor is proportional to the densities n of carriers therein:

$$\sigma = e(\mu_e n_e + \mu_h n_h) \quad (9.11)$$

where subscripts e and h refer to electrons and holes. For an intrinsic semiconductor under illumination, $n_e = n_h$, and the change in conductivity due to the optical signal can be shown to be

$$\Delta\sigma = \frac{e\eta\tau_{rec}F(\mu_e + \mu_h)}{xA} \quad (9.12)$$

where η is the quantum (absorption) efficiency, τ_{rec} is the recombination lifetime, F is the photon flux and x is the electrode spacing.

In the common situation where $\mu_e \gg \mu_h$, the current passed by the photoconductor may be expressed conveniently as

$$i_P = e\eta F \frac{\tau_{rec}}{\tau_e} \quad (9.13)$$

where τ_e is the transit time for electron drift between the electrodes. For each photon absorbed, the number of electrons that can drift between the electrodes before recombination occurs with the hole, and the gain of the photoconductor is, therefore, given by $G = \tau_{rec}/\tau_e$.

Careful design of the detector can achieve $G > 1$. For example, the structure should minimize the distance between electrodes both to minimize the transit distance and to maximize the electric field (and thus the drift velocity) for a given bias. Device geometry, therefore, usually consists of interdigitated electrodes deposited onto the semiconductor surface, as shown in Figure 9.13, allowing high gain whilst maintaining a reasonable active area. Measurement of the conductivity is achieved by monitoring the voltage across a small-load resistor in series with the detector. In this way, the biasing of the photoconductor remains constant and the output voltage is (approximately) proportional to the incident light intensity.

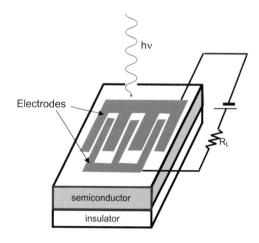

FIGURE 9.13 Schematic diagram of a photoconductive detector with an inter-digitated electrode structure.

Basic photoconductors can be produced extremely cheaply compared with photodiodes, since less care needs to be taken over material quality and contacting. Whilst offering a simple method to achieve signal gain, their detectivity is limited by generation–recombination noise (see Section 9.5.2). They, therefore, tend to be employed in imaging applications where large-scale arrays are needed but low sensitivity is acceptable, such as in photocopiers.

9.3.6 Intra-band Detectors or QWIPs

Intra-band electronic transitions—those that occur within the conduction band or within the valence band—can be utilized to permit photodetection using a doped semiconductor regardless of its bandgap. The threshold photon energy above which strong absorption occurs is determined by the energetic separation of the sub-bands or the energy required to eject a carrier from the quantum well altogether. This is determined partly by the properties of the materials used but also significantly by the width of the quantum wells. QWIPs usually consist of multiple quantum wells, which are doped n- or p-type to provide a source of carriers in the lowest confined level. Common well/barrier material combinations are GaAs/AlGaAs, SiGe/Si and AlGaInAs/InP. The same doping species is used in the contact regions, and the device is held under a bias of typically less than 1 V. Since the same intra-band transitions that are caused by photoabsorption can equally be caused thermally, QWIPs are invariably cooled. Detectivities approaching 10^{11} cm Hz$^{1/2}$ W^{-1} have been reported for a QWIP detecting light of ~ 10 µm wavelength [4]. This figure is some two orders of magnitude higher than the best pyroelectric detectors.

The most common application for QWIPs is in infrared imaging at wavelengths between 2 and 30 µm, for which they are fabricated into focal plane arrays (FPAs). Most commercial QWIP FPAs operate in the temperature range 20–80 K, cooled by Stirling closed-cycle coolers which are compact and light enough to be contained in a hand-held camera.

QWIPs that are designed to utilize a bound-to-continuum excitation are sensitive over a broad wavelength range with a sharp cut-off at long wavelength. Those that utilize a bound-to-bound transition have a narrow range of sensitive wavelengths ($\Delta\lambda \simeq 0.2\lambda$ is common). Two or more differently tuned bound-to-bound QWIPs can, therefore, be used in parallel to generate 'colour' images.

9.4 Thermal Detectors

Thermal detectors contain two principal elements: an absorption region, the temperature of which is a function of the incident optical power and a transducer to convert the temperature variation in the detector to an electrical signal. The latter takes several different forms in different types of detector. In this section, we will discuss the three most common of these: thermocouples and thermopiles; bolometers and thermistors; and pyroelectric detectors.

Thermal detectors are most commonly used for detection of optical radiation at wavelengths in the mid-infrared and longer, as the thermal detection process is slow and inefficient compared with the photoelectric processes described previously, and the latter are usually preferred within their range of sensitivity. However, the ability of thermal detectors to provide very uniform detectivity over a wide range of optical wavelengths has proved useful also for radiometric calibration purposes. In this chapter, we restrict ourselves to a brief introduction to the workings of thermal detectors. More detailed descriptions can be found in the references at the end of the chapter [5,6].

The change in temperature θ caused by incident radiation can be calculated by considering the thermal processes in a typical device. Figure 9.14 shows a schematic diagram that illustrates these processes. Heat is gained by the detector with the absorption of incident radiation and is lost through contact with a heat sink and through heat conduction by any electrical wires connected to the detector (radiation of heat at the surface is generally negligible by comparison).

The heat balance equation is, therefore,

$$H\frac{d\theta}{dt} = \eta P - G\theta \qquad (9.14)$$

where H is the heat capacity of the detector, G is the thermal conductivity of the electrical wires and the thermal link, P is

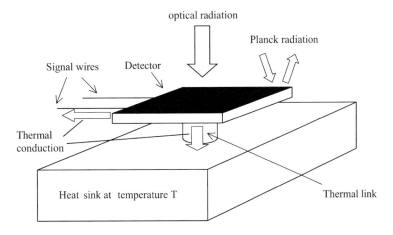

FIGURE 9.14 Heat balance in a thermal detector. The detector itself is coated with a 'black' surface for maximum absorption of radiation and is small in mass for minimum heat capacitance. Principal heat loss is to a heat sink through a thermal link and through any signal wires attached.

the incident radiation power and η is the absorptivity of the surface.

The temperature responsivity $S_\theta = d\theta/dP$ for a modulated incident signal is, therefore,

$$S_\theta = \frac{\eta}{G\sqrt{1+\omega^2 \tau_h^2}} \qquad (9.15)$$

where τ_h is the thermal time constant ($= H/G$), which limits the detector response time (pyroelectric detectors offer something of an exception to this—see Section 9.4.3). Under constant illumination, $\omega = 0$, and $\theta = \eta\, P/G$. Principal design considerations are to maximize θ_0 by maximizing η and to minimize τ_h by minimizing H. The former is achieved by coating the absorption region with a 'black' material for maximum absorption across a wide spectral range, whilst the latter is achieved by using a thin wafer of material with a low-specific heat capacity C (since $H = Cm$ for a body of mass m). The ultimate detectivity of thermal detectors is limited by fluctuations in the detector temperature. 'Temperature noise' will be discussed in Section 9.5.2.6.

9.4.1 Thermocouples and Thermopiles

A thermocouple is a thermoelectric transducer based on the Seebeck effect, which describes the current generated in an electrical circuit containing two or more different conductors when the junctions are held at different temperatures. A simple two-junction thermocouple is shown in Figure 9.15. The voltage measured by the voltmeter in the diagram is $V = S(T_1 - T_2)$, where S is the Seebeck coefficient. For temperature measurement, one of the junctions is held at known temperature while the other is held at the temperature to be measured. For improved sensitivity, junctions are often connected in series to make a thermopile. The very best thermopiles can provide detectivities as high as $10^9\,\text{W}^{-1}\text{Hz}^{1/2}$ and offer response times of a few milliseconds.

9.4.2 Bolometers and Thermistors

In bolometers and thermistors, the temperature change resulting from optical absorption is measured as a change in conductivity. In a metal, the conductivity decreases with temperature and the device is called a bolometer, whereas in a semiconductor, conductivity increases with temperature and the device is called a thermistor. Semiconductors offer temperature coefficients (α) of the order of $-4\%/\text{K}$, whilst metals offer coefficients of the order of $+0.5\%/\text{K}$ but metals have the advantage of being easier to fabricate into micrometre thin flakes which offer smaller thermal capacitance.

The voltage responsivity $S_V = dV/dP$ of a bolometer or thermistor is related to the temperature responsivity by

$$S_V = I\alpha R S_\theta \qquad (9.16)$$

where I is the device current and R is the device resistance.

The material to be used usually forms a thin flake that is coated directly with a highly absorbent 'black' material. Since it is a change in resistance that must be measured, the optical signal is generally chopped mechanically in front of the detector. Often a bridge circuit is used in which a second 'control' flake is placed close to the detecting flake but shielded from the radiation. This method serves to remove most of the dark current and reduces the effects of fluctuations in the ambient temperature on the measured signal. A typical measurement bridge circuit is shown in Figure 9.16.

Bolometers and thermistors offer time responses similar to those of thermopiles. Detectivities within an order of magnitude of the thermodynamic limit (see Section 9.5.2.6) are possible, and sensor flake sizes range from $1\,\text{cm}^2$ to as small as $0.01\,\text{mm}^2$.

9.4.3 Pyroelectric Detectors

The pyroelectric effect consists of a change in the surface charge of a material in response to a change in temperature. Since the detectors' electrical characteristic is that of a capacitor, only fluctuations in temperature can be measured and the optical signal must always be modulated. Materials are chosen for their high pyroelectric coefficient, p, and for their high Curie temperature, above which the electrical polarization is lost. An example of a popular high-p material is triglycine sulphate, which has $p \simeq 30\,\text{nC}\,\text{cm}^{-2}\,\text{K}^{-1}$ and a Curie temperature of 49°C. Also popular are ceramics such as lithium tantalate, LiTaO$_3$, which offers a higher Curie temperature of $T_C \approx 620°\text{C}$ combined with a pyroelectric coefficient of $p \approx 6\,\text{nC}\,\text{cm}^{-2}\,\text{K}^{-1}$.

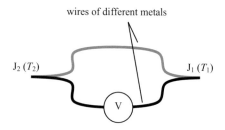

FIGURE 9.15 A two-junction thermocouple. The voltage is proportional to the temperature difference between the junctions.

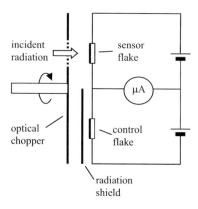

FIGURE 9.16 Measurement set-up using bolometers or thermistors. The micro-ammeter measures the difference between the currents passing through the two flakes. In addition, the optical signal is modulated using a chopper to minimize the background signal.

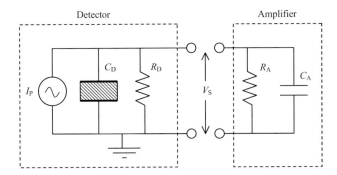

FIGURE 9.17 Equivalent circuit for an illuminated pyroelectric detector and amplifier input stage. Consideration of the complete circuit yields the frequency response of the system.

For a modulated temperature excess θ_ω, the pyroelectric detector produces an alternating current equal to

$$I_P = \omega p A \theta_\omega \qquad (9.17)$$

where ω is the angular frequency, p is the pyroelectric coefficient (in C cm^{-2} K^{-1}) and A is the device area.

The responsivity of the illuminated pyroelectric detector is found by considering its equivalent circuit, shown in Figure 9.17. The detector is represented as a capacitor C_D, a resistor R_D and an alternating current source I_P in parallel. The output, V_s, is usually amplified, whereby the amplifier input impedance must also be considered. The voltage responsivity of the complete detection circuit is found to be

$$S_V = \frac{\omega p A}{\sqrt{\left(1 + \omega^2 \tau_{RC}^2\right)}} S_\theta \qquad (9.18)$$

where τ_{RC} is the RC time constant of the equivalent circuit.

Equation (9.18) presents a very different frequency response to that of the other photothermal detectors. Assuming that $\tau_{RC} \ll \tau_h$, we find that (i) for $\omega \ll 1/\tau_h$, $S_V \alpha \omega$ reminding us that no signal occurs under constant illumination; (ii) in the frequency range $1/\tau_h \ll \omega \ll 1/\tau_{RC}$, the response is approximately flat; and (iii) for $\omega \gg 1/\tau_{RC}$, it falls off as $1/\omega$. The ability of pyroelectric detectors to respond to optical transients much faster than τ_h means that they are the only detectors capable of detecting nanosecond pulses into the far- and extreme-infrared, albeit with a much reduced responsivity.

9.5 Noise in Photodetection

Whenever photodetection is performed, as with all continuous physical measurements, the signal trace will contain a certain level of noise or random fluctuations in the detector output, which limits the accuracy that can be attained. The level of noise experienced depends upon a number of factors, and in many cases, steps can be taken to 'clean up' a noisy signal. Whilst some improvement can often be achieved by post-measurement data processing, better results will invariably be achieved through careful design and use of the measurement apparatus. In particular, we shall see that restriction of the measurement bandwidth can lead to a dramatic reduction in noise levels.

This section is intended as an introduction to the various sources of noise that may be encountered when performing photodetection. In each case, the physical source of the noise will be explained and a simple expression given with the aim of enabling the reader to estimate the magnitude of the effect as it pertains to the measurement in question.

Section 9.1.1 provides an introduction to the nomenclature commonly used in quantifying noise and in defining figures of merit with which to quantify photodetector performance. Sections 9.5.1–9.5.4 describe the various sources of noise that can arise in the photodetection process. These three sections reflect the sequential nature of the photodetection process, starting with the optical signal itself, through noise generated within common detector types and finishing with noise generated in the electrical detection circuit. Section 9.5.4 describes how to combine these noise contributions to generate an aggregate noise spectrum for a photodetection system. Finally, Section 9.5.5 provides a brief introduction to bandwidth-related methods of noise reduction.

9.5.1 Noise in the Optical Signal

Before examining noise generated by the detection process, let us consider the nature of the optical radiation incident on the detector.

9.5.1.1 Background Noise (Blackbody Radiation)

In any photodetection situation, effort must clearly be taken to minimize the detection of unwanted photons. However, this is not always possible to achieve perfectly since, according to Planck's theory, any surface at a finite temperature will radiate photons that can potentially be detected in addition to the desired signal. The optical power emitted per unit surface area and wavelength is

$$P_{bkgnd}(\lambda) = \frac{2\pi h c^2}{\lambda^5 \left(\exp[hc/\lambda kT] - 1\right)}. \qquad (9.19)$$

This relationship is plotted for several temperatures in Figure 9.18. For thermal equilibrium in a black box, the power radiated by any given area must be equal to the power incident on it. Using this approximation, we can estimate the optical power incident on a detector as being equal to the radiance by a surface of the same area at the temperature of the surroundings. In most practical situations, temperatures are a few hundred Kelvin or less and the effect is only significant for measurements at wavelengths in the mid-infrared or longer, for which emission from detector housing can be significant.

The background radiation incident on a detector leads to an additional 'dark' signal in the photoresponse—its effect still present when the signal source is switched off—the average of which can be subtracted from the total detector output signal. The signal is not constant, however, and is subject to random fluctuations which cannot be subtracted (because they are random!). These fluctuations ultimately limit the sensitivity of the

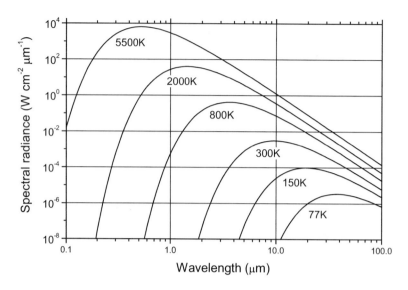

FIGURE 9.18 Blackbody spectral radiance at temperatures between 77 and 5500 K (approximate solar temperature). The radiance describes the background optical power incident on a detector situated in a black box at temperature T.

measurements. A detector exhibiting noise dominated by these fluctuations is said to be at the background-limited intrinsic performance limit. Further improvement can only be achieved by reducing the temperature (T in equation (9.19)) of as much of the detector field of view as possible, for example, by utilizing a cooled radiation shield. A shield should be chosen with an aperture corresponding closely to the spatial extent of the signal so as to limit the surface area over which 'uncooled' background radiation is detected.

9.5.1.2 Photon Noise

The behaviour of light as discrete 'particles' means that the arrival of light at a detector is not uniform but made up of pulses corresponding to the arrival of individual photons. Moreover, in most situations, the photons do not arrive at regular intervals and repeated observations of time duration T will reveal that the number of photon arrivals fluctuates about the mean value of $\langle n \rangle = PT/h\nu$, where P is the optical power and $h\nu$ is the photon energy. These fluctuations are known as photon noise and can be observed using a detector of sufficient detectivity.

In coherent light from a laser, the interval between photon arrivals is purely random (i.e. if we choose $T \ll h\nu/P$, the probability of a photon arriving in any time window is equal). The number of photon arrivals in a time window T is, therefore, governed by Poissonian statistics, and the variance in the photon number is equal to $\langle \delta n^2 \rangle = \langle n \rangle$.

Even with a perfect linear photodetector, therefore, in which each incident photon is converted into an exactly determined photocurrent, we can expect this noise to be present. In such a situation, the SNR is given by

$$\text{SNR} = \frac{\langle n \rangle^2}{\langle \delta n^2 \rangle} = \langle n \rangle \quad (9.20)$$

and so increases linearly with the photon flux or with the detection time T. Such a relationship between SNR and detection time indicates that photon noise is white noise (the maximum measurement bandwidth is $\Delta f = 1/T$).

For the vast majority of photodetection situations, photon noise constitutes an absolute lower limit on the achievable noise amplitude. Exceptions to this are only achieved by manipulation of the light itself, by techniques known as 'squeezing', which allow photon noise to be 'traded' between complementary measurements in two-dimensional amplitude/phase space [6–8].

9.5.2 Noise in the Photodetector

9.5.2.1 Photoelectron Noise

In most photodetectors, the quantum efficiency η is less than unity. Whether or not an individual photon is absorbed is arbitrary and so again Poissonian statistics apply. In the same way that the photon noise is proportional to the average photon flux, the photoelectron noise is proportional to the average photocurrent, I_{ph}.

$$\langle \delta i^2 \rangle = 2eI_{\text{ph}} \Delta f \quad (9.21)$$

where Δf is the bandwidth of the detector.

The linearity of the expression with Δf indicates that photoelectron noise, like photon noise, is also white noise.

9.5.2.2 Shot Noise

Shot noise is the general term for the noise that results from the varying number of photocarriers that contribute to the photocurrent. In a photodiode, it accounts for both photon and photoelectron noise, along with any other mechanisms that cause random loss of photogenerated electrons in the detector. One common mechanism is thermionic emission across a potential barrier, such as that encountered by photogenerated holes in a heterojunction design InGaAs/InP APD (see Section 9.3.4). The expression for the total shot noise is

Optical Detection and Noise

$$\langle \delta i^2 \rangle = 2eI_0 \Delta f \quad (9.22)$$

where I_0 is the average photocurrent, similar to I_{ph} but including these additional losses. Figure 9.19 shows how the random processes described result in shot noise in the detector photocurrent, on the assumption that each primary photoelectron generates a similar current pulse at the detector output.

The maximum SNR due to shot noise is found to be $I_0/2e\Delta f$ and is independent of any subsequent gain, as both signal and noise are amplified by the same factor.

9.5.2.3 Generation–Recombination (G–R) Noise

In a photoconductor, an additional source of shot noise is present, due to random fluctuations of the carrier density caused by the competing processes of thermal carrier generation and recombination. In contrast to the white noise sources encountered thus far, G–R noise reduces rapidly above a cut-off frequency corresponding to the recombination lifetime τ_{rec}

$$\langle \delta i^2 \rangle = \frac{4eGI_0}{1 + 4\pi^2 f^2 \tau_{rec}^2} \quad (9.23)$$

where G is the gain of the device. G–R noise can be reduced by cooling the detector, as the thermal generation rate is proportional to $\exp(-E_g/kT)$.

9.5.2.4 Gain Noise

In an amplifying photodetector with deterministic gain G, both signal and noise are amplified by this same factor and the SNR remains unchanged. In most common detectors, such as APDs and photomultipliers, the gain mechanisms are random in nature and so the current pulse generated by each absorbed photon will not be the same, as inferred in Figure 9.19, but will be subject to Poissonian fluctuations. This further variance, represented schematically in Figure 9.20, forms an additional source of noise, known as gain noise. It is characterized by the *excess noise factor*

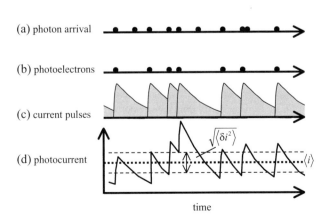

FIGURE 9.19 The origins of shot noise: (a) photons arriving at irregular intervals, some of which are absorbed to generate (b) photoelectrons; (c) each photoelectron results in a similar detector current pulse, which sum (d) to give the total photocurrent.

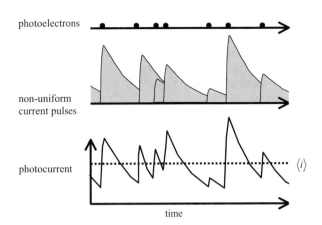

FIGURE 9.20 Gain noise. Random fluctuations in the gain mean that each photoelectron does not produce the same contribution to the current pulse. The SNR of the resulting photocurrent is smaller than that in Figure 9.19.

$$F = \frac{\langle G^2 \rangle}{\langle G \rangle^2} = 1 + \frac{\langle \delta G^2 \rangle}{\langle G \rangle^2} \quad (9.24)$$

where $\langle \delta G^2 \rangle$ is the variance in the gain. The total variance in the detector photocurrent due to the multiplied shot noise and gain noise is then simply $F\langle G \rangle$ times the shot noise given in equation (9.22).

In an APD, the excess noise factor is determined primarily by the relative contributions of electrons and holes to the multiplication process and

$$F = kG + \left(2 - \frac{1}{G}\right)(1-k) \quad (9.25)$$

where $k = \alpha_s/\alpha_p$ is the ratio of the impact ionization coefficients for the secondary and primary carrier types ($0 < k < 1$). The primary carrier type in this context depends upon the semiconductor material from which the avalanche gain region is made. For example, in Si and GaAs, electrons undergo impact ionization more readily than holes ($\alpha_e > \alpha_h$), and so most APDs made from these materials will use electrons as the primary carrier type and $k = \alpha_h/\alpha_e$. In InP, however, $\alpha_h > \alpha_e$, so holes are usually the primary carrier and $k = \alpha_e/\alpha_h$. Equation (9.25) reveals that the excess noise factor is lowest for $k \approx 0$, i.e. for a large difference between α_e and α_e.

Since the gain process in photomultipliers involves electrons alone, their excess noise factor is given by equation (9.25) with the substitution $k = 0$, whereby $F = 2 - 1/G$. The one situation in which gain noise can be neglected is that in which G is sufficiently large and the signal sufficiently weak, to perform single-photon counting. These digital measurements are subject only to shot noise and are, therefore, well suited to low-frequency photodetection where G–R noise or flicker (see next section) might otherwise dominate. (Note that, in the case of a photomultiplier, the gain distribution may manifest itself as an additional source of shot noise if discrimination of the pulse height from the detector causes some pulses to pass unrecorded.) An additional advantage of single-photon counting is the ability to measure fast optical pulses. Time-correlated single-photon counting (TCSPC) is a powerful technique in which the arrival time of a photon can be measured with picosecond accuracy relative to a synchronization signal, and the time dependence of the optical intensity can thus be acquired by repeating the measurement over many cycles. (For a detailed account of TCSPC, see Ref. [9].)

9.5.2.5 1/f or Flicker

At the low-frequency limit of analogue measurements, a further source of noise is universally dominant. It is often known simply as '1/f noise', as its intensity reduces uniformly with increasing frequency. 1/f noise exhibits equal power per increment in $\log(f)$ and is sometimes referred to as 'pink noise'. Its presence results in the unwelcome discovery that lower noise cannot necessarily be achieved by increasing the measurement integration time.

1/f noise can be found in a surprising range of physical systems, from cathode ray tubes to tidal movements, to inner city traffic flow! In semiconductor photodetectors, it is generally suspected to arise from carrier trapping by surface and bulk impurities and is particularly troublesome in devices with poor-quality ohmic contacts (for a review, [10]).

9.5.2.6 Temperature Noise

Under certain circumstances, noise can be measured that is due to random fluctuations in the temperature of the detector. This is rarely an issue in detectors based on photoelectric processes but is of crucial importance to photothermal detectors.

A theoretical lower limit in temperature fluctuations is imposed by the absorption and emission of Planck radiation, described in Section 9.5.2. The resulting temperature variance is given by statistical thermodynamics considerations:

$$\langle \Delta\theta^2 \rangle = \frac{4kT^2 G \Delta f}{G^2 + 4\pi^2 f^2 H^2} \quad (9.26)$$

where G and H are the thermal conductance and heat capacitance as defined in Section 9.4. At measurement frequencies well below the maximum response frequency of the detector, fmax = G/H, temperature noise is white and the noise power delivered to the detector is

$$\langle \Delta\phi^2 \rangle = G^2 \langle \Delta\theta^2 \rangle 4kT^2 G \Delta f. \quad (9.27)$$

A thermal detector is said to be 'ideal' when the thermal conductance is due primarily to Planck radiation, whereupon $G = 4\sigma e \varepsilon A T^3$, where σ is the Stefan–Boltzmann constant, ε is the emissivity and A is the detector area. This imposes a theoretical upper limit on the specific detectivity of photothermal detectors,

$$D^*_{max} = \left(8\pi\sigma\varepsilon k T^5 \right)^{-1/2}. \quad (9.28)$$

Assuming $\varepsilon = 1$ at 300 K, $D^*_{max} = 1.8 \times 10^{10}$ cm Hz$^{1/2}$ W^{-1}, while at 77K, $D^*_{max} = 5.2 \times 10^{11}$ cm Hz$^{1/2}$ W^{-1}. In real thermal detectors, these lower limits of thermal fluctuation are some way from being achieved, however, and the highest detectivities of uncooled detectors are those of the best pyroelectric detectors and bolometers, at around 10^9 cm Hz$^{1/2}$ W^{-1}.

9.5.3 Noise in the Measurement Circuit

9.5.3.1 Thermal Noise (Aka Johnson or Nyquist Noise)

The random thermal motion of electrons in a resistor leads to fluctuations in the current passing through it and thus also in the voltage measured across it. This affects all photodetection measurements, the most common sources being the load resistance in the measurement circuit and the output impedence of the photodetector itself.

The expression for the resulting current variance is derived from the Bose–Einstein statistics applied to electromagnetic radiation of frequency f:

$$\langle \delta i^2 \rangle = \frac{4hf}{R\left(\exp[hf/kT]-1\right)} \quad (9.29)$$

where k is Boltzmann's constant, T is the temperature in kelvin and R is the resistance in ohms. For detection frequencies low compared with kT/h (i.e. below about 10^{12} Hz at room temperature), thermal noise is white and $\langle \delta i^2 \rangle \approx 4kT\Delta f/R$. An important comparison can be made between the level of white noise caused by Johnson noise and that caused by shot noise. Equations (9.22) and (9.29) reveal that the Johnson noise level falls below the shot noise level if

$$kT < \frac{eI_0 R}{2}. \quad (9.30)$$

The Johnson noise in the load resistance seen by the photodetector is thus important in determining the dominant noise source. For example, in a photodiode with a dark current of 100 pA, equation (9.30) reveals that a load resistance greater than 500 MΩ would be required to achieve shot noise limited detectivity at a temperature of 300 K. Alternatively if we consider a fast photodiode terminated with a 50 Ω load, a photocurrent of at least 1 mA is required for the signal to be shot noise limited.

9.5.3.2 Amplifier Noise and Impedance Matching

Amplification of the photodetector signal is often essential. Here, we look at two common circuits in which amplification is used and consider their noise characteristics. The first, shown in Figure 9.21a, concerns a detector, represented as a current source generating current is and terminated to earth with a load resistor R_L, the voltage across which is amplified using a voltage amplifier. The total voltage noise at point A in the circuit is, in the absence of amplification, calculated by combining the Johnson noise in R_L with the detector current noise across R_L.

$$\langle \delta \Delta^2 \rangle_A = \langle \delta i_S^2 \rangle R_L^2 + 4kTR_L\Delta f. \quad (9.31)$$

If the signal at point A is fed into a voltage amplifier, we can represent the further noise contribution by two quantities—an

Optical Detection and Noise

input noise current i_n and an input noise voltage v_n (representing rms values)—in combination with a hypothetical noise-free amplifier G. The further variance in the voltage at the input to G, due both to and to passing through R_L, is then

$$\langle \delta v_G^2 \rangle = v_n^2 + (i_n R_L)^2 \quad (9.32)$$

If necessary, v_n and i_n can be measured by recording the noise level in the amplifier output under appropriate limiting input conditions (see Refs. [11,12]).

The factor by which the amplifier stage increases the effective input noise is given by the noise figure (NF):

$$\mathrm{NF} = 1 + \frac{\langle \delta v_G^2 \rangle}{\langle \delta v^2 \rangle_A}. \quad (9.33)$$

It is straightforward to show that if $\langle \delta v^2 \rangle_A$ is dominated by thermal noise in resistor R_L, then NF is minimized when $v_n^2 = i_n^2 R_L^2$. Under these conditions, the circuit is said to be impedance matched for minimum amplifier noise.

For photodiodes and photomultipliers, it is common to use a trans-impedance amplifier to convert the current output of the detector to a voltage. The idealized circuit is shown in Figure 9.21b. The ideal operational amplifier has an infinite input impedance and infinite gain, ensuring that all of is flows through the feedback resistor R_T and that the input is a 'virtual earth' held at 0 V. The output voltage is then simply $V_\mathrm{out} = -i_s R_T$. Johnson noise in the feedback resistor is the main source of noise, combining with the noise in the detector current to give a total voltage noise at the output equal to

$$\langle \delta V_\mathrm{out}^2 \rangle = \langle \delta i_s^2 \rangle R_T^2 + 4kTR_T \Delta f. \quad (9.34)$$

The circuit in Figure 9.21b offers a significant advantage of presenting a low load resistance to the detector due to the virtual earth at the input but without a corresponding Johnson noise penalty. Trans-impedance amplifiers are, therefore, widely used as pre-amplifiers, providing at the output a voltage source with low output impedance that can be connected to further amplification or processing components.

9.5.4 Combining Noise Sources

More than one source of noise may well be present in a measurement and it is important to know how to combine them to arrive at a figure for the total noise. In equations (9.31) and (9.34), variances from different sources have been added to give a 'total variance'. This is the appropriate procedure for noise from *uncorrelated* sources. Most noise sources to be encountered, and all of those discussed previously in this chapter, are uncorrelated and can be combined in this way. Note that adding the variances is equivalent to performing a 'sum of squares' on the rms noise amplitude.

Occasionally, noise sources will be encountered that are correlated. An example of this is electromagnetic pickup from a single source by two components in the detection circuit. In such cases, the rms noise amplitudes are added.

By combining the frequency dependence of the relevant noise sources, a noise spectrum is generated. Figure 9.22 shows a typical noise spectrum for a photoconductor, which is subject to flicker, G–R and thermal or shot noise.

In this situation, it is clear that the lowest noise will be witnessed for high-frequency signals. For a particular measurement bandwidth, the total noise power is the integral of the noise spectrum across the appropriate range of frequencies.

9.5.5 Bandwidth-related Noise Reduction Methods

Having optimized the signal and detector properties, the most common method of reducing the noise in a measurement involves manipulation of the measurement bandwidth.

The simplest example of this, for measurement of a constant signal, is the averaging of the detector output over time.

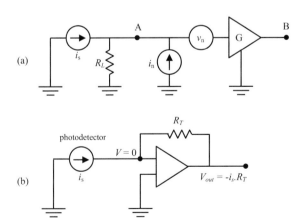

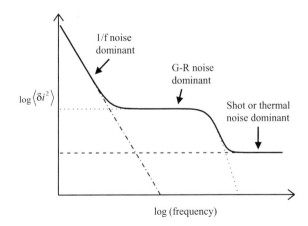

FIGURE 9.21 Equivalent circuits demonstrating the treatment of amplifier noise. (a) Shows how noise can be treated where a voltage amplifier is employed. The amplifier circuit (between points A and B) is represented by a noise-free gain G with an additional input noise current in and input noise voltage v_n (see the text). (b) Shows an idealized trans-impedance amplifier commonly used with photodiodes and photomultipliers. Here, the only noise contribution by the amplifying circuit is Johnson noise in the feedback resistor.

FIGURE 9.22 A typical frequency spectrum for noise in a photoconductor. At low frequencies (<1 kHz), flicker noise dominates; at intermediate frequencies (1 kHz < f < 1 MHz), generation–recombination noise dominates and at high frequencies (>1 MHz), shot or thermal noise dominates. Calculation of the total noise power in a measurement is performed by integrating the noise power spectrum over the measurement bandwidth.

The bandwidth of the measurement is thus reduced and much of the noise power is filtered out.

As Figure 9.22 illustrates, however, the detection process may contribute less noise to the detector output if the measurement is performed at higher frequency. To take advantage of this, we must modulate the optical signal at the desired frequency, then isolate this frequency component in our detection process. This can be achieved readily using an optical chopper coupled with a lock-in amplifier.

Such 'phase-sensitive detection' methods can be extremely useful in identifying small signals amongst large amounts of noise and have the added advantage that unwanted background signals can often be eliminated by careful application of the initial signal modulation (for a more detailed description, see Refs. [10,13]).

REFERENCES

1. Chuang S L 1995 *Physics of Optoelectronic Devices* (New York: Wiley) (Theory-oriented treatment of basic semiconductor detectors. Clear, friendly notation).
2. Sze S M 1981 *The Physics of Semiconductor Devices* (New York: Wiley) (General physics of semiconductor photodetectors).
3. Wood D 1994 *Optoelectronic Semiconductor Devices* (London: Prentice-Hall) (Good general reference, especially on pn and avalanche photodiodes).
4. Moon J, Li S S and Lee J H 2001 *Electron. Lett.* **37** 1249.
5. Budde W 1983 *Optical Radiation Measurements Vol 4: Physical Detectors of Optical Radiation* (New York: Academic) (Most rigorous and comprehensive of the texts. Excellent coverage of photothermal detectors. A little outdated in some areas).
6. Donati S 2000 *Photodetectors* (Englewood Cliffs, NJ: Prentice-Hall) (Good general reference on optical detection, good chapters on photomultipliers and semiconductor detectors. Also discusses non-demolitive detection and optical squeezing).
7. Bachor H 1998 *A Guide to Experiments in Quantum Optics* (Weinheim: Wiley) (Good practical explanation of squeezing and homodyne detection).
8. Loudon R 2000 *The Quantum Theory of Light* 3rd edn (Oxford: Oxford University Press) (Quantum optics bible).
9. O'Connor D V and Phillips D 1984 *Time-Correlated Single Photon Counting* (New York: Academic) (Everything you need to know about TCSPC. Concentrates on photomultipliers rather than semiconductor detectors).
10. Kogan Sh 1996 *Electronic Noise and Fluctuations in Solids* (Cambridge: Cambridge University Press) (Thorough treatment of noise sources in solids. Good section on $1/f$ noise).
11. Horowitz P and Hill W 1989 *The Art of Electronics* 2nd edn (Cambridge: Cambridge University Press) (Thorough treatment of noise in amplifier circuits. Excellent overall electronics handbook. Again, not much in the way of physics. Essential reading for circuit designers).
12. Hartley Jones M 1985 *A Practical Introduction to Electronic Circuits* (Cambridge: Cambridge University Press) (Electronics perspective, with good introduction to low noise amplifier circuits. Equations, but little physical explanation).
13. Jenkins T E 1987 *Optical Sensing Techniques and Signal Processing* (London: Prentice-Hall) (Excellent introductory text. Especially useful for post-detection signal processing, as few other photodetection texts cover this subject).

FURTHER READING

Battacharya P 1994 *Semiconductor Optoelectronic Devices* (New York: Prentice-Hall).

Good general introduction to different types of semiconductor photodetector and their performance characteristics.

Dereniak E L and Crowe D G 1984 *Optical Radiation Detectors* (New York: Wiley).

Excellent text on optical detectors.

Saleh B E A and Teich M C 1991 *Fundamentals of Photonics* (New York: Wiley).

Excellent chapter on semiconductor photodetectors and noise therein. Good explanation, equations and diagrams. Nothing on thermal detectors though.

Senior J M 1992 *Optical Fiber Communications* 2nd edn (London: Prentice-Hall).

Good coverage of semiconductor photodiodes for telecommunications.

10

Laser Safety

J. Michael Green and Karl Schulmeister

CONTENTS

10.1 Introduction 189
10.2 Laser Injuries to the Eyes and Skin 190
    10.2.1 Injury Mechanisms 190
    10.2.2 Principal Components and Operation of the Eye 190
    10.2.3 Laser Injuries to the Eye 191
    10.2.4 Retinal Injuries 192
10.3 Exposure Limits 192
    10.3.1 Establishing a Threshold Level 192
    10.3.2 Calculating MPE from Tables 193
        10.3.2.1 Assessing the Exposure Duration 193
        10.3.2.2 Dealing with Multiple Pulse Exposures 193
        10.3.2.3 Small and Extended Sources 194
    10.3.3 Nominal Ocular Hazard Distance 194
10.4 Safety in Product Design 195
    10.4.1 The Classification Scheme 195
    10.4.2 Optical Viewing Aids 195
    10.4.3 Engineering Safety Features on Laser Products 200
        10.4.3.1 Protective Housing 200
10.5 Safety in Practice 200
    10.5.1 Class-based User Guidance 200
    10.5.2 Application of Control Measures 202
    10.5.3 Personal Protective Equipment 202
        10.5.3.1 Eye Protection 202
        10.5.3.2 Skin Protection 202
    10.5.4 Accident Reports 202
10.6 Associated Hazards 203
10.7 Summary 203
References 203
Further Reading 204

10.1 Introduction

This chapter describes the hazard posed by laser radiation, how it is assessed and the principles of how to minimize the risk. The nature of the laser radiation hazard is well known, yet safety standards and guidance documents must continually evolve in response to the challenges of new laser applications and laser source developments; in this chapter, the major revisions to the international laser safety standard introduced in 2001 have been included.

The general trend in laser safety standards is a relaxation of what was historically a highly cautious approach to the setting of threshold limits and classifying of laser products. These subjects form the more complex and technical aspects of laser safety, yet from a user point of view, the subject remains an essentially practical one, dealing primarily with the design and selection of appropriate engineering controls, administrative controls and protective eyewear.

Wavelength is the principal determinant of the nature of the laser injury. This is particularly true for the eyes, and after a review in Section 10.2 of the wavelength dependence of absorption in the various components of the eye and the various injury mechanisms, Section 10.3 addresses the setting of threshold limits, with particular emphasis on recent developments in the international laser safety standard. The hazard classification scheme, the bedrock of manufacturer requirements and user guidance, is described in Section 10.4. Finally,

Section 10.5 reviews the principal control measures for safe laser use.

This chapter is intended only to provide a basic overview of the subject. It highlights the principal hazards and how to achieve safe use. However, it should not be regarded as a safety manual. In particular, the assessment of laser hazards, classification of laser products and the implementation of laser safety controls require detailed considerations beyond the scope of this chapter.

10.2 Laser Injuries to the Eyes and Skin

10.2.1 Injury Mechanisms

The full range of radiation wavelengths available from laser sources extends from the ultraviolet through to the far-infrared. Over this range, the skin is a strong absorber, thereby providing primary protection for all the bodily organs, except for the eyes. When absorption of radiation occurs, the temperature of the absorbing tissue rises, and if the temperature exceeds a critical temperature of approximately 60°C, proteins in cells denature and the cell dies. If the temperatures exceed 100°C, water in the tissue begins to boil and further temperature increases lead to a carbonization of the tissue. This is the *thermal* damage process. It is characterized by a sharp threshold for injury and is non-cumulative in that if thermal injury does not occur during the first approximately 10 s during which the temperature of the exposed tissue will be rising to some steady-state value, then it will not occur for prolonged or repeated exposure at that level.

For radiation in the ultraviolet and the blue/green end of the visible range of wavelengths, the thermal competes with the photochemical damage process. The capability of radiation to initiate photochemical effects is strongly wavelength-dependent and generally increases with decreasing wavelength. Chemical changes in the exposed tissue induced by this shorter wavelength radiation are cumulative, certainly over a working day, so the degree of damage depends on the radiant exposure (J m^{-2} on the exposed surface), i.e. the accumulated energy. Unlike the threshold for thermal damage, which varies with wavelength only in response to the changes in the absorption depth in the tissue, the photochemical 'action curve' has a strong wavelength dependence and photochemical injury can be the dominant mechanism, but only for prolonged or repeated exposures at levels that are too small to cause a sufficient temperature rise for thermal damage. Conversely, thermal damage will always be the dominant injury mechanism for short-time (less than 1 s) single exposures.

10.2.2 Principal Components and Operation of the Eye

There are many reference sources describing the structure and operation of the eye in sufficient detail for laser safety purposes. Figure 10.1 shows the structural elements of the human eye, which is equipped to transmit electromagnetic radiation in the restricted wavelength range of 400 nm (blue light) through to 1400 nm (near-infrared):

- through the cornea, which provides the primary means of focusing of the radiation;
- through the pupil at the centre of the iris, whose partial closure in response to bright light provides a degree of regulation;
- through the somewhat pliable crystalline lens, which under muscular control provides the remaining amount of focusing of the radiation;
- creating an image of the radiation source on the foveal region of the retina, where photoreceptors provide a signal to the brain from which the structure and colour of the image are interpreted.

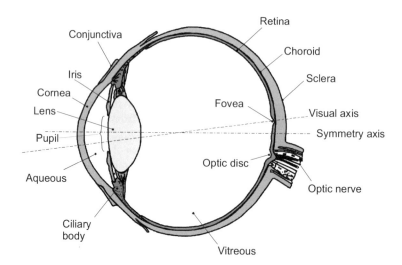

FIGURE 10.1 A schematic horizontal cross section of a human eye. Central sharp vision is only possible in the central part of the retina, the macula (yellow spot) and particularly in the even smaller centre of the macula, the fovea. The main refractive power of the eye is provided by the cornea, whilst the lens is needed for accommodation (imaging) of distant and close objects.

10.2.3 Laser Injuries to the Eye

The absorption of radiation incident on the eye has a strong and complex wavelength dependence. This is shown in figure 10.2, which plots the percentage absorption of radiation incident on the cornea within the different components of the ocular media. Various wavelength bands can be discerned over which the cornea, the anterior region (aqueous and lens) and the posterior (vitreous) become the major absorbing components. Between 0.4 and 1.4 µm, the accumulated absorption of these components is less than 100%, and the retina is therefore at risk of injury.

Table 10.1 delineates the principal types and locations of injury according to spectral band and exposure duration. Note that damage to the cornea or lens can result in serious, though

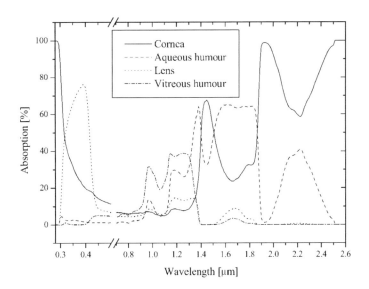

FIGURE 10.2 The spectral absorption of the ocular media for the eye. The break in the curves corresponds to the visible 0.4–0.7 µm region where there is very little absorption through to the retina. Of particular note is the strong UV and near-IR absorption of the lens and the 0.86–1.4 µm region (only) where the vitreous absorbs a significant fraction of the incident radiation. Corneal absorption of 100% at both wavelength extremes of the figure extends to longer and shorter wavelengths.

TABLE 10.1

Summary of Principal Ocular Injuries for Radiation from Ultraviolet to Infrared Wavelengths. The Cornea Has a Highly Effective Repair Mechanism for Minor Injuries, but for the Lens and Retina, the Injury is Permanent. Note That the Injury Mechanisms Apply to Both Laser and Non-Laser Sources, Except for Short Pulse (Typically Q-Switched Lasers with Pulse Durations Less Than 1 µs) Injury Mechanisms, which are Unique to Lasers

Wavelength Range	Tissue Affected	Single-Pulse Injury	Long Exposure (Several Seconds and More) cw and Repetitively Pulse Lasers
Ultraviolet (180–400 nm)	Cornea lens	Thermal damage dominates, leading to denaturation (clouding) of the cornea and (0.28–0.4 µm) lens.	Photochemical damage dominates, leading to photokeratitis ('arc-eye' or 'snow blindness') of the cornea and photochemical cataract (0.28–0.4 µm).
		Short high-power pulses can photoablate corneal tissue.	
Visible and near-infrared (400–1400 nm)	Retina	Thermal damage dominates, leading to a retinal burn (protein denaturation) with potential severe vision loss	400–550 nm (blue to green) may cause photochemical damage for prolonged exposure.
		Short pulses can cause photomechanical damage (i.e. rupture of tissue) leading to extensive retinal damage and bleeding into the inner eye.	Exposure to visible radiation is normally limited to less than 0.25 s by the natural aversion response to bright light. More generally, the saccadic eye movement reduces the retinal hazard for longer durations.
Far-infrared (1400 nm to 1 mm)	Cornea lens	Thermal damage dominates, leading to denaturation (clouding) of the cornea and (1.4–1.9 µm) lens.	Exposure for thermal damage is normally limited to a few seconds by reaction to pain due to heating of the cornea.
		Short high-power pulses can photoablate corneal tissue.	Long-term exposure can lead to infrared cataract (1.4–3 µm)

in principle correctable, vision loss. By contrast, injuries to the retina do not heal and are not repairable.

10.2.4 Retinal Injuries

Within the 'retinal hazard range' (0.4–1.4 μm), a laser beam entering the eye and passing through the pupil may be focused to a spot of the order of 20 μm diameter on the retina, resulting in a concentration of the exposure level at the cornea of up to approximately 100,000 times, i.e. the ratio of areas of a 7 mm diameter pupil to a 20 μm image. In practice, factors during laser exposure including pupil size and accommodation of the eye (i.e. the distance at which the eye is focused), combined with involuntary eye movements and, in the wavelength range of light (0.4–0.7 μm), aversion responses (including the blink reflex), lessen the retinal exposure. Nevertheless, quoted safe exposure limits at the cornea undergo a step function change of several orders of magnitude at 400 nm.

Figure 10.3 shows the overall transmission curve for the eye together with the retinal visual response curve. It shows the retina as sensitive only to wavelengths in the approximate range 0.4–0.7 μm, though a limited retinal sensitivity persists into the near-infrared (0.7–1.4 μm). For laser safety considerations, this sensitivity is regarded as insufficient to evoke an aversion response to hazardous levels of near-infrared radiation, and there are many cases of such near-infrared exposures causing damage without the victim being aware.

As confirmed by accident statistics [1], the near-infrared is the most hazardous wavelength range for lasers in general and near-infrared pulsed lasers are the most hazardous sources. A fundus photograph in figure 10.4 shows several types of thermal and thermo-mechanical retinal injuries in a rhesus monkey eye produced by a laser in this category. The white lesions result from thermally induced denaturation of proteins and are characteristic of exposures longer than 1 μs. The area of the injury is confined to the irradiated area, extended by thermal conduction. However, even minimal retinal lesions may result in permanent serious vision loss if the damage is located in the

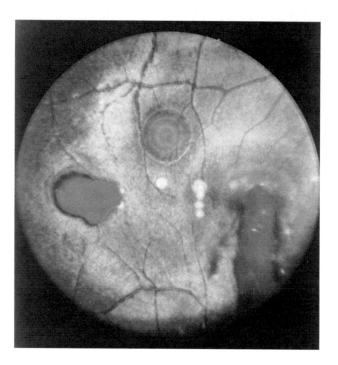

FIGURE 10.4 A range of injuries induced with a Nd:YAG laser on a monkey retina. The white spots in the centre are thermal burns, i.e. coagulation of retinal layers. With larger pulse energies, holes are produced in the retina which result in bleeding either into the vitreous or contained within the retina, both of which result in functional loss in the affected area. (Photograph courtesy of J Zuclich, TASC Litton, TX, USA.)

foveal region of the retina or the optic disc. For laser pulses of duration less than 1 μs, rapid heating of the absorbing tissue can cause a micro-explosion at the back of the eye, possibly leading to haemorrhaging and damage extending well beyond the irradiated area, the flow of blood into the vitreous humour greatly increasing the impairment of vision and the psychological impact of the injury on the victim. Examples of these injuries are illustrated in figure 10.4.

10.3 Exposure Limits

10.3.1 Establishing a Threshold Level

The level of exposure or irradiance which can be thought of as the border between safe and potentially harmful is called maximum permissible exposure (MPE). MPE values are mostly derived from animal experiments and a very limited number of human exposures. In such threshold experiments, a number of exposures are delivered to the eye or the skin for some fixed set of conditions (laser wavelength, exposure duration, spot size). After each exposure at a given energy level, the exposed site is examined for a detectable lesion. For obvious reasons, the exposed site can only be exposed once and therefore a range of sites per animal and also a number of animals have to be exposed. As a result, and in combination with experimental uncertainties, there is not a sharply defined threshold exposure value below which no lesions are found and above which all exposures lead to damage [2].

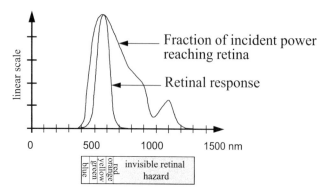

FIGURE 10.3 Semi-qualitative comparison of the photopic response curve and the transmission curve for the eye. The sensitivity of the eye in respect to visual stimulus is highest in the green, around 550 nm, and falls off towards red on one side and blue on the other. There is no sharp dividing line between visible and non-visible. Whereas the limits of the visible spectrum are defined as 380 and 780 nm [CIE], in laser safety, in the context of inducing an aversion response to bright light, they are taken as 400 and 700 nm.

Figure 10.5 is a typical dose–response curve, often called a 'probit plot'. The exposure dose at which 50% of the exposures lead to a lesion, ED-50, is generally referred to as the threshold, even though there is a finite probability for damage at energies somewhat below the ED-50. The MPE is set well below the ED-50 point to ensure that the probability of a detectable minimal lesion is practically zero.

The safety factor between ED-50 and MPE is often set at 10. However, where there is less variability and small experimental uncertainty, such as for corneal photochemical injury in the UV range, a safety factor less than 10 has been agreed by the international committee. To cover bands of wavelength and pulse duration, MPE values are expressed as single values or by a simple analytical expression, and as a result, safety margins greater than 10 are often encountered.

10.3.2 Calculating MPE from Tables

Internationally approved MPE values are to be found in the IEC 60825-1 standard [3]. Values expressed as an irradiance (W m^{-2}) or as a radiant exposure (J m^{-2}), averaged over an area of specified diameter or 'limiting aperture'. The importance of the limiting aperture is primarily in hazard assessment. For example, the ocular MPE for a 0.25 s intra-beam exposure to radiation in the 0.4–0.7 μm wavelength range is quoted as 26 W m^{-2}, and the corresponding limiting aperture size is 7 mm. In this case, multiplying the MPE by the area of the limiting aperture gives a result of 1 mW. If a 0.25 s laser exposure is not to exceed the MPE, then the power passing through a 7 mm diameter must be less than 1 mW, even if the actual beam diameter is less than this. Separate MPE and limiting aperture values are assigned for the eye and the skin, though they broadly coincide outside the retinal hazard range.

MPE values vary principally with wavelength, exposure duration and, in the retinal hazard range, retinal image size and are quoted as irradiance and radiant exposure levels at the cornea, even in the retinal hazard range. Annex B 'Biophysical considerations' of [3] provides a more comprehensive explanation of the biophysical considerations behind the derivation of MPE values.

10.3.2.1 Assessing the Exposure Duration

Current MPE values are defined for wavelengths between 180 nm and 1 mm and for exposure durations from 10^{-13} s to 30 000 s. For single-pulsed lasers, the appropriate exposure duration is the duration of the laser pulse, but for exposure to cw and repetitively pulsed lasers, the exposure duration has to be assessed. Table 10.2 lists 'typical' values according to potential exposure scenarios. Among the situations considered in this table is unexpected exposure to bright light. Such an exposure evokes natural aversion responses, including the blink reflex and turning away of the eyes and head, for which laser safety committees have established a figure of 0.25 s. At the other extreme, 30 000 s corresponds to almost 8 h, a full working day.

10.3.2.2 Dealing with Multiple Pulse Exposures

Assessing the MPE for exposure to a repetitively pulsed laser beam involves selecting the most restrictive of at least two approaches: (i) exposure to each pulse within the train taken separately and (ii) exposure to a cw beam of the same average power. This approach is satisfactory for wavelengths less than 400 nm, where the injury mechanism is photochemical. However, for wavelengths exceeding 400 nm, a third approach to the assessment of MPE is required to take into account the cumulative heating effect of multiple pulses, the chosen MPE being the most restrictive of the three. For a train of pulses of the same magnitude and duration, this third MPE is the

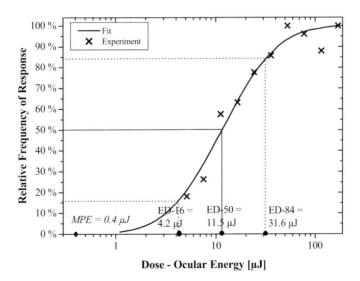

FIGURE 10.5 A typical dose–response curve as obtained in threshold experiments (here for a 850 nm laser, 180 ns pulse duration, minimal retinal spot size). For a given laser wavelength, pulse duration and spot size, the energy is varied. Due to variability of the threshold within the sampled population, a distribution of percentages of exposures which lead to a lesion results (in this example, the sample population consisted of a total of 191 exposures on four animals). Also shown on the plot is the exposure corresponding to the MPE; in this case, 3.5% of the ED-50 value to ensure that there is practically zero risk when exposed at this level [2].

TABLE 10.2

Suggested Exposure Times for Calculating MPE Values for cw and Repetitively Pulsed Lasers

Wavelength (nm)	Situation Assessed	Duration[a]	Comment
180–400	Eye and skin exposure	30 000 s	Accumulated photochemical injury over a full working day.
400–700	Accidental eye exposure	0.25 s	Aversion response to bright light
700–1400	Accidental eye exposure	10 s[b]	Involuntary eye movement
400–1400	Accidental skin exposure	10 s	Aversion response to local heating.
400–1400	Continuous intentional exposure of eyes and skin	30 000 s	Full working day
>1400 nm	Accidental exposure of skin and eyes	10 s	Aversion response to local heating.

[a] If the laser emission duration is less than the suggested duration then, of course, the former value should be used.

[b] For small sources. A larger duration is recommended where the image size corresponds to an angular subtense greater than 1.5 m rad (see discussion on apparent source size).

single-pulse MPE reduced by $N^{-0.25}$, where N is the number of pulses in the assessed duration of exposure. A more flexible approach, however, is to evaluate the MPE for a single total-on-time-pulse (TOTP).

The TOTP is a hypothetical pulse with an energy equal to the sum of the energy of all pulses and a duration given by summing their individual durations. The time frame recommended in the standard [3] for this summation is generally limited to 10 s (eye movement limit) but is longer for extended source viewing. However, if within this time frame, there are laser pulses of duration less than a specified wavelength-dependent value T_i, typically 1 ms or less, such pulses are for the sake of calculation given a duration T_i. This accounts for the time-independence of ocular MPE for exposure duration values below T_i.

10.3.2.3 Small and Extended Sources

The minimum image size that can be produced by the eye is estimated to be of order 20 μm diameter, a value strongly influenced by diffraction, aberrations and scattering. The cornea/lens optical system of a normal eye has a focal length of 17 mm, so the minimum spot corresponds to a linear angle of 1.2 mrad. In laser safety standards, a source is referred to as a 'point' or 'small' source if the angular size of a source as perceived by the eye during exposure (a value known as the 'angular subtense') is less than 1.5 mrad. This 'minimum angular subtense' value applies to intra-beam viewing of all but the poorest quality laser beams and can apply to the viewing of small near-monochromatic incoherent sources (e.g. LEDs).

Ocular MPE values in the 400–1400 nm wavelength range relate directly to exposure to 'point' or 'small' sources. Exposure conditions where the image size corresponds to an angular subtense greater than 1.5 mrad are referred to as 'extended source viewing' and apply to most conditions of viewing incoherent sources or diffuse reflections of laser beams, when viewed close enough for shape to be geometrically resolved by the eye (as illustrated in figure 10.6). Although the area of the retinal image has an α^2 dependence, where α is the angular subtense, thermal diffusion from the centre of the retinal image will be slower, and as a result, thermal retinal MPE values are quoted with a multiplication factor $\alpha/1.5$ (α in milliradians) for $\alpha \geq 1.5$ mrad. A different procedure is

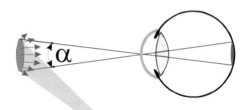

Angular subtense of apparent source

FIGURE 10.6 Definition of the angular subtense: the linear angle subtended by the retinal image. In this illustration, a laser beam is incident on a matt surface and the eye 'sees' the illuminated area as it would a conventional light source. More generally, the angular subtense of a source is determined by placing a lens at the desired point of measurement and dividing the image size in the focal plane by the focal length of the lens.

required to separately assess the photochemical retinal injury for cw exposure in the 400–600 nm wavelength range.

10.3.3 Nominal Ocular Hazard Distance

The minimum distance from the laser source at which the beam is so large that the MPE for ocular exposure is no longer exceeded is termed the nominal ocular hazard distance (NOHD). Class 1 and class 2 lasers (see Section 10.4) do not have a NOHD, but a well-collimated class 3B or class 4 laser could have a NOHD measured in kilometres. For indoor applications involving class 3B and class 4 lasers, the NOHD will often exceed the dimensions of the room or enclosure that it is in, and the calculation is of little value except if the laser radiation is highly divergent, for example, is expanding from focus or from the end of an optical fibre. NOHD calculations for diffuse reflections generally assume a Lambertian cosine law for angle of reflection and an inverse square law for irradiance, though for extended source viewing the calculation is complicated by the image size dependence of MPE as discussed in Section 10.3.2.

NOHD calculations are particularly useful in assessing the safety of outdoor applications of lasers, and in most cases, an approximate answer based on a simple conical expansion of the beam (i.e. expansion at the far-field divergence angle from the position of the beam waist) and a Gaussian beam profile is sufficient. Outdoor assessments may need to take account of the increased hazard created by the use of binoculars or telescopes, use of which may be hazardous at distances greater than the NOHD. Such considerations form part of the test procedures for laser classification and are addressed quantitatively in Section 10.4.2. The extended nominal ocular hazard distance (ENOHD) is defined on this basis.

10.4 Safety in Product Design

10.4.1 The Classification Scheme

The laser hazard classification scheme groups laser products according to increasing hazard potential and to recommended safety control measures such as interlocked access to laser areas (engineering controls) and restricted use of optical instruments (administrative controls). The IEC laser product safety standard [3], approved in Europe as EN 60825-1 and reproduced by national standards bodies in the EU member states (e.g. in the UK as BS EN 60825-1, in Germany as DIN EN 60825-1), specifies requirements for manufacturers on how to classify a laser product and the safety features and warning labels to be provided. At the present time, there is a convergence but not an equivalence between the IEC standard and the standard adopted in the USA [4].

In essence, the classification scheme divides laser products into four major classes.

> *Class 1: No risk to eyes or skin.* Lasers that are safe in normal operations under reasonably foreseeable conditions, including direct intra-beam viewing.
>
> *Class 2: Low risk to eyes. No risk to skin.* Lasers emitting visible radiation in the wavelength range 400–700 nm for which the natural aversion response to bright light (including the blink reflex) prevents retinal injury for direct intra-beam viewing. These lasers do, however, present a dazzle hazard.
>
> *Class 3: Medium risk to eyes. Low risk to skin.* Lasers for which intra-beam viewing is hazardous, but for which the viewing of diffuse reflections is normally safe. Natural aversion response to localized heating prevents serious skin injury.
>
> *Class 4: High risk to eyes and skin.* Lasers for which intra-beam viewing and skin exposure is hazardous and for which the viewing of diffuse reflections may be hazardous. These lasers are also a fire hazard.

Each class has associated with it an accessible emission limit (AEL), generally expressed in watts and joules, except class 4 for which there is no upper limit. The AEL is a maximum level of 'accessible' laser radiation for the class, according to prescribed measurement conditions, measurement detector positions and collection aperture sizes. The measurement procedures are intended to take account of many of the worst-case assumptions of exposure conditions for which the product could be used, i.e. exposure duration, closeness of viewing and the use of optical instruments.

The current scheme introduces relaxed forms of class 1, 2 and 3 in a systematic manner and, for the first time, addresses measurement conditions for classification more appropriate to high-divergence sources.

Table 10.3 summarizes the current classification scheme. A sub-division of classes has been included to reveal how, on specifics of warning signs, manufacturer-installed safety measures and (in Section 10.5) user guidance, a total of 7 classes and 11 'sub-classes' can be identified from the original four.

10.4.2 Optical Viewing Aids

Measurements for classification for class 1, 2, 3R and 3B in the wavelength range 302.5–4000 nm (the transmission window of silica) take into account the use of optical viewing aids, as illustrated in figure 10.7. For example, for outputs in the 400–1400 nm wavelength range (only), three measurement conditions apply:

i. the maximum output collected through a 50-mm aperture at 2 m from the product must not exceed the appropriate AEL. This measurement simulates the output that could enter the eye if a ×7 telescope or binocular of 50-mm acceptance aperture is used to view the laser output at this distance;

ii. the appropriate AEL must not be exceeded through a 7-mm aperture at 14 mm from the apparent source (this term is used to include virtual as well as real source positions). This measurement simulates the output that could enter the eye if a magnifier or eye loupe is used close to the laser source, but different measurement distances and aperture sizes are required to assess the photochemical and thermal hazard for extended sources;

iii. For class 1 and 2 only, the AEL must not be exceeded through a 7-mm aperture placed 100 mm away from the apparent source and simulates the output that could enter the eye in 'worst case' naked eye viewing, i.e. the shortest distance of accommodation. However, if (i) and (ii) are satisfied, then this condition is redundant.

Outside the retinal hazard range, the limiting aperture for MPE values reduces from 7 to 3.5 mm, and the collection apertures referred to in the measurement conditions are reduced accordingly. The reader should refer to the standard for full measurement details [3].

For class 1M and 2M laser products, defined as safe only for unaided viewing, (iii) only applies, with the caveat that the output collected under measurement conditions (i) and (ii) must not exceed the class 3B AEL. It follows that lasers with high divergence or large area outputs can be class 1M or 2M yet have total output powers greatly exceeding the class 1 and 2 AEL limits. Although these measurements for classification are designed to be performed by the manufacturers of lasers, they can usefully be exploited by users wishing to assess situations where optical viewing aids may be used.

TABLE 10.3

Qualitative Description of Laser Safety Classes

Class	Sub-Division	Meaning	Warning Label and Safety Features	Examples	Relationship of AEL to MPE	Typical AEL for cw Lasers	Comparison with 1993 Scheme
1	Intrinsic	Safe by virtue of the intrinsic low power of the laser, even with the use of optical instruments	No warning label. No additional safety features	Laser scanning ophthalmoscopes, low-power rangefinders, most LEDs	(i) Output does not exceed MPE (time base 30 000 s for UV sources and products intended to be intentionally viewed, otherwise 100 s).	10 mW in wavelength range 1.4–4 μm 40 μW for blue light	AELs are higher than previous class 1 by virtue of higher MPE values and some changes in limiting aperture sizes
	Engineering	Embedded laser products, safe by virtue of engineering controls e.g. total enclosure guarding, scan failure mechanism.	No warning label No additional safety features (but see requirements for protective housings in C6.4.3)	CD players, laser printers, some fully enclosed cutting, drilling, welding and marking machines	(ii) Output collected by 7 mm aperture 14 mm from apparent source does not exceed AEL. (iii) Output collected by 50 mm aperture 2 m from source does not exceed AEL.		
1M	Collimated	Well-collimated beam, output in range 302.5–4000 nm, with large diameter that is safe for unaided viewing but potentially hazardous when a telescope or binoculars are used.	LASER RADIATION DO NOT VIEW DIRECTLY WITH BINOCULARS OR TELESCOPES[a] No additional safety features	Some LIDAR and large aperture infrared optical test instruments	Relationship (i) and (ii) above are satisfied but not (iii).	As for class 1 but, additionally, for a collimated beam, output collected by 50 mm aperture 2 m from source less than class 3B AEL.	New, previously was non-visible part of class 3A
	High divergence	Output in range 302.5–4000 nm. Safe for unaided viewing but potentially hazardous when an eye loupe or magnifier is used.	LASER RADIATION DO NOT VIEW DIRECTLY WITH MAGNIFIERS[a] No additional safety features	Some low to medium power lasers with optical fibre outputs, raw beam laser diodes, line lasers, higher power infrared LEDs	Relationship (i) and (iii) above are satisfied but not (ii).	As for class 1 but additionally, for a high divergence beam output collected by 7 mm aperture 14 mm from source less than class 3B AEL	

(Continued)

Laser Safety

TABLE 10.3 (Continued)
Qualitative Description of Laser Safety Classes

Class	Sub-Division	Meaning	Warning Label and Safety Features	Examples	Relationship of AEL to MPE	Typical AEL for cw Lasers	Comparison with 1993 Scheme
2	–	Output in range 400–700nm. Safe for unintended exposure, even with the use of optical instruments, by virtue of natural aversion response to bright light.	DO NOT STARE INTO THE BEAM No additional safety features	Visible alignment laser, laser pointer	(i) Output does not exceed MPE (time base 0.25 s). (ii) Output collected by 7 mm aperture 14 mm from source does not exceed AEL. (iii) Output collected by 50 mm aperture 2 m from source does not exceed AEL.	1 mW for cw beam.	Unchanged
2M	Collimated	Well-collimated beam in range 400–700nm, with large diameter that is safe for unaided viewing by virtue of natural aversion response to bright light but potentially hazardous when a telescope or binoculars are used.	LASER RADIATION DO NOT STARE INTO THE BEAM OR VIEW DIRECTLY WITH BINOCULARS OR TELESCOPES[a] No additional safety features	Some visible LIDAR and large aperture optical test instruments	Relationship (i) and (ii) above are satisfied but not (iii).	As for class 2 but, additionally, for a collimated beam, output collected by 50 mm aperture 2 m from source less than class 3B AEL.	New, previously was visible part of class 3A, 5 mW power cap is lifted
	High divergence	High-divergence source with output in range 400–700nm. Safe for unaided viewing by virtue of natural aversion response to bright light but potentially hazardous when an eye loupe or magnifier is used.	LASER RADIATION DO NOT STARE INTO THE BEAM OR VIEW DIRECTLY WITH MAGNIFIERS[a] No additional safety features	Some visible lasers with optical fibre output, some visible LEDs and raw beam laser diodes.	Relationship (i) and (iii) above are satisfied but not (ii).	As for class 2 but, for a high-divergence beam, output collected by 7 mm aperture 14 mm from source less than class 3B AEL.	

(Continued)

TABLE 10.3 (*Continued*)
Qualitative Description of Laser Safety Classes

Class	Sub-Division	Meaning	Warning Label and Safety Features	Examples	Relationship of AEL to MPE	Typical AEL for cw Lasers	Comparison with 1993 Scheme
3R	Visible	Output in range 400–700 nm. Direct intra-beam viewing is potentially hazardous, but by virtue of natural aversion response to bright light the risk is lower than for class 3B.	AVOID DIRECT EYE EXPOSURE No additional safety features	Alignment lasers such as HeNe lasers and laser diodes, which are used widely in industry, science and medicine.	(i) Output does not exceed MPE (time base 0.25 s) by more than 5 times. (ii) Output collected by 7 mm aperture 14 mm from source does not exceed class 2 AEL by more than 5 times. (iii) Output collected by 50-mm aperture 2 m from source does not exceed class 2 AEL by more than 5 times.	5 times the limit for class 2, i.e. 5 mW.	Equivalent to the existing class 3B with less than 5 mW (often referred to as 3B star).
	Non-visible	Output in UV (302.5–400 nm) or IR (700 nm–1 mm). Direct intra-beam viewing is potentially hazardous, but the risk is lower than for class 3B.	AVOID DIRECT EYE EXPOSURE (700–1400 nm) AVOID EXPOSURE TO THE BEAM (outside the range 400–1400 nm) Emission warning device	Some low power IR range finders.	(i) Output does not exceed MPE (time base 30 000 s for UV sources and products intended to be intentionally viewed, otherwise 100 s) by more than 5 times. (ii) Output collected by 7 mm aperture 14 mm from source does not exceed class 1 AEL by more than 5 times. (iii) Output collected by 50 mm aperture 2 m from source does not exceed class 1 AEL by more than 5 times.	5 times the limit for class 1.	New, no equivalence in previous system.

(*Continued*)

Laser Safety

TABLE 10.3 (*Continued*)
Qualitative Description of Laser Safety Classes

Class	Sub-Division	Meaning	Warning Label and Safety Features	Examples	Relationship of AEL to MPE	Typical AEL for cw Lasers	Comparison with 1993 Scheme
3B	-	Medium power laser. Direct ocular exposure is hazardous, even taking into account aversion responses, but diffuse reflections are usually safe.	AVOID EXPOSURE TO THE BEAM Key switch, emission warning device, external interlock connection and beam stop	Large area illuminators (e.g. industrial holography) and laser displays.	Ocular MPE (10s) not exceeded at 130 mm from normal incidence viewing of beam-off diffuse (Lambertian) reflector.	500 mW for cw laser with a wavelength >315 nm.	Meaning unchanged, but 1993 class 3B included some lasers now moved to lower classes.
4	-	High-power laser. Direct exposure is hazardous to eye and diffuse reflection may also be hazardous. Skin and potential fire hazard.	AVOID EYE OR SKIN EXPOSURE TO DIRECT OR SCATTERED RADIATION Safety features as for 3B	Lasers for materials processing. High-power diode arrays. Professional laser displays.	No AEL.	No limit.	Unchanged.

[a] The general phrase 'optical instruments' can be used in place of 'binoculars or telescopes' or 'magnifiers'. However, generally only one or other group of optical instruments leads to an increase in the hazard for a given laser product. Therefore, at the discretion of the manufacturer, a specific wording can be added to the warning label. This distinction and information should help to place the hazard into perspective: for instance, the use of class 1M range finders (highly collimated large diameter beam) might have to be restricted in outdoor applications where binoculars are in frequent use (military base, bird sanctuary), while class 1M visual LED displays are safe to be viewed with binoculars.

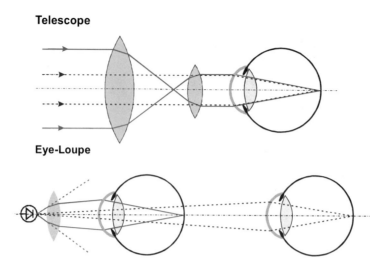

FIGURE 10.7 Use of viewing aids increasing the ocular hazard. Top: A telescope or binocular increases the hazard of a well-collimated beam if it has a diameter exceeding the pupil size. Bottom: An eye loupe or magnifier increases the hazard of highly divergent sources by allowing the source to be imaged at a closer distance (full lines) than would be the case for the unaided eye, where the closest distance of accommodation is about 100–200 mm (broken lines).

10.4.3 Engineering Safety Features on Laser Products

The product safety standard provides details of general requirements for warning labels and engineering safety features. These are summarized in Table 10.3. There are additional requirements for medical laser equipment [5] and optical fibre communication equipment [6] and for machines that use lasers to process materials [7,8].

10.4.3.1 Protective Housing

All classes of laser product are required to have a protective housing to limit human access to levels of laser radiation below the AEL assigned to the product. The design of this primary engineered safety feature on a laser product becomes particularly important for class 1 (embedded) laser products, particularly those enclosing class 4 laser sources.

Laser safety standards include requirements for labelling on access and service panels that form part of the protective housing and provide access to hazardous levels of laser radiation and, in addition, interlocking for such panels that are intended to be removed during normal use or maintenance.

Thin-walled enclosures will 'keep fingers out' and block any weakly scattered laser radiation, but so far as they are incapable of withstanding a high-power laser beam, they rely heavily on maintaining the direction of the beam path. Regular alignment checks and the maintenance of optics and the environment (e.g. vibration, humidity, temperature, dust) can therefore be important safety considerations in the context of high-power laser beams [9]. Indeed, the subject of laser guards, a 'machine' term but essentially simply a component of a protective housing, is the subject of a separate standard [8].

10.5 Safety in Practice

10.5.1 Class-based User Guidance

The laser hazard classification scheme is designed primarily as the basis for specifying manufacturer requirements, but as the basis for user guidance, it is widely recognized as insufficient, in particular, since severity of injury plays no part in the classification scheme. For example, most user guidance relates to the safe use of class 3B and 4 lasers, but class 3B lasers which threaten permanent visual impairment are in the same class as class 3B lasers for which the maximum injury is mild corneal irritation lasting a few days, and class 4 lasers capable of burning through firebricks in seconds are placed together with lasers capable of causing no more serious skin burn than could be inflicted by momentary contact with a soldering iron.

Section 10.5.2 addresses considerations by which appropriate control measures can be selected and imposed to reflect the severity and probability of injury based on wavelength, power and use of the laser. Nevertheless, there is a relatively small number of safety control measures to choose from to deal with the laser hazard, and these can generally be divided according to laser class, as summarized in Table 10.4. The particular layout of the table has been chosen to emphasize the progressive adding of appropriate control measures as the class of laser increases. Also, the table separates engineering controls (generally the most effective and therefore to be preferred, but also the most expensive and restrictive) from administrative controls (less effective, but lower cost and more flexible), and personal protective equipment. As described in Section 10.5.3, PPE is generally recognized as the measure of last resort and certainly not an equivalent alternative to the others.

TABLE 10.4

Summary of Recommended Control Measures by Laser Class

Class	SSub-division	Appropriate Controls During Normal Operation[a]		
		Engineering Controls	**Administrative Controls [NOHD, ENOHD]**	**Eye Protection**
1	Intrinsic	None.	None. [0,0]	None.
	Engineering	None.	Operation procedures. [0[a],0[a]]	None.[a]
1M	Collimated	Employ basic beam path design principles[b].	Safety training of laser user. Control use within NHZ of binoculars and telescope and specularly reflecting surfaces that may focus the beam. [0, >2 m]	None required. Filters should be fitted to binoculars or telescope within NOHA.
	High div.	None.	Control use of magnifiers close to source and specularly reflecting surfaces that may focus the beam.	None required. Filters should be fitted to magnifiers if used.
2	-	Employ basic beam path design principles[b].	Restrict use around activities where dazzle could cause an accident.	None required.
			Control use of specularly reflecting surfaces in vicinity of unenclosed beam path.	
			[0.0] for accidental exposure.	
2M	Collimated	Same controls as appropriate for class 1M (collimated) products plus dealing with dazzle hazard		
	High div.	Same controls as appropriate for class 1M (high-divergence) products		
3R	Visible	Same controls as appropriate for class 2 products, but some training of laser user is appropriate (to enforce the importance of avoiding intentional exposure.)		
		NOHD>0 but risk of injury during short-time exposure is very small.		
	Invisible	Except for dazzle hazard, same controls as appropriate for class 3R (visible) products plus:		
		Employ restricted beam path design principles[c].	Appoint a Laser Safety Officer Provide basic safety training to users and other personnel in the NHZ appropriate to the small but finite risk of injury.	Recommended.
3B	-	Same controls as appropriate for class 3R (visible) products plus:		
		Employ remote interlock connector.	Post warning signs at approaches to the hazard area.	
			Provide training to users and other personnel in the NHZ appropriate to the severity of potential injury.	
			Control access to key switch for laser use. Interpose the beam stop or attenuator when the beam is not required.	
4	-	Same controls as appropriate for class 3B products, plus those below:		
		Design exposed surfaces for power handling capability.	Potentially large NOHD	Skin protection recommended.
		Shroud diffuse reflections.		

[a] For embedded laser products, controls appropriate to the embedded laser may be required during maintenance and service operations and the NOHD is potentially large.

[b] Basic beam path design: locate horizontal beams above or below eye level, avoid upwardly directed beam paths and terminate beam path at end of its useful range.

[c] Restricted beam path design: beam enclosed and secondary beams blocked (for class 3B and 4) as far as reasonably practicable; robust, secure mounting of laser and beamline optics; open beam paths as short as reasonably practicable with a minimum number of directional changes, clear of surfaces producing hazardous reflections, not crossing walkways.

10.5.2 Application of Control Measures

The international safety standard presents control measures in the form of guidance (this includes parts of the IEC 60825 series that provide application-specific laser safety guidance [9–11]), leaving it to the user to decide if, for example, the 'should' in 'the remote interlock connector of class 3B and 4 lasers *should* be connected to ... a room door' should be interpreted as a 'shall'. Where, for example within a particular organization or profession, the conditions of laser use are well defined and a common approach to laser safety is desirable, then working codes based on the hazard classification system can be devised to take away some or all of the decision from the user. The most comprehensive example of working codes is the USAs ANSI Z136 series [12] with additional parts dealing with optical fibre communication systems (part 2), healthcare facilities (part 3), measurement (part 4), education (part 5) and outdoor lasers (part 6). This prescriptive approach can only work, however, where the scope of the code is restricted to specific laser applications and/or types of users and environments.

A more general and more flexible approach to laser safety is provided by risk assessment and is the basis by which safety in general is dealt within the various EU directives for both product and workplace safety. The approach involves first identifying potentially hazardous events and then, for each such event, assessing the likelihood of occurrence, the severity of harm and, thereby, the risk. For each unacceptable risk identified, and with reference to applicable standards and best practice guidelines, appropriate measures are identified and implemented, beginning with reasonably practicable engineering controls, followed by safe working practices and other administrative controls, until the risk is as low as reasonably practicable. Any residual risk is dealt with by the use of personal protective equipment.

The imposition of reasonable practicability and reference to applicable guidelines lead to different solutions depending on application and environment. For example, engineering controls are used in industrial laser processing (full enclosure of the laser radiation), whereas laser displays invariably rely heavily on administrative controls (crowd control) and many medical procedures require close proximity to open beams and therefore rely heavily on the use of laser safety eyewear. The flexibility often required for laser use in research may make full-beam enclosure impractical (though partial enclosure is often feasible and is strongly recommended), so a mix of controls including interlocked access to laser laboratories and training of laser operators in safe working procedures is generally required, including the use of laser safety eyewear.

10.5.3 Personal Protective Equipment

10.5.3.1 Eye Protection

The primary requirement to laser safety eyewear is a filter attenuation at the laser wavelength(s) sufficient to ensure that the transmitted radiation will not exceed the MPE at the maximum foreseeable laser exposure, usually but not necessarily chosen to be the exposure level at the laser aperture. The product standards for safety eyewear [13] specify a 30 000 s time basis for MPE values at UV wavelengths, to address cumulative photochemical damage from scattered light over a working day, otherwise a value of 10 s, based on head movement and other behavioural considerations. For providing protection at visible laser wavelengths whilst allowing sufficient visible transmission for beam alignment purposes, a related standard [14] specifies a time basis of 0.25 s.

Where laser safety eyewear is required, every reasonable means should be taken to ensure that it is properly used. This includes the choice of an appropriate model, i.e. high visible light transmission, a comfortable and appropriate design, convenient protective storage, clear labelling and regular maintenance. In Europe, this advice is embodied in the Personal Protective Equipment at Work Directive. This Directive also requires that all PPEs be type-tested and CE-marked. The eyewear product standards [13,14] include a test requirement for 'stability to laser radiation' (i.e. resistance to laser damage for 10 s exposure) which non-CE-marked eyewear would generally not meet.

Beam alignment is the most hazardous laser operation, but it also provides the greatest temptation for the user to remove protective eyewear, especially if the beam to be aligned is of a visible wavelength. Consequently, at the same time as selecting safety eyewear, it is important to consider the implications of its use. Beam visualization techniques involving fluorescent or scintillation cards and cameras and co-linear class 2 laser beams should always be investigated. Equally, depending on the visible transmission of the eyewear, it may be important to provide strong room lighting and to pay particular attention to trip hazards.

10.5.3.2 Skin Protection

Except for exposure to focused beams, skin injuries inflicted by lasers at the high end of class 3B and the low end of class 4 are generally regarded as minor (equivalent to a soldering iron burn), and skin protection is not worn. This practice can generally be justified on the grounds that the wearing of protective gloves, face shields and/or suits induces discomfort and can result in a greater accident as a result of the reduced ability of the wearer to carry out delicate operations. However, the wearing of gloves and other protective clothing should certainly be considered for open beam use of the higher power industrial lasers, and for work with open beam UV lasers in general, though, to the author's knowledge, proprietary laser-specified skin-wear is not generally available.

10.5.4 Accident Reports

Reports of laser injuries are notoriously unreliable and incomplete, and the available statistics are poor, with relatively few accidents reported and recorded per year [1]. The value of these reports lies primarily in identifying potentially hazardous activities and here they clearly imply that:

- research workers and service engineers are in the highest risk category for laser injuries;

TABLE 10.5

Summary of the Main Associated (i.e. Non-beam) Hazards and Their Control

Origin	Hazard	Typical Hazardous Situation	Typical Control Measures
Laser	High voltage	Laser head and power supply exposed during servicing.	Proper screening of exposed HV. Restricted access to qualified persons. Use of earthing stick.
	Explosion	During changes of high-pressure flashlamps in laser.	Gloves and face shield. Training.
	Collateral radiation	RF and UV accessible during servicing.	Proper screening combined with access restricted to service engineers.
	Mechanical	Unloading and positioning of laser power supplies.	Attachment points provided for use of lifting equipment by qualified persons. Training and use of gloves.
	Chemical	Dyes for dye lasers, halogen gas in excimer lasers, zinc selenide optics.	Procedures for storing, handling and disposing of hazardous materials. Training and use of gloves.
Process	Fume	Material removed during laser materials processing.	Fume extraction and filtration. Face mask and gloves worn during cleaning operations.
	Fire	Improperly terminated class 4 laser radiation.	Control of flammable materials and beam path. Provision of fire extinguisher.
	Secondary radiation	X-ray, UV and blue light during laser-workpiece/target interactions.	Enclosure of target area and monitoring of hazard. PPE for exposure to UV and blue light.
	Mechanical	Manipulators for laser and/or workpiece/target.	Guarding of traps. Restricted access to moving parts.

- the most hazardous laser activity is laser beam alignment;
- pulsed class 3B and 4 lasers inflict the worst injuries;
- the near-infrared (700–1400 nm) is the most hazardous part of the laser spectrum;
- the most common cause of accidents is an over-reliance on laser safety eyewear combined with poor enforcement in its use;
- the lowest power eye injuries inflicted by visible cw lasers have occurred at 3 mW and involve use by untrained people and deliberate viewing of several seconds.

10.6 Associated Hazards

In addition to the laser radiation hazard, there are other general hazards associated with laser use. Hazards can be divided into two categories: those generally associated with the laser source (including high voltage, high-pressure gases, collateral radiation) and those generally associated with the laser application (including fume, fire and mechanical hazards) though the list is not intended to be exhaustive or exclusive. Table 10.5 summarizes the main laser-related hazards and control measures but details go beyond the scope of this chapter.

Manufacturers of machines for laser materials processing should be aware that the international standard for these products [7] requires measures to deal with associated hazards, the fume hazard in particular.

10.7 Summary

Whilst the potential for injury exists, lasers have an excellent track record for safety. The vulnerability of the eye to laser injury has always been well appreciated, but as the nature of the injury has become more precisely defined, so the setting of safe exposure limits has become more complex, especially for retinal injuries. The classification scheme provides the basis for manufacturers' requirements and goes some way to assisting users on precautions, but the appropriateness of control measures in a given situation can really only be decided on the basis of a well-informed risk assessment.

REFERENCES

1. Rockwell Laser Industries web site http://www.rli.com provides up-to-date accident statistics.
2. Sliney D H, Mellerio J, Gabel V-P and Schulmeister K 2002 What is the meaning of thresholds in laser injury experiments? Implications for human exposure limits *Health Phys.* **82** 335–47.
3. IEC 60825-1: 2001 Safety of laser products. Part 1. Equipment classification, requirements and user's guide.
4. Federal Laser Products Performance Standard 21 CFR 1040.10 and 1040.11.
5. IEC 601-2-22: 1991 Medical laser equipment. Part 2: Particular requirements for the safety of diagnostic and therapeutic laser equipment.
6. IEC 60825-2: 2000 Safety of laser products. Part 2: Safety of optical fibre communication systems.
7. ISO 11553: 1996 Safety of machinery—Laser processing machines—Safety requirements (equivalent to EN 12626).

8. IEC 60825-4: 1997 Safety of laser products. Part 4: Laser guards.
9. Green J M 1989 High power laser safety *Opt. Laser Technol.* **21** 244–8.
10. IEC 60825-3: 1997 Safety of laser products. Part 5: Guidance for laser displays and shows.
11. IEC 60825-8: 1997 Safety of laser products. Part 8: Guidelines for the use of medical laser equipment.
12. ANSI Z136.1–2000 American National Standard for Safe Use of Lasers.
13. EN 207: 1998 Personal eye-protection—Filters and eye-protectors against laser radiation (laser eye-protectors).
14. EN 208: 1998 Personal eye-protection—Eye-protectors for adjustment work on lasers and laser systems (laser adjustment eye-protectors).

FURTHER READING

Henderson A R and Schulmeister K 2003 *Laser Safety* (Bristol: Institute of Physics).

Sliney D H 1995 (ed) Selected papers on laser safety *SPIE Milestone Series* MS-117 1995.

Sliney D H and Wolbarsht M L 1980 *Safety with Lasers and other Optical Sources* (New York: Plenum).

11

Optical Components: Section Introduction

Julian Jones

Chapters 12–15 are concerned with the optical and optomechanical components used to form and transport laser beams, building on the basic concepts of laser optics expressed in Chapters 2 and 5.

The topic of Leo Beckmann's Chapter 12 is the passive components of lenses and mirrors that are used to condition laser beams. His starting point is to contrast the properties of coherent laser beams with the familiar formalism of incoherent optics and the formation of images, developing the theory of Gaussian beams. Armed with the theoretical foundations, he goes on to consider the characterization of laser beams, introducing the M^2 parameter and more sophisticated means for describing the wavefront. The simplest components are those with plane surfaces, causing reflections, and whose properties can be modified by thin films, either singly or in multiples. The most familiar optical component is surely the lens; nevertheless, the characteristics required for laser beam transmission are different from those in imaging systems, for reasons of high spatial coherence and power density. Methods for lens design and multiple lens systems are described and extended to mirror systems. The section closes by considering thermal effects, optical specifications and manufacturing techniques.

In Chapter 13, Alan Greenaway considers the optical components used for dynamic control of lasers, in scanning and positioning the beam, changing its size and shape, and modulating it in time and space. Relevant components for scanning are mechanical, acousto-, electro- and magneto-optic and the use of diffractive optical effects. For controlling beam shape, spatial light modulators and various adaptive optical components are appropriate. Adaptive optics and phase conjugate reflectors are the subject of Michael Damzen and Carl Paterson's Chapter 14. Here are described the techniques first developed for astronomical imaging through the turbulent atmosphere for the correction of wavefront distortion. More recently, such techniques have been considered for intra-and extra-cavity use in lasers. They describe the relevant sensors (such as the Shack–Hartmann wavefront sensor), actuators (adaptive mirrors, faceplates, electrostatic membranes and biomorphs) and control systems. A special case of adaptive optics, in which there is no closed-loop control, is the phase-conjugate mirror, also capable of restoring distorted wavefronts to their original shape, and the uses of, for example, four-wave mixing, stimulated Brillouin scattering, photorefractivity and self-intersecting loops are all discussed.

Optical components are of little use unless they can be held in the right place (statically or dynamically) and kept there; such is the function of the optomechanical components that form the subject of Frank Leucke's Chapter 15, which considers the formalism for optomechanical design, practical approaches and testing, with special attention to precision positioning.

12

Optical Components

Leo H. J. F. Beckmann

CONTENTS

12.1 Introduction 207
12.2 Optical Design Aspects of Laser Optics 208
    12.2.1 What and Where Are the Object and Image? 208
    12.2.2 Size of the Image Waist 208
    12.2.3 Real Laser Beams 209
    12.2.4 Multiple Optical Elements and the Use of Ray Tracing 209
    12.2.5 Evaluation of Aberrations and Diffraction Patterns 209
12.3 Surface Phenomena and Thin Layer Coatings 209
    12.3.1 Reflection at the Surface of a Dielectric 209
    12.3.2 Vertical Incidence and Effect of a Single Thin Layer 210
    12.3.3 Multi-layer Coatings and Their Applications 210
12.4 Elementary Lens Forms 211
    12.4.1 The Singlet 211
    12.4.2 The Dialyte and the Achromat 211
    12.4.3 Lens Systems with More than Two Elements 213
12.5 Use of (Curved) Mirrors 214
12.6 Non-focusing Optical Laser-beam Handling and Relaying 214
12.7 Thermal Effects in Optical Materials 215
12.8 Specifying Optics for Laser Applications 215
    12.8.1 Tolerances 215
    12.8.2 Surface Imperfections: Shape Deviations 216
    12.8.3 Surface Imperfections: Surface Quality 216
    12.8.4 Communication of Specifications—ISO 10110 216
12.9 Manufacture of Optical Components 216
12.10 Summary and Conclusions 217
References 217

12.1 Introduction

This chapter will deal with the design, manufacture and handling of the components of optical systems used for manipulating laser radiation. Attention will focus on those aspects which are specific to laser-based optics. These occur mainly in two quite unrelated fields: (i) the propagation of laser beams due to spatial coherence; and (ii) the thermal aspects of laser optics caused by the (often) high power of laser beams. Commonalities and analogies between classical and laser-based optics, which are plentiful, will be addressed if needed for completeness or clarity.

As a general rule, optical components for laser applications have to meet high-quality standards and must be handled with care and the utmost cleanliness. At high-power densities, any organic contaminants on optical surfaces will readily carbonize, become unremovable and cause absorption and local heating resulting in thermal gradients and stress. Mechanical damage, such as scratches, may cause unwanted local power concentrations, thermal stress and rupture, aside from stray radiation which may reach sensitive parts of the equipment. Sub-standard or damaged anti-reflection coatings will make a part of the incident radiation pursue unintended optical paths with unpredictable results. Some but not all of these risks can be minimized by careful design of the overall system, including suitable cooling provisions to ensure thermal stability.

12.2 Optical Design Aspects of Laser Optics

12.2.1 What and Where Are the Object and Image?

In a classical radiation source (an incandescent lamp, etc.), each spot on the source emits radiation independently of all others and is the origin of a spherical wave, which propagates into a hemisphere. The centre of this diverging hemispherical wave locates the source as an object and its radius of curvature is proportional to the distance from the object. By means of suitable optics, the divergent wavefront can be made convergent in the direction of propagation to form a region of smallest extent called an image. As the object and image are located at the centres of the respective wavefronts, we may, and customarily do, consider just the object and image distances and their relations to the 'wavefront-bending-capability' of the optical system, also known as its optical power. This well-known relation can be written as follows [1]:

$$1/s + 1/s\phi = 1/f \quad (12.1)$$

where s is the object distance, s' is the image distance and $1/f$ is the optical power of the optical system (we refer to its reciprocal, the quantity f, as the 'focal length').

The radiation from a laser is spatially coherent and propagates accordingly (see Chapter 8). There is a point of minimum beam radius W_0, called the beam waist, but it does not have the characteristics of an object in the sense of classical optics: it does not form the centre point of a spherical wave. Instead, the wavefront of laser radiation is flat at the waist (like that from a classical object at infinity) and its radius of curvature varies non-linearly with the axial distance z from that waist according to

$$R(z) = z\left[1 + (z_R/z)^2\right] \quad (12.2)$$

while the beam radius expands according to

$$W(z) = w_0 \sqrt{\left[1 + (z/z_R)^2\right]}. \quad (12.3)$$

As both relations are non-linear in z, the simple relation (12.1) is clearly not applicable to laser beam waists. The difference is most dramatic if z is smaller than z_R and vanishes only for z much larger than z_R. z_R, the so-called Rayleigh range, with

$$z_R = \pi w_0^2/\lambda \quad (12.4)$$

marks the distinction between the near-field ($z<z_R$) and the far-field ($z>z_R$) and is the axial position, where $R(z)$ attains a minimum of $2z = 2z_R$.

Yet, intuitively, one would like to consider laser beam waists as 'objects' and their optical conjugates as 'images'. This can be formalized by writing equation (12.1) in expanded form:

$$1/(s+q) + 1/(s\phi + q\phi) = 1/f \quad (12.5)$$

where the beam parameters q and q' at the object and image waists, respectively, are purely imaginary [2]:

$$q = j\pi\omega_0^2/\lambda \quad q' = j\pi\omega_0'^2/\lambda. \quad (12.6)$$

As shown in Section 12.4.3, from this, we can derive an expression for the axial position of the image waist (s') as a function of the object waist position (s) and optical power ($1/f$) [1]:

$$1/\left[s + (z_R)^2/(s-f)\right] + 1/s\phi = 1/f. \quad (12.7)$$

If we rewrite these equations in normalized form [1]:

$$(s'/f) - 1 = \frac{(s/f) - 1}{\left[(s/f) - 1\right]^2 + (z_R/f)^2} \quad (12.8)$$

and compare this with the equally rewritten classical formula:

$$(s'/f) - 1 = \frac{1}{(s/f) - 1} \quad (12.9)$$

we see that the appearance of the term $(z_R/f)^2$ removes the pole at $s=f$ in the classical formula, the case for which the object is at the (forward) focal point and for which the image moves to infinity. Thus, the position of the image waist of a laser is always (!) finite, its maximum distance is

$$s\phi(\max) = f\left[1 + (f/z_R)/2\right] \quad (12.10)$$

for a source beam-waist distance of $s = f+z_R$. There is also no minimum distance between the source and image beam waists, unlike in classical optics. Finally, if the source distance becomes zero—which means a flat wavefront entering the optics and is the equivalent of a classical object at infinity—the image-waist distance is

$$s\phi(s \to 0) = f/\left[1 + (f/z_R)^2\right] \quad (12.11)$$

that is, the image waist is always closer to the imaging optics than the classical image. It is useful to note, however, that in most cases, the focal length is much smaller than the Rayleigh range, making the resulting deviation negligibly small for most practical considerations.

All these formulas for laser beams reduce to the simpler relations for classical optics when z_R approaches zero, which emphasizes that classical optics is merely a special case of the more general (=coherent) laser optics, where the Rayleigh range shrinks to zero.

12.2.2 Size of the Image Waist

From equations (12.3) and (12.4), it is seen that the divergence of a laser beam is related to its waist radius ω_o. Conversely, the size of an image waist ω_0' is related to the convergence of the beam in image space

$$\omega_0' = \lambda/(\pi \sin U'). \quad (12.12)$$

Optical Components

U' is the half-angle of the radiation cone that emerges from the optical system and is given by the beam radius $w(z)$ at the source-waist distance $z = s$ from the optical system (to be taken from (12.3)) and the image-waist distance s' (to be taken from (12.7)). ω'_0 is a minimum value. It assumes that virtually all radiant power from the source laser beam contributes to the (diffraction) image. This requires that the clear optical diameter of the imaging optics be substantially larger than $\omega(z)$, typically by a factor of 1.5, to avoid 'aperturing' (beam truncation). An even larger factor would theoretically be advantageous but is rarely effective in practice since some aperturing has always already taken place within the laser itself.

12.2.3 Real Laser Beams

While the previous formulas assume an ideal (or 'diffraction-limited') laser beam and ideal images, they can be extended to real beams, by multiplying the wavelength by a 'times diffraction-limited factor', M^2, introduced by Siegman [3] (see also Section C4). The quantity $M^2\lambda$ may be seen as an 'effective' wavelength. Real beams have a stronger divergence and a shorter Rayleigh range than an ideal laser beam of equal waist radius.

12.2.4 Multiple Optical Elements and the Use of Ray Tracing

The previous calculation of image-to-object relations applies to a single (thin) optical element and the distances from it. In practice, however, optical systems typically consist of a succession of optical components as needed to perform a given task. These relations can be expanded to cover this situation, but for all practical purposes, it is preferable to take an approach that is closer to the ray-tracing procedures which have been well developed for classical optical design. When dealing with laser beams, it is obviously necessary to take into account the specific rules of laser-beam propagation, equations (12.2) and (12.3). The laws of refraction (and reflection) apply unchanged. Details are discussed in Chapter C4.2.3.

12.2.5 Evaluation of Aberrations and Diffraction Patterns

Standard ray-tracing algorithms include the computation of the optical path difference, the difference between the optical path lengths along the ray and a reference (chief-)ray. By taking a large number of ray samples, typically arranged as a grid raster over the entrance pupil, one obtains a model of the actual wavefront in image space. The aberrations caused by the passage through the individual surfaces will be mapped into that emerging wavefront and will show up as deformations with respect to its ideal (i.e. spherical) shape. There are several ways in which such a computed wavefront can be evaluated. One is to express its shape by means of Zernike polynomials (see Chapter C1.3), another is to compute the pattern of the power density, which will be created by diffraction in some reference plane. The latter involves addition of the (amplitude) contributions from samples, properly distributed and weighted for the type of beam, and the amplitude distribution associated with its wavefront. Diffraction calculations give the most relevant information for practical purposes, in particular, for systems which approach the theoretical diffraction limit. Care has to be taken to ensure that the samples are taken over the full entrance aperture of the optical system and, thus, represent the true extent of the wavefront, which is admitted by the system's clear optical diameter. In practice, as already mentioned, this must well exceed the beam diameter to avoid undue beam aperturing. In the diffraction image, any beam aperturing will show up as a broadening of the central peak and a shift of power from that central peak to outer rings.

The calculation of the true diffraction pattern for a real beam requires full knowledge of the amplitude distribution within the source beam, which is usually known only for low-order modes. The characterization of a real beam by its M^2 value does not include that information and its diffraction calculation will yield the power distribution of a Gaussian beam, broadened in accordance with M^2 and including the effects of aberrations as applicable.

A treatment of laser beams in full accordance with this is not necessarily implemented in commercial optical design and analysis software, and the respective instruction manuals must be consulted for any details about the actually used methods of computation of Gaussian beams and eventual simplifying assumptions or shortcuts.

12.3 Surface Phenomena and Thin Layer Coatings

12.3.1 Reflection at the Surface of a Dielectric

When radiation travelling in air hits the surface of a dielectric material, it undergoes several important changes:

1. A part of it enters the dielectric, where it may be transmitted or absorbed.
2. Its speed is slowed down by a factor equal to the refractive index n of the material, and as a result, it is refracted.
3. The remainder of the incident radiation is reflected.
4. The distribution between entering and reflected radiation depends on the angle of incidence and on the polarization (see Chapter A5) and is governed by Fresnel's equations, which can be written as follows:

$$\frac{R_s}{E_s} = -\frac{\sin(I-I')}{\sin(I+I')} \quad \frac{R_p}{E_p} = \frac{\tan(I-I')}{\tan(I+I')}$$
$$\frac{E'_s}{E_s} = -\frac{2\sin I' \cos I}{\sin(I+I')} \quad \frac{E'_p}{E_p} = \frac{2\sin I' \cos I}{\sin(I+I')\cos(I-I')} \quad (12.13)$$

where E, R and E' are the amplitudes of the electric vectors of the incident, reflected and refracted radiation, respectively. The subscripts p and s denote the two planes of polarization; I and I' are the angles of incidence and of refraction. As is immediately seen from these equations, light of different polarization behaves quite differently, but the differences vanish at zero angle of incidence ($I = 0$, which also means $I' = 0$), and when I approaches 90° (grazing incidence, $\cos(I) = 0$), where all light is reflected. For p-polarized light, no light is reflected at the angle, where $\tan(I+I')$ passes through infinity, and for which $\tan(I) = n$, and since $\tan(I+I')$ changes sign at that angle, the reflected radiation also undergoes a phase shift of 180° beyond that angle.

The angle, for which $\tan(I) = n$, and where the refracted and the reflected rays form a right angle, is known as Brewster's angle. Since, as mentioned already, no p-polarized light is reflected under that angle, a (plano-parallel) plate of dielectric material placed under that angle will fully transmit p-polarized light and is then called a 'Brewster window'. This has important applications in laser technology: if included in a laser cavity, (i) it offers a loss-free means to contain a laser gas and (ii) it causes a higher cavity gain for p-polarized radiation and, thereby, incites the laser to oscillate exclusively in that plane of polarization, usually the preferred mode of operation.

For low-index glass, with a refractive index near 1.5, Brewster's angle is 56.3°. Since the tangent function is relatively steep in that range, Brewster's angle runs up to just 61 degrees for a refractive index of 1.8. To avoid losses (from stray radiation, etc.) at these angles, the surface quality requirements for Brewster windows are much higher than those for near-vertical incidence.

12.3.2 Vertical Incidence and Effect of a Single Thin Layer

In many situations, including practically all applications of lenses, radiation impinges (almost) vertically on the surface of an optical component. To obtain equations for vertical incidence ($I = 0$), we evaluate the sines of $(I - I')$ and $(I+I')$, set the cosine terms to 1, use Snell's law to replace $\sin(I)$ by $(N/n) \sin(I')$, multiply everything by n and divide by $\sin I'$ giving

$$R_s/E_s = R_p/E_p = (N-n)/(N+n) \quad (12.14)$$

where N is the refractive index of the component and n that of air, and as the reflected power R is given by the square of the electric vector, we find that

$$R = \left[(N-n)/(N+n)\right]^2 \quad (12.15)$$

Thus, at each interface between air ($n = 1$) and a component of refractive index N, a sizable percentage of the incident radiation is lost by reflection. The loss is already 4% per surface for low-index glasses ($N = 1.5$) and reaches 17% per surface for a material like ZnSe ($N = 2.4$).

In general optical imaging, (dielectric) reflection losses are not only a nuisance because they reduce image brightness but—mostly more important—because the reflected radiation will, to some extent, find its way to the image plane and act like stray light, which reduces image contrast. In high-power laser systems, radiation lost through reflection is particularly troublesome in that the reflected power will typically hit the interior walls of the lens housing, which is thereby heated up with unpredictable results (misalignment, stress, thermal damage, etc.).

Fresnel's equations in the simplified form for vertical incidence (12.15) also show the way in which to apply interference in order to reduce reflection. If the component (with refractive index N_s) is covered by a thin layer of a material with refractive index N_t, there will be two reflections:

1. at the interface between the optical component and the thin layer with

$$R_1 = \left[(N_s - N_t)/(N_s + N_t)\right]^2$$

2. at the interface between the thin layer and air with

$$R_2 = \left[(N_t - 1)/(N_t + 1)\right]^2.$$

The condition for the two reflections to become equal is easily found to be:

$$N_t = \sqrt{N_s} \quad (12.16)$$

If the (optical) thickness of the thin layer is $\lambda/4$, the radiation reflected at the interface between the thin layer and the component has travelled twice that distance, meets that reflected at the air–layer interface with a phase difference of $\lambda/2$ wavelength and is, thus, cancelled by interference.

Thus, a single thin layer can, in principle, completely suppress reflection losses but the necessary condition (12.16) is often difficult or even impossible to meet, as there is a lack of low-index materials: the lowest available are MgF_2 ($N = 1.38$) and kryolite ($N = 1.36$). Single layers are adequate for anti-reflection coating of materials with refractive indices above, say, $N = 1.8$, such as yttrium aluminium garnet (YAG), the most common host material for Nd lasers ($N = 1.82$) or ZnSe ($N = 2.4$), the most common material for optics for CO_2 laser radiation.

12.3.3 Multi-layer Coatings and Their Applications

Very substantial design freedom for reducing reflection losses is gained by expanding the thin film coatings on an optical component to two or more layers. In a typical situation, the component (N_s) is coated with a first layer of a high-index material (N_1) and a second layer of low-index material (N_2), chosen such that

$$N_s(N_2)^2 = (N_1)^2 \quad (12.17)$$

where both layers have an optical thickness of $\lambda/4$. This condition requires a high-index material of about 1.70 and a low-index material of 1.38 for a component of $N = 1.5$, both available.

In the cases discussed so far, thin-layer coatings were used to reduce reflection losses at only one wavelength—that of the laser radiation. The principle of manipulating reflection at the interface with a dielectric by means of thin-film coatings is, however, much more flexible and extremely powerful.

It allows for the provision of low or high reflection, in a small or wide band of wavelengths and, if designed for finite angles of incidence, to create surfaces with highly selective polarization behaviour. In essence, thin-film coatings exploit the wave nature of light, and as Fresnel's equations already suggest, there is an almost infinite variability of complexity, given the dependencies of the splitting of power at a dielectric interface on refractive index, angle and polarization. The design of such coatings, which go much beyond the examples given here, requires advanced computer methods. Suitable software is offered by several vendors. The subject, which developed somewhat late (by Smakula in 1936 [4]), is treated in a large body of literature and a number of textbooks. The relevant chapter in the *Handbook of Optics* [5] includes 532 literature citations.

The technology of thin-film coatings is difficult to master and has become the foundation of highly specialized enterprises whose viability critically depends on a substantial body of experience. Technical challenges arise in practically every single issue, as the adherence of thin films critically depends on the method of deposition, which must be performed in high vacuum, and the cleanliness of the substrate. Layer thicknesses have to be adhered to—and thus monitored—with high accuracy while the materials for the successive layers have to be chosen to not only meet the required refractive indices but also to avoid undue thermal stress caused by differences in the expansion coefficients.

In general, one has to expect that coatings of any type will be much more sensitive to damage by high laser energy densities (of short laser pulses) and high-power densities (of cw laser radiation) than the respective bulk materials. It is, therefore, mandatory to avoid focusing (high peak power) laser radiation onto optical surfaces, and this includes secondary foci ('ghost images') created by unwanted reflections from refractive optical surfaces.

12.4 Elementary Lens Forms

12.4.1 The Singlet

The singlet is a stand-alone lens made from a material which is transparent for the wavelengths of its intended use and of (mostly) homogeneous refractive index. It may have spherical or non-spherical surfaces, the former preferred for several reasons, including the ease of use and alignment and the cost of manufacturing to high precision. If r_1 and r_2 are the radii of curvature of the two lens surfaces, t is the centre thickness and N the refractive index, the lens power P, which is, by definition, the reciprocal of the focal length f, is

$$P = 1/f = (N-1)\left[1/r_1 - 1/r_2 + t\frac{(N-1)}{N(r_1 r_2)}\right] \quad (12.18)$$

The sign convention is such that light travels from left to right, and a radius is considered positive for a surface with the centre of curvature to the right. A lens of positive power makes an incoming light beam more convergent. If the lens is thin, i.e. $t = f$, as is usually the case, the thickness term becomes small and may often be neglected, giving

$$P = 1/f = (N-1)[c_1 - c_2] \quad (12.19)$$

where we have also replaced the surface radii r_1, r_2 by the respective curvatures $c_1 = 1/r_1$, $c_2 = 1/r_2$.

For a lens of given material (N), the sum of the curvatures is thus fixed by the required lens power (or focal length) and the only design parameter left free is the ratio of the curvatures, that is, the shape of the lens. This is typically taken so as to optimize the imaging quality, i.e. to minimize the 'aberrations' of the lens. If a (near) parallel beam of radiation (e.g. from a classical object at infinity or from a laser beam near the beam waist) is to be focused, the best lens shape depends on N and is approximately symmetrical (biconvex) for low-index materials ($N < 1.5$). It becomes plano-convex for $N = 1.8$ and is meniscus-shaped for higher values of N. The stronger curved surface is always in the direction of the incoming radiation. With increasing refractive index, the absolute values of the curvatures become smaller for a given lens power. This has the highly desirable result that the primary aberrations, spherical aberration and coma are also reduced. This means that the same level of performance is maintained at higher numerical aperture (NA) (which is the sine of the angle under which the beam radius is seen from the focal point) or lower 'f-number' (focal length divided by beam diameter). A factor of about five is gained by switching from a refractive index of 1.4–2.5 for a lens of equal focal length, which explains why optical designers prefer to work with high-index materials. As ZnSe, one of the very few materials which are highly transparent for CO_2 laser radiation, has a refractive index of 2.403, singlets made from ZnSe perform satisfactorily in many CO_2 laser applications. Details, examples and limits are discussed in Chapter C4.2.5.

Singlets of the plano-convex and biconvex type can be purchased as catalogue items from a number of vendors. Cost and—often more important—much time can be saved if a given requirement can be (nearly) met with a catalogue lens. The steps between available focal lengths are, however, rather large and materials are mostly limited to BK7 and quartz but occasionally include some high-index glasses for lenses with high apertures.

12.4.2 The Dialyte and the Achromat

Any combination of two lenses offers much extended design freedom over the singlet as there are four additional parameters: two curvatures, one refractive index and one lens separation. The classical use of this design freedom is to 'achromatize', i.e. to compensate the wavelength dependence of the refractive index (i.e. dispersion) of one lens by choosing a material with different dispersion for the second lens and arranging the curvatures for simultaneous correction of other aberrations. The subject is discussed extensively in all textbooks on optical design (see, for example, [6,7]). In a very general way, 'splitting' a single lens into two elements will reduce the aberrations of that single lens and, thus, can, for instance, greatly improve the monochromatic performance, and this is often the primary goal when handling (monochromatic) laser radiation. It should be noted that the rules for the best shape of a singlet almost never apply to combinations of lenses. Optimization by computer program, guided by the previous experience of the designer, is applied in practice.

For example, we consider the improvement gained by a two-lens arrangement (called a 'dialyte') over a single biconvex lens made from BK7 glass as discussed earlier (Figure 12.1). If the singlet (i), which has acceptable performance at NA = 0.062 (or f/8) is replaced by two identical biconvex lenses (each having half the optical power) (ii), the same performance is obtained at (a marginally better) NA = 0.071 (f/7.0). If replaced by two plano-convex lenses, arranged such that the curved surfaces are facing each other (iii), the performance is just a bit better and allows use at NA = 0.073 (f/6.8). If the two lenses have the same orientation, both with their curved sides facing the entering radiation (iv), the same performance is obtained at NA = 0.087 (f/5.7). If the second lens is given the optimum shape, which is a meniscus in case [e], the system is further improved to allow use at NA = 0.104 (f/4.8). This turns out to be the best that can be achieved with two positive (!) lens elements (with that refractive index). If, however, the design is forced to have a positive (biconvex) first, and a negative meniscus second, with a small air gap between the lenses [f], it becomes substantially better and can be used at NA = 0.15 (f/3.3). This type of dialyte is particularly useful for such laser applications, where the wavelength dependence of the optics is no issue.

Proper shaping (also known as 'bending') of lens elements is by far the most powerful method to achieve best performance (i.e. low monochromatic aberrations). Correcting the wavelength dependence of lens properties, however, can only be

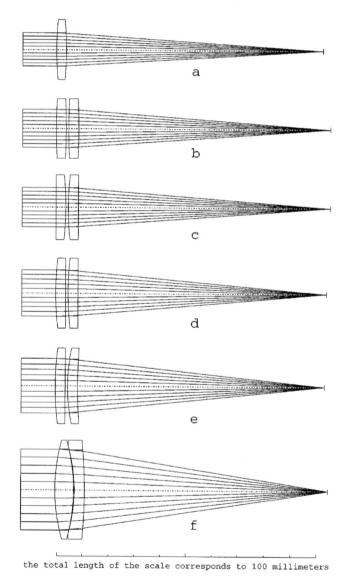

the total length of the scale corresponds to 100 millimeters

FIGURE 12.1 Comparison of six lens types of equal (diffraction-limited) performance. (a) Singlet at NA = 0.062 (f/8); (b) two equal symmetrically bi-convex lenses of half-power at NA = 0.071 (f/7); (c) two equal plano-convex lenses facing each other at NA = 0.073 f/6.8); (d) the same lenses as in (c) both facing the incoming beam at NA = 0.087 (f/5.7); (e) one plano-convex lens plus one positive lens of same power bent to optimum (= meniscus) shape at NA = 0.104 (f/4.8); (f) one positive, symmetrically bi-convex lens plus one negative lens of optimum shape in edge contact with the first lens at NA = 0.15 (f/3.3).

Optical Components

achieved by selecting suitable materials in conjunction with proper distribution of the optical power over these materials, whereas lens shape has very little effect.

A special form of the two-lens system, not a dialyte but an 'achromat', consists of a low-dispersion positive and a high-dispersion negative lens element, cemented together at the common interface. With properly selected materials (one popular combination for visible light uses the Schott glasses BK7 and SF2), the resulting system is corrected for spherical aberration and coma and has the same focal length at two wavelengths in the visible (e.g. C and F light) band. Cementing lenses has the advantage of creating a single mechanical unit where the mutual alignment between the elements has already been done by the manufacturer. It also removes two air–glass interfaces (and the need for anti-reflection coating of these surfaces). The use of cemented lenses must, however, be strongly discouraged for use with laser radiation of any appreciable power. The thermal properties of the constituents of an achromat tend to differ enough to create thermal stress upon exposure to high radiant powers, and this will readily break up the cemented assembly.

The well-known glass combinations, which allow simultaneous correction of colour and monochromatic aberrations for visible light, are not transferable to other spectral regions. When moving towards the near infrared, the differences in the dispersion between the optical glasses become progressively smaller [8]. Colour correction, if at all possible, then demands different glass combinations. A typical example is discussed in Chapter 27, Section 27.6.3.

Building a dialyte from catalogue lenses is only marginally successful. As the previous example showed, replacing a single lens by a pair of identical lenses (of, accordingly, double focal length) will give some, but not much, improvement in performance, and the best solution is not accessible with catalogue lenses, as it requires a very special negative lens element. If lowest aberrations are not mandatory, however, combinations of cleverly chosen catalogue lenses can help to achieve an otherwise not available focal length, since lens powers are, in a first approximation, additive.

12.4.3 Lens Systems with More than Two Elements

Much more complex designs than those already discussed are needed if a lens system has to provide imaging over an appreciable angular field onto a flat image plane, a requirement which is typical for all photographic applications but not normally for laser optical systems. In applications with direct focusing of a laser beam, we usually need to cover only small angular fields of less than, say, 5 mrad—just enough to make them insensitive to alignment errors. But there are exceptions: one is scanner optics (which often requires $f - \theta$ lenses) and another is the use of mask imaging for micromachining with excimer lasers. The latter subject is treated in Chapter C4.2.7, the former in C4.4. More complex designs than those already discussed are also needed to achieve very high numerical apertures and/or a distance between the last optical surface and the image, which is substantially larger than the focal length. The latter may be desirable to protect a lens from spatters and fumes from a laser-machining process (see Chapter 27, Section 27.5.2.2).

The art and science of designing optical systems for just about any application is highly developed. It is treated in a number of textbooks [6–9]. Excellent computer software for optical design and analysis is offered by a number of vendors

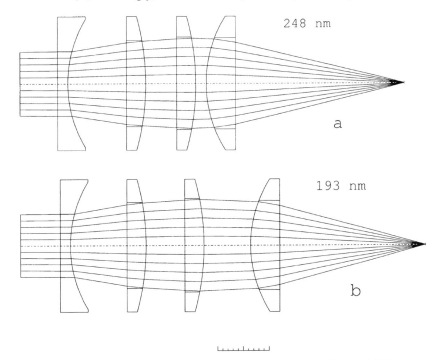

the total length of the scale corresponds to 10.0 millimeters

FIGURE 12.2 Example of a (UV) lens system assembled from (quartz) catalogue lenses, with lens distances adapted for optimum—diffraction limited—performance at wavelengths of 248 and 193 nm, respectively. For further details, see text.

and covers a broad price range. While such software is of great help to the educated designer (it alleviates the tediousness of manual calculations, provides excellent visualization of system layouts and (aberration) performance and always includes powerful optimization routines), it is no replacement for insight into the optical problem to be solved and the most adequate approach to its solution, which basically come from experience.

Designing multi-lens systems with the exclusive use of catalogue lenses is an interesting design challenge and can be quite successful. One will often end up with more lens elements than are necessary for a truly optimized system, but this should not be a major concern. Figure 12.2 shows, as an example, a lens for 30:1 imaging of a small object in excimer laser light of (i) 248 nm (a KrF-laser) and (ii) 193 nm (an ArF laser). Both versions use the same four (quartz) lenses, which can be found in most catalogues: one plano-concave, three plano-convex, of which two are identical. The respective focal lengths (for visible light) are −50, 100 and 50 mm. Only the lens separations are different (and have been used for optimizing performance). Both versions have excellent imaging qualities up to, respectively, NA = 0.22 (f/2.17) at 248 nm and NA = 0.25 (f/1.94) at 193 nm wavelength, albeit for small fields.

12.5 Use of (Curved) Mirrors

The optical power of a lens can also be provided by a curved mirror but with important differences, some advantageous, some not; mirrors function independently of wavelength (except for properties caused by surface coatings) and a single concave spherical mirror has much smaller aberrations than a comparable lens. In fact, a lens would have to be made of very high-index material (N = 4.4) to match the spherical aberration of a single concave spherical mirror of equal aperture! The problem with mirror optical systems lies in the overlap between the incoming and outgoing beams which, particularly in laser optics, requires the mirrors to be placed in some 'off-axis' arrangement, with the individual mirror elements tilted with respect to the optical axis to avoid obscuration. This is in sharp contrast to axially symmetrical mirror systems such as those used in astronomical instruments where a central obscuration is acceptable. Tilted non-flat optical surfaces are inevitably inflicted with off-axis aberrations, notably coma and astigmatism, which, in turn, must be corrected by additional (and again tilted) optical surfaces, resulting in more complex set-ups with at least three curved mirrors [10]. Thus, mirror optical systems lack the ease of alignment along a common optical axis, which makes lens-based systems attractive. This and the lower number of degrees of freedom have made mirror systems much less popular with optical designers. Extended mirror designs with non-spherical surface shapes do not really offer much relief: the improvement is typically limited to an extremely small angular field and at the expense of a large coma beyond. This is discussed for the off-axis paraboloid in Section C4.2.5.3.2. Not surprisingly, mirror systems are discussed much less extensively in textbooks on optical design than lenses. A systematic study of systems with three and four spherical mirrors, respectively, is presented in Refs. [11,12]. For use at very high laser power levels, mirrors are the only choice, as they offer excellent thermal stability due to the high thermal conductivity of the more common substrate materials (copper and silicon) and the ease with which efficient cooling can be implemented. These aspects are discussed in conjunction with high-power CO_2 laser applications in Chapter 27, Section 27.5.3.

12.6 Non-focusing Optical Laser-beam Handling and Relaying

Large distances often have to be bridged between the location of a high-power stationary laser installation and that of the workpiece on which a process has to be performed. This situation is typical for (and limited to) CO_2 lasers. Since the laser beam waist is usually located near the output window of the laser, the beam will diverge while travelling the distance to the workpiece which, for instance, in the case of large gantry-type workstations is large and variable (see Chapter C4.4). The beam radius at the location of the focusing optics, as well as its wavefront curvature, will vary accordingly (equations (12.2) and (12.3)). This is unwanted for two reasons: (i) it affects the axial position of the focus (equation (12.7)), and (ii) as the beam radius controls the numerical aperture of the image beam, the size of the focal radius (equation (12.12)) will not be constant over the working envelope. A numerical example is given in Table 12.1. Both drawbacks are avoided by relaying the primary beam so that an image waist is formed near the median of the range of distances to the workpiece and, thus, the range of waist distances with respect to the focusing optics. It is a good practice to combine this operation with creating an image waist of optimum radius and Rayleigh range. This will keep the distances to the focusing optics to well within the near-field and minimize focal distance variations as well as variations of the effective numerical aperture and resulting focal spot size. For the example of table 12.1, this would mean the use of a telescope which creates a waist of 12 mm radius at

TABLE 12.1

Effect of Lens-to-Waist Distance on the Focus of a Lens with 100 mm Focal Length. The Laser-Beam Characteristics are M^2 = 2; W_0 = 6 mm; Wavelength, 10.6 μm

Lens-to-Waist Distance (m)	Bram Radius on Lens (mm)	Diffraction Image Diameter (mm)
8.0	10.8	0.132
10.0	12.8	0.119
12.0	14.8	0.112

The diffraction image includes the effects of the varying beam truncation at the lens diameter of 36 mm

a distance of, say, 10 m. As a result, the focus remains constant within less than 1% over the same range of distances (±2 m).

Optical systems for performing this relaying function are, in essence, telescopes but with a capability of adjusting the exiting beam for (weak) convergence. In this application, the telescope must enlarge the beam enough to allow formation of a waist of suitable radius at the desired (forward) distance (c.f. equation (12.10)). It is often referred to as a 'beam expander'. Telescopes can, of course, be built with lenses or mirrors and can be of the 'Keplerian' or 'Galilean' type, i.e. with or without an intermediate focus. Unless needed for some special function (such as mode filtering), an intermediate focus is best avoided. Examples of telescopes based on (off-axis) mirrors and for use with (high power) CO_2-laser radiation are given in Section C4.2.5.1. Lens-based beam expanders are more commonly used in conjunction with shorter wavelength lasers. Beam adaptation—expansion or compression—in only one dimension is often needed for diode laser applications. This is discussed in Section C4.2.6.4.3.

Planar mirrors play an important role in beam positioning for whatever purpose (relaying, alignment, scanning a workpiece, etc.). The high speeds with which some laser-based processes (scribing, marking, etc., see Section D1) can be performed, call for equally high speed of beam deflection by mirrors (see Chapter C4.4). Low inertia and high mirror rigidity are then primary requirements, to be met by selection of (i) a suitable material with a high modulus of elasticity and low specific gravity and (ii) a mirror geometry that minimizes inertia. For best results—highest scan speeds, fast settling to the end position with minimum overshoot and accurate positioning—matching of the scan mirror to the drive motor is essential. Several vendors offer integrated (and optimized) mirror-drive assemblies for a range of mirror sizes.

By their very nature, weakly curved mirrors form an essential component of laser resonators (see Section A2). In this application, the local reflectivity of a mirror is occasionally adapted such that some desirable mode of oscillation is promoted within the resonator. If used in a folded cavity, (dielectric) mirrors will favour one direction of polarization, sufficient to induce a nearly completely (linearly) polarized output beam.

Mirror functions in systems for visible and near-infrared light are often implemented by means of reflecting prisms, usually exploiting total internal reflection for enhanced performance over surface mirrors. Reflective (and non-dispersing) prisms are often designed for multiple reflections. They, thus, allow not only for the direction of a beam to be changed but also for it to be deviated, displaced, reverted and/or inverted. The more common types of reflective prisms are treated in all textbooks on optics. A more complete yet concise overview can be found in the relevant chapter of the *Handbook of Optics* [13].

12.7 Thermal Effects in Optical Materials

The optical properties of materials are temperature-dependent. The refractive index typically varies of the order of 10^{-4}–10^{-5} per °C. As long as the temperature distribution within an optical component is homogeneous, this variation is of concern only in situations where an optical instrument has to meet very high performance standards (e.g. an aerial camera) or where very wide temperature ranges are envisaged (as in certain military equipment). Specific cases have been discussed by Rogers [14,15]. Quite a different problem arises in optical systems using high-power lasers. If a lens is exposed to high-power radiation and absorbs some of the incident radiation, the lens centre is heated but the heat will typically flow to the lens rim where it is eventually taken up by the lens mounting. This heat flow will, thus, cause a thermal gradient and, consequently, a radial gradient of refractive index in the lens material even under stationary conditions. Its optical effects are threefold:

1. a change of the lens shape due to stronger thermal expansion in the lens centre area,
2. an angle of refraction at the air–lens interfaces depending on ray height and
3. a propagation of radiation within the lens along an inwardly curved path, leading to a shortening of the focal length.

A quantitative computational model based on these effects reveals that, in any realistic situation, the third phenomenon accounts for the largest contribution. If the actual temperature rise in the lens centre is known, the resulting shortening of the focal length can be predicted reliably [16]. The effect is of substantial importance—and this is largely underestimated—for the ZnSe lenses used extensively for focusing CO_2 laser radiation. As a rule of thumb, the shortening of the focal length of a ZnSe lens will be about one Rayleigh range if, in thermal equilibrium, the lens centre runs 25 K warmer than the lens rim. Interestingly, the corresponding effect on aberrations is small; in fact, a thermal gradient reduces the spherical aberration of a single lens [17,18].

12.8 Specifying Optics for Laser Applications

12.8.1 Tolerances

The performance of optics in practice depends on many factors. While the parameters as designed will set an upper limit to what may be achieved under ideal circumstances, unavoidable manufacturing tolerances and imperfections cause real systems to perform below that limit. Thus, in the first place, the design of an optical system for a given application should always well exceed the acceptable minimum performance parameters by a reasonable margin to account for tolerances. Second, designs which, due to their basic principles, are less sensitive to tolerances should be preferred if possible. This typically involves a search for several, sometimes radically different, alternative possible solutions. Third, and usually quite complex, the tolerance budget has to be allocated to the individual components of a system to minimize the cost of manufacturing while staying within a given performance degradation. It is quite common for the effect of the tolerances of individual optical surfaces on the overall performance of the system to vary considerably, which makes it important to locate critical parts of a design and to cleverly balance the tolerances between all components. Modern commercial optical design software offers help for this task.

12.8.2 Surface Imperfections: Shape Deviations

Imperfections in an optical component can be divided into two parts: shape deviations and surface (and bulk material) quality imperfections. Shape deviations, also called form deviations, of an optical surface can be of different types. If the surface is perfectly spherical but its curvature deviates from the design value, the deviation is a 'sagitta error'. Deviations from the spherical shape, which are, in essence, higher-order shape deviations, are generally referred to as 'irregularities'. In practice, allowable surface shape tolerances are computed and specified in terms of sagitta errors, and the corresponding tolerances for irregularity are typically specified at a substantially lower value.

12.8.3 Surface Imperfections: Surface Quality

Optical surfaces are supposed to be 'optically smooth', which means that their microscopic profile varies gradually with steps well below the wavelength for which the optics is destined. This is achieved by a polishing process. Polishing must both remove the irregularities left by prior grinding of the surface and precisely shape the surface to conform to the specified curvature within the specified form tolerances. If the polishing ends prematurely, surface damage caused by the grinding process may still be present; alternatively, the polished surface may be damaged by contaminated polishing agents or by inadequate handling. The resulting faults are termed 'surface imperfections' if they are localized small individual pits, scratches, etc., and referred to as 'long scratches' if they exceed 2 mm in length, because of their visibility. All surface imperfections which are of the order of, or larger than, the wavelength cause losses by stray radiation, and in classical optical systems, the acceptability of such losses is the basis for setting a limit for the accumulated sizes of the surface imperfections. For laser applications, the situation is generally much more severe. Surface imperfections of wavelength size or beyond may cause local power concentration and thereby lower damage thresholds. It is, therefore, customary to require the highest standards for surface quality imperfections for optical components to be used with laser radiation, particularly, if pulsed laser operation with accordingly high peak powers is envisaged.

12.8.4 Communication of Specifications—ISO 10110

The communication of design data to a manufacturer is invariably by technical drawings which, to avoid misinterpretation, should be prepared according to a common standard. With the increasing globalization of the production of optical parts and systems during the last decades, a need was felt to agree upon an international standard. Issued by ISO, the International Organization for Standardization (Geneva, Switzerland), ISO 10110: (*Optics and Optical Instruments—Preparation of Drawings for Optical Elements and Systems*), has, since 1996, replaced the various national standards which were in effect before. ISO 10110 is now the preferred way of drawing optical elements and systems. Conformity to the standard must be explicitly embodied in the respective drawings.

ISO 10110 consists of 12 parts. An originally envisaged part 13 on laser damage thresholds has been cancelled but is still (and maybe confusingly) referred to in several other parts. Each part is a separate document and subject to revision. ISO 10110 is intended and has been written to standardize both the specification and measurement of the parameters of optical components.

Part 1 (General) explicitly regulates the drawing of optical components and systems, their dimensions and tolerances (linear dimensions always in millimetres, angular dimensions in degrees, minutes and seconds of arc). Parts 2, 3 and 4 are short documents describing how to specify three different types of bulk material imperfections: stress birefringence (part 2), bubbles and inclusions (part 3) and inhomogeneity and striae (part 4). Part 5 addresses the definition, specification, measurement and analysis of surface form tolerances, with particular attention to interferometric measurements and their interpretation. In part 6, the complex problem of specifying and indicating centring tolerances of single lenses and lens combinations is treated. Part 7 addresses surface imperfection tolerances and includes a choice of techniques for measurement or visual inspection. In part 8, surface textures as created by grinding or polishing are treated, whereas part 9 deals with (protective) surface treatment and (functional) coatings. Part 10 indicates how to optionally specify the data of an optical lens element in a table rather than on the drawing itself. Part 11 shows which tolerances apply, in case explicit indications are not given in a drawing. These tolerances then depend on the size of the component. Part 12 deals exclusively with aspheric surfaces, their mathematical description and presentation in drawings.

12.9 Manufacture of Optical Components

The manufacture of lenses from brittle materials—which include optical glasses and most crystals—involves two basic steps: grinding and polishing. Grinding, which may be preceded by some coarse machining of the part to approximate shape, is mostly done with a slush of loose grit of a hard material such as emery or carborundum. The actual mechanism is that the localized pressure exerted by a hard particle on the glass surface causes compressive stress in the glass surface area, which, upon release, lets a small part of the glass surface separate from the bulk. A ground surface, therefore, retains compressive stress in a surface layer of thickness roughly proportional to the size of the grinding particles and decaying exponentially. Grinding is typically done in steps using successively smaller particles as the intended final shape is approximated and the damage caused by the coarser particles is alleviated.

Polishing is the final process for generating an optical surface of the required smoothness and conformity to the specified shape. A measurement to determine surface conformity to the required shape by interferometric methods can only be done on a polished surface (see Chapter A5). Polishing must also remove the stressed part of the surface layer and as much more of this layer as is affected by damage (digs, scratches) from previous grinding. Polishing, in contrast to grinding, is a slow process, basically involving chemical interaction between the glass and water, enhanced by a polishing agent. The degree of polishing of a surface is characterized by one of four grades, which specify the number of micro-defects—irregularities of less than 1 μm size—per 10 mm of sampling length (ISO 10110-8). What is acceptable for optics for laser applications depends largely on the laser wavelength and on the peak power density, which will be encountered by the surface. The highest grade (P4 with less than three micro-defects per 10 mm of sampling length) is typically required for optical components for use with pulsed lasers in the near-infrared (Nd:YAG), visible (ruby) and ultraviolet (excimer lasers) spectral region. Much less stringent requirements apply to optics for the longer wavelengths of (cw) CO_2 lasers.

The quality of polish affects that of a coating, both in terms of adherence and blemishes and, therefore, strongly affects the onset of damage induced by laser irradiation. Coated optical surfaces are much more prone to damage by high-peak-power (pulsed) laser radiation than bulk materials, but, if produced by the best manufacturing techniques, coatings have approached the bulk thresholds to within one order of magnitude. Laser-induced damage to optical materials is a subject of extensive current research, and results are periodically reported in conferences, many of which have been organized or, at least, supported by the US National Bureau of Standards. Proceedings have been published by 1999 *SPIE* **3902**, 1998 **3578**, 1997 **3244**, 1994 **2428**, 1992 **1848**. Sixty-two 'classical' papers on the subject are found in Vol. MS24 of the SPIE Milestone Series [19].

The manufacture of mirrors is similar to that of lenses in that brittle substrate materials (glasses, zerodur, silicon) are used. Metal mirrors for high laser power applications based on copper bodies with integrated cooling channels are mostly made by single-point diamond turning on precision machines, a process which readily produces aspheric shapes.

12.10 Summary and Conclusions

When used with laser radiation, optical components perform similar tasks as in classical optics and, thus, benefit from the achievements in optical design and manufacturing, which have accumulated over centuries. The coherent nature of laser radiation has to be considered when planning and evaluating laser beam imaging. The possibly high (peak and or average) powers of laser radiation requires utmost care in the manufacturing and handling of optical components for laser applications and attention to achieving thermal stability in laser optics.

REFERENCES

1. Self S A 1983 Focusing of spherical Gaussian beams *Appl. Opt.* **22** 658–61.
2. Kogelnik H and Li T 1966 Laser beams and resonators *Appl. Opt.* **5** 1550–67.
3. Siegman A E 1990 New Developments in laser resonators *Proc. SPIE* **1224** 2–14.
4. Smakula 1936 German patent 685 767.
5. Dobrowolski J A 1995 Optical properties of films and coatings *Handbook of Optics* 2nd edn, ed M Bass (New York: McGraw-Hill).
6. Kingslake R 1978 *Lens Design Fundamentals* (New York: Academic).
7. Smith W J 1990 *Modern Optical Engineering* 2nd edn (New York: McGraw-Hill).
8. Shannon R R 1997 *The Art and Science of Optical Design* (New York: Cambridge University Press).
9. Smith W J 1989 Optical design *The Infrared Handbook* 3rd edn, ed W L Wolfe and G I Zissis (Washington, DC: Office of Naval Research) ch 8.
10. Beckmann L H J F and Ehrlichmann D 1994 Three-mirror off-axis systems for laser applications *Proc. Int. Optical Design Conf. (Rochester)* (Washington DC: OSA) pp 340–8.
11. Howard J M and Stone B D 2000 Imaging with three spherical mirrors *Appl. Opt.* **39** 3216–31.
12. Howard J M and Stone B D 2000 Imaging with four spherical mirrors *Appl. Opt.* **39** 3232–42.
13. Wolfe W L 1995 Nondispersive prisms *Handbook of Optics II* 2nd edn ed M Bass (New York: Mcgraw Hill).
14. Rogers P J 1996 The selection of IR optical materials for low-mass applications *Proc. SPIE* **2774** 301–10.
15. Rogers P J 1993 Optics in hostile environments *Proc. SPIE* **1780** 36–48.
16. Tangelder R J, Beckmann L H J F and Meijer J 1992 Influence of temperature gradients on the performance of ZnSe-lenses *Proc. SPIE* **1780** 294–302.
17. Beckmann L H J F 1994 Modelling of, and design for, thermal radial gradients in lenses for use with high power laser radiation *Proc. Int. Optical Design Conf. (Rochester)* (Washington, DC: OSA) pp 11–15.
18. Beckmann L H J F and Meijer J 1998 Modelling of and design for thermal radial gradients in lenses for use with high power laser radiation *Weld. World* **41** 97–104.
19. Wood R M (ed) 1990 *Laser Damage in Optical Materials (SPIE Milestone Series MS24)* (Bellingham, WA: SPIE).

13
Optical Control Elements

Alan Greenaway

CONTENTS

13.1 Introduction 219
13.2 Amplitude Modulation 219
    13.2.1 Electro-optic Modulators 219
    13.2.2 Acousto-optic Modulators 220
    13.2.3 High-power Beams 221
    13.2.4 Magneto-optic Isolators 221
13.3 Scanning and Positioning the Beam 222
    13.3.1 Mechanical Beam-directing Systems 222
    13.3.2 Acousto-optic and Electro-optic Scanners 223
    13.3.3 Diffractive Beam Steering 224
    13.3.4 Positioning the Beam 224
13.4 Controlling the Size and Shape of the Beam 224
13.5 Safe Disposal of Unwanted Beams 225
References 225
Further Reading 225

13.1 Introduction

The extra-cavity control of laser beams includes requirements for scanning and positioning the beam, changing the size and shape of the beam, modulation of the beam in both time and space (also in amplitude and phase) and, in this increasingly safety-conscious age, ensuring the safe disposal of stray beams. Each of these requirements will be considered briefly in this section.

In many cases the most suitable technology will depend on the power and wavelength of the laser beam that is to be controlled. Other considerations are cost and the environment in which the control function is to be exercised (for example, when used in a laboratory by experienced researchers a degree of control and access to the beam may be required that would be wholly unsuited to a commercial environment in which the operators may be much less aware of the potential risks).

As used in this section the term 'adaptive optics' will refer exclusively to linear optical systems and discussion will refer to extra-cavity uses of such systems. The treatment of beam positioning and scanning technologies here will include discussion of such applications as flat-bed scanners only briefly.

13.2 Amplitude Modulation

The most basic level of beam control is achieved through the use of a shutter. Most laser systems (but especially solid-state systems) operate most effectively when producing output at their design power output. The resonator design can, under these circumstances, take full account of optical defects and distortions due to the operating temperature of the laser. It follows that the use of a shutter is preferable to variation of the output power level (including power-down) in most situations. Various forms of shutter are available from a multitude of commercial sources, and these are widely used in safety interlock systems on lasers.

Amplitude modulators are also widely used with laser systems at all power levels. In the case of low-power lasers, the amplitude modulation is most often a simple means to ensure that the power level delivered to a detector can easily and quickly be varied. In this case a simple neutral density filter wheel or variable attenuator is generally satisfactory (but if these are of the reflective type care may be required with reflected beam). With low-power beams the use of absorptive filters or a polarizer–analyser system provides effective beam attenuators. In all cases it is advisable to locate filters between a laser source and any beam-conditioning optics (such as an optical fibre or a spatial filter) so that optical imperfections in these elements do not lead to any degradation of beam quality.

13.2.1 Electro-optic Modulators

Electro-optic modulators using the Kerr effect or the Pockels effect are available, although most modern electro-optic modulators are based on Pockels cells. These modulators exploit the voltage-dependent birefringence induced in some crystalline materials when they are subjected to an external electric field. The Pockels effect has a linear relationship and the electro-optic Kerr effect a quadratic relationship between the indicatrix of the material and the applied electric field. Birefringence is a difference in refractive index dependent on the alignment of the electric field vector with crystal lattice directions.

A linearly polarized light beam can be resolved into two orthogonal polarizations of equal strength. Suppose the beam enters a birefringent crystal along the crystalline optic axis (the axis of crystal symmetry denoted as the z-axis). If the linear polarization direction is at 45° to the other principal dielectric axes (x- and y-axes) in the crystal, one of the orthogonal polarization components aligns with the fast axis and the other with the slow axis within the birefringent material. As a result of the different refractive indices associated with the fast and slow axes, a phase change develops between the two polarization components as they propagate through the crystal and this, in turn, causes a rotation of the plane of polarization of the radiation.

In an electro-optic device the electro-optic element is positioned between a polarizer and analyser with a mutual orientation of 90°. In the absence of birefringent effects, no radiation will be passed by the analyser. However, if a voltage is applied that makes the crystal birefringent and the fast and slow axes of the induced birefringence are oriented at 45° to the input polarization plane, the plane of polarization is rotated as the beam passes through the crystal. If the plane of polarization is rotated by 90°, the polarization will align with the analyser and the radiation will pass through the system. If the voltage is removed and the plane of polarization is unchanged by passage through the crystal, the polarization will be orthogonal to the analyser and the radiation cannot pass through. If the applied voltage results in a rotation of the polarization plane between 0° and 90°, the fraction of the energy passed will follow a cosine-squared law.

Suitable materials for electro-optic modulators include lithium niobate, lithium tantalate (for visible and near-IR regions) and, for use in the thermal IR, cadmium telluride. Electro-optic modulators may use various orientations of the crystalline axes. Dependent on the orientation of the electric field relative to the direction of propagation of the light, these are referred to as longitudinal (electric field and propagation direction are parallel and aligned with the crystalline optic axis) or transverse (electric field and propagation direction are orthogonal).

With longitudinal modulators the voltage must be applied across the entire propagation path within the crystal and, to maintain the required field strength, high voltages (in the kilovolt range) must be used. For KDP (potassium dihydrogen phosphate), approximately 8.7 kV is required at a wavelength of 633 nm to retard one polarization component by half a wave compared to the other (thus to obtain an electrically programmable half-wave plate and rotate the plane of polarization by 90°). It is generally difficult to switch high voltages at high frequency and, thus, longitudinal devices are used for lower-frequency applications. An additional feature is that the radiation must enter and leave the crystal through transparent (or wire grid) electrodes required to apply the electric field.

As the applied voltage increases from zero to that required to achieve a half-wave delay, the polarization changes from the linear input to elliptical with the major axis aligned with the original polarization direction, through circular polarization, to elliptical polarization with the long axis aligned normal to the original direction of polarization and, finally, to linear polarization at 90° to the original direction. The interaction of these polarization states with the analyser is what provides the desired amplitude modulation. Longitudinal modulators are used when large acceptance angles and wide fields of view are required.

Transverse modulators, by comparison, use an orientation that requires the electric field to be applied along the z-axis which is now orthogonal to the direction of propagation. This construction makes it possible to achieve a long interaction length at a given field strength. Lower applied voltages are, therefore, required across the short dimension of long thin crystals, because the potential is applied across the material and orthogonal to the direction of propagation. Additionally, the input and output beams are no longer required to pass through the electrodes. Because of the long, thin structure only small acceptance angles are suitable. This design is ideally suited to modulators in waveguide configurations for integrated optics applications and has the added benefit of using lower electric voltages.

The voltage required to induce a given phase change between the polarization components (and thus rotation of the plane of polarization) increases with the wavelength of the radiation to be controlled.

Material exhibiting intrinsic birefringence can be used in electro-optic devices by cutting the active crystal into two segments and rotating the segments so that the fast and slow axes of the first segment align with the slow and fast axes in the second segment. Such an arrangement cancels the natural birefringence of the material and increases the range of materials that can be exploited to include birefringent materials such as ADP.

Pockels cells with extinction ratios (ratio between maximum and minimum transmission) of 10^3 are available and bandwidths of $>10^9$ Hz are available. This extinction ratio is not always adequate for use as a shutter but two Pockels cells may be used in series if required. Electro-optic devices offer a higher switching speed than magneto-optic and acousto-optic devices but are generally of smaller aperture and have the disadvantage of requiring relatively high-voltage drivers. If the polarizer and analyser in an electro-optic device are absorbing, it is clear that the energy deposited in these elements will limit the power that can be handled (Figure 13.1).

13.2.2 Acousto-optic Modulators

Acousto-optic modulators rely on the variation of refractive index with position that is produced by pressure (acoustic) waves in various media. Acousto-optic materials require good optical quality, good optical transmission and good acoustic properties, but suitable materials exist across the UV, visible and near IR (e.g. fused silica/quartz) and in the IR (germanium). A typical acousto-optic modulator will consist of a material with suitable elasto-optic properties and a carefully

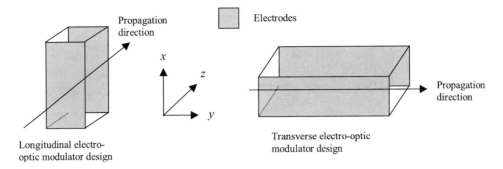

FIGURE 13.1 Schematic diagram showing axis directions and electrode location for longitudinal and transverse electro-optic modulators.

designed and coupled piezoelectric transducer bonded to one side of the material and an acoustic sink well coupled to the opposite side (to prevent acoustic reflection and the generation of standing waves).

The (acoustic) pressure waves give rise to a pattern of periodic variations in refractive index that is moving slowly compared with the electromagnetic wave. The light, therefore, sees a near-stationary Bragg grating structure and can be scattered by this grating (note, however, that Doppler effects lead to a slight frequency change, desirable, for example, in generating heterodyne carrier frequencies in some forms of interferometry).

For a given transducer frequency and a material with a given speed of sound, the acoustic wavelength (and thus the period of the pressure-induced refractive index changes) will be defined by the usual relationship between frequency, speed and wavelength. The angle at which an optical wave is scattered from the acoustically generated grating is dependent on the ratio of the wavelength of the light and the wavelength of the Bragg grating. The fraction of the light energy scattered by the Bragg grating will be dependent on the amplitude of the Bragg grating and that, in turn, is determined by the acoustic power coupled into the material from the transducer.

By keeping constant the frequency of excitation but varying the acoustic power, the energy in the scattered beam can be varied from zero (no acoustic signal) to some maximum level. Acousto-optic modulators, therefore, have a high extinction ratio. Acousto-optic materials are compared through a 'figure of merit' that relates the optical diffraction efficiency to the acoustic power coupled into the material. The 'figure of merit' is not the only important parameter and the speed of operation (modulation bandwidth) also needs to be considered.

The pressure-induced Bragg grating propagates through the acousto-optic cell at the speed of sound in the given material, and it follows that the reaction time of the device cannot significantly exceed the time taken for the acoustic wave to transit the optical beam diameter. The speed of operation of opto-acoustic devices can, therefore, be improved through the use of smaller light-beam diameters. However, by the principle of Fourier optics, the beam spread induced by the Bragg grating is dependent on the number of grating periods across the beam. This effect, and the power-density level that the material can support, may both limit the extent to which the bandwidth may be improved by focusing the beam.

Acousto-optic devices have the merit of requiring only low electrical power and delivering good extinction ratios (from 90% of input to zero) and being physically compact. They are intrinsically slower than electro-optic devices and have the disadvantage that the beam used has been deviated from its original direction (the undeviated beam is dumped).

13.2.3 High-power Beams

Amplitude modulators for higher-power systems may be based on the use of diffractive or polarizing effects to divide the beam, after which one beam is used as the required modulated output and the other beams are directed to beam dumps. For high-power applications, these approaches have the merit that the absorption of the beam energy occurs on the beam stop and not within the diffractive or polarizing element.

At their simplest, these elements can consist of a circular disc having phase gratings of various modulation depths in various segments around the disc. The grating diffracts the input beam into various diffraction orders 0, ±1, ±2,.... An aperture then passes one diffraction order and blocks the remaining orders. In a simple device of this form, it is generally the zero diffraction order (undeviated beam) that will be used downstream. These are low-cost elements that can be designed to specific user requirements and suitable materials for fabrication of the phase gratings can be found for wavelength ranges from the UV to the IR. In general, such elements would not be used where accurately calibrated attenuation ratios are required.

Variants on the same principle exploit polarization or acousto-optic effects to divide the input beam in amplitude and to dump all but one output beam safely whilst providing the required modulation on the remaining beam. In general, a larger dynamic range (i.e. a greater range in the extinction ratio) can be achieved through the use of a deviated beam as the output beam (since it is generally possible to switch the output energy to zero in the deviated beam). For all of these devices, maximum continuous power levels of a few hundred W cm^{-2} appear to be widely available.

If a sample of a high-power beam is required for laser diagnostic purposes, the use of a low-amplitude diffractive structure, or the Fresnel reflection from one or more accurately figured transparent wedges, may be used.

13.2.4 Magneto-optic Isolators

The distinctive fact that the rotation of the polarization plane of an optical beam induced by the magneto-optic effect does not reverse direction if the direction of propagation is reversed

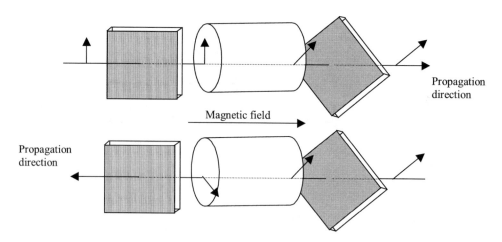

FIGURE 13.2 A vertically polarized beam propagating left to right is rotated clockwise by the Faraday rotator and passes unhindered through the analyser. A reflected beam of the same polarization passes through the Faraday rotator and is rotated anticlockwise, resulting in an orientation normal to the polarizer. The backscattered beam is thus blocked.

gives rise to the unique application using optical isolators. A reflected beam of initially weak amplitude re-entering a laser cavity can be amplified within the cavity and lead to serious effects in terms of the resonant properties of the laser cavity. Isolators offer a mechanism that can be used to prevent such unwanted feedback.

Many materials become optically active if they are immersed in an axial magnetic field. The rotation due to this Faraday effect depends on the Verdet constant, the length of the material and the modulus of the applied magnetic field. The vital feature here is that if the rotation of the polarization is clockwise when the light propagates in the same direction as the magnetic field vector, it is anticlockwise if the direction of propagation is reversed with respect to the magnetic field. Thus, a beam reflected back through the system suffers twice the rotation. The basic configuration of a Faraday isolator is shown in Figure 13.2.

The maximum achievable isolation from a Faraday isolator is likely to be limited by inhomogeneities in the crystal. It is, however, possible to square the attenuation of the backscattered beam by placing two isolators in series and arranging the magnetic field for each isolator to be reversed with respect to the other. Such an arrangement leaves the output polarization of the laser system plus isolators unchanged in addition to providing a high level of isolation.

13.3 Scanning and Positioning the Beam

In all scanning operations it should be recalled that the focus of the beam scanned by any device may vary significantly in quality across the scan when the output plane is flat.

13.3.1 Mechanical Beam-directing Systems

At the simplest level, a system for re-directing a laser beam may be required in a research laboratory to provide a safe method for sharing the output from a laser between several experiments or simply to provide a convenient and easily re-arranged coherent source. Flat mirrors for angular deflection can be obtained from a variety of commercial sources. These can range from simple mirrors with defined mechanical positions (e.g. Thorlabs' 'indexing optical mount' provides 15 positions to allow a laser to be shared at very low cost) to controllable scanning mirrors of various degrees of sophistication and cost. In each case, care is likely to be required to intercept and to contain all stray beams generated, not only in the positions required but also during transitions from one beam position to the next.

For raster scans it is normal to use two single-axis deflectors arranged so that the scan directions are orthogonal. This arrangement provides a high degree of flexibility in terms of the scan direction sequence. For repeated raster scans, rotating systems using rotating polygonal mirrors are useful and are widely used in the IR [2].

For applications that require an ability to actively re-position the beam under computer control and over relatively short periods of time, various systems are available. Fast-steering mirrors of the type used in military applications can provide steering of high-power laser beams at high frequencies and with a 'go to' capability. Such devices do not come cheaply—a high-performance model from Ball Aerospace may cost in excess of $50 000 (cheaper commercial devices are available through Newport Corp)—and are, therefore, not well suited to many applications where cost is an important factor. Such devices may offer kHz bandwidths and better than μrad resolution. Other, more modestly priced, devices can offer accurate 'go to' capability but generally with lower angular range, speed or power-handling ability.

The range of devices available includes galvanometer mirrors, voice-coil scanners, acousto-optic deflectors, electro-optic deflectors and various forms of adaptive optics mirrors. Mirrors with various forms of actuation are produced, the choice of actuation being determined by speed of operation, positional accuracy and repeatability required, cost, power-handling capability and angular deflection required. Moving magnet scanners are widely available and offer low-inertia, reliable and low-cost scanning systems capable of optical deflections of tens of degrees with good linearity and stability. Those available from Laser 2000, for example, have

Optical Control Elements

small-angle response times of a few tenths of a millisecond and can include position sensors to make them suitable for feedback control. Moving coil systems are likely to be more accurate but to have slower response. Repeatability ranging from a few 10s of μrad to a few μrad is offered over these scan ranges and, in both cases, optical scans over wide angles (up to 80°) are offered in the commercial literature. Gimballed mirrors driven by voice-coil or piezoelectric actuation with more modest range and price are available from a wide variety of sources.

Wavefront modulator systems used in adaptive optics (for example, deformable membrane mirrors can be used for scanning the beam over small deflections in addition to their use for changing the beam shape (this latter use will be discussed later). However, such adaptive optics wavefront modulators are generally designed for the shape deformation that they can produce. Since the amplitude available for wavefront deformation is very limited, it makes little sense in adaptive optics applications to compromise the wavefront shape change that can be achieved by using any of the limited modulator strokes for control of beam direction. For this reason, it is a general practice in adaptive optics to include within the system a 'tip-tilt' mirror of exactly the type that can be used for general laser-beam steering applications.

13.3.2 Acousto-optic and Electro-optic Scanners

When considering scanners, one is generally interested in the format of the raster scan that can be produced, i.e. how many resolvable spots can be produced. This is dependent both on the total scan angle that can be achieved and on the beam divergence (spot size). A second important consideration is the time taken to 'go to' a specific location within the scan area and the repeatability with which that spot can be re-acquired. For some applications, however (e.g. image display), the time required to complete the sequential scan may be of more interest.

A range of electro-optic and acousto-optic beam deflectors is in use. These devices are essentially the same as those discussed in the previous section and used as modulators and Q-switches. Electro-optic devices exploit the change in refractive index that occurs in electrically induced birefringence. The acousto-optic effect exploits the change in refractive index that occurs as a result of the passage of sound waves (i.e. pressure waves) through a material.

Acousto-optic diffraction occurs from the pressure-induced grating and is maximum when the diffracted beam emerges at exactly twice the Bragg angle relative to the undeviated position. This angle depends on the acoustic frequency and, hence, can be controlled electronically through the drive signal fed to the transducer. The Bragg angle is given by $\theta_B = \dfrac{\lambda}{2n\Lambda}$, where θ_B is the Bragg angle, λ is the optical wavelength, Λ is the acoustic wavelength and n is the optical refractive index (see, e.g. [5]). The principle is shown schematically in Figure 13.3.

Acousto-optic deflectors offer simple and fast control of the beam deflection for low electrical power input. The devices are much faster than mechanical scanners, simple in structure and offer a larger scan format than the electro-optic deflectors (but are slower in operation than electro-optical scanners). In some applications the small Doppler shift of the deflected beam may be a disadvantage.

Electro-optic deflectors use prisms of electro-optic material. Simple scanners may be based on the use of a single prism where the electric field is applied across the triangular ends of the prism to change the bulk refractive index and thus vary the beam deflection. However, a more common arrangement is to cement together two prisms whose optic axes (z-axis) are reversed in direction (Figure 13.4). By applying an electric field along the z-axis the increase in refractive index in the upper prism is balanced by an equal and opposite refractive index decrease in the lower prism. Thus, the upper ray is retarded with respect to the lower ray and the optical beam is deflected. Unfortunately, the deflection is rather small for

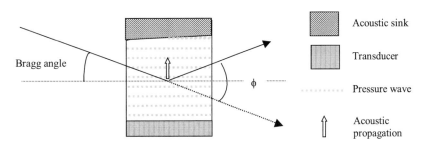

FIGURE 13.3 Illustrating the principles of acousto-optic deflection through angle ϕ.

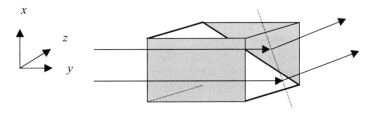

FIGURE 13.4 Schematic diagram of a double prism beam deflector. The upper and lower prisms have the orientation of their optic axis reversed. The shaded sides represent the electrodes across which the input voltage is applied.

modest voltages, and electro-optic deflectors are not as widely used as their acousto-optic counterparts.

The principal advantage of acousto- and electro-optic scanning systems is their speed, but this is delivered over small apertures and small scan angles (typically only a few degrees). It can be harder to control the properties of acousto-optic and electro-optic materials than to control mirror surface properties (especially emission) and for these reasons mechanical scanning systems tend to be more widely used in IR systems, although this is less true of laser-beam scanning applications.

In general, galvanometer systems are slowest but offer the highest format and largest overall scan angles. Acousto-optic scanners offer significantly higher access time but with smaller scan angles. Electro-optic scanners offer the fastest performance but the smallest scan formats.

13.3.3 Diffractive Beam Steering

Programmable diffractive optical elements can be used with laser (and other monochromatic) beams for beam steering and spatial multiplexing. This is particularly useful since small-amplitude devices can exploit the intrinsic modulo 2π nature of the diffraction process to achieve beam division and relatively wide-angle beam-steering performance. Low-cost, widely available components such as liquid crystal spatial light modulators with large spatial format can be exploited here, as can waveguide-type devices using electro-optic effects. Micro-machined mirrors, now widely used as display drivers, will offer an extremely flexible device for such applications when better control of the flatness of the individual mirror elements has been achieved—at present the curvature of the individual segments is an impediment to such application of these devices.

13.3.4 Positioning the Beam

When several sources (multiple lasers or both lasers and broadband sources) are to be used with a single optical system, the use of an optical waveguide (fibre) provides a method by means of which this may be achieved. The fibre end provides a well-defined source that may be left in position and energized sequentially with a wide variety of sources or energized with several sources simultaneously through the use of a cascade of fibre-optic couplers.

In general, the coherent interaction between the waveguide modes will mean that mono-mode fibres will be used here. Thus, the laser system may be used with a lens (usually a microscope objective) to couple the laser beam into the fibre core, and the laser and coupling optics may be safely boxed. Such a procedure has the advantage that all stray beams are confined within the box and that the only laser hazard then comes from the fibre output. It has the additional advantage that if the fibre used is mono-mode at the laser wavelength, the beam is also conditioned by the fibre. All modes except the fundamental fibre mode leak into the fibre cladding and ultimately out of the fibre, thus, at the expense of the optical efficiency, a well-defined Gaussian mode can be obtained

after mode stripping by even relatively short propagation lengths within the fibre. As a result, the output from a mono-mode fibre is a well-defined Gaussian mode irrespective of the quality of the laser beam itself. This represents a simple method to achieve a high-quality laser beam that may be coupled with precision into a laboratory experiment. The use of a mono-mode fibre is more effective, often has a lower cost and is easier to use than a conventional spatial filter based on a pinhole. Fibre-optic delivery systems also provide a mechanism for beam attenuation through the simple mechanism of deliberately misaligning the coupling optics at the fibre feed.

Within limits imposed by transport efficiency and modes supported, the use of fibres and fibre-couplers offers a convenient method for combining several wavelengths of laser output in a single source that may be conveniently and easily positioned in both location and angle. The different frequencies in the output beam are then easily selected by using shutters at the points where the lasers are coupled to their individual fibres. Pre-packaged couplers for injecting laser beams into both mono-mode and multi-mode fibres are commercially available, and many laser diodes are available with fibre pig-tails.

For situations where pure modal quality of the beam is less critical and some over-moded behaviour can be tolerated, the use of a pinhole filter offers an alternative to the use of fibres for spatial filtering and beam transport. The principal disadvantages associated with the use of a pinhole are the bulk optics required to couple the laser light through the pinhole filter, but the advantage can be more efficient use of the available laser output.

Applications using higher-power laser beams are generally less critical in terms of the beam quality, although the beam M^2 (i.e. the beam quality) can have a significant influence on the quality and efficiency of laser percussion drilling, for example. Here the use of fibres with large core diameter can provide a useful mechanism for transport of the beam. Fibres with smaller cores (e.g. mono-mode fibres) may suffer physical damage due to the high input brightness over the small core area).

13.4 Controlling the Size and Shape of the Beam

To expand (or to reduce) the beam diameter, zoom lens expanders suitable for both manual and computer control can be obtained. These systems can be capable of handling substantial power (500 MW cm^{-2} is quoted in the commercial literature) for operation over the 0.45–1.1 μm wavelength range. Such systems are of good optical quality (distortion less than $\lambda/4$) and offer either variable expansion or, with somewhat better optical performance, fixed expansion ratios. Designs using a combination of a negative lens and a positive lens may be preferred in high-power applications because the intense focused spot in a beam expander using two positive lenses can cause breakdown.

The beam shape (as opposed to diameter) can be controlled using linear adaptive optics, spatial light modulators of various descriptions [5] and diffractive optical elements. Linear adaptive optics technology [4] for controlling light has been developed primarily for military and astronomical applications. It was declassified by the USAF in May 1991. Linear adaptive optics is also used in laser applications, where it may be confused with non-linear optical effects, also described as 'adaptive optics' in the laser literature (e.g. [3]). Linear adaptive optics technologies are used for real-time measurement and correction of dynamic and stochastic aberrations due, for example, to turbulence in the earth's atmosphere or thermal blooming in the atmospheric propagation of high-power laser beams. Thermal blooming, caused by local heating of the atmosphere by a high-power beam, leads to defocusing and, in the presence of wind, causes the beam to swing upwind towards the higher density air. Used intra-cavity, an adaptive mirror can be used to modify the resonator properties to alter the output beam profile (usually to produce a super-Gaussian beam [1], i.e. one whose brightness profile matches a 'top-hat' more closely than the usual Gaussian beam modes). Used extra-cavity, adaptive optical wavefront modulators can produce variable focus control and/or small-angle beam redirection, functions of considerable use in laser-beam materials processing. By changing the relative phase across a laser beam, adaptive mirrors (or other wavefront modulation devices) can be used to produce particular beam patterns in the diffracted beam.

13.5 Safe Disposal of Unwanted Beams

Stray beams are a safety hazard and 'beam dumps' are available to dispose of them. Such systems may range from a simple curtain screen used in a laboratory to confine low-power stray beams to a work area used only by personnel aware of the hazards and to the use of specially designed conical systems designed to ensure the minimum back leakage of any 'dumped' beam. The latter are particularly useful where beam splitters are used to produce many beams, not all of which may be required during a particular experiment.

REFERENCES

1. Cherezova T Y, Chesnokov S S, Kaptsov L N and Kudryashov A V 1998 Super-Gaussian laser intensity output formation by means of adaptive optics *Opt. Commun.* **155** 99–106.
2. Montagu J and DeWeerd H 1996 *Optomechanical Scanning Applications, Techniques, and Devices (The Infrared and Electro-Optical Systems Handbook 3)* 2nd edn (Bellingham, PA: SPIE Optical Engineering Press).
3. Pepper D M 1986 Applications of optical phase conjugation *Sci. Am.* **254** 56–65.
4. Tyson R K 2000 *Introduction to Adaptive Optics* (Bellingham, PA: SPIE).
5. Yu F T S and Khoo I C 1990 *Principles of Optical Engineering* (New York: Wiley).

FURTHER READING

Davis C C 1996 *Lasers and Electro-Optics* (Cambridge: Cambridge University Press).
Written to be suitable for 'graduate students in electrical engineering and physics. It should also be useful to mechanical engineers or chemists who use lasers and electro-optic devices in their research'. Thorough, including much on crystal symmetries, tables of properties and many diagrams. Chapters 17 *Optical Fibres and Waveguides*, 18 *Optics of Anisotropic Media*, 19 *The Electro-Optic and Acousto-Optic Effects and Modulation of Light Beams* are particularly relevant here.

Smith S D 1995 *Optoelectronic Devices* (London: Prentice-Hall).
Written for 'graduate students in physics and engineering, as well as those working in research and in industry'. A balanced treatment with detailed comparisons of operational parameters for some specific (if now out of date) commercial devices. Chapter 7 *Fast optical modulators* is particularly relevant to this material.

Yariv A 1991 *Optical Electronics* 4th edn (Fort Worth, TX: Harcourt Brace Jovanovich College).
A standard text aimed at those 'interested in the generation and manipulation of optical radiation'. Contains much of interest on lasers. Chapter 9, *Electro-Optic Modulation of Laser Beams*, is particularly relevant to the material here. A thorough treatment with examples.

Yariv A and Yeh P 1984 *Optical Waves in Crystals* (New York: Wiley).
Written 'to present a clear physical picture of the propagation of laser radiation in various optical media and to teach the reader how to analyze and design electro-optical devices'. Chapters 4 *Electromagnetic Propagation in Anisotropic Media*, 7 *Electro-Optics*, 8 *Electro-Optic Devices*, 9 *Acousto-Optics* and 10 *Acousto-Optic Devices* are particularly relevant to the material presented here. Well explained and thorough with examples and tables of material properties.

The LEOT Laser Tutorial (http://www.dewtronics.com/tutorials/lasers/leot/) is a good source of background material on many topics discussed here. Course 4 Module 7 *Electro-Optic and Acousto-Optic Devices* is particularly relevant.

There is a wealth of introductory and general background materials available online in the websites of suppliers of these forms of instrumentation. Anyone interested should examine the sites of all major suppliers (a few were noted in the text here) in order to get up-to-date information on the newer commercially available devices.

14
Adaptive Optics and Phase Conjugate Reflectors

Michael J. Damzen and Carl Paterson

CONTENTS

14.1 Adaptive Mirrors 227
14.2 Wavefront Sensors, Reconstruction and Control 229
14.3 Non-linear Optical Phase Conjugation 229
    14.3.1 Four-wave Mixing 230
    14.3.2 Stimulated Brillouin Scattering 231
    14.3.3 Photorefraction 231
    14.3.4 Self-intersecting Loop Conjugators 232
References 232
Further Reading 232

Adaptive optics techniques, originally developed in the context of ground-based astronomy to compensate dynamically for optical aberrations arising from turbulence in the Earth's atmosphere, have found application in a number of aspects of laser system design. Intra-cavity adaptive optics allows fine-tuning of the laser cavity, for example, to compensate for thermal lensing effects and drift or to optimize efficiency automatically as operating parameters change. Extra-cavity adaptive optics systems are used widely in beam-shaping, beam-steering and temporal pulse-shaping applications.

At the core of an adaptive optical system is an active element such as a mechanically deformable mirror having a number of externally controllable actuators that are used to alter the shape of its optical surface. An error signal derived from elsewhere in the system is used to control this adaptive mirror in a closed feedback loop. The error signal, which may in fact consist of a number of different parallel measurements in spatially complex systems, can be obtained from a wavefront sensor or by monitoring some other property of the laser output such as power.

Phase-conjugate reflectors differ significantly from conventional adaptive mirrors in that they do not have externally controllable actuators. Therefore, requiring no external error signal, their practical use in adaptive laser systems is markedly different from conventional adaptive mirrors and will be treated separately.

14.1 Adaptive Mirrors

A key component to an adaptive optical system is the adaptive mirror, for which a number of different technologies are available. The suitability of a given type of mirror for the application is determined primarily by the spatial properties of the deformations that it can produce and by its temporal response.

To understand the basic principles involved, consider first one of the simplest types of adaptive mirror, the continuous faceplate mirror. This consists of a single plate with piston actuators (such as piezoelectric stacks) attached at regularly spaced points across the back surface. Using the actuators, the front surface of the mirror can be deformed in a controllable fashion (Figure 14.1a). Each actuator produces its own characteristic deformation to the mirror, referred to as an influence function, the amount of the deformation depending on the value of the control signal given to that actuator. Treating the mirror as a linear system, which, given the small deformations involved, is usually a very good assumption, for a mirror with N_A actuators, the total deformation of the mirror surface $\Phi_M(x,y)$ is given by summing the deformations due to each actuator,

$$\Phi_M(x,y) = \sum_{j=1}^{N_A} r_j(x,y) c_j \quad (14.1)$$

where $r_j(x, y)$ is the influence function of the jth actuator (with unit control signal) and c_j is the actual value of the control signal applied to that actuator. The influence functions determine how well the mirror can compensate for wavefront errors. Their shape depends on a number of factors including the plate mechanics, the actuator positioning geometry and the flexibility of the actuator-to-mirror mountings. The type of actuator technology also affects the influence functions: piezoelectric stacks control the vertical displacement, whereas voice-coil-type actuators apply a controlled force. The displacement is then determined by boundary conditions and plate mechanics. The influence functions overlap and, in general, will not be orthogonal; however, they are normally linearly independent

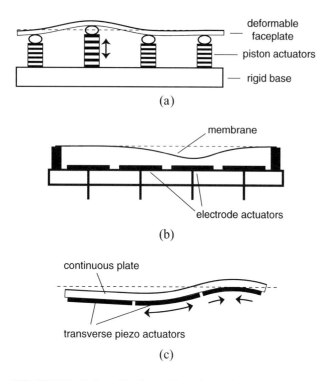

FIGURE 14.1 Deformable mirrors: (a) continuous faceplate; (b) electrostatic membrane and (c) bimorph.

and an orthogonal set of N_A mirror modes can be constructed from them. Expanding in terms of a suitable orthonormal basis, such as Zernike polynomials, $Z_j(x, y)$, the deformation can be written in matrix form as

$$\Phi_M = Mc. \qquad (14.2)$$

where M is called the influence matrix of the mirror and c is the control vector. Least-squares fitting can be used to calculate the control signals that should be applied to fit the mirror to the required shape,

$$c = M^\dagger \Phi_0 = [MM^T]^{-1} M^T \Phi_0 \qquad (14.3)$$

where $M^\dagger$ is the least-squares inverse of M. The difference between the least squares best-fit and the required deformation gives the fitting error, which is used to assess the suitability of a given mirror. For the continuous faceplate mirror with piezoelectric stack actuators, the influence functions are smooth and localized near each actuator. The number of actuators needed depends approximately on the space–bandwidth product of the deformations to be produced.

Also important are the dynamic range of the deformations, referred to as the stroke, and the temporal response. For the faceplate mirror, typical values are several micrometres for the stroke and around 1 kHz for the temporal response (limited by mechanical resonance of the device).

Other types of continuous devices include membrane and bimorph mirrors, which are both curvature-type mirrors. Electrostatic membrane mirrors consist of a thin membrane (e.g. silicon nitride) held under tension above an array of electrode pads (Figure 14.1b) [1,2]. Applying a voltage between the membrane and one of the electrode pads sets up an electrostatic attractive pressure, deforming the membrane. The deformations obey the membrane equation

$$\nabla^2 z(x,y) = -P(x,y)/T = -\varepsilon_0 V^2 / Td^2 \qquad (14.4)$$

where z is the deformation, T is the membrane tension, P is the electrostatic pressure due to the voltage V across the gap d between the membrane and the electrode, with the boundary condition of $z(x, y)=0$ at the fixed edges of the membrane. The influence functions are not localized at the actuators but extend across the whole of the mirror surface. As can be seen from equation (14.4), each actuator controls the local curvature of the mirror above that actuator, for which reason these devices are referred to as curvature mirrors. A consequence of this behaviour is that the stroke is strongly dependent on the spatial frequency of the deformation being produced, and the amplitude of the deformation varies as the inverse square of its spatial frequency.

Bimorph mirrors consist of a single plate with piezoelectric actuators bonded onto the back surface such that they can expand or contract laterally causing the plate to bend (Figure 14.1c) [3]. More complex sandwiching designs can be used to improve the stroke or make the mirror less susceptible to differential thermal expansion effects. As for the membrane device, each actuator alters the local curvature of the mirror so the influence functions are similar in nature to those of the membrane mirror. Bimorph mirrors with water-cooling channels have been used for high-power laser correction.

An alternative to the continuous mirrors are segmented devices. These use an array of separate rigid mirrors, each with its own actuators, which can be used to adjust the height (piston only) or both height and gradient (piston+tilt) of each mirror segment independently. The influence functions of this type of mirror are localized to individual segments and do not overlap. With this type of device, it is possible to introduce discontinuities in the mirror surface at the segment boundaries. This may be undesirable in many applications, requiring constraints to be placed on the mirror operation so that the edges of adjacent segments remain in phase, reducing the number of effective degrees of freedom of the mirror.

A number of different MEMS[1]-type mirrors are available, both segmented and monolithic. Although their actuator technologies, which include electrostatic, electrostatic comb and thermal bimorph, may differ from those already described, they are similar in operation, having similar types of influence functions. By their nature, MEMS devices are often considerably smaller, which may lead to more stringent limits to their optical power handling. Liquid crystal devices can be used as transmissive wavefront correctors, operating by changing the birefringent properties of a liquid crystal cell electrically. Pixellated devices can be modelled as analogous to segmented piston-only mirrors—the poor fitting error achievable with piston-only correctors being compensated by the large number of pixels available in liquid crystal devices. The so-called 'modal' devices which have more smoothly tapering influence functions are also available. The drawbacks of liquid crystal devices include the strong polarization dependence of the device and absorption.

---

[1] Micro-electromechanical system.

14.2 Wavefront Sensors, Reconstruction and Control

To control the deformable mirror in a closed-loop system requires an error signal, which is the purpose of the wavefront sensor. The basic principle will be illustrated with a commonly used wavefront sensor, the Shack–Hartmann (Figure 14.2). This consists of a lenslet array placed in the beam to be measured. Each lenslet produces its own focal spot, which for an on-axis collimated beam will lie on the optical axis of the lenslet. If, however, the beam is not on-axis and collimated, the position of each focus will be shifted laterally by an amount that depends on the local tilt of the wavefront entering the corresponding lenslet. By measuring the transverse displacement of the foci, one can obtain samples of the local gradient of the wavefront at each lenslet. Modelling the wavefront sensor as a linear system, the response of the sensor to an arbitrary input wavefront ϕ_0 is given by

$$s = S\phi_0. \qquad (14.5)$$

S is the wavefront sensor response matrix, the elements of which are calculated from the geometry of the sensor [4]. An estimate of the original wavefront can then be reconstructed from these samples using, for example, a least-squares inverse of S or optimal estimation [5]. In practice, it is possible to dispense with explicit wavefront reconstruction and estimate the required mirror control signals c directly from the sensor output,

$$c = [SM]^\dagger s \qquad (14.6)$$

For a closed-loop system, this has the advantage that the matrix product SM can be measured directly by varying each mirror actuator control in turn and observing the effect on the wavefront sensor output [6]. Since the wavefront sensor can only sense a finite number of wavefront modes, which for a Shack–Hartmann is twice the number of lenslets, the wavefront cannot be reconstructed completely from the sensor measurements. It is important, therefore, to choose the wavefront sensor geometry with regard to the mirror geometry to ensure that the deformations of the mirror are measurable [7].

There are many alternative wavefront sensor techniques, including wavefront curvature sensing, knife-edge detectors and interferometric techniques: the reader is referred to Hardy or Tyson in further reading for an overview. For some applications, it is possible to dispense with the wavefront sensor. Direct optimization methods rely on monitoring some property of the output (usually the on-axis intensity) while continually adjusting the mirror actuator controls. If a change makes the system worse, the direction of subsequent changes is altered. Optimization methods can work well for systems with small numbers of degrees of freedom; however, they tend to be significantly slower for more spatially complex cases, where they tend to require many iterations to converge. Modulating or dithering the corrective element at a higher frequency than that of the aberrations combined with synchronous detection of the output can be used to lock the system onto a maximum. Multi-dither approaches, where each actuator is modulated at a different frequency, have been used successfully for laser beam correction through turbulence. The number of channels is limited by the finite bandwidth of the mirror and the necessary separation of the dither frequencies.

14.3 Non-linear Optical Phase Conjugation

Non-linear optical phase conjugation was first demonstrated in the early 1970s [8]. In its most common form, it is a technique that uses non-linear optical processes to reverse precisely both the direction of propagation and phase variation of a coherent light wave. Phase conjugation produces true wavefront conjugation in that it preserves both the amplitude distribution while conjugating the phase distribution. It has the advantage over conventional adaptive optics, which relies on discrete active segmented control elements, of being a passive continuous wavefront reversal mechanism and without the requirement for mechanical, magnetic or electro-optical components or software processing. It, therefore, has a potential advantage in terms of resolution, speed, cost, weight and energy efficiency. There are, however, requirements on the input radiation, such as power and temporal coherence for specific non-linear phase-conjugation techniques that limit its use for certain applications (e.g. astronomy) but it can be used to advantage in laser applications.

The non-linear device that 'reflects' the input beam to produce the phase conjugate is often referred to as the phase-conjugate mirror. The non-linear mirror is a real-time processor of the electromagnetic optical field and can respond to the complex amplitude distribution and phase front of the incident field and adapt its reflective properties. Much of the fascination of this technique arises from its ability to restore an aberrated optical beam or image-bearing wavefront to its original, undistorted state (Figure 14.3). A practical example of this is the correction of phase aberrations in a laser amplifier. A high spatial quality (fundamental Gaussian mode) of a laser oscillator is degraded on a single pass through the amplifier by aberrations (e.g. thermally induced refractive index distortions in a solid-state laser rod due to intense pumping by the inversion mechanism). The double pass with a phase-conjugate mirror reflection recovers the high spatial quality of the incident beam as well as receiving a double-pass amplification in power by the amplifier. A polarization combination (polarizer and quarter-wave retardation plate or Faraday rotator) can

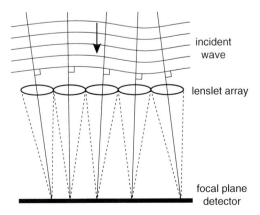

FIGURE 14.2 Shack–Hartmann wavefront sensor.

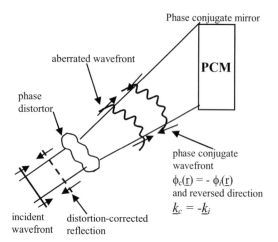

FIGURE 14.3 Phase conjugation as a means for distortion correction.

act as a simple means of extracting the amplified radiation and isolating the low-power laser oscillator. This was the first application used for phase conjugation [9], and the phase-conjugate mirror was based on the process of stimulated Brillouin scattering (SBS). For an introduction to non-linear optics.

A surprisingly large number of physical processes in non-linear optics have the ability to generate a phase conjugate signal [10–15]. The main mechanisms, illustrated in Figure 14.4, include:

i. four-wave mixing (FWM),
ii. stimulated scattering (especially SBS),
iii. photorefraction and
iv. self-intersecting loop schemes.

The two most generic phase-conjugating processes are FWM and SBS. FWM is a real-time holographic process that has the attraction that the beam to be conjugated can have low input power while SBS normally requires a high-power input and is especially attractive for high-power pulsed laser applications. FWM is the most versatile of the effects and also plays a major role in the phase-conjugation processes that occur in photorefraction and in self-intersecting loop schemes [10,12]. Other mechanisms also exist but have had more limited applicability. Three-wave mixing in a material with a second-order non-linear susceptibility has been shown, in principle, to produce phase conjugation but has a very limited phase-matching bandwidth, and this makes it impractical for beams with strong aberrations or divergent wavefronts. Another technique for phase conjugation is surface waves generated at the interface of a material from which the light reflects [11]. In this case, it is the counter-propagating surface waves excited by the incident light that produce the phase conjugation.

14.3.1 Four-wave Mixing [11,13–15]

The geometry of the FWM process is shown in Figure 14.4a, where three waves (E_1, E_2 and E_3) are incident onto a material with a third-order non-linear susceptibility $\chi^{(3)}$. Different nomenclature is used in the scientific literature for the numbering of the fields. Here, E_1 and E_2 are counter-propagating and

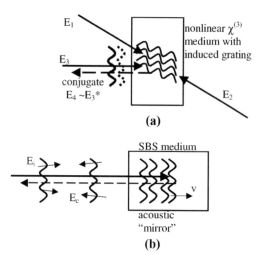

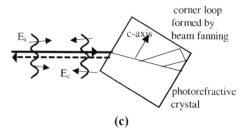

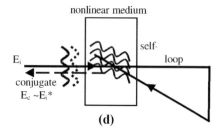

FIGURE 14.4 Non-linear phase conjugate reflectors include (a) FWM; (b) SBS; (c) photorefraction in corner-pumped geometry and (d) self-intersecting loop conjugators.

described as the pump fields and E_3 is the signal field to be phase-conjugated. In the non-linear medium, all three waves are coupled in such a manner that a new optical field, the conjugate wave E_4, is generated with a magnitude given by

$$E_4(\mathbf{r}) = \left[i\frac{\mu_0 c \omega}{2n} \chi^{(3)} L E_1 E_2 \right] E_3^*(\mathbf{r}) = r_C E_3^*(\mathbf{r}) \quad (14.7)$$

where $\mathbf{r}$ is the transverse spatial coordinate, L is the interaction length, $\chi^{(3)}$ is the third-order non-linear susceptibility of the medium, μ_0 is the permeability of free space, c is the speed of light in vacuum, n is the refractive index and ω is the angular frequency of the optical light. The term r_C is an effective amplitude reflectivity of the phase-conjugate mirror and is dependent on the material parameters, length of the interaction region and strength of pump fields E_1 and E_2.

If the pump fields are constant, e.g. uniform plane-waves, the generated field E_4 is seen to be the complex conjugate of the field E_3 with the phase conjugate wavefront. It is possible to control the conjugate reflectivity $R_C = |r_C|^2$ by varying the intensity of the pump beams and it is even possible to achieve amplified reflectivity $R_C > 1$, since the energy of the conjugate beam is derived from the pump beams. Various types of $\chi^{(3)}$ non-linearities can be exploited by the use of different materials and physical processes. The highest reflectivities produced to date have been greater than one million (10^6) using Brillouin enhanced FWM with pulsed radiation. The highest reflectivity for continuous-wave radiation has been in saturable gain FWM in a laser amplifier medium, where the radiation is resonant with the laser transition of the medium. High reflectivity can be used as a means of optical image amplification or for boosting weak signals. Many of the FWM processes are non-resonant and provide broad tunability in wavelength often covering the total visible spectrum and even near-UV and IR.

FWM can be thought of as 'real-time' holography—in FWM, the writing and reading processes can be virtually simultaneous and hence dynamic (e.g. atmospheric turbulence in communications) as well as static aberrations (e.g. in passive optical systems) can be compensated. In the case of transient processes, the response time may be a function of intensity of the writing beams, and in the case of photorefraction, the response time is inversely proportional to intensity.

In its fullest form, FWM can be considered as an optical multiplication since the conjugate beam is proportional to the product of the three input fields $E_4 \propto E_1 E_2 E_3^*$. An all-optical processor can be implemented to produce convolution and correlation of images using the Fourier-transforming property of lenses and the multiplicative property of FWM.

Exploitation can even be made of the finite response time of the phase conjugator to produce time-averaged interferometry effects and novelty filters. The novelty filter is effectively a movement detector of speeds faster than the response speed of the phase-conjugating device. In its basic form, it is a type of Michelson interferometer with phase-conjugate mirrors in each of the beam-splitter arms. For a static input image, perfect cancellation (destructive interference) results when the two conjugate beams recombine, resulting in zero signal at the output arm. Any movement in the image can be made visible at the output arm if the two conjugators have different speeds. Such a device has been considered for application in surveillance or robotic inspection. Since only the moving object is detected in a large scene, only a small and convenient amount of signal processing is required.

14.3.2 Stimulated Brillouin Scattering [11,8,9,13–15]

SBS is an alternative method of producing phase conjugation resulting from inelastic non-linear scattering of high-power radiation from coherent acoustic waves generated by electrostriction in the medium due to the intense optical field (Figure 14.4b). A backscattered field is produced which, under many experimental conditions, is found to be the phase conjugate of the incident field. Analysis indicates that when the backscattered field is the phase conjugate of the incident laser field, the growth experienced by the scattered field is greater than any other backscattered field configuration. Physically, one can visualize the acoustic wavefront forming a shape that reproduces the wavefront of the incident light wave. This concept of a flexible mirror has direct analogy to conventional adaptive optics but forming by passive non-linear optical means. The attraction of SBS compared to FWM is its self-conjugating property, requiring no high-quality additional pump beams.

Typical interaction geometries include focusing into the bulk of a cell containing the Brillouin medium or by launching into a hollow light-pipe filled with a Brillouin-active liquid (e.g. CS_2, n-hexane, acetone) or gas (e.g. CH_4, SF_6). Solids have been less frequently used, since their damage thresholds tend to be comparable to the thresholds for efficient SBS. Later, there has been renewed interest in solid materials in the form of multi-mode optical fibre whose long interaction length allows reduced power thresholds for SBS as low as ~1 W. In bulk geometries, very high intensities are required for SBS, and this means that it is most suitable for phase conjugation of high-power pulsed laser radiation. A major motivation of SBS phase conjugation has been directed towards the improvement of beam quality in high-power (solid-state, excimer and dye) laser systems. At high-energy and/or high-average-power operation, the laser gain medium becomes increasingly distorted by thermally induced refractive index distortions and lensing in solid-state and dye lasers and by discharge instabilities in gas excimer lasers. Phase-conjugation is a means to correct for these static and dynamic aberrations.

Other forms of non-linear inelastic scattering can also lead to phase-conjugation with the phase conjugate wave being selected by a higher growth rate than non-conjugate forms of scattering. The frequency shift of the SBS process determined by the acoustic frequency (acoustic phonon) is typically ~0.1–10 GHz; however, the shift in backward stimulated Raman scattering process due to the vibrational transition of the molecular system ('the optical phonon') can be as high as 10% of the optical frequency. The selection of the phase conjugate can be impacted by the difference in diffraction at the shifted frequency leading to an error in correction of aberrations when double-passing the distorting system.

14.3.3 Photorefraction [12]

The photorefractive effect is a change in the refractive index due to the optical transfer of charge in the medium leading to change in refractive index. The effect was first observed in lithium niobate ($LiNbO_3$) and was a detrimental effect due to its change in the propagation properties of the light. The basic mechanism is due to the photoexcitation of electrons (or holes) from donor (or acceptor) sites to the conduction (or valence) band, where they are free to migrate by drift or diffusion before recombining into an empty trap site. The dominant rate of charge photoexcitation is in the bright intensity regions with preferential trapping in dark regions. Hence, with an intensity interference pattern, the charge is redistributed to dark fringes. The imbalance of charge leads to a space-charge electric field E_{SC} that modulates the refractive index Δn through the electro-optic (Pockels) effect $\Delta n = 1/2 n^3 r_{eff} E_{SC}$, where n is the refractive index and r_{eff} is the effective electro-optic coefficient. Hence, good choices of photorefractive media exhibit high photoconductivity and have high electro-optic coefficient; examples include barium titanate ($BaTiO_3$), bismuth silicon oxide (BSO) and lithium niobate ($LiNbO_3$). Later, interest has been centred on photorefractive polymer materials where the composite polymer material can be engineered to have particular donor and trapping species, but they tend to require high externally applied electric fields and have high absorption.

An interesting effect known as two-beam coupling occurs when two beams intersect in the photorefractive medium. The non-local displacement of the charge induces a refractive index grating that is $\pi/2$ phase-shifted from the light interference pattern (assuming diffusion of charge). This leads to a self-diffraction where one beam is amplified at the expense of the other beam. The direction of energy transfer depends on the sign of the electro-optic coefficient, the orientation of the crystal (that needs to be poled) and the sign of the charge carriers. In highly non-linear crystals such as barium titanate, the gain of one beam can be by orders of magnitude and, indeed, so high that internal linear scattered light in the crystal can be strongly amplified, a phenomenon known as beam fanning, leading to depletion of the incident light. A new type of phase-conjugate mirror has been demonstrated in this case where the beam fanning is directed to a corner of the crystal acting as a retroreflector (Figure 14.4c). A loop of light is observed to evolve, and it acts as a counter-propagating pump beam to phase conjugate the input light beam. A phase-conjugate output with reflectivity of 60% or more has been demonstrated. The exact mechanism of the conjugation is rather complex but the internal fanning and beam development are self-organizing and have a preferential tendency to produce phase conjugation, which is the condition for highest spatial overlap and mutual coherence of beams. Like SBS, this process is self-conjugating without the requirement of external pump beams (as in FWM). The light required for phase conjugation in photorefractive crystals can be typically at the milliwatt level (or even lower), but the response time of the effect can be very slow. For barium titanate, the response is ~1 s at 1 W cm^{-2} intensity but speed is faster with higher intensity. For other crystals, e.g. BSO, the response is faster (~1 ms) but the non-linearity and phase conjugation effects are weaker.

14.3.4 Self-intersecting Loop Conjugators [10]

Any mechanism that can produce phase conjugation by FWM has the potential to produce a self-intersecting loop conjugator without requiring external pump beams. This geometry was first operated in photorefractive media where the incident beam, after transmission through the crystal, is re-injected to overlap with itself in the crystal (Figure 14.4d). In this case, it is the two-beam coupling initiated by beam fanning that causes a build-up of a backward phase-conjugate wave. Reflectivities of many tens of per cent can be achieved, and the threshold power is generally lower than that required for the corner-pumped phase conjugation. Later, a number of demonstrations of very high reflectivity loop conjugators have been made with saturable gain as the non-linear mechanism. Indeed, the self-intersecting input beam can induce a gain grating with sufficient diffraction efficiency to induce a ring laser oscillation in which the lowest-order backward spatial mode is the phase conjugate of the input. Reflectivities approaching 10^5 have been produced by this technique.

REFERENCES

1. Grosso R P and Yellin M 1977 The membrane mirror as an adaptive optical element *J. Opt. Soc. Am.* **67** 399–406.
2. Vdovin G and Sarro P M 1995 Flexible mirror micromachined in silicon *Appl. Opt.* **34** 2968–72.
3. Steinhaus E and Lipson S G 1979 Bimorph piezoelectric flexible mirror *J. Opt. Soc. Am.* **69** 478–81.
4. Hardy J W 1998 *Adaptive Optics for Astronomical Telescopes* (Oxford: Oxford University Press).
5. Wallner E P 1983 Optimal wave-front correction using slope measurements *J. Opt. Soc. Am.* **73** 1771.
6. Paterson C, Munro I and Dainty J C 2000 A low cost adaptive optics system using a membrane mirror *Opt. Express* **6** 175–85.
7. Roddier F 1994 The problematic of adaptive optics design *Adaptive Optics for Astronomy* eds D M Alloin and J M Mariotti (Dordrecht: Kluwer Academic) pp 89–111.
8. Ya Zel'dovich B, Pilipetsky N F and Shkunov V V 1985 *Principles of Phase Conjugation* (Springer Optical Sciences 42) (Berlin: Springer).
9. Nosach D et al. 1972 Cancellation of phase distortions in an amplifying medium with a Brillouin mirror *Sov. Phys.–JETP* **16** 435.
10. Damzen M and Crofts G 2002 Wave-mixing and phase conjugation in laser active media *Progress in Photorefractive Nonlinear Optics* ed K Kuroda (London: Taylor & Francis) ch 7.
11. Fisher R A (ed) 1983 *Optical Phase Conjugation* (New York: Academic).
12. Gunter P and Huignard J P (ed) 1989 *Photorefractive Materials and Their Applications* vols I and II (Berlin: Springer).
13. Pepper D M 1981 Applications of optical phase conjugation *Sci. Am.* January 56.
14. Ya Zel'dovich B et al 1972 Connection between the wave fronts of the reflected and exciting light in stimulated Mandel'shtam–Brillouin scattering *Sov. Phys.–JETP* **15** 109.
15. Yariv A 1978 Phase conjugate optics and real-time holography *IEEE J. Quantum Electron.* **QE-14** 650.

FURTHER READING

Fisher R A (ed) 1983 *Optical Phase Conjugation* (New York: Academic).

Hardy J W 1998 *Adaptive Optics for Astronomical Telescopes* (Oxford: Oxford University Press).

A comprehensive and detailed account of adaptive optics technologies and methods: although written for astronomy, most of the techniques are directly applicable to lasers.

Love G D (ed) 2000 *Proceedings of the 2nd International Workshop on Adaptive Optics for Industry and Medicine* (Singapore: World Scientific).

Contains many examples of adaptive optics applied to lasers.

Tyson R K 1998 *Principles of Adaptive Optics* 2nd edn (New York: Academic).

An overview of available technologies with an extensive bibliography.

15
Opto-mechanical Parts

Frank Luecke

CONTENTS

15.1 Introduction 233
15.2 Requirements and Specifications 233
15.3 System Considerations 234
    15.3.1 Position Description 234
    15.3.2 Mounting Accuracy 234
    15.3.3 Mounting Techniques 235
    15.3.4 Optimization 235
    15.3.5 Design for Manufacturability 235
    15.3.6 Testing 236
15.4 Materials and Finishes 236
15.5 Parts Configuration 236
    15.5.1 Visualization 236
    15.5.2 Distortion, Stress and Strain 236
15.6 Precision Positioning 237
    15.6.1 Stages 237
    15.6.2 Actuators 237
    15.6.3 Servo-actuator Systems 237
15.7 Closure 238
References 238
Further Reading 238

15.1 Introduction

Art is the triumph over chaos (John Cheever)
The art of opto-mechanical design is to

- put optical surfaces and elements into their proper—not necessarily static—positions;
- keep them there;
- protect them and their surroundings from distortion, damage and stray light and, in some cases,
- quantify their positions.

Further, it is usually necessary to achieve all this within cost, weight, size, time and numerous other constraints.
To accomplish these objectives requires

- understanding the overall system requirements,
- selecting appropriate materials,
- determining the configuration of the parts,
- anticipating a process for building the product at each phase from feasibility to maximum production and
- testing to assure the objectives are met.

Ideally, the designer pursues a global optimum but reality usually forces compromise among oft-conflicting goals. The design process is seldom simple or linear; most often, it is highly iterative as well as interactive with other parts of the bigger picture.

The objective of this chapter is to present some thoughts, information and further resources for creating an opto-mechanical design effectively and efficiently.

15.2 Requirements and Specifications

A good design starts with understanding the functional requirements, environmental context and administrative details of the proposed system as completely and accurately as possible. Although it may seem obvious, the importance of fully understanding what a system will be required to do as one starts to design that system cannot be overemphasized.

Functional requirements:

- What is this device supposed to do?
- How well must it perform: speed, resolution, repeatability and accuracy?

- How long must it continue to operate: operational cycles, number of years?
- How big or small can or should it be?
- How loud or quiet can or should it be?
- What mass and inertia are allowable?
- In what ways are aesthetics important? (Consider all five senses!)
- Which attribute is best to optimize and which attributes are bounds?
- What is the history of the technology? What are the nearest examples presently in existence? In what ways would they be acceptable for the current project and in what ways not? Can you examine, test and dissect some examples?
- Who will use it? Install it? Service it? What skill level will apply? What operator and service interfaces are needed?

Environmental context:

- Where will it be used?
- What conditions of temperature, humidity, pressure/vacuum, vibration/shock, chemical, biological, particulates, ambient light, orientation, etc., must it withstand during shipment and during operation?
- Are there any EMI, RFI or electrical or magnetic fields considerations?
- Are there any other special materials requirements?
- What energy and supplies are available for input and desired or tolerable for output?
- Are there any safety or health considerations?

Administrative details:

- Are any regulations or certifications applicable to the device, e.g. laser safety, European Compliance (CE), Underwriters' Laboratories Inc. (UL)?
- What is the schedule for feasibility, development and production?
- How many units are likely to be needed and when?
- Who is responsible for the rest of the system: optics, electronics and manufacturing?
- What resources are available—funding, skills and facilities—for concept, prototype and production?

Once the functional requirements, environmental context and administrative details of the proposed system are identified and understood as completely and accurately as possible, the creation and refinement of the opto-mechanical design can proceed through the usual steps of layout, analysis, prototyping, testing and preparation for production.

15.3 System Considerations

15.3.1 Position Description

Many systems consist of an assembly of essentially rigid optical elements that need to be held in a particular spatial relationship with each other. The position of a rigid body with respect to an external frame of reference can be described by six degrees of freedom (DOFs): in a Cartesian coordinate system, these are the X, Y, Z locations of a given point on the element and the rotational orientation α_x, α_y, α_z of the element with respect to the X, Y and Z axes. The origin of the coordinate system, i.e. the external frame of reference, can be set at any convenient location; typically the Z-axis is coincident with the optical axis of a linear system. A folded optical path may be described by using an axis-aligned coordinate system for each section with coordinate transforms at each fold. An external ground reference may be used. A polar or spherical coordinate system may be more suitable for some systems.

The term 'kinematic mounting' refers to an approach that provides exactly one constraint for each DOF. This concept was clearly described by Clerk Maxwell in 1876 [1]. Figure 15.1 shows under-constrained, kinematic and over-constrained assemblies. The block, shown separated for clarity, is assumed to be uniformly urged against all the supporting constraints by some method not shown. Many other examples of kinematic mounting appear in the literature [2–4].

15.3.2 Mounting Accuracy

Each element in an optical system must be in its correct position within some tolerance for the system to function properly. Table 15.1 is an example of an error budget for a fictitious simple camera showing the mounting accuracy required for

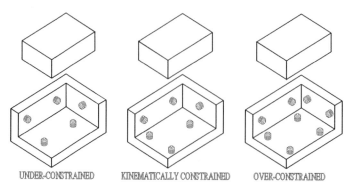

FIGURE 15.1 Under-constrained, kinematic and over-constrained assemblies of Kinematic mounting.

Opto-mechanical Parts

TABLE 15.1

Error Budget Example

	X (mm)	Y (mm)	Z (mm)	α_X (deg)	α_Y (deg)	α_Z (deg)
Lens	0.2	0.2	0.1	1	1	∞
Shutter	1	1	0.5	5	5	∞
Film	0.3	0.2	0.1	1	1	2

each optical element. Approximately, the same level of performance degradation would result from mis-positioning any element in any one direction by the amount shown. Many programs used to design lenses can supplement judgement and testing in quantifying the positioning tolerances. Tolerances for some DOFs may be unimportant (e.g. the rotation of a circularly symmetric lens about its optic axis), while others will be critical. The inter-relationship between a few elements in a sub-group may be more critical than the alignment of the group to the rest of the system; in such a case, it helps to add a separate error budget table for the sub-group. Usually, a system need not perform to specification with every element simultaneously at the worst extreme of misalignment; an error probability distribution may apply [4].

15.3.3 Mounting Techniques

The methods for placing and holding optical elements fall into three categories:

- deterministic, in which all DOFs of each optical element are adequately constrained by features on the opto-mechanical parts (this category includes innumerable configurations for mounting lenses, prisms, mirrors and other elements as shown in Refs. [3,56]);
- adjustable, if deterministic mounting cannot feasibly provide sufficient accuracy and finally,
- servo-actuated, in which elements are continuously and automatically positioned as needed. Such systems are described in more detail in Section 15.6.3.

The relationship between the cost and achievable accuracy of each of these approaches is shown in Figure 15.2. The cost of holding a tolerance increases as the tolerance is reduced. In some cases, it can be more economical to use a more complex approach, migrating from adjustment to servo (region A) or from deterministic mounting to adjustment (region B). However, including an adjustment or servo system in a product that does not require it (region C) would be a needless complication and a waste of money.

There are hundreds of approaches to holding lenses, mirrors, prisms, windows and other optical elements in place. For a laboratory or prototype system, devices available from suppliers such as those appearing in optical magazines and directories (e.g. *Photonics Directory*) may suffice. For more rigorous requirements and for systems to be produced in quantity, the references show many examples of specific opto-mechanical design elements [1,3,4,6,7]. Many more can be found by searching through patents and other literature. Dissecting available examples of similar devices can also be instructive.

15.3.4 Optimization

Optimization as a specific discipline is fundamental to the design of modern lens systems as well as many other devices. This discipline can also be used to advantage in certain aspects of opto-mechanical element design [5]. Although literal use of this tool may not be feasible when designing overall opto-mechanical systems due to the difficulty in creating and evaluating a mathematically tractable merit function, it still is helpful to think in optimization terms; e.g. 'Create a gizmo that is as lightweight as possible, while holding performance, cost, size and all other important variables within acceptable limits'.

15.3.5 Design for Manufacturability

A design has little value unless it is feasible to actually make and assemble the parts. The processes chosen are influenced

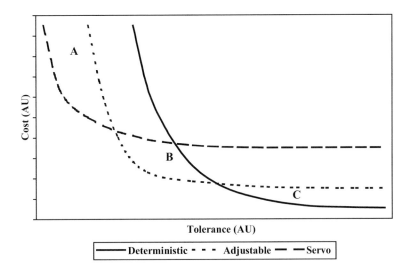

FIGURE 15.2 Cost of holding tolerances.

by the quantity and quality of each part needed. Basic model-shop equipment including manual mills and lathes may suffice for making one or two copies of relatively simple parts. Rapid prototyping can produce visually representative as well as in some cases completely functional complex parts [8]. Computer-numerically-controlled machining can economically produce small-to-medium quantities of parts, particularly those with complex but machinable shapes or large numbers of features. Production of high volumes must include consideration of other methods such as moulding to final or near-finished condition. The references cited in Section 15.5 describing the characteristics of a variety of materials also describe many parts-forming processes applicable to different materials.

The cost of producing products varies approximately with the negative root of the quantity to be produced: e.g. per-part cost may decline about tenfold when quantities increase 100-fold. This rule of thumb is far from exact or continuous, however, and the discontinuity seems greater in opto-mechanical devices than in many other fields: many optical products resemble prototypes even in volume production.

The considerations for part production apply to assembly and alignment as well. Although it may be appropriate for a PhD to spend hour after hour putting together a first article and coaxing it to life, this probably is not the best way to produce the 10 000th or even the 100th unit. A good design will eliminate as many assembly and alignment steps as possible and will make the remaining steps as easy as possible for a modestly skilled operator, robot or specialized machine to perform correctly.

When adjustments must be performed during assembly, it is better to have excellent manufacturing equipment perform the adjustment on the bench rather than building adjustment devices into each product.

15.3.6 Testing

It may be tempting to minimize development costs by omitting rigorous testing, letting customers discover if the product will work satisfactorily for a reasonable lifetime under the conditions to which it may be subjected. This could be acceptable for very limited production quantities as long as customers knowingly accept the risk. In the long run, though, it is better to verify that the product continues to meet its performance objectives, as outlined in Section 15.2, when subjected to the various environmental insults also listed in that section. For example, a pair of binoculars might need to maintain alignment, throughput and modulation transfer function after being dropped, sprayed with salt water, frozen, irradiated and vibrated. Of course, testing *must* be done in those cases requiring certification by regulatory agencies or industry standards organizations. Numerous suppliers can conduct the needed tests if facilities are not available in-house.

15.4 Materials and Finishes

Although black anodized 6061-T6 aluminium may be the most common material and finish now used to build opto-mechanical parts, innumerable other candidates are worthy of consideration [9,10]. The best choice depends on the combination of functional and aesthetic properties required. One very useful approach is to rank the candidate materials using a figure of merit based on the most important functional properties for the system [7]. One example of such a figure of merit is the stiffness-to-density ratio, which can affect the weight and performance of a simple system as well as the bandwidth of a complex, servo-actuated system because of the dependence of the acoustic propagation velocity, as mentioned in Section 15.6. The most appropriate figure of merit will depend on the specific system requirements and may involve a combination of thermal coefficient of expansion, electrical conductivity, chemical resistance and other factors. The materials palette includes many types of aluminium [11]; other non-ferrous as well as ferrous metals; plastics, elastomers and composites [12] and structural ceramics [13]. Similarly, many different finishes including paint, powder coating, anodizing, passivation and plating are available to meet the functional as well as aesthetic requirements of the product.

15.5 Parts Configuration

15.5.1 Visualization

Since the basic objective of opto-mechanical design is to hold optical elements in their proper positions, a good starting point for creating opto-mechanical parts is to first imagine the complete set of optical elements suspended in space, free of any constraints other than the mutual spatial relationship among the elements (and sometimes between the collection of elements and ground) required for acceptable performance. Ignore for a while the bits and pieces of hardware needed to hold the elements and just picture where the elements need to be. Once that image is clear, add to it the voids and barriers needed to accommodate the light paths. Only then, after contemplating this picture from all angles, use the remaining space as needed for parts to hold and move the optical elements.

A number of companies produce 3D mechanical computer-aided design (CAD) programs that can help in visualization and the more mundane processes of laying out, designing and documenting the mechanical parts. Each of the programs has some merits as well as some shortcomings. Unless the software already standard at your institution restricts the choice of CAD system, there is a daunting array of programs and features to choose from. There may be add-ons and interfaces to ease the process of combining mechanical and optical elements. Use caution in employing the ray-tracing functions of many general-purpose mechanical design programs, because they may be more oriented towards presenting a pleasing image than accurately portraying optical system performance.

15.5.2 Distortion, Stress and Strain

Nothing is perfectly rigid; all real objects are elastic. A force applied to an element will distort both the element providing the force and the one receiving it. Forces will be exerted by gravity, acceleration, vibration, mounting devices and thermal effects, among other agents.

A force applied to any optical element will distort its optical surfaces and will also change the bulk properties of transmissive elements. One example of surface effect is the gravity-induced change in figure of the primary mirror in a large telescope as it moves to view different parts of the sky; this also exemplifies situations in which the structural properties of the optical element itself must be considered. Volume effects include the changes in birefringence of a plastic block subjected to non-uniform external forces, a phenomenon exploited for visualizing stresses in models of structures.

The mechanical elements that are used to hold optics also distort under stress, moving the optical parts attached to them. Such movement must be considered in meeting the overall error budget.

Stress beyond the elastic limit will result in permanent deformation or breakage. It is best to avoid that level of stress in the parts as well as in the personnel on the project. The challenge in creating an acceptable design from a distortion viewpoint, therefore, lies in keeping the stresses and resultant strains within acceptable limits for both the optical and mechanical elements. For simple systems, basic static stress–strain analysis may be more than adequate. For more complex systems or situations, finite-element analysis (FEA) [4,5] can be useful for predicting the effects of applied loads, showing the magnitude of stress at various parts of the element and quantifying the resultant deflections. A skilled practitioner using FEA can help move a design towards optimal or acceptable much faster than would be possible by a 'cut and try' approach alone.

Most opto-mechanical systems will need to function properly over a range of temperatures. Most optical elements change size and shape with changes in temperature. The refractive index of most transmissive optical elements also changes with temperature. Both the sign and magnitude of these changes are dependent on the optical material. Most materials used to support optics expand with increasing temperature. The challenge then is to create a combination of materials and configuration in the support system that closely counteracts the change in optical properties. Many examples exist [5].

Vibration can also perturb an optical system. A system can be partially isolated from externally generated vibrations [14], and modifying the structural stiffness and damping within the system can make it less susceptible to the effects of vibrations from which it cannot readily be isolated. Since a system's natural frequency increases with increasing stiffness and decreasing mass, the figure of merit of stiffness-to-density ratio is applicable, although not necessarily in a simple way.

15.6 Precision Positioning

All optical elements of a system must be positioned to some level of precision. In many cases, deterministic mounting or one-time adjustments provide sufficient accuracy to meet the system specifications as outlined earlier. When they do not or parts of the system need to track moving external objects, a precision positioning mechanism must provide for motion along one or more of the six DOFs.

15.6.1 Stages

Stages nominally constrain motion to follow only the desired path(s). Several companies offer linear and rotary stages for laboratory and prototype applications, and a variety of approaches are built into end products. A well-designed device will allow motion only in the direction(s) of interest while maintaining adequate constraint in the remaining DOFs. A multi-direction-of-motion system ideally will have its coordinate system placed and aligned to minimize the interaction between the functions addressed by motion along each axis; i.e. its axes will be functionally orthogonal.

15.6.2 Actuators

Actuators are the devices that cause motion to occur. They are, in essence, variable links that replace one or more of the six fixed links in a kinematic mounting. Manually operated actuators include the ubiquitous micrometer-type devices. Powered actuator types include electrostatic, electrostrictive, magnetostrictive, piezoelectric, electret, hydraulic, pneumatic and electromagnetic. Electromagnetic actuators can be further subdivided by the magnet, coil and/or iron being the moving element. Each type of actuator has characteristics of moving force, holding force, resolution, range, cost, speed, size and power requirements. The choice of actuator will depend on matching its characteristics with the corresponding set of system requirements [5].

15.6.3 Servo-actuator Systems

Servo-actuator systems combine a sensor to detect the difference between an actual and desired position, a processor to interpret these data, an amplifier and an actuator to move parts to reduce the position error. In addition to the limits to performance imposed by the choice of actuator type, several other factors further limit high-speed servo system performance. The acoustic propagation velocity through the structure between the actuator and the payload imposes a fundamental limit to bandwidth. Simply put, the payload will not begin to move until some finite time after the actuator is stimulated. In an ideal system having a nominally solid bar connecting these two elements, this limit is proportional to $(KM^{-1})^{0.5}$, (where K = stiffness and M = mass). In a real system, this limit is usually analytically complex and is established by the combination of material properties and structural configuration. In addition to this fundamental limit, unintentional mechanical resonances can cause the servo loop to oscillate like the squeal in microphone-loudspeaker systems. FEA can help evolve a high-performance system design by predicting structural stiffness and resonant modes. Delays and inaccuracies in position error signal generation, signal processing and amplification may further limit system performance [15].

15.7 Closure

This chapter has touched on the major considerations involved in creating a good opto-mechanical design, emphasizing aspects rarely mentioned in other resources. Remember: it takes a good mechanical system to put optics in their place and keep them there!

REFERENCES

1. Evans C 1989 *Precision Engineering: An Evolutionary View* (Bedford: Cranfield).
2. Blanding D L 1999 *Exact Constraint: Machine Design Using Kinematic Principles* (New York: ASME).
3. Kamm L J 1993 *Designing Cost-Efficient Mechanisms* (Warrendale, PA: SAE).
4. Rothbart H A (ed) 1996 *Mechanical Design Handbook* 23rd edn (New York: McGraw-Hill).
5. Ahmad A (Editor-in-Chief) 1997 *Handbook of Optomechanical Engineering* (Boca Raton, FL: Chemical Rubber Company).
6. Smith W J 1966 *Modern Optical Engineering: The Design of Optical Systems* (New York: McGraw-Hill).
7. Smith S T and Chetwynd D G 1992 *Foundations of Ultraprecision Mechanism Design* (Amsterdam: Overseas Publishers Association).
8. Binstock L (ed) 1994 *Rapid Prototyping Systems: Fast Track to Product Realization* (Dearborn, MI: Society of Manufacturing Engineers).
9. Brady G S and Clauser H R 1991 *Materials Handbook* (New York: McGraw-Hill).
10. Lewis G 1990 *Selection of Engineering Materials* (Englewood Cliffs, NJ: Prentice-Hall); Lewis G 1990 *Photonics Directory* (Pittsfield, MA: Laurin Publishing).
11. Davis J R (ed) 1993 *Aluminum and Aluminum Alloys* (Materials Park, OH: ASM International).
12. Harper C A (Editor-in-Chief) 1996 *Handbook of Plastics, Elastomers and Composites* 3rd edn (New York: McGraw-Hill).
13. Schwartz M M 1992 *Handbook of Structural Ceramics* (New York: McGraw-Hill).
14. Slocum A H 1992 *Precision Machine Design* (Englewood Cliffs, NJ: Prentice-Hall).
15. Chang D K 1959 *Analysis of Linear Systems* (Reading, MA: Addison-Wesley).

FURTHER READING

Bouwhuis G, Braat J, Huijser A, Pasman J, van Rosmalen G and Schouhamer Immink K 1985 *Principles of Optical Disc Systems* (Bristol: Adam Hilger).

Includes an extensive section on the theory, considerations, approaches and limitations of servo-control mechanisms.

Goldberg N 1992 *Camera Technology: The Dark Side of the Lens* (San Diego, CA: Academic Press).

Presents a good example of the myriad details and design approaches to the multitude of functions incorporated into an everyday opto-mechanical device.

Moore W R 1970 *Foundations of Mechanical Accuracy* (Bridgeport: The Moore Special Tool Company).

Explains the great lengths taken in making and measuring the extremes of flat, straight, and round.

Oberg E, Jones F D, Horton H L and Ryffel H H 1988 *Machinery's Handbook* (New York: Industrial Press).

Includes extensive details of basic machine elements including screw threads.

Reiss R S 1992 *Instrument Design—A Collection of Columns and Features from OE Reports* (SPIE)

Describes the workings of a wide variety of familiar as well as not so familiar devices and designs.

16

Optical Pulse Generation: Section Introduction

Clive Ireland

The most commonly used categorization of laser output is into continuous wave (cw) or pulsed operation. This comes from the earliest days of laser research investigating the practical possibility of operating three- and four-level systems [1,2]. Subsequently, experiment confirmed that generating sufficient inversion for lasing in three-level systems requires high peak power pulse excitation. However, this is not to say that the majority of pulsed lasers in use today are based on three-level systems. Although there are more than 15 000 laser transitions reported in a broad range of media (the majority only providing weak output under powerful pulsed excitation), even for those that can be operated cw, the great majority finding practical use are employed with modulation or pulsed control of the beam [3].

There can be many application advantages of lasers producing pulsed output and, indeed, many that just would not work with a cw laser beam. In particular, pulsed operation allows a whole range of applications, including basic research, requiring radiation of high (or very high) peak power that is both spatially and temporally very well defined.

Today, pulsed lasers cover sources producing beams from quasi-cw through to the femtosecond and attosecond regimes. Here, for convenience, they are considered in three time domains; i.e. quasi-cw and modulated beams, beams comprising short pulses and beams of ultrashort pulses. Although this division might initially seem somewhat arbitrary, it has the merit of corresponding reasonably well to that of the different technologies employed to achieve pulsing across the full time domain.

In this section, the time regimes are broadly defined and the technologies used for laser pulse generation in each is described and explained. Where appropriate, examples are given of important types of pulsed laser and the performance that can be achieved. As noted earlier, pulsed performance impacts significantly on applications and this has resulted in a continuing strong R&D effort directed at enlarging the pulse regime envelope, particularly into the attosecond range.

REFERENCES

1. Schawlow A L and Townes C H 1958 *Phys. Rev.* **112** 1940.
2. Maiman T H 1960 *Phys. Rev. Lett.* **4** 564.
3. Weber M J (ed) 1999 *Handbook of Laser Wavelengths* (Boca Raton, FL: Chemical Rubber Company).

17

Quasi-cw and Modulated Beams

K. Washio

CONTENTS

17.1 Operation of Solid-state Lasers 241
    17.1.1 Lamp-Pumped Operation 241
    17.1.2 Diode-Pumped Operation 242
    17.1.3 Effects of Thermal Distortion in Solid-state Lasers 243
17.2 Operation of CO_2 Lasers 243
17.3 Examples of Quasi-cw or Modulated Beam Applications 244
References 245
Further Reading 246

Laser beams with reproducible and arbitrary controllable pulse shapes are often required for controlling process quality and increasing throughput in material processing. Laser beams with low-noise, single-longitudinal oscillation modes are often required for coherent optical measurements. This section deals with the topics of quasi-cw and modulated beams in the time domain where pumping control or gain dynamics predominantly determine the laser's temporal performance and omits other cases where intra-cavity active optical components such as Q-switches, modulators, etc., are used.

17.1 Operation of Solid-state Lasers

Solid-state lasers are optically pumped by means of either lamps or other lasers, e.g. diode lasers. Therefore, the temporal characteristics of such pumping sources, in addition to the gain dynamics in the laser media, greatly influence the performance of solid-state lasers. The relatively long lifetime of the upper laser level (around 0.24 ms for Nd:YAG, for example) limits full-modulation-depth highly repetitive pulsed operation, up to several kHz without utilizing intra-cavity optical modulators. Adequately long pre-lasing auxiliary excitation can relax the peak power requirement somewhat for pumping sources for fast short-pulse generation and help reduce the laser pulse delay and timing jitter.

17.1.1 Lamp-Pumped Operation

It has been shown that krypton-filled flashlamps can be efficiently operated at high repetition rates by using simmer current densities of the order of 10 A cm^{-2}. There are several advantages to be gained from using high simmer currents:

1. it helps prevent lamp misfiring;
2. it stops the lamp extinguishing;
3. it removes amplitude and timing jitter from the flashlamp output and
4. it improves the energy transfer from the pulse-forming circuit to the laser rod.

A flashlamp-pumped Nd:YAG laser with pulses exceeding 1500 pulse s^{-1} has been obtained [1].

The traditional way of producing high-power high-voltage current pulses was to use a pulse-forming network (PFN), consisting of capacitors and inductors in a network configuration. As the flashlamp impedance is not constant, perfect matching between the PFN and the flashlamp is only possible at one operating voltage. It is very difficult to obtain efficient operation for pulse lengths of more than 10 ms. PFNs are very limited in their range of pulse duration adjustment. With the emergence of high-power solid-state switching devices such as gate-turn-off thyristors, insulated-gate bipolar transistors and power MOSFETs, PFNs have almost been replaced by such high-power solid-state switching devices. There are a number of advantages in utilizing a transistorized power supply with feedback control of the flashlamp current in an Nd:YAG laser for material processing. These include the ability to tailor the pulse shape and modify the pulse parameter (height, width and rate) on a pulse-to-pulse basis. The current is controlled within a feedback control loop, which minimizes the output variation as a result of changes in the supply-line voltage or flashlamp impedance [2]. Various pulses, such as flat-top rectangular, ramp-up or ramp-down ones, with pulse durations in the range 0.1–20 ms are widely used for laser welding. Figure 17.1 shows a schematic diagram of a power supply with a lamp discharge current controller.

High-power lamp-pumped Nd:YAG lasers of 6 kW and 10 kW average output power with good beam quality have been developed by using a master oscillator power amplifier arrangement with multi-pump cavities for the oscillator and amplifiers [3]. By pulse modulation control of the power

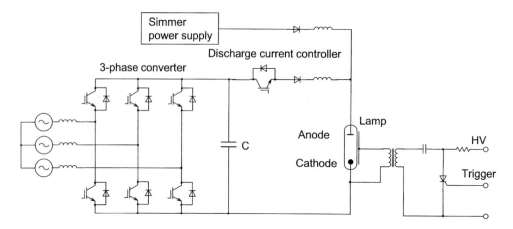

FIGURE 17.1 Schematic diagram of power supply with lamp discharge current controller.

supplies, peak laser powers three times the average power output and fast repetition modulation as high as 500 Hz have been achieved. For a 6-kW average-power Nd:YAG laser, a high peak power exceeding 18 kW has been obtained with a pulse width of 0.6 ms and a repletion rate of 500 Hz. For a 10-kW average power laser, a peak power exceeding 26 kW has been obtained with a pulse width of 8 ms. Figure 17.2 shows examples of waveforms from a 10-kW average-power Nd:YAG laser system [3].

17.1.2 Diode-Pumped Operation

Progress in high-power diode lasers has so far realized up to multi-kilowatt, high-brightness diode-pumped lasers for precision machining [4]. An average output power of up to 3.6 kW has been realized with good beam quality two to four times diffraction-limited, by pumping a zigzag Nd:YAG slab laser by stacked quasi-cw laser diode arrays. Typical diode pulse durations range from 0.1 to 1 ms. In multi-mode operation, 5 kW has been obtained with an optical efficiency of 35%.

The adoption of diode-laser pumping imparts a number of advantages to solid-state lasers. Excellent modulation controllability and fast excitation capability realize solid-state lasers with higher performances than those attained with lamp pumping.

The low-frequency dynamics of four-level lasers, such as Nd:YAG lasers, has been investigated by many researchers [5,6,8]. With gain or loss perturbation, the response of the homogeneously broadened, single-frequency solid-state laser can exhibit one sharp peak in the vicinity of the relaxation oscillation frequency together with spiking oscillation or regular pulses of high peak intensity can be obtained. Spiking behaviour becomes chaotic for multi-mode operation.

Gain switching has been shown to be an effective method for obtaining single-frequency short-pulse operation from a miniature monolithic Nd:YAG laser [6]. This method utilizes a low pumping power around 70 mW to select a high-gain single-longitudinal mode of the Nd:YAG laser near its lasing threshold with a long pedestal pulse of about 20 µs duration. This longitudinal mode is then driven by a gain-switching pulse to produce a single-frequency spiking relaxation oscillation. A short (less than 1 µs), high-power (335 mW) gain-switching diode-laser pulse drives the Nd:YAG well above threshold and produces the desired single-frequency emission. Figure 17.3 shows schematic diagrams of the pulse shapes for diode-laser pump power and gain-switched spiking Nd:YAG laser output power. With gain switching, the monolithic Nd:YAG laser generates single-frequency pulses with a bandwidth of less than 8 MHz (FWHM) at repetition rates up to 7 kHz. By the use of an Nd:glass laser amplifier, single-frequency Nd:YAG pulses can be boosted to powers exceeding 200 W.

In pulsed diode bars used to pump solid-state lasers, the temperature rise during the pulse results in a change in the centre wavelength, known as a chirp, and a reduction in efficiency and output power, known as sag [7]. Since many solid-state materials such as Nd:YAG have absorption features

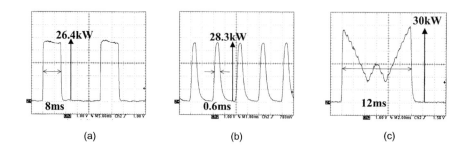

(a)　　　　　(b)　　　　　(c)

FIGURE 17.2 Waveform examples for a 10-kW average-power Nd:YAG laser system: (a) large-pulse energy operation with pulse repetition rate of 40 Hz; (b) 500 Hz, high-repetition rate operation; and (c) arbitrary pulse shape with 20 segments in a waveform.

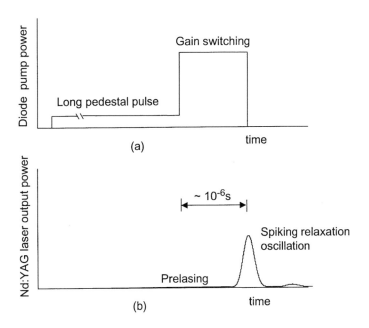

FIGURE 17.3 (a) Diode-laser pump power pulse shape; (b) gain-switched Nd:YAG laser output pulse shape.

only a few nanometres in width, efficient system operation requires a correspondingly narrow diode emission including the effects of wavelength chirp. An AlGaAs diode laser shows a shift of about 0.3 nm °C in the emission centre wavelength. In high-peak-power diode-pumped solid-state laser systems, diode chirp and sag during the pump can play a major role in determining the pump energy absorbed in the laser medium.

A substantial reduction in intensity noise has been demonstrated for diode-laser-pumped Nd:YAG lasers by application of electronic negative-feedback loops. A fraction of the laser light is detected with a photodiode, and the ac components are appropriately amplified to generate an error signal. This signal is then fed back to the pump diode. By carefully adjusting the parameters of the feedback loop, suppression of about 25 dB at low frequencies and more than 50 dB at the relaxation oscillation frequency is achieved [8].

17.1.3 Effects of Thermal Distortion in Solid-state Lasers

One of the big problems in solid-state lasers is thermal distortion of the laser medium and beam quality deterioration which are commonly seen in high-power or highly repetitive operation. Detailed interferometric measurements and comparisons have been made of the thermal distortions in xenon-flashlamp- and diode-pumped Nd:YLF laser rods [9]. In both cases, defocus and astigmatism were the dominant thermal distortions. The thermal distortions in the flashlamp-pumped rods were greater than the thermal distortion in the diode-pumped rod when pumped to the same small-signal gain. In the diode-pumped Nd:YLF laser system, a 5-mm-diameter × 54-mm-long Nd:YLF laser rod was used and was pumped with three banks of laser diodes along the length of the rod. The thermal relaxation time constant of their diode-pumped Nd:YLF rod was measured to be approximately 1.5 s.

17.2 Operation of CO_2 Lasers

Quasi-cw or modulated CO_2 laser beams are extensively used, particularly in materials processing. Among excitation schemes using dc, rf and excitation microwave, rf excitation now seems to be the most popular pumping scheme due to its excellent controllability and ease of incorporation into laser systems. Modulated laser operation has been easily realized by modulating the rf excitation directly. The rf frequencies used for CO_2 lasers vary from 100 kHz to about 100 MHz, depending on the type of laser systems. The average laser power can be controlled by rf phase modulation or varying rf amplitude in cw mode or by gating rf power in pulsed mode with the appropriate duty cycle and repetition frequency. The average power can be more widely controlled stably in pulsed mode than in cw mode. Figure 17.4 shows schematic diagrams of waveforms for gated rf power and CO_2 laser output pulses.

A fast axial flow CO_2 laser system using a 13.56 MHz rf discharge with an output power of up to 1500 W is well known from the early days (around 1986) [10,11]. Pulse repetition rates between 100 Hz and 12 kHz at duty cycles of 0.1–1 were attained [10]. The rf power was coupled capacitively through an Al_2O_3-plate into the discharge volume of 4 cm × 4 cm × 95 cm. With a maximally available rf-power of 50 kW, a power density of 33 W cm^{-3} has been achieved during pulsed operation [10].

An rf discharge with a frequency below 1 MHz is often called a silent discharge (SD). Transverse flow as well as fast-axial flow CO_2 lasers with SD excitation have been investigated [12,13]. When an SD excitation scheme is compared with other rf excitation schemes with higher frequencies (typically 13.56, 27, 40.68, 80.7 MHz, etc.) or a microwave excitation scheme (typically 2.45 GHz), it has the following major advantages: (i) high-efficiency high-power solid-state power sources are

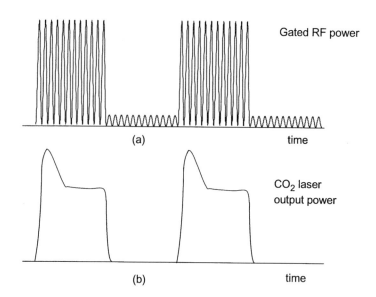

FIGURE 17.4 (a) Gated rf power waveform; (b) CO_2 laser output pulses.

readily available and (ii) the efficiency of the energy input is high because the power source and the discharge rod are easily coupled effectively. The performance characteristics of the TEM_{00} 2.5 kW SD-excited, transverse-flow CO_2 laser were investigated [12]. Each electrode was a square-shaped metal pipe covered with borosilicate glass 0.8 mm thick. The discharge space is 45 mm wide in the electric-field direction. The power source, which consisted of an inverting circuit switched by a transistor (MOSFET) and a transformer, provided 100 kHz, high-voltage, AC. Pulsed operation of the laser power has been achieved by modulating the applied voltage (around 10 kV) to the electrodes. When the frequency was low, flat-top pulses were obtained. When the frequency was high, the pulse shape lost its flat bottom.

The operation of a pulsed transverse-flow CO_2 laser system with rf excitation has been studied theoretically by computer simulation [14]. The pulse repetition frequency of the CO_2 laser system is restricted by the time delay between the electric field and the laser pulse and by relaxation oscillations. The time delay is roughly inversely proportional to the gas pressure. A pulse repetition frequency over some 30 kHz can be attained only if the excitation pulse period is equal to the delay time, but this operation is very sensitive to changes in the duty cycle. Above several 100 mbar, the discharge may become inhomogeneous and can break down into an arc discharge, which reduces the efficiency.

The response of rf-excited CO_2 lasers to pulsed modulation can be improved by increasing the gas pressure and decreasing the discharge gap distance inversely to keep the gas pressure–discharge gap distance product constant. A fast-axial-flow, rf-excited slab tube CO_2 laser incorporating a high-aspect-ratio rectangular-flow tube and a 5-mm discharge gap has been investigated [15]. The rf frequency was 27 MHz. This laser is expected to be free of thermal filamentation up to 800 mbar. In pulsed operation at pressures of around 400–540 mbar, a repetition frequency of 15 kHz has been obtained with the pulse duration reduced to 15–25 µs, suitable for processing low-thermal-conductivity materials. Diffusion-cooled, sealed CO_2 slab lasers are advantageous for obtaining efficient high-power operation with smaller inter-electrode distances.

The dependency voltage–current characteristics of a transverse rf-discharge-excited waveguide CO_2 laser have been investigated over a wide range of excitation frequencies (f) (100–160 MHz), for gas pressures of 40–100 Torr and inter-electrode distances (D) of 1–3 mm [16]. It has been found that there is a scaling law—of fD = constant—for such rf-excited CO_2 waveguide lasers.

The trend in rf-excited CO_2 lasers is towards sealed systems, and the power of these systems is increasing. There is a commercially available 'quasi-sealed' 3.5 kW laser (basically cw but it can be modulated). Sealed laser discharges have to be narrow to allow gas cooling by diffusion, and they should be excited at high frequencies to be efficient, to avoid arcs, etc. The narrower they get, the higher the required gas pressure becomes for stable operation.

A high-power, high-repetition-rate CO_2 laser system can be constructed with the aid of an external modulator. A 1-kW laser capable of full-depth modulation in excess of 200 kHz has been developed to satisfy a need within the flexographic printing industry [17]. The laser is based upon a diffusion-cooled CO_2 laser converted into an oscillator–acousto–optic–modulator–amplifier configuration. Switch-ON and switch-OFF both occur within 1 µs, easily achieving full-depth modulation at 200 kHz frequency. Full-depth modulation has been achieved up to 500 kHz.

17.3 Examples of Quasi-cw or Modulated Beam Applications

A wide variety of pulsed or modulated laser beams are required for high-quality materials processing such as drilling, cutting, welding and hardening.

In the drilling of microvias for epoxy-glass printed wiring boards (PWBs), for example, the biggest problem to be solved is carbonization of the epoxy resin. The thickness of the heat-affected layer can be reduced from 8 to 2.6 µm by decreasing the pulse duration from 500 to 50 µs and, again, to 1 µm by decreasing the pulse duration to 10–20 µs [18]. On the basis of these data, a new laser drilling system has been developed for PWBs to allow high-quality drilling at the maximum drilling rate of 500 holes s^{-1} with a pulse duration that can be set in the range of 16–250 µs [18].

In the cutting process, piercing is performed before cutting is started. To minimize the total piercing time, the laser power is preferably modulated to obtain the desired output waveform during piercing. Ramping is recommended when it is necessary to pierce the work gently to make the first hole. Laser power can be programmed to vary as a function of work-feed rate during cutting. The necessary data for power control can be stored in the laser controller unit. Pulsing is commonly used to reduce burning on corners or detailed work. Furthermore, pulsing is often used to obtain cut surface improvement during cutting. Using the correct pulsing parameters, it is possible to produce cut-edge striations which are very close together, and this, in turn, means that the cut surface quality improves because the roughness diminishes.

A mathematical analysis describing keyhole laser welding by pulsed beams has been investigated [19]. The effect of temporal pulse shape on weld dimensions has been examined over a range of power densities, pulse times and pulse frequencies. Several pulse types (e.g. top-hat, Gaussian, ramp-up and ramp-down) have been considered. Pulse shape had a significant effect on weld dimensions. The predicted width and penetration depth of weld with a ramp-up and ramp-down pulse were greater than those obtained with Gaussian and top-hat pulses.

The cause of penetration instability and counter measures was investigated for butt-welding A3003 aluminium alloys [20]. A slab Nd:YAG laser was used for carrying out the welding experiment in order to exclude instability factors such as a change in the laser beam quality. It was found that the addition of a high peak power, short-width pulse superimposed on the early part of the rectangular pulse is very effective in preventing the penetration instability of A3003 alloy.

The weld profile with cw processing commonly shows a wine-cup configuration because a large amount of laser energy is absorbed near the surface. However, when a pulse-modulated beam was used with an increased peak power of up to three times the average power, the weld profile was found to have a larger penetration depth with a wider width near the bottom than those obtained by cw beams [3]. Two- and threefold pulse enhancement seems to be very effective for multi-layer deep-penetration overlap welding with low thermal distortion of the workpiece.

REFERENCES

1. Corcoran V J, McMillan R W and Barnoske S K 1974 Flashlamp-pumped YAG:Nd^{+3} laser action at kilohertz rates *IEEE J. Quantum Electron.* **QE-10** 618–20.
2. Weedon T M W 1987 Nd–YAG lasers with controlled pulse shape *Proc. LAMP '87 (Osaka, May)* (Osaka: High Temperature Society) pp 75–80.
3. Ishida T, Togawa T, Morita H, Suzuki Y, Okino K, Takenaka H, Kubota K, Washio K and Yamane T 2000 6 kW and 10 kW high power lamp pumped MOPA Nd:YAG laser system *Proc. SPIE* **3888** 568–76.
4. Machan J, Moyer R, Hoffmans D, Zamel J, Burchman D, Tinti R, Holleman G, Marabella L and Injeyan H 1998 Multi-kilowatt, high brightness diode-pumped laser for precision laser machining *Technical Digest of Advanced Solid-State Lasers Topical Meeting (Coeur d'Alene, Idaho, February)* AWA2 (Washington, DC: OSA) pp 263–5.
5. Harrison J, Rines G A and Moulton P F 1988 Stable-relaxation-oscillation Nd lasers for long pulse generation *IEEE J. Quantum Electron.* **24** 1181–7.
6. Hadley G R, Owyoung A, Esherick P and Hohimer J P 1998 Numerical simulation and experimental studies of longitudinally excited miniature solid state lasers *Appl. Opt.* **27** 819–27.
7. Honea E C, Skidmore J A, Freitas B L, Utterback E and Emanuel M A 1998 Modelling chirp and sag effects in high-peak power laser-diode-bar pump sources for solid state lasers *Advanced Solid State Lasers (OSA TOPS 19)* ed W R Bosenberg and M M Fejer (Washington, DC: OSA) pp 326–32.
8. Freitag I, Tünnermann A and Welling H 1997 Intensity noise transfer in diode-pumped Nd:YAG lasers *Advanced Solid State Lasers (OSA TOPS 10)* ed C R Pollock and W R Bosenberg (Washington, DC: OSA) pp 380–5.
9. Skeldon M D, Saager R B and Seka W 1999 Quantitative pump-induced wavefront distortions in laser-diode- and flashlamp-pumped Nd:YLF laser rods *IEEE J. Quantum Electron.* **35** 381–6.
10. Hügel H 1986 rf-excitation of high power CO_2 lasers *Proc. SPIE* **650** 2–9.
11. Wollermann-Windgasse R, Akkermann F, Weick J and Brix W 1986 rf-excited high power CO_2-lasers for industrial material processing *Proc. SPIE* **650** 290–35.
12. Yasui K, Kuzumoto M, Ogawa S, Tanaka M and Yagi S 1989 Silent-discharge excited TEM_{00} 2.5 kW CO_2 laser *IEEE J. Quantum Electron.* **25** 836–9.
13. Kuzumoto M, Ogawa S, Tanaka M and Yagi S 1990 Fast axial flow CO_2 laser excited by silent discharge *IEEE J. Quantum Electron.* **26** 1130–4.
14. Offenhäuser F 1988 Theory and experiments on the power modulation of CO_2 lasers *IEEE J. Quantum Electron.* **24** 1289–96.
15. Markillie G A J, Baker H J, Betterton J G and Hall D R 1999 Fast-axial-flow CO_2 slab laser with a narrow-gap rf discharge operating at high pressure *IEEE J. Quantum Electron.* **35** 1134–41.
16. Vitruk P, Baker H J and Hall D R 1994 Similarity and scaling in diffusion-cooled rf-excited carbon dioxide lasers *IEEE J. Quantum Electron.* **30** 1623–34.
17. Wheatley D W 1996 A high power, high modulation bandwidth CO_2 laser (Proc. XI Int. Symposium on Gas and Chemical Lasers and High-Power Laser Conf. (25–30 August) *Proc. SPIE* **3092** 109–13.
18. Takeno S, Moriyasu M and Kuzumoto M 1997 Laser drilling of epoxy-glass printed circuit boards *Proc. ICALEO 1997* (Orlando, FL: Laser Institute of America) section B-ICALEO, pp 63–72.

19. Mahanty P S, Kar A and Mazumder J 1995 Effects of pulse shaping on keyhole laser welding: A mathematical analysis *Proc. ICALEO 1995* (Orlando, FL: Laser Institute of America) pp 999–1008.
20. Nakamura T, Togawa T, Okino K, Watanabe S and Washio K 1997 Seam welding of A3003 aluminum alloy using high-brightness pulsed slab YAG laser *Proc. ICALEO 1997* (Orlando, FL: Laser Institute of America) section G-ICALEO, pp 130–9.

FURTHER READING

Iffländer R 2001 *Solid-State Laser for Materials Processing* (Berlin: Springer).

Particularly covers the characteristics, design, construction and performance of high-power solid-state lasers for materials processing, including pump sources, power supplies and thermal management.

Koechner W 1999 *Solid-State Laser Engineering* 5th edn (New York: Springer).

Includes an explanation and illustrations for the characteristics, design, construction and performance of a wide variety of solid-state lasers for practical use, including pump sources and power supplies.

Metev S M and Veiko V P 1998 *Laser-Assisted Microtechnology* 2nd edn (Berlin: Springer).

This book introduces the principles and techniques of laser-assisted micro-processing of materials, including laser micro-shaping, laser melting, laser heat treatment, etc.

Powell J 1998 CO_2 *Laser Cutting* 2nd edn (London: Springer).

Wide coverage of CO_2 laser cutting for metals and non-metals with practical advice and processing parameters for pulsed laser cutting of mild steels, piercing of high-reflectivity metals, etc.

Steen W M 1991 *Laser Material Processing* (London: Springer).

This book provides an introduction to the industrial application of lasers in cutting, welding and the many new processes of surface treatment.

18 Short Pulses

Andreas Ostendorf

CONTENTS

18.1 Gain Switching 247
18.2 Q-switching 248
    18.2.1 Q-switched cw Pumped Lasers 249
    18.2.2 Methods of Q-switching 250
    18.2.3 Mechanical Q-switches 251
    18.2.4 Electro-optic Q-switches 252
    18.2.5 Acousto-optic Q-switches 252
    18.2.6 Saturable-absorber Q-switches 252
18.3 Cavity Dumping 253
18.4 Mode-locking 253
18.5 Master Oscillator with Power Amplifiers 253
18.6 Beam Characterization and Pulse Measurement 254
References 256
Further Reading 257

The pulse operation mode of lasers has been of great interest for laser developers and engineers for a long time. The following chapter provides an overview on the different principles of generating short laser pulses. In the case of concrete specifications for the pulse width, these are defined by the full width at half maximum (FWHM) method.

18.1 Gain Switching

Gain switching can be regarded as the most direct method to generate pulsed laser radiation. During gain switching, the pumping process is modulated which results in switching the amplification in the laser medium. In general, after the pumping process has been switched on, the population inversion starts to build up. When the critical inversion is reached, i.e. the gain becomes larger than losses or the loop gain reaches 1.0, the laser starts to oscillate. The oscillation continues until the pumping process is switched off, or until the losses become higher than the amplification. In contrast to the Q-switch mode, which is described in the following section, gain switching is primarily controlled by the pumping process [1]. Gain switching can make use of the transient spiking phenomena in the laser oscillator in order to achieve a high-peak-power pulse. If a laser medium is pumped at a very fast rate and the population inversion exceeds the threshold significantly before the oscillation starts, the laser reacts by relaxation oscillations, i.e. spiking occurs. If the pump pulse is not only fast but also stops directly after the first peak, a laser pulse is generated which consists of only one spike. The inversion after the first peak is below the threshold, and due to the absence of further pump power, there is no possibility for further emission.

Gain switching is well established for diode lasers, since for semiconductor lasers, the pumping current can easily be modulated. The maximum modulation frequency is mainly limited by the relaxation time constants provided by the active medium and the resonator. Due to the small resonator dimensions within semiconductor lasers, these time constants can become very short resulting in modulation frequencies up to the GHz range.

Another example of gain switching is lasers which are excited by diode lasers [2]. Diode lasers can also generate very high optical powers on a short-term basis and are thus suited for the build-up of high inversions. For diode-laser-pumped solid-state lasers, a spiking phenomenon can be observed which can be used for gain switching. The amplifier gain rises quickly to a high value because of the intense pumping level. This results in a high round trip gain and further relaxation oscillations. These oscillations quickly deplete the population inversion for that particular wavelength, spiking occurs and the lasing process stops. If, at that point, the pump diode current is also turned off, the laser switches itself off momentarily by using up all of its gain.

A representative of gain-switched gas lasers is also the transversally excited atmospheric pressure laser (TEA laser). For a TEA-CO_2 laser, a discharge voltage is applied to transversally arranged electrodes. If the voltage is pulsed at a duration shorter than 1 µs, instabilities in the discharge will be suppressed and gas pressures of the CO_2/N_2/He mixture can be increased up to 1 bar [3,4]. This leads to laser pulse energies

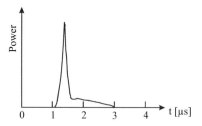

FIGURE 18.1 Typical pulse of a TEA-CO$_2$ laser.

up to 50 J per litre discharge volume at pulse widths of typically 100 ns. Laser pulses from a TEA laser are characterized by a main pulse of typically 100–500 ns width followed by a pulse tail of approx. 1 µs [5] (see Figure 18.1). The short main pulse is determined by gain switching. The tail is produced by the inversion recovery which is caused by impacts with N$_2$ and it is, therefore, dependent on the N$_2$:CO$_2$ ratio of the active gas.

Because gain switching in pulsed solid-state lasers is related to the spiking in the output, it is difficult to determine the pulse peak power which varies from shot to shot, although pulse energy and overall pulse width remain constant. For these reasons, specifications of pulsed solid-state lasers usually do not include the maximum output power. Instead, pulse energy and pulse width are specified. The peak output power may be approximated by dividing the energy of the output pulse by pulse width as with other pulsed lasers. In gain switching, laser pulse width and peak power are of the same order of the pump pulse. In Q-switching, which is described in the next section, this ratio becomes much more attractive. Moreover, the reproducibility and stability tend to be stochastic, which makes gain switching, in spite of its simplicity, unacceptable for many applications.

18.2 Q-switching

The output of a gain switched, pulsed solid-state laser is generally a train of irregular pulses—irregular in peak power, pulse width and repetition frequency. It is possible to remove these irregularities and at the same time greatly increase the peak power by a technique called Q-switching. Q-switched lasers normally emit only one giant pulse in an operational cycle. The pulse will typically have a time duration less than one microsecond down to several nanoseconds and a peak intensity between 10^6 and 10^9 W. This technique can also be applied to continuously pumped lasers in order to produce a train of Q-switched pulses with regular width, peak power and repetition rate [6]. Compared to gain switching, a Q-switch converts a relatively long pump pulse with low peak power into a very short pulse with high peak power over several orders of magnitude.

Q-switching is a mode of laser operation in which energy is stored in the laser active material during pumping in the form of excited atoms and suddenly released in a single, short burst. This is accomplished by changing the optical quality of the laser cavity. The quality factor Q is defined as the ratio of the energy stored in the cavity to the energy loss per cycle.

During the pumping process, the beam path in the Q-switched system is interrupted, resulting in a low Q-factor and preventing the onset of laser emission (Q ≈ 1). After a large amount of energy has been stored in the active medium, the beam path in the resonator is returned to proper alignment, and most of the stored energy emerges in a single, short pulse (Q≫1) [7–11].

In a Q-switched laser, a physical change is made in the feedback loop, i.e. in the resonant cavity which drastically lowers the effective reflectivity of the cavity during optical pumping. This prevents the system from oscillating and allows the population inversion to increase without an output being generated—resulting in an increase in amplifier gain and stored energy in the upper laser level. At the time of maximum population inversion (usually near the end of optical pumping), the quality factor Q is switched back to a high value, causing the system to emit laser radiation and, thereby, releasing its available, stored energy in one, short giant pulse [12]. A Q-switch is essentially a shutter placed between the active medium and the HR mirror. With this shutter closed, the HR mirror is blocked, preventing oscillation. When the amplifier gain reaches a predetermined value, the shutter is opened to increase the cavity quality.

For a further understanding of Q-switched lasers, the rate equations have to be solved before and after Q-switching. In the following, it is assumed that the Q-switch will take place at $t = t_Q$. The general rate equations for lasers are for the population inversion N:

$$dN/dt = 2R(t) - 2B\phi N - 2\tau_{sp}^{-1} N \qquad (18.1)$$

and for the photon density ϕ:

$$d\phi/dt = -\kappa(t)\phi + BN\phi \qquad (18.2)$$

with R(t) describing the pumping rate, B is the modified Einstein coefficient, τ_{sp} is the characteristic lifetime for spontaneous emission and κ representing the resonator losses. For the description of the Q-switch process, the spontaneous emission can be neglected. The pumping rate R(t) is equal to R for $t < t_P$ and equal to 0 for $t \geq t_P$. According to these definitions, $Q(t<t_Q) \approx 1$, $\kappa(t<t_Q) = \kappa_1 \gg 1$ and $Q(t \geq t_Q) \gg 1$, $\kappa(t \geq t_Q) = \kappa_2 \approx 0$.

Shortly before Q-switching, i.e. $t \approx t_Q$, the population inversion n_Q can be calculated from the steady-state case of equation (18.1) to

$$N(t_Q) = N_Q = \frac{\kappa_1}{B} \qquad (18.3)$$

which is much higher than the steady-state population inversion of a laser without Q-switch. This is achieved by keeping the photon density ϕ ($t<t_Q$) at a low level. There is nearly no stimulated emission, which reduces the population inversion within this mode of operation. Also from the steady-state case of equation (18.2), the resulting photon density shortly before the switching time can be calculated to

$$\phi(t_Q) = \phi_Q = \frac{R}{\kappa_1} \qquad (18.4)$$

which is much lower than the photon density of a laser without Q-switch. Shortly after the Q-switch at $t \geq t_Q$, the following behaviour of the population inversion can be calculated from equation (18.1)

$$dN/dt = -2B\phi_Q N_Q = -2R \tag{18.5}$$

or

$$N(t) = N_Q - 2Rt. \tag{18.6}$$

The photon density shortly after the Q-switch can be calculated from equation (18.2) to

$$d\phi/\phi = BN_Q dt \approx \kappa_1 dt \tag{18.7}$$

or

$$\phi(t) = \phi_Q e^{\kappa_1 t} \tag{18.8}$$

which shows that the photon density increases exponentially after the Q-switch. To calculate the pulse peak power, the rate equations (18.1) and (18.2) have to be solved after the Q-switch, i.e. with $R(t) = 0$. By taking the quotient of (18.1) and (18.2), time as a parameter can be eliminated.

$$2\frac{d\phi}{dN} = \frac{N_{\text{stat}}}{N} - 1 \tag{18.9}$$

with N_{stat} the solution of the rate equations in the steady-state case. Equation (18.9) can be integrated to give

$$2(\phi - \phi_Q) = (N_Q - N) - N_{\text{stat}} \ln(N_Q/N). \tag{18.10}$$

The maximum photon density can be derived from

$$d\phi/dt = 0 = (-\kappa_2 + BN)\phi \tag{18.11}$$

which results in

$$\phi_{\max} = \frac{1}{2} N_Q \left(1 - \frac{\kappa_2}{\kappa_1} + \frac{\kappa_2}{\kappa_1} \ln \frac{\kappa_2}{\kappa_1}\right). \tag{18.12}$$

The pulse peak power can be derived directly from the maximum photon density by $P_{\max} = \phi_{\max} h\nu \kappa_2$

Figure 18.2 provides a time scheme of a laser pulse generated by Q-switched operation of an optically pumped laser.

For Q-switched lasers both the amplifier gain and the deposited energy in the laser active medium reach much higher levels than for the normal gain switching mode. At a time near the end of the optical pumping ($t_4 = t_Q$), the losses are rapidly switched to its minimum value—causing the system to emit a giant laser pulse. During the time of the output pulse, the energy being put back into the laser active medium by external pumping can be neglected [13]. Laser emission starts when the loop gain passes 1.0 at time t_5. The laser pulse builds rapidly up depleting stored energy and reducing the amplifier gain. The peak of the laser output pulse occurs when the loop gain falls below the 1.0 level. After this, the output laser power starts to drop.

The pulse width may be controlled in two ways. In the first possibility, the Q-switch can be switched back to lower loop gain and terminate laser radiation before the stored energy has been used completely. This can shorten the laser pulse and reduces the pulse energy since some of the energy remains in the active medium. Later on, this energy is depleted through fluorescence. Alternatively, pumping can continue slightly beyond the laser pulse. This extra input energy contributes to fluorescence only. In some cases, the Q-switch may remain open with pumping ending during the Q-switch cycle. This usually results in a slightly longer output pulse duration.

The energy of the Q-switched output pulse is dependent on the repetition rate and can be up to 90% of the pulse energy available from normal mode operation. One factor contributing to this is a remainder of energy left in the active medium. Another is that fluorescence begins with pumping, and considerable energy is lost through spontaneous emission before the Q-switch opens. During normal mode operation, each atom in the active medium may participate in the emission process several times. In a Q-switched laser, each atom contributes only once. This further reduces laser efficiency.

The delay time between the onset of pumping and the opening of the Q-switch influences the efficiency of the laser significantly and should be optimized with respect to the lifetime of the upper level of the active medium. The amplifier gain and the stored energy rise from the beginning of pumping until an optimum has passed. Delaying Q-switch operation beyond this time will not result in more stored energy as energy is lost to fluorescence at the same rate as it is added by pumping. Obviously, the greater the fluorescent lifetime of the laser material, the more energy may be deposited in the active medium. All solid-state laser systems may be effectively Q-switched as their fluorescent lifetimes fall in the range above one microsecond (Nd:YAG 230 μs; Nd:YVO$_4$ 100 μs; Yb:YAG 980 μs; Nd:YLF 480 μs). Molecular gas lasers such as CO_2 are also often Q-switched. Ion lasers cannot be effectively Q-switched because the lifetime of their upper level is too short, allowing insufficient time to build up a large stored energy.

18.2.1 Q-switched cw Pumped Lasers

Solid-state and molecular gas lasers may be pumped continuously and repetitively Q-switched to produce a regular train of output pulses. The most common examples of this are cw pumped Nd:YAG lasers and cw CO_2 lasers with low output power. Since the pumping rate is much weaker for cw systems, the maximum stored energy is small, and the resulting peak power is lower (typically 10^4–10^6 watts).

Nd:YAG is by far the most common Q-switched, cw pumped laser system. Such systems can typically produce several thousand pulses per second without degradation of the pulse energy. Typical pulse widths for such a system range from several tens of nanoseconds to a few microseconds. If the pulse width should be further minimized, it is necessary to use a laser crystal with a larger emission cross section, e.g. Nd:YVO$_4$. Nd:YVO$_4$ lasers can deliver pulses with half the

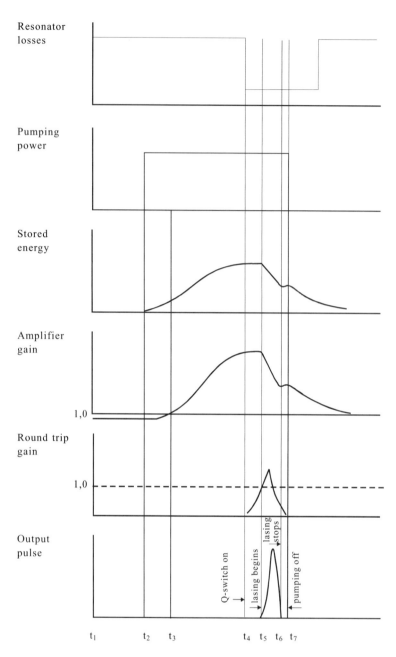

FIGURE 18.2 Time history of laser parameters during pulse pumping and Q-switching.

duration of Nd:YAG assuming that the same stored energy and resonator configuration is used.

18.2.2 Methods of Q-switching

Several techniques can be used for Q-switching of lasers. Each has its advantages, disadvantages and specific applications. The four most important Q-switch types are discussed in this section, illustrated in Figure 18.3, and can be characterized by the following values.

1. Dynamic loss is the maximum loss introduced in the optical cavity when the Q-switch shutter is closed. Ideally the dynamic loss should be 100% to ensure that lasing does not occur until the Q-switch is opened.

2. Insertion loss is the minimum loss introduced by the presence of the Q-switch in the open condition. Ideally this is zero, but most Q-switches include optical surfaces that introduce reflection and scattering losses.

3. Switching time is the time necessary for the Q-switch to open. Faster switching times result in shorter, high-peak-power pulses because the switch can become fully open before lasing has a significant effect on the population inversion. Slower switches allow significant amounts of the stored energy to be depleted before the Q-switch is fully opened. This lowers the

Short Pulses

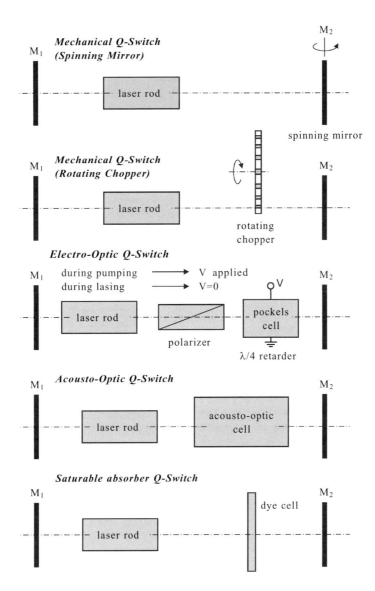

FIGURE 18.3 Principles of Q-switches.

maximum loop gain of the system and stretches out the output laser pulse.

4. Synchronization is an indication of how well the laser output can be timed with external events. Some Q-switches allow precise control of when the output pulse occurs. Others offer virtually no control at all.

The Q-switch can be carried out by different methods. The most commonly known methods are active switching by mechanical, electro-optical or acousto-optical Q-switches, where the quality control is an explicit function of time, and passive Q-switching by saturable absorbers, where the quality is controlled as a function of the photon density.

18.2.3 Mechanical Q-switches

Two representatives of this category are a light chopper and a spinning mirror. A light chopper is a spinning disc with a hole in it, or a spinning blade (like a fan blade). The chopper is inserted into the optical cavity between the laser rod and the maximum reflectivity mirror. This system provides 100% dynamic losses and 0% insertion losses. A mechanical chopper is relatively slow, however, in that it can Q-switch only a fraction of the beam area at a time as it is swept across the aperture. For this reason, mechanical chopper Q-switches are not practical or effective.

Spinning reflectors are used quite frequently in Q-switched systems where it is not necessary to closely synchronize the output to some other event. Usually, the maximum-reflectivity mirror is rotated so that the mirror is tilted out of alignment. The system is Q-switched when the mirror rotates back into alignment (it is in alignment once each revolution). Rotating mirror Q-switches offer 100% dynamic losses and 0% insertion losses. The Q-switch dynamics can be increased to a sufficient level by rotating the mirror at high speed (20 000 –60 000 rpm), or by various optical schemes to multiply the effect of the rotating element. Switching time is typically a few nanoseconds.

For both the chopper and spinning reflector, it is necessary to synchronize the firing of the flashlamps with the position of the spinning element, so that the pumping pulse occurs before the system is Q-switched. Synchronization of the output pulse in mechanical systems is poor. Rotating mirror Q-switches may be used with either pulsed or cw pumped lasers.

18.2.4 Electro-optic Q-switches

These devices usually require the placing of two elements into the reflecting cavity between the laser rod and the maximum reflecting mirror. These elements are a polarization filter (passive) and a polarization rotator (active). Producing a low-cavity feedback with these devices involves rotating the polarization vector of the laser beam inside the cavity so that it cannot pass through the polarization filter. When this polarization rotation is terminated, the cavity reflectivity is relatively high and the system will produce a giant pulse. Two of the electro-optic devices used in this application are Kerr cells and Pockels cells. The Pockels effect is a linear electro-optical effect, i.e. the refraction index change in the parallel and orthogonal direction is proportional to the applied voltage ($\Delta n \propto V_{el}$). For Pockels cells, usually non-linear crystals like KD*P, LiNbO$_3$ or LiTaO$_3$ are used for the visible and near-IR region, whereas CdTe is used for mid-IR. The Kerr effect is a non-linear electro-optical effect, i.e. the dependency is a square function of the applied voltage ($\Delta n \propto V_{el}^2$). Electro-optical Q-switches have high dynamic losses (99%) and acceptable insertion losses (<5%) because of the losses in the optical elements. Switching time is faster compared with competing technologies (mechanical and acousto-optic devices), typically less than a nanosecond, and synchronization is good. Electro-optic devices have some limitations, including their relatively high-voltage requirements and limited beam-diameter capabilities.

18.2.5 Acousto-optic Q-switches

This technique involves the use of a transparent element placed in the cavity between the laser rod and the maximum reflectivity mirror. This transparent device, when excited with intense, standing, acoustic waves by piezoelectric crystals, exhibits a diffraction effect on the intra-cavity laser beam and diffracts part of the beam out of the cavity alignment. This results in a relatively low feedback. When the acoustic wave is removed, the diffraction effect disappears, the cavity is again aligned and the system emits a giant pulse [14]. Acousto-optic devices have low insertion losses (typically less than 1%) and high dynamic losses (90% maximum). Switching time is slow at 100 ns or greater, and the synchronization is good. These devices are ideally suited for use with cw pumped systems or low-gain pulsed lasers.

18.2.6 Saturable-absorber Q-switches

A saturable absorber can be regarded as a passive switch and is the easiest method for a Q-switch [15–20]. The switching time is not externally triggered but defined by the intensity of the beam. Saturable absorbers are available as thin films on glass substrates or as liquids in glass cells. For Q-switching, the absorber is placed in the laser cavity between the amplifier and the maximum reflectivity mirror. It effectively absorbs the laser wavelength at low light intensities, presenting a very high cavity loss to the laser, and preventing lasing until the amplifier has been pumped to a very high gain state. When the irradiance from the active medium becomes intense enough, the energy that is absorbed optically pumps the absorber material, causing it to be transparent at the laser wavelength. The absorber cell is bleached and no longer represents a high cavity loss to the laser, i.e. the quality of the resonator increases. The absorption change is the equivalent of Q-switching in the laser, and it can occur in a period less than a nanosecond. In general, the behaviour of saturable absorbers can be characterized by

$$\alpha = \frac{\alpha_0}{1 + I/I_S} \quad (18.13)$$

with $\alpha_0 = \alpha(I \cong 0)$, the initial absorption coefficient and I_S, the saturation intensity. For $I = I_S$, the absorption coefficient drops to half of the level without irradiation. For $I \gg I_S$, both energy levels in a two-level system are occupied at the same rate, i.e. the absorption coefficient tends towards zero, the transmission coefficient becomes close to one. In this case, the absorber is bleached.

The initial absorption coefficient is determined such that the laser resonator at the time of maximum inversion in its active medium just reaches the threshold. For increasing intensities, the transmission coefficient T of the absorber grows, within a very short interval, from $T \ll 1$ to $T \cong 1$. After a relaxation time τ, which is between 1 ps and 1 μs dependent on the absorber material, the molecules fall back to their ground states. The initial absorption coefficient has to be selected so that the remaining intensity in the resonator will not produce a second switching.

Saturable absorber Q-switches provide very high dynamic losses (>99%) and insertion losses (a few per cent at most), and their switching time is fast. Due to their immanent properties, they need virtually no synchronization at all. Saturable absorber Q-switches can be used with pulse pumped systems only because a cw pumped laser never produces sufficient irradiance to bleach the absorber. The main problem with saturable absorbers is to find appropriate materials for the defined absorption wavelength, the saturation intensity, the relaxation time and the susceptibility towards UV-radiation, e.g. from flashlamps in solid-state lasers. For Nd-doped lasers, mostly Cr^{4+}:YAG crystals are used as satuarable absorber materials due to their high damage threshold of >500 MW cm^{-2}.

A good example of the use of saturable absorbers is in microchip lasers. Q-switched microlasers are simple, compact and reliable sources that can produce pulses of less than 300 ps duration with peak powers in excess of 25 kW [21,22]. These lasers can produce output pulses as short as large mode-locked lasers with peak powers as high as commercially available Q-switched systems and with a diffraction-limited output beam. Their main fields of application are range finding, micromachining, microsurgery, environmental monitoring, ionization spectroscopy and non-linear frequency generation [23]. In order to achieve a further decrease of complexity, passive Q-switches have also been applied to diode-pumped microchip lasers. The set-up is simple and compact consisting of a short piece of a solid-state gain medium Nd:YAG, bonded to a saturable-absorber crystal, for example Cr⁴⁺:YAG, each of 0.5 mm thickness [24]. Alternatively, the saturable absorber can be grown epitaxially onto the laser crystal. The facets at the interface between the gain medium and saturable-absorber crystal are coated dielectrically, such that the interface is totally transmitting at the oscillating frequency and highly reflecting at the frequency of the pump. The output face of the saturable absorber is coated to be partially reflecting at the oscillating frequency (reflectivity R) and provides the optical output from the device. As the resonator length is only a few hundred micrometres, the laser oscillates on a single longitudinal mode which allows Q-switching with pulse widths below 1 ns.

When pumped with one watt from a fibre-coupled diode, microchip lasers can even produce pulses as short as 218 ps, pulse energies as high as 14 mJ, time-averaged powers of up to 120 mW, with pulse repetition rates between 8 and 15 kHz. If semiconductor saturable-absorber mirrors are used as an end mirror with a Nd:YVO₄ laser, crystal pulses down to 37 ps can be realized at different wavelengths by Q-switching the microchip laser [25,26].

18.3 Cavity Dumping

As a result of a high population on the upper long-lifetime laser level, energy is stored in the active medium and subsequently emitted in the form of short laser pulses during Q-switching. In contrast, using cavity dumping, light energy is stored in the resonator. In this case, the loss of the cavity is switched from low to high. Cavity dumping is applied for laser media which have an upper laser level with a lifetime that is too short for Q-switching. For cavity dumping, the continuously pumped laser medium is placed between two highly reflective mirrors (see Figure 18.4). The intensity of the laser beam is highly amplified in the resonator. If, for example, an acousto-optic deflector, which is transparent during intensity build-up, is triggered in the resonator, almost the complete energy in the form of a short laser pulse can leave the resonator. For a cavity-dumping laser the acousto-optic deflector usually operates in an inactive state resulting in extremely high Q-values of the resonant cavity and very intense lasing within the cavity. If an electric pulse is applied to the device, nearly all the laser energies are dumped out of the resonant cavity in the form of a single optical pulse (see Figure 18.5). Typical pulse widths are of the order of 10 ns, i.e. a round-trip duration to dump out all photons. In order to achieve short pulses, the time in which the acoustic wave passes the beam diameter has to be as small as possible. For this reason, the beam is focused at the modulator.

Cavity dumping is used, e.g. to generate short laser pulses with a high peak power in continuously pumped argon lasers. The peak power of these laser pulses can be 50 times the

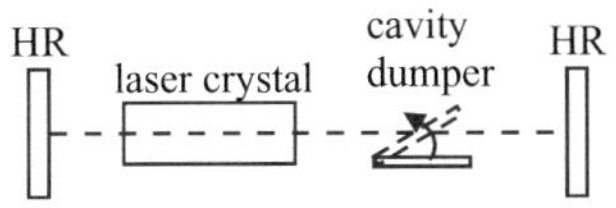

FIGURE 18.4 Principle of cavity dumping.

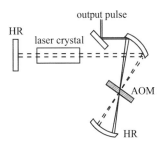

FIGURE 18.5 Cavity dumping by an acousto-optic modulator.

output power of the continuous wave mode. Q-switching is not useful in this case, since the lifetime of the upper laser level of these laser media is too short, meaning that a sufficiently high population inversion cannot be generated. Depending on the output coupling mechanism, cavity dumping makes it possible to select pulse repetition rates of up to several tens of MHz at pulse widths of 10–30 ns. The switching elements which are basically used for cavity dumping are electro-optic Pockels cells, acousto-optic devices and polarizing beam splitters. If used in combination with mode-locking, cavity dumping can generate sub-nanosecond pulses at repetition rates of several MHz without reducing the average laser power.

18.4 Mode-locking

While Q-switching can be used to generate pulses with high intensities in the ns range, mode-locking is used to generate ultrashort laser pulses with pulse duration in the ps to fs range.

Mode-locking can be used very effectively for lasers with a relatively broad laser transition bandwidth, and thus for lasers with a broad amplification profile, in which numerous longitudinal modes can oscillate simultaneously.

18.5 Master Oscillator with Power Amplifiers

If high-energy pulses with high brightness are required, pulse amplifiers have to be used. The generation of high-energy pulses is based on the serial combination of a master oscillator and a multistage power amplifier (master oscillator with power amplifier (MOPA)). The principle of the configuration is illustrated in Figure 18.6.

The oscillator generates the initial low-energy laser pulse with high-quality parameters, e.g. small beam divergence or narrow spectral width. The power amplifier with its large volume of active material then multiplies the energy up to a factor of 100. With multiple-stage amplifiers, even factors of 10^{12} can be reached. For a MOPA design, the pulse width and its influence on the amplification mechanism are of significant importance. If a Q-switched or mode-locked pulse with a pulse width shorter than the pumping rate and the spontaneous emission time of the amplifier enters the amplifier, energy, which was stored in the amplifying medium prior to the arrival of the pulse, will be extracted. As the input pulse passes through the rod, the atoms are stimulated to release the stored energy.

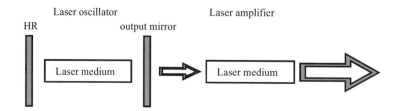

FIGURE 18.6 Schematic set-up of a MOPA configuration.

The amplifier can be described in this case by the two equations for the population inversion n and the photon flux ϕ:

$$\frac{dN}{dt} = -N \cdot c \cdot \sigma \cdot \phi \quad (18.14)$$

$$\frac{d\phi}{dt} = N \cdot c \cdot \sigma\phi - c \cdot \frac{d\phi}{dx} \quad (18.15)$$

where σ is the cross section for stimulated emission and c is the speed of light.

The rate at which the photon density changes is equal to the net difference between the generation of photons by simulated emission and the flux of photons leaving that region. The latter term in equation (18.15) is usually absent in basic oscillation processes of laser resonators (Figure 18.7).

The efficiency of amplifiers, i.e. the ratio of the output energy to the energy stored in the active medium, can be calculated by solving these equations. As a characteristic material-dependent parameter the saturation energy represents the energy which can be extracted from the amplifier divided by the small-signal gain. If the input signal, i.e. the input energy, is very small compared with the saturation energy, no saturation effects occur and the signal gains exponentially with the length of the amplifier medium. For higher energies, e.g. towards the end of the medium, the gain grows linearly. This implies that every excited state contributes its stimulated emission to the beam which obviously represents the most efficient conversion. Usually, lasers are designed to operate within this regime. The efficiency can be further increased by passing the active medium twice. Within a double-pass amplifier, a mirror at the output returns the beam and a $\lambda/2$ waveplate turns the polarization by 90° [27]. For Nd-doped glass laser materials, the saturation fluence is between 5 and 10 J cm^{-2} [28]. For Nd:YAG crystals, the saturation fluence is 0.2 J cm^{-2}, whereas ruby amplifiers provide a saturation fluence of 11 J cm^{-2}. Due to the inherent advantages of Nd:glass this medium has been chosen to be used for the realization of high-power systems for laser fusion experiments. By a superposition of 10 identical beam amplifier lines with pulse energies of 10 kJ each, a total pulse energy of 100 kJ have been realized by the NOVA laser system [29].

In very high-peak-power laser systems, the pulse is amplified until it begins to incur one of several non-linear problems associated with intense light.

At high optical intensities, the index of refraction becomes a function of intensity. The index of refraction determines the phase velocity of light and the optical path length experienced and so intense light begins to suffer phase delays, relative to weaker light. The accumulated phase-lag suffered by light in travelling through a medium is given by the B-integral:

$$B = \frac{2\pi}{\lambda} \int_0^L n_2(I) \cdot I(z)\, dz. \quad (18.16)$$

A beam with a non-uniform intensity distribution passing through a transparent medium suffers a delay of phase at the centre of the beam which differs from that at the beam edges. This alteration of the phase of the light wave is a distortion of the light beam which can seriously degrade the beam quality. If the B-value exceeds values of 3–5 filamentation, self-focusing occur. Besides self-focusing it is possible that a beam which is not perfectly smooth in its intensity profile will break up into beamlets and each of those may self-focus. Between whole-beam self-focusing and filamentation, the difference is only the relative growth rates of different spatial-frequency components of the beam and the initial amplitude of those components.

For mode-locked sub-ns pulses, however, the incident energy is far below the saturation energy as the high-peak-power would immediately destroy the optical components. For ultrashort laser pulses, the energy amplification usually is based on the chirped pulse amplification technique.

Regarding the pulse shape, the amplified pulse shapes differ significantly from the incident pulse shapes since the leading edge of the pulse experiences a higher gain. This usually leads to a distortion of the pulse shape which has to be taken into account by laser system designers and users alike.

18.6 Beam Characterization and Pulse Measurement

One of the most important parameters of a laser source is output power and, for pulsed lasers, pulse energy. Temporal measurement of the power of laser pulses down to several nanoseconds can be realized using fast diodes or oscilloscopes. The pulse energy can be calculated by applying the integral function $H = \int P(t)\, dt$. If the average power is measured by integral devices, the peak pulse power can roughly be estimated, if the repetition rate and pulse width are known.

All sensors can be subdivided into thermal detectors and sensors based on the internal and external photoelectric effect. For thermal photodetectors, the incident light heats up the sensor, and the temperature rise correlates with the laser pulse energy. The temperature difference can be measured using thermo-elements, temperature-dependent resistors or by using the pyroelectric effect.

In thermo-elements, the incident light of a laser pulse is completely absorbed by a measurement cone. The temperature rise is given by $\Delta T = H/C$, with H being the laser pulse energy and C is the heat capacity of the cone. ΔT is measured by thermo-elements and is transformed into $\Delta U = \alpha_{therm} \Delta T$, with α_{therm} being the thermal force in [V K^{-1}]. Pyroelectric sensors provide a spontaneous electric polarization $P_e(T)$. When the temperature alters, the effective surface charge changes and can be measured at a resistor R by $U(t) = R \cdot dQ/dt = A \cdot R \cdot dP_e/dT \cdot dT/dt$ with A representing the area of the sensor. Consequently, in pyroelectric sensors, voltages can only be measured when the temperature deviates. They are mainly used for determining the laser pulse energy, or as power meters. The energy can then be derived by calculating the time integral. Thermal detectors are nearly independent of the wavelength, which makes them a useful tool for infrared lasers. However, the temporal resolution is not very high. The rising time of pyroelectric sensors is in the range of a nanosecond, and the sensitivity is in the range of one µJ.

In addition to sensors based on thermal effects, detectors can exploit the internal photoeffect. Here, electrons are transferred by the incident radiation from the valence to the conduction band, resulting in a free electron and a free hole. The easiest solution is realized by a photo-resistor. Depending on the material, different band gaps and spectral sensitivities can be realized. To avoid free carrier generation by thermal effects, photoresistors are usually cooled. Their time resolution can be up to a few hundred ps. Photodiodes consist of pn junctions applied at a reverse bias. When light hits the depletion zone, free carriers are generated. Compared with photoresistors, the sensitivity of photodiodes is improved, and interference due to thermal effects can be neglected. Arrays of photodiodes can also measure the spatial and temporal distributions of laser pulses (e.g. CCD cameras).

Sensors which are based on the external photoeffect always operate in a vacuum. The external photoeffect is based on the generation of photoelectrons at a cathode by incident light. The efficiency and spectral sensitivity are strongly dependent on the material and the wavelength. The most simple device of these vacuum photodetectors is the vacuum diode. Here, the free electrons are accelerated and absorbed by an anode located at the opposite side of the diode. By using vacuum diodes, time resolutions down to 100 ps can be achieved. Higher sensitivities, compared to vacuum diodes, can be realized by photomultipliers. The primary photoelectrons from the cathode are accelerated and guided towards a dynode where secondary electrons are generated. By multiple impacts at numerous dynodes, amplification factors of 10^8 can be achieved. The amplification process can be built up within several ns, which makes photomultipliers unique sensors in terms of high amplification, broad bandwidth and low noise.

Characterization of the pulsed beam mainly consists of measuring the spatial, temporal and spectral properties, using appropriate methods. Whereas the spatial and spectral characterization of long and ultrashort laser pulses is identical to cw laser measurement methods (conventional beam profilers and spectrometers), the measurement of the temporal behaviour is more sophisticated for ultrashort pulses in the ps and fs regime. The measurement of ultrashort time intervals has been driven by the need for registration of fast events in chemistry, biology and electronics. Streak cameras and correlation methods are usually used to measure ultrashort time intervals. Both principles are based on mapping short time intervals in a spatial interval that can be easily measured using a camera or a microscope.

Streak cameras convert the time information, via the generation and deflection of free electrons, into spatial information. As the sensor can read out the information in two dimensions, the streak camera is able to record the time evolution of one spatial coordinate such as a line focus. Today, streak cameras cover the whole spectral range from the X-ray region to the near-infrared. The time resolution is limited to approximately 0.5 ps. The principle of streak cameras is described later [30]:

The incoming light passes through a one-dimensional slit and is imaged onto the photocathode of the tube. The number of electrons produced by the photoelectric effect is proportional to the number of incident photons and therefore to the intensity. They are accelerated towards the anode by an electric field in the tube. Before the electrons are imaged onto the phosphor screen, they pass a micro-channel plate (MCP) where they are amplified with a gain factor of 10^3–10^4. The HV sweep on the deflection plates is synchronized with the incident light pulse. Electrons generated at later times are more strongly deflected compared with electrons generated at earlier times. Therefore, they enter the MCP at different positions and can be detected on the phosphor screen at different x-coordinates.

For laser pulses considerably shorter than 1 ps streak cameras cannot be used [31]. For those short time intervals, autocorrelation and cross-correlation methods are usually used [32,33]. These methods do not resemble the incoming pulse directly; therefore, further mathematical transformations have to be applied.

The autocorrelation method is based upon a superposition of the direct incoming pulse E_1, and a delayed replica of the incoming pulse E_2 in a non-linear crystal. The main requirement for the non-linear effect is that it has to be instantaneous. The beam E_2 can be shifted in time by varying the beam path. The order of the autocorrelation method is determined by the order of the non-linear optical conversion in the non-linear crystal. The beams can either cross the crystal collinearly, which results in an interferometric autocorrelation, or under a certain angle, which results in an intensity autocorrelation. In both cases, however, a signal is recorded depending on the delay between the two incident pulses.

To determine the pulse width, the delay is gradually increased between successive laser shots and the higher order response signal is measured. This technique is called the multi-shot method. However, if the pulse-to-pulse fluctuations are significant, the technique provides a relatively high measurement error. Single-shot autocorrelators avoid these uncertainties [34]. In single-shot autocorrelators, the spatial information is also used, e.g. two line foci are superimposed in such a way that the temporal delay varies along the line focus [35]. The resulted signal can be measured by a CCD camera in one shot. An arbitrary point x_0 along the line focus is related to the temporal delay τ by:

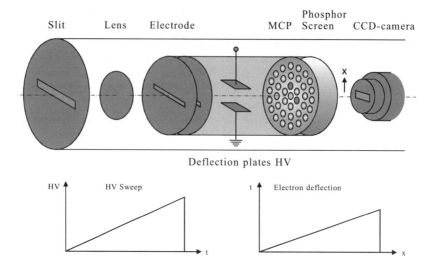

FIGURE 18.7 Principle set-up of a streak camera.

$$\tau = \frac{n \cdot x_0 \cdot \sin(\varphi/2)}{c} \quad (18.17)$$

where n is the refractive index of the non-linear medium, φ is the angle between the two incident beams and c is the speed of light. It can easily be seen that a smaller angle results in an increased time resolution. On the other hand, a certain angle facilitates the measurement as the resulting signal contains no portion of the incident beams and is, therefore, inherently background-free. For the described technique the measured spatial signal is directly proportional to the autocorrelation function. Under the assumption of a certain pulse shape, the pulse width can be determined from the measured autocorrelation signal. For rectangular pulse shapes, there is obviously no difference between the width of the autocorrelation function and the pulse width. For a Gaussian pulse shape, a correction factor of 0.7071 has to be adapted; for the hyperbolic secant shape, the correction factor is 0.6482. If the pulse shape is asymmetrical, an autocorrelation with even order will not reveal any information on the asymmetries as these functions are always symmetrical.

REFERENCES

1. Eggleston J M, DeShazer L and Kangas K 1988 Characteristics and kinetics of laser-pumped Ti:Sapphire oscillators *IEEE J. Quantum Electron.* **24** 1009–15.
2. Owyoung A, Hadley G R, Esherick R, Schmitt R L and Rahn L A 1985 Gain switching of a monolithic single-frequency laser-diode-excited Nd:YAG laser *Opt. Lett.* **10** 484.
3. Beaulieu A 1970 Transversely excited atmospheric pressure CO_2 laser *App. Phys. Lett.* **16** 504–5.
4. Dumanchin R and Rocca-Serra J 1970 High Power Density Pulsed Molecular Laser International Quantum Electronics Conference, Digest of Technical Papers (Kyoto) p 308.
5. Duley W W 1976 *CO2 Lasers: Effects and Applications* (New York: Academic).
6. Degnan J J 1988 Theory of the optimally coupled Q-switched laser *IEEE J. Quantum Electron.* **25** 214–20.
7. Collins R J and Kisliuk P 1962 Control of population inversion in pulsed optical masers by feedback modulation *J. Appl. Phys.* **33** 2009–11.
8. Hellwarth R W 1961 *Advances in Quantum Electronics* (New York: Columbia United Press).
9. McClung F J and Hellwarth R W 1962 Giant optical pulsations from ruby *J. Appl. Phys.* **33** 828–9.
10. McClung F J and Hellwarth R W 1963 Characteristics of giant optical pulsations from ruby *Proc. IEEE* **51** 46.
11. Stepke E T 1972 *The Key to Q-Switching Electro-Optical Systems Design.*
12. Wang C C 1963 Optical giant pulses from a Q-switched laser *Proc. IEEE* **51** 1767.
13. Wagner W G and Lengyel B A 1963 Evolution of the giant pulse in a laser *J. Appl. Phys.* **34** 2040.
14. Chesler R B, Karr M A and Geusic J E 1970 An experimental and theoretical study of high repetition rate Q-switched Nd:YAG lasers *Proc. IEEE* **58** 1899.
15. Degnan J J 1995 Optimization of passively Q-switched lasers *IEEE J. Quantum Electron.* **31** 1890–901.
16. Erickson L E and Szabo A 1967 Behaviour of saturable-absorber giant-pulse lasers in the limit of large absorber cross section *J. Appl. Phys.* **38** 2540–2.
17. Kafalas P, Masters J I and Murray E M E 1964 Photosensitive liquid used as a nondestructive passive Q-switch in a ruby laser *J. Appl. Phys.* **35** 2349.
18. Soffer B H 1964 Giant pulse laser operation by a passive, reversibly bleachable absorber *J. Appl. Phys.* **35** 2551.
19. Spaeth M L and Sooy W R 1968 Fluorescence and bleaching of organic dyes for a passive Q-switch laser *J. Chem. Phys.* **48** 2315.
20. Szabo A and Stein R A 1965 Theory of laser giant pulsing by a saturable absorber *J. Appl. Phys.* **36** 1562–6.
21. Zayhowski J J 1991 Q-switched operation of a microchip laser *Opt. Lett.* **16** 575–7.
22. Zayhowski J J and Dill C III 1992 Diode-pumped microchip lasers electro-optically Q-switched at high pulse repetition rates *Opt. Lett.* **17** 1201–3.
23. Zayhowski J J 1998 Passively Q-switched microchip lasers and applications *Rev. Laser Eng.* **26** 841–6.

24. Zayhowski J J and Mooradian A 1989 Single-frequency microchip Nd lasers *Opt. Lett.* **14** 24–6.
25. Braun B, Kaertner F X, Moser M, Zhang G and Keller U 1997 56 ps passively Q-switched diode-pumped microchip laser *Opt. Lett.* **22** 381–3.
26. Spuehler G J, Paschotta R, Fluck R, Braun B, Moser M, Zhang G, Gini E and Keller U 1999 Experimentally confirmed design guidelines for passively Q-switched microchip lasers using semiconductor saturable absorbers *J. Opt. Soc. Am. B* **16**(3) 376–88.
27. Michon M, Auffret R and Dumanchin R 1970 *J. Appl. Phys.* **41** 2739.
28. Krupke W F 1974 Induced-emission cross-section in neodymium laser glasses *J. Quantum Electron.* **10** 450–7.
29. Brown D C 1981 *High-Peak Power Nd:Glass Laser System* (Berlin: Springer).
30. Sarger L and Oberlé J 1998 *How to Measure the Characteristics of Laser Pulses Femtosecond Laser Pulses* ed C Rullière (Berlin: Springer) pp 177–201.
31. Sauerbrey R and Feurer T 1996 *Characterization of Short Laser Pulses Methods in Experimental Physics* vol 3 (New York: Academic).
32. Diels J-C, Fontaine J J, McMichael I C and Simoni F 1985 Control and measurement of ultrashort pulse shape (in amplitude and phase) with femtosecond accuracy *J. Appl. Opt.* **24** 1270–82.
33. Diels J-C, Fontaine J J and Rudolph W 1987 Ultrafast diagnostics revue *Phys. Appl.* **22** 1605–11.
34. Salin F, Georges P, Roger G and Brun A 1987 Single-shot measurement of a 52 fs pulse *Appl. Opt.* **26** 4528.
35. Collier J, Danson C, Johnson C and Mistry, C 1999 Uniaxial single shot autocorrelator *Rev. Sci. Instrum.* **70** 1599–602.

FURTHER READING

Diels J C and Rudolph W 1996 *Ultrashort Laser Pulse Phenomena: Fundamentals, Techniques and Applications* (Boston, MA: Academic).

Duling I N III 1995 *Compact Sources of Ultrashort Pulses* (Cambridge: Cambridge University Press).

Herrmann J and Wilhelmi B 1984 *Laser für Ultrakurze Lichtimpulse* (Berlin: Akademie).

Ifflände R 2001 *Solid-State Lasers for Materials Processing* (Berlin: Springer).

Koechner W 1999 *Solid-State Laser Engineering* (Springer Series of Optical Sciences 1) 5th edn (Berlin: Springer).

Ready J F 1978 *Industrial Applications of Lasers* (New York: Academic).

Rullière C 1998 *Femtosecond Laser Pulses* (Berlin: Springer).

Siegman A E 1989 *Lasers* (Mill Valley, CA: University Science Books).

Trebino R 2002 *Frequency-Resolved Optical Gating: The Measurement of Ultrashort Laser Pulses* (Dordrecht: Kluwer).

19

Ultrashort Pulses

Derryck T. Reid

CONTENTS

19.1 Theory of Ultrashort Pulse Generation and Mode-locking ... 259
    19.1.1 Active Mode-locking ... 259
    19.1.2 Passive Mode-locking ... 261
19.2 Sources of Ultrashort Pulses ... 262
    19.2.1 Dye Lasers ... 262
    19.2.2 Ti:sapphire Lasers ... 263
    19.2.3 Colour-centre Lasers ... 264
    19.2.4 Fibre Lasers ... 265
    19.2.5 Semiconductor Sources ... 266
    19.2.6 Other Common Solid-state Laser Sources ... 267
    19.2.7 Sources Based on Non-linear Frequency Conversion ... 267
    19.2.8 Sources of Amplified Ultrashort Pulses ... 268
19.3 Pulse Shaping and Dispersion in Optical Systems ... 268
    19.3.1 Linear Material Dispersion ... 268
    19.3.2 Non-linear Material Dispersion ... 270
    19.3.3 Other Sources of Dispersion ... 270
    19.3.4 Group-velocity Dispersion Compensation ... 271
    19.3.5 Fourier-transform Pulse Shaping ... 273
19.4 Diagnostic Techniques ... 273
    19.4.1 Direct Electronic Measurements ... 273
    19.4.2 Approximate Methods of Pulse-shape Measurement ... 274
    19.4.3 Exact Methods of Pulse-shape Measurement ... 275
19.5 Applications of Ultrashort Pulses ... 277
    19.5.1 Imaging ... 277
    19.5.2 Ultrafast Chemistry ... 278
    19.5.3 Semiconductor Spectroscopy ... 279
    19.5.4 Material Processing ... 279
    19.5.5 High Field Science ... 280
    19.5.6 Other Applications ... 281
References ... 281

19.1 Theory of Ultrashort Pulse Generation and Mode-locking

19.1.1 Active Mode-locking

In an inhomogeneously broadened laser system, optical frequencies within the emission bandwidth of the material can experience gain simultaneously at discrete frequencies corresponding to the longitudinal modes of the laser resonator. In the absence of any mechanism to coherently couple energy between these modes, the laser will operate with a continuous-wave (cw) output containing a narrow band of frequencies whose roundtrip cavity gain is highest. An alternative operating regime exists in which each mode is related to its neighbour by a fixed phase relationship, and it is this *mode-locked* condition that is used to generate a periodic sequence of ultrashort optical pulses.

To understand how mode-locking works, consider the frequency separation between adjacent longitudinal resonator modes which can be expressed as

$$\Delta\omega = 2\pi c/l \qquad (19.1)$$

where l is the optical roundtrip length of the resonator and c is the speed of light. If the nth longitudinal mode has an amplitude E_n then the total optical field can be denoted as

$$E(t) = \sum_n E_n \exp i[(\omega_o + n\Delta\omega)t + n\Delta\varphi] \quad (19.2)$$

where ω_o is the centre frequency of the output and $\Delta\varphi$ is the phase difference between adjacent modes. Figure 19.1 illustrates how a simple pulse can be produced by coherently adding together adjacent modes. In this example, three modes with equal amplitudes and a relative phase separation of $\Delta\varphi = 0$ are interfered (Figure 19.1a), and as further modes are added (Figure 19.1b), the pulse duration decreases.

Coupling of energy between adjacent cavity modes can be achieved in practice by modulating either the intra-cavity loss (*AM-mode-locking*) or the intra-cavity phase (*FM-mode-locking*). Consider first AM-mode-locking where an amplitude modulator with a modulation depth m is driven with a frequency Ω. If the unmodulated field of mode n is $E_n \exp(i\omega_n t)$, then the modulated field is given by

$$E = E_n \exp(i\omega_n t) \times (1 + m\cos\Omega t) \quad (19.3)$$

which can be written as

$$E = E_n \exp(i\omega_n t) + \frac{m}{2} E_n \exp(i(\omega_n + \Omega)t)$$
$$+ \frac{m}{2} E_n \exp(i(\omega_n - \Omega)t). \quad (19.4)$$

The second and third terms in equation (19.4) represent frequency sidebands added to the mode by amplitude modulation. When the modulation frequency is chosen so that $\Omega = \Delta\omega$, energy is coherently coupled from each cavity mode to its nearest neighbours and mode-locking is achieved. Phase modulation can be analysed in a similar way and the phase-modulated field can be written as

$$E = E_n \exp(i\omega_n t) \times \exp(im\sin\Omega t). \quad (19.5)$$

For small modulation depths ($m \ll 1$), the Taylor series approximation $\exp x \approx 1 + x$ can be used to give

$$E = E_n \exp(i\omega_n t) \times (1 + im\sin\Omega t) \quad (19.6)$$

which can be expressed as

$$E = E_n \exp(i\omega_n t) + \frac{m}{2} E_n \exp(i(\omega_n + \Omega)t)$$
$$- \frac{m}{2} E_n \exp(i(\omega_n - \Omega)t) \quad (19.7)$$

which indicates that phase-modulation can be used to achieve FM-mode-locking through the addition of spectral sidebands to each cavity mode.

A common practical amplitude modulator is an acousto-optic modulator in which a standing acoustic wave diffraction grating is formed twice every drive period, implying an electrical drive frequency of $\Delta\omega/2$ for mode-locking applications. For operation in the visible and near-infrared, fused silica is a common acousto-optic modulator material. Lithium niobate devices based on the Pockels effect are the most commonly used phase modulators and are attractive for mode-locked fibre-laser applications because fibre-pig-tailed waveguide devices are readily available.

The mode-locking mechanisms discussed so far are examples of *active mode-locking* in which the laser resonator is driven by an external reference frequency. Such a system is a classical forced oscillator, and for maximum efficiency, the external frequency must be accurately maintained at the resonant frequency of the laser cavity. Long-term thermal drift in

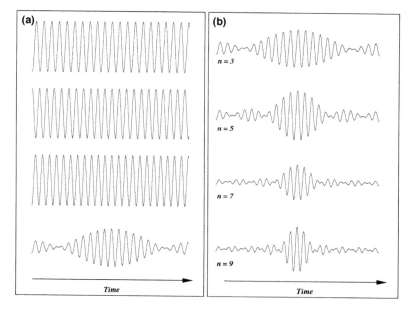

FIGURE 19.1 The principles of mode-locking: (a) three adjacent modes add coherently to produce a pulse; (b) shorter pulses require a larger number of n coupled modes.

the laser cavity length can make exact synchronization difficult, but a technique known as *regenerative mode-locking* can be applied to solve this problem. Using a fast photodiode detector, the mode-locked pulse sequence is sampled and the electrical signal bandpass-filtered to produce a clean sinusoidal waveform at the exact cavity frequency. With suitable filtering and amplification, the technique is self-starting because random mode-beating detected during cw operation is sufficient to produce the required drive frequency.

Active mode-locking has been employed in a wide variety of laser systems including argon ion [1], dye [2], erbium-doped fibre [3], semiconductor [4] and solid-state lasers [5]. Unlike passive mode-locking, no optical non-linearity is required which means that actively mode-locked lasers can be operated with low intra-cavity powers. The technique is particularly suitable for mode-locked telecommunication sources which need to be actively synchronized to an external clock source. Extremely high-frequency operation is possible using the technique of *harmonic mode-locking* in which the modulation frequency used is a multiple of the cavity frequency. In this way, sets of widely spaced *supermodes* are coupled together and interfere to generate a mode-locked pulse sequence with a repetition frequency equal to a high harmonic of the original cavity frequency. Lasers mode-locked using this method exhibit substantial high-frequency noise because the absence of strong coupling between independent sets of supermodes leads to random frequency-beating between them. A form of 'optical' active mode-locking exists in which gain modulation resulting from optical pumping using another mode-locked laser is the mode-locking mechanism. Known as *synchronously pumped mode-locking*, this technique has been used to obtain ultrashort pulses from oscillators incorporating gain media of laser or non-linear optical materials. The technique is most appropriate to laser gain materials which have a short-lived fluorescence lifetime and whose gain therefore exhibits a significant change when the pump light is modulated at high frequencies. Synchronous pumping has been used to generate mode-locked pulses from colour-centre lasers [6], dye lasers [7] and optical parametric oscillators (OPO) [8]. In mode-locked lasers, one uses the appropriate timing to achieve pulse compression through gain saturation. Mistiming the pump pulses changes the achieved pulse duration by orders of magnitude but has only a small effect on the output power. In OPOs, the exact matching of the master and slave oscillator pulse-repetition frequencies is critical for achieving high conversion efficiency at the desired wavelength, but the impact on the pulse duration is generally less noticeable although under certain circumstances substantial compression can result from careful control of the pump pulse timing [9].

19.1.2 Passive Mode-locking

In the active mode-locking technique, pulse shaping is determined directly by the performance (modulation depth, bandwidth, loss, etc.) of the optical modulator, and consequently, the generation of sub-picosecond duration pulses is difficult. To produce the shortest optical pulses, *passive mode-locking*, a technique based on exploiting intra-cavity optical non-linearities must be employed. The key difference between active and passive mode-locking methods is that, because passive mode-locking is a non-linear technique, the strength of the mode-locking action increases as the propagating pulses shorten and their peak intensities rise. One consequence of this non-linear behaviour is that, unlike active mode-locking which generally only locks together the phases of existing laser cavity modes, the passive technique commonly generates and mode-locks entirely new cavity frequencies which were not originally present during cw operation.

Passive mode-locking has been implemented using a wide range of different non-linear effects, but in every technique, the mode-locking process can be described in terms of *dynamic loss saturation* combined, in certain systems, with *dynamic gain saturation*. The principal mode-locking element is a *saturable absorber* whose loss decreases as the incident pulse intensity increases. A saturable absorber can take the form either of a physical component such as a solid-state semiconductor device [10] or a dye jet [11], or a virtual component (e.g. a Kerr lens) which replicates a saturable-absorber-like action using self-focusing or self-phase-modulation (SPM) effects. Saturable absorbers can be categorized as either *slow saturable absorbers* or *fast saturable absorbers*, depending on whether their recovery time is longer or shorter than the pulse duration, respectively. The pulse-shaping mechanisms are different in each case, and we will first describe mode-locking using a slow absorber. Most slow-saturable absorbers are physical devices, commonly dye molecules in solution [11] or semiconductor-doped glass [12], which rely on resonant excitation to produce an excited electronic state whose transmission is greater than that of the ground state. When an optical pulse is incident on a slow absorber, its leading edge is absorbed and creates an excited state which is relatively transparent to the trailing edge. This effect in isolation is not sufficient to generate ultrashort pulses and must be combined with dynamic gain saturation to achieve mode-locking. A laser medium with a gain relaxation time τ_g faster than the cavity roundtrip time, T_{cav}, but slower than the absorber recovery time, τ_a, must be used so that the gain recovers in time to amplify the next roundtrip pulse and the absorber recovers in time to be fully saturated by it. This principle is illustrated in Figure 19.2 which depicts the individual responses of the absorber and gain medium to an incident ultrashort pulse. A pulse propagating first through the absorber and then through the gain medium experiences a 'gain window' with a finite duration which strongly shapes the pulse. The situation is different when a gain medium with a slow relaxation time is used. Here, dynamic gain saturation cannot be used to shape the resonant pulses, and a fast saturable absorber is required to realize mode-locking. The time-dependent gain and loss profiles of the absorber and the gain medium are again depicted in Figure 19.2 and refer to an absorber with both a rapid response and recovery time. In this system, the dynamics of the absorber and gain medium no longer represent the limiting factor influencing the pulse duration, and other effects such as linear and non-linear dispersion become the dominant pulse-shaping mechanisms.

Pulse-shaping effects will be discussed in detail in Section 19.3, but it is useful to explain at this point the key mechanisms which are important in the context of passive mode-locking

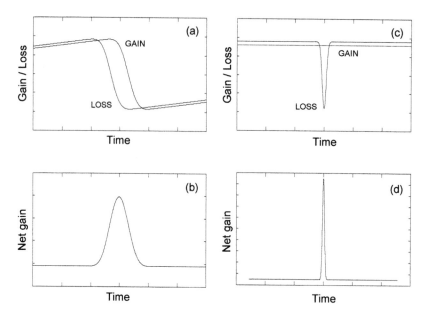

FIGURE 19.2 Saturable absorber action: (a) intra-cavity gain and loss profiles for a laser comprising a slow saturable absorber and a rapid relaxation time gain medium; (b) the net gain (gain − loss) for this system; (c) gain and loss profiles for a laser comprising a near-instantaneous saturable absorber and a slow relaxation time gain medium and (d) the net gain for this system.

using a near-instantaneous saturable absorber. Pulses propagating in a passively mode-locked laser experience strong soliton-like shaping effects associated with anomalous dispersion and SPM and can therefore be described in terms of a modified non-linear Schrödinger equation (NLSE) [13]. Unlike true solitons which propagate in a medium whose non-linearity and dispersion are distributed uniformly throughout the system, the pulses travelling within a mode-locked laser cavity generally experience dispersion and non-linearity within separate regions of the resonator. The non-linearity is localized within the gain medium while the (positive) dispersion arises here and within a separate (negative) dispersion-compensating component such as a prism-pair. This departure from the ideal soliton description means that passively mode-locked lasers operating under these conditions are better described as *solitary* lasers [14] rather than true soliton lasers [15]. The operating principles of such lasers have been analysed in detail by Krausz et al. [16], who have shown that, under the weak pulse shaping approximation, the output pulses can be described by a complex amplitude,

$$a_n(t) = \mathrm{sech}(t/\tau)\exp(i\varphi_n)\left(\frac{W}{2\tau}\right)^{1/2} \quad (19.8)$$

where n is the roundtrip number and φ_n is the linear phase given by,

$$\varphi_n = \varphi_o + nD/\tau^2 \quad (19.9)$$

and D is the roundtrip group-delay dispersion, W is the pulse energy, φ is the roundtrip non-linear phase shift and τ is the pulse duration given by,

$$\tau = 2D/\varphi W. \quad (19.10)$$

This result is identical to the solution of the NLSE for a soliton propagating in a uniform medium and justifies the common assertion that passively mode-locked lasers produce pulses with a $\mathrm{sech}^2(t)$ intensity profile. In actual mode-locked lasers, the situation is complicated by the effects of lumped dispersion/non-linearity, higher order linear dispersion and finite gain bandwidth which mean that spectral and temporal measurements of real lasers rarely conform to the ideal $\mathrm{sech}^2(t)$ solution [17]. Numerous implementations of active and passive mode-locking exist, and in this section, we have concentrated on outlining only the general operating principles. Such is the versatility of the passive mode-locking technique that the majority of practical ultrafast laser systems are now based exclusively on passive methods, with the exception of hybrid systems which employ multi-stage pulse compression schemes to generate femtosecond pulses from gain-switched or actively-mode-locked master oscillators. In the section which follows, we will review specific mode-locked laser systems and discuss the particular mode-locking effect in each case.

19.2 Sources of Ultrashort Pulses

19.2.1 Dye Lasers

Historically, the first practical passively mode-locked ultrafast lasers were those based on a gain medium of a thin (~10 μm) organic dye jet and incorporating a slow saturable absorber jet at a separate cavity focus as described in Ref. [18]. The most common combination was R6G in ethylene glycol (gain jet) and DOdcI in ethylene glycol (absorber jet) which produced pulses at a wavelength of 620 nm [10]. One notable refinement of this approach was the introduction of *colliding-pulse mode-locking* (CPM) [18] in which two counter-propagating pulses within a ring cavity are synchronized to arrive simultaneously at the absorber jet, so achieving greater saturation of the absorber and allowing the pulses to be transmitted with a loss significantly smaller than in the single-pulse case. CPM enabled the generation of 65 fs pulses directly from the laser [19] and represented the first demonstration of a system producing sub-100 fs pulses [18]. A further important refinement to the CPM dye laser was the compensation of intra-cavity group-velocity dispersion by incorporating a Brewster-angled 4-prism sequence [20] which resulted in pulse durations of 27 fs [21].

Throughout the 1980s, the CPM dye laser was the principal source for ultrafast laser science, and indeed (after amplification and compression), it was responsible for the record-breaking 6 fs duration pulses [22], which, for an entire decade, remained the shortest pulses ever produced. One drawback of the CPM configuration was the very low efficiency of the system, which required several watts of cw pump power from an Ar$^+$-ion laser to produce only around 10 mW of average mode-locked power. Synchronous pumping using an actively mode-locked ion laser allowed R6G dye lasers to produce mode-locked pulses without a saturable absorber and enabled the output power to be increased to 100 mW with pulse durations of around 200 fs [23]. Further developments included a hybridly mode-locked dye laser based on kiton red and malachite green and operating at 645 nm, which, as a commercial system, was available with average output powers of up to 300 mW and pulse durations of 150 fs [24]. All dye lasers are maintenance-intensive systems, and this, together with the hazardous nature of many of the chemicals used, led to their gradual replacement by more convenient solid-state alternatives, based principally on Ti:sapphire.

19.2.2 Ti:sapphire Lasers

The creation in 1986 [25] of a new broadband laser material—titanium-doped sapphire or Ti:sapphire—began a revolution in ultrafast laser sources which still continues today. Unlike other contemporary solid-state materials for mode-locked lasers such as Nd:YAG, Ti:sapphire is a *vibronic* gain medium which exhibits a strong electron-phonon coupling between the titanium ion transition and the host lattice. This vibronic coupling leads to a broad continuum of possible transition energies and results in a fluorescence bandwidth covering 670–1050 nm, making Ti:sapphire capable of supporting extremely short femtosecond pulses. Furthermore, Ti:sapphire can be grown as high-quality crystals, is mechanically strong with good thermal conductivity and has an absorption band in the blue/green which is compatible with the Ar$^+$-ion lasers formerly used to pump existing mode-locked dye oscillators. The long fluorescence lifetime of Ti:sapphire means that, unlike dye lasers, mode-locking requires either active intra-cavity modulation or a passive method using a fast saturable absorber. Various methods were reported including acousto-optic mode-locking [26], intra-cavity saturable absorbers in the form of a dye jet [27] or semiconductor-doped glass plate [12], additive-pulse mode-locking (APM) (see Section 19.2.3) [28–30], semiconductor saturable absorber mirrors (SESAMs) [31,32] and Kerr-lens mode-locking (KLM) [33]. Of the various approaches listed here, only the final two methods have prevailed, and these are now discussed in detail.

The first passively mode-locked Ti:sapphire lasers were based on APM in which mode-locking is achieved using interference between the intra-cavity pulse and a self-phase-modulated replica propagating in an auxiliary cavity. An important breakthrough was made when in 1990 it was reported that an APM Ti:sapphire laser continued to operate even when the auxiliary cavity was blocked [34]. This discovery was attributed to an entirely new mode-locking effect known as KLM or *self-mode-locking* in which the presence of a Kerr-lens within the gain medium during mode-locked-operation changes the mode focusing within the cavity so that, compared to cw operation, mode-locked pulses experience higher gain. Two configurations of KLM exist and are known as *soft aperture mode-locking* [35], where the presence of the Kerr-lens increases the gain by improving the overlap between the pump and laser modes, and *hard-aperture mode-locking* [36], in which a physical aperture such as a slit or the edge of a prism is adjusted to introduce greater loss for cw operation. Because KLM is based on a non-resonant non-linearity, Ti:sapphire lasers can be mode-locked at any wavelength in their gain bandwidth using this technique. Using specially designed broad-bandwidth mirror sets, femtosecond Ti:sapphire lasers have been demonstrated with continuous tunability across ~300 nm [37]. Independently tunable dual wavelength operation from a single mode-locked laser has also been reported [38] and enables two exactly synchronized pulse sequences to be produced which are suitable for spectroscopic pump-probe experiments [39]. Recent advances in controlling the intra-cavity group-velocity dispersion characteristics have led to the generation of sub-5 fs pulses directly from a Ti:sapphire oscillator [40] which utilize the entire gain bandwidth of the material. The principal drawback of the KLM method is that it is not self-starting because mode-locked operation must be seeded by an intense noise spike or other short fluctuation. Different starting methods have been applied successfully including mirror tapping, acousto-optic modulation [41] and mode-dragging [42]; once mode-locking has been initiated, it is generally stable until some external perturbation disturbs the intra-cavity beam.

An alternative method for obtaining femtosecond pulses has been pioneered largely in parallel with the KLM efforts and is based on using resonant non-linearities in semiconductors as fast saturable absorbers. The approach is to fabricate a semiconductor multiple-quantum-well (MQW) device whose bandgap has been engineered to match the laser wavelength and whose absorption can be saturated at high fluences. Initially, these devices were utilized in an APM geometry in which the MQW device was situated in an auxiliary cavity [43], but it was later realized that an equivalent monolithic configuration could be obtained by sandwiching the MQW layer between two high-reflectivity mirrors (see Figure 19.3) to form a Fabry–Pérot structure [44], and the resulting element was known as an anti-resonant Fabry–Pérot saturable absorber (A-FPSA). The first application of such a *SESAM* for mode-locking Ti:sapphire was reported in 1993 [45], where self-starting operation was demonstrated. A typical modern Ti:sapphire laser design is shown in Figure 19.4 and illustrates how a saturable absorber mirror can be integrated into the cavity. Further design enhancements such as reducing the finesse

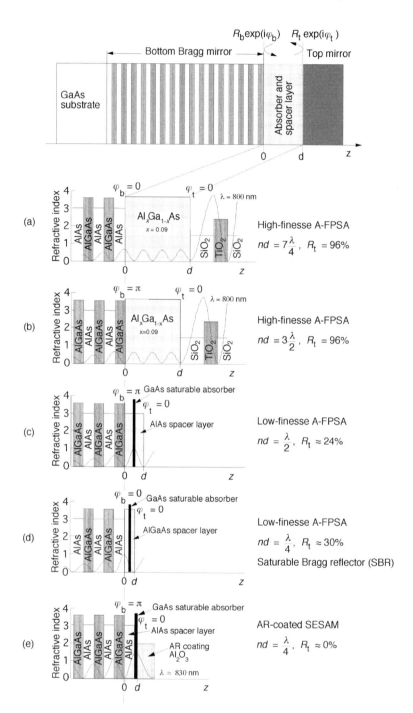

FIGURE 19.3 Different SESAM designs for a centre wavelength of about 800 nm which is suitable for a Ti:sapphire laser. The high-finesse A-FPSA was the first intra-cavity saturable absorber that stably cw-mode-locked solid-state lasers with long (>μs) upper state lifetimes. The saturable Bragg reflector is a special case of a low-finesse A-FPSA for which all layers are quarter-wave layers. (Image courtesy of Ursula Keller, Swiss Federal Institute of Technology.)

of the Fabry–Pérot cavity resulted in SESAMs with a broader wavelength response which enabled sub-10 fs self-starting operation.

19.2.3 Colour-centre Lasers

A new source of femtosecond pulses in the near-infrared was reported in 1984. Termed the *soliton laser* [15,47], the system was a synchronously pumped KCl:Tl^0 (1) colour-centre laser tunable from 1.4 to 1.6 μm and modified to allow direct generation of pulses as short as 50 fs, substantially shorter than the 8 ps achieved in synchronously pumped operation alone. Femtosecond operation was achieved by coupling a length of polarization-preserving anomalously dispersive fibre to the main laser cavity using a beam splitter and retroreflector arrangement. This coupled-cavity arrangement later became

Ultrashort Pulses

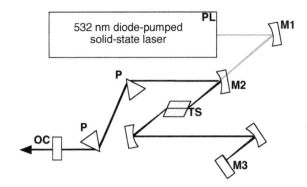

FIGURE 19.4 Typical configuration of a modern femtosecond Ti:sapphire laser. PL, pump laser; M1, pump-focusing mirror; M2, high-reflectivity cavity mirror (800 nm); TS, Ti:sapphire crystal, typically 2–10 mm long; M3, SBR/SESAM mirror; P, dispersion-compensating prisms; OC, output coupler.

known as APM and produced short pulses when the optical length of the fibre was adjusted to be an integral multiple of the main cavity length. In the original demonstration, operation was attributed to bright soliton generation in the fibre arm, but numerical modelling later showed [48] that mode-locking was enhanced even when the pulses returned from the fibre were temporally broadened, implying that solitonic effects were not necessary for femtosecond operation. This explanation was later confirmed experimentally when 260 fs pulses were generated from a KCl:Tl colour centre laser mode-locked using only normally dispersive fibre [49]. Pulses of 64 fs duration were subsequently produced from the same laser when erbium fibre, which exhibits enhanced non-linearity, was used [50]. This behaviour was consistent with theories which emphasized the importance of interference between the pulse travelling in the master cavity and the self-phase-modulated pulse returned by the non-linear fibre [51,52]. As Figure 19.5 illustrates, substantial pulse shortening can be achieved by this effect, even after only a single-pass. The long upper-state lifetime of the laser-active centres in the KCl:Tl gain medium allowed particularly stable APM operation, but mode-locking of this kind was also successfully employed in an actively stabilized NaCl:OH⁻ colour-centre laser to produce 110 fs pulses [53]. The need to maintain colour-centre crystals at liquid-nitrogen temperatures limited the convenience of these lasers, but until the advent of Ti:sapphire, they remained the only solid-state sources capable of femtosecond operation.

19.2.4 Fibre Lasers

Ultrafast fibre lasers, particularly those based on silica doped with rare-earth Er^{3+}, Nd^{3+} or Yb^{3+} ions, have enjoyed renewed attention as efforts to scale their average output powers and pulse energies to practical levels have proved successful. Four main mode-locking strategies have been applied to these systems: active mode-locking [54,55] which is generally restricted to picosecond generation; intensity-dependent feedback using a non-linear amplifying loop mirror (NALM) [56]; polarization APM [57–59] and semiconductor saturable absorber mode-locking [60]. The upper-state lifetime of rare-earth-doped fibres is long and therefore a fast saturable absorber mechanism is necessary for passive mode-locking, limiting the choice of mode-locking elements to those based on Kerr or semiconductor non-linearities.

The development of contemporary actively mode-locked fibre sources has concentrated on Er-doped lasers because of the compatibility of their output wavelengths with the 1.5 µm optical communication window in standard silica fibre. One common configuration is the sigma laser which comprises separate gain and phase-modulation sections and has produced 1.5 ps pulses at a repetition frequency of 10 GHz [61]. Passively mode-locked designs based on the non-linear loop mirror work [62] by exploiting the Kerr effect in an intracavity fibre Sagnac interferometer containing a gain section

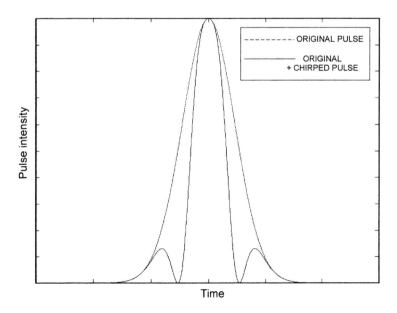

FIGURE 19.5 APM locking. Figure shows the reduction in pulse duration obtained by interfering a pulse with its self-phase-modulated replica. In this example, the replica has been self-phase-modulated with a peak phase shift of 2π.

[63]. With suitable control of the intra-cavity polarization, such a NALM can be set up to transmit light of high intensities but reflect low intensities because of the differential phase shift induced between the two interferometer arms. By adding the loop mirror to a unidirectional ring cavity, a *figure-of-eight laser* [64,65] is formed (Figure 19.6) which is capable of producing femtosecond soliton pulses with average powers of around 1 mW [66]. Drawbacks of the figure of eight laser include the presence of spectral sidebands at the shortest pulse durations [67] and a tendency for multiple pulse operations.

Mode-locking can also be achieved using a variation on the APM technique. An elliptically polarized pulse propagating inside the fibre can be resolved into right- and left-handed components of different intensities which accumulate a differential non-linear phase shift through the optical Kerr effect. The launch condition into the fibre is adjusted using a waveplate or polarization controller so that interference between the two circular polarization components at a polarizing beam splitter rejects the wings but transmits the centre of the pulse. The technique can also be understood in terms of non-linear polarization rotation in which Kerr-induced birefringence in the fibre rotates at the centre of the pulse differently from the wings so that only the pulse centre is transmitted by the polarizing beam splitter. Lasers based on *polarization-APM* can be operated in a soliton mode [68] or in a dispersive mode (*'stretched pulse'*) mode [69]. Notably, significant average output powers have been demonstrated from stretched-pulse-APM lasers, with up to 90 mW being reported at pulse energies of 2.25 nJ [70].

Semiconductor saturable absorber mode-locking has been demonstrated in fibre lasers using a variety of configurations including bulk InGaAsP on InP [71] and saturable Bragg reflectors based on InGaAs/InP MQWs on a AlAs/GaAs mirror structure [72]. Sub-picosecond pulse durations are readily available from these systems at repetition frequencies in the 10–100 MHz range. Another variation which has been demonstrated is mode-locking using a semiconductor optical amplifier (SOA) and 900 fs pulses at 10 MHz has been reported in a ring laser configuration incorporating a quantum-well diode amplifier. Fibre lasers mode-locked using semiconductor saturable absorbers are self-starting and exhibit stable mode-locking, normally at the fundamental cavity frequency. The narrow bandwidth of saturable absorber devices at near-infrared wavelengths makes the generation of sub-100 fs pulses difficult.

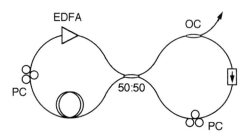

FIGURE 19.6 Schematic diagram of a figure-of-eight femtosecond fibre laser. EDFA, erbium-doped fibre amplifier gain section; OC, output coupler; PC, polarization controller.

19.2.5 Semiconductor Sources

Diode lasers and external-cavity lasers based on SOAs are of importance because of their ability to produce ultrashort pulses from a compact system at wavelengths of relevance to optical fibre communications. The two principal methods used to generate ultrashort pulses from such devices are mode-locking with an external cavity [73,74] or *gain-switching* (equivalently known as Q-switching) in a bow-tie waveguide structure [75]. Sub-picosecond pulse durations are now routinely produced using dispersion compensation and/or non-linear pulse compression techniques which, in fact, pre-empted parallel work in solid-state laser mode-locking [76], and by combining these strategies with Er:fibre amplification, 20 fs pulses have been produced [77]. The mode-locking and gain-switching techniques each have their individual advantages. Gain-switching provides a simple and compact approach but significant timing jitter and poor spectral quality can accompany pulses produced in this way. Mode-locked systems are physically larger than gain-switched devices because they require an external cavity and additional optical components, but shorter and higher quality pulses are achievable.

Laser action in electrically pumped semiconductor gain media is unique because the gain can be directly controlled by varying the drive current across the device. As one primary application of semiconductor lasers is as signal sources for optical fibre communications, the ability to modulate their output at high data-bit frequencies is particularly important. The gain-switching method works by biasing a laser diode with a constant dc current slightly below the threshold value and then, using a *bias tee*, superimposing a radio-frequency (rf) current derived from an electrical pulse synthesizer or sine-wave generator. For a short time each period, the rf current brings the diode above threshold and pulsed laser emission occurs, with the pulse duration being sensitive to the relative magnitudes of the dc and rf currents. Increasing the optical gain coefficient and the internal photon density and decreasing the photon lifetime inside the laser waveguide by minimizing facet reflectivity and separation are ways of increasing the maximum modulation frequency [78]. Maximum modulation frequencies of around 70 GHz have been reported by using short cavity lengths and strong optical confinement, as well as minimizing parasitic effects associated with the rf driver circuitry [79]. Gain-switched Fabry–Pérot diodes are commercially available offering pulse durations of 17 ps at modulation frequencies of 2 GHz [80]. One problematic characteristic of gain-switched laser diodes is their tendency to produce highly chirped output pulses. Modulating the device current leads to a modulation in the density of electrons and holes which, in turn, perturbs the positions of the Fermi levels lying beside the valence and conduction bands and modifies the oscillation wavelength of the laser. In this way, a non-linear frequency chirp is imprinted across the laser pulse as the drive current is modulated, and this effect places a limit on the minimum pulse duration that can ultimately be produced.

Mode-locked pulses have been produced from SOAs using the same range of techniques that are commonly applied to other laser media. Because mode-locking requires the mode from the SOA to be efficiently coupled into an external cavity, the SOA must be designed to operate with a fundamental transverse mode and this is achieved using flared-waveguide amplifier structures [81] with facets which are anti-reflection-coated to minimize etalon effects. Active mode-locking [82] has been demonstrated by using gain modulation near threshold [83], and passive mode-locking has been reported using a MQW, saturable Bragg [84] reflector which resulted in 5.6 ps pulses [85]. A theoretical analysis of active mode-locking in SOAs has been used to determine the practical limitations of these systems for producing sub-picosecond pulses [86], while similar studies of passively mode-locked semiconductor lasers have concentrated on understanding the roles of non-linearity, dispersion and pulse interaction within the cavity [87]. Active and passive techniques can be combined to implement *hybrid mode-locking* [88], and this method has been used to stabilize a passively mode-locked system which exhibited fourth-harmonic mode-locking due to gain saturation effects in the SOA [85]. A linear-cavity variation of the colliding pulse mode-locking technique has been applied in a monolithic form to semiconductor lasers [89], and the technique, which is implemented using laser diodes comprising separate saturable absorber and gain sections [90–92], is now an established method for directly mode-locking laser diodes without using an extended cavity.

19.2.6 Other Common Solid-state Laser Sources

Femtosecond Ti:sapphire oscillators are widely acknowledged as the modern standard for ultrafast laser systems, but other solid-state sources offer the potential for direct diode pumping, miniaturization and operation at wavelengths or average powers not accessible using Ti:sapphire. Alternative systems based on Nd or Cr ion transitions have also been shown to be capable of ultrashort picosecond or femtosecond operation, and a brief review of these sources will now be presented.

Laser materials based on Nd active ions exhibit emission bandwidths enabling sub-picosecond pulse generation, and Nd-based mode-locked lasers have existed in a variety of configurations for many years, based either on flashlamp excitation or direct diode-pumping at 808 nm. Actively mode-locked Nd:YAG mainframe systems produce a few picosecond duration pulses at repetition frequencies of around 80 MHz, and these have been commonly used for synchronous pumping of other ultrashort lasers [93]. The long upper-state lifetime of Nd:YAG and related materials means that attempts to passively modelock these lasers using saturable absorbers often result in the production of a mode-locked pulse sequence underneath a sub-µs-duration Q-switched envelope [94]. APM-locking has been used successfully to achieve cw-mode-locked operation from Nd:YLF [95] and other Nd-based lasers, and this technique produces pulse durations of around 1 ps. Passive mode-locking of Nd-based lasers using saturable absorber mirrors has been applied to eliminate Q-switching [96] and has enabled the demonstration of picosecond sources with repetition frequencies of 77 GHz [97] and average output powers of >4 W [98]. Femtosecond pulses with durations of 175 fs and an average power of 1 W have been produced using Nd:glass [99] which has one of the broadest fluorescence spectra of the Nd materials.

Cr-ion-doped crystals are promising broadband gain materials for generating femtosecond pulses and are particularly interesting because many have absorption bands which are suitable for pumping using readily available diode lasers. Crystals can be obtained with active ions of Cr^{2+}, Cr^{3+} and Cr^{4+} which, respectively, have transitions covering wavelengths in the 2.0 µm, 800 nm and 1.4 µm regions. Femtosecond mode-locked operation has been demonstrated in Cr^{4+}:YAG at 1.52 µm [100], Cr^{4+}:Mg_2SiO_4 (Cr:forsterite) at 1.25 µm [101,102], Cr^{3+}:$LiSrAlF_4$ (Cr:LiSAF) at 850 nm [103] and Cr^{3+}:$LiSrGaF_6$ (Cr:LiSGAF) [104]. Direct diode pumping of Cr-ion-based lasers is attractive because of the opportunities for producing low-noise, compact and efficient femtosecond sources, but the limited power available from high-quality pump diodes results in low intra-cavity powers which make KLM operation difficult. As a result, stable operation is better achieved using a physical saturable absorber, and SESAM/SBR mode-locking has been applied successfully to many of these lasers [105].

19.2.7 Sources Based on Non-linear Frequency Conversion

Laser sources alone are unable to provide ultrashort pulses continuously tunable from the ultraviolet to the mid-infrared, and large gaps exist at wavelengths where no broadband laser materials are available. Harmonic generation using Ti:sapphire can be used to access some parts of the visible [106] and ultraviolet [107] regions, but infrared wavelengths cannot be generated in this way and coverage is limited by the ability to tune the source itself. The non-linear production of infrared wavelengths requires parametric down-conversion, and as early as 1972, the synchronously pumped OPO was proposed as an effective means for generating ultrashort pulses in new wavelength regions [108]. Before the invention of the self-mode-locked Ti:sapphire laser, ultrafast pump sources with sufficiently high peak power to enable singly resonant operation of an OPO were not available. For this reason, the synchronously pumped parametric oscillator has its origins in frequency-unstable, doubly resonant configurations such as the one demonstrated by Piskarskas and pumped by the second harmonic of a picosecond cw-mode-locked Nd:YAG laser [109]. Although the output of a doubly resonant OPO is generally characterized by large amplitude and spectral instabilities, Ebrahimzadeh and co-workers exploited the low operating threshold associated with this configuration to demonstrate an all-solid-state oscillator based on KTP and pumped by 15 ps pulses from a frequency-doubled cw-mode-locked Nd:YLF laser [110].

The development of synchronously pumped OPOs had for many years relied on picosecond pump lasers because they represented the only sources capable of providing sufficiently large pulse energies. In 1989, Edelstein et al. became the first to demonstrate a high-repetition rate continuous-wave femtosecond OPO using an oscillator based on a thin crystal of KTP and pumped at the intra-cavity focus of a dye laser [111]. The singly resonant OPO produced 105 fs pulses at a repetition frequency of 100 MHz, and tuning was reported from 755–1040 nm (signal) and 1.5–3.2 µm (idler) [112–115]. The low output powers from the CPM-dye laser necessitated the alignment-critical intra-cavity pumping approach, so an important advance was made when Mak et al. demonstrated a KTP-based femtosecond OPO pumped extra-cavity by the output of a hybridly mode-locked dye laser operating at 645 nm [116]. This configuration allowed independent alignment of the laser and the OPO and was an important intermediary to the Ti:sapphire-pumped femtosecond OPO first demonstrated by Pelouch et al. [117] which produced near-infrared 57 fs pulses at average signal powers of 340 mW. OPO-based visible sources of femtosecond pulses have also been demonstrated, and access to the yellow/red region has been reported using intra-cavity frequency doubling of KTP [118] and RTA-based [119] oscillators. Direct oscillation in the visible was first demonstrated by Driscoll et al. in a β-barium-borate-based OPO pumped by the second harmonic of a self-mode-locked Ti:sapphire laser and resonant at 630 nm [120]. Pulses as short as 14 fs were reported from this oscillator [121] because of its unique phase-matching geometry [122]. New, highly non-linear materials based on the *quasi-phase-matching* and *periodic-poling* techniques emerged in the late-1990s and were soon applied as gain media in femtosecond [123] and picosecond [124] OPOs. Using materials such as PPLN [125] and PPRTA [126], oscillation from Ti:sapphire-pumped OPOs has been achieved at wavelengths extending to the long-wavelength edge of the transmission window at 6.8 μm [127].

19.2.8 Sources of Amplified Ultrashort Pulses

Modern ultrafast oscillators can produce pulses with peak powers of >1 MW [128,129], average powers of several watts [130] and cavity frequencies from 4 MHz [131] to 2 GHz [132], but many applications (see Section 19.5) require ultrashort pulses with further enhanced intensities and/or lower pulse-repetition frequencies. Additional amplification stages can be used to satisfy these requirements, and these are commonly implemented in Nd:glass for picosecond pulses and Ti:sapphire for femtosecond pulses. Other media are also becoming more widely used for amplifier configurations and include Yb:glass [133], Cr:LiSAF [134] and Er/Yb fibre [135]. In order to avoid damage and unwanted non-linear effects caused by high pulse intensities within the amplifier system, most practical sources now apply the *chirped-pulse amplification* (CPA) approach [136,137] in which a low-energy femtosecond seed pulse stretched to sub-ns durations to reduce its peak power is then amplified and is finally compressed to its original duration. Using the CPA technique, pulses with peak powers as high as 1.5 PW have been generated [138]. Two possible amplifier geometries are used which are categorized either as *multi-pass* or *regenerative* amplifiers. In the multi-pass configuration, the injected pulse is refocused many times through the same gain crystal using a system of mirrors. The gain crystal is pumped above the saturation fluence in order to extract the maximum stored energy from the medium per pass. The multi-pass geometry is more commonly used for high-average-power operation, and critical alignment is required to maximize the overlap between the pump mode profile and the multiple beams intersecting the gain medium. The alternative regenerative amplifier is essentially a stable laser resonator which includes a Pockels cell and a polarizing beam splitter to enable pulses to be switched into and out of the cavity. The gain per pass is considerably lower than in the multi-pass configuration, but regenerative systems produce superior beam quality and their performance can be optimized without affecting the beam pointing of the output. Commercially available Ti:sapphire amplifiers commonly use the regenerative method and are available with pulse energies of up to 1 mJ at repetition frequencies in the 1–300 kHz range.

Several factors affect the quality of the pulse amplification process and the ability of the system to produce output pulses with a similar duration and spectral bandwidth similar to the input pulses. *Gain narrowing* is a fundamental effect caused by the wavelength dependence of the amplifier gain material and, when the seed pulse is centred on the peak of the amplifier gain, leads to greater amplification in the centre of the pulse spectrum than in the wings. The consequent spectral narrowing caused by this effect leads to an increase in the pulse duration. In a similar way, *gain saturation* effects can lead to spectral broadening and introduce chirp on the pulse due to non-uniform amplification across the pulse spectrum. Strategies for minimizing spectral re-shaping include incorporating intra-amplifier frequency filters and using seed pulses whose wavelength is offset from the spectral centre of the amplifier gain. Proper control of second-, third- and fourth-order *group-velocity dispersion* inside an ultrafast amplifier is important and is critically dependent on the design and adjustment of the pulse compressor. Increasingly, exact pulse diagnostics such as FROG and SPIDER (see Section 19.4.3) are being used to characterize and minimize the high-order spectral phase errors which ultimately limit the maximum peak power which can be achieved.

19.3 Pulse Shaping and Dispersion in Optical Systems

19.3.1 Linear Material Dispersion

Chromatic dispersion, the variation of refractive index with wavelength, leads to a wavelength-dependent group velocity which is responsible for pulse broadening and break-up in optical systems. Dispersive effects in materials originate as a result of a frequency-dependent dielectric susceptibility:

$$P(\omega) = \varepsilon_o \chi^{(1)}(\omega) E(\omega) + \varepsilon_o \chi^{(2)}(\omega) E(\omega)^2 + \varepsilon_o \chi^{(3)}(\omega) E(\omega)^3 + \cdots \quad (19.11)$$

which is due to physical resonances (poles) in the electronic response to an applied *E*-field. Linear dispersion refers to the refractive index associated with the real part of the $\chi^{(1)}$ term, i.e.

$$n_o = \sqrt{1 + \text{Re}\{\chi^{(1)}\}}. \quad (19.12)$$

Dispersive effects are best analysed in terms of optical phase which is related to the refractive index by

$$\varphi(\omega) = \frac{n_o(\omega)\omega L}{c} \qquad (19.13)$$

where L is the medium length and c is the vacuum speed of light. The local variation of the spectral phase due to an optical medium can be represented as the Taylor series

$$\varphi(\omega) = \varphi(\omega_o) + (\omega - \omega_o)\frac{d\varphi}{d\omega}\bigg|_{\omega_o}$$
$$+ \frac{1}{2}(\omega - \omega_o)^2 \frac{d^2\varphi}{d\omega^2}\bigg|_{\omega_o} + \frac{1}{6}(\omega - \omega_o)^3 \frac{d^3\varphi}{d\omega^3}\bigg|_{\omega_o} + \cdots$$
$$(19.14)$$

where ω_o is the centre frequency of the incident light. When an ultrashort pulse propagates through the medium, each term in (19.14) is responsible for a distinct dispersive effect. An arbitrary pulse $E(t)$ with a centre frequency of ω_o can be represented as a carrier field modulated by a complex amplitude $A(t)$ so that

$$E(t) = A(t)\exp(i\omega_o t). \qquad (19.15)$$

The corresponding spectral field is given by the Fourier transform

$$e(\omega) = \frac{1}{\sqrt{2\pi}}\int_{-\infty}^{\infty} E(t)\exp(-i\omega t)\,dt. \qquad (19.16)$$

After propagating through a dispersive medium, the pulse acquires spectral phase according to

$$e(\omega)\rlap{/}{} = e(\omega)\exp(i\varphi(\omega)) \qquad (19.17)$$

and the resulting pulse in time is described by the reverse Fourier transform

$$E(t)' = \frac{1}{\sqrt{2\pi}}\int_{-\infty}^{\infty} e(\omega)'\exp(i\omega t)\,d\omega. \qquad (19.18)$$

The first term in equation (19.14) is the change in carrier phase after passing through the medium and (within the limits of the *slowly varying envelope approximation*) has no effect on the pulse. Within the second term, the time spent by the pulse inside the medium is given by the constant

$$\tau = \frac{d\varphi}{d\omega}\bigg|_{\omega_o} \qquad (19.19)$$

where τ is known as the *group delay*. In its entirety, the second term in equation (19.14) describes a linear phase ramp in frequency which corresponds (through the well-known Fourier shift theorem [139]) to a delay of the pulse in time.

The quadratic and higher order terms in equation (19.14) describe *group-delay dispersion*—the variation of the group delay with frequency—and have a direct influence on the shape of the pulse in time. The $(\omega - \omega_o)^2$ term is responsible for adding *linear chirp*, which refers to a time-varying uniform increase or decrease in the instantaneous optical frequency across the pulse. As an example (neglecting the carrier phase), consider a pulse with a Gaussian amplitude spectrum and quadratic spectral phase represented as

$$e(\omega) = \exp(-(a+ib)\omega^2) \qquad (19.20)$$

The pulse amplitude in time is given by the Fourier transform of equation (19.20) which, neglecting a constant normalization factor, has the form

$$E(t) \approx \exp(-at^2/4(a^2+b^2)) \times \exp(ibt^2/4(a^2+b^2)) \qquad (19.21)$$

The *full-width half-maximum* (FWHM) intensity duration of the pulse described by equation (19.21) is

$$\Delta\tau = 2\sqrt{\left(\frac{a^2+b^2}{a}\right)\ln 4} \qquad (19.22)$$

and illustrates that the presence of the spectral quadratic phase of either sign ($\pm b$) will broaden an ultrashort pulse from its unchirped ($b = 0$) duration which is given by

$$\Delta\tau_o = 2\sqrt{a\ln 4}. \qquad (19.23)$$

The unchirped duration given in equation (19.23) is related directly to the FWHM spectral intensity bandwidth of the pulse which is given by

$$\Delta\omega = \sqrt{\frac{\ln 4}{a}}. \qquad (19.24)$$

The quantity $\Delta\omega\Delta\tau$ is known as the *duration-bandwidth product* of the pulse and, as comparison of equations (19.23) and (19.24) confirms, has a constant value for unchirped pulses which is independent of their duration or spectral bandwidth. The duration-bandwidth product is, however, sensitive to the pulse shape: chirp-free Gaussian pulses have $\Delta\omega\Delta\tau = 2.77$ while $\text{sech}^2 t$ pulses have $\Delta\omega\Delta\tau = 1.98$. In experimental measurements of ultrafast lasers, the pulse shape is often assumed to be $\text{sech}^2 t$ and the chirp of the pulses is then estimated by comparing their measured duration-bandwidth product with the ideal chirp-free value. Practically, the duration-bandwidth product can therefore be treated as a figure-of-merit for ultrashort pulses.

It was mentioned earlier that linear chirp describes a time-varying uniform increase or decrease in the local carrier frequency across the pulse, and this is evident from the last term in equation (19.21) which corresponds to a temporal phase of

$$\phi(t) = bt^2/4(a^2+b^2) \qquad (19.25)$$

which leads to a time-dependent variation of the carrier frequency given by

$$\Omega(t) = -\frac{d\phi(t)}{dt} = -bt/2(a^2+b^2). \qquad (19.26)$$

All the wavelengths within an unchirped pulse ($b = 0$) travel together and, consequently, $\Omega(t) = 0$, and the local frequency across the entire pulse is the carrier frequency.

The quadratic term in the spectral phase expansion of equation (19.14) cannot introduce asymmetry into an originally symmetric pulse. In contrast, cubic spectral phase variations associated with the $(\omega - \omega_o)^3$ term are significant because, in the time-domain, they lead to sharpening of one edge of the pulse and broadening of the other, ultimately resulting in pulse break-up. Strategies for generating the shortest pulses from ultrafast lasers rely on controlling the net intra-cavity dispersion so that the spectral phase associated with one cavity roundtrip is constant across the entire spectral bandwidth of the pulse. Special optical components are used for controlling second-, third- and fourth-order spectral phase, and these are described in detail in Section 19.3.4.

19.3.2 Non-linear Material Dispersion

In our treatment of non-linear dispersion, we will only consider effects due to the $\chi^{(3)}$ term in (19.11) because, unlike those associated with the $\chi^{(2)}$ term (e.g. the Pockels effect), $\chi^{(3)}$ effects are present in all materials. Neglecting the $\chi^{(2)}$ term, (19.11) can be re-written as

$$P(\omega) = \varepsilon_o c(\omega) E(\omega) \quad (19.27)$$

where

$$\chi(\omega) = \chi^{(1)}(\omega) + \chi^{(3)}(\omega) E^2. \quad (19.28)$$

In a transparent medium, the refractive index due to this net susceptibility is

$$n = \sqrt{1 + \chi^{(1)} + \chi^{(3)} E^2} \quad (19.29)$$

which, by applying the Taylor series expansion $\sqrt{1+x} \approx 1 + x/2$, can be approximated as

$$n \approx n_o + \frac{\chi^{(3)} E^2}{2 n_o} \quad (19.30)$$

or simply

$$n \approx n_o + n_2 I. \quad (19.31)$$

The importance of the $\chi^{(3)}$ susceptibility is, therefore, that it causes the refractive index to become intensity-dependent, and significant changes in the refractive index can be induced by the high peak intensities of ultrashort pulses. Two principal effects—*SPM* and *self-focusing*—result from the non-linear refractive index, and both are essential in generating the shortest pulses from modern ultrafast lasers.

SPM is caused by the time-varying intensity profile of the pulse. The higher-intensity components induce a larger non-linear refractive index change and therefore experience a greater phase shift than the weaker components. Using equations (19.26) and (19.31), the resulting frequency shift across the pulse can be shown to be

$$\Omega(t) = -\frac{d\phi(t)}{dt} = \frac{\omega n_2 L}{c} \frac{dI(t)}{dt} \quad (19.32)$$

which describes a redshift of the pulse leading edge and a blue shift of the trailing edge. Only the pulse-centre frequency remains unchanged, and the result of SPM is, therefore, to broaden the pulse spectrum by redistributing energy from the centre to the wings. Figure 19.7 illustrates this effect for an initially chirp-free Gaussian pulse. Across the centre of the pulse, SPM leads to an approximately linear positive chirp which, with appropriate optics, can be corrected using an equal but opposite amount of negative second-order dispersion. The combination of spectral broadening using SPM and subsequent chirp removal using linear dispersion is the basis for optical pulse compression techniques which have been applied effectively to create pulses which exist for only a few carrier-wave periods [22140]. When the pulse propagates in an environment where SPM and linear dispersion are well balanced, a soliton [13,141], will evolve which, in the absence of loss, can propagate indefinitely without changing shape.

A spatial analogue to SPM is self-focusing, in which the variation of intensity across the beam profile of an ultrashort pulse leads to a positive *Kerr lens* which retards the centre of the wavefront more than the edges. As already mentioned in Section 19.2.2, by configuring a laser cavity to properly exploit this effect, the Kerr lens can be used as a versatile means of achieving passive mode-locking. A special case exists when beam divergence due to diffraction is exactly balanced by self-focusing. In analogy with temporal solitons, pulses propagating in this way are known as *spatial solitons* [142] and are observed under certain conditions of the beam diameter, pulse intensity and medium non-linearity. *Catastrophic self-focusing* occurs when the beam power exceeds a critical value [143] given by

$$P_{\text{crit}} = \frac{\pi \, \varepsilon_o c^3}{2 n_2 \omega^2}; \quad (19.33)$$

above this value, the beam diameter continually decreases until a focal point is formed whose intensity is large enough to damage the medium. Catastrophic self-focusing is of particular concern in ultrafast laser amplifiers and must be avoided by using large beam diameters and long pulses with peak powers below P_{crit}.

19.3.3 Other Sources of Dispersion

Although dispersion arising from the dielectric response of the propagation medium has the greatest effect on a pulse, other sources also contribute to the overall dispersion profile of an optical medium. The principal remaining source, *modal dispersion*, is relevant only to waveguides and arises because the wave propagating in the guide is split into a travelling forward component and a stationary transverse component [144]. Stronger confinement increases the size of the transverse wavevector component and reduces the size of the travelling component, leading to a lower group-velocity compared with propagation in an equivalent bulk material. In single-mode optical fibres, the contribution is significant and shifts the zero-dispersion wavelength from 1.27 μm (bulk silica) to 1.31 μm (typical single-mode silica fibre), and special techniques can be used to achieve even larger shifts [145]. Modal dispersion

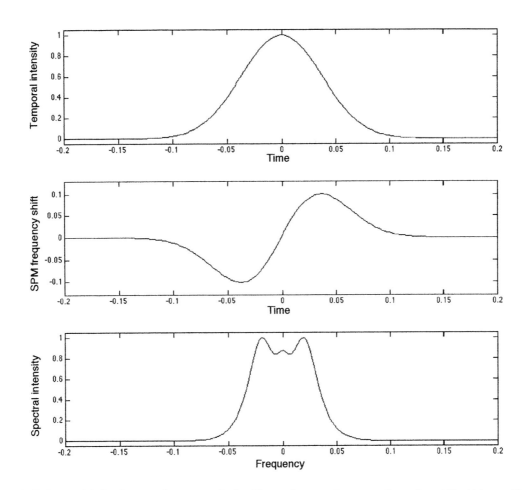

FIGURE 19.7 Self-phase modulation. Intense pulses propagating in a Kerr medium experience a non-linear phase shift which resembles linear chirp across the centre of the pulse.

can be engineered to dominate the dispersion characteristics of a waveguide, and this has been demonstrated recently in *photonic-crystal fibres* [146] which, although fabricated from pure silica, can be made to have zero group-velocity dispersion wavelengths as low as 650 nm by varying the core size and void distribution [147].

Dispersion also arises in multi-layer dielectric coatings, particularly high-reflectivity mirrors, and is associated with the wavelength dependence of the transmitted or reflected phase. In Bragg quarter-wave mirror structures, the reflected phase is flat at the centre of the stop-band but changes rapidly at the edges, leading to positive dispersion at shorter wavelengths and negative dispersion at longer wavelengths. Interestingly, Bragg mirrors can be understood as one-dimensional *photonic bandgaps* [148], and tunnelling of ultrashort pulses through a Bragg mirror has been observed at superluminal group velocities [149].

19.3.4 Group-velocity Dispersion Compensation

Obtaining the shortest pulses from a mode-locked laser requires the quadratic and higher-order terms in the roundtrip intra-cavity spectral phase to be zero. In practice, this means minimizing the material group-velocity dispersion of the gain medium by using a short but highly doped laser crystal and then identifying additional optical elements whose group-velocity dispersion is equal in magnitude but opposite in sign to that of the crystal. Optical components for compensating group-velocity dispersion fall into two categories: those which use bulk optics and geometry to achieve a wavelength-dependent path length, and others based on interference effects within a dielectric optical coating.

Pulse compression by dispersion compensation was originally considered by Treacy [150] who discussed in this and a later [151] paper on how pairs of diffraction gratings could be configured to compensate for positive quadratic spectral phase. Subsequent work by Fork et al. [20] showed that prism pairs could be used for the same purpose, thus avoiding the transmission loss associated with diffraction gratings. In both cases, the geometry of the systems is such that the group delay is wavelength-dependent and can be made greater for longer wavelengths than for shorter ones. Other authors [22,152,153] have refined the analytic expressions describing the quadratic and cubic spectral phase of prism and diffraction grating pairs, and their results are summarized in Table 19.1. Prism pairs and four-prism sequences have been extremely successful in enabling the generation of short pulses from mode-locked solid-state lasers, and 11 fs pulses were reported using this approach [154]. Dispersion compensation using prism pairs has fundamental limitations, and the simultaneous compensation

TABLE 19.1

Expressions for the Second- and Third-Order Dispersion of Common Optical Systems

Bulk material, thickness t 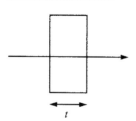	$\dfrac{d^2\varphi}{d\omega^2} = t\dfrac{\lambda^3}{2\pi c^3}\dfrac{d^2 n}{d\lambda^2}$ $\dfrac{d^2\varphi}{d\omega^3} = -t\dfrac{\lambda^4}{4\pi^2 c^3}\left(3\dfrac{d^2 n}{d\lambda^2} + \lambda\dfrac{d^3 n}{d\lambda^3}\right)$
Double-pass prism-pair, apex separation l, angular deviation within ray of β 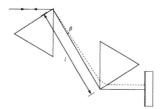	$\dfrac{d^2\varphi}{d\omega^2} = \dfrac{\lambda^3}{2\pi c^2}\dfrac{d^2 P}{d\lambda^2}$ $\dfrac{d^3\varphi}{d\omega^3} = -\dfrac{\lambda^4}{4\pi^2 c^3}\left(3\dfrac{d^2 P}{d\lambda^2} + \lambda\dfrac{d^3 P}{d\lambda^3}\right)$ $\dfrac{d^2 P}{d\lambda^2} = 4l\sin\beta\left[\dfrac{d^2 n}{d\lambda^2} + (2n - 1/n^3)\left(\dfrac{dn}{d\lambda}\right)^2\right] - 8lz\cos\beta\left(\dfrac{dn}{d\lambda}\right)^2$ $\dfrac{d^3 P}{d\lambda^3} = l\cos\beta\left[(24/n^3 - 48n)\left(\dfrac{dn}{d\lambda}\right)^3 - 24\dfrac{dn}{d\lambda}\dfrac{d^2 n}{d\lambda^2}\right]$ $\phantom{\dfrac{d^3 P}{d\lambda^3}} + l\sin\beta\left[\left(\dfrac{dn}{d\lambda}\right)^3(12/n^6 + 12/n^4 + 8/n^3 - 16/n^2 + 32n)\right]$ $\phantom{\dfrac{d^3 P}{d\lambda^3}} + (24n - 12/n^3)\dfrac{dn}{d\lambda}\dfrac{d^2 n}{d\lambda^2} + 4\dfrac{d^3 n}{d\lambda^3}$
Double-pass grating-pair, separation l, angle of incidence γ, line spacing d 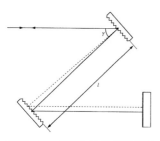	$\dfrac{d^2\varphi}{d\omega^2} = l\dfrac{\lambda^3}{\pi c^2 d^2}\left[1 - \left(\dfrac{\lambda}{d} - \sin\gamma\right)^2\right]^{-3/2}$ $\dfrac{d^3\varphi}{d\omega^3} = -6l\dfrac{\lambda^4}{c^3 d^2}\left(1 + \dfrac{\lambda}{d}\sin\gamma - \sin^2\gamma\right)\left[1 - \left(\dfrac{\lambda}{d} - \sin\gamma\right)^2\right]^{-5/2}$

of arbitrary second- and third-order spectral phase is difficult and involves identifying a suitable prism material, apex spacing and tip insertion [152].

Dispersion control using mirror coatings is not a new idea and, in 1986, was applied in CPM dye lasers to obtain 50 fs pulses by tuning the laser to a wavelength longer than the mirror centre wavelength. This configuration achieves negative dispersion from a standard Bragg mirror because, towards the long-wavelength edge of the mirror stopband, longer wavelengths penetrate further into the coating than the shorter ones before being reflected. The drawbacks of using a standard Bragg mirror design in this way are that the dispersion response is not readily controllable, and the longer wavelengths experience lower reflectivity and so the pulse spectral intensity is modified. A new mirror technology described as *chirped multilayer dielectric coatings* was reported in 1994 [155] and is based on the idea that a Bragg mirror with smaller layer periods at the surface and larger periods deeper in the coating will provide negative group-velocity dispersion because the longer wavelengths experience larger group delays within the coating than the shorter ones. Unlike the earlier approach using simple periodic quarter-wave Bragg layers, 'chirped mirrors' provide constant reflectivity across their stopband and, therefore, only modify the spectral phase of a pulse, leaving the intensity unchanged. The design of chirped mirrors has been refined to enhance their operating bandwidth [156] and to reduce undesirable modulations in the group-delay response [157]. The technique has been used to obtain 6.5 fs pulses directly from a Ti:sapphire oscillator [46, 158] and for compression of amplified pulses to 4.9 fs [140].

An alternative approach for obtaining negative group-velocity dispersion in a mirror structure is to exploit the frequency-dependent phase properties of a resonant Fabry–Pérot etalon [159]. A *Gires–Tournois (GTI) mirror* is an etalon fabricated within a dielectric Bragg mirror by including a layer with a thickness of $\lambda/2$ near the surface to form a weakly resonant cavity [160]. The group-delay across the GTI bandwidth increases approximately linearly with wavelength and can therefore be used for dispersion compensation. In comparison to other mirrors used for dispersion compensation, the GTI

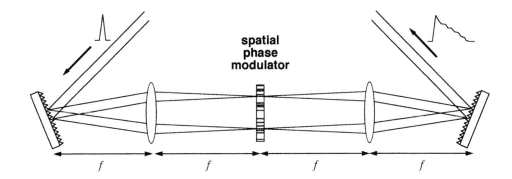

FIGURE 19.8 Femtosecond pulse shaping using a spatial light modulator to modify the spectral phase of a pulse at the Fourier plane of a dispersion-free grating stretcher/compressor.

provides larger amounts of negative dispersion, and a single GTI mirror can therefore provide enough dispersion to compensate the dispersion of all the other components in a laser cavity [161]. GTI mirrors have been applied in both picosecond [162] and femtosecond [163] mode-locked lasers.

19.3.5 Fourier-transform Pulse Shaping

Ultrashort pulses with tailored intensity and phase profiles are of interest for applications in high-energy amplifiers [164], *wavelength-division multiplexed* optical communications [165] and control of chemical reactions [166]. Femtosecond pulse shaping has been studied for a number of years (see the review in Ref. [167]), but only recently diagnostic tools and modulator components have become available which enable truly programmable pulse control. Shaping is commonly achieved by placing a spatial light-modulator such as a liquid-crystal display (LCD) array [168] at the Fourier plane of a dispersion-free pulse compressor [169] as illustrated in Figure 19.8. By modifying the intensity and phase of its spectral components, the pulse is shaped in the time-domain through the Fourier relation given in (19.18). Although LCD-based shaping is the most common approach, other modulation methods have been used successfully and include acousto-optic modulation [170] and deformable mirrors [171]. Pulse shaping is often carried out by using genetic algorithms to condition the modulation signal so that the output pulse maximizes some directly measurable quantity such as second-harmonic generation to optimize pulse peak power [172] or reaction yield in a photochemical process [173,174]. Feedback can also be applied using exact pulse measurement methods, and a system based on frequency-resolved optical gating (FROG) and a genetic algorithm has been reported.

19.4 Diagnostic Techniques

19.4.1 Direct Electronic Measurements

The principal motivation for using electronic devices such as photodiodes to measure the output from mode-locked lasers is to analyse the stability of the mode-locked pulse sequence in the radio-frequency (rf) domain. Direct pulse shape measurement using photodiodes is not possible because even the fastest devices and sampling oscilloscopes have finite response times with values in the picosecond range.

RF spectrum analysers are commonly used to determine the presence of multiple-pulsing, amplitude-noise and *phase noise*. Phase noise or *timing jitter* is a form of random noise found in all electronic or optical oscillators and, in mode-locked lasers, results in fluctuations in the pulse-arrival time. Although in a solid-state mode-locked laser the pulse timing error per cavity roundtrip time is generally very small (≤ 0.01 fs), phase noise prevents the arrival time of any particular future pulse from being predicted exactly. Applications, such as some pump-probe experiments, which rely on synchronization between two ultrafast sources, are particularly vulnerable to the presence of phase noise in the laser oscillator. In solid-state lasers, the major sources of phase noise are acoustic vibrations ($\approx 1-100$ Hz) and variations of the refractive index and gain of the laser material due to ripple on the pump laser. Amplitude variations of the intra-cavity mode-locked pulses can also lead to timing fluctuations through non-linear coupling between intensity and phase.

The rf power spectrum measured using a fast photodiode consists of a series of modes centred on the nth harmonics of the cavity frequency, f_R, with the nth mode broadened by a noise sideband S_n [175]:

$$P(\bar{f}) = \bar{P}^2 \sum [\delta(\bar{f} - nf_R) + S_n(\bar{f} - nf_R)]. \quad (19.34)$$

The noise sideband surrounding the nth harmonic can be expressed as an energy fluctuation term, S_E, whose size is independent of the harmonic number, and a timing fluctuation term, S_J, which dominates at high harmonics because of an n^2 dependence. Ignoring coupling between energy and time fluctuations, the nth sideband is

$$S_n(f) = S_E(f) + n^2(2\pi f_R)^2 S_J(f). \quad (19.35)$$

Measurements of the sideband intensity at different harmonics allow the energy and timing fluctuation terms to be separated [176] and the rms fluctuations within a finite rf band from f_{min} to f_{max} are obtained using

$$\sigma_{E(J)} = \sqrt{s \int_{f_{min}}^{f_{max}} S_{E(J)}(f)\, df}. \quad (19.36)$$

Quantitative measurements of amplitude and phase noise allow different mode-locking methods and laser sources to be compared, and the suitability of any source for an application requiring precise timing synchronization can be determined effectively using this technique.

19.4.2 Approximate Methods of Pulse-shape Measurement

Because direct electronic detection is not fast enough to characterize an ultrashort pulse—or even to estimate its duration—it is necessary to use all-optical methods to make the measurement. The ideal measurement device would act like a super-fast mechanical shutter or *gate* which could be opened periodically to sample the energy in a particular time-slice of every pulse. An *autocorrelator* is an optical analogue of this system in which the pulse is overlapped in a non-linear medium with its time-delayed replica and generates a mixing signal. In the idealized measurement device, an infinitesimally short gate would be able to sample the exact pulse intensity with time as its delay relative to the pulse was scanned. In autocorrelation measurements, the gate pulse is identical to the measurement pulse and results in a broadened signal which, although not a direct map of the pulse envelope, gives a useful estimate of the pulse duration.

Autocorrelation is commonly implemented using second-harmonic generation in a non-linear crystal to produce a mixing signal which has the form [177]

$$g_2(\tau) = \frac{\int_{-\infty}^{\infty} \left| \{E(t) + E(t-\tau)\}^2 \right|^2 dt}{2 \int_{-\infty}^{\infty} \left| \{E(t)\}^2 \right|^2 dt}. \quad (19.37)$$

The interference term in equation (19.37) leads to $g_2(\tau)$ being known as the *interferometric autocorrelation* profile. When $g_2(\tau)$ is recorded using a measurement system that has insufficient bandwidth to resolve the interference fringes, a modified signal is recorded which is known as the *intensity autocorrelation* profile and is given by

$$G_2(\tau) = 1 + \frac{2\int_{-\infty}^{\infty} I(t)I(t-\tau)\,dt}{\int_{-\infty}^{\infty} I(t)^2\,dt}. \quad (19.38)$$

Regardless of pulse shape, the autocorrelation functions described always exhibit a characteristic peak-to-background contrast ratio which for the interferometric signal is 8:1 and for the intensity signal is 3:1. A special situation exists when the second-harmonic light generated by each separate pulse (mixing with itself) is not detected, and in this case, a background-free intensity autocorrelation is recorded which is equal to the last term of equation (19.38).

Separate optical implementations are generally used for amplifier and oscillator sources because of the differences in pulse energy and repetition rate. To characterize pulses from an oscillator, the output is split using a scanning Michelson interferometer (see Figure 19.9) which then recombines the two beams by focusing into a second-harmonic crystal and detecting the frequency-doubled light using a photomultiplier or other linear detector. The same functionality can be obtained by mixing the pulses in a semiconductor whose bandgap energy, E_g, is related to the pulse photon energy by

$$E_g > E_{photon} > E_g/2 \quad (19.39)$$

and detecting the photocurrent due to two-photon absorption [178–180]. Scanning autocorrelators are most suitable for high-repetition-frequency (~ MHz) pulses because each part of the autocorrelation trace results from mixing between distinct pairs of pulses, and the measurement is averaged over a time

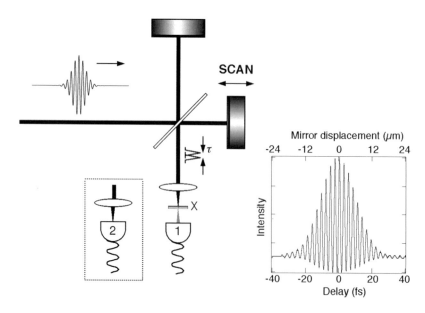

FIGURE 19.9 A scanning Michelson interferometer configured as an autocorrelator. The input pulse is split by the beam splitter into two replicas which subsequently leave the interferometer with a relative delay τ. A mixing signal is derived either by using a linear (single-photon) detector (1) to record second-harmonic generation in a non-linear crystal (X) or by focusing the pulses directly onto a non-linear (two-photon) detector (2).

Ultrashort Pulses

equal to many millions of pulse repetition periods. Mechanical scanning is inconvenient and slow for lower repetition-rate sources such as kHz ultrafast amplifiers, and for these systems, a single-shot autocorrelator [181] is used. Single-shot methods measure the background-free autocorrelation by using non-collinear SHG between two pulses, often by mixing opposite sides of a single beam as shown in Figure 19.10. When a sufficiently large mixing angle is used the spatial intensity distribution of the frequency-doubled light is proportional to the intensity autocorrelation.

Autocorrelation measurements do not allow the pulse duration to be determined exactly because the width of the autocorrelation function, although directly proportional to the pulse width, is also dependent on the shape of the pulses. Figure 19.11 reproduces the intensity and interferometric autocorrelations of five representative pulses with identical FWHM durations and demonstrates the difficulty of trying to infer pulse shape information from autocorrelation measurements. Practically, autocorrelation can be used to estimate the pulse duration from a laser by assuming a pulse shape—normally a Gaussian or $sech^2 t$ intensity profile—and using a known *deconvolution factor* to infer the pulse duration from the autocorrelation width. For Gaussian pulses $\Delta\tau_{ac}/\Delta\tau = 1.414$ and for pulses $\Delta\tau_{ac}/\Delta\tau = 1.543$.

19.4.3 Exact Methods of Pulse-shape Measurement

In many experiments, precise quantitative information about the intensity and phase profile of a pulse is invaluable. Exact pulse measurement methods have been used to tailor pulse chirp to a pre-determined value for quantum-control studies [177], to test the validity of competing laser theories [182] and to obtain pulse data for numerical simulations [183]. Throughout the 1990s, various new methods were developed which were able to provide exact shape and chirp information about ultrashort pulses from high-repetition-rate mode-locked oscillators or single-shot laser amplifiers. The approaches used can be classified either as two-dimensional methods which

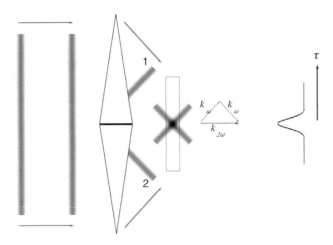

FIGURE 19.10 A single-shot autocorrelator. The input beam is split at a bi-prism and a second-harmonic signal is obtained by non-collinear phase-matching. At the top of the crystal, the trailing edge of pulse 1 mixes with the leading edge of pulse 2, while the opposite is true at the bottom of the crystal. In the centre of the crystal, the delay between pulses 1 and 2 is zero.

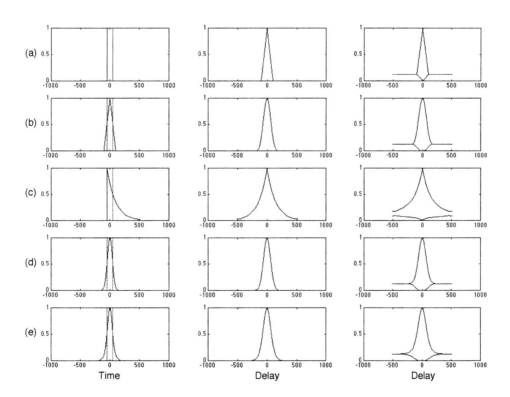

FIGURE 19.11 Illustration of the sensitivity of the second-order autocorrelation profile to pulse shape. Rows: pulses of different shapes but equal FWHM (broken lines) durations—(a) square, (b) triangular, (c) single-side exponential, (d) Gaussian and (e) $sech^2(t)$. Columns (left to right): temporal pulse intensity, background-free intensity autocorrelation, envelope of the interferometric autocorrelation.

measure the pulse spectrogram or sonogram signal in a hybrid time-frequency domain, or one-dimensional methods which measure one or more signals exclusively in either time or frequency. In this section, the principal implementations of each type of technique will be explained, and their advantages and limitations discussed.

Although approximate methods for determining the spectral phase of ultrashort laser pulses have existed for some time [177,184], the first truly exact method was FROG which was reported in 1993 by Trebino and Kane [185]. Based on measuring a *spectrogram*, FROG records the frequency spectrum of different time components of the pulse and uses a two-dimensional (2D) phase-retrieval algorithm (for which a unique solution exists) to obtain from this the pulse characteristics. FROG has been developed extensively since its original demonstration, and a wide variety of implementations has been introduced which allows the measurement of pulses from sources covering a diverse range of wavelengths, pulse durations and energies [186–189]. The mixing of two pulses—the 'probe' and the 'gate'—in a non-linear medium to obtain a cross-(or auto-) correlation signal is the basis of FROG. The FROG spectrogram is built up by measuring the spectrum of the correlation signal for a continuous range of positive and negative delays between the probe and gate pulses. The specific form of the FROG spectrogram depends on the non-linearity used to obtain the mixing signal, and common FROG implementations include second-harmonic generation FROG (SHG-FROG) [190], Kerr-effect polarization-gating FROG (PG-FROG) [191], third-harmonic generation FROG (THG-FROG) [192], difference-frequency generation (DFG-FROG) [193] and sum-frequency generation (SFG-FROG) [193]. Table 19.2 lists the field of the non-linear mixing signal, $E_{sig}(t, \tau)$, associated with the more popular FROG implementations. The FROG spectrogram is related to the mixing signal by

$$S_{FROG}(\omega,\tau) = \left| \int_{-\infty}^{\infty} E_{sig}(t,\tau)\exp(-i\omega t)\,dt \right|^2. \quad (19.40)$$

In some FROG geometries, notably SHG-FROG, the roles of the probe and gate pulses are interchangeable, and this leads to an ambiguity in the direction of time of the inferred pulse [194]. Figure 19.12 shows SHG-FROG traces calculated for some common pulse shapes and illustrates the temporal-symmetry of the measurement. Once the FROG spectrogram has been measured and calibrated, the pulse intensity and phase can be obtained using an iterative *retrieval algorithm* based on an initial guess pulse whose shape is then refined by applying prior knowledge of the non-linearity and by constraining the FROG trace of the retrieved pulse to match the experimental data. The details of FROG algorithms are beyond the scope

TABLE 19.2

Expressions for the Non-linear Mixing Field Associated with Common FROG Geometries

Second-harmonic	$E_{sig}(t, \tau) = E(t)E(t - \tau)$
Third-harmonic	$E_{sig}(t, \tau) = E2(t)E(t - \tau)$
Polarization gating	$E_{sig}(t, \tau) = E(t)|E(t-\tau)|2$

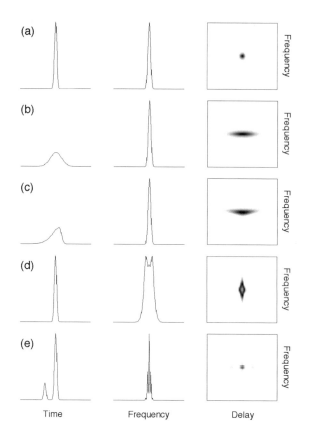

FIGURE 19.12 Examples of SHG-FROG traces for different pulse shapes. Rows: (a) transform-limited Gaussian pulse; (b) as (a) but with quadratic spectral phase added; (c) as (b) but with cubic spectral phase added; (d) as (a) but after SPM with a peak phase shift of π; (e) double pulse. Columns (left to right): temporal pulse intensity, pulse spectrum, SHG-FROG trace.

of this article but have been described fully by several authors [191,195,196]. With the availability of fast computers, algorithms retrieving in less than one second have been reported, allowing real-time implementations of FROG to be demonstrated [197].

The pulse *sonogram* can also be used for characterization purposes and is recorded by measuring the cross-correlation signal between different frequency-filtered components of the pulse [198] and a shorter (uncharacterized) reference. Commonly, the reference is the original pulse itself [199,200] but any available pulse which is synchronous with the pulse to be measured can be used [201]. The sonogram obtained using a zero-dispersion bandpass filter $g(\omega - \Omega)$ with a variable centre frequency Ω is

$$S_{SONO}(\Omega,t) = \left| \int_{-\infty}^{\infty} E(\omega)g(\omega - \Omega)\exp(i\omega t)\,d\omega \right|^2. \quad (19.41)$$

Sonogram measurements can be made using two-photon-absorption cross-correlation and, using waveguide detectors [202], can be both sensitive and broadband [203]. The requirement for a dispersion-free frequency-filter makes the application of the sonogram technique to sub-50 fs pulses difficult but, unlike SHG-FROG, sonogram measurements have no time ambiguity [200]. Recently, another method known as dispersion-propagation time-resolved optical gating (DP-TROG) has used multiple autocorrelation measurements to record a sonogram using a dispersive filter [204]. In general, 2D techniques are accessible to error-checking methods based on comparing the trace *marginals* (its integral over delay or frequency) with independently measurable quantities. The form and application of the trace marginals have been reported for the FROG [205], sonogram [213] and DP-TROG techniques [206].

Two-dimensional methods are characterized by a high-redundancy factor because the measurement of a pulse described by N intensity and N phase samples requires N^2 data points. Such oversampling is desirable because it makes the measurement more resistant to corruption by noise, but the acquisition of 2D data sets often necessitates mechanical scanning and complicated data-handling algorithms. It is well known that when a previously characterized reference pulse exists at the same wavelength as an unknown pulse, the interference signal between the pulses is sufficient for characterization [188,207]. The use of 2D methods was originally adopted because no well-characterized reference pulse generally exists but, in 1998, a one-dimensional (1D) interferometric method was reported [208] which used a form of spectral shearing interferometry to solve this problem. Known as spectral phase interferometry for direct electric-field reconstruction (SPIDER), the method eliminates the need for a reference pulse by interfering two time-delayed pulses with a relative frequency shift. The SPIDER signal is the interference spectrum between the two pulses,

$$S_{\text{SPIDER}}(\omega) = |e(\omega)|^2 + |e(\omega+\Omega)|^2 + 2|e(\omega)e(\omega+\Omega)|$$
$$\cos[\varphi(\omega+\Omega) - \varphi(\omega) + \omega\tau] \quad (19.42)$$

and, unlike 2D methods, both the time-delay, τ, and the frequency-shift, Ω, are fixed. In practice, the time-delay is achieved using an etalon or a fixed Michelson interferometer and the frequency shift, by sum-frequency mixing the delayed pulses with a quasi-monochromatic wave which is a highly chirped copy of the original pulse. SPIDER is attractive because it requires no moving parts and its 1D nature means that spatially resolved measurements of chirp can be recorded [209] but, for optimum performance, the technique requires careful selection of time and frequency shifts [208]. Using SPIDER, pulses as short as 6 fs have been measured [210].

The practicality of measuring pulses using non-phase-matched $\chi^{(3)}$ two-photon processes in semiconductors [211] instead of phase-matched $\chi^{(2)}$ frequency conversion in bulk non-linear crystals has motivated efforts to extract intensity and phase information from spectral and auto- or cross-correlation measurements alone. This approach is essentially 1D because measurements are made in either the time or frequency domain and the resulting data sets are vectors, not arrays. Various algorithms which use autocorrelation measurements for exact characterization have been reported [212–214], but complex pulse shapes are known to present problems for such techniques. A variation based on the cross-correlation between the measurement pulse and its frequency-chirped replica has been shown to cope successfully with asymmetric and double pulses [215,216]. The approach uses the spectrum with an initial random phase to estimate the complex spectral field amplitude $e(\omega)$ from which the temporal intensities of the pulse before and after a dispersive element are calculated. By minimizing the RMS error between the measured and calculated cross-correlations of the chirped and unchirped pulses, the algorithm converges to the correct pulse shape. The simplicity of techniques which require only measurements of the spectrum and correlation of the pulse is attractive, but the minimization algorithms can be sensitive to the initial guess phase and their convergence can be slow in comparison with iterative spectrogram or sonogram algorithms.

19.5 Applications of Ultrashort Pulses

19.5.1 Imaging

Beyond the optics research laboratory, many of the applications for low-energy ultrashort pulses concern novel imaging methodologies which, in comparison to conventional techniques, provide greater resolution, three-dimensional (3D) or four-dimensional (4D) sectioning, deeper penetration or other information. Biomedical imaging has been a rapidly developing area and was the first to adopt ultrafast sources for non-linear microscopy [217]. Two-photon laser-scanning fluorescence imaging uses a femtosecond laser to achieve confocal-like imaging without an aperture. Because two-photon fluorescence is only generated at the focus of the microscope objective, 3D resolution is obtained and phototoxic and photobleaching effects are minimized. The ability to use longer wavelengths also achieves greater depth penetration with reduced Mie scattering. Similar methods have been applied to image semiconductor devices by mapping the photocurrent across a chip resulting from two-photon absorption [218].

Medical imaging often involves tissues with good optical transmission but high optical scattering cross sections and short mean-free paths. Ultrashort pulse illumination has been combined with time-gated detection to measure only the unscattered *ballistic* light which travels directly through the sample and therefore carries the image information. Experiments by Alfano and others [219,220] have demonstrated the principle; although for biological samples thicker than a few millimetres, the intensity of the ballistic light becomes extremely weak and difficult to detect. Optical coherence tomography (OCT) [221] uses interferometric detection of back-scattered light to provide 3D sectioning of biological samples such as the cornea [222] and has also been implemented at video rates to create movies of living specimens [223]. OCT requires a light source with a short coherence length, and although it can be implemented with a broadband incoherent source, the best results have been obtained by using ultrashort laser pulses. Figure 19.13 shows an OCT image through hamster cheek tissue made using a 5 fs Ti:sapphire laser and illustrates the optical sectioning ability of the technique. Another related technique for medical imaging is time-gated fluorescence [224] measurement which, after excitation by a short laser pulse, has the potential to discriminate between cancerous and healthy tissue on the basis of the different fluorescence lifetimes in each region [225].

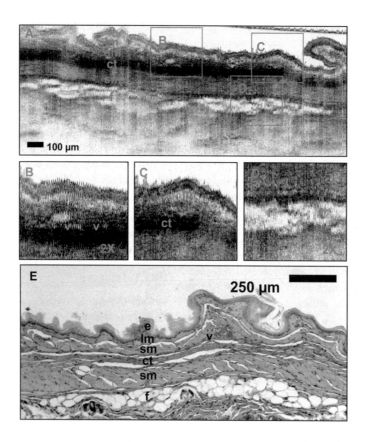

FIGURE 19.13 Optical section through a sample of hamster cheek tissue recorded by OCT using a 5 fs Ti:sapphire laser as the light source. The letters identify distinct microstructures such as blood vessels (v) and fat (f). (Image courtesy of James Fujimoto, MIT, Massachusetts.)

By implementing this technique in a whole-field configuration, 4D (x, y, z and t) imaging has been demonstrated [226].

Imaging with terahertz radiation is made possible by using femtosecond pulses to trigger a photoconductive antenna [227] or achieve optical rectification [228]. The unique transmission properties of terahertz images have been used to reveal water uptake within plants [229], penetrate optically opaque materials like paper [245] and image dental caries.

19.5.2 Ultrafast Chemistry

Many chemical processes, such as bond-breaking and formation; molecular collisions, rotation, vibration and fragmentation; isomerisation and ionization occur with characteristic timescales measurable in femtoseconds. Chemical reactions which are known to proceed from reactants to products *via* an excited transition state can be studied using ultrafast lasers. Commonly, an initial pulse is used to excite a molecule into the transition state and a second pulse is used to stimulate *laser-induced fluorescence* (LIF), ionization or *chemiluminescence* whose intensity describes the time-evolution of the product state. One of the first applications of such a scheme was to monitor the vibrational motion of I_2 molecules [230] which can be studied using LIF. Similar approaches have been used to investigate bond breaking in ICN [231], NaI [232] and $C_2F_4I_2$ [233]. Ultrafast laser techniques have also been used extensively to study the effect of the solvent on the dynamics of reactions which occur in solution [234].

A contemporary theme in femtosecond chemistry aims to use femtosecond pulses to drive reactions into particular product channels or prepare molecules in precise quantum states. Work by Kent Wilson and others has applied the FROG pulse measurement technique to prepare negatively chirped femtosecond pulses with the appropriate duration required to excite I_2 molecules so that, sometime after excitation, their vibrational wave packets had evolved to a state of minimum uncertainty ($\Delta p \Delta x = \hbar/2$) corresponding to a known atomic velocity, separation and momentum [177]. The equivalent positively chirped pulse was shown, in agreement with theory, to fail to produce localized wave packets. A related strategy has been used by Gerber and others to optimize the branching ratio of products resulting from the photodissociation of $CpFe(CO)_2Cl$ [235]. The experiment used a genetic algorithm controlling a pulse-shaper to maximize or minimize a feedback signal based on the ratio of $Fe(CO)_5^+/Fe^+$ ions by creating a pulse with an optimized phase profile.

In physical chemistry, X-ray diffraction is a long-established tool for determining atomic configurations, and a recent emphasis has concentrated on ways of recording how a crystal structure changes after perturbation by an intense laser pulse. *Femtosecond x-ray diffraction* (Figure 19.14) is a pump-probe technique in which an optical pulse vibrationally perturbs the atoms in a material, and a synchronous X-ray pulse arriving later at the sample is scattered to produce diffraction patterns [236]. Currently, the method is being developed to allow the analysis of lattice dynamics and thermalization in bulk

Ultrashort Pulses 279

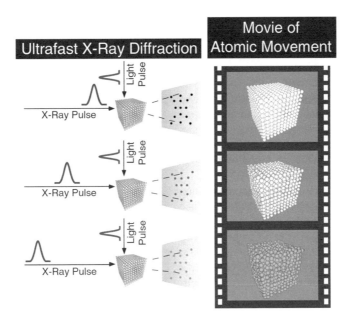

FIGURE 19.14 Intuitive representation of an ultrafast X-ray diffraction experiment. Ideally, after excitation with a short laser pulse, the transient structure of a dynamically evolving system is measured at each time delay. (Image courtesy of Craig W Siders, University of Central Florida.)

semiconductors and to determine the time-evolution of strain within semiconductor heterostructures, but potential exists for the technique to be applied to study complex biological molecules and protein conformations.

19.5.3 Semiconductor Spectroscopy

The ability to measure the dynamic behaviour of electrons in semiconductors is important because it enables the speed of modern electronic or opto-electronic devices to be optimized and provides the experimental data necessary to develop theoretical models of these materials. When a semiconductor of bandgap energy E_g absorbs a photon with an energy $\hbar\omega > E_g$, electrons are excited to the conduction band, leaving behind holes in the valence band. The scattering and radiative decay mechanisms by which these carriers return to equilibrium determine the electronic properties of the semiconductor. For example, electrons which subsequently become trapped by impurity levels may return to the ground state slowly, with the result that the sample conductivity remains high for a long time after the initial excitation. Intense illumination of the kind available using ultrafast lasers can cause the absorption of a semiconductor to become saturated because the Pauli's exclusion principle results in *band-filling*, and full saturation occurs when the populations of the excited and ground states are equal. This effect is enhanced by using (multiple) quantum wells in which strong spatial confinement of the carriers further restricts the allowed transitions to discrete energies. Absorption saturation effects lead to a corresponding change in the refractive index which has been exploited in optical logic devices [237]. The recovery of the absorption can be measured using the *pump-probe* technique in which a strong ultrafast pump pulse induces saturation by exciting carriers, and a time-delayed probe pulse then samples the transmission at a known time after excitation. The rate at which the saturation recovers provides information about the rate at which electrons and holes are scattered out of their initial excited states by non-radiative processes such as phonon emission or electron–electron scattering. Some implementations use a broadband probe pulse created by continuum generation to simultaneously sample absorption across a broad spectral range [238]. Radiative carrier recombination can be measured directly using *time-resolved photoluminescence* techniques [239]. The photoluminescence spectrum is a measure of the carrier energy distribution because the emitted photon energy is the sum of the carrier kinetic energy and the material bandgap energy. By gating the arrival time of the luminescence light using optical sum-frequency mixing, or an electrical instrument such as a streak camera, it is possible to record the time-evolution of the carrier energies and infer the routes by which electrons return to thermal equilibrium.

19.5.4 Material Processing

Conventional machining of mechanical components with sub-100 µm dimensions is difficult, and laser material processing provides a solution which enables the fabrication of features with sizes down to 1 µm. Laser machining is appropriate for metals, plastics, ceramics and glasses sometimes with CW lasers but often using pulsed radiation. When long pulses are used, thermal processes strongly influence the quality of the resulting feature; in thermal processing, heat is conducted away from the immediate optical absorption region into a wider *heat-affected zone*. In this region, local melting occurs and the ejection of liquid from the interaction region to its edge leads to recast and burr formation. In contrast, femtosecond laser machining is fundamentally non-thermal in nature, and the intense electric field of the pulses results in strong multi-photon absorption and a shallow, well-defined absorption region. The material is ablated by ionization and is removed as plasma, with the result that there is no heat-affected zone, negligible melting and no burr formation.

The favourable properties of femtosecond material processing have made it attractive for creating surgical implants [240] and waveguides [242]. Material removal is both precise and predictive and permits accurate 3D profiling of surfaces on the very small scale. A simple example of this capability is shown in Figure 19.15. Here, 40 fs laser pulses have been used to machine grooves of semi-circular cross section in alumina blocks then joined to form capillaries of 300 μm diameter and length 40 mm. Measurements have implied that femtosecond ablation proceeds by a multi-photon absorption process [241], and this strongly non-linear behaviour leads to a precisely defined ablation threshold below which no damage occurs. By using tight focusing, this effect has been used to produce structural changes within the bulk of a transparent material without affecting the surface [242]. Further, this process can be controlled to produce only a refractive index modification in plastics and glasses and may thus be utilized to produce a range of 3-D optical devices, including optical circuits, splitters, couplers, sensors, photonic crystals and optical storage.

19.5.5 High Field Science

The intense electromagnetic field available at the beam-focus of powerful ultrafast lasers enables extreme physical conditions to be achieved which are not obtainable in any other way. Using CPA, 440 fs duration pulses with an energy of 680 J have been demonstrated and focused to an intensity of 6×10^{20} W cm^{-2} [138]. The pulse peak power of 1.5 PW represents the highest laser pulse power ever achieved, and various experiments from laser fusion to experimental astrophysics can exploit the power densities available when the beam is focused on a target.

Intense laser pulses have been used to produce relativistic electrons [138] and collimated beams of high-energy protons [243]. The ion beams produced in this way may be used as short-pulse injectors for conventional particle accelerators or as the ignitor for fast-ignition inertial confinement fusion when an indirectly driven target is used. X-rays and γ-rays can be produced when laser-generated ion beams are accelerated by the field in the beam focus onto a solid target, causing bremsstrahlung radiation to be emitted. Laser-induced plasmas and ion beams have become a viable alternative to conventional synchrotron accelerator sources, and a review of the physics and applications of the plasmas is given in Ref. [244]. Efficient soft X-ray production using moderate energy femtosecond pulses is of interest because of the attractiveness of a 'table-top' source of collimated X-rays for medical and pump-probe experiments (see Section 19.5.2), and these have been produced by high-harmonic generation from a solid-target using a picosecond Nd:glass amplifier [245] and a gas-cell using a femtosecond Ti:sapphire amplifier [246].

Laser-induced nuclear reactions are made possible by the extreme power densities available at the focus of a high-power ultrafast amplifier chain. *Inertial confinement fusion* uses an intense pulse to create a shock wave which compresses a capsule of nuclear fuel to a sufficient density that fusion is initiated in the core of the material, and a thermonuclear burn wave travels outwards through the fuel releasing energy. The geometry of the fuel and the laser focusing arrangement is critical, and both conical [247] and spherical [248] geometries have been used. In the fast ignitor concept, the fuel is compressed by a pre-pulse and the thermonuclear burn wave is initiated at the outside of the fuel by a second shorter 'ignition' pulse. Nuclear fission is possible by using intense laser pulses to remove neutrons from atomic nuclei to create radioactive isotopes, and fission of uranium has been demonstrated using this approach. The short-lived isotopes which can be produced in this way may have applications in medical therapy.

Conditions at the focus of the most intense amplifier chains are comparable to those found in the centres of stars and

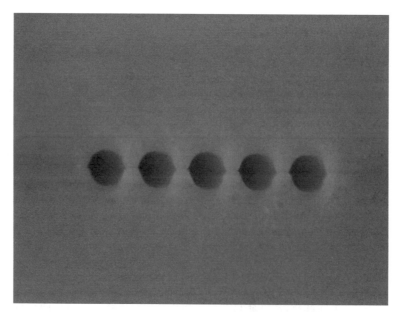

FIGURE 19.15 Cross-section of capillaries in alumina, of 300 μm diameter and 40 mm length. The capillaries are formed from two blocks joined together. The semi-circular grooves in each block are machined by 40 fs laser pulses. (Image courtesy of David R Jones, University of Strathclyde.)

planets, and therefore, these lasers are suitable sources with which to carry out experimental simulations of astrophysical events. Using targets of suitable composition, researchers have simulated conditions inside a star as it becomes a supernova [249] and inside a planet where the extreme densities result in novel states of matter with unusual properties not found elsewhere.

19.5.6 Other Applications

A range of other applications exist for ultrafast laser pulses but the space limitations within this article prevent a full discussion of these here. Amplified femtosecond and picosecond pulses are commonly used for frequency conversion by optical parametric amplifiers (see Section 19.2.7), particularly to access wavelengths in the visible and mid-infrared not available directly from laser materials [250]. In dense WDM telecommunications, the spectrum of a mode-locked femtosecond fibre laser has been sliced to create multiple wavelength channels synchronized to a common clock signal [251]. Attosecond pulse generation in the soft X-ray region will rely on femtosecond lasers as suitable optical pump sources [252], and frequency metrology will benefit from optical self-referencing schemes using phase-locked femtosecond pulses to relate optical frequency standards to conventional microwave reference oscillators [253]. High-power picosecond oscillators are being used to generate the primary colours needed for laser projection cinema [254] which will make possible the digital distribution of cinema films. This list is by no means exhaustive, and as ultrafast lasers become more compact, less expensive and more reliable, other applications beyond the scientific and telecommunication sectors can be expected to emerge.

REFERENCES

1. Ruddock I S and Illingworth R 1985 Cavity length optimization in acousto-optically mode-locked argon-ion lasers *J. Phys. E: Sci. Instrum.* **18** 121–3.
2. Marinero E E and Jasny J 1981 An interferometrically tuned and actively modelocked cw dye-laser *Opt. Commun.* **36** 69–74.
3. Bakhshi B and Andrekson P A 1999 Dual-wavelength 10-GHz actively mode-locked erbium fiber laser *IEEE Photon. Tech. Lett.* **11** 1387–9.
4. Dimmick T E, Ho P T and Burdge G L 1984 Coherent pulse generation by active modelocking of a GaAlAs laser in a Selfoc lens extended resonator *Electron. Lett.* **20** 831–3.
5. Albrecht G F 1982 Temporal shape-analysis of Nd-YLiF active modelocked/Q-switched oscillator *Opt Commun.* **41** 287–91.
6. Langford N, Smith K and Sibbett W 1987 Subpicosecond-pulse generation in a synchronously mode-locked ring color-center laser *Opt. Lett.* **12** 817–9.
7. May P G, Sibbett W and Taylor J R 1981 Subpicosecond pulse generation in synchronously pumped and hybrid ring dye-lasers *Appl. Phys. B* **26** 179–83.
8. Kafka J D, Watts M L and Pieterse J W 1995 Synchronously pumped optical parametric oscillators with LiB_3O_5 *J. Opt. Soc. Am. B* **12** 2147–57.
9. Lefort L, Puech K, Butterworth S D, Svirko Y P and Hanna D C 1999 Generation of femtosecond pulses from order-of-magnitude pulse compression in a synchronously pumped optical parametric oscillator based on periodically poled lithium niobate *Opt. Lett.* **24** 28–30.
10. Acobovitzveselka G R, Keller U and Asom M T 1992 Broad-band fast semiconductor saturable absorber *Opt. Lett.* **17** 1791–3.
11. Ippen E P, Shank C V and Dienes A 1972 Passive modelocking of the cw dye laser *Appl. Phys. Lett.* **21** 348–50.
12. Bilinsky I P, Prasankuma R P and Fujimoto J G 1999 Self-starting mode locking and Kerr-lens mode locking of a $Ti:Al_2O_3$ laser by use of semiconductor-doped glass structures *J. Opt. Soc. Am. B* **16** 546.
13. Agrawal G P 1995 *Nonlinear Fiber Optics* 2nd edn (New York: Academic) p 44.
14. Brabec T, Spielmann C and Krausz F 1991 Mode-locking in solitary lasers *Opt. Lett.* **16** 1961–3.
15. Mollenauer L F and Stolen R H 1984 The soliton laser *Opt. Lett.* **9** 13–15.
16. Krausz F, Ferman M E, Brabec T, Curley P F, Hofer M, Ober M H, Spielmann C, Wintner E and Schmidt A 1992 Femtosecond solid-state lasers *IEEE J. Quantum Electron.* **28** 2097–122.
17. Penman Z E, Schittkowski T, Sleat W, Reid D T and Sibbett W 1998 Experimental comparison of conventional pulse characterization techniques and second-harmonic-generation frequency-resolved optical gatin *Opt. Commun.* **155** 297–300.
18. Fork R L, Greene B I and Shank C V 1981 Generation of optical pulses shorter than 0.1 psec by colliding pulse mode locking *Appl. Phys. Lett.* **38** 671.
19. Fork R L, Shank C V, Yen R and Hirlimann C A 1983 Femtosecond optical pulses *IEEE J. Quantum Electron.* **19** 500–5.
20. Fork R L, Martinez O E and Gordon J P 1984 Negative dispersion using pairs of prisms *Opt. Lett.* **9** 150–2.
21. Valdmanis J A, Fork R L and Gordon J P 1985 Generation of optical pulses as short as 27 fs directly from a laser balancing, self-phase modulation, group-velocity dispersion, saturable absorption and saturable gain *Opt. Lett.* **10** 131–3.
22. Fork R L, Brito Cruz C H, Becker P C and Shank C V 1987 Comparison of optical pulses to 6 fs by using cubic phase compensation *Opt. Lett.* **12** 483–5.
23. Johnson A M and Simpson W M 1985 Tunable femtosecond dye laser synchronously pumped by the compressed second harmonic of Nd:YAG *J. Opt. Soc. Am. B* **2** 619.
24. Coherent Satori Hybrid Dye Laser (Kiton Red/Malachite Green).
25. Moulton P F 1986 Spectroscopic and laser characteristics of $Ti:Al_2O_3$ *J. Opt. Soc. Am. B* **3** 125.
26. Squier J, Salin F, Coe S, Bado P and Mourou G 1991 Characteristics of an actively mode-locked 2-psec Ti–sapphire laser operating in the 1 μm wavelength regime *Opt. Lett.* **16** 85–7.
27. Sarukura N, Ishida Y and Nakano H 1991 Generation of 50 fsec pulses from a pulse-compressed, cw passively mode-locked Ti–sapphire laser *Opt. Lett.* **16** 153–5.
28. Goodberlet J, Wang J, Fujimoto J G and Schulz P A 1989 Femtosecond passively mode-locked $Ti\text{-}Al_2O_3$ laser with a nonlinear external cavity *Opt. Lett.* **14** 1125–7.

29. Goodberlet J, Wang J, Fujimoto J G and Schulz P A 1990 Starting dynamics of additive-pulse mode-locking in the Ti-Al$_2$O$_3$ laser *Opt. Lett.* **15** 1300–1302.
30. French P M W, Williams J A R and Taylor J R 1989 Femtosecond pulse generation from a titanium-doped sapphire laser using nonlinear external cavity feedback *Opt. Lett.* **14** 686–8.
31. Keller U, 't Hooft G W, Knox W H and Cunningham J E 1991 Femtosecond pulses from a continuously self-starting passively-modelocked Ti:sapphire laser *Opt. Lett.* **16** 1022–4.
32. Keller U, Weingarten K J, Kartner F X, Kopf D, Braun B, Jung I D, Fluck R, Honninger C, Matuschek N and Der Au J A 1996 Semiconductor saturable absorber mirrors (SESAMS) for femtosecond to nanosecond pulse generation in solid-state lasers *IEEE J. Selected Topics Quantum Electron.* **2** 435.
33. Spence D E, Kean P N and Sibbett W 1991 60 fsec pulse generation from a self-mode-locked Ti–sapphire laser *Opt. Lett.* **16** 42–4.
34. Spence D E et al. 1990 *Proceedings of the Conference on Lasers and Electro-Optics, Optical Society of America* (Anaheim, CA) Paper CPD10.
35. Juang D-G, Chen Y-C, Hsu S-H, Lin K-H and Hsieh W-F 1997 Differential gain and buildup dynamics of self-starting Kerr lens mode-locked Ti:sapphire laser without an internal aperture *J. Opt. Soc. Am. B* **14** 2116.
36. Brabec T, Curley P F, Spielmann C H, Wintner E and Schmidt A J 1993 Hard-aperture Kerr-lens mode locking *J. Opt. Soc. Am. B* **10** 1029.
37. Xu L, Spielmann C and Krausz F 1996 Ultrabroadband sub-10 fs Ti:sapphire ring oscillator *Conference on Lasers and Electro Optics Europe Optical Society of America* (Hamburg) Paper CFF2.
38. Evans J M, Spence D E, Burns D and Sibbett W 1993 Dual-wavelength self-mode-locked Ti:sapphire laser *Opt. Lett.* **18** 1074.
39. Sohn J Y, Ahn Y H, Yee K J and Kim D S 1999 Two-color femtosecond experiments by use of two independently tunable Ti:sapphire lasers with a sample-and-hold switch *Appl. Opt.* **38** 5899.
40. Ell R, Morgner U, Kärtner F X, Fujimoto J G, Ippen E P, Scheuer V, Angelow G, Tschudi T, Lederer M J, Boiko A and Luther-Davies B 2001 Generation of 5 fs pulses and octave-spanning spectra directly from a Ti:sapphire laser *Opt. Lett.* **26** 373–5.
41. Kafka J D et al. 1991 *Optical Society of America Annual Meeting Optical Society of America* (San Jose, CA) Paper TUI4.
42. Negus D K et al. 1991 *Optical Society of America Topical Meeting on Advances in Solid-State Lasers Optical Society of America* (Hilton Head, SC).
43. Keller U, Knox W H and Roskos H 1990 Coupled-cavity resonant passive mode-locked Ti:sapphire laser *Opt. Lett.* **15** 1377.
44. Keller U, Miller D A B, Boyd G D, Chiu T H, Ferguson J F and Asom M T 1992 Solid-state low-loss intracavity saturable absorber for Nd:YLF lasers---an antiresonant semiconductor Fabry–Pérot saturable absorber *Opt. Lett.* **17** 505.
45. Mellish R, French P M W, Taylor J R, Delfyett P J and Florez L T 1993 Self-starting femtosecond Ti–sapphire laser with intracavity multiquantum-well absorber *Electron. Lett.* **29** 894–6.
46. Jung I D, Kartner F X, Matuschek N, Sutter D H, Morier-Genoud F, Zhang G, Keller U, Scheuer V, Tilsch M and Tschudi T 1997 Self-starting 6.5 fs pulses from a Ti:sapphire laser *Opt. Lett.* **22** 1009.
47. Mitschke F M and Mollenauer L F 1987 *Opt. Lett.* **12** 407.
48. Blow K J and Wood D 1988 *J. Opt. Soc. Am. B* **5** 629.
49. Kean P N, Zhu X, Crust D W, Grant R S, Langford N and Sibbett W 1989 *Opt. Lett.* **14** 39.
50. Zhu X, Kean P N and Sibbett W 1989 *Opt. Lett.* **14** 1192.
51. Mark J, Liu L Y, Hall K L, Haus H A and Ippen E P 1989 *Opt. Lett.* **14** 48.
52. Ouellette F and Piché M 1986 *Opt. Commun.* **60** 99.
53. Kennedy G T, Grant R S, Sleat W E and Sibbett W 1993 *Opt. Lett.* **18** 208.
54. Takara H, Kawanishi S, Saruwatari M and Noguchi K 1992 Generation of highly stable 20 GHz transform-limited optical pulses from actively mode-locked Er^{3+}-doped fibre lasers with an all-polarization maintaining ring cavity *Electron. Lett.* **28** 2095–6.
55. Harvey G T and Mollenauer L F 1993 Harmonically mode-locked fiber ring laser with an internal Fabry–Pérot stabilizer for soliton transmission *Opt. Lett.* **18** 107.
56. Duling I N 1991 All-fiber ring soliton laser mode locked with a nonlinear mirror *Opt. Lett.* **16** 539.
57. Hofer M, Ferman M E, Haberl F, Ober M H and Schmidt A J 1991 *Opt. Lett.* **16** 502.
58. Matsas V J, Newson T P, Richardson D J and Payne D N 1992 *Electron. Lett.* **28** 1391.
59. Tamura K, Haus H A and Ippen E P 1992 *Electron. Lett.* **28** 2226.
60. Zirngibl M, Stulz L W, Stone J, Hugi J, DiGiovanni D and Hansen P B 1991 *Electron. Lett.* **27** 1734.
61. Horowitz M, Menyuk C R, Carruthers T F and Duling I N 2000 Theoretical and experimental study of harmonically modelocked fiber lasers for optical communication systems *J. Lightwave Technol.* **18** 1565–74.
62. Doran N J and Wood D 1988 Nonlinear-optical loop mirror *Opt. Lett.* **13** 56.
63. Fermann M E, Haberl F, Hofer M and Hochreiter H 1990 Nonlinear amplifying loop mirror *Opt. Lett.* **15** 752.
64. Duling I N 1991 Subpicosecond all-fiber erbium laser *Electron. Lett.* **27** 544–5.
65. Richardson D J, Laming R I, Payne D N, Matsas V and Phillips M W 1991 Selfstarting, passively modelocked erbium fiber ring laser based on the amplifying Sagnac switch *Electron. Lett.* **27** 542–4.
66. Nakazawa M, Yoshida E and Kimura Y 1993 Generation of 98 fs optical pulses directly from an erbium-doped fiber ring laser at 1.57 µm *Electron. Lett.* **29** 63.
67. Kelly S M J 1992 *Electron. Lett.* **28** 806.
68. Tamura K, Nelson L E, Haus H A and Ippen E P 1994 Soliton versus nonsoliton operation of fiber ring lasers *Appl. Phys. Lett.* **64** 149–51.
69. Tamura K, Ippen E P, Haus H A and Nelson L E 1993 77 fs pulse generation from a stretched-pulse mode-locked all-fiber ring laser *Opt. Lett.* **18** 1080.

70. Lenz G, Tamura K, Haus H A and Ippen E P 1995 All-solid-state femtosecond source at 1.55 µm *Opt. Lett.* **20** 1289.
71. De Souza E A, Soccolich C E, Pleibel W, Stolen R H, Simpson J R and Digiovanni D J 1993 Saturable absorber modelocked polarization maintaining erbium-doped fiber laser *Electron. Lett.* **29** 447–9.
72. Collings B C, Bergman K, Cundiff S T, Tsuda S, Kutz J N, Cunningham J E, Jan W Y, Koch M and Knox W H 1997 Short cavity erbium/ytterbium fibre lasers mode-locked with a saturable Bragg reflector *IEEE J. Selected Topics Quantum Electron.* **3** 1065–75.
73. Mar A, Helkey R, Bowers J, Mehuys D and Welch D 1994 Mode-locked operation of a master oscillator power amplifier *IEEE Photon. Technol. Lett.* **6** 1067–9.
74. Goldberg L, Mehuys D and Welch D 1994 High power mode-locked compound laser using a tapered semiconductor amplifier *IEEE Photon. Technol. Lett.* **6** 1070–2.
75. Williams K A, Sarma J, White I H, Penty R V, Middlemast I, Ryan T, Laughton F R and Roberts J S 1994 Q-switched bow-tie lasers for high energy picosecond pulse generation *Electron. Lett.* **30** 320–1.
76. Silberberg Y and Smith P W 1986 Sub-picosecond pulses from a modelocked semiconductor laser *IEEE J. Quantum Electron.* **22** 759.
77. Matsui Y, Pelusi M D and Suzuki A 1999 Generation of 20 fs optical pulses from a gain-switched laser diode by a four-stage soliton compression technique *IEEE Photon. Technol. Lett.* **11** 1217–19.
78. Yariv A 1988 *Quantum Electronics* 3rd edn (New York: Wiley) p 259.
79. Vasil'ev P, White I H and Gowar J 2000 Fast phenomena in semiconductor lasers *Rep. Prog. Phys.* **63** 1997–2042.
80. For example, NTT device, Nlk1501.
81. Alphonse G 1997 *Conference on Lasers and Electro-Optics Optical Society of America* (Baltimore, MD).
82. Bowers J E, Morton P A, Mar A and Corzine S W 1989 Actively mode-locked semiconductor lasers *IEEE J. Quantum Electron.* **25** 1426–39.
83. Hou A S, Tucker R S and Eisenstein G 1990 Pulse compression of an actively modelocked diode laser using linear dispersion in fibre *IEEE Photon. Technol. Lett.* **2** 322–4.
84. Haus H A and Silberberg Y 1985 Theory of mode locking of a laser diode with a multiple-quantum-well structure *J. Opt. Soc. Am. B* **2** 1237.
85. Gee S, Alphonse G, Connolly J and Delfyett P J 1998 High-power mode-locked external cavity semiconductor laser using inverse bow-tie semiconductor optical amplifiers *IEEE J. Selected Topics Quantum Electron.* **4** 209–15.
86. Schell M, Weber A G, Bottcher E H, Scholl E and Bimberg D 1991 Theory of subpicosecond pulse generation by active modelocking of a semiconductor laser amplifier in an external cavity: limits for the pulsewidth *IEEE J. Quantum Electron.* **27** 402–9.
87. Koumans R G M P and van Roijen R 1996 Theory for passive mode-locking in semiconductor laser structures including the effects of self-phase modulation, dispersion, and pulse collisions *IEEE J. Quantum Electron.* **32** 478–92.
88. Morton P A, Bowers J E, Koszi L A, Soler M, Lopata J and Wilt D P 1990 Monolithic hybrid mode-locked 1.3 µm semiconductor lasers *Appl. Phys. Lett.* **56** 111–13.
89. Young-Kai C and Wu M C 1992 Monolithic colliding-pulse mode-locked quantum-well lasers *IEEE J. Quantum Electron.* **28** 2176–85.
90. Derickson D J, Helkey R J, Mar A, Karin J R, Wasserbauer J G and Bowers J E 1992 Short pulse generation using multisegment mode-locked semiconductor lasers *IEEE J. Quantum Electron.* **28** 2186–202.
91. Werner J, Melchior H and Guekos G 1988 Stable optical picosecond pulses from actively mode-locked twin-section diode lasers *Electron. Lett.* **24** 140–1.
92. Martins-Filho J F, Ironside C N and Roberts J S 1993 Quantum well AlGaAs/GaAs monolithic colliding pulse modelocked laser *Electron. Lett.* **29** 1135–6.
93. Sizer T, Mourou G and Rice R R 1981 Picosecond dye laser pulses using a cw frequency doubled Nd:YAG as the pumping source *Opt. Commun.* **37** 207.
94. Jeong T M, Chung C-M, Kim H S, Nam C H and Kim C-J 2000 Generation of passively Q-switched and mode-locked pulse from Nd:YVO$_4$ laser with Cr^{4+}:YAG saturable absorber *Electron. Lett.* **36** 633–4.
95. Malcolm G P A, Curley P F and Ferguson A I 1990 Additive-pulse mode locking of a diode-pumped Nd:YLF laser *Opt. Lett.* **15** 1303.
96. Keller U, Weingarten K J, Kartner F X, Kopf D, Braun B, Jung I D, Fluck R, Honninger C, Matuschek N and Aus Der Au J 1996 Semiconductor saturable absorber mirrors (SESAMs) for femtosecond to nanosecond pulse generation in solid-state lasers *IEEE Selected Topics Quantum Electron.* **2** 435–53.
97. Krainer L, Paschotta R, Moser M and Keller U 2000 77 GHz soliton modelocked Nd:YVO$_4$ laser *Electron. Lett.* **36** 1846–8.
98. Graf T, Ferguson A I, Bente E, Burns D and Dawson M D 1999 Multi-watt Nd:YVO$_4$ laser, mode locked by a semiconductor saturable absorber mirror and side-pumped by a diode-laser *Opt. Commun.* **159** 84–7.
99. Aus Der Au J, Loesel F H, Morier-Genoud F, Moser M and Keller U 1998 Femtosecond diode-pumped Nd:glass laser with more than 1 W of average output power *Opt. Lett.* **23** 271.
100. Sennaroglu A, Pollock C R and Nathel H 1994 Continuous-wave self-mode-locked operation of a femtosecond Cr^{4+}:YAG laser *Opt. Lett.* **19** 390.
101. Seas A, Petricevic V and Alfano R R 1992 Generation of sub-100-fs pulses from a cw mode-locked chromium-doped Forsterite laser *Opt. Lett.* **17** 937.
102. Sennaroglu A, Pollock C R and Nathel H 1993 Generation of 48 fs pulses and measurement of crystal dispersion by using a regeneratively initiated self-mode-locked chromium-doped Forsterite laser *Opt. Lett.* **18** 826.
103. Miller A, Likamwa P, Chai B H T and Van Stryland E W Generation of 150-fs tunable pulses in Cr:LiSrAlF$_6$ *Opt. Lett.* **17** 195.
104. Sorokina I T, Sorokin E, Wintner E, Cassanho A, Jenssen H P and Noginov M A 1996 Efficient continuous-wave TEM$_{00}$ and femtosecond Kerr-lens mode-locked Cr:LiSrGAF laser *Opt. Lett.* **21** 204.
105. Kopf D, Weingarten K J, Zhang G, Moser M, Emanuel M A, Beach R J, Skidmore J A and Keller U 1997 High-average-power diode-pumped femtosecond Cr:LiSAF lasers *Appl. Phys. B* **65** 235–43.

106. Ellingson R J and Tang C L 1992 High-repetition-rate femtosecond pulse generation in the blue *Opt. Lett.* **17** 343–5.
107. Rotermund F and Petrov V 1998 Generation of the fourth harmonic of a femtosecond Ti:sapphire laser *Opt. Lett.* **23** 1040–2.
108. Burneika K, Ignatavicius M, Kabelka V, Piskarskas A and Stabinis A 1972 *IEEE J. Quantum Electron.* **QE-8** 574.
109. Piskarskas A, Smil'gyavichyus V and Umbrasas A 1988 *Sov. J. Quantum Electron.* **18** 155.
110. Ebrahimzadeh M, Malcolm G P A and Ferguson A I 1992 *Opt. Lett.* **17** 183.
111. Edelstein D C, Wachman E S and Tang C L 1989 *Appl. Phys. Lett.* **54** 1728.
112. Wachman E S, Pelouch W S and Tang C L 1991 *J. Appl. Phys.* **70** 1893.
113. Tang C L, Wachman E S and Pelouch W S 1991 *Conference on Lasers and Electro-Optics Optical Society of America Paper* CFM5.
114. Wachman E S, Edelstein D C and Tang C L 1990 *Opt. Lett.* **15** 136.
115. Moon J A 1993 *IEEE J. Quantum Electron.* **29** 265.
116. Mak G, Fu Q and van Driel H M 1992 *Appl. Phys. Lett.* **60** 542.
117. Pelouch W S, Powers P E and Tang C L 1992 *Opt. Lett.* **17** 1070.
118. Ellingson R J and Tang C L 1993 *Opt. Lett.* **18** 438.
119. Reid D T, Ebrahimzadeh M and Sibbett W 1994 *J. Opt. Soc. Am. B.*
120. Driscoll T J, Gale G M and Hache F 1994 Ti:sapphire second-harmonic-pumped visible range femtosecond optical parametric oscillator *Opt. Commun.* **10** 638.
121. Gale G M, Cavallari M, Driscoll T J and Hache F 1995 Sub20fs tunable pulses in the visible from an 82MHz optical parametric oscillator *Opt. Lett.* **20** 1562.
122. Gale G M, Hache F and Cavallari M 1998 Broad-bandwidth parametric amplification in the visible---femtosecond experiments and simulations *IEEE Selected Topics Quantum Electron.* **4** 224–9.
123. Reid D T, Penman Z, Ebrahimzadeh M, Sibbett W, Karlsson H and Laurell F 1997 Broadly tunable infrared femtosecond optical parametric oscillator based on periodically poled $RbTiOAsO_4$ *Opt. Lett.* **22** 1397.
124. Butterworth S D, Pruneri V and Hanna D C 1996 Optical parametric oscillation in periodically poled lithium niobate based on continuous-wave synchronous pumping at 1.047 μm *Opt. Lett.* **21** 1345.
125. Myers L E, Eckardt R C, Fejer M M. Byer R L, Bosenberg W R and Pierce J W 1995 Quasi-phase-matched optical parametric oscillators in bulk periodically poled $LiNbO_3$ *J. Opt. Soc. Am. B* **12** 2102.
126. Karlsson H, Laurell F, Henriksson P and Arvidsson G 1996 Frequency doubling in periodically poled $RbTiOAsO_4$ *Electron. Lett.* **32** 556–7.
127. Loza-Alvarez P, Brown C T A, Reid D T, Sibbett W and Missey M 1999 High-repetition-rate ultrashort-pulse optical parametric oscillator continuously tunable from 2.8 to 6.8 μm *Opt. Lett.* **24** 1523.
128. Beddard T, Sibbett W, Reid D T, Garduno-Mejia J, Jamasbi N and Mohebi M 1999 High-average-power, 1 MW peak-power self-mode-locked Ti:sapphire oscillator *Opt. Lett.* **24** 163–5.
129. Xu L, Tempea G, Spielmann C, Krausz F, Stingl A, Ferencz K and Takano S 1998 Continuous-wave mode-locked Ti:sapphire laser focusable to 5×10^{13} W cm^{-2} *Opt. Lett.* **23** 789–91.
130. Liu Z L, Murakami H, Kozeki T, Ohtake H and Sarukura N 2000 High-gain, reflection-double pass, Ti:sapphire continuous-wave amplifier delivering 5.77 W average power, 82MHz repetition rate, femtosecond pulses *Appl. Phys. Lett.* **76** 3182–4.
131. Cho S H, Kärtner F X, Morgner U, Ippen E P, Fujimoto J G, Cunningham J E and Knox W H 2001 Generation of 90 nJ pulses with a 4-MHz repetition-rate Kerr-lens mode-locked $Ti:Al_2O_3$ laser operating with net positive and negative intracavity dispersion *Opt. Lett.* **26** 560–2.
132. Bartels A, Dekorsy T and Kurz H 1999 Femtosecond Ti:sapphire ring laser with a 2 GHz repetition rate and its application in time-resolved spectroscopy *Opt. Lett.* **24** 996–8.
133. Liu H, Biswal S, Paye J, Nees J, Mourou G, Hönninger C and Keller U 1999 Directly diode-pumped millijoule subpicosecond Yb:glass regenerative amplifier *Opt. Lett.* **24** 917–19.
134. Mellish R, Hyde S C, Barry N P, Jones R, French P M, Taylor J R, Van der Poel C J and Valster A 1997 All-solid-state diode-pumped Cr:LiSAF femtosecond oscillator and regenerative amplifier *Appl. Phys. B* **65** 221–6.
135. Boskovic A, Guy M J, Chernikov S V, Taylor J R and Kashyap R 1995 All-fibre diode-pumped, femtosecond chirped pulse amplification system *Electron. Lett.* **31** 877–9.
136. Fischer R A and Bischel W K 1975 Pulse compression for more efficient operation of solid-state laser amplifier chains II *IEEE J. Quantum Electron.* **11** 46.
137. Strickland D and Mourou G 1985 Compression of amplified chirped optical pulses *Opt. Commun.* **56** 219.
138. Pennington D M, Brown C G, Cowan T E, Hatchett S P, Henry E, Herman S, Kartz M, Key M, Koch J, MacKinnon A J and Perry M D 2000 Petawatt laser system and experiments *IEEE J. Selected Topics Quantum Electron.* **6** 676–88.
139. Riley K F 1987 *Mathematical Methods for the Physical Sciences* (Cambridge: Cambridge University Press) p 213.
140. Baltuska A, Wei Z, Pshenichnikov M S and Wiersma D A 1997 Optical pulse compression to 5 fs at a 1 MHz repetition rate *Opt. Lett.* **22** 102.
141. Dudley J M, Peacock A C and Millot G 2001 The cancellation of nonlinear and dispersive phase components on the fundamental optical fiber soliton---a pedagogical note *Opt. Commun.* **193** 253–9.
142. Akhmediev N N 1998 Spatial solitons in Kerr and Kerr-like media *Opt. Quantum Electron.* **30** 535–69.
143. Yariv A 1988 *Quantum Electronics* 3rd edn (New York: Wiley) p 487.
144. Pain H J 1987 *The Physics of Vibrations and Waves* 3rd edn (New York: Wiley) p 222.
145. Ainslie B J and Day C R 1986 *J. Lightwave Technol.* **4** 967.
146. Knight J C, Birks T A, Cregan R F, Russell P S and De Sandro P D 1998 Large mode area photonic crystal fibre *Electron. Lett.* **34** 1347–8.

147. Wadsworth W J, Knight J C, Ortigosa-Blanch A, Arriaga J, Silvestre E and Russell P S 2000 Soliton effects in photonic crystal fibres at 850 nm *Electron. Lett.* **36** 53–5.
148. Yablonovitch E 1994 Photonic crystals *J. Mod. Opt.* **41** 173–94.
149. Spielmann C, Szipöcs R, Stingl A and Krausz F 1994 Tunnelling of optical pulses through photonic band-gaps *Phys. Rev. Lett.* **73** 2308–11.
150. Treacy E B 1968 Compression of picosecond light pulses *Phys. Lett. A* **28** 34–5.
151. Treacy E B 1969 Pulse compression with diffraction gratings *IEEE J. Quantum Electron.* **5** 454–8.
152. Lemoff B E and Barty C P J 1993 Cubic-phase-free dispersion compensation in solid-state ultrashort-pulse lasers *Opt. Lett.* **18** 57–9.
153. Sherriff R E 1998 Analytic expressions for group-delay dispersion and cubic dispersion in arbitrary prism sequences *J. Opt. Soc. Am. B* **15** 1224–30.
154. Curley P F, Spielmann C, Brabec T, Krausz F, Wintner E and Schmidt A J 1993 Operation of a femtosecond Ti:sapphire solitary laser in the vicinity of zero group-delay dispersion *Opt. Lett.* **18** 54.
155. Yamashita M, Ishikawa M, Torizuka K and Sato T 1986 Femtosecond-pulse laser chirp compensated by cavity-mirror dispersion *Opt. Lett.* **11** 504.
156. Szipocs R, Ferencz K, Spielmann C and Krausz F 1994 Chirped multilayer coatings for broadband dispersion control in femtosecond lasers *Opt. Lett.* **19** 201.
157. Kartner F X, Matuschek N, Schibli T, Keller U, Haus H A, Heine C, Morf R, Scheuer V, Tilsch M and Tschudi T 1997 Design and fabrication of double-chirped mirrors *Opt. Lett.* **22** 831–3.
158. Jung I D, Kartner F X, Matuschek N, Sutter D H, Morier-Genoud F, Zhang G, Keller U, Scheuer V, Tilsch M and Tschudi T 1997 Self-starting 6.5 fs pulses from a Ti:sapphire laser *Opt. Lett.* **22** 1009.
159. Yariv A 1988 *Optical Electronics* 4th edn (Philadelphia, PA: Saunders College Publishing) p 117.
160. Gires F and Tournois P 1964 *C. R. Acad. Sci. Paris* **258** 6112.
161. For example: Robertson A, Ernst U, Knappe R, Wallenstein R, Scheuer V, Tschudi T, Burns D, Dawson M D and Ferguson A I 1999 Prismless diode-pumped mode-locked femtosecond Cr:LiSAF laser *Opt. Commun.* **163** 38–43.
162. Kuhl J, Serenyi M and Gobel E O 1987 Bandwidth-limited picosecond pulse generation in an actively mode-locked gas laser with intracavity chirp compensation *Opt. Lett.* **12** 334.
163. Heppner J and Kuhl J 1985, Intracavity chirp compensation in a colliding pulse mode-locked laser using thin-film interferometers *Appl. Phys. Lett.* **47** 453–5.
164. Suda A, Oishi Y, Nagasaka K, Wang P and Midorikawa K 2001 A spatial light modulator based on fused-silica plates for adaptive feedback control of intense femtosecond laser pulses *Opt. Express* **9** 2–6.
165. Patel J S and Silberberg Y 1995 Liquid crystal and grating-based multiple-wavelength cross-connect switch *IEEE Photon. Technol. Lett.* **7** 514–16.
166. Kohler B, Yakovlev V V, Che J, Krause J L, Messina M, Wilson K R, Schwentner N, Whitnell R M and Yan Y 1995 Quantum control of wave-packet evolution with tailored femtosecond pulses *Phys. Rev. Lett.* **74** 3360–3.
167. Weiner A M 1995 Femtosecond optical pulse shaping and processing *Prog. Quantum Electron.* **19** 161–237.
168. Weiner A M, Leaird D E, Patel J S and Wullert J R 1992 Programmable shaping of femtosecond optical pulses by use of 128-element liquid crystal phase modulator *IEEE J. Quantum Electron.* **28** 908–20.
169. Froehly C, Colombeau B and Vampouille M 1983 Shaping and analysis of picosecond light pulses *Prog. Opt.* **20** 65.
170. Hillegas C W, Tull J X, Goswami D, Strickland D and Warren W S 1994 Femtosecond laser pulse shaping by use of microsecond radio-frequency pulses *Opt. Lett.* **19** 737.
171. Chriaux G, Albert O, Wnman V, Chambaret J P, Flix C and Mourou G 2001 Temporal control of amplified femtosecond pulses with a deformable mirror in a stretcher *Opt. Lett.* **26** 169.
172. Brixner T, Oehrlein A, Strehle M and Gerber G 2000 Feedback-controlled femtosecond pulse shaping *Appl. Phys. Suppl. B* **70** S119–24.
173. Bergt M, Brixner T, Kiefer B, Strehle M and Gerber G 1999 Controlling the femtochemistry of $Fe(CO)_5$ *J. Phys. Chem. A.* **103** 10381–7.
174. Zeek E, Maginnis K, Backus S, Russek U, Murnane M, Mourou G, Kapteyn H and Vdovin G 1999 Pulse compression by use of deformable mirrors *Opt. Lett.* **24** 493.
175. Poppe A, Xu L, Krausz F and Spielmann C 1998 Noise characterization of sub-10-fs Ti:sapphire oscillators *IEEE Sel. Top. Quantum Electron.* **4** 179–84.
176. Von der Linde D 1986 Characterization of the noise in continuously operating modelocked lasers *Appl. Phys. B* **39** 201.
177. Diels J C M, Fontaine J J, Mcmichael I C and Simoni F 1985, Control and measurement of ultrashort pulse shapes (in amplitude and phase) with femtosecond accuracy *Appl. Opt.* **24** 1270.
178. Laughton F R, Marsh J H and Kean A H 1992 Very sensitive 2-photon absorption GaAs/AlGaAs wave-guide detector for an autocorrelator *Electron. Lett.* **28** 1663–5.
179. Takagi Y, Kobayashi T, Yoshihara K and Imamura S 1992 Multiple-shot and single-shot autocorrelator based on 2-photon conductivity in semiconductors *Opt. Lett.* **17** 658–60.
180. Reid D T, Padgett M, McGowan C, Sleat W E and Sibbett W 1997 Light-emitting diodes as measurement devices for femtosecond laser pulses *Opt. Lett.* **22** 233–5.
181. Gyuzalian R N, Sogomonian S B and Horvath Z G 1979 Background-free measurement of time behaviour of an individual picosecond pulse *Opt. Commun.* **29** 239.
182. Taft G, Rundquist A, Murnane M M, Kapteyn H C, DeLong K W, Trebino R and Christov I P 1995 Ultrashort optical waveform measurements using frequency resolved optical gating *Opt. Lett.* **20** 743.
183. Loza-Alvarez P, Reid D T, Faller P, Ebrahimzadeh M, Sibbett W, Karlsson H and Laurell F 1999 Simultaneous femtosecond pulse compression and second-harmonic generation in aperiodically poled $KTiOPO_4$ *Opt. Lett.* **24** 1071–3.
184. Treacy E B 1971 Measurement and interpretation of dynamic spectrograms of picosecond light pulses *J. Appl. Phys.* **42** 3848.

185. Kane D J and Trebino R J 1993 Single-shot measurement of the intensity and phase of an arbitrary ultrashort pulse by using frequency-resolved optical gating *Opt. Lett.* **18** 823.
186. Richman B A, Krumbugel M A and Trebino R 1997 Temporal characterization of mid-IR free-electron-laser pulses by frequency-resolved optical gating *Opt. Lett.* **22** 721–3.
187. Michelmann K, Feurer T, Fernsler R and Sauerbrey R 1996 Frequency resolved optical gating in the UV using the electronic Kerr effect *Appl. Phys. B-Lasers* **63** 485–9.
188. Fittinghoff D N, Bowie J L, Sweetser J N, Jennings R T, Krumbügel M A, DeLong K W, Trebino R and Walmsley I A 1996 Measurement of the intensity and phase of ultraweak, ultrashort laser pulses *Opt. Lett.* **21** 884–6.
189. Baltuska A, Pshenichnikov M S and Wiersma D A 1999 Second-harmonic generation frequency-resolved optical gating in the single-cycle regime *IEEE J. Quantum Electron.* **35** 459–78.
190. Paye J, Ramaswamy M, Fujimoto J G and Ippen E P 1993 Measurement of the amplitude and phase of ultrashort light pulses from spectrally resolved autocorrelation *Opt. Lett.* **18** 1946.
191. Trebino R and Kane D J 1993 Using phase retrieval to measure the intensity and phase of ultrashort pulses: frequency-resolved optical gating *J. Opt. Soc. Am.* A **10** 1101–11.
192. Tsang T, Krumbügel M A, DeLong K W, Fittinghoff D N and Trebino R 1996 Frequency-resolved optical-gating measurements of ultrashort pulses using surface third-harmonic generation *Opt. Lett.* **21** 1381–3.
193. Linden S, Kuhl J and Giessen H 1999 Amplitude and phase characterization of weak blue ultrashort pulses by downconversion *Opt. Lett.* **24** 569.
194. Delong K W, Trebino R and Kane D J 1994 Comparison of ultrashort-pulse frequency-resolved-optical-gating traces for three common beam geometries *J. Opt. Soc. Am.* B **11** 1595.
195. Delong K W, Fittinghoff D N, Trebino R, Kohler B and Wilson K 1994 Pulse retrieval in frequency-resolved optical gating based on the method of generalized projections *Opt. Lett.* **19** 2152–4.
196. Kane D J 1998 Real-time measurement of ultrashort laser pulses using principal component generalized projections *IEEE J. Selected Topics Quantum Electron.* **4** 278–84.
197. Kane D J 1999 Recent progress toward real-time measurement of ultrashort laser pulses *IEEE J. Quantum Electron.* **35** 421–31.
198. Chilla J L A and Martinez O E 1991 Direct determination of the amplitude and the phase of femtosecond light pulses *Opt. Lett.* **16** 39–41.
199. Wong V and Walmsley I A 1997 Ultrashort-pulse characterization from dynamic spectrograms by iterative phase retrieval *J. Opt. Soc. Am.* B **14** 944–9.
200. Reid D T 1999 Algorithm for complete and rapid retrieval of ultrashort pulse amplitude and phase from a sonogram *IEEE J. Quantum Electron.* **35** 1584–9.
201. Taira K and Kikuchi K 2001 Optical sampling system at 1.55 µm for the measurement of pulse waveform and phase employing sonogram characterization *IEEE Photon. Technol. Lett.* **13** 505–7.
202. Skovgaard P M W, Mullane R J, Nikogosyan D N and McInerney J G 1998 Two-photon photoconductivity in semiconductor waveguide autocorrelators *Opt. Commun.* **153** 78–82.
203. Reid D T, Thomsen B C, Dudley J M and Harvey J D 2000 Sonogram characterization of picosecond pulses at 1.5 µm using waveguide two photon absorption *Electron. Lett.* **36** 1141–2.
204. Koumans R G M P and Yariv A 2000 Pulse characterization at 1.5 µm using time-resolved optical gating based on dispersive propagation *IEEE Photon. Technol. Lett.* **12** 666–8.
205. Taft G, Rundquist A, Murnane M M, Christov I P, Kapteyn H C, DeLong K W, Fittinghoff D N, Krumbugel M A, Sweetser J N and Trebino R 1996 Measurement of 10 fs laser pulses *IEEE J. Selected Topics Quantum Electron.* **2** 575–85.
206. Cormack I G, Sibbett W and Reid D T 2001 Measurement of femtosecond optical pulses using time-resolved optical gating, Paper CTuN₃ *Conference on Lasers and Electro-Optics Optical Society of America* (Baltimore, MD).
207. Rothenberg J E and Grischkowsky D R 1987 Measurement of optical phase with subpicosecond resolution by time-domain interferometry *Opt. Lett.* **12** 99.
208. Iaconis C and Walmsley I A 1998 Spectral phase interferometry for direct electric-field reconstruction of ultrashort optical pulses *Opt. Lett.* **23** 792.
209. Gallmann L, Steinmeyer G, Sutter D H, Rupp T, Iaconis C, Walmsley I A and Keller U 2001 Spatially resolved amplitude and phase characterization of femtosecond optical pulses *Opt. Lett.* **26** 96–8.
210. Gallmann L, Sutter D H, Matuschek N, Steinmeyer G, Keller U, Iaconis C and Walmsley I A 1999 Characterization of sub-6 fs optical pulses with spectral phase interferometry for direct electric-field reconstruction *Opt. Lett.* **24** 1314.
211. Reid D T, Sibbett W, Dudley J M, Barry L P, Thomsen B and Harvey J D 1998 Commercial semiconductor devices for two photon absorption autocorrelation of ultrashort light pulses *Appl. Opt.* **37** 8142–4.
212. Peatross J and Rundquist A 1998 Temporal decorrelation of short laser pulses *J. Opt. Soc. Am.* B **15** 216.
213. Baltuska A, Pugzlys A, Pshenichnikov M S and Wiersma D A 1999 Rapid amplitude-phase reconstruction of femtosecond pulses from intensity autocorrelation and spectrum *Conference on Lasers and Electro-Optics Optical Society of America* (Baltimore, MD) Paper cwF22.
214. Yau, T-W, Jau Y Y, Lee C H, Wang J 1999 Photodiode-based phase-retrieval ultrafast waveform measurements *Conference on Lasers and Electro-Optics Optical Society of America* (Baltimore, MA) Paper cwF21.
215. Nicholson J W, Jasapara J, Rudolph W, Omenetto F G and Taylor A J 1999 Full-field characterization of femtosecond pulses by spectrum and cross-correlation measurements *Opt. Lett.* **24** 1774.
216. Nicholson J W, Mero M, Jasapara J and Rudolph W 2000 Unbalanced third-order correlations for full characterization of femtosecond pulses *Opt. Lett.* **25** 1801.
217. Denk W, Strickler J H and Webb W W 1990 2-Photon laser scanning fluorescence microscopy *Science* **248** 73–6.

218. Xu C and Denk W 1997, Two-photon optical beam induced current imaging through the backside of integrated circuits *Appl. Phys. Lett.* **71** 2578–80.
219. Yoo K M, Xing Q and Alfano R R 1991 Imaging objects hidden in highly scattering media using femtosecond second-harmonic-generation cross-correlation time gating *Opt. Lett.* **16** 1019.
220. Chen H, Chen Y, Dilworth D, Leith E, Lopez J and Valdmanis J 1991 Two-dimensional imaging through diffusing media using 150 fs gated electronic holography techniques *Opt. Lett.* **16** 487.
221. Huang D et al. 1991 Optical coherence tomography *Science* **254** 1178–81.
222. Drexler W, Morgner U, Ghanta R K, Kärtner F X, Schuman J S and Fujimoto J G 2001 Ultrahigh-resolution ophthalmic optical coherence tomography *Nat. Med.* **7** 502–7.
223. Boppart S A, Tearney G J, Bouma B E, Southern J F, Brezinski M E and Fujimoto J G 1997 Non-invasive assessment of the developing *Xenopus* cardiovascular system using optical coherence tomography *Proc. Natl Acad. Sci., USA* **94** 4256–61.
224. Jones R, Dowling K, Cole M J, Parsons-Karavassilis D, Lever M J, French P M and Hares J D 1999 Fluorescence lifetime imaging using a diode-pumped all-solid-state laser system *Electron. Lett.* **35** 256–8.
225. Cubeddu R, Ramponi R O, Liu W Q and Docchio F 1989 Time-gated fluorescence spectroscopy of the tumor localizing fraction of hpd in the presence of cationic surfactant *Photochem. Photobiol.* **50** 157–63.
226. Dowling K, Dayel M J, Hyde S C, Dainty C, French P M, Vourdas P, Lever M J, Dymoke-Bradshaw A K, Hares J D and Kellett P A 1998 Whole-field fluorescence lifetime imaging with picosecond resolution using ultrafast 10 KHz solid-state amplifier technology *IEEE J. Selected Topics Quantum Electron.* **4** 370–5.
227. Fattinger C and Grischkowsky D 1989 Terahertz beams *Appl. Phys. Lett.* **54** 490–2.
228. Luo M S C, Planken PC, Brener I, Roskos HG and Nuss MC 1994 Generation of terahertz electromagnetic pulses from quantum-well structures *IEEE J. Quantum Electron.* **30** 1478–88.
229. Mittleman D M, Jacobsen R H and Nuss M C 1996 T-ray imaging *IEEE J. Selected Topics Quantum Electron.* **2** 679–92.
230. Gruebele M, Roberts G, Dantus M, Bowman R M and Zewail A H 1990 Femtosecond temporal spectroscopy and direct inversion to the potential---application to iodine *Chem. Phys. Lett.* **166** 459.
231. Dantus M, Bowman R M and Zewail A H 1990 Femtosecond laser observations of molecular vibration and rotation *Nature* **343** 737.
232. Rose T S, Rosker M J and Zewail A H 1989 Femtosecond real-time probing of reactions. IV. The reactions of alkali halides *J. Chem. Phys.* **91** 7415–36.
233. Khundkar L R and Zewail A H 1990 Picosecond photofragment spectroscopy. IV. Dynamics of consecutive bond breakage in the reaction $C_2F_4I_2C_2F_4+2I$ *J. Chem. Phys.* **92** 231–42.
234. Fleming G R and Wolynes P G 1990 Chemical dynamics in solution *Phys. Today* **43** 36–43.
235. Assion A, Baumert T, Bergt M, Brixner T, Kiefer B, Seyfried V, Strehle M and Gerber G 1998 Control of chemical reactions by feedback-optimized phase-shaped femtosecond laser pulses *Science* **282** 919–22.
236. Cavalleri A, Siders C W, Sokolowski-Tinten K, Toth C, Blome C, Squier J A, Von der Linde D, Barty C P J and Wilson K 2001 Femtosecond X-ray diffraction *Opt. Photon. News* **12**(5) 29.
237. Jewell J L, Scherer A, McCall S L, Gossard A C and English J H 1987 *Appl. Phys. Lett.* **51** 94.
238. Shank C V, Fork R L, Leheny R F and Shah J 1979 Dynamics of photoexcited GaAs band-edge absorption with subpicosecond resolution *Phys. Rev. Lett.* **42** 112–15.
239. Shah J, Deveaud B, Damen T C, Tsang W T, Gossard A C and Lugli P 1987 Determination of intervalley scattering rates in GaAs by subpicosecond luminescence spectroscopy *Phys. Rev. Lett.* **59** 2222–5.
240. Nolte S, Momma C, Kamlage G, Chichnov B N, Tunnerman A, von Alvensleben F and Welling H 1998 *Conference on Lasers and Electro-Optics Optical Society of America* (San Francisco, CA) Paper CPD3.
241. Kruger J and Kautek W 1996 Femtosecond-pulse visible laser processing of transparent materials *Appl. Surf. Sci.* **96** 430–8.
242. Schaffer C B, Brodeur A, García J F and Mazur E 2001 Micromachining bulk glass by use of femtosecond laser pulses with nanojoule energy *Opt. Lett.* **26** 93–5.
243. Roth M, Cowan T E, Brown C, Christl M, Fountain W, Hatchett S, Johnson J, Key M H, Pennington D M, Perry M D and Phillips T W 2001 Intense ion beams accelerated by petawatt-class lasers *Nucl. Inst. Meth. Phys. Res.* **464** 201.
244. Umstadter D 2001 Review of physics and applications of relativistic plasmas driven by ultra-intense lasers *Phys. Plasmas* **8** 1774–85.
245. Norreys P A, Zepf M, Moustaizis S, Fews A P, Zhang J, Lee P, Bakarezos M, Danson C N, Dyson A, Gibbon P, Loukakos P, Neely D, Walsh F N, Wark J S and Dangor A E 1996 Efficient extreme UV harmonics generated from picosecond laser pulse interactions with solid targets *Phys. Rev. Lett.* **76** 1832–5.
246. Schnurer M, Cheng Z, Sartania S, Hentschel M, Tempea G, Brabec T and Krausz F 1998 Guiding and high-harmonic generation of sub-10 fs pulses in hollow-core fibers at 10^{15} W cm^{-2} *Appl. Phys. B* **67** 263–6.
247. Norreys P A, Allott R, Clarke R J, Collier J, Neely D, Rose S J, Zepf M, Santala M, Bell A R, Krushelnick K and Dangor A E 2000 Experimental studies of the advanced fast ignitor scheme *Phys. Plasmas* **7** 3721–6.
248. Rosen M D 1999 The physics issues that determine inertial confinement fusion target gain and driver requirements: A tutorial *Phys. Plasmas* **6** 1690–9.
249. Kane J, Arnett D, Remington B A, Glendinning S G, Bazan G, Drake R P, Fryxell B A 2000 Supernova experiments on the Nova laser *Astrophys. J. Suppl.* **127** 365–9.
250. Vodopyanov K L and Voevodin V G 1995 Type-I And Type-II $ZnGeP_2$ Travelling-wave optical parametric generator tunable between 3.9 And 10 μm *Opt. Commun.* **117** 277–82.

251. Boivin L, Wegmueller M, Nuss M C and Knox W H 1999 110 channels×2.35 Gb s^{-1} from a single femtosecond laser *IEEE Photon. Tech. Lett.* **11** 466–8.
252. Drescher M, Hentschel M, Kienberger R, Tempea G, Spielmann C, Reider G A, Corkum P B and Krausz F 2001 X-ray pulses approaching the attosecond frontier *Science* **291** 1923–7.
253. Holzwarth R, Udem T, Hänsch T W, Knight J C, Wadsworth W J and Russell P S 2000 Optical frequency synthesizer for precision spectroscopy *Phys. Rev. Lett.* **85** 2264–7.
254. Ruffing B, Nebel A and Wallenstein R 2001 High-power picosecond LiB$_3$O$_5$ optical parametric oscillators tunable in the blue spectral range *Appl. Phys. B* **72** 137–49.

20
Mode-locking Techniques and Principles

Rüdiger Paschotta

CONTENTS

- 20.1 Introduction 289
- 20.2 Basic Principles of Mode-locking 290
  - 20.2.1 Origin of the Term "Mode Locking" 290
  - 20.2.2 Active Mode-locking 291
  - 20.2.3 Passive Mode-locking 292
    - 20.2.3.1 Basic Principle 292
    - 20.2.3.2 Stability of the Circulating Pulse 292
    - 20.2.3.3 Start-up Phase 293
    - 20.2.3.4 Q-switching Instabilities 293
  - 20.2.4 Fundamental vs. Harmonic Mode-locking 294
  - 20.2.5 Frequency Combs 294
- 20.3 Saturable Absorbers for Mode Locking 295
  - 20.3.1 Parameters of Saturable Absorbers 295
  - 20.3.2 Semiconductor Absorbers 296
  - 20.3.3 Carbon Nanotubes and Graphene 297
  - 20.3.4 Laser Dyes 298
  - 20.3.5 Artificial Saturable Absorbers 298
- 20.4 Soliton Mode-locking 299
- 20.5 Mode-locked Solid-state Bulk Lasers 299
  - 20.5.1 Initial Remarks 299
  - 20.5.2 Picosecond Lasers 300
  - 20.5.3 Femtosecond Lasers 300
  - 20.5.4 High-power Operation 300
  - 20.5.5 High Pulse Repetition Rates 300
- 20.6 Mode-locked Fibre Lasers 301
- 20.7 Mode-locked Semiconductor Lasers 302
  - 20.7.1 Mode-locked Diode Lasers 302
  - 20.7.2 Mode-locked VECSELs 302
- 20.8 Modelling of Ultrashort Pulse Lasers 303
- References 304

20.1 Introduction

One of the most intriguing features of lasers is the possibility to generate ultrashort light pulses with durations on the picosecond or femtosecond scale. Such laser pulses find a wide range of interesting applications. Some of those directly make use of the ultrashort temporal durations. For example, with pump–probe measurement techniques, one can investigate in detail how a system evolves after the excitation with such a laser pulse; a second laser pulse, arriving at the sample with a variable short time delay, serves to probe the system. Another example is distance measurement with the time-of-flight technique, where a 1-ps pulse duration, for example, gives the potential for sub-millimetre accuracy even over large distances.

A wide variety of other applications make use of additional interesting aspects. For example, the typically rather high timing accuracy of ultrashort pulse trains makes them interesting as short-term timing references for optical clocks, and their coherence properties, which are the basis for so-called optical frequency combs, have opened a wide field of applications in optical metrology, particularly concerning measurements of time, frequencies and distances.

Another interesting aspect is that the concentration of a moderate amount of optical energy on extremely short timescales leads to enormously high peak powers. That in combination with the high focusability of laser beams allows one to generate enormously high optical intensities, which can be used for laser material processing (the commercially most relevant field of application) but also in some hot topics of fundamental scientific research.

In most cases, ultrashort light pulses are generated with mode-locking techniques, where one obtains regular trains of usually coherent ultrashort pulses with relatively low energy, rather than single pulses. Pulse repetition rates are typically in the region of many megahertz or even gigahertz. There are various techniques for picking a single pulse out of such a pulse train (or making a pulse train with much lower pulse repetition rate), and then amplifying such pulses to much higher energy levels. That area, however, goes beyond the scope of this chapter.

The chapter is meant as an overview on that field, concentrating on the most common techniques and the most common types of lasers. The latter are treated in Sections 20.5–20.7. Rather than providing a complete overview on achieved performance figures, the chapter discusses which aspects of mode-locking techniques and which additional physical effects are particularly relevant in certain lasers, and what impact that has on the performance limitations.

Mode-locked lasers are also called *ultrashort pulse lasers*, *picosecond lasers* or *femtosecond lasers*; somewhat odd is the term *ultrafast lasers*, since certain processes, but not the lasers themselves are ultrafast.

20.2 Basic Principles of Mode-locking

20.2.1 Origin of the Term "Mode Locking"

The term "mode locking" for techniques of ultrashort pulse generation resulted from an analysis of phenomena in the frequency domain. Each laser resonator has some number of modes, i.e. field configurations which are (in a certain sense) self-consistent when considering one complete resonator round trip; each mode has a certain resonance frequency. Typically, one considers only fundamental modes and ignores higher-order modes, having more complicated transverse intensity profiles and also different mode frequencies. For lasing on a single mode with constant power, the electric field corresponding to the laser output at a certain location is simply a sinusoidal oscillation. For simultaneous lasing on two modes, one obtains a beat note, i.e. an output with periodically modulated optical power. When combining several modes with equidistant frequencies, one can obtain a periodic train of well-separated short pulses (see Figure 20.1); however, that works only if the optical phases of the mode are "locked", i.e. if those modes maintain a certain phase relationship over long times: with random phases instead of that phase locking, one would obtain a complicated temporal evolution rather than clearly separated pulses.

Special measures (mode-locking techniques) are required to fulfil that phase condition. Note that the frequencies of fundamental modes in a real laser resonator are not exactly equidistant, so that the mentioned phase condition would normally be lost within a short time; one requires some additional locking effect which can enforce the phase locking at least if the natural tendency for dephasing is not too strong.

Although the one can easily understand the importance of the phase condition in the frequency domain, the workings of different mode-locking techniques are normally more easily understood in the time domain. They are explained in the following sub-sections.

Typically, mode-locking leads to a situation where a single ultrashort pulse circulates in the laser resonator. (In some situations, achieved with *harmonic* mode-locking, one obtains multiple circulating pulses.) Each time when a circulating pulse hits the output coupler mirror, part of its energy is transmitted by that mirror, and one obtains an output pulse. Therefore, one obtains a regular pulse train, with the pulse spacing determined by the round-trip time of the laser resonator. The pulse repetition rate (i.e. the inverse pulse spacing) typically has a value of tens or hundreds of megahertz, sometimes a few or even many gigahertz.

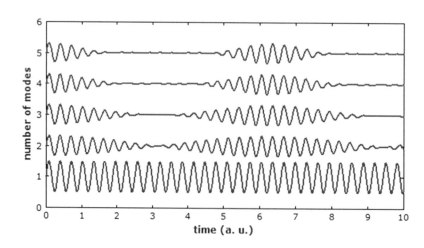

FIGURE 20.1 Synthesis of a pulse train by combination of an increasing number of fundamental modes in a laser resonator, where the relative phases are locked.

Normally, stable mode-locking is possible only when lasing on higher order transverse resonator modes can be safely suppressed, because those would have different round-trip phase shifts and times than the fundamental modes, making a multi-mode pulse decay quite quickly.

20.2.2 Active Mode-locking

The basic idea behind active mode-locking is to provide a periodic loss modulation by inserting an electrically controlled optical modulator into the laser resonator – typically close to an end mirror of a linear resonator. The electronic driver signal is in some way synchronized to the resonator round trips, such that the circulating pulse can each time pass the modulator with the minimum amount of loss. The loss modulation has two different effects:

- Any other radiation circulating in the resonator would pass the modulator at a time where the losses are substantially higher. It would thus have a lower net round-trip gain. As the circulating pulse in the steady state saturates the laser gain such that its energy stays constant, the other radiation would have a negative round-trip gain and thus die out sooner or later. Therefore, the modulator "protects" the circulating pulse against competing other radiation.
- The temporal wings of the circulating pulse, being slightly outside the temporal loss minimum, experience a slightly negative round-trip gain. To keep the pulse energy constant, the pulse centre must have a slightly positive round-trip gain. Those effects in combination cause some pulse shortening in each round trip (see Figure 20.2). In the start-up phase of the laser, this is what leads to rapid formation of a pulse. In the steady state, that shortening effect just compensates other effects which tend to make the pulse temporally longer. (Typically, the limited gain bandwidth and sometimes also chromatic dispersion can lead to temporal pulse broadening.)

As the pulse duration gets shorter, the effectiveness of the pulse-shortening effect of the modulator becomes weaker and weaker. On the other hand, pulse-broadening effects such as those arising from the limited gain bandwidth of the laser crystal or from chromatic dispersion become more effective. For those reasons, active mode-locking is not suitable for generating very short pulses, even when the modulator has a rather high modulation depth. This matter has been quantitatively analysed by Kuizenga and Siegman 1970, who derived an equation for the achievable pulse duration if the gain bandwidth is the limiting factor.

Somewhat shorter pulses are obtained by operating the modulator with the frequency which is an integer multiple of the round-trip frequency of the laser resonator, but that has the side effect of creating additional time windows, where additional pulses could circulate. For suppressing those, one can use the combination of the fundamental and a harmonic modulation frequency.

The modulator drive frequency must remain in quite precise synchronism with the round-trip frequency of the pulses. In principle, a modulator driven at a slightly deviating frequency can "pull" the pulses to that frequency, but such a regime quickly leads into instabilities, since the achievable "pulling force" for the pulse timing is quite weak. In many cases, sufficiently precise synchronism would be lost already by thermal drifts of the laser resonator or the driver electronics, and stable operation then requires some kind of feedback loop. It is possible to use a fixed (or even slightly variable) drive frequency of the modulator and automatically regulate the resonator length (e.g. via a piezo actuator attached to a resonator mirror). Alternatively, one may use regenerative mode-locking, where the drive frequency is automatically adjusted to follow fluctuations of the resonator length.

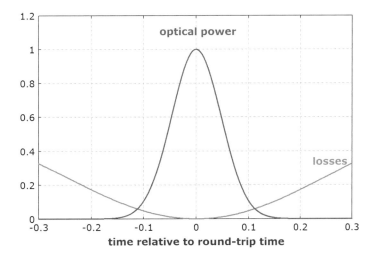

FIGURE 20.2 Temporal evolution of the pulse power and the losses at the modulator in an actively mode-locked laser. The temporal wings of the pulse experience higher losses than the central part and that leads to some pulse shortening. Under typical conditions with a low ratio of pulse duration to round-trip time, however, the pulse shortening effect may be quite weak.

20.2.3 Passive Mode-locking

20.2.3.1 Basic Principle

The inherent problem of active mode-locking concerning the achievable pulse duration is that the pulse-shortening effect obtained from the modulator rapidly becomes less effective as the pulses become shorter. That aspect is far more favourable for passive mode-locking, where the loss modulation is provided by a saturable absorber. Here, particularly the loss modulation on the leading front of the pulse becomes faster as the pulse gets shorter. Depending on the type of saturable absorber and the pulse duration, the loss recovery may be as fast as the decay of the pulse power – one would then have a *fast absorber* – or slower (for a *slow absorber*) with a fixed recovery time, which may be substantially longer than the pulse duration. Figure 20.3 shows the situation for a slow absorber.

Quite obviously, the use of saturable absorber allows for substantially stronger pulse-shortening effects even if the pulses become rather short. In the steady state, we may again have a situation where the pulse-shortening effect of the saturable absorber compensates pulse broadening effects, e.g. related to the finite gain bandwidth or to chromatic dispersion. However, there are also situations where the pulse duration is essentially determined by a balance of other effects (frequently, by soliton effects), while the saturable absorber has a negligible effect on the pulse duration. Its role is then only to initiate the pulse formation and to stabilize the circulating pulse. More details on soliton mode-locking are discussed in Section 20.4.

Similar to a modulator, a saturable absorber can also serve to suppress any other radiation circulating in the laser resonator. In that case, however, that suppression is not based on the other timing of the other radiation, but on the fact that the competing radiation is far less intense than the circulating pulse (e.g. in a situation close to the steady state) and thus not able to saturate the absorber; it thus experiences the full unsaturated loss in each resonator round trip.

Although very different kinds of saturable absorbers can be used for passive mode-locking (see Section 20.3), semiconductor saturable absorber mirrors (SESAMs) (Keller 1996) have become most common, since essentially all of their parameters can be optimized through the choice of materials and the device design. (Of course, such optimization also requires a solid understanding of which absorber parameters are most appropriate for the task.) Although expensive fabrication equipment is required, the fabrication cost can be small if many devices are fabricated together on a single semiconductor wafer. In most cases, a SESAM is used as an end mirror in a linear laser resonator. The laser resonator design must provide a suitable spot size of the beam on the SESAM, such that the saturation parameter (the ratio of pulse fluence to the saturation fluence of the absorber) has a reasonable value (typically of the order of 3–10).

20.2.3.2 Stability of the Circulating Pulse

In passively mode-locked lasers, the pulse duration may often become so short that the pulse bandwidth becomes quite substantial and is no longer small compared with the gain bandwidth. This results in a substantial gain disadvantage of the pulse against any temporally long background radiation also present in the laser resonator. That situation can only be stable if the losses caused by the saturable absorber for that background radiation are high enough to over-compensate the gain disadvantage of the pulses: essentially, one should assume that we need a situation where the round-trip gain for the circulating pulse is higher than that for any other radiation in the resonator. Interestingly, however, that condition is not absolutely necessary, as discussed in the following paragraph.

A particular concern for stability of the pulses arises for a slow absorber, having a recovery time which is several times longer than the pulse duration. Here, the net gain just behind the pulse can be positive because the losses are still close to their minimum, while the laser gain is nearly constant, e.g. in a typical solid-state laser, where the saturation energy of the gain medium is orders of magnitude higher than the pulse energy. According to the simple criterion formulated above, one should conclude that the pulses cannot be stable under such conditions. Indeed, that has been believed for many years,

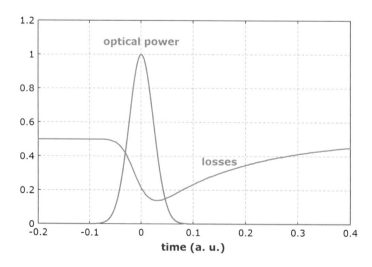

FIGURE 20.3 Temporal evolution of the pulse power and the losses at the (slow) saturable absorber in a passively mode-locked laser. Particularly, the leading temporal wing of the pulse experiences higher losses than the central part.

although practical experience actually showed that many lasers worked quite well even with pretty slow absorbers. An early explanation for that was limited to soliton mode-locked lasers (Kärtner and Keller 1995), because it was based on a mechanism which works only in such lasers, where chromatic dispersion causes a substantial temporal broadening of any noise behind the pulse. However, better than expected stability was observed even in simple picosecond lasers, where the effects of chromatic dispersion and the Kerr non-linearity are negligible. In fact, the same phenomenon of unexpected stability was also observed in numerical simulations and that finally led to the important finding that another effect, which had previously been overlooked, serves to stabilize the mode-locking: due to the decrease of losses during the pulse, the "center of gravity" of the pulse envelope is constantly shifted backwards, but the same does not happen for any noise behind the pulse. Therefore, the pulse is constantly shifted into that noise, which thus has only limited time to grow under the influence of the positive net gain. As a result, some significant amount of positive net gain can be tolerated without making the circulating pulse unstable. That allows for stable mode-locking even when the recovery time of the absorber is more than 10 times the pulse duration (Paschotta and Keller 2001). Of course, soliton effects and the like may provide some additional stabilization.

The previously mentioned analysis was done for the simplest situation, where the laser gain is temporally constant and only the limited gain bandwidth and the absorber action shape the pulse, while additional effects such as chromatic dispersion and the Kerr non-linearity are negligible. This is realistic for simple picosecond solid-state bulk lasers. Substantially more complicated situations arise for femtosecond bulk lasers and particularly for fibre lasers (see Section 20.6). Generally, the question of stability of a mode-locked laser, which can be a highly non-linear system, can be a highly non-trivial issue. It can depend on various details such as the required pulse energy and duration, the amount of chromatic dispersion and Kerr non-linearity, spectral filtering effects, etc.

20.2.3.3 Start-up Phase

In the start-up phase (directly after turning on the laser), there will always be some fluctuations of the laser power, e.g. arising from noise effects and from mode beating. Although the resulting peak power may be very small compared with the peak power of the pulse in the steady-state (because the final pulse is far shorter than initial power fluctuations), it causes some small amount of loss modulation at the saturable absorber. Even a small loss modulation of that type "favors" those parts of the circulating radiation which are more intense, since its net gain during each resonator round trip is slightly higher. Even if that process takes many thousands of round trips, for practical purposes, it often provides rapid self-starting of the mode-locking process. However, that self-starting feature may be inhibited by disturbing effects, particularly by parasitic reflections within the laser resonator. For example, even carefully anti-reflection-coated end faces of a laser crystal causes some amount of reflections, and those will be very disturbing unless the laser crystal is placed such that its surfaces are not exactly perpendicular to the laser beam. That rule also has to be observed for other optical surfaces in such a laser, except in some cases where an optical filtering effect from a sub-resonator needs to be employed, and the relevant optical surfaces have a small and stable distance from an end mirror.

For slow absorbers, the above-mentioned initial loss modulation is substantially stronger than for fast absorbers, essentially because for substantial absorber saturation, one does not require to reach the saturation energy within the short final pulse duration, but only within the much longer absorber recovery time. Self-starting mode-locking is thus tentatively more difficult to achieve, e.g. with Kerr lens mode-locking (KLM) (see Section 20.3.5), where the loss response is very fast. In many cases, such a laser will not automatically start mode-locking after being turned on, but will rather exhibit a noisy operation mode without ultrashort pulse generation. Mode-locking may then be started, e.g. by tapping one of the resonator mirrors, which introduces enhanced power fluctuations. Such requirements are obviously not ideal for commercial laser products.

Reliable self-starting of the passive mode-locking process is also tentatively more difficult to achieve in lasers with long laser resonators, leading to low pulse repetition rates. In such cases, the steady-state peak power of the pulse can be orders of magnitude higher than the power of initial random fluctuations. It is important to realize that one will usually have to optimize the absorber parameters such that they fit well to the steady-state conditions; the absorber will then have little effect on the initial phase.

20.2.3.4 Q-switching Instabilities

An inherent problem of passive mode-locking is that a saturable absorber influences the dynamical behaviour of a laser also on longer time scales. In particular, it tends to decrease the damping of the relaxation oscillations, as occurring in lasers with high saturation energies of the gain medium (e.g. solid-state lasers). If that effect becomes too strong, one obtains Q-switching instabilities, which usually lead to the regime of Q-switched mode-locking: the pulse energy is no longer stable, but undergoes strong oscillations, on a timescale of dozens, hundreds or more resonator roundtrips (Figure 20.4), depending on the system parameters. That regime is normally undesirable; it would in principle allow one to generate pulses of even substantially higher energy, but particularly when optimizing the parameters for a large enhancement of pulse energies, the operation becomes rather unstable. Therefore, one usually needs to adjust parameters such that one remains in the regime of stable mode-locking.

There is usually a well-defined transition between the regimes of stable mode-locking and the unstable regime. Frequently, a passively mode-locked laser is in the unstable regime below a certain value of the pump power, which is called the threshold power for stable mode-locking. In most cases, the value of that threshold power can be estimated based on a few system parameters, namely, the modulation depth of the absorber, its saturation parameter S, the saturation energy $E_{sat,g}$ of the gain medium and the intra-cavity pulse energy E_p: stability can be expected if the following condition is fulfilled: $E_p > E_{sat,g} \, \Delta R/S$ (Kärtner et al. 1995, Hönninger et al. 1999).

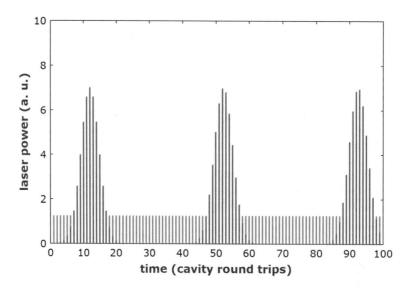

FIGURE 20.4 Evolution of pulses in the regime of Q-switched mode-locking (darker) and stable regular mode-locking (lighter).

(In the literature, the equation is often given in a modified form without the saturation parameter: $E_p^2 > E_{sat,g} E_{sat,a} \Delta R$, where $E_{sat,a}$ is the saturation energy of the absorber.) From that, one can conclude that the following measures are beneficial for avoiding Q-switching instabilities:

- One should avoid an unnecessarily strong modulation depth.
- The absorber saturation should be strong enough (e.g. $S > 4$ or even larger).
- One should have gain medium with low saturation energy (i.e. with high emission cross section at the laser wavelength and not too large beam area).
- A high intra-cavity pulse energy, e.g. achieved with a small output coupler transmission and small other intra-cavity losses, and with a long resonator, is also beneficial.

In some operation regimes, Q-switching instabilities are difficult to avoid and that may set limits to achievable performance parameters. In particular, this applies to lasers with very high pulse repetition rates, where it is difficult to achieve a high enough intra-cavity pulse energy, because that would correspond to a very high average power.

Note that Q-switching instabilities are not always easy to distinguish from other types of instabilities. For example, instabilities arising from a too slow saturable absorber may appear quite similarly in experimental tests.

20.2.4 Fundamental vs. Harmonic Mode-locking

In most mode-locked lasers, only a single ultrashort pulse circulates in the laser resonator. The obtained pulse repetition rate is then just the inverse of the round-trip time of the resonator. It has been mentioned above that shorter pulses can be obtained with active mode-locking if the modulator is driven at several times higher frequency, because that increases the last difference between the temporal wings and the power maximum of the pulse. At the same time, one obtains multiple temporal locations where a pulse can circulate and experience minimum losses at the modulator in every round trip. However, that does not automatically lead to clean harmonic mode-locking, i.e. multiple pulses with equal pulse energies and ideally perfect mutual phase coherence. For that, one requires additional measures which allow the circulating pulses to interact with each other such that their energies and phases are stabilized. That is typically achieved with some kind of sub-resonator within the laser resonator. A difficulty related to that concept is that the relative lengths of the involved resonators must be interferometrically stable.

Another motivation for harmonic mode-locking can be to achieve higher pulse-repetition rates. The resonators of fibre lasers, for example, can often not be made very short, because some length of active fibre is required in addition to various other components; on the other hand, multi-GHz pulse repetition rates are often required, for example, for telecom applications. One then requires harmonic mode-locking with a large number (hundreds or thousands) of pulses circulating in the laser resonator. Unfortunately, it is challenging to achieve long-term stability of operation under such conditions; at least, it requires a careful and relatively sophisticated device design. Therefore, a fundamentally mode-locked laser (possibly realized with a different technology) may be preferable.

20.2.5 Frequency Combs

The optical spectrum of a single ultrashort pulse shows a continuous distribution of power over the optical frequencies, the minimum (bandwidth-limited) width of which scales inversely with the pulse duration. For example, unchirped Gaussian-shaped pulses have a time–bandwidth product (the product of temporal and spectral width, both measured in terms of full width at half maximum) of ≈ 0.44. Such a pulse with 100-fs duration at 1064-nm wavelength, for example, has a FWHM bandwidth of 4.4 THz or 17 nm. For soliton pulses, the time–bandwidth product is smaller, ≈0.315.

Optical spectra of regular and virtually infinite pulse trains, as obtained from mode-locked lasers, look entirely different due to the mutual coherence of subsequent pulses. The spectrum resembles a *frequency comb*, i.e. it consists of discrete lines (Figure 20.5), which in the absence of any noise would have a zero width. The comb lines are exactly equidistant, and their spacing corresponds to the pulse repetition rate f_{rep}. The line frequencies can be formulated as $v_j = v_{ceo} + j \cdot f_{rep}$, where j is an integer index and v_{ceo} is the so-called carrier–envelope offset frequency; the latter is determined by the carrier-envelope offset phase shift which the circulating pulse experiences in each resonator round trip.

Interestingly, the two parameters v_{ceo} and f_{rep} precisely determine the frequencies of all (possibly many thousands) lines of the frequency comb. This effect is very important for a number of interesting applications in frequency metrology. For example, a single broadband frequency comb source with carefully stabilized and precisely measured values of v_{ceo} and f_{rep} can be used for accurately measuring the optical frequencies of any optical signals within its bandwidth: with a fast photodetector, one can measure the frequency of a beat note between such a signal and the closest comb line and calculate the signal frequency as the comb frequency plus or minus the beat note frequency. Conversely, with a precisely defined optical frequency standard (e.g. based on a laser source stabilized to a certain atomic or ionic optical transition) in combination with a high-precision radio frequency source one can very accurately determine the comb parameters.

Desirable properties of a frequency comb source are

- that it covers the optical frequency range of interest,
- that v_{ceo} can be measured accurately (typically with an *f-2f* interferometer, requiring a large spectral width of the laser),
- that it generates the required comb lines with sufficiently high optical powers,
- that the influences of noise (both quantum noise and technical noise sources) are as weak as possible and
- that the comb parameters v_{ceo} and f_{rep} can be rapidly adjusted, for example, within a feedback loop.

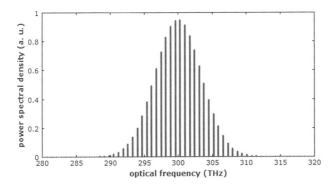

FIGURE 20.5 Optical spectrum of a frequency comb. In reality, the line spacing is normally much smaller, e.g. some hundreds of megahertz or some gigahertz.

20.3 Saturable Absorbers for Mode Locking

20.3.1 Parameters of Saturable Absorbers

In analytical or numerical models, for example, of mode-locked laser operation, saturable absorbers are often described by a couple of parameters of simple analytical absorber models:

- The *modulation depth* is the maximum amount by which the losses (in a certain wavelength range) can be reduced by saturation. In case of reflecting devices such as SESAMs, the modulation depth is normally referred to as ΔR (indicating the maximum increase of reflectivity).
- There are also often some *non-saturable losses*, which are normally undesirable but often to some extent unavoidable.
- For a fast absorber, the *saturation intensity* is the optical intensity required to reduce the saturable losses to half of their unsaturated value. For a given beam area on a saturable absorber, the *saturation power* is the saturation intensity times the beam area.

For a slow absorber, the *saturation fluence* is the fluence of a short pulse required to reduce the saturable losses to $1/e$ times their unsaturated value; however, the loss for the pulse itself is still higher ($(1-e^{-1})$ the initial loss), since that saturation value is only reached at the end (see Figure 20.6). For a given beam area on a saturable absorber, the *saturation energy* is the saturation fluence times the beam area.

- The *recovery time* is the time within which the loss reduction is reduced to $1/e$ times its initial value if there is no more saturating radiation. (At least in simple cases, there is an exponential decay of the saturation.)
- A *damage threshold* may be specified in terms of optical intensity or fluence, for example. Normal operation conditions are far below the damage threshold. However, there may also be accelerated aging for operation with high repetition rate pulse trains, e.g. in conjunction with excessive heating.

Of course, a saturable absorber must also be suitable for the specific wavelength of the pulses. If an absorber can be used within a certain wavelength range, its parameters as listed above may significantly vary within that range. For very broadband lasers, it can be a challenge to achieve a sufficiently large operation bandwidth of a saturable absorber.

Note that real absorbers will not always exhibit saturation characteristics as expected from simple absorber models. For example, they may not exhibit a simple exponential law for loss recovery after a pulse. Also, the reflectivity for pulses may not monotonically increase with increasing pulse fluence but eventually exhibit a roll-over, e.g. due to two-photon absorption. Therefore, a set of absorber parameters may not completely describe the actual characteristics. In addition, the transverse intensity variation of laser beam profiles, causing

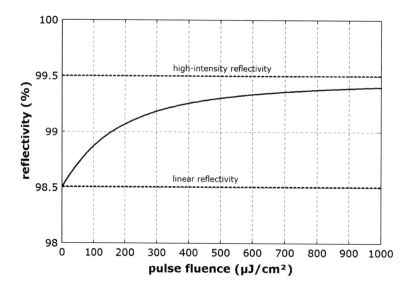

FIGURE 20.6 Saturation characteristics of a slow absorber mirror vs. pulse fluence with 1% modulation depth, 0.5% non-saturable losses and a saturation fluence of 100 μJ cm^{-2}. Shown is the reflectivity for short pulses (with a duration well below the recovery time) as a function of the pulse fluence. After a pulse with the saturation fluence, the saturable absorption is reduced to $1/e$ times the initial value, but the average loss for the saturating pulse itself is still higher.

a spatially variable degree of saturation, is often ignored in simple calculations.

The required values of absorber parameters depend very much on the circumstances:

- For many solid-state bulk lasers, a modulation depth of the order of 1% is fully sufficient, whereas fibre lasers often require a modulation depth of the order of 10%. For high-power operation of bulk lasers, one sometimes tries to work with less than 0.5% modulation depth.
- Non-saturable losses are no problem in some cases but can be critical for high-power operation (also for high pulse repetition rates) due to the involved heating. Tentatively, they are higher for absorbers with particularly fast loss recovery.
- Although a suitable saturation energy (assuming a slow absorber) can in principle be achieved for any saturation fluence just by choosing the appropriate beam area, it may be desirable to have a not too high saturation fluence because otherwise one requires inconveniently strong focusing, which may also cause thermal problems.
- An appropriate value for the saturation parameter (ratio of pulse energy to saturation energy) may be only 3 for some lasers, but it may have to be substantially higher in some cases, e.g. for suppressing Q-switching instabilities.

For successful use of saturable absorbers in mode-locked lasers, it is highly desirable to have their properties reliably measured (Haiml et al. 2004), since one can hardly derive absorber properties from their performance in mode-locked lasers. For that purpose, some standard measurements are often used:

- An optical reflection spectrum of a SESAM, for example (see the following section), gives information on the contained Bragg mirror structure.
- A pump–probe measurement, ideally performed with shorter and longer pulses, shows how the losses evolve over time; one can measure both the modulation depth and the recovery time.
- For a closer inspection of the saturation characteristics, one often measures the reflectivity or transmittivity of an absorber device as a function of the pulse energy (or fluence). From that, one obtains the saturation fluence. It is also possible to detect unusual features in the saturation characteristics, e.g. a rollover at high pulse fluences.
- In some cases, damage measurements are performed. Here, one may gradually increase the incident pulse fluence until damage or degradation is observed, e.g. by observation of the reflection characteristics or increased levels of scattered light. However, it still remains challenging to characterize absorbers in terms of their long-term degradation, since such measurements are intrinsically quite time-consuming. (Test with accelerated aging could in principle be a solution, but one would first need to reliably establish how much the aging is accelerated under certain conditions.)

20.3.2 Semiconductor Absorbers

Although early experiments with passive mode-locking where often based on other types of absorbers, SESAMs are nowadays the most widely used absorbers for passive mode-locking.

In the simplest case, a SESAM structure consists of a semiconductor Bragg mirror, on top of which a single absorber layer is placed, normally between some transparent layers (Figure 20.7).

The incident pulses are reflected on the Bragg mirror and experience some absorption losses essentially only in the absorber layer. Frequently, that absorber layer is so thin (e.g. 10 nm) that it acts as a quantum well. The absorption process moves some carriers from the valence band to the conduction band of the semiconductor material, and the absorption is substantially reduced once a substantial fraction of the carriers has been excited in that way. In the simplest case, recovery of the absorption would occur only through the return of carriers into the valence band; the recovery time would then simply be the carrier lifetime (which may, for example, be of the order of some tens or hundreds of picoseconds). However, particularly for femtosecond pulses, one frequently observes (e.g. in pump-probe measurements) a partial recovery of the absorption on a much shorter (sub-picosecond) time scale, which results from the thermalization of carriers in the conduction and valence bands. This effect can substantially support the use of such devices for the generation of femtosecond pulses, where a simple recovery with the recovery time of 50 ps, for example, would be too slow.

There are methods for making the recovery of SESAMs faster, as required for the generation of rather short laser pulses. One of them is to perform the growth of the semiconductor layers at significantly lower temperatures, so that a material with a higher density of certain microscopic defects is obtained. There are other methods to increase the defect density, e.g. the bombardment with high-energy heavy ions, which are then implanted into the material. Partly, an additional annealing process at higher temperature is required thereafter. It is common to those methods that while they can substantially speed up the loss recovery, they also tend to increase the level of non-saturable losses and possibly the tendency for short-term damage and long-term degradation. The optimization of such methods should thus be used to find a reasonable compromise between faster loss recovery and the negative side effects.

The choice of semiconductor materials largely depends on the required operation wavelength. For example, for the common wavelength region between 1 and 1.1 μm, one usually uses GaAs-based devices, i.e. with a GaAs substrate, a well lattice-matched AlGaAs-based Bragg mirror and an InGaAs-based absorber with higher indium content. The lattice mismatch caused by a high indium content tends to introduce material defects, which can be detrimental in terms of non-saturable losses and device lifetime but can also speed up the recovery, which can be beneficial for femtosecond pulse generation.

For a low saturation fluence, one typically places the absorbing layer in an anti-node of the optical field in the structure (i.e. at a location with high field intensities). If several absorbing layers are required for a higher modulation depth (e.g. for application in fibre lasers), one may place each one in a separate anti-node, or several of them (e.g. 3) close to one anti-node. The use of multiple absorbing layers increases the tendency for problems with deteriorated material quality due to the lattice mismatch.

Typically, the dimensions of the top layers are chosen such that the device is anti-resonant around the operation wavelength. One could achieve a lower saturation fluence with a resonant design, but this would have the disadvantage that the effective absorption would be lower at other wavelengths, possibly encouraging the laser to "escape" the absorption by some shift of wavelength. Also, the achievable values of the saturation fluence are usually small enough with anti-resonant designs – for example, of the order of 100 μJ cm^{-2} or even significantly lower. Reasonable beam radii on the SESAM are then required for the typical intra-cavity pulse energies.

Although SESAMs for 1-μm lasers have been best developed, devices for other wavelength regions such as 0.8, 1.5 and 2 μm have also been developed based on adaptive semiconductor materials (e.g. InP for 1.5-μm devices).

The fabrication of SESAMs requires sophisticated and expensive clean-room equipment for methods such as MOCVD, MOVPE or MBE. The development of a new type of SESAM (e.g. for a new wavelength region) can be time-consuming and quite expensive. However, once the fabrication technique is established, many SESAMs (each one being only, e.g. 10 mm^2 large) can be fabricated from a single processed wafer; the cost per device is then quite low, not contributing much to the overall production cost of a laser. Also, quite reproducible performance can be reached.

20.3.3 Carbon Nanotubes and Graphene

Although semiconductor-based saturable absorbers were already well established particularly for mode-docking applications, since 2004, there have been various reports on saturable absorbers based on carbon nanotubes (Set et al. 2004). Such nanotubes essentially consist of graphene layers which are wrapped up in microscopic dimensions to form rather thin tubes with diameters in the nanometre region. Depending on the detailed structure, they can have rather different optical properties, related to semiconducting, metallic or semi-metallic characteristics. The detailed composition, length

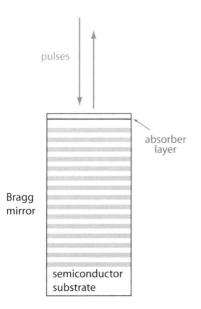

FIGURE 20.7 Structure of a simple SESAM (semiconductor saturable absorber layer).

spectrum, etc. depends on the growth method (e.g. a laser ablation technique followed by purification steps) and its details, which can thus be optimized to obtain certain optical properties. For saturable absorption, an excitonic absorption transition of semiconductor nanotubes, showing sub-picosecond recovery characteristics, can be used. Metallic nanotubes, which are also to some extent contained in the typically fabricated materials, might act as recombination centres for the semiconducting tubes, speeding up their loss recovery.

For a saturable absorber, one uses a thin layer of randomly oriented purified carbon nanotubes, which can be sandwiched between two glass plates, for example. If one of the plates contains a highly reflecting mirror, one obtains a reflecting saturable absorber device similar to a SESAM. It is also possible to place an absorbing layer between the ends of two optical fibres, resulting in a saturable absorber which can be operated in transmission within an all-fibre ring laser set-up. It is also possible to directly deposit carbon nanotube absorbers on fibre-end faces (Yamashita et al. 2004).

A disadvantage of using carbon nanotubes is that they are random in nature, and the influence of additional purification steps makes the fabrication overall not very reproducible. More controllable conditions are possible when using pure graphene sheets (Bonaccorso et al. 2010), which may either be used directly in transmission (e.g. after transfer to a fibre-end) (Bao et al. 2011) or in evanescent devices based on a side-polished fibre (Park et al. 2015) for obtaining a stronger absorption. Single or multiple graphene sheets may be used.

It is conceivable that graphene-based saturable absorbers will replace or complement more traditional semiconductor-based devices in some areas, particularly where the limitations of SESAMs are felt. For example, they may be useful for operation in some spectral regions for which well-developed SESAMs are not available. For mode-locking of fibre lasers, an advantage may be that a large modulation depth is possible (at least for evanescent devices) without negative side effects. There might also be other advantages, for example, concerning the lifetime, but the development is at a too early stage for a reliable assessment of the merits and limitations of that technological approach.

20.3.4 Laser Dyes

In the early days of femtosecond pulse generation, dye lasers were often used, in contrast to many solid-state laser gain media which provide a rather large gain bandwidth. The gain medium is typically a dye solution in the form of a dye jet exposed to pump light. It was then common to use a second dye jet (without optical pumping) as the saturable absorber.

When using the same dye for the saturable absorber as for the laser gain medium, one may be concerned that gain saturation would be just as strong as the loss saturation, so that one would overall not achieve a modulation of net gain as required for ultrashort pulse formation. However, one can simply use the absorber dye jet in a correspondingly smaller beam focus of the laser resonator, so that effectively the absorber's saturation energy becomes smaller than that of the gain medium.

Due to the inconvenience of using dye solutions, which are relatively maintenance-intensive and short-lived, and also cause hazards due to their poisonous and partially even carcinogenic nature, dye lasers were more and more replaced with solid-state lasers.

20.3.5 Artificial Saturable Absorbers

There are several techniques for making devices with reduced transmission losses for high optical powers, as required for mode-locking, but without employing any absorption. In most cases, they are based either on the Kerr non-linearity or on the $\chi^{(2)}$ non-linearity of some non-linear crystal material:

- For *KLM* (Salin et al. 1991), one exploits non-linear self-focusing of an intense laser beam, in most cases within the laser crystal. In case of hard aperture KLM, the losses at a subsequent optical aperture are reduced for a non-linearly focused beam, comparing with a low-power beam. In case of soft-aperture KLM, one exploits the higher laser gain for a more strongly focused beam, passing the laser crystal with increased overlap with the more strongly pumped centre. Such techniques (most often soft-aperture KLM) are frequently used for Ti:sapphire lasers, particularly when requiring the shortest possible pulse durations well below 10 fs. For reaching a sufficiently high sensitivity to the non-linear lens, such a laser usually has to be operated close to the stability limit of its resonator – which unfortunately makes it relatively sensitive to length changes, thermal lensing and other effects.

- *Additive-pulse mode-locking* is a method where self-phase modulation of light in an additional passive resonator is exploited. A pulse coming from the passive resonator has a phase profile such that the return of its energy into the main laser resonator is favoured for the intense centre of the pulse, but not for the temporal wings. That technique can also be fairly effective, but it requires interferometric stabilization of resonators.

- In fibre lasers, one often exploits *non-linear polarization rotation* in combination with some polarizing effect, e.g. in a Faraday isolator. Here, the polarization evolution and thus the overall power transmission through the polarizing element are power-dependent due to the fibre non-linearity. Unfortunately, the polarization evolution also depends on poorly controlled influences like temperature changes; therefore, this method is less suitable for industrial devices, where one often prefers polarization-maintaining fibres, excluding the use of non-linear polarization rotation.

- There are *non-linear fibre loop mirrors* of Sagnac type, where the effective reflectivity becomes power-dependent due to the power-dependent relative non-linear phase shifts of counter-propagating pulses in an asymmetric device. For example, the input pulse is split into two pulses, traveling through the fibre ring in opposite directions, and one of them, being amplified in an active fibre before going through a longer passive fibre, experiences higher non-linear phase shifts than the other one. Such operation principles can be realized with polarization-maintaining fibres, leading to environmentally stable devices.

- For some bulk lasers, one uses the combination of a *frequency-doubling non-linear crystal in combination with a dichroic mirror*. For high optical powers, much of the incident light is frequency-doubled, then effectively reflected on the mirror and converted back to the original wavelength in the backward path. For lower optical powers, leading to less efficient frequency doubling, the mirror exhibits higher reflection losses.

In most cases, the achieved non-linear response of artificial saturable absorbers is very fast, which can be beneficial for generating very short pulses. Another advantage is that a high modulation depth (as required, e.g. for fibre lasers) can often be achieved without negative side effects. On the other hand, the effective absorber parameters are often not well known, because they depend on poorly controlled influences.

A frequently encountered problem is that self-starting mode-locking is tentatively more difficult to achieve with artificial saturable absorbers – for example, with KLM. That is essentially a consequence of the fast absorber response, as explained in Section 20.2.3.3.

20.4 Soliton Mode-locking

In many mode-locked lasers, the pulse shaping effect of the saturable absorber is only one of several strong effects acting on the circulating pulse. In particular, chromatic dispersion and optical non-linearities (most frequently the Kerr non-linearity) are often quite important. They can have seriously performance-limiting effects but can, in some cases, also be utilized. This is particularly the case for soliton mode-locking. Here, one has substantial soliton effects in the laser resonator, i.e. an interplay of (typically anomalous) chromatic dispersion with the Kerr non-linearity: the two effects can combine such that in total there is neither a temporal broadening nor a change of the optical spectrum. For that balance, the non-linear and dispersive parameters of the laser resonator need to be adjusted such that the fundamental soliton condition is approximately fulfilled for the wanted operation conditions. The circulating pulse, which will then automatically acquire the needed temporal and spectral shape, may be mainly shaped by those soliton effects, and the saturable absorber is needed only for initiating the mode-locking and stabilizing the pulse in the steady state. In particular, the absorber needs to safely suppress the rising of any low-power background radiation, which may have a gain advantage over the pulses due to its potentially smaller bandwidth, i.e. due to its better spectral fit to the gain spectrum.

Originally, soliton effects have been employed in fibre lasers (see Section 20.6) (Mitschke and Mollenauer 1987). Here, however, a problem is that soliton pulse energies in fibres are usually very small, and the achievable pulse duration is also quite limited due to the onset of instabilities. Therefore, other mode-locking mechanisms have been developed for fibre lasers, which offer much better performance.

On the other hand, the technique of quasi-soliton mode-locking has become very important for solid-state bulk lasers (Kärtner and Keller 1995). Many of those can be designed such that the amount of intra-cavity chromatic dispersion (e.g. controlled via dispersive mirrors or a prism pair) and Kerr non-linearity are suitable for forming a quasi-soliton pulse; that is, also based on the balance of chromatic dispersion and Kerr non-linearity, although here those two effects are not continually spread in the laser resonator. For quasi-soliton mode-locking to work, one does not only need the appropriate relative strength of dispersion and non-linearity but also an appropriate absolute strength of both effects, quantified, e.g. with a peak non-linear phase shift of the order of 100 mrad per round trip. Soliton effects are then strong enough to substantially contribute to the formation of pulses with the desired pulse duration (usually in the femtosecond domain) and with clean temporal and spectral shapes. For substantially stronger non-linear phase shifts, the discrete distribution of the chromatic dispersion and non-linearity would lead to strongly disturbed pulses and correspondingly unstable pulse evolution.

In some bulk lasers, the amount of anomalous intra-cavity dispersion can be varied, e.g. through the insertion of a prism into the beam path. One then typically finds that the pulse duration (for a given pulse energy) is proportional to the overall amount of anomalous dispersion; it is hardly dependent on absorber parameters, the gain bandwidth, etc. However, the pulses become unstable as soon as they would get too short. At this point, the overall non-linear phase shift and/or the gain filtering in the gain medium gets too strong for stable operation. Therefore, details like the gain bandwidth and absorber parameters indirectly determine the minimum possible pulse duration.

20.5 Mode-locked Solid-state Bulk Lasers

20.5.1 Initial Remarks

Although a lot of initial work in the area of ultrashort pulse generation was made with dye lasers, solid-state lasers are nowadays dominating, since they are substantially more practical, not involving short-lived and poisonous liquid dye solutions. In this section, we consider only bulk lasers, i.e. lasers based on bulk pieces of some crystal or glass material; fibre lasers are discussed in the following section.

An important aspect of doped-insulator solid-state lasers is that the saturation energy of the gain medium is typically orders of magnitude larger than the achievable intra-cavity pulse energy. Therefore, gain saturation during a single pulse is negligible, and only the cumulative effect of many pulses can saturate the gain. This leads to a simple situation for a single pulse (temporally constant gain) but also to long-term gain dynamics (during many round trips) with relaxation oscillations – which in case of mode-locked lasers mean oscillations of the pulse energy. Through the effect of a saturable absorber, those oscillations can even become undamped for passively mode-locked lasers, which imply the so-called Q-switching instabilities (see Section 20.2.3.4). The need to maintain sufficient damping of the relaxation oscillations can introduce substantial constraints on the laser design and the achievable performance, at least for gain media with particularly low-emission cross sections.

20.5.2 Picosecond Lasers

It is relatively simple to realize a picosecond laser by incorporating a suitable saturable absorber (e.g. a SESAM) in the laser resonator (see Figure 20.8), which also contains a usual kind of laser crystal (e.g. Nd^{3+}:YAG or Nd^{3+}:YVO$_4$). Normally, such a device is used as an end mirror, and the resonator design is made such that the fundamental mode size on the absorber is suitable for achieving an appropriate strength of absorber saturation. Additional means such as dispersion compensation are not required in that operation regime. The pulse duration is often determined essentially by the balance of pulse shortening by the absorber and pulse broadening by the limited gain bandwidth. That means that an absorber with higher modulation depth allows the generation of shorter pulses; the pulse duration is inversely proportional to the square root of the modulation depth (Paschotta and Keller 2001), while the absorber recovery time has no significant influence on the pulse duration; it only needs to be short enough to obtain stable mode-locking. The pulse duration is also inversely proportional to the gain bandwidth.

20.5.3 Femtosecond Lasers

One might imagine that femtosecond pulses can be generated simply by using a laser crystal (or glass) with substantially broader gain bandwidth; there is a wide range of broadband crystalline and glass media with active ions such as Nd^{3+}, Yb^{3+}, Er^{3+}, Tm^{3+}, Cr^{2+}, Cr^{3+}, Cr^{4+} or Ti^{3+}. However, that alone would not work. First of all, pulse broadening by chromatic dispersion would become relevant. Even if chromatic dispersion would be totally removed by dispersion compensation (i.e. by inserting some dispersive elements with a dispersion opposite to that of other elements such as the laser crystal), self-phase modulation due to the Kerr effect would become problematically high due to the increased peak power of the circulating femtosecond pulse. However, there is a simple solution for both problems: quasi-soliton mode-locking (see Section 20.4) can be achieved

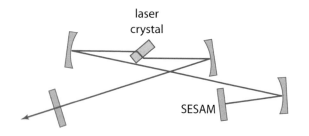

FIGURE 20.8 Set-up of a simple picosecond laser (without optical pump source).

if the overall chromatic dispersion per round-trip is adjusted to have a suitable value in the anomalous dispersion region. That works well, allowing for stable and clean pulse formation in a wide range of pulse durations, even where the pulse shortening action of the saturable absorber alone would be too weak. It turns out that the strength of the Kerr non-linearity, which can to some extent be adjusted, e.g. via the length of the laser crystal, is quite suitable down to pulse durations of the order of 10 fs.

For even shorter pulse durations, a number of challenges need to be met. First of all, one requires a gain material with very large gain bandwidth; the most successfully used material is Ti^{3+}:sapphire. Second, careful dispersion compensation is required, not only considering second-order dispersion but also dispersion of higher orders due to the large relevant frequency range; double-chirped dielectric mirror designs (Kärtner et al. 1997) are often used. Of course, any intra-cavity components such as laser mirrors and a saturable absorbers also need to have a very high reflection bandwidth. Finally, a rather fast absorber is required; for the shortest pulse durations (around 5 fs) (Sutter et al. 1999), KLM is usually used – sometimes in combination with a semiconductor absorber for reliable self-starting.

20.5.4 High-power Operation

For many years, average output powers of more than a few watts from mode-locked lasers were difficult to achieve. A breakthrough has then been obtained by passive mode-locking of thin-disk lasers, which have a number of attractive properties:

- the ability of fundamental mode operation even at very high power levels (due to well-managed thermal effects),
- the high power conversion efficiency and
- the relatively large gain bandwidth of the most frequently used gain material Yb^{3+}:YAG (and the potential to use some other crystal materials like Yb^{3+}:Lu$_2$O$_3$ or Yb^{3+}:Sc$_2$O$_3$ with even substantially larger gain bandwidth).

Not too long after the first demonstration with 16 W average output power (Aus der Au et al. 2000), much higher powers of ≈60 W (Innerhofer et al. 2003) and later even powers well above 200 W (Saraceno et al. 2014) have been demonstrated; pulse durations were in most cases somewhat below 1 ps. With KLM, substantially shorter pulses were achieved (Pronin et al. 2001).

20.5.5 High Pulse Repetition Rates

Typical pulse repetition rates of such lasers are between 50 and 500 MHz, corresponding to a length between ≈3 and 0.3 m of a linear laser resonator. For obtaining much higher repetition rates, the main challenge is not to build much smaller laser resonators but rather to avoid Q-switching instabilities (see Section 20.2.3.4). The highest pulse repetition rate of 160 GHz

(Krainer et al. 2002) has been achieved by using an Nd^{3+}:YVO_4 laser crystal (having a particularly high emission cross section and a reasonable gain bandwidth) and an optimized low-loss laser resonator containing a SESAM with quite low modulation depth. With Er:Yb:glass for emission in the 1.5-μm wavelength region, high pulse repetition rates are more interesting (for telecom applications) but also more difficult to achieve due to the much lower emission cross sections; 50 GHz have been obtained (Zeller et al. 2002).

20.6 Mode-locked Fibre Lasers

Fibre lasers are a special kind of solid-state lasers where the gain medium is an optical fibre doped with laser-active ions. Such active fibres have a number of attractive properties for ultrashort pulse generation, including a substantial gain bandwidth, good coverage of wide wavelength regions and the potentially low cost and ruggedness (provided that an all-fibre set-up it can be realized). However, their substantial non-linearity (resulting from the relatively long length of the gain medium and the very small mode size) and partly their chromatic dispersion often constitute substantial challenges. Since the introduction of many kinds of photonic crystal fibres (Russell 2003), there is at least a much greater variability of fibre parameters.

In principle, soliton mode-locking would appear to be the most natural technique for handling problems with dispersion and non-linearities, and it has indeed been demonstrated long ago (Mitschke and Mollenauer 1987). However, soliton pulse energies in fibres are typically very small (in the picojoule domain), which results in low output powers. The average intra-cavity powers are normally smaller than the gain saturation power, so that operation only slightly above threshold is required, which itself is a problem. Also, for pulse durations well below 1 ps, the soliton period quickly gets shorter than the practical length of the active fibre and the desirable resonator length, quickly leading to Kelly sidebands (Kelly 1992) and eventually to instabilities.

For those reasons, alternative methods for handling the dispersion and non-linearity have been developed, mostly based on various methods of dispersion management. One of those is to build stretched-pulse fibre lasers (Tamura et al. 1993) containing fibre segments with normal and anomalous dispersion; at most locations in the laser resonator, the pulses are then strongly chirped, leading to a reduced peak power and thus to a reduced non-linear phase shifts. With such methods, one can achieve pulse energies in the nanojoule domain.

Other types of dispersion management in fibre lasers have been developed which resulted in further improved performance. Most notably, it has been realized that while pulse propagation in the anomalous dispersion regime (with soliton effects) is subject to serious limitations in the context of strong non-linearities, attractive possibilities arise in the normal dispersion regime. In a normal dispersion fibre with laser gain, one can obtain self-similar propagation of parabolic pulses (Fermann et al. 2000), where the pulse duration and bandwidth grow together with the pulse energy, and wave-breaking effects are avoided. That principle has been utilized for mode-locked fibre lasers (Chong et al. 2008, Renninger et al. 2012); here, the laser resonator must contain a bandpass filter for "resetting" the pulse bandwidth after the gain medium (or some other location), which (for strongly chirped intra-cavity pulses) also resets the pulse duration. Pulse durations far below 1 ps in conjunction with multi-nanojoule output pulse energies are achieved. In a landmark experiment (Baumgartl et al. 2012), an unusually high average output power of 66 W in pulses which can be compressed to a duration below 100 fs has been demonstrated, based on a photonic crystal fibre with very large mode area in a laser resonator containing several bulk-optical elements.

While the operation principle of a soliton fibre laser (similar to a quasi-soliton bulk laser) can be relatively easily analysed, the pulse evolution in many other mode-locked fibre lasers is rather complicated (with strongly non-linear dynamics), and the dependence of the steady-state pulse parameters and the pulse stability on various device parameters can often be calculated only with numerical simulations. Therefore, the development particularly of high-performance mode-locked fibre lasers is substantially more difficult and time-consuming than for a bulk laser, which can be operated with much lower non-linearity and therefore based on much simpler operation principles.

Most mode-locked fibre lasers are passively mode-locked. Saturable absorbers are often SESAMs (usually with higher modulation depth than those used for bulk lasers) based on semiconductors, but other materials such as carbon nanotubes and graphene can also be used (see Section 20.3.3). Alternatively, one can use non-linear polarization rotation (see Section 20.3.5): in a not polarization-maintaining fibre, the Kerr non-linearity can lead to power-dependent polarization evolution, which in combination with a polarizing element can provide an artificial saturable absorber. While that method is versatile (e.g. applicable in any wavelength region without requiring special parts), it typically leads to lasers which are environmentally not very stable, because the polarization evolution is also affected by bending of the fibre, temperature changes, etc.

Certain specific practical advantages of fibre lasers in terms of stability and robustness (e.g. immunity to dirt and dust) can be achieved only with all-fibre set-ups, where the light is everywhere contained in a fibre, thus being well protected against dust. That, however, strongly limits the choice of available components; in particular, in many of the record performance figures were achieved only with set-ups containing a fibre together with bulk-optical elements and would be hard or impossible to realize with an all-fibre set-up. On the other hand, fibre laser set-ups containing bulk optics are often even less robust than pure bulk lasers, since they involve air-to-fibre interfaces which are particularly sensitive to dust and misalignment. (Launching a laser beam into a single-mode fibre typically requires micrometre precision.)

In order to overcome the power limitations of mode-locked fibre lasers, they are often combined with one or several fibre amplifiers. Active fibres can provide a high gain with a substantial gain bandwidth and high average power; therefore, they may appear to be ideal for ultrashort pulse amplification. However, particularly, the limitations arising from non-linearities are again quite severe. For example, a 100-nJ pulse with 100 fs duration, as can easily be generated in a mode-locked bulk laser, has a peak power of the order of 1 MW, while a peak power of only 1 kW already produces substantial non-linear phase shifts within a 1-m length of a fibre with standard mode area. Although the mode areas can be increased by roughly two orders of magnitude, that is not sufficient to avoid serious non-linear phase shifts in ultrashort pulse fibre amplifiers even for moderate pulse energies. In addition, there remains the hard limit set by non-linear self-focusing at a few megawatts peak power (independent of the mode area). To some extent, the principle of chirped pulse amplification can be used to mitigate such problems, but even then the peak power limitations are far more severe than for bulk devices.

20.7 Mode-locked Semiconductor Lasers

20.7.1 Mode-locked Diode Lasers

Semiconductor laser gain media, particularly when directly electrically pumped, obviously have important practical advantages and are in principle interesting for ultrashort pulse laser sources, which could be very compact, potentially cheap and would not require another laser as pump source. The gain bandwidth – typically, some tens of nanometres – is sufficient for pulse durations in the picosecond or high femtosecond regime, as is sufficient for many applications.

There are different ways of incorporating a mode-locking device into a diode laser. For active mode-locking (Bowers et al. 1989), different types of modulators can be implemented on a semiconductor platform; in the simplest case, one can just modulate the gain section. Alternatively, one can implement electro-absorption modulators, being quite similar to active gain sections.

Passive mode-locking of diode lasers is also possible by incorporating a saturable absorber section. Comparing with the gain section, such an absorber section usually does not require an electrical contact, and it should be prepared such that the carrier lifetime is reduced in that area, so that a sufficiently fast loss recovery is obtained. That can be achieved by irradiating the material with accelerated ions, for example. (Similar techniques are used for SESAMs.)

It is also possible to use a hybrid approach, where a saturable absorber section has an electrical contact through which the carrier density can be modulated with an external signal. That way, one can achieve the synchronization of the generated pulse train with an external signal while at the same time achieving shorter pulses than with active mode-locking only.

As the length of such a semiconductor device is typically quite small (e.g. a few millimetres), the laser resonator typically needs to be extended according to the desired pulse repetition rate; otherwise, extremely high repetition rates of possibly more than 1 THz are obtained (Arahira et al. 1996). In an extended-cavity laser, the semiconductor gain medium (including absorber sections and the like) is then just one part of a longer resonator, which may either contain free-space optical elements just as in a solid-state bulk laser (with a doped-insulator gain medium) or can be made with optical fibre. (In the latter case, one should nevertheless avoid the term *fibre laser* since the fibre is used only for forming the resonator not as the laser gain medium.) Mode-locked external cavity diode lasers with a fibre resonator are most suitable for optical fibre communications (Sato 2002, Jiang et al. 2005) as far as the performance requirements can be met.

A serious limitation for the achievable output pulse energy arises from the very small gain saturation energies of gain and saturable absorption in a semiconductor waveguide; more than some tens of picojoules is difficult to achieve. Particularly for pulse repetition rates well below 1 GHz (i.e. devices with a small duty cycle), the gain non-linearity limits the average output power to a value far below what would be possible in continuous-wave operation. This is remarkably different to ion-doped insulator gain media, which typically have saturation energies far larger than the pulse energies; they can thus easily store enough energy to amplify energetic pulses even at far lower repetition rates. Note that the saturation of gain and absorption in a semiconductor is also accompanied by significant non-linear phase changes (resulting in a kind of self-phase modulation), which must be dealt with, e.g. using certain chirp compensation techniques (Delfyett et al. 1992), at least in cases with relatively high pulse energies and/or short pulse durations. In that sense, boosting the output power with an external semiconductor optical amplifier (SOA) is less problematic; it can at least not disturb the mode-locking dynamics.

20.7.2 Mode-locked VECSELs

For a long time, only laser diodes were considered in the area of mode-locked semiconductor lasers. Only in the year 2000, the first mode-locked external-cavity surface-emitting semiconductor laser was reported (Hoogland et al. 2000). Such a mode-locked vertical external-cavity surface-emitting laser (VECSEL) (Keller and Tropper 2006) contains a semiconductor gain chip, which is most often optically pumped with radiation from a laser diode (e.g. from a diode bar), and is part of a laser resonator made of bulk optical elements (see Figure 20.9). The semiconductor gain chip is used as an end mirror or folding mirror. Due to its geometry, the term *semiconductor disk laser* has also become common. Indeed, there are strong similarities to thin-disk lasers based on doped-insulator laser crystals, including the potential for true power scalability, provided that an effective cooling method (e.g.

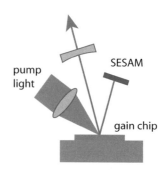

FIGURE 20.9 Set-up of a passively mode-locked optically pumped VECSEL.

based on a strongly thinned wafer mounted on a cooled metal plate) is applied.

Passive mode-locking is achieved by incorporation of a SESAM (see Section 20.3.2). Although such lasers are in many respects similar to more traditional mode-locked bulk lasers based on an ion-doped insulator gain medium, they differ from those in some important aspects:

- The semiconductor gain medium has a far lower gain saturation energy, which makes it suitable for passive mode-locking at very high pulse repetition rates without Q-switching instabilities. On the other hand, it is less suitable for low pulse repetition rates, where gain saturation could be excessive, unless the average power is reduced accordingly.
- Although the fabrication of such semiconductor gain chips requires sophisticated and expensive equipment, devices can be relatively cheap if many gain chips are fabricated together on a common wafer.
- Semiconductor gain media can be tailored to a wide range of wavelengths, including some wavelength regions which are difficult to access with ion-doped gain media. Also, other properties such as the gain bandwidth can be tailored by device design.

On the other hand, this type of device also substantially differs from mode-locked diode lasers:

- The area used on a surface-emitting semiconductor gain chip can be scaled up according to the required pulse energy (at least for optical pumping) while maintaining transverse single-mode emission due to the (comparatively long) external resonator. Therefore, far higher pulse energies are possible than with laser diodes based on small-area semiconductor waveguides (which would become multi-mode when increasing the beam area too much). Output pulse energies of hundreds of picojoules and multi-watt average output powers have been achieved early on (Aschwanden et al. 2005), and even much higher pulse energies should be possible when using even larger beam areas in conjunction with a sufficiently effective cooling scheme.
- Optical pumping substantially simplifies the design and fabrication requirements of the gain chip; it has so far been used in most cases. On the other hand, electrical pumping as for a diode laser would obviously be attractive since one would no longer need a separate pump laser diode and the pump optics; the overall power conversion efficiency could then also be higher. However, it is difficult to implement electrical pumping for larger beam areas; for example, one cannot work with an electrode ring as in monolithic small-area surface emitting lasers (VCSELs) because carrier diffusion for obtaining a reasonable transverse gain distribution works only over limited areas. For such reasons, there have been some demonstrations of electrically pumped mode-locked VECSELs (Barbarin et al. 2011) but usually with much lower performance than optically pumped devices.

In conclusion, mode-locked VECSELs constitute a new type of ultrafast laser with average output powers similar to those from traditional diode-pumped ion-doped insulator lasers (if optical pumping is employed) while being better suited for higher (multi-GHz) pulse repetition rates and for other wavelength regions. Pulse durations are often in the picosecond regime, but much shorter pulse duration down to 60 fs (Quartermann et al. 2009) has also been demonstrated.

In order to simplify the optical set-up, VECSEL gain chips with integrated saturable absorber (called MIXSELs) have been developed which despite of the additional design and fabrication challenges have reached attractive performance figures (Maas et al. 2007, Wittwer et al. 2012). For example, multi-watt average output powers in picosecond pulse trains are possible.

20.8 Modelling of Ultrashort Pulse Lasers

The generation of ultrashort pulses in mode-locked lasers often involves a complicated interplay of various physical effects, including potentially strong optical non-linearities. Although at least a qualitative understanding of the pulse formation and shaping processes is normally possible, a comprehensive and quantitative understanding is often highly desirable for successful implementation and optimization of such laser devices.

In some cases, pulse formation is dominated by the interplay of two or three essential effects, which can be satisfactorily modelled based on reasonably simple analytical equations. An example for this is the active or passive mode-locking of simple solid-state picosecond lasers as described in the beginning of Section 20.5, where, for example, the expected pulse duration, the required absorber recovery time and the Q-switching threshold can well be estimated with some simple formulas. Another example is quasi-soliton mode-locking in similar solid-state lasers, where the pulse duration is essentially determined by only a few parameters and the stability limit can at least be roughly estimated based on the calculated soliton period. The key parameters of such a laser design can thus be determined based on a relatively simple analysis, and if some guidelines are followed (e.g. concerning a reasonable amount of absorber saturation or of non-linear phase shifts in a quasi-soliton laser), the results are fairly predictable in a wide parameter region. Only when exploring extreme parameter regions (e.g. concerning output powers or pulse repetition rates), somewhat more sophisticated analysis is required.

Particularly in mode-locked fibre lasers, where strong effects of optical non-linearities and chromatic dispersion lead to strongly non-linear dynamics, the pulse shaping processes are so complicated that they can only be vaguely "explained" with simple physical arguments, and reliable predictions are not possible based only on a couple of simple equations. For example, the pulse duration, spectral width and spectral shape of the output of a fibre laser operating in the all-normal dispersion regime have a complicated dependence on the pump power and various device parameters (also on the distribution of chromatic dispersion in the resonator), and it is not possible to reliably predict performance figures and the stability of parameter regimes on a simple basis. In such cases, numerical simulations are often the only way to achieve a comprehensive understanding of the workings of such a laser, and thus to reach optimum performance and stability. Essentially, this approach involves the numerical simulation of pulse propagation in the laser over many resonator round trips, taking into account various physical effects related to the laser gain, the saturable absorber, chromatic dispersion, non-linearities, etc. Such a model can reveal under which circumstances the pulse propagation approaches a stable steady state and how the steady-state pulse parameters depend on various system parameters. Even in numerical simulations, it is sometimes difficult to identify suitable parameter sets, but at least that is much less difficult than doing the same exercise experimentally, where it is much harder to try out different configurations; also (perhaps more importantly) from experimental failures one obtains much less information concerning possible causes.

REFERENCES

Arahira S, Matsui Y and Ogawa Y 1996, "Mode-locking at very high repetition rates more than terahertz in passively mode-locked distributed-Bragg-reflector laser diodes", *IEEE J. Sel. Top. Quantum Electron.* 32 (7), 1211–1224.

Aschwanden A, Lorenser D, Unold H J, Paschotta R, Gini E and Keller U 2005, "2.1-W picosecond passively mode-locked external cavity semiconductor laser", *Opt. Lett.* 30 (3), 272–274.

Aus der Au J, Spühler G J, Südmeyer T et al. 2000, "16.2-W average power from a diode-pumped femtosecond Yb:YAG thin disk laser", *Opt. Lett.* 25 (11), 859–861.

Bao Q, Zhang H, Ni Z et al. 2011, "Monolayer graphene as a saturable absorber in a mode-locked laser", *Nano Res.* 4 (3), 297–307.

Barbarin Y, Hoffmann M, Pallmann W et al. 2011, "Electrically pumped vertical external cavity surface emitting lasers suitable for passive modelocking", *IEEE J. Sel. Top. Quantum Electron.* 17 (6), 1779–1786.

Baumgartl M, Lecaplain C, Hideur A, Limpert J and Tünnermann A 2012, "66 W average power from a microjoule-class sub-100 fs fiber oscillator", *Opt. Lett.* 37 (10), 1640–1642.

Bonaccorso F, Sun Z, Hasan T and Ferrari A C 2010, "Graphene photonics and optoelectronics", *Nature Photonics* 4 (9), 611–622.

Bowers J E, Morton P A, Mar A and Corzine S W 1989, "Actively mode-locked semiconductor lasers", *IEEE J. Quantum Electron.* 25 (6), 1426.

Chong A, Renninger W H and Wise F W 2008, "Properties of normal-dispersion femtosecond fiber lasers", *J. Opt. Soc. Am. B* 25 (2), 140–148.

Delfyett P J, Florez L T, Stoffel N et al. 1992, "High-power ultrafast laser diodes", *IEEE J. Quantum Electron.* 28 (10), 2203–2219.

Fermann M E, Kruglov V I, Thomsen B C, Dudley J M and Harvey J D 2000, "Self-similar propagation and amplification of parabolic pulses in optical fibers", *Phys. Rev. Lett.* 84 (26), 6010–6013.

Haiml M, Grange R and Keller U 2004, "Optical characterization of semiconductor saturable absorbers", *Appl. Phys. B.* 79, 331–339.

Hönninger C, Paschotta R, Morier-Genoud F, Moser M and Keller U 1999, "Q-switching stability limits of cw passive mode locking", *J. Opt. Soc. Am. B* 16 (1), 46–56.

Hoogland S, Dhanjal S, Tropper A C et al. 2000, "Passively mode-locked diode-pumped surface-emitting semiconductor laser", *IEEE J. Photon. Technol. Lett.* 12 (9), 1135–1137.

Innerhofer E, Südmeyer T, Brunner F et al. 2003, "60 W average power in 810-fs pulses from a thin-disk Yb:YAG laser", *Opt. Lett.* 28 (5), 367–369.

Jiang L A, Ippen E P and Yokoyama H 2005, "Semiconductor mode-locked lasers as pulse sources for high bit rate data transmission", *J. Opt. Fiber Commun. Rep.* 2, 1–31.

Kärtner F X, Brovelli L R, Kopf D, Kamp M, Calasso I G and Keller U 1995, "Control of solid-state laser dynamics by semiconductor devices", *Opt. Eng.* 34, 2024–2036.

Kärtner F X and Keller U 1995, "Stabilization of solitonlike pulses with a slow saturable absorber", *Opt. Lett.* 20 (1), 16–18.

Kärtner F X, Matuschek N, Schibli T et al. 1997, "Design and fabrication of double-chirped mirrors", *Opt. Lett.* 22 (11), 831–833.

Keller U and Tropper A C 2006, "Passively modelocked surface-emitting semiconductor lasers", *Phy. Rep.* 429 (2), 67–120

Keller U, Weingarten, K J, Kartner, F X et al. 1996, "Semiconductor saturable absorber mirrors (SESAMs) for femtosecond to nanosecond pulse generation in solid-state lasers", *IEEE J. Sel. Top. Quantum Electron.* 2, 435–453.

Kelly S M 1992, "Characteristic sideband instability of periodically amplified average soliton", *Electron. Lett.* 28 (8), 806–807.

Krainer L, Paschotta R, Lecomte S, Moser M, Weingarten K J and Keller U 2002, "Compact Nd:YVO$_4$ lasers with pulse repetition rates up to 160 GHz", *IEEE J. Quantum Electron.* 38 (10), 1331–1338.

Kuizenga D J and Siegman A E 1970, "FM and AM mode locking of the homogeneous laser – Part I: theory", *IEEE J. Quantum Electron.* 6 (11), 694–708.

Maas D J H C, Bellancourt A R, Rudin B et al. 2007, "Vertical integration of ultrafast semiconductor lasers", *Appl. Phys. B* 88, 493–497.

Mitschke F M and Mollenauer L F 1987, "Ultrashort pulses from the soliton laser", *Opt. Lett.* 12 (6), 407–409.

Park N H, Jeong H, Choi S et al. 2015, "Monolayer graphene saturable absorbers with strongly enhanced evanescent-field interaction for ultrafast fiber laser mode-locking", *Opt. Expres* 23 (15), 19806–19812.

Paschotta R and Keller U 2001, "Passive mode locking with slow saturable absorbers", *Appl. Phys. B* 73 (7), 653–662.

Pronin O, Brons J, Grasse C et al. 2011, "High-power 200 fs Kerr-lens mode-locked Yb:YAG thin-disk oscillator", *Opt. Lett.* 36 (24), 4746–4748.

Quarterman A H, Wilcox K G, Apostolopoulos V et al. 2009, "A passively mode-locked external-cavity semiconductor laser emitting 60-fs pulses", *Nature Photon.* 3, 729–731.

Renninger W H, Chong A and Wise F W 2012, "Pulse shaping and evolution in normal-dispersion mode-locked fiber lasers", *IEEE J. Sel. Top. Quantum Electron.* 18 (1), 389–398.

Russell P 2003, "Photonic crystal fibers", *Science* 299, 358–362.

Salin F, Squier J and Piché M 1991, "Modelocking of Ti:sapphire lasers and self-focusing: a Gaussian approximation", *Opt. Lett.* 16 (21), 1674–1676.

Saraceno C J, Emaury F, Schriber C et al. 2014, "Ultrafast thin-disk laser with 80 µJ pulse energy and 242 W of average power", *Opt. Lett.* 39 (1), 9–11.

Sato K 2002, "Semiconductor light sources for 40-Gb/s transmission systems", *J. Lightwave Technol.* 20 (12), 2035–2043.

Set SY, Yaguchi H, Tanaka Y and Jablonski M 2004, "Ultrafast fiber pulsed lasers incorporating carbon nanotubes", *IEEE J. Sel. Top. Quantum Electron.* 10 (1), 137–146.

Sutter D H, Steinmeyer G, Gallmann L et al. 1999, "Semiconductor saturable-absorber mirror-assisted Kerr lens modelocked Ti:sapphire laser producing pulses in the two-cycle regime", *Opt. Lett.* 24 (9), 631–633.

Tamura K, Ippen E P, Haus H A and Nelson L E 1993, "77-fs pulse generation from a stretched-pulse mode-locked all-fiber ring laser", *Opt. Lett.* 18 (13), 1080–1082.

Wittwer V J, Mangold M, Hoffmann M et al. 2012, "High-power integrated ultrafast semiconductor disk laser: multi-Watt 10 GHz pulse generation", *Electron. Lett.* 48 (18), 1144–1145.

Yamashita S, Inoue Y, Maruyama S et al. 2004, "Saturable absorbers incorporating carbon nanotubes directly synthesized onto substrates and fibers and their application to mode-locked fiber lasers", *Opt. Lett.* 29 (14), 1581–1583.

Zeller S C, Krainer L, Spühler G J et al. 2004, "Passively mode-locked 50-GHz Er:Yb:glass laser", *Electron. Lett.* 40 (14), 875–877.

21
Attosecond Metrology

Pierre Agostini, Andrew J. Piper, and Louis F. DiMauro

CONTENTS

21.1 Introduction 307
21.2 General Principles of Attosecond Pulse Characterization 309
21.3 Second-order XUV AC/FROG 310
21.4 Reconstruction of Attosecond Beating by Interference of Two-photon Transitions (RABBITT) 310
    21.4.1 Two-Colour IR-XUV Photoionization in the Perturbative Regime 310
    21.4.2 Spectral Amplitude 311
    21.4.3 Spectral Phase: RABBITT 311
    21.4.4 Rainbow RABBITT 312
21.5 Isolated Attosecond Pulses 312
21.6 Momentum Streaking 312
    21.6.1 Angular Streaking and Attoclock 313
21.7 Complete Reconstruction of Attosecond Beating (CRAB) 313
21.8 Phase Retrieval by Omega Oscillation Filtering (PROOF) 314
    21.8.1 Improved PROOF (iPROOF) 316
21.9 Comparison of RABBITT, Momentum Streaking, CRAB and PROOF 316
21.10 All-optical Method 316
21.11 Other Methods 317
21.12 Some Experimental Remarks 317
21.13 Principle Component Generalized Projection Algorithm 317
21.14 Conclusions and Outlook 318
References 319

21.1 Introduction

Extreme Ultraviolet (XUV) attosecond pulses have come of age with the 21st century and are now commonly used in dynamical studies on the timescale of electron motion in atoms (the atomic unit of time is 24 as). Both their generation and characterization have made great progress over the last decade or so, with the current state-of-the-art characterized by 53 ± 6 as and 43 ± 1 as pulses with a central energy of 100–150 eV (Li et al. 2017, Gaumnitz et al. 2017) (see Section 21.9), figure 21.1.

Such pulses have an optical period of ~25 as and extend over just a couple of cycles. They are basically the shortest light pulses ever produced, but sources which could reach 10 as are under study. It is obviously of utmost importance to be able to measure and characterize attosecond pulses as completely as possible, for the purpose of applying them to studies of physical or chemical processes, improving existing sources, or developing new ones.

Attosecond metrology is the ensemble of techniques and methods used to measure and characterize such light pulses. It is inseparable from the production of the pulses themselves, which is only briefly sketched in this introduction.

Currently, the main source of attosecond pulses is high harmonic generation (HHG), discovered in the late 1980s (Ferray et al. 1988). Conventional, perturbative, non-linear optics predicts that in an atom irradiated by an intense laser field a non-linear polarization source at the odd harmonic frequencies is generated. However, perturbative theory fails to account for the observed spectra (Kulander et al. 1993). A non-perturbative treatment of the interaction (L'Huillier et al. 1991,

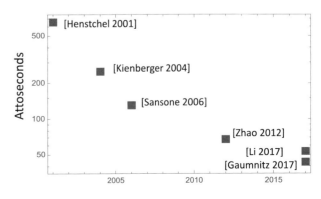

FIGURE 21.1 Evolution of sub-femtosecond pulses (see Zhao et al. 2012).

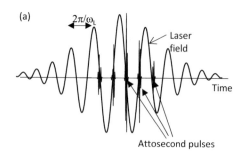
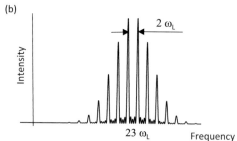

FIGURE 21.2 APT (a) and its spectrum (b) around the $23\omega_L$ energy.

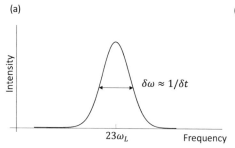
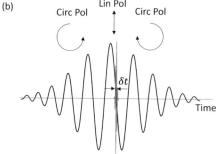

FIGURE 21.3 IAP (a) and its spectrum (b).

Kulander et al. 1993) predicts a sequence of XUV attosecond bursts, occurring twice per optical cycle according to the following simple model. First, atoms irradiated by a strong IR laser field emit an electron wave packet at each maximum of the driving field by tunnel ionization. The emitted wave packet is then accelerated back and forth, following the field, and moves essentially as a free electron except when its trajectory brings it in the vicinity of the nucleus where it experiences a strong acceleration and emits a burst of XUV light, thus generating a train of attosecond light pulses (Kulander et al. 1993) separated by half the laser optical period (Figure 21.2).

The corresponding spectrum comprises discrete peaks at only the odd harmonics (for symmetry reasons) of the laser frequency ω_L (Figure 21.2).

It extends quickly over the XUV or soft X-ray range: a typical IR photon is 1–1.5 eV and just 10 harmonics separated by twice this energy leads to a bandwidth of tens of eV. The highest harmonic frequency generated is proportional to the IR laser intensity, and harmonic orders of several hundred can be generated, albeit with a rather low efficiency around 10^{-7} (Lewenstein et al. 1994, Chang 2011).

In the beginning, the HHG spectrum was the only available information. Its structure suggested, by Fourier transform, the possibility of extremely short time-domain bursts separated by half the optical period of the driving laser (1.3 fs for a Ti:sapphire laser with a wavelength of 800 nm). That attosecond pulses which could be synthesized from the wide bandwidth of HHG was actually proposed at the beginning of the 1990s by Farkas and Toth (1992), but it was only after 2000 that techniques able to measure such pulses and actually confirm their existence were developed: RABBITT (Paul et al. 2001), Streaking (Drescher 2001)), CRAB (Mairesse and Quéré 2005), PROOF (Chini et al. 2010), based on photoelectron spectra analysis and the all-optical methods of Dudovich et al. (2006), to be described in the following sections. Later, it was discovered that the train of pulses could be reduced to a single pulse (isolated attosecond pulse (IAP)) through various optical tricks, for instance, polarization gating (Sansone et al. 2006), with a continuous, rather than discrete (Figure 21.3), spectrum.

An attosecond pulse clearly requires a high carrier frequency[1] if at least one period of the field must be contained within the pulse. This implies that the pulses are generated, characterized and used under vacuum, since air is strongly absorbing for XUV, and causes demanding constraints for the possible techniques (Figure 21.4).

Another difficulty lies in the relatively low intensity of the current attosecond pulses. This makes the popular technique of intensity autocorrelation (AC) difficult. Although not impossible, as demonstrated since the mid-2000s in a few exceptional set-ups (Nabekawa et al. 2005), it remains beyond the performance of most attosecond beamlines. In addition, neither FROG (Trebino 2000) nor SPIDER (Walmsley and Dorrer 2009) techniques, well established for visible/IR fs pulses, could be applied directly, owing to the general lack of intensity and the absence of essential optics, like beam splitters, in the XUV domain.

Thus, new complete characterization methods had to be developed. Most of the current techniques to be discussed hereunder start by making a photoelectron replica of the attosecond pulse, by photoionizing a target atom. The non-linearity,

---
[1] A 150 as period corresponds to a wavelength of 45 nm, well into the XUV (124-10 nm) range.

Attosecond Metrology

FIGURE 21.4 Typical attosecond RABBITT beam line entirely under vacuum. The target jet is inside the magnetic bottle spectrometer.

which is necessary to reveal the spectral phase, is provided by the interaction of the target atomic gas with a superposition of the XUV pulse to be measured and the driving IR pulse, via two-colour ionization, a non-linear process easily accessible in spite of the low intensity XUV HHG radiation (Figure 21.5) since the rate is proportional to both the IR and XUV intensities and thus can easily be made large enough. A part of the IR laser pulse generating the XUV spectrum is split off the main beam to drive the non-linearity. The IR pulse must be intense enough (10^{11}–10^{12} W cm^{-2} after focusing) but not so much as to induce significant ionization of the target itself.

The IR field is naturally synchronized with the HHG attosecond fields, with a well-defined frequency, allowing attosecond precision.

RABBITT (Paul et al. 2001) is tailored to attosecond pulse trains (APT) while Streaking (Drescher et al. 2001), CRAB (Mairesse and Quéré 2005), and PROOF (Chini et al. 2010) are applicable to IAP with a complex temporal structure. All these methods have in common the use of the IR optical cycle as the basic clock and an attosecond optical delay line to scan the delay between the two pulses with attosecond precision.

The purpose of this chapter is to describe the principles of these methods, as well as an all-optical method, and provide some details on their implementation. The intensity AC is briefly discussed in Section 21.3.

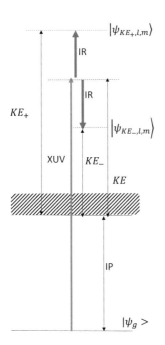

FIGURE 21.5 Two-colour XUV IR two-photon ionization.

21.2 General Principles of Attosecond Pulse Characterization

Very generally, a light pulse can be represented in the *time domain* by a temporal amplitude $|E(t)|$ and phase $\psi(t)$ Carrier-Envelope Phase (CEP) ψ_0 and a carrier frequency ω_0 (Walmsley and Dorrer 2009):

$$E(t) = |E(t)| e^{-i\psi(t)} e^{-i\psi_0} e^{-i\omega_0 t} \qquad (21.1)$$

A pulse is "chirped" when the temporal phase $\psi(t)$ is a quadratic function of time, e.g. $\psi(t) = \omega_0 t + b\, t^2$ (where b is the chirp parameter). Then, the chirp is called linear, since one can define a temporal frequency which is a linear function of time:

$$\omega(t) = \frac{d\psi}{dt} = \omega_0 + 2bt \qquad (21.2)$$

In the *frequency domain*, the same pulse is described by a spectral amplitude $|\tilde{E}(\omega)|$ and phase $\varphi(\omega)$:

$$\tilde{E}(\omega) = |\tilde{E}(\omega)| e^{-i\phi(\omega)} \qquad (21.3)$$

The two representations are strictly equivalent and related by a Fourier transform:

$$E(t) = \int_{-\infty}^{\infty} d\omega \, \tilde{E}(\omega) e^{i\omega t} \qquad (21.4)$$

In general, the appropriate time-bandwidth products are constant. Thus, for a pulse train and its discrete harmonic spectrum, or an IAP and its continuous spectrum, the duration of the attosecond burst is inversely proportional to the total bandwidth. The two situations have given rise to different solutions to the problem of their characterization. Note in addition that the determination of the CEP is usually not possible in any of the methods to be described and requires other solutions (Paulus et al. 2001) for fs pulses.

Since it is never possible to measure $E(t)$ directly, as the only available detectors have a response time much longer than the optical periods (typically ps), the measurement is done in the frequency domain. It is obvious from (21.3) to (21.4) that both the spectral amplitude and phase must be known to reconstruct the temporal function (21.1). The energy spectrum is easy to get with conventional equipment and integrating detectors, e.g. an optical spectrometer (or an electron spectrometer, after a photoelectron replica of the XUV pulse is made), but it is not sufficient to determine the pulse duration if the phase is unknown. The phase measurement is, however, more demanding and the methods used for fs visible or near-IR pulses, like FROG or SPIDER, are not easily adaptable to XUV. The use of cross correlations between the attosecond pulse and the optical cycle of the IR laser, which drives the HHG for the attosecond pulses, is at the heart of the methods to be described. The information is derived in all cases from a two-colour (XUV-IR) photoionization process. In RABBITT or PROOF, the IR intensity is kept low enough and induces only one extra absorption or emission photon process; the spectral phase is encoded in the amplitude of this extra photoionization. In Momentum Streaking or CRAB, the intensity must be high enough to significantly change the momentum of the photoelectrons and the phase is encoded in that change.

21.3 Second-order XUV AC/FROG

The simplest and most common technique for gathering some quantitative information about an ultrashort pulse is the intensity AC (Walmsley and Dorrer 2009): two replicas of the pulse, separated by a controllable delay, generate a second harmonic in a non-linear material, and the average power of the generated beam is recorded as a function of the delay. The width of the AC function provides an estimate of the pulse, provided one *assumes* a temporal profile (e.g. Gaussian). It was long thought that AC was beyond the reach of the usually too weak high harmonic sources. Progress in laser technology has actually allowed it in a few cases though. A two-colour cross-FROG based on two-colour, two-photon ionization, and an intensity AC for a monochromatic XUV (27.9 eV) fs pulse (Nabekawa 2005) (9th-order harmonic of the second harmonic of a Ti:sapphire laser) have been demonstrated. However, this remains a difficult solution for most attosecond beam lines.

Besides the low-available intensity, another obstacle is the lack of normal-incidence, wide-band, multi-layer XUV mirrors able to tightly focus the spectral bandwidth of attosecond pulses. One case has been demonstrated of an intensity AC using a spherical mirror reflecting a train of sub-femtosecond pulses with a 12-eV bandwidth (harmonics 7–15 of 800 nm) (Tzallas 2003) to ionize helium atoms. For more details and a discussion of AC's general limitations, the reader is referred to Walmsley and Dorrer (2009).

21.4 Reconstruction of Attosecond Beating by Interference of Two-photon Transitions (RABBITT)

This method is well adapted to the measurement of the spectral phase of discrete harmonics, i.e. when the spectral intensity is concentrated in narrow peaks separated by twice the driving laser frequency. It is based on a quantum interference which modulates the intensity of sidebands due to two-photon ionization as a function of a time delay introduced between the driving pulse and the attosecond pulses (Véniard et al. 1996). The first experimental demonstration was done by Paul et al. (2001), see also Muller (2002) and Toma and Muller (2003).

21.4.1 Two-Colour IR-XUV Photoionization in the Perturbative Regime (Figure 21.5)

Considering the harmonic spectrum as a series of narrow peaks of frequencies $(2q+1)\omega_L$, separated by twice the driving laser frequency, ω_L, the goal is to determine the phase relation between the contributing harmonics by considering them in pairs.

Because of the low intensity of the XUV, an individual harmonic only causes single-photon ionization, and each harmonic produces photoelectrons with a kinetic energy (KE) according to Einstein's equation for the photoelectric effect:

$$KE = \hbar\omega_{\text{XUV}} - IP \qquad (21.5)$$

where *IP* is the atom ionization potential. The *IR* field induces additional transitions in the continuum. Ionization with a harmonic photon can then be accompanied by absorption or emission of *IR* photons (Figure 21.5) which produce sidebands with kinetic energies:

$$KE_{\pm} = \hbar\omega_{\text{XUV}} \pm \hbar\omega_{IR} - IP = (2q+1)\hbar\omega_{IR} \pm \hbar\omega_{IR} - IP \qquad (21.6)$$

that is exactly in-between harmonics (Figure 21.6).

Attosecond Metrology

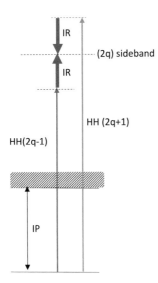

FIGURE 21.6 Quantum pathways in three-colour, two-photon ionization, basis of RABBITT (Veniard 1996) (equation 21.10).

These are two-photon processes involving a continuum-continuum transition since the XUV photon energy is larger than IP, the ionization potential. Its quantum transition amplitude is always complex: in second-order perturbation theory, the two-photon amplitudes between the ground state $|\psi_g\rangle$ and a continuum state of energy $KE_\pm$ and angular momentum quantum numbers l, m can be written as (using the rotating wave approximation):

$$M^-_{KE_-,l,m} = \int \frac{\langle \psi_{KE_-,l,m} | D^-_{IR} | \psi_{E'} \rangle \langle \psi_{E'} | D^+_{2q-1} | \psi_g \rangle dE'}{E_g + (2q-1)\hbar\omega - E'} \quad (21.7a)$$

$$M^+_{KE_+,l,m} = \int \frac{\langle \psi_{KE_+,l,m} | D^+_{IR} | \psi_{E'} \rangle \langle \psi_{E'} | D^+_{2q-1} | \psi_g \rangle dE'}{E_g + (2q-1)\hbar\omega - E'} \quad (21.7b)$$

where $D^\pm_{IR}$ and D^+_{2q+1}, D^+_{2q-1} are the dipole operators for the laser and harmonics, respectively.

The integral over the continuum involves a singularity which is handled through the rule:

$$\frac{1}{x+i\varepsilon} = \mathcal{P}\frac{1}{x} - i\pi\delta(x) \quad (21.8)$$

where $\mathcal{P}$ is the principal part, and $\delta(x)$ is the Dirac delta function. The singularity therefore imposes a complex part to the two-photon matrix element and, hence, an extra-phase (see equation (21.10) hereunder) in the process. For precise calculations of this phase in a particular case, see Toma and Muller (2002). For a general discussion of its physical interpretation in terms of "Photoionization Delays", see Dahlstrom et al. (2012) and Schoun (2015) and the rather complete reference lists therein. Thus, for each harmonic order $(2q+1)$, which would itself give rise to a photoelectron of energy KE (Figure 21.5), there are two "sidebands" with energies $2q\hbar\omega_{IR}$ and $(2q+2)\hbar\omega_{IR}$ corresponding to the emission or absorption, respectively, of an IR photon. If the IR intensity is kept low enough, higher order sidebands can be neglected. When a series of harmonics is present, each couple of consecutive orders harmonics shares a sideband, and the spectral phase is encoded in the oscillation of the sidebands amplitude as a function of an optical delay between the IR and XUV pulses. This is the principle of the RABBITT method.

21.4.2 Spectral Amplitude

Although (integrating) photon detectors exist for the XUV domain, it is more practical to detect photoelectrons, which are required for the spectral phase retrieval anyhow. It can be shown from perturbation theory that the amplitude to produce an electron with momentum $\vec{v}$ by an attosecond pulse is given by:

$$b(\vec{v}) = i\, d(\vec{v}) E_{XUV}(\omega) \quad (21.9)$$

where $d(\vec{v})$ is the dipole moment between the atom ground state and the continuum (Chang 2011). Under the condition that $d(\vec{v}) \approx ct$ (i.e. a flat continuum), $b(\vec{v})$ can be considered a replica of the attosecond pulse, amplitude and phase. In the case of an APT, the photoelectron energy spectrum is the replica of the high harmonic one. If $d(\vec{v})$ varies over the light bandwidth, this variation must be known to infer the XUV spectral amplitude.

21.4.3 Spectral Phase: RABBITT

In the low IR intensity limit (typically 10^{11} Wcm^{-2}), only the two nearest harmonics $(2q-1)$ and $(2q+1)$ contribute to each $2q$ sideband, through two interfering quantum paths (Figures 21.6 and 21.7). The second-order perturbation theory of Véniard et al. (1996) actually preceded the experiment by several years and showed that the amplitude of the sideband of order $2q$ depends on the delay τ, the phase difference between the harmonics adjacent to the sideband $\Delta\phi_q$, and the "atomic" term, $\Delta\phi_q^{me\,2}$ arising from the two-photon complex matrix element as shown in equation (21.8):

$$I_{2q}(\tau) \propto \cos(-2\omega_L \tau + \Delta\phi_q + \Delta\phi_q^{me}) \quad (21.10)$$

where $\Delta\phi_q = \phi_{2q-1} - \phi_{2q+1}$ and ω_L is the IR laser frequency. The $2\omega_L$ comes from the frequency separation between odd harmonics, and the phase of the function (21.10) encodes the sought information (Figures 21.6 and 21.7).

By varying τ and observing the oscillation of I_{2q} (see spectrogram Figure 21.8) and measuring its phase, one can derive $\Delta\phi_q$, $\Delta\phi_q^{me}$ is known. The delay must be scanned with attosecond precision. One usual solution is a wedged glass plate delay, introduced in the interferometer, with a typical resolution of ~5 as (Chirla 2011).

The electron spectrometer is usually a time-of-flight and often a magnetic bottle (Kruit and Read 1983), which has a large collection efficiency. The energy resolution must be sufficient to resolve the harmonics from the sidebands (i.e. $\hbar\omega_L$

---
[2] The derivative of ϕ_q^{me} with respect to energy is called the Wigner Delay (Dahlstrom 2012).

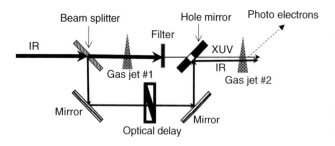

FIGURE 21.7 Principle of the RABITT set-up showing the basic Mach-Zehnder structure. The optical delay is only sketched.

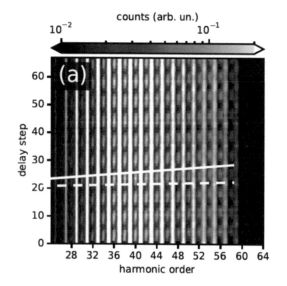

FIGURE 21.8 RABBITT spectrogram. Reproduced from Schoun (2015). The Ohio State University with permission.

from 1.55 eV at 800 nm (Ti:sapphire) to 0.62 eV at 2000 nm). An example is shown in Figure 21.9.

The RABBITT method has proved over the years to be not only an excellent way to characterize the APTs but also a remarkable tool for spectroscopy (see Dahlstrom et al. (2012), Schoun (2015), and references therein).

21.4.4 Rainbow RABBITT

In some cases, the phase varies over the sideband spectral width, and it is interesting to spectrally resolve smaller intervals (if permitted by the electron spectrometer resolution). Then instead of a phase averaged over the sideband, RABBITT provides more detailed information. This new variant, called Rainbow RABBITT, has been recently demonstrated and applied to the study of resonant two-photon ionization (Busto 2018).

21.5 Isolated Attosecond Pulses

The characterization methods in Sections 21.6–21.8 are more suited for an IAP. The principles and implementations of IAP emission are beyond the scope of this chapter, but in brief, the basic idea is to restrain the emission of the XUV burst to one cycle of the pump pulse by quenching the harmonic generation during the other cycles. Several techniques have been elaborated to reach this goal: spectral filtering (Kienberger 2004), polarization gating (Sansone et al. 2006), double optical gating (Mashiko et al. 2008), etc. The interested reader is referred to the original publications or Z. Chang's book (Chang 2011).

21.6 Momentum Streaking

The concept of the streak camera is well known: a photocathode converts the photon pulse into an electron pulse which is swept by a voltage ramp and detected on a 2D detector. The width of the image carries information on the pulse duration, assuming knowledge of the initial image width (ideally a thin slit) and of the streak speed. A typical streak camera time resolution is ps. The translation into the attosecond regime is obtained by using the IR electric field sub-cycle time dependence as a streaking ramp. The photoelectron replica of the XUV attosecond pulse is analysed in a spectrometer as a function of the delay, and the final momentum distribution allows to retrieve the pulse duration.

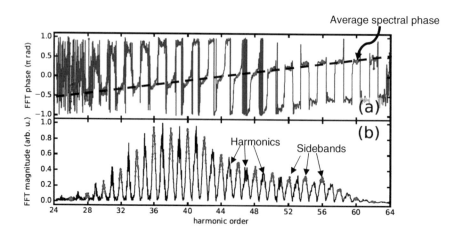

FIGURE 21.9 Example of RABBITT data. Lower box: photoelectron energy spectrum showing the odd harmonics and the even sidebands. Upper box: signal FT showing the average phase as well as the phase within each sideband. The slope of the dashed line is the GDD or attochirp. (Reproduced from Schoun (2015) with permission.)

Attosecond Metrology

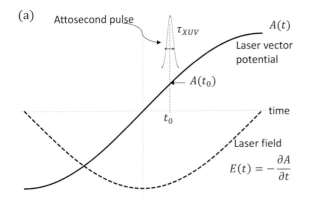
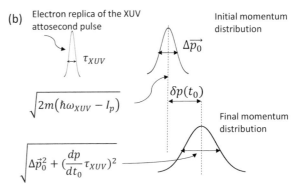

FIGURE 21.10 (a) Half-period of the IR laser field and vector potential. An attosecond pulse ionizes the target around time t_0 with a momentum distribution $\vec{p}_0(\omega)$ with width $\Delta \vec{p}_0$ shown in (b). After the end of the laser pulse, the momentum distribution is shifted by $\delta \vec{p}(t_0) = \vec{p}(\infty) - \vec{p}_0(t_0)$ and its width broadened to $\sqrt{\vec{p}_0^2 + \left(\dfrac{dp}{dt_0}\tau_{\text{XUV}}\right)^2}$.

In more details (Figure 21.10), let us assume an IAP of duration τ_{XUV}, which photoionizes a target atom gas, producing a photoelectron replica (see equation 21.8) with an initial momentum distribution centred on (m, electron mass, $\hbar$ the Planck constant/2π):

$$p_{in}(\omega_{\text{XUV}}) = \sqrt{2m(\hbar\omega_{\text{XUV}} - I_p)} \tag{21.11}$$

where ω_{XUV} is a Fourier component of the pulse and I_p is the atom ionization potential. Let Δp_0 be the initial width of that distribution (which would be $\sim 1/\tau_{\text{XUV}}$ for a Fourier-limited pulse). A direct time-domain determination of the pulse duration is possible by streaking the photoelectrons using an IR laser field (Figure 21.10), as first described in Kienberger et al. (2004).

If an electron is released in the IR field at time t_0, a classical calculation shows that its central momentum after the IR pulse is shifted by $-eA(t_0)$, where $A(t_0)$ is the laser vector potential at the instant of ionization by the XUV pulse, and its width broadened to:

$$\Delta p_f = \sqrt{\Delta p_0^2 + \left(\frac{dp}{dt_0}\tau_{\text{XUV}}\right)^2} \tag{21.12}$$

where $\dfrac{dp}{dt_0} = e\,E(t_0)$ is the streak speed (e, electron charge) (Chang 2011) which is maximum at the zero-crossing of the vector potential or peak of the IR pulse electric field.

The free electron is accelerated and can gain energy from the IR laser pulse (this is possible because the electron is "born" in the field, through the XUV photoionization process). Experimentally, t_0 is adjusted to the maximum streaking at the vector potential zero-crossing, i.e. the maximum of the field. The pulse duration τ_{XUV} is then derived from the measurements through (21.12). For a quantum theory of streaking, see (Kitzler et al. 2003).

Streaking requires a laser intensity strong enough to make the shift clearly identified against other broadening causes and typically is of order 10^{12} W cm^{-2} (Chang 2011).

21.6.1 Angular Streaking and Attoclock

An interesting variant of (linear) streaking is attosecond angular streaking (Eckle et al. 2008) in which a circularly polarized IR field is used to deflect the photoelectrons, so time is mapped into angle with an estimated resolution of ≈ 200 as. When using a close-to-circular polarization instead, the rotating electric vector gives a time reference which provides a very attractive tool for studying strong field ionization of atoms (Pfeiffer et al. 2011) but has not to our knowledge been applied to HHG attosecond pulses.

21.7 Complete Reconstruction of Attosecond Beating (CRAB)

Closely related to streaking, two methods use the whole spectrogram obtained by scanning the delay: CRAB and PROOF. CRAB, also dubbed FROG-CRAB, is due to Mairesse and Quéré (2005) and is inspired by Trebino's well-known FROG (Trebino 2000). As with Momentum Streaking, it uses photoionization by an XUV pulse in the field of an intense IR laser. CRAB can handle any attosecond pulse shape (train, isolated, or other). In the language of FROG, CRAB is a temporal *phase-gate* FROG. To see this, it is necessary to look at the quantum, *strong field approximation*, two-colour ionization theory (Mairesse and Quéré 2005, Quéré et al. 2005, Chang 2011). In the approximation where the IR field is not strong enough to cause ionization, it induces only a temporal phase to the electron wave packet put in the continuum by the XUV field. The transition amplitude from the ground state to a continuum final state with momentum $\vec{v}$ (in atomic units $(e = m = \hbar = 1)$ in the XUV and IR laser fields is given by Quéré et al. (2005):

$$a(\vec{v},\tau) = -i\int_{-\infty}^{\infty} dt\; e^{i\phi(t)} d(\vec{v}(t)) E_{\text{XUV}}(t-\tau) e^{i(W+I_p)t} \tag{21.13}$$

with

$$\phi(t) = -\int_t^\infty dt' \left(\vec{v}.\vec{A}(t') - \frac{A^2(t')}{2} \right) \quad (21.14)$$

As in Section 21.6, $\vec{A}(t')$ is the IR laser vector potential, $d(\vec{v}(t))$ is the dipole matrix element between the ground state and the continuum state with momentum $\vec{v}$, $W = \frac{v^2}{2}$ is the electron initial kinetic energy, Ip is the atom ionization potential, and $E_{XUV}(t - \tau)$ is the XUV pulse to be measured. The ionization probability, $|a(\vec{v}, \tau)|^2$, as a function of the momentum (or energy) and the delay τ, is, by definition, the 2D CRAB "trace", analogous to the FROG optical "spectrogram". This CRAB trace is obtained by recording the spectrum of the pulse to be characterized, $E(t)$, gated by the function $G(t)$ for different delays (Figure 21.11).

$$S(\omega, \tau) = |\int_{-\infty}^\infty dt \, G(t) E(t-\tau) e^{i\omega t}|^2 \quad (21.15)$$

In the case of CRAB, the gate is the pure temporal phase gate $G(t) = e^{i\phi(t)}$ (modulus = 1) if the momentum dependence in (21.14) can be ignored.

The momentum dependence can be removed by making the *central momentum approximation*, replacing $\vec{v}$ by $\vec{v}_0$ (the central momentum of the initial distribution). The approximation is valid if the *XUV spectrum width is much smaller than its central energy* and thus puts a limit to the shortest pulse that can be characterized. As an example, assuming a spectral bandwidth 50% of the centre energy $W_0 = \frac{1}{2} v_0^2$ the limit is, in attoseconds, $\frac{3650}{W_0(eV)}$, i.e. about 40 as for $W_0 = 90$ eV (Chang 2011).

CRAB is clearly related in its principle and set-up to RABBITT (it can be seen as an ensemble of interference patterns at different delays) and momentum streaking (it can be seen as a series of streaking at different delays), but it differs greatly from both in the way the amplitude and phase of the XUV attosecond pulse are extracted. The assimilation of the CRAB spectrogram to a FROG one allows for the use of the same retrieval algorithm. A fast algorithm allowing the retrieval of the unknown pulse in "real" time is called the Principal Components Generalized Projection Algorithm (PCGPA) (Kane 2008, Mairesse and Quéré 2005) and permits to reconstruct from the trace both the unknown pulse $E_{XUV}(t)$ (intensity and phase) and the gate function (blind CRAB analogue to blind FROG (Trebino 2000)), see Section 21.13.

The experimental conditions, namely, the shot noise and the IR intensity directly affect the precision of the CRAB measurement (Wang et al. 2009). Given the photoionization cross-section (~10^{-17} cm^2), the HHG photon flux (~10^8 photon/pulse), the jet pressure, and its dimensions, the detection probability is on the order of ~10^{-7} per shot and the number of detected electrons per shot is about 1. For a 50-electron count, simulations show that the CRAB retrieved pulse is within 5% of the actual one (Chang 2011). Looking at the CRAB measurement as a streaking, it can be shown (Chang 2011) that the minimum IR intensity (i.e. yielding a broadening comparable to the initial spectrum width) is of the order of 5×10^{13} W.cm^{-2} for a 90 as resolution at a carrier energy of 35 eV. Unfortunately, for most target atoms, such an intensity is sufficient for multiphoton ionization and must be avoided. In fact, it was shown (Wang et al. 2009) that for CRAB, an intensity of only 7.5×10^{11} W.cm^{-2} is required, and this makes the interference interpretation of CRAB more relevant.

Recently, CRAB has been successfully pushed to explore the water window (roughly 300–550 eV) soft X-ray regime for the first time. An IAP of 322 as duration, generated from HHG, was measured, and requirements on the streaking excursion and signal-to-noise ratios were established for this ultra-broadband regime (Cousin et al. 2017). Curiously, the spectral phase has no GDD in this measurement although a definite conclusion on the reason for this lack is still pending.

21.8 Phase Retrieval by Omega Oscillation Filtering (PROOF)

PROOF (Figure 21.12) was first proposed in 2010 (Chini et al. 2010) to get around some of the limitations of the other methods of characterizing IAPs, namely, the high IR intensity required in streaking and the central momentum approximation of CRAB.

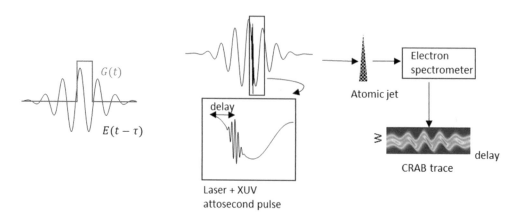

FIGURE 21.11 Principle of the CRAB method (right) compared with an amplitude gate FROG (left).

Attosecond Metrology

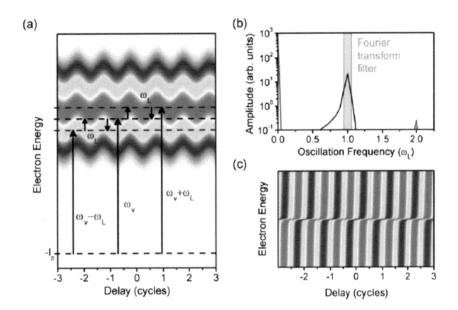

FIGURE 21.12 (a) PROOF trace (electron energy spectrum versus delay) and interfering quantum paths. (b) Fourier transform components at 0, ω_L and $2\omega_L$. (c) Spectrogram obtained by inverse Fourier transform of the filtered ω_L component, from which the spectral phase is extracted. (Reproduced from Chini (2010) http://creativecommons.org/licenses/by/4.0/.)

The theory starting point is of course very close to that of CRAB, i.e. equations (21.13) and (21.14). For an electric field $E_L(t) = E_0(t)\cos\omega_L t$ linearly polarized and detection of the photoelectron in the direction of the polarization, the phase of equation (21.14) writes, introducing $U_P(t) = \dfrac{A(t)^2}{4} = \dfrac{E_0(t)^2}{4\omega_L^2}$ the ponderomotive energy (average electron quiver energy):

$$\phi(t) = -\int_t^\infty dt\, \left(U_P(t) + \left(\frac{\sqrt{8WU_P(t)}}{\omega_L}\right)\cos\omega_L t\right) - \frac{U_P(t)}{2\omega_L}\sin 2\omega_L t \quad (21.16)$$

From equation (21.16), in the case of low-intensity IR field, i.e. small $U_P(t)$, the 0 and $2\omega_L$ component can be neglected and the phase reduces to:

$$\phi(t,v) \approx \frac{v\,E_0(t)}{2\,\omega_L^2}\cos\omega_L t\,\frac{v\,E_0(t)}{4\,\omega_L^2}\left(e^{-i\omega_L t} + e^{i\omega_L t}\right) \quad (21.17)$$

and, when the energy shift of the streaking is much less than the energy of the IR photon, that is

$$\frac{v\,E_0(t)}{2\,\omega_L} \ll WU_P(t) \quad (21.18)$$

$e^{i\phi(t)}$ can be approximated to:

$$e^{i\phi(t)} \approx 1 + i\phi(t) \quad (21.19)$$

such that (21.13) can be rewritten in the following form (Chang 2011):

$$a(\vec{v},\tau) = \int_{-\infty}^{\infty} dt \int_{-\infty}^{\infty} d\omega\; U(\omega) e^{i\varphi(\omega)} e^{i\omega(t-\tau)}$$

$$\times \left(1 + i\,\eta\left(e^{-i\omega_L t} + e^{i\omega_L t}\right)\right) e^{i\left(\frac{v^2}{2} + I_P\right)t} \quad (21.20)$$

where $\eta = \dfrac{v\,E_0}{4\,\omega_L^2} \ll 1$, and the unknown XUV pulse has been written in terms of its spectral amplitude $U(\omega)$ and phase $\varphi(\omega)$:

$$\varepsilon_{\text{XUV}}(t) = \int_{-\infty}^{\infty} U(\omega) e^{i\varphi(\omega)} e^{i\omega t} d\omega \quad (21.21)$$

For an arbitrary spectrum (i.e. $U(\omega) \neq ct$), the Fourier-filtered ω_L component amplitude of the normalized spectrogram writes:

$$S_{\omega_L}(v,t) = \eta\gamma(v)\sin[\omega_L\tau + \alpha(v)] \quad (21.22)$$

where the phase of the ω_L oscillation is given by:

$$\tan\alpha(v) = \frac{\sqrt{I(\omega_v + \omega_L)}\sin[\varphi(\omega_v) - \varphi(\omega_v + \omega_L)] - \sqrt{I(\omega_v - \omega_L)}\sin[\varphi(\omega_v - \omega_L) - \varphi(\omega_v)]}{\sqrt{I(\omega_v + \omega_L)}\cos[\varphi(\omega_v) - \varphi(\omega_v + \omega_L)]\sqrt{I(\omega_v - \omega_L)}\cos[\varphi(\omega_v - \omega_L) - \varphi(\omega_v)]} \quad (21.23)$$

and the modulation amplitude by

$$\gamma(v) = \frac{1}{I(\omega_v)}\left[I(\omega_v + \omega_L) + I(\omega_v - \omega_L) - 2\sqrt{I(\omega_v + \omega_L)I(\omega_v - \omega_L)}\cos[\varphi(\omega_v - \omega_L) - \varphi(\omega_v + \omega_L)]\right] \quad (21.24)$$

In (21.24) and (21.25), $I(\omega)$ is the spectral intensity of the XUV pulse at frequency ω. Although using a set-up very similar to CRAB/streaking, PROOF differs from those techniques in the phase encoding in the electron spectrogram and the method of phase retrieval. In a way analogous to the RABBITT quantum interference, the ω_L component arises from two-photon, XUV-IR, transitions from the ground state to the continuum states of energy W, as shown in Figure 21.12. The measured spectrum indeed shows these three components when taking the inverse Fourier transform of the spectrogram and the ω_L component can be filtered out for the analysis. The spectral phase can then be decoded from the ω_L component oscillation *at each energy*, in the spectrogram. Note that PROOF uses only those energy components of the spectrogram that are separated by one IR photon energy, which is much larger than a typical photoelectron spectrometer resolution. Retrieving the phase from multiple energies thus allows for determination of error bars.

As RABBITT, PROOF requires a small dressing laser intensity so the two-photon transitions can be treated *perturbatively*, and thus, only one-photon and two-photon transition pathways interfere. Nevertheless, interestingly, even at higher IR laser intensities, when the limited expansion of the exponential (21.19) is no longer valid, PROOF still works, possibly because the filtering of the ω_L component of the spectrogram eliminates the effects of higher order transition pathways, where under the same conditions for CRAB, the central momentum approximation would break down (Chini et al. 2010, Chang 2011). Experimentally, the intensity is then only limited by the production of spurious electrons by IR ionization of the target. Figure 21.13 shows a recent example of a PROOF measurement.

21.8.1 Improved PROOF (iPROOF)

iPROOF is a method directly inspired by PROOF proposed and demonstrated by Laurent et al., in (2013). The harmonic spectrum is again transformed into a photoelectron replica in a target gas, and the spectral phase of the XUV pulse is retrieved by a recursive semi-analytical procedure, rather than the iterative process of the PCPGA, which incorporates the physics of the photoionization process. The phase is evaluated by optimizing the fit with the theory (see Section 21.4.3). The demonstration is done by a HHG source driven by a superposition of two lasers at ω_L and $2\omega_L$ which generates odd and even harmonics (see Section 21.10) because of the asymmetry in the generating field, so in this case, another improvement derives from a sampling which is twice denser ($\hbar\omega_L$ rather than $2\hbar\omega_L$) than in the RABBITT method; however, an unexpected large phase shift between the odd and even harmonics is observed which has not received a clear explanation.

21.9 Comparison of RABBITT, Momentum Streaking, CRAB and PROOF

The four methods mentioned above have many common features since they all use spectrometry of photoelectrons produced by IR-HHG ionization of a target atom, but they also differ from each other both in the principle of measurement, retrieval procedures, types of attosecond pulses they can handle, etc. In the following table, some elements for a comparison and selection are given for quick reference (see the text for details).

21.10 All-optical Method

An interesting alternative phase retrieval method was established by Dudovich and coworkers (Dudovich 2006), which, like RABBITT, is well suited to a train of as pulses and is purely optical. The high harmonics are generated by a combined field composed of the fundamental driving field and its weak second harmonic. Even a weak second harmonic field ($<10^{-3}$) is sufficient to break the centro-symmetry of the HHG in centro-symmetric atoms, resulting in even-order harmonics depending on the relative phase of the two fields. The amplitudes of the even harmonics carry an information about the attochirp of the harmonic APT, i.e. the emission time of the different harmonics (Dudovich 2006).

By controlling the sub-cycle delay between the fundamental and second harmonic pulses (which is relatively easier than

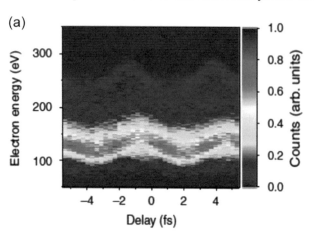
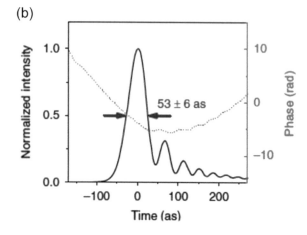

FIGURE 21.13 Example of PROOF trace (a), filtered ω_L component (b) temporal intensity and phase. (Reproduced from Li et al. (2017) http://creativecommons.org/licenses/by/4.0/.)

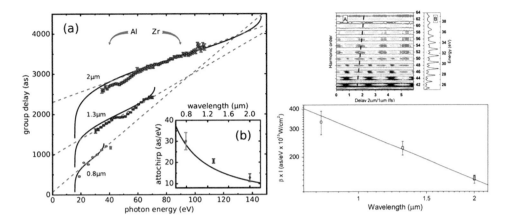

FIGURE 21.14 (a) RABBITT measurements at three wavelengths. The insert shows the GDD, or attochirp, as a function of the wavelength. (Reproduced from (Schoun 2015) with permission.) (b) measurements by the all-optical method. (Reproduced from Doumy et al. (2009).) http://creativecommons.org/licenses/by/4.0/). The results agree after scaling to the same intensity.

the XUV IR delay) and recording the resulting oscillations of the even order harmonics, it is possible to retrieve the group delays (or emission times) of the even-order harmonics and, by interpolation, that of the odd orders ones. In the limit of a sufficiently weak second harmonic beam, the group delay corresponds to the unperturbed values. As stressed by Dudovich et al. (2006), this method yields an *in-situ* value of the group delay dispersion (or attochirp), independent of possible subsequent compensation. However, this all-optical method relies on the modelling of the harmonic generation process. A comparison with RABBITT measurements of the attochirp versus the laser wavelength shows consistency of the two methods (Figure 21.14).

21.11 Other Methods

Besides those discussed above, several other procedures have been proposed. For the sake of completeness, they are cited here without any details: first, a refinement of PROOF and iPROOF called Scattering Wave PROOF (swProof) (Wei 2015); second, the Volkov Transform General Projection Algorithm (VTGPA) method (Keathley et al. 2016); and third Phase Retrieval of Broadband Pulses (Zhao et al. 2017). Each of these methods (as PROOF) does not rely on the central momentum approximation and can be applied to an IAP. VTGPA was recently demonstrated by measuring a 43 ± 1 as pulse, the shortest attosecond pulse on record at the time of writing (Gaumnitz et al. 2017). For a comparison of the theory and various approximations that delineate PROOF, iPROOF, and swPROOF, see Wei (2015).

21.12 Some Experimental Remarks

All the phase retrieval methods described above are interferometric measurements, which present particular challenges in the XUV. The lack of effective dichroic optics for the XUV and IR makes it that holed mirrors are commonly used for recombination (Figure 21.7). Thin silica wafers, and certain other semiconductor materials, set at Brewster's angle to the IR light can also be used to recombine the IR and XUV pulses, but the XUV reflectivity can vary dramatically with frequency depending on the semiconductor. For those phase-retrieval methods requiring photoelectron measurements from a second gas jet, the IR pulse must be focused to reach the required intensity (see Table 21.1). Then, the XUV pulse is also focused to a spot size within that of the IR at focus, ensuring that all the photoelectrons from the XUV have also interacted with the IR pulse. Metallic mirrors at grazing incidence can be used to focus a XUV pulses with bandwidths as high as 100 eV. Broadband multilayer mirrors are more effective closer to normal incidence but generally are unable to efficiently reflect as broad of bandwidths as grazing incidence metallic mirrors.

Each of these phase-retrieval methods requiring a second gas jet take a spectrogram as input, thus sufficient spectral and time delay resolution is essential to accurately retrieve the phases. However, what constitutes sufficient resolution depends on the attosecond pulse or pulse train to be measured and the phase retrieval method employed. The spectral resolution is dependent on three parameters of the electron time-of-flight spectrometer: flight tube length, magnitude of retarding potential, and timing resolution of the data acquisition electronics. The time delay resolution is dependent on the minimal step size of the delay stage and the phase stability of the interferometer itself. Active phase stabilization is often employed by co-propagating an additional laser with the interferometer, where the interference measured after recombination gives an error signal minimized by actuating the delay stage.

21.13 Principle Component Generalized Projection Algorithm

The PCGPA is an algorithmic method for extracting the spectral phases from a CRAB or PROOF spectrogram (Figure 21.15). The PCGPA assumes a FROG spectrogram of the form given in equation (21.15). The algorithm begins with guesses for the signal and gate fields sampled at constant time intervals to produce a pair of vectors. The outer product of these vectors

TABLE 21.1

Comparison between Attosecond Metrology Methods

	RABBIT	Momentum STREAKING	FROG-CRAB	PROOF	IPROOF
XUV-IR photoionization	Perturbative limit	Strong field limit	Strong field limit	Perturbative limit	Perturbative limit
IR intensity (W/cm^2)	Low 10^{11}–10^{12}	High 10^{12}–10^{13}	Medium high 7.5×10^{11}	Low 10^{11} But also works at higher intensity	Low 10^{11}–10^{12}
Principle	$2\omega_L$ sampling $2\omega_L$ filtering	Momentum distribution broadening	Blind FROG iteration	ω_L filtering	ω_L filtering and fit to perturbation theory
Retrieval	Inverse FT	Algebraic retrieval	PCGPA	PCGPA or Genetic Algorithm	Semi-analytical
Limitations	Only works for APT with harmonics separated by $2\omega_L$	IR intensity; Minimum pulse duration imposed by the broadening condition	Central Momentum Approximation	Neglects the atomic phase (Section 21.3)	Recursive phase retrieval can lead to compounding errors
Examples	Mairesse et al. (2003)	Drescher et al. (2001)	Mairesse and Quéré (2005)	Chini et al. (2010)	Laurent et al. (2013)

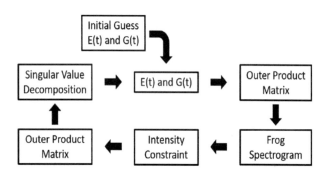

FIGURE 21.15 Flowchart of the PGCPA algorithm. The algorithm begins with guesses for $E(t)$ and $G(t)$, typically Gaussian pulses are used, to generate an initial outer product. For FROG-CRAB and PROOF, between a few hundred and a thousand iterations of the PCGPA are typically required for convergence (Mairesse and Quéré 2005, Chini 2012).

is then converted to a FROG spectrogram through a series of row/column manipulations and a Fast Fourier Transform (FFT) along the time axis. An intensity constraint is then applied to each point in the FROG spectrogram. In FROG-CRAB, the intensity at each pixel of the FROG spectrogram is set equal to the intensity of the corresponding pixel in the measured spectrogram. For PROOF, this same constraint is applied for the first few iterations of the PCGPA but is later replaced with a weighted sum of Fourier-filtered components of the measured spectrogram: those at the frequency of the gating field, twice that frequency, and the DC component (Chini 2012).

After applying the appropriate intensity constraint, a new outer product matrix is constructed by first applying an Inverse FFT along the frequency axis then reversing the row/column manipulations used previously to convert the outer product to the time domain. In the last step of each iteration, the pair of signal and gate vectors with the largest weight in the singular value decomposition of the outer product matrix is selected for the next iteration of the algorithm. To reduce the computational overhead, the singular value decomposition is often done only approximately using the power method (Trebino 2000). In each iteration, an error function of the generated FROG spectrogram is evaluated, and the algorithm breaks once the error reaches a minimum threshold. For FROG-CRAB, this error function is the least-squares error between the generated and measured spectrograms. For PROOF, it is the least-squares error between the same spectrograms after Fourier filtering each at the frequency of the gating field. For a more detailed description of the PCGPA, refer to Kane (2008) and Ch. 21 of Trebino (2000).

21.14 Conclusions and Outlook

Because of the low intensity and XUV wavelength of attosecond pulses from HHG, the metrology methods rely (in general) on the non-linearity of the XUV-IR photoionization of a target atom, for both APT and the IAP, electron spectrometry, and similar set-ups based on a Mach-Zehnder, or other, interferometer which provides the attosecond time steps. The IR pulse is the same one that generates the harmonics and is thus easily synchronized with the attosecond pulse. However, in all methods using photoelectrons from ionization of a target jet, the measurement is carried out in that jet. Propagation (forward or backward) from that position might involve corrections. The all-optical method of Dudovich et al. (2006) still needs corrections if the attosecond pulse is to propagate to another interaction region spatially separated from the generating jet. The uncertainty varying with the experimental conditions and for the shortest pulses so far was reported as 6 as and 1 as, respectively (Li et al. 2017; and Gaumnitz et al. 2017).

The methods described so far do not record the direction of the attosecond pulse electric field which is assumed to be known (and parallel to the polarization direction of the driving laser). New methods for complete temporal reconstruction of the electric field, including the direction, of few cycle fs pulses have been proposed and verified (Carpeggiani et al. 2017). PROOF analysis was recently applied to circularly polarized XUV attosecond pulses generated on a copper surface (Chen et al. 2016).

One remark to end this chapter: Figure 21.1 suggests that the evolution of the attosecond pulses is progressively slowing down and going to hit an "attosecond wall" just as the femtosecond pulses appeared to hit a femtosecond wall by the end of the 1990s. In the attosecond case, the wall is due to the attochirp (or Group Delay Dispersion) inherent to the HHG physics[3] (Kazamias and Balcou 2002). Approaches to solving this problem have been proposed (Mairesse et al. 2003) and shown to work, at least in one case (Lopez-Martens et al. 2005). Interestingly, a measurement in the soft X-ray regime of the water window (Cousin et al. 2017) appears to show a zero-attochirp; although at the time of writing, there is no clear interpretation of that result.

REFERENCES

Busto D, Barreau L, Isinger M, et al., (2018), Time-frequency representation of autoionization dynamics in helium, *Journ. J. Phys. B: At. Mol. Opt. Phys.* 51: 044002.

Carpeggiani P, Reduzzi1 M, Comby A, et al., (2017), Vectorial optical field reconstruction by attosecond spatial interferometry, *Nature Phot.* 11: 483.

Chang Z, (2011), *Fundamentals of Attosecond Optics* CRC Press, Boca Raton, Florida, USA.

Chen C, Tao Z, Hernández-García C, et al., (2016), Tomographic reconstruction of circularly polarized high-harmonic fields: 3D attosecond metrology, *Science. Adv.* 2: e1501333.

Chini M, (2012), Characterization and applications of isolated attosecond pulses, *PhD Dissertation University of Central Florida*. http://purl.fcla.edu/fcla/etd/CFE0004781.

Chini M, Gilbertson S, Khan S D, and Chang Z, (2010), Characterizing ultra-broadband attosecond lasers, *Opt. Express* 18: 13006.

Chirla R C, (2011), Attosecond pulse generation and characterization, *PhD Dissertation, The Ohio State University*. http://olc1.ohiolink.edu/record=b30121393~S0.

Cousin S L, Di Palo N, 1 Buades B, et al., (2017), Attosecond streaking in the water window: a new regime of attosecond pulse characterization, *Phys. Rev. X*. 7: 041030.

Dahlstrom J M, Guénot D, Klünder K K, et al., (2012), Theory of attosecond delays in laser-assisted photoionization, *Chem. Phys.* 414: 53.

Doumy G, Wheeler J, Roedig C, et al., (2009), Attosecond synchronization of high-order harmonics from midinfrared drivers, *Phys. Rev. Lett.* 102: 093002.

Drescher M, Hentschel M, Kienberger R, et al., (2001), X-ray pulses approaching the attosecond frontier, *Science* 291: 1923–1927 published online 15 February 2001; 10.1126/science.1058561.

Dudovich N, Smirnova O, Levesque J et al., (2006), Measuring and controlling the birth of attosecond XUV pulses, *Nature Phys.* 2: 781.

Eckle P, Smolarski M, Schlup P, et al., (2008), Attosecond angular streaking, *Nature Phys.* 4: 565.

Farkas G and Toth C, (1992), Proposal for attosecond light pulse generation using laser induced multiple-harmonic conversion processes in rare gases, *Phys. Lett. A* 168: 447–450.

Ferray M, L'Huillier A, Li X F, et al., (1988), Multiple-harmonic conversion of 1064 nm radiation in rare gases Journal of Physics B—Atomic, *Molecular and Optical Phys.* 21: L31.

Gaumnitz T, Jain A, Pertot Y et al., (2017), Streaking of 43-attosecond soft-X-ray pulses generated by a passively CEP-stable mid-infrared driver, *Opt. Express* 25: 305125.

Kane D J, (2008), Principal components generalized projections: a review, *Journ. Opt. Soc. Am. B* 25: A120.

Kazamias S and Balcou P, (2004), Intrinsic chirp of attosecond pulses: single-atom model versus experiment, *Phys. Rev. A* 69: 063416.

Keathley P D, Bhardwaj S, Moses J, Laurent G, and Kärtner F X, (2016), Volkov transform generalized projection algorithm for attosecond pulse characterization, *New J. Phys.* 18: 073009.

Kienberger R, Goulielmakis E, Uiberacker M et al., (2004), Atomic transient recorder, *Nature* 427: 817.

Kitzler M, Fabian C, Milosevic D, Scrinzi A, and Brabec T, (2003), Quantum theory of single subfemtosecond extreme-ultraviolet pulse measurements, *Journ. Opt. Soc. Am. B* 20: 591.

Kulander K C, Schafer K J, and Krause J L, (1993), Dynamics of short-pulse excitation, ionization and harmonic conversion, in *Super-Intense Laser-Atom Physics*, NATO *ASI Series II, Vol.12,* Piraux B, L'Huillier A and Rzazewski K Eds., Plenum, New York, p. 95.

Kruit P and Read F H, (1983), Magnetic field paralleliser for 2p electron-spectrometer and electron-image magnifier, *J. Phys. E* 16: 313.

Laurent G, Cao W, Ben-Itzhak I, and Cocke C L, (2013), Attosecond pulse characterization, *Opt. Expr.* 21: 16914.

Lewenstein M, Balcou Ph., Ivanov M Y, L'Huillier A, and Corkum P B, (1994), Theory of high-harmonic generation by low-frequency laser fields, *Phys. Rev. A* 49: 2117.

L'Huillier A, Schafer K J and Kulander K C, (1991), Theoretical aspects of intense field harmonic generation, *J. Phys. B At. Mol. Opt. Phys.* 24: 3315.

Li J, Ren X, Yin Y, et al., (2017), 53-attosecond X-ray pulses reach the carbon K-edge, *Nat. Commun.* 18: 1.

Lopez-Martens R, Varju K, Johnsson P, et al., (2005), Amplitude and phase control of attosecond light pulses, *Phys. Rev. Lett.* 94: 033001.

Mairesse Y, A. de Bohan A, Frasinski L J, et al., (2003), Attosecond synchronization of high-harmonic soft X-rays, *Science* 302: 1540.

Mairesse Y, and Quéré F, (2005), Frequency-resolved optical gating for complete reconstruction of attosecond bursts, *Phys. Rev. A* 71: 011401.

Mashiko H, Gilbertson S, Li C et al., (2008), Double optical gating of high-order harmonic generation with carrier-envelope phase stabilized lasers, *Phys. Rev. Lett.* 100: 103906.

Muller H G, (2002), Reconstruction of attosecond harmonic beating by interference of two-photon transitions, *Appl. Phys. B* 74: S17.

Nabekawa Y, Hasegawa H, Takahashi E J, and Midorikawa K, (2005), Production of doubly charged helium ions by two-photon absorption of an intense sub-10-fs soft X-ray pulse at 42 eV photon energy, *Phys. Rev. Lett* 94: 043001.

Paul PM, Toma E S, Breger P et al., (2001), Observation of a train of attosecond pulses from high harmonic generation, *Science* 292: 1689.

Paulus G G, Grasbon F, Walther H et al., (2001), Absolute-phase phenomena in photoionization with few-cycle laser pulses, *Nature* 414: 182.

---

[3] Pascal Salières Private Commun. (2017).

Pfeiffer A N, Cirelli C, Smolarski M, et al., (2011), Attoclock reveals natural coordinates of the laser-induced tunneling current flow in atoms, *Nature Phys.* 8: 76.

Quéré F, Mairesse Y, and Itatani J, (2005), Temporal characterization of attosecond XUV fields, *Jour. Mod. Opt.* 52: 339.

Sansone G, Benedetti E, Calegari F, et al., (2006), Isolated single-cycle attosecond pulses, *Science* 314: 443.

Schoun S B, (2015), Attosecond high-harmonic spectroscopy of atoms and molecules using mid-infrared sources, PhD Dissertation, The Ohio State University and references therein. https://etd.ohiolink.edu/pg_10?0::NO:10:P10_ACCESSION_NUM:osu1436853089.

Toma E S and Muller H G, (2002), Calculation of matrix elements for mixed extreme-ultraviolet–infrared two-photon above-threshold ionization of argon, *J. Phys. B---At. Mol. Opt. Phys.* 35: 3435.

Trebino R, (2000), *Frequency-Resolved Optical Gating: The Measurement of Ultrashort Laser Pulses*, Springer US: New York.

Véniard V, Taïeb R, and Maquet A, (1996), Phase dependence of (N+1)-color (N>1) ir-uv photoionization of atoms with higher harmonics, *Phys. Rev. A* 54: 721.

Walmsley I A and Dorrer C, (2009), Characterization of ultrashort electromagnetic pulses, *Adv Opt Photonics* 1: 308–437.

Wang H, Chini M, Khan S D et al., (2009), Practical issues of retrieving isolated attosecond pulses, *J. Phys. B At. Mol. Opt. Phys.* 42: 134007.

Wei H, Anh-Thu L, Morishita T, Yu C, and Lin C D, (2015), Benchmarking accurate spectral phase retrieval of single attosecond pulses, *Phys. Rev A* 91: 023407.

Zhao X, Wei H, Wu Y and Lin C D, (2017), Phase-retrieval algorithm for the characterization of broadband single attosecond pulses, *Phys. Rev A* 95: 043407.

Zhao K, Zhang Q, Chini M, Wu Y, Wang X and Chang Z, (2012), Tailoring a 67 attosecond pulse through advantageous phase mismatch, *Opt. Lett.* 37: 3891.

22

Chirped Pulse Amplification

Donna Strickland

CONTENTS

22.1 Introduction 321
22.2 CPA Basics 322
    22.2.1 Original CPA System 322
    22.2.2 Nd:glass and Ti:sapphire Systems 322
22.3 Dispersion Control 323
    22.3.1 Treacy Grating Compressor 323
    22.3.2 Martinez Grating Stretcher 324
    22.3.3 Offner Triplet 325
    22.3.4 Dispersion Compensation for Optical Elements in the Amplifier 325
    22.3.5 Grating Alignment Issues 326
22.4 Amplification to PW Level Power 326
    22.4.1 Energy Extraction from CPA Amplifier 326
    22.4.2 Energy Limitations 327
    22.4.3 Pulse Duration Limitations 327
    22.4.4 OPCPA 327
22.5 High-intensity Requirements 328
    22.5.1 Beam Quality 328
    22.5.2 ASE Issues 329
22.6 Concluding Remarks 329
References 329

22.1 Introduction

With the advent of the laser, optical waves were powerful enough for the first time to cause non-linear interactions with media [1]. Since then, the pursuit for higher power lasers has continued unabated. Non-linear interactions respond to instantaneous power rather than average power. For laser pulses, the peak instantaneous power is defined as the pulse energy over the pulse duration and so can be increased in one of two ways, increasing the energy or decreasing the pulse duration. Increasing the energy requires further amplification stages, which are necessarily larger to handle the higher energy. On the other hand, reducing the pulse duration is done at the oscillator stage and so the amplification system remains the same size. In the 1980s when Chirped Pulse Amplification (CPA) was invented, mode-locked dye lasers had achieved the shortest pulse durations of ~100 fs at wavelengths of ~600 nm [2]. Dye lasers though are low-storage-energy lasers and so could not achieve high peak powers. The high-energy lasers were solid-state lasers, typically Nd:glass used for laser fusion, at wavelengths of ~1060 nm and 10 kJ energy levels had been achieved [3]. The wavelengths were incompatible, but that was not the only reason that short pulses were not amplified in the large energy amplifiers. If a short pulse is amplified up to high energy in a solid-state laser, the peak power is sufficient for non-linear optics to occur in the lasing medium and cause self-focusing, where the beam collapses on itself [3]. The self-focusing causes damage to the laser rods, so high-energy amplification was limited to nanosecond long pulses.

A similar problem occurred decades earlier in the field of radar. When attempting to amplify shorter pulses to detect smaller objects at a further distance, the electronic amplifiers were damaged. Chirped radar was developed to overcome this problem [4]. The short pulses were stretched by dispersion lowering the peak power and then amplified to the required energy. The returning stretched pulse was compressed to give the correct image of the object. CPA works on the same principle. The short pulse from a mode-locked oscillator is first stretched sufficiently to maintain the peak power below the critical power for self-focusing in the amplifiers. The pulse can then be safely amplified to the required energy, and following amplification, the pulses can be recompressed giving ultra-intense pulses at the output of the laser system. The CPA scheme is depicted in Figure 22.1.

In this chapter, we will cover the basics of the CPA technique in Section 22.2, starting with the original CPA laser system [5],

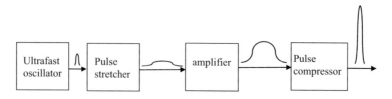

FIGURE 22.1 Schematic of basic CPA system showing the main components: short-pulse oscillator delivering a short pulse with small energy; pulse stretcher giving a long, low-power pulse, energy amplifier that boosts the energy but leaves the long pulse duration; pulse compressor, yielding a short pulse with high energy at the output.

which delivered just GW-level peak power, with pulse duration of 2 ps at a wavelength of 1064 nm. The first improvement to the original CPA system was better dispersion control to allow shorter pulse duration. Pulse stretching and compression and high-order dispersion control needed for optimized pulse duration will be discussed in Section 22.3. Currently, there are a number of CPA systems delivering petawatt (PW) level power around the world with several other systems under construction to go beyond 10 PW [6]. In Section 22.4, the various amplification systems will be described, and the issues limiting the energy and the gain bandwidth of the various systems will be discussed. For most high-intensity laser interactions, it is not sufficient to generate these high powers. In Section 22.5, various issues that limit the laser brightness and pulse contrast will be discussed. For a given power, maximum brightness occurs when the beam can be focused to a diffraction limited spot size. Ultra-high intensity interactions require excellent contrast between peak power and background signal. At the highest achieved focused intensities of $>10^{21}$ W cm^{-2}, a background that is 8 orders of magnitude below the peak can still lead to non-linear interactions and alter the medium before the central peak arrives. The chapter will conclude in Section 22.6 with a discussion of the current trends in CPA development.

22.2 CPA Basics

CPA solves the problem of how to get ultra-intense pulses to the application without any non-linear processes destroying the amplifier. A measure of the accumulated non-linear phase, ϕ_{NL}, in any medium is given by the B-integral [3],

$$B = \phi_{NL} = \frac{2\pi}{\lambda} \int n_2 I(z) \mathrm{d}z \quad (22.1)$$

where n_2 is the non-linear refractive index and $I(z)$ is the on-axis intensity as a function of propagation distance, z, through the medium. In the laser community, the intensity is defined as power per unit area (in the optics field, this is referred to as irradiance). Fluence is defined as energy per unit area, so that intensity is equal to the fluence per unit time. As will be discussed in Section 22.4.1, efficient energy extraction requires that a laser amplifier be seeded at the saturation fluence or higher. Therefore to maintain efficient amplification while reducing the intensity to decrease the B-integral, the pulse duration must be increased. At the time CPA was invented, the common rule was to keep the B-value below 5 to avoid damage of the laser rods from self-focusing. Now it is more typical in CPA systems to keep the B value even lower to minimize beam and pulse distortions.

22.2.1 Original CPA System

The original CPA laser [5] used a mode-locked Nd:YAG oscillator because it matched the 1064 nm wavelength of the high-energy-storage gain medium of Nd-doped silicate glass. Nd:YAG lasers could not generate the ultra-short pulses given by dye lasers. Since the duration of the pulses was 150 ps, self-phase modulation (SPM) in optical fibres was required to generate a broader spectrum that could be then compressed. Fibre optic pulse compression had already demonstrated that the pulse duration of mode-locked Nd:YAG lasers could be compressed to single picoseconds with the technique [7]. It was also known that optimum fibre compression occurred when the pulse duration was lengthened by material dispersion in the fibre [8]. The fibre in the first CPA system then served two purposes: generate a broad spectrum and be the pulse stretcher. The stretcher comprised a 1.4-km length of single-mode fibre. Over 2 W of average power was coupled into the fibre allowing the bandwidth to increase to 4 nm by SPM. This bandwidth allowed the pulse width to stretch to 300 ps by group velocity dispersion. The pulses were amplified to the mJ level in a regenerative amplifier using a standard cavity design, with a single Pockel cell to switch the pulses in and out of the cavity. The amplified pulses were compressed to 2 ps in a grating compressor that had been developed in 1969 by Treacy [9]. The Treacy grating compressor is discussed in detail in Section 22.3. The 2 ps pulse duration is greater than the transform limited pulse, showing that more work needed to be done to optimize the CPA technique beyond simply adding more amplifiers to reach higher energy.

22.2.2 Nd:glass and Ti:sapphire Systems

The original CPA concept could deliver pulses with picosecond durations and so to reach beyond the terawatt peak power level already achieved with nanosecond pulses and kilojoule energies, required amplifying the picosecond pulses to greater than the 1 J level. This required a large amplifier, which operated at low repetition rate. For this reason, in the late 1980s, further CPA development was mainly carried out in the large laser labs that were studying laser fusion with high-energy Nd:glass lasers. It was the concurrent development of the ultra-broadband solid-state laser medium of Ti:sapphire [10] that

Chirped Pulse Amplification

allowed small-scale terawatt level lasers to be developed. Soon after the development of Ti:sapphire as a laser gain medium, mode-locked lasers achieved pulse durations of less than 100 fs [11] and very quickly after that Ti:sapphire lasers were found to self-mode-lock [12], making it easier to generate 100 fs pulses than picosecond pulses. Ti:sapphire has a saturation fluence of 1 J cm^{-2} compared with the 5 J cm^{-2} of Nd:glass amplifiers so that larger systems would be needed to achieve the same energy as the glass system. However, since Ti:sapphire can support much shorter pulse durations down to ~30 fs, a smaller energy is required to achieve the same peak power as the high-energy, 1 ps Nd:glass systems. Ti:sapphire can therefore be used both as the short-pulse oscillator and as the amplifier gain medium. The very large gain bandwidth of Ti:sapphire can also produce mode-locked pulses at the 1054 wavelength corresponding to the gain peak of Nd-doped phosphate glass amplifiers allowing the mode-locked Ti:sapphire pulses also to be amplified to the highest energies. With the advent of mode-locked Ti:sapphire lasers, the worldwide development of CPA began.

22.3 Dispersion Control

The instantaneous power of a laser pulse is defined as energy per unit time, and for a laser pulse, the peak power P_0 is defined as the pulse energy, ε, divided by the effective pulse width τ_{eff}

$$P_0 = \frac{\varepsilon}{\tau_{eff}} \quad (22.2)$$

Since the total energy of the pulse is the instantaneous power $P(t)$ integrated over all time, the effective pulse duration is then given by

$$\tau_{eff} = \frac{1}{P_0} \int P(t) dt \quad (22.3)$$

Often the temporal width of laser pulses is defined by the full-width at half-maximum (FWHM). Even for a perfect Gaussian pulse, the FWHM duration is less than τ_{eff}. More importantly, actual laser pulses are not perfect Gaussian pulses and have energy in the wings of the pulse lowering the peak power. One of the crucial steps in CPA development is minimizing both the pulse duration and the energy in the wings of the pulse.

Because mode-locked lasers now deliver ultra-short pulses, there is no need to increase the bandwidth by SPM in an optical fibre, which leads to a non-linear phase dispersion that cannot be fully corrected. Stretching pulses in an optical fibre also leads to the problem of unmatched dispersion between the fibre stretcher and the grating compressor. This unmatched dispersion becomes more significant as the pulse duration decreases and the bandwidth increases.

The dispersion of the phase delay caused by transmission though an optical element, whether it is caused by angular dispersion of a grating or the frequency-dependent refractive index of a material, can be approximated using a Taylor series:

$$\phi(\omega) = \phi(\omega_0) + \left.\frac{\partial \phi}{\partial \omega}\right|_{\omega_0} (\omega - \omega_0) + \frac{1}{2}\left.\frac{\partial^2 \phi}{\partial \omega^2}\right|_{\omega_0} (\omega - \omega_0)^2 + \frac{1}{6}\left.\frac{\partial^3 \phi}{\partial \omega^3}\right|_{\omega_0} (\omega - \omega_0)^3 + \ldots \quad (22.4)$$

where $\phi(\omega)$ is the phase delay for frequency ω, near central frequency ω_0, $\phi(\omega_0)$ is the phase for the central frequency, $\left.\frac{\partial \phi}{\partial \omega}\right|_{\omega_0}$ is the group delay giving the time it takes for the peak of the pulse to travel through the element. The second-order dispersion term, $\frac{1}{2}\left.\frac{\partial^2 \phi}{\partial \omega^2}\right|_{\omega_0}$, is the group delay dispersion (GDD) which is the lowest order term that causes the pulse duration to stretch with propagation. If the bandwidth of the pulse is small, you can typically ignore the higher order terms, but as pulse durations get shorter, the higher order dispersion terms become important.

22.3.1 Treacy Grating Compressor

The Treacy grating compressor [9] is the one standard optical element that is used in almost every CPA laser system. Figure 22.2 shows a Treacy grating compressor, which is composed of two parallel gratings. One ray is traced through the compressor corresponding to a single wavelength within the bandwidth of the pulse. G is the perpendicular separation of the two parallel gratings. The separation of the gratings along the beam is given by b, which varies with wavelength, such that for the angle of incidence, γ, b = Gcos($\gamma - \theta$) and θ is the angle between the incident and diffracted ray. At the output of the grating compressor, all the rays corresponding to each wavelength are parallel, but spatially separated. Typically, a retroreflector is used at the output of a grating stretcher sending the light through a second pass of the compressor to spatially recombine all wavelengths back to a circular beam profile, while doubling the GDD of the compressor.

Treacy determined the phase delay through the compressor as a function of wavelength. Figure 22.3 shows two different rays for two different wavelengths, travelling through the compressor from input point A to the output plane containing points D and E of the output beam as well as point A. The phase delay between point A and the output plane is composed of two components. The first is the path length of the ray and the second component is the grating phase delay, which causes the different wavelengths to travel the different paths. The path

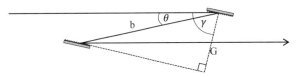

FIGURE 22.2 Schematic of Treacy grating compressor, two parallel gratings placed a perpendicular distance G apart. All rays leave the compressor parallel to the input beam. The displacement of the output beam perpendicular to the input beam is wavelength-dependent.

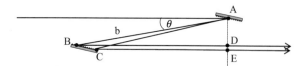

FIGURE 22.3 Schematic of two rays having different wavelength passing through Treacy grating compressor. The path length difference between the two rays is given by the difference of path ABD and ACE.

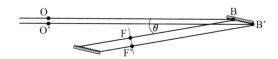

FIGURE 22.4 Schematic of two rays having the same wavelength from two different positions on the phase front of the input beam passing through Treacy grating compressor. The path length difference between the two rays is given by the difference of path ABF and O'B'F'.

length AB or AC is given by b, which is defined above for grating separation G and angle of incidence γ, which are common to both paths and angle θ, which is wavelength-dependent. The path length ABD or ACE is given by $b(1+\cos\theta)$. The phase delay ϕ_1 due to path length, p, is then:

$$\phi_1 = \frac{\omega}{c} p = \frac{\omega}{c} b(1+\cos\theta) \qquad (22.5)$$

The second component of the phase delay can be understood if you draw two rays having the same wavelength but originating at two different points in the same phase front of the beam, O and O' as shown in Figure 22.4. The phase fronts must remain perpendicular to the rays and so the phase must be constant across F and F'. You can see from the figure that the path lengths OBF and O'B'F' are different, and so the angular dispersion of the grating must compensate for this path difference. The grating adds a phase of -2π for each groove spacing, d, along B to B' giving a phase delay ϕ_2 at each grating of:

$$\phi_2 = -\frac{2\pi}{d} G \tan(\gamma - \theta) \qquad (22.6)$$

The total phase delay through a single pass of the grating compressor is then the addition of the two different phase delays: $\phi = \phi_1 + 2\phi_2$. With the angle θ being ω-dependent, this phase is a complicated function with respect to ω and so is usually written using a Taylor expansion. Treacy worked this out to second order and McMullen [13] first derived the expression to third order.

Below is a chart showing the expressions for second- and third-order dispersion terms for material dispersion and angular dispersion of a parallel grating pair. The material dispersion comes from successive derivatives of the optical path length, $\omega n z/c$ through a medium, where n is the refractive index, c is the speed of light and z is the length of the medium.

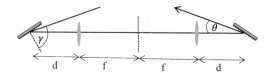

FIGURE 22.5 Schematic of Martinez grating stretcher; two anti-parallel gratings placed inside the focal points of a 1:1 telescope. The ray shown is for the central frequency. All rays leave the stretcher parallel to central frequency but displaced from it. The displacement of the output beam is wavelength-dependent.

positive second-order dispersion of the optical fibre stretcher, the positive third-order dispersion of both the stretcher and compressor add rather than subtract. This mismatched higher order dispersion leads to imperfect compression leaving the pulse duration longer than the transform limited duration. Not only do the higher order dispersion terms lead to longer pulses, they also cause the wings of the pulse or the background to rise. For this reason, these higher order terms must be corrected to get the shortest pulse durations with the highest contrast ratio between peak power and background signal.

22.3.2 Martinez Grating Stretcher

Soon after CPA was developed, Martinez [14] realized that gratings could also be used for compression of negatively chirped pulses. It was soon realized by Pessot, Maine and Mourou [15] that the Martinez grating system could be used as a stretcher in a CPA system. By using both a grating stretcher and compressor, all orders of dispersion can be matched, except for aberrations introduced by the telescope. In a Martinez grating stretcher, a telescope is used between the gratings as depicted in Figure 22.5. Martinez discusses the limitations imposed by aberations of the lenses [14]. Rather than the gratings being parallel they are placed at the mirror

Dispersion order	Material	Grating pair [13]
Second	$\left.\dfrac{\partial^2}{\partial \omega^2}\left(\dfrac{\omega n z}{c}\right)\right|_{\omega_0}$	$\dfrac{-4G}{c\omega_0}\left(\dfrac{2\pi c}{\omega_0 d}\right)^2 \cos^{-3}(\gamma-\theta)$
Third	$\left.\dfrac{\partial^3}{\partial \omega^3}\left(\dfrac{\omega n z}{c}\right)\right|_{\omega_0}$	$\dfrac{3G}{2c\omega_0}\left(\dfrac{2\pi c}{\omega_0 d}\right)^2 \left[\cos^2(\gamma-\theta)+\left(\dfrac{2\pi c}{\omega_0 d}\right)\sin(\gamma-\theta)\right]\cos^{-5}(\gamma-\theta)$

For a grating pair, the second- and third-order dispersion terms always have opposite sign, but the material dispersion is positive for both second- and third-order dispersion in the normal dispersion regime. Therefore, when the second-order negative dispersion of the grating compressor compensates the

image of each other. If the gratings are placed inside the focal region of the lenses, the pulse undergoes positive dispersion. For the gratings in the stretcher placed a distance d from a lens inside the focal distance, f, the corresponding distance, b, between the two gratings in a Treacy compressor is equal to 2(f–d). As in the case of the compressor, the output beam is spatially dispersed in one direction and so the beam is retro reflected back through the stretcher to once again double the dispersion and overlap the colors back to a circular beam. Also rather than using two gratings, a mirror can be placed in the focal plane at the center of the telescope and the beam is then passed four times off a single grating.

22.3.3 Offner Triplet

Of course lenses are not ideal and the first thing to note is that for large bandwidths, the chromatic aberrations of lenses cause large phase distortions corresponding to higher order dispersion terms. For this reason, the focusing elements in a grating stretcher are all reflective elements. Ideal lenses or spherical mirrors assume the rays are in the paraxial limit, but this is not the case for large bandwidths and large stretching ratios and so the Seidel aberrations of lenses also lead to non-linear phase distortions. Because the frequencies are mapped onto different rays, the aberrations of the rays correspond again to higher order frequency dispersion. In 1996, an Offner triplet [16] was first used to greatly reduce unwanted aberrations from the telescope in the stretcher of a CPA system [17]. An Offner triplet is composed of two spherical mirrors: one concave with radius of curvature R and a convex mirror with radius of curvature of R/2. The two mirrors are placed concentrically. The symmetry of the Offner triplet cancels all odd-order aberrations, leaving just spherical and astigmatism. If the object is placed at the centre of curvature, these two aberrations would also be eliminated. However, a single-grating stretcher needs to have the grating displaced from the centre leaving some spherical aberration. A stretcher with an Offner triplet is depicted in Figure 22.6. The two curved mirrors are placed concentrically about the centre of curvature C. The single grating is placed at distance, l, from C. Again, the ray trace shows that the output beam is parallel but displaced from the input beam. The amount of displacement varies with wavelength so the beam is again a line, and a retroreflector would be used to send the beam back through a double pass of the stretcher. This stretcher compensates for the dispersion of a compressor with a grating separation given by, $b = 2l$. It was shown that with a grating distance of 250 mm from C, for $R = 1024$ mm, the remaining spherical aberration only caused a small increase in the compressed pulse duration from 30 to 33 fs for a stretching factor of 10 000 [17].

The Offner triplet can be used to correct dispersion up to third order. For the shortest pulse durations (<100 fs), it is not sufficient to correct to third-order dispersion to obtain a nearly transformed limited pulse duration. There have been other stretcher designs developed that correct up to fifth-order dispersion [18,19]. More complete information about stretcher designs can be found in a review article written by Yakovlev [20].

22.3.4 Dispersion Compensation for Optical Elements in the Amplifier

The pulse compressor in a CPA system needs to compensate for more than just the dispersion of the stretcher. The amplifier system itself contains a number of optical elements; gain media, mirrors, Pockel cells. Each of these elements adds dispersion that must be compensated in the compressor. Compensation of this extra dispersion cannot be simply accomplished by increasing the separation of the compressor gratings, as that would only eliminate the second-order dispersion. It requires using a different angle of incidence for the stretcher and compressor to compensate the dispersion up to third order. This, however, requires an exact calculation of the dispersion of all elements. Also, the first amplifier is quite often a regenerative amplifier, and so if the number of round trips of this amplifier changes, the compressor would need to have a different alignment, not simply a change in grating separation.

Tournois in 1993 realized that by replacing the reflection gratings used in stretchers and compressors with a transmission grating placed directly on the face of a prism, the third-order dispersion could be eliminated when operated at a particular angle [21]. These diffractive elements are known as grisms. Having zero third-order dispersion does not help eliminate the third-order dispersion of the entire CPA system, but Kane and Squier showed in 1997 that the grism could be used at a different angle to have the same sign of dispersion for both second- and third-order dispersion [22]. They also showed that the grism compressor and stretcher could have the same ratio of second- to third-order dispersion as the material dispersion in the amplifier system. This ratio is independent of the material length, and so under this condition, the material dispersion can be eliminated with a grating compressor at the same angle as the stretcher, and the distance variation of the gratings eliminates the material dispersion up to third order.

Using grisms for compressors works well for the high repetition rate and lower energy (<1 mJ) systems, but cannot be used for the highest intensity CPA systems because of non-linear interactions in the grism. However, recently, grisms have been used as part of the stretcher where the peak power is low to help eliminate the third order of the amplification system [23].

In addition to optimizing the stretcher and compressor design, pulse shaping techniques are often used to correct for the higher order dispersion terms. Even for the picosecond pulse duration systems, at the highest intensity, it is not enough to cancel the dispersion up to third order because the background in the wings of the pulses would still be too high. The pulse shaping devices are placed before the amplification

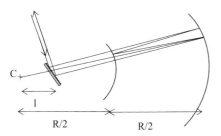

FIGURE 22.6 Schematic of grating stretcher with Offner triplet, consisting of a convex and a concave spherical mirror with radii of curvature of R/2 and R, respectively. One ray through the stretcher is depicted. All rays leave the stretcher parallel to the input beam but displaced. The displacement of the output beam is wavelength-dependent. The dispersion is determined by the angle of incidence and the distance l from the centre of curvature, C, of the two mirrors.

stages, where the power is low to pre-compensate for the high-order dispersion terms. The most commonly used pulse shaping device is the Acousto-Optic Programmable Dispersive Filter (AOPDF) developed and commercialized by Tournois [24].

It should also be pointed out here that for peak powers beyond 10 TW, the entire grating compressor must be kept in vacuum as the final intensity at the output is sufficient for non-linear interactions in air including self-focusing.

22.3.5 Grating Alignment Issues

In the previous section, we discussed the grating stretchers and compressors assuming that the grating angles could be set with perfect accuracy. Scientists at Rutherford Appleton Laboratory (RAL) studied the effects of the various alignment issues of the grating compressor [25]. Errors in compression result from imperfect matching of the compressor angle of incidence to that of the stretcher, and the non-parallelism of the two gratings in the compressor. All of these errors add together to lengthen the overall pulse duration and reduce the contrast ratio. The RAL group showed that either a 13 mrad misalignment between stretcher and compressor or just a 23 μrad non-parallelism of compressor gratings in a double-pass grating compressor would add 200 fs pulse durations for a 4-nm bandwidth at 1054 nm. These very tight angular constraints require interferometric techniques to optimally align the gratings. Yakovlev gives a good review of the various alignment measurement techniques [20].

22.4 Amplification to PW Level Power

There are three basic approaches to amplifying short pulses to 1 PW or greater power. The first two are CPA systems, using either high-energy storage gain media, typically Nd:glass, to reach kJ or greater energies, or broadband gain media, mainly Ti:sapphire that can amplify pulses with durations as short as 30 fs. The other scheme is Optical Parametric CPA (OPCPA) [26], which does not use laser amplification but rather the non-linear process of Optical Parametric Amplification (OPA). OPCPA has two advantages over CPA. OPCPA can amplify the broadest spectral bandwidths, and it generates the smallest background power.

22.4.1 Energy Extraction from CPA Amplifier

The two main parameters for pulsed amplification are the overall energy gain, G, and the efficiency of energy extraction. For cw laser beams, the interaction of the optical wave with the media is determined by the intensity of the wave, and the output is measured by power. For laser pulses, the duration of the interaction is given by the pulse duration and so the intensity can be integrated over this duration to give the energy fluence, F. The gain is then measured as the ratio of the output-to-input pulse energy. For amplification of pulses with durations that are short compared with either the radiative lifetime of the upper lasing level or the inverse of the pumping rate, the maximum energy that can be extracted by the seed pulse is the energy stored in the gain medium when the pulse arrives.

The solid-state gain media used in typical CPA systems have amplification described by a 4-level system. The stored energy, ε_{st}, in a simple 4-level system is given by:

$$\varepsilon_{st} = h\nu n_0 (al) \quad (22.7)$$

where $h\nu$ is the stimulated photon energy, n_0 is the initial population inversion density and (al) is the interaction volume, with l is the length of the gain medium and a is the cross-sectional area of the optical beam. For high extraction efficiency, most of this stored energy should be converted to the output energy of the laser pulse.

In the small signal limit, for low input fluence F_{in}, the gain is exponential,

$$G = \frac{F_{out}}{F_{in}} = e^{g_0 l} \quad (22.8)$$

where F_{out} is the output fluence and g_0 is the small signal gain:

$$g_0 = \sigma n_0 \quad (22.9)$$

where σ is the stimulated cross-section of the gain medium.

The gain is exponential because the population is not significantly altered by the small input fluence. Of course, if the population is not altered, very little of the stored energy is extracted. The population inversion only becomes significantly altered as the fluence reaches the saturation level, F_{sat}, and the gain is no longer exponential. F_{sat} is defined as the fluence that reduces the population n to $1/2 n_0$ through stimulated emission:

$$g = \sigma n = \frac{\sigma n_0}{1 + F\sigma/h\nu} = \frac{g_0}{1 + F/F_{sat}} \quad (22.10)$$

giving:

$$F_{sat} = h\nu/\sigma \quad (22.11)$$

As the fluence is increased, the gain is no longer exponential, since g is no longer constant with fluence. The gain, G, becomes linear with length when the input flux is greater than the saturation fluence. At these high inputs, the population inversion is greatly reduced, and so the energy is fully extracted. On the other hand, at high fluence, the gain is greatly reduced. This is why the initial amplification stage of a high-energy amplifier is either a regenerative or multi-pass amplifier, where the initial passes will see small-signal exponential gain, but the output pass has sufficient fluence to extract the full energy from the amplifier. The first stage of amplification then supplies the most orders of gain, and the final amplifiers bring the system to the required energy. Each amplification stage is correspondingly larger to maintain the same input fluence at each stage.

The inverse relationship between the saturation fluence and stimulated cross-section shows that high-gain lasing media such as dyes have low saturation fluence, which is why they are not used as high-energy amplifiers. The most energetic amplifiers have high saturation fluence levels. The maximum energy is limited by size that the gain medium can be manufactured and maintained high quality surfaces and uniform gain. However, for some gain media with very high saturation fluence levels, the energy may be limited by the damage fluence of the optics in the amplifier including the gain medium. Nd:glass has a high saturation fluence of 5 J cm^{-2}, which is still well below the damage level of the optics, which is why it is a standard high-energy amplifier gain medium.

22.4.2 Energy Limitations

The common element to all types of CPA is the Treacy grating compressor. Of all optical elements, gratings have the lowest damage thresholds. Also for optimum compression, the groove alignment and spacing must be highly accurate, which then puts constraints on the maximum size that gratings can be constructed. The limited size and damage threshold of the gratings then place the ultimate constraint of the maximum energy that can be delivered from any CPA system.

At the time of the original CPA system, gold-coated holographic gratings were used. These gratings had a damage threshold of just 50 mJ cm^{-2}. A decade later, Lawrence Livermore National Laboratory (LLNL) developed new grating technology that allowed the first PW laser system to be built [27]. They first developed multilayer dielectric (MLD) gratings, which proved to have very high efficiency and damage threshold but difficult to manufacture [28]. The initial PW laser actually used metallic gratings that LLNL developed, which had a damage threshold >0.4 J cm^{-2} and could be made to ~1 m in width [27]. The metallic gratings had a diffraction efficiency over 90% from 800 to 1100 nm [29] making them suitable for the shorter pulse Ti:sapphire CPA systems. The MLD gratings on the other hand have higher damage thresholds of 3 J cm^{-2} but can only support a bandwidth of the longer picosecond systems. The fluence is kept at least a factor of 2 below this damage threshold to ensure the longevity of the gratings. To increase the powers beyond the grating limit, a mosaic of gratings or tiled gratings are used to extend the size of the overall grating to support higher energies [30].

22.4.3 Pulse Duration Limitations

Whether using narrowband Nd:glass systems or broadband Ti:sapphire systems, gain narrowing limits the final pulse duration because over 10 orders of magnitude of gain is required to reach the PW power level. This amount of gain means that the amplified bandwidth is much less than the gain bandwidth of the amplifier. Assuming Gaussian pulses, with input spectral bandwidth of $\delta\lambda$ and Gaussian profile gain width, Δ, the output bandwidth of the pulse, $\Delta\lambda$, is given by [25]:

$$\frac{1}{(\Delta\lambda)^2} = \frac{\ln G_0}{(\Delta)^2} + \frac{1}{(\delta\lambda)^2} \quad (22.12)$$

where G_0 is the peak gain. For gains of greater than 10 orders of magnitude, the output bandwidth is about one quarter of the gain bandwidth.

The first solution to avoiding gain narrowing is to use a broader gain amplification system for the pre-amplifiers that give the first several orders of magnitude gain. The first PW laser not only used Ti:sapphire for the mode-locked oscillator but also for the pre-amplifiers to bring the energy to the ~50 mJ level before final amplification in Nd:glass [27]. The second PW laser system was developed at the University of Texas at Austin, and for this system rather than using Ti:sapphire as the pre-amplifiers, OPCPA was employed to deliver broadband spectra at the ~700 mJ level as seed pulses for the glass amplifiers [31].

The Texas PW system used a second solution to gain narrowing. In the amplification stage rather than using a single gain medium, the Texas system used two different glasses [31]. Nd-doped silicate glass, which has a gain peak at 1064 nm, was used for two stages of multi-pass amplification. The final amplification was done in a multi-pass Nd-doped phosphate glass amplifier, which has a gain peak at 1054 nm. By using the two different glasses for the final amplification stages, the gain bandwidth is increased from 30 to 40 nm [25].

In addition to the normal gain narrowing, CPA has added spectral narrowing because of gain saturation. When the gain is saturated, the front end of the pulse is amplified more than the trailing edge. However, in CPA, the spectrum is stretched over the pulse duration. Normally, the dispersion of the pulse stretcher is positive, and so the red side of the spectrum sees gain compared to the blue side. To overcome this type of gain narrowing, Kalashnikov and co-workers developed negatively and positively chirped pulse amplification (NPCPA) [32]. In NPCPA, the pulse is first negatively chirped with a Treacy grating compressor and then amplified, allowing the blue side of the spectrum to see higher gain. The pulses are then sent through a pulse stretcher, with positive dispersion that exceeds the original negative dispersion, leaving the pulse positively chirped for the final amplification that preferentially amplifies the red side of the spectrum before compression in a final Treacy grating compressor. By using NPCPA, the authors demonstrated an amplified bandwidth of 50 nm from a 60 nm seed with a Ti:sapphire amplification system. This is compared with an amplified spectrum of just 42 nm with conventional CPA with the same overall gain to reach the 100 mJ energy level.

22.4.4 OPCPA

Since the energy of the system is limited by optical damage, further increase in the peak power can only be achieved by reducing the pulse duration. Even with the broadest bandwidth gain medium of Ti:sapphire and using techniques such as NPCPA, the output laser bandwidth is still limited by gain narrowing. A different amplification technology is needed to increase the bandwidth. The non-linear process of OPA can amplify pulses with ultra-wide spectral bandwidth with proper phase matching. OPA is a second-order non-linear interaction of two fields having different frequencies (v_p, v_s) where v_p is known as the pump frequency because it is the higher energy photon and v_s is known as the signal frequency. Through second-order non-linear interactions, a generated field at frequency $v_p + v_s$ can be created through sum frequency generation or a field with frequency $v_p - v_s$ can be created from difference frequency mixing. In the case of sum frequency generation, the energy of both the pump and signal photons goes into the generated third higher energy frequency, but in the case of difference frequency generation, the pump energy is split between the signal field and the generated field at the lowest energy frequency, v_i, known as the idler. The electric field at the signal frequency then actually increases with difference frequency mixing, and this increase in the amplitude of the signal field is termed OPA. This energy conservation of OPA is depicted in Figure 22.7 showing the absorption of photon of frequency v_p and emission of photons at the two frequencies v_s and v_i.

However, energy conservation is not enough. Momentum must also be conserved through phase matching given by:

$$k_p = k_s + k_i \qquad (22.13)$$

where k_p, k_s, k_i are the wave vectors of the pump, signal, and idler waves, respectively. This phase-matching condition for three frequencies can be met by angle-tuning a birefringent non-linear mixing crystal. However, to amplify a broadband signal with a narrowband pump pulse, the following phase-matching condition must be met [33].

$$k_p = k_s + 2\pi\Delta v \left(\frac{1}{v_{gs}} - \frac{1}{v_{gi}} \right) \qquad (22.14)$$

where v_{gs} and v_{gi} are the group velocities of the signal and idler pulses, respectively, and Δv is the frequency bandwidth. This phase-matching condition can be met in one of two ways. For collinear pumping, the condition is met at the degeneracy point where $v_{gs} = v_{gi}$, or broadband phase-matching can occur with non-collinear pumping. With non-collinear OPA known as NOPA, the phase matching is accomplished by the idler frequencies dispersing at different angles for different wavelengths. It has been shown with NOPA that very large bandwidths of 2000 cm^{-1} can be amplified [34]. NOPA then makes an excellent choice for amplification of broad bandwidths to ultra-high powers.

As with CPA, the limit to the final output power is given by the damage fluence of the optics including the non-linear crystals. The intensities must also be kept low enough in the mixing crystals so that the third-order non-linearities do not occur significantly and add temporal or spatial phase distortions. The highest energy beam is necessarily the pump beam, and so its fluence is kept a factor of at least two below the damage fluence. The high-energy pumps are again from the solid-state lasers that are most efficient for narrow band, longer pulses of ~1 ns. The OPA would not be efficient if the long ns pump

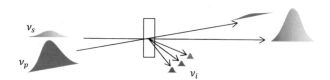

FIGURE 22.8 Schematic of OPCPA using non-collinear phase matching. The long, high-energy pump, v_p, mixes with the lower energy, chirped, stretched signal v_s. At the output, the energy from the pump pulse is depleted and the chirped signal is amplified. The idler frequencies, v_i, are dispersed.

pulses were mixed with the much shorter, broad bandwidth pulses. To maximize the OPA efficiency, the signal pulses must be stretched to match the pulse duration of the pump. The signal pulses are therefore chirped and this CPA scheme is known as OPCPA. NOPA is typically used in OPCPA as the idler beam is not needed, and so the angular dispersion of the idler beam is not an issue. The basic amplification process of OPCPA is depicted in Figure 22.8, where the energy from the highest energy photons goes into the chirped signal and the lowest energy idler photons. Note that the figure does not show the oscillator and pulse stretcher for the seed, nor the final compression stage that is needed to achieve the high-intensity pulses, as these are the same as for conventional CPA.

Not only does OPCPA offer higher bandwidths than conventional CPA, it also has less background ASE as will be discussed in the next section. For this reason, OPCPA is used for the shortest pulse systems and as previously noted, OPCPA is also used as the front end of conventional CPA systems to achieve ultra-broadband seed pulses for the final amplification to large energies.

22.5 High-intensity Requirements

The performance of a high-intensity laser is given by two main qualifiers: the laser brightness and peak-to-background power ratio. Laser brightness is defined as power per unit area per unit solid angle. To maximize the brightness, CPA laser systems must maximize the energy and minimize the pulse duration while maintaining a high-quality beam. The beam quality is affected by the flatness and roughness of the optical elements, thermal conditions, and non-linear phase distortions across the beam. To achieve the highest intensities, adaptive optics is employed. The background signal comes from imperfect dispersion control leading to wings in the pulses and Amplified Spontaneous Emission (ASE). Non-linear techniques have been developed to reduce the background signal.

22.5.1 Beam Quality

The peak laser intensity, I_0, is given by:

$$I_0 = \frac{\varepsilon}{\tau_{\text{eff}} A_{\text{eff}}} = \frac{P_0}{A_{\text{eff}}} = \frac{1}{\tau_{\text{eff}}} F_0 \qquad (22.15)$$

where A_{eff} is the effective area of the beam and F_0 is the peak fluence given by:

FIGURE 22.7 Energy-level diagram for OPA, where the pump photon energy, hv_p, is absorbed and photons at energies hv_s and hv_i are emitted.

$$F_0 = \frac{\varepsilon}{A_{\text{eff}}} \quad (22.16)$$

The energy is given by integration of the spatially varying fluence $F(x,y)$ over the entire (x,y) plane perpendicular to the wave vector. The effective area is then determined by:

$$A_{\text{eff}} = \frac{1}{F_0} \int F(x,y) \, dx \, dy \quad (22.17)$$

As in the case of the effective duration of the pulse, the beam diameters along x and y given by the FWHM sizes are not typically a good indicator of the area of the beam. Spatial phase distortion causes spreading of the focal spot size beyond the diffraction limit.

Most CPA lasers are high repetition rate, low energy systems where the amplification takes place in a regenerative amplifier. The cavity of a regenerative amplifier ensures high quality of the beam for these low-power systems. The higher energy systems rely on larger amplifiers, which require larger optics. The larger dimensions make it more difficult to maintain high-quality phase fronts across the beam, both because of aberrations in the focusing elements and the roughness and flatness across all the optical surfaces. In addition to the imperfections of the optical elements, there will also be thermal phase distortions. Since these imperfections in beam quality cannot be completely eliminated, adaptive optics is used to correct for the spatial distortions. With adaptive optics, intensities of 2×10^{22} W cm^{-2} have been achieved with a power of 300 TW using an $f/1$ parabolic mirror [35].

OPCPA has a couple of advantages over conventional CPA to maintain better beam quality. The first is that there is very small thermal loading in the non-linear crystal as ideally all the energies deposited by the pump beam are carried away in the signal and idler beams. In reality, there is always some scattering and absorption in the crystal, but this is significantly smaller than the energy deposited as heat from the pumps in conventional laser amplification. The other advantage in OPCPA is that the aberrations from a pump beam do not get imprinted onto an unaberrated seed beam, but rather onto the generated idler beam [33].

22.5.2 ASE Issues

Another advantage OPCPA has over conventional CPA is the reduction in ASE. Ross and co-workers made a comparison between the focused intensity of ASE for conventional CPA and for OPCPA [33]. For conventional lasers, the intensity of the ASE is given by

$$I_{\text{ASE}} = \frac{F_S}{16 f^2 \tau_{\text{Rad}}} \frac{\Delta \lambda_{\text{ASE}}}{\Delta} G_0 \quad (22.18)$$

where F_s is the saturation fluence of the gain medium, f is the f-number of the focusing optic, τ_{Rad} is the radiative lifetime of the upper state, and $\Delta \lambda_{\text{ASE}}$ is the bandwidth of the ASE.

The ASE from OPA comes from total gain, G_0, of one noise photon per mode and the intensity of the ASE is given by

$$I_{\text{ASE}} = \frac{\pi}{4 f^2} \frac{h \nu \Delta \nu}{\lambda^2} G_0 \quad (22.19)$$

where $\Delta \nu$ is the gain bandwidth of the OPCPA.

In both cases, the intensity of the ASE is dependent on the total gain of the system. Since the seed pulses from the oscillators for any of these systems have comparable energy of ~1 nJ regardless of the pulse duration, shorter pulse systems then exhibit less ASE because they require less energy gain to reach the same peak power. However, in conventional amplifiers, the broadband amplification media of Ti:sapphire also have the shorter radiative lifetime and so suffer higher ASE than the Nd:glass amplifiers for the same gain. On the other hand, OPCPA, which can deliver the shortest pulse durations, exhibits the least ASE for the same gain and so considerably smaller ASE at the shortest pulses for the same peak power. Ross and co-workers [33] derived that I_{ASE} from a Ti:sapphire laser achieving 100 TW of power with 30 fs pulse duration would be ~1.3×10^{12} W cm^{-2}, whereas a 1 PW, 30 fs OPCPA system would only generate $I_{\text{ASE}} \sim 7 \times 10^{11}$ W cm^{-2}.

The background coming from ASE and imperfect pulse compression must be removed for the highest intensity laser systems. High-energy CPA systems use high-gain regenerative amplifiers as the front end and low gain, high-energy storage amplifiers as the final stage. Since most of the gain is in the pre-amplification stage, most of the ASE is also produced in the front end. To eliminate this, double CPA has been developed [36]. For double CPA, the pulses are compressed after pre-amplification and then undergo a non-linear process that passes the highest intensities near the peak of the pulse and blocks the lower intensity background. The cleaned pulses are then sent for final amplification in the high-energy, low-gain amplifiers. For double CPA then, only the back ground from the final amplification stages remains at the output of the system.

22.6 Concluding Remarks

Not all CPA development works are about increasing the peak power. The repetition rate and average power are also important for a number of applications. For the highest average power, fibre CPA has been developed. Fibre CPA is distinctly different from conventional CPA since self-focusing is not an issue as the beam is guided in the fibre. Fibre CPA is concerned with controlling the temporal phase through the amplification process to generate the shortest pulses. It is beyond this chapter to discuss the issues and technological solutions for fibre CPA, but they are well covered in a review by Fermann and Hartl [37]. The majority of CPA lasers are high repetition rate, solid-state systems with energy less than a milli-joule. It was a high-repetition rate CPA system that was first developed for a commercial application in surgical ophthalmology [38]. However, it is the push to go beyond 10 PW in peak power that is driving new CPA technologies. Although a number of these technologies have been discussed in this chapter, a more complete review of the world-wide effort in developing these powerful laser systems is covered by Colin and co-workers [6].

REFERENCES

1. Franken P A, Hill A E, Peters C W and Weinreich G 1960 Generation of optical harmonics *Phys. Rev. Lett.* **7** 118–119.

2. Fork R L, Greene B I, and Shank C V 1981 Generation of optical pulses with shorter than 0.1 ps by colliding pulse mode-locking *Appl. Phys. Lett.* **38** 671–672.
3. Holzrichter J F 1985 High-power solid-state lasers *Nature* **316** 309–314.
4. Cook C E 1960 Pulse compression – key to more efficient radar transmission *Proc. IRE* **48** 310–316.
5. Strickland D and Mourou G 1985 Compression of amplified chirped optical pulses *Opt. Commun.* **56** 219–221.
6. Danson C, Hillier D, Hopps N, and Neely D 2015 Petawatt class lasers worldwide *High Power Laser Sci Eng* **3** e3.
7. Kafka J D, Kolner B H, Baer T M, and Bloom D M 1984 Compression of pulses from a continuous wave mode-locked Nd:YAG laser *Opt. Lett.* **9** 505–506.
8. Tomlinson W J, Stolen R H, and Shank C V 1984 Compression of optical pulses chirped by self-phase modulation in fibers *J. Opt. Soc. Am. B.* **1** 139–140.
9. Treacy E B 1969 Optical pulse compression with diffraction gratings *IEEE J. Quantum Electron.* **5** 454–458.
10. Moulton P F 1986 Spectroscopic and laser characteristics of Ti-AL_2O_3 *J. Opt. Soc. Am. B* **3** 125–133.
11. French P M W, Williams J A R, and Taylor J R 1989 Femtosecond pulse generation from a titanium-doped sapphire laser using nonlinear external cavity feedback *Opt. Lett.* **14** 686–688.
12. Spence D E, Kean P N, and Sibbett W 1991 60-fsec pulse generation from a self-mode-locked Ti:sapphire laser *Opt. Lett.* **16** 42–44.
13. McMullen J D 1979 Analysis of compression of frequency chirped optical pulses by a strongly dispersive grating pair *Appl. Opt.* **18** 737–741.
14. Martinez O E 1987 3000 times grating compressor with positive group velocity dispersion: application to fiber compensation in 1.3-1.6 μm region *IEEE J. Quantum Electron* **23** 59–64.
15. Pessot M, Maine P and Mourou G.1987 1000 times expansion/compression of optical pulses for chirped pulse amplification *Opt. Commun.* **62** 419–421
16. Offner A 1973 Unit power imaging catoptric anastigmat U.S. Patent 3748015A.
17. Cheriaux G, Rousseau P, Salin F Changaret J P Walker B, and Dimauro L F 1996 Aberration-free stretcher design for ultrashort-pulse amplification *Opt. Lett.* **6** 414–416.
18. White W E, Patterson F G, Combs R L, Price D F, and Shepherd R L 1993 Compensation of higher-order frequency-dependent phase terms in chirped-pulse amplification systems *Opt. Lett.* **18** 1343–1346.
19. Lemoff B E and Barty C P J 1993 Quintic-phase-limited spatially uniform expansion and recompression of ultrashort optical pulses *Opt. Lett.* **18** 1651–1653.
20. Yakovlev I V 2014 Stretchers and compressors for ultrahigh power laser systems *Quantum Electron.* 44 393–414.
21. Tournois P 1993 New diffraction grating pair with very linear dispersion for laser-pulse compression *Electron. Lett.* **29** 1414–1415.
22. Kane S and Squier J 1997 Fourth-order-dispersion limitation of aberration-free chirped pulse amplification systems *J. Opt. Soc. Am. B* **5** 1237–1244.
23. Li S, Wang G, Liu Y, Xu Y, Li Y, Liu X, et al. 2017 High-order dispersion control of 10-peatwatt Ti:sapphire laser facility *Opt. Express* **25** 17488–17498.
24. Tournois P 1997 Acousto-optic programmable dispersive filter for adaptive compensation of group delay time dispersion in laser systems. *Opt. Commun.* **140** 245–249.
25. Ross I N, Trentelman M, and Danson C N 1997 Optimization of chirped-pulse amplification Nd:gass laser *Appl. Opt.* **36**, 9348–9358.
26. Dubeitis A, Jonasauskas G, and Piskarskas A 1992 Powerful femtosecond pulse generation by chirped and stretched pulse parametric amplification in BBO crystal *Opt. Commun.* **88** 437–440.
27. Perry M D, Pennington D, Stuart B C, Tietbohl, G, Britten J A, Brown C, et al. 1999 Petawatt laser pulses *Opt. Lett.* **24** 160–162.
28. Perry M D, Boyd R D, Britten J A, Decker D E, Shore B W, Shannon C, and Shults E 1995 High-efficiency multilayer dielectric diffraction gratings *Opt. Lett.* 20 940–942.
29. Britten J A, Perry M D, Shore B W, and Boyd R D 1996 Universal grating design for pulse stretching and compression in the 800–1100-nm range *Opt. Lett.* **21** 540–542.
30. Kessler T J, Bunkenburg J, Huang H, Kozlov A, and Meyerhofer D D 2004 Demonstration of coherent addition of multiple gratings for high-energy chirped-pulse-amplified lasers *Opt. Lett.* **29** 635–637.
31. Gaul E W, Martinez M, Blakeney J, Jochmann A, Ringuette M, Hammond D et al. 2010 Demonstration of a 1.1 petawatt laser based on a hybrid optical parametric chirped pulse amplification/mixed Nd:glass amplifier *Appl. Opt.* **49** 1676–1681.
32. Kalashnikov, M P, Osvay K, Lachko I M, Schonnagel H, and Sandner W 2005 Suppression of gain narrowing in multi-TW lasers with negatively and positively chirped pulse amplification *App. Phys. B*, **81**, 1059–1062.
33. Ross I N, Matousek P, Towrie M, Langley A J, and Collier J L 1997 The prospects for ultrashort pulse duration and ultrahigh intensity using optical parametric chirped pulse amplifiers *Opt. Commun.* **144**, 125–133.
34. Shirakawa A and Kobayashi T 1997 Noncollinearly phase-matched femtosecond optical parametric amplification with a 2000 cm^{-1} bandwidth *Appl. Phys. Lett.* **72** 147–149.
35. Yanovsky V, Chvykov V, Kalinchenko G, Rousseau P, Planchon T, Matsuoka T et al. 2008 Ultra-high intensity-300-TW laser at 0.1 Hz repetition rate *Opt. Express* **16**, 2109–2114.
36. Kalashnikov M P, Risse E, Schonnagel H, Husakou, A Herrmann J, and Sandner W 2004 Characterization of a nonlinear filter for the front-end of a high contrast double-CPA Ti:sapphire laser *Opt. Express* **12** 5088–5097.
37. Fermann M E and Hartl I 2013 Ultrafast fiber lasers *Nat. Photonics* **7** 868–874.
38. Juhasz T, Loesel F H, Kurtz R M, Horvath C, Bille J F, and Mourou G 1999 Corneal refractive surgery with femtosecond lasers *IEEE J. Quantum Electron.* **5** 902–910.

23
Optical Parametric Devices

M. Ebrahimzadeh

CONTENTS

- 23.1 Introduction ... 331
- 23.2 Non-linear Frequency Conversion ... 332
  - 23.2.1 Optical Parametric Generation ... 333
  - 23.2.2 Optical Parametric Gain ... 334
  - 23.2.3 Optical Parametric Amplification ... 335
- 23.3 Phase-matching ... 335
  - 23.3.1 Birefringent Phase-matching ... 337
  - 23.3.2 Quasi-phase-matching ... 340
- 23.4 Optical Parametric Devices ... 341
- 23.5 Optical Parametric Oscillators ... 342
  - 23.5.1 Continuous-wave OPOs ... 344
    - 23.5.1.1 Steady-state threshold ... 344
    - 23.5.1.2 Conversion Efficiency ... 345
  - 23.5.2 Pulsed OPOs ... 347
    - 23.5.2.1 Nanosecond OPOs ... 347
    - 23.5.2.2 Picosecond and Femtosecond OPOs ... 349
- 23.6 OPO Design Issues ... 349
  - 23.6.1 Non-linear Material ... 349
  - 23.6.2 Pump Laser ... 351
- 23.7 Continuous-wave OPO Devices ... 352
- 23.8 Nanosecond OPO Devices ... 353
- 23.9 Synchronously Pumped OPO Devices ... 355
  - 23.9.1 Picosecond OPO Devices ... 355
  - 23.9.2 Femtosecond OPO Devices ... 356
- 23.10 Summary ... 359
- References ... 359

23.1 Introduction

Since its invention in 1960 [1], the *laser* has become an indispensable tool that has transformed optical science and technology over the past four decades. The important characteristics of laser light with regard to intensity, spectral purity, spatial coherence and directionality have led to the establishment of a new branch of optical science and technology now referred to as *photonics*, and many new applications in previously unexplored areas of pure and applied science have become possible with the advent of the laser.

The laser, however, is not without its limitations. By virtue of its unique coherence properties, light emission from the laser can occur over only a confined band of frequencies determined by the energy gap in the gain material, and emission over extended spectral bands is generally difficult. It is possible to develop lasers with more extended spectral emission by using gain materials with broadened energy levels (or energy *bands*). In this case, transitions from the upper to the lower energy bands can occur over a correspondingly wide band of frequencies, resulting in laser emission over a broad spectral range. Examples of this type of *tunable* laser are the liquid dye or solid-state vibronic lasers such as the Ti:sapphire laser. However, even in such cases, the maximum spectral coverage available to most prominent tunable lasers (e.g. Ti:sapphire) is limited to at best 300–400 nm. Moreover, the restricted availability of suitable laser gain materials has confined the wavelength coverage of existing tunable lasers mainly to the visible and near-infrared spectrum, as shown in Figure 23.1. These limitations have left substantial portions of the optical spectrum, particularly in the infrared, inaccessible to lasers, and alternative methods for the generation of coherent light in these important regions have had to be devised. One of the most effective techniques to address the spectral limitations of conventional lasers is to exploit *non-linear optics*. In

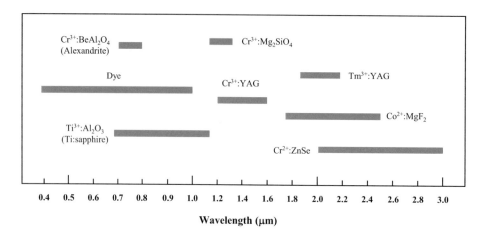

FIGURE 23.1 Spectral coverage of the most prominent tunable lasers, including the traditional dye laser and a number of solid-state vibronic lasers. The tuning range corresponding to the dye laser requires up to 20 different dyes. The vertical scale has no significance.

particular, *optical parametric generation, amplification* and *oscillation* in a non-linear material have long been recognized as versatile techniques for the generation of widely tunable radiation in spectral regions inaccessible to traditional lasers. The aim of this chapter is to provide an overview of optical parametric devices, including their basic operating principles, material and pump source requirements, and the latest developments in the field. While many of the fundamental principles are universal to parametric devices of all types, the main focus of the discussion will be on *optical parametric oscillators* (OPOs), which represent the majority of the systems developed to date.

23.2 Non-linear Frequency Conversion

The potential of non-linear optical techniques as a means of providing coherent light in new spectral regions was recognized soon after the invention of laser, and many of the early ideas were formulated more than forty years ago [2,3]. Non-linear optical phenomena are based on a fundamentally different principle from conventional lasers for the generation of coherent radiation. While the process of light generation in a laser is a direct consequence of the transitions (absorption, spontaneous and stimulated emission) between the energy levels in the gain material, non-linear processes rely on an alternative mechanism for light generation, namely, electric dipole oscillations in a dielectric medium.

Non-linear optical processes can readily be understood in terms of a simple classical picture [4]. When a dielectric material is subjected to an external optical field, the constituent dipoles within the material can be set into oscillations through interaction with the oscillating electric field of the incoming light wave. The dipole oscillations can result in the emission of new light waves with an intensity and frequency determined by the dipole moment and the dipole oscillation frequency. In the case when the input optical intensity is small, the dipole displacement follows a linear dependence on the input field strength. The result is dipole oscillation at the same frequency as that of the input optical wave. This situation corresponds to the regime of *linear optics*. On the other hand, when the input optical intensity is large, the dipole displacement from rest will become non-linear towards the higher field strengths, as shown in Figure 23.2. This is the regime of *non-linear optics*. The exact form of the non-linear response will depend on the structural symmetry of the dielectric material[1]. However, regardless of the crystallographic structure, the dipoles will oscillate not only at the input frequency but also over an infinite range of frequencies, above and below the input frequency.[2] The resulting light emission from the oscillating dipoles will therefore also be over an infinite band of frequencies. In the photon picture, this process is equivalent to the generation of large numbers of new photons with energies extending over an infinite range, below and above the input photon energy, but subject to the conservation of energy.

The potential of non-linear optical processes is thus immediately clear—they provide a mechanism for the generation of new frequencies (wavelengths) from an already existing input frequency. In other words, they provide a convenient technique for *frequency conversion* of light from an old to a new spectral range. This is one of the most fundamental principles of non-linear optics, first suggested more than four decades ago [2,3].

Non-linear optical processes can take a variety of forms. The most important in the context of frequency conversion are second harmonic generation (SHG), sum- and difference-frequency mixing (SFM and DFM) and, central to this discussion, optical parametric generation and amplification. The underlying physical principles responsible for all such processes are essentially the same and relate to the second-order susceptibility $\chi^{(2)}$ in a dielectric crystalline medium. In such a medium, the induced polarization, $\vec{P}$, due to an input optical field, $\vec{E}$, is in its most general form given by

---

[1] In a centrosymmetric material, only the odd harmonics of the input wave are generated through dipole oscillations, whereas in a non-centrosymmetric material all frequency harmonics, odd and even, are emitted.

[2] Mathematically, the range of frequencies emitted by the nonlinear oscillating dipoles can be viewed as the Fourier components of the input frequency. In a centrosymmetric material, only odd Fourier components of the input wave are generated, whereas in a non-centrosymmetric material all frequency components are available.

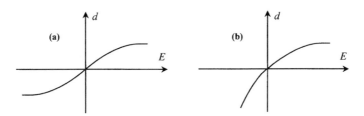

FIGURE 23.2 Dipole displacement from rest, d, as a function of the input electric field amplitude, E, for (a) centrosymmetric and (b) non-centrosymmetric dielectric materials. The regime of linear optics corresponds to the small and linear displacements of the dipole at low values of field. The regime of non-linear optics corresponds to the high values of field, where the dipole response is distorted.

$$\vec{P} = \varepsilon_0 [\chi^{(1)} \cdot \vec{E} + \chi^{(2)} \cdot \vec{E} \cdot \vec{E} + \chi^{(3)} \cdot \vec{E} \cdot \vec{E} \cdot \vec{E} + \ldots] \quad (23.1)$$

where $\chi^{(1)}$ is the linear susceptibility and $\chi^{(2)}$, $\chi^{(3)}$, ... are the non-linear susceptibilities of the medium. The linear susceptibility is related to the refractive index through $\chi^{(1)} = n^2 - 1$ and is responsible for the linear optical properties of the material such as refraction, dispersion, absorption and birefringence. On the other hand, the second-order susceptibility $\chi^{(2)}$ is the component that gives rise to the processes of SHG, SFM and DFM, linear electro-optic (Pockels) effect and, most importantly in the context of this discussion, optical parametric generation. The third-order non-linear susceptibility $\chi^{(3)}$ is responsible for the phenomena of third harmonic generation, optical bi-stability, phase conjugation and the optical Kerr effect including self-focusing and self-phase-modulation. The susceptibilities $\chi^{(1)}, \chi^{(2)}, \chi^{(3)}, \ldots$ are tensors of the second, third, fourth and higher rank, respectively. In writing equation (23.1), we have used a tensor notation, because most crystalline media that exhibit second-order non-linearity are optically anisotropic, so that $\vec{P}$ and $\vec{E}$ are generally not parallel in such media.

For the onset of non-linear optical processes, an essential prerequisite is clearly a large input optical intensity. In general, non-linear optical effects can be observed only when the input electric field approaches the electric field strength binding the material dipoles themselves (of the order of $\sim 10^8$–10^9 V cm^{-1}). The attainment of such high electric field strengths requires optical intensities as large as $\sim 10^{13}$ W cm^{-2}, which can be provided only by a laser. Moreover, the magnitude of the susceptibility tensors decreases rapidly with increasing rank of non-linearity, so that higher-order non-linear effects become increasingly difficult to induce and require higher input intensities. Thus, it is not surprising that the observation of non-linear optical effects only became possible after the invention of the laser.

In this discussion, we are concerned with the second-order non-linear susceptibility $\chi^{(2)}$ and focus on the implications of this property only in relation to the parametric process. More extensive treatments of other $\chi^{(2)}$ processes and higher-order non-linear effects can be found in numerous texts and review articles in the literature [5–8]. In its most general form, the $\chi^{(2)}$ tensor has 27 elements, but in practice, many of the components vanish under certain symmetry conditions, so the total number of independent components is generally far fewer. The tensor is non-zero only in non-centrosymmetric media that lack inversion symmetry in their crystalline structure. In centrosymmetric materials, $\chi^{(2)}$ and all other even-order susceptibilities reduce to zero, and so second-order non-linear frequency conversion processes, including parametric generation, are not attainable in such crystals. In the definition of the induced polarization through equation (23.1), the units of $\chi^{(2)}$ are m V^{-1} in MKS units, which is the system of units used throughout this chapter.

23.2.1 Optical Parametric Generation

As highlighted above, non-linear frequency conversion processes can take a variety of forms. For light emission over extended spectral regions, optical parametric generation [9–11] is the process of primary interest. It corresponds to the most fundamental non-linear effect where, in the classical picture, a single input *pump* frequency is converted to a broad range of lower optical frequencies in a new spectral range, as illustrated in Figure 23.3a. In other words, it provides the mechanism for tunable light generation in a new and longer wavelength range

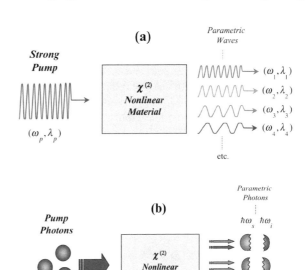

FIGURE 23.3 (a) Simple classical wave picture and (b) the photon picture of the optical parametric generation. In (a), strong dipole oscillations induced by an intense optical pump field result in the generation of new optical waves at lower frequencies. In (b), the pump photons are spontaneously broken up into a pair of constituent photons (signal and idler), whose energies add up to the energy of the pump photon, due to energy conservation.

using an existing laser pump source at a fixed wavelength. The process is thus highly effective for the generation of tunable coherent radiation in spectral regions where conventional lasers are not available. In the photon picture, Figure 23.3b, the process is equivalent to the spontaneous break-up of each high-energy pump photon into two constituent parts of lower energy (termed *signal* and *idler*), subject to the conservation of energy. Since energy must be conserved, the sum of energies of the constituent photons has to equal the pump photon energy. Given the large number of pump photons, there are statistically an infinite number of ways in which the break-up can occur, so that an infinite range of signal and idler energies (frequencies) will be emitted during the process. However, of the infinite number of generated signal and idler photons, only a single pair determined by the *phase-matching* condition and satisfying the energy conservation condition will be emitted in practice. The concept of phase-matching is discussed in more detail in Section 23.3.

Analytically, the parametric generation process can therefore be treated by considering the interaction of three optical fields at frequencies ω_3, ω_2 and ω_1, such that $\omega_3 = \omega_2 + \omega_1$. The field at ω_3 corresponds to the intense input optical pump field, giving rise to a pair of signal and idler fields at ω_2 and ω_1, respectively. The generated field at the higher frequency, ω_2, say, is usually designated as the signal, while the field at the lower frequency, ω_1, is designated as the idler, but variations in this nomenclature are also frequently used in the literature. Although more rigorous treatments of the parametric generation process can be found elsewhere [3], here we restrict discussion to the essential features of the treatment and focus on the main results of the analysis.

At a fundamental level, the parametric process can be described by considering Maxwell's wave equation for the propagation of three optical fields, the pump, signal and idler, at respective frequencies ω_3, ω_2 and ω_1, in a non-centrosymmetric dielectric exhibiting second-order susceptibility, $\chi^{(2)}$. The propagation of the three optical fields in such a medium involves the solution of Maxwell's non-linear wave equation with the second-order non-linear polarization as the source term, namely

$$\frac{\partial^2 E}{\partial z^2} = \mu\varepsilon\frac{\partial^2 E}{\partial t^2} + \mu\frac{\partial^2 P^{(2)}}{\partial t^2} \quad (23.2)$$

where $P^{(2)} = \varepsilon_0\chi^{(2)}E^2$ is the second-order polarization and E is the electric field of the propagating wave, given by

$$E = \tfrac{1}{2}[E(z)e^{i(kz-\omega t)} + \text{cc}] \quad (23.3)$$

with $E(z)$ representing the complex field amplitude. In writing equation (23.2), we have ignored non-linear polarizations terms higher than second order. We have also reverted to a scalar notation for convenience and taken the propagation to be along the z-axis. In addition, we have assumed that the medium is lossless, non-conducting and non-magnetic, as is generally the case in practice. Since the parametric process involves the interaction of three optical fields, the total field E will comprise three harmonic waves representing the pump, signal and idler, so that

$$E = E_1(\omega_1) + E_2(\omega_2) + E_3(\omega_3) \quad (23.4)$$

To simplify the analysis further, we neglect the effects of focusing and assume that the optical fields are infinite uniform plane waves. We also assume that the fields are monochromatic and neglect any effects due to double refraction.

The parametric interaction process can then be understood by seeking solution to the non-linear wave equation (23.2). This is done by separating equation (23.2) into three components at the three different frequencies ω_1, ω_2 and ω_3, each of which must separately satisfy the wave equation. Then, by considering the three separate wave equations at each frequency and assuming that the field amplitudes vary only slowly over distances compared with a wavelength, after some manipulation, we obtain the variations of three field amplitudes with propagation as

$$\frac{\partial E_1(z)}{\partial z} = i\kappa_1 E_3(z)E_2^*(z)e^{i\Delta kz} \quad (23.5a)$$

$$\frac{\partial E_2(z)}{\partial z} = i\kappa_2 E_3(z)E_1^*(z)e^{i\Delta kz} \quad (23.5b)$$

$$\frac{\partial E_3(z)}{\partial z} = i\kappa_3 E_1(z)E_2(z)e^{-i\Delta kz} \quad (23.5c)$$

where $\kappa_j = (\omega_j d_{\text{eff}}/n_j c)$, with $j = 1, 2, 3$, n is the refractive index, $\Delta k = k_3 - k_2 - k_1$ is the *phase-mismatch* parameter and $d_{\text{eff}} = \chi_{\text{eff}}^{(2)}/2$ is the *effective non-linear coefficient* representing the appropriate combination of the non-linear tensor elements taking part in the parametric process. These are the *coupled-wave equations* governing the parametric interaction of the pump, signal and idler ($\omega_3 \rightarrow \omega_2 + \omega_1$) in a dielectric medium exhibiting second-order non-linear susceptibility.

The coupled-wave equations are the starting point in the analysis of a wide range of non-linear optical effects and apply universally to any three-wave mixing process involving the second-order susceptibility, where the frequencies of the fields satisfy $\omega_3 = \omega_2 + \omega_1$. Similar equations to (23.5a)–(23.5c) apply to other mixing processes including SFM ($\omega_1 + \omega_2 \rightarrow \omega_3$), DFM ($\omega_3 - \omega_2 \rightarrow \omega_1$) and SHG ($\omega + \omega \rightarrow 2\omega$) [8]. We notice from equations (23.5a) to (23.5c) that the amplitudes of the three optical fields are coupled to one another through d_{eff}. Physically, this coupling provides the mechanism for the exchange of energy among the interacting fields as they propagate through the non-linear material. The direction of energy flow in a given three-wave mixing process depends on the relative phase and the intensity of the input fields.

23.2.2 Optical Parametric Gain

In practice, the parametric process is initiated by a single intense pump field at frequency ω_3 at the input to a non-linear crystal. This field, which is provided by a laser, in turn mixes, through the non-linear susceptibility, with a signal field at ω_2 to give rise to an idler field at $\omega_1 = \omega_3 - \omega_2$. The idler field so generated in turn mixes back with the pump to produce additional signal, and the re-generated signal re-mixes with the pump to produce more idler. Under suitable phase-matching conditions (see Section 23.3), the process can continue in this way until power is gradually transferred from the strong pump to the initially weak signal and idler fields. The generated signal and idler fields can therefore grow to macroscopic levels by draining power from the input pump field as they propagate through the non-linear crystal.

In the absence of a coherent source of signal and idler beams at the input, the initial supply of photons at ω_1 and ω_2 for mixing with the input pump is provided by the spontaneous break-up of pump photons through *spontaneous parametric fluorescence*. This process, also referred to as *parametric noise* or *parametric luminescence*, may be viewed to arise from the mixing of the zero-point flux of the electromagnetic field at the signal and idler frequency, quantized within the volume of the crystal, with the incoming pump photons, through the non-linear polarization [12,13]. The effective zero-point flux at the signal and idler is obtained by allowing one half-photon of energy at both ω_2 and ω_1 or one photon of energy at either frequency to be present in each black-body mode of the quantizing volume.

We can obtain the gain and amplification factor for the growth of the signal and idler fields in the parametric process from the solution of the coupled wave equations (23.5a)–(23.5c). The general solution is beyond the scope of the present discussion and can be found elsewhere [3]. However, if it is assumed that the input pump field does not undergo strong depletion with propagation through the medium, then $\partial E_3/\partial z = 0$ in equation (23.5c). The coupled-wave equations are reduced to two, with E_3 independent of z in both equations (23.5a) and (23.5b). Subject to the initial condition of no input idler field, $E_1(z=0) = 0$, and finite input signal, $E_2(z=0) \neq 0$, the fractional gain in signal intensity with propagation through the non-linear crystal is obtained as

$$G_2(\ell) = \frac{I_2(z=\ell)}{I_2(z=0)} = 1 + \Gamma^2 \ell^2 \frac{\sinh^2[\Gamma^2\ell^2 - (\Delta k \ell/2)^2]^{1/2}}{[\Gamma^2\ell^2 - (\Delta k\ell/2)^2]} \quad (23.6)$$

where ℓ is the interaction length, $I = nc\varepsilon_0 EE^*/2$ is the intensity or flux (in W m^{-2}) and Γ is the *gain factor* defined as

$$\Gamma^2 = \frac{8\pi^2 d_{\mathrm{eff}}^2}{c\varepsilon_0 n_1 n_2 n_3 \lambda_1 \lambda_2} I_3(z=0). \quad (23.7)$$

Here, n and λ are the refractive index and wavelength of the respective waves, $I_3(z=0)$ is the input pump intensity and d_{eff} has units of m V^{-1}. From the same analysis, an expression similar to equation (23.6) may be derived for the growth of the idler field from its initial zero value at the input to the non-linear crystal [7]. The case of non-zero input idler as well as signal can also be treated similarly [14], and results analogous to equation (23.6) can be derived for the amplification of the generated fields. It is sometimes useful to express the gain factor in the form

$$\Gamma^2 = \frac{8\pi^2 d_{\mathrm{eff}}^2}{c\varepsilon_0 n_0^2 n_3 \lambda_0^2}(1-\delta^2) I_3(z=0) \quad (23.8)$$

where δ is the *degeneracy factor* defined as

$$1+\delta = \frac{\lambda_0}{\lambda_1} \quad 1-\delta = \frac{\lambda_0}{\lambda_1} \quad 0 \leq \delta \leq 1 \quad (23.9)$$

where $\lambda_0 (=2\lambda_3)$ is the degenerate wavelength and n_0 is the refractive index at degeneracy, with $n_0 \sim n_1 \sim n_2$. The factor δ is a measure of how close the signal and wavelengths are to degeneracy. It is clear from equation (23.8) that parametric gain has a maximum value at degeneracy, where $\delta \sim 0$, and decreases for operation away from degeneracy as $\delta \to 1$.

23.2.3 Optical Parametric Amplification

It can be seen from equations (23.6) to (23.7) that the magnitude of non-linear gain in the parametric process depends on material parameters such as the refractive index, interaction length and non-linear coefficient, the signal and idler wavelengths as well as the input pump intensity. It is also seen that amplification is strongly dependent on the phase-mismatch parameter, Δk. For $\Delta k = 0$, the generated fields experience maximum gain, whereas the growth of the parametric waves is severely hampered by an increase in the magnitude of Δk. In practice, therefore, it is imperative to ensure maximum gain and amplification for the parametric waves by setting $\Delta k = 0$. This is achieved through the technique of phase-matching, as discussed in more detail in Section 23.3. Under this condition, $\Delta k \sim 0$ and equation (23.6) is reduced to

$$G_2(\ell) = \cosh^2(\Gamma \ell) \quad (23.10)$$

Parametric devices often operate under different gain conditions, depending on the magnitude of the gain factor, Γ. In the low-gain regime, corresponding to $\Gamma\ell \lesssim 1$, equation (23.6) can be simplified to

$$G_2(\ell) \approx 1 + \Gamma^2 \ell^2 \quad (23.11)$$

On the other hand, in the high-gain regime, corresponding to $\Gamma\ell \gg 1$, equation (23.6) can be approximated by

$$G_2(\ell) \approx \tfrac{1}{4} e^{2\Gamma\ell} \quad (23.12)$$

It can be seen from equations (23.11) to (23.12) that, under the phase-match condition, the single-pass intensity gain has a quadratic dependence on $\Gamma\ell$ in the low-gain limit, whereas it increases exponentially with $2\Gamma\ell$ in the high-gain limit. Experimentally, the low-gain limit corresponds to parametric generation when using continuous-wave (cw) or low- to moderate-peak-power pulsed pump sources. This regime is pertinent to most parametric devices. On the other hand, the high-gain limit corresponds to pumping with high-intensity, pulsed and amplified laser pump sources.

23.3 Phase-matching

From the preceding discussion, it is clear that in the presence of a strong input pump, the signal (and idler) field in the parametric process can experience growth with propagation through the crystal, provided the phase-matching condition, $\Delta k = 0$, is satisfied. However, because of normal dispersion, optical fields at different frequencies cannot generally travel with the same phase velocity and maintain synchronism in propagating through the material due to different refractive indices. The degree of phase velocity synchronism among the pump, signal and idler is determined by the phase-mismatch parameter $\Delta k = k_3 - k_2 - k_1$. Under normal conditions, $\Delta k \neq 0$, so that the optical waves at different frequencies slip out of phase after travelling a short distance through the medium. This distance is known as the *coherence length*, given by $\ell_c = \pi / |\Delta k|$, and is illustrated in Figure 23.4. With propagation beyond a coherence length, the intensities of the generated waves begin to fall back to zero, and with further propagation, the waves step in and out of phase periodically, with a period determined by the coherence length. As a result, the optical waves cannot maintain any phase unison beyond a coherence length. The pump and the parametric fields exchange energy back and forth, and no net transfer of energy from the pump to the generated waves can occur in travelling through the medium. The coherence length is thus a measure of the maximum interaction length over which amplification of the parametric fields can be sustained in the presence of dispersion. It is the distance after which the relative phase of the pump, signal and idler waves slips by π, and the generated intensities begin to fall back to zero. Typical values of ℓ are a few to tens of microns in most non-linear materials. Given the direct dependence of $G_2(\ell)$ on the interaction length, ℓ, using equation (23.6), it is clear that over negligible distances corresponding to a coherence length, only trivial gains are available.

The general functional dependence of $G_2(\ell)$ on Δk may be derived from equation (23.6) and can be found in a number of reference texts [7,8,14,15]. In most situations of practical interest, however, we are concerned with small gains, for which $\Gamma^2 \ell^2 \ll (\Delta k \ell / 2)^2$. Under this condition, the fractional single-pass signal gain, given by equation (23.6), is modified to

$$G_2(\ell) \approx 1 + \Gamma^2 \ell^2 \left[\frac{\sin(\Delta k \ell / 2)}{(\Delta k \ell / 2)} \right]^2 \quad (23.13)$$

which has a sinc² dependence on the $\Delta k \ell / 2$, as shown in Figure 23.5. As expected, the gain has a maximum value at $\Delta k = 0$ and is reduced with increasing Δk, dropping to zero when $\Delta k = 2\pi / \ell$.

Clearly, for any meaningful growth of the parametric waves, phase velocity synchronism must be maintained over interaction lengths significantly longer than the coherence length. In fact, in order to exploit the full length of the non-linear crystal, the necessary length for velocity synchronism must be comparable or longer than the crystal length. This requirement can be met by exploiting the technique of phase-matching, which permits the condition $\Delta k = 0$ to be satisfied, hence allowing the coherence length to become infinite, and phase velocity synchronism between the interacting fields to be maintained indefinitely. Under this condition, the signal and idler fields undergo amplification over the full interaction length of crystal by continually draining power from the pump, as shown in Figure 23.6, and coherent macroscopic output can be extracted at ω_2 and ω_1.

The phase-match condition, $\Delta k = 0$, may also be usefully expressed as $n_3 \omega_3 - n_2 \omega_2 - n_2 \omega_2 = 0$, which highlights one of the most attractive features of the parametric process—its

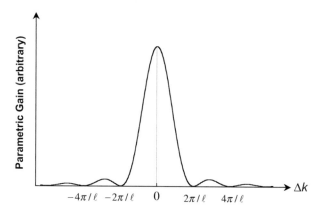

FIGURE 23.5 Dependence of parametric gain on the phase-mismatch parameter $\Delta k \ell / 2$ in the small gain limit, $\Gamma^2 \ell^2 \ll (\Delta k \ell / 2)^2$.

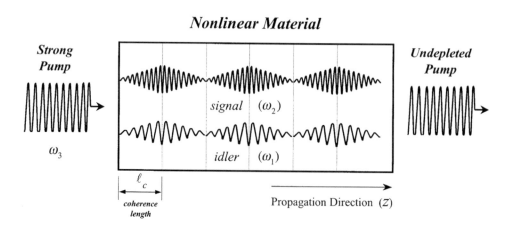

FIGURE 23.4 Wave velocity mismatch and coherence length. The signal and idler wave amplitudes undergo oscillations in propagating through the material, without experiencing any growth. The pump amplitude is unaffected by the generation and propagation of the parametric waves. The amplitudes of the generated waves are grossly exaggerated relative to that of the pump.

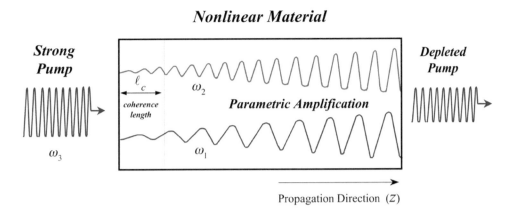

FIGURE 23.6 Optical parametric amplification. Phase-matching provides the velocity synchronism for a particular pair of signal and idler waves, out of the infinite number of waves generated, resulting in amplification. Power transfer occurs from the pump to the parametric waves, resulting in a monotonic increase in the signal and idler intensity and a drop in the transmitted pump intensity with propagation through the material. The amplitudes of the generated waves are grossly exaggerated relative to that of the pump.

tunability. If for a given pump frequency, ω_3, the refractive indices n_3, n_2 and n_1 are made to vary through, for example, the change in crystal angle or temperature, the phase-match condition will be satisfied for a new pair of signal and idler frequencies, ω_2 and ω_1. As a consequence, the generated frequencies will be 'forced' to shift to new values for which phase-matched amplification is available. In addition to *angle-* and *temperature-tuning*, there are also other tuning methods including *pump-tuning* and, in the case of *quasi-phase-matched* interactions, *grating-tuning* (see Section 23.3.2).

23.3.1 Birefringent Phase-matching

The traditional method for achieving phase-matching is to use the birefringence of optically anisotropic media to compensate for dispersion [16,17]. Most non-centrosymmetric crystals in which second-order non-linear processes are attainable are optically anisotropic and exhibit the phenomenon of birefringence. In such media, the index of refraction (and hence the phase velocity) for a wave at a given frequency depends on its state of polarization as well as the direction of propagation. It can be shown that for an arbitrary propagation direction in a birefringent crystal, two orthogonal linear polarization states are permitted [18]. This means that an optical wave at a given frequency can exhibit two different phase velocities depending on its state of polarization. Alternatively, two optical waves at different frequencies may propagate with the same phase velocity in a birefringent crystal by suitable choice of orthogonal polarization vectors and propagation direction. It is this important property that can be used to obtain phase velocity synchronism between optical waves at different frequencies in non-linear processes, including parametric generation. The technique is known as *birefringent phase-matching* (BPM).

Detailed discussion of crystal optics and the general procedures for determining the allowed polarization states and phase-matching directions in BPM are not within the scope of this treatment and can be found in other reference texts [6,8,19]. Here, we focus on the essential elements of the approach and provide a brief summary of procedures involved for the attainment of BPM.

The orientation and phase velocities of the two allowed linear polarization states in a birefringent crystal can be obtained from the so-called *normal index surface*, which uniquely describes the optical properties of the particular crystal. This surface is defined in such a way that the index of refraction (or the phase velocity) of a wave propagating in a given direction is equal to the distance between the intersection of the wave normal with the surface in that direction and the origin. The principal axes of the surface are defined by the principal refractive indices n_x, n_y and n_z, which lie along the principal dielectric axes of the crystal, ε_x, ε_y and ε_z. Because of dispersion, the length of the principal axes of the normal index surface, and thus its shape, varies with the frequency of the optical radiation.

With regard to their optical properties, birefringent crystals may be classified into two distinct categories: optically *uniaxial* crystals in which $n_x = n_y \neq n_z$ and optically *biaxial* crystals in which $n_x \neq n_y \neq n_z$. In uniaxial crystals, one may find a unique direction along which the two allowed orthogonal polarizations have the same refractive index (i.e. travel with the same phase velocity). In biaxial crystals, two such directions may be identified. These directions define the *optic axes* of the crystal. In uniaxial materials, the optic axis coincides with the z-axis of the index surface, while in biaxial materials, the two optic axes generally lie symmetrically about the z-axis, in the xz-plane. In both crystal classes, it is in principle possible, by an appropriate choice of polarizations and propagation direction, for the optical waves at different frequencies to achieve phase-matching with $\Delta k = 0$.

In uniaxial crystals, of the two allowed polarization waves, one experiences the same refractive index regardless of the direction of propagation in the medium. This wave is known as the *ordinary wave* (or o-wave). If θ is the angle between the wave normal direction and the z-axis of the normal index surface (Figure 23.7), then the refractive index of the o-wave is a constant for all θ, given by $n_o(\theta) = n_o$. The wave with orthogonal polarization to the o-wave, however, experiences a refractive index that varies with the propagation direction in the crystal. This wave is referred to as the *extraordinary wave* (or e-wave). The refractive index of the e-wave, $n_e(\theta)$, varies from

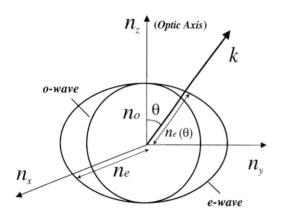

FIGURE 23.7 Normal index surfaces, representing the refractive indices for the o-wave and the e-wave in a uniaxial crystal. The surfaces correspond to a positive uniaxial crystal.

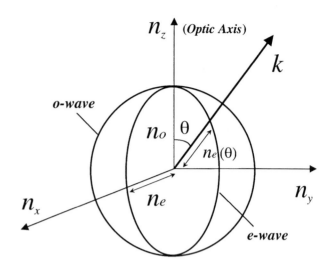

FIGURE 23.8 Normal index surfaces for a negative uniaxial crystal.

n_o to n_e as θ varies from 0° to 90°. The refractive indices n_o and n_e, therefore, represent the principal axes of the normal index surfaces for the uniaxial crystal. The normal index surface for the o-wave is a sphere, whereas that for the e-wave is an ellipsoid of revolution about the z-axis, as shown in Figure 23.7.

We may distinguish between two classes of uniaxial crystals: *positive* uniaxial for which $n_e > n_o$ and *negative* uniaxial for which $n_e < n_o$. The normal index surfaces for a positive crystal are those shown in Figure 23.7, and the surfaces for a negative crystal are shown in Figure 23.8. In either case, it is often possible to find a direction along which suitably polarized pump and parametric waves have refractive indices (and phase velocities) that satisfy the phase-match condition $\Delta k = k_3 - k_2 - k_1 = 0$. In a positive crystal, two phase-matching schemes are possible, both involving the propagation of the pump as an o-wave, with either the signal (or idler) or both as e-waves. In a negative crystal, the pump is polarized as an e-wave, with either the signal (or idler) or both polarized as o-waves. The case where the signal and idler have the same polarization is referred to as *type I* (or *parallel*) phase-matching. When the signal and idler have orthogonal polarizations, it is known as *type II* (or *orthogonal*) phase-matching. Table 23.1 summarizes the possible phase-matching schemes for parametric generation in uniaxial crystals.

We may also identify two classes of biaxial crystal depending on whether $n_x < n_y < n_z$ or $n_x > n_y > n_z$. In both cases, the normal index surface for the allowed polarizations does not follow a simple spherical or spheroidal symmetry as in a uniaxial crystal, but has a more complex bilayer structure with four points of interlayer contact through which the two optic axes pass [6]. The determination of the phase-match condition and allowed polarizations for an arbitrary propagation direction (θ, ϕ) is thus considerably more complex, since the refractive indices of not only one but both allowed polarizations in general vary with propagation direction. Therefore, one cannot strictly consider the allowed polarizations as an o-wave or an e-wave, as in uniaxial crystals. The situation may, however, be greatly simplified if propagation is confined to one of the principal planes xy, xz or yz. In this case, the refractive index of one of the allowed waves becomes independent of the direction of propagation, so that one may think of this wave as an o-wave in analogy with the uniaxial case. The wave with orthogonal polarization to the o-wave has a refractive index that varies with the propagation direction in the phase-match plane and may thus be treated as an e-wave. The situation thus simplifies to that in a uniaxial crystal.

As an example, if we consider propagation in the xy-plane ($\theta = 90°$), for a wave polarized along the z-axis, the refractive index does not vary with propagation angle ϕ. This wave can, therefore, be considered as an o-wave with a refractive index $n_1(\phi) = n_z = n_o$. However, the refractive index of a wave with an orthogonal polarization vector in the xy-plane varies with the phase-match angle ϕ. This wave can then be treated as an e-wave with a refractive index $n_2(\phi) = n_{xy} = n_e(\phi)$, say, where n_{xy} varies from n_y to n_x as ϕ varies from 0° to 90°. It thus becomes possible to obtain phase-matching in

TABLE 23.1

Possible BPM Schemes and Corresponding Field Polarizations for Parametric Generation ($\omega_3 \rightarrow \omega_2 + \omega_1$) in Uniaxial Crystals. E Represents the Electric Field Vector of the Optical Waves.

Crystal Class	$\omega_3 \rightarrow \omega_2 + \omega_1$			
	Positive Uniaxial		**Negative Uniaxial**	
Phase-matching scheme	Type I	Type II	Type I	Type II
Field polarizations	$E_o \rightarrow E_e + E_e$	$E_o \rightarrow E_o + E_e$	$E_e \rightarrow E_o + E_o$	$E_e \rightarrow E_o + E_e$
		$E_o \rightarrow E_e + E_o$		$E_e \rightarrow E_e + E_o$

The subscripts o and e refer to *ordinary* and *extraordinary* polarization, respectively.

Optical Parametric Devices

the *xy*-plane in the same way as in a uniaxial crystal using similar type I and type II schemes. Phase-matching in the other principal planes *xz* and *yz* can also be treated in a similar way. Whether the phase-matching scheme in a particular plane corresponds to a positive or negative uniaxial crystal depends on relative indices of the biaxial crystal. The possible phase-matching schemes for parametric generation in the principal planes of the two classes of biaxial crystals with $n_x<n_y<n_z$ and $n_x>n_y>n_z$ are summarized in Tables 23.2a and 23.2b, respectively. More general treatments of BPM and the calculation of phase-matching directions in biaxial crystals can be found elsewhere [6,20].

It is therefore clear that by the judicial choice of wave polarizations and propagation direction in a particular crystal, it is

TABLE 23.2

(*a*) Possible BPM Schemes and Corresponding Field Polarizations for Parametric Generation ($\omega_3 \to \omega_2+\omega_1$) in the Three Principal Planes of Biaxial Crystals with $n_x<n_y<n_z$.

Crystal Class	$\omega_3 \to \omega_2+\omega_1$ Biaxial ($n_x<n_y<n_z$)			
	xy-plane ($\theta=90; 0<\phi<90$) (Negative uniaxial)		*yz*-plane ($\phi=90; 0<\theta<90$) (Positive uniaxial)	
Phase-matching scheme	Type I	Type II	Type I	Type II
Field polarizations	$E_{xy} \to E_z+E_z$	$E_{xy} \to E_z+E_{xy}$ $E_{xy} \to E_{xy}+E_z$	$E_x \to E_{yz}+E_{yz}$	$E_x \to E_x+E_{yz}$ $E_x \to E_{yz}+E_x$
	xz-plane ($\phi=0; 0<\theta<\theta_0$) (Negative uniaxial)		($\phi=0; \theta_0<\theta<90$) (Positive uniaxial)	
Phase-matching scheme	Type I	Type II	Type I	Type II
Field polarizations	$E_{xz} \to E_y+E_y$	$E_{xz} \to E_y+E_{xz}$ $E_{xz} \to E_{xz}+E_y$	$E_y \to E_{xz}+E_{xz}$	$E_y \to E_y+E_{xz}$ $E_y \to E_{xz}+E_y$

The angles θ and ϕ are the polar and azimuthal angles, measured from the principal optical axes *z* and *x*, respectively. The angle θ_0 is the angle between the optic axis and the *z*-axis, measured in the *xz*-plane [6]. *E* represents the electric field vector of the optical waves. The subscripts represent the polarization of the electric vector, with *x*, *y*, *z* representing polarization along the respective axes and *xy*, *xz* and *yz* designating polarization in the respective planes.

TABLE 23.2

(*b*) Possible BPM Schemes and Corresponding Field Polarizations for Parametric Generation ($\omega_3 \to \omega_2+\omega_1$) in the Three Principal Planes of Biaxial Crystals with $n_x>n_y>n_z$.

Crystal Class	$\omega_3 \to \omega_2+\omega_1$ Biaxial ($n_x<n_y<n_z$)			
	xy-plane ($\theta=90; 0<\phi<90$) (Positive uniaxial)		*yz*-plane ($\phi=90; 0<\theta<90$) (Negative uniaxial)	
Phase-matching scheme	Type I	Type II	Type I	Type II
Field polarizations	$E_z \to E_{xy}+E_{xy}$	$E_z \to E_z+E_{xy}$ $E_z \to E_{xy}+E_z$	$E_{yz} \to E_x+E_x$	$E_{yz} \to E_x+E_{yz}$ $E_{yz} \to E_{yz}+E_x$
	xz-plane ($\phi=0; 0<\theta<\theta_0$) (Positive uniaxial)		($\phi=0; \theta_0<\theta<90$) (Negative uniaxial)	
Phase-matching scheme	Type I	Type II	Type I	Type II
Field polarizations	$E_y \to E_{xz}+E_{xz}$	$E_y \to E_y+E_{xz}$ $E_y \to E_{xz}+E_y$	$E_{xz} \to E_y+E_y$	$E_{xz} \to E_y+E_{xz}$ $E_{xz} \to E_{xz}+E_y$

The angles θ and ϕ are the polar and azimuthal angles, measured from the principal optical axes *z* and *x*, respectively. The angle θ_0 is the angle between the optic axis and the *z*-axis, measured in the *xz*-plane [6]. *E* represents the electric field vector of the optical waves. The subscripts represent the polarization of the electric vector, with *x*, *y* and *z* representing polarization along the respective axes and *xy*, *xz* and *yz* designating polarization in the respective planes.

possible to obtain phase-velocity synchronism in the parametric generation process using the BPM and to achieve amplification of the generated waves.

23.3.2 Quasi-phase-matching

An alternative technique for the attainment of phase velocity synchronism in non-linear processes is *quasi-phase-matching* (QPM) [21,22]. The technique was first proposed as early as four decades ago when the original ideas on non-linear frequency conversion were first formulated [3]. However, its practical implementation did not become possible until the development of reliable fabrication methods in ferroelectric materials [22].

In the QPM process, the orientation of the electric dipoles in the non-linear material is periodically reversed by 180° along the pump propagation direction (Figure 23.9). A practical method to achieve such domain reversal is through *periodic poling* of ferroelectric materials during the fabrication process by applying a high electrical field (several kV) across the crystal using patterned electrodes. The result of periodic poling is that the radiated optical waves from the oscillating dipoles in consecutive domains become out of phase by π in both time and space. The domain reversal is equivalent to a periodic change in the sign of the effective non-linear coefficient between $+d_{eff}$ and $-d_{eff}$. If the poling period, Λ, is made to correspond to two coherence lengths for the non-linear interaction ($\Lambda = 2\ell_c$), then the phase of the generated waves is periodically reversed by π for every two coherence lengths. This periodic re-adjustment of phase preserves a constructive relative phase between the pump and parametric waves (albeit in a quasi-continuous manner) as they propagate through the material. This prevents the waves from slipping out of phase after only one coherence length, as would be the case under normal dispersion.

The end result of periodic poling is that the generated waves can undergo quasi-continuous growth as they propagate with the pump through the material, as illustrated in Figure 23.10. A qualitative comparison of the generated intensity between QPM and BPM, as well as the non-phase-matched process with propagation through the crystal, is shown in Figure 23.11. It can be seen that, in the QPM process, the growth of optical waves is not monotonic, as it is in BPM. However, because QPM does not rely on a prescribed combination of wave polarizations for phase-matching, it can provide access to the highest non-linear tensor coefficients in the material, which are not generally available under BPM. This flexibility more than compensates for the quasi-monotonic amplification of the generated waves, so that the QPM process can ultimately provide substantially higher gains for the same crystal length and operating conditions than BPM. In addition, the direction of propagation in QPM can also be freely chosen to provide optimized phase-matching geometries such as *non-critical phase-matching* (NCPM), where the effects of *spatial walk-off* are minimized (see Section 23.6.1). This is again not possible in BPM, where the choice of propagation direction in the crystal is dictated by the birefringence properties of the material and the type of phase-matching permitted. The NCPM capability is particularly important in the presence of low parametric gains, since it allows for the use of long crystal interaction lengths, to maximize device efficiency. Another important merit of QPM is that it can be implemented in optically isotropic as well as anisotropic materials, as it does not rely on birefringence for phase-matching. With regard to tuning capability, QPM also has an over-riding advantage over BPM in that it can be freely engineered to provide maximum spectral coverage for the generated waves. By fabricating the correct poling (or *grating*) period, it is in principle possible to generate any desired wavelengths within the entire transparency range of the material.

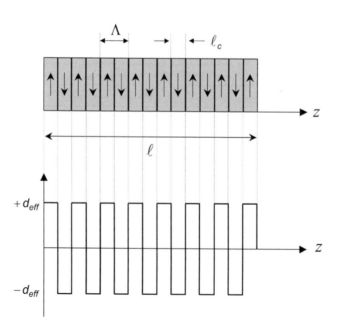

FIGURE 23.9 QPM through periodic poling. The orientation of the electric dipoles is periodically flipped through the material, resulting in a corresponding modulation in the effective non-linear coefficient.

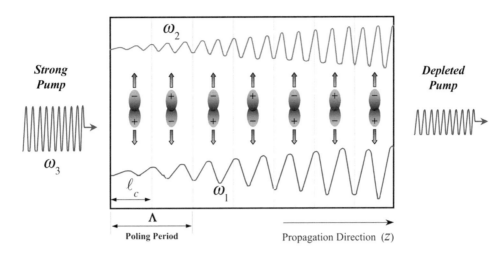

FIGURE 23.10 Optical parametric amplification under QPM. The generated waves undergo phase re-adjustment every coherence length, preserving a constructive relative phase, and thus experiencing quasi-continuous growth with propagation through the material. Power transfer occurs from the pump to the parametric waves, resulting in the depletion of the transmitted pump. The amplitudes of the generated waves are grossly exaggerated relative to that of the pump.

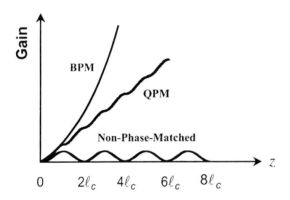

FIGURE 23.11 Qualitative comparison of gain under BPM, QPM, and non-phase-matched interaction with propagation through the non-linear crystal. The plot assumes the same effective non-linear coefficient for QPM as for BPM. The vertical scale, hence the relative gain, is arbitrary. The intensity of the non-phase-matched interaction is grossly exaggerated relative to that of BPM and QPM.

The phase-matching condition under QPM can be expressed in a similar way to BPM but is modified by the periodic phase re-adjustments introduced by the poling process. As a result, the phase-mismatch parameter under QPM is given by $\Delta k_Q = k_3 - k_2 - k_1 - k_Q$, where $k_Q = 2\pi/\Lambda$ is the so-called *grating vector* corresponding to the periodic grating. The effective non-linear coefficient in QPM is also similarly modified to take account of the periodic phase reversal of the optical waves. The form of the coefficient depends on the order of the QPM process and the poling cycle [21,22] and is given by $d_Q = G_m d_{\text{eff}}$, where d_{eff} would be the effective coefficient for the same wave polarizations and propagation direction in the absence of the periodic grating. The factor G_m is given by

$$G_m = \frac{2}{m\pi}\sin(m\pi D) \tag{23.14}$$

where m is the order of QPM and $D = l/\Lambda$ is the poling duty factor, with l being the length of the reversed domain. The QPM order, m, represents the periodicity of the poling in terms of coherence length. For an mth-order process, the poling period is given by $\Lambda = 2ml_c$, so that for a first-order grating ($m = 1$), the domains are reversed every coherence length, for a second-order grating ($m = 2$) every two coherence lengths, and so on. The existing analyses of parametric amplification and gain used in Section 23.2 equally apply to QPM processes. For a given QPM order, grating period and poling cycle, the appropriate expressions for Δk_Q and d_Q may be substituted into equations (23.7) and (23.6) to obtain the gain factor and growth in signal intensity.

For a first-order QPM process ($\Lambda = 2l_c$) and a 50% poling duty cycle, the effective non-linear coefficient is maximized to $d_Q = (2/\pi)d_{\text{eff}}$. This is therefore the configuration of most practical interest, as it minimizes device threshold and maximizes output power and efficiency. Higher-order processes and poling duty factors other than 50% are of interest for novel applications including spectral and spatial tailoring, temporal pulse shaping, expansion and compression, tailoring of phase-matching line-shape and bandwidth modification in frequency conversion processes [21–25]. The unique flexibility and important advantages of QPM have now established this technique as a highly effective approach to the development of a new class of frequency conversion sources, particularly parametric devices [26–28], with exceptional versatility and unprecedented performance capabilities with regard to wavelength coverage, output power and efficiency, and spectral and temporal characteristics (see Sections 23.7–23.9).

23.4 Optical Parametric Devices

From the preceding discussion, it may be concluded that on the attainment of the phase-match condition ($\Delta k = 0$) in the parametric process, the generated waves will experience optical gain and amplification, and coherent macroscopic output can be obtained at the signal and idler wavelengths. To obtain practical estimates of the magnitude of parametric gain under different pumping conditions, we can consider the main operating regimes of parametric devices. These are summarized in Table 23.3, where the gain factor, Γ, and the fractional single-pass intensity gain, $G_2(\ell)$, have been calculated under the phase-matched condition ($\Delta k = 0$) for different pumping regimes, using equations (23.8) and (23.10). We have taken typical values for pump pulse energy, duration, power, focused beam waist radius (ω_3) and crystal length (ℓ) appropriate to each mode of operation in practice. We have also assumed a typical non-linear coefficient of $d_{eff} \sim 3$ pm V^{-1} and have considered degenerate operation at 2 μm ($\lambda_2 \sim \lambda_1 \sim 2$ μm) with $n_3 \sim n_0 \sim 1.5$, for convenience.

We can see from Table 23.3 that the net percentage single-pass parametric gain varies from a mere 0.8% under cw pumping with a relatively powerful 5 W laser to as much as $\sim 3.4 \times 10^{31}$ when using high-energy ultrashort pump pulses. The latter would correspond to the use of high-intensity, mode-locked and amplified laser systems based on, for example, regenerative amplification and cavity-dumping schemes. In the intermediate regime of pumping with Q-switched nanosecond pulses, net optical gains of ~70% are available, whereas with ultrashort pulses of relatively low energy, corresponding to typical cw mode-locked lasers (e.g. Ti:sapphire), single-pass intensity gains of the order of ~1300% are expected. Given that the initial signal intensity at the crystal input is provided by spontaneous parametric noise, it is clear that any meaningful amplification to macroscopic levels over a single pass of the crystal is not practicable, except in the high-energy ultrashort pump pulse regime where large exponential gains are available. This is the configuration corresponding to *optical parametric generator* (OPG) and *amplifier* (OPA) devices. In all other operating regimes, discernible output can only be made available by enclosing the non-linear material within an optical resonator to provide feedback at the generated parametric wave(s). This is the configuration corresponding to OPOs, which is the most common architecture for parametric devices. In both architectures, there is an operation threshold associated with the device, which in practice would correspond to the detection and extraction of coherent output at the signal and idler wavelengths. In this treatment, we focus mainly on a description of OPO devices and their operating characteristics. While many of the design criteria and operating principles are equally applicable to OPG and OPAs, more extensive discussion of these single-pass devices can be found elsewhere in the literature [29].

23.5 Optical Parametric Oscillators

Of the different types of parametric devices, a large majority conform to the low- to moderate-gain operating limit. This is the regime most frequently encountered in practice when commonly available laser pump sources and non-linear materials are deployed. From the previous discussion, it is seen that the small-signal gains available in such devices are of the order of 0.1%–1000% for typical non-linear materials and pumping intensities. Such gains are clearly insufficient to achieve macroscopic amplification of parametric waves from noise in a single pass through the non-linear crystal. Therefore, parametric devices of this type are operated in OPO configuration, as in a conventional laser, by enclosing the non-linear gain medium within an optical cavity to provide feedback at the generated waves. In this way, the amplification of parametric waves is achieved to macroscopic levels by successive transits through the non-linear crystal, and a coherent output can be extracted from the oscillator.

The potential of OPOs derives from their exceptional wavelength flexibility, which allows access to spectral regions unavailable to conventional lasers. The OPO can readily provide widely tunable radiation across substantial portions of the optical spectrum by a suitable choice of non-linear material and laser pump source. Figure 23.12 shows the wavelength tuning range of a number of OPO devices developed to date. The spectral range available to several conventional tunable lasers (as in Figure 23.1) is also shown for comparison. It is seen that many wavelength regions unavailable to lasers are readily accessible to OPOs. Moreover, spectral regions far more extensive than any tunable laser can be accessed by a single device using one non-linear crystal. In addition to its spectral versatility, the OPO can be configured as a highly

TABLE 23.3

The Parametric Gain Factor and the Fractional Single-Pass Intensity Gain for Different Pumping Regimes, Calculated from (23.8) and (23.10). The Calculations are Based on Typical Experimental Values for Pump Laser and Non-linear Material Parameters for Each Operating Regime and Assume Phase-Matched Interaction ($\Delta k = 0$) and Near-Degenerate Operation at ~2 μm

	CW	Pulsed		
		Q-Switched	Mode-Locked	Mode Locked Amplified
Pump pulse energy	—	10 mJ	15 nJ	10 μJ
Pump pulse duration	—	10 ns	100 fs	200 fs
Peak pump power	5 W	1 MW	150 kW	50 MW
Focused waist radius (w_3)	20 μm	1 mm	15 μm	15 μm
Peak intensity (I_3)	400 kW cm^{-2}	30 MW cm^{-2}	20 GW cm^{-2}	7 TW cm^{-2}
Crystal length (ℓ)	10 mm	10 mm	1 mm	1 mm
$\Gamma\ell$	0.09	0.77	1.99	37
$G_2(\ell)$	1.008	1.72	13.99	34×10^{31}

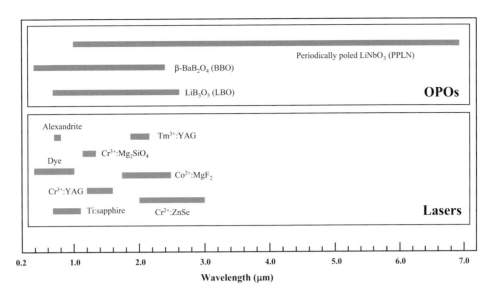

FIGURE 23.12 Comparison of the spectral coverage of prominent conventional tunable lasers (in figure 23.1) with that of a number of OPO devices demonstrated to date. The vertical scale has no significance.

compact device, has a simple tuning mechanism and offers high efficiencies in converting the input pump energy into useful output. It also has a practical solid-state design unlike, for example, tunable dye lasers employing liquids as the gain medium. Another important characteristic of OPOs is their temporal flexibility, which allows these devices to operate across all temporal regimes. This property is a consequence of the instantaneous nature of electronic polarization, the origin of non-linear gain. In contrast to conventional lasers, where the generation of the shortest optical pulses is limited by the upper-state lifetime of the laser transition, OPOs can provide output in all temporal regimes, from the cw to ultrafast femtosecond timescales, by the suitable choice of laser pump source.

As in a conventional laser oscillator, the OPO is characterized by a *threshold* condition, defined by the pumping intensity at which the growth of the parametric waves in one complete round trip of the optical cavity just balances the total loss in that round trip. However, in contrast to the laser, amplification of the generated waves in an OPO occurs only along the pump beam, so that optical gain is generally available only in one direction, unless the pump beam is returned to the crystal by back-reflection (see Section 23.5.1). The gain is thus single pass, whereas the losses occur in both passes of the non-linear crystal. Once the threshold has been surpassed, coherent light at macroscopic levels can be extracted from the oscillator.

In order to provide feedback in an OPO, a variety of resonance configurations may be deployed by the suitable choice of mirrors forming the optical cavity. The principal cavity geometries are summarized in Figure 23.13. The mirrors may be highly reflecting at only one of the parametric waves (signal or idler, but not both), as in Figure 23.13a, in which case the device is known as a *singly resonant oscillator* (SRO). This configuration is characterized by the highest operation threshold. In order to reduce the threshold, alternative resonator schemes may be employed where additional optical waves are resonated in the optical cavity. These include the *doubly resonant oscillator* (DRO), Figure 23.13b, in which both the signal and idler waves are resonant in the optical cavity, and the *pump-resonant* or *pump-enhanced* (PE) SRO, Figure 23.13c, in which the pump as well as one of the generated waves (signal or idler) is resonant in the optical cavity. In an alternative scheme, the pump may be resonated together with both parametric waves, as in Figure 23.13d, in which case the device is known as a PE-DRO or a *triply resonant oscillator* (TRO). Such schemes effect substantial reductions in threshold from the SRO configuration, with the TRO offering the lowest operation threshold. On the other hand, the threshold reduction is achieved at the expense of increased spectral and power instability in the OPO output arising from the difficulty in maintaining resonance for more than one optical wave in a single optical cavity. For this reason, the SRO offers the most direct route to the attainment of high output stability and spectral control without stringent demands on the frequency stability of the laser pump source. On the other hand, multiple resonant oscillators (DRO, PE-SRO and TRO) require active stabilization techniques to achieve output power and spectral stability, with the TRO representing the most difficult configuration in practice. In addition, the implementation of practical OPOs in multiple resonant cavities can only be achieved by using stable, single-frequency pump lasers and such devices also require more complex protocols for frequency tuning and control than the SRO [30].

An alternative technique to obtain major reductions in the threshold pump power requirement in cw SROs is the use of intra-cavity pumping [31]. The *intra-cavity* SRO (IC-SRO) takes advantage of the large circulating intensities within the pump laser oscillator itself to reach the operation threshold. In the scheme, shown in Figure 23.13e, the OPO is placed inside the cavity of the pump laser to allow access to high-circulating pump intensities, which would otherwise not be available external to the laser. The technique has proved highly effective in bringing the operation of cw SROs within the reach of commonly available pump lasers and birefringent non-linear materials and enabling the development of practical OPO devices with minimal requirements on pump power and frequency stability.

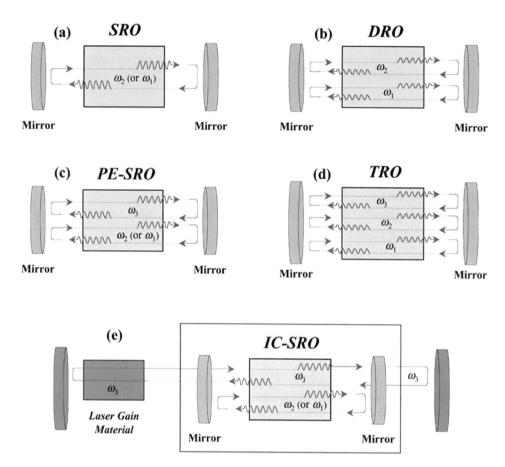

FIGURE 23.13 Principal cavity configurations for OPOs.

More detailed descriptions of the different resonance and pumping schemes for OPOs and analytical treatment of tuning mechanisms, spectral behaviour, frequency control and stabilization can be found elsewhere [30]. Here, we focus on the main results of the analysis and provide a summary of the main operating parameters for OPOs in the different resonance configurations and temporal regimes from the cw to ultrafast femtosecond timescales.

Because of the high operation threshold, successful operation of SROs in conventional external pumping configurations using commonly available non-linear materials generally necessitates large input pumping intensities, which are often beyond the reach of many existing cw laser sources. Such high intensities can, however, be made readily available with pulsed lasers operating in nanosecond, picosecond or femtosecond regimes. As a result, OPOs operating in the cw regime generally require multiple resonant cavities based on the DRO and PE-SRO to reach threshold, whereas pulsed OPOs can be readily implemented in the SRO configuration. On the other hand, practical implementation of cw OPOs in the SRO configuration is readily attainable using the intra-cavity pumping approach.

23.5.1 Continuous-wave OPOs

This class of OPO devices conforms to a steady-state operating regime as in a cw laser. The successful operation of the OPO is achieved by increasing the cw pumping intensity until the optical gain at the parametric waves overcomes the parasitic losses in the cavity, and the threshold is surpassed. Once above threshold, steady-state oscillation is established and parametric down-conversion from the pump to the signal and idler waves occurs, resulting in macroscopic output from the device. Starting from the coupled-wave equations (23.5a)–(23.4c), analytical expressions for threshold pumping intensity and conversion efficiency in cw OPOs can be obtained under steady-state conditions by considering the balance between parametric gain and parasitic loss within the OPO cavity [30].

23.5.1.1 Steady-state threshold

The analysis is based on the same assumptions as used in the derivation of the coupled-wave equations and parametric gain in Section 23.2. All three optical waves, the pump, signal and idler, are assumed to be infinite plane waves with uniform intensity across the beams, and the effects of focusing and double refraction are ignored. The interaction is also assumed to be phase-matched ($\Delta k = 0$), with the pump making a single pass through the non-linear crystal, so that the gain is only in the forward direction. In addition, the analysis assumes that the OPO cavity losses at the resonant wave(s) are small, which is generally true in practice. Pump depletion is also considered to be small, a condition which is true close to threshold. Both these assumptions allow E_3 to be constant throughout the length of the non-linear crystal, making $\partial E_3/\partial z$ = constant in the third coupled-wave equation (23.5c), hence simplifying the analysis.

Under these assumptions, steady-state oscillation thresholds can be obtained for both SRO and DRO cavities. For the SRO, the threshold pumping intensity is derived as

$$I_{th} = \alpha_2 \left(\frac{c\varepsilon_0 n_1 n_2 n_3 \lambda_1 \lambda_2}{8\pi^2 \ell^2 d_{eff}^2} \right) \quad (23.15)$$

where $\alpha_2 \ll 1$ is the total fractional round-trip power loss for the resonant wave (in this case, the signal). The loss at the idler wave is taken as 100% ($\alpha_1 = 1$), corresponding to an ideal SRO. For the DRO, the threshold condition is similarly obtained as

$$I_{th} = \frac{\alpha_1 \alpha_2}{4} \left(\frac{c\varepsilon_0 n_1 n_2 n_3 \lambda_1 \lambda_2}{8\pi^2 \ell^2 d_{eff}^2} \right) \quad (23.16)$$

where α_2 and α_1 are the total fractional round-trip power losses for resonant signal and idler waves, respectively. Both signal and idler losses are assumed to be small ($\ll 100\%$), so that α_2 and α_1 are both $\ll 1$. It can be seen from equations (23.15) and (23.16) that a DRO with signal and idler losses, α_1 and α_2, will have a threshold which is a factor $\alpha_1/4$ lower than a SRO with the same loss at the resonant signal and 100% loss at the idler. Therefore, a DRO with an idler loss of 2% will have a pump power threshold that is 200 times lower than the equivalent SRO. The threshold advantage of the DRO is therefore immediately clear.

In both SRO and DRO configurations, the pump threshold can be substantially reduced by returning the pump beam back to the crystal after its first pass in the forward direction [32]. The double-pass pumping ensures that optical gain is available in both directions, as in a laser oscillator, hence increasing the round-trip optical gain relative to loss. For a SRO, the threshold is reduced by a factor of two using this technique. Further reductions in threshold by as much as four times can also be obtained in a SRO if, in addition to the pump, the non-resonant wave is also double-passed through the crystal. However, to achieve the maximum four-fold reduction, it is necessary to maintain the appropriate relative phase between the pump and the parametric waves on reflection. In a DRO, the threshold reduction can be as much as four times, but this again requires that the correct relative phase among the three waves be preserved on reflection.

For a PE-SRO, the threshold condition can be derived in a similar manner to SRO, but by considering gain in both directions through the non-linear crystal, as the pump is now also resonant within the optical cavity. On resonance, the internal pump power threshold for the PE-SRO will, therefore, be the same as the incident threshold of the SRO with double-pass pump. The external threshold pumping intensity can, however, be shown to be [30]

$$I_{th} = \alpha \left(\frac{c\varepsilon_0 n_1 n_2 n_3 \lambda_1 \lambda_2}{16\pi^2 \ell^2 d_{eff}^2} \right) \left[\frac{(1-\sqrt{r_1 r_2} tt')^2}{t_1} \right] \quad (23.17)$$

where $\alpha \ll 1$ is the round-trip power loss of the parametric wave (signal or idler) that is resonant together with the pump in the cavity. Hence, α is equal to α_2 or α_1, depending on whether the resonant wave is the signal or the idler. r_1 and t_1 are the pump power reflectivity and transmission of the input mirror through which the pump is coupled into the cavity and r_2 is the pump power reflectivity of the output mirror. t and t' are the effective round-trip pump transmission of the crystal to linear and non-linear loss (due to parametric conversion), respectively. Clearly, for $r_1 = 0$ and $t_1 = 1$, (23.17) reduces to the SRO threshold condition (23.15). It can also been seen from equation (23.17) that resonating the pump has the effect of reducing the SRO threshold by a factor $(1-\sqrt{r_1 r_2} tt')^2/t_1$. To achieve maximum threshold reduction, this factor must thus be minimized through the interplay and optimization of the controllable parameters r_1, r_2, t_1 and t. The non-linear transmission of the crystal, t', must also be maximized through the use of long crystal lengths, and optimization of phase-matching and focusing parameters in order to achieve the lowest threshold.

In an IC-SRO, where the OPO is placed within the pump laser cavity, the steady-state threshold corresponds to a value of circulating pump intensity within the laser cavity, which is equal to the external pump intensity for an externally pumped SRO with double-pass pumping. The required external input power to the laser will in turn be determined by the internal slope efficiency of the pump laser in the presence of the SRO within the cavity [30,31].

23.5.1.2 Conversion Efficiency

The steady-state analysis of cw OPOs can be similarly extended to operation above threshold to determine the maximum conversion efficiency from the pump to the parametric waves in the different resonance configurations [30]. The plane-wave analysis is again based on the same assumptions as used in the threshold calculations. These include small cavity losses, optimum relative phase for parametric generation and phase-matched interaction. For a SRO with a single pass of the pump, the maximum conversion efficiency can be derived from the relation

$$I_{out} = I_{in} \cos^2 \left[\left(\frac{I_{in} - I_{out}}{I_{th}} \right)^2 \right] \quad (23.18)$$

where I_{in} and I_{out} are the steady-state pump intensities at the input and output of the non-linear crystal, respectively, and I_{th} is the threshold pumping intensity defined by equation (23.15). Form equation (23.18), we can find the condition for maximum conversion efficiency in the SRO. This occurs when the pump is completely depleted in passing through the crystal, so that $I_{out} = 0$. This results in the relation

$$I_{in} = \left(\frac{\pi}{2} \right)^2 I_{th} = 2.47 I_{th} \quad (23.19)$$

which implies that in the SRO a maximum efficiency of 100% is attainable when the input pump intensity is 2.47 times that at threshold. The analysis, as presented, holds up only to the point where the pump intensity just reaches zero at the output face of the non-linear crystal. If the input pump intensity is further increased, then the pump wave will be totally depleted at some point within the crystal. In the remaining length of the crystal, the signal and idler waves are then able to regenerate the pump through SFM. As a result, pump intensity once again now exits the crystal, and the SRO conversion efficiency is reduced. There is hence an optimum value of the pump intensity, at 2.47 times threshold, for which 100% efficiency is obtained. Above and below this value, lower conversion efficiencies are available.

Interestingly, it can also be shown that if the pump is double-passed through the SRO, a conversion efficiency of 100% is still attainable but will require the same input pump intensity, namely, 2.47 times the single-pass threshold intensity [32]. This is despite the two-fold reduction in SRO threshold under double-pass pumping. The analysis of conversion efficiency in the SRO leads to the same results, (23.18) and (23.19), regardless of whether the SRO is configured in a standing wave or ring cavity. This is because the non-resonant wave is not present on the return pass through the crystal, and so regeneration of the pump wave from mixing between the signal and idler waves does not arise here. Since the non-resonant wave is generated from zero at the input of the crystal, the appropriate relative phase for maximum conversion efficiency is acquired among the optical waves from the start and, under phase-matching, is maintained throughout the process.

In a DRO, the analysis of conversion efficiency must include the effects of pump wave regeneration, which can limit the maximum attainable efficiency [32,33]. In a standing-wave cavity with a single-pass pump only in the forward direction, the pump wave is regenerated through the mixing of signal and idler waves on their backward propagation through the non-linear crystal. If it is assumed that all waves maintain the optimum relative for maximum conversion efficiency, then the pump intensity at the output of the crystal is derived as

$$I_{out} = I_{in} - \left[2\sqrt{I_{th}} \left(\sqrt{I_{in}} - \sqrt{I_{th}} \right) \right] \quad (23.20)$$

where I_{in} is the steady-state pump intensity at the input of the crystal, and I_{th} is the threshold pump intensity defined by equation (23.16). From the expression, we see that the maximum conversion efficiency in a DRO (corresponding to $I_{out} = 0$) occurs at four times threshold, instead of the 2.47 times in the SRO. The maximum efficiency is also now limited to only 50% due to pump wave back-conversion. It can be readily shown that the back-converted pump intensity is given by

$$I_{BC} = \left(\sqrt{I_{in}} - \sqrt{I_{th}} \right)^2 \quad (23.21)$$

from which we can calculate the transmitted intensity of the pump under steady-state operation. If we define the down-converted pump intensity as $I_{DC} = I_{in} - I_{out}$, then the transmitted intensity can be obtained from the relation $I_{out} = I_{in} - I_{BC} - I_{DC}$. Substituting for I_{BC} and I_{DC} using equations (23.20) and (23.21), we find $I_{out} = I_{th}$. This is an interesting result as it implies that, under steady-state operation above the threshold, the transmitted pump intensity through the DRO is clamped at its threshold value. In other words, the DRO acts as an optical intensity or power limiter for the pump and regardless of the magnitude of the input pump intensity above threshold, the transmitted intensity remains fixed at I_{th}. Any increase in the input intensity above the threshold value has the effect of increasing the back-generation of the pump, without a rise in either the down-converted or the transmitted pump intensity.

The effects of back-generation in a DRO can, however, be eliminated by allowing the pump to be present with the parametric waves on the return pass through the crystal using double-pass pumping [32]. Under this condition, 100% conversion efficiency becomes attainable in the DRO. As noted earlier, double passing of the pump also has the effect of reducing the DRO threshold by as much as four times. However, unlike the SRO, the required pump intensity for 100% conversion efficiency in the DRO is four times the new and lowered threshold. At the same time, the attainment of the maximum four-fold reduction in DRO threshold requires that the optimum relative phase among the pump, signal and idler is preserved on the return pass through the crystal. This may, in practice, be best achieved by interferometric control of the reflected pump wave using a separate mirror.

On the other hand, it is also possible to achieve 100% conversion efficiency in the DRO without double passing of the pump, but by using a travelling-wave rather than a standing-wave cavity for the OPO. In this geometry, the pump wave always accompanies the signal and idler waves in every pass of the non-linear crystal, so that no back-generation of the pump can occur. The steady-state analysis of conversion efficiency in a travelling-wave DRO is similar to a standing-wave DRO, but by ignoring back-generation of the pump. The result of the analysis is the following expression for the steady-state output pump intensity at the exit of the crystal [30]

$$I_{out} = \left(\sqrt{4I_{th}} - \sqrt{I_{in}} \right)^2 \quad (23.22)$$

where I_{in} is the pump intensity at the input of the crystal and I_{th} is the threshold intensity. From the above, it is thus clear that the maximum DRO conversion efficiency (corresponding to $I_{out} = 0$) still occurs at four times threshold. However, if we consider the down-converted pump intensity, $I_{DC} = I_{in} - I_{out}$, we find from equation (23.22) that

$$I_{DC} = 4\sqrt{I_{th}} \left(\sqrt{I_{in}} - \sqrt{I_{th}} \right) \quad (23.23)$$

which implies that a DRO conversion efficiency of 100% is now attainable at four times threshold because of the absence of pump back-conversion. It is, however, important to note that back-conversion of the pump can still occur if the pump is fully depleted within the crystal. In this case, the signal and idler fields can mix in the remaining length of the crystal to regenerate the pump, hence reducing the conversion efficiency. This would occur when the input pump intensity increases above the optimum value of four times threshold. The back-conversion of the pump within the crystal can also arise in the SRO, where 100% depletion of the pump is possible.

In a PE-SRO, the conversion efficiency may be analysed by considering the down-converted pump intensity, defined as [30]

$$I_{DC} = (1 - t')I_{cc} \quad (23.24)$$

where I_{cc} is the circulating pump intensity within the PE-SRO cavity at threshold. It is equal to the incident threshold of the SRO with double-pass pump. The down-converted intensity may then be shown to be

$$I_{DC} = I_{cc}\left[1-\left(\frac{1}{r_1 r_2 t}\right)\left(1-\sqrt{\frac{t_1 I_{ext}}{I_{cc}}}\right)^2\right] \quad (23.25)$$

where I_{ext} is the external pump power incident on the input mirror and all other parameters as defined previously (see Section 23.5.1.1). For a perfect input mirror with $r_1+t_1 = 1$, the optimum transmission of the input mirror can be found from equation (23.25) to be $t_1^{opt} = I_{ext}/I_{cc}$. The optimum down-converted pump intensity can similarly be obtained as

$$I_{DC}^{opt} = I_{cc}\left[1-\left(\frac{1}{r_2 t}\right)\left(1-\frac{I_{ext}}{I_{cc}}\right)\right] \quad (23.26)$$

We can see from this that, for zero parasitic loss of pump through the output mirror and no linear transmission through the crystal ($rt = 1$), the optimum down-converted intensity in the PE-SRO under optimized input coupling will equal the external pump intensity at the input. In other words, the PE-SRO is capable of 100% conversion efficiency from the external pump to the generated parametric waves in the absence of any linear parasitic loss of pump from the cavity.

For an IC-SRO, the analysis of conversion efficiency leads to the following relation for down-converted pump intensity [30,31]

$$I_{DC} = \sigma_{max}(I_{in} - I_{th})\left(1-\frac{I_{th}^L}{I_{th}}\right) \quad (23.27)$$

where σ_{max} is the output slope efficiency of the pump laser with optimum output coupling for a given parasitic loss. I_{in} is the external input intensity to the pump laser; I_{th}^L and I_{th} are the external input intensities corresponding to the pump laser and IC-SRO threshold, respectively. From equation (23.27), the condition for maximum down-converted intensity can be obtained as

$$I_{th} = \sqrt{I_{th}^L I_{in}} \quad (23.28)$$

Under this condition, the total down-converted power can be shown to be

$$(I_{DC})_{max} = \sigma_{max}\left(\sqrt{I_{in}} - \sqrt{I_{th}^L}\right)^2 \equiv (I_{out}^L)_{max} \quad (23.29)$$

which also identically describes the condition for the maximum attainable output intensity from the pump laser, $(I_{out}^L)_{max}$, at a given input intensity, when subject to optimum output coupling. This implies that when the pump laser and SRO input intensity thresholds satisfy equation (23.28), then 100% of the maximum output intensity potentially extractable from the pump laser itself can be down-converted into the parametric waves. In other words, equation (23.28) describes the condition for optimization of the non-linear coupling loss presented to the pump laser by the IC-SRO to achieve 100% down-conversion at a given input power. Under this condition, the IC-SRO can be viewed as an optimum output coupler to the pump laser. Increasing the external input intensity to the pump laser leads to the down-converted intensity from the IC-SRO approaching the maximum available output from the optimally out-coupled pump laser, reaching 100% efficiency when the input intensity reaches a value $I_{in} = I_{th}^2/I_{th}^L$. Above this value, the efficiency gradually tails off as the IC-SRO begins to over-couple the laser field. Back-conversion of the pump is generally not relevant in influencing the efficiency of IC-SRO, since one of the generated waves is non-resonant in the OPO cavity and the pump also always accompanies the parametric waves in every pass of the crystal. In addition, full-pump depletion within the crystal is not possible because the single-pass conversion is at best only a few per cent (equivalent to the loss presented to the pump laser by an optimized output coupler). Equation (23.28) provides a practical design criterion for the attainment of maximum power conversion in the IC-SRO for a given available input power. In practice, however, it is often difficult to achieve the maximum 100% efficiency, but down-conversion efficiencies in excess of 90% can be obtained experimentally through suitable design of the pump laser and ICSRO threshold parameters, as described in Section 23.7.

23.5.2 Pulsed OPOs

As highlighted earlier, the SRO cavity configuration in which only one of the generated waves is resonant represents the simplest and most practical device architecture for an OPO, because of its minimal demands on pump beam coherence, high passive stability and reduced complexity in spectral and temporal behaviour. On the other hand, the SRO yields the highest threshold of all cavity configurations, so that the attainment of oscillation threshold using conventional external pumping schemes is generally beyond the reach of commonly available cw pump lasers, particularly when conventional birefringent materials are employed. Given this limitation, practical operation of SROs has historically necessitated the use of pulsed lasers, where the high available peak powers readily permit device operation without resorting to intra-cavity pumping or any form of pump enhancement. Such devices can be readily implemented in the nanosecond regime by deploying Q-switched pump lasers or in the ultrafast picosecond and femtosecond regime by using mode-locked laser sources. In the following discussion, we provide an overview of the main operating characteristics of pulsed OPOs, with an emphasis on the SRO configuration, as the majority of practical devices developed to date conform to this mode of operation.

23.5.2.1 Nanosecond OPOs

Because of their high peak powers, nanosecond pulsed lasers represent the most viable choice of pump for SROs. Combined with the widespread availability of such laser sources, OPOs operating in the nanosecond pulsed regime have traditionally been the most extensively developed of all parametric devices. The basic operation principles in nanosecond OPOs are the same as in cw oscillators. However, the steady-state threshold analysis used earlier in the treatment of cw OPOs is not strictly applicable here, because the instantaneous nature of parametric gain does not allow steady-state conditions to prevail within the finite duration of the pump pulse. This situation is markedly different from that in a conventional laser, where under pulsed excitation, the gain storage capability of the gain medium generally permits amplification after the pump pulse excitation for a time determined by the upper-state lifetime of the transition. In the parametric process, the electronic susceptibility is effectively instantaneous, so that the non-linear medium has no gain-storage capacity. This means that coherent amplification of the parametric wave(s) has to occur in the presence of the pump and that no gain is available outside the temporal window of the pump pulse. Given the finite duration of the pump pulse, only a limited number of round trips for the parametric wave(s) can be made available through the non-linear crystal in OPO cavities of practical length, thus preventing the establishment of steady-state conditions. Therefore, nanosecond-pulsed OPOs generally operate in a transient regime, and a modified model taking account of the dynamic behaviour of OPO is necessary for the analysis of such devices.

By using a time-dependent gain analysis, it is possible to derive the threshold condition in pulsed SROs in terms of pump energy fluence (energy/area) [34]. The model assumes a Gaussian temporal profile for the incident pump pulse and a Gaussian spatial distribution for the pump and the resonant signal, with an idler wave that is unconstrained by the optical cavity. The model also assumes a single-pass pump and includes the effects of mode overlap and spatial walk-off. The threshold condition is then derived from the solution of the coupled-wave equations (23.5a)–(23.5c) in the limit of low pump depletion and zero input idler field, by allowing the resonant signal wave to be amplified from the initial parametric noise power to a detectable level by successive transits of the SRO cavity. By defining the SRO threshold as a signal energy of ~100 µJ, corresponding to a signal power to parametric noise power of $\ln(P_s^{th}/P_n) = 33$, the threshold energy fluence is derived as

$$J_{th} \cong \frac{2.25}{\gamma g_s \zeta}\tau\left[\frac{L}{2\tau c}\ln\frac{P_s^{th}}{P_n} + 2\alpha_2 \ell + \ln\frac{1}{\sqrt{R}} + \ln 2\right]^2 \quad (23.30)$$

where τ is the $1/e^2$ intensity half-width of the pump pulse, L is the cavity length, α_2 is the signal field amplitude loss coefficient within the crystal, R is the mirror reflectivity coefficient and ℓ is the crystal length. The parameter γ is a modified gain coefficient, which is related to the gain factor Γ defined in equation (23.7) by

$$\gamma = \frac{\Gamma^2}{I_3(z=0)} = \frac{8\pi^2 d_{eff}^2}{c\varepsilon_0 n_1 n_2 n_3 \lambda_1 \lambda_2}. \quad (23.31)$$

The factor g_s is the spatial coupling coefficient describing the mode overlap between the resonant signal and the pump field. It is defined as

$$g_s = \frac{\omega_3^2}{\omega_3^2 + \omega_2^2} \quad (23.32)$$

where ω_3 and ω_2 are the Gaussian mode electric field radii of the pump and signal, respectively. The parameter ζ is the effective parametric gain length given by

$$\zeta = \ell_w \text{erf}\left(\frac{\sqrt{\pi}}{2}\frac{\ell}{\ell_w}\right) \quad (23.33)$$

where ℓ_w is the *walk-off length* defined as

$$\ell_w = \frac{\sqrt{\pi}}{2}\frac{\omega_3}{\rho}\sqrt{\frac{\omega_3^2 + \omega_2^2}{\omega_3^2 + \omega_2^2/2}} \quad (23.34)$$

where ρ is the double-refraction angle (see Section 23.6.1). The walk-off length, ℓ_w, is closely related to the aperture length, ℓ_a (see Section 23.6.1), and accounts for spatial walk-off between the Gaussian pump and signal fields. In the absence of spatial walk-off, $\ell_w \to \infty$, the crystal length ℓ becomes the effective parametric gain length.

The main conclusions of the time-dependent model relate to the strong dependence of pulsed OPO threshold on the characteristic *rise time* of the oscillator. This is a measure of the time required for the parametric gain to build up from noise to oscillation threshold [35]. In order for the OPO to switch on, it is essential that the oscillator rise time is shorter than the pump pulse duration. For efficient operation, however, the rise time must be minimized so that the parametric waves are rapidly amplified above the threshold level in a time significantly shorter than the pump pulse interval. Therefore, the rise time represents an effective loss for pulsed OPOs, with a direct impact on oscillation threshold. To exploit maximum gain, it is essential that the parametric waves are amplified in as short a time interval as possible in the presence of the pump pulse. In practice, this can be achieved by minimizing the OPO cavity length to allow the maximum number of round trips over the pump pulse interval. This is also evident from the direct dependence of J_{th} on the cavity length L in equation (23.30). In nanosecond OPOs, cavity lengths of a few centimetres or shorter are often practicable, yielding as many as one hundred round trips for pump pulses of 1–50 ns duration. As can be verified from equation (23.30), the pulsed OPO threshold can be further reduced by using longer pump pulse durations, τ (and maintaining the same peak intensity), and by lowering the intra-cavity parasitic losses through minimizing α and maximizing R. Above threshold, the operation of pulsed OPOs is characterized by the extraction of energy from the pump pulse and saturation of non-linear gain, in a similar manner to cw OPOs discussed earlier. This is often observed in the form of depletion towards the trailing edge of pump pulse. The signal (and idler) pulse is then amplified with a characteristic delay from the pump pulse by a time interval determined by the OPO rise time [36–38].

We can obtain a measure of the threshold fluence in pulsed SROs by considering a typical example based on a 10 mm long KTP crystal, pumped by 10 ns pulses from a Q-switched Nd:YAG laser 1.064 µm. We assume type II NCPM (see Section 23.6.1), so that the full crystal length is utilized for the non-linear interaction. We take a practical cavity length $L = 20$ mm, mirror reflectivities $R \sim 99.5\%$, an effective non-linear coefficient $d_{eff} \sim 3$ pm V^{-1}, refractive indices $n_3 \sim n_2 \sim n_1 \sim 1.5$ and mode radii $\omega_3 \sim \omega_2$. For near-degenerate operation with $\lambda_1 \sim \lambda_2 \sim 2$ µm and for $\alpha \sim 2\%$ and $\rho = 0$, substitution of these parameters into equations (23.31)–(23.34) and then into equation (23.30) yields a threshold pump energy fluence $J_{th} \sim 1.2 \, \text{J cm}^{-2}$. This translates into a threshold pump pulse energy of $E_{th} \sim 37 \, \text{mJ}$ for a focused pump mode radius $\omega_3 \sim 1 \, \text{mm}$. The threshold energy can, of course, be substantially reduced by using tighter focusing. For example, the use of a pump beam waist of $\omega_3 \sim 0.5 \, \text{mm}$ will lead to a nearly fourfold reduction in threshold energy to ~9 mJ, whereas reducing the focused radius to $w_3 \sim 0.1 \, \text{mm}$ will result in nearly a ten-fold threshold reduction to ~0.4 mJ. It should, however, be noted that considerations of material damage in the presence of nanosecond pulses present a practical upper limit to the minimum focused pump spot radii that can be used in such devices (see Section 23.6.1). Therefore, the optimum choice of experimental parameters for the attainment of minimum operation threshold in pulsed OPOs is often compromised by material damage threshold. Material damage remained a major limitation to the operation of early devices, even at moderate power levels, and for many years hampered the development of practical nanosecond OPOs.

23.5.2.2 Picosecond and Femtosecond OPOs

The high peak pulse intensities available with mode-locked picosecond and femtosecond pulses can facilitate the attainment of sufficient non-linear gain in an OPO to overcome threshold, even in SRO configurations exhibiting the highest operation threshold. Moreover, the low-energy fluence associated with ultrashort pulses results in increased damage tolerance for the material, thereby enabling reliable operation of the OPO at high average powers using high-power input pump sources. On the other hand, unlike cw and pulsed nanosecond devices, the operation of ultrafast OPOs is based on the principle of *synchronous pumping*. As noted in Section 23.5.2.1, due the instantaneous nature of non-linear polarization, optical gain in an OPO is available only in the presence of the pump pulse, during which macroscopic amplification of the parametric waves from noise to coherent output must take place. However, in contrast to nanosecond pulses, the temporal window in picosecond and femtosecond pulses is too narrow to allow a finite number of cavity round trips for the build-up of the parametric waves over the pump pulse duration, even for practical OPO cavity lengths as short as a few millimetres. To overcome this difficulty, the OPO resonator length is matched to the length of the pump laser, so that the round-trip transit time in the OPO cavity is equal to the repetition period of the pump pulse train. In this way, the resonated parametric pulses experience amplification with consecutive coincidences with the input pump pulses as they make successive transits through the non-linear crystal. In practice, the technique is practical only at relatively high pulse-repetition rates (>50 MHz). At lower repetition frequencies, the required OPO cavity lengths for synchronous pumping become too long and cumbersome to be useful in practice.

In general, synchronously pumped OPOs may be classified into either cw or pulsed devices. In cw oscillators, the input pump radiation comprises a continuous train of ultrashort pulses, whereas in pulsed devices, the pump consists of trains of ultrashort pulses contained within a nanosecond or microsecond envelope. With regard to their operating characteristics, cw synchronously pumped OPOs may be treated as steady-state devices in the same way as cw OPOs, but with the peak pump pulse intensity determining the non-linear gain. As such, the steady-state analysis of cw OPOs is similarly applicable to cw synchronously pumped OPOs by using the peak pulse intensity as the incident pump intensity. On the other hand, the operating dynamics of pulsed synchronously pumped OPOs is similar to nanosecond OPOs, where a transient analysis taking account of rise-time effects is necessary to adequately describe the OPO behaviour [84]. In pulsed synchronously pumped OPOs, the non-linear gain is similarly determined by the peak pulse intensity, but rise-time effects arising from the finite duration of the pulse envelope lead to an additional loss mechanism. In either case, however, further effects pertinent to the operation of synchronously pumped OPOs including *temporal walk-off* and group-velocity dispersion (see Section 23.9.2) also have to be included in the model. In picosecond OPOs, where pulse durations of >1 ps are involved, such temporal effects can generally be ignored for practical crystal lengths of up to 20 mm. However, they become increasingly important in femtosecond OPOs, where pulses of 100 fs or shorter are involved (see Section 23.9.2).

The analysis of synchronously pumped OPOs under steady-state and transient operation has been performed for plane waves [39] and for cw synchronously pumped SROs with Gaussian beams [40]. One of the main conclusions of the analyses is that in the absence of group velocity dispersion and temporal walk-off, the output pulses from a synchronously pumped OPO are always shorter than the input pump pulses, with the parametric pulses broadening towards the pump pulse length with increasing pump depletion. They also show that it is possible to obtain conversion efficiencies as high as ~70% in such devices with a suitable choice of design parameters. The analysis of cw synchronously pumped SROs also shows that, for high efficiency and short output pulses, the oscillator should be designed to operate with a peak pump intensity that is approximately twice the steady-state threshold value. Thus, for a given pump intensity and non-linear material parameters, the output-coupling mirror can be adjusted to allow this condition to be met. Near threshold, the effects of temporal walk-off can be almost perfectly compensated for by fine-tuning the cavity length.

23.6 OPO Design Issues

The principal design criterion for optimum operation of parametric devices of all types, including OPOs, is the maximization of parametric gain through the suitable choice of non-linear material and laser pump sources, optimization of phase-matching, cavity design and focusing parameters.

23.6.1 Non-linear Material

The selection of non-linear material for OPOs is governed by a number of universal factors that are equally applicable to all types of parametric devices. A fundamental material parameter is a broad transparency and the ability to be phase-matched in the wavelength range of interest. As evident from equation (23.7), it is also important for the material to have a large effective non-linear coefficient, d_{eff}, for the attainment of maximum gain and highest efficiency. Other material requirements include a high optical damage threshold, good optical quality and low absorption loss at the pump, signal and idler wavelengths, mechanical, chemical, and thermal stability, and availability in bulk form and sufficiently large size.

In addition to the fundamental material parameters, there are also a number of general material requirements that are desirable for optimum operation of OPOs. These include favourable phase-matching geometries, small double-refraction angles and spatial walk-off and, in the case of femtosecond OPOs, low temporal walk-off and group velocity dispersion. It is also desirable for the material to display large tolerances to possible deviations in the spectral and spatial quality of the pump beam, which result in an increase in the magnitude of the phase mismatch, Δk, from zero. Such deviations arise in practice from the finite spectral bandwidth and divergence of the pump beam. The tolerance of the non-linear crystal to such effects is measured in terms of the so-called *spectral* and *angular acceptance bandwidths*, which can be calculated from the rate of change of phase mismatch with the wavelength spread ($\Delta\lambda$) and angular divergence ($\Delta\theta$) at the pump, using a series expansion of Δk [6]. The acceptance bandwidths can then be obtained by solving for the quantities $\Delta\lambda \cdot \ell$ and $\Delta\theta \cdot \ell$, using the boundary condition $|\Delta k \ell| \sim 2\pi$. For a given crystal length, ℓ, the acceptance bandwidths set an upper limit to the maximum allowable pump linewidth, and the divergence before parametric gain is severely degraded. Equivalently, for a given pump linewidth and angular divergence, the acceptance bandwidths determine the maximum length of the non-linear crystal that can effectively contribute to the non-linear gain. Since the crystal length follows an inverse relationship with the maximum allowable pump bandwidth and divergence, for shorter crystal lengths, larger deviations in pump beam quality can be tolerated and *vice versa*. For materials that exhibit temperature-dependent refractive indices and, hence, a temperature-tuning capability, one can similarly define a *temperature acceptance bandwidth*, $\Delta T \cdot \ell$, which is a measure of the sensitivity of phase-matching and, thus, parametric gain, to changes in the crystal temperature. As a general rule, for the attainment of maximum non-linear gain and minimum OPO threshold, it is advantageous to use materials which exhibit large spectral, angular and temperature acceptance bandwidths.

Another important material parameter is double-refraction, which leads to spatial walk-off among the interacting fields. Under this condition, the strength of non-linear coupling can rapidly diminish and the interaction becomes ineffective after a finite length through the medium, known as the *aperture length*, and given by $\ell_a = \omega_3 \sqrt{\pi}/\rho$, where w_3 is the pump beam waist radius and ρ is the double-refraction angle. It is clear that for longer aperture lengths, it is desirable to use non-linear crystals with small double-refraction and larger pump beam waists. In pulsed nanosecond OPOs, the high peak intensities available from Q-switched pump lasers allow for the use of large beam waists. However, in cw and synchronously pumped OPOs, it is often necessary to use tightly focused beams because of the substantially lower available pumping intensities.

In the operating regime of cw and synchronously pumped OPOs, therefore, it is imperative to use materials that possess a NCPM capability, which allows phase-matched interaction along a principal optical axis with no spatial walk-off. Given the generally low intensities available from cw- and mode-locked pump lasers, the NCPM geometry allows tight focusing of the pump beam to achieve the necessary intensities without the deleterious effects of beam walk-off. Therefore, NCPM is generally a highly important material property in the context of cw and synchronously pumped OPOs, particularly picosecond OPOs, where long interaction lengths are imperative for the attainment of threshold. This requirement is somewhat less stringent when deploying high-power (multi-watt) pump sources for picosecond OPOs, or in femtosecond OPOs, where the high peak powers and short interaction lengths of the material allow for the use of critical phase-matching (CPM) in collinear or non-collinear schemes (see Section 23.9.2). However, even in these cases NCPM still offers the advantage of zero spatial walk-off and hence higher non-linear gain. Another important advantage of NCPM is that under this geometry the angular acceptance bandwidth of the material is maximized, because of the small sensitivity of refractive indices to beam propagation angle along a principal optical axis. Therefore, the NCPM geometry is also highly beneficial when using pump beams of poor spatial quality.

In addition to these requirements, one of the most important properties that can ultimately limit operation of parametric devices in practice is material damage threshold. Because of the high pumping intensities in the parametric process, the materials must withstand power densities typically >10 MW cm^{-2} and often many times this value. Optical damage may be caused by several different mechanisms and can be of various types. The most predominant type is the irreversible damage, which can take the form of surface or bulk damage. As a general trend, material damage tolerance is found to decrease for increasing pump energy fluence and with shorter pumping wavelengths. As a result, material damage becomes more important in nanosecond OPOs than in synchronously pumped and cw devices and also in the presence of ultraviolet and visible radiation. Another type of optical damage that is prevalent in PPLN is a reversible form, which results from illumination by visible radiation. This type of damage, known as photorefractive damage, manifests itself as inhomogeneities in the refractive index, which can lead to beam distortions, non-uniformities and instabilities in the output beam. The induced damage can be reversed by ultraviolet illumination of the crystal or by elevating its temperature.

The relative importance of the various material parameters discussed is critically dependent on the particular mode of operation. In Table 23.4, we summarize the key material requirements for each OPO mode of operation, from the cw to the synchronously pumped femtosecond regime. The effective non-linearity, transparency and potential tuning range of a number of prominent birefringent and periodically poled materials used in the development of OPOs are also summarized in Figure 23.14. A more comprehensive survey of a wide range of birefringent non-linear materials can be found elsewhere [6].

Optical Parametric Devices

TABLE 23.4
The Relative Significance of the Key Material Parameters for Different Modes of OPO Operation, from cw and Pulsed Nanosecond to cw Synchronously-Pumped Picosecond and Femtosecond Regimes. The Universal Material Requirements of Transparency, Phase-Matchability, High Optical Quality and Low Absorption Loss, and Mechanical and Chemical Stability are Essential for all Modes of Operation

		OPOs		
	CW	Nanosecond	Picosecond	Femtosecond
High d_{eff} (> 3 pm V^{-1})	Essential	Desirable	Essential	Desirable
Low spatial walk-off	Essential	Desirable	Essential	Desirable
NCPM	Essential	Desirable	Essential	Desirable
Long-interaction length (>10 mm)	Essential	Desirable	Essential	Undesirable
High damage threshold	Desirable	Essential	Desirable	Desirable
Low group velocity dispersion	Unnecessary	Unnecessary	Desirable	Highly desirable
Low temporal walk-off	Unnecessary	Unnecessary	Desirable	Highly desirable
Temperature phase-matching	Highly desirable	Unnecessary	Highly desirable	Unnecessary
Angle phase-matching	Generally not permitted	Desirable	Generally not permitted	Unnecessary
Large phase-matching acceptance bandwidths	Desirable	Desirable	Desirable	Desirable

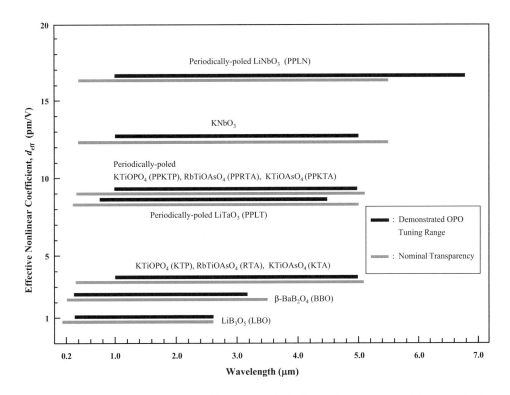

FIGURE 23.14 Characteristics of a number of important birefringent and periodically poled non-linear materials used in the development of OPOs. The indicated spectral coverage may not in all cases correspond to a single device but is potentially available to the material in different device configurations.

23.6.2 Pump Laser

As with the non-linear material, the choice of laser pump source is also governed by a number of universal factors relating to the maximization of non-linear gain. A principal requirement is set by phase-matching to access the wavelength regions of interest, and the pump wavelength should obviously be within the transparency range of the material. The pump source must also be of sufficient intensity, as evident from (23.7), to provide appreciable non-linear gain for the OPO to reach operation threshold. This, in turn, necessitates a sufficiently low beam divergence to allow high focused intensities and depth of focus within the non-linear material. Considerations of phase-mismatch, Δk, place further demands on the spectral and spatial coherence of the pump beam. As can be seen from equation (23.6), parametric gain is strongly dependent on the phase-mismatch parameter, Δk. Maximum gain occurs for $\Delta k = 0$, while increases in the magnitude of Δk result in severe reductions in non-linear gain. Despite the use of phase-matching techniques, the attainment of perfect phase-matching ($\Delta k = 0$) is generally not possible in practice, because of finite spectral bandwidth and spatial divergence of the pump beam. This leads to an increase in the magnitude of Δk, with the net result that the parametric gain is severely reduced from its peak value at $\Delta k = 0$. The maximum allowable pump bandwidth and divergence can be calculated from considerations of phase-mismatch [41]. As discussed in Section 23.6.1, such limitations to pump beam quality can be equivalently considered in terms of the spatial and spectral acceptance bandwidths for the non-linear process. To maintain high parametric gains, it is important to employ laser pump sources of narrow linewidth and low beam divergence.

23.7 Continuous-wave OPO Devices

Of the different types of OPO devices, the development of practical OPOs in the cw regime has been traditionally more difficult than pulsed and synchronously pumped devices because of the substantially lower non-linear gains available under cw pumping. The development of novel birefringent and periodically poled materials, new innovations in resonance and pumping techniques, and the advent of stable, high-power solid-state laser pump sources has led to important breakthroughs in cw OPOs in the past few years, particularly for the infrared spectral range at wavelengths >1 µm. Because of low spatial walk-off, NCPM capability, moderate non-linear optical coefficients, an extended transparency window to 5 µm and the birefringent material KTP and its arsenate isomorphs, KTA and RTA, have shown promise as candidates for cw OPOs. Configured in IC-SRO cavities and pumped by Ti:sapphire lasers, devices based on KTP and KTA have been shown to be practical sources of cw radiation, providing total output powers of up to 1.46 W at up to 90% conversion efficiency, with as much as 840 mW of idler power available in the 2.4–2.9 µm spectral range [42,43]. In alternative to TRO and PE-SRO resonance configurations, low-threshold operation of cw OPOs has also been achieved using semiconductor diode lasers as the pump source [44,45]. By using KTP and RTA under NCPM and GaAlAs diode lasers at ~800 nm, pump power thresholds <100 mW and tuning in the 2.1–2.7 µm range have been demonstrated through the tunability of the diode laser pump source. Because of a lack of significant temperature tuning in KTP and its arsenate analogues, wavelength tuning in cw OPOs based on these materials is generally achieved through static pump tuning while maintaining the NCPM condition. With the available laser sources based on cw Ti:sapphire, semiconductor and Nd:YAG lasers, the potential infrared spectral coverage of cw OPOs based on the KTP family of crystals under NCPM is limited to ~3.5 µm.

In contrast, the most important advances in cw OPOs have been brought about by periodically poled materials. In particular, the advent of PPLN with high effective non-linearity and long interaction lengths has brought the threshold of cw SROs within the reach of cw solid-state lasers and enabled the operation of these devices at unprecedented power levels and conversion efficiencies in simple external pumping schemes. Using a 50-mm long PPLN crystal pumped with a high-power diode-pumped Nd:YAG laser, infrared output powers of as much as 3.6 W have been generated from a cw SRO over a spectral range from 3.25 to 3.95 µm, for 13.5 W of input pump power [46]. Wavelength tuning was achieved through temperature tuning or by using different grating periods incorporated onto the single crystal. Alternative tuning schemes include the use of PPLN with fanned gratings, where the grating period is progressively varied across the crystal [47]. With the advent of PPLN, the operation of cw SROs in external pumping configuration has also extended to high-power semiconductor diode pump lasers [48]. Pumped by an InGaAs laser at 925 nm in a master oscillator-power amplifier configuration and a 38-mm PPLN crystal, SRO operation thresholds of 1.7 W and idler output powers of 480 mW were generated over a tuning range of 2.03–2.29 µm, for 2.5 W of pump power. Appropriate synchronization of pump tuning and cavity length scanning results in a wide, continuous, single-frequency tuning range. Later, high-power fibre lasers have been successfully used to pump cw PPLN-based SROs [49]. Using an Yb-doped cw fibre laser tunable over 1.031–1.100 µm and delivering 8.3 W of pump power into a 40 mm PPLN crystal, mid-IR idler powers of up to 1.9 W over a spectral range of 2.98–3.7 µm have been generated, with a SRO power threshold at 3.5 W.

At the same time, the use of intra-cavity pumping schemes in combination with PPLN has, for the first time, permitted cw SRO operation with minimal operation threshold based on commonly available, low-power, diode-pumped solid-state lasers [50]. Using a Nd:YVO$_4$ laser pumped by a 1 W diode laser, practical mid-IR output powers of 70 mW over a spectral range from 3.16 to 4.02 µm have been generated from an IC-SRO in a simple, compact, all-solid-state design, with a diode pump power threshold of only 310 mW. Operation of mid-IR cw OPOs based on PPLN has also been achieved in PE-SRO configurations. Pumped by a miniature diode-pumped single-frequency Nd:YAG laser, 140 mW of idle power over the range 2.29–2.96 µm has been obtained from a PE-SRO, for 800 mW of pump power. The device had an external pump power threshold of 250 mW, and wavelength tuning was available by a combination of temperature and grating period tuning. Later, the operation of a mid-IR PPLN PE-SRO pumped by a cw single-frequency Ti:sapphire laser and tunable from 4.07 to 5.26 µm was also reported [51]. By using a twin-cavity arrangement, mode-hop-free tuning of the single-frequency idler over 10.8 GHz was demonstrated by fine-tuning the pump laser over 12.3 GHz.

The high optical non-linearity of PPLN and long available interaction lengths has also permitted the development of PE-SRO devices pumped directly by single-stripe semiconductor diode lasers. Operation of a device based on a 50-mm PPLN crystal and pumped at ~810 nm with a single-mode extended-cavity AlGaAs laser has been achieved with a power threshold of only around 25–30 mW [52]. This device could provide up to 4 mW of unidirectional idler power, with tuning coverage from 2.58 to 3.44 µm. Using alternative DRO configurations, similar cw OPOs based on PPLN have also been demonstrated by the use of solitary, single-stripe diode lasers

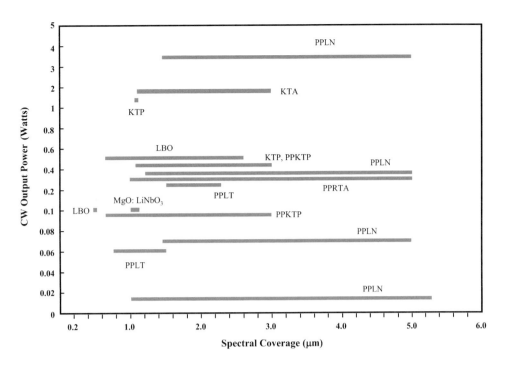

FIGURE 23.15 Survey of tuning range and output power of a number of cw OPOs demonstrated to date. The indicated tuning limits correspond to the potential coverage available to the particular crystal in the given phase-matching and pumping configuration, although the experimental tuning range may not be continuous or may have been limited by the available mirrors and crystal coatings. The output powers correspond to the maximum combined signal and idler power generated within the demonstrated tuning range.

[53,54]. These devices can operate with pump powers of less than 20 mW and can provide practical mid-IR output powers of up to 5 mW over a spectral range 2.2–3.7 µm. It is important to note that the potential tuning range available to all PPLN-based cw OPOs under Ti:sapphire, Nd:YAG or direct diode laser pumping extends across a continuous range of ~1.5–5 µm, limited by the increased absorption in the material beyond ~5 µm. By suitable choice of OPO mirror and crystal coatings, it is possible to access this entire tuning range with one device incorporating a single PPLN crystal.

At present, substantial progress continues in the development of cw OPOs based on PPLN. These devices have now been developed to the point where they can provide widely tunable, stable, single-frequency coherent radiation at practical power levels in the mid-IR. However, effective frequency control and power stabilization of PPLN devices, particularly under high-intensity pumping regimes in SRO configurations, are impeded by the photorefractive effect, thermal phase-mismatching and green-induced infrared absorption; practical operation of these devices requires careful control of these parameters. In this context, alternative periodically poled materials, particularly phosphates and arsenates of KTP, represent attractive material candidates for mid-IR cw OPOs, because of the absence of such detrimental effects, an equally extended transparency range to >5 µm, and lower coercive fields for poling. However, the shorter available interaction lengths (typically <20 mm) and lower effective non-linear coefficients lead to higher operation thresholds than in equivalent PPLN devices. This precludes SRO operation in external pumping configurations even with high-power multi-watt laser pump sources, and so other approaches based on the DRO, PE-SRO and IC-SRO are the most viable route to the development of mid-IR cw OPOs based on these materials [55,56]. Currently, the KTP family of periodically poled materials also offer more limited tuning flexibility than PPLN, because of the lack of significant temperature tuning and the difficulty in incorporating several different grating periods onto a single crystal due to limited apertures. However, the continuing advances in poling and fabrication technology are expected to pave the way for the practical development of stable and widely tunable cw OPOs for the mid-IR based on these materials. Figure 23.15 provides a summary of the spectral coverage and power performance of several cw OPOs developed to date.

23.8 Nanosecond OPO Devices

Advances in OPO devices have traditionally been most deliberate in the nanosecond regime, where the high peak pump intensities can be readily made available, even for SRO operation, with the use of Q-switched laser sources delivering pulses between 1 and 100 ns. Indeed, the first operation of an OPO was achieved with a pulsed nanosecond laser [57] and the majority of the early devices conform to this mode of operation. The use of this device configuration was also driven by the easier accessibility to such pulsed lasers in the early days of non-linear optics. This operating regime has continued to represent the most widespread and extensively developed of all OPO device architectures.

In contrast, practical operation of nanosecond OPOs is more demanding on material damage tolerance than other device configurations, because of the relatively high optical energy associated with nanosecond pulses. Material damage is generally exacerbated at higher pump energy fluence, and this was perhaps the single most important factor hampering the practical development of early nanosecond OPOs that followed the demonstration of the very first device in 1965 [57]. With the emergence of a new generation of non-linear materials in the 1980s, exhibiting unprecedented damage thresholds, the development of truly practical nanosecond OPOs became a reality for the first time. Of the new materials, BBO, LBO and KTP have been the prime candidates for visible and near-IR parametric generation. Other non-linear crystals such as KNB and the arsenate isomorphs of KTP, namely, KTA, CTA, RTA, have been of particular interest for wavelength generation in the difficult 3–5 µm spectral range in the mid-IR, because of their longer infrared transmission beyond 5 µm.

The majority of nanosecond OPOs have been pumped by the well-established Q-switched Nd:YAG laser [58–68]; some in diode-pumped all-solid-state configurations [69–71]. Due to their high average power capability, ultraviolet excimer lasers have also been used as optical pumps [72–74]. The upper limit to the conversion efficiency and output pulse energy in nanosecond OPOs is generally set by optical damage to the mirror and crystal coatings, which can be caused either by the pump pulses or by the circulating signal intensities. This can be minimized by using novel pumping and crystal configurations [60], longer pump pulses and shorter cavity lengths. For enhancement of spectral and spatial coherence and output efficiency, novel design concepts and new resonator architectures have also been effectively deployed [75–80]. These have enabled the generation of optical pulses with high spectral purity, excellent spatial beam quality and broad spectral tunability using nanosecond OPOs. These devices have now been established as viable sources of coherent light pulses from ~400 nm in the near-UV to ~5 µm the mid-IR. Optical energies from typically a few mJ to as much as 200 mJ can now be made available from such OPOs over a range of pulse durations from 1 to 50 ns. For wavelengths beyond 5 µm, classical materials such as $ZnGeP_2$ have proved highly effective. By using a *cascaded* pumping approach, spectral coverage to ~12 µm in spectrally narrow output pulses has been achieved [81,82]. The important advances in this area have also led to the commercial realization of several OPO devices based on the nanosecond pumping approach. The development of wide-aperture periodically poled materials PPKTP and PPRTA also promises further progress in high-energy OPOs with more flexible operating parameters [83–85]. The large non-linear gain of PPLN has also enabled fibre laser pumping of nanosecond OPOs [86] and the development of new types of single-pass parametric generators using nanosecond pulse pumping [87]. Figure 23.16 provides a summary of the main performance characteristics of several nanosecond OPOs developed to date.

At the same time, the availability of PPLN, PPKTP and its arsenate isomorphs such as PPRTA and PPKTA, with large effective non-linearity (d_{eff} ~8–16 pm V^{-1}, see Figure 23.14) and long interaction lengths (up to 60 mm in PPLN), has enabled the realization of pulsed OPOs at considerably lower pump pulse energies than practicable with birefringent materials. Combined with the advances in pump laser technology, this has facilitated the development of nanosecond OPOs using high-repetition-rate or long-pulse pump sources with relatively low pump pulse energies (few µJ to few mJ) [88,89]. While avoiding material damage due to the lower energy fluence, this approach has permitted the successful development of OPOs that can deliver a continuous train of nanosecond pulses at

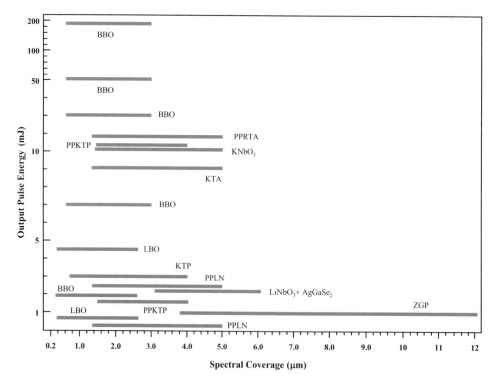

FIGURE 23.16 Survey of the tuning range and output pulse energy of several pulsed nanosecond OPOs developed to date. The indicated tuning limits correspond to the potential tuning range available to the particular crystal in the given phase-matching and pumping geometry, although not necessarily demonstrated. The indicated energies correspond to the maximum pulse energy obtained over the demonstrated tuning range.

Optical Parametric Devices

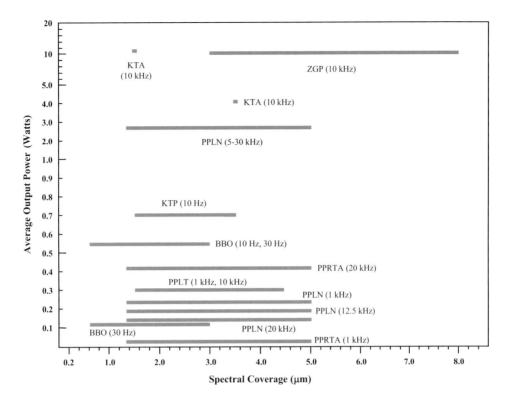

FIGURE 23.17 Survey of performance characteristics of high-repetition-rate nanosecond OPOs. The indicated tuning limits correspond to the potential tuning range available to the particular crystal in the given phase-matching and pumping geometry, although not necessarily demonstrated. The indicated average powers correspond to the maximum power generated over the demonstrated tuning range.

high repetition rates up to 30 kHz, with average powers as high as 10 W. Almost all devices in this operating regime have been pumped by the Nd:YAG laser or its variations, many in diode-pumped all-solid-state configurations. The main operating characteristics of a number of high-repetition-rate nanosecond OPOs are summarized in Figure 23.17.

23.9 Synchronously Pumped OPO Devices

As noted in Section 23.5.2.2, synchronously pumped OPOs may be operated either as steady-state cw or pulsed (quasi-cw) devices. For most practical applications, cw operation is the desired configuration because the output consists of a truly continuous train of identical pulses. In pulsed oscillators, the amplitude, intensity and duration of the output pulses can vary across the pulse envelope, and the output does not constitute a truly repetitive pulse train. However, because of the significantly higher peak powers available from pulsed mode-locked than the equivalent cw mode-locked pump lasers, the operation of synchronously pumped OPOs is more readily attainable under pulsed conditions, particularly in SRO configurations of practical interest. For this reason, most of the early synchronously pumped OPOs based on birefringent materials have been pulsed oscillators. With the availability of high-power mode-locked laser sources and novel non-linear materials, operation of cw synchronously pumped OPOs has become a practical reality, circumventing the need for pulsed pumping or the use of DRO schemes. We, therefore, focus on a description of synchronously pumped OPOs in the cw SRO configuration, as this represents the most practical and stable mode of operation for picosecond and femtosecond OPOs.

23.9.1 Picosecond OPO Devices

As with nanosecond OPOs, the majority of picosecond OPOs demonstrated to date have been based on the cw mode-locked Nd laser or its variations as the pump source. These include all-solid-state cw picosecond SROs based on KTP [90] and LBO [91,92], pumped by coupled-cavity mode-locked Nd:YLF lasers, providing output pulses with durations of 1–5 ps at repetition rates of up to 130 MHz. The LBO-based picosecond SRO can generate continuously tunable output from <650 nm to >2.5 μm with a single crystal, while the KTP device is tunable over 900 nm to 1.2 μm. Average output powers of tens of milliwatts at efficiencies of up to 30% are routinely available from these devices. High-power operation of cw picosecond SROs has also been achieved with the use of flashlamp-pumped cw mode-locked neodymium lasers. By using the fundamental output of a Nd:YLF laser at 1.053 μm, total average output powers of up to 2.8 W have been obtained from a KTP picosecond SRO for 14 W of pump, with 800 mW in the idler beam near 3.2 μm [93]. This device generated signal pulses of 12 ps duration for 40 ps input pump pulses at 76 MHz repetition rate. A similar mid-IR oscillator was demonstrated by using temporally compressed pulses from a Nd:YAG laser at 1.064 μm to pump a KTP SRO [94]. Average output powers of up to 350 mW in pulses of 2–3 ps at 75 MHz were generated for 4 W of pump power. Later, the operation of a cw picosecond SRO at unprecedented power levels was achieved by using a diode-pumped, cw mode-locked Nd:YVO$_4$ oscillator-amplifier laser system at 1.064 μm to pump the crystal of KTA in a NCPM geometry [95]. With the pump radiation consisting of 7 ps pulses at 83 MHz repetition rate and an average power of 29 W, the OPO delivered as much as 6.4 W of idler output at 3.47 μm in the mid-IR. The combined signal and idler output power from this OPO was as high as 21 W, corresponding to an external efficiency in excess of 70%. High-repetition-rate operation of the KTA OPO at frequencies as high as 1.33 GHz was also demonstrated. Picosecond pulses in the visible have also been generated in LBO using the third harmonic of a cw mode-locked Nd:YLF laser [96]. Average powers of 275 mW in the blue spectral range 453–472 nm were obtained in 15 ps pulses at 75 MHz repetition rate.

The operation of cw picosecond SROs has also been extended to the Kerr-lens mode-locked (KLM) Ti:sapphire laser as the pump source. This approach is attractive because the tunability of the Ti:sapphire pump source allows wavelength tuning in the OPO without resorting to angle phase-matching. This minimizes intra-cavity reflection losses caused by crystal rotation, thus maintaining maximum efficiency across the available tuning range. By using type II NCPM in KTP, picosecond pulses in the 1–1.2 μm and 2.3–2.9 μm spectral range have been generated by tuning the Ti:sapphire laser over 720–853 nm [97]. Average output powers of up to 700 mW in 1.2 ps pulses at 82 MHz repetition rate were produced for 1.6 W of pump. The combination of tunable Ti:sapphire laser with temperature-tuned LBO under type I NCPM has also been shown to provide picosecond pulses with continuous tunability over 1–2.4 μm [98]. This SRO could generate average powers of up to 325 mW for 1.2 W of pump in pulses of ~1–2 ps durations at 81 MHz repetition rate. For wavelength coverage further into the infrared, the arsenate isomorphs of KTP, namely KTA and RTA, represent excellent material candidates because of their extended transparency beyond that of KTP and LBO as well as their NCPM capability under Ti:sapphire pump tuning. The use of type II NCPM in KTA has enabled the generation of picosecond pulses out to 3.6 μm, with average powers of up to 400 mW in 1–3 ps pulses [99]. The extension of tuning range of Ti:sapphire-pumped picosecond OPOs to the visible has also been demonstrated by intra-cavity frequency doubling of signal pulses. By using LBO as the non-linear material, visible pulses of ~1 ps duration in the 584–771 nm spectral range have been obtained at 320 mW average power with this technique [100].

The availability of periodically poled materials has also brought about new opportunities for the development of cw picosecond SROs with improved performance characteristics. The large non-linearity of these materials combined with NCPM capability and broad infrared transparency has enabled the implementation of picosecond SROs with minimal pump power requirement, high output efficiency, practical output powers and extended tunability into the difficult mid-IR spectral range. In particular, compact all-solid-state SROs based on PPLN or PPRTA and pumped by mode-locked Nd:YLF and Ti:sapphire lasers have been shown to be versatile sources of picosecond pulses with tunability across the entire 1.5–6 μm range [101–103]. These devices exhibit average pump power thresholds as low as 10 mW and can provide signal output powers of up to ~400 mW in pulses of ~1–5 ps duration, with >10 mW of idler available beyond 5 μm.

Power scaling of cw picosecond SROs to unprecedented levels has become possible by using flashlamp-pumped Nd lasers in combination with PPLN. By using a picosecond Nd:YLF pump laser, an average mid-IR output power of 2.4 W in pulses of ~45 ps at ~76 MHz repetition rate was generated in PPLN, with a mid-IR tuning range from ~2.15 to ~2.6 μm [104]. The SRO generated a total output power of ~5 W for 7.4 W of input pump power. Major enhancements in the mid-infrared output power and photon conversion efficiency have been shown to be possible by using intra-cavity DFM techniques [105]. In a separate experiment, the use of 80 ps pulses at 76 MHz from a high-power Nd:YAG laser has permitted the generation of more than 4 W of mid-IR idler power from a PPLN picosecond OPO, with a tunable range over 2.2–2.8 μm [106]. With an average input pump power of 18 W, the OPO provided a combined signal and idler output power in excess of 12 W at ~65% extraction efficiency. Operation of cw picosecond OPOs has later also been extended to novel mode-locked semiconductor pump lasers operating at GHz repetition rates. By using an InGaAs oscillator-amplifier system delivering 7.8 ps pulses at 2.5 GHz, up to 78 mW of mid-IR idler power over a tunable range of 1.99–2.35 μm was generated from a PPLN OPO for 900 mW of input pump power [107]. Later, operation of a cw picosecond SROs has been achieved at repetition rates as high as 10 GHz by using novel all-solid-state Nd lasers in combination with monolithic PPLN [108]. Figure 23.18 provides a summary of the performance characteristics of cw picosecond OPOs developed to date.

23.9.2 Femtosecond OPO Devices

OPOs operating in the femtosecond time domain represent the latest class of parametric devices. The high peak intensities available from ultrashort femtosecond pulses are particularly suited to the exploitation of the small non-linear gain in parametric devices, enabling the practical development of femtosecond OPOs in cw SRO configurations. However, because of the short temporal duration (~100 fs) and large spectral content (~10 nm) of femtosecond pulses, additional effects such as group velocity dispersion, temporal walk-off and spectral acceptance bandwidths play an important role in the operation of these devices. Group velocity dispersion often leads to pulse broadening in femtosecond OPOs, while temporal walk-off can degrade non-linear gain or even modify the temporal characteristics of the output pulses. The spectral acceptance bandwidth of the crystal can also reduce gain as well as set a lower limit to the minimum attainable pulse duration from the OPO. The influence of temporal walk-off and spectral bandwidth can be minimized by using short crystal lengths of typically 1–2 mm, while the effects of group velocity dispersion can be overcome by the inclusion of dispersion compensation in the OPO cavity, as in conventional femtosecond lasers. The choice of non-linear crystal, therefore, requires a trade-off with the crystal length. The need for high optical non-linearity for low operation threshold must be traded against the desire for large phase-matching bandwidth and low temporal walk-off for optimum design. While crystal lengths of 1–2 mm are too short

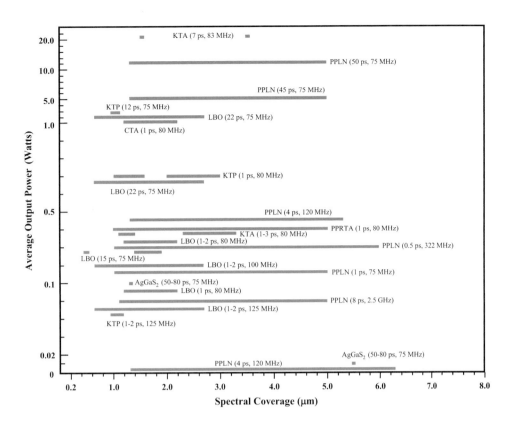

FIGURE 23.18 Survey of the performance characteristics of several cw picosecond OPOs demonstrated to date. The average output powers represent the maximum combined power in the signal and idler beams. The indicated tuning limits correspond to the potential tuning range available to the particular crystal in the given phase-matching geometry and pumping configuration, although the experimental tuning range may have been limited by the available mirrors and crystal coatings. The pulse durations correspond to the measured signal or idler pulse width within the tuning range.

to provide sufficient gain in picosecond and nanosecond oscillators, the substantially higher peak intensities available with femtosecond pulses can adequately compensate for this shortfall in gain. At the same time, the high in-crystal peak intensities (a few GW cm^{-2}) can introduce higher-order non-linear effects such as self-phase-modulation in femtosecond OPOs. Such effects, which are not generally present in picosecond and nanosecond OPOs, often lead to spectral broadening and chirping of the output pulses and become more pronounced with longer crystal lengths. The spectral broadening due to self-phase-modulation can, however, be exploited for subsequent compression of output pulses. Therefore, in addition to the general phase-matching and material requirements for parametric devices, the design of femtosecond OPOs involves consideration of the pertinent temporal and spectral effects.

The first successful operation of a cw femtosecond OPO was achieved in a 1.4 mm crystal of KTP, pumped by 170 fs pulses from a colliding-pulse mode-locked dye laser at 620 nm [109]. To access the high peak intensities necessary for oscillation, the KTP crystal was pumped at the intra-cavity focus of the dye laser in a non-collinear phase-matching geometry. The SRO produced 220 fs signal pulses at milliwatt average power levels in the near-IR. The approach was subsequently extended to an externally pumped SRO where increased signal powers of up to 30 mW were generated [110].

Soon after, the availability of the KLM Ti:sapphire laser provided a laser source capable of providing cw femtosecond pulses at far higher intensities and many times above SRO threshold. This laser has since become the primary pump source for femtosecond OPOs, enabling the development of a KTP-based device capable of producing total average output powers of up to 750 mW at wavelength < 2 μm in the near-IR using non-collinear CPM or collinear NCPM schemes [111–113]. Pulse durations of <100 fs at repetition rates of 75–85 MHz are routinely available from these devices. The non-collinear CPM technique was soon extended to other KTP-based systems [114], as well as to KTA [115,116], CTA [117,118], RTA [119] and KNB [120], providing femtosecond pulses into the 2–5 μm spectral range using pump and angle tuning schemes. Total average powers of hundreds of milliwatts and mid-IR output powers of tens of milliwatts in 100–200 fs pulses have been obtained from these devices. By using alternative collinear NCPM schemes in combination with Ti:sapphire pump tuning, the operation of a femtosecond near-IR OPO based on KTP was achieved, generating 40 fs output pulses [121], while a similar device based on RTA was reported as having a pump power threshold as low as 50 mW [122]. This device provided mid-IR pulses in the 2–3 μm spectral range at 50–100 mW of average power. A femtosecond OPO using type I temperature-tuned NCPM in LBO was also demonstrated, providing ~500 mW in ~100 fs pulses from 1.1 to 2.4 μm range [123].

The spectral range of Ti:sapphire-pumped femtosecond OPOs has also been extended to the visible by using two different approaches. The first relies on SHG of near-IR signal pulses internal to the OPO cavity. By using BBO as the doubling crystal, output powers of up to 240 mW in 70–100 fs pulses have been generated over a spectral range of 580–660 nm in KTP- and RTA-based OPOs [124,125]. Intracavity SFM between the pump and signal pulses in BBO has also been used in a KTP OPO to provide sub-100 fs pulses in the blue spectral range from 426 to 483 nm at ~200 mW average power [126]. Internal SHG and parametric generation has also been simultaneously achieved in an OPO based on a single KTP crystal, providing visible femtosecond pulses in the 530–580 nm range with up to 200 mW average power [127]. The second approach to visible pulse generation has been the use of frequency-doubled output of the Ti:sapphire laser at ~400 nm to directly pump a visible OPO based on BBO as the non-linear crystal [128]. This approach provided 30 fs signal pulses with average powers of up to 100 mW from 566 to 676 nm. Subsequently, femtosecond pulses with durations as short as 13 fs were generated from such a device [129].

Later, the introduction of periodically poled non-linear materials has led to further advances in femtosecond OPOs. Devices based on PPLN, PPRTA and PPKTP have been shown to be highly versatile sources of femtosecond pulses, offering minimum pump thresholds, high output powers and vast spectral coverage from a single device. Operation of femtosecond OPOs based on periodically poled materials has been demonstrated over a wide spectral range from ~1 to ~6.8 µm using non-collinear CPM or collinear NCPM schemes combined with temperature, grating or angle-tuning [130–134]. Because of the large non-linear gains available, these oscillators exhibit pump power thresholds typically well below 100 mW, enabling the use of all-solid-state Ti:sapphire lasers as the pump source [130,132]. These devices can readily provide practical powers in excess of 100 mW in the 1–4 µm range, with milliwatt-level output attainable in the difficult spectral regions beyond 5 µm. Output pulse durations of 100–200 fs at ~80 MHz repetition rate are typically available. Because of the large non-linear gain bandwidth of periodically poled crystals, combined with the short interaction lengths used in femtosecond OPOs, devices based on such materials have been shown to be particularly flexible in providing extensive wavelength coverage only through cavity length tuning of the OPO [131–133]. This provides a highly convenient method for wavelength tuning and without the need for pump, angle, grating or temperature tuning. This tuning mechanism occurs because the cavity length detuning introduces a loss at the signal wavelength by reducing the synchronism between the pump and signal pulses. To maintain synchronism and optimize gain, the signal shifts to a more favourable wavelength with a group velocity that satisfies a constant round-trip time. This cavity length tuning, which was observed in the first demonstration of a femtosecond OPO [109], is a useful mechanism for tuning the output wavelength, often by as much as 50 nm in birefringent crystals.

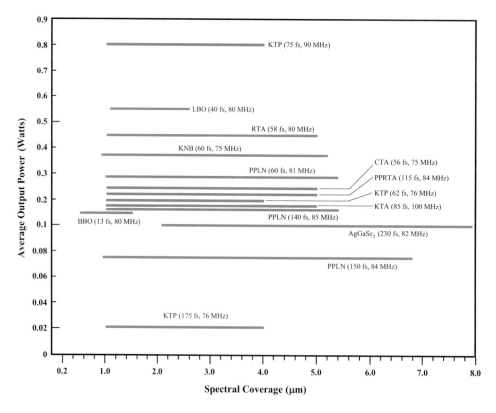

FIGURE 23.19 Survey of the performance characteristics of Ti:sapphire-pumped cw femtosecond OPOs demonstrated to date. The indicated tuning limits correspond to the potential tuning range available to the particular crystal in the given phase-matching geometry and pumping configuration, although the experimental tuning range may have been limited by the available mirrors and crystal coatings. The output powers correspond to the maximum combined signal and idler power generated within the demonstrated tuning range. The pulse durations correspond to the measured signal or idler pulse width within the tuning range.

In periodically poled materials, the large phase-matching bandwidths can provide cavity length tuning over hundreds of nanometres, limited by the bandwidth of OPO mirrors.

Femtosecond pulses in the 5–10 µm spectral range have also been generated by using a cascaded, two-stage pumping arrangement, where the output of a Ti:sapphire-pumped CTA OPO has been used as the pump for a second oscillator based on the classical non-linear material AgGaSe$_2$ [135]. The wavelength flexibility of the primary OPO provides a suitable pump wavelength above the absorption edge of the material, which avoids material absorption and at the same time facilitates the required phase-matching condition for mid-IR generation. Using this technique, femtosecond pulses in the 4–8 µm spectral range with average powers of up to 35 mW and pulse durations of 300–640 fs have been obtained at 82 MHz repetition rate. Successful implementation of this technique, however, necessitates the use of high-power input laser pump sources. Later, operation of Ti:sapphire-pumped femtosecond OPOs at repetition rates as high as 1 GHz has been achieved [136,137], and the use of mode-locked fibre lasers as pump sources for femtosecond OPOs has been demonstrated [138]. Figure 23.19 provides a summary of the performance characteristics of several Ti:sapphire-pumped cw femtosecond OPOs demonstrated to date.

23.10 Summary

This chapter has provided a description of optical parametric devices, with particular emphasis on OPOs operating from the cw to the ultrafast femtosecond timescales. The discussion has included the basic elements of non-linear frequency conversion, the main operating principles of parametric devices, and a survey of the developments in OPOs. A key element in the rapid advancement of OPO technology has been the emergence of new non-linear materials, replacing the more traditional non-linear crystals, as well as the development of novel resonator and pumping techniques. The advent of PPLN, PPRTA and other periodically poled materials as well as birefringent materials such as BBO, LBO, KTP and its arsenate isomorphs has led to the realization of practical OPOs covering spectral regions from <400 nm in the near-UV to >12 µm in mid-IR across all temporal timescales and with unprecedented performance characteristics.

In cw operation, OPOs have been shown to be capable of providing output powers of up to ~4 W and wavelength coverage to >5 µm, while in pulsed nanosecond operation, pulse energies as high as 200 mJ and spectral access to >12 µm have been achieved. In picosecond operation, average output powers of more than 20 W have been generated and tuning to >6 µm demonstrated. In the femtosecond regime, average powers in excess of 1.5 W, spectral coverage to ~8 µm and pulse durations as short as ~13 fs have been obtained from OPO devices. With the continuing progress in non-linear materials, laser pump sources, novel phase-matching, pumping and resonator techniques and the need for new coherent light sources in new spectral regions, particularly the mid-IR, the potential for further advancement of OPO device technology remains strong.

REFERENCES

1. Maiman T H 1960 *Nature* **187** 493.
2. Franken P A, Hill A E, Peters C W and Weinreich G 1961 Generation of optical harmonics *Phys. Rev. Lett.* **7** 118–19.
3. Armstrong J A, Bloembergen N, Ducuing J and Pershan P S 1962 Interaction between light waves in a nonlinear dielectric *Phys. Rev.* **127** 1918–39.
4. Ebrahimzadeh M 2003 Parametric light generation *Phil. Trans. R. Soc. A* in press.
5. Shen Y R 1984 *Principles of Nonlinear Optics* (New York: Wiley).
6. Dimitriev V G, Gurzadyan G G and Nikogosyan D N 1991 *Handbook of Nonlinear Optical Crystals* (Berlin: Springer).
7. Boyd R W 1992 *Nonlinear Optics* (New York: Academic).
8. Sutherland R L 1996 *Handbook of Nonlinear Optics* (New York: Marcel Dekker).
9. Kingston R H 1962 Parametric amplification and oscillation at optical frequencies *Proc. IRE* **50** 472–3.
10. Kroll N M 1962 Parametric amplification in spatially extended media and application to the design of tuneable oscillators at optical frequencies *Phys. Rev.* **127** 1207–11.
11. Akhmanov S A and Khokhlov R V 1963 Concerning one possibility of amplification of light waves *Sov. Phys. J. Exp. Theor. Phys.* **16** 252–4.
12. Louisell W H, Yariv A and Siegman A E 1961 Quantum fluctuations and noise in parametric processes *Phys. Rev.* **124** 1646.
13. Yariv A 1989 *Quantum Electronics* (New York: Wiley).
14. Smith R G 1976 *Optical Parametric Oscillators (Lasers)* ed A K Levine and A J DeMaria (New York: Marcel Dekker) pp 189–307.
15. Harris S E 1969 Tunable optical parametric oscillators *Proc. IEEE* **57** 2096–113.
16. Giordmaine J A 1962 Mixing of light beams in crystals *Phys. Rev. Lett.* **8** 19–20.
17. Maker P D, Terhune R W, Nissenoff M and Savage C M 1962 Effects of dispersion and focusing on the production of optical harmonics *Phys. Rev. Lett.* **8** 21–2.
18. Born M and Wolf E 1984 *Principles of Optics* (Oxford: Pergamon).
19. Ebrahimzadeh M and Ferguson A I 1993 *Novel Nonlinear Crystals in Principles and Applications of Nonlinear Optical Materials* (London: Chapman and Hall) pp 99–142.
20. Hobden M V 1967 *J. Appl. Phys.* **38** 4365.
21. Fejer M M, Magel G A, Jundt D H and Byer R L 1992 Quasi-phase-matched second harmonic generation: tuning and tolerances *IEEE J. Quantum Electron.* **28** 2631–54.
22. Myers L E, Eckardt R C, Fejer M M, Byer R L, Bosenberg W R and Pierce J W 1995 Quasi-phase-matched optical parametric oscillators in bulk periodically poled LiNbO$_3$ *J. Opt. Soc. Am. B* **12** 2102–16.
23. Arbore M A, Marco O and Fejer M M 1997 Pulse compression during second-harmonic generation in aperiodic quasi-matching gratings *Opt. Lett.* **22** 865–7.
24. Imeshev G, Galvanauskas A, Harter D, Arbore M A, Procter M and Fejer M M 1998 Engineerable femtosecond pulse shaping by second-harmonic generation with synthetic quasi-phase-matching gratings *Opt. Lett.* **23** 864–6.

25. Beddard T, Ebrahimzadeh M, Reid D T and Sibbett W 2000 Five-optical-cycle pulse generation in the mid-infrared from an optical parametric oscillator based on aperiodically-poled lithium niobate *Opt. Lett.* **25** 1052–54.
26. Byer R L and Piskarskas A S (ed) 1993 Optical parametric oscillation and amplification *J. Opt. Soc. Am. B* **10** 1656–791, 2148–243.
27. Bosenberg W R and Eckardt R C (ed) 1995 Optical parametric devices *J. Opt. Soc. Am. B* **12** 2084–322.
28. Ebrahimzadeh M, Eckardt R C and Dunn M H (ed) 1999 Optical parametric devices and processes *J. Opt. Soc. Am. B* **16** 1477–597.
29. Danielius R, Piskarskas A, Stabinis A, Banfi G P, Di Trapani P and Righini R 1995 Travelling-wave parametric generation of widely tunable, highly coherent femtosecond light pulses *J. Opt. Soc. Am. B* **10** 2222–31.
30. Ebrahimzadeh M and Dunn M H 2000 Optical parametric oscillators *Handbook of Optics* vol IV (New York: McGraw-Hill) pp 2201–72.
31. Ebrahimzadeh M, Turnbull G A, Edwards T J, Stothard D J M, Lindsay I D and Dunn M H 1999 Intracavity continuous-wave singly resonant optical parametric oscillators *J. Opt. Soc. Am. B* **16** 1499–511.
32. Bjorkholm J E, Ashkin A and Smith R G 1970 *IEEE J. Quantum Electron.* **QE-6** 797.
33. Siegman A E 1962 Nonlinear optical effects: an optical power limiter *Appl. Opt.* **1** 739–44.
34. Brosnan S J and Byer R L 1979 Optical parametric oscillator threshold and linewidth studies *IEEE J. Quantum Electron.* **QE-15** 415–31.
35. Pearson J E, Ganiel U and Yariv A 1972 Rise time of pulsed parametric oscillators *IEEE J. Quantum Electron.* **QE-8** 433–40.
36. Bjorkholm J E 1968 Efficient optical parametric oscillation using doubly and singly resonant cavities *Appl. Phys. Lett.* **13** 53–5.
37. Kreuzer L B 1968 High-efficiency optical parametric oscillation and power limiting in $LiNbO_3$ *Appl. Phys. Lett.* **13** 57–9.
38. Bjorkholm J E 1971 Some effects of spatially nonuniform pumping in pulsed optical parametric oscillators *IEEE J. Quantum Electron.* **QE-7** 109–18.
39. Becker M F, Kuizanga D J, Phillion D W and Siegman A E 1974 Analytic expressions for ultrashort pulse generation in mode-locked optical parametric oscillators *J. Appl. Phys.* **45** 3996–4005.
40. Cheung E C and Liu J M 1991 Theory of a synchronously pumped optical parametric oscillator in steady-state operation *J. Opt. Soc. Am.* **B 7** 1385–401; Cheung E C and Liu J M 1991 Efficient generation of ultrashort, wavelength-tunable infrared pulses *J. Opt. Soc. Am. B* **8** 1491–506.
41. Ebrahimzadeh M, Henderson A J and Dunn M H 1990 *IEEE J. Quantum Electron.* **26** 1241–52.
42. Colville F G, Dunn M H and Ebrahimzadeh M 1997 Continuous-wave, singly resonant intracavity parametric oscillator *Opt. Lett.* **22** 75–9.
43. Edwards T J, Turnbull G A, Dunn M H and Ebrahimzadeh M and Colville F G 1998 High-power, continuous-wave, singly resonant intracavity optical parametric oscillator *Appl. Phys. Lett.* **72** 1527–9.
44. Scheidt M, Beier B, Knappe R, Boller K-J and Wallenstein R 1995 Diode-laser-pumped continuous-wave KTP optical parametric oscillator *J. Opt. Soc. Am. B* **12** 2087–94.
45. Scheidt M, Beier B, Boller K-J and Wallenstein R 1997 Frequency-stable operation of a diode-pumped continuous-wave $RbTiOAsO_4$ optical parametric oscillator *Opt. Lett.* **22** 1287–9.
46. Bosenberg W R, Drobshoff A, Alexander J I, Myers L E and Byer R L 1996 93% pump depletion, 3.5 W continuous-wave, singly resonant optical parametric oscillator *Opt. Lett.* **21** 1336–8.
47. Powers P E, Kulp T J and Bisson S E 1998 Continuous tuning of a continuous-wave periodically poled lithium niobate optical parametric oscillator by use of a fan-out grating design *Opt. Lett.* **23** 159–61.
48. Klein M E, Lee D H, Meyn J-P, Boller K-J and Wallenstein R 1999 Singly resonant continuous-wave optical parametric oscillator pumped by a diode laser *Opt. Lett.* **24** 1142–4.
49. Gross P, Klein M E, Walde T, Boller K-J, Auerbach M, Wessels P and Fallnich C 2002 Fiber-laser-pumped continuous-wave singly-resonant optical parametric oscillator *Opt. Lett.* **27** 418–20.
50. Stothard D J M, Ebrahimzadeh M and Dunn M H 1998 Low-pump-threshold, continuous-wave, singly resonant optical parametric oscillator *Opt. Lett.* **23** 1895–7.
51. Turnbull G A, McGloin D, Lindsay I D, Ebrahimzadeh M and Dunn M H 2000 Extended mode-hop-free tuning using a dual-cavity, pump-enhanced optical parametric oscillator *Opt. Lett.* **25** 341–3.
52. Lindsay I D, Petridis C, Dunn M H and Ebrahimzadeh M 2001 Continuous-wave pump-enhanced singly-resonant optical parametric oscillator pumped by an external-cavity diode laser *Appl. Phys. Lett.* **78** 871–3.
53. Lindsay I D, Turnbull G A, Dunn M H and Ebrahimzadeh M 1998 Doubly-resonant continuous-wave optical parametric oscillator pumped by a single-mode laser diode *Opt. Lett.* **23** 1889–91.
54. Henderson A J, Roper P M, Borschowa L A and Mead R D 2000 Stable, continuously tunable operation of a diode-pumped doubly resonant optical parametric oscillator *Opt. Lett.* **25** 1264–6.
55. Edwards T J, Turnbull G A, Dunn M H and Ebrahimzadeh M 1998 Continuous-wave, singly-resonant optical parametric oscillator based on periodically-poled $RbTiOAsO_4$ *Opt. Lett.* **23** 837–9.
56. Edwards T J, Turnbull G A, Dunn M H and Ebrahimzadeh M 2000 Continuous-wave, singly-resonant optical parametric oscillator based on periodically-poled $KTiOPO_4$ *Opt. Exp.* **6** 58–63.
57. Giordmaine J A and Miller R C 1965 Tunable coherent parametric oscillation in $LiNbo_3$ at optical frequencies *Phys. Rev. Lett.* **14** 973–6.
58. Cheng L K, Bosenberg W R and Tang C L 1988 Broadly tunable optical parametric oscillation in β-BaB_2O_4 *Appl. Phys. Lett.* **53** 175–7.
59. Fan Y X, Eckardt R C and Byer R L 1988 *Appl. Phys. Lett.* **50** 2014.
60. Bosenberg W R, Cheng L K and Tang C L 1989 Ultraviolet optical parametric oscillation in fi-BaB_2O_4 *Appl. Phys. Lett.* **54** 13–15.

61. Fan Y X, Eckardt R C, Byer R L, Chen C and Jiang A D 1989 *IEEE J. Quantum Electron.* **25** 1196.
62. Bosenberg W R, Pelouch W S and Tang C L 1989 High-efficiency and narrow-linewidth operation of a two-crystal β-BaB$_2$O$_4$ optical parametric oscillator *Appl. Phys. Lett.* **55** 1952–4.
63. Bosenberg W R and Tang C L 1990 Type II phase matching in β-barium borate optical parametric oscillator *Appl. Phys. Lett.* **56** 1819–21.
64. Kato K 1991 Parametric oscillation at 3.2 μm in KTP pumped at 1.064 μm *IEEE J. Quantum Electron.* **27** 1137–40.
65. Hanson F and Dick D 1991 Blue parametric generation from temperature-tuned LiB$_3$O$_5$ *Opt. Lett.* **16** 205–7.
66. Bosenberg W R, Cheng L K and Bierlein J D 1994 Optical parametric frequency-conversion properties of KTiOAsO$_4$ *Appl. Phys. Lett.* **65** 2765–7.
67. Gloster L A W, Jiang Z X and King T A 1994 *IEEE J. Quantum Electron.* **30** 2961.
68. Urschel R, Fix A, Wallenstein R, Rytz D and Zysset B 1995 *J. Opt. Soc. Am. B* **12** 726.
69. Cui Y, Dunn M H, Norrie C J, Sibbett W, Sinclair B D, Tang Y and Terry J A C 1992 *Opt. Lett.* **17** 646.
70. Cui Y, Withers D E, Rae C F, Norrie C J, Tang Y, Sinclair B D, Sibbett W and Dunn M H 1993 *Opt. Lett.* **18** 122–4.
71. Marshall L R, Kasinski J and Burnham R L 1991 Efficient optical parametric oscillation at 1.6 μm *Opt. Lett.* **16** 1680.
72. Komine H 1988 Optical parametric oscillation in beta-barium borate crystal pumped by an XeCl excimer laser *Opt. Lett.* **13** 643–5.
73. Ebrahimzadeh M, Henderson A J and Dunn M H 1990 An excimer-pumped β-BaB$_2$O$_4$ optical parametric oscillator tunable from 354 nm to 2.370 μm *IEEE J. Quantum Electron.* **26** 1241–52.
74. Ebrahimzadeh M, Robertson G and Dunn M H 1991 *Opt. Lett.* **16** 767.
75. Bosenberg W R and Guyer D R 1993 *J. Opt. Soc. Am. B* **10** 1716.
76. Fix A, Schroder T, Wallenstein R, Haub J G, Johnson M J and Orr B J 1993 *J. Opt. Soc. Am. B* **10** 1744.
77. Boon-Engering J M, van der Veer W E and Gerritsen J W 1995 Bandwidth studies of an injection-seeded β-barium borate optical parametric oscillator *Opt. Lett.* **20** 380–2.
78. Haub J G, Johnson M J, Powell A J and Orr B J 1995 Bandwidth characteristics of a pulsed optical parametric oscillator: application to degenerate four-wave mixing spectroscopy 1995 *Opt. Lett.* **20** 1637–9.
79. Urschel R, Bader U, Borzutsky A and Wallenstein R *J. Opt. Soc. Am. B*
80. Ribet I, Drag C, Lefebvre M and Rosencher E 2002 *Opt. Lett.* **27** 255.
81. Vodopyanov K L, Ganikhanov F, Maffetone J P, Zwieback I and Ruderman W 2000 *Opt. Lett.* **25** 841.
82. Ganikhanov F, Caughey T and Vodopyanov K L 2001 *Opt. Lett.* **25** 818.
83. Karlsson H, Olson M, Arvidsson G, Laurell F, Bader U, Borsutzky A, Wallenstein R, Wickstrom S and Gustafsson M 1999 *Opt. Lett.* **24** 330–2.
84. Hellstrom J, Pasiskevicius V, Karlsson H and Laurell F 2000 *Opt. Lett.* **25** 174.
85. Feve J-P, Pacaud O, Boulanger B, Menaert B, Hellstrom J, Pasiskevicius V and Laurell F 2001 *Opt. Lett.* **26** 1882.
86. Britton P E, Offerhaus H L, Richardson D J, Smith P G R, Ross G W and Hanna D C 1999 *Opt. Lett.* **24** 975–7.
87. Missey M J, Dominic V, Powers P E and Schepler K L 1999 *Opt. Lett.* **24** 1227–9.
88. Myers L E and Bosenberg W R 1997 *IEEE J. Quantum Electron.* **33** 1663–72.
89. Bader U, Bartschke J, Klimov I, Orsutzky A B and Wallenstein R 1998 *Opt. Commun.* **147** 95–8.
90. McCarthy M J and Hanna D C 1992 *Opt. Lett.* **17** 402–4.
91. Hall G J, Ebrahimzadeh M, Robertson A, Malcolm G P A and Ferguson A I 1993 *J. Opt. Soc. Am. B* **10** 2168–79.
92. Butterworth S D, McCarthy M J and Hanna D C 1993 *Opt. Lett.* **18** 1429–31.
93. Grasser C, Wang D, Beigang R and Wallenstein R 1993 *J. Opt. Soc. Am. B* **10** 2218–21.
94. Chung J and Siegman A E 1993 Singly resonant continuous-wave mode-locked KTiOPO$_4$ optical parametric oscillator pumped by a Nd:YAG laser *J. Opt. Soc. Am. B* **10** 2201–10.
95. Ruffing B, Nebel A and Wallenstein R 1998 All-solid-state cw mode-locked picosecond KTiOAsO$_4$ (KTA) optical parametric oscillator *Appl. Phys. B* **67** 537.
96. Wang D, Grasser C, Beigang R and Wallenstein R 1997 The generation of tunable blue ps-light-pulses from a cw mode-locked LBO optical parametric oscillator *Opt. Commun.* **138** 87–90.
97. Nebel A, Fallnich C, Beigang R and Wallenstein R 1993 Noncritically phase-matched continuous-wave mode-locked singly resonant optical parametric oscillator synchronously pumped by a Ti:sapphire laser *J. Opt. Soc. Am. B* **10** 2195–200.
98. Ebrahimzadeh M, French S and Miller A 1995 Design and performance of a singly resonant picosecond LiB$_3$O$_5$ optical parametric oscillator synchronously pumped by a self-mode-locked Ti:sapphire laser *J. Opt. Soc. Am. B* **12** 2180–91.
99. French S, Ebrahimzadeh M and Miller A 1996 High-power, high-repetition-rate picosecond optical parametric oscillator for the near- to mid-infrared *Opt. Lett.* **21** 131–3.
100. French S, Ebrahimzadeh M and Miller A 1996 High-power, high-repetition-rate picosecond optical parametric oscillator tunable in the visible *Opt. Lett.* **21** 976–8.
101. Kennedy G T, Reid D T, Miller A, Ebrahimzadeh M, Karlsson H, Arvidsson G and Laurell F 1998 Broadly tunable mid-infrared picosecond optical parametric oscillator based on periodically-poled RbTiOAsO$_4$ *Opt. Lett.* **23** 503–5.
102. Lefort L, Peuch K, Ross G W, Svirko Y P and Hanna D C 1998 Optical parametric oscillation out to 6.3 μm in periodically poled lithium niobate under strong idler absorption *Appl. Phys. Lett.* **73** 1610–12.
103. Phillips P J, Das S and Ebrahimzadeh M 2000 *Appl. Phys. Lett.* **77** 469.
104. Finsterbusch K, Urschel R and Zakarias H 2000 Fourier-transform-limited, high-power picosecond optical parametric oscillator based on periodically poled lithium niobate *Appl. Phys. B* **70** 741–6.

105. Dearborn M E, Koch K, Moore G T and Diels J C 1998 Greater than 100% photon-conversion efficiency from an optical parametric oscillator with intracavity difference-frequency mixing *Opt. Lett.* **23** 759–61.
106. Hoyt C W, Sheik-Bahae M and Ebrahimzadeh M 2002 High-power picosecond optical parametric oscillator based on periodically poled lithium niobate *Opt. Lett.* **27** 1543.
107. Robertson A, Klein M E, Tremont M A, Boller K-J and Wallesnstein R 2000 2.5-GHz repetition-rate singly resonant optical parametric oscillator synchronously pumped by a mode-locked diode oscillator amplifier system *Opt. Lett.* **25** 657.
108. Lecomte S, Krainer L, Paschotta R, Dymott M J P, Weingarten K J and Keller U 2002 *Opt. Lett.* **27** 1714.
109. Edelstein D C, Wachman E S and Tang C L 1989 Broadly tunable high repetition rate femtosecond optical parametric oscillator *Appl. Phys. Lett.* **54** 1728–30.
110. Mak G, Fu Q and van Driel H M 1992 Externally pumped high repetition rate femtosecond infrared optical parametric oscillator *Appl. Phys. Lett.* **60** 542–4.
111. Fu Q, Mak G and van Driel H M 1992 High-power, 62-fs infrared optical parametric oscillator synchronously pumped by a 76-MHz Ti:sapphire laser *Opt. Lett.* **17** 1006–8.
112. Pelouch W S, Powers P E and Tang C L 1992 Ti:sapphire-pumped, high-repetition-rate femtosecond optical parametric oscillator *Opt. Lett.* **17** 1070–2.
113. Dudley J M, Reid D T, Ebrahimzadeh M and Sibbett W 1994 Characteristics of a noncritically phase-matched Ti:sapphire-pumped femtosecond optical parametric oscillator *Opt. Commun.* **104** 419–30.
114. McCahon S W, Anson S A, Jang D-J and Boggess T F 1995 Generation of 3–4 mm femtosecond pulses from a synchronously pumped, critically phase-matched $KTiOPO_4$ optical parametric oscillator *Opt. Lett.* **22** 2309–11.
115. Powers P E, Ramakrishna S, Tang C L and Cheng L K 1993 Optical parametric oscillation with $KTiOAsO_4$ *Opt. Lett.* **18** 1171–3.
116. Reid D T, McGowan C, Ebrahimzadeh M and Sibbett W 1997 Characterization and modelling of a noncollinearly phase-matched femtosecond optical parametric oscillator based on KTA and operating beyond 4 µm *IEEE J. Quantum Electron.* **33** 1–9.
117. Powers P E, Tang C L and Cheng L K 1994 High-repetition-rate femtosecond optical parametric oscillator based on $CsTiOAsO_4$ *Opt. Lett.* **19** 37–9.
118. Holtom G R, Crowell R A and Cheng L K 1995 Femtosecond mid-infrared optical parametric oscillator based on $CsTiOAsO_4$ *Opt. Lett.* **20** 1880–2.
119. Powers P E, Tang C L and Cheng L K 1994 High-repetition-rate femtosecond optical parametric oscillator based on $RbTiOAsO_4$ *Opt. Lett.* **19** 1439–41.
120. Spence D E, Wielandy S and Tang C L 1996 High average power, high repetition rate femtosecond pulse generation in the 1–5 µm region using an optical parametric oscillator *Appl. Phys. Lett.* **68** 452–4.
121. Dudley J M, Reid D T, Ebrahimzadeh M and Sibbett W 1994 Characteristics of a noncritically phase-matched Ti:sapphire-pumped femtosecond optical parametric oscillator *Opt. Commun.* **104** 419–30.
122. Reid D T, Ebrahimzadeh M and Sibbett W 1995 Noncritically phase-matched Ti:sapphire-pumped femtosecond optical parametric oscillator based on $RbTiOAsO_4$ *Opt. Lett.* **20** 55–7.
123. Kafka J D, Watts M L and Pieterse J W 1995 Synchronously pumped optical parametric oscillators with LiB_3O_5 *J. Opt. Soc. Am. B* **12** 2147–57.
124. Ellingson R J and Tang C L 1993 High-power, high-repetition-rate femtosecond pulses tunable in the visible *Opt. Lett.* **18** 438–40.
125. Reid D T, Ebrahimzadeh M and Sibbett W 1995 Efficient femtosecond pulse generation in the visible in a frequency-doubled optical parametric oscillator based on $RbTiOAsO_4$ *J. Opt. Soc. Am. B* **12** 1157–63.
126. Shirakawa A, Mao H W and Kobayashi T 1996 Highly efficient generation of blue-orange femtosecond pulses from intracavity-frequency-mixed optical parametric oscillator *Opt. Commun.* **123** 121–8.
127. Kartaloglu T, Koprulu K G and Aytur O 1997 Phase-matched self-doubling optical parametric oscillator *Opt. Lett.* **22** 280–2.
128. Driscoll T J, Gale G M and Hache F 1994 Ti:sapphire second-harmonic-pumped visible range femtosecond optical parametric oscillator *Opt. Commun.* **110** 638–44.
129. Gale G M, Hache F and Cavallari M 1998 Broad-bandwidth parametric amplification in the visible: femtosecond experiments and simulations *IEEE J. Sel. Top. Quantum Electron.* **4** 224–9.
130. Burr K C, Tang C L, Arbore M A and Fejer M M 1997 Broadly tunable mid-infrared femtosecond optical parametric oscillator using all-solid-state-pumped periodically poled lithium niobate *Opt. Lett.* **22** 1458–60.
131. Reid D T, Penman Z, Ebrahimzadeh M, Sibbett W, Karlsson H and Laurell F 1997 Broadly tunable infrared femtosecond optical parametric oscillator based on periodically poled $RbTiOAsO_4$ *Opt. Lett.* **22** 1397–9.
132. Reid D T, Kennedy G T, Miller A, Sibbett W and Ebrahimzadeh M 1998 Widely tunable near- to mid-infrared femtosecond and picosecond optical parametric oscillators using periodically poled $LiNbO_3$ and $RbTiOAsO_4$ *IEEE J. Sel. Top. Quantum Electron.* **4** 238–48.
133. McGowan C, Reid D T, Penman Z E, Ebrahimzadeh M, Sibbett W and Jundt D T 1998 Femtosecond optical parametric oscillator based on periodically poled lithium niobate *J. Opt. Soc. Am. B* **15** 694–701.
134. Kartaloglu T, Koprulu K G, Aytur O, Sundheimer M and Risk W P 1998 Femtosecond optical parametric oscillator based on periodically poled $KTiOPO_4$ *Opt. Lett.* **23** 61–3.
135. Marzenell S, Beigang R and Wallenstein R 1999 Synchronously pumped femtosecond optical parametric oscillator based on $AgGaSe_2$ tunable from 2 µm to 8 µm *Appl. Phys. B* **69** 423–8.
136. Zhang X P, Hebling J, Bartels A, Nau D, Kuhl J, Ruhle W W and Giessen H 2002 *Appl. Phys. Lett.* **80** 1873–5.
137. Jiang J and Hasama T 2003 *Opt. Commun.* **220** 193–202.
138. O'Connor M V, Watson M A, Shepherd D P, Hanna D C, Price J H V, Malinowski A, Nilsson J, Broderick N G R, Richardson D J and Lefort L 2002 *Opt. Lett.* **27** 1052.

24

Optical Parametric Chirped-Pulse Amplification (OPCPA)

László Veisz

CONTENTS

24.1 Introduction 363
24.2 Comparison of OPCPA and Lasers 364
24.3 Theory 365
    24.3.1 Parametric Amplification 365
    24.3.2 Phase-matching 367
    24.3.3 Saturated Regime – Intensity, Phase and CEP 370
    24.3.4 Simulations 372
24.4 OPCPA Architecture 373
24.5 OPCPA in Practice 375
    24.5.1 Broad Spectral Coverage 375
    24.5.2 Few-cycle Pulse Duration 376
    24.5.3 Ultrahigh Power Systems 377
24.6 Optical Parametric Synthesizers 377
24.7 Summary 378
References 378

24.1 Introduction

After the invention of laser amplification, shorter and shorter light pulses with increasing intensity were generated. This direct amplification reached its limits due to non-linear effects and damage in the amplifier medium triggered by the high intensities. The chirped-pulse amplification (CPA) technique (Chapter 22 'CPA' by Donna Strickland; Strickland and Mourou, 1985) mitigated these difficulties and permitted to reach much higher laser energies and intensities. Furthermore, this made intense lasers available for a broad range of applications, therefore their inventors, Donna Strickland and Gerard Mourou, obtained the Nobel Prize in 2018. A certain spectrum supports a minimum (so-called Fourier limited) pulse duration, which is a natural lower limit defined by the Fourier transformation. The next limit in laser development towards even shorter pulses is posed by the laser materials providing finite gain bandwidth and therefore longer pulses.

Optical parametric chirped-pulse amplification (OPCPA) (Dubietis, Jonušauskas and Piskarskas, 1992) is an approach to generate ultrashort light pulses down to the few optical cycle duration with high energy and intensity as illustrated in Figure 24.1. OPCPA is the combination of two established techniques (Dubietis, Butkus and Piskarskas, 2006). The first of them is CPA, which involves the temporal stretching (chirping) of weak ultrashort pulses, their amplification and compression. To remove the bandwidth limitations, a second technique is applied to amplify in CPA, which is called the optical parametric amplification (OPA) (Manzoni and Cerullo, 2016). OPA is a second-order non-linear interaction that leads to an energy transfer from a strong narrowband pump pulse to a weak, broadband seed pulse in a non-linear optical crystal. OPA is an alternative to laser amplification offering much

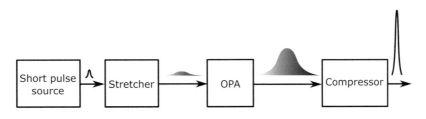

FIGURE 24.1 Schematics of an OPCPA system. A short pulse source (oscillator, OPA or laser amplifier) provides low energy pulses typically with high repetition rate for amplification. A stretcher elongates them in time by a large factor (10^3–10^4) and a broad bandwidth OPA amplifies them typically at a reduced repetition rate. The pulses are temporally shortened in a compressor matched to the stretcher.

broader gain bandwidth and therefore shorter pulses (Ross et al., 1997; Cerullo and De Silvestri, 2003; Manzoni and Cerullo, 2016) or new spectral regions unavailable for lasers.

In this chapter, we first discuss the general properties of OPA/OPCPA, then their theoretical description and related numerical simulations. It is followed by the description of typical OPCPA system architecture. Thereafter, the application of OPCPA in different wavelength ranges and its potential to generate few-cycle pulses and high peak power are reviewed. Then, the optical parametric synthesizers (OPS) are shortly described that further broaden the spectral gain bandwidth towards or even beyond an octave supporting quasi-single optical cycle or even sub-cycle pulses. A short summary concludes the chapter with an outlook into the future of OPCPA.

24.2 Comparison of OPCPA and Lasers

OPCPA (and also OPA) as an alternative light amplification technique has certain advantages over as well as challenges compared with conventional laser systems. Certainly, the benefits are the main motivation to use OPCPA for specific applications and therefore will be discussed in detail later. The main advantages are:

- *Broader bandwidth.* OPA has a significantly broader gain bandwidth than lasers, up to an octave in special cases. Correspondingly, it can amplify light pulses with durations down to a few (1–3) optical cycles, i.e. 3–9 fs around 800 nm central wavelength, see Figure 24.2.
- *Spectral tunability.* If the whole gain bandwidth is not necessary, the spectrum and central wavelengths have a broad tuning range with minor changes in the OPA setup.
- *Spectral region.* Spectral gain is achievable not only in limited ranges in the visible or near-infrared (NIR) typical for lasers but also in regions where no laser gain medium is available, covering continuously the former ranges as well as the mid-infrared. This is mainly limited by the transmission range of the non-linear crystal.
- *Better temporal intensity contrast.* There is no unwanted background such as amplified spontaneous emission (ASE) outside of the pump temporal window as the gain is limited in time to the pump pulse duration, which is typically short (ps range). Many large petawatt-scale laser systems have an OPCPA front end to utilize this property.
- *Higher gain.* Single-pass gain values of 10^2–10^4 can be easily reached in OPCPA. This reduces the system complexity and size as multi-pass configurations are not required.
- *Low thermal load in amplifier crystal.* The gain medium does not heat up as the upper energy level is virtual and there is no energy deposition in the non-linear optical crystal in contrary to lasers, where a real upper energy level and population inversion is used. The non-linear crystals are transparent at the pump, signal and idler wavelengths.

- *Scalability to high energy, repetition rate and average power.* The absence of relevant thermal effects supports the increase of pulse energy and repetition rate and thus power scaling of OPCPA systems.
- *CEP conservation.* A *carrier-envelope phase* (CEP), defining the position of the carrier wave relative to the envelope as shown in Figure 24.2, stabilized source remains stabilized after saturated amplification as the CEP is influenced in the same way by the OPA phase for slightly different pump intensities.

To realize an OPCPA system, the technology of CPA lasers can partially be utilized, however, there are some challenges:

- *Stretching and compression.* The larger spectral bandwidth of OPCPA systems makes them much harder to temporally stretch and compress due to two factors. First, the throughput of these components for a broad bandwidth is reduced, and second, the higher order (3.-5.) spectral phase needs to be controlled to obtain a few-cycle pulse after the OPCPA.
- *Pump laser.* Requirements for the pump laser are high. Optimal parameters are 1–100 ps duration, high energy, high temporal and pointing stability, flat-top beam profile, high average power and low non-linearities (B-integral).
- *Temporal synchronization.* The pump and seed need to be synchronized with a much better accuracy than the pump duration, typically a few percent of it. This is increasingly difficult towards 1 ps pump duration and calls for all-optical-synchronization and slow-timing jitter stabilization.
- *Contrast preservation.* Suppression or control of the amplification of unwanted background within the temporal gain window of the pump is needed. This background originates from pre-pulses or ASE from the short-pulse laser source or optical parametric fluorescence (super-fluorescence) from the OPCPA amplifier.

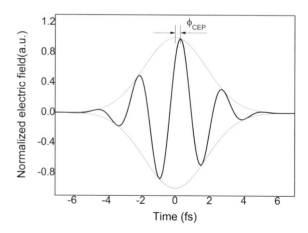

FIGURE 24.2 Normalized field envelope and electric field of a 3 fs (intensity FWHM) pulse at 750 nm central wavelength with a carrier-envelope phase (CEP) of 45°.

OPCPA

- *Sensitivity.* OPCPA, especially when providing a gain bandwidth that supports few-cycle pulse duration, is very sensitive to certain input parameters. Among others the accurate day-to-day alignment of the pump and seed beam geometry and non-linear crystal direction with significantly higher accuracy than 0.05° are needed.

24.3 Theory

In the following an overview of relevant theoretical aspects of OPA is given. It starts with the basic equations and non-saturated amplification regime, and it is followed by the phase-matching and its realization, the saturated regime of amplification and the most complete description using numerical simulations.

24.3.1 Parametric Amplification

In this sub-chapter the theoretical basics of OPA/OPCPA processes will be described. OPA as illustrated in Figure 24.3a is a second-order non-linear interaction that transfers energy from a high energy, high frequency, typically narrow bandwidth pump to a weak, lower frequency, broadband seed. During this process, an idler beam is generated, which did not exist before. This interaction takes place in a specially cut non-linear optical crystal. The amplified seed is called signal. The energy conservation for the parametric process as shown in Figure 24.3b has the form

$$\hbar\omega_p = \hbar\omega_s + \hbar\omega_i \tag{24.1}$$

where $\omega_j, j = s, i, p$ are the angular frequencies of the signal, idler and pump waves, and $\hbar$ is the reduced Planck constant. Here, $\omega_p > \omega_s \geq \omega_i$ corresponding to the parametric amplification, where a pump photon decays into a signal and an idler photon in the quantum picture. When $\omega_s = \omega_i = \omega_p/2$, the process is called degenerate and has a broad gain bandwidth.

The momentum conservation, which is also called *phase-matching condition*, is

$$\hbar\mathbf{k}_p = \hbar\mathbf{k}_s + \hbar\mathbf{k}_i \tag{24.2}$$

where $\mathbf{k}_j, j = s, i, p$ are the wave vectors of the signal, idler and pump. Phase-matching is a very important concept in OPA and will be further discussed later. The dispersion relation in non-magnetic materials ($\mu_r = 1$), as assumed here, has the form

$$\omega_j/k_j = c_0/n_j = c_0/\sqrt{\varepsilon_j} \tag{24.3}$$

where $j = s, i, p$ and c_0 is the speed of light in vacuum, n_j is the index of refraction and ε_j is the relative permittivity or dielectric constant of the medium at the corresponding frequencies.

In the following, the theoretical description of the OPA process will be presented limited to the collinear case along the z-axis. However, we will also discuss non-collinear phase-matching later. The electric field of the linearly polarized plane waves propagating into z-direction is described as

$$E_j(z,t) = \mathrm{Re}\{A_j(z,t)\exp[i(\omega_j t - k_j z)]\} \tag{24.4}$$

where $A_j(z,t), j = s, i, p$ are the complex electric field envelopes of signal, idler and pump and $i = \sqrt{-1}$ when it is not in the index. Furthermore, the induced polarization in the medium is written in the form

$$P = P_{\mathrm{lin}} + P_{\mathrm{nl}} \tag{24.5}$$

The linear term for monochromatic waves and isotropic medium is $P_{\mathrm{lin},j}(z,t) = \varepsilon_0 \chi_j^{(1)} E_j(z,t)$, where $\chi_j^{(1)} = \varepsilon_j - 1$ is the electric susceptibility at the corresponding frequencies. However, the dispersion, i.e. frequency dependence of $\chi^{(1)}$ within the signal, idler or pump bandwidth, is important for broadband pulses, which corresponds to the memory effect of the material (Akhmanov, Vysloukh and Chirkin, 1992), i.e. the polarization depends also on former values of the electric field. In this case, the polarization of the medium is expressed as

$$P_{\mathrm{lin},j}(z,t) = \varepsilon_0 \int_0^\infty \chi^{(1)}(\tau) E_j(z, t-\tau) \mathrm{d}\tau \text{ or}$$

$$P_{\mathrm{lin},j}(z,\omega) = \varepsilon_0 \chi^{(1)}(\omega) E_j(z,\omega) \tag{24.6}$$

This term includes the dispersion of the medium. Approximations of different orders of this linear part originate from Taylor expansion of the slowly varying $A_j(t)$ and also

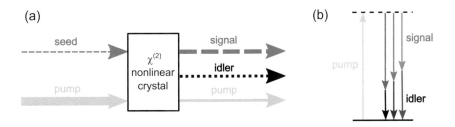

FIGURE 24.3 Schematics of OPA. (a) A high energy, high frequency, typically narrow bandwidth pump beam and a weak, lower frequency, broadband seed beam enter a non-linear crystal, where the pump amplifies the seed, which is then called signal, and generates an idler beam. (b) Energy conservation of broadband OPA process. The upper level is virtual. In the quantum picture, a pump photon decays into a signal and an idler photon corresponding to the energy conservation.

contain the corresponding frequency derivative of the dielectric constant. The non-linear polarization is also divided into different orders in the electric field, but only the second order is relevant for OPA, which is written in the general form as

$$P_{\text{nl}}(z,t) = \varepsilon_0 \chi^{(2)} E(z,t) E(z,t) \tag{24.7}$$

where $\chi^{(2)}$ is the relevant component of the second order non-linear susceptibility tensor (Baumgartner and Byer, 1979), and in general, $E(z,t)$ is the sum of the 3 (s, i, p) fields. Using equations (24.4)–(24.7) in the wave equation, Taylor expanding the complex amplitude and applying the slowly varying-envelope approximation, the *coupled non-linear wave equations* for $A_j(z,t)$ $j=s$, i, p in the first-order approximation can be obtained (Akhmanov, Vysloukh and Chirkin, 1992):

$$\frac{\partial A_s}{\partial z} + \frac{1}{v_{gs}} \frac{\partial A_s}{\partial t} = -i \frac{\omega_s d_{\text{eff}}}{n_s c_0} A_i^* A_p \exp(-i\Delta kz) \tag{24.8}$$

$$\frac{\partial A_i}{\partial z} + \frac{1}{v_{gi}} \frac{\partial A_i}{\partial t} = -i \frac{\omega_i d_{\text{eff}}}{n_i c_0} A_s^* A_p \exp(-i\Delta kz) \tag{24.9}$$

$$\frac{\partial A_p}{\partial z} + \frac{1}{v_{gp}} \frac{\partial A_p}{\partial t} = -i \frac{\omega_p d_{\text{eff}}}{n_p c_0} A_i A_s \exp(i\Delta kz) \tag{24.10}$$

where $\Delta k = k_p - k_s - k_i$ is the phase mismatch, $v_{gj} = \left(\partial k / \partial \omega\right)^{-1}\Big|_{\omega_j}$ is the *group velocity* of $j=s$, i, p, $d_{\text{eff}} = \chi^{(2)}/2$ is the effective non-linear coefficient and * is the complex conjugate. The first term on the left-hand sides describes propagation in z-direction, the second the dispersion in first order and the right hand sides contain the non-linear coupling between the waves including the influence of non-perfect phase-matching.

If these equations are transformed to $z \rightarrow z'$, $t \rightarrow \tau - z/v_{gp}$, the coefficient of the first temporal derivative terms will change from $1/v_{gj}$ to $1/v_{gj} - 1/v_{gp}$ $j=s$, i, p, which is 0 for the pump in equation (24.10) and contains the signal-pump $\left(1/v_{gs} - 1/v_{gp}\right)$ and idler pump $\left(1/v_{gi} - 1/v_{gp}\right)$ group velocity mismatch (GVM) for equations (24.8) and (24.9). These describe how signal and idler move away from the pump, which is called *temporal walk-off*. If this is larger than the pump pulse duration, the pulses are split and the amplification is seriously influenced. However, typical GVM is on the order of 100 fs/mm and so it is not relevant for OPCPA, only for OPA.

The second-order approximation of the above equations contains on the left-hand side $+\frac{k_{2j}}{2i} \frac{\partial^2 A_j}{\partial t^2}$ terms, where $k_{2j} = \frac{\partial^2 k}{\partial \omega^2}\Big|_{\omega_j}$ is the group velocity dispersion, which is important in certain cases as we will see.

The zeroth-order approximation of the equations (24.8)–(24.10) contains the phase velocities (c_0/n_j) instead of the group velocities (v_{gj}) in front of the first temporal derivatives, i.e. does not contain dispersion. The case of monochromatic waves is also relevant, where the complex amplitude does not depend on time $A_j(z)$, i.e. the equation does not contain the temporal derivative (second term on the left-hand side). The first-order approximation with equal group velocities ($v_g = v_{gs} = v_{gi} = v_{gp}$) after the transformation $z \rightarrow z'$, $t \rightarrow \tau - z/v_g$ has the same form for $A_j(z,\tau)$ (as the longitudinal coordinate z does not change, we will use z instead of z' in the following)

$$\frac{\partial A_s}{\partial z} = -i \frac{\omega_s d_{\text{eff}}}{n_s c_0} A_i^* A_p \exp(-i\Delta kz) \tag{24.11}$$

$$\frac{\partial A_i}{\partial z} = -i \frac{\omega_i d_{\text{eff}}}{n_i c_0} A_s^* A_p \exp(-i\Delta kz) \tag{24.12}$$

$$\frac{\partial A_p}{\partial z} = -i \frac{\omega_p d_{\text{eff}}}{n_p c_0} A_i A_s \exp(i\Delta kz) \tag{24.13}$$

where A_j depends only on z for monochromatic waves or z and τ for broadband waves when using the first-order approximation. It is instructive to investigate this case and neglect pump depletion $[A_p(z) = \text{const.}]$, assume a low seed $[I_s(0) \ll I_p(0)]$ and no idler $[I_i(0) = 0]$ input. Using the temporal (cycle-averaged) intensity $I_j = \frac{1}{2} n_j c_0 \varepsilon_0 |A_j|^2$, we get a solution for the intensities in the form

$$I_s(z) = I_s(0)\left[1 + \frac{\Gamma^2}{g^2} \sinh^2(gz)\right] \tag{24.14}$$

$$I_i(z) = I_s(0) \frac{\omega_i}{\omega_s} \frac{\Gamma^2}{g^2} \sinh^2(gz) \tag{24.15}$$

$$I_p(z) = I_p(0) \tag{24.16}$$

where the parametric gain is the expression in the square brackets in equation (24.14), and the square of the parametric (small-signal) gain coefficient for perfect phase-matching is

$$\Gamma^2 = \frac{2\omega_s \omega_i d_{\text{eff}}^2 I_p(0)}{n_s n_i n_p \varepsilon_0 c_0^3} \tag{24.17}$$

and the *(small-signal) gain coefficient*

$$g = \sqrt{\Gamma^2 - \left(\frac{\Delta k}{2}\right)^2} \tag{24.18}$$

Equations (24.14) and (24.15) indicate that the number of generated signal and idler photons is the same, and the transferred energy to signal and idler have a ratio of ω_s/ω_i.

For perfect phase-matching ($\Delta k = 0$) and high gain ($\exp(2\Gamma L) \gg 1$), the signal intensity increases as

$$I_s(z) = \frac{1}{4} I_s(0) \exp(2\Gamma z) \tag{24.19}$$

The *parametric gain* in this case is

$$G = \frac{1}{4} \exp(2\Gamma L) \tag{24.20}$$

It should be noted that the first-order approximation, i.e. equations (24.8)–(24.10), with perfect phase-matching and no temporal walk-off effect $\left(v_{gs} = v_{gi} = v_{gp} = v\right)$

in the frame of reference moving with this group velocity ($z' = z$, $\tau = t - z/v$) has a similar solution to equation (24.19) (Akhmanov, Vysloukh and Chirkin, 1992), which shows that this result is more general and valid for pulsed pump and seed as well.

There are a few important consequences of the former results:

- Equations (24.11) and (24.12) can be interpreted as idler is generated at the beginning from seed and pump. Then, this idler produces with the pump more signal, which leads to further idler generation. This feedback mechanism leads to general exponential growth of signal until the pump depletion stops it.
- Signal and idler intensities grow very fast (exponentially) with propagation in the non-linear medium. Thus, small changes in the crystal thickness have large effects in the overall (small-signal) gain. Other non-linear effects such as second harmonic generation or sum-frequency generation depend only quadratically on the crystal length, which makes OPA to act as a real amplifier. The phase-matching is also influenced by this as discussed in the next chapter.
- This growth depends also strongly [$\exp(\text{const} \cdot I_p^{1/2})$] on the pump intensity. As a pulsed pump has a similar solution to equations (24.14)–(24.16), the gain changes temporally and is therefore limited to the pump pulse duration or a slightly even shorter time window. This is a very important property utilized to enhance the temporal intensity contrast of ultrahigh power lasers on a high dynamic range.
- Under realistic conditions, the parametric gain of a single pass reaches large values ($\gg 100$), much larger than typical laser amplifiers or optical parametric oscillators (Chapter 23 'Optical Parametric Devices' by Majid Ebrahimzadeh).
- The damage threshold intensity of materials and thus also non-linear crystals increases with decreasing laser pulse duration. Therefore, shorter pump pulses can have higher intensity in the amplifier and apply thinner crystals to reach the same gain. Furthermore, the gain bandwidth also increases as indicated by equation (24.18), because higher phase-mismatch can be tolerated.

To get familiar with OPCPA, we will investigate two examples. The first example is a 532-nm pumped OPCPA at a signal wavelength of 800 nm, which corresponds to an idler wavelength is 1588 nm. In a 5-mm thick beta-barium borate or BBO ($n_s \approx n_i \approx n_p \approx 1.66$, $d_{\text{eff}} \approx 2.1$ pm/V for $\theta = 23.8°$) and at a pump intensity of $I_p = 5$ GW/cm^2, the parametric small-signal gain coefficient is $\Gamma = 1.07$ mm^{-1} and the corresponding parametric gain is about 10^4. That means the signal (and idler) energy is increased by a factor of approximately 3 after every 500 µm in the crystal, except at the very beginning and at the end of the crystal, where the gain is still small or saturation takes place, respectively. In OPAs, the pump intensity can be significantly higher 10–100 GW/cm^2; therefore, shorter crystals can be used. A second example is a 1064-nm pumped OPCPA at a signal wavelength of 1550 nm and an idler wavelength of 3400 nm. In a 5-mm thick Type II KTP ($n_s \approx n_i \approx n_p \approx 1.73$, $d_{\text{eff}} \approx 2.8$ pm/V for $\theta = 45°$) at $I_p = 5$ GW/cm^2, the parametric small signal gain is 166, or for 8-mm crystal length, it is the same as before (10^4). This lower gain originates mainly from the longer signal and idler wavelengths.

The strong dependence of parametric gain on the pump laser intensity necessitates very stable pump lasers for non-saturated amplifiers. For example, 1% pump laser intensity fluctuation in the former example generates 5.5% fluctuation in the signal. In practice, most amplifiers operate in saturation which reduces this strong dependency. However, their correct analytical description is very challenging and numerical simulations are needed.

It can be shown that the solution of equations (24.11)–(24.12) for the phase with neglecting pump depletion, assuming a small initial signal [$I_s(0)$] intensity, no initial idler intensity, input pump and seed phases of $\varphi_p(0)$ and $\varphi_s(0)$ is:

$$\varphi_i(z) = \varphi_p(0) - \varphi_s(0) - \frac{\pi}{2} - \frac{\Delta k z}{2} \qquad (24.21)$$

$$\varphi_s(z) = \varphi_s(0) - \frac{\Delta k z}{2} + \arctan\left[\frac{\Delta k}{2g}\tanh(gz)\right] = \qquad (24.22)$$

$$= \varphi_s(0) - \frac{\Delta k z}{2}$$
$$+ \arctan\left[\frac{\Delta k}{2\sqrt{\Gamma^2 - (\Delta k/2)^2}}\tanh\left[\sqrt{\Gamma^2 - (\Delta k/2)^2}\, z\right]\right]$$
$$(24.23)$$

It indicates that the phase of the signal, which is identified with the spectral phase with broadband pulses, changes if the phase mismatch is not zero. Therefore, we investigate the phase mismatch and if it can be made zero in a broad spectral width. Later, the phase will be studied again in the more general case of pump depletion. It should be noted that certain conclusions about the phase with pump depletion are also valid here.

24.3.2 Phase-matching

The *phase-mismatch* is defined for collinearly propagating pump, signal and idler [Figure 24.4a] as

$$\Delta k = k_p - k_s - k_i \qquad (24.24)$$

Equations (24.14) and (24.18) indicate that it plays an important role in the parametric amplification as it quickly reduces gain, maximum pump depletion and therefore efficiency and also influences the phase of the waves. In the former example of 532-nm pumped OPCPA at 800 nm signal wavelength and $L = 5$ mm thick BBO crystal, the gain coefficient times the thickness reduces from its large maximum value of 5.4 to 0 when $\Delta k \approx 2.1$ rad mm^{-1}, i.e. $\Delta kL \approx 10.7$ rad. Therefore, it is necessary to fulfil the phase-matching condition $\Delta k = 0$, or momentum conservation, equation (24.2). However, the energy conservation [see equation (24.1)] and the reformulated phase-matching condition

$$\omega_p = \omega_s + \omega_i, \quad n_p \omega_p = n_s \omega_s + n_i \omega_i \quad (24.25)$$

cannot be fulfilled, taking into account the relation $\omega_p > \omega_s \geq \omega_i$ and the fact that most transparent, isotropic materials in the visible and NIR spectral region have normal dispersion ($\partial n/\partial \omega > 0$), correspondingly, n_p is the largest.

There are two common solutions to this problem:

- Utilize birefringent materials, which have an index of refraction depending on polarization and propagation direction of the waves.
- Apply quasi-phase-matching, where periodic modulation of the sign of the non-linear coefficient of the crystal leads to an associated wave vector that helps to satisfy the phase-matching condition even if the three waves are collinear and have the same polarization. This provides a net energy flow from the pump to the signal and idler waves, and allows utilizing orientations of the crystal in which the effective non-linear coefficient is the highest. Quasi-phase-matching is utilized in some mid-IR OPCPA systems, especially for the first amplifier stages due to the relative small crystal sizes.

We shortly summarize the relevant points about birefringent phase-matching, for more information about it or quasi-phase-matching, see (Chapter 23 'Optical Parametric Devices' by Majid Ebrahimzadeh) or (Manzoni and Cerullo, 2016).

Birefringent crystals have two characteristic linear polarization states. In uniaxial crystals ($n_x = n_y = n_e \neq n_z = n_o$), the two states have polarizations perpendicular to the plane formed by their wave vector and z-axis of the crystal (ordinary wave), or parallel to this plane (extraordinary wave). The ordinary wave has an index of refraction independent of the propagation direction (n_o), while the index of refraction of the extraordinary wave has a value between n_e and n_o and depends on the propagation direction, although only on the angle between its k vector and the z-axis (θ). A uniaxial crystal is called positive if $n_o < n_e$ and negative if $n_o > n_e$. In biaxial crystals ($n_x \neq n_y \neq n_z$), the index of refraction of both polarization states depends on the propagation direction (θ, ϕ) and the phase-matching is more complex. Therefore, we restrict ourselves to uniaxial crystals, and the reader is referred to (Dmitriev, Gurzadyan and Nikogosyan, 1999) for biaxial crystals.

As a relevant consequence of birefringence, the Poynting vector is not parallel to the wave vector of the extraordinary wave, i.e. the pulse front containing the energy propagates in a different direction than the wave vector, which is normal to the wavefront. This is called (spatial) *walk-off*, and the angle between Poynting vector and wave vector is the *walk-off angle*, which is typically a few degrees.

Birefringent phase-matching is termed Type I if the signal and idler have the same polarizations and Type II if their polarization is crossed. As an example, BBO is a negative uniaxial crystal and has n_o and n_e values of 1.674 and 1.555 at 532 nm and 1.646 and 1.539 at 1588 nm (Polyanskiy, 2019), respectively. The difference between ordinary and extraordinary indices is much larger than the difference due to dispersion, which allows realizing birefringent phase-matching. Type I situation corresponds to ooe, that is, ordinary signal, ordinary idler and extraordinary pump polarization.

As mention before, one of the biggest advantages of OPA is its broad gain bandwidth. We express the full width at half maximum (FWHM) bandwidth for a collinear phase-matching with plane waves and monochromatic pump beam (Danielius et al., 1993). The Taylor expansion of the phase mismatch is

$$\Delta k = \Delta k_0 + \left.\frac{\partial \Delta k}{\partial \omega}\right|_0 \Delta \omega + \left.\frac{\partial^2 \Delta k}{\partial \omega^2}\right|_0 \Delta \omega^2 + \cdots \quad (24.26)$$

where index 0 indicates the central frequency of the corresponding pump, signal or idler waves, Δk_0 is equation (24.24) at the central frequencies, $\left.\frac{\partial \Delta k}{\partial \omega}\right|_0 = \left.\frac{\partial k}{\partial \omega}\right|_{\omega_i} - \left.\frac{\partial k}{\partial \omega}\right|_{\omega_s} = \frac{1}{v_{gi}} - \frac{1}{v_{gs}}$ is the GVM between signal and idler, and $\left.\frac{\partial^2 \Delta k}{\partial \omega^2}\right|_0 = \left.\frac{\partial^2 k}{\partial \omega^2}\right|_{\omega_i} + \left.\frac{\partial^2 k}{\partial \omega^2}\right|_{\omega_s}$ is the sum of the group velocity dispersion of the signal and idler. Normal birefringent (or quasi-) phase-matching results in $\Delta k_0 = 0$. The FWHM gain bandwidth is obtained in this case, by inserting the first-order expansion of the phase-mismatch into equation (24.18) and using equation (24.14), as

$$\Delta \omega_{\text{FWHM}} = \frac{4\sqrt{\ln 2}}{|1/v_{gi} - 1/v_{gs}|}\sqrt{\frac{\Gamma}{L}} \quad (24.27)$$

where L is the crystal thickness and Γ is defined by equation (24.17).

Using the previous example of a 532 nm pumped BBO at 800 nm signal and 1588 nm idler, the estimated FWHM bandwidth for collinear phase-matching is 12.4 nm, which is not very broad and support only 75 fs pulse duration.

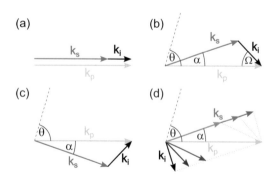

FIGURE 24.4 Illustration of phase-matching in (a) collinear, (b–d) non-collinear geometry. k_p, k_s and k_i are wave vectors of pump, signal and idler, correspondingly. Dashed black line is the crystal axis in a negative uniaxial crystal (for example, BBO) and θ is the phase-matching angle, while α is the non-collinear angle between pump and signal, and Ω is the angle between pump and idler. (b) and (d) are illustrating tangential phase-matching scheme, while (c) is Poynting vector walk-off compensation scheme. (a–c) illustrate monochromatic or narrow bandwidth, and (d) broad bandwidth radiation. In (d), different wave vector lengths correspond to a broad signal spectrum, which generate an angularly dispersed idler, i.e. the different idler frequencies propagate in different directions.

To achieve a broader bandwidth, the first derivative should also be zero. In this case, equation (24.27) loses validity and the second-order expansion should be used to get

$$\Delta\omega_{\text{FWHM}} = \frac{4\sqrt[4]{\ln 2}}{\sqrt{|k_{2i} + k_{2s}|}} \sqrt[4]{\frac{\Gamma}{L}} \quad (24.28)$$

where $k_{2j} = \left.\frac{\partial^2 k}{\partial \omega^2}\right|_{\omega_j}$ for $j = s, i$ are the group velocity dispersion as before. This is naturally reached at degeneracy, which is defined as $\omega_s = \omega_i = \omega_p/2$, and therefore, the group velocities of signal and idler are also the same. Using the example of a 532 nm pumped BBO at 1064 nm signal and idler, the estimated FWHM bandwidth for collinear phase-matching is 92.1 nm, which is significantly broader than the non-degenerate case and supports 18 fs pulses.

A possible route to broaden the phase-matching bandwidth is to apply a broadband pump. If it is carefully angularly dispersed, the bandwidth can be increased even further (Baltuška, Fuji and Kobayashi, 2002b). This works well for small OPA beams; however, it is not practical for OPCPA as there are no high-quality, high-energy and broadband pump lasers, and it is challenging to angularly disperse large beams.

In general, to make the derivative zero and broaden the phase-matching bandwidth, an additional degree of freedom is required. A common solution is *non-collinear phase-matching*, i.e. introducing an extra-control degree, the internal non-collinear angle α (measured in the crystal) between the pump and seed beams. It is illustrated in Figure 24.4b. An important consequence of a broadband seed is the angularly dispersed idler as shown in Figure 24.4d, which is a consequence of energy and momentum conservations. The phase mismatch contains vectors that can be split into parallel and perpendicular components to the pump wave vector

$$\Delta k_{\parallel} = k_p - k_s \cos\alpha - k_i \cos\Omega, \quad \Delta k_{\perp} = k_s \sin\alpha - k_i \sin\Omega \quad (24.29)$$

where Ω is the angle between pump and idler wave vectors, which is determined by α and the k vectors that can also depend on θ. θ is chosen for a certain α that these equations are zero at the central frequency. The first-order expansion around the central signal frequency is

$$\Delta k_{\parallel} = -\left.\frac{\partial k}{\partial \omega}\right|_{\omega_s} \cos\alpha\, \Delta\omega + \left.\frac{\partial k}{\partial \omega}\right|_{\omega_i} \cos\Omega\, \Delta\omega - k_i \left.\frac{\partial \Omega}{\partial \omega}\right|_{\omega_i} \sin\Omega\, \Delta\omega \quad (24.30)$$

$$\Delta k_{\perp} = \left.\frac{\partial k}{\partial \omega}\right|_{\omega_s} \sin\alpha\, \Delta\omega + \left.\frac{\partial k}{\partial \omega}\right|_{\omega_i} \sin\Omega\, \Delta\omega + k_i \left.\frac{\partial \Omega}{\partial \omega}\right|_{\omega_i} \cos\Omega\, \Delta\omega \quad (24.31)$$

α is determined that also these quantities are zero, which provides an important relation

$$v_{gs} = v_{gi} \cos(\alpha + \Omega) \quad (24.32)$$

As a consequence, a broader phase-matching bandwidth, where the first derivative of the phase mismatch is also zero, requires matching of the projected idler group velocity in the direction of the signal. This demands that the idler has a higher group velocity than the signal.

In practice, the optimal non-collinear and phase-matching angles are determined by plotting θ vs. wavelength for different α values and searching for the α that provides the same θ in the broadest signal spectral range. This is plotted in Figure 24.5 for Type I phase-matching in BBO for a pump wavelength of 532 nm. The choice of $\theta = 23.8°$ and $\alpha = 2.3°$ provides the broadest spectrum from 700 to 1000 nm. At this

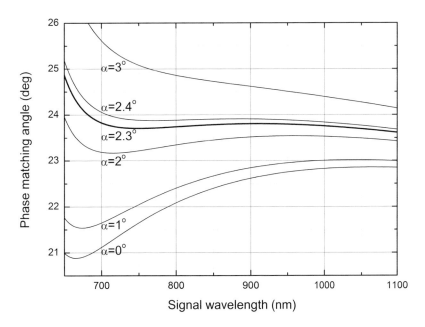

FIGURE 24.5 Phase-matching angle vs. signal wavelength for different non-collinear angles (α) for Type I phase-matching in BBO and a pump wavelength of 532 nm. The combination $\theta = 23.8°$ and $\alpha = 2.3°$ provides a broad phase-matching bandwidth from 700 nm to beyond 1000.

optimal parameter combination, the projection of idler group velocity in the signal direction equals relatively well with the signal group velocity.

In a negative uniaxial crystal like BBO and Type I phase-matching (ooe, s, i, p polarization), there are two possible geometries to realize non-collinear geometry: the *tangential phase-matching* scheme (Figure 24.4b), where the seed propagates between the pump and the optical axis ($\theta - \alpha$ to axis) and the *Poynting vector walk-off compensation* scheme (Figure 24.4c), where the seed propagates on the other side of the pump ($\theta + \alpha$ to axis). While the Poynting vector walk-off compensation scheme as its name suggests compensates for the spatial walk-off as the non-collinear propagation shifts the signal beam in the walk-off direction of the extraordinary wave. A disadvantage of this scheme is that the signal propagation direction is similar to its phase-matching direction for second harmonic generation, and thus, a significant amount of second harmonic is produced creating a hole in the fundamental spectrum. To avoid second harmonic generation the tangential scheme is practical. However, the pump and signal beams overlap only shorter during propagation as non-collinearity and walk-off are adding and not compensating each other. This limits the scheme to Pointing vector walk-off compensation in µJ-scale OPAs with 100s of µm spot sizes in the amplifier, but this shift is negligible in OPCPA stages with high energy (mJs) and correspondingly larger pump diameter (>1 mm).

Using as an example, a Type I OPA in 5 mm BBO with 532 nm extraordinary pump, the walk-off angle is 3.3° and the spatial walk-off is 288 µm. While in this case, the optimal non-collinear angle is 2.3° and the corresponding shift at the end of the crystal is 200 µm. In tangential scheme, there is 488 µm shift at the end of the crystal, while it is 88 µm in Pointing vector walk-off compensation scheme. However, if the pump diameter is a few mm as usual in OPCPA, both are negligible.

The phase-mismatch is determined by equation (24.29) with given θ, α and crystal thickness as a function of the signal wavelength. The phase mismatch vector shows in the idler direction. Normally, the propagation direction can be taken along the pump (z-axis) using only the parallel component of Δk to the pump as the difference is completely negligible compared with the full Δk along the idler or the corresponding component along the signal. When the phase mismatch is obtained, the spectral gain in the non-saturated regime is calculated using equations (24.14), (24.17) and (24.18). This is a very practical estimation of the spectral gain to evaluate a given phase-matching geometry. Even if it is based on non-saturated calculations, the bandwidth agrees well with the saturated case and the measurements. The calculated spectral gain for our example of 532 nm wavelength, 5 GW cm² intensity pump, 5-mm thick Type I BBO and 2.31° internal non-collinear angle is shown in Figure 24.6. It indicates a large predicted gain in the 700–1050 nm signal range.

However, this configuration is sensitive to the alignment, only 0.05° change in the phase-matching angle between crystal axis and pump wave vector direction modifies the spectral gain significantly. As the phase-matching in this geometry

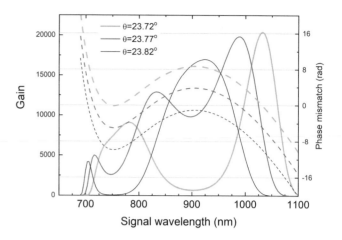

FIGURE 24.6 Calculated spectral gain and phase-mismatch for non-collinear OPCPA with 2.31° non-collinear angle, three different phase-matching angle and 5 mm Type I BBO, 5 GW/cm² pump intensity and 532 nm wavelength in the non-saturated regime. The spectral gain is very sensitive to the alignment of the OPCPA and 0.05° change significantly modifies it.

depends on the angle between crystal axis and signal wave vector, the gain is also similarly sensitive to the non-collinear angle. Nevertheless, if the phase-matching angle is adjusted with the change of the non-collinear angle so that angle between crystal axis and signal wave vector is approximately constant, i.e. the pump propagation direction is changed instead of the signal, a slightly lower sensitivity is observed with the non-collinear angle (approx. 0.15–0.2°). As a conclusion, a precise daily alignment of the OPCPA is necessary to generate few-cycle pulses.

24.3.3 Saturated Regime – Intensity, Phase and CEP

The former analytical results show how the amplitude of various waves changes during OPA in the non-saturated regime (no pump depletion) and how it is influenced by phase-matching. Now, the intensity and phase evolution in the saturated regime are investigated. There is an analytic solution of equations (24.11)–(24.13) describing the monochromatic approximation of the coupled non-linear wave equations (and even for the first-order approximation when all three waves have equal group velocities) when they are separated into amplitude and phase terms by substituting $A(z) = B(z)\exp[i\varphi(z)]$ (Armstrong et al., 1962). Assuming perfect phase-matching and an initial zero idler intensity that corresponds to OPA, the solution provides the *intensities* in the form (Baumgartner and Byer, 1979)

$$I_s(z) = I_s(0) + \frac{\omega_s}{\omega_p} I_p(0)\left[1 - \mathrm{sn}^2\left(\frac{z-z_0}{l}, \gamma\right)\right] \quad (24.33)$$

$$I_i(z) = \frac{\omega_i}{\omega_p} I_p(0)\left[1 - \mathrm{sn}^2\left(\frac{z-z_0}{l}, \gamma\right)\right] \quad (24.34)$$

$$I_p(z) = I_p(0)\mathrm{sn}^2\left(\frac{z-z_0}{l}, \gamma\right) \quad (24.35)$$

where sn(x, m) is Jacobi elliptic function sn, $l^{-1} = \sqrt{\left(1 + \frac{\omega_p}{\omega_s} \frac{I_s(0)}{I_p(0)}\right)} \Gamma$ and for small input seed compared with the pump $l^{-1} = \Gamma$, where Γ is the small-signal gain coefficient for perfect phase-matching [see equation (24.17)], $z_0/l = k(\gamma) \approx \frac{1}{2} \ln\left[32\left(1 + \frac{\omega_s}{\omega_p} \frac{I_p(0)}{I_s(0)}\right)\right]$, $K(\gamma)$ is the complete elliptic integral of the first kind, and $\gamma = \sqrt{\frac{1}{1 + \frac{\omega_p}{\omega_s} \frac{I_s(0)}{I_p(0)}}} \approx 1 - \frac{1}{2} \frac{\omega_p}{\omega_s} \frac{I_s(0)}{I_p(0)}$, where the last approximation is valid for small input seed. l is practically a growth length of the non-saturated amplifier, and the saturation is reached within a few growth lengths (z_0/l) depending on the initial seed to pump intensity ratio. Saturation is reached at z_0, and thereafter, the energy is transferred back to the pump from signal and idler within another distance of z_0 where the initial conditions are reached, i.e. the energy transfer has a period of $2z_0$. Certainly, in the low saturated limit, equations (24.33)–(24.35) converge to the former non-depleted results.

This solution is plotted in Figure 24.7a for our example of Type I BBO with pump intensity of $I_p = 5$ GW cm^2, for wavelengths of 532, 800 and 1588 nm corresponding to pump, signal and idler, respectively, and $\Gamma = 1.07$ mm^{-1} as estimated before, plus the initial pump to seed intensity ratio is 1000. The pump is completely depleted after 4.7 mm, which corresponds to 3 growth length (=1/Γ) and regains its energy after the double distance. If the initial pump and seed intensities are the same, all pump energy is still transferred to the signal and idler, but this saturation length and thus also the period of energy regain are shorter as shown in Figure 24.7b.

There is no analytical solution of the first-order approximation [equations (24.8)–(24.10)] with different group velocities or the second-order of the coupled non-linear wave equations in the saturated regime even if the phase-matching is perfect. A very simple estimation for small input seed intensity gives the maximum pump depletion due to non-perfect phase-matching as

$$\frac{I_{p\min}}{I_p(0)} = \left(\frac{\Delta k}{2\Gamma}\right)^2 \quad (24.36)$$

where $I_{p\min}$ is the pump intensity at maximal depletion. It should be noted that as the gain coefficient is also reduced by phase-mismatch, the depletion is reached at a later position in the crystal than for perfect phase-matching.

The spectral as well as temporal *phase* plays, especially for short pulses, an important role. This is because the change of the phase during the OPA process can influence the compressibility of the pulses in OPCPA and also the CEP.

Using the same separation of the coupled non-linear wave equations (24.11)–(24.13) into amplitude and phase terms $(A(z) = B(z)\exp[i\varphi(z)])$ as for intensities, the following results are obtained for the phases in the saturated regime with zero initial idler (Ross et al., 2002)

$$\varphi_s(z) = \varphi_s(0) - \frac{\Delta k}{2} \int \frac{f}{f + \gamma_s^2} dz \quad (24.37)$$

$$\varphi_i(z) = \varphi_p(0) - \varphi_s(0) - \frac{\pi}{2} - \frac{\Delta k z}{2} \quad (24.38)$$

$$\varphi_p(z) = \varphi_p(0) - \frac{\Delta k}{2} \int \frac{f}{1-f} dz \quad (24.39)$$

where $f = 1 - I_p(z)/I_p(0)$ is the fractional pump depletion, and $\gamma_s^2 = \omega_p I_s(0)/[\omega_s I_p(0)]$. Here, the initial conditions of pump and seed phase are given, but the idler phase self-adjusts itself at the beginning to $-\pi/2$ to maximize the idler amplitude growth. These phases are identified with the spectral phase in the broadband case as a certain wavelength acquires it during amplification.

These equations support the following important consequences:

- Phase of the amplified signal is independent of the initial phase of the pump, which has the advantage that any phase degradation in an independent high-energy pump laser is not inherited by the amplified signal. For example, CEP degradation due to highly non-linear and thermal effects in the pump.

- Change of the phase of all three waves is proportional to phase mismatch. It means that imperfect phase-matching ($\Delta k \neq 0$) is not only reducing the gain but also influencing the phase of amplified waves.

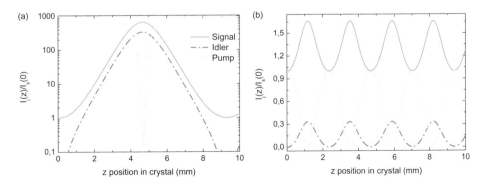

FIGURE 24.7 Intensities in a saturated OPA amplifier. (a) Calculated intensity of j=signal at 800 nm (gray solid), idler at 1588 nm (black dash-dotted) and pump at 532 nm (light gray dashed) along the BBO non-linear crystal in the saturated regime using monochromatic waves and perfect phase-matching. All intensities are normalized to the seed (initial signal) intensity. Initial intensity of pump $I_p(0) = 1000 I_s(0)$ and idler: $I_i(0) = 0$. (b) Same parameters and colours as before except initial intensity of pump $I_p(0) = I_s(0)$.

- Pump depletion influences the phase of signal and pump and alters them compared with their initial values, but it does not influence the idler phase.
- The phase of the amplified signal (see Eq. 24.22) is proportional to $+\Delta k$, which slightly compensates the phase mismatch $-\Delta k z/2$ and increases slightly the phase-matching bandwidth (Herrmann et al., 2010b).
- Idler phase is the difference of initial pump and signal phases, which provides a mean for CEP stabilization based on difference frequency, also called passive CEP stabilization. To this end, the pump and the signal should have the same CEP, for example, both of them are generated from (possibly broadened) fundamental or alternatively from second harmonic (Baltuška, Fuji and Kobayashi, 2002a). This is a very popular technique to CEP stabilize OPAs (Fuji et al., 2006).
- Change of signal CEP, originating from fluctuations in saturation triggered by pump intensity fluctuations, is a very small effect as the Δk has positive as well as negative values within the spectrum, which is averaged out over the whole spectrum (Witte and Eikema, 2012). Therefore, it is a secondary effect and OPA preserves the CEP stability (Baum et al., 2003). However, the spectral phase fluctuations generate a small broadening of a few-cycle pulse after compression.

24.3.4 Simulations

To account for dispersion including group velocity dispersion and higher orders and saturation or even further effects such as spatial walk-off and diffraction numerical simulations are necessary. Normally, the split-step Fourier algorithm is applied to solve the coupled non-linear wave equations (Fisher and Bischel, 1975; Arisholm, 1997). It splits the non-linear material into many small steps. In each of them, it simulates the dispersion in the frequency domain without any approximation using the material dispersion and the non-linearities in the time domain. If the steps are small enough, the result converges to the exact solution. Every non-linearity has to be defined with extra field or even extra non-linear equations in these simulations. An alternative approach still uses the split-step Fourier method but combines all electric fields in an ordinary and an extraordinary component and describes dispersion, diffraction and walk-off in the temporal and spatial Fourier domain. Furthermore, it considers all second-order non-linearities, if necessary also self-phase modulation, in the time domain using 2+1 dimensional simulations (Lang et al., 2013). By this way, unexpected mixing products and temporal, spatial, cascaded and parasitic effects are automatically taken into account.

Using the code Chi2D (Lang et al., 2013) based on this latter technique, simulations were performed for our typical parameters: 532 nm pump, 80 ps FWHM pump pulse duration, and 9.3 mJ energy, super-Gaussian beam profile with 1 mm radius, and 5 GWcm^{-2} peak intensity, Gaussian seed with a central wavelength of 850 and 240 nm FWHM bandwidth supporting 4.25 fs Fourier limited pulses, a linear chirp to stretch the seed to 26 ps FWHM duration, 3.6 µJ energy, and super-Gaussian beam profile with 1.2 mm radius, corresponding to a pump to seed intensity ratio of 1000, Type I BBO. The results are plotted in Figure 24.8. The signal intensity increases until the pump depletes at 6.5 mm, where the saturated parametric gain is 420, and then it transfers energy back to the pump. This distance

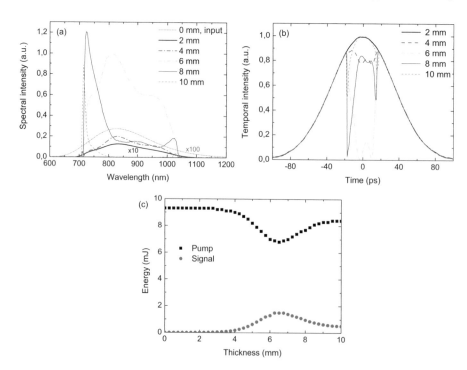

FIGURE 24.8 Simulation of saturated OPCPA with dispersion, diffraction and walk-off using Chi2D. (a) Signal spectrum evolution with area scaled with energy and normalized at 6 mm, (b) pump temporal profile evolution within the crystal, (c) signal and pump energy along the BBO crystal. (Courtesy of Peter Fischer and Tino Lang.)

is a bit longer than in the analytical result. However, at the edges of the phase-matching bandwidth, the gain is lower and saturation is reached later; therefore, these edges – around 720 and 1030 nm – are still increasing when the back conversion already reduces the middle part. The pump laser reaches significant depletion in the middle of the Gaussian temporal beam profile. The maximal signal energy is 1.5 mJ corresponding to 16% conversion efficiency, which are significantly lower than the ideal ω_s/ω_p. This is due to the Gaussian shape of the temporal pump intensity, the ratio of the pump to signal pulse duration, intensity differences in transverse pump beam profile and different signal wavelengths as well as spectrally dependent phase-mismatch. The temporal intensity profile of the pump and the pulse duration ratio make only a small time window and thus reduced pump energy usable for amplification, and all three other reasons influence the gain coefficient in equations (24.17) and (24.18) that cause saturation at various positions in the crystal. As a conclusion, the simple analytical model is a good approximation, but for optimal OPCPA, amplifiers' simulations are necessary.

24.4 OPCPA Architecture

In the previous chapters, OPCPA was described with theoretical and numerical methods to understand its basic properties. We turn our attention to the experimental architecture of an OPCPA system and discuss also how the former principles are implemented. The general OPCPA structure is illustrated in Figure 24.1. The main parts are (i) short pulse source, (ii) stretcher, (iii) parametric amplifier stages and (iv) compressor. An experimental realization of this concept (Herrmann et al., 2009) is depicted in Figure 24.9.

The *front end* is a Titanium:sapphire (Ti:sa) laser. It contains a Ti:sa oscillator generating about 4 nJ, 5–6 fs pulses with 80 MHz repetition rate. To synchronize the pump laser and the seed, all-optical synchronization is used (Ishii et al., 2006), by splitting the oscillator pulses into two and seeding the pump laser and also the OPCPA with them. As the pump laser is Nd:YAG with 1064 nm wavelength and the oscillator has low spectral intensity there, a non-linear photonic crystal fibre is used to shift the oscillator spectrum to 1064 nm (Teisset et al., 2005). Then, the pump generates 80 ps pulses with 10 Hz and about 1 J energy in two parallel arms that are frequency-doubled in one LBO crystal to provide 1 J at 532 nm. The other part of the oscillator energy is seeding a 9-pass, 1 kHz amplifier. The pulses after the multi-pass are compressed in a prism compressor to 20–25 fs with 0.8 mJ energy. However, the spectrum is narrow as typical for lasers and not suitable to seed a broadband OPCPA. Therefore, these pulses are broadened in a neon-filled hollow-core fibre to reach a Fourier limit of sub-6 fs with 0.35 mJ and significant spectral components between 700 and 1000 nm, which is optimal for the OPCPA even if broader spectrum can be achieved from the fibre. Optionally, the beam can be coupled into a cross-polarized wave generation stage with a throughput of less than 10% to increase the temporal ASE contrast by 4 orders of magnitude.

The next component is the *stretcher* containing prisms and gratings in a special configuration, called grism stretcher. The reason of this special hardware is connected to the dispersion management of the OPCPA system plus the significance of higher order (>2) spectral phase terms to obtain a well-compressed few-cycle pulse at the end and explained by the compression. This grism stretcher extends the pulses to 50 ps, which is the group delay between 700 and 1000 nm. Transmission is about 10%. Behind the stretcher, an acousto-optic programmable dispersive filter (Dazzler) compresses the pulses to about 33 ps and has transmission of about 10%. The Dazzler can shape the spectral amplitude and phase before amplification to obtain a well-compressed pulse at the end of the system.

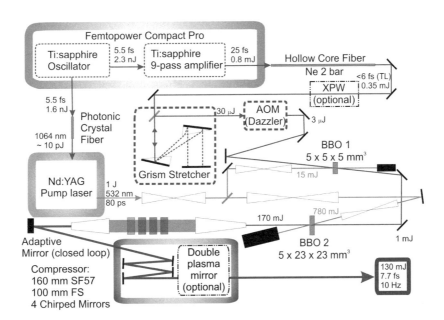

FIGURE 24.9 Schematics of a few-cycle OPCPA system, old version of the Light Wave Synthesizer 20. XPW: cross-polarized wave generation; AOM: acousto-optic modulator (Herrmann et al., 2009).

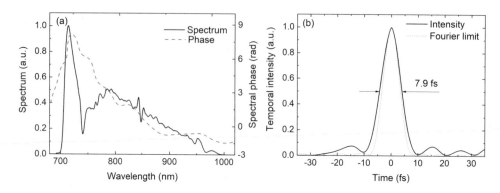

FIGURE 24.10 (a) Spectrum and spectral phase, and (b) evaluated and Fourier-limited (7.5 fs) temporal intensity of the old version of LWS-20 OPCPA system with 7.9 fs pulse duration (Herrmann et al., 2009).

Two single-pass *parametric amplifier* stages follow in the setup and enhance the energy from ~3 μJ to 150–170 mJ in 5-mm thick Type I BBO crystals using non-collinear geometry with a phase-matching angle of $\theta = 23.8°$ and a non-collinear angle of $\alpha = 2.3°$. It should be noted that the accurate values of these angles depend on the dispersion relation and has relevance only for simulations. The pump beam profile is relay imaged to the OPA stages from the second harmonic crystal – indicated by the double triangles in the 532 nm pump beam path in Figure 24.9 – and has a super-Gaussian shape to increase conversion efficiency into the signal. The parametric gain approaches 1000 in the first and >100 in the second stage, taking into account saturation pumped by 15 and 780 mJ, respectively. Most systems in the visible and NIR range use this configuration of green-pumped BBO.

The *dispersion management* (stretching and compression) of the OPCPA is based on a high transmission (75%–80%) *compressor* including 4 pieces of bulk glasses (2 × 80 mm SF57, and 2 × 50 mm quartz) with anti-reflex coating and 4 chirped mirrors. Conventional grating compressors would have a maximum transmission of 50% (Witte et al., 2006). The drawback of a bulk compressor is the special higher (>2) order spectral phase of the material that has to be compensated well to get few-cycle pulse duration. Conventional grating stretcher is not capable of compensating the third and fourth orders well, and a combination of conventional design of a 'grating compressor' and a 'prism compressor' (Adachi et al., 2008) or even better a grism stretcher is needed (Tan et al., 2018). These solutions generate a negatively chirped pulse (blue part comes first) for amplification. Certain geometric parameters in the stretcher design provide some extra control over the higher order spectral phase and thus a good compression. By this way, compression of a spectrum in the region 700–1000 nm and a corresponding sub-8-fs pulse duration is feasible as shown in Figure 24.10. The spectral range with shorter pump lasers is typically slightly broader due to the higher applicable pump intensity.

The high-dynamic range *temporal contrast* is the ratio of the intensity at a given time instant to the peak intensity of the pulse. It is relevant to keep this below 10^{-8}–10^{-12} – depending on peak intensity – to avoid early unwanted plasma generation that expands until the main pulse arrives in experiments especially with solid targets (Veisz, 2010). OPCPA with short pump pulse duration is an ideal tool to enhance the contrast outside

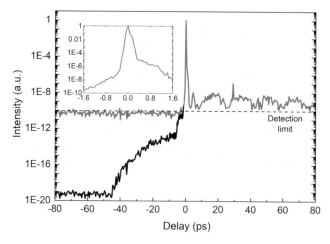

FIGURE 24.11 High-dynamic range temporal contrast of the old version of the LWS-20 OPCPA system. The gray line is measured and the black one is evaluated from independent measurements of contrast enhancement by the XPW, short-pump OPCPA and plasma mirror. For details, see Mikhailova et al. (2011).

of the pump window by a factor corresponding to the overall gain of the system. Care must be taken that ASE from the seed is suppressed enough, for example, by crossed-polarized wave generation as indicated in Figure 24.9. Implementing beyond XPW also a plasma mirror at the end of LWS-20 system (Mikhailova et al., 2011) an extreme estimated contrast value of 19 orders of magnitude has been reached as shown in Figure 24.11. This contrast exceeds all requirements even for a 10-PW laser.

The discussed OPCPA is one specific example, which has typical structure and properties to other systems working in the visible and NIR regime. Some comments about relevant general properties follow.

General remarks:

- Typical pump pulse duration is in the 1–100 ps interval due to multiple reasons. First, as the optical parametric process is instantaneous, the seed and pump should have approximately similar durations. However, the dispersion management for broadband spectra is challenging. Stretching beyond 100 ps and compression to few cycles with good transmission, especially for the compressor, is not possible. Second, as mentioned before, the contrast is enhanced outside of the pump temporal window. To this end a 1 ps short pump is optimal and a ns pump does not provide any advantage.

- Correspondingly, the seed and pump should be well synchronized to a few percent of the pump duration or even better. This can be solved electronically between two independent laser oscillator for seed and pump (Witte et al., 2006) or in the above described all-optical way with a common oscillator for both (Ishii et al., 2006). Dominantly this last solution is used nowadays.

- The stretched seed duration is limited by two effects. If it is too short the temporal overlap with the pump is not optimal and the transferred energy and conversion efficiency are low. If the seed is too long not the whole seed is amplified only that part of it, which temporally overlaps with the pump selectable with the delay. As the seed is chirped the different times correspond to different frequencies, and consequently, only a portion of the seed spectrum is amplified increasing the Fourier limit of the amplified pulse. Optimum ratio for a Gaussian pump is about 0.35–0.4 to have broad gain bandwidth and still have high energy and conversion efficiency (Witte and Eikema, 2012). However, by sacrificing bandwidth, both conversion efficiency and contrast can be improved (Moses et al., 2009). It is important to mention that a flat-top temporal intensity pump profile is the best for both efficiency and bandwidth (Fülöp, Zs. Major, et al., 2007a).

- Similarly to the temporal shape, a flat-top beam pump profile in transverse direction increases the conversion efficiency and also stability with simultaneous saturation (Waxer et al., 2003). It is necessary to relay image the pump beam to the OPCPA stages to maintain its original flat-top profile.

- The contrast improvement using an 80 ps pump is visible in Figure 24.11. The black line indicates about 7 orders of magnitude between −50 and −10 ps, which corresponds to the gain (3×20^5 when XPW reduces the seed energy) plus the parametric fluorescence or super-fluorescence degradation ($\times 50$ compared with the XPW cleaned ASE from the front end) produced by the pump laser at −10 ps. If XPW is not used, the contrast improvement by the pump is reduced to the gain and the contrast is generally worse before ($\times 10^4$ degradation, which is the cleaning of XPW) as well as during the pump ($\times 200$ degradation, which is the difference between ASE and parametric fluorescence). Shortly, the short pump helps to increase the contrast by at least a factor of the parametric gain.

- Limited contrast originates from amplified incoherent light by optical parametric fluorescence (Tavella et al., 2005), which is seeded by vacuum fluctuations (Homann and Riedle, 2013). It can be suppressed by ps pump duration at least in the first stage, using fundamental Gaussian mode, and shifting low-transmission stretcher and acousto-optic modulator behind the first amplifier stage, where also the fluorescence background is reduced. Certainly a high seed energy also helps that the seeding photon number significantly exceeds vacuum fluctuations (Tavella, Marcinkevičius and Krausz, 2006). An alternative front end to the presented one in Figure 24.9 utilizes white light generation in sapphire, difference frequency generation (DFG) for passive CEP stabilization, another white light generation to get a broad seed spectrum and amplification in two noncollinear OPA stages (Budriunas, Stanislauskas and Varanavičius, 2015). The OPAs certainly have very short duration and improve the contrast significantly.

24.5 OPCPA in Practice

In the following, we discuss typical features of OPCPA systems and how they are realized in practice. Among others, the spectral coverage of OPCPA systems, the generation of broad bandwidth supporting few-cycle pulse duration and their application in ultrahigh power lasers will be scrutinized.

24.5.1 Broad Spectral Coverage

A unique feature of parametric amplification that it covers the full optical range from visible, NIR, to mid-infrared (mid-IR) (Wilhelm, Piel and Riedle, 1997; Brida et al., 2010). This is also utilized in OPCPA systems by applying different non-linear crystals, pump lasers and system architectures. The spectral operating range of some typical OPCPA systems is shown in Figure 24.12. Even if this is an incomplete overview of the existing systems, it reflects well the fact that almost from 400 up to 9000 nm, the whole spectrum is covered by OPCPA. Upper and lower wavelength limits are defined by the transmission of crystals. Further increase of the upper border is expected, while the lower limit is defined by the bandgap and associated absorption, which is harder to improve. A short review of the typical ranges follows.

OPCPA in the *NIR and mid-IR* (8–20 in the caption of Figure 24.12) utilizes as typical pump radiation the fundamental of diode-pumped picosecond Nd or Yb doped lasers with a wavelength around 1 μm, having high quality regarding energy and pointing stability, beam profile, temporal shape, etc. Longer wavelengths use Ho:YLF with 2 μm wavelength (Sanchez et al., 2016).

The amplification takes place at low (<100 μJ) energies typically in non-linear crystals such as periodically poled lithium niobate (PPLN) (Chalus et al., 2010) and MgO-doped PPLN (Rigaud et al., 2016), which is limited to small sizes, however, with energies towards a mJ potassium titanyl phosphate (KTP) (Mücke et al., 2009), potassium titanyl arsenate (KTA) (Mero et al., 2018), LN (Deng et al., 2012) and zinc germanium phosphide (ZGP) (Sanchez et al., 2016) are used. A general property of OPCPA in this range is passive CEP stabilization via intra-pulse DFG of a shorter wavelength seed (Baltuška, Fuji and Kobayashi, 2002a). As the wavelength is longer, natural fluctuations do not influence CEP so much and better stability

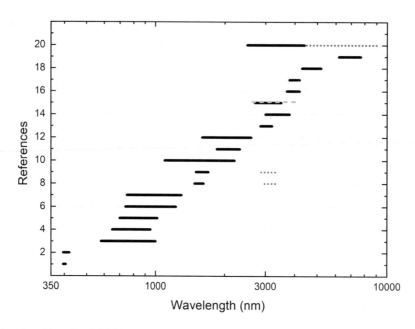

FIGURE 24.12 Spectral region of broadband OPCPA systems, covering the visible, NIR and Mid-IR ranges. Continuous black: signal spectrum; dotted gray: idler for collinear configuration or angular dispersion compensated; dashed light gray: tunability range of central wavelength. References on vertical axis 1: (Kurdi et al., 2004); 2: (Wnuk, Stepanenko and Radzewicz, 2010); 3: (Rivas et al., 2017; Harth et al., 2012); 4: (Adachi et al., 2008; Fülöp, Zs. Major, et al., 2007b); 5: (Herrmann et al., 2009; Witte et al., 2006; Prinz, Schnitzenbaumer, et al., 2018; Ahrens et al., 2016; Budriūnas et al., 2017); 6: (Stanislauskas et al., 2018); 7: (Kessel et al., 2018); 8: (Mero et al., 2018); 9: (Rigaud et al., 2016); 10: (Ishii et al., 2011); 11: (Hong et al., 2014); 12: (Deng et al., 2012); 13: (Chalus et al., 2010); 14: (Mayer et al., 2014); 15: (Thiré et al., 2018a); 16: (Andriukaitis et al., 2011); 17: (Wang et al., 2018); 18: (Bock et al., 2018); 19: (Sanchez et al., 2016); 20: (Liang et al., 2017).

can be reached (Thiré et al., 2018b). Furthermore, many systems utilize collinear geometry, which preserves the idler from angular dispersion and it is easily utilized after compression (Andriukaitis et al., 2011).

OPCPA in the *visible and NIR* range (3–7 in the caption of Figure 24.12) typically is pumped by the second or third harmonic of a 1 μm Nd or Yb doped lasers. Tests to use the second harmonic of Ti:sa as pump have been made as well (Fülöp, Zs. Major, et al., 2007b; Adachi et al., 2008). The non-linear crystals are BBO or LBO, which provide a high non-linear coefficient and can be produced also in larger sizes. The spectral range is 650–1100 nm or less – depending on the BBO thickness – for second-harmonic and 550–700 nm for third-harmonic pumped stages. The CEP stabilization is more challenging as the wavelength range overlaps with the spectrum of conventional oscillators, and DFG would generate much longer wavelength components. Still some front-end architectures exist to realize passive-CEP stabilization with DFG (Budriūnas et al., 2017). Most systems utilize the traditional CEP stabilization of the oscillators and secondary feedback loops (Prinz, Schnitzenbaumer, et al., 2018). As mostly non-collinear geometry is applied, the idler is angularly dispersed and not used as an independent source.

OPCPA in the *ultraviolet (UV) – blue part of visible* (Kurdi et al., 2004; Wnuk, Stepanenko and Radzewicz, 2010) – is more challenging. The main reason is the significant linear and non-linear (two-photon) absorption of the pump that reduces the pump intensity and also damages the crystal. Experiments around 400 nm signal wavelengths were demonstrated in BBO and UV pump (Kurdi et al., 2004; Wnuk, Stepanenko and Radzewicz, 2010), but there are no working systems in this range yet.

24.5.2 Few-cycle Pulse Duration

The most prominent feature of OPCPA is the capability of amplifying ultra-broad bandwidth that supports pulse durations of only a few optical cycles. The spectral gain bandwidth is determined by the crystal transmission range and the possible phase-matching configuration. This corresponds to 3–9 fs in the visible, while even the shortest conventional CPA laser pulses are 20 fs or longer. In the mid-IR regime, it seems to be more challenging to get a very short pulse as typical systems have more than 3 optical cycles, while certain experiments require shorter than 2 such as generation of isolated attosecond pulses in gas or relativistic plasma media. Nonetheless, 1.5 cycles duration at 2.1 μm has already been achieved (Deng et al., 2012). However, in the visible-NIR, the common BBO and second harmonic pump architecture routinely provide sub-3 cycles. Pump pulses with shorter duration (around 1 ps) utilize higher intensity in the non-linear crystals and thus support slightly larger gain bandwidth (Kessel et al., 2018; Prinz, Schnitzenbaumer, et al., 2018) with correspondingly slightly reduced pulse duration, compared with longer pump duration (around 100 ps). On the other hand, all-reflective optics is necessary for a 1 ps pump to avoid non-linearities along with a sophisticated synchronization system beyond all-optical synchronization (Schwarz et al., 2012; Prinz, Häfner, et al., 2018). Furthermore, the amplifier stages should be in vacuum for high pump energies (Kessel et al., 2018). Under special circumstances, OPCPA can even support an octave spanning spectrum from a single stage (Ishii et al., 2011), using 800 nm pump wavelength, which can be compressed below two optical cycles (Ishii et al., 2012).

24.5.3 Ultrahigh Power Systems

Due to the excellent contrast generated by OPCPA, it is often used in combination with conventional laser amplifiers, typically around 800 nm or 1 μm central wavelength. OPCPA and laser amplifiers in a hybrid infrastructure with the same pump beam for both provide high gain, high conversion efficiency and high pre-pulse contrast (Jovanovic, Ebbers and Barty, 2002; Hama et al., 2007; Zhou et al., 2007). Different front-end light sources have been developed based on OPCPA (Dorrer et al., 2007; Musgrave et al., 2010; Liang et al., 2018). Some of them even for sub-10-fs pulse duration with moderate (10^{12}) (Papadopoulos et al., 2017) and extreme (10^{19}) contrast (Mikhailova et al., 2011). Correspondingly, many large high-energy laser facilities are utilizing OPCPA front ends (Danson et al., 2005; Waxer et al., 2005; Gaul et al., 2010; Kiriyama et al., 2012). Furthermore, OPA (no stretching)-based contrast improvement has also been realized for 500 fs pulses (Shah et al., 2009) and even for few-cycle pulses with passive CEP stabilization (Budriunas, Stanislauskas and Varanavičius, 2015). This approach reduces the gain window and improves the contrast practically in the whole relevant temporal range. OPCPA provides also a broad and typically energetic seed pulse, which is less influenced by gain narrowing in the final laser amplifiers, especially if the OPCPA pump is shaped in time, and thus increases the final bandwidth and shortens the pulse duration (Lee et al., 2018).

High-energy (>10 J) OPCPA beyond the 100 TW power level has also been demonstrated (Chekhlov et al., 2006; Lozhkarev et al., 2007). These use KDP non-linear crystals for the final amplifier as only these can be produced in large enough size. Correspondingly, the pump is the second harmonic of a μm wavelength laser and the signal wavelength is 910 or 1050 nm. However, the broad bandwidth capability of parametric amplification is not utilized with the 40–80 fs pulse durations. The highest peak power obtained from a pure OPCPA system is 560 TW to date (Lozhkarev et al., 2007).

24.6 Optical Parametric Synthesizers

As shown, the OPCPA technology supports the amplification of few-cycle pulses. The question arises whether the pulse duration can be made even shorter under realistic conditions such as the application of available high-quality pump laser at 1 μm wavelength. The recent development of field synthesis provides a positive answer to this question (Manzoni et al., 2015). The basic principle of field synthesis is illustrated in Figure 24.13. The idea is the coherent combination of two or more pulses with complementary spectra and well-defined spectral phase.

One way to attain this combination is the parallel synthesis, where the different spectral regions are separated, independently amplified, compressed and combined again. Another way is the sequential or serial synthesis, where the spectral components are propagating together and they are consecutively amplified in different stages before compression. The advantage of parallel synthesis is the limited spectral range that is treated in the amplifier and compressor of one arm,

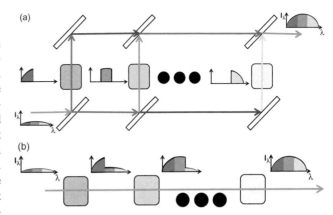

FIGURE 24.13 Principle of coherent field synthesis in (a) parallel and (b) serial (or sequential) configuration. The three points indicate possible extra stages. If the amplification is based on OPA, the systems are called OPS.

which makes it easier to deal with ultra-broad (up to multi-octave) spectra. Utilizing a narrower gain bandwidth and optics for a narrower spectral range with higher throughput are enough, and the chirped mirror technology often used for compression also supports only limited bandwidth (Kärtner et al., 1997; Pervak et al., 2009). On the other hand, coherent synthesis requires very low jitter between the parallel arms, much below one optical cycle, and stable CEP and compensated spectral phase for all channels. This is very challenging and calls for high repetition-rate systems to keep the changes between the shots small. Up to an octave spectral bandwidth, serial synthesis is obviously well suited.

The approach of coherent field synthesis for the generation of spectra spanning about an octave or even more has been applied with fibre lasers (Krauss et al., 2010), a combination of a hollow-core fibre and a special chirped mirror compressor (Wirth et al., 2011) and OPAs (Manzoni et al., 2012). A special type of synthesizer is the frequency domain optical parametric amplifier (Schmidt et al., 2014), where OPA stages amplifying different spectral regions are placed in the Fourier plane of a 4f zero-dispersion shaper.

Up to now, only very few OPCPA systems use the principle of synthesis. These OPS systems with parallel synthesis (Huang et al., 2011) can provide gain bandwidth beyond an octave and promise multi-octave light transients with high average power (Fattahi et al., 2014). The principle of serial OPS has been demonstrated (Herrmann et al., 2010a) providing almost an octave-spanning spectrum, and applied on the μJ level (Harth et al., 2012) and 80 mJ level (Rivas et al., 2017), with a compression below two-optical cycles (<5 fs). As an example the description of this last system follows.

The spectral evolution of the synthesized waveform in a 4-stage serial OPS is plotted in Figure 24.14 (Rivas et al., 2017). This system is an upgraded version of the Light Wave Synthesizer 20, which is shown in Figure 24.9. While the original system has an OPCPA stage with second harmonic (532 nm) pumping and amplification in the 700–1000 nm range, the upgrade contains another non-collinear stage using BBO in different geometry (phase-matching and non-collinear angle) pumped by the third harmonic (355 nm) of the Nd:YAG laser.

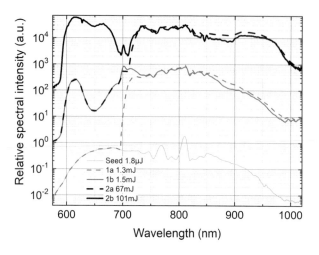

FIGURE 24.14 Spectral evolution after the different amplifiers in a serial OPS (Rivas et al., 2017). Stages marked with 'a' ('b') are pumped with 532 nm (355 nm) pump wavelength and amplified in different regions.

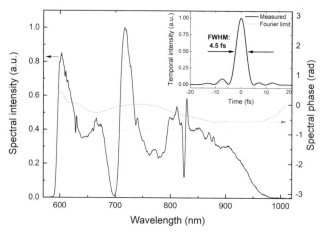

FIGURE 24.15 Spectrum and spectral phase of the few-cycle serial OPS system, LWS-20 (Rivas et al., 2017). Inset: corresponding measured and Fourier-limited temporal intensities. The FWHM pulse duration is 4.5 fs, which corresponds to 1.8 optical cycles.

This stage amplifies in the 580–700 nm range and correspondingly extends the spectrum that is approaching a full octave. A second pair of amplifiers with 532 and 355 nm pumping boosts the energy and maintains the broad spectrum at the same time. To get the broadest spectrum and thus the shortest possible pulse, the two spectral regions are slightly shifted away from each other, which generates a small hole between them at around 700 nm.

Figure 24.15 shows the amplified spectrum on a linear scale and the corresponding temporal characterization. The spectral phase is optimized with the help of chirp scan and an iterative algorithm. The pulse after compression has 70–75 mJ energy and a FWHM duration of 4.5 fs within 0.1 fs of its Fourier limit (see inset in Figure 24.15) showing the high-quality compression. Furthermore, it is much shorter than the original only 532 nm pumped part, compared with Figure 24.10, which clearly indicates the stable coherent combination of the two spectral parts. Further, development of this system is expected to deliver a CEP-stable 100 TW-scale 4-fs OPS.

24.7 Summary

This chapter has given an overview about the OPCPA technology, its theoretical description, its experimental architecture, OPCPA systems in practice and their further development towards OPS. In summary, OPCPA became a well-established technology in the last 27 years, which not only provides alternative light sources to CPA lasers but also exceeds their capabilities in certain areas. It is unique in ultra-broadband spectral amplification to support few-cycle pulse duration, much beyond any direct laser amplifier. Optical parametric synthesis can produce single-cycle or even sub-cycle temporal waveforms and tailor their pulse shape at will. OPCPA systems available today cover almost the whole spectral range from about 400 nm to 9 μm, and the development is still ongoing. Some OPCPA systems also provide ultra-high peak power and promise significant enhancement by combining high energy and short pulse duration. In the future, it is expected that high average power systems will be widely used. Furthermore, the combination of extreme bandwidth beyond an octave and high energy is envisioned to deliver petawatt peak powers with sub-cycle pulse duration.

REFERENCES

Adachi, S. et al. (2008) '5-fs, multi-mJ, CEP-locked parametric chirped-pulse amplifier pumped by a 450-nm source at 1 kHz', *Optics Express*, 16(19), p. 14341. doi: 10.1364/OE.16.014341.

Ahrens, J. et al. (2016) 'Multipass OPCPA system at 100 kHz pumped by a CPA-free solid-state amplifier', *Optics Express*, 24(8), p. 8074. doi: 10.1364/OE.24.008074.

Akhmanov, S. A., Vysloukh, V. A. and Chirkin, A. S. (1992) *Optics of Femtosecond Laser Pulses*. New York: Springer.

Andriukaitis, G. et al. (2011) '90 GW peak power few-cycle mid-infrared pulses from an optical parametric amplifier', *Optics Letters*, 36(15), p. 2755. doi: 10.1364/OL.36.002755.

Arisholm, G. (1997) 'Second-order nonlinear interactions in birefringent media', *Journal of the Optical Society of America B*, 14(10), pp. 2543–2549. doi: 10.1364/JOSAB.14.002543.

Armstrong, J. A. et al. (1962) 'Interactions between light waves in a nonlinear dielectric', *Physical Review*, 127(6), pp. 1918–1939. doi: 10.1103/PhysRev.127.1918.

Baltuška, A., Fuji, T. and Kobayashi, T. (2002a) 'Controlling the carrier-envelope phase of ultrashort light pulses with optical parametric amplifiers', *Physical Review Letters*, 88(13), p. 133901. doi: 10.1103/PhysRevLett.88.133901.

Baltuška, A., Fuji, T. and Kobayashi, T. (2002b) 'Visible pulse compression to 4 fs by optical parametric amplification and programmable dispersion control', *Optics Letters*, 27(5), p. 306. doi: 10.1364/OL.27.000306.

Baum, P. et al. (2003) 'Phase-coherent generation of tunable visible femtosecond pulses', *Optics Letters*, 28(3), p. 185. doi: 10.1364/ol.28.000185.

Baumgartner, R. and Byer, R. (1979) 'Optical parametric amplification', *IEEE Journal of Quantum Electronics*, 15(6), pp. 432–444. doi: 10.1109/JQE.1979.1070043.

Bock, M. et al. (2018) 'Generation of millijoule few-cycle pulses at 5 μm by indirect spectral shaping of the idler in an optical parametric chirped pulse amplifier', *Journal of the Optical Society of America B*, 35(12), p. C18. doi: 10.1364/josab.35.000c18.

Brida, D. et al. (2010) 'Few-optical-cycle pulses tunable from the visible to the mid-infrared by optical parametric amplifiers', *Journal of Optics*, 12(1), p. 013001. doi: 10.1088/2040-8978/12/1/013001.

Budriūnas, R. et al. (2017) '53 W average power CEP-stabilized OPCPA system delivering 5.5 TW few cycle pulses at 1 kHz repetition rate', *Optics Express*, 25(5), p. 5797. doi: 10.1364/OE.25.005797.

Budriunas, R., Stanislauskas, T. and Varanavičius, A. (2015) 'Passively CEP-stabilized frontend for few cycle terawatt OPCPA system', *Journal of Optics (United Kingdom)*, 17(9). doi: 10.1088/2040-8978/17/9/094008.

Cerullo, G. and De Silvestri, S. (2003) 'Ultrafast optical parametric amplifiers', *Review of Scientific Instruments*, 74(1 I), pp. 1–18. doi: 10.1063/1.1523642.

Chalus, O. et al. (2010) 'Six-cycle mid-infrared source with 3.8 μJ at 100 kHz', *Optics Letters*, 35(19), pp. 3204–3206.

Chekhlov, O. V. et al. (2006) '35 J broadband femtosecond optical parametric chirped pulse amplification system', *Optics Letters*, 31(24), p. 3665. doi: 10.1364/ol.31.003665.

Danielius, R. et al. (1993) 'Traveling-wave parametric generation of widely tunable, highly coherent femtosecond light pulses', *Journal of the Optical Society of America B*, 10(11), p. 2222. doi: 10.1364/JOSAB.10.002222.

Danson, C. N. et al. (2005) 'Vulcan petawatt : Design, operation and interactions at 5×10^{20} Wcm2 C.N.,' *Laser and Particle Beams*. Umea University Library, 23, pp. 87–93.

Deng, Y. et al. (2012) 'Carrier-envelope-phase-stable, 1.2 mJ, 1.5 cycle laser pulses at 2.1 μm', *Optics Letters*, 37(23), pp. 4973–4975.

Dmitriev, V. G., Gurzadyan, G. G. and Nikogosyan, D. N. (1999) *Handbook of Nonlinear Optical Crystals*. Springer-Verlag Berlin Heidelberg.

Dorrer, C. et al. (2007) 'High-contrast optical-parametric amplifier as a front end of high-power laser systems', *Optics letters*, 32(15), pp. 2143–2145. doi: 10.1364/OL.32.002143.

Dubietis, A., Butkus, R. and Piskarskas, A. P. (2006) 'Trends in chirped pulse optical parametric amplification', *IEEE Journal of Selected Topics in Quantum Electronics*, 12(2), pp. 163–172. doi: 10.1109/JSTQE.2006.871962.

Dubietis, A., Jonušauskas, G. and Piskarskas, A. (1992) 'Powerful femtosecond pulse generation by chirped and stretched pulse parametric amplification in BBO crystal', *Optics Communications*, 88(4–6), pp. 437–440. doi: 10.1016/0030-4018(92)90070-8.

Fattahi, H. et al. (2014) 'Third-generation femtosecond technology', *Optica*, 1(1), p. 45. doi: 10.1364/OPTICA.1.000045.

Fisher, R. A. and Bischel, W. K. (1975) 'Numerical studies of the interplay between self-phase modulation and dispersion for intense plane-wave laser pulses', *Journal of Applied Physics*, 46(11), pp. 4921–4934. doi: 10.1063/1.321476.

Fuji, T. et al. (2006) 'Parametric amplification of few-cycle carrier-envelope phase-stable pulses at 21 μm', *Optics Letters*, 31(8), p. 1103. doi: 10.1364/OL.31.001103.

Fülöp, J. A., Major, Zs., et al. (2007a) 'Shaping of picosecond pulses for pumping optical parametric amplification', *Applied Physics B: Lasers and Optics*, 87(1), pp. 79–84. doi: 10.1007/s00340-006-2488-3.

Fülöp, J. A., Major, Zs., et al. (2007b) 'Short-pulse optical parametric chirped-pulse amplification for the generation of high-power few-cycle pulses', *New Journal of Physics*, 9, p. 438. doi: 10.1088/1367–2630/9/12/438.

Gaul, E. W. et al. (2010) 'Demonstration of a 1.1 petawatt laser based on a hybrid optical parametric chirped pulse amplification/mixed Nd:glass amplifier', *Applied Optics*, 49(9), p. 1676. doi: 10.1364/AO.49.001676.

Hama, Y. et al. (2007) 'Control of amplified optical parametric fluorescence in hybrid chirped-pulse amplification', *Springer Series in Optical Sciences*, 132(2), pp. 527–533. doi: 10.1007/978-0-387–49119-6_68.

Harth, A. et al. (2012) 'Two-color pumped OPCPA system emitting spectra spanning 1.5 octaves from VIS to NIR', *Optics Express*, 20(3), p. 3076. doi: 10.1364/OE.20.003076.

Herrmann, D. et al. (2009) 'Generation of sub-three-cycle, 16 TW light pulses by using noncollinear optical parametric chirped-pulse amplification', *Optics Letters*, 34(16), p. 2459. doi: 10.1364/OL.34.002459.

Herrmann, D et al. (2010a) 'Approaching the full octave: noncollinear optical parametric chirped pulse amplification with two-color pumping.', *Optics Express*, 18(18), pp. 18752–18762. doi: 10.1364/OE.18.018752.

Herrmann, D et al. (2010b) 'Investigation of two-beam-pumped noncollinear optical parametric chirped-pulse amplification for the generation of few-cycle light pulses', *Optics Express*, 18(5), p. 4170. doi: 10.1364/OE.18.004170.

Homann, C. and Riedle, E. (2013) 'Direct measurement of the effective input noise power of an optical parametric amplifier', *Laser & Photonics Reviews*, 7(4), pp. 580–588. doi: 10.1002/lpor.201200119.

Hong, K. et al. (2014) 'Multi-mJ, kHz, 2.1 μm optical parametric chirped-pulse amplifier and high-flux soft X-ray high-harmonic generation', *Opt. Lett.*, 39(11), pp. 3145–3148.

Huang, S. W. et al. (2011) 'High-energy pulse synthesis with sub-cycle waveform control for strong-field physics', *Nature Photonics*, 5(8), pp. 475–479. doi: 10.1038/nphoton.2011.140.

Ishii, N. et al. (2006) 'Seeding of an eleven femtosecond optical parametric chirped pulse amplifier and its Nd 3+ picosecond pump laser from a single broadband Ti: sapphire oscillator', *IEEE Journal on Selected Topics in Quantum Electronics*, 12(2), pp. 173–180. doi: 10.1109/JSTQE.2006.871930.

Ishii, N. et al. (2011) 'Carrier-envelope-phase-preserving, octave-spanning optical parametric amplification in the infrared based on BiB_3O_6 pumped by 800 nm femtosecond laser pulses', *Applied Physics Express*, 4(2), pp. 3–5. doi: 10.1143/APEX.4.022701.

Ishii, N. et al. (2012) 'Sub-two-cycle, carrier-envelope phase-stable, intense optical pulses at 1.6 μm from a BiB_3O_6 optical parametric chirped-pulse amplifier', *Optics Letters*, 37(20), pp. 4182–4184.

Jovanovic, I., Ebbers, C. a and Barty, C. P. J. (2002) 'Hybrid chirped-pulse amplification', *Optics Letters*, 27(18), pp. 1622–1624. doi: 10.1364/OL.27.001622.

Kärtner, F. X. et al. (1997) 'Design and fabrication of double-chirped mirrors', *Optics Letters*, 22(11), p. 831. doi: 10.1364/OL.22.000831.

Kessel, A. et al. (2018) 'Relativistic few-cycle pulses with high contrast from picosecond-pumped OPCPA', *Optica*, 5(4), p. 434.

Kiriyama, H. et al. (2012) 'Temporal contrast enhancement of petawatt-class laser pulses', *Optics Letters*, 37(16), p. 3363. doi: 10.1364/ol.37.003363.

Krauss, G. et al. (2010) 'Synthesis of a single cycle of light with compact erbium-doped fibre technology', *Nature Photonics*, 4(1), pp. 33–36. doi: 10.1038/nphoton.2009.258.

Kurdi, G. et al. (2004) 'Optical parametric amplification of femtosecond ultraviolet laser pulses', *IEEE Journal on Selected Topics in Quantum Electronics*, 10(6), pp. 1259–1267. doi: 10.1109/JSTQE.2004.837706.

Lang, T. et al. (2013) 'Impact of temporal, spatial and cascaded effects on the pulse formation in ultra-broadband parametric amplifiers.', *Optics Express*, 21(1), pp. 949–959. doi: 10.1364/OE.21.000949.

Lee, H. W. et al. (2018) 'Spectral shaping of an OPCPA preamplifier for a sub-20-fs multi-PW laser', *Optics Express*, 26(19), p. 24775. doi: 10.1364/oe.26.024775.

Liang, H. et al. (2017) 'High-energy mid-infrared sub-cycle pulse synthesis from a parametric amplifier', *Nature Communications. Springer US*, 8(1), pp. 1–9. doi: 10.1038/s41467-017-00193-4.

Liang, X. et al. (2018) 'Broadband main OPCPA amplifier at 808 nm wavelength in high deuterated DKDP crystals', *Optics Letters*, 43(23), p. 5713. doi: 10.1364/ol.43.005713.

Lozhkarev, V. V. et al. (2007) 'Compact 0.56 petawatt laser system based on optical parametric chirped pulse amplification in KD*P crystals', *Laser Physics Letters*, 4(6), pp. 421–427. doi: 10.1002/lapl.200710008.

Manzoni, C. et al. (2012) 'Coherent synthesis of ultra-broadband optical parametric amplifiers', *Optics Letters*, 37(11), p. 1880. doi: 10.1364/OL.37.001880.

Manzoni, C. et al. (2015) 'Coherent pulse synthesis: towards subcycle optical waveforms', *Laser and Photonics Reviews*, 9(2), pp. 129–171. doi: 10.1002/lpor.201400181.

Manzoni, C. and Cerullo, G. (2016) 'Design criteria for ultrafast optical parametric amplifiers', *Journal of Optics*. IOP Publishing, 18(10), pp. 1–33. doi: 10.1088/2040-8978/18/10/103501.

Mayer, B. W. et al. (2014) 'Mid-infrared pulse generation via achromatic quasi-phase-matched OPCPA', *Optics Express*, 22(17), p. 20798. doi: 10.1364/OE.22.020798.

Mero, M. et al. (2018) '43 W, 155 μm and 125 W, 3.1 μm dual-beam, sub-10 cycle, 100 kHz optical parametric chirped pulse amplifier', *Optics Letters*, 43(21), p. 5246. doi: 10.1364/ol.43.005246.

Mikhailova, J. M. et al. (2011) 'Ultra-high-contrast few-cycle pulses for multipetawatt-class laser technology', *Optics Letters*, 36(16), p. 3145. doi: 10.1364/ol.36.003145.

Moses, J. et al. (2009) 'Highly stable ultrabroadband mid-IR optical parametric chirped-pulse amplifier optimized for superfluorescence suppression,'*Optics Letters*, 34(11), pp. 1639–1641.

Mücke, O. D. et al. (2009) 'Self-compression of millijoule 1. 5 μm pulses', *Optics Letters* 34(16), pp. 2498–2500.

Musgrave, I. et al. (2010) 'Picosecond optical parametric chirped pulse amplifier as a preamplifier to generate high-energy seed pulses for contrast enhancement.', *Applied optics*, 49(33), pp. 6558–62. doi: 10.1364/AO.49.006558.

Papadopoulos, D. N. et al. (2017) 'High-contrast 10 fs OPCPA-based front end for multi-PW laser chains', *Optics Letters*, 42(18), p. 3530. doi: 10.1364/OL.42.003530.

Pervak, V. et al. (2009) 'Double-angle multilayer mirrors with smooth dispersion characteristics', *Optics Express*, 17(10), p. 7943. doi: 10.1364/oe.17.007943.

Polyanskiy, M. N. (2019) *Refractive Index Database*. Available at: https://refractiveindex.info/.

Prinz, S., Häfner, M., et al. (2018) 'Active pump-seed-pulse synchronization for OPCPA with sub-2-fs residual timing jitter: erratum', *Optics Express*, 26(5), p. 5512. doi: 10.1364/oe.26.005512.

Prinz, S., Schnitzenbaumer, M., et al. (2018) 'Thin-disk pumped optical parametric chirped pulse amplifier delivering CEP-stable multi-mJ few-cycle pulses at 6 kHz', *Optics Express*, 26(2), p. 1108. doi: 10.1364/OE.26.001108.

Rigaud, P. et al. (2016) 'Supercontinuum-seeded few-cycle mid-infrared OPCPA system', *Optics Express*, 24(23), pp. 26494–26502. doi: 10.1523/JNEUROSCI.4261–09.2009.

Rivas, D. E. et al. (2017) 'Next Generation Driver for Attosecond and Laser-plasma Physics', *Scientific Reports*, 7(1), p. 5224. doi: 10.1038/s41598-017-05082-w.

Ross, I. N. et al. (1997) 'The prospects for ultrashort pulse duration and ultrahigh intensity using optical parametric chirped pulse amplifiers', *Optics Communications*, 144(1–3), pp. 125–133. doi: 10.1016/S0030-4018(97)00399-4.

Ross, I. N. et al. (2002) 'Analysis and optimization of optical parametric chirped pulse amplification', *Journal of the Optical Society of America B*, 19(12), p. 2945. doi: 10.1364/JOSAB.19.002945.

Sanchez, D. et al. (2016) '7 μm, ultrafast, sub-millijoule-level mid-infrared optical parametric chirped pulse amplifier pumped at 2 μm', *Optica*, 3(2), p. 147. doi: 10.1364/assl.2016.aw4a.6.

Schmidt, B. E. et al. (2014) 'Frequency domain optical parametric amplification', *Nature Communications*, 5(May), pp. 1–8. doi: 10.1038/ncomms4643.

Schwarz, A. et al. (2012) 'Active stabilization for optically synchronized optical parametric chirped pulse amplification', *Optics Express*, 20(5), p. 5557. doi: 10.1364/oe.20.005557.

Shah, R. C. et al. (2009) 'High-temporal contrast using low-gain optical parametric amplification.', *Optics Letters*, 34(15), pp. 2273–2275. doi: 10.1364/OL.34.002273.

Stanislauskas, T. et al. (2018) 'Towards sub-2 cycle, Several-TW, 1kHz OPCPA system based on Yb:KGW and Nd:YAG lasers', *CLEO 2018*, p. STu4O.1. doi: 10.1364/cleo_si.2018.stu4o.1.

Strickland, D. and Mourou, G. (1985) 'Compression of amplied chirped optical pulses', *Optics Communications*, 56(3), pp. 219–221.

Tan, J. et al. (2018) 'Dispersion control for temporal contrast optimization', *Optics Express*, 26(19), p. 25003. doi: 10.1364/oe.26.025003.

Tavella, F. et al. (2005) 'High-dynamic range pulse-contrast measurements of a broadband optical parametric chirped-pulse amplifier', *Applied Physics B: Lasers and Optics*, 81(6), pp. 753–756. doi: 10.1007/s00340-005-1966-3.

Tavella, F., Marcinkevičius, A. and Krausz, F. (2006) 'Investigation of the superfluorescence and signal amplification in an ultrabroadband multiterawatt optical parametric chirped pulse amplifier system', *New Journal of Physics*, 8(6). doi: 10.1088/1367-2630/8/10/219.

Teisset, C. Y. et al. (2005) 'Soliton-based pump–seed synchronization for few-cycle OPCPA', *Opt. Express*, 13(17), p. 6550. doi: 10.1007/s00340-005-1929-8.36.

Thiré, N. et al. (2018a) 'Highly stable, 15 W, few-cycle, 65 mrad CEP-noise mid-IR OPCPA for statistical physics', *Optics Express*, 26(21), p. 26907. doi: 10.1364/OE.26.026907.

Thiré, N. et al. (2018b) 'Highly stable, 15 W, few-cycle, 65 mrad CEP-noise mid-IR OPCPA for statistical physics', *Optics Express*, 26(21), p. 26907. doi: 10.1364/OE.26.026907.

Veisz, L. (2010) 'Contrast improvement of relativistic few-cycle light pulses', in Duarte, F. J. (ed.) *Coherence and Ultrashort Pulsed Emission*. InTech Publishing. doi: 10.5772/12932.

Wang, P. et al. (2018) '2.6 mJ/100 Hz CEP-stable near-single-cycle 4 μm laser based on OPCPA and hollow-core fiber compression', *Optics Letters*, 43(9), p. 2197. doi: 10.1364/OL.43.002197.

Waxer, L. J. et al. (2003) 'High-conversion-efficiency optical parametric chirped-pulse amplification system using spatiotemporally shaped pump pulses', *Optics Letters*, 28(14), p. 1245. doi: 10.1364/ol.28.001245.

Waxer, L. J. et al. (2005) 'High-energy petawatt capability for the omega Laser', *Optics and Photonics News*, 16(7), p. 30. doi: 10.1364/OPN.16.7.000030.

Wilhelm, T., Piel, J. and Riedle, E. (1997) 'Sub-20-fs pulses tunable across the visible from a blue-pumped single-pass noncollinear parametric converter', *Optics Letters*, 22(19), p. 1494. doi: 10.1364/OL.22.001494.

Wirth, A. et al. (2011) 'Synthesized light transients', *Science*, 334, pp. 195–200. doi: 10.1126/science.1210268.

Witte, S. et al. (2006) 'A source of 2 terawatt, 2.7 cycle laser pulses based on noncollinear optical parametric chirped pulse amplification', *Optics Express*, 14(18), p. 8168. doi: 10.1364/OE.14.008168.

Witte, S. and Eikema, K. S. E. (2012) 'Ultrafast optical parametric chirped-pulse amplification', *IEEE Journal on Selected Topics in Quantum Electronics*, 18(1), pp. 296–307. doi: 10.1109/JSTQE.2011.2118370.

Wnuk, P., Stepanenko, Y. and Radzewicz, C. (2010) 'High gain broadband amplification of ultraviolet pulses in optical parametric chirped pulse amplifier', *Optics Express*, 18(8), p. 7911. doi: 10.1364/oe.18.007911.

Zhou, X. et al. (2007) 'An 11-fs, 5-kHz optical parametric/Ti: sapphire hybrid chirped pulse amplification system', *Applied Physics B: Lasers and Optics*, 89(4), pp. 559–563. doi: 10.1007/s00340-007-2844-y.

25

Laser Beam Delivery: Section Introduction

Julian Jones

The process of conveying a laser beam from the laser to the point at which the beam is to be applied is *beam delivery*, the subject of this section. Delivery may be either by conventional optics in free space (Chapter 27), or in waveguides (Chapter 28). In either case, the point of delivery must be controllable and accurately defined in space, either statically or dynamically—hence the related topics of positioning and scanning, the subjects of Chapter 29.

The basic principles of beam delivery—the focusing, shaping and manipulation of the beam relative to the workpiece—are expounded in Chapter 26 by Duncan Hand. Focusing and shaping are dependent on the properties of the laser beam, the required beam profile at the workpiece and whether the beam is to be imaged onto the workpiece and scanned or projected via a mask. The laser wavelength and power densities set the choice of materials to be used (and whether fibre-optic beam delivery would be an option).

Chapter 27 (Leo Beckmann) is concerned with the design of delivery systems using free-space optics and builds on the foundations of Chapter 11 in which the optical components to be used were described. Much depends on the wavelength of the laser. For CO_2 optics at the infrared wavelength of 10.6 μm, the choice of transmissive optical materials is limited, with ZnSe a popular choice and artificial diamond a modern alternative in some situations: reflective optics are appropriate too. For the Nd:YAG and diode lasers of the visible/near-infrared range, more familiar materials are appropriate and fibre delivery (see Chapter 28) is a promising option. Diode lasers create special problems because their output beam is of relatively low quality, elliptical and can be astigmatic: more complex optics are required when an array of diode lasers is to be coupled or even a stack of arrays. Excimer lasers have UV wavelengths, requiring careful material choice and have low beam quality, favouring techniques of mask projection combined with beam homogenization.

For Nd:YAG lasers, fibre-optic beam delivery has become the technique of choice in many situations and such is the topic of Chapter 29 (Duncan Hand). For the 1.06 μm wavelength of the Nd:YAG laser, the familiar solid-core fused silica fibre, of step or graded core refractive index profile, guiding by total internal reflection, is most commonly used. However, materials such as chalcogenide, fluoride and germanate glasses are all used to extend operation to longer wavelengths—as are the crystalline materials of the silver halide class (polycrystalline) and sapphire (single crystal). The most successful mid-infrared wavelength fibres currently in use are those based on hollow fibres, where the wave-guiding mechanism is relatively lossy but the radiation need not pass through significant depths of the waveguide materials. Causing great excitement are a new class of optical fibres, often called micro-structured, photonic crystal or 'holey' fibres, which rely on their structure rather than material properties in order to guide radiation. One type of photonic crystal fibre, the *photonic bandgap*, is capable of providing guidance in a hollow-core structure with orders of magnitude less attenuation than a conventional hollow-core fibre. Fibre systems, whilst flexible, have distinct limitations in beam quality, power and energy-handling capacity, all of which are discussed.

Jürgen Koch's Chapter 30, on scanning and positioning, is essentially *dynamic* beam delivery, covering the basic actuation mechanisms: mechanical, electromagnetic, electro-optic, magneto-optic, acousto-optic and piezoelectric—building on Chapter 13. The chapter covers the basic design issues in positioning—motion, accuracy, repeatability and drift and dynamic effects (essentially, the speed of operation)—as well as discussing practical systems with several axes of motion. Positioning is mainly concerned with stepping between fixed positions: scanning is more dynamic, progressing from stepping through vector scanning (stepping at rates beyond system bandwidth) to raster scanning. Scanning is often based on moving mirrors of which the galvanometer and polygon are special cases. More rapid, although with limited amplitude, are piezoelectric scanners and faster still are those based on acousto-optics.

26
Basic Principles

D. P. Hand

CONTENTS

26.1 Beam Manipulation ... 385
26.2 Materials for Transmissive and Reflective Optics ... 385
26.3 Beam Quality ... 386
26.4 Beam Requirements at the Workpiece ... 387
26.5 Attenuation ... 388
26.6 Optical Damage ... 388
26.7 Safety ... 388
26.8 Summary ... 389
References ... 389

The beam delivery of laser light is an important consideration for any high-power laser application, whether for manufacturing or medical applications. Such lasers are normally large and heavy and so cannot be easily moved. In addition, in many applications such as welding in automotive manufacture or ship-building, the workpiece is also of considerable size and mass. This means that the ideal solution is to manipulate the laser *beam* relative to the workpiece.

In addition to beam manipulation issues, most processes require a focused beam and so the beam delivery system must provide a correctly focused beam at, or close to, the surface of the workpiece. This can be achieved using either transmissive or reflective optics. In some cases, simple focusing of the laser beam is not sufficient; instead different and more complex beam shapes at the workpiece may be required. For example there is evidence that in processes such as 'laser direct casting' and 'concrete scabbling', alternative beam shapes are more efficient [1]. Also, with excimer lasers a 'mask-imaging' system is normally used, due to the poor beam quality and low coherence.

The high capital cost of high-power lasers means that there is often a requirement for 'multiplexing' arrangements, where a single laser beam can either be switched 'time-shared' or split 'power-shared' into different paths to different machines. This is particularly useful where workpiece fixturing and positioning is more time-consuming than the laser processing itself or where it is important to process more than one point at a time on a component to ensure symmetrical heat input and, therefore, prevent distortion problems.

Beam delivery, therefore, encompasses optics and associated mechanics for beam manipulation, multiplexing, focusing and beam shaping. The main lasers used for industrial applications are Nd:YAG (1.06 µm) and CO_2 (10.6 µm), and so I will concentrate on beam delivery for these lasers here. However, where appropriate I will also address specific issues with other lasers currently used, including excimer (248, 193 nm), diode (~800, 980 nm), copper vapour (510.6, 578.2 nm), femtosecond Ti:sapphire (~800 nm), and frequency-multiplied versions of Nd:YAG and copper vapour.

26.1 Beam Manipulation

There are three main ways of manipulating the beam: (i) flying optics, (ii) articulated arms and (iii) fibre optic delivery. Examples of these are shown in Figures 26.1–26.3, respectively. The choice of technique depends on the laser wavelength, peak power and the application. For example, CO_2 lasers for cutting flat plates typically use a flying optics configuration, since suitable fibre optics are not available for this wavelength and movement is only required in two dimensions. Alternatively, for the delivery of high peak power, short pulse Nd:YAG light for dentistry applications, an articulated arm is normally used, due to the damage problems experienced with optical fibres at such high peak powers and the requirement for true 3D manipulation. Fibre optics are used in many applications, however, including the delivery of high cw Nd:YAG power for laser welding. Indeed, Nd:YAG welding lasers are often *only* supplied with fibre optic beam delivery.

26.2 Materials for Transmissive and Reflective Optics

The choice of material for beam delivery optics is strongly dependent on wavelength. With visible and NIR lasers (including Nd:YAG), a number of different optical glasses may be used, with fused silica particularly useful due to its very low

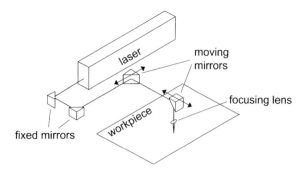

FIGURE 26.1 Flying optics configuration, suitable for moving laser beam around surface of large workpiece.

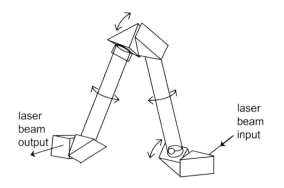

FIGURE 26.2 Articulated arm for flexible delivery of high-power laser beams.

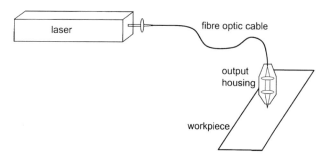

FIGURE 26.3 Fibre optic delivery system.

absorption and suitability for the UV (excimer or frequency-tripled Nd:YAG) as well as the visible and NIR parts of the spectrum.

With CO_2 lasers however, standard optical glasses cannot be used due to the long wavelength (10.6 μm). A number of other materials can be used instead, with the most popular being ZnSe and GaAs. ZnSe, in particular, has low absorption and, hence, good thermal properties; also its good transmission in the visible means that it can be used with on-axis monitoring alignment systems, which view through the laser-focusing optics. GaAs has particular application as a final focusing lens, with its high thermal conductivity reducing the likelihood of thermal run-away damage if spatter from the workpiece deposits onto the surface. An alternative possibility is artificial diamond.

For high-power CO_2 applications (>5 kW), reflective rather than transmissive optics are normally used due to their higher damage resistance. Copper and silicon mirrors are popular, since they offer advantages in terms of reflectivity (>99.3% for copper) and thermal conductivity. Copper is also easy to machine, with diamond machining used to produce mirror surfaces directly. This greatly simplifies the production of aspheric surfaces for focusing the beam–the most common design being the off-axis parabola. In welding and heat-treating applications, there is no high-pressure coaxial gas flow to protect the optics from spatter (unlike cutting and drilling) and so molybdenum is sometimes used for the final focusing optic, due to its extremely high resistance to laser damage [2].

26.3 Beam Quality

In general, high-power lasers do not produce a single transverse mode, due to thermal effects in the laser gain medium. Heating gives rise to a non-uniform temperature profile and, hence, thermal lensing. Conventional lenses are often placed within the laser resonator to compensate for this lensing effect but residual effects remain, which lead to the generation of higher order modes. This is particularly an issue for lamp-pumped solid-state lasers, where only a small percentage of the broadband pump light has the correct wavelength to pump the laser, with the remainder being converted into heat. This means that a typical Nd:YAG welding laser will have a highly multi-mode beam with a beam profile which is roughly 'top-hat'. Spatial filtering can be used to reduce the number of modes; however, this also reduces the average power. Narrow-band pumping can significantly increase efficiency and, hence, reduce heating effects, and so high-power laser diode arrays are being increasingly used as pump sources for Nd:YAG lasers. The capital cost of these arrays is significantly higher than flash lamps; however, they do have advantages in terms of maintenance cost, with typical lifetimes of 10 000 h compared with 1000 h for flash lamps. With gas lasers, thermal lensing effects are significantly reduced since the dependence of refractive index on temperature is several orders of magnitude lower than with solid-state lasers; also, novel resonator designs such as the hybrid unstable resonator [3] have been used to remove heat more efficiently.

The modal characteristics of a laser beam are important because they directly affect its 'focusability', or beam quality. This depends on both the wavelength and the transverse modes produced by a laser. For a given laser beam, the product of the focused spot-size (ω_0) and the focusing angle (α_0) is a constant (see Figure 26.4):

$$\omega_0 \alpha_0 = M^2 \frac{\lambda}{\pi}. \qquad (26.1)$$

The M^2 parameter [4] is often quoted as a measure of the beam quality of a particular laser. A 'perfect' laser beam has a single mode with a fundamental Gaussian profile, which has an M^2 value of 1. In reality, even a single-mode laser will have an M^2 value of ≥1.1, and a high-power Nd:YAG welding laser may have an M^2 value of 100 or more. However, the M^2 parameter

Basic Principles

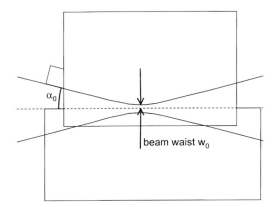

FIGURE 26.4 Focused laser beam, showing beam waist ω_0 and focusing angle α_0

is not the only term used to describe beam quality; in some countries (particularly in Germany), it is more common to use the beam propagation factor K to describe beam quality, where

$$K = \frac{1}{M^2}. \qquad (26.2)$$

In addition, many laser manufacturers describe beam quality using the 'beam parameter product', normally $w_0\alpha_0$, expressed in mm mrad. This has the advantage of incorporating the effect of wavelength but the disadvantage that sometimes focus *diameter* $2\omega_0$ and *full-angle* $2\alpha_0$ are used for this product instead of focus *radius* w_0 and *half-angle* a_0, giving a factor of four difference in the numerical value quoted.

There is generally a trade-off between average power and beam quality, particularly in solid-state lasers, due to thermal lensing effects as discussed earlier; or at least a trade-off between beam quality and cost. It is, therefore, clearly undesirable that any beam delivery system should reduce the beam quality. There are three main ways in which beam quality can be lost: (i) lens aberrations, (ii) thermal distortions in mirrors and lenses and (iii) mode coupling in optical fibres. Lens aberrations can be avoided by using appropriate lens designs. Thermal distortions are a particular problem with CO_2 optics but can be minimized by using effective cooling, e.g. water cooling of mirrors. However, there is always some loss of beam quality (or laser power) when a multi-mode laser beam is coupled into an optical fibre. In order to achieve coupling of light into the fibre, a focused spot radius of typically 90% of the fibre core radius is used. This will 'grow' to 100% after a short distance within the fibre. Furthermore, mode coupling occurs in the fibre, due to imperfections and bending, which means that the output divergence angle will always be greater than the focusing angle at the input [5]. For a typical fibre beam delivery system using 600 μm diameter optical fibre, a reduction in beam quality of about 16% would be expected for an input M^2 value of ~100.

26.4 Beam Requirements at the Workpiece

There are two basic types of processing: (i) a focused spot is moved relative to the workpiece (e.g. Figures 26.1 and 26.3);

and (ii) a mask-imaging system (Figure 26.5). CO_2, Nd:YAG, copper vapour and diode lasers all use the first technique, whereas excimer lasers use the second. With mask imaging, the main issues are beam uniformity and optical aberrations, with such aberrations being particularly critical in high-resolution photolithographic applications in the electronics industry.

Any focused spot process has particular requirements in terms of focused spot size, depth of field, beam shape and, in some cases, polarization. As shown in equation (26.1), the spot size varies with both the M^2 parameter and wavelength for a given focusing angle. It is important to realize that a small spot size carries the penalty of a short depth of field, which has a quadratic dependence on spot size, for a given wavelength and M^2 parameter. Although there are significant differences between the required focused beam for different processes, there are some general principles which can be defined.

First, there is a requirement for a large stand-off (several tens of mm) between the focusing optics and the workpiece. This is to prevent damage to these optics during processing. Such damage can occur due to direct thermal radiation from the process, reflected laser light or debris from the process (e.g. spatter from laser drilling) contaminating the lens and, hence, causing runaway thermal damage due to subsequent heating from the transmitted laser beam. Gas jets are normally used to help prevent such contamination, which can be either co-axial (as used in cutting) or off-axis (commonly used for welding)—see Figure 26.6.

Second, many processes (e.g. precision cutting) require a small focused spot (tens to hundreds of micrometres in diameter, dependent on the process) at the surface of the workpiece. This combination of small focused spot and large stand-off means that relatively large diameter lenses must be used, particularly if the beam quality is poor, and so 50 mm diameter optics are commonly employed, although some systems are based around 25 mm diameter optics. The high numerical aperture used means that aberrations must be considered. Spherical aberration is the main problem, which is commonly addressed in the visible-NIR by using achromatic doublets; aspheric optics are used at CO_2 wavelengths. Although achromatic doublets are optimized to deal with chromatic aberration, they also provide good correction for spherical aberration, are relatively low cost and have the additional advantage of

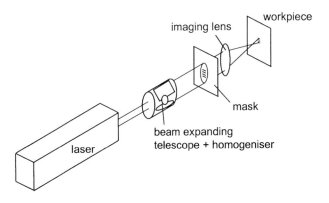

FIGURE 26.5 Mask projection arrangement.

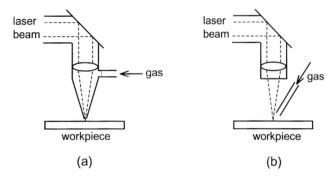

FIGURE 26.6 Laser processing head with (a) co-axial gas assist and (b) off-axis gas shielding.

providing a high-quality imaging route for the on-axis camera systems which are often used for workpiece set-up and monitoring.

The final issue is depth of field. This is normally defined as being twice the Rayleigh range and is the distance over which the laser beam area (and hence intensity) is within a factor of two of its value at the beam waist. However, such a variation in intensity would have a significant effect on most high-power laser processes, so the effective depth of field is somewhat less than this. For example, a 10% change in intensity occurs at a distance after the waist of roughly one-third of the Rayleigh range. Some processes, such as deep-hole drilling, require a long depth of field [6], whilst with others, it is simply important to minimize the tolerancing requirements of the beam-positioning system. For example, a typical Nd:YAG beam used for drilling cooling holes in turbine blades has an M^2 parameter of 20 and a focused spot diameter of 0.5 mm, giving a Rayleigh range of 9 mm. This Rayleigh range is necessary since holes of several mm depth are required, due to the combination of material thickness and the shallow angle of the holes relative to the surface of workpiece. A typical Nd:YAG beam used for welding, meanwhile, may have an M^2 value of 125, with a focused spot diameter of 0.6 mm and Rayleigh range of 2.1 mm.

In processes which do not require a standard focused beam but instead demand a special beam shape at the workpiece, other types of optics may be necessary. For example, diffractive optical elements (DOEs) may be used, which can be designed to produce an almost infinite variety of beam shapes [1]. Such DOEs have, for example, been applied to useful effect in the laser scabbling of concrete. Other techniques, including kaleidoscope-type reflective cavities, have been developed for beam homogenization, as described in C4.2.5.4. Another optical element which has been applied to great effect is the 'dual-focus' lens [7]. This produces a beam which is particularly suited to deep-section cutting, with the outside of the beam focusing at the workpiece surface and the central part a few mm below the surface.

26.5 Attenuation

An ideal beam delivery system would be perfectly transparent. In reality that can never be the case, primarily due to reflections from the surfaces of the optical elements but also due to absorption, particularly in optical fibres, and incomplete reflection from mirrors. The reflectivity of a transmissive optic is dependent on its refractive index. For NIR, visible and UV wavelengths (Nd:YAG, copper vapour, diode, excimer), fused silica or BK7 optics are normally used, which have a refractive index of 1.46 and so each surface reflects about 3.5%. For CO_2 laser light, however, ZnSe is commonly used. It has a refractive index of 2.4, and hence, 17% of the beam is reflected at each surface. Anti-reflection coatings are used to considerably reduce these surface reflectivities; however, they reduce the damage threshold of these optics and so cannot always be used with very high-power beams. Mirror reflectivities are normally very high, up to 99.9%, and so do not significantly affect the overall attenuation of the beam delivery system. Metal mirrors have slightly lower reflectivities but are still typically >99%.

Attenuation in large-core fused silica optical fibres is typically 2.5 dB km^{-1} at 1064 nm (see Figure 26.5), which means that it only becomes significant when lengths of 200 m or more are used and so is not normally an issue with industrial applications. However, the absorption increases at shorter wavelengths, giving a loss of 2.4 dB in 10 m at 266 nm (frequency-tripled Nd:YAG). The dominant effect with fibres is end-face reflection; here, it is not practical to use anti-reflection coatings due to the high-intensity incident on the fibre-end face which would quickly result in damage. In addition, some light is normally lost when coupled into the fibre, as the focused spot from the laser is not a perfect 'top-hat' intensity distribution and so a small percentage (typically 1%–2%) is not coupled into the core.

For a typical Nd:YAG fibre optic delivery system, consisting of two mirrors and a single doublet lens at the input end of the 10 m fibre, with two doublets at the output to image the light onto the workpiece, the overall loss is about 11% if the lenses have anti-reflection coatings, rising to about 26% if uncoated.

26.6 Optical Damage

Beam delivery systems for industrial applications must be able to deliver light with sufficient power to melt and vapourize a wide range of materials. This has the potential to cause catastrophic damage to the various optical elements in the delivery system. The threshold for damage at the surface of a material is usually lower than in the bulk, due to small imperfections at or near the surface, dependent on the quality of the polish. Care must be taken to keep all surfaces clean, as any dust can be burnt onto the surface resulting in localized heating and, hence, damage. The risk of damage is reduced if the beam size is as large as possible, although, in some cases, the damage has been shown to scale with diameter rather than area [8]. Thin film optical coatings, such as those used to make mirrors or anti-reflection coatings, significantly reduce the damage threshold.

26.7 Safety

Safety must always be a key consideration, since the high power (average and/or peak) which must be delivered could easily cause damage if even a small fraction was directed somewhere other than intended—hazards include eye damage, skin burns and fire. Delivery systems are designed to prevent the beam from being accidentally intercepted, by enclosing the beam either within tubing or in an optical fibre. In addition, optical fibre cables for high-power laser light normally incorporate some kind of continuity sensor, to detect if fibre breakage occurs. One such example is a fine wire coiled around the fibre along its length, whose continuity is continually checked. If the fibre breaks, the light which emerges will very quickly burn through the wire, breaking the continuity. Sensors have also been developed to detect any misalignment of the laser beam relative to the input end of the fibre by monitoring the power coupled into the fibre cladding [9].

26.8 Summary

Laser beam delivery is an intrinsic part of any high-power laser processing system, in order to transport, focus and manipulate the beam as required. The components of a beam delivery system include lenses, mirrors and optical fibres, together with appropriate safety enclosures or continuity monitors and process gas delivery. The particular components used (and the material from which they are made) primarily depend on the laser wavelength, peak power and beam quality required for a particular application.

REFERENCES

1. Taghizadeh M R, Blair P, Balluder K, Waddie A J, Rudman P and Ross N 2000 Design and fabrication of diffractive elements for laser material processing applications *Opt. Lasers Eng.* **34** 289–307.
2. Danielewicz E, von Der Ahe T and King S 2000 *Reflective Optics Add Focus* Industrial Laser Solutions April
3. Jackson P E, Baker H J and Hall D R 1992 CO_2 large-area discharge laser using an unstable-waveguide hybrid resonator *Appl. Phys. Lett.* **54** 1950.
4. Sasnett M W 1989 Propagation of multimode laser beams— The M^2 factor *The Physics and Technology of Laser Resonators* ed D R Hall and P E Jackson (Bristol: Adam Hilger).
5. Kuhn A, Blewett I J, Hand D P and Jones J D C 2000 Beam quality after propagation of Nd:YAG laser light through large-core optical fibers *Appl. Opt.* **39** 6754–60.
6. Kudesia S S, Rodden W S O, Hand D P and Jones J D C 2001 Effect of beam quality on single pulse laser drilling *Proc. ICALEO 2001 Jacksonville, USA* (Orlando FL: Laser Institute of America).
7. Powell J, Tan W K, Maclennan P, Rudd D, Wykes C and Engstrom H 2000 Laser cutting stainless steel with dual focus lenses *J. Laser App.* **12** 224–31.
8. Wood R M 1997 Laser induced damage thresholds and laser safety levels. Do the units of measurement matter? *Opt. Laser Technol.* **29** 517–22.
9. Boechat A A P, Su D and Jones J D C 1992 Bidirectional cladding power monitor for fiberoptic beam delivery systems *Meas. Sci. Technol.* **3** 897–901.

27

Free-space Optics

Leo H. J. F. Beckmann

CONTENTS

- 27.1 Introduction 392
- 27.2 Laser Beam Propagation and Its Optical Consequences 392
  - 27.2.1 Beam Size 393
- 27.3 Computation of Laser Optical Systems 393
  - 27.3.1 Location of the Laser Beam Image 393
    - 27.3.1.1 Multiple Optical Elements and the Use of Ray Tracing 394
  - 27.3.2 Focal Spot Size 395
  - 27.3.3 Effect of Aberrations on Image Size and Shape 395
  - 27.3.4 Beam Aperturing Requirements and Effects 396
  - 27.3.5 Low Beam Quality Sources 396
  - 27.3.6 Non-rotationally Symmetrical Laser Beams 396
- 27.4 General Practical Guidelines of Optics for Laser Applications 397
- 27.5 Optics for CO_2 Laser Systems 397
  - 27.5.1 Beam Transport 397
    - 27.5.1.1 Articulated Arms 398
  - 27.5.2 Lenses for Focusing CO_2 Laser Radiation 399
    - 27.5.2.1 The Single (ZnSe) Lens 399
    - 27.5.2.2 Extending the Limits of the Single Lens 399
    - 27.5.2.3 Use of Zoom Optics 399
    - 27.5.2.4 Thermal Effects in ZnSe Lenses 400
    - 27.5.2.5 Use of Artificial Diamond 401
  - 27.5.3 Mirror Systems for Use with CO_2 Lasers 401
    - 27.5.3.1 Angular Field Considerations 401
    - 27.5.3.2 The Off-axis Paraboloid 401
    - 27.5.3.3 Coma-corrected Optics 403
    - 27.5.3.4 General Remarks on Aspherical Mirrors 403
  - 27.5.4 CO_2 Laser Beam Integration for Homogeneous Illumination 403
  - 27.5.5 Power Distribution Shaping by Phase Modulation 403
- 27.6 Optics for Lasers Operating in the Visible or Near-IR 404
  - 27.6.1 Optical Handling of Fibre-delivered Nd:YAG Laser Radiation 404
  - 27.6.2 Lens Design 404
  - 27.6.3 Colour Correction 405
  - 27.6.4 Optics for Diode Lasers 405
    - 27.6.4.1 Classes of Diode Lasers and Applications 405
    - 27.6.4.2 Diode Laser Optical Output Properties 407
    - 27.6.4.3 Optical Handling of Single-diode Laser Beams 407
    - 27.6.4.4 Single-diode Laser-focusing Optics 407
    - 27.6.4.5 Optics for Applications of Diode Laser Arrays 408
    - 27.6.4.6 Stacks of Diode Laser Arrays 409
- 27.7 Optics for Excimer and Other UV Lasers 409
  - 27.7.1 Material Aspects 410
  - 27.7.2 Beam Homogenization 410
  - 27.7.3 Optics for Imaging a Mask 410

  27.7.3.1 Design Examples 410
  27.7.3.2 Photolithography 412
27.8 Optics for Other Laser Sources 412
27.9 Conclusions 413
References 413
Further Reading 415

27.1 Introduction

The subject of this chapter is free-space optics, that is, the optical means and devices which transfer and transform the radiation output from a laser to a workpiece in such a way that it activates the desired processes. Thus it entails both radiation beam transport (relaying) and beam manipulation. The latter is mostly in the form of focusing to a spot of some desired size and shape but an application may also require a radiation beam transformation, which produces a homogeneous irradiation over a certain area of the workpiece. The optical methods and devices for achieving these goals vary considerably with the type of laser used and, more specifically its wavelength, mainly due to restrictions set by wavelength-dependent material properties.

The unique properties of laser radiation, in particular, its spatial coherence in conjunction with monochromaticity and scalability to high power levels, have given lasers a firm, often exclusive, place in applications ranging from information processing, transport and storage to thermal machining and material processing. The usefulness of lasers of medium and high power stems largely from the exceptionally high radiance of the laser radiation source. A (rotationally symmetrical) laser which delivers a radiation beam of radius w_0, and has a radiating surface $A = \pi\omega_0^2$ at the point of smallest beam radius, or beam 'waist', will radiate its power P of wavelength λ into a cone with a divergence half-angle $\vartheta = M^2\lambda/(\pi\omega_0)$, where M^2 is the beam quality, which is equal to or greater than 1. The corresponding (small) solid angle is $\Omega = \pi\vartheta^2$, so that the radiance $N = P/(A\Omega)$ of the laser is found to be $N = P(M^2\lambda)^{-2}$, a noteworthy result in that M^2 and λ are the only variables besides the power P and radiance depends on the squares of both the beam quality and the wavelength. The radiance found by this simplified calculation is an average quantity—the actual radiance is even higher in a central peak or for pulsed lasers.

Since radiance is a field quantity, invariant to (loss and aberration-free) optical manipulation, optical systems allow the transformation of laser radiation to (very high) power density (or irradiation), $P/A' = N\Omega'$, by suitably choosing the solid angle Ω' at which the laser radiation impinges on a surface A'. This solid angle can—and most often will—be much greater than that into which the laser radiates. A parameter, more commonly used in optics, is the numerical aperture (NA), $NA = n \sin U'$, where U' is the half-angle of the radiation cone, which emerges from the optical system. It is related to the solid angle (in air, and for a Lambertian absorber) by $\Omega' = \pi(NA)^2$, so that

$$P/A' = \left[\pi P/(M^2\lambda)\right](NA)^2 \qquad (27.1)$$

It should be noted that, with the exception of numerical aperture, all parameters in this equation are fixed by the choice of laser source, so that the NA is the only quantity available to the system designer for controlling the power density on the workpiece and, thus, the process.

For a numerical example, consider a CO_2 laser of 1 kW with a beam quality of $M^2 = 2$. Its radiance is approximately 2.2×10^8 W cm^{-2} sr^{-1}, and it is noteworthy that this level corresponds to blackbody radiation at a temperature of more than 100 000 K. It allows power densities well in excess of 1 MW cm^{-2} to be achieved as is necessary for such processes as the cutting and drilling of metals, even with optics of moderate NA.

In many respects, laser radiation behaves similarly to non-coherent radiation from classical radiation sources and the design of devices for its manipulation can, therefore, often benefit from classical optical expertise which has accumulated over more than a century but there are important differences and hitherto unknown challenges and traps. Aside from coherence and its optical consequences, treated in the next section, it is the power of a laser beam, the high power density—whether wanted or unwanted—it creates in foci and the associated thermal effects, which require careful systems consideration. Lasers which deliver high-power outputs span a wide range of wavelengths, from ultraviolet (less than 200 nm) to infrared (more than 10 μm), and their optical manipulation faces serious, unavoidable and partly unsolvable problems rooted in basic materials properties, specific to the appropriate wavelength region. The wide range has also consequences with respect to the manufacturing tolerances of the optics, those for the ultraviolet region being roughly 50 times more severe—and costly to achieve—than for the infrared. The following sections will address these problems in conjunction with the laser types which have found broad application in materials processing or otherwise, notably the CO_2 laser, the Nd:YAG laser, the diode laser and the family of excimer lasers.

27.2 Laser Beam Propagation and Its Optical Consequences

Spatial coherence is probably the most characteristic and most consequential property of a laser beam. It causes a beam, which simultaneously has a Gaussian beam profile, to propagate such that the beam radius $\omega(z)$ along the axis of propagation expands in a non-linear fashion [1]:

$$\omega(z) = \omega_0 \left\{1 + [\lambda z/(\pi\omega_0^2)]^2\right\}^{1/2} \qquad (27.2)$$

so that the beam boundaries do not form straight lines as in classical optics. Instead, the beam radius attains a minimum of ω_0 at $z = 0$, known as the beam waist and fans out to a radiation cone with a limiting half-angle (also called the half-beam divergence):

$$\vartheta(z \to \infty) = \lambda/(\pi\omega_0). \quad (27.3)$$

The axial distance over which the beam has grown to √2 times its size at the waist is known as the Rayleigh range and

$$z_R = \pi\omega_0^2/\lambda. \quad (27.4)$$

At the site of the waist, the beam wavefront is flat and its radius of curvature, $R(z)$, changes along the axis of propagation:

$$R(z) = z\left[1 + (z_R/z)^2\right] \quad (27.5)$$

These relations differ from those for classical ray optics, yet the concept of optical rays retains some usefulness when treating laser beams, provided the rays are strictly seen as the normal to the wavefront of the beam. The well-established methods for optical design and analysis, which are essentially based on geometrical ray tracing, can, with some modifications, be applied to optics for laser-beam handling. The modifications concern the non-linear change in the wavefront curvature, as given by equation (27.5), when the beam propagates from one optical surface to the next but the difference, and thus, the necessary correction is small except for near-telescopic situations. The laws of refraction and reflection apply unchanged.

The axial region between the beam waist and the Rayleigh range is referred to as the near-field. It is in this region that the difference between laser and classical optics, notably the curvature of the wavefront, is greatest. At distances much greater than the Rayleigh range, called the far-field, the laser-beam wavefront curvature approaches that described by classical theory. As the differences are all tied to the Rayleigh range, classical optics may well be considered to be a special—or limiting—case of general (coherent) optics for the Rayleigh range becoming zero.

The analogy between coherent and incoherent optics may be extended to the generation of an image, which is, by definition, the formation of a diffraction pattern from a converging wavefront at some point in space and which may be simulated by calculating the (Huygens) diffraction integral. Quantitative differences result from differences in the amplitude and phase distribution over the wavefront, however, and these affect the shape and size of the diffraction pattern as well as the axial point of minimum size (or maximum energy concentration). Most marked is the absence of outer rings in the case of a unapertured coherent beam with a Gaussian profile, as contrasted to the Airy pattern which results from non-coherent illumination. Indeed, the Gaussian profile of the laser beam is preserved throughout all optical-beam manipulations by aberration-free optics [1].

27.2.1 Beam Size

The previous relations apply to purely Gaussian or fundamental mode (TEM_{00}) laser beams and have to be modified for real beams, which are characterized by higher and mixed modes [2].

Before elaborating on real beams, the question of beam-size definition has to be addressed. While non-coherent beams are usually limited by hard aperture stops to well-defined beam widths, the definition of that quantity for a laser beam is much less obvious. For pure Gaussian (TEM_{00}) beams, the diameter at which the power has dropped to $1/e^2$ of its central peak has been quite generally used to describe the beam width but for higher and mixed-mode beams; this is no longer a meaningful quantity and must, therefore, be replaced by a more rigorous and fundamental definition [3,4]. A suitable and generally applicable description of the beam diameter is based on the second moments of its power distribution function, and this has been adopted for both the relevant ISO and CEN standards [5]. Using this beam size definition, the previous formulas for Gaussian beams can also be applied to real beams, if the wavelength λ is replaced by the quantity

$$M^2\lambda = \lambda/K \quad (27.6)$$

which could be looked at as an 'effective' wavelength. M^2 is called the beam quality or, more specifically, the 'times diffraction-limited factor' and was introduced by Siegman [6]. The ability to express the most important performance parameters of any real laser by a single additional quantity—where (in accordance with the previously mentioned standards) $1/K$ and M^2 may be used interchangeably—is the main reason and advantage of its introduction, but it must be borne in mind that it does not describe the shape of the power distribution within the beam and, consequentially, the focal spot.

The beam quality is directly related to the product of the beam-waist radius (W_0) and the half-beam divergence (ϑ):

$$\vartheta\omega_0 = M^2\lambda/\pi. \quad (27.7)$$

This quantity is also known as the beam parameter product [7]; it is invariant throughout the beam propagation, as long as aberration-free optics are involved and it attains a minimum for a 'diffraction-limited', i.e. purely TEM_{00} beam, for which $M^2 \to 1$. In fact, the relation (27.7) is typically used to determine M^2 from measured values of the beam-waist radius and beam divergence [8]. This definition of the beam parameter product is not adhered to consistently throughout the literature and occasionally the product of beam-waist diameter and the full-beam divergence is used instead.

27.3 Computation of Laser Optical Systems

27.3.1 Location of the Laser Beam Image

In the following sections, we describe how to calculate the position and dimensions of laser beam waists when transformed by lenses. In geometrical optics, an object (loosely something that can be imaged) is considered to be the point of origin of a spherical wavefront, the radius of which equals the distance from the object. Given a lens (or, more generally, an optical imaging system) of focal length f, object (s) and image (s') distances from the (thin) lens is given by the well-known 'lens formula' [9]:

$$1/s + 1/s' = 1/f \qquad (27.8)$$

The waist of a laser beam is not an object in the same sense, as the wavefront varies with distance in accordance with equation (27.5), but the lens formula can be applied in an expanded form:

$$\frac{1}{s+q} + \frac{1}{s'+q'} = \frac{1}{f} \qquad (27.9)$$

where the beam parameters q, q', at the object and image waists, respectively, are purely imaginary [1]:

$$q = j\pi\omega_0^2/\lambda \qquad q' = j\pi\omega_0'^2/\lambda \qquad (27.10)$$

Equating the imaginary parts, one obtains

$$\frac{s-f}{s'-f} = \frac{\omega_0^2}{\omega_0'^2}. \qquad (27.11)$$

Equating the real parts results in

$$(s-f)(s'-f) = f^2 - f_0^2 \qquad (27.12)$$

with

$$f_0 = \pi\omega_0\omega_0'/\lambda. \qquad (27.13)$$

From this, we can derive an expression for the position of the image waist (s') as a function of object-waist position (s) and focal length [9]:

$$\frac{1}{s + b^2/(s-f)} + \frac{1}{s'} = \frac{1}{f} \qquad (27.14)$$

with

$$b = \pi\omega_0^2/\lambda. \qquad (27.15)$$

It will be noted that b is the Rayleigh range of the source beam. The non-linear dependence of s' on s, together with the variation of beam diameter with distance s from the beam waist (equation (27.2)), affects both the focal position and the numerical aperture of the focused beam in systems where the laser is stationary, and flying optics are used to concentrate its radiation on the workpiece (such as in gantry-type machining stations). To minimize these dependencies, it is advisable to laterally expand the source beam, which effectively increases the source-beam Rayleigh range and, thus, reduces the contribution of s. If this is not done, the resulting variation of focal point distance from the lens may need to be compensated for at the optical focusing head.

If the expression (27.14) is rewritten in normalized form [9]:

$$(s'/f) - 1 = \frac{[(s/f) - 1]}{[(s/f) - 1]^2 + (b/f)^2} \qquad (27.16)$$

and compared with the equally rewritten geometrical lens formula

$$(s'/f) - 1 = \frac{1}{[(s/f) - 1]} \qquad (27.17)$$

it is clear that the appearance of the term $(b/f)^2$ removes the pole at $(s/f) = 1$, the case for which the object is at the (forward) focal point and which causes the image of a classical object to move to infinity. Instead, the image position (s') of a laser beam waist is always finite: its maximum distance from the lens is

$$s'(\max) = f\left[1 + (f/b)/2\right] \qquad (27.18)$$

for a source-beam-waist distance of $s = f + b$. Furthermore, unlike the case of classical geometrical objects where there is a minimum object-to-image distance of $4f$, there is no minimum separation between input (source) and output (image) beam waist for Gaussian beams.

The magnification m, i.e. the ratio of the image- and source-waist radii, is found to be

$$m = \omega_0'/\omega_0 = \frac{1}{\{[1 - (s/f)]^2 + (b/f)^2\}^{1/2}}. \qquad (27.19)$$

As is easily seen, equations (27.14), (27.16) and (27.19) reduce to the corresponding geometrical optics formulae when $b \to 0$, which again emphasizes that classical geometrical optics is the special case at the $b \to 0$ limit.

It is interesting to discuss another special case: in many applications, the focusing lens will be situated at or near (an image of) the source waist and $s \to 0$. Then

$$s' = \frac{f}{1 + (f/b)^2} \qquad (27.20)$$

that is, the image waist is always closer to the lens than the geometrical image. Since the focal length is, however, in most cases much smaller than the Rayleigh range, the deviation is usually very small. In other words, if a laser beam is focused by an optical system located well within the near-field, the position of the focus will be near the geometrical focal point.

27.3.1.1 Multiple Optical Elements and the Use of Ray Tracing

The previous discussion has tacitly assumed a single (thin) optical element and the distances from it. In practice, however, we will often encounter optical systems with a succession of optical components as needed to perform a given task. For this situation, the previous treatment with its use of the imaginary beam parameters q and q', although mathematically quite elegant, is less suitable. In practice, we prefer an approach which is closer to the well-established methods of ray tracing but where a 'ray' is strictly seen as the normal to a wavefront and ray propagation is computed in accordance with equation (27.2). Such a calculation will involve the following steps:

1. Calculation of the wavefront size w and radius of curvature $R(z)$ at the point of intersection with the first optical surface, using the laser beam waist radius and

waist distance. These data result in a 'ray' of height ω and slope $\omega/R(z)$ on the first optical surface.

2. Refraction of the ray, which results in the slope of the exiting ray (there is no change of the ray height). Conversion to the corresponding wavefront curvature of the exiting beam.
3. Calculation of the waist size (ω'_0) and position (z') of this exiting beam from that wavefront curvature and size.
4. Calculation of the wavefront curvature and size for that beam at the intersection with the next optical surface and conversion to a 'ray' of appropriate slope and height.
5. Repetition of steps (2) through (4) for as many optical surfaces as there are in the system, ending with a step (3), which yields the size and position of the image waist.

The formulas for calculating the beam waist size and position from a point along the axis, where the wavefront has radius w and radius of curvature R, as needed for step (3), are [10]:

$$\omega_0 = \omega/\{[1 + (\pi\omega 2/\lambda R)^2]\}^{1/2} \quad (27.21)$$

$$z = R/[1 + (\lambda R/\pi\omega^2)^2]. \quad (27.22)$$

For use in a computer program, it is preferable to replace these formulas by the equivalent expressions:

$$\omega_0 = \{R^2(\lambda/\pi)^2/[\omega^2 + R^2(\lambda/\pi)^2/\omega^2]\}^{1/2} \quad (27.23)$$

$$z = R[1 - \omega_0^2/\omega^2] \quad (27.24)$$

This procedure is easily extended to include real beams of a given beam quality M^2 by multiplying λ with M^2. It should further be noted that the calculations can be (and often are) carried out with 'paraxial formulas' for the refraction part, which means approximating the sine of an angle by the angle itself and the cosine by 1. The final result is the ideal image waist size and position, i.e. neglecting aberrations and beam aperturing (to be discussed later).

27.3.2 Focal Spot Size

As in geometrical optics, the size and energy distribution within the focus of a laser beam are governed by diffraction. With aberration-free optics, the size of a real laser focal spot, actually the radius ω'_0 of an image beam waist, for a laser of beam quality M^2 is

$$\omega'_0 = M^2 \lambda / [\pi(NA)] \quad (27.25)$$

and since λ and M^2 are determined by the choice of the laser used, the only parameter left to the user for control of the spot size is the NA of the focusing optics—quite in analogy to classical optics. It is also apparent that a laser of superior beam quality (lower M^2) allows one to achieve a desired spot size at lower NA. This is important for three reasons: (i) a higher NA generally requires more sophisticated optics, which (ii) has typically a shorter back focus (i.e. the last optical surface will be closer to the workpiece). It (iii) also has a shorter depth of focus (i.e. the radiation cone has a wider angle at the image). The latter is quantified by the Rayleigh range (R_i) of the image,

$$R_i = \pi(\omega'_0)^2/(\lambda M^2) \quad (27.26)$$

or

$$R_i = \omega'_0/(NA) \quad (27.27)$$

which shows that, for any given spot size ω'_0, the Rayleigh range will be inversely proportional to both M^2 and to the NA which is required to achieve that spot size.

27.3.3 Effect of Aberrations on Image Size and Shape

The aberrations of a non-ideal optical system will, in general, cause a deformation of the wavefront with respect to its—originally spherical—shape. The equally occurring (small) changes in the amplitude distribution over the wavefront are normally—and in by far the most cases justifiably—neglected. The effect of wavefront deformations in laser beams is quite analogous to that in non-coherent optics: a reduction in the power density in the central peak of the diffraction image, associated with a shift of optical power to surrounding rings. There is also some broadening of the central peak.

A quantitative comparison between the diffraction images obtained from non-coherent and coherent wavefronts relies on a suitable definition of the image boundary. For a non-coherent diffraction image, it is the diameter of the first dark ring of the well-known Airy pattern [11, p 182 ff] which provides a suitable measure of the image diameter but the diffraction pattern of a perfect Gaussian wavefront is Gaussian itself, has no such ring features and (theoretically) extends to infinity. The generally accepted [5] diameter definition of laser-beam images—actually image waists—again uses the second moments of the power distribution function, completely analogous to the definition of beam diameters. This value corresponds to the $1/e^2$ diameter for the limiting case of a TEM_{00} beam; that diameter is $2\lambda/(\pi NA)$ or $0.64\lambda/(NA)$. We note that the diameter of the first dark ring of the Airy pattern (for homogeneous illumination by non-coherent radiation) is $1.22\lambda/(NA)$, seemingly much bigger for equal NAs but it must be borne in mind that focusing a Gaussian beam will require that the free diameter of the optics be larger than that of the beam to avoid beam aperturing and associated image degradation (to be discussed shortly). Thus, the difference in NA essentially offsets the numerical factors of these expressions. The diffraction image of a laser beam is, however, more compact than the Airy pattern with its extended ring structure and more than 99% of the energy falls within two diffraction diameters as long as the aberrations are small.

What constitutes a 'small' aberration is somewhat arbitrary [12]. In non-coherent optics, it is customary to consider an optical system to be 'diffraction limited' if the wavefront error due to, for instance, spherical aberration is less than one-quarter wavelength, which causes a reduction in the Strehl ratio (the peak intensity in relation to the theoretical maximum) to 80%. Much smaller (1/14–1/20 wavelength) random peak-to-peak wavefront errors, such as those resulting from manufacturing tolerances, cause a deterioration of the same order of magnitude [13] and make the 80% limit, which was originally proposed by Lord Rayleigh [14], widely accepted. There is no such commonly agreed acceptance limit in laser applications, and since peak power density is quite often the quantity which determines the effectiveness of a laser process (for example cutting and drilling), low aberration optical systems, with a Strehl ratio greater than 0.9, are preferable. The effect of spherical aberration on the diffraction image of a Gaussian beam—essentially a shift of power into an outer ring structure—is shown in Figure 27.1.

27.3.4 Beam Aperturing Requirements and Effects

The already mentioned requirement that the free diameter of a focusing optics well exceeds the beam diameter arises for two reasons: One, of particular importance for high-power beams [15], is to ensure that no appreciable portion of the incident radiation falls on the lens-holder and is converted to heat hence causing unforeseen thermal problems. The second, which is valid for all power levels, is a change of shape of the central maximum of the diffraction image with a corresponding reduction in the peak power density. The increase in focal diameter from aperturing a TEM_{00} beam at 1.5 times the $1/e^2$ diameter amounts to about 10% for an aberration-free lens. As a practical rule, the free optical diameter for laser-beam handling should be at least 1.5—and preferably 2.0—times the beam diameter. Lower factors are acceptable for very low-quality beams. Although an even larger factor would be theoretically beneficial, it is hardly useful in practice, as some aperturing has always already occurred within the laser source. Formulas describing the effect of beam aperturing on the far-field beam divergence have been derived by Drége [16].

27.3.5 Low Beam Quality Sources

For low-quality laser beams—$M^2 > 10$—as is typical for most high-power solid-state lasers, the focal spot size approaches that of the classical image of an extended object. This is easily demonstrated for the case of the radiation output from, for instance, an Nd:YAG laser, which is delivered via an optical fibre of diameter D. The fibre output forms a source of radiant surface

$$A = \pi D^2 / 4 \qquad (27.28)$$

and if the beam quality is M^2, it will radiate all its power into a cone of half-divergence

$$\sin V = \frac{\lambda M^2}{\pi D/2}. \qquad (27.29)$$

Now let this (divergent) radiation beam enter an optical system of magnification m, then the exiting radiation cone will have a half-beam divergence of $\sin V' = (1/m) \sin V$ and the geometrical image diameter will be $d = mD$. The diffraction image of a Gaussian source, however, would have a diameter of

$$\delta = \frac{2\lambda}{\pi \sin V'} = \frac{2\lambda m}{\pi \sin V} = mD/M^2 \qquad (27.30)$$

and is seen to be smaller by a factor $1/M^2$. In other words, the diffraction contribution is only a fraction $1/M^2$ of the image size.

27.3.6 Non-rotationally Symmetrical Laser Beams

Throughout most of the previous discussion, it has been tacitly assumed that the laser beams are essentially rotationally symmetrical, since this is the preferred shape for most applications and generally striven for by laser manufacturers. A number of real lasers, notably diode lasers, however, emit beams of different width and/or beam quality in two transverse, mutually perpendicular, directions. Focusing such beams generally

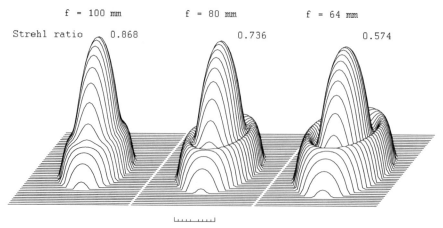

FIGURE 27.1 Diffraction patterns of foci of TEM_{00}-laser radiation obtained with lenses of increasingly shorter focal length and associated spherical aberration. Logarithmic vertical scale, 0.002 threshold. Spherical aberration reduces peak intensity and broadens the image as power is shifted to outer rings.

results in more or less elliptical spots, and in accordance with the previous formulas, adaptation of the NA in one direction will be needed to form a circular spot (see Section 27.6.4.3).

27.4 General Practical Guidelines of Optics for Laser Applications

As in any other application of optics, the optical layout is dictated by system requirements but there are additional and important considerations specific to laser applications, mostly related to the typically high (peak) powers. For lasers operating outside the visible and near-infrared spectrum, applications rely on few, often costly, transmissive materials, suitable for use as lenses or windows. Mirror optics can be used rather universally at any wavelength and throughout the laser optical train but obviously cannot reproduce the function of a (pressure) window.

Lenses are often preferred over mirrors because they allow straight optical paths and, thus, make the mechanical layout, assembly and alignment easier. Mirror systems, however, are always 'off-axis', causing folded paths, and the mechanics must occasionally accommodate odd angles as determined by the optical design.

At very high powers, mirrors are the only choice: metal-based mirrors can withstand very high powers and power densities (up to, say, 10^5 W cm^{-2}), if channels for circulating cooling water are integrated into the mirror body, a widely applied technique. Metal-based mirrors are typically manufactured by computer-controlled single-point diamond machining, and the generation of various non-spherical shapes is well within the capability of that technique, though often at higher cost.

When designing an opto-mechanical system for laser powers in excess of, say, 500 W, due attention has to be given to thermal problems, which derive from stray radiation, residual absorption in transmissive materials and thermal expansion and resultant stress. Forced cooling ensures thermal stability and avoids thermal runaway, an effect which otherwise can occur if absorption increases with temperature. Since laser power is often applied abruptly, the risk of thermal shock and rupture is to be considered, and this favours the use, if possible, of materials with low thermal expansion and high thermal conductivity. Lens surfaces exposed to possible contamination from process fumes or even spatters are particularly vulnerable and are often protected by an (expendable) window made from lower-cost material. Purging with a well-directed stream of a clean shielding gas is helpful and is implemented more easily for optics with a comparatively long back focus (the distance between the last lens surface and the focal spot), and corresponding lens designs of the 'retrofocus' type are, therefore, preferable [17]. Thermal aspects also put restrictions on the optical design of lens systems in general: the refractive surfaces, even if coated for maximum transmission, will always have some residual reflection. In complex lens systems consisting of several lenses at appreciable distances, such residual reflections may cause a focus in or, even worse, on the surface of one of the earlier lens elements. This situation, which will inevitably cause lens damage, must be avoided at all cost, in particular for applications with pulsed lasers, due to their inherently very high peak powers.

It is a good practice in optical design to strive to achieve a specified performance with a minimum number of optical elements. For laser applications, this aspect is emphasized by a often high cost of the materials, the need to minimize the loss of (expensive) laser power and the complexity of cooling. What helps is that many laser focusing optics need to perform only over a small angular field, typically less than one degree and this often permits relatively unsophisticated designs with low element counts. An exception is the optics for excimer laser applications, which use mask projection techniques rather than single-point imaging. These must cover angular fields of several degrees and require excellent field flatness. The problems are compounded by the short wavelength, the scarcity and low refractive index of the available materials and the requirement for a relatively long back focus.

Finally, the user of laser optics has to face the fact that laser optics are easily damaged, be it due to low hardness of the material, burn-in of accumulated surface contaminations or irreversible changes within the material, a known effect in excimer laser optics used at high power levels. Cleanliness and special care in handling laser optics, in particular lenses, are absolutely essential.

27.5 Optics for CO_2 Laser Systems

The relatively high power efficiency of the CO_2 laser, combined with scalability to high powers (greater than 10 kW), with only a gradual decrease in beam quality from, typically, $M^2 < 2$ at powers below 3 kW to $M^2 > 5$ well above 10 kW, has given this type of laser a strong position as a 'work horse' for materials processing, such as cutting, welding and surface treatment. Its long wavelength, 10.6 µm, is a drawback in that there are very few environmentally stable optical materials which have high transmission—i.e. low absorption losses—at that wavelength, none of them even near-perfect, a fact which has frustrated hopes for the feasibility of beam transport through an optical fibre (short hollow waveguides are a limited alternative [18]). At the higher power levels, beam handling, including focusing, is by means of mirror systems rather than lenses. The limit is somewhat elastic, depending on the effort spent on lens cooling and ranges between 3 and 5 kW.

27.5.1 Beam Transport

High-power CO_2 lasers are bulky and lend themselves to stationary installation, while applications such as cutting and welding of large workpieces often require bridging long distances to the point of processing. When relaying a CO_2 laser beam from the source to the focusing optics near the point of application, two aspects have to be considered. One is the position of the source beam waist: this is typically located near the out-coupling window of the laser and, thus, often far away from the location of the focusing optics, at which the beam may have expanded to an undesirably large diameter. The second is the beam divergence in the far-field (equation (27.2)), which causes a variation the beam diameter with varying distance from the source, as typically occurs with gantry-type installations. The answer to these problems is the use of a telescope, operating as a beam expander and, if adjustable from slightly converging to slightly diverging, as a means for relocating the beam waist to an axial position near the focusing optics, together with an adaptation of the beam size. CO_2 laser beam telescopes are mostly based on the Galilean type, using a positive–negative combination of optical elements, usually mirrors, because it avoids an intermediate focus and is shorter (Figure 27.2). If the mirrors are paraboloidal, the in-going and out-going beams are parallel (Figure 27.2a). If spherical mirrors are applied, the angles of deflection must be different (Figure 27.2b) to compensate for astigmatism [19]. Waist-position control is achieved by varying the distance between the two mirrors. This is, of course, associated with a lateral beam shift, which must be compensated by a third mirror, often arranged so that the unit has a total deflection of exactly 90° (Figure 27.2c). The use of adaptive mirrors for relocating a source waist has also been proposed [20,21].

27.5.1.1 Articulated Arms

An alternative approach to beam positioning in multiple dimensions is the use of an 'articulated arm', a hinged tubular structure with mirrored rotational joints and sufficient degrees of freedom of motion, to allow it to passively follow a three-dimensional path plus angular orientation if gripped at the end by, for instance, an industrial robot. The necessary very precise alignment, and freedom from mechanical play, is difficult to achieve (and maintain) in practice, and the required large number of mirrors and their cooling demands are further drawbacks. Applications of CO_2 laser systems requiring articulated arms have now been largely superseded by installations based on fibre-optically transported Nd:YAG laser radiation, but since the latter approach is typically more costly at power levels in the multi-kilowatt range, the use of articulated arms for positioning CO_2 laser beams and foci by industrial robots is still a viable option.

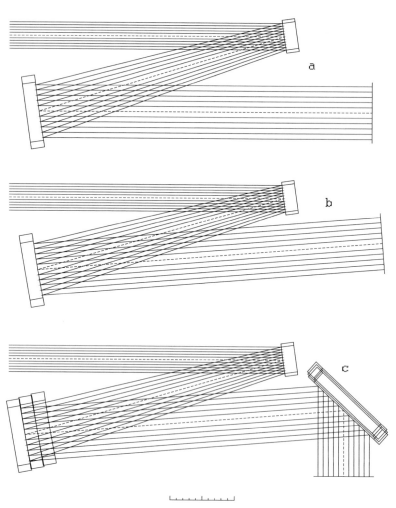

the total length of the scale corresponds to 100.00 millimeters

FIGURE 27.2 Galilean telescopes for beam expansion and/or waist relocation of CO_2 laser beams, using off-axis paraboloidal (a) or spherical mirrors (b). Note different tilt angles in the latter case to compensate for astigmatism. Mirror separation may be adjusted to control the location of the image waist, and the associated lateral beam shift is compensated by synchronously moving a third, flat, mirror, often arranged for a total deflection of 90° (c).

27.5.2 Lenses for Focusing CO_2 Laser Radiation

27.5.2.1 The Single (ZnSe) Lens

CO_2 laser radiation can be conveniently focused by means of lenses at power levels up to, say, 3 kW and, with proper measures for forced cooling, at even higher power levels. In applications such as cutting and welding, which are by far the most common applications of this type of laser, the lens simultaneously serves as a pressure window for the cutting or shielding gas used with these processes. These lenses are almost exclusively made from ZnSe, as this is the only environmentally stable material, which, in pure form—referred to as 'laser grade'—has the required low absorption. GaAs has also been considered in the past (for a comparison of the properties, see Reedy [22]). A new and upcoming possibility is the use of artificial diamond, to be briefly discussed in Section 27.5.5. High-quality ZnSe is available from several vendors with a bulk absorption of less than 0.0005 cm^{-1}. Due to the relatively high refractive index of ZnSe (2.4028), it is mandatory to apply anti-reflection coatings, and in most situations, it is the quality of the coating which determines the overall absorption of laser radiation by a ZnSe lens. For a good new lens, this may run as low as 0.15% of the incident radiation.

The relatively high refractive index of ZnSe helps the lens designer: it keeps spherical aberration low and, thus, allows diffraction-limited performance to be achieved with a single lens up to useful NAs. The lens shape for which spherical aberration is minimum is a meniscus with the stronger curved side directed towards the incident beam [23, p 25 ff]. Figure 27.3 shows the performance of such a lens with a focal length of 80 mm over a range of $(1/e^2)$ beam diameters in comparison with an aberration-free lens. The diameter of the focus will be somewhat larger than that obtained with an aberration-free lens, but the deviation is below 10% for an 18 mm beam ($f/4.4$, Strehl ratio 0.8) and gradually grows to about 20% at, say $f/2.8$, where the Strehl ratio has dropped to 0.6, which means that the peak intensity at the focus is only 60% of what it would be with perfect optics. If a (cheaper) plano-convex lens is used instead of the meniscus lens, the performance drops by another 15%.

In practice, where peak intensity is often the controlling factor for process speed, a plano-convex ZnSe lens is adequate up to NA = 0.1 ($f/5$).

While the above numbers were evaluated for a TEM$_{00}$ laser beam, they apply equally well to low-order mixed mode beams such as a combination TEM$_{01*}$ (ring shaped) plus TEM$_{00}$ beam, a beam type often generated by medium-power fast axial flow CO_2 lasers.

27.5.2.2 Extending the Limits of the Single Lens

When a higher NA at the focal spot is desired, for instance, to cut very thin material at maximum speed, there are several options to shorten the focal length: one is the use of diffractive optics, an approach obviously favoured by the relatively long wavelength, and diffractive optical elements with the required high diffraction efficiency are available commercially. A second option is to use an aspherical rather than a spherical lens surface shape. Both these techniques have the disadvantage of a relatively small back focus: the distance between the last lens surface and the workpiece is typically only 90% of the focal length of a single lens. It is, therefore, much more desirable to use dual lens systems of the retrofocus type [17], exemplified by Figure 27.4, which can be considered to combine a beam widening and focusing function and creates a long back focus albeit at the expense of a larger lens diameter. This lens type allows excellent aberration correction, with diffraction-limited performance extended to NA greater than 0.25 ($f/1.9$) at the lens rim.

27.5.2.3 Use of Zoom Optics

The NA is the quantity of choice for adapting the focal spot size, on one hand, and the depth of focus (the Rayleigh range), on the other, to the needs of a process. Thus, a zoom lens system which allows a continuous adjustment of the NA can optimize processes like sheet metal cutting by selecting an appropriate NA for a given material thickness. Such zoom systems are comparable to zoom lenses which are widely used

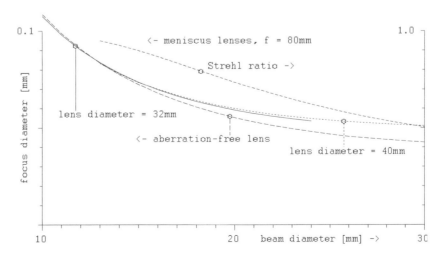

FIGURE 27.3 Performance of ZnSe meniscus lenses for focusing TEM$_{00}$ CO_2 laser radiation in comparison with an aberration-free lens, showing broadening of the $(1/e^2)$ focal spot and reduction of the peak intensity (Strehl ratio). Deterioration becomes noticeable at a beam diameter of 16 mm ($f/5$) but is acceptable up to 18 mm ($f/4.4$). The aberration-free lens used as a reference is the one shown in figure 27.4.

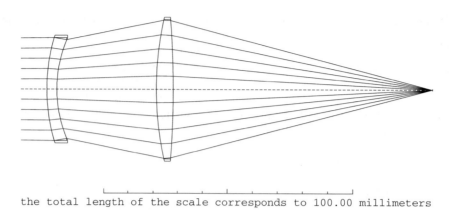

FIGURE 27.4 2 element ZnSe lens of the retrofocus type for aberration-free focusing of CO_2-laser radiation at high NAs (up to 0.25 or f/1.96 at the lens rim). Note that the back focus (100 mm) is substantially larger than the focal length (80 mm).

in photography, except that here they are operated at constant entrance pupil diameter rather than exit pupil diameter, and the NA varies accordingly—which is here the reason for zooming. They are also much simpler since they only need to function near the axis: A range of 3:1 in focal lengths (and NA) can be covered with optics using only two lenses (Figure 27.5) with one aspheric surface or three lenses—all spherical [24]. Performance is diffraction-limited. Note that the coverage of the depth of focus is nearly one order of magnitude as the Rayleigh length goes by the square of the focal length. The zoom lens approach has hardly been exploited commercially but deserves more attention as its feasibility and usefulness have been well demonstrated.

27.5.2.4 Thermal Effects in ZnSe Lenses

The unavoidable residual absorption of CO_2 laser power in ZnSe lenses, of which the largest part normally occurs at the surfaces due to imperfect coatings or contamination, causes lens heating and a radial temperature gradient, since the lens is heated at the centre and cooled by heat conduction to the lens rim and mounting [25–27]. In a new and perfectly clean lens, the effect is small enough to be neglected in practice, but it increases strongly if lens surfaces are contaminated and/or damaged. A radial temperature gradient is associated with a radial gradient of the refractive index, and the primary effect of this is that radiation paths inside the lens are bent inwardly, thereby shortening the focal length. If the actual temperature increases at the centre of the lens is known, the effect can be computed. A simple model which describes the situation in thermal equilibrium—reached typically after half a minute—has been shown to match experimental data within about 10% [28,29]. A more sophisticated model based on finite-element calculations leads to similar results and also covers the time dependency of the effect [30]. In an example situation of a (contaminated) lens, which absorbs 20 W (i.e. 1% of a 2 kW beam), the centre of the lens will, in thermal equilibrium, run about 25 K warmer than the lens rim, and the focal shift is of

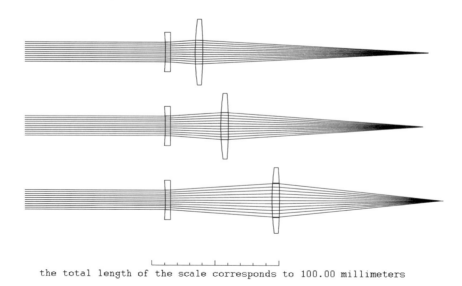

FIGURE 27.5 Zoom lens for focusing CO_2 laser radiation at variable focal lengths for adaptation to material thickness in laser cutting. The lens is of the retrofocus type and has only two (ZnSe) lens elements, surface 3 is (weakly) aspheric. Note the relatively long back focus at the position of shortest focal length (below). The lens covers a factor of >3 in focal length (and spot diameter) which corresponds to nearly one order of magnitude in Rayleigh range and, thus, depth of focus.

the order of one Rayleigh range. For an application like cutting, this is normally beyond acceptable limits, and as a rule of thumb, the limit of the lifetime of a ZnSe lens is reached when it absorbs more than 10 W.

Interestingly, the warm-up of the lens centre hardly affects its aberrations; in fact, the spherical aberration is even slightly reduced [28]. Lens heating and its associated focal shift have had surprisingly little attention in the literature on laser materials processing and are probably largely underestimated as a source of poor process reproducibilty.

27.5.2.5 Use of Artificial Diamond

Diamond has long been recognized as an excellent optical material for use at wavelengths in excess of 10 μm, but natural diamond is scarce in large enough sizes and prohibitively expensive. Artificial diamond, synthesized by a chemical vapour deposition (CVD) process, has become available with a quality and size which allows its use in industrial CO_2 laser systems. Due to the manufacturing process, CVD-diamond comes in plates of up to 2 mm thick and up to, say, 120 mm in diameter. Highly transparent samples are sold under the trade name DIAFILM and are polycrystalline with grain sizes below 0.1 mm, and a predominant crystallographic orientation of (1 1 0) perpendicular to the growth plane. The material closely approaches the properties of natural diamond type IIb, which is the purest form, and likewise exhibits intrinsic multiphonon absorption bands in the medium infrared region near 5 μm. The wings of these bands are responsible for a residual absorption at the CO_2 laser wavelength, where the absorption coefficient is typically 0.05 cm^{-1}, which is close to the theoretical limit. Although this coefficient is substantially higher than that of ZnSe, there are two main factors that make (CVD-)diamond superior for high-power CO_2 laser applications:

1. the thermal conductivity is more than two orders of magnitude higher than that of ZnSe (2000 versus 17 Wm^{-1} K^{-1});
2. the temperature dependence of the refractive index is only about one-sixth (10 *versus* 57 × 10^{-6} K^{-1}).

In addition, the high strength of diamond allows its use at much lower thickness for a given mechanical strength requirement. The total absorption—bulk plus surface coatings—is, thus, somewhat higher than that of the corresponding ZnSe component, but the optical effects of heat absorbed in a lens (or window) previously discussed in Section 27.5.2.4 are much reduced with diamond, as the higher thermal conductivity keeps the thermal gradients low and the lower temperature coefficient of the refractive index reduces the optical effect of the remaining gradient. Theoretical modelling has shown a gain by a factor between 40 and 240 depending on lens (or window) thickness [31]. Diamond has about the same refractive index as ZnSe (2.38 at 10.6 μm), which likewise makes it attractive for lens applications, but since the material is best available in plates typically 1 mm thick or less, its actual use in CO_2 laser systems has so far been predominantly for outcoupling windows in high-power CO_2 lasers above, say, 3 kW.

27.5.3 Mirror Systems for Use with CO_2 Lasers

The unavoidable residual absorption of CO_2 laser radiation by lenses makes mirror systems an attractive and at very high power levels, the only feasible alternative for radiation focusing. Since there is no pressure window function as with lenses, the use of all-mirror systems also requires alternative nozzle designs for applying cutting or shielding gas. Owing to the excellent thermal conductivity of the relevant mirror substrate materials and efficient cooling geometries, which are essentially independent of the mirror size, reflective optics can withstand high overall power levels and local power densities up to 10^5 W cm^{-2}. At large beam diameters, typically associated with high-power beams, mirrors also offer substantial cost advantages over refractive optics.

The two most common substrate materials for CO_2-laser mirrors are silicon and copper. Copper has a higher thermal conductivity by a factor of 2.5, which favours its use at very high power levels but its thermal expansion is more than six times higher so that, under equal thermal loading, silicon will deform less [32] and is preferred for reflectors inside a CO_2 laser. Both materials are normally used with reflection-enhancing coatings, which can raise reflectivity up to >99.5%. Suitability for precision-machining and integration of cooling channels into the mirror body have made (oxygen-free) copper the most popular substrate material for mirrors in CO_2 laser applications. Its main drawback is the low hardness and susceptibility to scratching during cleaning. In applications where a mirror is likely to be easily contaminated by spatters or fumes and may require frequent cleaning, a material of higher durability and hardness is preferable. These properties are offered by (uncoated) molybdenum, albeit at the expense of reduced reflectivity (typically 98%) [33]. In view of their application at high power levels, it is common practice to have cooling water channels integrated into the mirror body with a geometry that minimizes surface distortion by unavoidable thermal gradients.

27.5.3.1 Angular Field Considerations

Since laser beam focusing normally involves a very small angular field centred about the optical axis, it is tempting to consider only the axial aberrations of such systems. It must be borne in mind, however, that mirror optics for laser-beam handling are of necessity applied in some off-axis arrangement (Figure 27.6) in order to avoid central obscuration and that such systems are notoriously difficult to align such that they are truly operating with their optical axis parallel to that of the ingoing laser beam. Consequently, the optical system should tolerate a sensible angular deviation from the theoretical axis of at least, say, 2 mrad. Only if the optical aberrations at such a field angle, notably coma, are negligible, can an optical system be considered fit for general applications.

27.5.3.2 The Off-axis Paraboloid

The only axially corrected one-mirror focusing system, the paraboloid, is particularly troublesome in this respect. Due to the impossibility of fulfilling the optical sine condition, the

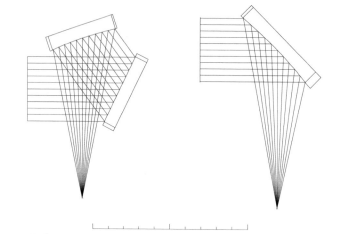

FIGURE 27.6 Off-axis paraboloids for 90° beam deviation.

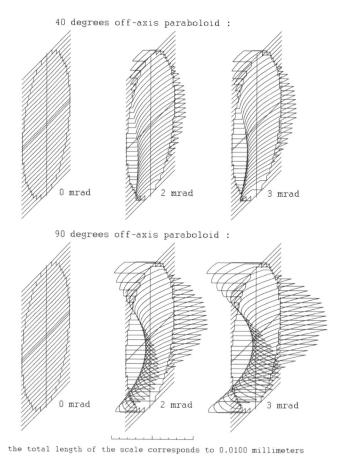

FIGURE 27.7 Phase-front distortion of off-axis paraboloid mirrors used under 40° and 90° off-axis (figure 27.6) for beams entering under angles of 0, 2 and 3 mrad with respect to the optical axis. The phase distortions are saddle-shaped and represent segments of the off-axis comatic phasefront which increase linearly with the angle but they result in astigmatic images.

paraboloid suffers from severe coma, when used at an angle to the (theoretical) axis [11, p 95 ff]; the effect rapidly increases when an off-axis section of the paraboloid is used, which is the only way to avoid obscuration (Figure 27.6). Figure 27.7 shows the deformations of the wavefront at angular deviations from the optical axis of 2 and 3 mrad for paraboloids used under 40° and 90° off-axis, respectively. It will be noted that the wavefront deformation is of a saddle shape, quite similar to that caused by astigmatism, but it is actually a segment of the unsymmetrical comatic wavefront and its magnitude is proportional to the field angle as is characteristic for (third-order) coma terms [10, p 118 ff]. From calculations of the diffraction image, it can be concluded that a 40° off-axis paraboloid requires alignment to better than 2 mrad, and a 90° off-axis paraboloid to better than half a milliradian. Thus, in spite of its apparent simplicity, the one-mirror 90° off-axis paraboloid is the least recommendable solution, while the two-mirror folded variety with the paraboloid under about 40° is just acceptable for practice.

27.5.3.3 Coma-corrected Optics

There is no cure for coma of single mirrors other than the use of designs involving two or more curved surfaces. This has, of course, been known by optical scientists for a long time and has led to well-known designs such as, for instance, the Ritchey–Chretien system which dominates modern astronomical telescopes. Unfortunately, this and related systems do not lend themselves well to off-axis derivatives in that (i) the NA of an off-axis cut-out will be much too small for most laser applications and (ii) the back focus will be much shorter than the focal length, as is true in all systems with a concave–convex sequence of mirrors, regardless of their shape. Thus, for use with lasers, the opposite configuration, a convex–concave sequence of mirrors, i.e. the well-known Schwarzschild design [34], is a better starting point for creating an off-axis laser focusing system. One family of designs derived on this basis uses four mirrors—three spherical and one flat—in a variety of arrangements with the focus either in line with the incoming laser beam or under 90°. Such systems have negligible aberrations over a field of more than one degree which makes them truly alignment-tolerant [35].

27.5.3.4 General Remarks on Aspherical Mirrors

Some general remarks about the use of aspheric mirrors may be in order. While in the past, aspheric surfaces were known to be costly to produce by classical grinding and polishing techniques and, thus, likely to be avoided by designers, mirrors using copper substrates are currently manufactured by single-point diamond turning which allows machining of aspherical shapes of high precision, albeit at sensibly higher cost for off-axis segments. There are two aspects that make aspheric mirrors less attractive than spherics:

1. An off-axis section of an aspheric has no rotational symmetry, which makes shape measurement and verification, and positioning in a mechanical mount, quite demanding and time consuming, unless sophisticated measuring tools are available.
2. All aspheric mirrors contribute strong coma, which drastically limits the angular field of use. This was discussed earlier for the off-axis paraboloid but it applies equally well to the ellipsoid and the hyperboloid, which are candidates in conjunction with divergent or convergent beams. Contrary to common belief, off-axis aspheric surfaces are no universal problem-solvers and should be used selectively and with caution.

27.5.4 CO_2 Laser Beam Integration for Homogeneous Illumination

CO_2 laser applications such as cutting and welding use beam focusing to produce small circular spots. A quite different radiation pattern is desirable for applications in various forms of surface treatments such as transformation hardening, cladding and alloying. In these applications, the surface is scanned with a spot of—preferably—rectangular shape and a flat power profile, in order to create a stroke of approximately homogeneously elevated temperature. The required power densities are in the 10^4–10^5 W cm^{-2} range and typical stroke widths range from a few millimetres for lasers in the low kilowatt class to several centimetres for laser powers above 10 kW. Several techniques have evolved for creating rectangular flat-topped beam profiles, referred to as 'beam homogenization' or 'beam integration'. These methods split the original beam into smaller rectangular parts and overlap them to form the desired rectangular pattern in a given plane. This can, for instance, be achieved by a mirror having rectangular flat areas [36], the size of, say, 1/10–1/50 of the beam area, or by just folding two halves of the beam [37]. If these areas are properly tilted with respect to each other, the beamlets reflected from such a mirror overlap and add up in one plane. Note that the recombination of beam parts stemming from the same source invariably leads to interference which shows up as a rather disturbing modulation of the power density within the integrated spot. A way to bring this modulation down to acceptable limits is to have the beamlets recombine at angles as large as are compatible with the system configuration, which at least raises the spatial frequency of the interference pattern [38].

Another approach to beam integration uses a kaleidoscope-type rectangular reflective cavity, where different parts of the original wavefront undergo different reflections so that the outcome is a more or less randomized sum. More reflections result in better randomization but also higher losses due to the high angles of incidence. In-coupling of the radiation by means of an axicon is helpful to minimize the number of internal reflections needed to achieve a given homogeneity [39]. The effective randomization of the optical path lengths reduces the modulation due to interference to an unproblematic level.

A pre-determined irradiation pattern over a given stroke width can also be achieved by rapid scanning or oscillation of a more or less focused beam in a direction perpendicular to the (slow) motion of the beam over the workpiece, creating an irradiation profile by time-averaging. To avoid excessive temperature fluctuations, the scan frequency must be very high, of the order of 1000 Hz for transformation hardening, somewhat lower for processes, where a melt pool is formed. The method is quite flexible and with little extra effort allows variation of the stroke in width and temperature profile by scanning with non-linear speed, but the latter is difficult to combine with high repetition rates.

27.5.5 Power Distribution Shaping by Phase Modulation

A radically different concept for beam shaping, be it beam homogenization or other, makes use of phase modulation, exploiting in effect the spatial coherence of the laser beam and manipulating the phase distribution of the original beam so that the power distribution in the diffraction image attains a predetermined shape and profile. Although known for some time as a viable principle [40–43], the method has been slow to be implemented in practice, partly because of technical hurdles that need to be overcome and partly because the theoretical understanding and modelling have so far been limited to TEM_{00} beams. Computer simulations have shown that phase modulation has the potential to shape a three-dimensional power distribution in the focal region in such a way that an axially elongated, thread-like region is formed, within which the power density exceeds a certain value (Beckmann, unpublished). Although that value will be a little less than that at the peak of a diffraction-limited image, it stays high over an extended axial range due to the formation of multiple waist-like concentrations along the axis. This particular behaviour is obtained with a very simple phase element, having a single phase step of $\pi(= 1/2$ wavelength) somewhere near the $1/e^2$ diameter of the (TEM_{00}) beam. The exact diameter of that phase step—the amount of outer beam radiation which interferes with the π phase difference with the inner part—determines the exact nature of the three-dimensional power pattern. It is mandatory that beam aperturing is virtually absent. The method still awaits experimental verification.

27.6 Optics for Lasers Operating in the Visible or Near-IR

A common advantage of all lasers which operate in the visible or near-infrared spectral region, that is, between, say, 0.40 and 1.5 µm, is that they can make use of the wide range of optical glasses of high quality, precision of definition in terms of refractive index and dispersion, stability and workability, which have been available for many decades. At least some of the more popular glasses have low residual absorption in this spectral region. It must be noted though that the data are, by tradition, given in, for instance, the Schott glass catalogues, as bulk transmissions for samples, respectively, 5 and 25 mm thick and with a precision of only three figures. More precise data would be highly desirable for (cw) laser applications in view of the ever increasing powers at which, for instance, Nd:YAG-lasers have become available over the last decade. Quite a different matter is the handling of radiation from Q-switched solid-state lasers, where power densities well above 10^{10} W cm^{-2} are routine and the possibility of laser damage to optical glasses has to be considered. There are two mechanisms that cause internal damage to optical glasses at high peak powers: one is due to absorption by foreign particles in the glass, such as platinum, which entered during the manufacturing process; the other is self-focusing of the laser radiation due to an increase in the effective refractive index with the local power density as characterized by the non-linear refractive index n_2 of the glass. A further concern is surface damage, which is largely independent of the glass type, i.e. its chemical composition, but related to the polishing agent and, of course, surface cleanliness. Schott [44] published a list of optical glasses for which damage-related quantities have been measured. Low-index glasses typically perform better. Data for fused silica were reviewed in Ref. [45].

27.6.1 Optical Handling of Fibre-delivered Nd:YAG Laser Radiation

The transport of radiation by an optical fibre has important advantages from a systems standpoint as it allows laser machining to be combined with positioning by (standard) industrial robots. It is also beneficial in that the fibre output constitutes a homogeneous radiation source with sharply defined boundaries. It is customary to combine the fibre output with collimating optics into one hermetically sealed unit, protecting the fibre end from contamination and ensuing catastrophic failure. The beam quality at the fibre output of a (lamp-pumped) state-of-the-art laser may typically run around $M^2 = 100$ but the advantages of the (collimated) fibre output readily outweigh that low beam quality in typical applications such as welding. Diode-laser-pumped high-power (>2 kW) Nd:YAG lasers have become available with beam qualities $M^2 < 50$ at the fibre output. Focusing the radiation onto the workpiece should be by (near) diffraction-limited optics, if the sharpness of the fibre boundary and its homogeneous (= top-hat) power profile are to be preserved in the image, which is usually the preferred mode of operation [46].

27.6.2 Lens Design

The design of focusing optics for visible- and near-infrared lasers follows essentially the general design rules and practice known from classical geometrical optics [23,47,48], with noteworthy exceptions: cemented lenses are better avoided where appreciable powers are to be handled and the glass catalogues should be checked for (bulk) optical transmission at the wavelength of use. To be on the safe side when designing optics for high-power Nd:YAG lasers above, say, 2 kW, the use of quartz glass of the kind specifically made for that spectral region by fusion of natural quartz [49,50] has merits, as it greatly reduces the risk of thermal stress and rupture. A typical design is shown in Figure 27.8. It is diffraction-limited at a focal length of 177 mm for an optical diameter of 50 mm, corresponding to a NA of 0.14 (f/3.5) at the lens rim. Table 27.1 shows what can be achieved with higher refractive index materials. All these lenses are air-spaced doublets with the positive element in front and the separation precisely set by edge contact (Figure 27.8). More lens elements will be needed for higher NAs; an example is shown in Figure 27.9. The basic three-element lens (Figure 27.9a) has a 100 mm focal length, an optical diameter of 50 mm and is diffraction limited up to a NA of 0.25 (f/2.0). Note that this system has a negative front lens followed by two positive lenses, a configuration which maximizes the back focus and, at the same time, minimizes spherical aberration. Addition of a (near aplanatic) lens element (Figure 27.9b) shortens the focal length to 70 mm, increases the NA to 0.36 (f/1.3), while maintaining diffraction-limited performance (Strehl ratio >0.95). The back focus is much shorter, however, and use of a protective window made from a less expensive material is

TABLE 27.1

Air-Spaced Doublet Lenses for Focusing Collimated Nd:YAG Radiation, Made from Different Materials. All Lenses Are of the Type Exemplified by Figure 27.8, with the Lens Elements in Edge Contact. Optical Diameter is 50 mm. The Lens Systems Are Optimized (and Diffraction Limited) for 1060 nm and Obviously not Colour-Corrected. They Nevertheless also Show Diffraction-Limited Performance at the HeNe Laser Wavelength (633 nm) for the Indicated Focal Offset, a Feature That is Helpful for Testing and Adjustment

Glass	Refractive Index	Focal Length [mm]	Back Focus [mm]	NA	F-number	Focal Offset [mm]
SiO_2	1.44967	177.0	156.3	0.1412	3.505	−0.295
BK7	1.50669	163.0	142.2	0.1534	3.222	−0.274
LF5	1.56594	152.0	131.0	0.1645	3.000	−0.336
SF2	1.62766	143.0	121.9	0.1748	2.816	−0.361
SF15	1.67516	137.0	115.8	0.1825	2.694	−0.378
SF14	1.73313	131.0	109.7	0.1908	2.572	−0.397

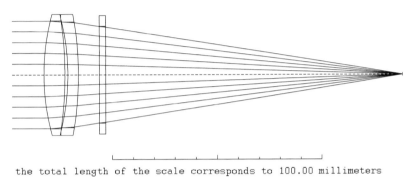

the total length of the scale corresponds to 100.00 millimeters

FIGURE 27.8 Air-spaced doublet for aberration-free focusing of collimated high-power Nd:YAG radiation at NAs up to 0.16 (f/3.1). The two lenses are in edge contact to ease manufacturing. This type of lens can be made from different materials (see Table 27.1) and covers a wide spectral range, albeit at varying focal position. A protective window, typically made from low-cost glass, is optional.

highly recommended. Note that, at this high NA, such a window has to be accounted for in the design of the last lens.

27.6.3 Colour Correction

System requirements may demand that a lens for handling Nd:YAG laser radiation be colour-corrected to have the same focal position for the wavelength of a pilot laser emitting visible light, often a HeNe laser. In this case, the classical and well-established methods for correcting colour for visible light optics by a suitable choice of glasses come into play but have to be modified for the change in wavelength region. The dispersion of the high index—and, in the visible, high dispersion—glasses, becomes progressively smaller when moving towards the near-infrared [51, p 142 ff]. Thus, expertise and experience about glass combinations which allow colour correction in the visible spectrum and simultaneously result in low monochromatic aberrations cannot be transferred to the longer wavelengths [52]. Figure 27.10 shows an example. For use in the visible spectrum (C, d, F light), the positive lens elements are made from BK7, the negative lens from SF2, a popular choice. To correct for use with Nd:YAG and HeNe laser light, however, the negative lens is better made from SF10, a glass of much higher index and higher dispersion for visible light but a good match for the longer wavelengths. The system is diffraction-limited (up to NA = 0.155 or f/3.2) and has exactly the same focal position for both laser wavelengths but there is strong (and uncorrectable) secondary colour, i.e. the focal position deviates sensibly at wavelengths in-between. Full-colour correction is not always desired: a well-chosen amount of chromatic aberration can be used to image different radiation sources at different axial positions back into the fibre for process control in laser welding [53].

27.6.4 Optics for Diode Lasers

Diode lasers differ from all other laser types in several important aspects. They (i) are inherently simple devices in that they convert electrical energy directly to (laser) light in a properly configured semiconductor crystal, which makes them very suitable for (low-cost) mass production and (ii) allows high energy conversion efficiencies but they (iii) are not scalable to high output powers from a single diode. The process of generating laser light in a diode does (iv) create a beam with different properties in two mutually perpendicular directions and, due to the very small size of the laser cavity, with (v) small waist sizes and accordingly large far-field divergence. The last three aspects govern all optics for diode laser applications.

27.6.4.1 Classes of Diode Lasers and Applications

There are three major areas of diode laser applications and corresponding classes of laser diodes and associated optics:

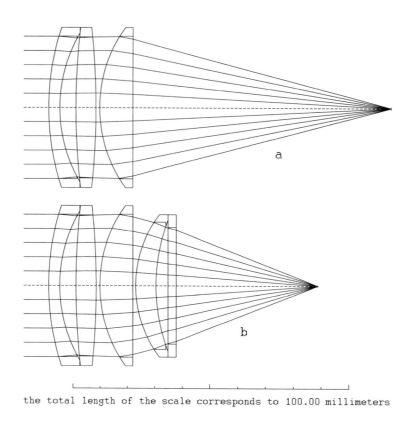

the total length of the scale corresponds to 100.00 millimeters

FIGURE 27.9 High-aperture focusing optics for high-power Nd:YAG radiation using SiO_2. The basic lens (a) has 3 elements and 100 mm focal length for a NA of up to 0.25 (f/2.0). Adding an aplanatic lens (b) shortens the focal length to 70 mm for a NA of up to 0.36 (f/1.3), but at the expense of a much shorter back focus, so that concurrent use of an (expendable) protective window is recommended. Note that the aberrations caused by this window at high aperture have to be taken care of in the design of the added lens element. To ease manufacturing, lenses 1 and 2 are in edge contact, lens 2 is symmetrically biconvex and lens 3 has a flat rear surface.

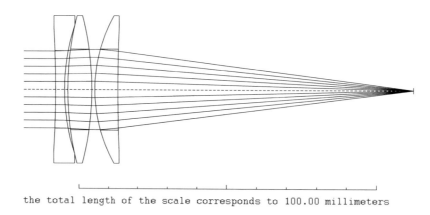

the total length of the scale corresponds to 100.00 millimeters

FIGURE 27.10 Three-element lens simultaneously corrected for Nd:YAG and HeNe laser radiation. The lens uses BK7 glass in the two positive lenses and SF10 for the negative front element to match the change of dispersion when shifting from visible light to the near-infrared. For use in the visible spectral region, the front lens would be made from SF2 with near identical surface curvatures.

1. telecommunication via optical fibres, which is mainly done at wavelengths in the 1.3–1.5 μm range and requires diodes of medium power and highest reliability;
2. optical information read-out and storage, which applies low-power (< 100 mW) diodes at near-infrared (CD) and red (DVD) wavelengths with a projected switch-over in the near future to shorter wavelengths; and
3. a cluster of applications which demand the highest available powers from individual diodes or from (stacks of) linear arrays of diodes, for such widely differing applications as pumping of solid-state lasers, material processing and a number of medical uses.

Due to their compactness, low power consumption and low operating voltages, diode lasers have also largely replaced gas lasers such as the HeNe laser in various types of sensors (e.g. bar code readers) and are gradually displacing ion lasers in applications such as printing, but these applications fall into one or the other of the previous classes.

27.6.4.2 Diode Laser Optical Output Properties

Optically speaking, the diode laser is an astigmatic light source with, in general, different beam properties in two orthogonal directions [54]. In the direction perpendicular to the diode junction, the beam quality is high, with M^2 typically running less than 1.4, and close to 1 at low powers, and the source waist width w_s is typically less than 1 μm, so that in accordance with equation (27.3), the corresponding half-beam divergence can exceed 0.5 rad. This direction is, therefore, usually referred to as the 'fast' direction. In the other direction, the beam quality and beam width vary widely, depending on diode size and power. For low-power diodes, with outputs below, say, 100 mW, a beam quality close to 1 is achieved in this direction as well but high power diodes may show a $M^2 > 50$ and a waist width ω_p up to 0.25 mm, with a typical half-beam divergence of 0.08 rad. This direction is therefore called the 'slow' direction. Radiances of high-power diode lasers well exceed 10^7 W cm^{-2}sr^{-1}.

27.6.4.3 Optical Handling of Single-diode Laser Beams

The strong divergence of the diode laser beam in the 'fast' direction (if not in both) suggests the need to intercept that radiation by some optics as close to the diode emitter as possible. In the most simple case—coupling of the diode laser output straightaway into a single-mode fibre—a single, small ball lens is often used, which offers by far the most economic, albeit optically imperfect solution. Methods for calculating and optimizing the coupling efficiency have been described by Ratowsky [55] and Wilson [56]. In all other cases, an aspheric lens with a short focal length is typically used for collimating the beam. Alignment is critical because of the very limited angular field of such lenses. Since a diode laser beam nearly always has strongly differing divergence in the directions parallel and perpendicular to the diode junction, the collimated beam is, in general, elliptical and will, for most applications, require equalization of the beam width by beam expansion in one direction so that subsequent focusing yields a circular spot. Figure 27.11 shows how this is achieved with a pair of prisms. Alternatively, the required beam shaping can be performed in a single step by a properly designed thick lens [57] with non-spherical surfaces that can be produced by plastic moulding. This is currently the preferred approach for use in high-volume consumer products (CD readers), which use diode lasers of 785 nm wavelength with beam divergences of 7° and 20° (FWHM), respectively, and require beam transformation to a wavefront of about 3 mm diameter with less than 1/50 wavelength distortion.

27.6.4.4 Single-diode Laser-focusing Optics

As with all other laser types, the focusing requirements, especially the NA, depend on the application. In optical storage, which is the largest diode laser application by numbers, diffraction-limited performance is required at NA = 0.45 for standard CD read-out and NA = 0.50 for writing. Since the focusing lens forms part of the servo-controlled optical head, its mass must be as small as possible, so a single aspheric lens is used exclusively. Several options are shown in Figure 27.12. The design has to take care of the thickness (= 0.12 mm) and thickness tolerance, of the (refractive) CD substrate ($n = 1.58$)

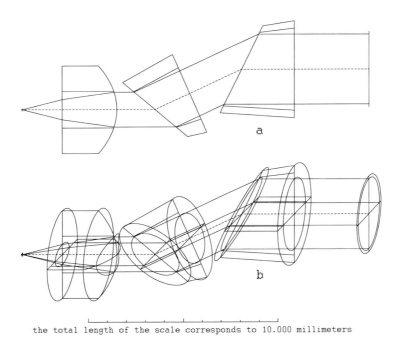

the total length of the scale corresponds to 10.000 millimeters

FIGURE 27.11 Collimation and one-dimensional beam expansion for diode laser beam equalization using a prism pair in the collimated beam: (a) meridional cross section; (b) 3D a view showing elliptical input and circular output cross section.

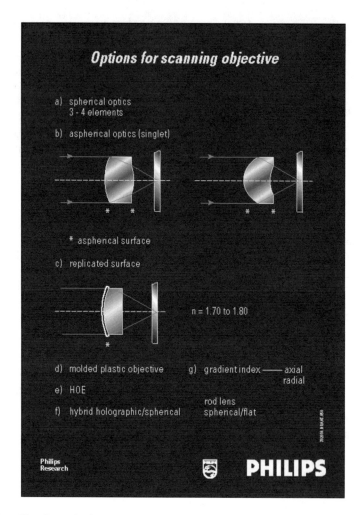

FIGURE 27.12 Options for the collimating optics for use in optical storage devices. (Courtesy J Braat, Philips Research Laboratories, Eindhoven, The Netherlands.)

in the convergent beam [58]. To ensure a stable product, this lens (option d of Figure 27.12) is typically manufactured from a basic plano-convex spherical glass lens of high refractive index, with a thin plastic aspheric shell moulded onto the convex surface. Higher density optical data storage (on the DVD, Figure 27.13) uses a somewhat shorter wavelength (635–670 nm) and a substantially higher NA = 0.60, in order to generate a smaller diffraction spot, imposing correspondingly higher demands on the quality of the optical components. For the next generation of optical storage, a wavelength of 400–405 nm and a yet higher NA = 0.85 are envisaged.

27.6.4.5 Optics for Applications of Diode Laser Arrays

A single-diode laser is only scalable to powers around a few watts, and for higher output powers, a series of diodes is arranged in a linear array, also referred to as a diode bar, with a length of typically 10 mm [59]. The individual diodes have widths varying between 0.5 mm and some tens of micrometres, and they are multi-mode in the 'slow' direction; the divergence of each emitted beam will typically be 10°–25° (FWHM). As the thickness of the diode junction is as small as in any single diode, the beam quality in the 'fast' direction is high (typically $M^2 < 1.4$) as is the divergence. For just about all applications, it is a common practice to place a collimating optic for the fast direction very close (<0.2 mm) to the emitting surface. This is mostly in the form of a cylindrical lens of short focal length, with a non-circular cross section and a planar surface at the side of the diode emitter. Because of the relatively high thermal load, such lenses are made from (high-index) glass. Lenses of this type, with focal lengths in the 1–2 mm range, are available commercially. Alignment is critical and unless both the diode bar and the cylinder lens are strictly straight within, say, a few micrometres, the centre lines of the individual collimated beams will not lie in one plane, a deviation referred to as 'smile'.

Behind the fast-axis collimator, the beam divergence will be reduced to a few milliradians but the beams fan out in the slow direction. They also start to overlap in that direction at some distance. Depending on the fast-axis collimator focal length, the beams can, in principle, be collimated in the slow direction before they overlap but it should be noted that this requires an array of cylindrical lenses with the lens distances exactly matched to the positions of the diodes in the bar. Alternatively, and depending on the application, the beams can be jointly manipulated with common optics. Without any further imaging, they are, for instance, suitable for side-pumping of a solid-state laser [60]. When it is desired to concentrate the total

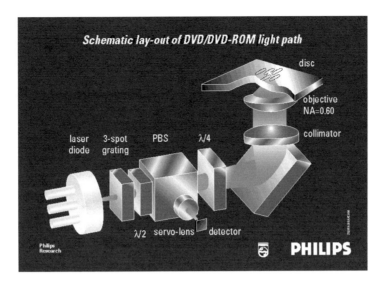

FIGURE 27.13 Schematic lay-out of the readout optics for the DVD optical storage system. (Courtesy J Braat, Philips Research Laboratories, Eindhoven The Netherlands.)

optical power from the bar into one (circular) spot or an optical fibre, one has to bear in mind that the beam quality of the individual diodes is low in the slow and high in the fast direction. Thus, to combine them into one common beam with reasonably equal quality in both directions, they should be added (concatenated) in the fast direction. This is, however, not the direction in which they are lined up behind the fast axis collimator, and for concatenation in the proper (=fast) direction, the individual beams have to be rearranged, i.e. effectively rotated by 90°. A number of optical schemes for this purpose have been proposed in the patent literature [61].

27.6.4.6 Stacks of Diode Laser Arrays

In practice, diode laser bars as described earlier are mounted (soldered) onto a heat sink which has provisions for water cooling and forms a package of, typically, 2 mm in height. Such packages, complete with collimating cylinder lenses, can be stacked to form larger units with output powers of hundreds of watts. It should be noted that collimation in the slow direction is virtually impossible for a stack since the bars are not in line within the required small tolerances. Stacks of diode laser bars are suitable for pumping (high-power) solid-state lasers and for direct materials processing, the latter so far limited to applications which require power densities below, say, 10^5 Wcm^{-2}, such as soldering or surface heat treatment. With increasing power output from the individual bars and through the use of more sophisticated schemes for combining diode laser outputs, such as polarization and wavelength multiplexing, the achievable power densities are expected to rise to levels which allow key-hole-based welding of metals, within just a few years.

27.7 Optics for Excimer and Other UV Lasers

Excimer lasers and their applications differ strongly from all other lasers in several respects. The radiation spans the whole ultraviolet region, the exact wavelength being determined by the gas combination used. The output is in the form of short pulses (typically 10–20 ns), pulse energies run up to, say, 1 J and create peak powers of tens of megawatts with average power levels of tens of watts. In typical commercial excimer lasers, beam quality is poor, with M^2 well in excess of 100, so an excimer laser beam behaves largely like (collimated) incoherent radiation. Direct focusing does not make much sense but the capability for high-precision machining, which results from the short (UV) wavelength, can nevertheless be exploited by using the laser for illuminating a mask, which is imaged on the workpiece so as to replicate the mask pattern. Although the mask can be in direct contact with the workpiece, it is more common to project a de-magnified image onto the workpiece. This then requires essentially diffraction-limited optics with a very flat image plane of, typically, several square millimetres for simple applications (simultaneous drilling of small holes in a small area) and up to some square centimetres for higher end systems. Due to the high photon energy of excimer laser radiation, the interaction with matter, in particular with organic matter (plastics), is largely non-thermal at the shorter wavelengths and referred to as ablation; it allows the machining of sharply defined features in the micrometre and submicrometre range without causing thermal damage. There is a threshold of laser energy density for the ablation process (typically less than 0.5 J cm^{-2} for plastics, higher for metals and ceramics). This limits the image area over which the radiation may be spread out and, indirectly, sets a lower limit for the NA at the mask image. Material removal is minute (several 10 μg per pulse) but care has to be taken that the resulting debris does not deposit on the imaging lens. A long back focus of, say, 70 mm and preferably more is, therefore, mandatory. Finally, the typical excimer laser beam is rectangular and inhomogeneous in at least one direction, so it requires homogenization before it is fit for mask illumination. The technique of homogeneously illuminating a mask and imaging the mask pattern makes, in general, only inefficient use of the laser light. When the mask pattern is simple and fixed, the laser energy can better be concentrated on the open parts of the mask [62].

These comments also largely apply to handling of radiation from other lasers operating in the ultraviolet, such as frequency-multiplied solid-state lasers.

27.7.1 Material Aspects

The ultraviolet region below, say, 300 nm is known for the very small number of available transparent optical materials. SiO_2 in the form of fused synthetic silica is most widely used, as are some of the crystalline fluorides, notably CaF_2 and LiF and, to a lesser extent, MgF_2, as it suffers from birefringence. At the shorter UV-wavelengths, the bulk absorption coefficient of fused silica depends on the fabrication process. Manufacturers offer certified material with an absorption coefficient of <0.0009 cm^{-1} at 248 nm (for KrF laser applications) and <0.003 cm^{-1} at 193 nm (for ArF laser applications) [63]. The high peak powers and high photon energy of excimer laser pulses cause irreversible changes [64], the mechanisms of which are not yet fully understood [65]. Thus, optics for excimer lasers have a limited lifetime. Under prolonged irradiation by, for instance, KrF excimer laser pulses, fused silica develops an absorption band around 210 nm [66,67], which affects the transmission at both the wavelengths of the ArF and the KrF excimer lasers. Although this absorption vanishes for low intensities within minutes after the irradiation, it re-occurs at the previous level if high-energy pulses are again applied [66]. Excimer laser irradiation also increases the density and, thus, the refractive index of fused silica, depending upon the fluence and number of pulses but the index change is only of the order of 10^{-6}–10^{-5} and, thus, not harmful in most applications.

All UV-transmitting materials have low refractive indices, so single lenses are rarely adequate. The spherical aberration of a single lens is strongly dependent on the refractive index [23, p 25 ff], and optical systems virtually always have multiple lens elements. SiO_2 and CaF_2 have a sufficient difference in dispersion to allow for colour correction at two wavelengths. Durable anti-reflection coatings are available for both materials [68,69].

27.7.2 Beam Homogenization

A typical beam homogenizer for excimer lasers has two mutually perpendicular arrays of cylindrical lenses, which slice the original beam into four to six strips and diverge and—by means of a common lens—redirect each strip of radiation so that they all overlap with the desired size in one common plane, which then becomes the plane of the mask (Figure 27.14). Diffractive micro-lenses have also been proposed for this purpose [70]. The homogenized beam tends to diverge behind the mask, and a (low-power) field lens is needed to direct the radiation into the opening of the lens system that images the mask onto the workpiece. In fact, the field lens has a dual function: it ensures, on one hand, that all laser radiation is utilized and, on the other, that the entrance pupil of the imaging lens is properly filled so that the lens can perform to its full diffraction-limited capability corresponding to the aperture for which it has been designed.

27.7.3 Optics for Imaging a Mask

Most of the requirements which have to be met by optics for imaging a mask for excimer laser machining have already been mentioned. For processing of plastics, they translate typically to a de-magnifying lens (or mirror system) with a NA at the image of about 0.1 (f/5). For machining of edges well perpendicular to the surface over a sizeable field and to reduce sensitivity to axial alignment, (near) telecentricity in image space is advantageous and often mandatory. To achieve a flat image plane over a sizeable field calls for optics with a low Petzval sum, achievable only with lens system types with some length, and this, in turn, causes problems with foci from reflections at later surfaces (see Section 27.4). This seriously limits the design options.

Optics for UV radiation are costly to manufacture because all tolerances tend to go down proportionally to the wavelength. This, and the already mentioned limited lifetime, justifies every effort to keep the number of optical elements as small as possible.

27.7.3.1 Design Examples

Figure 27.15 shows an example of an 'economic' lens with only three elements for use in micro-machining of small areas (less than 2 mm circular): de-magnification is 10, object-to-image distance 1048 mm and the back focus a comfortable 88 mm. The system is corrected for 248 nm (KrF) excimer laser radiation and, for ease of focusing and alignment, simultaneously for HeNe laser radiation. The two outer lenses are made from CaF_2, the inner (negative) lens from fused silica. The lens has sensible field curvature, and this limits the Strehl ratio in the best image plane for the total field to not more than 0.33 for a

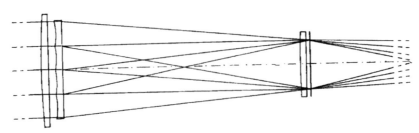

FIGURE 27.14 Beam homogenization system for excimer laser radiation (schematic diagram). The beam is split by means of an array of cylindrical lenses into four lightly diverging beamlets which are overlapped by a common lens in the plane of the mask, which is then imaged onto the workpiece (not shown). Two mutually perpendicular cylindrical lens arrays provide homogenization in both directions of the beam.

Free-space Optics

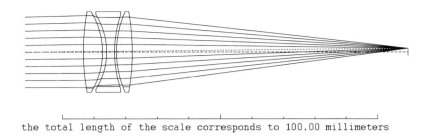

FIGURE 27.15 Three-element lens simultaneously corrected for 248 nm (KrF) excimer and HeNe laser radiation for 10:1 mask imaging on a workpiece for micro-machining with feature sizes down to about 2 μm over an area of a few millimetres diameter. Note the long back focus intended to protect the lens against debris from the process.

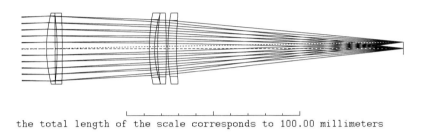

FIGURE 27.16 Five-element lens of the Petzval type for 4:1 mask imaging for micro-machining with 248 nm KrF excimer laser radiation. The lens covers an image field of 7 mm diameter with better than 1.25 μm resolution and is near-telecentric in image space. The positive–negative lens pairs are in edge contact.

NA of just over 0.11 (*f*/4.7). The image of an edge has a steepness, defined here as the distance over which the energy rises from 15% to 85% of the total, of better than 2 μm over the total field, and the overall performance, in practice, has been quite satisfactory. It should be mentioned, though, that the design is difficult to manufacture as the strongly curved surfaces imply tight tolerances for lens separation and centration.

Figure 27.16 shows a more sophisticated system with five lenses, made exclusively from SiO_2 and thus not colour-corrected, and optimized for 248 nm (KrF laser). De-magnification is only 4 and the long object-to-image distance, a system requirement in this case, actually 891.7 mm, produces a long back focus of 131.8 mm. An image field of up to 7 mm diameter is covered with an NA of 0.11 (*f*/4.7). The system is very well corrected with a Strehl ratio of >0.8 over most of the field and near telecentric. The steepness, as previously defined, is better than 1.25 μm. The system is basically of the Petzval lens type (with the last lens functioning as an aplanatic amplifier) which is known for its excellent definition over a limited angular field. To ease manufacturing, all lens curvatures are weak and the components of the two positive–negative lens groups are in edge contact. As Figure 27.17 demonstrates, the only critical reflected image (from surface 8) is well outside the front lens group. Although designed and optimized for operation at 248 nm, the lens is surprisingly perfect at the wavelength of the HeNe laser, yet with a much longer back focus of 154.7 mm, and this can be exploited to (i) mark the focal position with a properly convergent HeNe laser beam or (ii) check for manufacturing errors with visible (HeNe) light. Figure 27.18 shows photographs of these two lens systems.

A final example, shown in Figure 27.19, is again a lens for the 248 nm KrF laser with, again, five elements (all SiO_2) and ×4

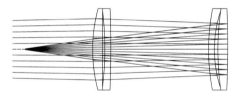

FIGURE 27.17 Focus caused by (residual) reflected radiation at surface 8 of the lens shown in figure 27.16. Such foci must lie outside the lens elements if the lenses are to be used at the high (peak) power levels needed for micro-machining with excimer lasers.

demagnification but with an even larger field of 14 mm diameter (10 mm square). The object-to-image distance, 376.2 mm, is much shorter than in all previous examples, yet combined with a long back focus of 109.5 mm. The lens is (perfectly) telecentric in image space, and fully diffraction limited at NA = 0.106 (*f*/4.7), with a Strehl ratio >0.9 and the previously defined steepness a nearly constant 1.2 μm over the whole field. Efforts to reduce manufacturing costs led to (i) a plane rear surface of the first lens, (ii) edge contact between the second and the third lens and (iii) identical and symmetrical biconvex lenses 4 and 5. The lens is again diffraction-limited for HeNe laser light as well (on-axis only). Reflections of concern are from surfaces 6, 8 and 10, the last just outside the troublesome region (Figure 27.20).

An interesting mirror system for 1:1 imaging is shown in Figure 27.21. It dates back to the early days of photolithography [71] and has a remarkable set of properties, achieved with only two spherical mirrors in a concentric set-up. The Petzval sum is zero and, thus, the image field essentially flat, at least in the sagittal direction, with some field curvature in the meridional direction and resulting astigmatism. Thus, imaging is best

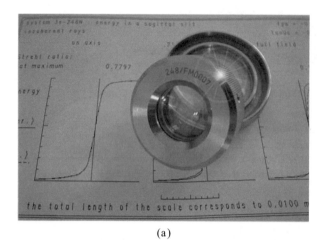

(a)

(b)

FIGURE 27.18 Photograph of the mask imaging lens systems of (a) Figure 27.15 and (b) Figure 27.16 (Courtesy OPTEC s.a., Hornu, Belgium).

over a strip-shaped (or better, an annular segment) image field of, say, 6×30 mm in the example layout, at NA = 0.1 (f/5), the strip extending in the sagittal direction. The convex secondary mirror acts as aperture stop, and the system is telecentric in both object and image space. The large distances between the (concave) mirror and the object and image planes, respectively, allow to freely re-arrange their positions with flat folding mirrors, if so desired. Derivatives of this system are presently under study for use at 157 nm (F_2 excimer laser).

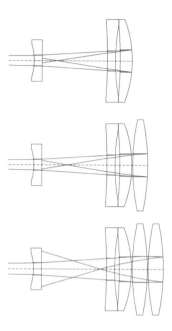

FIGURE 27.20 Foci caused by (residual) reflected radiation at surfaces 6, 8 and 10 of the lens shown in Figure 27.19. Note that the reflection from surface 10 barely meets the requirement.

27.7.3.2 Photolithography

Photolithography, a key technology for the fabrication of semi-conductor-integrated circuits, ('chips') also uses UV (excimer laser) radiation to image a mask on, in this case, a semiconductor wafer but, in contrast to laser machining, the laser power density is small and geared to a chemical change in the photo-resist which covers the surface. As there is no material removal, the back focus may be short (beneficial for the designer) but there are ever increasing demands on resolution, NA and field size and flatness, which puts photolithographic optics into a class of its own and outside the scope of this treatment.

27.8 Optics for Other Laser Sources

The previously described lasers and their optics are representative for the three main spectral regions and their respective materials limitations. There is virtually only one material for handling CO_2 laser radiation; when shifting to shorter infrared

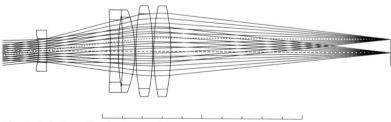

FIGURE 27.19 Five-element retrofocus-type lens for 4:1 mask imaging for micro-machining with 248 nm KrF excimer laser radiation. The lens covers a flat image field of 13 mm diameter with better than 1.2 µm resolution and is perfectly telecentric in image space. The back focus is 109.5 mm for a mask-to-image distance of only 376.2 mm. To ease manufacturing, lenses 2 and 3 are in edge contact, lenses 4 and 5 are identical and symmetrical, and lens 1 has a flat rear surface.

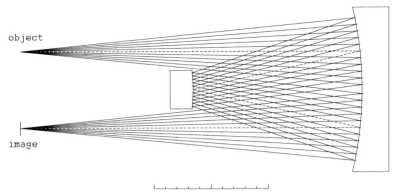

FIGURE 27.21 Two-mirror 'Offner'-type 1:1 imaging system. The system is telecentric in both object and image space and remarkably free of aberrations over a sizeable field. The mirrors are concentric and there is ample space for folding the system to match to mechanical constraints.

wavelengths of, say, about 5 µm, several materials with excellent transmission and durability become usable, most of them crystalline such as Al_2O_3 (sapphire), MgO, ZnS and spinel; these materials also have favourable (= high) refractive indices and thermal conductivities. Yet, the CO laser which operates in this wavelength region and could, in principle, benefit from these facts has not attained any sensible market shares, clearly for other than optical (or materials) reasons.

Ion lasers are visible light sources with excellent beam quality, and this fact has accounted for their popularity in applications which need that property, such as scanners. Needless to say, optical systems must be of a corresponding (i.e. diffraction-limited) quality level. Design aspects correspond to those discussed before for the more widespread visible and near-infrared lasers (27.6). There is a tendency to replace ion lasers with diode lasers wherever possible, for reasons of efficiency and life expectancy.

The copper vapour laser which has become available at powers of 200 W also has high beam quality which makes it a candidate for precision machining [72], allowing focal diameters down to 10 µm at a wavelength favourable for both the radiation–material interaction and optical handling. Optics have to reckon with high peak powers resulting from the short pulse duration.

Most of the lasers considered in 27.5 through 27.7 are typical high-power lasers, so possible thermal problems with the optics require adequate attention. Other lasers rarely exceed 10 W average output, and the thermal aspects are much more relaxed. With the great effort spent on ultrashort laser pulses down to the femtosecond region, however, extremely high peak powers become increasingly common. Little is known, and less published, about the performance of optics at these extremes. Theory predicts a different diffraction pattern [73].

27.9 Conclusions

Free-space optics play a key role in achieving optimum—even adequate—results in a laser application. To select the right optics for each task within a laser system is far from trivial and requires a full understanding of the applicable optical design principles, materials properties and power-related thermal aspects [74]. Optical expertise which has accumulated over centuries has not always found its way into the laser community and this suggests a need for a stronger interaction between optical engineers, laser system designers and laser users [75].

REFERENCES

1. Kogelnik H and Li T 1966 Laser beams and resonators *Appl. Opt.* **5** 1550–66.
2. Jabczynski J 1992 Quasi-geometrical model of partially coherent beam propagation in real axially symmetric optical systems *Proc. SPIE* **1780** 584–9.
3. Siegman A E, Sasnett M W and Johnston jr T F 1991 Choice of clip levels for beam width measurements using knife-edge techniques *IEEE J. Quantum Electron.* **27** 1098–104.
4. Siegman A E 1991 Defining the effective radius of curvature for a nonideal optical beam *IEEE J. Quantum Electron.* **27** 1146–8.
5. CEN/TC 123 N79, 1996 Optics and optical instruments, Lasers and laser related equipment, vocabulary and symbols, Secretariat DIN Aussenstelle Pforzheim, Westliche 56, D-75172 Pforzheim.
6. Siegman A E 1990 New developments in laser resonators *Proc. SPIE* **1224** 2–14.
7. Borik S, Wittig K and Zoske U 1994 Zur Bedeutung des Strahlparameterprodukts für das Propagationsverhalten von Hochleistungslaserstrahlen *Laser Optoelektron.* **26** 50–7.
8. Johnston T F Jr 1998 Beam propagation (M^2) measurement made as aesy as it gets: the four-cuts method *Appl. Opt.* **37** 4840–50.
9. Self S A 1983 Focusing of spherical Gaussian beams *Appl. Opt.* **22** 658–61.
10. Welford W T 1986 *Aberrations of Optical Systems* (Bristol: Adam Hilger).
11. Schroeder D J 1987 *Astronomical Optics* (San Diego, CA: Academic).
12. Borik S 1994 Einfluss optischer Komponenten auf die Fokussierbarkeit *Laser Optoelektron.* **26** 58–63
13. Smith W J 1989 Optical design *The Infrared Handbook* ed W L Wolfe and G I Zissis 3rd printing (Washington, DC: Office of Naval Research) ch 8.
14. Rayleigh L 1879 *Phil. Mag.* **8** 403.

15. Yura H T and Rose T S 1995 Gaussian beam transfer through hard-aperture optics *Appl. Opt.* **34** 6826–8.
16. Drége E M, Skinner N G and Byrne D M 2000 Analytical far-field divergence angle of a truncated Gaussian beam *Appl. Opt.* **39** 4918–25.
17. Beckmann L H J F and De Meijere J L F 1988 Retrofocus optics for the focusing of CO_2-laser radiation; a comparison between designs using spherical optics and aspherics *Proc. SPIE* **1020** 192–5.
18. Matuura Y, Miura D and Miyagi M 1999 Fabrication of copper oxide-coated hollow waveguides for CO_2 laser radiation *Appl. Opt.* **38** 1700–3.
19. Hello P and Man C N 1996 Design of a low-loss off-axis beam expander *Appl. Opt.* **35** 2534–6.
20. Bea M, Giesen A and Huegel H 1994 Gezielte Steuerung der Fokusgeometrie durch gekoppelte adaptive systeme *Laser Optoelektron.* **26** 43–9.
21. Jarosch U K 1996 Adaptive metal mirror for high power CO_2 lasers *Proc. SPIE* **2774** 457–67.
22. Reedy H E and Herrit G L 1988 Comparison of GaAs and ZnSe for high power CO_2 laser optics *Proc. SPIE* **1020** 180–91.
23. Smith W J 1992 *Modern Lens Design* (New York: McGraw-Hill).
24. Beckmann L H J F and Märten O 1992, Zoom lens designs for use in sheet metal cutting by high power CO_2-lasers *Proc. SPIE* **1780** 765–76.
25. Miyamoto I, Nanba H and Maruo H 1990 Analysis of thermally induced optical distortion in lens during focusing high power CO_2 laser beam *Proc. SPIE* **1276** 112–21.
26. Miyamoto I, Horiguchi Y and Maruo H 1990 Novel shaping optics of CO_2 laser beam: LSV optics—principle and applications *Proc. SPIE* **1276** 202–16.
27. Tangelder R J, Beckmann L H J F and Meijer J 1992 Influence of temperature gradients on the performance of ZnSe-lenses *Proc. SPIE* **1780** 294–302.
28. Beckmann L H J F 1994 Modelling of, and design for, thermal radial gradients in lenses for use with high power laser radiation *Proc. Int. Optical Design Conf. (Rochester)* p 11–15.
29. Beckmann L H J F and Meijer J 1998 Modelling of and design for thermal radial gradients in lenses for use with high power laser radiation *Welding in the World* **41** 97–104.
30. Weck M, Hermanns C, Ostendarp H and Wermeyer K 1994 Betriebsverhalten transmissiver Optiken bei der Laser-Materialbearbeitung *Laser Optoelektron.* **26** 67–72.
31. Godfried H P, Coe S E, Hall C E, Pickles C S J, Sussmann R S, Tang X and van der Voorden W K L 2000 Use of CVD diamond in high-power CO_2 lasers and laser diode arrays *Proc. SPIE* **3889** 553–63.
32. Herrit G L and Reedy H E 1989 Advanced figure of merit evaluation for CO_2 laser optics using finite element analysis *Proc. SPIE* **1047** 33–42.
33. Plansee 1987 Wolfram- und Molybdänlaserspiegel (datasheet) Metallwerk Plansee GmbH, A-6600 Reutte.
34. Erdös P 1959 Mirror anastigmat with two concentric spherical surfaces *J. Opt. Soc. Am.* **49** 877–82.
35. Beckmann L H J F and Ehrlichmann D 1994 Three-mirror off-axis systems for laser applications *Proc. Int. Optical Design Conf. (Rochester)* pp 340–8.
36. Henning T, Scholl M, Unnebrink L, Habich U, Lenert R and Herziger G 1997 Beam shaping for laser materials processing with non-rotationally symmetric optical elements *Proc. SPIE* **3092** 126–9.
37. Armengol J, Lupon N and Laguarta F 1997 Two-faceted mirror for active integration of coherent high-power laser beams *Appl. Opt.* **36** 658–61.
38. Schneider M 1998 Laser cladding *PhD Thesis* (Enschede: Univerity of Twente).
39. Beckmann L H J F 1990 A small-computer program for optical design and analysis written in 'C' *Proc. SPIE* **1354** 254–61.
40. Bett T H, Danson C N, Jinks P, Pepler DA, Ross I N and Stevenson R M 1995 Binary Phase zone-plate arrays for laser-beam spatial-intensity distribution conversion *Appl. Opt.* **34** 4025–36.
41. Dickey F M and Holswade S C 1996 Gaussian laser beam profile shaping *Opt. Eng.* **35** 3285–95.
42. Kuittinen M, Vahimaa P, Honkanen M and Turunen J 1997 Beam shaping in the nonparaxial domain of diffractive optics *Appl. Opt.* **36** 2034–41.
43. Zhang G Q, Gu B Y and Yang G Z 1995 Design of diffractive phase elements that produce focal annuli: a new method *Appl. Opt.* **34** 8110–15.
44. Schott 1988 Resistance of optical glasses to short laser pulses, Technical Information—Optical Glass, No 21, 4/1988 Schott Glaswerke, Postfach 2480, 55014 Mainz, Germany.
45. Milam D 1998 Review and assessment of measured values of the nonlinear refractive- index coefficient of fused silica *Appl. Opt.* **37** 546–50.
46. Verboven P E 1995 Beam delivery for Nd:YAG lasers *Opt. Eng.* **34** 2683–6.
47. Kingslake R 1978 *Lens Design Fundamentals* (New York: Academic).
48. Smith W J 1990 *Modern Optical Engineering* 2nd ed. (New York: McGraw-Hill).
49. Heraeus 1994 Infrasil 301, 302 and 303 POL Information, POL-O/439M-E, 06/94 and: HOQ 310, POL Information, POL-O/424M-E, 06/94, Heraeus Quarzglas GmbH & Co KG, Division POL, PO Box 1554, 63405 Hanau, Germany.
50. Heraeus 2000 Homosil 101 and Herasil 102, POL Information, POL-O/439M-E, Heraeus Quarzglas GmbH & Co KG, Division POL, PO Box 1554, 63405 Hanau, Germany.
51. Shannon R R 1997 *The Art and Science of Optical Design* (New York: Cambridge University Press).
52. Walker B H 1995 Lens design for the near IR… correction of primary chromatic aberration *Appl. Opt.* **34** 8072–3.
53. Haran F M, Hand D P, Peters C and Jones J D C 1997 Focus control system for laser welding *Appl. Opt.* **36** 5246–51.
54. Sun H 1997 Measurement of laser diode astigmatism *Opt. Eng.* **36** 1082–7.
55. Ratowsky R P, Long Y, Deri R J, Chang K W, Kallman J S and Trott G 1997 Laser diode to single-mode fiber ball lens coupling efficiency: full-wave calculation and measurements *Appl. Opt.* **36** 3435–8.
56. Wilson R G 1998 Ball-lens coupling efficiency for laser-diode to single-mode fiber: comparison of independent studies by distinct methods *Appl. Opt.* **37** 3201–5.

57. Braat J 1995 Design of beam-shaping optics *Appl. Opt.* **34** 2665–70.
58. Braat J 1997 Influence of substrate thickness on optical disk readout *Appl. Opt.* **36** 8056–62.
59. Muller N G, Weber R and Weber H P 1995 Output beam characteristics of high-power continuous-wave diode laser bars *Opt. Eng.* **34** 2384–9.
60. Du K, Zhang J, Quade M, Liao Y, Falter S, Baumann M, Loosen P and Poprawe R 1998 Neodymium:YAG 30 W cw laser side pumped by three diode laser bars *Appl. Opt.* **37** 2361–4.
61. see, for example, US 5,784,203; UK 2319630; DE 19752416.8.
62. Holmer A-K and Hård S 1995 Laser-machining experiment with an excimer laser and a kinoform *Appl. Opt.* **34** 7718–23.
63. Heraeus 1999, Quartz Glass for Excimer Laser Applications, Application Note, June 1999, Heraeus Quarzglas GmbH & Co KG, Division POL, PO Box 1554, 63405 Hanau, Germany.
64. Schenker R and Oldham W 1998 Damage-limited lifetime of 193 nm lithography tools as a function of system variables *Appl. Opt.* **37** 733–8.
65. Rothschild M 1993 Optical materials for excimer laser applications *Opt. Photon. News* May 8–15.
66. Leclerc N, Pfleiderer C, Hitzler H, Wolfrum J, Greulich K O, Thomas St Fabian H, Takke R and Englisch W Transient 210-nm absorption in fused silica induced by high-power UV laser irradiation *Opt. Lett.* **16** 940–2.
67. Thomas T and Kuehn B 1996 KrF Laser induced absorption in synthtic fused silica *Proc. SPIE* **2966** 56–64.
68. Krajnovich D J, Kulkarni M, Leung W, Tam A C, Spool A and York B 1992 Testing of the durability of single-crystal calcium fluoride with and without antireflection coatings for use with high-power KrF excimer lasers *Appl. Opt.* **31** 6062–75.
69. Laux S, Mann K, Granitza B, Kaiser U and Richter W 1996 Antireflection coatings for UV radiation obtained by molecular-beam deposition *Appl. Opt.* **35** 6216–18.
70. Nocolajeff F, Hård S and Curtis B 1997 Diffractive microlenses replicated in fused silica for excimer laser-beam homogenizing *Appl. Opt.* **36** 8481–9.
71. Offner A 1975 New concepts in projection mask aligners *Opt. Eng.* **14** 131–7.
72. Hartmann M, Koch J, Lang A, Schutte K and Bergmann H W 1997 Industrial aspects of precision machining with copper vapour lasers *Proc. SPIE* **3097** 260–6.
73. Aleshkevich V, Kartashov Y and Vysloukh V 1999 Diffraction and focusing of extremely short optical pulses: generalization of the Sommerfeld integral *Appl. Opt.* **38** 1677–81.
74. Klein C A 1997 Materials for high-power laser optics: figures of merit for thermally induced beam distortions *Opt. Eng.* **36** 1586–95.
75. Beckmann L H J F and Ehrlichmann D 1995 Optical systems for high-power laser applications: principles and design aspects *Opt. Quantum Electron.* **27** 1407–25.

FURTHER READING

There are no specific books on the subject of this chapter, but useful information can be found in most texts on Laser materials processing, such as:

Caird Luxon J T and Parker E 1992 *Industrial Lasers and Their Application* (Englewood Cliffs, NJ: Prentice-Hall).
Crafer R C and Oakley P J 1993 *Laser Processing in Manufacturing* (London: Chapman and Hall).
Dickey F M and Holswade S C (ed) 2000 *Laser Beam Shaping* (New York: Marcel Dekker).
Herziger G and Loosen P 1993 *Werkstoffbearbeitung mit Laserstrahlung* (München: Hanser).

The design of optical systems in general is treated in a number of textbooks, such as those cited for this chapter, but without addressing laser-specific subjects, with the exception of Smith (1990, 1992) and Welford (1986) (in a 2-page appendix). The subject of power distribution shaping of laser beams (see section B3.2.5.5) has been treated in detail in a book:

28

Optical Waveguide Theory

George Stewart

CONTENTS

28.1 Introduction 417
28.2 Basic Types of Optical Waveguides 417
28.3 Planar and Rectangular Guides 418
    28.3.1 Planar Guides 418
    28.3.2 Two-dimensional Guides 421
    28.3.3 Numerical Methods for Waveguide Analysis 422
28.4 Optical fibres 423
    28.4.1 Description of the Modes and Fields in Optical Fibres 424
    28.4.2 Modal Birefringence and Polarization-maintaining Fibres 425
28.5 Propagation Effects in Optical Fibres 426
    28.5.1 Attenuation in Optical Fibres 426
    28.5.2 Dispersion in Optical Fibres 427
        28.5.2.1 Inter-modal Dispersion 427
        28.5.2.2 Chromatic Dispersion 428
        28.5.2.3 Polarization-mode Dispersion 429
    28.5.3 Non-linear Effects in Optical Fibres and Solitons 429
28.6 Mode-coupling 430
28.7 Conclusion 432
References 432
Further Reading 433

28.1 Introduction

Over the last 40 years or so, much effort has been devoted to advancing the theory and practice of optical waveguides, mainly driven by the optical fibre data and communication systems market. In addition to the optical fibres themselves, a number of optical devices, such as diode lasers, couplers, wavelength multiplexers/demultiplexers, external modulators and photonic integrated circuits are formed on, or make use of, optical waveguide structures and their development has demanded a thorough understanding of waveguide theory. Equally important, but on a smaller scale, research and development of optical sensors for physical, chemical and biomedical sensing, including microfluidic and lab-on-a-chip devices, has also provided a stimulus for the development of new types of waveguide devices and theoretical models to predict their performance.

In this chapter, after presenting a brief review of the various types of optical waveguides, we outline the key principles and parameters which describe and define the operation of optical waveguides and fibres. The ways in which propagation through optical fibres affects the properties of the guided waves are discussed, including dispersion and non-linear effects. Power transfer between propagating waves is essential to the operation of a number of optical devices, and the fundamentals of coupling theory are reviewed. In summary, the theory given provides the essential foundation for understanding the operation of a wide variety of optical components based on waveguide and optical fibre technology.

28.2 Basic Types of Optical Waveguides

The simplest form of optical waveguide is the three-layer *planar* or *slab guide* shown in Figure 28.1a consisting of a central guiding layer sandwiched between lower index layers. If the layers on either side are of equal index, it is known as a *symmetric* planar guide; otherwise, it is an *asymmetric* guide. The lower layer (on which the waveguide may be formed in practice) is called the *substrate* and the top layer the *superstrate*. A further distinction arises depending on the index distribution within the guide. If the layers are of uniform index, then the guide is referred to as a *step-index* structure, whereas if the index varies (usually within the central layer), it is known as a *graded-index* guide. As discussed later, the

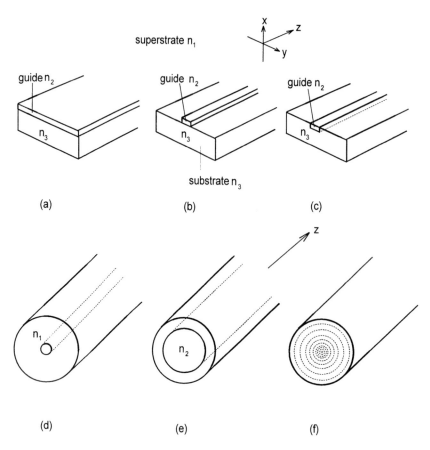

FIGURE 28.1 Various types of optical waveguide: (a) planar, (b) ridge, (c) embedded, (d) single-mode fibre, (e) multi-mode step-index fibre and (f) graded-index fibre.

central (wave-guiding) layer is typically a few micrometres in thickness for single-mode operation.

Typical *rectangular* or *two-dimensional (2D)* guides are illustrated in figures 28.1b and c. Here, the central (wave-guiding) layer is confined to a narrow channel or strip a few micrometres in dimension. Depending on the fabrication technology employed, the 2D guide may have a ridge or rib structure or may be embedded within a planar substrate. As with planar guides, the 2D guide is classified as step or graded index in structure.

Optical fibre waveguides, shown in Figure 28.1d–f, consist of a circular core surrounded by a *lower-index* cladding. Typically, standard *single-mode* fibres have a core/cladding diameter of 9/125 µm, whereas *multi-mode* fibres have dimensions of 50/125 µm or 62.5/125 µm and may be step or graded index in construction. In terms of transmission data rates, *single-mode* fibres are superior, followed by the graded-index multi-mode type.

The above examples constitute the simplest form of guiding structures, and in practice, many more complex guiding structures have been designed to meet specific requirements. For example, multi-layer planar or coated fibre guides may be used to tailor the optical power distribution throughout the structure [1,2] or facilitate phase-matching in non-linear optics while a variety of index profiles in optical fibres have been used to modify the dispersion properties [3]. Other examples include holey fibre where the core is solid silica with small air holes in the cladding region to effectively lower its index. In photonic band-gap fibre, the core is hollow and guidance is achieved by photonic band-gap effects [4]. Here however we mainly concentrate on the basic wave-guiding structures of Figure 28.1, where ray optics and/or fairly straightforward analytical solutions of Maxwell's equations are possible which allows a clear understanding of the fundamental principles and parameters of optical wave-guiding.

28.3 Planar and Rectangular Guides

28.3.1 Planar Guides

The simple planar guide, with a ray optics approach, provides a useful starting point for understanding the key properties of optical waveguides and fibres in general. If we first consider a light ray incident on the boundary between two materials of index n_1 and n_2, where $n_1 < n_2$, as shown in Figure 28.2a, then if θ exceeds the critical angle, θ_c, given by $\sin\theta_c = n_1/n_2$, total internal reflection (TIR) occurs. In TIR, the light actually penetrates a short distance beyond the boundary, i.e. the field does not abruptly drop to zero at the boundary but decays exponentially on the lower index side. This exponential field is called the *evanescent field* and has an associated penetration (or $1/e$) depth. As a result, light which has undergone TIR is phase-shifted from the incident light by an amount $2\phi_1$ where

Optical Waveguide Theory

$$\tan\phi_1 = \xi \frac{\sqrt{\sin^2\theta - \sin^2\theta c}}{\cos\theta} \quad (28.1)$$

In equation (28.1), $\xi = 1$ for s-polarization (i.e. an electric field perpendicular to the plane of incidence) and $\xi = n_2^2/n_1^2$ for p-polarized light.

As noted earlier, a planar optical waveguide is formed by a higher index layer sandwiched between regions of lower index, and so light rays may be trapped in the core layer by TIR whenever $\theta > \theta_c$ as shown in Figure 28.2b. However, for guiding to occur, a standing-wave pattern must be established across the guide since the wave must be confined between the boundaries. Put in another way, the round-trip phase change in the transverse (x-direction, see Figure 28.2c) must satisfy

$$-\left[\frac{2\pi}{\lambda_x} \times 2d\right] + 2\phi_1 + 2\phi_3 = \pm 2m\pi \quad (28.2)$$

where m is an integer and λ_x is the spacing of the wavefronts in the x-direction, $\lambda_x = \lambda/\cos\theta = \lambda_0/(n_2\cos\theta)$. Hence

$$k_0 d n_2 \cos\theta = m\pi + \phi_1 + \phi_3 \quad (28.3)$$

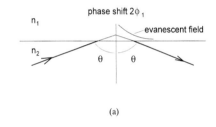

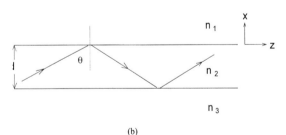

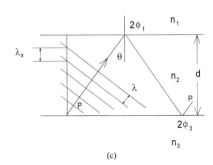

FIGURE 28.2 (a) Expanded view of total internal reflection (TIR), (b) light trapped in guide by TIR and (c) transverse resonance condition for guided modes.

where $m = 0, 1, 2,...$ is the mode order, with $m = 0$ for the fundamental mode and $k_0 = 2\pi/\lambda_0$.

Since m is an integer, this condition implies that only a *discrete* set of values of θ are allowed, $\theta_0, \theta_1, \theta_2, ..., \theta_m$ and each allowed value of θ corresponds to a certain transverse standing-wave pattern or *guided mode*. (Compare standing waves on a string, fixed at both ends, giving rise to different modes of vibration.) The modes which arise from s-polarized rays are transverse-electric (TE) modes because the electric field (but not the magnetic field) is entirely transverse to the propagation direction (z-direction) of the mode. Similarly, p-polarized rays give rise to transverse-magnetic (TM) modes. Hence, in general, a multi-mode guide will support two classes of modes, $TE_0, TE_1, TE_2, ...$, etc. and $TM_0, TM_1, TM_2, ...$, etc. Since ϕ is polarization-dependent, corresponding TE and TM modes have slightly differing values of θ and, hence, propagation constant (this is *waveguide birefringence*).

A very useful definition for describing the propagation of a guided mode is its *effective-index* value. With reference to Figure 28.1c, the phase velocity of the ray in the guide is c/n_2, but since the ray zig-zags at an angle θ, the wavefronts of the guided mode propagate in the z-direction with a phase velocity of $c/(n_2 \sin\theta_m) = c/n_e$, where $n_e = n_2 \sin\theta_m$ is the effective index of the mode. Because $\theta > \theta_c$ for all guided modes, the allowed range of n_e is $n_2 > n_e > (n_1$ and $n_3)$ and the propagation constant of the mode, $\beta_m = k_0 n_e$. With this definition, the eigenvalue equation for the modes (28.3) can also be written in the form:

$$k_0 d \sqrt{n_2^2 - n_e^2} = m\pi + \phi_1 + \phi_3. \quad (28.4)$$

The condition $n_e = n_1$ or $n_e = n_3$ (whichever is the greater) corresponds to the *cut-off* condition for a mode, since at that point the ray is travelling at the critical angle at one of the boundaries. This condition may be used in equation (28.4) to determine the cut-off thickness (or cut-off wavelength, if the thickness is known) given the other waveguide parameters, for a particular mode of order m. To illustrate, table 28.1 gives typical cut-off values calculated from equation (28.4) for a film of index $n_2 = 1.5$ on a quartz substrate ($n_3 = 1.45$) with air superstrate ($n_1 = 1$) at a wavelength of 1 μm.

The cut-off thickness for a mode is the minimum thickness which will support that mode. Using the data in table 28.1 as an example, if the thickness is in the range $0.51 < d < 0.58$ μm, then the guide will support the TE_0 mode only. Data of this type are used to design waveguides for single-mode operation or for operation in a selected number of modes, by choosing the appropriate guide dimensions.

At this point, let us consider in some detail the symmetric slab (which provides a very simple planar model for an optical fibre) and introduce a parameter known as the *V-number* or normalized frequency. The V-number is a very important parameter for determining the number of modes supported by a guide. By analogy with an optical fibre, the V-number for the planar (slab) guide may be defined as

$$V_{\text{slab}} = k_0 \left(\frac{d}{2}\right) \sqrt{n_2^2 - n_1^2}. \quad (28.5)$$

(Note that $d/2$ is the half-width of the planar guide which is analogous to the fibre radius a as used later in the definition of the fibre V-number).

In addition, it is customary to define a normalized propagation constant, or b-parameter, which lies in the range $0<b<1$, as follows:

$$b = \frac{n_e^2 - n_1^2}{n_2^2 - n_1^2} \quad (28.6)$$

With these definitions, equation (28.4) can be recast in a form which applies to any symmetric slab guide:

$$2V\sqrt{1-b} = m\pi + 2\tan^{-1}\xi\sqrt{\frac{b}{1-b}}. \quad (28.7)$$

Figure 28.3 shows universal b–V curves (or dispersion curves) plotted from equation (28.7) with $\xi = n_2^2/n_1^2$ chosen as 1.2 to show waveguide birefringence. Note that mode cut-off occurs at $b = 0$, giving the cut-off value for a mode of order m as $V_C = m\pi/2$ and that TE and TM modes have the same cut-off points in a symmetric guide. Also for single-mode operation, $0<V<\pi/2$, so that in theory, the fundamental mode is always supported in a symmetric guide (compare the asymmetric slab as illustrated in table 28.1 where TE and TM have different cut-off points and the fundamental mode is cut-off below a certain value).

Waveguides which are made by diffusion or ion-exchange processes generally have a graded-index core. In terms of ray optics, we can visualize the ray as being continuously refracted and travelling in a curved path through the graded index medium as illustrated in Figure 28.4. By applying the same procedure of requiring the round-trip phase change in the transverse direction to be an integer number of 2π, the following equation may be derived [5] for the modes of a graded-index guide:

$$k_0 \int_0^{d_m} \sqrt{n^2(x) - n_e^2}\, dx = m\pi + \frac{1}{4}\pi + \phi \quad (28.8)$$

where $n(x)$ is the index profile of the guide, d_m is the mode depth given by $n(d_m) = n_e$ and 2ϕ is the usual phase shift at the superstrate interface. Note that if the guide is buried into the substrate (e.g. a symmetric parabolic profile with maximum index below the surface), then $2\phi = \pi/2$. For certain profile forms, such as a linear or parabolic variation in index, the integral in equation (28.8) may be evaluated in closed form [5,6].

So far, all the results discussed are derived on the basis of a simple ray-optics model of a planar waveguide. A more rigorous method is to obtain the field solutions by application of Maxwell's equations. The guided modes are assumed to propagate in the z-direction as described by a travelling-wave term of the form: $\exp j(\omega t - \beta_m z)$. For a planar guide, the transverse-field distribution is assumed to vary in only the x-dimension so that the Maxwell wave equation is simplified by the assumption that the $\partial^2/\partial y^2$ term is zero. From the one-dimensional wave equation, one can then obtain the transverse field component (for example, E_y for TE modes) in the form of either a decaying exponential for the cladding regions or as a co-sinusoidal distribution in the guiding core. The other field components may then be obtained from E_y by application of Maxwell's curl equations. When continuity conditions are applied on tangential field components at the boundaries of the guide the same eigenvalue equation (28.4) is obtained for the guided modes. For TE modes in a step-index planar guide, the TE field component, E_y, can thus be obtained in the form:

Substrate $(x<-d)$: $E_y = \pm C \dfrac{\cos\phi_3}{\cos\phi_1}\exp\{\gamma_3(x+d)\}$

Core $(-d<x<0)$: $E_y = \dfrac{C}{\cos\phi_1}\cos\{k_x x + \phi_1\}$

Superstrate $(x>0)$: $E_y = C\exp\{-\gamma_1 x\}$

(28.9)

where + is for even-mode orders and – is for odd-mode orders, C is an arbitrary constant, γ is the decay constant of the evanescent field, $\gamma_i = k_0\sqrt{n_e^2 - n_i^2}$ and $k_x = k_0\sqrt{n_2^2 - n_e^2}$. Note that in these equations, the field is matched at the boundaries $x = 0$ and $x = -d$ and all the fields should be multiplied by the travelling-wave term: $\exp j(\omega t - \beta_m z)$. The constant C is the field

TABLE 28.1

Cut-off thickness, d_C, for several modes of a slab guide (parameters in text)

Mode	TE0	TM0	TE1	TM1	TE2	TM2
dC (μm)	0.51	0.58	1.81	1.89	3.11	3.19

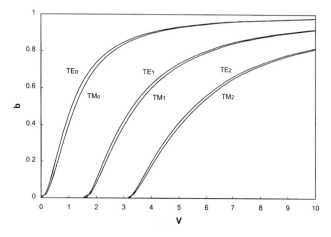

FIGURE 28.3 Universal dispersion curves for a symmetric planar guide.

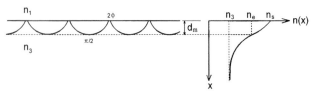

FIGURE 28.4 Graded-index slab guide.

Optical Waveguide Theory

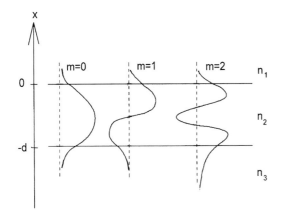

FIGURE 28.5 Field distribution of modes in a planar step-index slab guide.

amplitude at the $x = 0$ boundary and may be obtained from the total power per unit width, P_w, coupled into the planar guide by computing the integral $P_w = \frac{1}{2} n_e c \varepsilon_0 \int_{-\infty}^{+\infty} E_y^2(x) dx$ using equation (28.9) for the electric field in the various layers (c is the speed of light in vacuum and ε_0 is the permittivity constant).

Figure 28.5 shows the field distribution, E_y, for the first few TE modes of an asymmetric step-index slab guide. Note the cosine distribution of the field in the core, forming a standing-wave pattern, with an increasing number of cycles for increasing mode order. The cosine distribution is connected smoothly (through the continuity conditions as noted earlier) to the evanescent field tails which penetrate the substrate and superstrate. The evanescent field penetration depth, defined by $d_p = 1/\gamma$, increases with mode order and is larger in the substrate for $n_3 > n_1$. Each of these field patterns represent a particular way in which the light is guided by the structure and, hence, is described as a *mode* of the guide.

28.3.2 Two-dimensional Guides

Planar structures are generally impractical for guiding light because of diffraction spreading of the beam in the plane, so 2D or channel guides are required for the manufacture of integrated optic guided wave devices such as Y-junctions, branching elements, couplers, interferometers and modulators [7]. However, the theoretical description and analysis of 2D guides are complex, and in general, a rigorous electromagnetic solution is required. There are some approximate methods that can be applied, mainly based on the extension of planar waveguide concepts, which we briefly consider here.

First of all, Marcatili [8] provided an approximate electromagnetic solution in closed form for the general structure shown in Figure 28.6a under the assumption of small index differences and rigorous field matching along the sides only (not in the shaded corners where the field is relatively negligible). Under these conditions, one can regard the light as travelling nearly axially down the guide and consequently the modes are essentially TEM in character with electric field polarization along either the x- or y-axis. The field distribution in the x–y plane (transverse to the propagation direction) may then be constructed in the form of a product: $E(x, y) = E_1(x) E_2(y)$, where $E_1(x)$ and $E_2(y)$ have the form of planar-type solutions as in equation (28.9). For example, for the y-polarized mode, we would write:

in the core: $E_y = C \cos(k_x x + \phi_x) \cos(k_y y + \phi_y)$

in the region: $x > 0$: $E_y = C \cos(\phi_x) \cos(k_y y + \phi_y) \exp\{-\gamma_1 x\}$
(28.10)

and so on.

Following on from the Marcatili's approach, the mode parameters may be obtained by the *effective index method* [5,9,10] which makes use of planar mode conditions. To understand this, Figure 28.6b shows a simple ray-optics picture of a mode in the 2D guide where the ray undergoes TIR in the sequence: A (bottom) → B (left side) → C (top) → D (right side) → A' (bottom), and so on. Viewed from above or from the side, the ray describes a zig-zag path as in a planar guide and the 2D guide appears as a combination of two planar guides. In the effective-index method, the two planar guides are linked

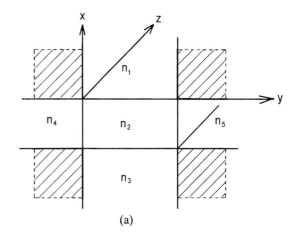

(a)

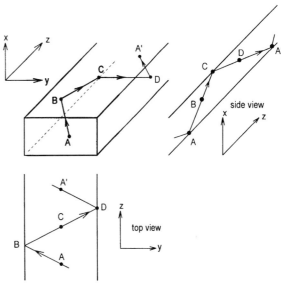

(b)

(*Continued*)

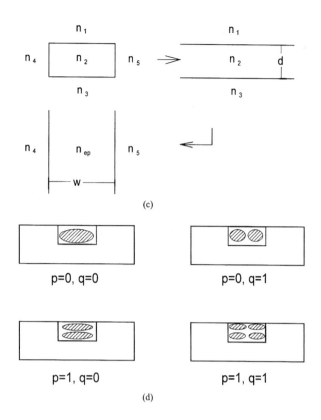

FIGURE 28.6 (CONTINUED) (a) Model for the general analysis of 2D guides, (b) ray paths in a 2D guide, (c) effective index method for 2D guides and (d) field and intensity distribution of modes in a 2D guide.

by using the effective index calculated from the first guide to replace the core index of the second. The procedure is as follows (see Figure 28.6c).

i. Let the long dimension of the guide $\to \infty$ to obtain the first planar guide. Use the characteristic equation (28.4) for a planar guide (or the b-V characteristic from equation (28.7) if the guide is symmetrical) to obtain the effective index, n_{ep}, for each of the allowed modes with mode order $p (= 0, 1, 2, \ldots)$.

ii. Construct the second planar guide in the other direction as shown, but replacing n_2 with each value of n_{ep} in turn. Again calculate the effective index n_{epq} for each allowed mode with mode order q ($= 0, 1, 2, \ldots$) from the planar guide equation (28.4).

The set of calculated effective indices, n_{epq}, provides an approximate description of the set (p, q) of 2D guided modes. Field parameters, $\gamma_1, \gamma_3, \gamma_4, \gamma_5$ and k_x, k_y may also be calculated from the effective-index values. Figure 28.6d shows the typical field distribution for several modes of the 2D guide.

Note that in applying the method, the appropriate mode polarization in the planar guides must be used. For example, for 2D modes polarized along the y-axis, TE modes of the first planar guide would be used with TM modes of the second (*vice versa* for x-polarized 2D modes). The effective-index method can also be applied to graded-index structures [5], in which case the eigenvalue equation (28.8) would be used.

28.3.3 Numerical Methods for Waveguide Analysis

Analytical solutions for four-layer planar isotropic guides are available [2,11], but for complex multilayer wave-guiding structures, it is usually necessary to revert to numerical methods to derive the wave-guiding properties. Useful reviews of the various numerical techniques are given in references [12,13]. Typical problems for numerical analysis include obtaining the eigenmodes of a guiding structure (normal mode-solving), finding the evolution of the electric field as the wave propagates through the structure (beam propagation) or analysing devices with forward and backward travelling waves [14]. Several of the more important methods include:

i. Matrix method [1,16,17]
ii. Finite difference method (FDM) [12–15]
iii. Finite element method (FEM) [10,12–14]
iv. Finite difference time domain (FDTD) method [13,14]
v. Beam propagation method (BPM) [10,13,14]

We briefly review these methods below.

The matrix method is particularly useful for analysing multi-layer and/or graded index structures. A graded index structure is approximated by a large number of layers over which the refractive index is assumed constant. Additionally, any of the layers may be defined to have real and imaginary components of its refractive index so that the effect of loss on the propagation of the guided modes may be modelled. The basic procedure for a planar structure is as follows [1,16]:

i. Introduce a temporary semi-infinite medium with a refractive index higher than that of the guiding layer(s) so that plane waves in this medium at appropriate angles will excite the guided modes of the structure. (The perturbation effect of this medium is removed in the final solution by letting its distance from the guiding structure $\to \infty$.)

ii. In each layer, $i = 1 \ldots n$, the electric field is written as the sum of downward, E_i^+ and upward E_i^- propagating waves. The field in one layer is related to that in the adjacent layer through the boundary conditions, expressed by a 2×2 matrix, S_i, based on the Fresnel reflection and transmission coefficients (r_i and t_i) at the interface and phase shift, $\delta_i = k_0 n_i d_i \cos\theta_i$, dependent on the layer thickness, index and ray angle. This leads to the matrix equation relating the fields in the layers [16]:

$$\begin{bmatrix} E_1^+ \\ E_1^- \end{bmatrix} = S_1 \begin{bmatrix} E_2^+ \\ E_2^- \end{bmatrix} = S_1 S_2 \begin{bmatrix} E_3^+ \\ E_3^- \end{bmatrix}$$

$$= S_1 S_2 S_3 \ldots S_{n-1} \begin{bmatrix} E_n^+ \\ E_n^- \end{bmatrix} \quad (28.11)$$

where $S_i = \dfrac{1}{t_i}\begin{bmatrix} e^{j\delta_i} & r_i e^{j\delta_i} \\ r_i e^{-j\delta_i} & e^{-j\delta_i} \end{bmatrix}$

iii. The real and imaginary components of the propagation constants of the guided modes are obtained by plotting the excitation efficiency against the incident angle of the plane wave introduced into the semi-infinite medium above. By fitting the peaks that occur in this curve to a Gaussian function, the propagation constant is obtained from the peak position and the loss from the peak width.

In general, the method is fast, involving the multiplication of simple 2×2 matrices, and does not require iterative procedures. The method may also be applied to optical fibres [17].

The basis of the FDM for waveguide analysis is to represent the field in the transverse plane by discrete values at the elements of a rectangular grid transverse to the propagation direction. The field distribution is then obtained by the process of iteration using standard finite difference relations for the partial derivatives in the wave equation which may be written, for example, for the quasi-TE modes of a waveguide as:

$$\frac{\partial^2 E_x}{\partial x^2} + \frac{\partial^2 E_x}{\partial y^2} + k_0^2 \left\{ n^2(x, y) - n_e^2 \right\} E_x = 0 \quad (28.12)$$

where $n(x,y)$ is the transverse index distribution and the guided modes are assumed to have a time and z-dependence of: $\exp\{j(\omega t - k_0 n_e z)\}$

Similar to the FDM, the FEM also uses a mesh in the transverse plane, but here, the field in each element is described by trial functions with unknown coefficients. A functional, derived from the wave equation, is used for the analysis, and when the variation of this functional is set to zero, a matrix eigenvalue equation results from which the solutions are obtained. The FDTD method includes propagation of the guided mode fields by direct integration of Maxwell's equations using finite difference relations for the partial derivatives in both space and time.

The BPM is the most popular method for waveguide analysis and is the basis of many commercial software packages. A full account of the method is beyond the scope of this chapter, and the reader is referred to reference [13] and the book on BPM by Pedrola (2016) in the list of further reading for a detailed description of the procedures and its applications. Briefly, however, the BPM starts with an approximation to the exact wave equation which is then solved numerically by, for example, the FDM. The simplest approximate form for the wave equation can be obtained under the assumptions of (i) a scalar field (polarization effects neglected) and (ii) a slow variation of the field envelope in the propagation direction. These assumptions lead to the basic BPM equation [13]:

$$\frac{\partial u}{\partial z} = \frac{j}{2k_0 \bar{n}} \left[\frac{\partial^2 u}{\partial x^2} + \frac{\partial^2 u}{\partial y^2} + k_0^2 \left\{ n^2(x, y, z) - \bar{n}^2 \right\} \right] \quad (28.13)$$

where $n(x, y, z)$ is the refractive index distribution of the guiding structure, $\bar{n}$ is a reference refractive index and the scalar field is represented as: $u(x, y, z) e^{j(\omega t - k_0 \bar{n} z)}$

To solve equation (28.13) by the FDM, the transverse plane is divided into a rectangular grid and the field represented by discrete values at each element of the grid. Similarly, the propagation direction, z, is divided into discrete planes. Knowing the field values in one z-plane, the values at the next plane may be computed using finite difference approximations for the partial derivatives in equation (28.13) in the Crank-Nicholson scheme [13].

If the above assumptions for the simple BPM are not valid for a particular problem, then more advanced versions of the BPM may be used. Where polarization effects are important, the BPM may be extended by starting from the vector wave equation to include transverse field components in the x- and y-directions, leading to a set of coupled equations for the slowly varying fields, $\partial u_x / \partial z$ and $\partial u_y / \partial z$, in the z-direction. The other assumption (slow variation of the field envelope in the z-direction) involves neglecting a term, $\partial^2 u / \partial z^2$, in the derivation of equation (28.13). This restriction is relaxed in the so-called wide-angle BPM where the effect of this term is included through various approximations.

28.4 Optical Fibres

Optical fibres are the most widely used form of waveguide because of their flexibility, low cost, small dimensions and ability to transport optical data over long lengths with low loss (<0.5 dB km^{-1}). As noted earlier, depending on the core size, optical fibres are either multi-mode (supporting perhaps many thousands of modes) or single mode in operation. Single-mode fibres are the most important in communication systems because of their high information-carrying capacity but multi-mode fibres have found several important applications in sensor technology, such as in evanescent field sensors and in sensors exploiting mode-coupling through micro- or macro-bending. Graded-index multi-mode fibres may also be used for lower capacity or shorter length communication links.

Three important parameters often quoted for fibres are the numerical aperture, $NA = n_0 \sin\theta_0 = \sqrt{n_2^2 - n_1^2}$, V-number, $V = k_0 a \sqrt{n_2^2 - n_1^2}$ and birefringence, $B = (n_{ex} - n_{ey})$.

The numerical aperture (see Figure 28.7) defines the cone of light accepted into guided modes by the fibre and is particularly useful for multi-mode fibres. The V-number, similar

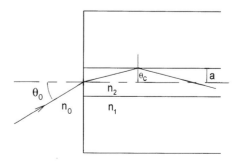

FIGURE 28.7 Optical fibre parameters and the numerical aperture.

to that defined for planar waveguides, may be used to determine the number of modes supported by the fibre and to determine the range for single-mode operation. The birefringence defines the difference in effective index between the two polarization states and will be discussed later.

28.4.1 Description of the Modes and Fields in Optical Fibres

Under the assumption of a core surrounded by an infinite cladding, the electric and magnetic fields of guided waves in optical fibres, as well as the eigenvalue equations for the modes, may be obtained from application of Maxwell's wave equations to the cylindrical structure [3,10]. In general, the fibre supports several types of modes classified as transverse electric (TE) modes where $E_z = 0$, transverse magnetic (TM) modes where $H_z = 0$ and hybrid modes (HE and EH) where both E_z and H_z are non-zero. The fundamental mode is designated as HE_{11} mode. The problem with this approach is that the mode description and the associated eigenvalue equations are rather complicated mathematically and not easy to apply in practical applications.

A much more useful description can be obtained by realizing that most fibres used in practice have a relatively small index difference between core and cladding,[1] i.e. $\Delta = (n_2 - n_1)/n_1 \ll 1$. Under this condition of weak guidance, modes propagate nearly axially down the fibre and field components along the propagation direction (z-direction) are, therefore, small. Simplified solutions for these modes, where the fields are essentially transverse in nature, were first derived by Snyder [18] and designated *Linearly Polarized* or *LP* modes by Gloge [19]. For polarization along either the x- or y-axis, the TE field has the form (see Figure 28.8)

$$E_{core} = E_l \frac{J_l(ur/a)}{J_l(u)} \cos(l\phi)$$

$$E_{core} = E_l \frac{K_l(\omega r/a)}{K_l(\omega)} \cos(l\phi)$$

(28.14)

where l is an integer ($l = 0, 1, 2, \ldots$), J_l is a Bessel function of the first kind of order l and K_l is the modified Bessel function of order l, E_l is the field at the core/cladding interface and the parameters u and ω are defined by:

$$u = k_0 a \sqrt{(n_2^2 - n_e^2)} \quad \omega = k_0 a \sqrt{(n_e^2 - n_1^2)} \quad V^2 = u^2 + \omega^2$$

with n_e the effective index of a mode as defined for planar guides. Note that these equations are multiplied by $\exp\{j(\omega t - \beta z)\}$ for mode propagation along the z-axis.

Bessel and modified Bessel functions are shown in Figure 28.9. Note the comparison with planar guides where the field in the guiding layer is described by a sinusoidal function (compare the Bessel function for the fibre-core field) and in the substrate by an exponentially decaying field (compare the modified Bessel function for the fibre-cladding field).

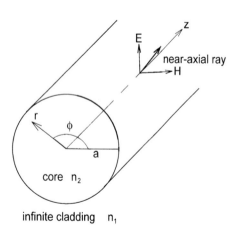

FIGURE 28.8 Fibre coordinates for describing LP modes.

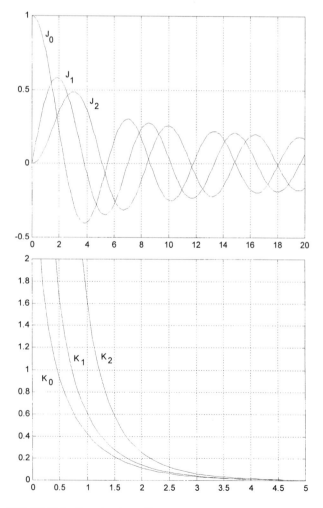

FIGURE 28.9 Bessel and modified Bessel functions.

---

[1] Sometimes Δ is defined as: $\Delta = (n_2^2 - n_1^2)/2n_1^2$ but the two definitions are the same in weakly guiding fibre where the core and cladding refractive indices are similar.

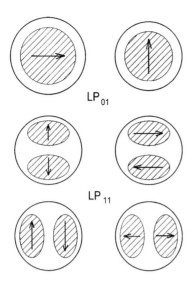

FIGURE 28.10 Intensity patterns for several LP modes.

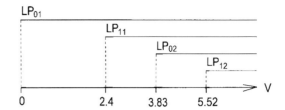

FIGURE 28.11 Mode cut-off points.

The various modes are given the designation LP_{lm} where m is also an integer, $m = 1, 2, \ldots$, which indicates the zero of the Bessel function at which the mode is cut-off as explained later. The fundamental mode is LP_{01} which corresponds to the HE_{11} mode from the exact theory. Apart from the fundamental, all the LP modes are, in fact, combinations of several exact modes which essentially become degenerate (i.e. have nearly the same propagation constant) under the weakly guiding approximation; for example, $HE_{21} + TE_{01}$ (or TM_{01}) $\rightarrow LP_{11}$. Figure 28.10 shows the intensity distribution and polarization for several LP modes. Note that l determines the number of field zeros in the azimuth direction through the $\cos(l\phi)$ term while m determines the number of zeros in the radial direction through the number of zero crossings of the Bessel function.

As in the case of planar waveguides, the electric and magnetic fields must satisfy boundary conditions at the core/cladding interface which leads to a characteristic or eigenvalue equation for the allowed modes. As Figure 28.10 shows, the various modes correspond to standing-wave patterns in the plane transverse to propagation. For LP modes, the characteristic equation for the allowed modes is [19]:

$$u \frac{J_{l-1}(u)}{J_l(u)} = -\omega \frac{K_{l-1}(\omega)}{K_l(\omega)}. \quad (28.15)$$

This equation is very useful for determining mode cut-off conditions. The cut-off is the point when the mode just ceases to be guided and occurs when the mode effective index equals the cladding index, $n_e = n_1$ or, equivalently, $\omega = 0$ which means that $V = u$. From equation (28.15), this gives the general cut-off condition as $J_{l-1}(V) = 0$. The cut-off conditions for the various modes are therefore (see Figure 28.11):

LP_{0m} modes: since $l = 0$, the cut-off condition becomes: $J_{-1}(V) = -J_1(V) = 0$

LP_{01} $m = 1 \rightarrow$ first zero crossing of $J_1 \Rightarrow V = 0$

LP_{02} $m = 2 \rightarrow$ second zero crossing of $J_1 \Rightarrow V = 3.832$

etc.

LP_{1m} modes: since $l = 1$, the cut-off condition becomes: $J_0(V) = 0$

LP_{11} $m = 1 \rightarrow$ first zero crossing of $J_0 \Rightarrow V = 2.405$

LP_{12} $m = 2 \rightarrow$ second zero crossing of $J_0 \Rightarrow V = 5.52$

etc.

From this, the condition for only the LP_{01} mode to be guided is: $0 < V < 2.405$. This condition is very important in fibre design to determine the appropriate range of wavelengths or core diameters for single-mode operation. Two-mode operation (LP_{01} and LP_{11} only) occurs over the range $2.405 < V < 3.832$.

As in the case of planar guides, the characteristic equation (28.15) may be used to obtain b–V dispersion curves for the various modes with mode cut-off at $b = 0$. Gloge [19] provides an approximate solution for the characteristic equation and presents the b–V curves for a large range of LP modes. For more complex fibre geometries, the numerical methods discussed earlier may also be used to obtain the dispersion characteristics and field distribution in core and cladding.

28.4.2 Modal Birefringence and Polarization-maintaining Fibres

As noted earlier, the modal birefringence, B, is the difference in effective indices between the orthogonally polarized modes, $B = (n_{ex} - n_{ey})$. In a perfectly symmetric circular core fibre, the x- and y-polarized modes would be identical so the propagation constants or effective indices would be equal and $B = 0$. In practice, the lack of perfect symmetry in conventional single-mode fibres means that two nearly-degenerate modes are propagated, with principal x- and y-axes defined by the asymmetry of the cross section (through geometrical, composition or strain-induced factors) and B ranges from about 10^{-5} to 10^{-6} with a lowest value of $\sim 10^{-9}$. In addition, perturbations arising during fibre manufacture or from twists and bends in operation will cause the principal axes to vary randomly along the fibre length. Coupling between the polarization states will therefore occur, and such fibres cannot maintain a fixed polarization state for more than a few metres [10,20].

In certain applications, it is sometimes desirable to maintain a fixed polarization state on propagation through the optical fibre path, for example, in interferometric optical fibre sensors or to avoid noise arising from random fluctuations in the output polarization state. Polarization-maintaining (PM) fibres with $B \sim 10^{-4}$ can be manufactured by making use of either geometrical birefringence, as in the elliptical core fibre, or strain-induced birefringence by incorporating stress elements in the fibre as illustrated in Figure 28.12.

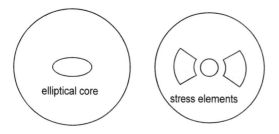

FIGURE 28.12 Polarization-maintaining fibres.

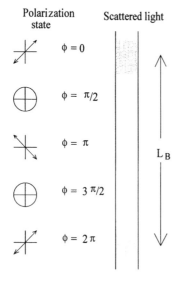

FIGURE 28.13 Evolution of polarization state along a PM fibre.

With PM fibre, if linearly polarized light is launched into one of the polarization eigenmodes (i.e. polarized along the *x*- or *y*-axis), then the PM fibre will maintain the polarization state over a considerable distance (depending on the value of *B*). If, however, the input polarization direction is not parallel to either the *x*- or *y*-axis, both eigenmodes will be launched into the fibre. Since the *x*- and *y*-eigenmodes have different effective indices (or phase velocities), the phase difference between the modes will change along the length of the fibre, and the polarization state will evolve from linear to elliptical to linear as illustrated in Figure 28.13. After a certain distance, called the beat length, L_B, the phase difference will be 2π and the original polarization state will be recovered. The beat length is thus given by

$$\Delta\phi = (\beta_x - \beta_y)L_B = 2\pi$$

giving the beat length $L_B = \lambda_0/B$. The beat length can be observed visually in a PM fibre from the periodic variation of the scattered light along the fibre length. The Rayleigh scattered light can be thought of as originating from dipole radiators. The radiation pattern of a dipole has a minimum along the dipole axis and a maximum transverse to the axis, so, if the fibre is viewed from any particular direction, the scattered light will vary periodically as the dipole orientation changes due to the changing polarization state. Observation of the beat length provides a useful way of measuring the birefringence of a fibre.

28.5 Propagation Effects in Optical Fibres

The most widespread use of optical fibres is for data and communication systems involving transmission of light pulses at rates of up to hundreds of gigabits per second. Consequently, there is a need to consider the effects of guided wave propagation on the data pulses. We consider here three physical factors that can modify or distort the data pulses as a result of propagation through the fibre, namely,

- attenuation which causes a reduction in the pulse intensity leading to reduced signal-to-noise ratios.
- dispersion which broadens the pulses, leading to possible overlap of pulses and errors in detection.
- non-linear effects which can modify the pulse shape and also cause cross-talk between neighbouring channels in wavelength-division-multiplexed (WDM) systems.

28.5.1 Attenuation in Optical Fibres

Figure 28.14 illustrates the attenuation characteristic of silica fibres, showing the operational wavelengths around 850, 1300 and 1550 nm of first-, second- and third-generation optical fibre communication systems, respectively. The shape of the curve arises from several factors [3,21].

On the short wavelength side, the attenuation is dominated by Rayleigh scattering and, to a lesser extent, by the tail of the UV absorption band. Rayleigh scattering has a strong wavelength dependence of the form $1/\lambda^4$ and arises from index variations in the glass over distances that are small in relation to the wavelength. Index variations occur because of fluctuations in the density and composition of the glass (especially since dopants are added) and from inhomogeneities during manufacture. UV absorption occurs when photons excite electrons from the valence to the conduction band of the silica material, and the tail of this band extends into the 1 µm region adding a small contribution to the loss. On the long wavelength side, the loss characteristic is dominated by the tail of IR absorption bands from the fundamental molecular vibrations in the 7–12 µm region of bonds such as Si–O, Ge–O, B–O and P–O. Various impurities and dopants also contribute to the loss. Water in the form of the hydroxyl group (OH) bonded into the silica structure gives overtone absorption lines at 0.72, 0.95 and 1.38 µm and combination lines (with silica vibrations) at 0.88, 1.13 and 1.24 µm. Other sources of loss are ionic impurities of transition metals including Cr, Fe, Cu, Ni, Mn and V.

Finally, it should be noted that in the practical use of fibres, other factors may contribute, to a greater or lesser extent, to the attenuation experienced by guided light in the fibre. If the fibre is bent beyond a certain critical radius, *macro-bending* losses become significant as a result of radiation from the evanescent wave tail on the outer curved side of the fibre. *Micro-bending* losses from mode-coupling occur when a fibre is subjected to periodic or repetitive variations in curvature along the fibre axis (micro-bends) which may happen during manufacture, cable installation or in service.

Optical Waveguide Theory

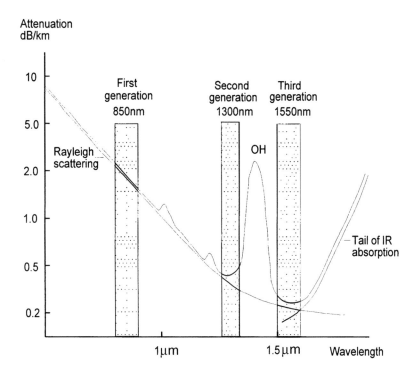

FIGURE 28.14 Loss characteristic for silica fibres.

28.5.2 Dispersion in Optical Fibres

In general, dispersion in optical fibres arises from three factors, namely, inter-modal dispersion, chromatic (or intra-modal) dispersion and polarization-mode dispersion. The effect on pulse broadening from the different dispersion factors can be combined according to the relation [3,21]

$$\Delta t_{tot} = \sqrt{\Delta t_i^2 + \Delta t_c^2 + \Delta t_P^2}$$

but usually only one (or possibly two) factors are significant, depending on the fibre type and the application.

28.5.2.1 Inter-modal Dispersion

The greatest dispersion occurs with the step-index multi-mode fibre where inter-modal dispersion is dominant. Here, the pulse power at the fibre input is distributed over the (very) large number of modes supported by the fibre. From a simple ray-optics description, the ray experiences multiple reflections as it travels along the fibre, and the path length increases with mode order. Hence, each mode has its own transit time through the fibre link so there is a spread in the arrival times of the pulse power from each mode at the output. The two extreme cases are illustrated in Figure 28.15a which shows the lowest-order mode making a near direct transit in the shortest time and the highest-order mode taking the longest time. The pulse broadening can be estimated from the difference in transit time of these two modes (ignoring skew rays). For the fastest (axial) mode, the transit time is $n_2 L/c$, whereas for the extreme meridional ray (at the critical angle θ_c), it is $n_2 L/(c \sin\theta_c)$. Hence, the inter-modal pulse broadening is:

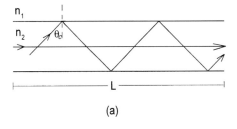

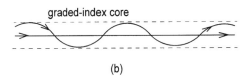

FIGURE 28.15 (a) Fastest and slowest modes in a step-index fibre and (b) ray paths in a graded-index fibre.

$$\Delta t_i = \frac{L}{c}\frac{n_2}{n_1}(n_2 - n_1) \approx \frac{nL\Delta}{c} \quad (28.16)$$

where $\Delta = (n_2 - n_1)/n_1$ is small. For $\Delta \approx 0.01$, then $\Delta t_i \simeq 50\,\text{ns}\,\text{km}^{-1}$. However, a useful definition is the rms pulse broadening for inter-modal dispersion [21], given by $nL\Delta/(2\sqrt{3}c)$ giving a typical value of ~14 ns km^{-1}.

Inter-modal dispersion can be dramatically reduced by using a graded-index core fibre with a parabolic or near-parabolic index profile as illustrated in Figure 28.15b. The transit times for the various rays are equalized by using the fact that off-axis rays experience a lower index in the outer regions of the core where the group velocity is higher, thus compensating for the longer path length. For the best case with optimum index profile, the rms pulse broadening can be reduced by a factor of $\sim\Delta/10$ compared with the step-index case [22], giving a value of ~ 14 ps km^{-1}. In practice, slight deviations of the profile from its ideal shape due to manufacturing difficulties can greatly increase this value. Note that with graded-index fibre, both inter-modal and chromatic (intra-modal) dispersion may be significant. Although graded-index fibres have improved dispersion properties over the step-index type, they are not used for high-data-rate systems because of the superior performance of single-mode fibres as discussed in the following section.

28.5.2.2 Chromatic Dispersion

Single-mode fibres are used in high-data-rate communication systems due to their superior performance as a result of no inter-modal dispersion but the effects of chromatic dispersion on pulse broadening must still be considered. Because of the wavelength spread (linewidth) of the source, the pulse is composed of a range of wavelength components which disperse on propagation through the fibre due to their different group velocities. Because of the great importance of single-mode fibres in communication systems, chromatic dispersion is considered here in some detail.

Before examining dispersion in single-mode fibres, consider first the propagation of a pulse through a (dispersive) medium where the refractive index, $n(\lambda_0)$, depends on wavelength. A pulse travels a distance L in a time, $t = L/v_g$, where v_g is the group velocity given by $v_g = d\omega/d\beta$ and β is the propagation constant, $\beta = k_0 n = (2\pi/\lambda_0) n = (\omega/c) n$, so

$$v_g = \left(\frac{d\beta}{d\omega}\right)^{-1} = \left(\frac{n}{c} + \frac{\omega}{c}\frac{dn}{d\omega}\right)^{-1} = \frac{c}{N_g} \quad (28.17)$$

where the group index N_g is given by

$$N_g = \left\{n + \omega \frac{dn}{d\omega}\right\} = \left\{n - \lambda_0 \frac{dn}{d\lambda_0}\right\}. \quad (28.18)$$

The transit time (or *group delay*) for the pulse is therefore $t = (L/c) N_g$. If the pulse is generated by a source with a wavelength spread of $\Delta\lambda = (\lambda_2 - \lambda_1)$, then the transit time for the pulse will also have a spread in values because the group index is wavelength-dependent. The pulse broadening as a result can thus be approximated by

$$\Delta t = (t_2 - t_1) = \frac{L}{c}\left\{N_g(\lambda_2) - N_g(\lambda_1)\right\} \cong \frac{L}{c}\frac{dN_g}{d\lambda_0}\Delta\lambda. \quad (28.19)$$

Using equation (28.18) for the group index gives the result:

$$\Delta t = -\frac{L}{c}\frac{\Delta\lambda}{\lambda_0}\left[\lambda_0^2 \frac{d^2 n}{d\lambda_0^2}\right]. \quad (28.20)$$

This expression shows that the pulse broadening depends on the relative spectral linewidth of the source and the chromatic dispersion coefficient $\lambda_0^2 (d^2 n/d\lambda_0^2)$. Note that pulse broadening occurs irrespective of the sign of Δt (positive $\Rightarrow$ longer wavelengths have longer transit times, negative $\Rightarrow$ longer wavelengths have shorter times). The dispersion is often expressed in terms of a parameter, D, with units of ps nm^{-1} km^{-1} through the definition:

$$D = -\frac{1}{c\lambda_0}\left[\lambda_0^2 \frac{d^2 n}{d\lambda_0^2}\right] = -\frac{\lambda_0}{c}\frac{d^2 n}{d\lambda_0^2}. \quad (28.21)$$

This definition means that pulse broadening is simply given by: $\Delta t = DL\Delta\lambda$. Figure 28.16 shows typical values for the material dispersion parameter D_m for conventional silica optical fibres. Note that the coefficient is negative for $\lambda_0 <\sim 1270$ nm and positive for $\lambda_0 >\sim 1270$ nm.

Consider now the case of pulse propagation in a waveguide. The propagation constant of a guided mode is $\beta_g = k_0 n_e = (2\pi/\lambda_0) n_e$, where the effective index, n_e, is a function of λ_0 and so equation (28.20) describes pulse broadening where the dispersion coefficient is now $\lambda_0^2 (d^2 n_e /d\lambda_0^2)$. This dispersion coefficient may, in principle, be derived from the characteristic equation for the modes of a guide (for example equation (28.4) for planar guides), taking into account the wavelength dependence of the refractive indices.

For this purpose, it is convenient to consider the dispersion in terms of two contributions, namely, material dispersion and waveguide dispersion. As already noted, material dispersion arises from the wavelength dependence of the refractive indices of the glasses making up the guide, whereas waveguide dispersion arises from the nature of guided modes (the b–V curves shown earlier are essentially waveguide dispersion curves).

Consider the most important case that of dispersion in a single-mode fibre [3, 10, 21]. Assuming a small index difference between core and cladding (as for LP modes), $\Delta = (n_2 - n_1)/$

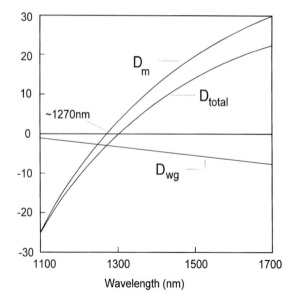

FIGURE 28.16 Material, waveguide and total dispersion for a conventional single-mode fibre.

$n_1 \ll 1$, then from equation (28.6), the effective index can be written as

$$n_e \cong n(b\Delta + 1)$$

and, hence

$$\frac{\mathrm{d}n_e}{\mathrm{d}\lambda_0} \cong \frac{\mathrm{d}n}{\mathrm{d}\lambda_0} + n\Delta \frac{\mathrm{d}b}{\mathrm{d}\lambda_0} \quad (28.22)$$

where $n = n_1 \approx n_2$, and it has been assumed that the dispersion of n_1 and n_2 is similar so that Δ is not a function of λ_0.

The dispersion parameter can now be expressed as a sum of material and waveguide dispersion effects in the form:

$$D = -\frac{\lambda_0}{c}\frac{\mathrm{d}^2 n_e}{\mathrm{d}\lambda_0^2} \approx -\frac{\lambda_0}{c}\frac{\mathrm{d}^2 n}{\mathrm{d}\lambda_0^2} - n\Delta \frac{\lambda_0}{c}\frac{\mathrm{d}^2 b}{\mathrm{d}\lambda_0^2} = D_\mathrm{m} + D_\mathrm{wg} \quad (28.23)$$

where D_m is the material dispersion as given earlier and D_wg is the waveguide dispersion given by:

$$D_\mathrm{wg} = -n\Delta \frac{\lambda_0}{c}\frac{\mathrm{d}^2 b}{\mathrm{d}\lambda_0^2} = -\frac{n\Delta}{c\lambda_0} V \frac{\mathrm{d}^2 (bV)}{\mathrm{d}V^2} \quad (28.24)$$

where the approximation $V \cong k_0 a n\sqrt{2\Delta}$ is used. (Note that for the approximation of equation (28.23) a term involving $\mathrm{d}n/\mathrm{d}\lambda_0$ is ignored [21].)

Figure 28.16 shows typical material, waveguide and total dispersion values for conventional single-mode fibre. Note that waveguide dispersion is negative over the range shown. Hence, material and waveguide dispersion are of opposite sign beyond ~1270 nm, and the total dispersion is zero around 1300 nm. The point of zero total dispersion can, however, be shifted to longer wavelengths by modifying the waveguide dispersion through the use of special refractive index profiles for the fibre core and cladding [3] to give *dispersion-shifted fibres*. In this way, the point of minimum dispersion can be made to coincide with the minimum loss region around 1550 nm. Alternatively, special profiles can also be used to flatten the dispersion minimum over a wider range to give *dispersion-flattened fibres*.

A problem can arise in WDM communication systems using near-zero dispersion fibre. In WDM systems, a number of closely spaced wavelength channels (grid interval 100 GHz or ~0.8 nm) are sent along the same fibre to multiply its information-carrying capacity. With zero dispersion fibre, nearby wavelength channels have similar group velocities and the consequent phase-matching between them enhances cross-phase modulation and four-wave mixing effects [23,24], giving rise to severe cross-talk between channels. Rather than using zero-dispersion fibre, an alternative solution is to employ *dispersion compensation* whereby the (low) positive dispersion of the transmission fibre is compensated at appropriate intervals by shorter lengths of *dispersion-compensating fibre* which has a large negative dispersion factor. In this way, the total net dispersion is zero but there is always dispersion present at any point to inhibit phase-matching between channels.

28.5.2.3 Polarization-mode Dispersion

For high bit-rate long-haul systems using fibres operating around the zero-dispersion point or where compensation techniques are employed to minimize chromatic dispersion, *polarization mode dispersion* (PMD) becomes the limiting factor on the maximum bit rate from dispersion. The origin of this dispersion, which is typically of the order of 0.1–1.0 ps km$^{-1/2}$, can be explained as follows [3,24].

As we noted in Section 28.4.2, although a perfectly circular fibre has zero birefringence, in practice factors such as geometric irregularities, stress variations and other perturbations arising during fibre manufacture result in a small birefringence. This is further exacerbated by twists, bends or pinching of the fibre during operation. As a result of the birefringence, the fibre has two principal polarization states and an input pulse is split into the two states which propagate with slightly differing group velocities, giving rise to dispersion. In fact, only a short, straight and undisturbed length of fibre can be described in this way with uniform birefringence and fixed principal axes of polarization. In practice, the fibre is more accurately modelled as a chain of birefringent segments 'spliced' at random angles with respect to their principal axes. At the 'splices', mode-coupling occurs when the output polarization modes of one segment are decomposed into the polarization states of the next. Environmental factors (such as temperature or wind effects for aerial cables) cause fluctuation both in the birefringence and in the degree of mode coupling. Hence, unlike chromatic dispersion which is relatively stable, the polarization mode delay varies randomly in time and a time average value is used to characterize the PMD (which accounts for the square-root dependence on the fibre length). The PMD also varies with wavelength, which means that different channels in a WDM system experience different amounts of pulse spreading.

In describing the PMD at a particular wavelength and time, a pair of *principal states of polarization* (PSP) can be defined at the input and output which correspond to the fast and slow modes of propagation of the fibre link. The polarization transformation between input and output states is independent of wavelength over a small bandwidth (average size referred to as the *PSP bandwidth*). The dispersion performance of a channel is then dependent on the relative intensities launched into these fast and slow modes and is worst when both are equally excited. PMD may be mitigated by electronic equalization or by optical methods for bit rates greater than 40 GBit/s [24]. With the optical method the output signal is split into its fast and slow components and compensation for PMD is applied by delaying the fast component. Since the PMD has a time-varying nature, the delay required for the fast component must be estimated in real time from the properties of the fibre optic link.

28.5.3 Non-linear Effects in Optical Fibres and Solitons

Light pulses propagating through optical fibres also experience several non-linear effects [24] which can impair performance, especially in WDM communication systems. The most important are *self-phase modulation* (SPM), *cross-phase modulation* (XPM), *four-wave mixing* (FWM) and *stimulated Raman scattering* (SRS). Briefly, SPM converts power changes in a propagating wave to phase changes in the same wave; XPM converts power fluctuation in one channel of a WDM system to phase fluctuations in the neighbouring channels; FWM generates new frequency components in WDM systems which may coincide with existing channels and cause severe cross-talk; and SRS results in transfer of power from shorter to longer wavelengths through interaction with vibrational modes of the silica molecules.

Consider SPM, which arises from the Kerr non-linearity where the refractive index has a (weak) dependence on the optical intensity, I, according to the relation

$$n = n_0 + n_{nl}I \qquad (28.25)$$

where n_{nl} is the non-linear index coefficient, $n_{nl} \simeq 3 \times 10^{-16}$ cm^2 W^{-1}.

Following from equation (28.25), the increasing intensity on the rising edge of a pulse will induce a positive dn/dt, whereas the falling edge will cause a negative dn/dt. This time variation in index produces a frequency change across the pulse (frequency chirp) with the leading edge reduced in frequency and the trailing edge increased.

If the pulse is travelling in a fibre where the dispersion parameter, as defined earlier, is negative (i.e. the normal dispersion regime where longer wavelengths have a greater group velocity), then the group velocity will be increased for the leading edge and reduced for the trailing edge. The effect is to redistribute the pulse power from the centre to the sides, broadening the pulse and impairing the system performance. However, if the pulse is travelling in a fibre where the dispersion parameter is positive (anomalous dispersion regime), the leading edge will be reduced in group velocity and the trailing edge increased. The result is that the pulse will be compressed, which may compensate for the normal dispersion broadening.

The latter case is the situation which leads to the existence of *temporal* soliton pulses [10,24], where compression from SPM balances dispersion broadening and the pulse shape is retained without broadening over an indefinitely long transmission path. In fact, propagation of temporal solitons along an optical fibre occurs in two forms—the fundamental soliton maintains its original shape indefinitely whereas periodic solitons change shape but return periodically to their original shape, with a typical period of 100 km. Soliton pulses are typically of 10–50 ps in duration with a few milliwatts of peak power, but only became viable with the advent of the optical fibre amplifier to maintain the required power levels. One of the key problems in the practical application of solitons for very high data-rate systems has been controlling timing jitter which arises from random fluctuation in the soliton's central frequency (from the Gordon–Haus [25] effect or from soliton collisions in WDM systems) which is transformed into timing jitter through dispersion.

28.6 Mode-coupling

Coupling of power between the modes of two optical waveguides is a very important phenomenon and is widely used to make integrated- and fibre-optic components such as power dividers, wavelength selective couplers and modulator devices.

Consider the situation shown in Figure 28.17 where two single-mode waveguides (planar, 2D or fibre guides), with mode propagation constants of β_1 and β_2 (in isolation), are placed side by side. If β_1 and β_2 are equal or closely matched and the separation, s, between the waveguides is reduced so that the evanescent field of one guide is able to penetrate the other, power is transferred periodically between the guides over their interaction length. The minimum length required for the transfer of maximum power from one guide to the other is called the *coupling length, L*.

The reason for this behaviour can be understood by examining the properties of the structure as a single, composite (five-layer) waveguide. If Maxwell's wave equations are applied to the whole structure and appropriate field-matching conditions are applied at each of the boundaries, then the allowed modes are described by a characteristic (or eigenvalue) equation for a five-layer system. For the case considered, we would find that the structure supports two modes, referred to as symmetric and anti-symmetric *super-modes*, illustrated in Figure 28.18. The propagation constants of these super-modes are: $\beta_s = \beta_{av} + \kappa'$ (slow mode) and $\beta_s = \beta_{av} - \kappa'$ (fast mode), respectively, where $\beta_{av} = (\beta_1 + \beta_2)/2$ and κ' is a constant to be defined later.

Suppose now that at some point $z = 0$, power is launched into guide 1 with no power in guide 2. Since the only way that light can be guided by the structure is in super-mode(s), the input field requires both super-modes to be excited with phases such that their field summation corresponds to power in guide 1 and zero power in guide 2, as illustrated in Figure 28.19. As the super-modes propagate along the z-axis, their phase difference, $(\beta_s z - \beta_a z)$, increases with z due to their different propagation constants. After a distance L where $(\beta_s - \beta_a) L = \pi$ and

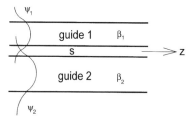

FIGURE 28.17 Waveguide coupling through the evanescent field of the guided modes.

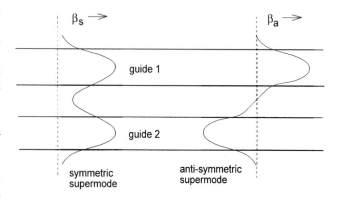

FIGURE 28.18 Symmetric and anti-symmetric super-modes of a five-layer guiding structure.

Optical Waveguide Theory

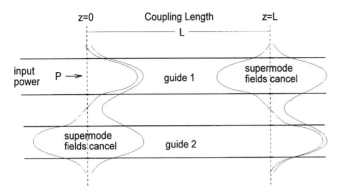

FIGURE 28.19 Periodic power transfer between two guides from the addition of the super-mode fields.

the phase difference has reached a value of π, the summation of the fields gives maximum power in guide 2 and minimum in guide 1. Using the values given for the propagation constants, β_s and β_a, the coupling length is, therefore

$$L = \frac{\pi}{2\kappa'}. \quad (28.26)$$

Note that, after another coupling length, the phase difference becomes 2π and so maximum power is back in guide 1. The power is thus cyclically transferred between the guides over their interaction length.

A very useful approximate technique for the theoretical analysis of mode coupling is based on the concept of *weak coupling* which leads to the *coupled mode equations* [10,26–28]. Returning to Figure 28.17, if the coupling between the guides is weak, the (transverse) field distribution of each guide in isolation will only be slightly perturbed by the presence of the other guide. Hence, the field distribution, ψ, of the whole structure can be approximated by

$$\psi(x,y,z) \approx A_1(z)\psi_1(x,y)\exp(-i\beta_1 z) + A_2(z)\psi_2(x,y)\exp(-i\beta_2 z) \quad (28.27)$$

where ψ_1 and ψ_2 are the normalized transverse field distributions of guides 1 and 2 in isolation, and $A_1(z)$ and $A_2(z)$ indicate that the field amplitudes vary with distance z along the guides due to the coupling.

Under these assumptions, ψ, ψ_1 and ψ_2 must all satisfy the wave equation since they all represent solutions of a guiding structure. When these functions are substituted into the wave equation (with some approximations which are valid for weak coupling), the following two relationships are obtained [26]:

$$\frac{dA_1}{dz} = -i\kappa_{12} \cdot A_2 \exp(-i\Delta\beta z)$$
$$\frac{dA_2}{dz} = -i\kappa_{21} \cdot A_1 \exp(+i\Delta\beta z) \quad (28.28)$$

where $\Delta\beta = (\beta_2 - \beta_1)$ and $\kappa = \sqrt{\kappa_{12}\kappa_{21}}$ is the *coupling coefficient* which depends on the overlap between the fields ψ_1, ψ_2 of the two guides.

Equations (28.28) are known as the *coupled-mode equations* and show that variations in the amplitude in one guide are linked to the amplitude in the other guide through the coupling coefficient. Note that if $\kappa = 0$ (i.e. no interaction between the guides), then the amplitudes in each guide, A_1 and A_2, remain constant along the z-direction, as expected.

For the particular case illustrated in Figure 28.19 where at $z = 0$, power, P, is launched into guide 1 with no power in guide 2, the solution of the coupled mode equations gives the power in each guide as a function of z:

$$P_1(z) = P\left[1 - \frac{1}{1+\delta}\sin^2\kappa'z\right]$$
$$P_2(z) = P\left[\frac{1}{1+\delta}\sin^2\kappa'z\right] \quad (28.29)$$

where $\kappa' = \kappa\sqrt{(1+\delta)}$ and $\delta = \left(\frac{\Delta\beta}{2\kappa}\right)^2$.

Equations (28.29) reveal the dependence of the coupling length and the maximum power transferred to guide 2 on the degree of mismatch, δ, between the guides. Note that the power fraction transferred is $1/(1+\delta)$ at a coupling length of $\pi/2\kappa'$. For efficient power transfer in a directional coupler, δ must be small, i.e. $\Delta\beta \ll 2\kappa$. Figure 28.20 shows examples of two cases: (a) phase-matched case where $\beta_1 = \beta_2$ and $\delta = 0$; and (b) mismatched case where $\beta_1 \neq \beta_2$.

The coupling coefficient, κ, is determined from the overlap of the fields of the individual guides [10,27] and may be calculated either analytically or numerically if the field distributions of the individual guides are known [29]. For example, using the field distributions given in equation (6.9), the coupling coefficient for the TE$_0$ mode of the two planar guides illustrated in Figure 28.21 is:

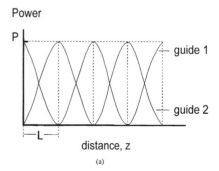

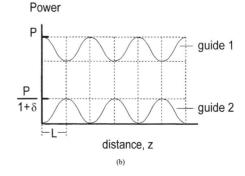

FIGURE 28.20 Power in each guide for a directional coupler: (a) phase-matched case and (b) mismatched case.

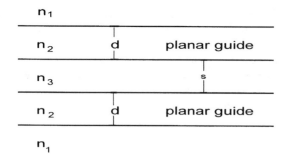

FIGURE 28.21 Coupling between two single-mode planar waveguides.

$$\kappa = \frac{2\gamma_3 \exp(-\gamma_3 s)}{\beta_0 d_e \left[1+\left(\frac{\gamma_3}{k_x}\right)^2\right]} \quad (28.30)$$

where

$$\gamma_3 = k_0\sqrt{n_e^2 - n_3^2} \quad k_x = k_0\sqrt{n_2^2 - n_e^2} \quad \text{and} \quad d_e = d + \frac{1}{\gamma_1} + \frac{1}{\gamma_3}.$$

These principles underpin the operation of a variety of integrated and fibre optic devices based on directional coupling. For example, by designing the guides with an appropriate interaction length in relation to the coupling length, fibre or integrated optic couplers with different power-splitting ratios may be manufactured. Also, since the coupling is stronger at longer wavelengths, as seen by the exp $(-\gamma_3 s)$ term in equation (28.30), wavelength splitters can be designed by choosing an interaction length which corresponds to the coupling length (or a multiple of it) for λ_1 but not for λ_2, so that maximum cross-coupling occurs at λ_1 and minimum at λ_2. Active devices are made by using directional couplers in electro-optic materials such as lithium niobate. The index change from the applied voltage is used to match or mismatch the propagation constants of the two guides, thus forming a modulator or switch.

This discussion has dealt with coupling between modes which are matched or nearly matched in propagation constants. It is also possible to couple modes which have substantially different propagation constants if periodic coupling is introduced [30] by, for example, a diffraction grating. This can be seen from equation (28.28) by introducing a periodic coupling coefficient of the form: $\kappa_{21}\kappa_{12}^* = \kappa \exp(-ik_c z)$. Combining this with the exponential terms in equation (28.28) gives the terms: $\exp \pm i(\Delta\beta - k_c)$ so that 'matching' now occurs when $(\Delta\beta - k_c) = 0$ where $k_c = (2\pi/\Lambda)$ and Λ is the periodicity of the perturbation. This phenomenon may cause unwanted coupling between the modes within a waveguide or may be exploited in certain types of fibre optic sensors and narrow-linewidth fibre Bragg gratings for use in optical transmission networks [31,32].

28.7 Conclusion

This chapter has outlined the fundamental concepts and principles involved in guiding light in various types of structures. Knowledge of these principles and the various parameters that have been defined is essential in the analysis and design of optical waveguides and waveguide-based components. Further detailed information on the theory of dielectric waveguides, integrated optic components and their applications in fibre optic systems may be found in the list of further reading.

REFERENCES

1. Stewart G and Culshaw B 1994 Optical waveguide modelling and design for evanescent field chemical sensors *Opt. Quant. Elect.* (invited paper in special issue on waveguide modelling), **26** 249–59.
2. Stewart G, Muhammad F A and Culshaw B 1993 Sensitivity improvement for evanescent wave gas sensors *Sens. Actuat.* **B11** 521–4.
3. Keiser G 2014 *Optical Fiber Communications* 5th ed. (New York: McGraw-Hill).
4. Russell P St J 2006 Photonic-crystal fibers *IEEE J. Lightw. Techn.*, **24** 12, 4729–4749.
5. Lee D L 1986 *Electromagnetic Principles of Integrated Optics* (New York: Wiley) pp 116–35.
6. Stewart G, Millar C A, Laybourn P J R, Wilkinson C D W and De La Rue R M 1977 Planar optical waveguides formed by silver-ion migration in glass *IEEE J. Quant. Electron.* **QE-13** 192–200.
7. Syms R and Cozens J 1992 *Optical Guided Waves and Devices* (London: McGraw-Hill) pp 217–49 and 253–6
8. Marcatili E A J 1969 Dielectric rectangular waveguide and directional coupler for integrated optics *Bell Syst. Tech. J.* **48** 2071–102.
9. Hocker G B and Burns W K 1977 Mode dispersion in diffused channel waveguides by the effective index method *Appl. Opt.* **16** 113–18.
10. Okamoto K 2006 *Fundamentals of Optical Waveguides* 2nd ed. (New York: Academic Press).
11. Sun M J and Muller M W 1977 Measurements on four-layer isotropic waveguides *Appl. Opt.* **16** 814–5.
12. Chiang K S 1994 Review of numerical and approximate methods for the modal analysis of general optical dielectric waveguides (invited paper) *Opt & Quant Elect.* **26** 113–34.
13. Scarmozzino R Gopinath A Pregla R and Helfert S 2000 Numerical techniques for modeling guided-wave photonic devices *IEEE J. Sel. Top Quant. Elect.* **6** 150–62.
14. Doerr C R and Kogelnik H 2008 Dielectric waveguide theory (invited paper) *IEEE J. Lightwave Technol.* **26** 1176–87.
15. Stern M S 1988 Semivectorial polarised finite difference method for optical waveguides with arbitrary index profiles *IEEE Proc. J.* **135** 56–63.
16. Ghatak A K Thyagarajan K and Shenoy M R 1987 Numerical analysis of planar optical waveguides using matrix approach *IEEE J. Lightwave Technol.* **5** 660–7.
17. Shenoy M R, Thyagarajan K and Ghatak A K 1988 Numerical analysis of optical fibers using matrix approach, *IEEE J. Lightwave Technol.* **6** 1285–91.
18. Snyder A W 1969 Asymptotic expressions for eigenfunctions and eigenvalues of dielectric optical waveguides *IEEE Trans. Microwave Theory Technol.* **MTT-17** 1130–38.
19. Gloge D 1971 Weakly guiding fibers *Appl. Opt.* **10** 2252–8.
20. Kaminow I P 1981 Polarization in optical fibers *IEEE J. Quant. Elect.* **QE-17** 15–22.

21. Senior J M 2009 *Optical Fiber Communications Principles and Practice* 3rd ed. (Pearson).
22. Olshansky R and Keck D B 1976 Pulse broadening in graded index optical fibres *Appl. Opt.* **15** 483–91.
23. Willner A E 1997 Mining the optical bandwidth for a terabit per second *IEEE Spect.* **34** 32–41.
24. Ramaswami R and Sivarajan K N 2002 *Optical Networks: A Practical Perspective* (San Francisco, CA: Morgan Kaufmann Publishers, Academic Press).
25. Gordon J P and Haus H A 1986 Random walk of coherently amplified solitons in optical fiber transmission *Opt. Lett.* **11** 665–7.
26. Yariv A 1985 *Optical Electronics* 3rd ed (New York: Holt-Saunders) pp 413–49.
27. Ghatak A K and Thyagarajan K 1989 *Optical Electronics* (Cambridge: Cambridge University Press) pp 447–54, 609–12.
28. Marcuse D 1971 The coupling of degenerate modes in two parallel dielectric waveguides *Bell Syst. Tech. J.* **50** 1791–816.
29. Marcuse D 1987 Directional couplers made of non-identical asymmetric slabs *IEEE J. Lightwave Technol.* **5** 113–18.
30. Miller S E 1969 Some theory and applications of periodically coupled waves *Bell Syst. Tech. J.* **48** 2189–219.
31. Othonos A and Kalli K 1999 *Fibre Bragg Gratings: Fundamentals and Applications in Telecommunications and Sensing* (London: Artech House).
32. Kashyap R 1999 *Fibre Bragg Gratings* (San Diego, CA: Academic).

FURTHER READING

Doerr C R and Kogelnik H 2008 Dielectric waveguide theory *IEEE J. Lightwave Technol.* **26** 1176–1187.
This invited paper gives a broad illustrative overview of the theory of dielectric waveguides with a historical perspective. Topics briefly reviewed include planar, rectangular and cylindrical guides, waveguide bends and couplers, holey fibres, photonic crystal fibres and numerical methods for waveguide analysis.

Hunsperger R G 1995 *Integrated Optics: Theory and Technology* 4th ed. (Berlin: Springer)
Presents the basic theory of waveguides and couplers but concentrates on components and technology and describes in detail the construction and operation of a number of devices including couplers, modulators and lasers.

Keiser G 2014 *Optical Fiber Communications* 5th ed. (New York: McGraw-Hill).
Gives a comprehensive account of all aspects of the design and practice of modern fibre communications systems and networks, including a readable account of fibre theory, signal degradation, non-linear effects and measurements in optical fibres.

Lee D L 1986 *Electromagnetic Principles of Integrated Optics* (New York: Wiley).
As the title suggests, this book presents the electromagnetic theory of planar and rectangular guides and mode coupling, but also includes optical fibres. Examples of integrated optics devices are given as well as the basic fabrication techniques.

Mynbaer D K and Scheiner L L 2001 *Fibre-optic Communications Technology* (Upper Saddle River, NJ: Prentice-Hall).
Provides a useful starter text for learning about fibre optic communications technology. Describes the operation and characteristics of key fibre components with specific, commercial examples. The book is also suitable for technician training in fibre optic systems.

Okamoto K 2006 *Fundamentals of Optical Waveguides* 2nd ed. (New York: Academic Press)
Provides a comprehensive theoretical account of the operation of optical waveguides and fibres.

Pedrola G L 2016 *Beam Propagation Method for Design of Optical Waveguide Devices* (Chichester: John Wiley & Sons
Gives a comprehensive account of the beam propagation method for numerical analysis of optical waveguides including vectorial and 3-D beam propagation techniques. The analysis of a number of integrated photonic devices is presented including couplers, interference devices, arrayed waveguide gratings and waveguide lasers.

Ramaswami R and Sivarajan K N 2002 *Optical Networks: A Practical Perspective* (San Francisco, CA: Morgan Kaufmann Publishers, Academic Press).
As the title suggests, this book give a comprehensive physical description of all the issues involved in the propagation of signals in optical fibres for high data-rate systems including the properties of the components involved and the engineering design of optical fibre networks.

Reider G A 2016 *Photonics* (Springer International Publishing Switzerland)
Chapter 5 of this book on dielectric waveguides gives a theoretical description of the operation of planar guides and optical fibres along with an overview of a number of integrated optical components based on dielectric waveguides, including couplers, switches, grating devices and modulators.

Senior J M 2009 *Optical Fiber Communications Principles and Practice* 3rd Edition (Pearson)
Similar in content and level to Keiser.

Snyder A W and Love J D 1983 *Optical Waveguide Theory* (London: Chapman and Hall).
A well-known and classic textbook in the area. Part I of the book gives a full treatment of ray optics in multi-mode fibres which is not so relevant for modern optical communications systems, but Parts II and III provide a comprehensive account of the electromagnetic analysis of optical waveguides, including analysis of bends and perturbations in waveguides and mode-coupling between guides.

Syms R and Cozens J 1992 *Optical Guided Waves and Devices* (London: McGraw-Hill)
Provides a fairly descriptive account of optoelectronic devices including both fibre- and integrated-optic components

as well as semiconductor devices with illustrations and applications. The background theory for understanding waveguide and component operation is also presented.

Zappe H P 1995 *Introduction to Semiconductor Integrated Optics* (Norwood, MA: Artech House).

Gives a comprehensive account of semiconductor properties and fabrication technology as applied to integrated optics and describes the construction and operation of a number of important optical components in semiconductor materials including waveguides, modulators, lasers and detectors.

29
Fibre Optic Beam Delivery

D. P. Hand

CONTENTS

29.1 Fibre Operation 435
    29.1.1 Total Internal Reflectance Fibres 437
    29.1.2 Hollow Waveguide Fibres...438
29.2 Fabrication..438
    29.2.1 Fused Silica Optical Fibres...438
    29.2.2 Chalcogenide, Fluoride and Germanate Glasses 439
    29.2.3 Crystalline Optical Fibres 439
    29.2.4 Hollow Waveguide Optical Fibres..439
    29.2.5 Micro-structured Optical Fibres..440
29.3 Implementation 440
29.4 Beam Division and Combination 440
29.5 Limitations..441
    29.5.1 Beam Quality and Profile 441
    29.5.2 Thermal Damage...441
    29.5.3 Pulsed Laser Damage 442
    29.5.4 Non-linear Effects 442
    29.5.5 Mechanical Damage 442
29.6 Summary and Future Directions...442
References..443

Fibre optics are a particularly attractive means of high-power laser beam delivery, offering the highest degree of flexibility and, hence, full directional control of light, coupled with transmission over long distances if required (hundreds of metres). If integrated into suitable robots and positioning systems, they provide multi-axis processing (important for complex workpieces) and remote working capabilities and allow one laser to be easily time-shared between many different workstations.

The most widely used type of optical fibre is made from fused silica. The telecommunications industry has driven the development of these fibres, which have since been adapted for use in laser materials processing. These fibres have excellent transmission throughout the visible and near-infrared (NIR) part of the spectrum, and are ideally suited to the 1.06 µm wavelength of Nd:YAG lasers, with losses as low as 1–2 dB km^{-1}. Unfortunately, they are not suitable for the 10.6 µm wavelength produced by CO_2 lasers where they are opaque. Alternative fibres do exist for such longer wavelengths but these have more severe limitations in terms of power handling, attenuation, beam profile preservation and available lengths. However, they are used in a limited range of power delivery applications, in particular, for laser surgery.

One example process where fibre optic delivery is often used is laser welding in the automotive industry, e.g. for welding car body parts. The 3D shapes being welded require a highly manoeuvrable laser beam, and so a fibre is ideal. Typically, in such an application, 2–4 kW of cw laser power is required, focused down to a spot of 0.3–0.5 mm diameter. A suitable beam from a Nd:YAG laser can be readily delivered through a 600 µm core diameter fused silica optical fibre. Indeed, many high-power Nd:YAG welding lasers are *only* available with fibre optic delivery. Nd:YAG laser drilling, meanwhile, requires high peak power pulses *and* a high beam quality (necessitating a smaller diameter optical fibre, 200 µm) which means that optical fibre beam delivery cannot normally be used due to optical damage problems.

29.1 Fibre Operation

An optical fibre is an example of an optical waveguide. Most optical waveguides work on the principle of total internal reflection (TIR), where the light is guided within a high-index core region surrounded by a lower index cladding (Figure 29.1). Light which is incident on the core/cladding interface at an angle θ greater than the critical angle θ_c is 100% reflected. The angle θ_c is dependent on the refractive index difference between the core and cladding, according to the equation $\sin \theta_c = n_1/n_2$. It is often more useful to express this angle in terms of the fibre's numerical aperture (NA). Light which is incident

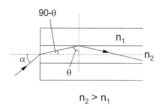

FIGURE 29.1 Step-index fibre optic.

on the front face of the fibre at an angle less than that defined by the fibre's NA will satisfy $\theta > \theta_c$ and so be guided by the fibre. Optical fibres based on this principle are normally made from fused silica, with a suitable dopant used to alter the refractive index of either the core or cladding region. Fused silica is suitable over a wide wavelength spectrum, from UV at 200 nm up to NIR at 2 μm but other glasses have been used for transmission of mid-IR light, for which fused silica is opaque. Plastic optical fibres are also available, as are plastic-clad silica fibres but these cannot be used with large optical powers due to their low damage thresholds.

It is also possible to provide guidance using Fresnel reflection (rather than TIR), which is the technique normally used with hollow fibres for CO_2 laser light. An example of such a fibre is shown in Figure 29.2, consisting of a very fine glass capillary tube with an internal dielectric coating to provide a high reflectivity at the appropriate wavelength. These hollow fibres are described in more detail later, but for an in-depth review, see Harrington [1].

Both TIR and hollow waveguide fibres guide a number of discrete waveguide modes. The number of modes depends on the guide's dimensions and NA. If the core region is made sufficiently small (~7 μm at 1.064 μm for a 0.11 NA fibre), only one mode can propagate and the fibre is said to be single mode.

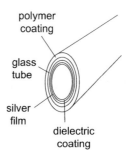

FIGURE 29.2 Hollow glass fibre optic.

Single-mode fibres are normally used in telecommunications applications, where intermodal dispersion would otherwise cause problems. For power delivery, however, multi-mode fibres with much larger core diameters are normally used. This clearly reduces the intensity in the core and, hence, the likelihood of optical damage. In addition, high-power laser beams, particularly from solid state lasers such as Nd:YAG, are normally highly multi-moded due to thermal lensing effects in the laser. It is impossible to couple such a beam efficiently into a single-mode fibre; instead only the lowest order mode will be coupled efficiently. A typical silica fibre used for high-power Nd:YAG beam delivery has a core diameter between 200 and 1000 μm, with an NA of 0.22, and can support many tens to hundreds of thousands of modes.

Finally, there is a new type of optical fibre. This is the microstructured fibre, often called a 'photonic crystal fibre' or PCF. PCF is an optical fibre with an ordered array of air holes running along its length—typical fibre cross sections are shown in Figure 29.3. There are two main categories of PCF: 'index-guiding' PCF in which a solid core is surrounded by a cladding of the same material but including an array of air holes (Figure 29.3a); air-guiding 'photonic bandgap fibres' where a hollow core is

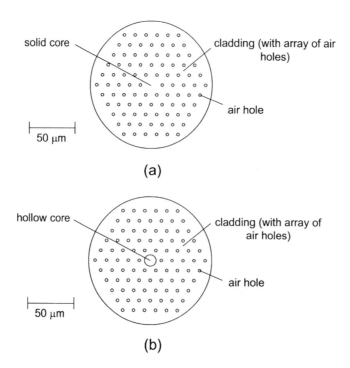

FIGURE 29.3 Photonic crystal fibre (a) index-guiding and (b) air-guiding 'photonic bandgap fibre'.

surrounded by a cladding with an array of holes (Figure 29.3b). With the index-guiding PCF structure, the core index is greater than the average index of the cladding, which allows guiding in much the same way as conventional optical fibres (by a modified form of TIR). Photonic bandgap fibres, meanwhile, use the special optical properties of the *periodic pattern* of the air holes to provide guidance by Bragg reflection.

The performance of these various types of optical fibre is now described and summarized in Table 29.1.

29.1.1 Total Internal Reflectance Fibres

Total internal reflectance (TIR) fibres can be either step-index or graded-index, with refractive index profiles as shown in Figure 29.4. Graded-index fibres were developed for telecommunications, since they provide multi-mode operation with greatly reduced intermodal dispersion. They have only been used to a very limited extent for high-power beam delivery. Step-index fibres, meanwhile, have been specifically developed for power delivery applications. These normally have a pure silica core to maximize the damage threshold and a fluorine-doped cladding to provide the requisite refractive index difference. Telecommunication fibres, by contrast, normally have a germanium-doped core and pure silica cladding. The attenuation spectrum is dependent on the OH content of the fibre, and typically, manufacturers offer both 'low OH' and 'high OH'

fibre. Attenuation spectra are shown in Figure 29.5, courtesy of Polymicro Technologies, Inc. High OH fibres are most suitable for the UV/visible part of the spectrum, whereas low OH fibres are used for the longer wavelength end of the visible and the NIR. In most power delivery applications, it is unusual to use more than about 10 m of fibre. This means that an attenuation of ~0.05 dB m^{-1} is acceptable (10% loss in 10 m).

Other materials have been developed for TIR optical fibres, specifically aimed at the transmission of longer wavelength light. These can be either glasses or crystalline materials. The glasses are suitable for the mid-IR, whereas only crystalline fibres can be used at the important CO_2 wavelength of 10.6 μm.

The main glasses used are fluorides (e.g. fluorozirconate and fluoroaluminate), chalcogenides [2] (sulphide, selenide or telluride) and germanates. Fluoride fibres transmit light from 1 to 4 μm, with attenuation < 1 dB m^{-1}, and so are suitable for delivery of Er:YAG laser light at 2.94 μm, used for medical applications. Germanate fibres, however, offer a much higher damage threshold (20 J pulse from Er:YAG), even though their attenuation is slightly higher. Chalcogenide fibres, meanwhile, may be used to transmit light from 1 to 6 μm, with an attenuation of 0.2–1 dB m^{-1}. Typical sizes available are 200–300 μm core diameter, with an NA of 0.3–0.5. These fibres are useful with CO lasers, which emit a number of lines between 5 and 6 μm. The maximum power which can be transmitted, however, is limited to about 10 W at present.

TABLE 29.1

Details and Performance of the Various Types of Optical Fibre

Fibre Type	Operating Wavelengths (μm)	Power Handling	Typical Core Diameter (μm)	Attenuation (Dependent on Wavelength)	Usable Length	Typical Beam Quality (M2)
Total internal reflectance fibres						
Fused silica (large core multimode)	0.2–2	> 4 kW cw in 600 μm 55 J/pulse in 400 μm (1 ms pulse)	100–1000	<0.1 dB/m from 0.5 to 1.9 μm	Hundreds of metres	20–300
Chalcogenide	1–6		200–300	0.2–1 dB	Hundreds of metres	15–200
Fluoride	1–4	970 mJ in 200 μs	70–600	< 0.1 dB/m from 2 to 3 μm	Hundreds of metres	20–200
Germanate	1–3	20 J/pulse		0.1–1 dB/m	Hundreds of metres	
Silver halide	4–16	100 W cw (1000 μm ∅)	500–1000	0.5–2 dB	15 m	20–200
Sapphire (solid)	0.5–3.5	1 J/pulse in 400 μm	150–400	0.25–2 dB/m	3 m	15–250
Sapphire (hollow)	10–16	1.9 kW (water cooled)	250–1000	1–10 dB/m (dependent on ∅)		
Hollow waveguide fibres	Internal coating optimized for wavelength					
Metal tube	1–25	2.7 kW cw	700–1000	0.1–10 dB/m (dependent on and ∅ bend radii)	Few metres	
Plastic tube	1–25	65 W cw (1800 μm ∅)	800–1800	0.1–2 dB/m		
Glass tube	1–25	1 kW cw (water cooled) 75 mJ in 7 ns	250–1000	0.1–0.3 dB	13 m	
MICROSTRUCTURED OPTICAL FIBRES						
Fused silica	0.2–2		5–150	Down to 0.05 dB/m	Hundreds of metres	

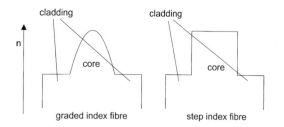

FIGURE 29.4 Refractive index profiles of graded and step index fibre optics.

Two crystalline materials are commonly used for optical fibres: silver halide (polycrystalline) and sapphire (single crystal). Silver halide fibres can be used to transmit light in the mid-to-far IR, with an attenuation at 10.6 μm as low as 0.5 dB m^{-1} [3]. These fibres do not normally have a cladding region; instead, the polycrystalline silver halide core is simply surrounded by air, and hence, they have a high NA of 0.7. The fibre must be held within an opaque tube, in order to prevent UV light from reaching the photosensitive core which causes premature failure. Diameters of 200–1000 μm are typical and the damage threshold is about 100 W cw with the 1000 μm fibre. The fibres are mechanically quite weak. Single-crystal sapphire fibres, meanwhile, can be used for wavelengths of up to 3.5 μm and are mechanically very strong but are limited to lengths of 3 m. Sapphire can also be used to guide 10.6 μm light; although in this case, it must be fabricated into a hollow tube. Anomalous dispersion means that sapphire has a refractive index of less than one between 10 and 16.7 μm [4], and so light is guided in the air core by TIR. These fibres have significant loss (about 10 dB m^{-1} for a 200 μm core fibre) but this varies with 1/R, with the attenuation for 1000 μm fibre ~1 dB. Hollow sapphire fibres have been used with CO_2 lasers for surgical applications and, with a water-cooled jacket, have delivered 1.9 kW of laser power [5].

29.1.2 Hollow Waveguide Fibres

Hollow waveguide fibres are made by depositing a high-reflectivity layer inside a glass [6], metal [7] or plastic [8] tube. For example, hollow glass fibre lengths of up 13 m have been manufactured with bore sizes varying from 250 to 1300 μm. One interesting feature of hollow fibres is the differential modal attenuation, where higher order modes are strongly attenuated. This means that the output will consist of only a few low-order modes and, hence, has a high beam quality *but* a large speckle pattern due to intermodal interference, which changes as the fibre is bent [9]. This can be a significant problem for many laser materials processing applications, where a more uniform spot is required. Matsuura et al. [10] have developed a silver halide beam homogenizer at the output end of the fibre in order to address this, but this, of course, will reduce the beam quality. The attenuation of higher order modes in hollow waveguide fibres means that bend loss becomes significant, since bends cause coupling to higher order modes [9]. Indeed, bend loss has been shown to vary with the inverse of the bend radius [1]. The loss also depends on the bore diameter, with a loss of 2 dB m^{-1} for a 250 μm diameter bore dropping to 0.25 dB m^{-1} for a 530 μm bore for 10.6 μm light in a hollow glass waveguide [11]. Designs based on a glass tube generally have the best performance, due to the superior interior surface quality in comparison with metal or plastic tubes.

The significant attenuation of hollow waveguide fibres means that water cooling is essential for very high power transmission. The largest CO_2 laser power reported to have been delivered through a water-cooled hollow glass waveguide with a bore of 700 μm is 1040 W [5].

29.2 Fabrication

29.2.1 Fused Silica Optical Fibres

Fused silica optical fibres are made by first manufacturing a very high purity glass rod or tube called the *preform* (typically 10–25 mm in diameter and 60–120 cm long) which is then pulled into a very thin filament in a drawing furnace. There are two main types of preform manufacture process used: one where the glass is deposited in layers and in the other the glass for the core (rod) and cladding (tube) are made separately and subsequently combined. The layered technique uses vapour phase deposition and allows complex radial refractive index profiles to be constructed and is the process normally used for small core fibres, such as the single-mode fibre used in telecommunications. The rod-in-tube technique, meanwhile, is quicker (radial heating is applied to the tube in order to collapse it onto the rod) and is particularly suitable for the large-core step-index fibres used for power delivery (200–1000 μm core diameter). However, it is more likely to create inherent problems such as core/cladding interface irregularities.

Vapour phase deposition involves an oxidation process where highly pure vapours of $SiCl_4$ react with oxygen to form silica (SiO_2) particles. Other vapours, such as $GeCl_4$, are included

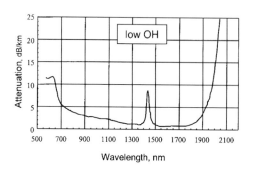
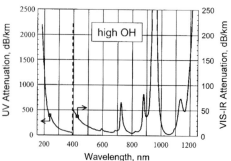

FIGURE 29.5 Attenuation spectra of low OH and high OH fibre optics. Figures courtesy of Polymicro Technologies, Inc.

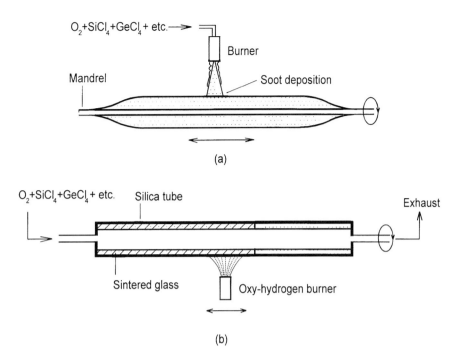

FIGURE 29.6 Manufacture of a fibre preform by vapour phase deposition. (a) Outside vapour phase oxidation (OVPO). (b) MCVD.

in the gas mixture to adjust the refractive index of the deposited layers through the Ge content. The deposition process can be performed in two basic ways, namely, external or internal. In the external lateral deposition process [12] as illustrated in Figure 29.6a, the glass particles (or soot) are formed on the outside of a rotating graphite or ceramic mandrel which is also translated in the lateral direction. The desired index profile is built up layer by layer. After deposition, the mandrel is removed and the preform is vitrified at high temperature to form a clear glass preform. (The small central hole from the mandrel disappears during the fibre drawing process). An alternative external technique is axial deposition [13], or VAD (vapour phase axial deposition), where both core and cladding layers are deposited simultaneously on the end of a fused silica seed rod using two oxy-hydrogen burners. Because of axial deposition, the preform can be made in continuous lengths. An internal deposition method, the modified chemical vapour deposition (MCVD) process [14], is illustrated in Figure 29.6b. Here, the constituent vapours, along with oxygen, flow through the inside of a rotating silica tube, and as the particles are deposited, they are sintered to glass by the oxy-hydrogen flame traversing back and forth along the tube. In plasma-activated CVD [15], the oxy-hydrogen flame is replaced with a moving microwave resonator which generates plasma within the tube (held at 1100°C) and activates the chemical reactions. The process produces a clear glass preform directly without the need for sintering.

29.2.2 Chalcogenide, Fluoride and Germanate Glasses

These fibres are normally fabricated using rod-in-tube techniques similar to that previously described for large-core fused silica fibres; although in some cases, extrusion techniques are used [16].

29.2.3 Crystalline Optical Fibres

There are both polycrystalline and single-crystal fibres which are made in different ways. Silver halide fibres are polycrystalline and are made into fibres using a hot extrusion process, where a single-crystal billet or preform is placed in a heated chamber and the fibre extruded through a diamond or tungsten die at a temperature of about half the melting point, resulting in a polycrystalline structure with a grain size of about 10 μm. Sapphire fibres, meanwhile, are single crystal and are made by melting some or all of the starting sapphire material and slowly drawing a fibre, either from a melt through a capillary tube or directly from a sapphire, using a CO_2 laser for local melting.

29.2.4 Hollow Waveguide Optical Fibres

A number of different techniques have been used in the fabrication of hollow waveguide fibres. The smoothness of the internal surface is particularly important, with a very smooth surface essential to prevent coupling to higher order modes and, hence, high losses.

a. *Metal tube waveguides.* In 1983, Miyagi et al. [17] developed a technique where a dielectric layer and then a metallic film are deposited on the outside of an aluminium tube. The metal film is subsequently electroplated with nickel, before etching away the aluminium tube. A different technique was developed by Bhardwaj et al. [7] in 1993 which used a silver tube as a starting point. This tube is first internally etched to make it smooth, before depositing an AgBr film on the inside using wet chemistry.

b. *Plastic tube waveguides.* Croitoru et al. [18] have developed a technique where a silver film is deposited on the inside of Teflon and polyethylene tubing, which is then over-coated with AgI using wet chemistry. Haan and Harrington [19] have used similar techniques to deposit Ag/AgI films inside polycarbonate tubing.

c. *Hollow glass waveguides.* Hollow glass waveguide fibres are normally based on fused silica. A fused silica tube is heated and pulled down to a small diameter in the same way as a TIR optical fibre and a polyimide coating is deposited on the outside for mechanical protection. A conventional electroless plating technique is then used to deposit a silver film internally. A dielectric layer is then deposited on top of the silver. One technique is to flow iodine through the tube, which reacts with the silver surface and, hence, forms a thin uniform dielectric layer of AgI [20]. An alternative is to deposit a suitable polymer layer inside the tube, by flowing a polymer solution through and subsequently heating [21].

29.2.5 Micro-structured Optical Fibres

Micro-structured optical fibres are made by stacking fused silica capillary tubes in order to give the required cross section. In the 'index-guiding' case, the core is defined by substituting one of the tubes with a solid rod, whereas in the 'photonic bandgap' (hollow core) case, a tube with a larger bore diameter is used. These stacks are then heated and fused together to form a preform, from which the optical fibre is pulled.

29.3 Implementation

Focusing optics are necessary to couple the light from the laser into the fibre. These are chosen to give a focal spot diameter only slightly smaller than that of the fibre core. However, with some lasers, the beam quality can change with operating parameters, which can lead to an increased focal spot size, so this must be taken into account. One way to avoid this is to use an imaging system where the cone angle rather than spot size changes with beam quality. This can be achieved, for example, by imaging an aperture within the laser resonator.

With solid TIR fibres, the end-faces must be very flat and clean to avoid damage on coupling the light in and so are usually prepared by careful polishing to give a scratch-free surface. With small diameter fibres (outer diameter <220 μm), however, high-quality end-faces can be produced simply by cleaving with a special diamond tool—this technique is much quicker and easier and is the standard one used with telecommunications.

At the output end of the fibre, the light is imaged onto the workpiece with typical magnification ratios between 1:1 and 1:2. A standard arrangement is to use a pair of lenses, both operating at infinity—see Figure 29.7. The fibre end-face is placed at the focal point of the first (collimating) lens, giving an image at the focal point of the second (focusing) lens with a magnification that is simply dependent on the ratio of the two focal lengths. The lenses are normally held in a rigid assembly

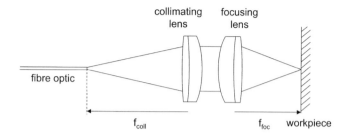

FIGURE 29.7 Imaging optics used at the output end of a fibre optic delivery system.

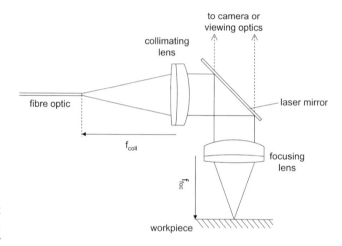

FIGURE 29.8 Combined high-power laser imaging and viewing optics at the output end of a fibre optic delivery system.

(output housing), and the fibre is held in a connector which plugs into the back of this. For many machining operations, high-pressure gases are used to assist with material removal and these are introduced through a nozzle coaxial with the laser beam, which is also integrated into the output housing.

For alignment and inspection purposes it is often beneficial to image the workpiece co-axially with the laser beam. This can be achieved by placing a laser turning mirror or beam splitter between the two lenses (Figure 29.8). For this reason, although the laser light has a very narrow bandwidth, achromats are often used in the output housing, thus allowing white light imaging to be used. In addition, they are well corrected for spherical aberration and are available at reasonable cost.

The main problem which can arise at the fibre output end is one of back-reflection of laser light from the workpiece. This is re-focused by the optics onto the fibre end and can cause the fibre and mounting to heat up and even damage when using high-average powers. To avoid this, the fibre output housing that contains these optics is often placed at an angle of perhaps 10° to the workpiece normal.

29.4 Beam Division and Combination

Fibre optic beam splitters are often used with single-mode optical fibres, in order to route light down more than one path and for (re-)combination of light from separate paths. These are normally fused tapered couplers, where a tapered section is used to expand the optical field out of the core, in order that coupling can occur between the two fibres [22]. It is possible to make multi-mode versions but they have high loss and so are easily damaged with high powers. An alternative design is to use two polished 'half-couplers' in close contact, but again, losses are likely to be too high for high-power applications. If a high-power laser is required to be coupled into a number of fibres, it is, therefore, normal to split it into separate beams using bulk optics before coupling into the optical fibres. Three techniques are commonly used: (i) high reflectivity mirrors partially inserted into the beam; (ii) partial reflectors; or (iii) diffractive optical elements. In addition to splitting light into several fibres, there are some very high-power applications where two or three lasers have been combined into a single fibre. Such a device has been developed [23] to combine light from three 4 kW Nd:YAG lasers.

29.5 Limitations

Despite the advantages of fibre optic beam delivery described here (flexibility, multiplexing), the implementation of fibre optics for delivery is often limited by optical damage and degradation of beam quality and/or beam profile. Changes in beam quality or profile are strongly dependent on the type of fibre (TIR or hollow waveguide), its dimensions and the coupling optics used at the input end. Optical damage, meanwhile, is dependent both on the peak power and the temporal profile of the laser; for long pulses (milliseconds up to cw), the main damage problems are thermal, associated with the high average laser power, whereas with short pulses (nanoseconds), damage occurs at much lower average powers and is limited by non-linear processes in the fibre such as self-focusing.

29.5.1 Beam Quality and Profile

It is important to consider how the properties of the beam are preserved through the fibre. As indicated in the introduction, there is really no problem with silica fibres for low beam quality applications such as welding, where the loss of beam quality on transmission through the fibre is very small. Unfortunately, for applications requiring higher beam quality, such as cutting, drilling and marking, degradation of both beam quality and profile can be a severe problem. Beam quality degradation arises partly because such beams do not match well to the mode structure of a step-index fibre (whereas the 'top-hat' profile typical of low beam quality is quite suitable) and also because mode coupling occurs as the light propagates through the fibre.

A rough measure of beam quality is the M^2 parameter, which is proportional to the product of beam waist and divergence angle. The beam waist at the fibre output will always be equal to the fibre radius, irrespective of the spot size at the input of the fibre. Similarly, even if a low NA beam is launched into the fibre, the process of mode coupling (e.g. when the fibre is bent) will cause the NA of the guided beam to increase, giving an increased divergence at the fibre exit [24]. Thus, the quality of the output beam is always less than that of the input. Consequently, high-quality beams must be guided by small core fibres.

Light propagates through a fibre in a number of discrete modes and laser light output from these modes will interfere at the output of the fibre, creating a 'speckle pattern'. With a low beam quality, large core fibre (600 μm diameter), there are so many modes excited (> 75 000) that these speckles are too small to be resolved from the perspective of the process. However, the smaller fibres required by higher beam quality lasers have significantly fewer modes, and so the speckle interference can become quite severe (and random!) to the extent of affecting the process. Such problems have been observed both with silica fibres for Nd:YAG precision machining [25] and hollow waveguide fibres for CO_2 marking [9]. One way of improving this is to excite equally all the modes within a particular NA, which can be achieved by increasing the fibre length or introducing controlled bends. A better solution is to restrict the fibre core size sufficiently so that only one mode will propagate; such fibres are described as 'single mode' and are commonly used in telecommunications applications. Unfortunately, however, to obtain efficient coupling from the laser into the fibre, the laser must have an $M^2 \sim 1$. As a demonstration, light delivered through a 200 m length of single-mode fibre has been used to cut 0.2 mm thick stainless steel sheet at 4 mm s^{-1} [26].

29.5.2 Thermal Damage

With hollow waveguide fibres, thermal damage can occur simply due to the attenuation of the fibre, since a small fraction of the light is lost and converted to heat on each Fresnel reflection. Since higher order modes are attenuated more quickly, this heating effect is normally greatest close to the input end. In addition, care must be taken to ensure that the focused laser beam does not strike the tube end-face, so an aperture is normally placed just in front of the fibre to ensure that light can only go *inside* the tube. An alternative solution proposed by Matsuura et al. [27] is a launching coupler consisting of a tapered section of hollow waveguide. In order to allow high-power transmission, hollow waveguide fibres are often water cooled [5], and in this way, average powers of up to 1040 W have been transmitted through a hollow glass fibre.

In TIR fibres, by contrast, the guiding mechanism is lossless, and fused silica fibres have very low intrinsic absorption and so no heating problems occur due to guided light. Problems can arise, however, on coupling the light into the fibre. The laser must be accurately focused onto the fibre core: it is important not to overfill either the core diameter or the NA of the fibre. If this is allowed to happen, a proportion of the light is coupled into the cladding, and damage can then occur by absorption in a material in optical contact with the cladding, for example, the glue used to mount the fibre. Provided the focusing condition can be satisfied, however, there is the potential to significantly increase the average power transmitted, compared with that used in current applications. Experiments with an 8 μm core diameter fibre [26] have shown that the maximum long-pulse (0.1 ms) power which can be transmitted is 300 W. If this is scaled up to a more typical 600 μm fibre, and assuming that the power handling capacity scales with area, it implies that a power level of 1.7 MW is transmissible. It may, however, be necessary to take into account the optical damage $1/d$

scaling law described by Wood [28] (in some regimes, optical damage threshold scales with $1/d$ rather than $1/d^2$ as might be expected) but even so a transmitted power of at least 23 kW should be possible.

With both hollow and TIR fibres, very sharp bends can also lead to damage. Sharp bends in a TIR fibre will couple light into the cladding, whereas in a hollow fibre, they will couple light to higher order modes, which are then strongly attenuated. Appropriate cabling designs are, therefore, used to prevent this.

29.5.3 Pulsed Laser Damage

Optical damage in silica fibres is much more of a limiting factor with short pulses (tens of ns). In this case the damage threshold can be orders of magnitude lower than expected from measurements with silica blocks—nominally the same material! Rather than the damage occurring at the fibre end-face, it typically occurs a few mm inside the fibre and is believed to be associated with self-focusing. The damage threshold can be significantly increased if the laser light is coupled to a greater range of fibre modes, reducing the likelihood of self-focusing [29]. This may be achieved by using a diffractive optical element (DOE), which is designed to reduce controllably the laser beam quality and match it to that of the fibre. By doing this, the damage threshold may be significantly increased and coupled with careful end-face polishing, the damage threshold becomes similar to that predicted by measurements on fused silica blocks—7.3 GW cm^{-2}, corresponding to 51 J cm^{-2} in a 7 ns pulse [30]. However, in a 200 µm fibre, this is still only equivalent to 16 mJ.

With the longer pulses (ms) used for laser percussion drilling, a different kind of damage is observed, which originates at the input end-face. The threshold for this damage has been shown to be related to the cleaving and polishing processes used to prepare the fibre end, which can leave surface scratches and sub-surface damage. Kuhn et al. [31] developed a technique using a CO_2 laser to 'anneal' this damage and demonstrated an increase in damage threshold intensity of almost two orders of magnitude. This means that it is now possible to deliver suitable pulse energy and beam quality through 400 µm diameter fibres for drilling holes of a few hundred micrometres diameter in a thickness of a few millimetres.

29.5.4 Non-linear Effects

In addition to the self-focusing effect described earlier, two other non-linear effects, stimulated Raman scattering (SRS) and stimulated Brillouin scattering (SBS), are also observed in fused silica optical fibres with a high peak power. For a full explanation see *Nonlinear Fiber Optics* by Agrawal [32]. Damage is not a problem with either of these processes; instead, the light is strongly coupled to other wavelengths. Brillouin scattering is an interaction of the laser beam with an acoustic wave, which couples the power into a frequency-shifted beam travelling in the opposite direction to the laser, thus greatly attenuating the light emerging from the output end of the fibre. Raman scattering, meanwhile, is the scattering of a photon from a molecule, with a change in vibrational energy level of the molecule and a consequent increase in the wavelength of the light. With SRS, the scattered beam normally travels in the same direction as the original beam. If a sufficient length of fibre is used, further scattering of the scattered light to longer wavelengths is observed, resulting in a spectrum with a number of broad peaks [33]. Since the scattered light propagates in the same direction as the original laser beam, SRS is not such a problem as SBS in wavelength-sensitive applications, as the power will still emerge from the far end of the fibre. However, the focusing optics at the output end must be free from chromatic aberration over the range of wavelengths produced.

The intensity thresholds for both SRS and SBS processes are dependent on the fibre composition, with a linear dependence on the fibre length. SBS is also dependent on the line width and pulse width of the laser. SBS has a lower threshold with cw laser light, whereas SRS dominates for short pulses—with Nd:YAG pulses, SRS has been shown to dominate even with a pulse length of 0.1 ms in a single-mode fibre [26]. In current fibre optic delivery systems, however, the fibre length is normally quite short, e.g. 10 m, so the threshold intensities are very high, $\sim 8.6 \times 10^8$ W cm^{-2}. With a 600 µm core diameter fibre, this corresponds to a peak power of 2.4 MW, much lower than the 4 kW cw laser power transmitted for welding applications. Based on these results, even for a drilling application where 30 kW peak power pulses are required through a 200 µm diameter fibre, SRS is only likely to become an issue if the fibre length exceeds 90 m.

29.5.5 Mechanical Damage

A fused silica optical fibre is very strong provided that there are no scratches on its surface. To prevent such mechanical damage, a coating is placed on the fibre as it is pulled from the preform, before it comes into contact with any other object. Such a coated fibre can be bent to a very tight radius without damage (50 mm with a 1000 µm diameter fibre). Hollow glass waveguides are also based on fused silica and so have similar properties. To make the fibres suitably robust for an engineering environment, the fibres are encased inside a suitable armoured cable structure.

29.6 Summary and Future Directions

Fibre optics have found wide acceptance for high power laser beam delivery and, in applications such as Nd:YAG laser welding, have become standard components. Hollow waveguide fibres are also becoming a practical proposition for delivery of high-power CO_2 light, with low-loss glass hollow waveguides now available. However, there is a range of applications where current fibres are still unsuitable, including welding using high-power (multi-kW) CO_2 lasers and Nd:YAG micromachining where high-power short pulses and high beam quality are required. These applications are still limited by damage problems.

One exciting advance which could have implications for the delivery of short pulse Nd:YAG light is the development of micro-structured or PCF optical fibres, as described in Figure 29.1. In particular, the air-guiding version holds promise as a means of delivering high peak power Nd:YAG laser pulses with good beam quality. New versions of hollow waveguide fibres designed for operation at 1.064 μm also have potential for this application, with low loss delivery of 158 mJ, 10 ns pulses already demonstrated [21,34].

There have been significant developments in hollow waveguide fibres in the mid-IR, and research in this area remains very active. Current fibres have a loss which is acceptable for many applications but their sensitivity to bending remains a problem—I would suggest that the major challenge that remains to be addressed with these fibres, therefore, is that of reducing differential modal attenuation, which is the primary reason for this sensitivity.

REFERENCES

1. Harrington J A 2000 A review of IR transmitting, hollow waveguides *Fiber Integrated Opt* **19** 211–17.
2. Nishii J, Morimoto S, Inagawa I, Iizuka R, Yamashita T, Yamagashi T 1992 Recent advances and trends in chalcogenide glass fiber technology: a review *J. Non-cryst. solids* **140** 199–208.
3. Artjushenko V G, Butvina L N, Vojtsekhovsky V V, Dianov E M, Kolesnikov J G 1986 Mechanisms of optical losses in polycrystalline KRS-5 fibers *J. Lightwave Technol.* **LT-4** 461–5.
4. Gregory C C, Harrington J A 1993 Attenuation, modal, polarization properties of $n < 1$, hollow dielectric waveguides *Appl. Opt.* **32** 5302–9.
5. Nubling R K, Harrington J A 1996 Hollow-waveguide delivery systems for high-power, industrial CO_2 lasers *Appl. Opt.* **34** 372–80.
6. Abel T, Hirsch J, Harrington J A 1994 Hollow glass waveguides for broadband infrared transmission *Opt. Lett.* **19** 1034–6.
7. Bhardwaj P, Gregory O J, Morrow C, Gu G, Burbank K 1993 Performance of a dielectric-coated monolithic hollow metallic waveguide *Mater. Lett.* **16** 150–6.
8. R Dahan, J Dror, N Croitoru 1992 Characterization of chemically formed silver iodide layers from hollow infrared guides *Mater. Res. Bull* **27** 761–6.
9. Su D, Somkuarnpanit S, Hall D R, Jones J D C 1995 Hollow core waveguide for high quality CO_2 laser beam delivery: exploitation of bend-induced mode coupling *Opt. Commun.* **114** 255–61.
10. Matsuura Y, Miyagi M, German A, Nagli L, Katzir A 1997 Silver-halide fiber tip as a beam homogenizer for infrared hollow waveguides *Opt. Lett.* **22** 1308–10.
11. Matsuura Y, Abel T, Harrington J A 1995 Optical properties of small-bore hollow glass waveguides *Appl. Opt.* **34** 6842–7.
12. van Dewoestine R V and Morrow A J 1986 Developments in optical waveguide fabrication by the outside vapor deposition *Proc. IEEE J. Lightwave Technol.* **4** 1020–5.
13. Murata H 1986 Recent developments in vapor phase axial deposition *IEEE J. Lightwave Technol.* **4** 1026–33.
14. Nagel S R, MacChesney J B and Walker K L 1985 *Modified Chemical Vapor Deposition Optical Fiber Communications: Fiber Fabrication* vol 1, ed T Li (New York: Academic)
15. Lydtin H 1986 PCVD: A technique suitable for large scale fabrication of optical fibers *IEEE J. Lightwave Technol.* **4** 1034–8.
16. Itoh K, Miura K, Masuda M, Ikwakura M and Yamagashi T 1991 Low-loss fluorozirco-aluminate glass fiber *Proc. 7th Int. Symp. Halide Glass* pp 2.7–2.12 (North Holland: Elsevier).
17. Miyagi M, Hongo A, Aizawa Y and Kawakami S 1983 Fabrication of germanium-coated nickel hollow waveguides for infrared transmission *Appl. Phys. Lett.* **43** 430–2.
18. Croitoru N, Dror J and Gannot I 1990 Characterization of hollow fibers for the transmission of infrared radiation *Appl. Opt.* **29** 1805–9.
19. Haan D J and Harrington J A 1999 Hollow waveguides for gas sensing and near-IR applications specialty fiber optics for medical applications *Proc. SPIE* **3596** 43–9.
20. Matsuura Y, Abel T and Harrington J A 1995 Optical properties of small-bore hollow glass waveguides *Appl. Opt.* **34** 6842–7.
21. Abe Y, Matsuura Y, Shi Y, Wang Y, Uyama H and Miyagi M 1998 Polymer-coated hollow fiber for CO_2 laser delivery *Opt. Lett.* **23** 89–90.
22. Payne F P, Hussey C D and Yataki M S 1985 Modelling fused single-mode-fibre couplers *Electron. Lett.* **21** 461–2.
23. Olivier C A, Hilton P A and Russell J D 1999 Materials processing with a 10kW Nd:YAG laser facility *Proc. ICALEO '99* pp D233–41 (Orlando FL: Laser Inst. of America).
24. Kuhn A, Blewett I J, Hand D P and Jones J D C 2000 Beam quality after propagation of Nd:YAG laser light through large-core optical fibers *Appl. Opt.* **39** 6754–60.
25. Hand D P, Su D, Naeem M and Jones J D C 1996 Fibre optic high quality Nd:YAG beam delivery for materials processing *Opt. Eng.* **35** 502–6.
26. Hand D P and Jones J D C 1998 Single-mode fibre delivery of Nd:YAG light for precision machining applications *Appl. Opt.* **37** 1602–6.
27. Matsuura Y, Hiraga H, Wang Y, Kato Y, Miyagi M, Abe S and Onodera S 1997 Lensed-taper launching coupler for small-bore, infrared hollow fibers *Appl. Opt.* **36** 7818–21.
28. Wood RM 1997 Laser induced damage thresholds and laser safety levels. Do the units of measurement matter? *Opt. Laser Technol.* **29** 517–22.
29. Sweatt W C and Farn M W 1993 Kinoform/lens system for injecting a high power laser beam into an optical fiber *Proc. SPIE* **2114** 82–6.
30. Maier R R J, Hand D P, Kuhn A, Blair P, Taghizadeh M R and Jones J D C 1999 Fibre optic beam delivery of nanosecond Nd:YAG laser pulses for micro-machining *Proc ICALEO '99, Laser Microfabrication Conf. (San Diego, CA)* pp 204–18 (Orlando FL: Laser Inst. of America).
31. Kuhn A, French P, Hand D P, Blewett I J, Richmond M and Jones J D C 2000 Preparation of fibre optics for the delivery of high-energy, high-beam-quality Nd:YAG laser pulses *Appl. Opt.* **39** 6136–43.
32. Agrawal G P 1989 *Nonlinear Fiber Optics* (London: Academic) chs 8 and 9.
33. Cohen L G and Lin C 1978 A universal fiber-optic (UFO) measurement system based on a near-IR fiber Raman laser *IEEE J. Quantum Electron.* **QE-14** 855–9.
34. Sato S, Ashida H and Arai T 2000 Vacuum-cored hollow waveguide for transmission of high-energy, nanosecond Nd:YAG laser pulses and its application to biological tissue ablation *Opt. Lett.* **25** 49–51.

30
Positioning and Scanning Systems

Jürgen Koch

CONTENTS

30.1 Introduction 445
30.2 General Requirements 445
30.3 Positioning Systems 446
    30.3.1 Motion Devices 446
    30.3.2 Kinematics 447
    30.3.3 Measuring Devices 448
    30.3.4 Advanced Positioning Systems and Optical Set-ups 448
    30.3.5 Influences on Beam Propagation 448
    30.3.6 Comparison of 3D-positioning Systems 449
30.4 Scanning Systems 449
    30.4.1 Scanning Methods 450
    30.4.2 Optical Configurations of Scanning Systems 450
    30.4.3 Polygonal Scanners 451
    30.4.4 Galvanometer Scanners 451
    30.4.5 Performance and Accuracy of Scanning Galvanometers 452
    30.4.6 Mirrors in Oscillatory Scanning Systems 453
    30.4.7 Piezoelectric Devices 453
    30.4.8 Acousto-Optic Deflectors 455
30.5 Conclusion 457
Acknowledgements 457
References 457
Further Reading 457

30.1 Introduction

In the context of industrial material processing, the 'positioning system' incorporates the optical and mechanical techniques for focusing, directing and controlling a high-powered laser beam, focused to a spot, to the coordinates of a specific location [position] on a workpiece. Either the beam or the workpiece is moved to achieve this objective; both can also be moved relative to each other to provide a more compact or a more flexible system. The 'scanning system' comprises the optical, acousto-optical (AO), electro-optical (EO) and optomechanical techniques for controlling and scanning a laser beam, focused to a spot, across a surface that is usually one- or two-dimensional, such as a bar code or a display screen. Such systems are referred to as beam delivery systems. Besides the positioning and scanning systems, the free-space and fibre optics mentioned in previous chapters are also part of beam delivery systems but predominantly their purpose is to locate the laser beam at a workstation and not at a workpiece.

30.2 General Requirements

The general requirements for positioning and scanning systems derive from the previous definitions. Pointing out the character of each system reveals its main property: positioning a laser beam on the coordinates of a specific location demands precision, while laser beam scanning necessitates speed. According to the specific application, these two properties are characterized by different parameters—see Table 30.1. Furthermore, both systems involve motion, so the dynamical behaviour caused by certain forces also affects these parameters. Further demands are associated with the type of drive used in the positioning system or are related to the method of steering a laser beam in scanning systems. In addition, specific requirements are associated with particular applications. Table 30.1 lists several parameters classified according to the main application. This table is not exhaustive. Each developer of a beam delivery system for a specific application has to bear in mind the important parameters, but engineers may use it as a checklist.

TABLE 30.1

Considerations for Beam Delivery System

Purpose of Application	Main Parameters
Motion	Traverse path, angle of deviation, velocity.
Accuracy	Stepwidth, resolution, accuracy of position, drift hysteresis.
Dynamic	Accuracy of path, acceleration, deceleration, rise time, resonance frequency, transient response, heating.
Repetition	Repeatability, drift.
Forces	Load and stress, stiffness.

It should be noted that different parameters cannot be chosen arbitrarily, because they are mostly mutually dependent. Furthermore, positioning and scanning systems are usually equipped with specific controllers, which principally influence the system performance by their settings. For a specific application, optimization of these settings is necessary to achieve the best performance. Thereby, the whole assembly has to be taken into account. Given the number of possible different set-ups, specific types of controllers cannot be mentioned in this chapter, but general remarks related to this subject are presented where necessary.

An example taken from laboratory practice will illuminate some aspects regarding the general requirements for beam delivery systems for laser material processing—some of this has been reported in Ref. [1]:

> A feasibility study on the manufacture of multiple-hole metal sheets, which form part of an active system for the reduction of aerodynamic resistance in aircrafts, has been carried out. The application's requirements were to drill thousands of holes perpendicular to the surface of flat aluminium or titanium 1 mm thick in an area of up to 200 mm×200 mm with a distance of 0.5 mm between adjacent holes. Each hole had to have a diameter of 50 μm and processing time should not exceed 20 ms, including positioning.
>
> A high-frequency (pulse repetition rate of approximately 10 kHz), short-pulsed (pulse width 30 ns) copper vapour laser (CVL) system was applied to perform the task. Hence, the drilling of a single hole had to be done by percussion drilling, i.e. the use of several consecutive pulses for each drilling and, hence, the positioning of each hole had to be done step by step.

This example—we will come back to it later—demonstrates only a few of the possible requirements placed upon a beam delivery system by a specific application.

First, *the laser system*: drilling holes with a diameter of 50 μm necessitates the use of a laser source with high beam quality and the pulses should also be of high energy, according to the amount of material that has to be ablated. For the processing time, a highly repetitive system has to be chosen. Nowadays, pulsed Nd:YAG systems would fit very well, but at the time, the investigations were carried out the copper vapour laser (CVL) was the best choice. More details about laser systems are given in part B of this handbook.

Second, *workpiece positioning*: if we neglect the necessity of obtaining the focus, i.e. placing the focused spot of the laser beam on the surface of the workpiece, the whole problem is two-dimensional. For this, taking into account the maximum size of the metal sheets, a coordinate table would do but, considering the distance between the adjacent holes and the processing time allowed for each drilling, a device with high speed—not velocity, but high values of acceleration and deceleration—is required. In addition, wear can play a role because of the many thousand repetitive positionings. Finally, vibrations caused by acceleration and deceleration have to be avoided.

As we can see, developing beam delivery systems starts with the laser source. The laser selected for an application limits the choice of usable positioning or scanning devices and *vice versa*. The wavelength, power, energy, mode of operation—continuous wave (cw) or pulsed—pulse duration and beam width of the designated laser have to be considered. Using mirror-based deflection systems, a wide range of different laser sources can be selected for many applications. Only the optical materials available for mirrors restrict our choice. The same applies to transmissive optics such as AO ones. The limitations on workpiece handling are even fewer because the positioning of the workpieces is independent of the laser. Hence, in the following, different types of beam delivery systems for specific laser sources are not treated in detail, rather certain devices are introduced and their advantages and limitations discussed.

30.3 Positioning Systems

Positioning in the context of laser material processing means controlling and directing the focused spot of a laser beam to the coordinates of a specific location on a workpiece. Either the beam or the workpiece is moved to achieve this objective—both can also be moved relative to each other to provide a more complex or a more flexible system. Hence, the main purpose of a positioning system is motion, whether it be translation or rotation.

30.3.1 Motion Devices

The technique most commonly used in laser material processing, e.g. cutting, welding, structuring and micromachining, to perform translation is mechanical in type. Usually, a rotation is translated into linear motion by a ball screw. For this purpose, the rotation is achieved by mounting an electromagnetic drive to the shaft. The accuracy of such systems is determined by the precision of the screw shaft and the ball bearings used to fix and guide the shaft. Mounting two such axes together, one perpendicular to the other, facilitates two-dimensional motion. These systems are called coordinate tables due to the possibility of addressing a specific position by its coordinates (X and Y). The same set-up can be realized with linear drives, where magnetic forces move the axes. The use of magnetic drives solves an

TABLE 30.2

Comparison of the Accuracy of Different Driving Techniques

System	Drive Mechanism/ Bearing	Field Size (mm²)	Position Accuracy Full Stroke (μm) at	Velocity (mm s⁻¹)	Angular Deviation (μrad)	Resolution (μm)
[3]	Ball screw/hydrostatic	200 × 200	0.05	_b	2.4	0.005
LPKF HS 8 GP[a]	Moving coil/air bearing	200 × 200	<2	up to 300	—	0.25
LPKF XY 60 G[a]	Ball screw/air bearing	600 × 600	±3	155	24	0.5

[a] Taken from supplier's catalogue.
[b] According to its use as a high-precision device, Kami did not report on velocity investigations. Besides, an increase in accuracy can also be achieved by sensor technology adapted to the specific device—see further reading [4,5].

additional problem for two-dimensional motion: a parallel set-up of the two axes overcomes the problem of a different load on each axis if they are mounted one on the other. The system is driven by moving coils and is also equipped with air bearings. A comparison of the two different driving techniques reveals the specific advantages and disadvantages.

> Back to the example. Different techniques were used to move the metal sheets linearly:
> At first, the stepping for positioning each hole was achieved by a screw-shaft-driven coordinate table. After optimization of the motion parameters, the positioning of a single hole already exceeded the required processing time by a factor of five. In addition, repetitive positioning of many thousand holes resulted in wear of the screw shaft, in turn causing unreasonable deviations in position.
> In the next step, a magnetically driven motion system with a parallel set-up for the axes was used. Hence, the stepping time was reduced by half but the drillings appeared somewhat elliptical. A close investigation of the whole system revealed that vibrations in one axis of the coordinate table's base frame, caused by high values of the acceleration and deceleration in combination with a weakness in the substructure in one direction, were responsible for the anomaly in shape. By just changing the main direction of motion, circular holes were once again processed but time was still a problem.

The differences in positioning time result from different driving forces. A mechanical system is very stiff because of the use of a screw shaft and ball bearings. Hence, high velocity can be achieved but as a serial assembly—the Y-axis is mounted on the X-axis—a large mass has to be moved and friction becomes noticeable. Hence, acceleration and deceleration, the main parameters for stepping small distances, are limited. In addition, repetitive positioning often results in wear of the screw shaft or bearings.

Using a parallel set-up of magnetically driven axes with air bearings reduces friction. Thus, wear is not observable; moreover, higher values of acceleration and an increase in deceleration are achieved. However, due to the mechanically contact-free assembly, the system shows no ideal stiffness—see also Ref. [2]. Hence, the load on the axes has to be taken into account because the resulting transient response, i.e. the behaviour of the drive until it reaches the desired position, depends on the mass mounted on the system. Furthermore, the noticeable vibrations at least demonstrate that at elevated rates of acceleration and deceleration, the environment also has to be taken into account.

Table 30.2 presents a comparison of the achievable accuracy of different driving techniques—hydrostatic ball screw, moving coil with air bearings and a ball screw with air bearings—for use in micromachining, e.g. with excimer lasers, Q-switched Nd:YAG lasers and ultrashort-pulsed laser systems.

Coordinate tables are not limited to linear motion: curved lines can also be realized, dividing the path into small consecutive parts of linear motion, thus limiting the accuracy of curved lines to the step width of the system, i.e. the smallest range of motion possible. Hence, applications necessitating rotation only employ a rotation axis. The power transmission of these systems is performed directly, as in ball-screw-driven devices, or indirectly with cogwheels, chains or belts.

Finally, the combination of several axes facilitates processing in three dimensions. However, the motion sequences necessary to locate a distinctive point in space have first to be determined.

30.3.2 Kinematics

In general, the positioning of an optic, hence a laser beam, for 3D processing tasks requires a minimum of five degrees of freedom regarding motion. The definition of a particular point in a workspace needs at least three coordinates of translation (usually called the X-, Y- and Z-axes) and another two of rotation (B- and C-axes) with respect to an orientation at that point—compare Figure 30.1. Differences in the realization of 3D motions are correlated to the type of movement chosen (translation, rotation) and to the sequence in which the axes move.

The gantry systems with translation axes for the main directions of motion as shown in Figure 30.1 are an example for a 3D assembly with a Cartesian solution. In a Cartesian coordinate system, it is possible to compute all the axial positions necessary to locate a point in space. The coordinate axes are linearly independent.

In contrast, industrial robots which use a combination of rotation axes, as shown in Figure 30.2, demand a minimum of six rotation axes to achieve the same performance as that of gantry systems. Furthermore, the coordinate axes of the kinematic chain are not linearly independent due to the

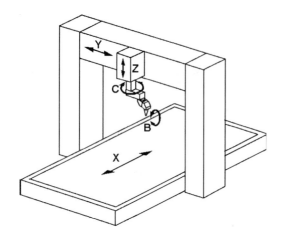

FIGURE 30.1 Gantry system (reproduced by permission of Trumpf, Germany).

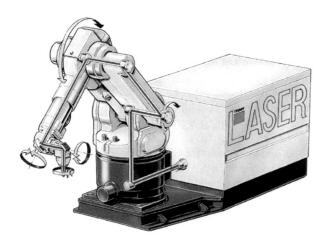

FIGURE 30.2 Robot (reproduced by permission of Trumpf, Germany).

superposition of non-Cartesian coordinate systems. The same point in space can be accessed with the elbow up or elbow down. Thus, increased effort has to put into the controlling tools—see Ref. [6]—and position measuring becomes inevitable.

30.3.3 Measuring Devices

The measurement of positions in linear devices, such as gantry systems and coordinate tables, is carried out by linear measuring scales. The advantage of these measuring systems is that the actual value of the position is measured directly. For very precise requirements, e.g. in micromachining with excimer lasers, the measuring scale is made of glass due to its lower thermal expansion coefficient. When linear motion is performed by a ball screw, encoders are usually applied. Their purpose is to transform the number of rotations into linear scaling and *vice versa*. Therefore, the measurement of the position of the axes is indirect. Comparable systems are used in rotation axes. Errors due to the low stiffness of the axis, fetch in the power train or overload of the axis are not registered. Moreover, the errors add up when chain-linking several axes, for example to achieve advanced systems for three-dimensional processing.

30.3.4 Advanced Positioning Systems and Optical Set-ups

The introduction of relative motion between workpiece and tool—here a laser beam—can be accomplished by moving the workpiece or the tool; both can also be moved relative to each other.

Gantry systems feature large mass as well as vibration-restricted constructions. Therefore, high processing and positioning velocities are achieved, even when considering the motion of large mass. A set-up with a fixed workpiece—examples 1 and 3 in Figure 30.3—denotes a definite position for the workpiece, but a complex beam delivery system with movable optical parts—also called flying optics—has to be used to locate the radiation at the optics needed for processing. Moving the workpiece during processing results in a reduction in the specifications for the beam delivery system and saves space but these are mostly limited to small and lightweight components. For precision processing, a hybrid solution where the workpiece moves along one direction—example 5 in Figure 30.3—is more convenient. Figure 30.3 also demonstrates some assembly examples for 2D and 3D machining, equipped with different numbers of axes.

Fixed workpieces can also be accessed by industrial robots. Such robots are available with internal beam delivery systems, as shown in Figure 30.2, or external beam supplies, comparable to articulated arms thus increasing flexibility. The internal beam supply demands careful adjustment of the optical components, because the beam has to be directed exactly along the rotation axes. Therefore, fibre optics are in use where applicable.

A current method to widen workspaces is to combine translation and rotation motion units. Hence, additional axes, for example, an assembly of industrial robots on linear motion devices and workpiece holders which are rotate-and-pivot units, allow processing at areas with difficult access but beam delivery and system control are more expensive.

Furthermore, the growing use of process control elements, e.g. [7], controlling accuracy during processing [8], optical focus [7] and feedback control systems, increases the precision of positioning systems and is steadily enhancing the complexity of beam delivery systems. Already available means—like adaptive optics—designed for this purpose can also be used.

30.3.5 Influences on Beam Propagation

The dimensions of the workspace and, therefore, the length of free-space beam propagation (the sum of the X-, Y- and Z-directions) influence the focability of the beam. In particular, the movable parts in the flying optics need careful adjustment, because they can greatly modify the length of beam propagation, whereby the quality of the processing varies across the workspace. Affected by the divergence, the beam width changes and, hence, the irradiance distribution which depends on the position of the working optics along the beam axis. Deviations in the position of the focus relative to the working optics can be corrected with telescopes, adaptive optics or mechanical systems for compensating the beam propagation length.

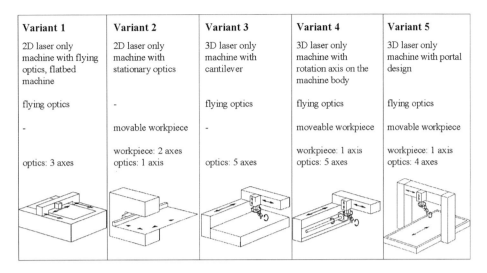

FIGURE 30.3 Overview on gantry systems (reproduced by permission of Trumpf, Germany).

TABLE 30.3

Comparison of 3D-Positioning Systems

Assembly	Accuracy of Position (mm)	Accuracy of Path (mm)	Investment Costs (€)
Robot	±0.05–±0.2	±0.05–±1.4	50 000–70 000
Gantry system	±0.03–±0.05	±0.1–±0.2	>200 000

30.3.6 Comparison of 3D-positioning Systems

Hoffmann compared the accuracy and investment cost of 3D positioning systems, e.g. those used for hard-soldering in the automobile industry [9]—see Table 30.3.

Differences in investment costs derive from the complexity and accuracy of the various systems. For example, robots, such as those shown in Figure 30.2, compared with gantry systems are a less expensive option but they do not give such good guidance. In addition, robots are more sensitive to misalignment and the loss in laser power is higher because of the necessary number of mirrors. The main advantage of robots is higher flexibility in positioning. With these systems, even working areas which are difficult to access can be reached, especially if the less expensive internal beam delivery is exchanged for an external beam supply.

30.4 Scanning Systems

In general, two different groups of scanning systems can be distinguished. In reflective systems, the laser beam scans by means of mirrors, which are usually mounted to torsion or rotation drives. Galvanometer scanners, polygonal scanners and piezoelectrically driven trepanning heads belong to this group. In contrast, transmissive systems, such as AO and EO devices, deflect the laser beam while it is being transmitted through a crystal. The optical properties of the crystal are changed by means of acoustic or electric waves applied to the crystal. Some essential aspects of AOs are mentioned in the corresponding section of this chapter. EOs are neglected here as they are rarely used in beam delivery systems for laser material processing. However, they are considered in other literature.

In principle, positioning systems can be applied to scanning operations. However, the nature of scanning processes is a high velocity with an adapted accuracy. So, some positioning devices may also be used in scanning systems and vice versa. For this to happen, only the accent has to be changed. Piezoelectric actuators are the prime example here, as will be shown later. In general, scanning devices are designed to meet a particular task. Reflective scanning systems are usually used in industrial laser material processing.

The benefit of scanning devices for use in industrial applications, such as materials processing or micro-structuring, can be shown with the example of perforating metal sheets, mentioned previously.

> After several optimization steps, first with a ball-screw-driven coordinate table, and then applying a moving-coil driving mechanism, the time necessary to position each hole was still a problem. It was finally solved by the implementation of a fast laser beam deflection system. To be more precise a galvanometer scanner due to its large aperture for beam deflection equipped with a telecentric $f-\theta$ lens was chosen to position the holes accurately in an area 25 mm×25 mm and to achieve appropriate focusing of the radiation. After processing such a patch, the metal sheet was translated to be worked on the next patch. That way, the pre-determined requirement (a processing time of less than 20 ms, including positioning) was fulfilled.

This example shows that scanning systems can be successfully applied to laser material processing. Hence, as their use is not restricted to small step distances, several scanning methods are available.

30.4.1 Scanning Methods

Three different scanning methods are used in laser material processing according to the various applications (Figure 30.4a–c).

The perforation of sheet metal in our example was finally carried out by *step scanning*: the laser beam is active at pre-defined positions and during positioning operations laser processing is suppressed—see Figure 30.4a.

In *raster scanning*, the scanning system sweeps the laser beam at a constant velocity over a fixed angular range in a repetitive manner, thus scanning a full image with a series of parallel lines. At the end of each line, the laser beam is turned off and the scanning device returns to the starting position. Where applicable, raster scanning is done bi-directionally, i.e. processing from the left to the right and *vice versa*—see the lower part of Figure 30.4b.

In *vector scanning*, the motion is separated into many small steps, indicated by the arrows drawing an 'H' in Figure 30.4c. A continuous track on the workpiece surface is made when the command waveform's frequency exceeds the scanning system's bandwidth. This technique enables writing (with a laser beam), e.g. letters or symbols, on the object surface when the scanning device moves the beam during laser operation. A new starting position is reached while the laser beam is deactivated (Figure 30.4c).

The various scanning methods can be performed by transmissive devices as well as by reflective ones. However, in laser material processing, the laser beam is usually focused to a spot to perform the processing tasks. Hence, a lens is part of the scanning system's set-up. Due to the position of the focusing lens relative to the scanning device, two optical configurations can be specified.

30.4.2 Optical Configurations of Scanning Systems

The focal spot size necessary for laser material processing can be achieved by two types of optical configurations: a pre-objective one (Figure 30.5), which is also applicable to AO devices and not restricted to reflective systems as depicted here; and a post-objective one (Figure 30.6), which depends on the relative position of the scanning device with respect to the focusing objective lens.

In the pre-objective configuration, the laser beam scans before it is focused by the objective lens, while in post-objective systems the beam is focused first. When necessary, an

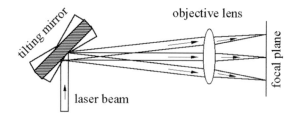

FIGURE 30.5 Pre-objective configuration.

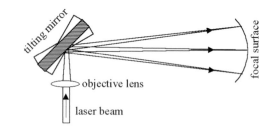

FIGURE 30.6 Post-objective configuration.

additional lens for field correction is applied, as is the case with wire marking. Here, a mask is projected by means of a telescope, settled before a rotating mirror, and a specific lens corrects the focal plane afterwards, in order to perform sharp images of the bar codes, for example, when the wire is moved during processing.

The pre-objective configuration allows only limited angles due to the aperture of the objective; the post-objective configuration, in principle, can reach a full circle.

Two-dimensional scanning is achieved by the use of a second scanning device and the optical configuration can be pre-objective, as shown by the sketch in Figure 30.7, or post-objective; a pre–post-objective configuration is also possible, when the objective lens is positioned between the two devices. The most common arrangement is one in which both scan units are in a pre-objective configuration. In general, $f-\theta$ lenses are applied to achieve a maximum flat field size h [10].

$$h = 2F\tan(\Theta) \quad (30.1)$$

with F the focal distance and Θ the optical scan half-angle.

As previously indicated, scanning operations are performed by transmissive or reflective devices. Hence, the means for deflecting a laser beam depend on the type of device chosen.

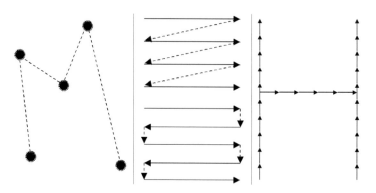

FIGURE 30.4 Step (a), raster (b) and vector (c) scanning.

Positioning and Scanning Systems

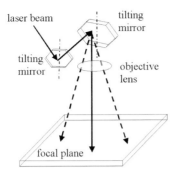

FIGURE 30.7 Pre-objective 2D scanning system.

Moreover, if using mirrors, the different types of driving systems have to be considered.

In industrial applications where large apertures due to the beam width of the applied laser systems, e.g. CO_2 lasers, are important, as in the heat treatment of metals, regular prismatic polygonal scanners or pyramid polygonal systems [11] move the laser beam across the workpiece's surface. Galvanometer-driven systems are used in laser welding applications, e.g. in the automotive industry [12], and are also employed for microstructure surfaces [13]. Furthermore, piezoelectric actuators are applied to drilling operations, e.g. in trepanning heads [14].

30.4.3 Polygonal Scanners

Polygonal scanners deflect the laser beam by means of a faceted mirror. Usually regular prismatic polygonal scanners, with the mirror facets placed on the periphery of a cylindrical wheel, are used but pyramid polygonal scanners, i.e. ones in which the mirror facets are placed on the surface of a truncated cone (Figure 30.8), are applied as well, for example for heat treatment applications, although they are difficult to manufacture [11]. The advantage of the pyramid polygonal scanner compared with the regular prismatic scanner is that the scan width can be varied due to the height chosen for the incident beam on the cone.

Two-dimensional scanning is achieved by inserting an additional beam-deflecting device, e.g. a galvanometer scanner. Hence, it is obvious that polygonal scanners are limited to raster scanning, as used in laser projection systems and laser-marking applications (Figure 30.9).

The performance of polygonal scanners, i.e. the scan frequency (representing the number of repetitive scans), depends

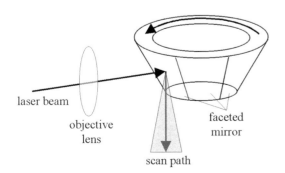

FIGURE 30.8 Diagram of a pyramid polygonal scanning system (post-objective configuration).

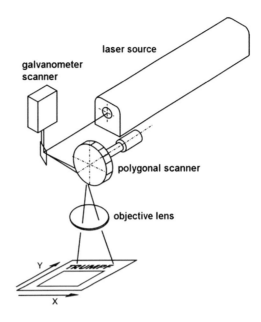

FIGURE 30.9 Regular prismatic polygonal scanners in a 2D assembly (pre-objective configuration) (reproduced by permission of Trumpf, Germany).

on the rotation drives of these scanners. Usually, they provide high scan frequencies f according to the rate N of the rotation axis and the number m of facets on the polygonal mirror:

$$f = Nm \tag{30.2}$$

Power transmission is performed directly, when the mirror wheel is fixed to the shaft of the axis, or indirectly, e.g. by belt.

Another way to perform raster scanning is by the use of scanning galvanometers, but these are not limited to raster scanning.

30.4.4 Galvanometer Scanners

The scanning galvanometer is an electromagnetic actuator that moves a flat mirror (see Figure 30.7) in order to deflect a laser beam in a controlled way and with high speed. Different galvanometer drives are used for laser beam scanning for industrial processes.

- *Resonant galvanometers* are driven at their resonant frequency, which limits their use to high-speed raster scanning. Drawing arbitrary paths or step scanning is not possible with these systems.
- *Moving magnet galvanometers* feature very low rotor momentum of inertia, in connection with the torque generated by them. Hence, they offer high accelerations, and in combination with efficient position sensors, they are applicable to all scan methods mentioned earlier.
- *Moving coil galvanometers* are mainly used due to their high and stable torque when the beam diameter exceeds 30 mm, and therefore, the mirrors have to be very large.

In any case, because of the high values of the tangential acceleration performed by a galvanometer, its shaft is mounted on high-precision, pre-loaded ball bearings. The shaft rotation has to be very smooth with very low friction to minimize positioning and tracking errors, at all ranges of working temperature. Indeed, some types of inertia effects cannot be neglected, as Figure 30.10 shows. In step scanning, the time of response depends on the desired angle of deviation (jump angle) as expected and also on the diameter of the aperture. Larger apertures are used with larger beam widths; therefore, the mirrors applied to the scanner have to be larger—being heavier, the momentum of inertia is, thus, increased. Obviously, at smaller apertures (up to approximately 11 mm) when the mirrors are small and lightweight, the friction from the drive and the inertia of the rotation axis prevail over the inertia effects of the mirrors.

The limitation of the measurement of the response time to a maximum jump angle of 25° in Figure 30.10 indicates the maximum mechanical deflection angle preferred in industrial material processing tasks.

A galvanometer drive for scanning purposes is usually equipped with a position sensor intimately linked to the shaft. The motor moves by means of a feedback amplifier controlled by the position sensor signal. In this way, the galvanometer shaft (which means the laser beam deflection) can be accurately positioned by injecting an electrical signal into the amplifier input, or if the electrical signal is time-dependent, the laser beam moves precisely following the applied signal. However, system performance and the accuracy of the galvanometer scanners depend on their driving force. The parameters reflect the oscillatory character of the electromagnetic drive.

30.4.5 Performance and Accuracy of Scanning Galvanometers

Usually, the performance and accuracy of scanning galvanometers are dependent on the scanning method used for a particular application. In addition, the focal distance of the objective lens also influences accuracy. Therefore, in the following, only a general view of these parameters with respect to the applied scanning method is given. It should be noted that the definitions of the parameters are valid for any scanning system but some are only specific to the use of galvanometers.

The performance parameter for step scanning is the *step time*, measured from the initial motion of the scanning device to a pre-defined position, within a settling tolerance, known as the transient response of the device—see Figure 30.11.

Besides heat, *dither*, i.e. a slight vibration of the shaft, caused by the noise of the power supply affects the accuracy of the position. Therefore, the power supplies used are specially designed for low noise output.

In raster scanning, the *scan frequency* is the important parameter which represents the number of repetitive scans. The retrace time, which is the time necessary to reposition the system from the end of one scan to the beginning of the next, limits the scan frequency.

Figure 30.12 describes the performance with vector scanning. Generally, a *tracking error* occurs due to acceleration of the scanning device. To minimize the tracking error, state-of-the-art systems offer a programmable 'head time' to overcome the acceleration effect in a particular application.

According to the desired accuracy in the position or path, three effects—overshoot, cross-scan error (also known as across-scan error or wobble) and in-scan error (also known as along-scan error or jitter)—should also be mentioned. Figure 30.13 displays the latter two errors in brief. Overshoot

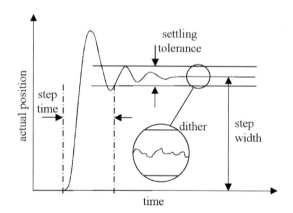

FIGURE 30.11 Step scan performance.

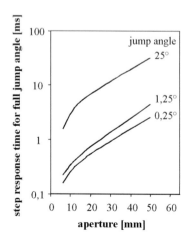

FIGURE 30.10 Step performance versus aperture size (reproduced by permission of Arges, Germany).

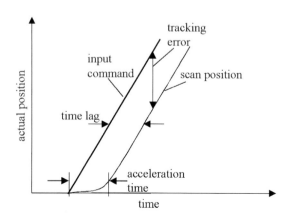

FIGURE 30.12 Vector scan performance.

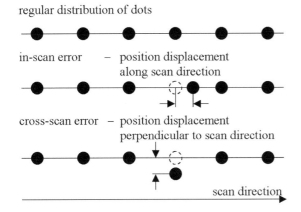

FIGURE 30.13 Position errors in scanning applications.

describes a movement that is farther than necessary. Hence, some applications overcome deceleration effects, such as an increase in the overlap of consecutive pulses at the end higher performance or accuracy but only when the overshoot is programmable by an additional 'tail time' and appropriate laser control. Cross-scan error and in-scan error are two positioning errors, perpendicular and parallel to the direction of motion, respectively. These arise from heating effects, calibration failures and power supply noise.

It should be mentioned that the performance of galvanometer scanners is generally affected by heating in dynamic applications. Therefore, most systems are available in a temperature-stabilized design where the system is heated to a value above the ambient temperature. However, working within a fixed coordinate system requires calibration of the scanning devices before a new process is started.

The *resolution* of galvanometer scanners as well as that of piezoelectric drives depends on the lenses used with the system; accuracy and linearity are also linked to this. In addition, the position sensor applied to the system is a very important item that has a considerable influence on the response time and accuracy.

The accuracy and electromechanical characteristics of a galvanometer–amplifier assembly enable high-speed positioning of the deflecting mirror with a resolution of the order of a micro-radian and a reaction time of a microsecond (Figure 30.10), if we take into account the previously mentioned limitations. However, the lateral resolution depends on the focal distance of the lenses used for focusing the laser beam. Employing equation (30.1), the lateral resolution r can be approximated by exchanging the optical scan half-angle α with half the minimum step width $\Delta\alpha$:

$$r = 2F\tan(\Delta\alpha). \quad (30.3)$$

To drive the motors, specialized software which runs on a PC is used. The input to this software is a datafile of a drawing. The motors are driven by means of an electronic card (path generator) that also switches the laser on or off.

Another aspect which has to be considered with respect to galvanometer scanners and their oscillatory nature and high values of acceleration is its mirrors.

30.4.6 Mirrors in Oscillatory Scanning Systems

Both the static and dynamic characteristics of scanning mirrors are important, especially in galvanometer scanners, where a flat mirror is fixed to a shaft. Inadequate mirror flatness increases the laser beam divergence. In addition, the mirror flatness can be degraded by rapid rotation of the mirror. In high-speed scanning systems, the mirror rotates with very high tangential accelerations (180×10^3 rad s^{-2}) and frequencies up to 500 Hz. Such accelerations strongly stress the mirror and elastically distort it or even make it vibrate. Therefore, the stiffness of the mirror is important. It has to be high enough to minimize distortion of the mirror. Furthermore, appropriate mirror stiffness prevents vibrations of the mirror as it rotates. Otherwise, the feedback electronic system is affected adversely, and especially when the vibrations match one of the mirror's resonant frequencies, the mirror will be destroyed. However, the mirror dimensions should not limit the laser beam diameter over the full angular field needed for the application.

In a pre-objective configuration, the width of the mirror along the rotation axis is determined by the width of the laser beam. The minimum height H of the mirror—perpendicular to the rotation axis—can be estimated with the assumptions that a laser beam of a width W completely fills the mirror at the desired maximum of the angle of rotation α:

$$H = W/\cos \alpha. \quad (30.4)$$

However, to overcome the diffraction effects caused by the mirror, both of its dimensions—parallel and perpendicular to the rotation axis—usually exceed the minimum requirements by 10%.

30.4.7 Piezoelectric Devices

Among many other applications, piezoelectric devices are also employed in high-resolution positioning and dynamic applications, e.g. in trepanning heads, they provide the laser beam with circular motion by moving a single mirror [14] but their use is limited to the generation of Lissajou patterns, comparable to resonantly excited 2D scanners [15].

According to Piezojena's handbook [16], there are many parameters which have to be taken into account with piezoelectric devices. In the following, some additional aspects to those already mentioned, concerning scanning and positioning applications, are described. For further information see [16].

The driving force of piezoelements is the piezoelectric or inverse piezoelectric effect.

A voltage applied to piezoelectric material (PZT or lead–zirconium–titanate) changes the dimensions of the material, thereby generating motion—the inverse piezoelectric effect. Hence, the driving force of a piezoelectric actuator is *expansion*. The relative expansion $S = \Delta l/L_0$ (without any external forces) of a piezoelement is proportional to the applied electrical field strength. Typical values for ceramic materials are $S = 0.1\%–0.13\%$ (field strength $E = 2$ kV mm^{-1}):

$$\Delta l/L_0 = S = d \times E. \quad (30.5)$$

Here S is the relative stretch (dimensionless); $d=d_{ij}$ is the piezomodule or parameter of the material; E (= U/d_s) is the electrical field strength, where U is the applied voltage and d_s is the thickness of a single disk.

The maximum expansion that can be achieved by using normal stacks, i.e. an assembly of several discs of piezoelectric material, is up to 300 µm. The length of such a stack will be 300 mm! Typical piezostacks have a motion of 20–60 µm.

According to equation (30.5), the maximum expansion will rise with increasing voltage, but the relation is not linear as predicted by the equation. Because of their ferroelectric nature, piezoelements show typical hysteresis behaviour. A butterfly-like curve reflects the inherent hysteresis. To reduce the hysteresis effect, usually unipolar voltage is applied, thus leading to the characteristic expansion curve of piezoelements (shown in Figure 30.14).

The typical width of hysteresis is 10%–15% of the full motion. Working in a small voltage range, the hysteresis is also smaller.

Apart from hysteresis, the piezoelectrical effect as a solid-state effect has a very high resolution. Investigations carried out with typical piezoelements, e.g. PX 38 from Piezosysteme Jena, show a resolution of 1/100 nm. Therefore, the resolution is limited by the noise characteristics of the power supply.

Another characteristic of piezoelectric actuators is a short-dimensional stabilization known as *creep*. A step change in the applied voltage will produce an initial response in a fraction of a millisecond, followed by a smaller change in a much longer time scale as shown in Figure 30.15. Creep will be within 0.5%–0.8% in the next decade.

The creep depends on the expansion, Δl, of the ceramic material, on the external loads, and on time. The creep can also be shown to have a logarithmic dependence on time:

$$\Delta l(t) = \Delta l_{0.1} \times (1 + \gamma lg(t/0.1\ s)) \quad (30.6)$$

Here, $\Delta l_{0.1}$ is the stroke 0.1 s after the end of the voltage's rise time.

In this case, a value of 0.015 is reached for γ. The value of γ depends on the material, the construction and the environmental conditions (e.g. forces). When the motion (voltage) stops, the creep takes a few seconds longer to stop.

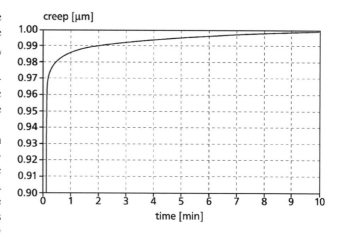

FIGURE 30.15 Typical creep of a PZT (reproduced by permission of Piezosysteme Jena, Germany).

When working with periodic signals, as is the case in step scanning, the *repeatability* of a position will not be affected by creep. Because of the strong time dependence of the motion, creep occurs in all oscillations of the same order.

The periodic oscillation of a mirror mount is shown in Figure 30.16. The full tilting angle is approximately 380 arcsec. In the figure, only a 10 arcsec section (302–312 arcsec) is displayed—the repeatability is better than 0.1 arcsec.

A result of this experiment is that high repeatability within the system is reached without the use of closed-loop control. For such experiments, the repeatability is only determined by the quality of the power supply.

Taking scanning operations into account, the dynamic properties also have to be considered. Piezoactuators are oscillating mechanical systems characterized by their *resonant frequency* f_{res}. The resonant frequency is determined by the stiffness and the mass distribution (effective mass moved) within the actuator. Therefore, the resonant frequency of a complete system can deviate considerably from the resonant frequency of a single actuator. This is an important fact, for example, when using a mirror for fast tilting.

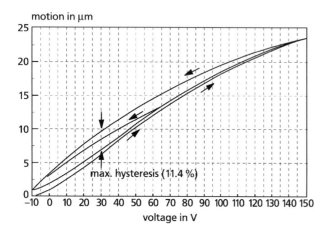

FIGURE 30.14 Hysteresis curve at unipolar voltage (reproduced by permission of Piezosysteme Jena, Germany).

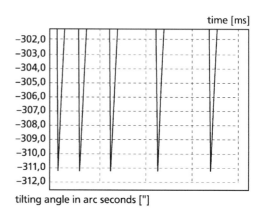

FIGURE 30.16 Repeatability of a mirror mount with periodic motion (reproduced by permission of Piezosysteme Jena, Germany).

Reputable suppliers provide data sheets not only with the resonant frequency but also with the effective mass. Knowing the effective mass, it is possible to estimate the resonant frequency f_{res}^1 with an additional mass (using formula (30.7)):

$$f_{res}^1 = 1/2\pi \times \sqrt{(c_T/(m_{eff}+M))} = f_{res}^0 \times \sqrt{(m_{eff}/(m_{eff}+M))}. \tag{30.7}$$

Here, f_{res}^0 is the resonant frequency of the single actuator, m_{eff} is the mass off the actuator system and M is the additional mass, e.g. a mirror.

The *rise time*, previously mentioned in general, can be treated in a special way for piezoactuators. The shortest rise time, t_{min}, an actuator needs for expansion is determined by its resonant frequency f_{res}:

$$t_{min} = (3 f_{res})^{-1} \tag{30.8}$$

If a short electrical pulse is applied to an actuator, the actuator expands within its rise time t_{min}. The piezoelement's resonant frequency will be simultaneously excited. Therefore, it begins to oscillate with a damped oscillation relative to its position. A shorter electrical pulse can result in a higher super-elevation but the rise time will not be shortened.

A typical response to a short electrical excitation of a piezoactuator (PAHL 40/20 from Piezosysteme Jena) is shown in Figure 30.17. Although the excitation pulse has a duration of approximately 8 μs, the rise time of the actuator is only 20 μs. This value agrees with a resonant frequency of 16 kHz.

While working in the dynamic regime, compressive stress and tensile forces act on the piezoactuators. The compressive strength of piezoactuators is very high, but they are very sensitive to tensile forces. However, both forces occur to the same order while moving dynamically. Typical tensile and compressive stress forces for piezoactuators without pre-loads are shown in Table 30.4.

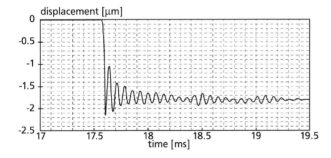

FIGURE 30.17 Response of a piezoelement (reproduced by permission of Piezosysteme Jena, Germany).

TABLE 30.4
Typical Values of Compressive Stress and Tensile Forces on Piezoelements during Dynamic Operation

Compressive Stress Forces (kg cm^{-2})	Tensile Force (kg cm^{-2})
~9000	~50

According to this, *dynamic forces* have to be considered while in dynamic motion. They also appear without external loads! Actuators without pre-loads can only be used for small motions in special cases. Hence, it is necessary to use pre-loaded actuators for dynamic applications.

30.4.8 Acousto-Optic Deflectors

AO deflectors are completely different from mirror-based scanning devices with respect to the physical mechanism for beam deflection. The mirror systems mentioned before reflect the radiation, AO deflectors deviate the beam while it is being transmitted through a crystal. According to the optical materials used in the AO deflectors, they can be applied to every application that necessitates beam manipulation but the parameters differ appropriately with the material.

The principle in brief is as follows. An acoustic wave, generated by an rf signal, is applied to a piezoelectric transducer, bonded to a suitable crystal, which promotes a phase grating in the crystal, while it travels through the crystal at the acoustic velocity of the material with the acoustic wavelength being dependent on the frequency of the rf signal. Any incident laser beam will be diffracted by this grating, generally giving a number of diffracted beams.

Using the *Bragg regime* (Figure 30.18), beam deflection can be achieved. In this case, at one particular angle of incidence, only one diffraction order is produced—all the others are annihilated by destructive interference. Most AO devices operate in the Bragg regime, e.g. modulators and deflectors—the common exception being AO mode-lockers and Q-switches.

For specific use as deflectors, the resolution, efficiency and bandwidth are of interest in most applications. A high *resolution* is provided by the use of large-sized crystals (up to 30 mm or more) in order to work with large beam diameters, decrease optical divergence and increase resolution. The *static resolution*, r, is defined as the number of distinct directions that the diffracted beam can have. The centre of two consecutive points will be separated by the laser beam diameter (at $1/e^2$) in the case of a TEM$_{00}$ beam:

$$r = (\pi/4) \times \Delta f \times (d/v) \tag{30.9}$$

where Δf is the AO frequency range (f_{min}–f_{max}); this is also called the *bandwidth*. d describes the beam diameter at an intensity $1/e^2$ and v is the acoustic velocity.

To avoid first- and second-order overlapping, the bandwidth is limited to an octave, and therefore, the maximum scan angle is also limited.

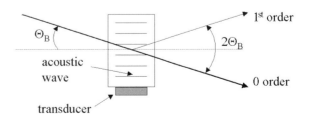

FIGURE 30.18 Bragg regime.

Due to the use of AO devices in scanning applications as well, the *access time* t_a—a synonym for transient response but more common when referring to AOs—of the deflector is of interest, too. This corresponds to the time necessary for the acoustic wave to travel through the laser beam and, thus, to the time needed by the deflector to move from one position to another:

$$t_a = d/v \qquad (30.10)$$

where d is the beam width and v is the acoustic velocity of the material.

Hence, beam width and acoustic velocity work arbitrarily. Therefore, a high static resolution is achieved by the use of a large-sized crystal with a high access time. It should be noted that the acoustic velocity depends on the temperature of the material. Therefore, even the AO characteristics are influenced by heating.

A deflector is often characterized with the access time and bandwidth product ($t_a \times \Delta f$). Typical values are summarized in Table 30.6.

When the field of frequencies no longer consists of discrete values, as in step scanning, but continuous sweeping as in raster scanning, it is necessary to define the *dynamic resolution* r_d, which takes account of the 'gradient' of the frequencies.

In the case of linear frequency sweeping, at $z=0$ (i.e. on entry into the crystal), the frequency f is equal to

$$f = f_0 + (\Delta f/T) \times 1 \qquad (30.11)$$

where f_0 is the frequency at the beginning of the sweeping and T is the time to sweep from f_{min} to f_{max}. At z,

$$f = f_0 + (\Delta f/T) \times t - (\Delta f/T) \times (z/v). \qquad (30.12)$$

The angle of deviation, $\Theta = \lambda f/v$, is now a function of the distance z and of time t. For z and $z+dz$, the angle of deviation is not the same. Focusing occurs in only one plane of the diffracted beam. It is important to note this cylindrical lens effect, which intervenes during sequential sweeping, e.g. in televisions with a raster scan, printing and so on.

An equivalent cylindrical focal length is given by

$$F_{cyl} = \alpha^2 (v/\lambda \times (df/dt)^{-1}) \qquad (30.13)$$

with df/dt is the frequency modulation slope and α is a parameter depending on the beam profile (1 for a rectangular shape, ~1.34 for TEM_{00}).

The *efficiency*, i.e. the amount of irradiance at a distinct angle of deviation with respect to the irradiance of the incident laser beam, usually shows a dependence on the frequency as can be seen in Figure 30.19 (dotted line). However, some applications require a quasi-constant efficiency, on all bandwidths. This can be obtained by decreasing the width of the ultrasonic beam but with a resulting loss in maximum efficiency, except for an anisotropic interaction. In this case, it is possible to match the Bragg condition at specific interaction angles with two synchronous frequencies. Therefore, the deflection angle range can be broadened, providing a good efficiency—see Figure 30.19 (full line).

Table 30.5 gives some examples of AO materials including specific parameters, and Table 30.6 shows some specific properties for TeO_2 with respect to different beam diameters.

The values displayed in Tables 30.5 and 30.6 are taken from consumer information supplied by AA Acousto-optics, France.

Due to their optical properties, AO devices affect the incident radiation. Hence, optical influences, e.g. on the polarization, should be mentioned but are no matter for this chapter. Additional information can be found in other literature, for example [17].

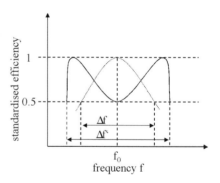

FIGURE 30.19 Efficiency of AO deflectors.

TABLE 30.5

Selected Properties of AO Materials

Material	Refractive Index	At λ (nm)	Max cw Laser Power Density (W cm^{-2})	Acoustic Velocity (m s^{-1})	M^2 (AO Figure of Merit) ($\times 10^{-15}$ S^{3}kg^{-1})	At λ (nm)
Ge	4	10 600	5	5500	180	10 600
Doped glass	2.09	633	1	3400	24	633
$Ge_{33}As_{12}Se_{55}$	2.59	1064	1	2520	248	1064
As_2S_3	2.46	1150	1	2600	433	633
$PbMoO_4$	2.26/2.38	633	0.5	3630	36	633
TeO_2	2.26	633	5	4200	34	633
TeO_2	2.26	633	5	620	1200	633
SiO_2 (fused silica)	1.46	633	>100	5960	1.5	633
SiO_2 (fused silica)	1.46	633	>100	3760	0.5	633

TABLE 30.6

Property Dependency of TeO_2 Regarding Maximum Aperture[a]

Aperture Ø (mm)	Efficiency (%)	Access Time T_a (µs)	Bandwidth Δf (MHz)	$T \times \Delta f$	Deflection Angle Θ (mrad)
1.7[b]	70	2	50	100	48
4.2[b]	70	5	50	250	48
6.7[b]	70	8	50	400	48
1.7[c]	40	2	50×50	100×100	48×48
4.2[c]	40	5	50×50	250×250	48×48
6.7[c]	40	8	50×50	400×400	48×48

[a] Spectral range (nm) depends on the specific anti-reflection coating: 514–532, 633–670, 780–820 or 360–1100.
[b] Single axis.
[c] Dual axes.

30.5 Conclusion

Given the broad distribution of laser applications, whether it be household, medicine, measuring techniques or material processing, it is not possible to mention all aspects linked to positioning and scanning systems. Limiting ourselves to industrial material processing has successfully reduced the work. However, the responsibility of finding appropriate solutions for beam delivery systems lies in the hands of those involved in developing them, aided by their imagination and knowledge of the possibilities. Hopefully, the latter has been amplified here.

Acknowledgements

The author would like to thank Dr Braun (Fraunhofer IPM), Fr Schmid (Trumpf GmbH), Dr Jovanovic (Pegasus Optik), Mr Hartmann (Arges), Dr Blank (LPKF Motion Control), Mr Elbinger (Piezosysteme Jena), Mr Höppe (Reis-robotics) and Crisel Instruments for support, assistance and discussion.

REFERENCES

1. Koch J, Wolf M and Sepold G 1999 Mikrobohrungen in Dünnblech-hergestellt mit dem Laser *wt werkstattstechnik* **89** 492–4.
2. Weck M, Kruger P and Brecher C 2001 Limits for controller settings with electric linear direct drives *Int. J. Machine Tools Manufacture* **41** 65–88.
3. Kami Y, Yabuya M and Shimizu T 1995 Research and development of an ultraprecision positioning system *Nanotechnology* **5** 127–34.
4. Chang S B, Wu S H and Hu Y C 1997 Submicrometre overshoot control of rapid and precise positioning *Precision Eng.* **20** 161–70.
5. Petiot J-F, Chedmail P and Hascoet J-Y 1998 Contribution to the scheduling of trajectories in robotics *Robot. Computer-Integrated Manufacturing* **14** 237–51.
6. Chen J S and Dwang I C 2000 A ballscrew drive mechanism with piezo-electric nut for preload and motion control International *J. Machine Tool Manufacture* **40** 513–26.
7. Hand D P, Fox M D T, Haran F M, Peters C, Morgan S A, McLean M A, Steen W M and Jones J D C 2000 Optical focus control system for laser welding and direct casting *Opt. Lasers Eng.* **34** 415–27.
8. Liu J S, Li L J and Jin X Z 1999 Accuracy control of three-dimensional Nd:YAG laser shaping by ablation *Opt. Laser Technol.* **31** 419–23.
9. Hoffmann P, Schwab J and Kugler P 2002 Laserstrahlhartlöten-Systemtechnik, Anwendungen und Potenziale *Laserstrahlfügen: Prozesse, Systeme, Anwendungen, Trends Strahltechnik* Band 19, ed G Sepold and T Seefeld (Bremen: BIAS) pp 315–24.
10. Grey D J 1997 Design of high performance CO2 laser beam scanning objectives *Lasers in Materials Processing (Proc. SPIE 3097)* ed L H J F Beckmann (Bellingham, WA: SPIE).
11. Kim J-D, Jung J-K, Jeon B-C and Cho C-D 2001 Wide band laser heat treatment using pyramid polygon mirror *Opt. Lasers Eng.* **35** 285–97.
12. Hornig 2002 Laserstrahlbearbeitung bei BMW: Anwendungen und Trends *Laserstrahlfügen: Prozesse, Systeme, Anwendungen, Trends Strahltechnik* Band 19, ed G Sepold and T Seefeld (Bremen: BIAS) pp 351–65.
13. Rizvi N and Apte P 2002 Developments in laser micro-machining technique *J. Mater. Process. Technol.* **127** 206–10.
14. Koch J and Lang A 1997 Feinstbohren mit dem Kupferdampflaser *Präzisionslaserstrahl-fertigungstechnik für den Maschinenbau* ed M Geiger et al. (Erlangen: Bayerisches Laserzentrum) pp 127–44.
15. Schenk H, Dürr P, Kunze D, Lakner H and Kuck H 2001 A resonantly excited 2D-micro-scanning-mirror with large deflection *Sensors Actuators A* **89** 104–11.
16. Piezosysteme 2001 *Piezoline part of catalogue (Jena: Piezosysteme Jena GmbH)* which is also available on the internet: www.piezojena.com.
17. Blomme E, Gondek G, Katkowski T, Kwiek P, Sliwinski A and Leroy O 2000 On the polarization of light diffracted by ultrasound *Ultrasonics* **38** 575–80.

FURTHER READING

Agullo-Lopez F and Cabrera J M 1994 *Electrooptics: Phenomena, Materials and Applications* (San Diego, CA: Academic).
Introduces the fundamental concepts and phenomena of electro optics at an undergraduate and graduate level. Gives a

broad overview of inorganic materials, including ceramics and novel single crystals.

Beiser L 2003 *Unified Optical Scanning Technology* (New York: Wiley) ISBN 0-471-31654-7.

A cohesive view of the expanding field of optical scanning, in a single compact volume.

Chen J S and Dwang I C 2000 A ballscrew drive mechanism with piezo-electric nut for preload and motion control *International J. Machine Tool Manufacture* **40** 513–26.

Another scientific presentation. Introduces a long stroke, high-precision positioning system performed by combination of a ball screw drive unit additionally equipped with a piezo-electric nut. Considers stick-slip motion at low speed and mentions friction as well.

Chang S B, Wu S H and Hu Y C 1997 Submicrometre overshoot control of rapid and precise positioning *Precision Eng.* **20** 161–70.

Written from a scientific point of view. An optimized controller for high-precision positioning of ball guide tables is introduced. Considers several possible errors, like stick-slip motion and overshoot.

Goutzoulis A and Pape D 1994 *Design and Fabrication of Acousto-Optic Devices* (New York: Marcel Dekker)

Offers in-depth discussions on all aspects of acousto-optic deflectors, modulators and tenable filters-emphasizing hands-on procedures for design, fabrication and testing.

Montague J and DeWeerd H 1996 Optomechanical scanning application, techniques, and devices *The Infrared and Electro-Optical Systems Handbook* 2nd edn, vol 3, ed J S Accetta and D L Shumaker (Bellingham, WA: SPIE).

A good review on scanning techniques and applications.

n.n. 1996 *The Fascinating World of Sheet Metal* ed TRUMPF GmbH & Co. (Dietzingen: Trumpf GmbH)

Overview on sheet metal with general information related to laser processing of sheet metal. Includes CO_2- and Nd:YAG high-power lasers as well as workpiece positioning and beam delivery systems. Also available on the Internet: www.trumpf.com.

Petiot J-F, Chedmail P and Hascoet J-Y 1998 Contribution to the scheduling of trajectories in robotics *Robot. Computer-Integrated Manufacturing* **14** 237–51.

Written from a scientific point of view. Considers a general problem that comes along with robotics, the travelling salesman, with brief mention of commonly used solutions. Introduces a new approach to optimization. With detailed reference listing.

Waynant R and Edinger M 2000 *Electro-Optics Handbook* (New York: McGraw).

Master the latest electro-optical sources, materials, detectors and applications. Mentions a wide array of gas, solid state, semiconductor, dye, chemical and free electron lasers.

31

Laser Beam Measurement: Section Introduction

Julian Jones

To characterize a laser beam requires complete knowledge of its mutual coherence function—the power (or energy) distribution and the relative phase distribution over a single transverse plane—achieved by complex measurement and subsequent analysis. In Chapter 32, Brooke Ward describes a useful approximation to the ideal, based on measuring the beam diameter and divergence at a set of transverse planes along the axis of the laser beam, considering how the diameter is to be defined (using a second-moment technique) for different types of beams; how the measurements are to be made, using sensor arrays or scanning slits.

Intrinsic to the characterization of a beam is the ability to be able to measure optical power or energy. In Chapter 33, the authors consider various practical detector devices that are required to cover the range of wavelengths of the lasers described elsewhere in this Handbook, extending from ultraviolet to infrared. In the ultraviolet, there remains a place for the vacuum-tube photomultiplier, although photodiodes (notably of the Schottky type), using semiconductor materials such as Si, SiC, GaP and GaN, are useful. More recently, semiconductor artificial diamond has come forward as a promising material on account of its very wide bandgap. For many purposes, the silicon photodiode is ubiquitous for measurements in the visible wavelength range. Infrared detection poses more of a challenge. In the nearest wavelength range, say up to the 1.55 µm typical of optical communications, InGaAsP approaches the ideal. At longer wavelengths, where the thermal quantum approaches the optical quantum energy, more complex structures (such as quantum wells) and different materials are needed, such as the lead salts in the mid-infrared; perhaps HgCdTe is the most flexible material at these wavelengths. Thermal detectors, or *bolometers*, have their place for the detection of the longest wavelengths and development continues, e.g. with the silicon micro-bolometer.

International standards exist for the measurement of the radiant power and energy of lasers. Techniques for the transfer of these standards are discussed by Robert Tyson in Chapter 34, spanning the wide wavelengths (IR to UV) and pulse lengths (fs to cw) required for practical lasers, noting the standard calibration categories, and describing some standard test configurations. Returning to the complete characterization of a laser beam in terms of the irradiance and phase distributions in a single transverse plane and which are often written as $I(x, y)$ and $\phi(x, y)$, Bernd Schäfer in Chapter 35 reviews the relevant international standards ISO13694 and ISO15367. The irradiance (power density) is measured with array or scanning detectors. The phase distribution is measured with wavefront sensors, such as those used in adaptive optics (Chapter 14), e.g. the Shack–Hartmann sensor, or by using an interferometer (Chapter 3). Alternatively, the complete mutual coherence function can be measured using an optical arrangement similar to that used for producing Young's fringes. Finally, temporal characterization of ultrashort laser pulses is reviewed in Chapter 36.

32
Beam Propagation

B. A. Ward

CONTENTS

32.1 Introduction ... 461
32.2 Beam Types ... 461
32.3 Beam Diameter ... 461
32.4 Practical Measurements ... 462
32.5 Propagation Characteristics ... 463
32.6 Beam Transformation by a Lens ... 464
32.7 Alternative Measurement Methods ... 464
32.8 Propagation of Astigmatic Beams ... 464
32.9 Summary ... 465
References ... 465

32.1 Introduction

The propagation behaviour of a highly coherent laser beam can be predicted from a detailed knowledge of both the power/energy density distribution and the relative phase distribution in a single transverse plane of the beam. Complete characterization of a more general beam requires knowledge of the mutual coherence function combined with complex measurement and analysis procedures. A much simpler and equally useful procedure that can be used to predict much of the propagation behaviour of laser beams, as well as some incoherent broadband beams of light, involves measurement of the beam diameter and divergence. The measurements are made using second-order moment analysis of the transverse power/energy density distribution at a number of positions along a beam. This section of the handbook is devoted to describing this latter technique. In what follows, the word 'irradiance' is used to mean the power/energy density in a transverse plane of a beam.

32.2 Beam Types

The analytic methods adopted in this section can usually be applied to beams of radiation whose full divergence angle is less than 30°. The simplest type is the circular cross-section beam whose transverse irradiance distribution is radially symmetric. This is a *stigmatic* beam. If the radial symmetry exists at all planes along a beam, it is classified as *intrinsically stigmatic*. If, on the other hand, the beam is circular at only one or two locations along the beam then the beam is classified as astigmatic.

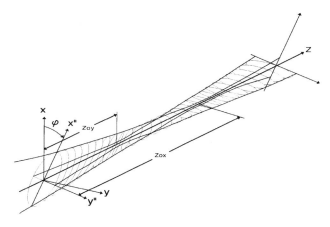

FIGURE 32.1 Beam axis coordinate system showing an astigmatic irradiance distribution and axially separated orthogonal waists at Z_{0x} and Z_{0y}.

If the axes of radial asymmetry of a propagating beam remain parallel to each other then the beam is classified as *simple astigmatic* (see Figure 32.1). If the axes of asymmetry rotate in the transverse plane of a propagating beam then that beam is classified as *general astigmatic*.

32.3 Beam Diameter

There are a number of alternative definitions [1] of the diameter of a beam of electromagnetic radiation. Many are based on selecting the diameter where the irradiance is a given fraction of that at the peak. Others are based on the diameter of the contour enclosing a fraction of the total beam power/energy. Perhaps, the most common definition is known as the

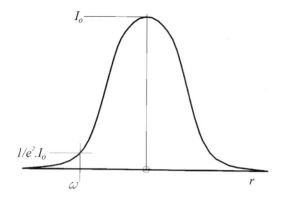

FIGURE 32.2 The $1/e^2$ radius of a lowest order Gaussian irradiance distribution.

$1/e^2$ diameter (Figure 32.2). This is based on the diameter of the fundamental TEM_{00} single Gaussian mode emitted by a stable resonator where the radial irradiance distribution takes the form

$$I = I_0 \cdot \exp\left(-2\left(\frac{r}{\omega}\right)^2\right)$$

and ω is the radius where the irradiance is $1/e^2$ of the peak irradiance I_0.

The fundamental Gaussian mode can be compared with the statistical normal probability or Gaussian error distribution

$$P = P_0 \exp\left(-\frac{1}{2}\left(\frac{r}{\sigma}\right)^2\right)$$

where σ is the standard deviation of the distribution. The relationship results in the convenient definition of the $1/e^2$ diameter to be that of the contour that contains 87% of the total beam power. It is referred to as the second moment diameter, $d_{87} \equiv d_\sigma = 2\omega = 4\sigma$.

This statistical definition of beam diameter is convenient since it does not require identification of a peak irradiance and is not restricted to Gaussian beam forms.

Evaluation of the beam diameter at location z along the beam axis in the direction of the transverse x and y axes is performed by taking the first moments of the irradiance distribution $I(x, y, z)$ to determine the centroid $\bar{x}$, $\bar{y}$ of the beam.

$$\bar{x} = \frac{\sum_x \sum_y x \cdot I(x, y, z)}{\sum_x \sum_y I(x, y, z)} \quad \bar{y} = \frac{\sum_x \sum_y y \cdot I(x, y, z)}{\sum_x \sum_y I(x, y, z)}. \tag{32.1}$$

The diameters of a beam are identified as 'second moment' diameters by using σ to suggest the similarity with statistically determined 'standard deviation' of a variable. The orthogonal diameters are given by:

$$d_{\sigma x}(z) = 4\sigma_x(z) \quad \text{and} \quad d_{\sigma y}(z) = 4\sigma_y(z) \tag{32.2}$$

where the second moments of the power density distribution of the beam at location z are given by:

$$\sigma_x^2(z) = \frac{\sum_x \sum_y (x-\bar{x})^2 \cdot I(x, y, z)}{\sum_x \sum_y I(x, y, z)} \equiv \langle x^2 \rangle$$

$$\sigma_y^2(z) = \frac{\sum_x \sum_y (y-\bar{y})^2 \cdot I(x, y, z)}{\sum_x \sum_y I(x, y, z)} \equiv \langle y^2 \rangle \tag{32.3}$$

32.4 Practical Measurements

The previous expressions have been written with summation symbols rather than integrals to underline the fact that the irradiance data will most likely be acquired by array detectors, and the analysis involved is simple processing of digitized pixel values. However, a number of precautions and approximations have to be made before second moment measurements can be used for propagation predictions.

The raw data collected from an array detector will require processing to reduce the uncertainty in beam diameter measurement that can be introduced by optical noise and sharp-edge diffraction effects. Sources of optical noise include stray light from pump lamps or discharge fluorescence as well as background illumination effects. A high background level can lead to a 'baseline offset' that can cause large errors in the evaluation of beam characteristics.

There are a number of methods for reducing the influence of optical noise. The simplest technique is to record, calculate and subtract an average background illumination pattern from the power density distribution that includes the beam pattern of interest as well as the interfering background. The background pattern can be obtained by carefully blocking and absorbing just the beam under study as it leaves the laser and recording the rest. The averaging procedure is necessary to reduce any statistical fluctuations that might be present in the background pattern. However, it is possible that some of the background light is coherent and capable of producing speckle interference patterns when combined with the main laser beam. If this is the case, more sophisticated compensation techniques may be required [2].

One result of proper background subtraction will be the presence of negative noise values in the corrected distribution. These negative values should be included in the further evaluation procedures to compensate for the presence of residual positive noise signals.

A further uncertainty is introduced by the fact that equations 32.1 and 32.2 imply that the summations have to be carried out over an infinitely wide aperture. This means that the measurements are sensitive to widely scattered light. Beams passing through sharp-edged apertures will suffer diffraction and a certain amount of large-angle scattering and may not be amenable to second moment measurements.

Infinite measurement apertures are obviously impractical, but a solution does exist that preserves the usefulness of second moment width measurements [2]. The total number of

Beam Propagation

pixels used should cover an area whose width is more than four times that of the beam diameter. There should be sufficient number of pixels across the beam to permit an accurate estimate to be made of the second moment width. It is suggested that there should be more than 400 kpixels in a suitable array detector. The technique involves making an initial estimate of the beam widths using all available pixels. The summations for both the centroid and width estimates are then repeated, but the pixels included are limited to those that are within a rectangle whose sides are three times the estimated beam widths along each axis and centred on the beam centroid. The procedure results in a new estimate of the beam width which is then used to repeat the process. The iteration continues until the width estimate converges to a steady value. In practice, five to seven cycles are generally sufficient to produce a value that is precise, reproducible and useful for propagation prediction. The procedure is summarized by a modification of equation 32.3 so that:

$$\sigma_x^2(z) = \frac{\sum_{y_1}^{y_2} \sum_{x_1}^{x_2} (x-\bar{x})^2 \cdot I(x, y, z)}{\sum_{y_1}^{y_2} \sum_{x_1}^{x_2} I(x, y, z)} \equiv \langle x^2 \rangle$$

$$\sigma_y^2(z) = \frac{\sum_{y_1}^{y_2} \sum_{x_1}^{x_2} (y-\bar{y})^2 \cdot I(x, y, z)}{\sum_{y_1}^{y_2} \sum_{x_1}^{x_2} I(x, y, z)} \equiv \langle y^2 \rangle.$$

(32.4)

The summations are carried out over a rectangle parallel to the *x*- and *y*-axes where

$$x_1 = \bar{x} - \frac{3}{2} d_{\sigma x} \quad x_2 = \bar{x} + \frac{3}{2} d_{\sigma x}$$
$$y_1 = \bar{y} - \frac{3}{2} d_{\sigma y} \quad y_2 = \bar{y} + \frac{3}{2} d_{\sigma y}.$$

(32.5)

This procedure for the analysis of beam diameter can also be applied to some incoherent and broadband beams as well as monochromatic laser beams. It has been used for the examination of a number of beam types. For many sources, it is found to provide a reproducible estimate of beam width and demonstrates useful precision. However, the absolute accuracy of the beam propagation parameters derived using this method may not be high enough for some of the more complex beams. Nevertheless, the procedure appears to provide a robust and transportable protocol for optical beam analysis that might be used in different laboratories to examine the same source and to give similar results.

32.5 Propagation Characteristics

Investigations of the propagation behaviour of an arbitrary optical beam are reviewed in [3–5]. They have revealed that two of the most useful parameters for predicting the diameters of a propagating beam are the dimensions of the beam waist $d_{\sigma 0}$ and the far-field divergence of the beam Θ_σ. If the diameter of the beam is measured using the second moment of its irradiance distribution, it will vary in a hyperbolic manner and be symmetrical about the beam waist [6].

The usefulness of the waist width and far-field divergence measurements stems from the fact that their product is invariant when passing through an aberration-free optical system of sufficient aperture to avoid detectable truncation of the beam. The far-field divergence can be determined by inserting an aberration-free lens (focal length *f*) on the axis of the beam under investigation and measuring the second moment diameter $d_{\sigma f}$ of the irradiance distribution in the focal plane of the transforming lens. The far-field divergence is $\Theta_\sigma = d_{\sigma f}/f$. The far-field is that part of a beam that is a significant distance (>5 Rayleigh lengths) from the waist of the beam (Figure 32.3)

Locations close to the beam waist are said to be in the near-field. A useful parameter for describing distances from a beam waist is the Rayleigh length $z_R = d_{0\sigma}/\Theta_\sigma$ (synonymous with the Rayleigh range. It is the distance from the waist over which the cross-sectional area of the beam has doubled).

Figure 32.3 indicates the 87% enclosed power/energy contour. This contour will have a symmetrical shape with an hyperbolic form $d_{\sigma x}^2(z) = a + bz + cz^2$. This fact indicates a convenient method for measuring the main propagation parameters of a beam. The second moment diameter is measured at ten or more positions around the waist, and statistical procedures are used to determine the most probable hyperbola coefficients [7]. The location and diameter of the beam waist are then given by:

$$z_0 = \frac{-b}{2c} \quad \text{and} \quad d_{\sigma 0} = \sqrt{a - \frac{b^2}{4c}}. \quad (32.6)$$

The other propagation constants, i.e. divergence, propagation ratio and Rayleigh length, are given by

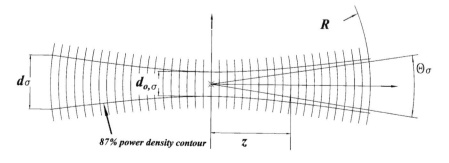

FIGURE 32.3 Characteristics of a beam of electromagnetic radiation.

$$\Theta_\sigma = \sqrt{c} \quad M^2 = \frac{\pi}{4\lambda}\sqrt{a \cdot c - \frac{b^2}{4}}$$

$$\text{and} \quad z_R = \frac{1}{c}\sqrt{a \cdot c - \frac{b^2}{4}}. \tag{32.7}$$

The propagation ratio M^2 is a quantity popularly known as the 'beam quality factor'. It is now called the 'beam propagation ratio'. Its value is based on the fact that the product of the waist diameter and far-field divergence of a beam is an invariant of propagation. In other words, this value is not changed by passing the beam through an aberration-free optical system. The smallest value of the diameter/divergence product is that achieved by the fundamental TEM$_{00}$ Gaussian beam and has the value $\frac{4}{\pi}\lambda^2$, where λ is the wavelength of the monochromatic radiation. M^2 is the ratio of the diameter/divergence product of a beam being characterized to this theoretical minimum product.

32.6 Beam Transformation by a Lens

When a beam enters an aberration-free lens, the emerging beam can also be described by a hyperbola. The new hyperbola will have its own Rayleigh length, waist diameter and asymptotic divergence (Figure 32.4). Prediction of the width and position of the waist of the laser beam after it has passed through a lens is relatively simple.

The expressions for the diameter and location of the transformed beam are best considered in terms of the displacement of the input and output beam waists from the front and rear focal planes of the lens, respectively (see Figure 32.4).

The relationship between the propagation characteristics of the input and output beams is summarized as follows:

$$\text{if } d_{02} = V_1 \cdot d_{01} \text{ where } V_1^2 = \frac{f^2}{x^2 + z_{R1}^2} \tag{32.8}$$

$$\text{then } y = V_1^2 \cdot x \quad \Theta_2 = \frac{1}{V_1} \cdot \Theta_1$$

$$\text{and } z_{R2} = V_1^2 \cdot z_{R1}. \tag{32.9}$$

It should be noted that as the Rayleigh length of the input beam approaches zero (i.e. the waist becomes a point source), then the relationship between the locations of the input and output waists of the beams approaches the Newtonian expression for geometrical image formation, $x \cdot y = f^2$. This is an interesting result since it enables examination of the relationship between the image-forming properties in a beam and the contrasting behaviour of propagation envelopes.

32.7 Alternative Measurement Methods

Sensor arrays provide the greatest detail and analysis capability when measuring beam properties. However, simpler detectors and technologies do exist. An effective method for measuring the power content diameter of a stigmatic beam is the use of a sequence of variable circular apertures to locate that diameter that contains (say) 87% of the total beam power. Alternative methods that can accommodate astigmatic beams are based on the scanning slit or knife edge sensor. These instruments, their limitations and correction methods are discussed in detail in the international standard [2]. They are very useful devices for characterizing good quality beams, but mixed beam modes and sharp-edge diffraction effects can lead to significant measurement uncertainties.

It is possible to test the measurement capability of an instrument when examining a beam of unknown quality. The technique is based on measuring the propagation characteristics of the raw subject beam and using them to predict the location and size of the new waist produced by insertion of a low-aberration (long focal length) transforming lens. The location and size of the transformed waist can then be confirmed by measurements taken with the same instrument on the actual transformed beam.

32.8 Propagation of Astigmatic Beams

Most of the previous discussion has been aimed at the propagation properties of stigmatic or simple astigmatic beams. In fact, if the ratio $\sigma_x/\sigma_y \leq 1.15$, then the beam may be considered circular with a diameter $d_\sigma = 2\sqrt{\sigma_x^2 + \sigma_y^2}$ at the measurement location. If the azimuth angle φ (Figure 32.1) varies by no more than 10° over a distance of two Rayleigh lengths, the beam may be simple astigmatic. In this case, an effective

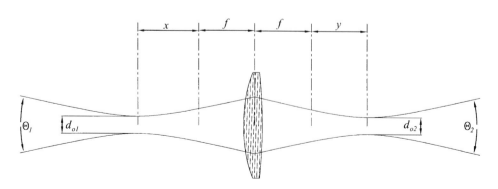

FIGURE 32.4 Beam transformation by a lens. Waist distances are measured from the focal planes of the lens.

beam propagation ratio can be defined to be $M_{\text{eff}}^2 = \sqrt{M_x^2 \cdot M_y^2}$. However, if neither of these two conditions applies, the beam is probably general astigmatic.

Analysis and propagation prediction of a general astigmatic beam is obviously more complex than the study of simple astigmatic beams. Nevertheless, the measurement and evaluation principles are virtually the same for both beam types. The concept of second moment measurements is now extended to include the 'mixed moments' of the spatial and divergence properties of the beam. For example, the spatial mixed moment is:

$$\sigma_{xy}^2(z) = \frac{\sum_{y_1}^{y_2} \sum_{x_1}^{x_2} (x - \bar{x})(y - \bar{y}) \cdot I(x, y, z)}{\sum_{y_1}^{y_2} \sum_{x_1}^{x_2} I(x, y, z)} \equiv \langle xy \rangle. \tag{32.10}$$

This measurement can be combined with the second moment diameter measurements to reveal the mean azimuth angle φ of the irradiance distribution in the laboratory coordinate system. The alternative nomenclature used in equations (32.4) and (32.10) for second moment measurements is now used to simplify the equations.

$$\varphi = \frac{1}{2} \arctan \left[\frac{2\langle xy \rangle}{\langle x^2 \rangle - \langle y^2 \rangle} \right] \tag{32.11}$$

There are ten second moment values that can be used to specify the propagation properties of a general astigmatic beam. Those moments can be used to form an ABCD matrix that can be used to compute the changes in the beam characteristics as it passes through an idealized optical system [8]. The form of the matrix and its application is summarized in considerable detail in the forthcoming revision of the international standard on beam propagation [2]. The overall beam matrix has the form:

$$P = \begin{pmatrix} \langle x^2 \rangle & \langle xy \rangle & \langle x\Theta_x \rangle & \langle x\Theta_y \rangle \\ \langle xy \rangle & \langle y^2 \rangle & \langle y\Theta_x \rangle & \langle y\Theta_y \rangle \\ \langle x\Theta_x \rangle & \langle y\Theta_x \rangle & \langle \Theta_x^2 \rangle & \langle \Theta_x \Theta_y \rangle \\ \langle x\Theta_y \rangle & \langle y\Theta_y \rangle & \langle \Theta_x \Theta_y \rangle & \langle \Theta_y^2 \rangle \end{pmatrix} \tag{32.12}$$

In spite of the complexity of the definition of a general astigmatic beam, it is still possible to define a beam propagation ratio as an invariant of propagation

$$M_{\text{eff}}^2 = \frac{4\pi}{\lambda} (\det(P))^{1/4} \tag{32.13}$$

Other invariants of propagation through a homogeneous system of spherical optical elements are the intrinsic astigmatism and the twist parameter. This first parameter results from a complex combination of the ten second moment properties of the beam. The twist parameter $t = \langle x\Theta_y \rangle - \langle y\Theta_x \rangle$ is a parameter that can be changed by propagation through cylindrical lenses.

32.9 Summary

Many of the measurement techniques and analysis methods discussed in this section are specified and described in the latest revised international standard on beam parameter measurements [2] published in 2005. This standard was last reviewed and confirmed in 2013. Therefore, this version remains current.

REFERENCES

1. ISO 11145:2001 Optics and optical instruments—Lasers and laser related equipment—Vocabulary and symbols. Available as BS EN ISO 11145:2001 from the British Standards Institute.
2. Revised ISO 11146 Lasers and laser related equipment—Test methods for laser beam widths, divergence angle and beam propagation factor—Part 1: Stigmatic and simple astigmatic beams. Part 2: General astigmatic beams. Part 3: Alternative test methods and geometrical laser beam classification and propagation. 2005.
3. Nemes G and Serna J 1997 The ten physical parameters associated with a full general astigmatic beam *4th Int, Workshop on Laser Beam and Optics Characterization* ed A Giesen and M Morin (Munich: IFSW/VDI Messe) pp. 92–105.
4. Sasnett M W and Johnston T F 1991 *Conf. on Laser Beam Diagnostics (Proc. SPIE)* **1414** 1–32.
5. Siegman A E 1993 *Laser Beam Characterization* ed P M Mejias, H Weber, R Martinez-Herrero and A Gonzalez-Urena (Madrid: SEDO) pp. 1–22.
6. Siegman A E 1990 New developments in laser resonators *Optical Resonators (Proc SPIE)* **1224** 2–14.
7. Ward B A 1993 *Laser Beam Characterization* P M Mejias, H Weber, R Martinez-Herrero and A Gonzalez-Urena (Madrid: SEDO) pp. 62–4.
8. Belanger P A 1991 Beam propagation and the ABCD ray matrices *Opt. Lett.* **16** 196.

33

Laser Beam Management Detectors

Alexander O. Goushcha and Bernd Tabbert

CONTENTS

- 33.1 Introduction 467
- 33.2 Position-sensitive Detectors 467
  - 33.2.1 Resistive Charge Division Sensors 468
  - 33.2.2 Strip Detector and Quadrant Detector 468
- 33.3 Multi-pixel Arrays 469
  - 33.3.1 Photodiode Arrays 469
  - 33.3.2 Charge-coupled Devices as Tracking/Image Sensors 470
  - 33.3.3 Complementary Metal Oxide Semiconductor Detectors 470
- 33.4 Fast Response Photodetectors 471
- 33.5 Photodetectors with the Intrinsic Amplification 472
- 33.6 Colour-sensitive Detectors 472
- 33.7 Detectors for the UV Spectral Range 473
- 33.8 Detectors for the IR Spectral Range 473
- 33.9 Common Customizations 474
  - 33.9.1 Active Area Sizes, Shapes and Apertures 474
  - 33.9.2 Coatings and Filters 474
- 33.10 Photodetector Integration 475
  - 33.10.1 Packaging 475
  - 33.10.2 Hybrids and Detectors for Fibre-coupled Lasers 475
- 33.11 Future Directions in Detectors Technology and Applications 476
- References 477
- Further Reading 477

33.1 Introduction

Photodetectors are widely used in laser applications to monitor laser power, beam alignment and shape, laser pulse characteristics and other parameters of radiation generated by laser systems. The principles of the reliable detection of light, in any region of the spectrum, depend on careful consideration of a large number of factors. The importance of signal level, noise and signal-to-noise ratio cannot be underestimated, and these general considerations are discussed in detail elsewhere (see Chapter 9 of this Handbook), together with details of the operating principles of many generic detectors. Some specifics of managing and detection the short laser pulses are discussed in Chapters 18, 19, 21 and 36 of this handbook and the details on design and operation of detectors for laser energy and power measurements are presented in Chapters 21, 34, and 35 of the Handbook.

This chapter focuses on features of the photodetectors used for various beam management applications – position sensing, beam tracking, laser guiding, remote sensing, spot cross-section analysis, imaging and other applications. We will start with briefly discussing design specifics of photodetectors for position sensing and beam tracking. Then we will provide some useful insights into detection of short laser pulses, specifics of the detectors with intrinsic amplifications, detection of multi-colour optical beams and features for the UV and IR light detection. Lastly, we will continue with an overview on how to find the right customization and discuss the latest trends in the development of the next-generation photodetectors.

33.2 Position-sensitive Detectors

Large-area position-sensing detectors (PSDs) can detect and record the position of incident light beams. Many industrial manufacturers and laboratories use PSDs in their daily work. PSDs are able to characterize lasers and align optical systems during the manufacturing process. In conjunction with lasers they can be used for industrial alignment, calibration, analysis of machinery, articulated robotic beam delivery systems, quality control of lasers, pointing stability measurements and other various positioning and beam monitoring systems. They provide outstanding resolution, fast response, excellent linearity for a wide range of light intensities and simple operating circuits.

PSD is a photoelectric device that converts an incident light spot into continuous position data. PSDs come in two types: Lateral effect or resistive charge division sensors and Quadrant or the dual-axis detectors. The purpose of these two types is to sense the position of the beam centroid in the X-Y plane orthogonal to the optical axis.

The other type of photodetectors used for position sensing is multi-pixel arrays. Such arrays are used also for optical beam profiling, remote sensing and other imaging applications. In this section, we will discuss properties of the most widely used PSDs and multi-pixel arrays and concentrate primarily on the semiconductor devices.

33.2.1 Resistive Charge Division Sensors

The lateral effect semiconductor PSDs take advantage of the resistive division of charges generated by light in the bulk of the detector. They typically offer less positional sensitivity, but much greater dynamic range and faster response time than multi-pixel array detectors. Resistive charge division sensors can be designed as either photoconductive or photovoltaic devices. Considering the most widely used lateral effect PSDs based of a photodiode structure, the active sensing region in PSD is formed by a p-n junction. Electrodes may be arranged on both the anode and cathode of the photodiode so that the photodiode internal resistance forms a series resistance for current between the electrodes.

As an example, semiconductor resistive charge division sensor consists of a segment of photodetective semiconductor (moderately doped p^+-layer on the top of n-type substrate) with two terminals for signal output and a terminal for application of back-bias voltage – see Figure 33.1a. When the absorbed photons produce charge carriers (electrons and holes) in the semiconductor bulk, the charge carriers move to the appropriate electrode (anode or cathode). However, this means the carriers have to first pass through the resistive semiconductor layer. The photocurrent at each electrode is inversely proportional to the distance x between that electrode and the centroid of the incoming light beam. Thus, the location of the beam centroid can be accurately determined by using the ratio of the signals from the different electrodes.

Position information using such devices is derived by comparing the signal outputs from each terminal. For example, for a four-terminal, two-dimensional PSD shown schematically in Figure 33.1b, the position may be calculated from:

$$x = \frac{D_x}{2} \frac{I_D - I_B}{I_D + I_B} \quad \text{and} \quad y = \frac{D_y}{2} \frac{I_C - I_A}{I_C + I_A} \quad (33.1)$$

where I_A, I_B, I_C and I_D denote photocurrents collected by the respective electrodes and D_x, D_y is the active area size in x and y dimensions, respectively.

Position sensors of this type have several advantages over, e.g. quadrant detectors or multi-pixel arrays. Because there are no gaps in the active area, the size of the image is not subject to constraints on a minimum diameter. A second advantage is that position information is available as long as the image falls somewhere on the detector's active area. Among the disadvantages is that frequency response of such detectors tends to be lower than for a conventional semiconductor detector of the same size, because of the series resistance that the photocurrent encounters at the detector surface. The other drawback is the pincushion distortion effect that causes non-linearity in the location determined using equation 33.1 versus the actual centroid of the impinging light beam. Such pincushion distortion occurs due to non-uniformity of the electric field distribution across the active area of the device.

33.2.2 Strip Detector and Quadrant Detector

Another way to measure position of the laser beam is obtained by dividing the large-area photodiode into many small, strip-like regions and to read them out separately – see Figure 33.2. The measurement precision depends mainly on the strip spacing and the readout method.

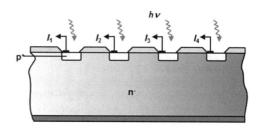

FIGURE 33.2 Cross-sectional schematics of a semiconductor diode strip detector.

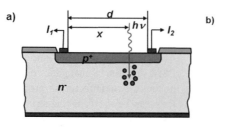
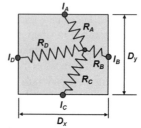

FIGURE 33.1 Cross section of the lateral effect PSD with two terminals (a) and schematic representation of different resistive paths on the four-terminal lateral effect PSD.

A quadrant photodiode is variation of a strip detector, the segments of which are the four quadrants that have separate connectors for a signal output. Each quadrant functions and behaves as a regular *pin* or avalanche photodiode (APD). Quadrant photodiodes are widely used in the tracking systems. Position information is derived from the relative signal output from each segment. When the focused image is centred on the quadrant detector, each segment will receive the same amount of optical radiation, and all four signal outputs will be equal (Figure 33.3, right). As the image moves on the detector surface (corresponding to an angular change of the object being tracked or imaged), more radiation falls on one of the segments and less on the opposite segment (Figure 33.3, centre and left).

The up-down position of the spot is characterized with the relative amplitudes of $(I_A + I_B) - (I_C + I_D)$. Similarly, the term $(I_A + I_D) - (I_B + I_C)$ gives information about the left-right position:

$$x = \frac{(I_A + I_D) - (I_B + I_C)}{I_A + I_B + I_C + I_D} \text{ and } y = \frac{(I_A + I_B) - (I_C + I_D)}{I_A + I_B + I_C + I_D}$$

(33.2)

Among the important factors that must be considered when designing a position sensing system based on a quadrant or other type of multi-element detector is the relationship between the diameter of the light spot and the dimensions of the detector. If the light spot is smaller than the spaces between the detector elements, then there will be a dead zone where beam tracking will be lost completely. This can be an issue in diffraction-limited applications such as data storage. Conversely, the light spot must be smaller than the overall detector dimensions or the signal will be truncated by the edges of the detector, resulting in erroneous position feedback.

The relationship between spot size and detector dimensions is often determined by the trade-off between spatial resolution and range. Tracking range increases with spot size, because once the spot is entirely located within one quadrant, further tracking is impossible. However, positional resolution is inversely proportional to the spot size. This is because a given displacement of a small spot produces a much bigger differential signal than the same movement in a larger spot.

In most tracking and alignment applications, a major parameter of concern is the ability of the detector to supply a precise description of the location of the focused image on the face of the detector. In other words, the change in output signal in relation to a change in the position of the spot is imperative. "Sensitivity" is used in this case to describe the ratio

$$\text{Sensitivity} = \frac{\text{Change in output signal}}{\text{Change in position}} \quad (33.3)$$

33.3 Multi-pixel Arrays

Multi-pixel arrays have a virtual monopoly in the fields of light detection and ranging (LIDAR), 3D laser radars (LADAR), laser vibrometers, medical imaging, etc. Among the types of arrays used in different applications are 2D *pin* photodiode and APD arrays. What makes *pin* and APD arrays unique is their ability to provide a fast access to any single pixel of the array, high dynamic range, high-frequency bandwidth (up to 1 GHz and even more) and possibility to provide a high-speed parallel read-out regime.

33.3.1 Photodiode Arrays

State-of-the-art *pin* photodiode arrays with high quantum efficiency in virtually the whole visible and near-infrared (NIR) spectral range were designed using both the front-illumination and back-lit configurations (Holland et al. 1997). In the front-illumination configuration, the light sensitive area of each pixel of the array is formed on the light impinging side of the semiconductor chip, and the electrical connections to each pixel are made on the same side of the chip. Despite the obvious advantage of ease of manufacturing, such arrays have a significant drawback – the requirement to trace signal from each pixel to the edge of the chip to connect to a downstream electronics, which consumes significant portion of the front surface of the chip.

Back-illuminated *pin* photodiode arrays allowed to overcome this problem. Such arrays feature the bonding pads located on the side of the chip that is opposite to the light entrance side (Figure 33.4), allowing creating arrays with virtually any number of pixels without "dead" spaces between the adjacent pixels. However, for high-resolution applications, the approach with separated pixels each requiring dedicated signal line proved to be not very promising. Instead, different

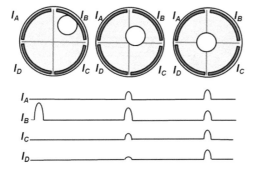

FIGURE 33.3 Formation of output signals from the four electrodes of the quadrant detector dependently on the optical spot location. White circle outlines the beam position.

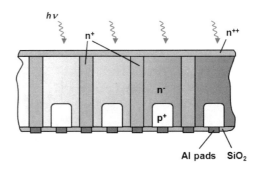

FIGURE 33.4 Schematic cross-section of a back-illuminated pin photodiode array built on a thin silicon substrate.

types of arrays based on charge transfer devices were developed and we will discuss them next.

33.3.2 Charge-coupled Devices as Tracking/Image Sensors

Charge-coupled device (CCD) is an array of metal-oxide-semiconductor (MOS) capacitors placed close together. A simplified structure and band diagram of the ideal p-type MOS capacitor in thermal equilibrium and at the so-called flat-band (FB) conditions is given in Figure 33.5a and b. Under this condition, the semiconductor band bending is absent (the electric field strength throughout the semiconductor is zero and potential level shift Ψ_S at the oxide-semiconductor interface is zero) and the applied to the gate electrode voltage $V_G = V_{FB} = \phi_m - \phi_s$ corresponds to the difference in the metal (ϕ_m) and semiconductor (ϕ_s) work functions.

Applying a bias to the semi-transparent gate electrode $V_G \neq V_{FB}$ allows shifting the potential level Ψ_S of the oxide-semiconductor interface, which drives the MOS capacitor into different regimes. There can be three distinctly different modes of operation dependently on the gate voltage polarity and amplitude: accumulation ($\Psi_S<0$, Figure 33.5c), depletion ($0<\Psi_S<\Psi_B$, Figure 33.5d) and inversion ($\Psi_S>\Psi_B$, Figure 33.5e). The Fermi level E_F remains flat at any regime of operation because of zero current flow. It means that in either depletion of inversion mode, the potential well is created in the semiconductor next to the oxide interface, facilitating non-equilibrium minority carriers' collection close to the interface (Sze 1981).

Under illumination, the photo-generated minority carriers (electrons for the p-type semiconductor) tend to collect in the potential well Ψ_{S1} close to the semiconductor-oxide interface. If the second MOS capacitor is placed next to the illuminated cell on the same semiconductor chip, then the carriers accumulated in the potential well Ψ_{S1} of the illuminated cell will have the opportunity to leak to the potential well Ψ_{S2} of the second, neighbour cell if the potential well Ψ_{S2} is deeper than Ψ_{S1}. If a one-dimensional array of MOS capacitors is placed on a single semiconductor chip, the accumulated photo-generated carriers in the wells can be sequentially transferred from one cell to the other. Applying the proper sequence of gate voltages allows transferring minority carriers across the array, thereby delivering electrical signal proportional to the amount of photons absorbed by each pixel to the output register. The simplified description given above illustrates the basic idea behind the functioning of MOS capacitor-based imaging sensors with charge transfer. The detailed description of the design, structure and operation of CCD is available in a large number of publications (Janesick 2001, Magnan 2003, Sze 1981). Contemporary cooled CCD can show the photon-limited performance and operate in a single-photon counting mode.

33.3.3 Complementary Metal Oxide Semiconductor Detectors

Many of the newer imaging devices use a different chip, one known as a complementary metal oxide semiconductor (CMOS) chip. In a CMOS sensor, each pixel has its own charge-to-voltage conversion, includes MOS transistors providing buffering and addressing capability and may also include the digitizing circuit. The price the designers pay for these additional functions is a reduced area available for the front-side light collection.

Similar to CCD, CMOS detector also performs four-step process: charge generation, charge collection, charge transfer and charge measurement. However, unlike CCD, the first three steps in CMOS sensor are performed within each individual pixel.

To convert two-dimensional spatial information into the serial stream of electrical signals, electronic scan circuits of CMOS photodetector read out each pixel sequentially. First, the vertical scan circuit selects a row of pixels by setting a high DC voltage on all gates of the MOS switches for that row. Next, the horizontal scan circuit selects the pixels in one particular column using the same technique. As a result, only one pixel in the two-dimensional matrix has a high DC voltage on both the row and the column switch, which electronically selects it for read-out. After the pixel dumps its information

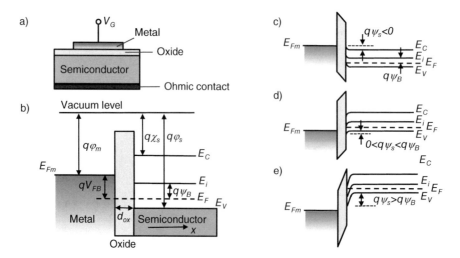

FIGURE 33.5 Cross section of a MOS capacitor (a) and simplified energy-band diagram of the ideal MOS capacitor at the gate bias $V_G=V_{FB}$. (b) Right panels show schematically band diagrams for the accumulation (c), depletion (d) and inversion (e) modes.

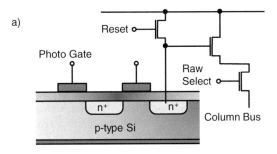

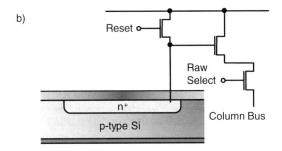

FIGURE 33.6 Typical designs of CMOS sensor pixels: photogate architecture (a) and photodiode architecture (b).

into the output stage, it is reset to begin a new integration, and the read-out process progresses to the next pixel in the row. Such organization, in contrast to CCD architecture, provides random access to each pixel and avoids multiple charge transfers over long distances of semiconductor material. Figure 33.6 shows schematically the most common pixel architectures for a CMOS sensor (Magnan 2003, Sze 1981).

The quantum efficiency of CMOS detectors may be not as high as that of CCD sensors because CMOS fabrication process requires low-resistivity wafers and relatively low power supply levels, which limits the depletion width of the semiconductor bulk to 1–3 μm. The spectral sensitivity range of CMOS sensors is usually narrower than that of CCD detectors since the absorption length increases towards the red spectral range.

33.4 Fast Response Photodetectors

Among main requirements for designing of fast photodiodes is minimizing the collection time of carriers created by light. To achieve it, one needs to minimize carriers' transit time by decreasing the thickness of the semiconductor's active layer, without compromising the optical absorbance. There are several ways to achieve it, one of which is to build resonantly enhanced structures and the other is to decouple the photodiode absorption length from the direction of non-equilibrium carriers drift.

A resonant cavity-enhanced photodetector relies on the idea of constructive interference of a Fabry–Perot cavity to enhance the optical field inside the photodetector bulk at specific wavelengths (Ünlu et al. 1995). Such resonant cavities can be formed using a buried backside reflector and the air/semiconductor top interface. The backside reflector can be of various types; the scattering reflector (Chen and Chou 1997) and distributed Bragg reflector (Emsley et al. 2002) are common examples.

For indirect-band gap semiconductors, it is difficult to achieve both high bandwidth and high efficiency with a vertical design since the length of the absorption region is proportional to the carrier transit time. Decoupling of the absorption length from the motion of non-equilibrium carriers is another efficient method to improve the speed and quantum efficiency of Si- and Ge-based photodiodes. Such decoupling is achieved in the lateral surface *pin* photodiodes on silicon-on-insulator (SOI) substrates, which exploits the idea of a grating coupler on top of the photodiode (Csutak et al. 2002). The coupler promotes the optical beam propagation in a thin, waveguide-type surface layer of *Si* – see Figure 33.7.

In top-illuminated, direct-band gap semiconductor photodiodes, a reduction of the carriers' transit time is limited by the surface recombination processes, which decreases the responsivity. The side-illuminated photodiodes, in which the absorption length is decoupled from the carriers' transit time, overcome the efficiency-bandwidth trade-off of top-illuminated photodiodes (Wake et al. 1991). However, they are not able to operate at high optical power levels, since carrier generation occurs in a very small volume at the diode input facet.

One of the solutions to that challenge is to design evanescently coupled photodiodes monolithically integrated with a waveguide (Takeuchi et al. 2000). The evanescent coupling optimizes the distribution of non-equilibrium carriers along the absorption layer. Due to more uniform light absorption, devices that utilize evanescent coupling can achieve several times higher saturation current than a traditional side-illuminated device. An example of evanescently coupled photodiode is shown schematically in Figure 33.8. The structure is usually based on a classical *pin* photodiode with a very thin (<0.5 μm) undoped InGaAs absorbing layer of 20–30 μm length and a p^+-doped InP layer on the top. The light is delivered from the side and propagates along the undoped diluted waveguide layer, which is a stack of several InP/InGaAsP sandwiches. The bandgaps and the thickness of n^+-doped InGaAsP optical matching layers are chosen in a way that they provide a gradual increase of the refractive index from the diluted waveguide towards the thin absorbing layer. As a result of the evanescent wave coupling, the optical wave when propagating along the waveguide is gradually transferred through the optical matching layers to the absorbing layer. The quantum efficiency can be achieved higher than 95% with the bandwidth of over 50 GHz, which should be beneficial for many applications.

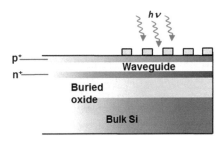

FIGURE 33.7 A schematic example of a *pin* photodiode with the waveguide-grating coupler.

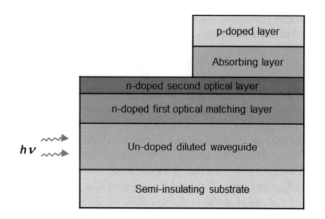

FIGURE 33.8 A schematic representation of the photodiode with a waveguide coupler.

33.5 Photodetectors with the Intrinsic Amplification

To improve the sensitivity, spatial and temporal resolution of optical signal detection, it is advantageous to use photodetectors with intrinsic amplification. There are a few types of photodetectors that are usually considered within this family – phototransistors, APDs, multi-channel plates (MCP), and PMTs are the most common examples. Concerning laser light management and detection, APDs are the prime choice in a vast majority of applications due to their superior performance, quality, form-factor and cost effectiveness. The basic properties of APDs were discussed in Chapter 9 (see also Dereniak and Boreman 1996, Sze 1981). Below we analyse some properties and design consideration important for laser applications.

The choice of APD for each specific application is determined primarily by the spectral range of interest, required frequency bandwidth and acceptable level of noise. Dependently on the required spectral range, the APD design relies on a number of different semiconductor materials available to this technology. The following materials have proved to be appropriate to fabricate high-performance APDs:

- *Silicon.* The spectral range from 400 to 1100 nm. The most commonly used design takes advantage of the reach-through structure, in which the depletion width propagates through the whole thickness of the device.
- *Germanium.* For the wavelengths of up to 1.65 μm. However, since the bandgap in Ge is lower than in Si, the noise is considerably higher and limits the Ge-based APDs application.
- *GaAs-based devices.* The designers usually use the heterostructures like $GaAs/Al_{0.45}Ga_{0.55}As$, and the large increase in gain occurs due to the avalanche effect in GaAs layers. $GaAs/Al_{0.45}Ga_{0.55}As$ structures are used below 0.9 μm. Applying InGaAs layers allows to extend the sensitivity range to ~1.4 μm.
- *InP-based devices.* For the wavelengths range 1.2–1.6 μm. The example is the double heterostructure with lattice matched layers n^+-InP/n-GaInAsP/p-GaInAsP/p^+-InP, in which either of carriers may create an avalanche. The other example is APD with separated absorption and multiplication regions p^+-InP/n-InP/n-InGaAsP/n^+-InP, which is similar to the Si reach through devices. The absorption occurs in the relatively wide InGaAsP layers, and avalanche multiplication of the minority carriers proceeds in the n-InP layer.

For the beam tracking, positioning and some other applications requiring ultra-low light-intensity detection with a relatively high-frequency bandwidth, the APDs operating in a single-photon counting mode (Geiger mode) promise undisputable advantages (see Chapter 9 for description of the Geiger mode of APD operation). A relatively novel type of avalanche photodetector with Geiger mode operation, known as silicon photomultiplier (SiPM) or multi-pixel photon counter, was developed by several groups, see, e.g. Buzhan et al. (2003). SiPM is an array of microcell APDs operating in the Geiger mode. A typical SiPM may contain several hundreds to thousands of microcells coupled to a common signal output terminal. The size of each microcell is usually ten to hundred square microns. The number of microcells defines the photon-counting dynamic range of SiPM. In each microcell, an arriving photon can trigger an avalanche flow of carriers, leaving the surrounding cells untriggered and ready to record other arriving photons. The output signal of the SiPM is the analogue sum of all individual cell signals. As long as the number of instantly impinging photons is less than approximately half the total number of microcells, it can be assumed that a given microcell will be hit by one photon at a time only. Under this condition, the sensor output will depend linearly on the input light flux.

The typical gain of a SiPM is $> 10^6$, providing a signal of a few nanovolts over a 50 Ω load resistor for a single photoelectron. The electronics noise of SiPMs is negligibly small due to the very high gain, in contrast to standard APDs, where the gain is typically about 100. The main source of noise, which limits the SiPM's single photon resolution, is the dark rate, originating from charge carriers thermally created in the sensitive volume.

The overall conversion factor from impinging photons to the number of detectable photoelectrons is called photon detection efficiency (PDE). PDE is normally lower than quantum efficiency (QE) η. The maximum achievable PDE of a SiPM is the product of the following factors: the geometrical efficiency ε_{Geom} (which is the fraction of total SiPM area occupied by active pixels); the Geiger efficiency (usually close to 100%); wavelength-dependent transport of impinging photons into the sensitive volume of the SiPM; and the intrinsic quantum efficiency η of Si. Typical SiPMs can achieve a peak PDE of 50% to 60% in both the yellow-to-green and the blue-to-near-UV ranges.

33.6 Colour-sensitive Detectors

The conventional way for detecting optical beams with multiple wavelengths relies on the idea that the light of different wavelengths can be detected separately. The variety of

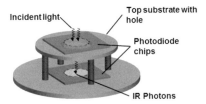

FIGURE 33.9 A stacked two colour or sandwich detector.

methods exploits two main principles: one uses colour overlay filters on top of the photodetectors active area and the other selects colours using the inherent property of the semiconductor to absorb light of different wavelength on a different depth.

In the first type of colour sensors, colour detection is performed by using different bandpass filters, and a corresponding number of individual photodetectors fabricated on the same substrate. The filters allow transmission of one of the selected wavelengths, and the corresponding photodiode measures the intensity of the incident light at that wavelength. A drawback to such methods is that the use of several sensors with their corresponding filters can become complicated and take up excessive space when forming a colour-sensing array.

Another type of multiple wavelength sensors employs more than one sensor in a vertically oriented arrangement. The operation of these devices is based on the intrinsic filtering property of semiconductors that results from the variation of absorption coefficient with the wavelength of light. For example, the blue, green and red photodiode sensors may be formed by the corresponding *p/n* junctions arranged one below the other underneath the Si sensor surface. The detector layers are individually connected to pixel sensor read-out circuits.

In an alternative design, for the detection of multi-wavelength optical beam the stack of two or more photodetectors may be applied (Figure 33.9). The top-level detector absorbs light in the shorter wavelength spectral range, and the bottom-level detector absorbs light of the longer wavelengths. In particular, the top-level detector may be a thin-layer Si photodiode absorbing primarily within the spectral range below ~950 nm, whereas the bottom detector may be an InGaAs photodiode to detect light in the spectral range above ~900 nm. The advantage of this type of structures is a possibility to tune independently the performance of each channel.

33.7 Detectors for the UV Spectral Range

Detection in the ultraviolet relies on the high energy of the individual photons. The photoelectric effect and the photomultiplier tube (PMT) detectors are still used for this application. The photocathode needs to be appropriate to the region of the spectrum of interest, such that the energy of the photon is greater that the work function of the material forming the cathode. Typical spectral responses of photocathode materials are given in Chapter 9. Care must also be taken to design appropriately the window of PMT due to the high absorption coefficient of many materials in the UV.

Solid-state semiconductor detectors for the UV rely on the same internal process as visible and NIR detectors and, generally, offer a higher quantum efficiency than photoemissive devices (though do not necessarily offer the overall sensitivity of PMT). UV photons have high energies, a wavelength of 200 nm corresponds to a photon energy of ~7.25 eV. This energy implies that any of the solid-state detectors used in the visible or near-infrared can also be used for the UV. However, because of a high value of the absorption coefficient for these materials in the UV, light is absorbed very close to the surface. In addition, the thin passivation layer that is produced on the surface often has high absorption in the UV, and moreover, extended exposure to UV may produce radiation defects degrading the detector's performance. These drawbacks have to be addressed in designing UV-sensitive semiconductor detectors (Razeghi and Rogalski 1996).

Among the most popular materials and solutions used for UV detectors are as follows:

- Silicon carbide (SiC) with the peak responsivity at around 280 nm and reasonably good sensitivity within the range from ~220 to 380 nm. Both photoconductive and photovoltaic structures have been developed with SiC.
- Gallium phosphide that is often used for the blue/green light detection. Its sensitivity also extends into the UV down to ~200 nm.
- The nitride III–V semiconductor materials also have large bandgaps suggesting suitability for application in the UV. Gallium nitride is the most developed and is used in the 240–380 nm spectral range. Other III–V nitrides have been studied in detail, for example, aluminium nitride (AlN) with a bandgap of 6.2 eV at 300 K and AlGaN that allows tailoring the peak responsivity by adjusting the content of Al and Ga (Moustakas and Paiella 2017).
- Semiconductor diamond as a UV detector is attractive as it has the largest bandgap and has good UV optical properties. In addition, it has high thermal conductivity and a small dielectric constant, bringing together properties that are ideal for UV optical detectors. However, getting high-quality devices did not allow good production yield.
- Schottky (metal-semiconductor) barrier devices can be used as photovoltaic detectors for many discussed above materials. The Schottky barrier creates a depletion layer between the metal contacting the semiconductor and the semiconductor material (see also chapter 9 and Dereniak and Boreman 1996, Sze 1981). Thus, the sensitive region of the device is brought very close to the surface, allowing secure absorption and photoelectric conversion of high-energy photons. The drawback of the Schottky barrier photodetectors is a high noise level.

33.8 Detectors for the IR Spectral Range

IR detectors cover the wavelength from approximately 1 to 14 μm. The radiation up to the lower end of this range can be satisfactorily detected by standard Si, Ge and InGaAs semiconductor devices. Thermal detectors are also used in this region of the spectrum, with the associated trade-off in sensitivity and response time. Many semiconductor materials appropriate for radiation up to 14 μm (corresponding to photon energy of 0.089 eV). However, special consideration must be exercised to take into account the background radiation as the black body radiation peaks around 10 μm at 300K. In fact, IR detectors may be considered as the background-limited infrared photodetectors (BLIP). Therefore, the total power of the background radiation establishes a sensitivity threshold independently of the actual detector. In practical applications, the IR detectors need to be cooled to minimize the contribution of the background radiation.

The lead salts (PbS, PbSe and PbTe) and indium antimonide-based (InSb) detectors have intrinsic bandgaps appropriate to detection in the mid-wavelength IR (MWIR) spectral range (typically 3–8 μm). Commercial PbSe-based detectors function well without cryogenic cooling, but they suffer from low specific detectivity (D^* less than 10^9 cm*Hz$^{1/2}$/W). The state-of-the-art PbSe detector had achieved a detectivity of 4×10^{10} cm*Hz$^{1/2}$/W and a noise-equivalent power (NEP) of 2 pW/Hz$^{1/2}$ at a wavelength of 3.8 microns. In InSb detectors, controlling the ratio of In to Sb allows building intrinsic, either n-type or p-type semiconductor detector. For the same operating condition, the InSb photodetector has a better specific detectivity D^* value than the detectors based on lead salts. The time response is also faster for InSb-based detectors.

Among very useful materials for MWIR detection is mercury cadmium telluride, $Hg_{1-x}Cd_xTe$, or MCT (Rogalski 2012). MCT is a semiconductor detector with an intrinsic bandgap and is characterized with a reasonable high D^* ranging from 10^{10} to 10^{11} cm×Hz$^{1/2}$/W but requires cooling. The spectral responsivity can also be tailored, as the bandgap energy varies with the ratio of mercury to cadmium. At $x=0.2$, the bandgap is 0.1 eV and the sensitive range extends out to 14 μm. Since the lattice constant changes very little as the composition is changed, it is also possible to make heterostructures based on MCT. However, difficulty of growing structures with uniform x makes the devices expensive. MCT detectors approach BLIP performance at higher temperature compared with the extrinsic semiconductor materials as they operate as intrinsic devices. MCT is widely used in the designing of multi-element photodetector arrays.

Since nineties, significant progress allowed developing IR detectors based on quantum well structures (Levine 1993). The quantum wells are designed such that the energy required to promote a carrier from a lower energy level of the well to a second bound state near the top of the well is close to the energy of the incident photons. Biasing of the device allows the photoexcited electrons to tunnel from the well to the continuum of the conduction band creating a photocurrent. The basic design principles and properties of such detectors are discussed in Chapter 9.

33.9 Common Customizations

Some laser applications may need a photo detector that is optimized regarding one or more features. Solid-state photodetectors based on semiconductors allow significant level of customization. Since wafer fabrication processes are mostly independent of the features of the design layout, it may be advantageous to use a custom design for each particular application.

33.9.1 Active Area Sizes, Shapes and Apertures

Aside from the choice of a semiconductor material, the active area size and shape are among the most important design features of a photodetector. Larger detector areas collect more light but also exhibit higher dark noise and capacitance. For low-noise and high-speed applications, the designs with small sensitive active areas are commonly used. For silicon photodiodes, active areas can range from a fraction of to tens of square millimetres. Devices made of other materials are typically limited to smaller designs. InGaAs detectors, for instance, rarely exceed about 25 mm^2 due to defects in the base material that cause low yield for large photodiodes. SiC or GaN are typically limited to a about a millimetre diameter of active area also because of the crystalline defects and wafer uniformity issues.

For many photoconductive applications, the designs capable of high-performance operation at high-voltage bias conditions are important. Since the probability of voltage breakdown and highly increased dark current is much higher around the sharp corners of the p/n junction, the circular shaped active areas are preferential for the photoconductive operations at high reverse bias.

The active area size of semiconductor detectors is one of the determining factor of the device capacitance. Therefore, to achieve a fast response time, the active area size has to be minimized. For example, to design Si photodetector with below 1-ns response time, the active area size should be smaller than 0.5 mm in diameter.

For some applications, it is imperative to ensure that there is no signal detected if light is falling on regions of the photodetector outside its active area. For the semiconductor photodetectors, all metal-covered surfaces provide protection against non-useful light sensitivity. But semiconductor chip surfaces covered with metal often produce the unwanted light reflection. Therefore, other materials like black polyimide or a stack of Chromium with Chrome oxide (Cr/CrOx) may be used to create apertures on the chip. Such coatings may provide both, less than 5% reflectance and better than OD3 transmittance blocking within the required spectral range.

33.9.2 Coatings and Filters

Semiconductor photodetectors often use a protective passivation layer (typically a thin layer of silicon oxide or silicon nitride or their combination) which also serves as anti-reflection coating. This layer can reduce the front surface reflection of the active area down to less than 1% but only for a part of the spectrum that is narrow compared with the spectral range of most common semiconductor materials. More sophisticated multilayer coatings can be applied allowing coarse wavelength selection for simple applications like low-cost RGB sensors.

For better spectral filtering, dielectrically coated glass sheets can be attached to the active area of a detector or be installed as part of a package window assembly.

33.10 Photodetector Integration

Photodetectors for specific applications do not work alone. In most of the cases, they are coupled to signal measuring and processing electronics or even integrated with other optoelectronic and/or electronic devices to provide useful input for a higher-level processing systems. This section describes some important properties and features of semiconductor photodetectors' assembly and integration.

33.10.1 Packaging

Regardless of the type of package of a detector assembly, the active area has to be aligned to other optical components in the system for most applications. Within the semiconductor surface, the alignment tolerance is typically better than 5 μm. The semiconductor chip then is placed onto a substrate, either a printed or flex circuit board, a ceramic, a metal header, a lead frame or another semiconductor chip. The tolerance for this placement depends on the tooling used during the attachment process. Manual placement will result in tolerance between ±75 and ±150um depending on materials and operator skill. With contemporary die placement, equipment tolerances better than 10 um can be achieved under high-volume production conditions.

Optoelectronic packages can be divided in hermetic and non-hermetic types. Over the years many different solutions have been developed. Below, we will outline the basic types only.

Hermetic packages have typically a cavity that contains the photodetector chip and, in many cases, other active and passive components. The cavity is sealed by a lid or cap that has to provide optical access to the active area. True hermeticity in the interpretation of aerospace standards is achieved if all interfaces are either soldered, brazed, welded or consist of a glass or ceramic to metal seal. One of the still popular package types is "transistor outline" packages or the TO-cans. These cans consist of a cap and a header made of a nickel–cobalt ferrous alloy (Kovar) or steel with a sealed borosilicate, quartz or sapphire window. Since no epoxy is involved in the sealing process, these packages generally exhibit the best hermiticity and lowest leak rates. Moreover, if all materials are chosen carefully, these packages can withstand temperatures in excess of 200°C and other stress conditions like shock, vibration and temperature cycling. However, the storage and operating temperature of such devices may be limited to lower values due to the use of epoxies and low-temperature solders for the attachment of components inside the package. Figure 33.10 shows typical configurations for simple devices assembled in TO-type packages.

Epoxy seals, over-moulded components or plastic encapsulated microcircuits are generally not considered hermetic. The materials and processes commonly used to assemble such components can be ruggedized to some extent as, for instance, for the use in many modern automotive parts. Examples of photodetectors in non-hermetic packages are shown in Figure 33.11.

33.10.2 Hybrids and Detectors for Fibre-coupled Lasers

Many applications require bringing the initial signal processing circuitry as close as possible to the output terminals of the photodiode chip. This not only allows more compact packaging of the overall assembly but may improve the noise and frequency bandwidth of the assembled photodetector. The simplest version of a hybrid is a photodiode chip with

FIGURE 33.11 1D photodiode arrays assembled in non-hermetic packages: (a) array on single chip assembled on a substrate; (b) array of individual chips placed on a substrate.

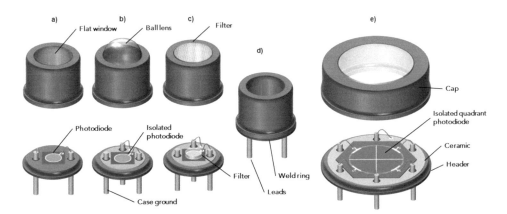

FIGURE 33.10 Examples of semiconductor photodiodes assembled in hermetically sealed TO packages: (a) two-leads package with a flat window, (b) three-leads package with isolated photodiode and a light focusing lens, (c) photodiode with the attached filter and a window cap with a filter, (d) sealed simple TO package and (e) quadrant detector in a large TO can.

the transimpedance amplifier (TIA) assembled in either TO or surface mount package. The purpose of a transimpedance circuit is to convert the input current from a photodiode into an output voltage V_out that can be approximated as $V_\text{out} = I_{ph}R_f$, where I_{ph} and R_f are the photodiode photocurrent and feedback resistance of TIA, respectively (see Figure 33.12 for an example electrical schematics). The TIA gain R_f and the 3dB cut-off frequency f_{3dB} are defined as:

$$R_f = V_\text{out}/I_{ph} \quad \text{and} \quad f_{3dB} = \sqrt{GBP/2\pi R_f C_\text{total}} \quad (33.4)$$

where GBP is the Gain-Bandwidth-Product of the operational amplifier and C_total is the total capacitance of the circuit, equal to the sum of the photodiode capacitance and input capacitance of the operational amplifier and feedback capacitance C_f. Obviously, increasing TIA gain R_f causes a decrease in the cut-off frequency value f_{3dB}. Therefore, selection of operational amplifier and photodiode chip is critical to provide a desired photodetector performance.

To improve frequency bandwidth and still ensure high gain of the photodetector, the hybrid may include two or more amplification stages. The first stage maybe TIA with a relatively low feedback resistance that ensures high-frequency bandwidth of the first stage. The second amplification stage may be a simple voltage amplifier also having a relatively low feedback resistance. The resulting frequency bandwidth of such hybrid will be rather high and determined by the 3dB cut-off frequency of the first amplification stage, whereas the total gain determined by the product of gains of the two stages will also be high.

Multi-element photodetector arrays can also be packaged in hybrids with TIA or even higher level processing components. It is common to add comparators, analogue-digital converters and even ASIC (application-specific integrated circuit) chips in the same package with a primary photodetector to save space and improve performance parameters. However, with increased functionality of a hybrid other considerations like heat dissipation have also to be taken into account.

In applications that require detection of optical signals delivered via fibres, a special type of hybrid assemblies called optical receivers is usually used. An example of the receiver assembled in a ceramic package is shown schematically in Figure 33.13. A single-mode optical fibre with the core diameter of ~10 μm and cladding diameter of several hundreds of micrometres is fed through the input port of the package. The core of the fibre

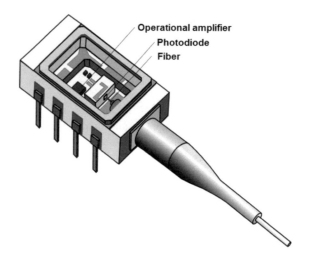

FIGURE 33.13 Schematic representation of an optical receiver with the top lid removed, optical fibre fed through the input port of the ceramic package, and fibre core aligned with the fast photodiode.

is brought close to the photodiode (~50 μm or less) and aligned with its centre to obtain maximum photocurrent from the photodiode. The package may contain TIA and other components to bring the output signal to a required level.

33.11 Future Directions in Detectors Technology and Applications

Recent advances in optoelectronics and photonics challenged researchers developing optical detectors with improved performance and facilitated application of new materials and principles in their design. Significant efforts were employed in the field of short-wavelength IR (SWIR, approximately 0.8–3 μm), MWIR (3–8 μm) and long-wavelength IR (LWIR, 8–14 μm) detectors (see Tan and Mohseni 2018 for the most recent review). Impressive advances were reached with the structures employing quantum wells, quantum dots (QD) and nanowires. Among new emerging photodetectors, graphene-based and other 2D structures with carrier multiplication and devices utilizing photonic crystals (PC) to couple incoming light to the absorbing medium were developed. We will discuss some representative examples in this sections.

PCs are periodically arranged structures with the periodicity scale proportional to the wavelength of light. In analogue to the periodically arranged atomic structures forming crystalline atomic lattice with electronic energy zones and bandgaps, the structure of PC is characterized with a forbidden bandgap (also known as a photonic stopband, PSB) where light cannot propagate or escape from it. If light with the wavelength within the range of PSB enters PC, it will experience multiple Bragg's reflection, which creates an effect of light retardation inside PC and may eventually stop light against further propagation through PC. Incorporating nanocrystals or other nanoscale materials within the pores of PC or optically coupling such nanoscale materials with PC in an appropriate alternative way allows enhancing interaction of light with those embedded materials due to multiple light-scattering effects in PC,

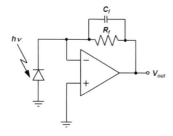

FIGURE 33.12 Typical photodiode transimpedance amplifier with C_f and R_f the feedback capacitance and resistance, respectively.

creating thereby favourable pre-requisites for an efficient wavelength-selective photo detection.

The effect of PC cavity enhancement of absorption was used to tune the graphene photodetector's responsivity by coupling of the graphene sheet to a planar PC prepared using SOI wafer with 450 nm pitch-etched cavities. As is known, graphene as an atomic layer material is characterized by rather low absorbance. However, integrating graphene with PC boosts graphene's absorption allowing to achieve high responsivity at ~1550 nm (Shiue et al. 2013). A comprehensive review on optoelectronics in silicon-based PC (including photodetectors) for MWIR is given in the most recent article (Lin et al. 2018).

A promising solution for SWIR and MWIR detectors was proposed recently based on the hybrid structures of graphene integrated with QD layers (Koppens et al. 2014). The design of such photodetectors is based on a multilayer structure, in which the top layers may perform a wavelength selection role. After passing these top layers, the MWIR optical beam enters the layer with QD in which light absorption and initial charge separation occur. The carriers of one polarity (electrons) are trapped by QD, whereas the other polarity carriers (holes) are transferred to the bottom graphene layer. This causes electrostatic perturbation of the graphene conductance and leads to a photo-gating effect changing the graphene resistance through capacitive coupling and allowing charge carriers in graphene to recirculate many times due to extremely high carrier's mobility in the graphene while the other polarity carriers remain trapped by QD. This structure is in fact the analogue of an efficient phototransistor with high photoconductive gain produced due to charge trapping in the QD layer. The devices of such type promise very high value of specific detectivity D^* of over 10^{14} cm×Hz$^{1/2}$/W. In addition, selecting of QD material and size allows fine-tuning the spectral sensitivity range from slightly above 1 μm for QD made of PbS to virtually any wavelength within the interval from 2 to above 6 μm for lead salts PbSe$_x$Te$_{1-x}$ with varied values of x.

REFERENCES

Buzhan, P., Dolgoshein, B., Filatov, L. et al. 2003. Silicon photomultiplier and its possible applications. *Nucl. Instr. and Meth. A* 504: 48–52.

Chen, E. and Chou, S.Y. 1997. High-efficiency and high-speed silicon metal-semiconducto-metal photodetectors operating in the infrared. *Appl. Phys. Lett.* 70: 753–755.

Csutak, S.M., Dakshina-Murthy, S. and Campbell, J.C. 2002. CMOS-compatible planar silicon waveguide-grating-coupler photodetectors fabricated on silicon-on-insulator (SOI) substrates. *IEEE J. Quant. Elect.* 38: 477–480.

Dereniak, E.L. and G.D. Boreman. 1996. *Infrared Detectors and Systems*. New York: John Wiley & Sons.

Emsley, M.K., Dosunmu, O. and Ünlu, M.S. 2002. High-speed resonant-cavity-enhanced silicon photodetectors on reflecting silicon-on-insulator substrates. *IEEE Photonics Technol. Lett.* 14: 519–521.

Holland, S.E., Wang, N.W. and Moses, W.W. 1997. Development of low noise, back-side illuminated silicon photodiode arrays. *IEEE Trans. Nuc. Sci.* 44: 443–447.

Janesick, J. 2001. *Scientific Charge Coupled Devices*. Bellingham: SPIE Press.

Koppens F. H. L., Mueller T., Avouris P., Ferrari A. C., Vitiello M. S. and Polini M. 2014. Photodetectors based on graphene, other two-dimensional materials and hybrid systems. *Nature Nanotechnology* 9 (10): 780–793.

Levine, B.P. 1993. Quantum-well infrared photodetectors *J. Appl. Phys.* 74: R1–81.

Lin H., Luo Z., Gu T. et al. 2018. Mid-infrared integrated photonics on silicon: a perspective. *Nanophotonics* 7(2): 393–420.

Magnan, P. 2003. Detection of visible photons in CCD and CMOS: a comparative view. *Nuclear Instr. Meth. Phys. Res. A* 504: 199–212.

Moustakas, T.D. and Paiella, R. 2017. Optoelectronic device physics and technology of nitride semiconductors from the UV to the terahertz. *Rep. Prog. Phys.* 80: 106501.

Rogalski, A. 2012. History of infrared detectors. *Opto-Electronics Review* 20: 279–308.

Razeghi, M. and Rogalski, A. 1996. Semiconductor ultraviolet detectors *J. Appl. Phys.* 79: 7433–7473.

Shiue R.J., Gan X., Gao Y. et al. 2013. Enhanced photodetection in graphene-integrated photonic crystal cavity. *Appl. Phys. Lett.* 103: 241109.

Sze, S.M. 1981. *Physics of Semiconductor Devices*. New York: John Wiley & Sons.

Takeuchi, T., Nakata, T., Makita, K. and Yamaguchi, M. 2000. High-speed, high-power and high-efficiency photodiodes with evanescently coupled graded-index waveguide. *Electron. Lett.* 36: 972–973.

Tan, C.L. and Mohseni, H. 2018. Emerging technologies for high performance infrared detectors, *Nanophotonics.* 7(1): 169–197.

Ünlu, M.S. and Strite, S. 1995. Resonant cavity enhanced photonic devices. *J. Appl. Phys.* 78: 607–639.

Wake, D., Spooner, T.P., Perrin, S.D. and Henning, I.D. 1991. 50 GHz InGaAs edge-coupled PIN photodetector. *Electron. Lett.* 27: 1073–1075.

FURTHER READING

Dereniak, E.L. and D.G. Crowe. 1984. *Optical Radiation Detectors*. New York: John Wiley & Sons.

Ferrari, A.C., Bonaccorso, F., Fal'ko, V. et al. 2015. Science and technology roadmap for graphene, related two-dimensional crystals, and hybrid systems. *Nanoscale.* 7: 4598.

Graeme, J.G. 1995. *Photodiode Amplifiers: OP AMP Solutions*. New York: McGraw-Hill.

Lutz, G. 1999. *Semiconductor Radiation Detectors. Device Physics*. Berlin: Springer.

Rieke, G. 1996. *Detection of Light*, Cambridge: Cambridge Univ. Press.

Wood, D. 1994. *Optoelectronic Semiconductor Devices*. New York: Prentice Hall.

34
Laser Energy and Power Measurement

Robert K. Tyson

CONTENTS

34.1 Measurement Technique Selection 479
34.2 Test Configuration 480
34.3 Pyroelectric Sensors 480
34.4 Thermopile Sensors 480
34.5 Laser Absorbers 481
34.6 Semiconductor Photodiode or Optical Sensors 481
34.7 Displays 481
References 481
Further Reading 481

To fully characterize a laser one must measure the energy and the power or, more specifically, the *radiant energy* in joules (J) and *radiant power* or *energy flux* in J s^{-1}. Due to the nature and diversity of lasers, which operate from the deep ultraviolet to the infrared, from continuous output to picosecond pulses and from microwatts to megawatts, energy-measuring devices encompass a wide range of possibilities [1–3]. Detectors are sensitive to the spectrum of the beam and the speed of energy deposition in their conversion of energy to a useful readable or recordable quantity that can be interpreted as energy or power.

A detector is any device in which optical radiation produces a measurable physical effect. Laser energy measurement devices are divided into general groups: calorimetric, photoelectric, photochemical and mechanical. Detector methods are covered in detail in Chapter 9. A specific method is chosen depending upon the duration of the energy pulse, the spectral response of the detector and the dynamic range of the incident energy. Absolute power and energy measurement and laser dose (energy density) measurement is dependent upon the accurate calibration of the detectors. Calibration, data collection and data interpretation are three necessary steps of the *measurement* process.

34.1 Measurement Technique Selection

The choice of measurement technique is governed by the application and generally when the measurement technique closely resembles the application the measurement is most useful. Lasers must meet standards dictated by their application in manufacturing, electronics, medicine, communications and the military. The first characteristic dictating the measurement is whether the laser is continuous wave (cw) or pulsed. In terms of modern energy measurement, applications for cw lasers include optical (lightwave) communications, laser-based medical instrumentation, materials processing, photolithography, data storage and laser safety equipment. Accurate measurement is required for pulsed lasers such as ultraviolet excimer lasers (optical lithography for semiconductors, corneal sculpting for photorefractive keratectomy (PRK) or laser *in situ* keratomileusis (LASIK), and micro-machining of small structures) and deep-UV lasers used for lithography.

These applications have various power levels and wavelengths and can be grouped into standard calibration categories [3].

- cw laser power below 2 W,
- cw laser power at 1064 nm above 2 W,
- cw laser power at 10.6 µm,
- pulsed laser energy (Q-switched YAG at 1064 nm) and
- pulsed laser energy (KrF excimer at 248 nm and ArF at 193 nm).

For any of the methods, despite their variation in employed physical principles, the measuring device must respond in a predictable fashion to the actual incident energy [4]. This responsivity R is expressed as the ratio

$$\Re = C\frac{\text{output}}{\text{power incident}} \quad (34.1)$$

where the output is the measurable voltage, current, resistance, pressure, etc., of the detector and C is a calibration factor. Because C can be, and almost always is, dependent upon wavelength, pressure, temperature, quantum efficiency, frequency, detector area, load resistance or noise, there are always important steps in calibrating the detectors and determining their useful ranges all the while relating them to radiometric standards.

Whenever the desired range of operation for a device is exceeded, it is necessary to employ electronic or optical techniques to keep the responsivity known [2]. Sometimes it is necessary to match the angular response of detectors with integrating spheres or diffusers. When the source energy or power exceeds the linear region of a detector, calibrated neutral density filters can be used to bring the incident energy onto the detector into a usable regime. Filters can also be used to match the spectral response of the detector to the spectrum of the incident radiation. The addition of choppers for attenuation or ac amplification is often employed. In all cases of photoelectric measurement of power and energy, it is necessary to maintain calibration and international standardization [5].

34.2 Test Configuration

For laser power and energy measurement, it is necessary to maintain repeatable and calibrated environments. For example, isoperibol (constant temperature environment) calorimeters are used in many cases. Solid-state and thermal detectors are used with well-characterized transfer standards (calibrated against primary standards). Ultra-high accuracy can be achieved using a cryogenically cooled electrical substitution laser power meter [3].

Laboratories often use beam-splitter-based calibration systems that allow various power and energy detectors. The critical parameters such as absorptivity and transmittance are well characterized at all the operating wavelengths. Small angles of incidence are used to minimize polarization effects.

Optical fibre power meters are calibrated using a system in which the test meter and the laboratory standard are both exposed to a stable diode laser source. For many commercial and standards laboratory measurements, the reported results include both the test data and the calibration data.

Power of the laser is measured in Watts (nW, mW, W, etc.). Optical power output is the continuous power output of a continuous wave (CW) laser or the average power of a pulsed or modulated laser. Energy (E) of a pulsed laser is the laser's peak power (P_{Peak}) multiplied by t, the pulse duration, $E = P_{Peak} \times t$. The average power of a pulsed laser (P_{Avg}) is the pulse energy (E) multiplied by the laser pulse repetition rate (PRR) given in Hz. $P_{Avg} = E \times PRR$. As an example, a laser with a 10 ns pulse width and an energy of 10 mJ per pulse with a repetition rate of 10 pulses per second have a peak power of 1 MW and an average power of 100 mW. The response time of a sensor is the time from initial pulse detection to the point where the read voltage is achieved. The integration time is the time from initial detection to reset. The time from the reset signal (or end of the pulse) to zero response is dependent upon thermal relaxation time of the detector and the RC time constant of the electronics. For all detectors, the total accuracy of the measurement depends upon: power level, energy level, wavelength, pulse rate, linearity, uniformity, calibration error, damage occurred, optical noise and electromagnetic interference.

Measurement systems are usually comprised of three subsystems, a sensor, a metre (analogue or digital) and various conditioners that make corrections such as differences between the calibration wavelength and the actual wavelength of the laser. Calibration to NIST, PTB or other standards is absolutely necessary and routinely done.

The most common power and energy measurement technologies are pyroelectric sensor (that can only measure pulsed lasers), thermopiles (that can measure pulsed or CW lasers) and semiconductor photodiodes (optical sensors that can possibly measure both pulsed and CW lasers). A recently developed technology uses radiation pressure for the measurement of extremely high-power (~ 500 kW) lasers [ref Laser Focus World Sep 2016, p. 13]. Using high-precision weighing techniques, the force of a beam reflecting from a mirror is linearly proportional to the beam power multiplied by the cosine of its incidence angle. This technique is not practical for low powers because the force cannot be measured, but for powers above 10 kW, the force can be measured. In a recent demonstration, a 500 kW laser produced a photon pressure of 330 milligrams ±0.002%.

34.3 Pyroelectric Sensors

Average power is calculated by measuring the PRR and multiplying by the pulse energy. Many manufacturers have systems that can measure pulsed lasers up to 25 kHz. Pyroelectric sensors use a ferroelectric crystal that has a permanent electrical polarization. Incident light heats the crystal, changing the dipole moment and causing current to flow. The current can charge a capacitor in parallel with the crystal to create a voltage proportional to the pulse energy and then the capacitor switched (short-circuited) to be ready for the next pulse. Because pyroelectrics respond only to change of temperature, the source must be pulsed or modulated. The response time of a pyroelectric sensor is much faster than a thermopile and can make a stable measurement within several hundred microseconds. Pyroelectric sensors work well with pulsed lasers with 100 nJ to >10 J pulses in the 0.15–12 µm wavelength range.

34.4 Thermopile Sensors

Thermopile sensors absorb incident laser radiation and convert it into heat. The heat flow into a heat sink is held at ambient temperature by air-convection cooling or water-cooling (calorimeters). The temperature difference is converted into an electrical signal by a thermopile junction. They can measure CW lasers and average power in pulsed lasers, or for long pulses (>1 ms), they can integrate the energy over time. Thermopile sensors are often chosen for CW lasers with power levels from tens of mW to the kW level. In the infrared region beyond 1.8 µm, the spectral limit of a Germanium photodiode, thermopiles are useful. Thermopiles often take a few seconds to achieve a stable measurement.

Thermopiles work well with CW average power or pulsed average power from 200 µW to >5 kW in the 0.15–12 µm wavelength range, or with long pulses (>1 ms) with 1 mJ to >300 J in the integrated energy mode.

34.5 Laser Absorbers

On thermopile detectors the absorbing coating is deposited on the substrate where thermocouples lay. In calorimeters the absorber coats the water-cooled elements used as heat exchangers. One of the major concerns about the absorber is its damage threshold, defined as the power density (W cm^{-2}) beyond which there is a variation of >1% measurement of laser power. The damage can occur in multiple ways. For very short pulses (<100 ns), the damage process is ablative. For long laser pulses (>10ms), the pulse is long enough to allow a diffusion of heat within the absorber and the damage is a thermal effect. For all absorbers, the absorption coefficient is wavelength-dependent and should be as high as possible with an aperture large enough to contain the entire beam and with the lowest possible reflection at the incident angle.

Surface absorbers are thin layers of refractory materials deposited on substrates that can easily transfer heat like high-conductivity metals. They are used for long pulses (>300 μs) or CW lasers. Radiation is entirely absorbed in the few-micron thin layer then released as heat to the thermopile.

Volume absorbers are used for short-time pulse lasers (< 1ms) where a gradual absorption of radiation penetrates into the material to a depth of 0.5–2 mm. Various types of glasses and ceramics are used as volume absorbers. For very high powers a super-hard coating is used. They are applicable for wavelengths from 0.25 to 11 μm and have damage thresholds up to 12 kW cm^{-2} CW and 40J cm^{-2} pulsed [ref Laserpoint].

34.6 Semiconductor Photodiode or Optical Sensors

Photodiode sensors convert incident photons into electrical current. They are quite stable and they have high sensitivity and low noise and are useful at low light levels (nW to low mW). They generally have fast response time, but their spectral range is limited to the response of the semiconductor detector. Optical sensors work with CW lasers from 10 nW to 50 mW in the 250 nm to 1800 nm wavelength range. Furthermore they work with pulsed lasers from 10 pJ to 800 nJ in the 325–1700 nm wavelength range. The sensors generally saturate above 1 mW cm^{-2} so attenuating filters (which must be stable and accurately characterized) from the detector window.

34.7 Displays

Power or energy displays can either be analogue (needles) or digital. Digital displays can be more accurate and are useful for computer interfacing; however analogue displays can be easily converted to digital for such interfacing. IEEE-488, RS-232 and USB interfaces are commercially available with a large supply of commercially available laboratory or field software.

For power-only meters, the important performance characteristic is the noise floor which restricts low light-level use. For energy-only meters, the important characteristic is the maximum PRR.

If four sensors, of any type, are placed together in a quad-cell, the power and energy sensor can be used to determine the centroid position of the beam. With an array of many sensors, the entire beam shape and distribution of power can be recovered.

For measuring scanned beams, such as barcode scanners, or uncollimated beams, the sensor can be inside an integrating sphere to capture the total integrated radiation.

Calibrations are often made at low powers and then by knowing any non-linearities or zero offsets, the measurement can easily be accurate to ±5% at higher powers [ref Ophir].

REFERENCES

1. Budde W 1983 *Physical Detectors of Optical Radiation* (New York: Academic).
2. Heard H G 1968 *Laser Parameter Measurements Handbook* (New York: Wiley).
3. National Institute of Standards and Technology (NIST), 2001, Electronics and Electrical Engineering Laboratory, Optoelectronics Division Programs, Activities, and Accomplishments, NISTIR 6602, January.
4. Grum F and Becherer R J 1979 *Optical Radiation Measurements* vol 1 (New York: Academic).
5. Commission Internationale de l'Eclairage (CIE) 1982 *Methods for Characterizing the Performance of Radiometers (Photometers)* (Publ. No. 53) (Paris: CIE).

FURTHER READING

Acetta J S and Shumaker D L (ed) 1993 The Infrared and Electro-Optical Systems Handbook (Ann Arbor, MI: ERIM) and (Bellingham, WA: SPIE)

Bode D E 1976 Infrared detector technology at the Santa Barbara Research Center *Proc. Electro-opt. Syst. Des. Conf.* p 63.

Dowell M L 2001 Choosing the right detector is key to accurate beam power measurements SPIE OE Magazine **56**.

Lehman J and Li X 1999 A transfer standard for optical fiber power metrology *Opt. Photon. News* **10** 44f–h.

Leong K H, Holdridge D J and Sabo K R Characteristics of power meters for high-power CO$_2$ lasers 1994 *J. Laser Appl.* **6** 231–6.

Putley E H 1980 *Thermal Detectors Optical and Infrared Detectors* ed R J Keyes (Berlin: Springer) ch 3.

Vayshenker I, Li X and Scott T R 1994 Optical power meter calibration using tunable laser diodes. *Proc. Natl Conf. Stud. Lab.* (Jul 31–Aug 4, Chicago, IL) pp 337–52.

West E D and Schmidt L B 1987 A system for calibrating laser power meters for the range 5–1000 W Natl Bur. Stand. (US) Technical Note 685.

35

Irradiance and Phase Distribution Measurement

B. Schäfer

CONTENTS

35.1 Basic Concepts and Definitions 483
35.2 Principal Measurement Set-Up 483
35.3 Irradiance Distribution Measurement..484
    35.3.1 Scanning Devices 484
    35.3.2 Camera-based Systems..484
35.4 Phase Distribution Measurement 484
    35.4.1 Hartmann–Shack Wavefront Sensor 484
    35.4.2 Interferometers 484
    35.4.3 Phase Retrieval from Intensity Transport Equations 485
35.5 Coherence Measurement..485
References..486
Further Reading 486

This section concerns the measurement of the irradiance distribution $I(x, y)$ [1] and the phase distribution $\phi(x, y)$ [2] of laser beams in xy-planes perpendicular to the direction of beam propagation. The topic intrinsically includes aspects of spatial coherence measurement which are, therefore, considered in the last section.

35.1 Basic Concepts and Definitions

In this context irradiance terms, i.e. either the relative power density or the relative energy density, depending on whether cw or pulsed laser beams are considered. A convenient phase definition, which applies to fully and partially coherent beams, is based on the time-averaged Poynting vector distribution $S(x, y)$ according to [3]

$$\frac{S_\perp}{|S|} = \frac{\lambda}{2\pi}\nabla_\perp \phi(x, y) \quad (35.1)$$

where λ denotes the mean wavelength of light and $\nabla_\perp$ is the two-dimensional gradient.[1] Defined in this way, $\phi(x, y)$ represents the deterministic part of the fluctuating phase associated with partially coherent fields and for coherent beams it coincides with the familiar definition as a surface of constant phase. However, equation (35.1) is consistent only if $\nabla \times S = 0$.

Whenever this condition is violated as, e.g. for some higher-order Gauss–Laguerre modes, an adequate description of the beam requires the full Poynting vector distribution.

35.2 Principal Measurement Set-Up

The principal experimental arrangement for electronic laser beam analysis (Figure 35.1) consists of a beam sampler for producing a high-quality low-energy replica of the original beam, a filter set for further attenuation, an optional re-imaging optics in order to adjust the beam diameter and sensor aperture, an irradiance or phase distribution sensor and a readout device, normally a personal computer system equipped with a frame grabber and beam analysis software.

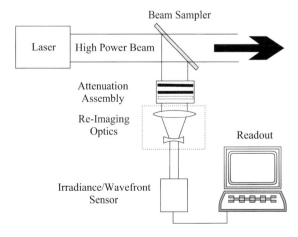

FIGURE 35.1 Set-up for electronic irradiance and phase distribution measurement.

---
[1] The quantity $\omega(x, y) = \lambda \phi(x, y)/2\pi$ is often called the wavefront of the beam. If there is no risk of confusion, both terms—phase and wavefront—will be used simultaneously.

35.3 Irradiance Distribution Measurement

35.3.1 Scanning Devices

Mechanical scanning devices are commonly based on a rotating drum or a moveable opaque screen, with slits, knife edges or pinholes mounted on it and a single-element detector positioned behind (Figure 35.2). These methods work only with cw beams but the requirements on beam attenuation are rather modest, as only a small part of the total beam power hits the detector. The spatial resolution can be as good as 0.5 μm but only a few (~10) one-dimensional profile scans are delivered at a time, thus yielding quite limited off-axis information.

35.3.2 Camera-based Systems

Camera-based systems deliver the complete two-dimensional irradiance distribution, covering a wavelength range from the far-IR to the extreme UV [4]. They are suitable for cw as well as for pulsed beams and, by means of fast electronic shutters, they can pick up single pulses from a 10 kHz pulse train. In general, very strong beam attenuation is needed to account for their high sensitivity. Without re-imaging, the spatial resolution is limited by the pixel size, which ranges between 5 and 15 μm for CCD detectors and approximately 30 μm for pyroelectric cameras. For optimum re-imaging conditions, the detector diameter d_{Det} should accommodate the beam profile in a way that less than 1% of the total beam power is lost by clipping [5]. The corresponding magnification, M, of the beam yields the effective resolution, δ:

$$\delta = \frac{d_{Det}}{\sqrt{N_{tot}} M}.$$

The total number of pixels, N_{tot}, varies between 128×128, say for pyroelectric cameras up to 4k×4k for digital CCD cameras. Qualitative profile inspection as well as low-accuracy beam parameter estimation can be performed by standard 8-bit cameras, but higher demands require 12- or even 16-bit dynamic resolution. A special problem occurring for vidicon-type cameras and for some interline transfer CCD chips is a spatially varying baseline offset which must be compensated by subtraction of a background map if reliable quantitative results are desired.

35.4 Phase Distribution Measurement

35.4.1 Hartmann–Shack Wavefront Sensor

The Hartmann principle [6–10] is based on an orthogonal or hexagonal array of lenses (Hartmann–Shack) or pinholes (Hartmann), which divides the incoming beam into a large number of sub-rays (Figure 35.3). The total irradiance and the position of the individual spots are monitored with a position-sensitive detector placed at a distance f behind the array. The displacement of the spot centroid x^c with respect to a plane-wave reference position x^r measures, for some sub-aperture (i, j), the local phase gradient distribution according to[2]

$$\left(\begin{array}{c} \partial\phi/\partial x \\ \partial\phi/\partial y \end{array} \right)_{ij} = \frac{2\pi}{f \cdot \lambda} \left(\begin{array}{c} x^c - x^r \\ y^c - y^r \end{array} \right)_{ij} \quad (35.2)$$

Obviously, the accuracy in phase estimation depends on the quality of the reference wavefront and on how accurate the centroid estimation is performed. The latter is not only influenced by detector properties as pixel size and dynamic range but suffers from diffraction-induced cross-talk between neighbouring sub-apertures. However, cross-talk can be reduced efficiently in several ways [13,14], so relative rms wavefront errors as low as $\lambda/20$ at 633 nm are achievable even with quite moderate effort [13]. If compared with interferometrical techniques, the spatial resolution (~50 μm) is lower by an order of magnitude, but on the credit side, Hartmann–Shack systems are capable of sensing white light, they are suited for pulsed laser beams and there is no phase ambiguity.

35.4.2 Interferometers

In principle, there are many interferometer designs which are capable of laser beam phase determination. In this section two

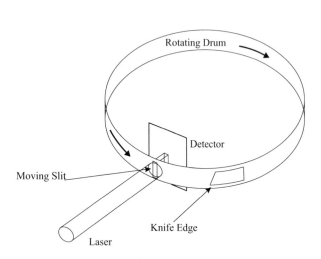

FIGURE 35.2 Principal arrangement of a scanning laser beam profiler.

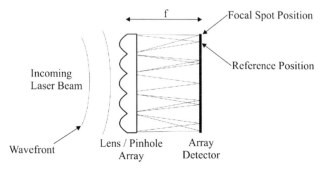

FIGURE 35.3 Principle of the Hartman and Hartmann-Shack wavefront sensors (see the text).

---

[2] The wavefront reconstruction from the gradient data is performed by standard techniques of linear theory and the reader is referred to [11] and [12] as well as reference therein for a detailed analysis.

particular attractive approaches utilizing the method of self-referencing are briefly considered.

The point-diffraction interferometer (PDI) is a common path interferometer exhibiting a simple design [15] (Figure 35.4). A laser beam is directed onto a neutral density filter containing a small pinhole. Diffraction at this pinhole creates a virtually undisturbed spherical reference wave which interferes with the beam. Furthermore, a configuration working in reflection rather than transmission is also possible. Except for a quadratic lag, the interference pattern in the detector plane gives directly the phase distribution with reference to the pinhole position. Obviously, the spherical reference wave has to interfere with the whole beam, so the latter must be globally coherent.

The principle of a lateral shearing interferometer [16] consists of amplitude division of the beam and introduces a small relative amount of transverse shear, s, which is produced, for example, by dual-frequency gratings [17] or precision etalons (Figure 35.5). Interference of the two beams produces fringes which represent, in the limiting case of small shear, curves of constant phase gradient. Complete gradient information requires shear in two orthogonal directions. The extraction of the phase differences Δf from the detector signal N_d^{ij} at pixel (i,j),

$$N_d^{ij} = \int_{pixel} dx\, dy\, I(x,y) + I(x+s_s, y+s_y)$$
$$+ 2\sqrt{I(x,y)I(x+s_s, y+s_y)}\, g(x,y,s_x,s_y,\tau)$$
$$\times \cos(\Delta\phi(x,y,s_x,s_y,\tau)) \quad (35.3)$$

with g denoting the modulus of the complex degree of coherence and $c\tau$ is the optical path difference (OPD) between both sheared beams, is performed either by ac modulating the phase with angular frequency ω (ac-shearing interferometer) or by introducing three or four fixed delays (three- or four-bin shearing interferometer) in one of the beams before detection. Finally, the phase gradient is obtained as

$$\frac{\partial \phi^{ij}}{\delta x} \approx \frac{\Delta\phi(x_i, y_j, s_x, 0)}{s_x} \quad \frac{\partial \phi^{ij}}{\partial y} \approx \frac{\Delta\phi(x_i, y_i, 0, s_y)}{s_y}. \quad (35.4)$$

From equation (35.3), it is evident that a reliable phase gradient measurement requires a certain amount of temporal and spatial beam coherence. However, these restrictions are much weaker than for the PDI, and even white light can be sensed by some modifications of the principal design [18]. Regarding the measurement accuracy, commercial devices yielding rms wavefront errors of $\lambda/50$ at 633 nm are already available.

35.4.3 Phase Retrieval from Intensity Transport Equations

The law of energy conservation applied to equation (35.1) yields with $|S(x,y)| \approx I(x,y)$:

$$\nabla \times (I \times \nabla \phi) = 0 \quad (35.5)$$

and, finally, the paraxial approximation gives the intensity transport equation:

$$k \cdot \frac{\partial I(x,z)}{\partial z} = -\nabla_\perp \cdot (I(x,z) \cdot \nabla_\perp \phi(x,z)) \quad (35.6)$$

It is, therefore, possible to determine the phase distribution from the derivative of the irradiance distribution in the direction of beam propagation by solving the intensity transport equation [19,20]. The advantage of this approach is that only irradiance distribution measurements in two or three neighbouring z-planes are needed and that there are no requirements on beam coherence. However, a partial differential equation has to be solved, and real-time operation requires the synchronization and simultaneous evaluation of several detectors.

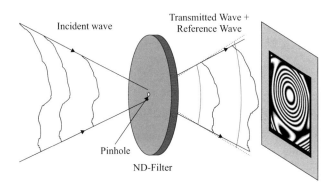

FIGURE 35.4 Basic set-up of the point diffraction interferometer (see the text).

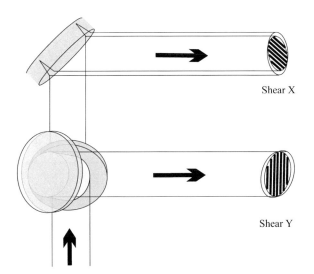

FIGURE 35.5 Operation principle of an etalon-based lateral shearing interferometer (see the text).

35.5 Coherence Measurement

A possible way of determining the coherence properties of laser beams, which is still frequently employed in practice, is Young's experiment employing pairs of slits or pinholes in an opaque screen. Indeed, each term in the expression for the mutual coherence function Γ

$$\Gamma(x_1,y_1,x_2,y_2;\tau) = \sqrt{I(x_1,y_1)}\sqrt{I(x_2,y_2)}$$
$$\times \underbrace{g(x_1,y_1,x_2,y_2;\tau)\exp[i\phi(x_1,y_1,x_2,y_2;\tau)]}_{\gamma(x_1,y_1,x_2,y_2;\tau)} \quad (35.7)$$

has its experimental Young counterpart [21]. In particular, the coherence function g determines the contrast in the centre of the interference pattern but, unfortunately, as g depends on four variables, even highly automated measurements become very tedious. A much more efficient way to determine the beam coherence uses a lateral shearing interferometer [22,23]. It is striking from equation (35.3) that the variation of the local fringe contrast with the applied shear enables the evaluation of g across the whole beam profile, and moreover, the temporal coherence can be estimated from the variation of the fringe visibility with the OPD between both beams.

REFERENCES

1. ISO/DIS 13694, Optics and optical instruments—Lasers and laser-related equipment—Test methods for laser beam power (energy) distribution.
2. ISO/CD 15367-1, Lasers and laser related equipment—Test methods for determination of the shape of a laser beam wavefront—Part I: Terminology and fundamental aspects.
3. Paganin D and Nugent K A 1998 Noninterferometric phase imaging with partially coherent light *Phys. Rev. Lett.* **80** 2586–9.
4. Hoover R B and Tat M W (ed) 1994 X-ray and UV detectors SPIE 2278.
5. ISO/FDIS 11146, Lasers and laser related equipment—Test methods for laser beam parameters—Beam widths divergence angle and beam propagation factor.
6. Flath L, An J, Brase J, Carrano C, Brent Dane C, Fochs S, Hurd R, Kartz M and Sawvel R 2000 Development of adaptive resonator techniques for high power lasers *Proc. 2nd Int. Workshop on Adaptive Optics for Industry and Medicine* ed G D Love (Durham) (Singapore: World Scientific) pp 163–8.
7. Hartmann J 1900 Bemerkungen über den Bau und die Justierung von Spektrographen, *Z. Instrumentenkunde* **8** 2.
8. Neal D R, Alford W J, Gruetzner J K and Warren M E 1996 Amplitude and phase beam characterization using a two-dimensional wavefront sensor *SPIE* **2870** 72–82.
9. Schäfer B and Mann K 2000 Investigation of the propagation characteristics of excimer lasers using a Hartmann–Shack sensor *Rev. Sci. Instrum.* **71** 2663–8.
10. Shack R B 1971 Lenticular Hartmann screen *Opt. Sci. Center Newsletters* **5** 15.
11. Tyson R K 1998 *Principles of Adaptive Optics* 2nd edn (New York: Academic).
12. Press W H, Teukolsky S A, Vetterling W T and Flannery P B 1992 *Numerical Recipes in C* 2nd edn (Cambridge: Cambridge University Press).
13. Laude V, Olivier S, Dirson C and Huignard J P 1999 Hartmann wave-front scanner *Opt. Lett.* **24** 1796–8.
14. Mansel J D, Byer R L and Neal D R 2000 Apodized microlenses for Hartmann wavefront sensing *Proc. 2nd Int. Workshop on Adaptive Optics for Industry and Medicine* ed G D Love (Durham) (Singapore: World Scientific) pp 203–8.
15. Smartt R N and Steel W H 1975 Theory and application of point-diffraction interferometers *Japan. J. Appl. Phys.* **14** 351–6.
16. Bates W J 1947 A wavefront shearing interferometer *Proc. Phys. Soc.* **59** 940–50.
17. Wyant J C 1973 Double frequency grating lateral shearing interferometer *Appl. Opt.* **12** 2057–60.
18. Wyant J C 1974 White light extended source shearing interferometer *Appl. Opt.* **13** 200–2.
19. Streibel N 1984 Phase imaging by the transport equation of intensity *Opt. Commun.* **49** 6–10.
20. Teague M R 1983 Deterministic phase retrieval: a Greens function solution *J. Opt. Soc. Am.* **73** 1434–41.
21. Born M and Wolf E 1982 *Principles of Optics* 6th edn (New York: Academic).
22. Kawata S, Hikima I, Ichihara Y and Watanabe S 1992 Spatial coherence of excimer lasers *Appl. Opt.* **31** 385–96.
23. Omatsu T, Kuroda K and Takase T 1992 Time resolved measurement of spatial coherence of a copper vapor laser using a reversal shear interferometer *Opt. Commun.* **87** 278–86.

FURTHER READING

Dickey ad S C Holswade (New York: Marcel Dekker).
Written from a practitioners point of view. Includes a lot of valuable advices for experimenters engaged in the field of laser beam profile and laser beam parameter measurement.

Geary JM 1995 *Introduction to Wavefront Sensors (Tutorial Texts in Optical Engineering TT18)* (Bellingham, WA: SPIE Optical Engineering Press).
A tutorial text with main emphasis on the optical principles and the hardware implementation of wavefront sensors as well as their application in different fields of optical engineering.

Mandel L and Wolf E 1994 *Optical Coherence and Quantum Optics* 2nd edn (Cambridge: Cambridge University Press).
A standard textbook on optical coherence theory including a comprehensible mathematical introduction to stochastic processes and a couple of applications.

Roundy CB 2000 Current technology of beam profile measurement *Laser Beam Shaping* ed FM

Tyson RK 1998 *Principles of Adaptive Optics* 2nd edn (New York: Academic).
Thorough treatment of wavefront sensing, wavefront sensors and wavefront reconstruction, considering both theoretical and practical aspects of the field. Many references.

36

The Measurement of Ultrashort Laser Pulses

Rick Trebino, Rana Jafari, Peeter Piksarv, Pamela Bowlan, Heli Valtna-Lukner,
Peeter Saari, Zhe Guang, and Günter Steinmeyer

CONTENTS

36.1	Ultrashort Laser Pulses	488
	36.1.1 Measuring the Spectrum	489
36.2	The Spectrum and One-dimensional Phase Retrieval	489
36.3	The Intensity Autocorrelation	490
	36.3.1 The Autocorrelation and One-dimensional Phase Retrieval	490
36.4	Autocorrelations of Complex Pulses	492
36.5	Autocorrelations of Noisy Pulse Trains	492
36.6	Third-order Autocorrelations	493
36.7	The Autocorrelation and Spectrum—in Combination	495
36.8	Interferometric Autocorrelation	496
36.9	Cross-correlation	496
36.10	Autocorrelation Conclusions	498
36.11	The Time-frequency Domain	498
36.12	Frequency-resolved Optical Gating (FROG)	499
36.13	FROG and the Two-dimensional Phase-retrieval Problem	500
36.14	FROG Beam Geometries	500
36.15	The FROG Algorithm	500
36.16	The RANA Approach	503
36.17	Properties of FROG	504
36.18	Single-shot FROG	504
36.19	Near-single-cycle Pulse Measurement	505
36.20	FROG and the Coherent Artefact	507
36.21	XFROG	508
36.22	Very Simple FROG: GRENOUILLE	509
36.23	Measuring Two Pulses Simultaneously	512
36.24	Error Bars	513
36.25	Other Self-referenced Methods	514
36.26	Spectral Interferometry	514
36.27	Advantages and Disadvantages of Spectral Interferometry	515
36.28	Crossed-beam Spectral Interferometry	516
36.29	Practical Measurement Weak and Complex Pulses: SEA TADPOLE	517
36.30	Measuring Very Complex Pulses in Time: MUD TADPOLE	519
36.31	Single-shot MUD TADPOLE	519
36.32	SPIDER	521
36.33	Spatiotemporal Pulse Measurement	524
36.34	Spatially Resolved Spectral Interferometry: One Spatial Dimension	524
36.35	Fibre-based Scanning Spatiotemporal Pulse Measurement: Two and Three Spatial Dimensions	525
36.36	Spatiotemporal Measurement Examples: Focusing Pulses	526
36.37	Other Spatiotemporal-measurement Methods	527
36.38	Spatiotemporal Measurement on a Single Shot: STRIPED FISH	527
36.39	Conclusions	531
Acknowledgements		531
References		531

36.1 Ultrashort Laser Pulses

Ultrashort light pulses are the shortest events ever created. Pulses as short as tens of attoseconds (10^{-18} s) have been generated, and it is now routine to generate pulses less than 100 fs long. In order to use them effectively and to determine how to make them shorter, less structured, and more stable from pulse to pulse, it is important to be able to measure them.

This task seems particularly difficult because, in order to measure an event in time, it would seem that one would need a shorter event with which to time it. For example, to resolve the action of a bubble popping requires a strobe light shorter than the time it takes for the bubble to pop. Then to measure the strobe light intensity vs. time requires a detector with an even shorter response time, etc. So, one could reasonably conclude that measuring the *shortest* event is impossible.

We will return to this issue later, but for now, we must first ask precisely what it is that we are trying to measure about a pulse. And that is its electric field as a function of time and space and which can potentially be a complicated function of time and space. We will temporarily ignore the field's spatial dependence and concentrate on the pulse's dependence on time. We can write the pulse electric field as:

$$\mathcal{E}(t) = \frac{1}{2}\sqrt{I(t)} \exp\{i[\omega_0 t - \phi(t)]\} + c.c. \qquad (36.1)$$

where t is time in the reference frame of the pulse, ω_0 is a carrier angular frequency on the order of $10^{15} s^{-1}$ for visible and near-IR light, and $I(t)$ and $\phi(t)$ are the temporal intensity and phase of the pulse.

As usual, "c.c." means *complex conjugate* and is required to yield a real pulse field. But we will use the equivalent *analytic signal* representation and ignore the complex-conjugate term, yielding a complex pulse field and simplifying the mathematics significantly. The *c.c.* can be re-added anytime. As a result, the *complex amplitude*—the quantity generally desired in a measurement because the centre frequency, ω_0, is easy to measure using a spectrometer—is:

$$E(t) \equiv \sqrt{I(t)} \exp[-i\phi(t)]. \qquad (36.2)$$

The pulse field in the frequency domain, $\tilde{E}(\omega)$, is the Fourier transform of the time-domain field and is usually separated into its spectral intensity $S(\omega)$ and phase $\varphi(\omega)$:

$$\tilde{E}(\omega) = \sqrt{S(\omega)} \exp[-i\varphi(\omega)]. \qquad (36.3)$$

Note that the temporal phase, ϕ, and spectral phase, φ, are both called "phi", but they are different (ϕ is a function of time and φ is a function of frequency), so we have used different Greek characters to distinguish them. See Figure 36.1 for a schematic drawing of a simple, common pulse, and these quantities for it.

Before we begin discussing methods for measuring these quantities, a few words on the philosophy of such devices are in order. First, a pulse-measurement device should be able to completely measure pulses such as the simple one above but also much more complex pulses as well. This is because, if a measurement technique *can* only measure simple pulses, it

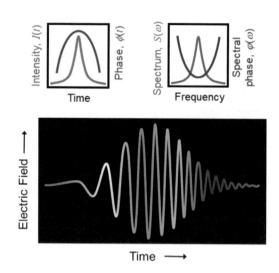

FIGURE 36.1 Upper plots: Intensity and phase vs. time (upper left) and the spectrum and spectral phase vs. frequency (upper right). The curves plotted here are for a positively "chirped" pulse, whose darkest curves (black) in each figure precede the curves that appear lighter (lower plot). In the latter plot, the electric field of a positively chirped white pulse is plotted using the colour of its instantaneous frequency at the particular time. From www.frog.gatech.edu.

will only measure simple pulses. In other words, when presented with a complex pulse, a technique that can only measure simple pulses would yield an incorrect simple pulse, rather than the correct complex pulse. This is unacceptable. Second, the measurement device should itself be simple, with a minimal number of components and knobs, so that non-experts in the field of pulse measurement should be able to use it. Also, tweaking a knob of a measurement device often changes the resulting measurement—a highly undesirable situation—so the device should not yield results that are sensitive to its alignment. The best way to accomplish this is for the device to have minimal knobs in the first place. Third, the technique should have a minimum of "ambiguities" (incorrect pulses that have the same measured trace as the correct pulse). Although we will find that ambiguities are unavoidable, some are *trivial* ambiguities, that is, they can easily be calculated and/or removed and which we can live with. Unfortunately, in some techniques, there will be *non-trivial* ambiguities, which can be neither calculated nor removed and so are unacceptable. Finally, the device should also yield some sort of feedback as to whether the measurement is correct or not and whether the assumptions of the measurement are satisfied.

Another issue is that often a device must average over many, possibly very differently shaped, pulses. But, by definition, devices can only provide one result. This is, of course, an impossible situation, in which no one answer can be correct, and so provides an excellent litmus test for pulse-measurement devices. In this case, how well does the device's "measured pulse" approximate an actual typical pulse? Does the device at least yield the correct pulse length? And critically, does the device provide an indication of the instability? Unfortunately, there is no such thing as a "pulse-shape stability meter", so this task necessarily also falls to the pulse-measurement technique.

While many pulse-measurement techniques have been proposed, very few meet these criteria. This article will focus mainly on devices and general techniques that have these important properties. It will also include some that do not, so the reader can see how things can go wrong.

This chapter will also distinguish between "self-referenced" techniques and those that require a well-characterized reference pulse to make a measurement. In principle, it would seem that self-referenced pulse measurement is impossible, as, by definition, a shorter event seems required to measure an event in time, and the pulse is only as short as itself, not shorter. However, this argument was shown to be a mere myth in 1991 by Rick Trebino, and it is, in fact, not only possible, but quite easy for a pulse to completely measure itself in time [1]. Indeed, self-referenced techniques with all the above important properties exist.

Then, once a pulse has been measured using only itself (and, of course, some optical components), that pulse can then be used as a reference pulse in other, much more difficult pulse-measurement problems, such as the measurement of a very weak pulse or a pulse's complete *spatiotemporal* intensity and phase.

Our standards for techniques that use a reference pulse for such difficult tasks will be much lower with regard to pulse-shape-stability measurement. This is because, once established by a self-referenced technique as having a stable shape, a pulse train can be relatively certain to remain stable when simply propagating around an optical table. Of course, if a technique that requires a reference pulse can also confirm pulse-shape stability in multi-shot measurements, all the better. Indeed, for measuring very complex pulses generated from a stable pulse train using, say, an optical fibre in which high-order non-linear-optical processes occur, pulse-shape stability cannot be counted on, and a technique that indicates the presence of instability is critical, even though it uses a reference pulse.

We will begin with a discussion of self-referenced techniques because, without a reference pulse, techniques that require one are useless.

36.1.1 Measuring the Spectrum

Of the above-mentioned four quantities, it has only been generally possible to measure the pulse spectrum, $S(\omega)$. Spectrometers perform this task well and are readily available. The most common spectrometer involves diffracting a collimated beam off a diffraction grating and focusing it onto a camera. But interferometers also work (see Figure 36.2).

Fourier-transform spectrometers operate in the time domain and measure the transmitted integrated intensity from a Michelson interferometer vs. delay τ, which is often called the light's *field autocorrelation* or its *interferogram*:

$$\Gamma^{(2)}(\tau) = \int_{-\infty}^{\infty} E(t)\, E^*(t-\tau)\, dt, \quad (36.4)$$

neglecting constant terms. The interferogram's Fourier transform (vs. delay τ) is simply the spectrum, a result known as the *Autocorrelation Theorem*:

$$\left|\tilde{E}(\omega)\right|^2 = \mathscr{F}\left\{\int_{-\infty}^{\infty} E(t)\, E^*(t-\tau)\, dt\right\}, \quad (36.5)$$

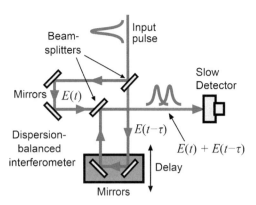

FIGURE 36.2 Experimental layout for a Fourier-transform spectrometer. Note that this Michelson interferometer design is "dispersion balanced" (has the same amount of glass, and hence dispersion, in each path), which is required for high-quality fringes in the presence of broadband light. From www.frog.gatech.edu.

where $\mathscr{F}$ indicates the Fourier transform (here, with respect to τ).

Thus, all spectrometers, whether diffraction-grating or Fourier-transform devices, yield the spectrum, and only the spectrum.

36.2 The Spectrum and One-dimensional Phase Retrieval

It is actually more interesting than it may seem to ask what information is, in fact, available from the spectrum. Obviously, if we have only the spectrum, what we lack is precisely the spectral phase. Sounds simple enough. Why belabour this point?

Here is why: suppose we have some additional information, such as the knowledge that we are measuring a *pulse*? What if we know that the pulse intensity vs. time is definitely zero outside a finite range of times? Or, at least asymptotes quickly to zero as $t \to \pm\infty$? This is a great deal of additional information, and it is interesting to ask how much this additional information allows us to limit the possible pulses that correspond to a given spectrum.

Whatever the additional information, this class of problems is called the *one-dimensional phase-retrieval problem* for the obvious reason that we have the spectral magnitude and we are trying to retrieve the spectral phase using this additional information.

It turns out that it is highly *ill-posed*. In other words, there are many, often infinitely many, ambiguities—pulses that correspond to a given spectrum and that satisfy any additional constraints such as those mentioned above. The one-dimensional phase-retrieval problem is unsolvable in almost all cases of practical interest, even when additional information is included.

Specifically, even with this additional information, there are obvious, or "trivial", ambiguities [2]. Clearly, if the complex amplitude $E(t)$ has a given spectrum, then adding a phase shift, yielding $E(t)\exp(i\phi_0)$, changes nothing. Same with a translation, $E(t-t_0)$. Not to mention the complex-conjugated mirror image, $E^*(-t)$, which corresponds to a time reversal. Most researchers, however, can live with these trivial ambiguities, hence the name. They are known, simple, and usually can be removed if additional information is available. Are there other, more difficult or even impossible to remove, "nontrivial" ambiguities?

Unfortunately, nearly always yes. In two classic papers written in 1956 and 1957, E. J. Akutowicz showed that knowledge of the spectrum in conjunction with the additional knowledge that $E(t)$ is of finite duration—often called *finite support*—is still insufficient to uniquely determine $E(t)$ [3,4]. Indeed, he showed that infinitely many pulse fields usually satisfy these constraints. For example, a Gaussian spectrum can have any amount of linear chirp and so can correspond to an intensity vs. time that is also Gaussian, but with any pulse width. And, of course, it can have almost any higher-order phase distortion, as well. Indeed, the number of possible pulses that correspond to a given spectrum is not just infinity; it is a *higher-order infinity*.

36.3 The Intensity Autocorrelation

Recall that, to measure such short events, we appear to need a shorter event, but we do not have one. The shortest event we have is the event itself. The *intensity autocorrelation*, $A^{(2)}(\tau)$, is a technique based on using the pulse to measure itself [5–13]. It involves splitting the pulse into two, variably delaying one with respect to the other, and spatially overlapping the two pulses in some instantaneously responding non-linear-optical medium, such as a second-harmonic-generation (SHG) crystal (see Figure 36.3). A SHG crystal produces light at twice the frequency of input light with a field that is given by

$$E_{sig}^{SHG}(t,\tau) \propto E(t)E(t-\tau). \quad (36.6)$$

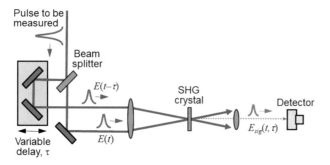

FIGURE 36.3 Experimental layout for an intensity autocorrelator using second-harmonic generation. A pulse is split into two: one is variably delayed with respect to the other and the two pulses are overlapped in an SHG crystal. The SHG pulse energy is measured vs. delay, yielding the autocorrelation trace. Other effects, such as two-photon fluorescence and two-photon absorption can also yield the autocorrelation, using similar beam geometries [9,11,12,14]. In this figure and others, a lens is shown focusing the beam into the crystal, but, for extremely short pulses, when propagation through glass unacceptably distorts the pulse, a curved mirror can be used in its place. Also, because intensity autocorrelation is not interferometric, compensation for dispersion from the beam-splitter is not needed. From www.frog.gatech.edu.

where τ is the delay between the two pulses. This field has an intensity (the squared magnitude of the electric field) that is proportional to the product of the intensities of the two input pulses

$$I_{sig}^{SHG}(t,\tau) \propto I(t)I(t-\tau). \quad (36.7)$$

Because detectors (even streak cameras) are typically too slow to time-resolve $I_{sig}^{SHG}(t,\tau)$, this measurement necessarily yields an integral over time:

$$A^{(2)}(\tau) = \int_{-\infty}^{\infty} I(t)I(t-\tau)\, dt. \quad (36.8)$$

Equation (36.8) is the definition of the intensity autocorrelation, or, for short, simply the autocorrelation. It is different from the *field* autocorrelation (equation (36.4)), which provides only the information contained in the spectrum.

It is clear that an (intensity) autocorrelation yields some measure of the pulse length because no second harmonic intensity will result if the pulses do not overlap in time; thus, a relative delay of one pulse length will typically reduce the SHG intensity by about a factor of two.

Figure 36.4 shows some simple pulses and their intensity autocorrelations. Because the intensity autocorrelation only attempts to provide a measure of the pulse intensity vs. time $I(t)$ and makes no attempt to measure the phase, only intensities are shown.

36.3.1 The Autocorrelation and One-dimensional Phase Retrieval

We can learn more about the autocorrelation by applying the Autocorrelation Theorem to equation (36.8), yielding:

$$\tilde{A}^{(2)}(\omega) = \left|\tilde{I}(\omega)\right|^2. \quad (36.9)$$

where $\tilde{I}(\omega)$ is the Fourier Transform of the intensity vs. time (note that it is not the spectrum, $S(\omega)$). In words, the Fourier transform of the autocorrelation is the mag-squared Fourier transform of the intensity. In other words, if we know the autocorrelation of an intensity, we have the magnitude, but not the phase of the Fourier transform of the quantity we wish to find, $I(t)$.

If this sounds familiar, it should. *It is another one-dimensional phase-retrieval problem!*

Thus, autocorrelation also suffers from trivial and non-trivial ambiguities. Figure 36.5 gives examples of different pulse intensities that have the same autocorrelation.

The approach taken by users of autocorrelation has been to assume a simple pulse shape, such as a Gaussian or a hyperbolic secant squared, and divide the width of their autocorrelation trace by the relevant factor computed for that pulse shape to obtain a possible pulse length. Although highly simplified models of ultrafast lasers yielded such theoretical pulse lengths, this has little to no theoretical justification in practice, but no better approach was available until the early 1990s, and that approach did not involve simple autocorrelation.

Measurement of Ultrashort Laser Pulses

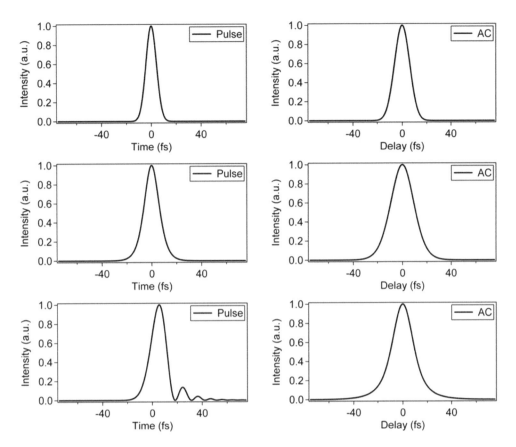

FIGURE 36.4 Examples of theoretical pulse intensities and their intensity autocorrelations (AC). Left: Intensities vs. time. Right: The intensity autocorrelation corresponding to the pulse intensity to its left. Top row: A 10fs Gaussian intensity. Middle row: A 7fs sech^2 intensity. Bottom row: A pulse whose intensity results from cubic spectral phase. Note that, because the pulse is measuring itself, the autocorrelation loses details of the pulse, and, as a result, all of these pulses have similar autocorrelations. From Ref. [1].

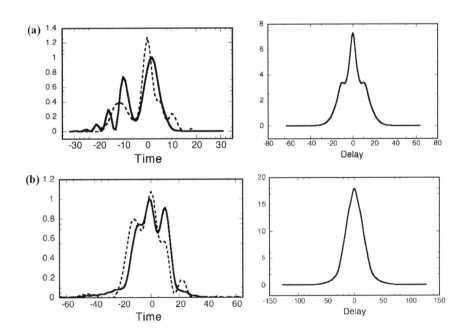

FIGURE 36.5 (a) Left: Two pulse intensities that yield numerically equivalent autocorrelations. Right: Their intensity autocorrelation. Both scales are in arbitrary units. (b) Left: Two additional pulse intensities that yield numerically equivalent autocorrelations. Right: Their intensity autocorrelation. In this case, despite their structured shapes, the intensity autocorrelations of these pulses never vary by more than the thickness of the above curve from a Gaussian. From Ref. [1].

36.4 Autocorrelations of Complex Pulses

Nowhere does the lack of power of the autocorrelation to reveal structure in a pulse reveal itself more than in the measurement of complicated pulses, where, unfortunately, there happens to be a great deal of structure waiting to be revealed. In fact, for complex pulses, it can be shown that, as the intensity increases in complexity, the autocorrelation actually becomes *simpler* and approaches a simple shape of a narrow spike on a pedestal, *independent of the intensity structure* (see Figure 36.6) [15].

The narrow central spike, called the *coherence spike* or *coherent artefact* of approximate width τ_c (the coherence time), sits on top of a broad *pedestal* or *wings* of the approximate pulse-length τ_p.

In this case, when only a single complex pulse (or a train of identical complex pulses) is to be measured, we call this spike the *single-pulse* coherent artefact.

36.5 Autocorrelations of Noisy Pulse Trains

Even relatively simple pulses can yield autocorrelations comprising a coherence spike on a pedestal if the measurement averages over a noisy train of them, in which their shape varies [16]. Consider, for example, double pulses. Figure 36.7 shows some double pulses and their autocorrelations, which have three bumps.

Now, when a laser double-pulses, it typically does so quite randomly. It will often emit a train of double pulses with different, random separations for each double-pulse in the train. Since a typical ultrafast laser emits pulses at a very high repetition rate (100 MHz), and most autocorrelators are multi-shot devices anyway (measuring the SHG energy for only one delay at a time), the autocorrelator will necessarily average over the autocorrelations of many such pulses.

This will also produce a trace that contains two components, a narrow central coherence spike sitting on top of a broad pedestal, whose height will typically be much less than the value of 1/2 we saw in the last section. Clearly, the coherence spike is a rough measure of the individual pulses within the double-pulse, and the pedestal indicates the distribution of double-pulse separations. Again, while it would be tempting to try to derive the pulse length from the coherence spike—especially now that the pedestal seems so weak in comparison—the true pulse length is related, not to the coherence spike, but to the pedestal.

Now consider a related problem: averaging over a train of pulses with randomly varying complex intensity pulse shapes.

This situation also yields precisely the same autocorrelation trace!

Unfortunately, in the 1960s, when autocorrelation was first introduced, researchers, desiring to claim a shorter (more exciting) pulse, neglected the background and confused the coherent artefact for the pulse length. The correct interpretation of such traces was provided by Fisher and Fleck [17].

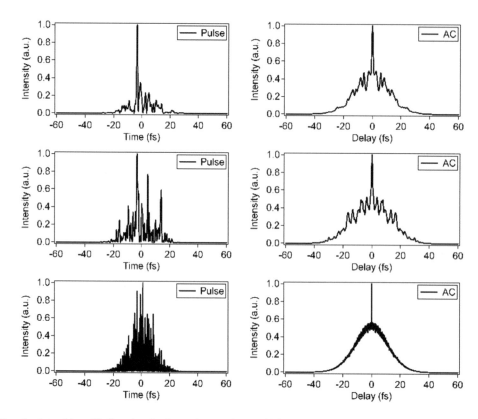

FIGURE 36.6 Complex intensities with Gaussian slowly varying envelopes with increasing amounts of intensity structure (left) and their autocorrelations (right). As the pulse increases in complexity (from top to bottom), the autocorrelation approaches the simple coherence-spike-on-a-pedestal shape, independent of the pulse intensity structure. Note that, as the structure increases in complexity, the coherence spike narrows along with the structure, while the pedestal continues to reveal the approximate width of the envelope of the intensity. From Ref. [1].

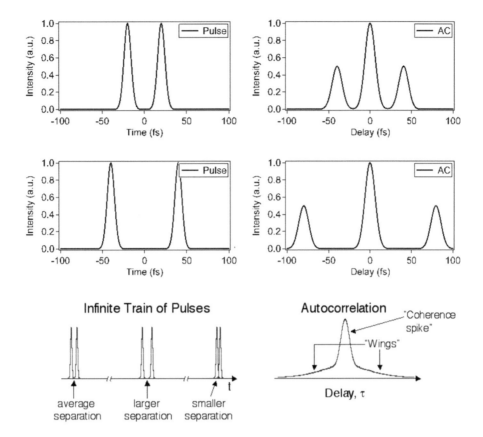

FIGURE 36.7 Examples of theoretical double-pulse intensities and their intensity autocorrelations. Left: Intensities vs. time. Right: The intensity autocorrelation corresponding to the intensity to its left. Top row: Two pulses (10fs Gaussians) separated by four pulse lengths. Second row: The same two pulses separated by eight pulse lengths. Third row: A train of double pulses with varying separation. A multi-shot autocorrelation measurement (third row, right) averages over many double-pulses. Note that the structure has washed out in the autocorrelation due to the averaging over many double pulses in the train, each with its own separation. The pulse length is better estimated by the width of the pedestal than by the width of the coherence spike. From Ref. [1].

This has been considered an embarrassing mistake, but it was not at all obvious at that time that such traces would result in a technique that yielded an always positive quantity. Coherence is generally considered to only be a property of waves, which go negative and which yield fringes that can cancel out in phase-sensitive measurements.

On the other hand, making this mistake *now*—when everyone has known about it for fifty years—would indeed be quite embarrassing! So, the reader is cautioned to be very careful in interpreting traces generated by this obsolete technique.

To see just how careful one must be, consider the case of partial mode-locking, a common problem in ultrashort-pulse lasers, which can lead to highly unstable and complex pulse shapes. Here we consider only slightly complex random pulse shapes consisting of a stable 12 fs flat-phase Gaussian component plus a longer random component. Autocorrelations are shown in Figure 36.8 for a stable 12 fs pulse train and also for two such random pulse trains. Note that the autocorrelation approaches the same shape as for complex and double pulses (bottom row). Note also that, when the coherence spike is only a factor of two or so shorter than the base (middle row), it blends in with the base and cannot be distinguished from it, yielding an autocorrelation trace that looks like that of a simple stable pulse train, when it in fact yields a pulse that is considerably shorter than the actual pulse. Thus, even for only slightly random pulse shapes and long before it approaches the classical spike-on-a-pedestal shape, autocorrelation can lead to significant errors in the pulse length—by a factor of 2 or more.

Because this case involves multiple pulses, we call it the *multi-pulse* coherent artefact. In autocorrelation, the two effects are essentially indistinguishable, but this will not be the case for techniques that will be discussed later.

So, to repeat, even when the pedestal is weak, do not make the mistake of misidentifying the coherence spike as an indication of the pulse length! But also recognize that any, even an innocent-looking autocorrelation, trace could in fact be the sum of a base and coherent artefact and yield pulses that are significantly shorter than in fact are present.

36.6 Third-order Autocorrelations

The inadequacies of the (second-order) intensity autocorrelation have not been lost on those who use it. As a result, several improvements emerged over the years, and one simple advance was the *third-order intensity autocorrelation*, or just the *third-order autocorrelation* [18–22].

Figures 36.9–36.11 give examples of beam geometries that yield third-order autocorrelations.

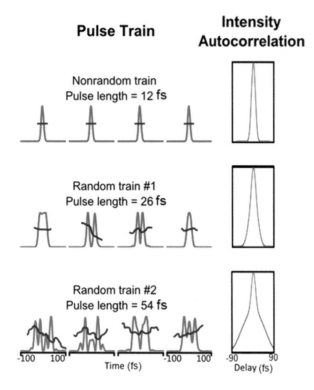

FIGURE 36.8 Examples of theoretical random complex intensities (light gray) and phases (dark gray, but the phase is irrelevant for autocorrelation) and their intensity autocorrelations. Top: Stable simple intensities vs. time and their autocorrelation, which correctly yields the pulse length. Middle: The sum of the stable pulse train of the top row and a random component and its corresponding intensity autocorrelation. Bottom row: The sum of the stable pulse train of the top row and a highly random, more complex component and its corresponding intensity autocorrelation. Note that the structure has washed out in the middle and bottom autocorrelations due to the averaging over many double pulses in the train. Also, note that the middle-row autocorrelation's coherent artefact is long enough that it plus the background look like an autocorrelation of a simple stable pulse and would easily be confused for one. However, this yields a factor of almost two shorter pulses than is actually present. From www.frog.gatech.edu.

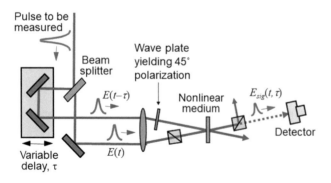

FIGURE 36.9 Experimental layout for a third-order intensity autocorrelator using polarization gating. A pulse is split into two: one (the "gate" pulse) has its polarization rotated by 45° and is variably delayed, and the other (the "probe" pulse) passes through crossed polarizers. Then, the two pulses are overlapped in a piece of glass. The 45°-polarized gate pulse induces birefringence in the glass, which slightly rotates the polarization of the probe pulse causing it to leak through the polarizers if the pulses overlap in time. The leakage pulse energy is measured with respect to delay, producing the third-order autocorrelation trace. From www.frog.gatech.edu.

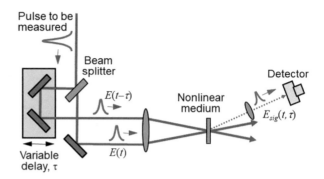

FIGURE 36.10 Experimental layout for a third-order intensity autocorrelator using self-diffraction. The pulse is split into two, delayed, and recombined in a third-order non-linear medium, as in the previous figure, but here, the pulses induce a grating in the glass, which diffracts one of the pulses into a new direction off to the side, at $2\mathbf{k}_1 - \mathbf{k}_2$. This also produces a third-order autocorrelation trace. From www.frog.gatech.edu.

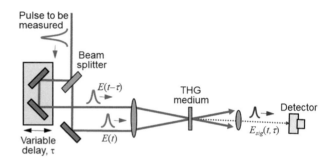

FIGURE 36.11 Experimental layout for a third-order intensity autocorrelator using third-harmonic generation (THG). The pulse is split into two, delayed, and recombined in a third-order non-linear medium, as in the previous figure, but here, the pulses yield the third harmonic. This also produces a third-order autocorrelation trace. An alternative approach for generating a THG autocorrelation is to perform SHG in one of the above beams after the beam splitter (removing any remaining input wavelength), and then perform sum-frequency generation (SFG) between that beam and the other. From www.frog.gatech.edu.

A polarization-gating (PG) autocorrelator produces a field given by:

$$E_{\text{sig}}^{\text{PG}}(t,\tau) \propto E(t)|E(t-\tau)|^2, \quad (36.10)$$

where $E(t)$ is the vertically polarized pulse and $E(t-\tau)$ is the delayed field of the 45°-polarized pulse. This yields a signal field with three factors of the field, hence the notion of third-order. It then yields a signal intensity that is proportional to three factors of the intensities of the two input pulses:

$$I_{\text{sig}}^{\text{PG}}(t,\tau) \propto I(t)\, I^2(t-\tau). \quad (36.11)$$

Again, detectors are too slow to time-resolve the rapidly varying intensity, $I_{\text{sig}}^{\text{PG}}(t,\tau)$, so this measurement produces a measured quantity, which is the time integral of $I_{\text{sig}}^{\text{PG}}(t,\tau)$:

$$A^{(3)}(\tau) = \int_{-\infty}^{\infty} I(t)\, I^2(t-\tau)\, dt. \quad (36.12)$$

This result is the third-order autocorrelation.

The other geometries yield different signal fields, but they all yield the same result. For example, self-diffraction (Figure 36.10) has a signal field given by:

$$E_{\text{sig}}^{\text{SD}}(t,\tau) \propto E^2(t)\, E^*(t-\tau), \qquad (36.13)$$

which has the signal pulse intensity:

$$I_{\text{sig}}^{\text{SD}}(t,\tau) \propto I^2(t)\, I(t-\tau). \qquad (36.14)$$

And, its integrated intensity is:

$$A^{(3)}(\tau) = \int_{-\infty}^{\infty} I^2(t) I(t-\tau)\, dt, \qquad (36.15)$$

which, with a simple change of variables, $t \to t - \tau$, yields the same result, except for a reflection about the vertical axis.

Similarly, THG has a signal field:

$$E_{\text{sig}}^{\text{THG}}(t,\tau) \propto E^2(t)\, E(t-\tau), \qquad (36.16)$$

which has the signal pulse intensity:

$$I_{\text{sig}}^{\text{THG}}(t,\tau) \propto I^2(t)\, I(t-\tau), \qquad (36.17)$$

whose time-integral is also the third-order autocorrelation.

Third-order autocorrelations have also been generated using other non-linear-optical effects, such as three-photon fluorescence [20].

Because $I(t)$ and $I(t-\tau)$ appear in the third-order autocorrelation *asymmetrically* (only one is squared), a third-order autocorrelation is symmetrical only if the intensity that produces it is. The asymmetry is not overwhelming, but it can be sufficient for determining whether a satellite pulse is a pre-pulse or a post-pulse, which is important in ultrahigh-intensity settings, where a pre-pulse can have enough intensity to damage the sample before the main pulse arrives [23].

Third-order non-linearities are weaker than second-order ones and hence require more pulse energy. As a result, they do not work well for unamplified pulses from typical ultrafast laser oscillators. But they are useful for amplified pulses and UV pulses (for which SHG cannot be performed and where third-order non-linearities are stronger).

Third-order autocorrelation also does not uniquely determine the pulse intensity. And, the third-order autocorrelation of a complicated pulse or a train of random pulses is similar to the second-order autocorrelation of such a pulse or train: a coherence spike on top of a broad pedestal [15].

36.7 The Autocorrelation and Spectrum—in Combination

If the autocorrelation by itself does not determine the intensity, and the spectrum by itself does not determine the field, why not just use *both* measures in combination and see what the two quantities *together* yield?

Unfortunately, for ultrashort laser pulses, we do not have the spectrum and the *intensity*. We have the spectrum and the *autocorrelation*. Of course, as we have seen, the autocorrelation does not uniquely determine the intensity. So, this process can at best yield only *a* possible pulse field, not *the* pulse field. Indeed, for very complicated pulses, because the autocorrelation contains so little information, this procedure is doomed to fail.

Even for simple pulses, no analytical work has been performed on this topic (it is mathematically very difficult). But Chung and Weiner [24] have performed numerical computations and found numerous non-trivial ambiguities, in addition to the obvious direction-of-time ambiguity. Figure 36.12 shows one of the many ambiguities they found.

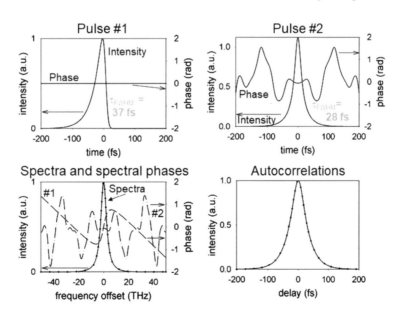

FIGURE 36.12 Two pulses (top row) with different intensities and phases, which yield numerically identical autocorrelations (bottom right) and spectra (bottom left). The spectral phase of both pulses is shown as the dashed curves at bottom left. From Ref. [1].

There are several variations on this theme. While no one has taken the time to evaluate them as Chung and Weiner have for the earlier scheme, they may work in some simple cases, but it is doubtful that they perform much better in general. And even if they yield a pulse consistent with the autocorrelation and spectrum, it may or may not be the correct pulse—all have many non-trivial ambiguities.

Occasionally, researchers use "regularization", an algorithmic approach that attempts to find the simplest pulse consistent with the given traces. The assumption that the shortest pulse is the correct one is dubious at best, so this approach is not helpful in this case or any other. Regularization should play no role in pulse measurement.

36.8 Interferometric Autocorrelation

The spectrum and the autocorrelation can essentially be combined into a single data trace in a method called the *interferometric autocorrelation*, often also called *phase-sensitive autocorrelation* or *fringe-resolved autocorrelation (FRAC)*. It was introduced by Jean-Claude Diels in 1983 [25–31], and quickly became popular. It involves measuring the second-harmonic energy vs. delay from an SHG crystal placed at the output of a Michelson interferometer (see Figure 36.13). In other words, it involves performing an autocorrelation measurement using collinear beams, so that the second harmonic light created by the interaction of the two different beams combines coherently with that created by each individual beam. As a result, interference occurs due to the coherent addition of the several beams, and interference fringes occur vs. delay, as well as a background out to ±∞ due to the SHG from the individual beams. This is in contrast to intensity autocorrelation, which is also often referred to as the *background-free autocorrelation* when interferometric autocorrelation is also being discussed.

The expression for the interferometric autocorrelation trace is:

$$I_{\text{FRAC}}(\tau) = \int_{-\infty}^{\infty} \left| \left[E(t) + E(t-\tau) \right]^2 \right|^2 \, dt \quad (36.18)$$

$$= \int_{-\infty}^{\infty} \left| E(t)^2 + 2E(t)E(t-\tau) + E(t-\tau)^2 \right|^2 \, dt. \quad (36.19)$$

Note that, if the $E(t)^2$ and $E(t-\tau)^2$ terms were removed from the above expression, only the cross term, $2E(t)E(t-\tau)$ would remain, yielding the usual expression for background-free autocorrelation. These new terms, integrals of $E(t)^2$ and $E(t-\tau)^2$, are due to SHG of each individual pulse. And their interference, both with each other and with the cross term, yields the additional information in the interferometric autocorrelation that is not present in intensity autocorrelation. Multiplying out all the terms yields four terms: a constant, the intensity autocorrelation, a term related to the interferogram, and the interferogram of the second harmonic. Unfortunately, the mathematics involved is complex, and interferometric autocorrelation does not yield the pulse or even its intensity and must also be curve-fit to a guessed field. Example interferometric autocorrelation traces are shown in Figure 36.14.

Chung and Weiner also shed some light on the issue of how well the interferometric autocorrelation determines pulses by calculating interferometric autocorrelation traces for the pairs of pulses that yielded ambiguities in the spectrum/autocorrelation approach. And, they found that the resulting traces of the pairs of pulses had very similar interferometric autocorrelation traces. See Figure 36.15.

In short, interferometric autocorrelation yields essentially the autocorrelation and the spectrum. But it does so in one measurement and so is considerably more convenient and convincing.

On the other hand, Diels and co-workers showed that, once a field has been fit to an interferometric autocorrelation trace, the direction of time could be determined by including a second interferometric autocorrelation measurement—actually a fringe-resolved *cross*-correlation—in which some glass is placed in one of the interferometer arms. This breaks the symmetry and yields an asymmetrical trace. Then, assuming that the dispersion of the glass is known, Diels and co-workers showed that the two traces could be used to completely determine the pulse field in some cases. Again, however, no study has been published on this algorithm's performance, and this approach is rarely used.

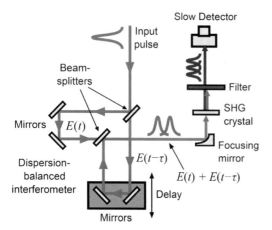

FIGURE 36.13 Experimental layout for interferometric autocorrelation. The correct setup uses a dispersion-balanced Michelson interferometer, as shown here. A curved focusing mirror is shown, rather than a lens, to emphasize its application to extremely short pulses, for which propagation through glass must be minimized. From www.frog.gatech.edu.

36.9 Cross-correlation

Occasionally, a shorter event is available to measure a pulse. In this case, a cross-correlation can be performed (see Figure 36.16). The cross-correlation, $C^{(2)}(\tau)$, is given by:

$$C^{(2)}(\tau) = \int_{-\infty}^{\infty} I(t)\, I_g(t-\tau)\, dt, \quad (36.20)$$

where $I(t)$ is the unknown intensity and $I_g(t)$ is the gate pulse (shorter event) intensity vs. time.

Measurement of Ultrashort Laser Pulses

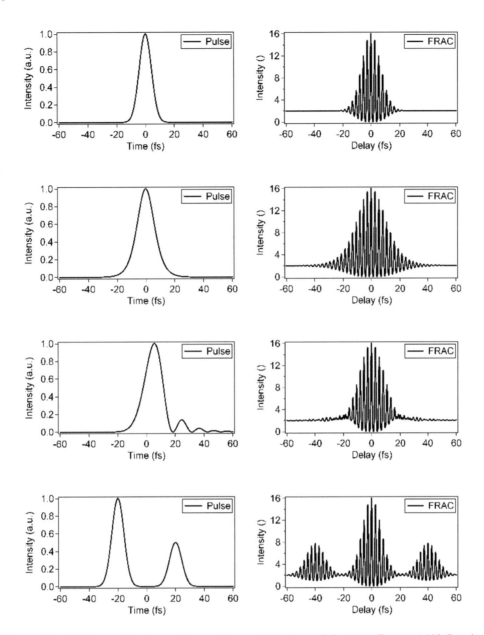

FIGURE 36.14 Pulses and their interferometric autocorrelation or fringe-resolved autocorrelation traces. Top row: A 10fs Gaussian intensity. Second row: A 7fs sech^2 intensity. Third row: A pulse whose intensity results from 3rd-order spectral phase. Note that the satellite pulses due to third-order spectral phase, which were invisible in the intensity autocorrelation, actually can be seen by looking carefully in the wings of the interferometric autocorrelation trace. Fourth row: A double pulse. From Ref. [1].

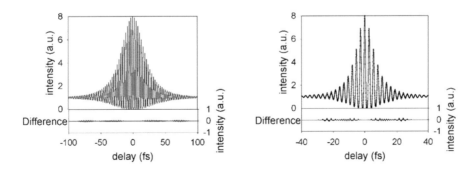

FIGURE 36.15 Left: interferometric autocorrelation traces of the pair of pulses from figure 36.12. The difference between the two interferometric autocorrelation traces is plotted below. Right: interferometric autocorrelation traces of the same pulses, but shortened by a factor of 5. Note that, in both cases, the two interferometric autocorrelation traces are very similar. Note also that the interferometric autocorrelation traces are even more difficult to distinguish as the pulse lengths decrease. From Ref. [1].

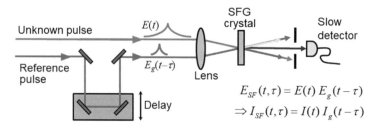

FIGURE 36.16 A cross-correlator. A shorter reference pulse can gate a longer one and yield the intensity vs. time of the unknown pulse. From www.frog.gatech.edu.

When a much shorter gate pulse is available, the cross-correlation yields the intensity precisely (but yields no information about the phase). This can easily be seen by substituting $\delta(t)$ for $I_g(t)$, which yields $I(t)$ precisely. In fact, it is not even necessary to know the gate pulse—just that it is much shorter. The problem is that there is not often a delta-function pulse lying around the lab.

36.10 Autocorrelation Conclusions

Despite drawbacks, ambiguities, and generally unknown information content, the autocorrelation and spectrum remained the standard measures of ultrashort pulses for over a quarter of a century, largely for lack of better methods. They allowed rough estimates for pulse lengths and time-bandwidth products (TBPs), and they helped researchers to make unprecedented progress in the development of sources of ever-shorter light pulses. However, the drawbacks of autocorrelation and its relatives began to severely limit progress in the 1990s, when the pulse spectral phase became the limiting factor in the generation of shorter pulses. Fortunately, complete-intensity-and-phase measurement techniques quickly became available. As a result, autocorrelation and its relatives are now obsolete and no longer appropriate measures of pulses. They have been discussed here for historical reasons and because they are the building blocks from which more modern techniques are constructed.

36.11 The Time-frequency Domain

Having considered the spectrum and autocorrelation in detail, the next step in any effort to develop a self-referenced pulse-measurement technique is to consider the spectrum of the autocorrelation. Interestingly, this extension will solve the problem nicely. It is remarkable that this simple idea was not considered in detail until 1991 [32]. Perhaps, it was because its mathematics appears complex; that it solves the problem so completely certainly is counterintuitive, as we shall see Ref. [1]. Perhaps, it was because it involves a hybrid domain: the *time-frequency domain* [33,34]. Measurements in the time-frequency domain involve both temporal *and* spectral resolution simultaneously. This intermediate domain has received much attention in acoustics and applied mathematics research but, at the time, had received only scant use in optics and then only for qualitative measurements. One notable exception, however, was the idea of spectrochronography [35], which, although not widely used, was the principle and work that partially inspired the work that follows.

In reality, everyone is actually quite familiar with the time-frequency domain. A well-known example of it is the *musical score*, which is a plot of a sound wave's short-time spectrum vs. time. Specifically, this involves breaking the sound wave up into short pieces and plotting each piece's spectrum (vertically) as a function of time (horizontally). So, the musical score is a function of both time and frequency. See Figure 36.17.

The musical score is actually not a bad way to look at a waveform. For simple waveforms containing only one note at a time, it graphically shows the waveform's instantaneous frequency, ω, vs. time, t, and, even better, it has additional information on the top indicating the approximate intensity vs. time (e.g., fortissimo, "ff", or pianissimo, "pp"). Of course, the musical score can handle complex waveforms, like symphonies, too.

The mathematically rigorous version of the musical score is the spectrogram, $\Sigma_g(\omega, \tau)$: [36]

$$\Sigma_g(\omega,\tau) \equiv \left| \int_{-\infty}^{\infty} E(t)\, g(t-\tau) \exp(-i\omega t)\, dt \right|^2, \quad (36.21)$$

where $g(t-\tau)$ is a variable-delay gate function, and the subscript on the Σ indicates that the spectrogram uses the gate function, $g(t)$. Figure 36.18 is a graphical depiction of the spectrogram, showing a linearly chirped Gaussian pulse and a rectangular gate function, which gates out a piece of the pulse. For the case shown in Figure 36.18, it gates a relatively weak, high-frequency region in the trailing part of the pulse, etc. The spectrogram is the set of spectra of all gated chunks of $E(t)$ as the delay, τ, is varied.

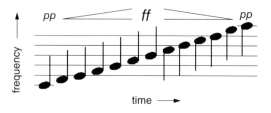

FIGURE 36.17 The musical score is a plot of an acoustic waveform's frequency vs. time, with information on top regarding the intensity. Here the wave increases in frequency with time. It also begins at low intensity (pianissimo), increases to a high intensity (fortissimo), and then decreases again. Musicians call this waveform a "scale", but ultrafast laser scientists refer to it (roughly) as a "linearly chirped pulse". From Ref. [1].

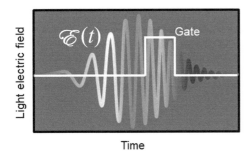

FIGURE 36.18 Graphical depiction of the spectrogram. A gate function gates out a piece of the waveform (here a linearly chirped Gaussian pulse), and the spectrum of that piece is measured or computed. The gate is then scanned through the waveform and the process repeated for all values of the gate position (i.e., delay). From www.frog.gatech.edu.

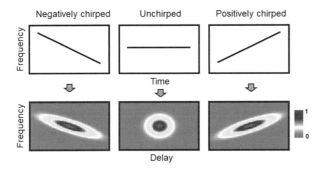

FIGURE 36.19 Spectrograms (bottom row) for linearly chirped Gaussian pulses (top row), all with the same spectrum, but different amounts of chirp and using a Gaussian gate pulse. The spectrogram, like the musical score, reflects the pulse instantaneous frequency vs. time. It also yields the pulse intensity vs. time: notice that the shortest pulse (centre) has the narrowest spectrogram in time. And looking at the spectrogram sideways yields the group delay vs. frequency, and its height yields the spectrum. From www.frog.gatech.edu.

The spectrogram is a highly intuitive display of a waveform. Some examples of it are shown in Figure 36.19, where it can be seen that the spectrogram intuitively displays the pulse instantaneous frequency vs. time. And, the pulse intensity vs. time is also evident in the spectrogram. Indeed, acoustics researchers can easily directly measure the intensity and phase of sound waves, which are many orders of magnitude slower than ultrashort laser pulses, but they often choose to display them using a time-frequency-domain quantity like the spectrogram. Importantly, knowledge of the spectrogram of $E(t)$ is sufficient to essentially completely determine $E(t)$ [36,37] (except for a few trivial ambiguities, such as the absolute phase, which are typically of little interest in optics problems).

36.12 Frequency-resolved Optical Gating (FROG)

As in autocorrelation, it will be necessary to use the pulse to measure itself and thus gate the pulse with itself. And to make a spectrogram of the pulse, it is necessary to spectrally resolve the resulting gated piece of the pulse. The Frequency-Resolved-Optical-Gating (FROG) technique measures such an "autospectrogram" of the pulse and was the first method (and arguably so far the only method) to completely solve the pulse-measurement problem, having all the required characteristics mentioned earlier [32,38–45].

In its simplest form, FROG is any autocorrelation-type measurement in which the autocorrelation signal beam is spectrally resolved [32,40,41]. In other words, instead of measuring the autocorrelation signal energy vs. delay, which yields an autocorrelation, FROG involves measuring the signal *spectrum* vs. delay.

As an example, consider a PG autocorrelation geometry. Ignoring constants as usual, this third-order autocorrelator's signal field is $E_{sig}(t, \tau) = E(t)\,|E(t-\tau)|^2$. Spectrally resolving yields the Fourier transform of the signal field with respect to time, and detectors, of course, measure the squared magnitude, so the PG FROG signal trace is given by:

$$I_{\text{FROG}}^{\text{PG}}(\omega,\tau) = \left| \int_{-\infty}^{\infty} E(t)\,|E(t-\tau)|^2 \exp(-i\omega t)\,dt \right|^2. \quad (36.22)$$

Note that the PG FROG trace is a spectrogram in which the pulse *intensity* gates the pulse *field*. In other words, the pulse gates itself.

So why is $E(t)$ obtained from its FROG trace?

First, consider $E_{sig}(t, \tau)$ to be the one-dimensional Fourier transform with respect to τ (*not* t) of a new quantity that we will call $\bar{E}_{sig}(t, \Omega)$:

$$E_{sig}(t,\tau) = \int_{-\infty}^{\infty} \bar{E}_{sig}(t,\Omega)\,\exp(-i\Omega\tau)\,d\Omega. \quad (36.23)$$

Since this Fourier transform involves τ, and not t, we are using a bar, rather than a tilde, on top of the Fourier-transformed functions here.

Now, it is important to note that, once found, $E_{sig}(t, \tau)$ or $\bar{E}_{sig}(t,\Omega)$ easily yields the pulse field, $E(t)$. Here is why: if we know $\bar{E}_{sig}(t,\Omega)$, we can inverse-Fourier-transform to obtain $E_{sig}(t, \tau)$. Then, we can simply evaluate the signal field at $t = \tau$: $E_{sig}(t, t) = E(t)\,|E(0)|^2$. Since $|E(0)|^2$ is merely a multiplicative constant, and we only care about the pulse shape, then, as far as we are concerned, $E_{sig}(t, t) = E(t)$. Thus, to measure $E(t)$, it is sufficient to find $\bar{E}_{sig}(t,\Omega)$.

Now substitute the above equation for $E_{sig}(t, \tau)$ into the expression for the FROG trace, which yields an expression for the FROG trace in terms of $\bar{E}_{sig}(t,\Omega)$:

$$I_{\text{FROG}}^{\text{PG}}(\omega,\tau) = \left| \int_{-\infty}^{\infty}\int_{-\infty}^{\infty} \bar{E}_{sig}(t,\Omega)\,\exp(-i\omega t - i\Omega\tau)\,dt\,d\Omega \right|^2. \quad (36.24)$$

Here, we see that the measured quantity, $I_{\text{FROG}}^{\text{PG}}(\omega,\tau)$, is the squared magnitude of the two-dimensional Fourier transform of $\bar{E}_{sig}(t,\Omega)$.

The above expression is true, but it certainly does not look helpful. We just took a difficult-looking one-dimensional integral-inversion problem and turned it into an impossible-looking two-dimensional integral-inversion problem. Indeed, in order to solve integral equations, the goal is usually to *reduce* the number of integral signs, not increase it.

But looking more closely at equation (36.24), it is more elegant than at first glance. From it, we see that the measured FROG trace yields the magnitude, but not the phase, of the two-dimensional Fourier transform of the desired quantity $\bar{E}_{\text{sig}}(t,\Omega)$. If we had the two-dimensional phase, we would be done because we would have $\bar{E}_{\text{sig}}(t,\Omega)$ in its entirety. So, the problem is then to find the phase of $\bar{E}_{\text{sig}}(t,\Omega)$.

It turns out that this inversion problem is well known. It is called, quite reasonably, the *two-dimensional phase-retrieval problem* [1,2].

36.13 FROG and the Two-dimensional Phase-retrieval Problem

Now, earlier, we discussed the *one-dimensional* phase-retrieval problem, and we concluded that it was ill-posed and so a poor choice for a pulse-measurement technique. Almost certainly, it would seem that the two-dimensional analogue of a one-dimensional piece of mathematical bad news can only be worse news.

Quite unintuitively, however, the two-dimensional phase-retrieval problem has been shown to have an essentially *unique solution* and is a *solved* problem when certain additional information regarding $\bar{E}_{\text{sig}}(t,\Omega)$ is available, such as that it has finite support (that is, zero outside a finite range of values of t and Ω) [2,46–48]. This interesting and useful fact follows from the fact that the Fundamental Theorem of Algebra, which holds for polynomials of one variable, fails for polynomials of two variables.

The two-dimensional phase-retrieval problem, when finite support is the case, has only the usual "trivial" ambiguities. If $\bar{E}_{\text{sig}}(t,\Omega)$ is the solution, then the ambiguities are as follows:

1. an absolute phase factor $\exp(i\phi_0)\bar{E}_{\text{sig}}(t,\Omega)$
2. a translation: $\bar{E}_{\text{sig}}(t-t_0,\Omega-\Omega_0)$
3. inversion: $\bar{E}_{\text{sig}}^*(-t,-\Omega)$

In addition, there is an extremely small probability that another, non-trivial, ambiguous solution may exist, but no such ambiguity has ever been found in FROG.

In FROG, we actually do not have finite support because the relevant function is $\bar{E}_{\text{sig}}(t,\Omega)$, and its extent along the t axis is essentially that of $E(t)$, and its extent along the Ω axis is essentially that of the $\tilde{E}(\omega)$. Since no function can be finite in extent in both time and frequency, $\bar{E}_{\text{sig}}(t,\Omega)$ does not have finite support.

However, FROG has another, much stronger constraint. We know that, for polarization gating, for example, $E_{\text{sig}}(t,t) = E(t)|E(t-t)|^2$. This is a very strong constraint—the *mathematical form* that the non-linear-optical signal field can have. Hence, we refer to this constraint as the *mathematical-form constraint* or *non-linear-optical constraint*. Other versions of FROG, which use other non-linear-optical processes, have analogous constraints that are slightly different. For example, in SHG FROG, $E_{\text{sig}}(t,t) = E(t)E(t-t)$.

This additional information turns out to be sufficient, and thus, the problem is solved [32]. Indeed, it is solved in a particularly robust manner, with many other advantageous features, such as feedback regarding the validity of the data [43,49,50]. FROG was recently rigorously proven to yield unique solutions, except for trivial ambiguities [51].

What are these trivial ambiguities? Clearly, from the above list, FROG does not measure the absolute phase, φ_0, in the Taylor expansion of the spectral phase. Also, translations in time are not measured. Physically, this is because FROG involves the pulse gating itself, so there is no absolute time reference, that is, FROG does not measure the pulse arrival time, which corresponds in the frequency domain to φ_1, the first-order term coefficient in the spectral-phase Taylor series. The mathematical-form constraint removes the direction-of-time, or inversion, ambiguity in all but one FROG variation (SHG). Also, some FROG versions have single-parameter ambiguities in the relative phases of well-separated pulses in time. Finally, no known technique is able to measure the relative phases of well-separated modes in frequency.

36.14 FROG Beam Geometries

Because FROG is a spectrally resolved autocorrelation, every non-linear-optical process that can be used to make an autocorrelator can also be used to make a FROG. The interested reader is referred to Rick Trebino's book, *Frequency-Resolved Optical Gating: The Measurement of Ultrashort Laser Pulses* for more information.

Here, we consider only the most common FROG geometries. SHG FROG achieves the best signal-to-noise ratios because it is the strongest (lowest order) non-linearity and its signal beam is a different colour, so scattered light is easily filtered. As a result, it is the most commonly used method. Its apparatus and those of other common FROG geometries are shown in Figures 36.20–36.22.

Figure 36.23a and b show theoretical FROG traces for the most common beam geometries. Pulses shown are all Gaussian-intensity pulses and include the Fourier-transform-limited (flat-phase) pulse, a pulse with negative chirp, a pulse with positive chirp, and a pulse with self-phase modulation, which does not change the pulse's intensity but distorts its spectrum. Note the symmetrical SHG FROG traces, which yield a trivial ambiguity in the direction of time. This ambiguity can be removed by inserting a piece of (dispersive) glass and making a second measurement; only one direction of time is consistent with both measurements.

36.15 The FROG Algorithm

There are many different FROG pulse-retrieval algorithms. They all start with an initial guess for the field $E(t)$, usually random noise (see Figure 36.24); unlike common minimization schemes, it is not necessary to start with a good guess (although we will see shortly that a good initial guess yields a highly reliable algorithm). A signal field $E_{\text{sig}}(t,\tau)$ is then generated using the relevant expression for the FROG trace for the

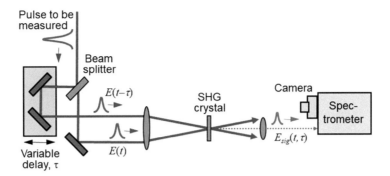

FIGURE 36.20 Experimental apparatus for SHG FROG. From www.frog.gatech.edu.

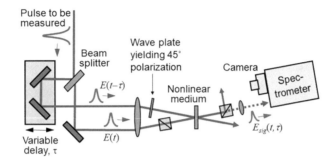

FIGURE 36.21 PG FROG apparatus. (www.frog.gatech.edu).

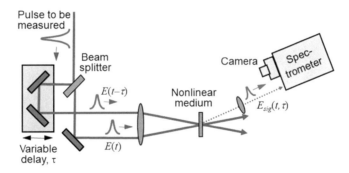

FIGURE 36.22 Self-diffraction FROG. From www.frog.gatech.edu.

beam geometry being used. This field is then Fourier transformed with respect to t in order to generate the signal field $E_{sig}(\omega,\tau)$ in the frequency domain. The measured FROG trace $I_{FROG}(\omega,\tau)$ is then used to generate an improved signal field $E'_{sig}(\omega,t)$ by realizing that the squared magnitude of $E_{sig}(\omega,\tau)$ should be equal to $I_{FROG}(\omega,\tau)$, so this step involves simply replacing the magnitude of $E_{sig}(\omega,\tau)$ with the square root of the measured trace to generate $E'_{sig}(\omega,t)$. $E'_{sig}(\omega,t)$ is then transformed back into the time domain by applying an inverse Fourier transform (IFT). In the last step of the cycle, the modified signal field $E'_{sig}(t,\tau)$ is used to generate a new guess for $E(t)$. And, the process is repeated. Ideally, each iteration of the algorithm generates a better signal field, which eventually approaches the correct complex electric field.

The algorithm that has made FROG a technique that can quickly measure virtually every imaginable pulse is the *generalized projections (GPs) algorithm* [1,45], which, in the absence of noise, generally converges to the correct solution, even for very complex pulses, with an accuracy only limited by the host-computer system's numerical precision. It is also very versatile: it can be modified for any non-linear-optical interaction, even some slow ones [52], used to measure a pulse.

GPs [1,47] are frequently used in phase-retrieval problems, from which it was borrowed for FROG. It is also commonly used in many other problems, from X-ray crystallography to the training of artificial neural networks. Indeed, it is one of the few algorithmic methods than can be *proven* to converge when reasonable conditions are met.

The essence of the GPs technique in FROG is graphically displayed in Figure 36.25. Consider it a Venn diagram in which the entire figure represents the set of all complex functions of two variables, i.e., potential signal fields, $E_{sig}(t,\tau)$. The set of signal fields satisfying the data constraint is indicated by the lower elliptical region, while those satisfying the mathematical-form constraint are indicated by the upper elliptical region. The signal-pulse field satisfying both constraints corresponds

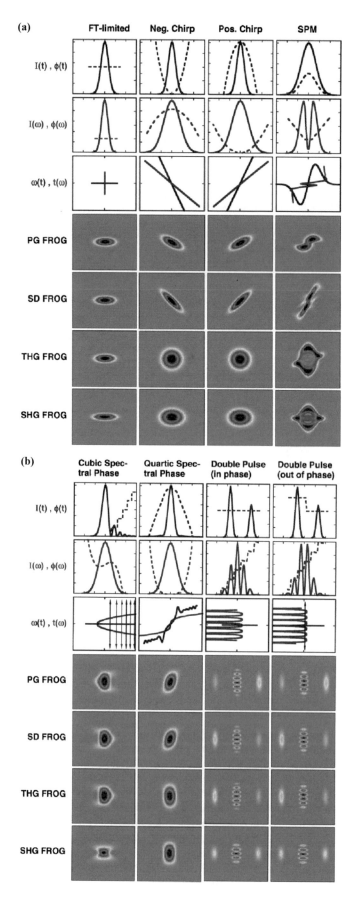

FIGURE 36.23 (a) Comparison of traces for common ultrashort pulse distortions for the most common FROG beam geometries. From Ref. [1]. (b) Comparison of traces for common ultrashort pulse distortions for the most common FROG beam geometries for additional pulses. From Ref. [1].

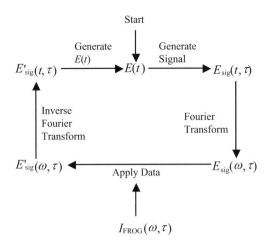

FIGURE 36.24 Schematic of a generic FROG algorithm. From Ref. [1].

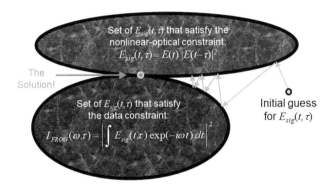

FIGURE 36.25 GPs applied to PG FROG. The two equations are considered as constraints on the function, $E_{sig}(t,\tau)$, which, when found, yield $E(t)$, the pulse field. Moving to the closest point in one constraint set and then the other yields convergence to the solution. Although the mathematical form constraint for PG FROG is shown, other FROG geometries can be treated analogously. From www.frog.gatech.edu.

to the intersection of the two elliptical regions and is the solution, uniquely yielding the pulse field, $E(t)$.

The solution is found by making function *projections*, which have simple geometrical analogues. We begin with an initial guess at an arbitrary point in signal-field space, which typically satisfies neither constraint. In the first iteration, we make a projection onto one of the constraint sets, which consists of moving to the point in that set (in function space) closest to the initial guess. From this point, we then project onto the other set, moving to the point in that set closest to the first iteration. This process is continued until the solution is reached.

When the two constraint sets are convex, i.e., all line segments connecting two points in each constraint set lie entirely within the set, convergence is guaranteed. Unfortunately, the constraint sets in FROG are not convex. When a set is not convex, the projection is not necessarily unique, and the computed projection is called a *GP*. The technique is then called GPs, and convergence cannot be guaranteed. Although it thus is conceivable that the algorithm may stagnate (that is, not converge), this approach is in practice fairly robust in FROG problems. In a recent study of the GP algorithm, convergence occurred performance for simple pulses in the presence of

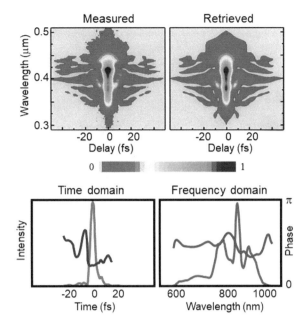

FIGURE 36.26 Upper left: Measured SHG FROG trace of a 4.5 fs pulse from a Ti:Sapphire oscillator designed to generate extremely short pulses. Upper right: Computed SHG FROG trace of the retrieved pulse. Lower: Retrieved intensity (gray) and phase (black) vs. time and retrieved spectrum (gray) and spectral phase (black). Note the excellent agreement between the measured and retrieved traces, indicating an accurate measurement. From Ref. [53] and www.frog.gatech.edu.

noise about 75% of the time on the first guess. The use of additional guesses yields convergence most of the time. Convergence occurs for very complex pulses (TBP ~ 100) in the presence of noise about half of the time. Since pulses are rarely that complex, and no other self-referenced technique has even been proposed that can measure such complex pulses unless a reference pulse is available, this is more than sufficient for essentially all cases.

When it was originally introduced in the early 1990s, the FROG algorithm had a well-deserved reputation for being slow. Convergence for even simple pulses could take a minute or more, especially for algorithms that combined multiple algorithms to absolutely ensure convergence. In a field devoted to *ultrafast* phenomena, this was quite frustrating. However, computers have sped up so much that the same code now requires less than 0.03 seconds on an inexpensive Windows personal computer to converge for even a somewhat complex pulse requiring a 64×64 trace. For perspective, this is less than the time required simply to *plot* the resulting pulse. In other words, the development of faster FROG algorithms at this time is pointless, unless it is for pulses with TBPs in excess of, say, ~20.

A measurement using SHG FROG is shown in Figure 36.26 [53].

36.16 The RANA Approach

When the GP algorithm fails, one can rerun the algorithm multiple times, but it would be far preferable to have a pulse-retrieval algorithm that converges every time. As a result,

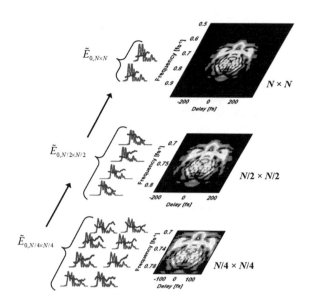

FIGURE 36.27 Schematic of the RANA approach. $\tilde{E}_0$: initial guesses; $N \times N$: array size. From www.frog.gatech.edu.

such an algorithmic approach has recently been developed. It is called the Retrieved-Amplitude N-grid Algorithm (RANA) approach [54–56]. It is based on the usual GP algorithm (or any other FROG algorithm), but it also uses a much better initial guess—the *actual spectrum*, which can actually be extracted directly from the measured FROG trace. This is possible because the frequency marginal (the integral of the trace with respect to delay) is known to be the autoconvolution of the spectrum for SHG FROG, which—unlike the autocorrelation!—can be inverted to yield the spectrum using the simple fact that the Fourier transform of a finite-width function (e.g., the spectrum) is continuous and has continuous derivatives.

It also uses smaller, coarser grids for initial iterations and so operates very quickly as a result. Finally, because it operates so quickly, it can use multiple initial guesses, all with the correct spectrum. It achieves 100% convergence for even extremely complex pulses with TBPs of 100 on tests of over 25 000 pulses and in the presence of significant noise in the measured traces (see Figure 36.27) [54–56].

36.17 Properties of FROG

It can accurately be said that, while autocorrelation yields a blurry black and white image of the pulse, FROG yields a high-resolution full-colour view of it. Indeed, the pulse intensity and phase may be estimated simply by looking at the experimental FROG trace, or the RANA iterative algorithm may be used to reliably retrieve the precise intensity and phase vs. time or frequency.

FROG has many useful features. It is very accurate; no approximations are made regarding the pulse. All that must be assumed in FROG is a nearly instantaneously responding medium (SHG crystals work very well in second order, as do pieces of glass in third order), and even that assumption has been shown to be unnecessary, as the medium response can be included in the pulse-retrieval algorithm [52]. Similarly, any known systematic error in the measurement may also be modelled in the algorithm [49,50], although this is not generally necessary, except for extremely short pulses (< 10 fs) or for exotic wavelengths. And, systematic error can often be removed by pre-processing the measured trace [50]. Also, unlike many other proposed ultrashort pulse measurement methods, FROG completely determines the pulse with essentially infinite temporal resolution [40,49]. It does this by using the time domain to obtain long-time resolution and the frequency domain for short-time resolution. As a result, if the pulse spectrogram has sufficient resolution and is entirely contained within the measured trace boundaries, then there can be no additional long-time pulse wings (because the spectrogram is essentially zero for off-scale delays, and there is no fine detail vs. frequency), and there can be no additional broad spectral wings (because the spectrogram is essentially zero for off-scale frequency offsets, and there is no fine detail vs. time). Interestingly, this extremely high temporal resolution can be obtained by using delay increments that are as large as, or even larger than, the time scale of the structure. Again, this is because the short-time information is obtained from large frequency-offset measurements. Thus, as long as the measured FROG trace contains all the non-zero values of the pulse FROG trace, the result is rigorous. (Of course, the trace typically only falls asymptotically to zero for delays and frequency offsets of $\pm\infty$, but these low values outside the measured trace do not significantly affect the retrieved pulse.)

Another useful and important feature that is unique to FROG is the presence of feedback regarding the validity of the measurement data. The FROG trace is a time-frequency plot, that is, an $N \times N$ array of points, which is then used to determine N intensity points and N phase points, that is, $2N$ points. There is thus a significant over-determination of the pulse intensity and phase—there are many more degrees of freedom in the trace than in the pulse. As a result, the likelihood of a trace composed of randomly generated points corresponding to an actual pulse is very small. Similarly, a measured trace that has been contaminated by systematic error is unlikely to correspond to an actual pulse. Thus, convergence of the FROG algorithm to a pulse whose trace agrees well with the measured trace virtually assures that the measured trace is free of systematic error.

In practice, FROG has been shown to work very well in the IR [57,58], visible [59], UV [60–62] and X-ray [63,64]. It has been used to measure pulse lengths from tens of attoseconds [64] to several ns [65]. It has measured pulses from aJ [66,67] to J in energy. And, it can measure simple near-transform-limited pulses to extremely complex pulses with TBPs in excess of 1000 [68]. It can use nearly any fast non-linear-optical processes that might be available. If an autocorrelator can be constructed to measure a given pulse, then making a FROG is straightforward since measuring the spectrum of it is usually easy.

36.18 Single-shot FROG

For high-repetition-rate ultrashort-pulse lasers, there often is not much variation from pulse to pulse. As a result, a spectrum may be obtained by scanning a monochromator in time

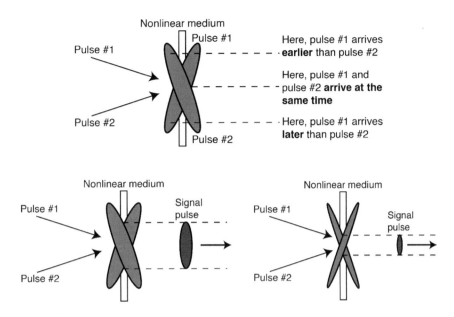

FIGURE 36.28 Crossing beams at a large angle causes the delay to vary across the non-linear medium, thus mapping delay onto transverse position. The length of the pulse is indicated by the width of the non-linear-optical signal beam. From www.frog.gatech.edu.

or by leaving the shutter open on a camera or diode array at a spectrometer output, averaging over many pulses. Similarly, the delay in an autocorrelator or FROG may often be scanned in time with confidence that the pulse has not changed during the scan. But not always.

Worse, some amplified laser systems have considerably lower repetition rates, and significant pulse-to-pulse variations are expected. In this latter case, a single-shot method is required. It is easy to obtain a single-shot spectrum, simply by opening the spectrometer camera shutter for only a single laser pulse. Single-shot autocorrelation or FROG, however, is not immediately obvious because the delay must somehow be scanned during a single pulse.

Fortunately, this problem is easily solved, and a single-shot autocorrelation or FROG trace is obtained by mapping the delay onto transverse position at the non-linear medium and spatially resolving the autocorrelation signal by imaging the non-linear medium onto a camera. This involves crossing the two beams in the non-linear-optical crystal at a large angle, so that, on the left, one pulse precedes the other, and, on the right, the other precedes the one (see Figure 36.28) [13,69,70]. In this manner, the delay ranges from a negative value on one side of the crystal to a positive value on the other. Usually, we focus with a cylindrical lens or mirror, so the beams are line-shaped at the crystal, and the range of delays is greater. Spectral resolution occurs later in the perpendicular direction.

36.19 Near-single-cycle Pulse Measurement

By definition, the shortest pulse for a given wavelength is only a single cycle long (~2.7 fs for an 800 nm wavelength pulse). Measuring such incredibly short—and incredibly broadband—pulses can be very challenging. Propagation through essentially all glass must be avoided to avoid distortions to the pulse due to dispersion in the glass over the pulse's broad spectrum. Also achieving SHG for all pulse wavelengths requires a very thin (~10 μm thick) crystal, which is challenging, no matter which technique is used. Fortunately, such crystals are now common. And because the pulse can propagate through any amount of glass after the crystal, such thin crystals can be placed on substrates of any thickness for added durability. Finally, the SHG efficiency varies significantly across the entire spectrum of the pulse. But this can easily be corrected for [50].

Geometrical distortions can also potentially plague the measurement, causing the trace to broaden unacceptably. For example, in the above figures, notice that, if the delay was to be scanned in the usual multi-shot manner using a delay stage, a range of delays would be sampled on any given shot, yielding a somewhat broadened autocorrelation or FROG trace, and the minimum amount of resulting pulse lengthening is on the order of a fraction of one cycle [50]. The presence of this "transverse geometrical smearing" means that all of the beam geometries shown above will fail to accurately measure a near-single-cycle pulse. In addition, other geometrical distortions can also occur in crossed-beam geometries (for more information on this effect, refer to Rick Trebino's book, *Frequency-Resolved Optical Gating: The Measurement of Ultrashort Laser Pulses*). As a result, multi-shot FROG geometries in which the delay is scanned and the beams crossed at an angle should *not* be used for near-single-cycle measurements. There are simple alternative FROG geometries that avoid this problem, but many authors, unaware of this fact, have erroneously concluded that FROG cannot measure such short pulses, so the reader should be cautioned not to draw such conclusions.

One way to avoid this geometrical smearing in measurements of near-single-cycle pulses was pioneered by Steinmeyer and co-workers [71] and is to use collinear beams, that is, a spectrally resolved interferometric autocorrelator, in what is called Interferometric FROG (IFROG) (see Figures 36.29 and 36.30). The collinear beams completely avoid the problem of

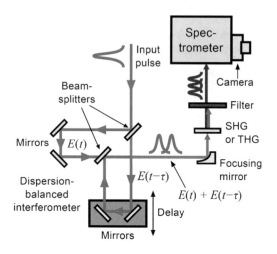

FIGURE 36.29 Schematic for IFROG for measuring near-single-cycle pulses. Note that a curved focusing mirror replaces the usual lens to avoid propagation through glass. Also, the beam-splitters are thin, so the dispersion introduced by their glass is also minimal and also designed to be identical for each beamlet yielding a "dispersion-balanced" interferometer—important for yielding correct, symmetrical fringes. From www.frog.gatech.edu.

geometrical distortions. IFROG involves propagation through only a thin beam-splitter and the non-linear-optical crystal and so is ideal for near-single-cycle pulses. The interference fringes present in this case arguably increase its sensitivity to spectral-phase distortions, which are more likely in such short pulses. However, the IFROG algorithm requires additional development before IFROG should be considered reliable.

Alternatively, single-shot FROG geometries work very well for near-single-cycle pulse measurements because the delay vs. position is measured and used for the delay range, so no transverse geometrical smearing can occur. Indeed, single-shot crossed-beam SHG FROG has no geometrical distortions at all (it is free of another effect that can be an issue for extremely short pulses called longitudinal geometrical smearing) and so is also ideal for performing such measurements, especially if an actual single-shot measurement is required due to a very low repetition rate or the presence of pulse-train instability (something IFROG cannot be used for). The best such approach was pioneered by Akturk and co-workers using the beam geometry shown in Figure 36.31, which has no glass in the beam until the SHG crystal (after which glass does not affect the result) [72]. A measured pulse using it is shown in Figure 36.32.

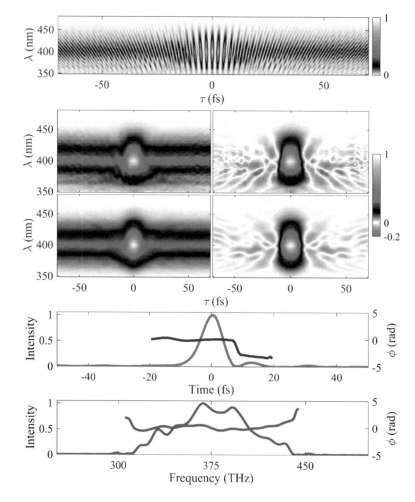

FIGURE 36.30 Near-single-cycle pulse measured using IFROG. Top row: measured trace. Note the fringes. Middle row: Measured and retrieved Fourier-filtered trace (with fringes removed). Bottom rows: retrieved pulse in the time and frequency domains. From www.frog.gatech.edu.

Measurement of Ultrashort Laser Pulses 507

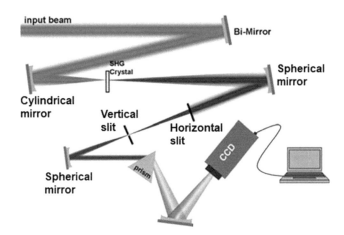

FIGURE 36.31 Single-shot, single-cycle SHG FROG set-up of Akturk, et al. The "bi-mirror" consists of two mirrors next to each other, each at a slightly different angle, which split the beam into two beams, each propagating at a slightly different angle, yielding crossing beams. Note that a curved focusing mirror replaces the usual lens to avoid propagation through glass. The pulse propagates through no glass at all before 5µm-thick crystal (and so is not distorted at all), essential for such extremely short pulses. This single-shot beam geometry can be used to measure a single pulse or to average over many pulses. The mirror after the prism should actually be a lens in order to avoid aberrations in the spectrometer. From Ref. [72].

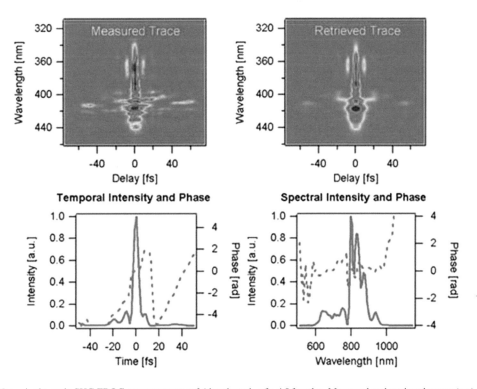

FIGURE 36.32 Near-single-cycle SHG FROG measurements of Akturk et al., of a 4.9 fs pulse. Measured and retrieved traces (top) and the measured intensity and phase vs. time (lower left) and frequency (lower right). From Ref. [72].

36.20 FROG and the Coherent Artefact

We have seen that autocorrelation performs quite poorly when presented with a complex pulse, variable-separation or variable-relative-phase double pulses, or trains of pulses with varying complex shapes. In these cases, it yields an inconvenient and misleading coherent artefact. Interestingly, FROG and its variations not only measure much more information but also deal very well with the coherent artefact.

First, as the pulse to be measured becomes very complex, the FROG trace also becomes very complex, not simpler, as in autocorrelation. This is good: the pulse information is not lost in its measurement. And, FROG's performance has been studied for complex pulses with TBPs as large as 100, even in the presence of noise, and it does very well [73]. In other words, FROG does not suffer from a single-pulse coherent artefact.

For variable-separation or variable-relative-phase double pulses, FROG yields traces that clearly indicate the presence of a double pulse, but frequency fringes in the centre lobe (see

Figure 36.33) tend to wash out [74]. However, because such a trace does not correspond to a possible FROG trace, the FROG algorithm correctly yields a double pulse. Unfortunately, in this case, it often yields an incorrect relative height for the two pulses. Fortunately, this is easily corrected using additional information in the trace [74].

Finally, when a FROG measurement must average over a train of randomly shaped complex pulses, it also yields a trace with a spike in it that cannot correspond to a trace of a single pulse. Fortunately, significant discrepancies between the measured and retrieved traces result and make it clear that the pulse train is unstable. See Figure 36.34.

While it was well known that discrepancies between measured and retrieved FROG traces could indicate pulse-shape instability [68], the above very specific simulations were not performed until a few years ago [74–78], when they resolved a mystery that had remained unresolved for two decades. While multiple-initial-guess FROG simulations always showed very reliable algorithm convergence [45,73], even in the presence of considerable noise, experimental FROG measurements often showed discrepancies between measured and retrieved traces, even after trying multiple initial guesses. Such discrepancies were routinely ascribed to algorithm non-convergence or FROG-device misalignment by those performing the measurements. However, with the above results, it is now clear that such discrepancies are generally due to pulse-shape instability. And, such pulse-shape variations can not only occur from pulse to pulse but also from place to place in the beam and so can occur even in single-shot measurements, although this latter effect requires more study.

In other words, FROG has a multi-pulse coherent artefact, but FROG's 2D traces and data redundancy allow FROG to take advantage of it for the identification of pulse-shape instability or its absence.

36.21 XFROG

When a well-characterized reference pulse is available and can be used to gate the unknown pulse in a FROG set-up, the resulting measurement is referred to as cross-correlation FROG, or XFROG [79]. The resulting trace is precisely a spectrogram. XFROG is the most reliable version of FROG and so is essentially always advised when a well-characterized reference pulse is available. In a study of the XFROG algorithm for pulses with TBPs as large as 100, it retrieved the correct pulses with a 100% reliability, even in the presence of significant noise in the trace [73].

Indeed, where XFROG truly excels is for measuring extremely complex pulses. It is the only known method for measuring ultra-broadband supercontinuum, whose TBP can be as high as 10 000. Shown below in Figure 36.35 are XFROG measurements of a train of supercontinuum pulses averaged

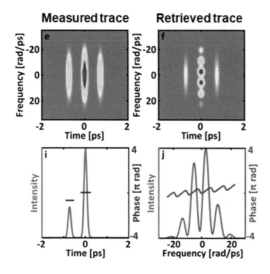

FIGURE 36.33 Simulations of SHG FROG measurements of double pulses with random relative phases and separations. Note that the central lobe's fringes, which ordinarily yield the pulse relative phase, have washed out due to the random relative phase. So, the relative phase is no longer an important parameter. But the pulse relative heights are wrong. However, the side lobes are too weak, indicating this. From www.frog.gatech.edu.

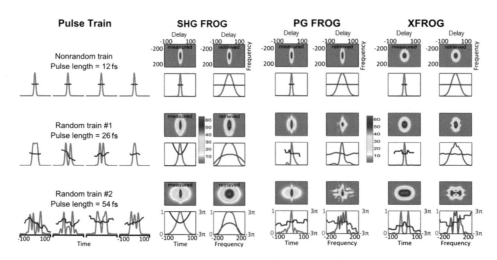

FIGURE 36.34 FROG traces and measurements of non-random (top row) and random (middle and bottom rows) trains of pulses. Note that FROG yields major discrepancies between the measured and retrieved traces when the pulse train is random. This is an excellent indication of pulse-shape instability. From www.frog.gatech.edu.

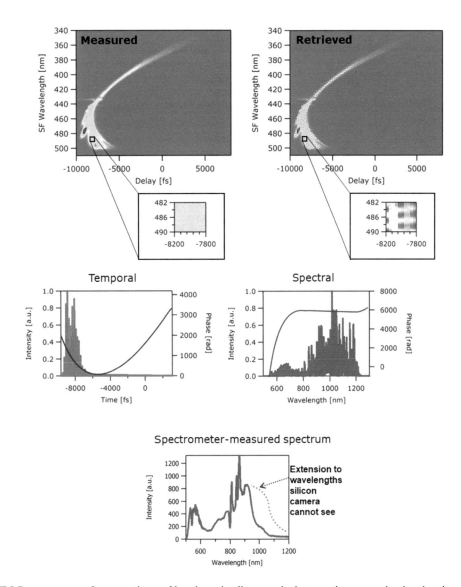

FIGURE 36.35 FROG measurement of supercontinuum. Note the major discrepancies between the measured and retrieved traces because the pulse train is random. The FROG-measured spectrum is more accurate than the spectrometer-measured spectrum. From www.frog.gatech.edu.

over 10^{11} laser shots. Notice that the measured and retrieved traces disagree in their fine structure—a clear indication of pulse-shape instability. This conclusion was confirmed by single-shot spectral measurements [68]. These measurements showed that the supercontinuum was the most complex and unstable pulse train ever generated (or measured).

XFROG is also useful for measuring pulses at difficult-to-detect frequencies, such as in the mid-IR (~5 – ~40 μm), where cameras are very expensive and typically still have only ~10^4 pixels. Using a near-IR reference pulse, the non-linear interaction can be chosen to produce the XROG trace at UV or visible frequencies so that a standard Si detector can be used. Such measurements have been done using sum or difference frequency generation between the mid-IR and a near-IR reference pulse in crystals [80,81]. Similarly, to use XFROG to measure mid-IR pulses with bandwidths spanning several octaves, four-wave mixing up-conversion, a $\chi^{(3)}$ process, can be performed in a gas such as air [82]. In these cases, measuring the spectrum vs. delay of the new visible or UV pulse yields an XFROG trace from which the mid-IR pulse can be retrieved, provided that the reference near-IR pulse is known.

In addition, by use of the optical-parametric-amplification non-linearity, gain can also be achieved along with the usual temporal gating. Gains of up to 10^6 are possible, allowing the measurement of extremely weak pulses. In this manner, continuum pulses with energies of aJ were measured [67].

36.22 Very Simple FROG: GRENOUILLE

FROG yields much more information than is available from autocorrelators, in particular, the full intensity and phase of the pulse vs. time and frequency.

But accuracy was initially the goal, not simplicity.

Indeed, FROG adds a spectrometer to an autocorrelator. A simple grating-lens home-made spectrometer that introduces no additional sensitive alignment degrees of freedom can be appended to an autocorrelator to make an excellent FROG, but

FROG still inherits the autocorrelator's moderate complexity, size, cost, maintenance, and alignment issues. How can we simplify FROG?

Figure 36.36 shows an SHG FROG device and a significantly simpler version of it.

This simpler device, like its other relatives in the FROG family of techniques, has a frivolous amphibian name: GRating-Eliminated No-nonsense Observation of Ultrafast Incident Laser Light E-fields (GRENOUILLE, which is the French word for "frog"). It and its operating principles are shown in more detail in Figures 36.37–36.39 [1,83].

It first involves replacing the beam splitter, delay line, *and* beam combining optics with a *single* simple element, a Fresnel biprism [83] (see Figure 36.37), which accomplishes all these tasks by itself. Second, in seemingly blatant violation of the SHG phase-matching-bandwidth requirement, it also involves replacing the thin SHG crystal with a *thick* SHG crystal (see Figure 36.38), which not only gives considerably more signal (signal strength scales as the approximate square of the thickness) but also simultaneously replaces the spectrometer.

How does it work? Consider the Fresnel biprism first (see Figure 36.37). It is a prism with an apex angle close to 180°, whose refraction crosses the two resulting beamlets at an angle—exactly what is required in conventional single-shot autocorrelator and FROG beam geometries, in which the relative beam delay is mapped onto horizontal position at the crystal. But, unlike conventional single-shot geometries, beams that are split and crossed by a Fresnel biprism are *automatically aligned* in space and in time, a significant simplification. Then, as in standard single-shot geometries, the crystal is imaged onto a camera, where the signal is detected vs. position (i.e., delay) in, say, the horizontal direction.

FROG also involves spectrally resolving a pulse that has been time-gated by itself. GRENOUILLE combines both of these operations in a single *thick* SHG crystal [83,84]. As usual, the SHG crystal performs the self-gating process: the two pulses cross in the crystal with variable delay. But, in addition, the thick crystal has a relatively small phase-matching bandwidth, so the phase-matched wavelength produced by it varies with angle (see Figure 36.38). Thus, the thick crystal also acts as the dispersive element of a *spectrometer*. The ability of a thick non-linear-optical medium to act as a low-resolution spectrometer was realized many years ago [85,86], but pulses then were longer and more narrowband, so its rediscovery and

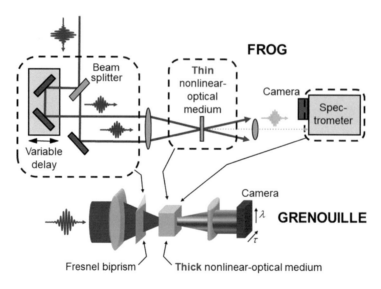

FIGURE 36.36 SHG FROG device and its simpler version, GRENOUILLE. While SHG FROG is a fairly simple device, there are a few components of it that can be replaced with even simpler ones. From www.frog.gatech.edu.

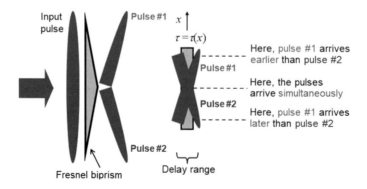

FIGURE 36.37 The Fresnel biprism (a prism with a near-180° apex angle) refracts each half of the pulse by different amounts, crossing them at the crystal. This maps delay onto transverse position and yields a single-shot autocorrelation, which, when spectrally resolved in the perpendicular direction and imaged onto a camera, yields a single-shot SHG FROG device. From www.frog.gatech.edu.

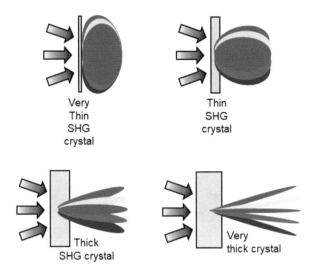

FIGURE 36.38 SHG crystals of various thicknesses illuminated by converging broadband light and polar plots of the generated shades vs. crystal exit angle. Note that the very thin crystal (ordinarily required in pulse-measurement techniques) generates the second harmonic of all colours in the forward direction. The very thick crystal, on the other hand, generates a small range of wavelengths in each direction and, in fact, acts like a dispersive element used in a spectrometer. Note also that the thick crystal generates considerably more SH in the relevant directions. From www.frog.gatech.edu.

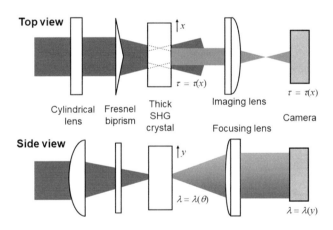

FIGURE 36.39 Side and top views of the GRENOUILLE beam geometry. Typically, the anamorphic lens between the crystal and camera has a focal length of f in the vertical direction to map angle and wavelength out of the crystal onto vertical position at the camera and $f/2$ in the horizontal direction to image delay at the crystal onto horizontal position at the camera. Note that the beam becomes a vertical line just before the camera, a convenient place for a slit to filter out any extraneous beams, ensuring optimal signal-to-noise ratio. From www.frog.gatech.edu.

use in pulse measurement had to wait until pulse bandwidths are increased and pulse lengths are significantly decreased.

Two additional cylindrical lenses complete the device (see Figure 36.39). The first cylindrical lens must focus the beam into the thick crystal tightly enough to yield a range of crystal incidence (and hence exit) angles large enough to include the entire spectrum of the pulse. After the crystal, a cylindrical lens then maps the crystal exit angle onto position at the camera, with wavelength a near-linear function of (vertical) position.

GRENOUILLE has many advantages over FROG. It has few elements and so is inexpensive and compact, with no alignment required. It naturally operates single shot. And, it is considerably more sensitive than other pulse-measurement devices. Furthermore, since GRENOUILLE produces (in real-time, directly on a camera) traces identical to those of SHG FROG, it yields the full pulse intensity and phase (except the direction of time and other trivial ambiguities). In addition, several feedback mechanisms on the measurement accuracy that are already present in the FROG technique also work with GRENOUILLE, allowing confirmation of—and confidence in—the measurement. But best of all, GRENOUILLE is extremely simple to set up and align: it involves no beam-splitting, no beam-recombining, and no scanning of the delay, and so has *zero* sensitive alignment degrees of freedom.

Figure 36.40 shows two measurements using GRENOUILLE and the good agreement obtained between it and FROG. The GRENOUILLE signal strength was ~1000 times greater than that of a standard single-shot SHG FROG device and also greater than that of an equivalent autocorrelator.

GRENOUILLE can measure much longer pulses; indeed, it works even better for such pulses. Use of a pentagonal SHG crystal even eliminates the Fresnel biprism, yielding a device with only three elements. Using such a device, pulses up to 20 ps have been measured.

Extremely short pulses, on the other hand, are more challenging for GRENOUILLE. Such pulses lengthen in the biprism and first lens, but simple theoretical back-propagation of the pulse through these elements remedies this. Alternately, an all-reflective GRENOUILLE can be built, using a "Fresnel bi-mirror" [87] allowing the measurement of pulses as short as 19 fs. GRENOUILLE also measures spatiotemporal distortions [88,89]. But for shorter, and hence broader-band, pulses, the crystal group-velocity dispersion begins to approach the crystal group-velocity mismatch, and a GRENOUILLE cannot be designed for such a measurement.

So GRENOUILLE cannot be extended to measurements of near-single-cycle pulses. But another variation of it can. Indeed, this variation can measure near-single-cycle pulses at odd wavelengths, which the two approaches for measuring such extremely short pulses that we described in an earlier section, cannot do. Even near-single-cycle *visible* light pulses can be difficult to measure if they are extremely broadband or include the violet, where SHG cannot be performed. Also, the SHG non-linear coefficient typically varies significantly over such a broad spectrum, complicating such measurements. To avoid these issues, a concept somewhat analogous to that of GRENOUILLE, but using a third-order induced-grating beam geometry, solves the problem. The result is a cross-correlation FROG (XFROG) device using a longer, intense reference pulse in a $\chi^{(3)}$ non-linear medium [90]. Specifically, the transient grating formed by two crossed intense reference pulses not only acts as the temporal gate but simultaneously as a spectrally dispersive element in a single-shot all-reflective FROG because diffraction, especially from nearly counter-propagating beams, involves significant angular dispersion [91]. The transient grating can be generated in various non-linear media, but a very thin slab of glass is best to avoid geometrical distortions (due to the nearly counter-propagating

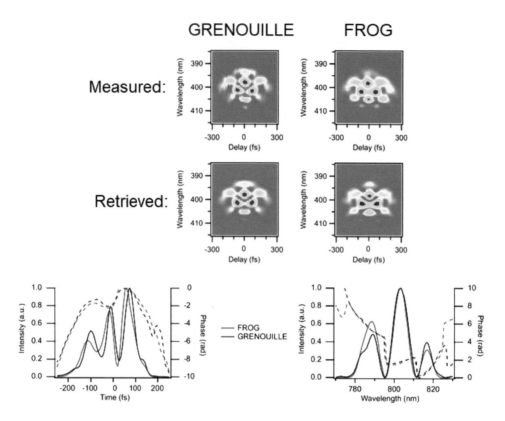

FIGURE 36.40 Comparison between GRENOUILLE and (multi-shot) FROG measurements of a complex test pulse. Top: measured and retrieved FROG and GRENOUILLE traces. Bottom: retrieved intensities (solid lines) and phases (dashed lines) for the time and frequency domains, respectively. From www.frog.gatech.edu.

reference beams) and group-velocity dispersion and its associated pulse-distortions.

36.23 Measuring Two Pulses Simultaneously

Quite frequently, labs require two different-colour pulses for an application, such as an excite-probe chemistry experiment. Or a pulse at one wavelength is generated by a laser, and another pulse at a new, more useful wavelength is generated from it for the desired application. One could build two FROGs, one for each wavelength, but a single device would certainly be preferred. This can most easily be accomplished using a set-up called Double Blind FROG. This name comes from the mathematical problem called Blind Deconvolution, in which two unknown functions are determined from their convolution. For one-dimensional functions, this is obviously not possible, but, unintuitively, it is often possible for two-dimensional problems. Alas, despite the fact that FROG pulse retrieval is a two-dimensional problem, it is not possible in FROG. However, if *two* traces are generated, it not only becomes possible, but it proves to be very reliable and the corresponding apparatus not too complex.

This is Double Blind FROG [92,93]. It works by one pulse gating the other in a PG XFROG beam geometry, while, simultaneously, the other gates the one in the same geometry but viewed from a 45° angle. See Figure 36.41.

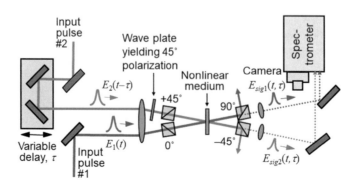

FIGURE 36.41 Beam geometry for the measurement of two different pulses using Double Blind FROG. Note that each pulse gates the other, yielding two separate PG FROG traces. From Ref. [1].

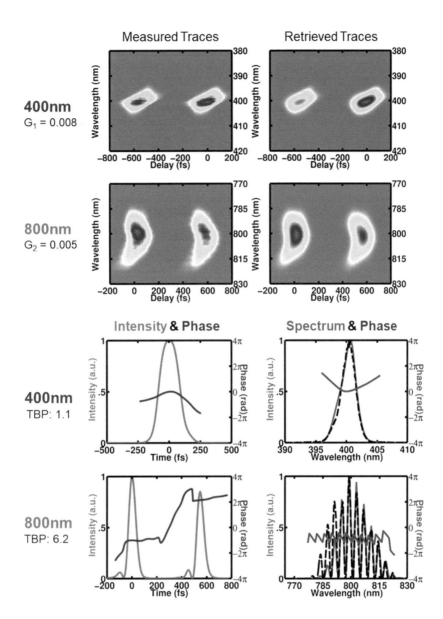

FIGURE 36.42 Top: Measured and retrieved traces for the measurement of two different-shades (black and dark gray) pulses using Double Blind FROG. The *G* values are the rms differences between the measured and retrieved traces, here revealing a very good measurement. Bottom: Measured pulses retrieved for the measurement of two different-shades (black and light gray) pulses using Double Blind FROG whose traces are shown in the previous figure. Dashed lines indicate the independently measured spectra using a spectrometer. From Ref. [94].

The pulse-retrieval algorithm uses the XFROG algorithm with one pulse treated as the known reference pulse on odd-numbered iterations, and the other treated as the known reference pulse on even-numbered iterations. This technique and its accompanying algorithm work extremely well and can measure extremely short pulses due to the PG beam geometry's extremely large bandwidth. It can also measure pulses at two completely different wavelengths. See Figure 36.42.

36.24 Error Bars

In any measurement, it is important to be able to know how accurately the quantity of interest has been measured. In the days of autocorrelation, it made no sense to even try to do this in view of the vast number of unknown nontrivial ambiguities present in the measurement and the fact that it was not clear precisely what was actually being measured. However, with FROG's precisely measured quantities and the massive redundancy in its measured trace, it has become, not only relevant but also possible and straightforward to place error bars on the measured intensity and phase in both domains. This approach, developed by statisticians for all curve-fitting applications, is a general approach for placing error bars of retrieved parameters. It is called the "bootstrap" method because it involves making no additional measurements and can be accomplished simply from the measured trace [95,96]. Applied to FROG measurements, it yields results like those shown in Figure 36.43.

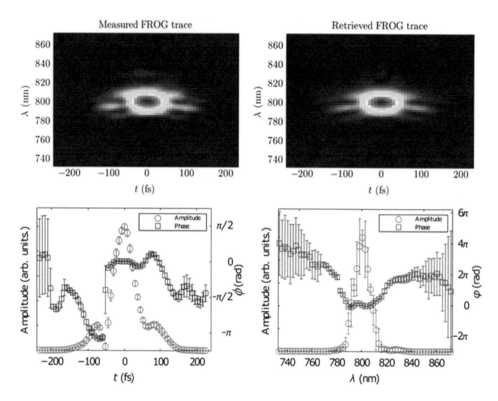

FIGURE 36.43 Measured trace and retrieved pulse including error bars. Plots courtesy of Kent Wootton, Stanford Linear Accelerator Center.

36.25 Other Self-referenced Methods

FROG was the first and remains the most powerful and popular pulse-measurement technique, so it has been described in detail here (and it is the subject of an entire book [1]). Since FROG's introduction, however, many additional methods have emerged. Many are simply variations on autocorrelation and so are obsolete. Most are interferometric, and so will be considered in the next section, specifically devoted to such approaches.

Many, however, are not known to be reliable as yet or can only measure very simple pulses. Others are very complex and/or very expensive. Some, for example, require the use of a pulse-shaper, which is a very expensive piece of equipment, and something whose performance an additional, separate pulse-measurement device is required to check.

Many others, as we will see, measure *only* the coherent artefact. While lasers are more stable now than in the 1960s, many are *not* stable, and simple misalignment, a rickety platform, or an overly powerful or unstable pump laser can turn an otherwise stable laser into an unstable one. Also, lasers at the edge of technology, such as those emitting near-single-cycle pulses, are often unstable and so require a check on their output pulse stability. A technique that only measures the coherent artefact cannot distinguish a stable train of short simple pulses from an unstable train of long complex pulses. Not being able to distinguish the most desirable situation from the least desirable one is the worst-case scenario for any type of measurement device. Such methods are unacceptable and so should not even be considered. They will not be treated here, except to illustrate this point.

Before we continue, it is important to point out that simple standards for pulse-measurement techniques have been established [77], but few technique developers have as yet subjected their techniques to them. These standards are straightforward and obviously important and include, for example, checking for trivial and non-trivial ambiguities and the coherent artefact. Unfortunately, these simple checks remain to be performed for many techniques, so such techniques will not be considered here. In particular, interferometric methods are likely to measure only the coherent artefact.

36.26 Spectral Interferometry

FROG and its variations work well for a wide variety of pulses, but they fail for very weak pulses due to the required non-linear-optical process if a more powerful reference pulse is not available. Also, for very complex pulses, FROG works, but the required data collection can be very slow. In these cases, spectral interferometry (SI) (see Figure 36.44), first introduced by Froehly and co-workers [97], is often a better choice. SI is, in principle, very simple, and it can measure a light pulse's intensity and phase for very complex (i.e., large TBP) pulses, reaching TBP values of ~65 000, or very weak pulses with energies of zeptojoules or less.

The catch is that SI is not self-referenced. It requires a temporally or spectrally completely characterized reference pulse (in intensity and phase) whose spectrum contains that of the unknown pulse. Fortunately, weak and/or complex pulses usually do not exist "in a vacuum." The processes that create them, whether a pulse-shaper or propagation through a

Measurement of Ultrashort Laser Pulses

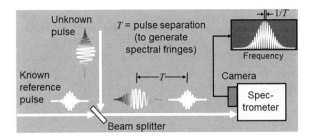

FIGURE 36.44 Basic SI experimental set-up. The reference and unknown beams are recombined collinearly and then sent into a spectrometer. Because the pulses expand in time inside the spectrometer, they interfere even in the presence of a delay, yielding spectral fringes. From www.frog.gatech.edu.

plasma or biological tissue, generally begin with a much stronger, simpler pulse with a spectrum that contains, and often is the same as, that of the unknown pulse as the input. Thus, a more intense, simpler reference pulse, which can be measured using FROG, is often available to use to measure the more difficult pulse. The combination of FROG and SI is often called TADPOLE or Temporal Analysis by Dispersing a Pair Of Light E-fields [98].

SI involves simply measuring the spectrum of the sum of the unknown pulse and a known. The resulting spectral interferogram is given by:

$$S_{SI}(\omega) = S_{ref}(\omega) + S_{unk}(\omega) + 2\sqrt{S_{ref}(\omega)}$$
$$\times \sqrt{S_{unk}(\omega)}\cos[\varphi_{unk}(\omega) - \varphi_{ref}(\omega) + \omega T], \quad (36.25)$$

where $S_{ref}(\omega)$ and $S_{unk}(\omega)$ are the reference and unknown pulse spectra, $\varphi_{ref}(\omega)$ and $\varphi_{unk}(\omega)$ are their spectral phases, and T is the delay between the two pulses.

The spectral phase is encoded in the phase of the interference fringes and is easily retrieved, provided that the reference pulse spectrum and spectral phase are known and its spectrum includes that of the unknown pulse.

The most common method for retrieving a pulse from its SI trace takes advantage of the delay between the reference and unknown beams. The spectral interferogram is Fourier transformed with respect to frequency to a "pseudo-time" domain ("pseudo" because it is only being used for data reduction, and the resulting quantity has no simple physical meaning). The delay is chosen so that, in the Fourier domain, the different components of the interferogram are well separated along the pseudo-time axis. As shown in Figure 36.45, the first two terms in equation (36.25), the "DC" terms, are centred at $t=0$. The cosine separates into its positive and negative components, called the "AC" terms, each centred at $\pm T$. Only one of the two "AC" terms is retained and is then translated to $t=0$ and inverse-Fourier transformed. What results is the product of the reference and unknown fields, $2\sqrt{S_{ref}(\omega)}\sqrt{S_{unk}(\omega)}\cos[\varphi_{unk}(\omega) - \varphi_{ref}(\omega)]$, which, after dividing out by the reference field, yields the unknown field spectrum and spectral phase. When using this retrieval algorithm, the approach is often called Fourier-Transform Spectral Interferometry (FTSI) [98].

36.27 Advantages and Disadvantages of Spectral Interferometry

SI uses an inherently single-shot geometry, and the retrieval algorithm is fast and direct. Also, SI is a heterodyne method because what is measured is squared magnitude of the sum of the reference and unknown (spectral) fields. This means that because the key measured quantity is the cross term of this squared magnitude, which is a product of the two fields, a strong reference field can effectively amplify a weak unknown field. Choosing the reference pulse to be M times more intense than the unknown pulse produces peak-to-peak spectral fringes that are $4\sqrt{M}$ times as intense as the spectrum of the unknown pulse; for $M=100$, the fringes are 40 times more powerful than the unknown pulse intensity in the absence of the reference pulse. As a result, for a 100 MHz rep-rate pulse train, SI was able to measure a pulse with only 42 zJ of energy or 1/5 of a photon per laser pulse [98], and pulses weaker by a factor of 1000 should be measurable using it, provided that the pulse train is stable. See Figure 36.46 for these results.

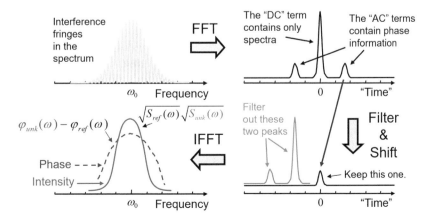

FIGURE 36.45 FTSI concept. The spectral interferogram (top, left) is Fourier-transformed from ω to a pseudo-time domain, t. A sufficiently large delay T is needed so that there are three well-separated nonzero regions (top right). The two side-bands are the phase-containing component of the data, so either one of these is retained (bottom, right), translated to $t=0$, and inverse-Fourier transformed, yielding the unknown pulse's spectrum and spectral phase (bottom, left). From www.frog.gatech.edu.

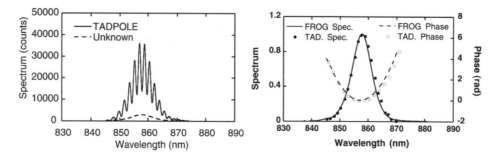

FIGURE 36.46 Left: Spectral interferogram for a 42 zJ pulse, and the spectrum of the pulse (here labelled "Unknown"). Right: Retrieved spectrum and spectral phase. From Fittinghoff et al. [98].

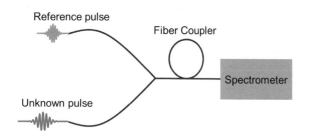

FIGURE 36.47 STARFISH: This very convenient experimental set-up for SI uses a fibre spectrometer, a fibre coupler, and two fibres to introduce the reference and unknown pulses. From www.frog.gatech.edu.

Notice that the spectral fringe contrast is easily visible even though the unknown pulse is much weaker than the reference.

It should always be remembered that SI measures only the coherent artefact. As for other methods, this is fine when there is perfect stability, as in the case above, when the coherent artefact is equal to the pulse itself. As a result, it is very important to measure the reference pulse using a technique that establishes its stability, such as FROG. Then a 100% SI fringe visibility confirms the stability of the unknown pulse as well (except for unequal beam intensities and/or misalignments that reduce the fringe visibility). This can be said for all variations of SI and also for holographic pulse-measurement techniques to be discussed later. Fortunately, this is easily accomplished when the reference pulse is strong and hence measurable with, for example, FROG.

Also, SI requires perfectly collinear beams, with perfectly matched spatial modes. This makes SI very difficult to align, despite its seemingly trivial apparatus. Fortunately, there are simpler alternatives. A recent implementation of SI using optical fibres, known as STARFISH, solves most of these problems [99]. See Figure 36.47. The reference and unknown pulses are coupled into two, equal-length fibres whose outputs are combined with a fibre coupler and sent into the spectrometer. This ensures collinear beams and matched modes. If the fibres have equal lengths, both the reference and unknown pulses see identical dispersions, which cancel out and so can be ignored. In practice, it is not possible to make the lengths exactly the same, but the small residual difference can be measured and then subtracted from all measurements. The fibres are extremely helpful, since the beams can now move around without affecting the alignment of the spectral interferometer.

The most serious drawback of SI, however, is the required spectral fringes and the resulting loss of spectral resolution due to the Fourier filtering [100]. Due to the need to separate the various regions in the pseudo-time domain, the reconstructed pulse has only a fraction of the original temporal range or equivalently a fraction of the spectrometer's spectral resolution. Typically, at best only 1/5 of the spectral resolution is retained. This severely limits SI's ability to measure even moderately complex pulses.

36.28 Crossed-beam Spectral Interferometry

To avoid this loss of spectral resolution, a variation of SI instead makes spectrally resolved *spatial* interference fringes. This approach involves a very convenient experimental set-up and also solves SI's alignment problems and yields spatial resolution in one dimension [101–105]. It has been shown in Figure 36.48.

The SI spectrometer's usual linear detector is replaced with a camera. Now the idea is to make the interference fringes along the spatial dimension instead by setting the reference-unknown pulse relative delay to 0 and crossing the two beams

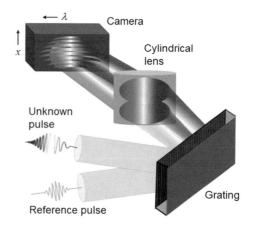

FIGURE 36.48 Crossed-beam spatial interferometry. The reference and unknown beams cross at the camera of a spectrometer, or equivalently at the slit of an imaging spectrometer. Mirrors in one of the two beam paths are tilted to make the beams cross. The spatial interference of the two beams at each frequency within the pulse is recorded. From this, $E_{unk}(\omega)$ can be reconstructed with the full resolution of the spectrometer. It also yields spatial resolution in one dimension (across the fringes). From www.frog.gatech.edu.

Measurement of Ultrashort Laser Pulses

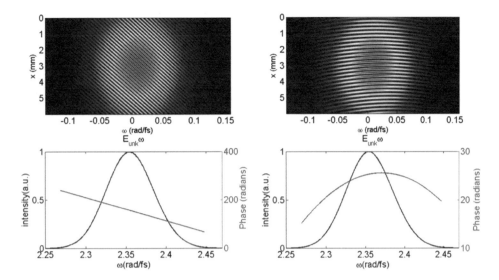

FIGURE 36.49 Simulations of crossed-beam spatial interferograms for two different unknown pulse shapes, assuming a reference pulse with a flat spectral phase and the same spectrum. Notice that the shape of the fringes is proportional to the spectral phase difference. From www.frog.gatech.edu.

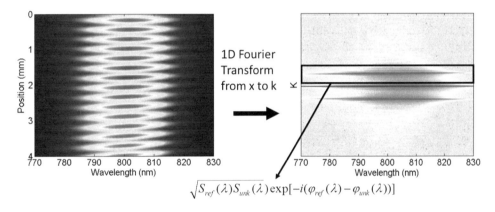

FIGURE 36.50 Fourier-filtering along the spatial axis. Take a 1-D Fourier transform to ky where the data separate into three bands. Use either of the side bands, which contain the phase information, and inverse-Fourier transform to obtain $E_{ref}(\omega) E_{unk}(\omega)^*$ without any loss of spectral resolution. From Ref. [103].

at a small angle. The crossing-angle plays the same role as the delay in usual SI. After Fourier transforming, now along the *spatial dimension*, the "AC" and "DC" terms separate due to their different *spatial frequencies*. This leaves the data along the frequency axis unchanged so that the unknown pulse is reconstructed with the spectrometer's full spectral resolution.

The equation for a spectrally resolved spatial interferogram is shown below:

$$S(\omega, x) = S_{ref}(\omega) + S_{unk}(\omega) + 2\sqrt{S_{ref}(\omega)}\sqrt{S_{unk}(\omega)}$$
$$\times \cos[\varphi_{unk}(\omega) - \varphi_{ref}(\omega) + 2kx\sin\theta]. \quad (36.26)$$

In equation (36.26), θ is the crossing half angle, k is the wavenumber, x is the vertical dimension in which the beams are crossing, and z is the propagation direction. Figure 36.49 shows such interferograms for two different pulse shapes, assuming an identical spectrum for the reference and unknown pulses and flat spectral phase for the reference. What is especially convenient about these interferograms is that the curvature of the interference fringes is proportional to the spectral phase difference. So, qualitatively, the spectral phase of the pulse can be immediately seen by the eye. This has been used for real-time adjustments of a pulse compressor [99]. Curve-fitting has been applied to very accurately extract the unknown pulse from the spatially resolved spectral interferogram [103,106], and it has the advantage that, if some spatial information is present, there is no resolution loss along this dimension.

Figure 36.50 shows the pulse-retrieval algorithm for crossed-beam spectral interferometry.

36.29 Practical Measurement Weak and Complex Pulses: SEA TADPOLE

Spatially Encoded Arrangement for Temporally Dispersing a Pair of Light E-fields (SEA TADPOLE) [103,107] is another experimentally simple version of crossed-beam spectral interferometry that uses optical fibres to ensure perfect mode-matching and a lens to ensure collimated beams (see Figure 36.51) entering a simple imaging spectrometer. Like STARFISH, it avoids

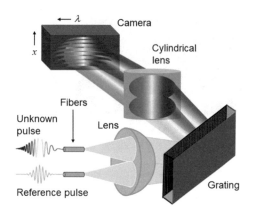

FIGURE 36.51 SEA TADPOLE experimental set-up. From www.frog.gatech.edu (see Ref. [103]).

alignment problems. But a great advantage of it is that it is also able to use the entire spectrometer resolution.

In SEA TADPOLE, the reference and unknown beams are coupled into equal-length fibres. At their output ends, the fibres are mounted with a small gap between them, and the diverging beams emerging from them are collimated with a single spherical lens. Because of the fibres' displacement from the optic axis, the beams are collimated, but one propagates downward, and the other upward. Eventually they cross, resulting in *spatial*, rather than spectral, interference fringes. As a result, the pulses can (and should) be coincident in time, so no additional spectral resolution is required. A camera is placed at the beam's crossing point. In the other dimension, a diffraction grating and a cylindrical lens map each wavelength to a different camera position (that is, act as a spectrometer). The Fourier-filtering algorithm described earlier is used to reconstruct the pulse, but now the filtering occurs in the spatial domain, rather than the spectral domain.

SEA TADPOLE is very easy to use in practice, collinear beams are unnecessary, and a low-resolution home-made spectrometer is used. It has measured complex pulses with TBPs as high as [100,102] with the only limitation being the spectrometer. An example of a SEA TADPOLE measurement of the spectrum and phase added by a pulse shaper is shown in Figure 36.52. Note the intuitive shape of the fringes, mirroring the actual spectral phase of the pulse.

The bottom left plot of the spectra reveals a very interesting feature of such measurements. The spectrum in green was that obtained from the SEA TADPOLE measurement. The spectrum in blue was obtained by simply blocking the reference beam and measuring the unknown pulse spectrum using

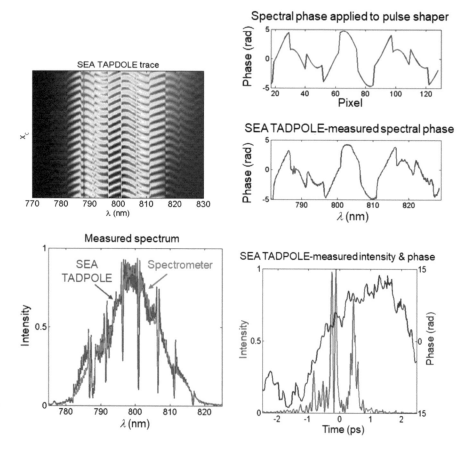

FIGURE 36.52 SEA TADPOLE measurement of a shaped pulse with a TBP of 30. The measured interferogram is shown at the top left. The top, right plots are the extracted spectral phases introduced by the pulse-shaper, compared with that of applied, showing good agreement, and having the same shape as the interference fringes. The temporal intensity and phase are shown at the bottom, right. The spectrum retrieved from the SEA TADPOLE measurement is shown in black (lower left). Also shown in the power left plot, in gray, is the spectrum measured directly with the SEA TADPOLE spectrometer, obtained simply by blocking the reference beam. Note that the SEA TADPOLE-measured spectrum is better resolved. Data and plots are taken from Ref. [102].

the same spectrometer, but now acting as a standard spectrometer, rather than a spectral interferometer. Note that, despite the use of the same spectrometer, the SEA TADPOLE-measured spectrum has considerably higher spectral resolution!

This is due to the fact that SEA TADPOLE measures the spectral *field*, not the spectral *intensity*, and convolutions of the spectrometer point-spread function with potentially negative functions, like the spectral field, do not broaden as much as always positive functions, like the spectral intensity [102].

36.30 Measuring Very Complex Pulses in Time: MUD TADPOLE

At each pixel on the detector of an SI spectrometer, the pulses stretch in time and have the duration of the inverse of the spectrometer spectral resolution. So, they temporally overlap for this amount of time. Because two pulses can only interfere when temporally overlapping, the task of measuring longer, more complex pulses using any version of SI, including SEA TADPOLE, would, in principle, require a higher-resolution spectrometer, in order to further broaden the reference pulse. If the unknown pulse is longer than the inverse of the spectrometer resolution, then a *single* reference pulse will not be able to interfere with all of it. The key to overcoming this limitation, that is, to beating the spectrometer's resolution, is to use *multiple*, delayed reference pulses. This is the key idea of MUltiple Delay TADPOLE (MUD TADPOLE) [108–110]. This idea is illustrated in Figure 36.53.

This idea is straightforward to implement in a multi-shot scanning set-up [108–110] and was accomplished using SEA TADPOLE for the individual measurements. Then, in the time domain, these temporal pieces are concatenated to yield the entire unknown pulse, which will now be resolved with a spectral resolution given by the inverse of the scanning range. Figure 36.54 illustrates this process.

MUD TADPOLE has succeeded in measuring a pulse with a TBP of 65 000 (see Figure 36.55), the most complex pulse ever measured.

The limiting factor in multi-shot MUD TADPOLE is the dynamic range of the camera. The camera only sees fringes from the temporal region of the unknown pulse that temporally overlaps with the reference pulse. All other temporal regions yield only relatively constant background. For long-enough unknown pulses, the fringes will no longer be discernable.

The main disadvantage of MUD TADPOLE is that it is multi-shot and so is not applicable to pulse trains in which every pulse is different. Also, the measurable spectral width is still limited by the spectrometer, and as with other SI-based approaches, the reference-pulse spectrum must contain the unknown pulse spectrum.

36.31 Single-shot MUD TADPOLE

It is straightforward to extend the multiple-reference-pulse idea to a single-shot measurement geometry. It turns out to be possible using a trick that is often employed for single-shot pulse measurement, tilting the pulse in order to achieve the required large range of delays. And, the variably delayed reference pulses overlap with the unknown pulse at different angles, yielding different fringe spacings. In other words, the fringe spacing codes for the delay [111]. See Figure 36.56.

TBP's as large as 4500 have been measured on a single shot with this device [111], a significant improvement over all other approaches. The main limitation is not yet the dynamic range of even an inexpensive eight-bit camera, but the pulse-front tilt and hence the delay range producible by a diffraction grating.

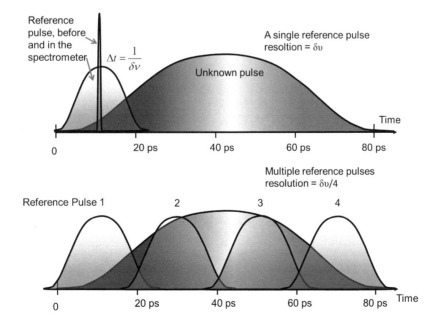

FIGURE 36.53 In a spectrometer, two pulses can only interfere with each other for as long as they overlap, which is determined by the inverse of the spectral resolution. N reference pulses could then be used to extend the spectral resolution by N times, as depicted here. The colour in the unknown pulse represents the instantaneous frequency, showing that it has a complex spectral phase. From www.frog.gatech.edu.

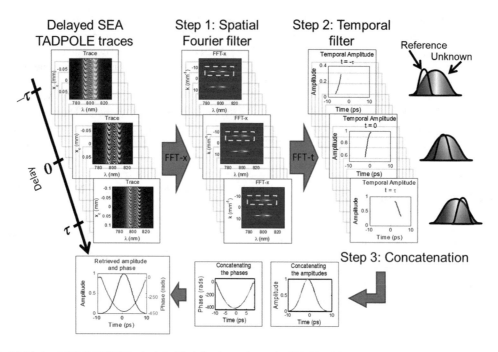

FIGURE 36.54 Multi-shot MUD TADPOLE principle. The left column shows the SEA TADPOLE traces measured at a few different reference pulse delays. $E(t)$ is extracted from these just as in SEA TADPOLE (middle and right columns). The concatenation, done in the time domain, is illustrated at the bottom. The individual temporal segments illustrate individual measurements. The final concatenated, reconstructed pulse is shown at the bottom, left. Figure from Ref. [108].

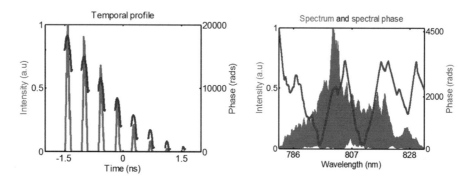

FIGURE 36.55 Reconstructed pulse with a TBP of 65 000 from a MUD TADPOLE measurement. Data from Ref. [109].

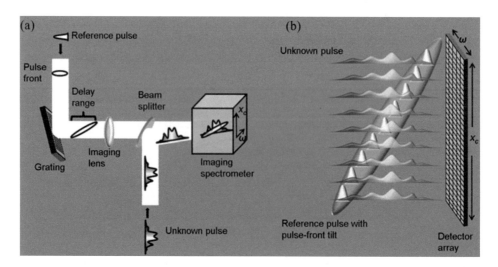

FIGURE 36.56 Single-shot MUD TADPOLE. The reference pulse is reflected off a diffraction grating, acquiring pulse-front tilt, and then imaged into the camera. This continuously varies the reference and unknown pulse relative delay across the y-axis. Figure from Ref. [110].

With other, more dispersive optics, or just larger diffraction gratings and cameras, single-shot MUD TADPOLE should be scalable to even more complex pulses.

36.32 SPIDER

Frustrated by the slow convergence of the FROG algorithm in the 1990s, researchers sought a non-iterative, that is, "direct" self-referenced intensity-and-phase pulse-measurement technique. They looked to SI, which has a direct pulse-retrieval algorithm. Unfortunately, in its usual configuration, SI cannot be self-referencing because it measures only phase differences, and if the unknown pulse interferes with itself, the spectral-phase difference necessarily vanishes or is at most the deliberately introduced delay. Either way, it can yield no pulse phase information.

However, *non-linear-optical* variations of SI were introduced that overcame this limitation, allowing the measurement of pulses without the need for a reference pulse. The first and most popular has been Spectral Phase Interferometry for Direct Electric-field Reconstruction (SPIDER) [112] (see Figure 36.57), and another is self-referenced spectral interferometry [113].

SPIDER involves performing SI on the pulse and a *frequency-shifted* (by $\delta\omega$) version of itself. In the ideal, simplest case, the SPIDER trace is given by:

$$S_{\text{SPIDER}}(\omega) = S(\omega) + S(\omega + \delta\omega) + 2\sqrt{S(\omega)}$$
$$\times \sqrt{S(\omega + \delta\omega)} \cos[\varphi(\omega + \delta\omega) - \varphi(\omega) + \omega T]. \quad (36.27)$$

Again, in the ideal limit, the SPIDER phase (the quantity inside the cosine) can be approximated by:

$$\Delta\varphi_{\text{SPIDER}} \equiv \varphi(\omega + \delta\omega) - \varphi(\omega) + \omega T \approx \delta\omega \frac{d\varphi}{d\omega} + \omega T. \quad (36.28)$$

The derivative, $d\varphi/d\omega$, is just the group delay, $\tau(\omega)$, of the pulse vs. frequency. Consequently, any deviations of the fringe phase from linearity are due to variations in the pulse group delay, and direct algorithms are available to retrieve this information.

SPIDER does not measure the spectrum and instead uses a separate independent measurement of it, which is generally considered an advantage because, for near-IR pulses, it does not require accurate measurements of the spectral intensity in the UV as SHG FROG does (although such values can easily be corrected in FROG measurements using the known spectrum, also not measured in the UV). SPIDER's ambiguities are similar to those of FROG: the absolute phase, the pulse arrival time, and relative phases of multiple pulses and multiple modes. And like most FROG versions, it does not have an ambiguity in the direction of time (Figure 36.58).

The SPIDER phase $\Delta\varphi_{\text{SPIDER}}(\omega)$ between the two pulses can be analytically retrieved from the measured $S_{\text{SPIDER}}(\omega)$ interferogram by extracting the phase of its IFT (with a specified finite frequency range). This is known as the Takeda algorithm [114] for numerical phase demodulation. As in standard spectral interferometry, it relies on a Fourier filtering approach. Then, it isolates one of the modulation side-bands, e.g., the positive modulation sideband, from which one can simply extract $\Delta\varphi(\omega) = \omega T$ by applying the complex logarithm (or arg function) to this modulation term. An alternative method for reconstructing the complex-valued sideband is given by the Hilbert transform [115], and a third method is the use of

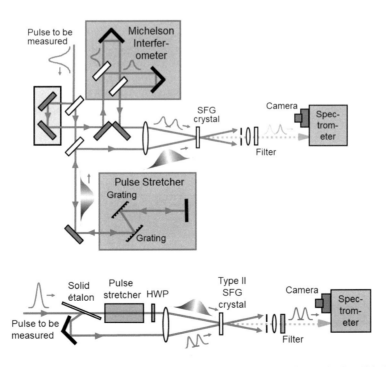

FIGURE 36.57 SPIDER apparatus. Top: As originally proposed. Bottom: Experimentally simplified version in which the grating pulse stretcher has been replaced with a simple block of glass whose dispersion stretches the pulse. Also, a simple etalon replaces the Michelson interferometer. The latter design is used mainly for few-cycle pulses. Plots from www.frog.gatech.edu.

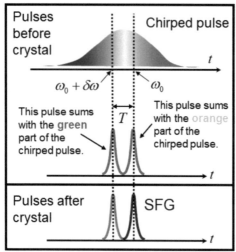

FIGURE 36.58 SPIDER principle. The undistorted double pulse performs sum-frequency-generation with two different temporal regions and hence two different wavelengths of the chirped pulse. This yields two pulses, one wavelength-shifted and delayed compared with the other. As a result, performing spectral interferometry with them yields a phase that is ideally the derivative of the spectral phase. Image from www.frog.gatech.edu.

wavelets [116]. All approaches ultimately isolate the underlying spectral phase difference, removing the constant group delay corresponding to the interference of identical pulses, and any deviation from a linear relation vs. ω indicates the SI of two pulses with different phases.

While analytical pulse retrieval in SPIDER is very fast and may look very robust at first sight, there is nevertheless a hidden caveat: isolated application of equation (36.28) to a measured SPIDER trace typically does not lead to meaningful results without a calibration measurement. This calibration measurement is performed using the identical beam splitting scheme for generating the two unchirped replicas but replacing the sum-frequency arrangement with second-harmonic generation. This second measurement results in an interferogram $S'_{\text{SPIDER}}(\omega)$, which is processed as above to yield $\tau'(w)$, which then needs to be subtracted from the previously measured $\tau(\omega)$ to remove the constant group delay as well as any spectral calibration errors from the measurement.

Because SPIDER is usually used to measure very broadband pulses, higher-order corrections are typically required to directly obtain the spectral phase of a pulse from equation (36.28) without subtracting the reference phase. As these corrections affect the SPIDER measurement in the same way as the reference measurement, one can completely avoid a careful wavelength calibration of the spectrograph by the reference subtraction.

Ignoring any possible issues with the spectrograph calibration, the remaining source of error in a SPIDER measurement is the spectral shear $\delta\omega$. To avoid the need for frequent recalibration of the set-up, it is preferred to implement both the pulse separation T and the dispersion of the stretched pulse using solid-state components, e.g., employing an etalon and a glass block, respectively, as shown in Figure 36.57 (bottom).

This arrangement provides two replica pulses and one chirped pulse with nearly identical pulse energies and has been shown to provide the optimum efficiency in the sum-frequency process [117].

While the widest bandwidth for SHG is obtained with type-I phase matching, the SFG mixing of a narrowband signal in the chirped arm and the broadband replica signal in the other is best performed with type-II phase matching. Given the same bandwidth of the pulse under test, it is therefore much easier to obtain the necessary phase-matching bandwidth than in SHG-based methods, and as only the fringe pattern phase is extracted. Also, because it only measures the spectral phase, SPIDER is also relatively immune to spectral variations of conversion efficiency. These properties have made SPIDER popular for the characterization of few-cycle pulses. Finally, it should be noted that type-II phase matching requires insertion of an additional half-wave retarder in the chirped-pulse arm, as also shown in Figure 36.57 (bottom).

For optimum performance of a SPIDER setup, it is mandatory to understand the proper choice of the pulse separation in one SPIDER arm and the chirp in the other, which both enter into the frequency shear between the two pulses. One generally demands that the shear be much smaller than the spectral bandwidth of the pulse under test and that the separation T be much larger than the duration of the pulses under test. A typical choice for the separation for pulses in the 10 fs range is ~1 ps and a chirp of ~10 000 fs^2. The relatively large dispersion is sufficient to stretch a 10 fs pulse to 1 ps duration, so that there is sufficient temporal overlap between the chirped pulse and both replicas. The resulting fringe frequency is ~1 THz, which allows effective removal of noise in the Fourier filtering process and enables dense sampling of the phase in the spectral domain. Finally, the pulse delay is about 100 times larger than the pulse duration. Longer pulses, e.g., in the 100 fs range require an increase of both parameters. Unfortunately, in this case, a limitation is provided by the spectral resolution of the available spectrometer, which then, in turn, limits the maximum practical value of the separation T. Additionally, the dispersion also must be massively increased to allow temporal stretching of the pulses to this value, which will typically require grating sequences rather than glass blocks.

While the basic SPIDER set-up discussed so far is probably the most widespread and commercially available version of SPIDER, there exist additional variants of SPIDER, which improve on some problems of the method. The first improvement is zero-additional phase SPIDER (ZAP-SPIDER) [118], in which the role of the replicas and the chirped pulse are inverted, and two chirped pulses with a separation T interact with one short replica that can be generated by a surface reflection. Consequently, the latter does not experience any temporal stretching in beam splitter substrates or in an etalon, which makes ZAP-SPIDER the method of choice for the ultraviolet range. If the SPIDER set-up uses two chirped pulses, one can also replace the dispersive element by two narrowband interference filters, which filter out different spectral portions from the input pulse [119]. This replacement is one of the central ideas of Two-Dimensional Spectral Shearing Interferometry (2DSI) [120], which additionally exploits the concept of spatial shearing interferometry, which is also utilized in the spatially encoded arrangement for SPIDER (SEA-SPIDER) [121], analogous to SEA TADPOLE. While traditional SPIDER only requires a one-dimensional detector array in the Fourier plane of the spectrograph, both 2DSI and SEA-SPIDER rely on two-dimensional interferograms and require area-scan rather than line-scan cameras and an imaging spectrograph. This increased experimental complexity pays off in the averaging effect from the multiple parallel interferograms recorded in the individual lines of a CCD camera. Finally, another, less related, method is self-referenced SI [113], which involves performing SI between the pulse field and its cube (which is usually shorter), and has been used on occasion to measure amplified pulses.

A SPIDER-measured pulse is shown in Figure 36.59.

All versions of SPIDER, however, have the same significant drawback: in the presence of pulse-shape instability, *they measure only the coherent artefact* [74–78]. This is because SPIDER is inherently interferometric; random components in the pulse yield fringes that cancel out of the measurement, yielding only an approximately constant background, which is necessarily ignored in SPIDER pulse retrieval. This leaves only the stable component to yield SPIDER fringes and hence the retrieved pulse (see Figure 36.60). Unfortunately, this background cannot be distinguished from identical background due to benign misalignment effects that are unavoidable in SPIDER devices. In addition, as shown in the simulations of Figure 36.60, a mere 2% background corresponded to an under-estimate of the pulse length by a factor of 2.2, and 10% corresponded to an under-estimate of a factor of 4.5. As a result, like other techniques that only measure the coherent artefact, *SPIDER cannot distinguish a stable train of short simple pulses from an unstable train of long complex pulses.* Consequently, in general, SPIDER usually only yields a lower bound on the pulse length. Also, it does not see satellite pulses at all when their relative phase is random [74], which is often the case when lasers double-pulse due to over-pumping—one of the main reasons for making pulse measurements. Unfortunately, readily available devices, such as spectrometers, also cannot distinguish between these diametrically opposed types of pulses. For example, spectral fringes due to unstable double-pulses or variable complex pulses also cancel out in measured spectra. In general, as mentioned earlier, the task of determining a pulse train's shape stability falls to the pulse-measurement technique.

Single-shot measurements could help, but variations from place to place in the beam could give rise to similar effects, and SPIDER is rarely used for single-shot measurements.

There is a way to be certain that a SPIDER measurement is of a stable pulse train and is accurate, and that is, if it has a 100% fringe visibility. Unfortunately, SPIDER measurements rarely achieve this, and the question remains as to whether the background is due to benign misalignment issues or instability.

On the other hand, SPIDER has been extensively tested and compared with FROG for well-known stable lasers. But the above issue necessarily presents a difficult dilemma for other researchers who use SPIDER: because there is no feedback regarding the measurement quality, and it measures only the coherent artefact, it is prone to yielding shorter pulses in the presence of otherwise undetected instability. So, it is not possible to know which SPIDER measurements are valid and

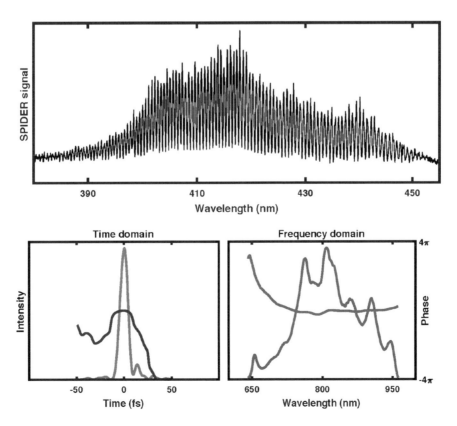

FIGURE 36.59 SPIDER measurement. Top: Measured SPIDER trace. Bottom: Retrieved pulse. From www.frog.gatech.edu (see Ref. [122]).

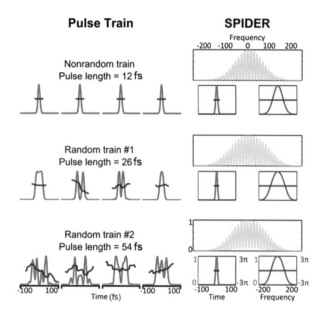

FIGURE 36.60 Simulated SPIDER measurements of a stable and two unstable trains of pulses (the same trains used for analogous studies for autocorrelation and FROG earlier). Note that SPIDER only measures the coherent artefact. It cannot distinguish a stable train of short simple pulses from an unstable train of long complex pulses. In general, SPIDER yields only a lower bound on the pulse length. From Ref. [78].

which have under-estimated the pulse lengths, instability, and/or complexity of the pulses due to the presence of pulse-shape instability or unstable double-pulsing. It is true that standard ultrafast lasers are more stable than ever, but many are not, especially extremely short pulses at the edge of technology, for which SPIDER is usually used. This issue is one that requires additional research and remains a dark cloud over all SPIDER measurements. As a result, SPIDER's popularity is rapidly waning.

So, the question is: when are pulse measurements using SPIDER appropriate? It appears only to be suited for measuring pulses with durations between 10 and 100 fs, with very limited complexity, and that are somehow known to have stable pulse shapes. SPIDER variants have been used to measure some of the shortest pulses to date [123], but such measurements are not as convincing as one would like in view of SPIDER's measurement of only the coherent artefact and that such measurements provide only a lower bound on the pulse length. SPIDER especially should not be used to measure pulses from newly developed, possibly misaligned, or poorly constructed lasers, all of which have uncertain stability. SPIDER should not be used to check for possibly unstable double-pulsing. SPIDER also should not be used for pulse-length claims in the absence of 100% fringe visibility or an independent stability confirmation. And, it is not suited for the characterization of irregular pulse trains, which are also more prone to instability [76]. On the other hand, it is ideal for measuring the group-delay dispersion of a pulse compressor, for which the average spectral phase is the desired quantity.

36.33 Spatiotemporal Pulse Measurement

The propagation of an ultrashort pulse is fundamentally *spatiotemporal*, meaning that its electric field cannot usually be separated into a product of purely temporal and spatial fields due to unavoidable spatiotemporal couplings [124]. Even the simple cases of diffraction of ultrashort pulses from a circular aperture [125] or merely the focusing of them [101,104,126–131] result in complex spatiotemporal structures. As a result, measuring the pulse vs. time, averaging over spatial coordinates (x and y), or measuring the pulse vs. x and y, averaged over time is not always useful. Complete-spatiotemporal-resolution measurements, on the other hand, enable direct visualization of these phenomena, so that they can be studied and understood in greater detail.

Complete spatiotemporal pulse-measurement techniques determine $E(x, y, z, \omega)$ or, equivalently, $E(x, y, z, t)$. No method directly measures all this information, of course. Fortunately, it is not necessary to measure the z-dependence because it can be obtained from a diffraction integral, unless some information is missing from the measurement of $E(x, y, t)$, in which case, additional information must be obtained.

Most multi-shot spatiotemporal-measurement methods are based on spectral interferometry (SI Which we describe in the next section) [132,133].

36.34 Spatially Resolved Spectral Interferometry: One Spatial Dimension

It is straightforward to simply perform standard SI, using the other dimension of the camera to obtain spatial information of the beam in one spatial direction (see Figure 36.61). Such an approach has been used to measure beams with spatiotemporal couplings, such as the pulse-front distortions caused by lenses [134,135]. The drawback is that a pulse cannot be measured directly at a focus, where it is too small, so the beam had to

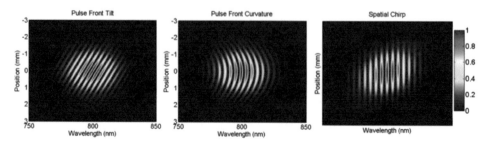

FIGURE 36.61 Calculated spatially resolved spectral interferograms of pulses with three different spatiotemporal couplings. Left: pulse front tilt. Centre: pulse front curvature. Right: spatial chirp. It is assumed that the transform-limited reference pulse is free of spatiotemporal couplings [124].

be first re-collimated, by propagating a second time through the lens. The advantage is that $E(y, \omega)$ can be retrieved from a single camera frame.

Spatially resolved spectral interferometry was also recently adapted for spatiotemporal measurements of a pulse with a huge pulse front tilt of ~89.9°. This was done using an etalon as the spectrometer's dispersive element, rather than a diffraction grating [136].

Analogously, spectrally resolved spatial interferometry can provide similar information [137–139].

36.35 Fibre-based Scanning Spatiotemporal Pulse Measurement: Two and Three Spatial Dimensions

SEA TADPOLE and STARFISH both naturally extend to spatiotemporal measurement in gathering spatial information by scanning a small fibre tip or probe through the light beam of interest. Using such methods, SI has been integrated into a scanning microscope to collect spatiotemporal information from a small sample [106]. Such methods have the same constraints as SI; for example, they require a well-characterized reference pulse that contains the spectrum of the unknown pulse, which now must also be spatially uniform. A significant benefit of using an optical fibre for introducing the unknown beam into a measurement device is that it provides spatial resolution—as small as a few nanometres. The fibre collects only the light from a small spatial sample. The tip of the fibre can then be freely scanned through three-dimensional space, measuring $E(\omega)$ at each fibre position, yielding the spatiotemporal electric field of the unknown pulse $E_{unk}(x, y, z, \omega)$ with high spatial, temporal, and spectral resolutions.

In scanning SEA TADPOLE (see Figure 36.62), a single-mode fibre in the unknown arm is mounted on an x, y, z translation stage, so that the spectrum and spectral phases of the unknown pulse are determined at multiple fibre positions in space [135,140,141]. The robustness of SEA TADPOLE was nicely illustrated by 3D-scanning measurements using a supercontinuum laser source [148].

Because of the high sensitivity of SEA TADPOLE, even very large beams can be measured, in which only a tiny fraction of the pulse energy is coupled in the small fibre tip. Also, a SEA TADPOLE measurement at a single fibre position is proportional to the product of the reference and unknown fields, so the much more intense reference pulse effectively amplifies the unknown pulse, making such a measurement even more sensitive.

Improving The spatial resolution of scanning SEA TADPOLE to <1 μm by uses a Near-Field Scanning Optical Microscopy (NSOM) fibre probe, rather than a single-mode fibre [142]. NSOM fibres have been used in the past to measure the spatial intensity distribution of tightly focused continuous-wave lasers. Using an NSOM probe with an aperture diameter of 500 nm, the complete electric field of focused pulses with numerical apertures (NAs) as high as 0.44 and features in their intensity of <1 μm have been measured with SEA TADPOLE [101]. NSOM probes have also been used in SEA TADPOLE to measure the refractive index and group velocity in a waveguide structure by collecting the local evanescent field into the fibre probe [143]. In addition, NSOM probes have been combined with Fourier transform spectral interferometry to characterize the local plasmons excited by ultrashort laser pulses in gold nanostructures [143].

Scanning in SEA TADPOLE typically cannot be performed stably enough to avoid a loss of the absolute *spatial* phase $\varphi_0(x, y, \omega_0)$. In some cases, for example, to measure

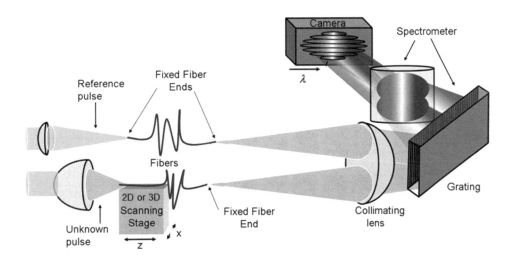

FIGURE 36.62 Scanning SEA TADPOLE experimental set-up. The laser beam is split into the reference and unknown arms in a Mach-Zehnder interferometer. The spatially uniform reference pulse goes through a delay line and is focused into one of the single-mode fibres (for nearly unlimited bandwidth the so-called endlessly single-mode photonic crystal fibres can be used). $E(t)$ of the reference pulse can be characterized by sending it with a flip mirror to a FROG device. In the unknown arm of the interferometer, the light is directed to the optical system being studied (the "experiment"), and the resulting field is sampled at every x, y, z point of interest by the other single-mode fibre. The outputs of the fibres are placed at the entrance of an imaging spectrometer with a few mm separation between them, so after the collimating lens, they cross at a small angle at the camera which records the 2D spectrally resolved spatial interference pattern. The interference is measured for each fibre position and $E(\omega)$ is reconstructed from each interferogram yielding $E(x, y, z, \omega)$. From Ref. [104].

spatiotemporal couplings or to see the pulse front, the spatial phase is not of interest. However, there are many benefits to having this additional spatial information. If it is known, the field $E(x, y, \omega)$ only needs to be measured at one value of z, and then it can be numerically propagated to any other plane using a diffraction integral.

Fortunately, it has been demonstrated that this information is still actually contained in the measured SEA TADPOLE data as long as $E(x, y, \omega)$ is measured for at least two values of z [144]. Using a simple *phase-diversity* approach, which involves Fresnel-transforming back and forth and replacing the intensity or spectrum by the known intensity or spectrum (often called a Gerchberg-Saxton-like phase-retrieval algorithm), the absolute spatial phase of a pulse can be very accurately numerically recovered. In addition, the complete spatiotemporal field at all locations is more accurately determined by this process. See Figure 36.63.

36.36 Spatiotemporal Measurement Examples: Focusing Pulses

The first motivation for scanning SEA TADPOLE was to measure the spatiotemporal field of focused pulses propagating using various lenses, including simple plano-convex lenses, aspheric lenses made of moulded PMMA, achromatic doublets, and microscope objectives and lenses with numerical apertures as high as 0.44 [101,133]. Figure 36.64 shows one such measurement exhibiting the so called "fore-runner pulse" resulting from the combination of diffraction at the edge of the lens and chromatic aberration [129–131]. In this measurement, a 500 nm NSOM probe was used to resolve the sub-micron spatial features.

Another application of SEA TADPOLE has been to measurements of superluminal Bessel-X pulses, first demonstrated

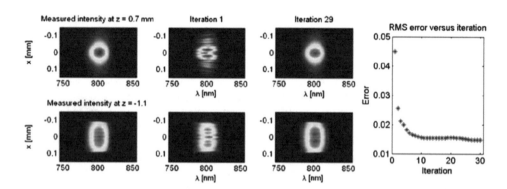

FIGURE 36.63 Spatial phase retrieval results. First column: measured spatio-spectral amplitudes at z_1 (top) and z_2 (bottom). Second column: spatio-spectral intensities obtained by propagating the field at z_2 to z_1 (top) and *vice versa* (bottom) using the measured spatial phase after only one iteration. Third column: same as the previous, but using the retrieved spatial phase after 29 iterations. Note the excellent agreement between the first and last columns. From Ref. [144].

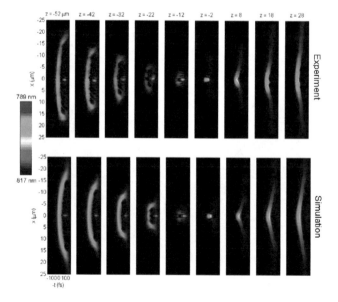

FIGURE 36.64 Measured $E(x,z,t)$ for a pulse focused with a 0.44 NA aspheric lens. Each box shows $E(x,t)$ at a certain distance from the focus (z) written above the box. The variation of grayscale tone within every frame in the plot is the instantaneous frequency of the pulse as indicated by the colour bar. Here, the colour also varies along the x-direction due to the severe chromatic aberrations that are present. The combination of overfilling the lens and chromatic aberrations results in the additional "fore-runner" pulse ahead of the main pulse, for $z < 0$ [101].

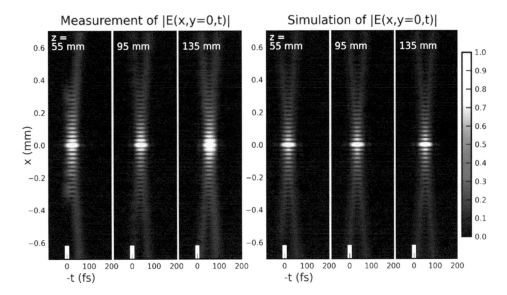

FIGURE 36.65 Left: the measured spatiotemporal amplitude of the Bessel-X pulse at three different distances after the axicon. Right: the corresponding simulations. Amplitude is indicated by the grayscale and is normalized for each field to have a maximum of 1. The white bar on the time axis emphasizes the zero of time—a reference frame moving along with the reference pulse at the speed of light [140].

in Ref. [145]. SEA TADPOLE yields highly accurate measurements of the group velocity. Although the superluminal speed of Bessel-X pulses had been measured before [146], the full spatiotemporal field of a Bessel-X pulse had never been directly recorded with simultaneous high spatial and temporal resolutions [140]. Figure 36.65 shows the spatiotemporal field $E(x, z, t)$ from these measurements, side-by-side with the corresponding simulations. These results can be considered "snapshots in flight" of the Bessel-X pulse.

Note that the central maximum of the pulse has a width of only ~20 μm and does not significantly diverge over 8 cm of propagation, a well-known and important characteristic of Bessel pulses. In comparison, a Gaussian beam of the same waist size would have expanded by 26 times. The superluminal velocity of the pulse is apparent in these plots because the reference pulse, determining the centre of the time-windows, propagates at the speed of light (and is indicated by the small white bars on the snapshots and labelled by $t=0$). The measured group velocity of this Bessel-X pulse is within 0.001% of the theoretically predicted value, $1.00012c$. Similarly, SEA TADPOLE has been used to study accelerating and decelerating Bessel pulses [147] and subluminal pulsed Bessel beams generated by diffractive axicons [141,148] as well as diffraction on openings [135,141].

36.37 Other Spatiotemporal-measurement Methods

Other methods have been developed for measuring the complete spatiotemporal intensity and phase on a multi-shot basis. One such technique, also with nanometre-scale spatial resolution, is called nanoFROG [149]. It involves making a FROG trace for the beam at a point in space using a nanometre-sized non-linear crystal placed at that point. This yields the intensity and phase vs. time for that point, with the usual trivial FROG ambiguities. While they comprise a larger set of ambiguities than the spatial phase ambiguity that arises in SEA TADPOLE, they should also be removable using the phase-diversity approach described earlier [144].

36.38 Spatiotemporal Measurement on a Single Shot: STRIPED FISH

Although the aforementioned complete spatiotemporal pulse-measurement techniques work well, and some even have sub-micron spatial resolution, they are all inherently multi-shot. But many applications—in particular, high-intensity extremely-low-rep-rate pulse measurements—require single-shot operation.

In attempting to take complete spatiotemporal measurement to three or even four dimensions, an obvious observation must be made. There are three spatial dimensions and only one temporal dimension, so it makes more sense to adapt a *spatial*-measurement technique to spatiotemporal measurement than to adapt a *temporal*-measurement technique. Arguably, the most common and powerful spatial intensity-and-phase technique is holography. Thus, it would seem reasonable that spectrally resolved digital holography could be used to measure the complete spatiotemporal intensity and phase of ultrashort pulses, indeed, on a single shot.

The difference between the spatiotemporal measurement problem and standard holography is that holograms usually only yield the spatial intensity and phase of a *monochromatic* beam, while pulses are *broadband*. Also, the relative phases of the various frequencies of the pulse.

So the best approach is to record three-dimensional unknown field information, $E_{unk}(x, y, \omega)$, using multiple two-dimensional digital holograms, each at a different frequency. By using a digital camera with high pixel count and small pixel size, it is possible to illuminate different regions of the imaging sensor with digital holograms recorded at different frequencies.

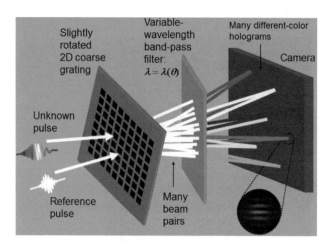

FIGURE 36.66 Conceptual schematic of STRIPED FISH. From Ref. [152].

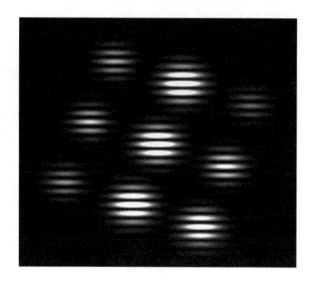

FIGURE 36.67 Conceptual schematic of a measured STRIPED FISH trace. Axes are simply the two spatial coordinates of the camera frame. From www.frog.gatech.edu (see Ref. [152])..

And in order to know the relative phases of the many different-frequency holograms, the reference pulse intensity and phase must be measured in time/frequency [150]. It must also have known spatial properties and so must be spatially filtered. Finally, as usual, the device should be simple and usable by those not skilled in pulse measurement. A device with these characteristics is STRIPED FISH (Spatially and Temporally Resolved Intensity and Phase Evaluation Device: Full Information from a Single Hologram) [150,151]. At the time of this writing, STRIPED FISH is the only technique ever proposed to measure the complete spatiotemporal intensity and phase on a single shot.

A schematic of the STRIPED FISH apparatus is shown in Figure 36.66. It comprises a very simple set-up of only a coarse two-dimensional diffractive optical element (DOE), an interference band-pass filter (IBPF), imaging optics, and a camera. It uses a previously spatially smoothed and temporally characterized reference pulse, accomplished at an earlier point using a spatial filter and a FROG measurement of the spatially filtered pulse. These operations can be performed on a replica of the unknown pulse, making the technique self-referenced. The pulse to be measured and the known reference pulse cross at a small vertical angle on the DOE, which simultaneously generates multiple divergent pairs of beams at different angles.

The DOE is also rotated slightly, so the horizontal propagation angle is different for each beam pair. Because the IBPF's transmission wavelength varies with horizontal incidence angle, it wavelength-filters each pair of beams to be essentially monochromatic and with different centre wavelengths. The beam pairs then overlap at the camera, generating an array of quasi-monochromatic holograms, each at a different wavelength. The spatiotemporal information of the unknown pulse is contained within multiple holograms, which are recorded simultaneously on the camera frame. See Figure 36.67.

In the recorded STRIPED FISH trace, for each hologram on the camera, a Fourier-filtering algorithm (similar to that of SEA TADPOLE) is applied to obtain the unknown field $E_{unk}(x, y, \omega)$ at that frequency. Once the unknown fields at all frequencies are obtained, an IFT vs. frequency is performed to convert the spatio-spectral field $E_{unk}(x, y, \omega)$ to yield the spatiotemporal pulse profile $E_{unk}(x, y, t)$. The pulse-retrieval algorithm is shown in Figure 36.68.

Also, once $E_{unk}(x, y, \omega)$ is obtained, diffraction integrals can be used to propagate the field into different z-locations as well.

The STRIPED FISH trace, comprising multiple spectral digital holograms, is quite informative. After spatial filtering, the reference pulse contains no spatiotemporal coupling and the spatio-spectral information of the unknown pulse is encoded in the STRIPED FISH trace [153]. Specifically, for a certain frequency, the unknown pulse's spatial structure is contained within each hologram: the spatial amplitude is represented by the intensity distribution and the spatial phase by the fringe shape across that hologram. Likewise, for each location, the unknown pulse's spectral information is reflected by multiple holograms: the spectral amplitude is represented by the intensity variations and the spectral phase is indicated by the fringe shifts among different holograms.

To illustrate these effects, simulated STRIPED FISH traces are shown in Figure 36.69 for the cases of temporal double pulses and spatial double pulses. From the traces (Figure 36.69a and b), we can clearly see the spectral intensity variations among different holograms. Also, from their fringe-shifting (e.g. the upper left hologram shows a shifted fringe pattern compared with the lower right one), we know that the spectral phase of the unknown pulse is varying with respect to different holograms, or different frequencies.

Similarly, the spatial effects are demonstrated by implementing a pair of spatial double pulses. Two spatial pulses are assumed to be propagating in the same direction, the left of which is half the amplitude (therefore quarter intensity) as the right one. To show the spatial phase variation, a π-phase jump was introduced between the two component pulses. As shown in Figure 36.69c and d, the left pulse appears dimmer than the right one, indicating different spatial amplitudes. Also, in the middle of each hologram, we observe a fringe discontinuity due to the introduced spatial phase jump.

Thus, it is possible to quickly identify various significant spatial-spectral (or spatiotemporal) structures of the unknown

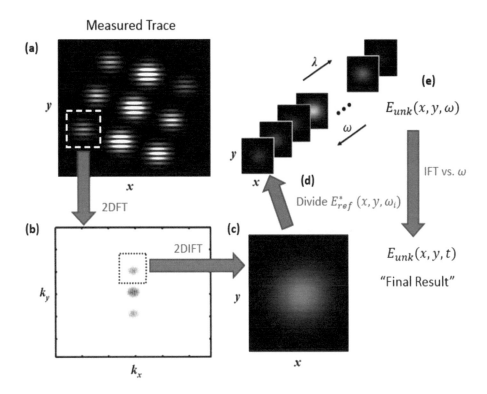

FIGURE 36.68 Illustration of the STRIPED FISH retrieval algorithm. Amplitudes are plotted for complex quantities. (a) Multiple holograms of different frequencies are recorded on the camera. (b) A hologram of a certain frequency ω_i is selected. A two-dimensional Fourier transform (2DFT) is taken over spatial dimensions x and y. (c) The oscillating interference term is extracted and inversely 2DFTed into the spatial domain, obtaining a product term $E_{unk}(x, y, \omega_i) E_{ref}^*(x, y, \omega_i)$. (d) Dividing the product term by the conjugated reference field $E_{ref}^*(x, y, \omega_i)$ yields the unknown spatial field at frequency ω_i, $E_{unk}(x, y, \omega_i)$. (e) Performing steps b through d for every hologram yields $E_{unk}(x, y, \omega)$; then, $E_{unk}(x, y, t)$ is obtained by an IFT into the time domain. From Ref. [151].

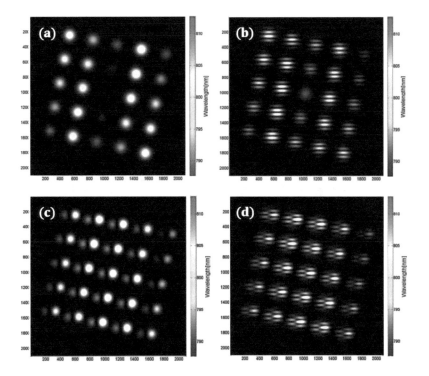

FIGURE 36.69 STRIPED FISH traces for double pulses. (a) Pattern of the temporal double pulses, with equal intensities and π phase jump between the two component pulses in the absence of the reference pulse. (b) The STRIPED FISH holograms of the temporal double pulses. (c) Pattern of the spatial double pulses, with the left pulse of one-fourth intensity as the right pulse in the absence of the reference pulse. A π-phase jump was introduced between the two pulses. (d) The STRIPED FISH holograms of the spatial double pulses. From Ref. [153].

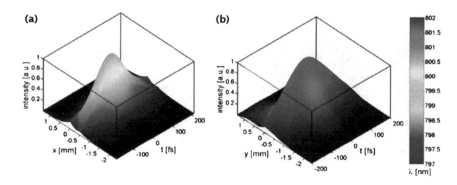

FIGURE 36.70 STRIPED FISH measurement of a spatially chirped pulse. (a) Measured intensity along x and t. (b) Measured intensity along y and t. Gray shades is the instantaneous frequency. From Ref. [151].

pulse even before complete retrieval, by observing the recorded STRIPED FISH trace on camera.

STRIPED FISH has been used to measure spatiotemporally complex ultrashort pulses. A measurement of simple spatially chirped pulse is shown in Figure 36.70, where the measured pulse intensity is plotted vs. x–t and y–t. The height and brightness denote the field intensity, and the colour reflects the instantaneous wavelength. It is clear that the pulse shows a spatial frequency chirp along the x-direction, whereas it has essentially no chirp along the y-direction.

A typical STRIPED FISH setup can generate ~30 holograms at different frequencies. Each hologram typically is a ~300 × ~300 array of spatial pixels. As a result, the maximum measurable space-TBP is on the order of 1 000 000.

If pulses more complex in time must be measured, more holograms but smaller in size could be generated. However, given by the practical bandwidth of the bandpass filter, STRIPED FISH is limited in its temporal range. The longest measurable pulse is the reciprocal of this spectral width, so pulses as long as ~10 ps can be measured using the narrowest-band available filters. This is usually more than adequate for measuring the most ultrahigh-intensity pulses. Alternatively, the temporal-range limitation can be overcome if a stable pulse train is to be measured, and scanning in one dimension (the delay) can be performed. In this case, the longest possible pulse length is the delay range, which can be ~ns in length. Snapshots of movies of pulses from multi-mode optical fibres measured using this approach are shown in Figure 36.71.

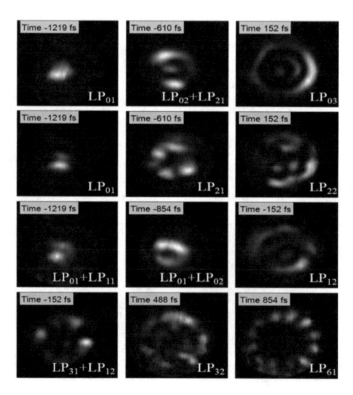

FIGURE 36.71 Measured delay-scanned STRIPED FISH results of output pulses from multi-mode optical fibres with different fibre-coupling situations. Each row contains three snapshots of the output pulses measured by delay-scanned STRIPED FISH for a different coupling arrangement. Top row: Centred coupling situation. Second row: small offsets in both x- and y-direction situation. Third row: small offset in only the x-direction. Fourth row: Large offset. All the snapshots show the spatial intensities (by brightness) and frequencies present at different times. Every snapshot has one or two dominant LP modes, which are labelled in the snapshots. From Ref. [154].

Such complex pulses are best displayed using movies based on spectrograms due to the massive amount of information measured about the pulse. This plotting scheme avoids artefacts in plotting instantaneous frequency, e.g., when the pulse contains all of its spectrum, the pulse appears green (instantaneous frequency) instead of white (what it should be) [152,154–156].

36.39 Conclusions

The field of ultrashort-pulse measurement has progressed spectacularly in the past three decades. It has rapidly evolved from a position well behind that of pulse generation to now leading it in most areas. In 1990, only blurry black and white images were measurable and then only for simple pulses. Worse, such pulse measurements contained misleading artefacts. Today, it is possible to measure high-definition full-colour spatiotemporal images of most pulses, including those as complex in time and in space-time as have ever been intentionally generated. What remains is to take advantage of these powerful methods to better understand the pulses that arise in laboratory settings and the processes that they can be used to study.

Acknowledgements

Text and figures in this article have been reprinted with permission from the authors from various articles (cited in the text), including R. Trebino, R. Jafari, S. A. Akturk, P. Bowlan, Z. Guang, P. Zhu, E. Escoto, and G. Steinmeyer, "Highly Reliable Measurement of Ultrashort Laser Pulses," *J. Appl. Phys.* 128, 171103–171101 (2020) with the permission of AIP Publishing, and also from Rick Trebino's book (*Frequency-Resolved Optical Gating: The Measurement of Ultrashort Laser Pulses*), as well as the partially written second edition of this book, and finally also from the Trebino group website, and they retain all copyrights, including this work as a whole. They grant permission to CRC to reprint this work in its entirety in both the digital and physical versions of the Laser Handbook. The Georgia Tech authors would like to acknowledge that this work was supported in part by the National Science Foundation under Grant ECCS-1609808. It should be noted that Rick Trebino owns a company that sells pulse-measurement devices.

REFERENCES

1. R. Trebino, *Frequency-Resolved Optical Gating: The Measurement of Ultrashort Laser Pulses* (Kluwer Academic Publishers, Boston, MA 2002).
2. H. Stark, ed., *Image Recovery: Theory and Application* (Academic Press, Orlando, FL 1987).
3. E. J. Akutowicz, "On the Determination of the Phase of a Fourier Integral, I," *Trans. Amer. Math. Soc.* **83**, 179–192 (1956).
4. E. J. Akutowicz, "On the Determination of the Phase of a Fourier Integral, II," *Trans. Amer. Math. Soc.* **84**, 234–238 (1957).
5. K. L. Sala, G. A. Kenney-Wallace, and G. E. Hall, "CW Autocorrelation Measurements of Picosecond Laser Pulses," *IEEE J. Quant. Electron.* **16**, 990–996 (1980).
6. D. M. Rayner, P. A. Hackett, and C. Willis, "Ultraviolet Laser, Short Pulse-Width Measurement by Multiphoton Ionization Autocorrelation," *Rev. Sci. Inst.* **53**, 537–538 (1982).
7. E. S. Kintzer and C. Rempel, "Near-Surface Second-Harmonic Generation for Autocorrelation Measurements in the UV," *App. Phys. B* **42**, 91–95 (1987).
8. O. L. Bourne and A. J. Alcock, "Ultraviolet and Visible Single-Shot Autocorrelator Based on Multiphoton Ionization," *Rev. Sci. Inst.* **57**, 2979–2982 (1986).
9. J. I. Dadap, G. B. Focht, D. H. Reitze, and M. C. Downer, "Two-Photon Absorption in Diamond and Its Application to Ultraviolet Femtosecond Pulsewidths Measurement," *Opt. Lett.* **16**, 499–501 (1991).
10. G. J. Dixon, "Advanced techniques measure ultrashort pulses," Laser Focus World **33**, 99–102, 104–105 (1997).
11. M. H. R. Hutchinson, I. A. McIntyre, G. N. Gibson, and C. K. Rhodes, "Measurement of 248-nm, Subpicosecond Pulse Durations by Two-Photon Fluorescence of Xenon Excimers," *Opt. Lett* **12**, 102–104 (1987).
12. M. H. R. Tünnermann, H. Eichmann, R. Henking, K. Mossavi, and B. Wellegehausen, "Single-Shot Autocorrelator for KrF Subpicosecond Pulses Based on Two-Photon Fluorescence of Cadmium Vapor at λ = 508 nm," *Opt. Lett* **16**, 402–404 (1991).
13. R. Wyatt and E. E. Marinero, "Versatile Single-Shot Background-Free Pulse Duration Measurement Technique for Pulses of Subnanosecond to Picosecond Duration," *Appl. Phys.* **25**, 297–301 (1981).
14. J. A. Giordmaine, P. M. Rentzepis, S. L. Shapieo, and K. W. Wecht, "Two-Photon Excitation of Fluorescence By Picosecond Light Pulses," *Appl. Phys. Lett.* **11**, 216–218 (1967).
15. R. Trebino, E. K. Gustafson, and A. E. Siegman, "Fourth-Order Partial-Coherence Effects in the Formation of Integrated-Intensity Gratings with Pulsed Light Sources," *J. Opt. Soc. Amer. B* **3**, 1295–1304 (1986).
16. A. Birmontas, R. Kupris, A. Piskarskas, V. Smil'gyavichyus, and A. Stabinius, "Determination of the duration of fluctuating picosecond optical pulses," *Sov. J. Quant. Electron.* **12**, 792–794 (1982).
17. R. A. Fisher and J. J. A. Fleck, "On the Phase Characteristics and Compression of Picosecond Pulses," *Appl. Phys. Lett.* **15**, 287–290 (1969).
18. J. Etchepare, G. Grillon, and A. Orszag, "Third Order Autocorrelation Study of Amplified Subpicosecond Laser Pulses," *IEEE J. Quant. Electron.* **19**, 775–778 (1983).
19. J. Janszky and G. Corradi, "Full Intensity Profile Analysis of Ultrashort Laser Pulses Using Four-Wave Mixing or Third Harmonic Generation," *Opt. Commun.* **60**, 251–256 (1986).
20. N. Sarukura, M. Watanabe, A. Endoh, and S. Watanabe, "Single-Shot Measurement of Subpicosecond KrF Pulse Width by Three-Photon Fluorescence of the XeF Visible Transition," *Opt. Lett* **13**, 996–998 (1988).
21. H. Schulz, H. Schuler, T. Engers, and D. von der Linde, "Measurement of Intense Ultraviolet Subpicosecond Pulses Using Degenerate Four-Wave Mixing," *IEEE J Quant Electron* **25**, 2580–2585 (1989).

22. R. Fischer, J. Gauger, and J. Tilgner, "Fringe Resolved Third-Order Autocorrelation Functions," *Proc. Am. Inst. Phys. Conf.* **172**, 727–729 (1988).
23. P. F. Curley, G. Darpentigny, G. Cheriaux, J. P. Chambaret, and A. Antonetti, "High dynamic range autocorrelation studies of a femtosecond Ti: sapphire oscillator and its relevance to the optimisation of chirped pulse amplification systems," *Opt. Comm.* **120**, 71–77 (1995).
24. J.-H. Chung and A. M. Weiner, "Ambiguity of ultrashort pulse shapes retrieved from the intensity autocorrelation and power spectrum," *IEEE J. Sel. Top. Quant. Electron.* **7**, 656–666 (2001).
25. J. C. Diels, J. J. Fontaine, and F. Simoni, "Phase sensitive measurements of femtosecond laser pulses from a ring cavity," in *Proceedings of the International Conference on Lasers* (STS Press, McLean, VA, 1983), pp. 348–355.
26. J. C. Diels, J. J. Fontaine, I. C. McMichael, and F. Simoni, "Control and Measurement of Ultrashort Pulse Shapes (in Amplitude and Phase) with Femtosecond Accuracy," *Appl. Opt.* **24**, 1270–1282 (1985).
27. J. C. Diels, "Measurement techniques with mode-locked dye laser," in *Ultrashort Pulse Spectroscopy and Applications, Proceedings of SPIE* (SPIE Press, Bellingham, 1985), pp. 63–70.
28. C. Yan and J. C. Diels, "Amplitude and Phase Recording of Ultrashort Pulses," *J. Opt. Soc. Am. B* **8**, 1259–1263 (1991).
29. J. C. M. Diels, J. J. Fontaine, N. Jamasbi, and M. Lai, "The Femto-nitpicker," in *Conference on Lasers & Electro-Optics*, (1987).
30. S. P. Le Blanc, G. Szabo, and R. Sauerbrey, "Femtosecond Single-Shot Phase-Sensitive Autocorrelator for the Ultraviolet," *Opt. Lett* **16**, 1508–1510 (1991).
31. G. Szabo, Z. Bor, and A. Muller, "Phase-Sensitive Single-Pulse Autocorrelator for Ultrashort Laser Pulses," *Opt. Lett* **13**, 746–748 (1988).
32. R. Trebino and D. J. Kane, "Using Phase Retrieval to Measure the Intensity and Phase of Ultrashort Pulses: Frequency-Resolved Optical Gating," *J. Opt. Soc. Amer. A* **10**, 1101–1111 (1993).
33. L. Cohen, "Time-Frequency Distributions—A Review," *Proc IEEE* **77**, 941–981 (1989).
34. L. Cohen, *Time-Frequency Analysis* (Prentiss-Hall, Englewood Cliffs, NJ, 1995).
35. A. Freiberg and P. Saari, "Picosecond Spectrochronography," *JQE Publications Office* **QE-19**, 622–630 (1983).
36. R. A. Altes, "Detection, Estimation, and Classification with Spectrograms," *J. Acoust. Soc. Amer.* **67**, 1232–1246 (1980).
37. S. H. Nawab, T. F. Quatieri, and J. S. Lim, "Signal Reconstruction from Short-Time Fourier Transform Magnitude," *IEEE Trans Acoustics, Speech, and Signal Proc* **ASSP-31**, 986–998 (1983).
38. R. Trebino and D. J. Kane, "The Dilemma of Ultrashort-Laser-Pulse Intensity and Phase Measurement and Applications," *IEEE J. Quant. Electron.* **35**, 418–420 (1999).
39. R. Trebino, K. W. DeLong, D. N. Fittinghoff, J. N. Sweetser, M. A. Krumbügel, and D. J. Kane, "Measuring Ultrashort Laser Pulses in the Time-Frequency Domain Using Frequency-Resolved Optical Gating," *Rev Sci Inst* **68**, 3277–3295 (1997).
40. D. J. Kane and R. Trebino, "Single-Shot Measurement of the Intensity and Phase of an Arbitrary Ultrashort Pulse By Using Frequency-Resolved Optical Gating," *Opt. Lett.* **18**, 823–825 (1993).
41. D. J. Kane and R. Trebino, "Characterization of Arbitrary Femtosecond Pulses Using Frequency Resolved Optical Gating," *IEEE J Quant Electron* **29**, 571–579 (1993).
42. K. W. DeLong and R. Trebino, "Improved Ultrashort Pulse-Retrieval Algorithm for Frequency-Resolved Optical Gating," *J. Opt. Soc. Amer. A* **11**, 2429–2437 (1994).
43. K. W. DeLong, R. Trebino, and D. J. Kane, "Comparison of Ultrashort-Pulse Frequency-Resolved-Optical-Gating Traces for Three Common Beam Geometries," *J. Opt. Soc. Amer. B* **11**, 1595–1608 (1994).
44. K. W. DeLong, R. Trebino, J. Hunter, and W. E. White, "Frequency-Resolved Optical Gating With the Use of Second-Harmonic Generation," *J. Opt. Soc. Amer. B* **11**, 2206–2215 (1994).
45. K. W. DeLong, D. N. Fittinghoff, R. Trebino, B. Kohler, and K. Wilson, "Pulse Retrieval in Frequency-Resolved Optical Gating Based on the Method of Generalized Projections," *Opt. Lett* **19**, 2152–2154 (1994).
46. V. V. Kotlyar, P. G. Seraphimovich, and V. A. Soifer, "Regularized Iterative Algorithm for the Phase Retrieval," *Optik* **94**, 96–99 (1993).
47. J. R. Fienup, "Phase Retrieval Algorithms: A Comparison," *Appl. Opt.* **21**, 2758–2769 (1982).
48. D. Peri, "Optical Implementation of a Phase Retrieval Algorithm," *Appl. Opt.* **26**, 1782–1785 (1987).
49. K. W. DeLong, D. N. Fittinghoff, and R. Trebino, "Practical Issues in Ultrashort-Laser-Pulse Measurement Using Frequency-Resolved Optical Gating," *IEEE J. Quant. Electron.* **32**, 1253–1264 (1996).
50. G. Taft, A. Rundquist, M. M. Murnane, I. P. Christov, H. C. Kapteyn, K. W. DeLong, D. N. Fittinghoff, M. A. Krumbügel, J. N. Sweetser, and R. Trebino, "Measurement of 10-fs Laser Pulses," *IEEE J. Sel. Top. Quant. Electron.* **2**, 575–585 (1996).
51. T. Bendory, P. Sidorenko, and Y. C. Eldar, "On the Uniqueness of FROG Methods," *IEEE Signal Proc. Lett.* **24**, 722–726 (2017).
52. K. W. DeLong, C. L. Ladera, R. Trebino, B. Kohler, and K. R. Wilson, "Ultrashort-Pulse Measurement Using Noninstantaneous Nonlinearities: Raman Effects in Frequency-Resolved Optical Gating," *Opt. Lett.* **20**, 486–488 (1995).
53. A. Baltuska, M. S. Pshenichnikov, and D. A. Wiersma, "Amplitude and Phase Characterization of 4.5-fs Pulses by Frequency-Resolved Optical Gating," *Opt. Lett.* **23**, 1474–1476 (1998).
54. R. Jafari, T. Jones, and R. Trebino, "100% Reliable Algorithm for Second-Harmonic-Generation Frequency-Resolved Optical Gating," *Opt. Exp.* **27**, 2112–2124 (2019).
55. R. Jafari and R. Trebino, "Highly Reliable Frequency-Resolved Optical Gating Pulse-Retrieval Algorithmic Approach," *IEEE J. Quant. Electron.* **55**, 1–7 (2019).
56. R. Jafari and R. Trebino, "Extremely Robust Pulse Retrieval from Even Noisy Second-Harmonic-Generation Frequency-Resolved Optical Gating Traces," *IEEE J. Quant. Electron.* **56**, 1–8 (2020).
57. B. A. Richman, M. A. Krumbügel, and R. Trebino, "Temporal Characterization of Mid-IR Free-Electron-Laser Pulses by Frequency-Resolved Optical Gating," *Opt. Lett.* **22**, 721–723 (1997).

58. B. Kohler, V. V. Yakovlev, K. R. Wilson, J. Squier, K. W. DeLong, and R. Trebino, "Phase and Intensity Characterization of Femtosecond Pulses from a Chirped-Pulse Amplifier by Frequency-Resolved Optical Gating," *Opt. Lett* **20**, 483–485 (1994).

59. T. S. Clement, A. J. Taylor, and D. J. Kane, "Single-Shot Measurement of the Amplitude and Phase of Ultrashort Laser Pulses in the Violet," *Opt. Lett* **20**, 70–72 (1995).

60. D. J. Kane, A. J. Taylor, R. Trebino, and K. W. DeLong, "Single-Shot Measurement of the Intensity and Phase of a Femtosecond UV Laser Pulse Using Frequency-Resolved Optical Gating," *Opt. Lett* **19**, 1061–1063 (1994).

61. K. Michelmann, T. Feurer, R. Fernsler, and R. Sauerbrey, "Frequency Resolved Optical Gating in the UV Using the Electronic Kerr Effect," *Appl Phys B (Lasers and Optics)* **B63**, 485–489 (1996).

62. S. Backus, J. Peatross, M. M. Murnane, and H. C. Kapteyn, "16 fs Ultraviolet Pulse Generation in Air: 1 µJ Pulses at 257 nm at 1 kHz," *Opt. Lett* **21**, 665–667 (1996).

63. I. Thomann, A. Bahabad, X. Liu, R. Trebino, M. M. Murnane, and H. C. Kapteyn, "Characterizing Isolated Attosecond Pulses from Hollow-Core Waveguides Using Multi-Cycle Driving Pulses," *Opt. Expr.* **17**, 4611–4633 (2009).

64. F. Quere, Y. Mairesse, and J. Itatani, "Temporal Characterization of Attosecond XUV Fields," *J. Modern Opt.* **52**, 339–361 (2005).

65. P. Bowlan and R. Trebino, "Complete Single-Shot Measurement of Arbitrary Nanosecond Laser Pulses in Time," *Opt. Expr.* **19**, 1367–1377 (2011).

66. S.-D. Yang, A. M. Weiner, K. R. Parameswaran, and M. M. Fejer, "Ultrasensitive Second-Harmonic Generation Frequency-Resolved Optical Gating by a Periodically Poled LiNbO3 Waveguides at 1.5 µm," *Opt. Lett* **30**, 2164–2166 (2005).

67. J. Zhang, A. P. Shreenath, M. Kimmel, E. Zeek, R. Trebino, and S. Link, "Measuring of the Intensity and Phase of Attojoule Femtosecond Light Pulses Using Optical-Parametric-Amplification Cross-Correlation Frequency-Resolved Optical Gating," *Opt. Expr.* **11**, 601–609 (2003).

68. X. Gu, L. Xu, M. Kimmel, E. Zeek, P. O'Shea, A. P. Shreenath, R. Trebino, and R. S. Windeler, "Frequency-Resolved Optical Gating and Single-Shot Spectral Measurements Reveal Fine Structure in Microstructure-Fiber Continuum," *Opt. Lett.* **27**, 1174–1176 (2002).

69. A. Brun, P. Georges, G. LeSaux, and F. Salin, "Single-Shot Characterization of Ultrashort Light Pulses," *J. Phys. D.* **24**, 1225–1233 (1991).

70. S. Szatmári, F. P. Schäfer, and J. Jethwa, "A Single-Shot Autocorrelator for the Ultraviolet with a Variable Time Window," *Rev. Sci. Inst.* **61**, 998–1003 (1990).

71. G. Stibenz and G. Steinmeyer, "Interferometric Frequency-Resolved Optical Gating," *Opt. Exp.* **13**, 2617–2626 (2005).

72. S. Akturk, C. D'Amico, and A. Mysyrowicz, "Measuring Ultrashort Pulses in the Single-Cycle Regime Using Frequency-Resolved Optical Gating," *J. Opt. Soc. Amer. B* **25**, A63–A69.

73. L. Xu, E. Zeek, and R. Trebino, "Simulations of Frequency-Resolved Optical Gating for Measuring Very Complex Pulses," *J. Opt. Soc. Am. B* **25**, A70–A80 (2008).

74. M. Rhodes, Z. Guang, and R. Trebino, "Unstable Multiple Pulsing Can Be Invisible to Ultrashort Pulse Measurement Techniques," *Appl. Sci.* **7**, 40–54 (2017).

75. M. Rhodes, M. Mukhopadhyay, J. Birge, and R. Trebino, "Coherent Artifact Study of Two-Dimensional Spectral Shearing Interferometry," *J. Opt. Soc. Am. B* **32**, 1881–1888 (2015).

76. M. Rhodes, G. Steinmeyer, J. Ratner, and R. Trebino, "Pulse-Shape Instabilities and their Measurement," *Laser & Photon Rev* **7**, 557–565 (2013).

77. M. Rhodes, G. Steinmeyer, and R. Trebino, "Standards for Ultrashort-Laser-Pulse-Measurement Techniques and their Consideration for Self-Referenced Spectral Interferometry," *Appl. Opt.* **53**, D1–D11 (2014).

78. J. Ratner, G. Steinmeyer, T. C. Wong, R. Bartels, and R. Trebino, "The Coherent Artifact in Modern Pulse Measurements," *Opt. Lett.* **37**, 2874–2876 (2012).

79. S. Linden, H. Giessen, and J. Kuhl, "XFROG-a New Method for Amplitude and Phase Characterization of Weak Ultrashort Pulses," *Physica Status Solidi B* **206**, 119–124 (1998).

80. D. Reid, P. Loza-Alvarez, C. Brown, T. Beddard, and W. Sibbett, "Amplitude and Phase Measurement of Mid-Infrared Femtosecond Pulses by Using Cross-Correlation Frequency-Resolved Optical Gating," *Opt. Lett.* **25**, 1478–1480 (2000).

81. B. Richman, M. Krumbügel, and R. Trebino, "Temporal Characterization of Mid-IR Free-Electron-Laser Pulses by Frequency-Resolved Optical Gating," *Opt. Lett.* **22**, 721–723 (1997).

82. A. Lanin, A. Voronin, A. Fedotov, and A. Zheltikov, "Time-Domain Spectroscopy in the Mid-Infrared," *Sci Rep* **4**, 6670 (2014).

83. P. O'Shea, M. Kimmel, X. Gu, and R. Trebino, "Highly Simplified Device for Ultrashort-pulse Measurement," *Opt. Lett* **26**, 932–934 (2001).

84. C. Radzewicz, P. Wasylczyk, and J. S. Krasinski, "A Poor Man's FROG," *Opt. Commun.* **186**, 329–333 (2000).

85. A. G. Akmanov, A. I. Kovrigin, and N. K. Podsotskaya, Frequency Discrimination in Laser Radiation, *Radio Eng. Electron. Phys,* **14**, 1315–1317 (1969).

86. D. H. Auston, Nonlinear spectroscopy of picosecond pulses, *Opt. Commun.* **3**, 272–276 (1971).

87. S. Akturk, M. Kimmel, P. O'Shea, and R. Trebino, "Extremely Simple Device for Measuring 20-fs Pulses," *Opt. Lett.* **29**, 1025–1027 (2004).

88. S. Akturk, M. Kimmel, P. O'Shea, and R. Trebino, "Measuring Pulse-Front Tilt in Ultrashort Pulses Using GRENOUILLE," *Opt. Expr.* **11**, 491–501 (2003).

89. S. Akturk, M. Kimmel, P. O'Shea, and R. Trebino, "Measuring Spatial Chirp in Ultrashort Pulses Using Single-Shot Frequency-Resolved Optical Gating," *Opt. Expr.* **11**, 68–78 (2003).

90. H. Valtna-Lukner, F. Belli, A. Ermolov, F. Kottig, K. F. Mak, F. Tani, J. C. Travers, and P. S. J. Russell, "Extremely Broadband Single-Shot Cross-Correlation Frequency-Resolved Optical Gating Using a Transient Grating as Gate and Dispersive Element," *Rev. Sci. Instrum.* **88**, 073106 (2017).

91. R. Trebino and A. E. Siegman, "Frequency Bandwidths in Nondegenerate N-Wave-Mixing Interactions and Induced-Grating Geometries," *Opt. Commun.* **56**, 297–302 (1985).

92. T. C. Wong and R. Trebino, "Single-Frame Measurement of Complex Laser Pulses Tens of Picoseconds Long Using Pulse-Front Tilt in Cross-Correlation Frequency-Resolved Optica Gating," *J. Opt. Soc. Am. B* **30**, 2781–2786 (2013).
93. T. C. Wong, J. Ratner, V. Chauhan, J. Cohen, P. M. Vaughan, L. Xu, A. Consoli, and R. Trebino, "Simultaneously Measuring Two Ultrashort Laser Pulses on a Single-Shot Using Double-Blind Frequency-Resolved Optical Gating," *J. Opt. Soc. Am. B* **29**, 1237–1244 (2012).
94. T. C. Wong, J. Ratner, and R. Trebino, "Simultaneous measurement of two different-color ultrashort pulses on a single shot," *J. Opt. Soc. Am. B* **29**, 1889–1893 (2012).
95. Z. Wang, E. Zeek, R. Trebino, and P. Kvam, "Determining Error Bars in Measurements of Ultrashort Laser Pulses," *J. Opt. Soc. Am. B* **20**, 2400–2405 (2003).
96. Z. Wang, E. C. Zeek, R. Trebino, and P. Kvam, "Beyond Error Bars: Understanding Uncertainty in Ultrashort-Pulse Measurements in the Presence of Ambiguity," *Opt. Expr.* **11**, 3518–3527 (2003).
97. C. Froehly, A. Lacourt, and J. C. Vienot, "Time Impulse Response and Time Frequency Response of Optical Pupils," *Nouvelle Revue D'Optique* **4**, 183–196 (1973).
98. D. N. Fittinghoff, J. L. Bowie, J. N. Sweetser, R. T. Jennings, M. A. Krumbügel, K. W. DeLong, R. Trebino, and I. A. Walmsley, "Measurement of the Intensity and Phase of Ultraweak, Ultrashort Laser Pulse," *Opt. Lett* **21**, 884–886 (1996).
99. B. Alonso, M. Miranda, Í. J. Sola, and H. Crespo, "Spatiotemporal Characterization of Few-Cycle Laser Pulses," *Opt. Exp.* **20**, 17880–17893 (2012).
100. C. Dorrer and M. Joffre, "Characterization of the Spectral Phase of Ultrashort Light Pulses," *Comptes Rendus de l'Academie des Sciences, Serie IV (Physique, Astrophysique)* **2**, 1415–1426 (2001).
101. P. Bowlan, U. Fuchs, R. Trebino, and U. D. Zeitner, "Measuring the Spatiotemporal Electric Field of Tightly Focused Ultrashort Pulses with Sub-Micron Spatial Resolution," *Opt. Expr.* **16**, 13663–13675 (2008).
102. P. Bowlan, P. Gabolde, M. A. Coughlan, R. Trebino, and R. J. Levis, "Measuring the Spatio-Temporal Electric Field of Ultrashort Pulses with High Spatial and Spectral Resolution," *J. Opt. Soc. Am. B* **25**, A81–A92 (2008).
103. P. Bowlan, P. Gabolde, A. Shreenath, K. McGresham, R. Trebino, and S. Akturk, "Crossed-Beam Spectral Interferometry: A Simple, High-Spectral-Resolution Method for Completely Characterizing Complex Ultrashort Pulses in Real Time," *Opt. Expr.* **14**, 11892–11900 (2006).
104. P. Bowlan, P. Gabolde, and R. Trebino, "Directly Measuring the Spatio-Temporal Electric Field of Focusing Ultrashort Pulses," *Opt. Expr.* **15**, 10219–10230 (2007).
105. J. P. Geindre, P. Audebert, A. Rousse, F. Falliés, J. C. Gauthier, A. Mysyrowicz, A. Dos Santos, G. Hamoniaux, and A. Antonetti, "Frequency-Domain Interferometer for Measuring the Phase of a Femtosecond Pulse Probing a Laser-Produced Plasma," *Opt. Lett.* **19**, 1997–1999 (1994).
106. S. D. Gennaro, Y. Sonnefraud, N. Verellen, P. Van Dorpe, V. V. Moshchalkov, S. A. Maier, and R. F. Oulton, "Spectral Interferometric Microscopy Reveals Absorption by Individual Optical Nanoantennas from Extinction Phase," *Nat. Commun.* **5**, 3748 (2014).
107. P. Bowlan and R. Trebino, "Extreme Light Pulse-Front Tilt and its Application to Single Shot Measurement of Picosecond to Nanosecond Laser Pulses," U.S. Patent No. 8,953,166 (2012).
108. J. Cohen, P. Bowlan, V. Chauhan, and R. Trebino, "Measuring Temporally Complex Ultrashort Pulses Using Multiple-Delay Crossed-Beam Spectral Interferometry," *Opt. Expr.* **18**, 6583–6597 (2010).
109. J. Cohen, P. Bowlan, V. Chauhan, P. Vaughan, and R. Trebino, "Measuring Extremely Complex Pulses with Time-Bandwidth Products Exceeding 65,000 Using Multiple-Delay Crossed-Beam Spectral Interferometry," *Opt. Expr.* **18**, 24451–24460 (2010).
110. J. Cohen, P. Bowlan, and R. Trebino, "Extending Femtosecond Metrology to Longer, More Complex Laser Pulses in Time and Space," *IEEE J. Sel. Top. Quant. Electron.* **18**, 218–227 (2012).
111. J. Cohen, P. Bowlan, V. Chauhan, P. Vaughan, and R. Trebino, "Single-Shot Multiple-Delay Crossed-Beam Spectral Interferometry for Measuring Extremely Complex Pulses," *Opt. Commun.* **284**, 3785–3794 (2011).
112. C. Iaconis and I. A. Walmsley, "Spectral Phase Interferometry for Direct Electric-Field Reconstruction of Ultrashort Optical Pulses," *Opt. Lett.* **23**, 792–794 (1998).
113. T. Oksenhendler, S. Coudreau, N. Forget, V. Crozatier, S. Grabielle, R. Herzog, O. Gobert, and D. Kaplan, "Self-Referenced Spectral Interferometry," *Appl. Phys. B: Lasers Opt.* **99**, 7–12 (2010).
114. M. Takeda, H. Ina, and S. Kobayashi, "Fourier-Transform Method of Fringe-Pattern Analysis for Computer-Based Topography and Interferometry," *J. Opt. Soc. Am.* **72**, 156 (1982).
115. S. L. Hahn, *Hilbert Transforms in Signal Processing* (Artech House, Boston, 1996).
116. J. Bethge and G. Steinmeyer, "Numerical Fringe Pattern Demodulation Strategies in Interferometry," *Rev. Sci. Instrum.* **79**, 073102 (2008).
117. G. Stibenz and G. Steinmeyer, "Optimizing Spectral Phase Interferometry for Direct Electric-Field Reconstruction," *Rev. Sci. Instrum.* **77**, 073105 (2006).
118. P. Baum, S. Lochbrunner, and E. Riedle, "Zero-Additional-Phase SPIDER: Full Characterization of Visible and Sub-20-fs Ultraviolet Pulses," *Opt. Lett.* **29**, 210–212 (2004).
119. T. Witting, D. R. Austin, and I. A. Walmsley, "Improved ancilla preparation in spectral shearing interferometry for accurate ultrafast pulse characterization," *Opt. Lett.* **34**, 881–883 (2009)
120. J. R. Birge, R. Ell, and F. X. Kärtner, "Two-Dimensional Spectral Shearing Interferometry for Few-Cycle Pulse Characterization," *Opt. Lett.* **31**, 2063–2065 (2006).
121. E. M. Kosik, A. S. Radunsky, I. A. Walmsley, and C. Dorrer, "Interferometric Technique for Measuring Broadband Ultrashort Pulses at the Sampling Limit," *Opt. Lett.* **30**, 326–328 (2005).
122. J. Hyyti, E. Escoto, G. Steinmeyer, and T. Witting, "Interferometric Time-Domain Ptychography for Ultrafast Pulse Characterization," *Opt. Lett.* **42**, 2185–2188 (2017).
123. T. Balciunas, C. Fourcade-Dutin, T. W. G. Fan, A. A. Voronin, A. M. Zheltikov, F. Gerome, G. G. Paulus, A. Baltuska, and F. Benabid, "A Strong-Field Driver in the Single-Cycle Regime Based on Self-Compression in a Kagome Fibre," *Nature Commun.* **6**, 6117 (2015).

124. S. Akturk, X. Gu, P. Bowlan, and R. Trebino, "Spatio-Temporal Couplings in Ultrashort Laser Pulses," *J. Opt.* **12**, 093001 (2010).
125. Z. Bor and Z. L. Horvath, "Distortion of Femtosecond Pulses in Lenses. Wave Optical Description," *Opt Commun.* **94**, 249–258 (1992).
126. Z. L. Horvath and Z. Bor, "Focusing of Femtosecond Pulses Having Guassian Spatial-Distribution," *Opt. Comm.* **100**, 6–12 (1993).
127. Z. L. Horvath and Z. Bor, "Behavior of Femtosecond Pulses on the Optical-Axis of a Lens - Analytical Description," *Opt. Comm.* **108**, 333–342 (1994).
128. Z. L. Horvath, K. Osvay, and Z. Bor, "Dispersed Femtosecond Pulses in the Vicinity of Focus," *Opt. Comm.* **111**, 478–482 (1994).
129. M. Kempe and W. Rudolph, "Impact of Chromatic and Spherical Aberration on the Focusing of Ultrashort Light Pulses by Lenses," *Opt. Lett.* **18**, 137–139 (1993).
130. M. Kempe and W. Rudolph, "Femtosecond Pulses in the Focal Region of Lenses," *Phys. Rev. A* **48**, 4721–4729 (1993).
131. U. Fuchs, U. D. Zeitner, and A. Tunnermann, "Ultra-Short Pulse Propagation in Complex Optical Systems," *Opt. Expr.* **13**, 3852–3861 (2005).
132. T. Tanabe, H. Tanabe, Y. Teramura, and F. Kannari, "Spatiotemporal Measurements Based on Spatial Spectral Interferometry for Ultrashort Optical Pulses Shaped by a Fourier Pulse Shaper," *J. Opt. Soc. Am. B* **19**, 2795–2802 (2002).
133. Y. Teramura, M. Suekuni, and F. Kannari, "Two-Dimensional Optical Coherence Tomography Using Spectral Domain Interferometry," *J. Opt. A: Pure Appl. Opt.* **2**, 21–26 (2000).
134. M. Lõhmus, P. Bowlan, P. Piksarv, H. Valtna-Lukner, R. Trebino, and P. Saari, "Diffraction of Ultrashort Optical Pulses from Circularly Symmetric Binary Phase Gratings," *Opt. Lett.* **37**, 1238–1240 (2012).
135. P. Saari, P. Bowlan, H. Valtna-Lukner, M. Lõhmus, P. Piksarv, and R. Trebino, "Basic diffraction phenomena in time domain," *Opt. Express*, **18**, 11083–11088 (2010).
136. P. Bowlan and R. Trebino, "Extreme Pulse-Front Tilt from an Etalon," *J. Opt. Soc. Am. B* **27**, 2322–2327 (2010).
137. D. Meshulach, D. Yelin, and Y. Silberberg, "Real-Time Spatial–Spectral Interference Measurements of Ultrashort Optical Pulses," *J. Opt. Soc. Am. B* **14**, 2095–2098 (1997).
138. A. Börzsönyi, A. P. Kovács, M. Görbe, and K. Osvay, "Advances and Limitations of Phase Dispersion Measurement by Spectrally and Spatially Resolved Interferometry," *Opt. Commun.* **281**, 3051–3061 (2008).
139. D. J. McCabe, A. Tajalli, D. R. Austin, P. Bondareff, I. A. Walmsley, S. Gigan, and B. Chatel, "Spatio-Temporal Focusing of an Ultrafast Pulse through a Multiply Scattering Medium," *Nat. Commun.* **2**, 447 (2011).
140. P. Bowlan, H. Valtna-Lukner, M. Lõhmus, P. Piksarv, P. Saari, and R. Trebino, "Measurement of the Spatio-Temporal Field of Ultrashort Bessel-X Pulses," *Opt. Lett.* **34**, 2276–2278 (2009).
141. M. Lõhmus, P. Bowlan, R. Trebino, H. Valtna-Lukner, P. Piksarv, and P. Saari, "Directly Recording Diffraction Phenomena in the Time Domain," *Lith J. Phys.* **50**, 69–74 (2010).
142. E. Betzig, M. Isaacson, and A. Lewis, "Collection Mode Near-Field Scanning Optical Microscopy," *Appl. Phys. Lett.* **51**, 2088–2090 (1987).
143. J. Trägårdh and H. Gersen, "Combining Near-Field Scanning Optical Microscopy with Spectral Interferometry for Local Characterization of the Optical Electric Field in Photonic Structures," *Opt. Expr.* **21**, 16629–16638 (2013).
144. P. Bowlan and R. Trebino, "Using Phase Diversity for the Measurement of the Complete Spatiotemporal Electric Field of Ultrashort Laser Pulses," *J. Opt. Soc. Am. B* **29**, 244–248 (2012).
145. P. Saari and K. Reivelt, "Evidence of X-Shaped Propagation-Invariant Localized Light Waves," *Phys. Rev. Lett.* **79**, 4135 (1997).
146. I. Alexeev, K. Kim, and H. Milchberg, "Measurement of the Superluminal Group Velocity of an Ultrashort Bessel Beam Pulse," *Phys. Rev. Lett.* **88**, 073901 (2002).
147. H. Valtna-Lukner, P. Bowlan, M. Lõhmus, P. Piksarv, R. Trebino, and P. Saari, "Direct Spatiotemporal Measurements of Accelerating Ultrashort Bessel-Type Light Bullets," *Opt. Expr.* **17**, 14948–14955 (2009).
148. P. Piksarv, H. Valtna-Lukner, A. Valdmann, M. Lõhmus, R. Matt, and P. Saari, "Temporal Focusing of Ultrashort Pulsed Bessel Beams into Airy–Bessel Light Bullets," *Opt. Expr.* **20**, 17220–17229 (2012).
149. J. Extermann, L. Bonacina, F. Courvoisier, D. Kiselev, Y. Mugnier, R. Le Dantec, C. Galez, and J.-P. Wolf, "Nano-FROG: Frequency Resolved Optical Gating by a Nanometric Object," *Opt. Expr.* **16**, 10405–10411 (2008).
150. P. Gabolde and R. Trebino, "Single-Shot Measurement of the Full Spatiotemporal Field of Ultrashort Pulses with Multispectral Digital Holography," *Opt. Expr.* **14**, 11460–11467 (2006).
151. P. Gabolde and R. Trebino, "Single-Frame Measurement of the Complete Spatio-Temporal Intensity and Phase of Ultrashort Laser Pulse(s) Using Wavelength-Multiplexed Digital Holography," *J. Opt. Soc. Am. B* **25**, A25–A33 (2008).
152. Z. Guang, M. Rhodes, M. Davis, and R. Trebino, "Complete Characterization of a Spatiotemporally Complex Pulse by an Improved Single-Frame Pulse-Measurement Technique," *J. Opt. Soc. Am. B* **31**, 2736–2743 (2014).
153. Z. Guang, M. Rhodes, and R. Trebino, "Numerical Simulations of Holographic Spatiospectral Traces of Spatiotemporally Distorted Ultrashort Laser Pulses," *Appl. Opt.* **54**, 6640–6651 (2015).
154. P. Zhu, R. Jafari, T. Jones, and R. Trebino, "Complete Measurement of Spatiotemporally Complex Multi-Spatial-Mode Ultrashort Pulses from Multimode Optical Fibers Using Delay-Scanned Wavelength-Multiplexed Holography," *Opt. Expr.* **25**, 24015–24032 (2017).
155. Z. Guang, M. Rhodes, and R. Trebino, "Measurement of the Ultrafast Lighthouse Effect Using a Complete Spatiotemporal Pulse-Characterization Technique," *J. Opt. Soc. Am. B* **33**, 1955–1962 (2016).
156. M. Rhodes, Z. Guang, J. Pease, and R. Trebino, "Visualizing Spatiotemporal Pulse Propagation: First-Order Spatiotemporal Couplings in Laser Pulses," *Appl. Opt.* **56**, 3024–3034 (2017).

Index

Note: **Bold** page numbers refer to tables and *italic* page numbers refer to figures

aberrations, effect of 395–6
absorption saturation effects 279
ACC *see* automatic current control (ACC)
accessible emission limit (AEL) 195
accident reports 202–3
achromat 211–13
acoustic modulation 118
acousto-optic deflection 223, *223*
acousto-optic deflectors 253, 455–6
acousto-optic modulators 118, 220–1
acousto-optic scanners 223–4
acousto-optic programmable dispersive filter (AOPDF) 326
acousto-optic Q-switches 118, 165, 252
active area sizes, shapes and apertures 474
active mode-locking technique 167, 259–61, 267, 291, 292, 302
actuators 237
adaptive mirrors 227–8, 398
'adaptive optics' 219
adaptive optical wavefront modulators 225
adaptive optics techniques 227
additive-pulse mode-locking 298
advanced positioning systems and optical set-ups 448
AEL *see* accessible emission limit (AEL)
A-FPSA *see* anti-resonant Fabry–Perot saturable absorber (A-FPSA)
airy disc intensity distribution 155, *156*
Alexandrite 162
all-optical method 316–17
allowed transitions 7
alternative measurement methods 464
amplified spontaneous emission (ASE) 328, 329, 364
amplified ultrashort pulses 268
amplifier
 energy extraction from CPA 326–7
 ideal operational 187
 laser 14
 noise 186–7
 to oscillator 33–4
amplifier–oscillator connection 4–5
amplitude
 coupling coefficient 126
 drift 168
 modulation 219–22
 noise 168
amplitude–wavefront representation 79
analogue-mode operation 179
angular acceptance bandwidths 350
angular field considerations 401
angular spectrum
 decomposition method 79
 propagation 79–81
anomalous dispersion 438
anti-reflection coatings 388
anti-resonant Fabry–Perot saturable absorber (A-FPSA) 263

AO cavity dumping configurations *166*
AOPDF *see* acousto-optic programmable dispersive filter (AOPDF)
APC *see* automatic power control (APC)
APDs *see* avalanche photodiodes (APDs)
aperture length 350
APM 265
application-specific integrated circuit (ASIC) chips 476
APT *see* attosecond pulse trains (APT)
arbitrary elliptical polarization 65
articulated arms 398
artificial diamond 401
artificial saturable absorbers 298–9
ASE *see* amplified spontaneous emission (ASE)
ASIC chips *see* application-specific integrated circuit (ASIC) chips
aspherical mirrors, general remarks on 403
astigmatic beams, propagation of 464–5
astigmatic wavefront distortion 163
astigmatism in collimators 163
asymmetric laser cavity 48, *48*
atomic systems 5
attenuation 388
attosecond metrology methods 307–8, 317, **318**
 all-optical method 316–17
 Complete Reconstruction of Attosecond Beating (CRAB) 313–14, 316
 general principles of attosecond pulse characterization 309–10
 isolated attosecond pulses 312
 momentum streaking 312–13, 316
 Phase Retrieval by Omega Oscillation Filtering (PROOF) 314–16
 principle component generalized projection algorithm 317–18
 Reconstruction of Attosecond Beating by Interference of Two photon Transitions (RABBITT) 310–12, 316
 second-order XUV AC/FROG 310
 some experimental remarks 317
attosecond pulse characterization 309–10
attosecond pulse generation 281
attosecond pulse trains (APT) 309
autocorrelation method 255
 of complex pulses 492
 conclusions 498
 measurements 275
 of noisy pulse trains 492–3
 and spectrum 495–6
 theorem 489
autocorrelation Theorem 490
autocorrelator 274
automatic current control (ACC) 168
automatic power control (APC) 168
"autospectrogram" 499
avalanche breakdown 149, 180

avalanche gain 179
avalanche photodiodes (APDs) 179–81, 469

background-free autocorrelation 496
background-limited infrared photodetectors (BLIP) 474
back-tracing beams 154
'bands' 6
bandwidth-related noise reduction methods 187–8
beam
 adaptation 215
 aperturing requirements and effects 396
 axis coordinate system 461, *461*
 characterization and pulse measurement 254–6
 concentration 155
 controlling size and shape of 224–5
 delivery systems 388, 445, 446
 diameter 461–2
 divergence 46
 division and combination 440–1
 homogenization system 410, *410*
 manipulation 385
 requirements at workpiece 387–8
 scanning and positioning 222–4
 transformation by lens 464
 transport 397–8
 types 461
 visualization techniques 202
beam-conditioning optics 219
beam propagation
 alternative measurement methods 464
 beam diameter 461–2
 beam transformation by lens 464
 beam types 461
 influences on 448
 practical measurements 462–3
 propagation characteristics 463–4
 propagation of astigmatic beams 464–5
beam propagation methods (BPM) 93
'beam propagation ratio' 464
beam quality 107–8, 328–7, 386–7
 factor 464
 profile 441
Beer's law 84
Bessel functions 424, *424*
Bessel-X pulse 527
biaxial crystal 64, *64*
bimorph mirrors 228
biomedical imaging 277
'birefringence' 64
birefringence axes 64
birefringent crystals 368
birefringent filters 163
birefringent phase-matching (BPM) 337–40, 368
black body radiation 11, *11*
blackbody spectral radiance *184*

537

Blind Deconvolution 512
BLIP *see* background-limited infrared photodetectors (BLIP)
bolometers 182
Boltzmann's constant 176
"bootstrap" method 513
Bose–Einstein statistics 186
BPM *see* Beam propagation methods (BPM); birefringent phase-matching (BPM)
Bragg quarter-wave mirror structures 271
Bragg grating 221
Bragg regime 455, *455*
'Brewster window' 210
Brewster's angle 210
broad spectral coverage 375–6
broad-band signals 78
broader bandwidth 364

camera-based systems 484
'Canada balsam' 66
carbon nanotubes 298
  graphene 297–8
carrier-envelope offset frequency 295
carrier-envelope phase (CEP) 309, 364
Cartesian coordinate system 106, 234, 447
cascaded pumping approach 354
catastrophic self-focusing 270
cavity dumping 165–6, 253, *253*
CCD *see* charge-coupled devices (CCD)
cemented lenses 213
central momentum approximation 314
CEP *see* carrier-envelope phase (CEP)
chalcogenide 439
  fibres 437
charge-coupled devices (CCD) 470
chemical vapour deposition (CVD) process 401
chirp compensation techniques 302
'chirped mirrors' 272
chirped multilayer dielectric coatings 272
chirped pulse amplification (CPA) 268, 302, 321
  amplification to PW level power 326–8
  basics 322–3
  dispersion control 323–6
  high-intensity requirements 328–9
  technique 363
chirped radar 321
chromatic dispersion 268, 299, 428–9
chromium with chrome oxide (Cr/CrOx) 474
circular birefringence 68, 72
class-based user guidance 200, **201**
classical electromagnetic theory 59
classical electron oscillator model 31
classical oscillator explanation for stimulated emission 31–2
cleavage planes 63
closed-loop system 229
CMOS *see* complementary metal oxide semiconductor (CMOS)
CO_2 lasers
  beam integration for homogeneous illumination 403
  operation of 243–4
  radiation, lenses for 399–401
'Coefficient of finesse' 56
coefficients
  Einstein A and B 11–13

third-order non-linear optical 138, **138**, 139, *139*, 140, **141**
threshold gain 36
'coherence' function 58
coherence length 47
coherence measurement 485–6
coherence time 47
coherent Gaussian beams 153
coherent-wave interference 52–3
coil galvanometers 451
colliding-pulse mode-locking (CPM) 262, 263
colour correction 405
colour-centre lasers 264–5
colour-sensitive detectors 472–3
coma-corrected optics 403
comb function 81–2
commercial helium–neon lasers 161
commercial numerical modelling software 95
commercial PbSe-based detectors function 474
common solid-state laser sources 267
communication of specifications 216
complementary metal oxide semiconductor (CMOS) 470–1
Complete Reconstruction of Attosecond Beating (CRAB) 313–14, 316
complete-intensity-and-phase measurement techniques 498
complex susceptibility, gain and absorption 30–1
computer-numerically-controlled machining 236
concave spherical curvature 44
condensed matter, energy bands in 18
conduction
  band 6
  band filling 141
confocal interferometer 116, *116*
confocal unstable resonators 112
continuous mathematical formulation 80
continuous-wave OPO devices 344, 352–3
contrast preservation 364
control measures, application of 202
conventional electroless plating technique 440
'conventional' electronics 4
conventional lenses 386
conventional monochromatic source 47
conventional optical design 91
conventional two-mirror cavity 159
conventions and notation 99
conversion efficiency 345–7
copper vapour laser (CVL) 413, 446
coupled mode equations 431
coupled non-linear wave equations 366
coupled wave equations 334, 335
coupling efficiency 126
coupling length 430
coupling loss 126–7
CPA *see* Chirped Pulse Amplification (CPA)
CPM *see* colliding-pulse mode-locking (CPM)
CRAB *see* Complete Reconstruction of Attosecond Beating (CRAB)
Cr/CrOx *see* Chromium with Chrome oxide (Cr/CrOx)
Cr-ion-based lasers 267
Cr-ion-doped crystals 267
cross-correlation 496–8
crossed-beam spatial interferometry 516, *516, 517*

crossed-beam spectral interferometry 516–17
crystal optics 63–5
crystalline optical fibres 439
cubic spectral phase variations 270
curvature mirrors 228
curved mirror–waveguide combination *126*
CVD process *see* chemical vapour deposition (CVD) process
CVL *see* copper vapour laser (CVL)
cylindrical coordinates 106–7

damage threshold 295
deconvolution factor 275
deformable mirrors 228, *228*
degrees of freedom (DOFs) 234
density of states 11
detectivity (D) 173
detectors technology and applications 476–7
dialyte 211–13
dielectric waveguides 87–9
diffraction-limited lens 155
diffractive beam steering 224
diffractive optical element (DOE) 388, 442
diode lasers 163–4, 247, 383
  and applications, classes of 405–7
  arrays 408–9
  optical output properties 407
  optics for 405–9
diode-laser-pumped solid-state lasers 247
diode-pumped operation 242–3
direct electronic measurements 273–4
direct (single cycle) field ionization 149
direct optimization methods 229
direct pulse shape measurement 273
direct-gap semiconductors 173
dispersion 271, 429
  control 323–30
  management 374
  for optical elements in amplifier 325–6
  sources of 270–1
dispersion-flattened fibres 429
dispersion-propagation time-resolved optical gating (DP-TROG) 277
dispersion-shifted fibres 429
distant mirrors 130–1
dither 452
DOE *see* diffractive optical element (DOE)
DOFs *see* degrees of freedom (DOFs)
doped-insulator solid-state lasers 299
Doppler broadening 16–18, 26, 27
  distribution of Lorentzian lineshapes 18, *19*
double refraction 64
double-blind FROG 512
doubly resonant oscillator (DRO) 343, 345, 346
1D-positioning systems, comparison of 449
DP-TROG *see* dispersion-propagation time-resolved optical gating (DP-TROG)
DRO *see* doubly resonant oscillator (DRO)
Drude model 140
2DSI *see* Two-Dimensional Spectral Shearing Interferometry (2DSI)
dual case I waveguide lasers 129
dual-beam interference 53
dual-beam interferometer 53
duration-bandwidth product 269
dye lasers 162, 262–3, 298, 321
dynamic gain saturation 261
dynamic loss saturation 261

Index 539

dynamic resolution 456
dynode materials 174

effective index method 421
'eigenstates' 64
Einstein A and B coefficients 104
    relationship between 11–13
Einstein relations 12, 13, 20
Einstein's equation 310
electromagnetic actuators 237
electromagnetic radiation 51, 463, *463*
electromagnetic wave 52, *52*
electron oscillator model 30, *30*
    radiative transition 28–33
electron spectrometer 311
'electron–hole pair' 176
electro-optic devices 252, 220
electro-optic effect 72, *72*, 72–3
electro-optic modulators 219–20
electro-optic q-switches 118, 165, 252
electro-optic scanners 223–4
electrostatic membrane mirrors 228
electrostriction 137
elementary lens forms 211–14
elementary trigonometry 52
ellipse 62–3
elliptical polarization state 65, 69
energy bands 6
    condensed matter 18
energy density
    and intensity, relation between 8–10
    spectrum, monochromatic 9, *9*
energy extraction from CPA amplifier 326–7
'energy level' 5
energy limitations 327
engineering safety features on laser
    products 200
EO cavity dumping configurations *166*
Epsilon-near-zero non-linearities 142
epsilon-near-zero wavelength 142
error bars 513
error budget example 234, **235**
etalon-based lateral shearing
    interferometer 485, *485*
etalons 117, 160
evanescent field 418
excess noise factor 185
excited states 5
exposure duration assessing 193
extended-source interference 58, *59*
extra-cavity adaptive optics systems 227
eye
    principal components and operation of 190
    protection 202

fabrication 438–40
Fabry–Perot etalon 55, 160
Fabry–Perot interferometer 41, 55, 57, 146
Fabry–Perot laser cavity 38, 46
Fabry–Perot resonator system 33, 34, *34*
Faraday effect 73, 74, 222
Faraday magneto-optic effect 73, *73*, 74
Faraday magneto-optic isolator 74
far-field divergence 463
fast Fourier transform (FFT) 78, 79, 318
fast photodiode detector 261
fast response photodetectors 471
fast saturable absorbers 261

FDP *see* finite-difference propagator (FDP)
FDTD *see* finite difference time domain
    (FDTD) 423
FEA *see* finite-element analysis (FEA)
feedback electronic system 453
femtosecond lasers 300
femtosecond OPO devices 349, 356–9
femtosecond pulse shaping 273, *273*
femtosecond x-ray diffraction 278
Fermat's principle 145
Fermi energy 173
Fermi's golden rule 149
few-cycle pulse duration 376
few-mode theory 127
FFT *see* fast Fourier Transform (FFT)
fibre lasers 118, 265–6, 298
fibre operation 435–8
fibre optic beam delivery
    beam division and combination 440–5
    fabrication 438–40
    fibre operation 435–8
    implementation 440
    limitations 441–2
fibre optic beam splitters 440
Fibre optic delivery system 224, *386*
fibre optic pulse compression 322
fibre optics 385, 442
fibre-based scanning spatiotemporal pulse
    measurement 525–6
fibre-coupled lasers, hybrids and detectors
    for 475–6
field autocorrelation 489
figure-of eight laser 266
filamentation 145
finite-difference method 79
finite-difference propagator (FDP) 79
finite difference time domain (FDTD) 423
finite-element analysis (FEA) 237
first-order theory 126
flashlamp-pumped rods 243
fluoride 439
    fluoride fibres 437
flying optics 448
focal plane arrays (FPAs) 181
focal spot size 395
'focused dynode chain' 175
Folded lasers 131–2
folded-waveguide CO_2 designs 132
forbidden transitions 92
"fore-runner pulse" 526
Fourier beam propagation methods 114
Fourier-filtering algorithm 518
Fourier optics 221
Fourier-transform 15, 81, 82
    autocorrelation 490
    operator 82
    pulse shaping 273
    spectrometers 489, *489*
    techniques 108, 112
Fourier-Transform Spectral Interferometry
    (FTSI) 515
Four-wave mixing (FWM) 230–1
FPAs *see* focal plane arrays (FPAs)
FRAC *see* fringe-resolved
    autocorrelation (FRAC)
Frantz–Nodvik Solution 85–6
'free spectral range' 57
free electron 149, 255, 313

free-electron metal 140
free-space beam propagation 448
free-space laser resonators 97–9, 121
    axial modes 116–17
    distortion effects 115–16
    fibre laser resonators 118
    frequency selection and frequency
        stability 117–18
    Gaussian beams 99–104
    higher-order modes of stable resonators
        105–8
    mode-matching 109–10
    plane parallel resonators 110–11
    stable resonators 104–5
    temporal resonator characteristics 118
    unstable resonators 111–15
free-space optics
    computation of laser optical systems 393–7
    general practical guidelines of optics for
        laser applications 397
    laser beam propagation and its optical
        consequences 392–3
    optics for CO_2 laser systems 397–404
    optics for excimer and other UV lasers
        409–12
    optics for lasers operating in
        visible 404–9
    optics for other laser sources 412–13
frequency and amplitude stabilization 167
frequency combs 294–5
frequency domain 310
frequency stability 99, 160–2
frequency-resolved optical gating (FROG)
    499–500
    algorithm 500–3
    beam geometries 500
    and coherent artefact 507–8
    properties of 504
    single-shot 504–5
    spectrogram 276
    two-dimensional phase-retrieval
        problem 500
Fresnel biprism 510, *510*
Fresnel condition 46
Fresnel diffraction 78
Fresnel's equations 209–11
fringe-resolved autocorrelation (FRAC) 496
FROG *see* frequency-resolved optical
    gating (FROG)
FTSI *see* Fourier-Transform Spectral
    Interferometry (FTSI)
full-width half-maximum (FWHM) 269,
    323, 368
Fundamental Theorem of Algebra 500
fundamental transverse distribution 98
fundamental *vs.* harmonic mode-locking 294
fundamental waveguide mode 128, *128*
fused silica 436
fused silica optical fibres 438–9, 442
FWHM *see* full-width half-maximum
    (FWHM)

GaAs-based devices 472
gain and non-linear media 84–6
gain narrowing 268
gain noise 185–6
gain saturation effects 268
gain-guided laser resonators 99

gain-switched gas lasers 247
gain-switching 242, 247–8, 266
gallium phosphide 473
galvanometer
　driven systems 451
　Scanners 449, 451–3
　System 224
Gantry systems 447, 448, *448, 449*
gas jets 387
gas-ion laser 159
Gaussian beam 84, 98–100, 126, 209
　characterization **101**
　diameter 156
　formula 154
　modes 121
　propagation equations 84
Gaussian beamlets 91
Gaussian dependence 24
Gaussian distribution 16, *16*
Gaussian error distribution 462
Gaussian function 17
Gaussian intensity distribution 155, *156*
Gaussian irradiance distribution 462, *462*
Gaussian modes 103, 167
　electric field 348
Gaussian optics 118
Gaussian pulse shape 256
Gaussian resonant modes 103
Gaussian spatial distribution 348
Gaussian spectrum 490
Gaussian-shaped function 16
Gauss–Laguerre modes 483
GDD *see* group delay dispersion (GDD)
Geiger efficiency 472
generalized projections (GPs) algorithm 501
Generation–Recombination (G–R) noise 185
geometrical distortions 505
geometrical rays 80
'geometrically unstable', resonators 98
germanate fibres 437
germanate glasses 439
germanium 472
Gires–Tournois (GTI) mirror 272
GPs algorithm *see* generalized projections
　　(GPs) algorithm
G–R noise *see* Generation–Recombination
　　(G–R) noise
graded-index fibres 428, 437
graded-index multi-mode fibres 423
graded-index slab guide 420, *420*
graphene-based saturable absorbers 298
grating alignment issues 326
grating vector 341
GRating-Eliminated No-nonsense Observation
　　of Ultrafast Incident Laser Light
　　E-fields (GRENOUILLE) 509–12
grinding 216
ground state 5
group delay 269
group delay dispersion (GDD) 269, 323
group-velocity dispersion 268, 271–3
GTI mirror *see* Gires–Tournois (GTI) mirror
gyroscopes 61

'half-wave' plate 65
Hankel transforms 81
hard-aperture mode-locking 263
hard-edged apertures 112–15

harmonic mode-locking 261, 290, 294
　fundamental *vs.* 294
Hartmann–Shack wavefront sensor 484
hazard classification scheme 189
heat balance equation 181
heat-affected zone 279
helium–neon gain envelope 161, *162*
helium–neon lasers 156, 161, 162, *162,* 168
hermetic packages 475
Hermite–Gaussian modes 98, 106, 107, *107,*
　108, 158
high field science 280–1
high Fresnel diffraction 79
high pulse repetition rates 300–1
high spatial frequencies 79
high-dynamic range temporal contrast 374
higher-order modes of stable resonators 105–4
higher-order transverse modes 98
high-frequency noise 167
high-frequency permittivity 140
high-gain pulsed lasers 117
high-intensity requirements 328–9
highorder harmonic generation 149–50
high-power beams 221
high-power laser systems 210
high-power operation 300
high-power picosecond oscillators 281
high-precision weighing techniques 480
high-speed scanning systems 453
hole burning 26
hollow dielectric waveguides 122
hollow glass waveguides 440
hollow sapphire fibres 438
hollow waveguide fibres 438, 441
hollow waveguide optical fibres 439–40
homogeneous broadening mechanisms 15–16
homogeneous broadening of group 18, *18*
homogeneous line broadening 14
homogeneous line shape function 20, 23, 25, 27
homogeneous media, propagation in 81–3
homogeneous systems 22–4
homogeneously broadened laser 24, 38, *39, 40*
homogeneously broadened system 21, 24
hot-electron non-linearities 140, 142, *142*
Huygens' principle 52
hybrid mode-locking 267
hybrid ray-tracing methods 91

IC-SRO *see* intra-cavity SRO (IC-SRO)
ideal monochromatic radiation 115
ideal operational amplifier 187
IFROG *see* interferometric FROG (IFROG)
impedance matching 186–7
Improved PROOF (iPROOF) 316
indirect-band gap semiconductors 471
indium antimonide-based (InSb) detectors 474
industrial laser processing 202
industrial robots 447
inertial confinement fusion 280
infinite measurement apertures 462
InGaAs detectors 474
inhomogeneous broadening 16–18
　optical frequency amplification with 21–2
inhomogeneous systems 24–8
inhomogeneously broadened 24
　amplifier 26, *27*
　laser 38, 40, *41–2*
injection locking 118, 158

injury mechanisms 190
InP-based devices 472
InSb detectors *see* indium antimonide-based
　　(InSb) detectors
intense laser pulses 280
intensity autocorrelation 274, 490–1
intensity transport equations 485
inter-band absorption 176
interference
　coherent-wave interference 52–3
　interferometers 53–7
　between partially coherent waves 57–9
　practical examples 59–62
　wave coherence 51–2
interferometers 53–7, 484–5
interferometric autocorrelation 274, 496
interferometric FROG (IFROG) 505, 506
inter-modal dispersion 427–8
internal beam delivery systems 448
internal deposition method 439
International Organization for
　　Standardization (ISO) 216
intra-band absorption 176
intra-band detectors 181
intra-band electronic transitions 181
intra-cavity adaptive optics 227
intra-cavity chromatic dispersion 299
intra-cavity DFM techniques 356
intra-cavity dispersion 299
intra-cavity mode-locked pulses 273
intra-cavity polarization 266
intra-cavity pumping approach 268
intra-cavity SRO (IC-SRO) 343
ion beams 280
ion lasers 413
iPROOF *see* Improved PROOF (iPROOF)
IR spectral range, detectors for 473–4
irradiance distribution measurement 484
ISO *see* International Organization for
　　Standardization (ISO)
isolated attosecond pulses 312

Kerr effect 72, 252, 265, 300
Kerr lens 270
Kerr lens mode-locking (KLM) 118, 293,
　　298, 356
Kerr non-linearity 299, 300
kinematic mounting 234, 237
kinematics 447–8
KLM *see* Kerr lens mode-locking (KLM)
KLM Ti:sapphire laser 357

Laguerre–Gaussian functions 157
Laguerre–Gaussian modes 106
lamb dip 41
lamp-pumped operation 241–2
laser absorbers 481
laser amplifier 14
laser applications, optics for 215–16, 397
laser beam control
　beam quality 167–8
　beam shape and astigmatism in diode
　　lasers 163–4
　Gaussian beam with simple lenses 154–7
　laser applications 153
　Q-switching, mode-locking and cavity
　　dumping 164–7
　single axial mode operation 159–60

Index 541

spatial filtering 169
transverse modes and mode control 157–9
tunable operation 162–3
laser beam management detectors
 colour-sensitive detectors 472–3
 common customizations 474–3
 detectors for IR spectral range 473–4
 detectors for UV spectral range 473
 detectors technology and applications 476–7
 fast response photodetectors 471
 multi-pixel arrays 469–71
 photodetector integration 475–7
 photodetectors with intrinsic amplification 472
 position-sensitive detectors (PSDs) 467–9
laser beams 241
 delivery 383, 389
 image 393–4
 measurement 459
 propagation and optical consequences 392–3
laser class **201**
laser dyes 298
laser energy and power measurement
 displays 481
 laser absorbers 481
 measurement technique selection 479–80
 pyroelectric sensors 480
 semiconductor photodiode or optical sensors 481
 test configuration 480
 thermopile sensors 480
laser frequency stabilization scheme 160, *161*
laser injection locking 158, *158*
laser injuries, to eye 191–2
laser machining 279
laser modelling software 91–5
laser modes 44–6
laser optical systems, computation of 393–7
laser oscillation 34
laser principles 1–2
 amplifier to oscillator 33–4
 amplifier–oscillator connection 4–5
 characteristics of laser radiation 43–7
 coherence properties 47–8
 collection of particles in thermal equilibrium 11
 electron oscillator model of radiative transition 28–33
 energy levels of atoms, molecules and condensed matter 5–6
 inhomogeneous broadening 16–18
 optical frequency amplification with homogeneously broadened transition 18–22
 optical frequency amplifiers and line broadening 14–16
 optical frequency oscillation—saturation 22–8
 optical resonators, amplifying media 34–6
 oscillation frequency 37–43
 power output of laser 48–9, *50*
 relationship between Einstein A and B coefficients 11–13
 spontaneous and stimulated transitions 6–10
laser radiation 43

beam divergence 46
 characteristics of 43
 hazard 203
 laser modes 44–6
 linewidth of 46–7
laser safety
 associated hazards 203
 classes, qualitative description of 195, **196–9**
 exposure limits 192–5
 to eyes and skin 190–2
 safety in practice 200–3
 safety in product design 195–200
laser speckle 47–8
laser systems 446
 angular spectrum propagation 79–81
 dielectric waveguides 87–9
 finite-difference propagation 79
 gain and non-linear media 84–6
 integration of geometrical and physical optics 86–7
 laser modelling software 91–5
 numerical analysis for 77–8
 propagation in homogeneous media 81–3
 reflecting wall waveguides 90–1
 representation of optical beams 78
 split-step method 79
laser-induced fluorescence (LIF) 278
laser-induced nuclear reactions 280
laser-induced plasmas 280
laser-pumped laser resonators 115
laser-pumped laser systems 167
lateral shearing interferometer 485
law of energy conservation 485
lens design 404–5
lens systems 213–14
lens-makers formula 155
level degeneracy 12–13
LIF *see* laser-induced fluorescence (LIF)
light regulation 168
'lightning rod effect' 140
light-scattering processes 147, **147**
$LiNbO_3$ *see* lithium niobate ($LiNbO_3$)
line broadening 14–16
line selection 132
linear absorption 149
linear adaptive optics 225
linear material dispersion 268–70
linear optical properties 140
linear optics 332
linear waveguide resonator 121, *122*
lineshape function 7–8, *8*
linewidth factors in laser 46, *46*
linewidth of laser radiation 46–7
liquid crystal devices 228
lithium niobate ($LiNbO_3$) 231
lithium niobate devices 260
Littrow prisms 162, 163
Lommel functions 83
longitudinal laser modes 159
longitudinal modes 44
Lorentzian factor 27
Lorentzian homogeneous lineshapes 24
Lorentzian lineshapes
 doppler-broadened distribution of 18, *19*
 function for natural broadening 15, *15*
 normalized Gaussian and 17, *17*
loss line 38

low beam quality sources 396
lower optical powers 299
lowest-lying energy band 6
low-index glass 210
low-order axisymmetric Laguerre–Gaussian resonator modes 157, *157*
low-order Hermite–Gaussian resonator modes 157, *157*
low-power diodes 407
low-saturation fluence 297

Mach–Zehnder interferometer 53, *53, 54,* 60, 147
magnet galvanometers 451
magneto-optic effects 67, 73, 74
magneto-optic isolators 221–2
manufacturability, design for 235–6
Martinez grating stretcher 324, *324,* 324–5
mask projection arrangement 387, *387*
master oscillator with power amplifiers (MOPA) 253–4, *254*
material processing 279–80
maximum permissible exposure (MPE) 192, 193, 202
Maxwell wave equation 420
Maxwell–Boltzmann distribution 16
Maxwell–Boltzmann statistics 12
Maxwell's equations 78, 418
McIntyre function 179
MCP *see* micro-channel plate (MCP); multichannel plate (MCP)
MCVD *see* modified chemical vapour deposition (MCVD)
measurement technique selection 479–80
measuring devices 448
mechanical beam-directing systems 222–3
mechanical damage 442
mechanical Q-switches 251–2
mechanical scanning devices 484
medical imaging 277
metal tube waveguides 439
metal-based mirrors 397
metallic nanotubes 298
metal-oxidesemiconductor (MOS) capacitors 470
Michelson interferometers 53, *54,* 59, *59,* 60, 231, 274, *274,* 496
Michelson's stellar interferometer 60, *60*
micro-bending losses 426
micro-channel plate (MCP) 255
microlasers 5
micro-structured optical fibres 440
mid-wavelength IR (MWIR) spectral range 474, 477
mirror optics 397
mirror systems, use with CO_2 lasers 401–3
MLD gratings *see* multilayer dielectric (MLD) gratings
modal birefringence 425–6
'modal' devices 228
modal dispersion 270
mode beating 42–3, 167
mode control 158
 phase-conjugate mirrors 158–9
mode coupling 126–7
mode losses 126–7
mode-locked diode lasers 302, 303
mode-locked fibre lasers 301–3

mode-locked lasers 117, 290, 295, 296, 299, 323
mode-locked pulses 266
mode-locked semiconductor lasers 302–3
mode-locked solid-state bulk lasers 299–301
mode-locked systems 266
mode-locked VECSELs 302–3
mode-locking 118, 166, 253, 259–62, 260, 266
  basic principles of 290–4
  fibre lasers 298
  mechanisms 260
  modelling of ultrashort pulse lasers 303–4
  mode-locked fibre lasers 301–2
  mode-locked semiconductor lasers 302–3
  mode-locked solid-state bulk lasers 299–301
  saturable absorbers for 295–9
  soliton mode-locking 299
  techniques 290
  techniques and principles 289
modern ultrafast oscillators 268
modes of oscillation 125, 133
modes of propagation 133
modified Bessel functions 424, 424
modified chemical vapour deposition (MCVD) 439
modulator drive frequency 291
molecular orientation 137, 137
momentum streaking 312–13, 316
monochromatic energy density spectrum 9, 9
monochromatic radiation field 18, 20, 20
MOPA see master oscillator with power amplifiers (MOPA)
MOS capacitors see metal-oxidesemiconductor (MOS) capacitors
motion devices 446–7
mounting accuracy 234–5
mounting techniques 235
MPE see maximum permissible exposure (MPE)
MQW device see multiple-quantum-well (MQW) device
MUD TADPOLE 519
multichannel plate (MCP) 175
multi-direction-of-motion system 237
multi-dither approaches 229
multi-element photodetector arrays 476
multi-layer coatings and their applications 210–11
multilayer dielectric (MLD) gratings 327
multi-lens systems 214
multi-longitudinal-mode oscillation 41, 43
multi-mode fibres 418
multi-mode laser oscillation 37–42
multi-mode resonator model 127, 132
multi-mode theory 127, 132
multi-photon absorption 148–9, 149
multiphoton ionization/dissociation 149
multi-pixel arrays 469–71
multiple optical elements 209, 394–5
multiple pulse exposures 193–4
multiple-quantum-well (MQW) device 263
multiple-wave interference 54, 55
multi-pulse coherent artefact 493
multi-shot method 255
mutual coherence function 58, 58

'mutual correlation function' 58
MWIR spectral range see mid-wavelength IR (MWIR) spectral range

NALM see non-linear amplifying loop mirror (NALM)
nanoFROG 527
nanosecond OPO Devices 347–9, 353–5
natural broadening 14–15
NCPM see noncritical phase-matching (NCPM)
Nd-doped silicate glass 327
Nd:glass system 322–3
near-single-cycle pulse measurement 505–6
negative electron affinity (NEA) photocathodes 174
negative uniaxial crystal 338, 338, 370
negatively and positively chirped pulse amplification (NPCPA) 327
NEP see noise-equivalent power (NEP)
nicol prism 66, 66
NLSE see non-linear Schrodinger equation (NLSE)
NOHD see nominal ocular hazard distance (NOHD)
noise
  measurement circuit 186–7
  optical signal 183–4
  in photodetector 184–6
  sources 187
noise-equivalent power (NEP) 172
noisy pulse trains, autocorrelations of 492–3
nominal ocular hazard distance (NOHD) 194–3
non-centrosymmetric crystals 337
non-coherent optics 395
non-collinear phase-matching 328, 328, 369
noncritical phase-matching (NCPM) 340
non-focusing optical laserbeam handling and relaying 214–15
non-Gaussian beams 156
non-Gaussian laser beams 156–7
non-interactive laser cavity 33, 33
non-linear amplifying loop mirror (NALM) 265
non-linear crystals 354
non-linear dependence 394
non-linear effects 442
  optical fibres and solitons 429–30
non-linear envelope equation 150
non-linear fibre loop mirrors 298
non-linear frequency conversion 332–5
  sources based on 267–8
non-linear inelastic scattering 231
non-linear Mach–Zehnder interferometer 147, 147
non-linear material 270, 349–50
nonlinear mode-pulling 43
non-linear optical effects 78, 135, 150
non-linear optical materials 138, **138–9**
non-linear optical phase conjugation 229–32
non-linear optical processes 137, 332, 333
non-linear optical self-action effects 150
non-linear optics 135, 331, 332
  basic concepts 135–6
  mechanisms of optical non-linearity 136–8
  multi-photon absorption 148–9
  non-linear optical materials 138, **138–9**

optical bistability 146–7
optical parametric oscillation 143–4
optical phase conjugation 144–5
optical solitons 145–6
optical switching 147
optically induced damage 149
optics in plasmonic materials 140–2
second- and third-harmonic generation 142–3
self-focusing of light 145
stimulated light scattering 147–8
strong-field effects and highorder harmonic generation 149–50
non-linear polarization rotation 298
non-linear Schrodinger equation (NLSE) 146, 262
non-linear techniques 328
non-linear-optical variations 521
non-reciprocal effect 73
non-resonant electronic response 137
non-rotationally symmetrical laser beams 396–7
non-saturable losses 295, 296
non-uniform intensity distribution 254
normal index surface 337
NPCPA see negatively and positively chirped pulse amplification (NPCPA)
nuclear fission 280
Nyquist sampling frequency 82, 83

occupation number 11
OCT see optical coherence tomography (OCT)
off-axis paraboloid 401–3
offline effects 86
offner triplet 325
one-dimensional (1D) interferometric method 277
one-dimensional phase retrieval 489, 490–1
one-lens approach 109
one-photon absorption processes 149, 149
OPA see optical parametric amplification (OPA)
OPCPA see Optical Parametric Chirped Pulse Amplification (OPCPA)
OPD see optical path difference (OPD)
OPG see optical parametric generator (OPG)
OPOs see optical parametric oscillators (OPOs)
OPS see optical parametric synthesizers (OPS)
'optical' active mode-locking 261
'optical activity' 68
optical beams, representation of 78
optical bistability 146–7
optical coherence tomography (OCT) 59, 277
optical components 205
  elementary lens forms 211–14
  manufacture of 216–17
  non-focusing optical laserbeam handling and relaying 214–15
  optical design aspects of laser optics 208–9
  optics for laser applications 215–16
  surface phenomena and thin layer coatings 209–11
  thermal effects in optical materials 215
  use of (curved) mirrors 214
optical configurations of scanning systems 450–1
optical control elements

Index

amplitude modulation 219–22
  controlling size and shape of beam 224–5
  Safe disposal of unwanted beams 225
  Scanning and positioning the beam 222–4
optical damage 388
optical damage threshold of fused silica 149, **150**
optical design aspects of laser optics 208–9
optical detection and noise
  noise in photodetection 183–8
  nomenclature and figures of merit 172–3
  photoemissive detectors 173–6
  semiconductor detectors 176–81
  thermal detectors 181–3
optical elements in amplifier 325–6
optical fibre communication systems 202
optical fibre power meters 480
optical fibre-based telecommunications applications 179
optical fibres 71, *71*, 87
  attenuation in 426
  dispersion in 427–9
  and solitons, non-linear effects in 429–30
  waveguides 418
optical frequency
  amplification with inhomogeneous broadening 21–2
  amplifiers 14–16, 18
  combs 289
  with homogeneously broadened transition 18–20
  oscillation—saturation 22–8
optical isolators 222
optical materials, thermal effects in 215
optical non-linearities 136–8, 140–1, 299
optical parametric amplification (OPA) 335, 336, *337*, 340, *341*, 342, 363
Optical Parametric Chirped Pulse Amplification (OPCPA) 326–9, 363
  architecture 373–5
  and lasers, comparison of 364–5
  optical parametric synthesizers 377–8
  in practice 375–7
  theory 365–73
optical parametric devices 331, 341–2
  continuous-wave OPO devices 352–3
  nanosecond OPO Devices 353–5
  non-linear frequency conversion 332–4
  OPO design issues 349–52
  optical parametric devices 341–2
  optical parametric oscillators 342–9
  phase-matching 335–41
  synchronously pumped OPO devices 355–9
optical parametric gain 334–5
optical parametric generation 333–4
optical parametric generator (OPG) 342
optical parametric oscillation 143–4
optical parametric oscillators (OPOs) 143, 261, 332, 342–9
  design issues 349–52
optical parametric synthesizers (OPS) 364, 378
optical path difference (OPD) 485
optical phase conjugation 144–5, 158
optical polarization 71
optical power 208
optical pulse generation 239
optical rays 77
optical reflection spectrum 296

optical resonators 97, 124
  containing an amplifying media 34–6
optical sensors 481
optical signal, noise in 183–4
optical solitons 145–6
optical spectrum of frequency comb 295, *295*
optical switching 147
optical viewing aids 195
optical wave 71
optical waveguide theory 121
  basic types of optical waveguides 417–18
  mode-coupling 430–2
  optical fibres 423–6
  planar and rectangular guides 418–23
  propagation effects in optical fibres 426–30
optical-fibre gyroscope 60, 61
optical-fibre Mach–Zehnder interferometer 53, *54*, 60, *61*
optics
  for CO_2 laser systems 397–404
  for excimer and other UV lasers 409–12
  for imaging mask 410–12
  for laser applications 397
  for lasers operating in visible 404–9
  for other laser sources 412–13
  plasmonic materials 140–2
optimal non-collinear angle 369
optimization 235
optimum re-imaging conditions 484
optoelectronic packages 475
opto-electronic techniques 53
opto-mechanical parts
  closure 238
  materials and finishes 236
  parts configuration 236–7
  precision positioning 237
  requirements and specifications 233–4
  system considerations 234–6
orthogonal astigmatic system 103
orthogonal linear polarizations 70, 71
orthogonal polarizations 58, 65
oscillation frequency 37–43
  mode beating 42–3
  multi-mode laser oscillation 37–42
oscillatory scanning systems 453

parametric amplification 365–7
parametric generation process 334
parametric interaction process 334
parametric luminescence 335
parametric noise 335
'paraxial formulas' 395
partial differential equation 485
partially coherent waves 57–9
passive mode-locking 167, 261–2, 292–4, 302, 303
passive optical resonators 97
passive Q-switching 165
Pauli's exclusion principle 18, 279
PC *see* photonic crystals (PC)
PCF *see* photonic crystal fibre (PCF)
PCGPA *see* Principal Components Generalized Projection Algorithm (PCGPA)
PDE *see* photon detection efficiency (PDE)
PDI *see* point-diffraction interferometer (PDI)
periodic-poling techniques 268
personal protective equipment 202
perturbative theory 307

PFN *see* pulse-forming network (PFN)
PG autocorrelator *see* polarization-gating (PG) autocorrelator
phase distribution measurement 484–5
phase modulation, power distribution shaping by 403–4
phase noise 273
Phase Retrieval by Omega Oscillation Filtering (PROOF) 314–16
phase-conjugate mirrors 144, 159, *159*
  mode control with 158–9
  reflection 229
phase-conjugate reflectors 227
phase-matching 335–41, 367–70
  angles 369
  condition 334, 365
  mirrors 128
  scheme 339
  techniques 352
phase-retrieval methods 317
phase-sensitive autocorrelation 496
'phase-sensitive detection' methods 188
phonons 15
photodetection processes 172, *172*
photodetectors 467
  integration 475–6
  intrinsic amplification 472
photodiodes 187, 255
  arrays 469–70
  transimpedance amplifier 476, *476*
photoelectric absorption 176
photoelectric effect 473
photoelectron noise 184
photoemissive detectors 172
photoemissive effect 173–4
photolithography 412
photomultiplier tube (PMT) detectors 473
photomultipliers 174–6, 187
photon detection efficiency (PDE) 472
photon noise 184
photon-counting applications 180
photonic bandgap 383, 440
  fibres 436, 437
photonic crystal fibre (PCF) 271, 301, 436, *436*
photonic crystals (PC) 476
photonic stopband (PSB) 476
photonics 4, 331
photon–particle collision processes 10, *10*
photons 4
photorefraction 231–2, 350
photorefractive effect 137–8, *138*
  photorefractive polymer materials 231
photovoltaic mode 178
picosecond lasers 300
picosecond OPO devices 349, 355–6
picosecond pulse duration systems 325
piezoactuators 454
piezoelectric actuators 449, 451
piezoelectric devices 453–5
piezoelectrical effect 454
pin photodiodes 176–8, *178*, 471, *471*
pixellated devices 228
planar guides 418–21
planar mirrors 215
Planck's constant 85
Planck's hypothesis 11
Planck's theory 183
plano-convex lens 399

plasma-activated CVD 439
plastic optical fibres 436
plastic tube waveguides 439
PM fibres *see* polarization-maintaining (PM) fibres
PMT detectors *see* photomultiplier tube (PMT) detectors
Pockels cells 220
Pockels effect 252, 220
Poincaré sphere 69–71, *70*
point-diffraction interferometer (PDI) 485
polarization
    analysis 68–71
    circular birefringence 68
    control with waveplates 65, *65*
    crystal optics 63–5
    eigenmodes 64
    ellipse *62*, 62–3
    holding property 72
    material interactions 63
    optics, applications of 71–4
    polarizing prisms 66–8
    prisms 66–8
    retarding waveplates 65–6
polarization-APM 266
polarization-gating (PG) autocorrelator 494
polarization-maintaining (PM) fibres 425–6, *426*
polarization-mode dispersion 71, 429
polarization-optical phenomena 69
polishing 215, 216
polygonal scanners 451
'ponderomotive energy' 150
ponderomotive non-linearities 141
position sensors 468
positioning systems 446–9
position-sensitive detectors (PSDs) 467–9
positive uniaxial crystal 65
positive-branch unstable resonators 113
post-objective configuration 450, *450*
Pound–Driever locking 118
power distribution shaping by phase modulation 403–4
power output of laser 48–9, *50*
poynting vector walk-off compensation scheme 370
practical measurement weak and complex pulses 517–19
practical measurements 462–3
pre-objective configuration 450
pre-objective 2D scanning system 450, *451*
Principal Components Generalized Projection Algorithm (PCGPA) 314
principal measurement set-up 483
principal refractive indices 64
principal states of polarization (PSP) 429
printed wiring boards (PWBs) 244
prismatic polygonal scanners 451
Programmable diffractive optical elements 224
Prominent conventional tunable lasers 342, *343*
PROOF *see* Phase Retrieval by Omega Oscillation Filtering (PROOF)
propagation characteristics 463–4
protective housing 200
PSDs *see* position-sensitive detectors (PSDs)
PSP *see* principal states of polarization (PSP)
pulse duration limitations 327

pulse repetition rate 290
pulse shape measurement 274–5
    approximate methods of 274–5
    exact methods of 275–7
pulse-broadening effects 291
pulsed beam 255
pulsed laser damage 442
pulsed Nd:YAG systems 446
pulsed OPOs 347–9
pulse-forming network (PFN) 241
pulse-measurement techniques 488
pulse-retrieval algorithm 513, 517
pulse-shaping 273
    dispersion in optical systems 268–73
    effects 261
    mechanisms 261
pump laser 351–2, 364
pump radiation 118
pump-probe
    measurement 296
    technique 279
PW level power, amplification to 326–8
PWBs *see* printed wiring boards (PWBs)
pyramid polygonal scanning system 451, *451*
pyroelectric detectors 182–3
pyroelectric sensors 480

QPM *see* quasi-phase-matching (QPM)
Q-switched CW pumped lasers 249–50
Q-switched laser 78, 118
Q-switched microlasers 252
Q-switched mode-locking 293
Q-switched solid-state lasers 404
Q-switching 98, 248–53, *251*
    instabilities 293–4, 300
    methods of 250–1
    scheme 167
    threshold 303
quadrant detector 468–9
quadrant photodiode 469
qualitative description of laser safety classes **196–9**
qualitative profile inspection 484
quantitative computational model 215
quantum electronics 4
quantum interference effects 145
quantum mechanical zero-point fluctuations 147
quantum of energy 4
quantum well infrared photodetector (QWIP) technology 172, 176, 181
quantum-size effects 141
'quarter-wave' plate 65
quartz 68
quasi-CW and modulated beams
    applications 244–5
    operation of CO_2 lasers 243–4
    operation of solid-state lasers 241–3
quasi-phase-matching (QPM) 268, 340–1, 368
quasi-soliton mode-locking 299, 300
QWIP technology *see* quantum well infrared photodetector (QWIP) technology

RABBITT *see* Reconstruction of Attosecond Beating by Interference of Twophoton Transitions (RABBITT)
radiative carrier recombination 279

radiative transition, electron oscillator model of 28–33
Rainbow RABBITT 312
RAL *see* Rutherford Appleton Laboratory (RAL)
Raman scattering process 231
RANA *see* Retrieved-Amplitude N-grid Algorithm (RANA)
raster scanning 450
ray tracing 209, 394–5
ray transfer matrices 101–3, **104**
ray-based methods 91
Rayleigh range 393
Rayleigh scattered light 426
real laser beams 209
real resonators 98
'real-time' holography 231
reciprocal rotation 73
reconstruction and control 229
Reconstruction of Attosecond Beating by Interference of Twophoton Transitions (RABBITT) 309–12, *311, 312*, 316, *317*
recovery time 295
reflective scanning systems 449
regenerative mode-locking 261
resistive charge division sensors 468
'resonant' coupling 72
resonant frequency 454
resonant galvanometers 451
resonator modes 121
    concept of 124–5
    degeneracies 132–3
    waveguide modes 125–6
resonator stability diagram 105, *105*
responsivity 173
retarding waveplates 65–6
'retinal hazard range' 192
retinal injuries 192
Retrieved-Amplitude N-grid Algorithm (RANA) 503–4
rigrod analysis for waveguide lasers 129–30
Rigrod equation 129
Rigrod modelling 127
ring dye laser 160
ring laser cavity 160, *160*
ring resonators 118
Ritchey–Chretien system 403
rod-in-tube techniques 438, 439
rotating mirrors 165
rotational quantum number 5
Rutherford Appleton Laboratory (RAL) 326

safe disposal of unwanted beams 225
sapphire 438
    sapphire fibres 439
saturable absorbers 118, 165, 261, *262*, 292, 295, 298, 301
    mode-locking 295–9
    Q-switches 252–3
saturated regime 370–2
saturation energy 295
saturation fluence 295
saturation intensity 24, 295
SBS *see* stimulated Brillouin scattering (SBS)
scan frequency 452
scanning autocorrelators 274
scanning devices 484

Index 545

scanning galvanometers, performance and accuracy of 452–3
scanning systems 445, 449
 acousto-optic deflectors 455–6
 galvanometer scanners 451–2
 mirrors in oscillatory scanning systems 453
 optical configurations of 450–1
 performance and accuracy of scanning galvanometers 452–3
 piezoelectric devices 453–5
 polygonal scanners 451
 scanning methods 450
Scattering Wave PROOF (SWProof) 317
Schott glass catalogues 404
Schottky barrier devices 473
Schottky diode detectors 178–9
Schottky photodiode 178, 179
SD see silent discharge (SD)
SEA TADPOLE see Spatially Encoded Arrangement for Temporally Dispersing a Pair of Light E-fields (SEA TADPOLE)
second harmonic-generation (SHG) 142–3, 332, 490
second-order dispersion of common optical systems 272, **272**
second-order non-linear optical materials 138, **138**
second-order non-linear optical processes 137
second-order perturbation theory 311
second-order XUV AC/FROG 310
Seebeck effect 182
self-diffraction FROG 500, 501
self-focusing of light 145
self-mode-locking 263
self-phase modulation (SPM) 270, 271, 322
self-referenced methods 489, 514
self-starting mode-locking 293
self-trapping 145
semiconductor absorbers 296–7
'semiconductor detectors' 172
semiconductor diode strip detector 468, 468
semiconductor disk laser 302
semiconductor optical amplifier (SOA) 266, 302
semiconductor photodetectors 474
semiconductor photodiode 481
semiconductor saturable absorber layer (SESAM) 292, 296–7, 297
semiconductor sources 266–7
semiconductor spectroscopy 279
semiconductor-based saturable absorbers 297
sensor arrays 464
servo-actuator systems 237
SESAM see semiconductor saturable absorber layer (SESAM)
Shack–Hartmann wavefront sensor 229, 229
SHG see second harmonic-generation (SHG)
shielded plasma frequency 140
short pulses
 beam characterization and pulse measurement 254–6
 cavity dumping 253
 gain switching 247–8
 master oscillator with power amplifiers 253–4

mode-locking 253
Q-switching 248–53
shot noise 184–5
SiC see silicon carbide (SiC)
signal frequency 327
signal-to-noise ratio (SNR) 167, 172
silent discharge (SD) 243
silica fibres, optical damage in 442
silicon 472
silicon carbide (SiC) 473
silicon photomultiplier (SiPM) 472
silicon-on-insulator (SOI) substrates 471
silver halide fibres 438, 439
simple amplifier with feedback 4, 4
simple energy-level diagram 6, 6
simple phase-diversity approach 526
simple picosecond laser, set-up of 300, 300
single broadband frequency comb 295
single (ZnSe) lens 399
single lens, limits of 399
single longitudinal mode 160
single thin layer, vertical incidence and effect of 210
single-cavity mode 40, 41
single-crystal sapphire fibres 438
single-diode laser beams, optical handling of 407
single-diode laser-focusing optics 407–8
single-frequency laser 160
single-grating stretcher 325
single-lens mode-matching 109, 110
single-mode fibres 418, 428, 436
single-mode model 130
single-mode optical fibres 270, 476
single-mode theory 127
single-mode waveguide resonators 127–9
single-photon ionization 310
single-shot autocorrelators 255
single-shot MUD TADPOLE 519–21
singleshot pulse measurement 519
Singlet 211
singly resonant oscillator (SRO) 343, 345
SiPM see silicon photomultiplier (SiPM)
skin protection 202
slow saturable absorbers 261
small-angle scalar propagation 78
Snell's law 66, 210
SNR see signal-to-noise ratio (SNR)
SOA see semiconductor optical amplifier (SOA)
soft aperture mode-locking 263
soft collision 16
soft-edged apertures 115
software, validation of 95
SOI substrates see silicon-on-insulator (SOI) substrates
solid TIR fibres 440
solid-state bulk lasers 296
solid-state devices 4
solid-state gain media 326
solid-state lasers 249
 operation of 241–3
 thermal distortion in 243
solid-state photodetectors 474
solid-state semiconductor detectors 473
solitary lasers 262
soliton laser 264
soliton mode-locking 299, 301
soliton pulse energies 299

spatial coherence 48, 392
spatial dimension 517
spatial filtering 169, 169, 386
spatial frequencies 517
spatial light-modulator 273
spatial solitons 146, 270
Spatially and Temporally Resolved Intensity and Phase Evaluation Device: Full Information from a Single Hologram (STRIPED FISH) 528, 530
Spatially Encoded Arrangement for Temporally Dispersing a Pair of Light E-fields (SEA TADPOLE) 517–19, 518, 525
spatially resolved spectral interferometry 524–5
spatiotemporal measurement
 examples 526–7
 methods 527
 pulse 524
 on single shot 527–31
spatiotemporal pulse-measurement techniques 524
spatio-temporal soliton 146
Special Theory of Relativity 59
specific detectivity 173
spectral amplitude 311
spectral hole burning 117
spectral interferometry 514–15, 525
 advantages and disadvantages of 515–16
Spectral Phase Interferometry for Direct Electric-field Reconstruction (SPIDER) 277, 521–4
spectral radiant intensity 92
spectral region 364
spectral tunability 364
spectrometer 510
spectrum and one-dimensional phase retrieval 489–90
SPIDER see Spectral Phase Interferometry for Direct Electric-field Reconstruction (SPIDER)
spinning reflectors 251
split-step Fourier algorithm 372
split-step method 79, 87
SPM see self-phase modulation (SPM)
spontaneous and stimulated transitions
 lineshape function 7–9
 relation between energy density and intensity 8–10
 spontaneous emission 6–7
 stimulated absorption 10
 stimulated emission 104
spontaneous emission 6–7, 7, 11–13, 21, 86
spontaneous light-scattering processes 147
spontaneous parametric fluorescence 335
spontaneous transitions, ratio of 13
'spun preform' technique 71
SRO see singly resonant oscillator (SRO)
SRS see stimulated Raman scattering (SRS)
stable lasers 105
stable mode-locking 291
stable resonators 98, 98
 higher-order modes of 105–6
stable single-frequency operation 118
stable-resonator axial-mode spectral separation 116–17

standard lens equation 154
standard single-mode fibres 418
STARFISH 516
Stark broadening 16
Stark effect 16
start-up phase 293
static resolution 455
steady-state threshold 344–5
Stefan–Boltzmann constant 186
step scan performance 452, *452*
stigmatic beam 461
stimulated absorption 10, *10*
stimulated Brillouin scattering (SBS) 144, 148, *148*, **148**, 159, 230, 231, 442
stimulated emission 104, 10
  classical oscillator explanation for 31–2
stimulated emission process 8, *8*
stimulated Raman scattering (SRS) 147, 148, 442
stimulated transitions, ratio of 13
Stirling closed-cycle coolers 181
Stokes parameters 69
streak cameras 255
strip detector 468–9
STRIPED FISH *see* Spatially and Temporally Resolved Intensity and Phase Evaluation Device: Full Information from a Single Hologram (STRIPED FISH)
strong-field effects 149–50
sub-Doppler spectroscopy techniques 118
sub-femtosecond pulses 307, *307*
sub-picosecond pulse durations 266
suitable phase-matching conditions 335
suitable saturation energy 296
surface absorbers 481
surface imperfections 216
SWIR detectors 477
SWProof *see* Scattering Wave PROOF (SWProof)
symmetric confocal interferometer 116
symmetrical mirror systems 214
synchronous pumping 167, 263, 349
synchronously pumped mode-locking 261
synchronously pumped OPO devices 355–9

tangential phase-matching scheme 370
Taylor expansion 324
TBPs *see* time-bandwidth products (TBPs)
TCSPC *see* time-correlated single-photon counting (TCSPC)
TE modes *see* transverse-electric (TE) modes
TEA laser *see* transversally excited atmospheric pressure laser (TEA laser)
TEA-CO_2 laser 248, *248*
telecommunication fibres 437
temperature acceptance bandwidth 350
temperature noise 182, 186
temporal coherence 47
temporal soliton 145
temporal synchronization 364
temporal walk-off 366
temporal-measurement technique 527
terahertz radiation 278
test configuration 480
Texas PW system 327
theoretical double-pulse intensities 492, *493*

theory of longitudinal modes 159–60
Thermal blooming 225
Thermal damage process 190, 441–2
thermal distortion in solid-state lasers 243
thermal effects
  optical materials 215
  ZnSe lenses 400–1
thermal equilibrium, collection of particles in 11
thermal noise 186
thermal photodetectors 254
thermistors 182
thermocouples 182
thermopiles 182
  detectors 481
  sensors 480
THG *see* third-harmonic generation (THG)
thin film optical coatings 388
thin-film coatings 211
third-harmonic frequency 143
third-harmonic generation (THG) 142–3, 493, *494*
third-order autocorrelations 493–5
third-order dispersion of common optical systems 272, **272**
third-order intensity autocorrelation 493
third-order non-linear optical coefficients 138, **138**, 139, **139**, 140, **141**
third-order non-linearities 495
three spatial dimensions 525–6
threshold gain coefficient 36
threshold kinetic energy 179
threshold level, establishing 192–3
threshold photon energy 181
threshold population inversion 36
TIA *see* transimpedance amplifier (TIA)
tilted mirrors 131–2
time-bandwidth products (TBPs) 498, 519
time-correlated single-photon counting (TCSPC) 185
time-frequency domain 498–9
time resolved photoluminescence techniques 279
TIR fiber *see* total internal reflectance (TIR) fiber
Ti:sa laser *see* titanium:sapphire (Ti:sa) laser
Ti:sapphire amplification system 327
Ti:sapphire lasers 263–4, 298
Ti:sapphire oscillators 267
Ti:sapphire systems 322–1
titanium:sapphire (Ti:sa) laser 162, 373
tolerances 215–16
total internal reflectance (TIR) fibres 66, 418, 435–8, 441, 442
totalon-time-pulse (TOTP) 194
TOTP *see* totalon-time-pulse (TOTP)
traditional methods of modelling 91–5
transimpedance amplifier (TIA) 476
trans-impedance amplifiers 187
transmissive and reflective optics 385–6
transmissive systems 449
transversally excited atmospheric pressure laser (TEA laser) 247
transverse beam development 115, *116*
transverse geometrical smearing 505
transverse Kerr effect 72
transverse modes 46
transverse modulators 220

transverse-electric (TE) modes 419
Treacy grating compressor 323–4
triply resonant oscillator (TRO) 343
truncation 155–6
tunability 132
two mirror resonators 104–5
two pulses simultaneously, measuring 512–13
two spatial dimensions 525–6
two-colour IR-XUV Photoionization 310–11
two-dimensional (2D) guides 418, 421–2
two-dimensional phase-retrieval problem 500
two-dimensional scanning 450, 451
two-dimensional spectral shearing interferometry (2DSI) 522
two-lens mode-matching 109–10
two-photon laser-scanning fluorescence imaging 277
Twyman–green interferometers 59–60, *60*

ultrafast chemistry 278–9
ultrafast laser techniques 278
ultra-high accuracy 480
ultra-high intensity interactions 322
ultrahigh power systems 377
ultrashort laser pulses 488–9
  advantages and disadvantages of spectral interferometry 515–16
  autocorrelation and spectrum 495–6
  autocorrelation conclusions 498
  autocorrelations of complex pulses 492
  autocorrelations of noisy pulse trains 492–3
  cross-correlation 496–8
  crossed-beam spectral interferometry 516–17
  error bars 513
  fibre-based scanning spatiotemporal pulse measurement 525–6
  frequency-resolved optical gating (FROG) 499–500
  GRENOUILLE 509–12
  intensity autocorrelation 490–1
  interferometric autocorrelation 496
  measuring the spectrum 489
  measuring two pulses simultaneously 512–13
  measuring very complex pulses in time 519
  near-single-cycle pulse measurement 505–6
  practical measurement weak and complex pulses 517–19
  RANA approach 503–4
  self-referenced methods 514
  single-shot MUD TADPOLE 519–21
  spatially resolved spectral interferometry 524–5
  spatiotemporal measurement examples 526–7
  spatiotemporal measurement methods 527
  spatiotemporal measurement on single shot 527–31
  spatiotemporal pulse measurement 524
  spectral interferometry 514–15
  spectrum and one-dimensional phase retrieval 489–90
  SPIDER 521–4
  third-order autocorrelations 493–5
  time-frequency domain 498–9

Index 547

XFROG 508–9
ultrashort light pulses 290
ultrashort pulses
　applications of 277–81
　diagnostic techniques 273–7
　illumination 277
　lasers, modelling of 303–4
　measurement 531
　pulse shaping and dispersion in optical systems 268–73
　sources of 262–7
　theory of ultrashort pulse generation and mode-locking 259–2
uniaxial crystals 337
unstable resonators 98, 99, 119
Unwanted beams, safe disposal of 225
UV spectral range, detectors for 473

vacuum ultraviolet (VUV) 4
VAD see vapour phase axial deposition (VAD)
valence band 6
valence electrons 18
vapour phase axial deposition (VAD) 439
vapour phase deposition 438
VECSEL see vertical external-cavity surface-emitting laser (VECSEL)
vector scanning 450, 452, *452*
'venetian blind' configuration 174

verdet constant 73
vertical external-cavity surface-emitting laser (VECSEL) 302, *302*
vibrational energy states 5
vibronic state 6
vibrot state 6
virtual source techniques 111
visualization 236
V-number 419
Voigt profile 18
Volkov Transform General Projection Algorithm (VTGPA) method 317
volume absorbers 481
VTGPA method see Volkov Transform General Projection Algorithm (VTGPA) method
VUV see vacuum ultraviolet (VUV)

wafer fabrication processes 474
wall waveguides 90–1
wave coherence 51–2
wave properties 51
wavefront sensors 229
'wavefront-bending-capability' 208
wavefront modulator systems 223
waveguide analysis, numerical methods for 422–93
waveguide coupling matrices 121

waveguide laser resonators
　first-order theory and its limits 127–30
　propagation in hollow dielectric waveguides 122–4
　real waveguide resonators 130–3
　waveguide resonator analysis 124–7
waveguide lasers, rigrod analysis for 129–30
waveguide mode expressions 122–4
waveguide resonator analysis 122
wavelength-division multiplexed optical communications 273
'well-behaved' systems 77
'white' energy density spectrum 115, *115*
Wollaston prism 66, *66*, 67
workpiece positioning 446

XFROG 508–9
X-ray diffraction 278
XUV IR two-photon ionization 309, *309*

Young's fringes 52
Young's modulus 47

Zernike polynomials 209, 228
zero incident wave—oscillation 36
zero-dispersion fibre 429
zero-point energy 11
zoom optics, use of 399–400

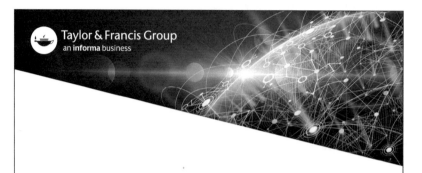